Natural Climate Variability On Decade-to-Century Time Scales

Climate Research Committee
Board on Atmospheric Sciences and Climate
Commission on Geosciences, Environment, and Resources
National Research Council

NATIONAL ACADEMY PRESS
Washington, D.C.   1995

Library of Congress Catalog Card Number 96-67828
International Standard Book Number 0-309-05449-4

Additional copies of this report are available from:

National Academy Press
2101 Constitution Avenue, NW
Box 285
Washington, DC 20055
800-624-6242
202-334-3313 (in the Washington Metropolitan Area)

B731

COVER: *The Eye of the Storm*, the oil painting reproduced on the cover of this book, is the work of Ilana Cernat of Bat-Yam, Israel. Dr. Cernat is linked to the world of intermediate-scale climate change through her son Michael Ghil, a dynamicist who has contributed substantially to our understanding of the interactions that affect climate variability. *The Eye of the Storm* (1989) is one of several of her paintings that express her concern for the future, particularly what sort of world we will be leaving to the generations to come. A lawyer by training and profession, Dr. Cernat began studying painting in her teens. Her work has been exhibited in Romania, Hungary, Israel, and the United States, and hangs in collections in other countries as well.

Editorial Committee for
Natural Climate Variability on Decade-to-Century Time Scales

Climate Research Committee

ERIC J. BARRON (*Chair*), Pennsylvania State University, University Park
DAVID S. BATTISTI, University of Washington, Seattle
RUSS E. DAVIS, Scripps Institution of Oceanography, University of California, San Diego
ROBERT E. DICKINSON, University of Arizona, Tucson
THOMAS R. KARL, NOAA National Climatic Data Center, Asheville, North Carolina
JEFFREY T. KIEHL, National Center for Atmospheric Research, Boulder, Colorado
CLAIRE L. PARKINSON, Goddard Space Flight Center, National Aeronautics and Space Administration, Greenbelt, Maryland
STEVEN W. RUNNING, University of Montana, Missoula
KARL E. TAYLOR, Lawrence Livermore National Laboratory, Livermore, California

Ex Officio Members

W. LAWRENCE GATES, Lawrence Livermore National Laboratory, Livermore, California
DOUGLAS G. MARTINSON, Lamont-Doherty Earth Observatory, Columbia University, Palisades, New York
EDWARD S. SARACHIK, University of Washington, Seattle
SOROOSH SOROOSHIAN, University of Arizona, Tucson
PETER J. WEBSTER, University of Colorado, Boulder

Staff

WILLIAM A. SPRIGG, Director
MARK HANDEL, Senior Program Officer
THERESA M. FISHER, Administrative Assistant

Board on Atmospheric Sciences and Climate

Commission on Geosciences, Environment, and Resources

Preface

One objective of the Climate Research Committee of the National Research Council's Board on Atmospheric Sciences and Climate is to promote progress in climate research through the organization of workshops and the preparation of reports. Over the course of a number of meetings and study sessions, the Committee began to recognize that climate variability on decade-to-century time scales had received significantly less attention than was warranted by its relevance to society. Much of climate research has focused either on seasonal-to-interannual climate variability, or on potential long-term climate change in response to human activities. These two research topics are of critical importance, but we cannot adequately address climate and climate change without also considering decade-to-century-scale variability.

Consequently, the CRC sponsored a workshop on Natural Climate Variability on Decade-to-Century Time Scales, which was held September 21-25, 1992. The conference was organized by CRC members Kirk Bryan, Michael Ghil, Doug Martinson, and Lynne Talley. It was generously sponsored by NOAA, NASA, DOE, and NSF, through the Climate Systems Modeling Program of the University Corporation for Atmospheric Research, with later assistance from EPA, USDA, the USAF Office of Scientific Research, and the Office of Naval Research. Much of the success of this workshop can be credited to the eagerness of members of the scientific community to participate and to contribute timely papers on the diverse topics that constitute this important subject.

This volume is built around the papers presented at the workshop. The organizers wished to obtain the broadest coverage of the topic possible at one meeting, recognizing that the collection of papers would be representative rather than complete and that their findings would not always be unambiguous. A commentary on each paper by its discussion leader has therefore been included, as well as an edited version of the ensuing discussion; both of these highlight uncertainties and offer additional information.

While in Irvine, the CRC selected (from among their and their panels' ranks) an expert in each of the fields represented—atmospheric and ocean observations, atmospheric and ocean models, coupled models, and proxy climate indicators—to serve as editor and first reviewer of the papers in the appropriate section of the workshop volume. To provide a context for the papers and appropriate references to germane material not covered at the workshop, the members of this editorial committee contributed an essay introducing each section. The editors, chaired by Doug Martinson, were responsible as well for initially drafting the conclusions and recommendations, and their efforts in producing this seminal volume are greatly appreciated by the CRC. We are also indebted to Ellen Rice not only for editing the papers and discussions but for seeing the volume (and its contributors) through the entire preparation, review, and production process.

The Climate Research Committee's goal was to produce a reference document on natural climate variability on decade-to-century time scales that would encourage greater interaction among the disciplines involved and between modelers and observationalists. We hope that this volume will serve as a foundation for new and stronger research programs on climate variability on these scales, which in turn will yield a broader and sounder understanding of the climate system and enhance the predictive capability that serves our society's needs.

Eric J. Barron, Chairman
Climate Research Committee

Table of Contents

Executive Summary

Human activities increasingly influence all aspects—biological, chemical, and physical—of the planet on which we live. To better understand what is being affected and how, the scientific method requires us to partition the immense problem at hand into slightly more manageable pieces. One of those pieces is the natural climate variability on which any human-induced change is superimposed. Climate variability, with or without anthropogenic change, represents one of the most fundamental issues of scientific and social interest today. The purpose of the workshop held by the NRC's Climate Research Committee in September 1992 (at the National Academies' Beckman Center in Irvine, California) was to define natural climate variability on the time scale of a few human generations. This volume reflects not only the proceedings of that workshop but considerable intervening work, both by the invited authors and their colleagues, and by anonymous reviewers and the book's editorial committee.

THE CHALLENGE

Natural climate variability on decade-to-century time scales is best defined in terms of the bio-chemical-physical system that must be studied, the principal components of that system, the mechanisms active within each component, and the interactions between components. The main components of the earth system are the atmosphere, oceans, land surface, snow and ice at the surface of both oceans and land, and biota near the interfaces of atmosphere, ocean, and land. The natural mechanisms include radiative transfer,

the planetary-scale circulation of the atmosphere and oceans, photochemical processes, and biogeochemical cycles of trace gases and nutrients. The major interactions between the components of the climate system so defined are given by the exchanges of energy, momentum, water, and trace constituents, which take a large number of specific forms. For instance, ice-albedo feedback affects radiative transfer in the atmosphere and the heat exchange between it and the underlying high-latitude surfaces, and evaporation-wind stress affects the feedback between the tropical atmosphere and oceans.

Our current understanding of the climate system on these time scales is based on insufficient observations and imperfect models. Historically, both observations and models have addressed only one component of the system; the best (but still unsatisfactory) data sets and models are those available for the atmosphere, followed in order by those for the oceans and, more recently, the snow and ice, land surface, and biota. Fortunately, sophisticated global observation systems and model studies are now addressing all these components. Credible results have been obtained with coupled ocean-atmosphere models in the last decade for the interannual variability of the tropical Pacific Ocean and the overlying atmosphere. Similar results for the longer time scales and global components are only starting to become available at the time of this writing.

The community owes much to the remarkable foresight of the individuals and institutions whose persistence is responsible for the sets of long-term observations that are available to us today. Studies of climate variability on the interannual time scale have since been greatly stimulated by the cooperative efforts of dynamic meteorologists and physical oceanographers, sustained by long-term scientific coordination and by relatively healthy and stable governmental support, both national and international. The effort required to shed light on the global system's variability on decade-to-century time scales is proportionately greater, and atmospheric scientists, other oceanographers, glaciologists, biologists, and ecologists will be needed to join those meteorologists and oceanographers thus far engaged. The problem is further complicated by the fact that both natural and anthropogenic effects occur on the time scales of interest, and are hard to separate. It is thus important to establish an understanding of the natural variability, so that it can serve as a baseline against which possibly anthropogenic effects can be gauged. The NRC workshop papers were contributions toward this goal.

FINDINGS FROM THE WORKSHOP PAPERS

Essays at the beginning of each section of this volume discuss our progress in atmospheric observations, atmospheric modeling, ocean observations, ocean modeling, coupled systems, and climate proxy data. They set the stage for the 42 papers included. The preliminary results of these explorations of natural climate variability summarized in this volume are interesting and encouraging; the Climate Research Committee has extracted four main findings from them.

First, the relatively short instrumental record of climate (the last 50 to 100 years) does not represent a stationary or steady record. The papers in this volume show that climate's natural propensity for change has manifested itself through periodic variations, sudden shifts, gradual changes, and changes in variability. Furthermore, such changes do not appear to be unique to this century, as proxy records such as ice cores and tree rings attest. Climate fluctuations over the past few millennia or so will need to be analyzed in greater detail to establish a baseline against which future variations can be gauged.

Second, we are not yet certain why these changes in climate occur. Models must be used to test our hypotheses and to increase our understanding of the climate system. Models of the atmosphere, the ocean, and the coupled atmosphere-ocean system are beginning to yield insights into the causes of natural climate variations. For example, recent ocean modeling studies suggest that significant changes in the deep-water circulation may occur over time scales of decades to centuries, and that these changes might critically affect climate. A second ocean model finding is that the thermohaline circulation can oscillate between quasi-steady 'equilibrium' modes.

Third, the Climate Research Committee feels that systematically combining observations and models, and ensuring the long-term continuity and sufficient quality of the data,

will be critical to the assessment of climate variability and of the models that are used for climate simulation and prediction. The observations permit us to initialize, force, and diagnose models, providing reassurance that we are simulating the real world. Models not only serve as the measure of our understanding and the means of predicting, but are now good enough to help guide observation, monitoring, and data-management programs.

Finally, additional data are needed to supplement and expand the currently sparse and sporadic record of past natural climate variability. Proxy data, historical records, current operational data, research data, and model simulations can all contribute significantly to our understanding. In some cases they are available but under-utilized; in others they must be obtained through special programs or refinement of existing collection programs. Consistent data quality and uniform data-management practices are essential, and all climate data should be standardized and made available to researchers worldwide.

RECOMMENDATIONS

The comfort and livelihood of future generations on this planet demand the vigorous pursuit of new insights into the characteristics of decade-to-century-scale climate variability. Significant benefits will be realized if this research yields reliable prediction capabilities. Such benefits have already been achieved for prediction on interannual time scales. In countries such as Peru, successful predictions of El Niño have enabled fisheries and agricultural communities to introduce adaptive measures in order to minimize its negative impacts. Evidence to date suggests that the magnitude of climate change is often proportional to the time scale over which it occurs, so that over the longer scales the potential benefits of accurate prediction could be even greater. The four recommendations below for making credible prediction methods a reality are discussed in detail in Chapter 7.

1. Criteria must be established to ensure that key variables are identified and future observations are made in such a way that their results will yield the most useful data base for future studies of climate variability on decade-to-century time scales. Assurances of continuity and high data quality are essential.

2. Modeling studies must be actively pursued, using a variety of models, in order to improve our skill in simulating and predicting the climate state, and to assess potential modes of variability. Closer links between models and observational studies will be necessary. Continuing efforts must be made to assimilate the data sets in dynamical models, so that ultimately the added value of dynamical consistency and dynamical interpolation may be realized.

3. Records of past climate change, particularly those reflecting the pre-industrial era, must be actively sought out and refined as a source of valuable new data on the natural component of climate variability.

4. Climate data must be made freely available to researchers worldwide; data from many sources contribute to the solution of research problems.

1 INTRODUCTION

Variations in the earth's climate have had considerable impact on society—particularly agriculture, fisheries, water resources, and recreation—throughout recorded history. Such natural climate variability must be identified, quantified, and understood if ways are to be found to minimize its negative consequences and maximize its positive ones. In addition, human activities could significantly alter this natural variability, and indeed may already have done so. If we are to make informed decisions about our own future, it is essential that we assess the climate's sensitivity to a variety of factors, particularly on the decade-to-century time scales that are of most concern to human beings.

Many information sources, including instrumental records, visual observations, and paleoclimate data, bear witness to substantial variability in the earth's climate on time scales from years to centuries. The notion of a stationary climate on these time scales has thus become untenable. While variability in the modern climate regime is small relative to the formidable changes that characterize transitions from glacial to interglacial periods, the rate of change is often similar or even greater.

The relatively short instrumental record of climate clearly does not represent a steady background against which future variations can be gauged. Human-induced change will be difficult to assess unless the long-term natural variability of the climate system can be characterized. Natural variations with time scales of decades to centuries may well be masking anthropogenic climate changes that have already been effected, and will continue to do so. We must be able to recognize natural variability and its results if we are to make reasoned estimates as to whether a particular climate perturbation or trend is likely to have been induced by human activities, or simply represents a natural variation.

These very basic issues give rise to four critical questions concerning climate change on time scales of decades to centuries:

(1) Can we characterize the climate system's variability on these scales, over both space and time?

(2) Can the causes of such climate variability be isolated?

(3) Can such climate changes be predicted?

(4) Can changes induced by human activities be distinguished from natural variability on these time scales?

ASSESSING DECADE-TO-CENTURY-SCALE VARIABILITY

To assess climate variability on intermediate scales, one must have some idea of what it, and the forcings, look like. Several possible forms of climate variation can be seen in Figure 1. For example, climate variability may involve periodic change (Figure 1a), which is similar in nature to a daily or annual cycle, but in this case has a cycle lasting tens to hundreds of years or longer. The climate may also undergo a sudden shift (Figure 1b) from its current state to a different state, possibly one characterized by significantly colder or warmer conditions. It may show a steady warming or cooling until a stable state is reached (Figure 1c). And last, the climate may maintain what appears to be a steady state, when characterized by a specific variable such as mean annual temperature, but variations in some other measure, such as seasonal temperature, diurnal range of temperature, snow or ice coverage, or storm frequency, may indicate

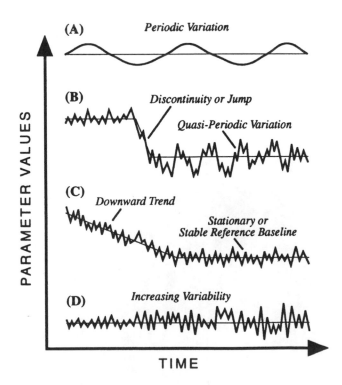

FIGURE 1   Types of climate variations. (From Marcus and Brazel, 1984; reprinted with permission of the Office of the State Climatologist for Arizona.)

that significant change has taken place (Figure 1d). Many of the papers in this volume contribute insights and data on the form of such climate variations.

The identification of the characteristics of climate variability involves several issues. Differentiating climate "change" from climate "variability" is a matter of the time scale. What appears to be a trend, in a single decade's recording, may reveal itself as fluctuating variability over a period of a century. The Dust Bowl period in the central United States in the 1930s represents a short, decadal-scale natural variation in climate for a specific region. In contrast, the "little ice age", which lasted from the 1400s to the 1800s, represents a variation on a time scale of centuries. (During the little ice age the decadal variability was typical of that observed during modern conditions.) An additional complication arises because variation that occurs on one time scale may influence changes on other time scales. The well-documented El Niño / Southern Oscillation phenomenon, which is characterized by variation over 4- to 7-year periods, seems to be affected by longer-time-scale variations and, in turn, to modulate the variations themselves.

Even after the time scale of interest has been defined, it may be difficult to recognize and classify natural variability as evidenced in modern instrumental observations, because anthropogenic change may already have contaminated them significantly. Limitations on data continuity (loss of records, changes in instrumentation, gaps, lack of data quality assurance) are also a serious handicap. Longer records, especially those based on proxy indicators of climate (i.e., indirect measures of climate such as the width of annual growth cycles in trees and corals), represent an excellent potential additional data set for estimating or extracting natural variability.

An additional difficulty in assessing variability is that change and variation are often characterized by strong spatial dependency. That is, while some region or regions of the earth may be experiencing significant climate variation over a given period, other regions may show virtually no change. Similarly, a specific variable (e.g., global temperature) may show a marked change between decades, while another (e.g., global pressure fields) may not show any significant difference. Where and how climate is measured can influence the findings and conclusions.

Our understanding of the causes of decade-to-century-scale variability is limited, in part because there are many possible causes, typically disguised by complex interactions. They include inherent variability (deterministic or random within individual components of the earth system), internal variability associated with the coupled ocean-atmosphere-cryosphere system, and forced variability such as solar variation and volcanic aerosol loading of the atmosphere. One key challenge is to isolate (if possible) the signatures of the various possible mechanisms so that we can discriminate between cause and effect as we examine the climate record.

Also, a greater understanding of the magnitude of the forcing required to produce observed variations will allow us to focus on the mechanisms most likely to contribute to climate variability. It seems likely that several mechanisms, operating in concert, are responsible for the variations apparent in the observed climate record.

Understanding the mechanisms that produce natural variability will require a hierarchy of climate models, including coupled models that are capable of addressing the interactions between the components of the earth system. The development of models that will yield simulations or predictions that can be verified and validated against the observational record, is another major challenge. Models that can describe the nature of decade-to-century-scale variations will serve not only as a measure of our understanding, but as a tool to increase this understanding. They provide our primary opportunity to predict climate variability, although ultimately prediction may be achieved through a variety of means. Given specified forcing scenarios, models may provide viable climate-response scenarios. But statistical characterization is another possible tool. For example, it would be useful to know that a particular extreme climate event (e.g., major regional flooding) tends to occur in clusters over a several-decade period rather than irregularly and infrequently.

Making the distinction between natural variations and human-induced changes—and ultimately predicting future changes—will require more complete characterization of the climate system's variability in both space and time, a greater understanding of the causes of decade-to-century-scale climate variability and the mechanisms by which it is produced, and the development of more advanced predictive models. The papers included in this volume reveal significant progress in identifying the behavior of many key components of the climate system, their climate signatures, their internal modes of variability, and their interactions with different components of the earth system. The coupling of observations and models, together with the availability of both long-term, consistent, and high-quality observations and proxy data records, will be critical to this assessment of climate and its change.

PURPOSE AND STRUCTURE OF THIS VOLUME

The importance of climate variability to society motivated the organization of a workshop by the Climate Research Committee (CRC) of the National Research Council's Board on Atmospheric Sciences and Climate. Not only have climate variations on decade-to-century time scales received significantly less attention than seasonal and interannual climate variations, or even the glacial/interglacial periods, but for nearly two decades natural variations have been overshadowed by much-publicized concerns about the possibility of long-term changes caused by increases in atmospheric concentrations of greenhouse gases. Both the workshop and the present volume cover a wide range of topics relevant to climate variability; they include the characteristics of the atmosphere and ocean environments as well as the methods used to describe and analyze them, such as proxy data and numerical models. The papers in this volume clearly demonstrate the range, persistence, and magnitude of natural variability as represented by many different climate indicators over the decade-to-century time scale.

This book is not a simple "workshop proceedings". Not only have the 42 papers included been refereed and edited, but each paper is followed by a brief critique by the workshop discussion leader and by a condensed version of the lively discussion that followed each presentation. In addition, each of the major sections is introduced by an essay that provides a perspective for the reader, and completes the picture sketched by the individual papers. Chapter 2 of this volume presents the atmospheric side of natural climate variability. It begins with papers dealing with observational data, and then discusses current atmospheric modeling. Similarly, Chapter 3 presents ocean observations first, and then ocean models. Papers about coupled atmosphere and ocean models appear in Chapter 4, and Chapter 5 introduces a variety of sources of proxy data, from lake beds to ice cores. Chapter 6 contains the conclusions the CRC has drawn from the papers, commentaries, and discussions included. The introductory essays, which outline the significance of the papers included, in conjunction with Chapters 1 and 6 constitute a portrait of our current understanding of many aspects of climate variability on decade-to-century

time scales. Chapter 7 presents the committee's recommendations for the direction of future research.

Despite the volume's breadth, it is still not comprehensive. Most of the current research effort has involved the oceans and the atmosphere. Other, less studied components, such as the cryosphere, land-surface processes, and biogeochemical cycles, may also play substantial roles in natural variability. For example, the cryosphere, through snow cover, sea ice, and land ice, can influence climate over subseasonal to millennial time scales. It has been implicated in a number of important climate processes, including ice–albedo feedbacks, thermohaline circulation, and abrupt climate change. Land-surface processes influence the hydrological cycle and surface albedo. The biosphere affects climate on a variety of time scales through its influence on surface moisture fluxes, albedo, and the carbon cycle. The roles of these components require considerable attention still; they are under-represented in current studies. The CRC nonetheless feels that this book will serve as a foundation upon which research into climate variation on decade-to-century time scales can build. The ultimate goals— determining the characteristics of natural climate variability, predicting climate changes on decade-to-century scales, and assessing the effects of human activities—will be realized only through the cooperation of scientists and national and international agencies in all the fields represented.

REFERENCE

Marcus, M.G., and S.W. Brazel. 1984. Climate Changes in Arizona's Future. Arizona State Climate Publication No. 1, Office of the State Climatologist, Arizona State University, Tempe.

2

THE ATMOSPHERE

Introduction

Humankind lives at the bottom of the sea of air, and climate change is perceived by us mainly as a change in the overall conditions of this sea's lower layers. The atmosphere lies at the heart of decade-to-century climate change: It filters the sun's rays as they reach the surface of the earth and as they are reflected again into outer space, and it is the principal medium of exchange of heat, water, trace gases, and momentum between the other components of the climate system—oceans, land surface, snow, ice masses, and the biosphere. The atmosphere and ocean are intimately coupled within the climate system, and are governed by similar physical laws. But the atmosphere has been explored in greater detail—in terms of available observations and of existing models—than any other component of the climate system. Thus it is natural to review the results of this exploration first, as we begin our examination of climate variability.

Understanding of natural phenomena proceeds through a sequence of observations, experiments, and models. Given the complexity of the climate system, laboratory experiments can reproduce only very incompletely the system's major aspects, and have not been included in the present volume. Atmospheric observations have led, in past centuries, to very simple, purely descriptive models of atmospheric motions. In the second half of this century, advanced computer models of the atmosphere have simultaneously benefited from an increase in the number and quality of observations, and stimulated vast field programs designed to verify model results and yield the new details necessary for improving the models. The separate treatment of observations and models in this chapter and the next is, therefore, only a matter of expository convenience.

The oldest instrumental records of atmospheric temperature—and, to some extent, precipitation—extend about 300 years into the past. The coverage and density of these measurements have grown more or less continuously, with a dramatic increase occurring in the 1940s and 1950s. This permits us to make a fairly informed assessment of past interannual variability, but we have considerably less confidence about interdecadal changes and only little or indirect information on the century-to-century time scale.

The Atmospheric Observations section below starts with a paper by H.F. Diaz and R.S. Bradley that addresses head-on the question of how different the climate of this century has been from those of previous ones. Proceeding from temperatures (which tend to be more uniform in space and time) to precipitation (which is considerably less so), S.E. Nicholson looks at the socio-economically critical issue of African rainfall variability on interannual and decadal time scales. J. Shukla provides complementary insight on the initiation and persistence of drought in the Sahel.

The role of snow cover in the radiation balance at the surface makes it an important player in climate variability; this role is discussed by J.E. Walsh. Variability and trends of both liquid and solid precipitation over North America are reviewed by P.Ya. Groisman and D.R. Easterling. Returning to temperature, a careful study of the difference between trends in daily temperature maxima and minima is presented by T.R. Karl and his colleagues. C.D. Keeling and T.P. Whorf then analyze the decadal oscillations in global temperatures and in atmospheric carbon dioxide. These oscillations are at the heart of understanding natural variability on this time scale; they are also covered later in this section in the essay introducing atmospheric modeling.

The Southern Hemisphere has less instrumental coverage, in space and time, than the Northern Hemisphere. While many of the earlier papers in the section concentrate on the latter, D.J. Karoly describes the observed variability in the atmospheric circulation south of the equator.

Atmosphere-ocean interaction is involved in the last two papers of this section. C. Deser and M.L. Blackmon review atmospheric climate variations at the surface of the North Atlantic, while D.R. Cayan and his associates study a general-circulation model simulation of the Pacific Ocean, driven by observed surface fluxes.

Atmospheric models have a relatively long tradition in the climate community, going back to the 1950s. They span a spectrum, from simple radiative-convective models in one vertical dimension, through energy-balance models in one and two horizontal dimensions, to fully three-dimensional general-circulation models. These models are of interest in their own right as important tools for investigating climate change, either by themselves or coupled to models of other climate subsystems. They also provide an instructive example of the creation of a full suite of models for intercomparison and validation; their development is only now being followed, more or less closely, by the modeling enterprise in oceanography, hydrology, and other disciplines contributing to the climate-change enterprise.

Modeling and observations are closely linked by the reciprocal problems of simulating the observed variability and validating the existing models. The Atmospheric Modeling section is thus appropriately begun by T.M.L. Wigley's and S.C.B. Raper's paper on modeling and interpreting paleoclimate data, with an emphasis on the greenhouse effect. G.R. North and K.-Y. Kim apply classic time-series analysis techniques to climate-signal detection. R.S. Lindzen then considers a few themes in these two papers from complementary viewpoints.

A number of causes for climate variability on time scales of decades to millennia are reviewed by D. Rind and J.T. Overpeck. They emphasize the modeling approach to a study of these causes, while J.M. Wallace addresses similar issues by analyzing the climate record.

The two sections of this chapter are introduced by essays, one by T.R. Karl and the other by M. Ghil. They provide an overview of the current state of the fields of atmospheric observations and atmospheric modeling, reviewing those topics not covered by the workshop and offering a perspective on the papers included. Following each paper, a commentary by the discussion leader and a condensation of the spirited discussion that took place at the workshop shed additional light on our current knowledge in both areas.

The ocean is examined in Chapter 3, and atmosphere-ocean interaction is explored further in Chapter 4. Also of interest are the proxy records of climate variability, such as tree rings and coral reefs, which are covered in Chapter 5. The sponsoring committee's conclusions, which draw on the material in this chapter and in the other three, appear in Chapter 6.

Atmospheric Observations

THOMAS R. KARL

INTRODUCTION

During the 1970s, the 1980s, and into the 1990s atmospheric scientists have accelerated research aimed toward identifying and explaining the presence of decade-to-century scale climate fluctuations. The motivation for special consideration of atmospheric climate fluctuations on these longer scales has its roots in the numerous major climate events whose occurrence is so well documented in the instrumental record. Probably the best-known of these fluctuations occurred about 1970, when rainfall suddenly decreased over the Sahel; the causes of this jump are discussed in this section by Nicholson (1995) and Shukla (1995). Other examples include the Dust Bowl years in North America during the 1930s, the diminished intensity of tropical storms affecting the East Coast of the United States during the 1960s, 1970s, and much of the 1980s, and the wet weather of the 1970s and 1980s over much of the United States.

As access to climate records has broadened during the past few decades, researchers such as Hurst (1957), Mandelbrot and Wallis (1969), Mitchell (1976), Douglas (1982), and Lorenz (1986) have presented evidence suggesting that the notion of a static climate is no longer tenable, even on less-than-geological time scales. We have come to realize that 30 years of data, the length of time that has been used to compute temperature and precipitation "normals" (Court, 1968), is inadequate to define climate. It does not provide us with sufficient information either to minimize adverse climate impacts within such sectors as energy, water supply, transportation, environmental quality, construction, agricul-

ture, etc., or to maximize the availability of the climate-governed resources, such as water and energy, on which both natural and man-made systems depend.

Several climate fluctuations that have affected the United States serve to illustrate this point. Beginning about 1975, the interannual variability of mean winter temperatures and total precipitation averaged across the contiguous United States substantially increased (Figures 1a and 1b); this increase persisted at least through 1985. In contrast, the interannual variability had been very low during the previous 20 years. The mean temperature increased dramatically over the United States during the 1930s, coincident with a large decrease in summer precipitation (Figures 1c and 1d), before returning to more typical conditions. The wetness of the 1970s and the first half of the 1980s, which is clearly evident in Figure 1e, resulted in record-setting high lake levels and caused considerable economic damage and human suffering (Changnon, 1987; Kay and Diaz, 1985). Over the past decade another climate fluctuation has been evident over the United States; as Figure 1f shows, temperatures increased in a jump-like fashion during the early 1980s. Interestingly, the temperature discontinuity that is apparent between January and June is not reflected by the temperatures during the second half of the year.

Due to the limited span of the instrumental climate record, evidence for climate fluctuations on decade-to-century time scales is biased toward higher frequencies. (The situation is much worse in the oceans; as Wunsch (1992) points out, the absence of comprehensive oceanographic data has led many oceanographers to focus on identi-

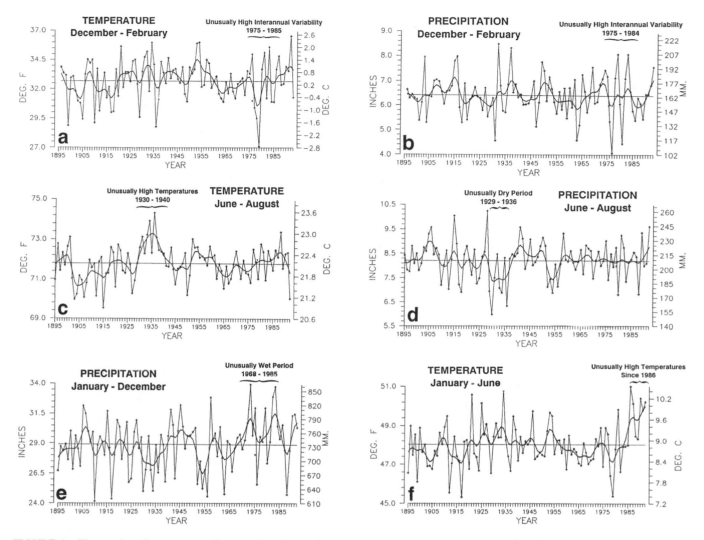

FIGURE 1   Time series of area-averaged seasonal and annual mean temperature and total precipitation for the contiguous United States. The smoothed curve is a nine-point binomial filter. The horizontal line reflects the mean over the period of record.

fying a mean climate state.) Nonetheless, there is evidence that climate fluctuations on longer time scales occur throughout the world (Jones and Briffa, 1995; Diaz and Bradley, 1995). As Karoly (1995) shows, there is a dearth of the data that would permit the identification of decadal-scale circulation variability in the Southern Hemisphere, but analysis of the longer-term surface data available suggests that significant climate fluctuations persist at least through decadal time scales.

The forcing agents of many climate fluctuations may have their origin in either anthropogenic or natural factors, or both. These fluctuations are important in terms both of their socio-economic and biophysical effects and of our need to distinguish between natural climate variations and anthropogenic climate changes. In analyzing the climate record, the dangers of "data dredging" must be kept in mind. As access to climate data increases, the chance of

finding trends and variations that appear to be significant also increases. For this reason decade-to-century-scale climate fluctuations have been referred to as the "gray area of climate change" (Karl, 1988). For instance, Keeling and Whorf (1995) find evidence for decadal fluctuations in the global temperature record that can be reproduced by assuming the existence of two oscillations with small differences in frequency that beat on time scales of about 100 years. The search for an explanation of this statistical result is a good example of the challenge presented by the existence of these decadal fluctuations.

IDENTIFYING CLIMATE FLUCTUATIONS

The instrumental record of atmospheric and related land and marine observations is fragmentary until at least the middle of the nineteenth century. Moreover, virtually all of the observations used to study climate fluctuations are derived from observing programs that were developed not for this purpose, but rather to support day-to-day weather forecasting. Since the day-to-day variability of the atmospheric system in much of the world is far larger than the decadal climate variability, identification of multi-decadal climate fluctuations is often hampered by the poor quality and lack of homogeneity of the observations. For example, Groisman and Easterling (1995), in their study of the variability and trends of precipitation over North America, painstakingly document the many discontinuities and false jumps in the climate record that arise from changes in observing practices. These changes, if ignored, can and do overwhelm important long-term precipitation fluctuations.

Other atmospheric quantities besides precipitation are also affected by problems in measurement. In this section, Robinson (1995) points out the large disparity in snow-cover extent between a NASA data set using microwave measurements and a NOAA data set using visible imagery. Karl et al. (1995) devote considerable discussion to jumps and trends introduced into the surface temperature record by such factors as urban heat islands, changes in irrigation practices, differences in instrumentation, and station relocations. Cayan et al. (1995) use surface marine observations to calculate the latent and sensible heat fluxes that are used as forcing agents of the sea surface temperature field in the North Pacific. Their analysis includes the basin-wide climate jump that began about 1976-1977. To account for known systematic biases related to trends in the wind field and the sea surface and marine air-temperature and moisture lapse rates (Ward, 1992; Cardone et al., 1990; Wright, 1988; Ramage, 1987), the global trend of each of these quantities is removed. As mentioned in Zebiak's commentary on Cayan's paper, it is unfortunate that such adjustments to the data are necessary.

Often the corrections and adjustments that must be applied to a climate record are of such magnitude that it is difficult to be confident that the resulting time series adequately reflects the climate fluctuations. For example, Figure 2 illustrates the significant adjustments required to calculate global temperature fluctuations since the nineteenth century. Through the use of other data bases (e.g., changes in snow cover, alpine glaciers, or sea level) and the isolation of various components of the surface temperature record (e.g., marine air temperature, sea surface temperature, and land temperature), it is possible to gain more confidence in the adjustments applied. Each of the data sets has distinctly different problems related to long-term homogeneity and data quality that require independent adjustment procedures. When these independent data sets provide a physically consistent picture of decade-to-century climate fluctuations, their agreement can be very compelling. In fact, the analysis of quasi-periodic oscillations of surface winds, pressure, and ocean and marine air temperatures

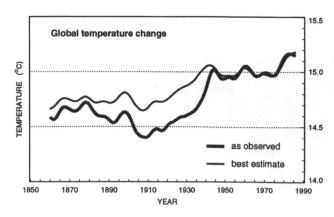

FIGURE 2   Smoothed global surface temperature variations, as derived from the original observations without adjustments for inhomogeneities in the climate record, compared to the same observations after adjustments for inhomogeneities as described by the Intergovernmental Panel on Climate Change (IPCC, 1990, 1992).

by Deser and Blackmon (1995) relies entirely on physical consistency among independent variables.

In order to improve our ability to discern climate fluctuations on decade-to-century time scales within the existing climate record, some federal agencies such as NSF, DOE, and NOAA are supporting data archeology efforts. Data archeology is the process of seeking out, restoring, correcting, and interpreting data sets. Such efforts are critical to identifying and understanding longer-scale climate fluctuations, since they often turn up information that reveals a pattern not otherwise obvious.

The Comprehensive Ocean-Atmosphere Data Set (COADS), which is just one of several major data-archeology efforts, provides a good example of the type of benefits that can be expected from such efforts. Now, several years after the project's inception (Woodruff et al., 1987), scientists are identifying decadal-scale variations and changes in many climate elements previously thought to be too uncertain to document with any confidence (London et al., 1991; Flohn et al., 1990; Parungo et al., 1994).

Interestingly, up to the present, satellite data have not figured prominently in the analysis of multi-decadal climate variability. Their short observing history and a lack of temporal homogeneity have hampered efforts to use them. Certainly, there are notable exceptions, such as NOAA's snow and sea-ice products and the Spencer and Christy (1990) microwave sounding-unit data. However, major temporal inhomogeneities in data sets, like those identified in the International Satellite Cloud Cover Project (ISCCP; see Klein and Hartmann, 1993), threaten the usefulness of much of the multi-decadal satellite data. The two major challenges over the next few decades for the United States will to be to ensure that global-change satellite-research projects such as EOS provide homogeneous data over their planned lifetimes, and to lay the foundation for converting them into a long-term operational observing systems.

Future observations will require more attention to data quality, homogeneity, and continuity if we are to understand the nature of decade-to-century-scale climate variability. Currently under consideration is a Global Climate Observing System (GCOS)[1] that includes observations of terrestrial, atmospheric, and oceanic aspects of climate. Since GCOS would be built around existing observing networks and environmental problems, it is essential that scientists effectively convey long-term climate-monitoring needs to the agencies sponsoring space-based and conventional observing systems. Among the climate-monitoring issues that should be addressed in the near future are stability of network sites, intercomparability of instruments, and increased sampling in data-sparse regions. Two concerns for both present and future observations are the better collection and documentation of metadata on observing instruments and practices as well as processing algorithms, and improved data-archiving practices and data-management systems. All these will increase the value of climate monitoring for decade-to-century-scale research; most essential, of course, is assurance of a long-term commitment to observational objectives.

THE FORCING AGENTS OF CLIMATE FLUCTUATIONS

The climate record provides the key to developing, refining, and verifying our hypotheses regarding the forcing agents responsible for climate fluctuations, be they anthropogenic or natural, internal or external to the climate system, global or regional, or persisting one or many decades. Diaz and Bradley (1995) suggest that the many decade-to-century climate fluctuations evident in both observations and proxy records may indeed have natural origins, since similar fluctuations seem to occur in the climate record both before and after humans became capable of modifying climate.

Often physically based models are the best means of testing our hypotheses about the cause of the fluctuations, as described in the modeling sections of this volume, but in addition they can be effectively used to help discern the physical consistency of apparent climate fluctuations and change. Cayan et al. (1995) demonstrate this approach; they use an ocean general-circulation model to help explain the climate jump over the North Pacific, as documented by Trenberth (1990; Trenberth and Hurrell, 1995, in this volume). Cayan et al. show how the atmosphere and ocean

can act together to maintain decadal-scale climate fluctuations on a large spatial scale.

Potentially important factors in explaining climate fluctuations on decade-to-century time scales are land-surface and atmospheric feedback effects. Walsh (1995) provides ample evidence that snow cover has important feedback effects on the climate system on short (daily and interannual) and long (thousands of years) time scales, but on the time scales of interest to us the impact of snow is still not well understood. However, Nicholson (1995) and Shukla (1995) present evidence that the feedback between land-surface characteristics and the atmosphere has led to the prolonged drought in the Sahel. Karl et al. (1995) document an asymmetric increase of the mean maximum and minimum temperatures over many portions of the global land mass. Although they cite a number of potential causes of this multi-decadal trend, such as increases of anthropogenic atmospheric sulfate aerosols (Charlson et al., 1992) and greenhouse gases, empirical evidence suggests that, at least in some regions, observed increases in cloud cover play an important role in modulating the surface temperature. The forcing responsible for the increase in cloud cover remains unknown.

There are many important aspects of climate forcings and associated responses that cannot be covered here. Of particular relevance are the known changes in solar irradiance associated with the sunspot cycle. Recently Friis-Christensen and Lassen (1991) have used the length of the sunspot cycle to explain the decadal fluctuations and trends of land temperatures. Although a linear response of the surface temperature to the sunspot cycle length implies some unlikely responses of the temperature record in the early part of the time series (Kelly and Wigley, 1992), it is clear that changes in solar irradiance must continue to be monitored as a potential source of global climate fluctuations or change.

Recently, Elliott et al. (1991) have found evidence for an increase of tropospheric water vapor since 1973, leading to an enhanced greenhouse effect. However, the many types of observing and data-processing inhomogeneities in the upper-air moisture record make it immensely difficult to separate spurious trends and discontinuities from the true climate signal, even when the signal may be as large as a 10 percent increase in specific humidity.

A critical atmospheric quantity affecting surface temperature is the variability of cloud cover—its type, height, and spatial distribution. Using the ISCCP data base, Hartmann et al. (1992) showed the net forcing of various cloud types as a function of season and latitude. A recent study by Rossow (1995), however, reveals that this data set has serious biases that make it inappropriate for decadal-scale climate assessments. Meanwhile, conventional in situ data, analyzed on a national basis by a number of researchers, show a widespread general increase in total cloud cover

---

[1] GCOS will develop a dedicated observation system designed specifically to meet the scientific requirements for monitoring the climate, detecting climate change, and predicting climate variations and change (ICSU/UNEP/UNESCO/WMO, 1993).

over the past several decades (Dutton et al., 1991; Henderson-Sellers, 1990; Karl and Steurer, 1990; London et al., 1991; McGuffie and Henderson-Sellers, 1989; Parungo et al., 1994). Clearly, a much more focused effort is required to better understand decadal and multi-decadal fluctuations of cloud cover, since they often have very important feedback effects on other climate quantities. For instance, cloud feedbacks work differently during the day from the way they do at night. They therefore differ at high and low latitudes, making generalizations speculative.

Also, there is no discussion in this volume of the world's glaciers, in spite of their known tendency to fluctuate markedly on decade-to-century time scales. Long-term trends of climate change are integrated by mountain glaciers, and during the past century mountain glaciers all over the world have been declining (IPCC, 1990). The World Glacier Monitoring Service (1993) summarized these changes over the past several decades, and the USGS Satellite Image Atlas of the World discusses observed variations of hundreds of glaciers over the past several centuries.

CONCLUSION AND RECOMMENDATIONS

Identification and explanation of the forcings and feedbacks responsible for decade-to-century-scale climate fluctuations are essential to distinguishing between natural and anthropogenic impacts on the climate system, as well as to developing any predictive skill with respect to these phenomena. Further progress in understanding this gray area of climate change will depend upon our ability to address three topics. First, we must be able to document climate fluctuations without spurious discontinuities and trends. Second, we must ensure that we are adequately monitoring potential forcings and feedbacks internal and external to the climate system, so that we will have the data necessary for testing hypotheses about their operation. Last, data analysts must work closely with modelers (and vice versa) to test the hypotheses we formulate on a variety of climate models, ranging from simple one-dimensional climate models to complex coupled ocean-atmosphere general-circulation models. The re-analysis modeling projects of the United States (Kalnay and Jenne, 1991) and the European community are likely to enhance our confidence in both how and why climate has varied on decade-to-century time scales. (It must be emphasized, however, that the value of any re-analysis effort can be jeopardized by the use of data that are biased or, most troublesome, inhomogeneous in time.) Only by advancing in all three of these areas will we be able to overcome our ignorance about multi-decadal climate fluctuations and make predictions with any degree of assurance.

Documenting Natural Climatic Variations: How Different is the Climate of the Twentieth Century from That of Previous Centuries?

HENRY F. DIAZ[1] AND RAYMOND S. BRADLEY[2]

ABSTRACT

Changes in decadal-mean surface temperature and its variance for different land areas of the Northern Hemisphere are examined. In the last 100 years, changes in surface air temperature have been greatest and most positive in the period since about 1970. Interannual variability, particularly at the largest spatial scales has also increased, although it differs according to regions. The most unusual decade of the last 100 years in the contiguous United States may have been the 1930s, although that of the 1980s is probably a close second, and in some regions perhaps the most anomalous. Both the 1930s and the 1980s experienced significant warming together with enhanced climatic variability.

To incorporate a longer-term perspective than is obtainable from the modern instrumental record, we used summer-temperature reconstructions based on tree-ring records, and on $\delta^{18}O$ ratios extracted from different ice-core records. Although none of these records is a simple or even direct temperature proxy, we present them as general indicators of prevailing environmental temperatures. The data were averaged by decades in order to focus on decadal-scale variability. With the exception of the data from tropical ice cores, the proxies indicate that the recent decades were not very unusual, either in regard to the mean or in terms of increased variability. While seasonal and annual temperature changes in the last two decades have been rather large in most areas of the Northern Hemisphere, the available paleoclimate evidence suggests that in many areas there have been decadal periods during the past several centuries in which reconstructed temperatures were comparable to those of the 1970s and 1980s, with climatic variability as large as any recorded in recent decades. Natural variability on decadal time scales is comparatively large—typically about half as large as the interannual variance.

[1]NOAA Environmental Research Laboratories, Boulder, Colorado
[2]Department of Geosciences, University of Massachusetts, Amherst, Massachusetts

INTRODUCTION

Considerable effort has been expended over the last 25 years to improve our knowledge and understanding of climate processes and mechanisms associated with changes in climate (Saltzman, 1983; Trenberth, 1992). A major impetus for much of the recent climate research has been the global and regional climate projections made with physical climate models (the so-called general-circulation models, or GCMs) for a doubling of carbon dioxide concentration in the atmosphere.

Most assessments of the climatic impact of increases in the "greenhouse" gases (CO_2, methane, nitrous oxide, chlorofluorocarbons) have concluded, more or less consistently, that global temperatures should rise (under a doubling of the most abundant of these greenhouse gases, CO_2) from 1.5°C to 4.5°C (NRC, 1982; Bolin et al., 1986; IPCC, 1990; Schlesinger, 1984, 1991). Such changes, should they occur, will be superimposed on the natural variability of the climate, which is quite large at all time scales (Karl et al., 1989). Depending on a number of assumptions, the amount of global warming that should have been realized by now appears to be more consistent with the lower end of the above estimates (Bloomfield, 1992).

Karl et al. (1991b) performed a variety of tests using output from three different GCMs to ascertain, given their different climate sensitivities, when a statistically significant greenhouse signal might be detected in the central United States. They concluded that it was likely that a greenhouse signal had been masked, to date, by natural climate variability, and that it would likely take another two to four decades before the greenhouse signal in temperature and precipitation might be unambiguously detected in this region.

We should note the possibility that other human-induced factors may be acting to counteract the radiative effects of increased greenhouse-gas concentrations in the atmosphere. Recent work has indicated that sulfate aerosols, which act to increase the planetary albedo (Charlson et al., 1992; Hansen and Lacis, 1990; Wigley, 1989, 1991) may have counteracted to some degree the enhanced infrared warming (see Michaels and Stooksbury, 1992). These and other factors, whether anthropogenic in origin or not (e.g., changes in vulcanism), will undoubtedly "muddy" the climate picture and may make it more difficult to identify unequivocally a contemporary greenhouse signal.

In this paper, we consider two aspects of climate-change detection that we feel bear strongly on the issue of natural versus anthropogenic climate variability. One aspect of the problem is, in effect, at what point one rejects the null hypothesis (no climate change) and accepts the premise that the climate of the last few decades belongs to a different sample population. The difficulty there arises because we are evaluating a relatively short observational record with the knowledge that the climate has fluctuated in the past few

centuries by an amount that may be of the same magnitude as the fluctuations observed in the recent record (see Bradley and Jones, 1992).

A second aspect of the problem of climate change versus natural variability, which we consider here, is temporal changes in that variability, i.e., in the variance. Changes in climatic variability are important, since they are likely to have a greater effect on a society's ability to mitigate and adapt to climatic changes than a slow alteration in the mean climate patterns. We have examined a variety of observational records of various lengths (typically 100 to 150 years), which are representative of different spatial scales (from hemispheric to regional basins). We will focus on decadal time scales, since one could argue that climatic changes will be better gauged at time scales that to some extent average out the high-frequency climatic "noise" associated with seasonal variability and other air-sea processes operating on annual time scales (e.g., the El Niño/Southern Oscillation system). We compare these instrumental series with a suite of high-resolution climate proxy records—namely, tree rings and oxygen isotopes extracted from ice cores—spanning the last 300 to 800 years. Our aim is to give the reader some feeling for the range of this intermediate-frequency (decade-to-century-scale) climatic variability, as well as some appreciation of the uncertainties inherent in the existing records (instrumental, historical, and paleoenvironmental).

We note that the "fingerprint" climate-change-detection technique (Barnett, 1986; Barnett and Schlesinger, 1987; Barnett et al., 1991) provides a very useful methodology for testing a hypothesis (the GCM climate-change projections) against observations. We will attempt here to highlight some aspects of climatic variability at decadal and longer time scales and will re-emphasize the conclusion of Barnett et al. (1991), who noted that the existence of a high degree of "unexplained" interdecadal climatic variability (see also Ghil and Vautard, 1991, and Karl et al., 1991b) will greatly complicate the detection of a greenhouse climate-change signal.

THE MODERN (INSTRUMENTAL) RECORD

Characteristics of Decadal-Scale Changes of Area-Mean Temperature Indices

The available data suggest that during the past century decadal-scale changes in global-scale mean annual temperature are on the order of 0.1°C to 0.3°C (see Folland et al., 1990). Figure 1 illustrates the annual and seasonal temperature changes for the Northern Hemisphere land areas. The plotted data, from Jones et al. (1986a) with subsequent updates, are consecutive 10-year area-weighted averages from 1891-1900 through 1981-1990 of gridded temperature anomalies referenced to the 1951-1970 period. On this "dec-

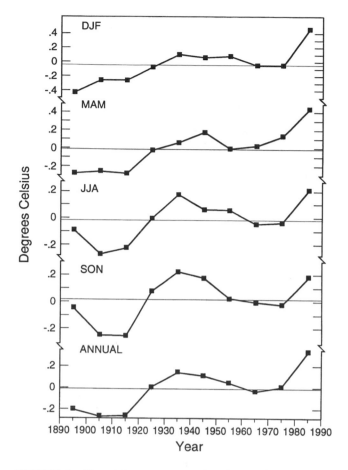

FIGURE 1 Decadal-mean seasonal and annual temperature anomalies (in °C) for land areas of the Northern Hemisphere. Data is from Jones et al. (1986a), with subsequent updates.

to be strongly driven by ocean-atmosphere heat exchanges resulting from changes in the ocean's thermohaline circulation.

Because of the changes in the earth's radiative balance imposed by human activities during the past century, climatologists have been trying to detect climatic-change signals that hitherto have typically been associated with time scales of millennia. The climate record available for such studies, even with the aid of high-resolution paleoenvironmental records, is sufficient to define only continental to hemispheric changes on decadal time scales, and perhaps regional (less than about 10^5 km^2) changes on century time scales.

In the United States, perhaps the singular climatic event of this century was the severe, widespread, and persistent drought of the 1930s. This event has been widely chronicled in both scientific (e.g., Skaggs, 1975) and popular literature. Other climatic events that persisted for more than one year and affected relatively large areas of the United States were the drought of the 1950s in south central areas of the United States (Chang and Wallace, 1987), and the drought of the 1960s in the Northeast (Namias, 1966). Periodic drought of variable duration has affected all areas of the contiguous United States to varying degrees (Diaz, 1983). Indeed, multi-year climate anomalies are a characteristic feature of the climate of the United States and of other parts of the world (Karl, 1988; Folland et al., 1990).

In Figure 2, decadal-mean seasonal (December-February for winter, March-May for spring, etc.) and annual temperature anomalies for various areas of the Northern Hemisphere are examined. Although Eurasia and North America (Figure 2a) have broadly similar seasonal trends over this period, differences can be noted in all seasons as well as in the annual averages. Figure 2b illustrates decadal temperature changes for the contiguous United States for the last century. The data used are from the adjusted divisional values (solid curve) described in Karl et al. (1984b). For comparison, the equivalent Jones et al. (1986a) gridded values are given as the dashed curve in Figure 2b. The decadal averages for the lower 48 states exhibit much smaller long-term trends than those noted for the continental-scale regions. Figures 2c and 2d illustrate the decadal temperature changes recorded in subregions of the United States, namely, the eastern and western halves of the United States (divided roughly along the 100th meridian), the state of Colorado, and a single state climatic division in Colorado where the city of Boulder is located. Again we note regional differences that can depart substantially from the temporal behavior of the larger spatial averages.

We have tabulated the mean temperature change from one calendar decade to the next over the 100-year period 1891 to 1990 for each of the above regions. No particular pattern emerges, except that at the largest space scales, the greatest decade-to-decade warming occurs mostly from the 1970s to the 1980s, whereas the time of the largest such

adal" time scale, the warming of the 1980s relative to the 1970s (0.32°C for the hemisphere) is comparable to the warming that occurred in the 1920s relative to temperatures in the 1910s (0.26°C). Some seasonal differences can be noted, with northern summer and fall showing the least warming trend and winter and spring showing the greatest warming trend over the last 100 years.

The twentieth century contains many examples of decadal-scale climatic variations of sufficient magnitude to have significantly affected societies. The two-decade-long drought in the African Sahel is perhaps the best-known example. The words "climate change" have become overused, and climatic variations from seasonal to multi-year are often lumped together. Until the mid-1970s, "climate change" was used to refer to climatic changes on time scales of 10^4 to 10^6 years (essentially, Milankovitch time scales). Variations of less than 10 years were generally considered to be part of the natural climate noise associated with interannual variations of the various components of the climate system and their nonlinear interactions, whereas variations on time scales of 10 to 10^3 years are now thought

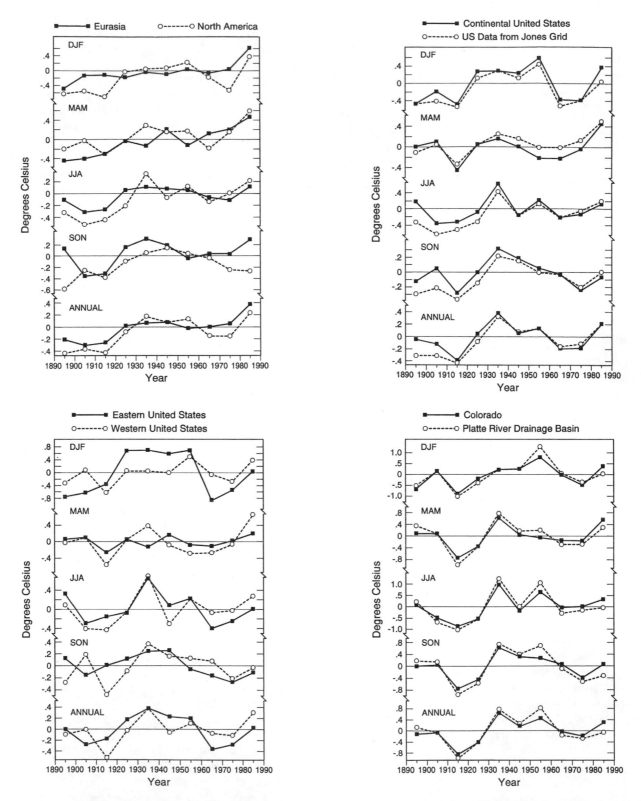

FIGURE 2   Decadal-mean seasonal and annual temperature anomalies (in °C). (a) For the Eurasian (solid line) and North American (dashed line) land masses. (b) For the contiguous United States, using data from the National Climatic Data Center described in Karl et al. (1984b) (solid line) and from the gridded temperature anomaly fields of Jones et al. (1986a) (dashed line). (c) For the eastern (solid line) and western (dashed line) United States. (d) For the state of Colorado (solid line) and the Platte River climatic division of Colorado (dashed line).

changes varies for the smaller regional averages, with most areas of the contiguous United States experiencing their greatest decadal changes during the first half of this century.

Recent studies have documented some rather pronounced and relatively rapid shifts in the climate system that occurred on decadal time scales. In particular, the atmospheric circulation in the extratropical North Pacific underwent a major shift in the mid-1970s that lasted over 10 years (Trenberth, 1990; Ebbesmeyer et al., 1991). This change was accompanied by a tendency for climate patterns in the tropical Pacific to exhibit conditions reminiscent of El Niño/Southern Oscillation (ENSO) warm-event conditions—weaker trades, warmer equatorial sea surface temperature (SST), anomalous rainfall patterns, etc.—relatively more frequently than during previous decades. Gordon et al. (1992) have also reviewed some of the relatively abrupt changes in water properties and surface climate in the Atlantic Ocean that have occurred in the past several decades, including sudden changes in SST anomalies in the South Atlantic compared to those of the North Atlantic, and the so-called "great salinity anomaly" in the North Atlantic from about the late 1960s to the early 1980s (Dickson et al., 1988). Certainly the sudden decrease of rainfall in the Sahel of Africa, a feature that has lasted for two decades, is a good example of how the climate can undergo significant, sudden, and prolonged change over relatively large spatial scales.

Clearly, increased knowledge of the behavior of climatic variability, from the interannual through decadal and century time scales, is needed to improve our assessments of any future changes in climate from regional to global scales. Indeed, as was noted by Ghil and Vautard (1991), the ability to distinguish a warming trend from natural variability is critical for separating out the greenhouse-gas-induced signal.

As we noted at the outset, an important element of climate that is too often overlooked is the variance or the characteristic variability of climatic means. In determining climate impacts, the importance to society of short- to medium-term climatic instability (i.e., climate fluctuations occurring on interannual to interdecadal time scales) is perhaps equal to or greater than that of slow changes in the background mean. Below we have analyzed some aspects of low-frequency changes in the temperature variance of different regional means.

Changes in Temperature Variability

We have examined contemporaneous variation of two measures of temperature variability, the standard deviation in running 15-year segments, and decadal means of the root-mean-square differences between successive yearly values of seasonal and annual regional temperature anomalies. The latter index is defined as

$$\mathrm{Ind}V \equiv \left[(N-1)^{-1} \sum_{i=1}^{N-1} (x_{i+1} - x_i)^2 \right]^{1/2} . \qquad (1)$$

Since the x_i are deviations from a reference mean and have approximately zero mean,

$$\mathrm{Ind}V = [2\sigma^2(1 - r_1^2)]^{1/2} , \qquad (2)$$

where σ^2 is the series variance, and r_1 is the autocorrelation of the time series with a lag of 1 year. We will use this measure, rather than the standard deviation, as the key index of interannual variability. Note that this index, $\mathrm{Ind}V$, amplifies changes in the high-frequency part of the variance spectrum.

Figures 3 and 4 illustrate the changes in the interannual variability of the regional groupings discussed earlier. The curves differ from one season to another and from one region to another. For the largest continental-sized regions, there appears to be an increase in interannual variability in the last couple of decades. At smaller spatial scales, there does not seem to be a consistent trend; instead, the interan-

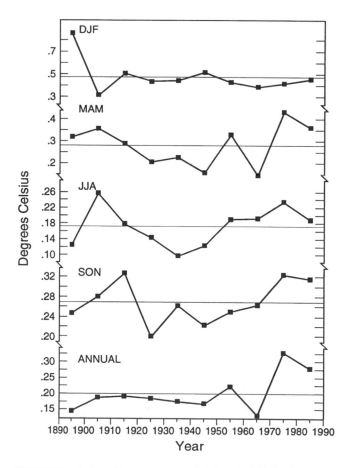

FIGURE 3   Index of interannual variability (in °C) for the Northern Hemisphere land areas. Values correspond to decadal means of the root-mean-square difference between successive seasonal and annual temperature anomalies over the last 100 years.

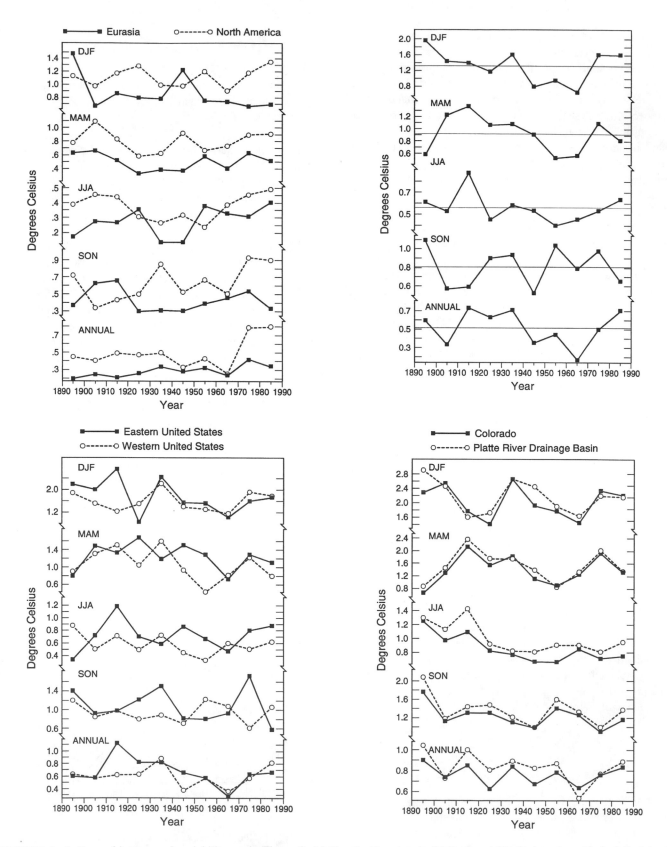

FIGURE 4 Indices of interannual variability, as in Figure 3. (a) For the Eurasian (solid line) and North American (dashed line) land masses. (b) For the contiguous United States, using data from the National Climatic Data Center. (c) For the eastern (solid line) and western (dashed line) United States. (d) For the state of Colorado (solid line) and the Platte River climatic division of Colorado (dashed line).

nual variability wanes and waxes over the last 100 years. The corresponding curves for the standard deviation (not shown) reveal similar patterns, with increases in variance at the largest spatial scales with time but no consistent trends in the standard deviation of the smaller regions. As one would expect, there is a well-defined log-linear relationship between the magnitude of the interannual variability and the size of the area comprised by the various indices (see Figure 5). The inherently higher variability at smaller regional scales should be kept in mind when interpreting the paleoclimate record, which is based on limited regional samples.

The question of whether a warming climate will exhibit increased short-term variability is still open. Previous studies have been generally inconclusive on this subject (van Loon and Williams, 1978; Diaz and Quayle, 1980). Certainly, there is no consistent relationship between interdecadal changes in average temperatures and interdecadal changes in variability. Climatic variability may increase, decrease, or stay the same, in response to an arbitrary change in decadal mean temperatures. On the basis of the above results, we can say that, considering surface temperature variations during the last century, there has been a recent increase in variability at large spatial scales (continental to hemispheric-scale averages). This increase in variability is also concurrent with relatively large temperature increases over those areas. However, this is not the case for earlier warming episodes.

THE PRE-INSTRUMENTAL PERIOD

Prior to the late 1800s, reliable quantitative data on climate variations are sparse, and conclusions about the magni-

tude of these changes on large spatial scales necessarily contain significant degrees of uncertainty. There is ample evidence that multi-decadal temperature changes occurring over regions the size of Europe and China have been at least of order 1°C (Bradley and Jones, 1992). This suggests that it is quite plausible, if not likely, that temperatures as warm as those prevailing since the end of the Second World War may have been experienced at regional spatial scales of about 10^5 to 10^6 km^2 for periods of one to several decades during some portions of the past thousand years under what amounts to natural (i.e., with negligible human influences) conditions.

Below we present some examples of long-term variability in a set of climate-sensitive paleotemperature records. The climate indices used here are discussed in several chapters of Bradley and Jones (1992). A listing of them, together with the source references and periods of record, appears in Table 1; each of them has been assigned a number for use in later tables comparing their data. Because of its very long record, the instrumental temperature series for central England (Manley, 1974) is considered in this section for comparison with other paleoclimate indices.

Tree-Ring Indicators

There is an extensive body of work relating climate variations to growth changes in particularly climate-sensitive trees (e.g., Fritts, 1976; Fritts et al., 1979; Hughes et al., 1982; Cook and Kairiukstis, 1990). We have selected a suite of climate-sensitive tree-ring records to study possible changes in climatic variability. Two main considerations were applied in the selection of these indices. First, we chose

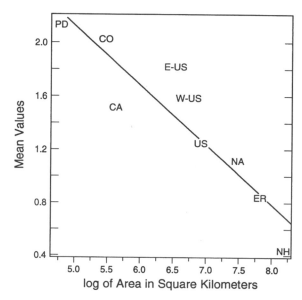

FIGURE 5 Interannual variability (in °C) as a function of the regional area. Values shown are for northern summer (left panel), and winter (right panel). Area is plotted as $\log_{10}$ values.

TABLE 1   Temperature Indices Used in This Study

Type of Record	Time Coverage	Reference	Index No.*
Ice-core $\delta^{18}O$			
Agassiz Ice Cap (Ellsmere Is.)	A.D. 1349-1977	Fisher and Koerner (1983)	1
Devon Island Ice Cap	A.D. 1512-1973	Koerner (1977)	2
Camp Century (Greenland)	A.D. 1176-1967	Johnsen et al. (1972)	3
Milcent (Greenland)	A.D. 1176-1967	Hammer et al. (1978)	4
Quelccaya Ice Cap (Peru)	A.D. 1481-1981	Thompson et al. (1986)	14
Dunde Ice Cap (China)	A.D. 1606-1987	Thompson et al. (1990)	15
Temperature reconstructions from tree rings			
Western U.S. (annual) summer	A.D. 1602-1961	Fritts (1991)	5
Eastern U.S. (annual) summer	A.D. 1602-1961	Fritts (1991)	6
Western U.S. & SW Canada (annual) summer	A.D. 1602-1961	Fritts (1991)	7
Northern Treeline (North America) (annual)	A.D. 1601-1974	D'Arrigo & Jacoby (1992)	9
U.S. & SW Canada (April-September)	A.D. 1600-1982	Briffa et al. (1992b)	8
Northern Scandinavia (April-August)	A.D. 500-1980	Briffa et al. (1992a)	12
Northern Urals (June-July)	A.D. 961-1969	Graybill and Shiyatov (1992)	13
Tasmania, Australia (November-April)	A.D. 900-1989	Cook et al. (1992)	16
Patagonia, Argentina (December-April)	A.D. 1500-1974	Boninsegna (1992) & Villalba et al. (1989)	17
Rio Alerce, Argentina (December-February)	A.D. 870-1983	Villalba (1990)	18
Instrumental			
Central England Temperature (annual)	A.D. 1660-1987	Manley (1974)	
Dec-Jan-Feb series			10
June-July-Aug series			11

*These numbers are used in text and later tables to identify the various indices.

climate-sensitive tree-ring records that have been thoroughly documented in the refereed scientific literature. Second, we chose the longest of those records in order to maximize the temporal coverage, and tried, as far as possible, to select samples from representative geographical areas.

A number of problems are inherent in any climate reconstruction. These problems are discussed in detail in the original published papers; however, we will briefly highlight here the more critical ones as they may affect the results presented below. Because we have analyzed changes in variability, temporal changes in the composition of the tree-ring network used for the reconstructions will, in general, affect the high-frequency variance, and to some degree the low-frequency variance as well. In interpreting the temperature reconstructions shown here, we have taken into account these potential sources of biases as much as possible.

Regardless of whether one uses the width of the annual

growth rings or the maximum latewood density to reconstruct a particular climate variable (generally, growing-season temperature and/or precipitation), a process of standardization is required to account for different rates of tree growth as a function of age. This standardization is achieved by fitting a growth curve to the series of annual values, usually a cubic smoothing spline (see Cook and Kariukstis, 1990), and taking residuals about the smoothed curve. The degree of smoothing and the functional form of the growth curve partly determine the spectral properties of the residual series. In order to develop a climate reconstruction from these records, it is necessary to formulate a transfer function to convert the tree-growth index into, say, a temperature index. The procedure usually involves a calibration phase, in which a set of regression coefficients is derived that convert the tree-growth parameter into a climatic estimate, and a verification phase (see, e.g., Cook and Kariukstis). Verification is usually done against an independent set of predictors. The reader is urged to review the reference sources listed in Table 1 for specific details regarding each paleotemperature reconstruction.

Another feature of long-term reconstructions from tree rings that should be borne in mind is that the composition of tree-ring samples that make up each yearly value is variable. In general, an index value for a given site is composed of a number of tree samples from several individual trees. These may vary in age, and in some cases may come from both dead as well as living trees. The number of samples making a particular ensemble average will vary with time, although this is more generally the case near the beginning and end of a particular chronology. These extraneous factors can introduce "noise" and spurious fluctuations into the reconstructions. Nevertheless, the tree-ring indices considered here, while suffering to varying degree from the above-named shortcomings, comprise a high-quality set of proxy temperature records that are of value to compare.

Figures 6 through 8 show decadal-mean values of various tree-ring reconstructions, generally representing summer (or growing-season) mean temperature. For the most part, these tree-ring reconstructions do not sample the 1980s; nevertheless, it is clear from the degree of interdecadal variability that it would be hard to point to any particular recent period as being unique or exceptional. The November-to-April

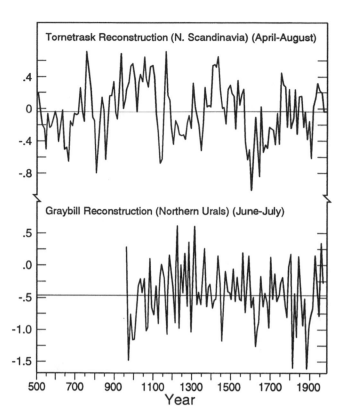

FIGURE 7 Decadal mean values of reconstructed summer-temperature anomalies for Scandinavia and the northern Urals. Units in °C.

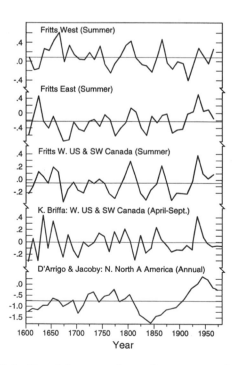

FIGURE 6 Decadal mean values of reconstructed summer-temperature anomalies for various North American regions. The top three graphs are in standardized units, the bottom two are in °C. See Table 1 for source information.

temperature reconstruction for Tasmania (Cook et al., 1992) does indicate that the last two decades of that record (1970s and 1980s) exhibit significantly greater tree growth (Figure 8) and hence higher reconstructed mean growing-season temperatures than previously, but it would seem premature to conclude that a "climatic change" has taken place in this region of southern Australia. We also note that, unlike the North American and Scandinavian reconstructions, this one is based on tree samples from only one location.

Some unique features are evident for a few of the temperature indices. For example, the curve of D'Arrigo and Jacoby (1992) is a reconstruction of annual mean temperature based on several sites along the northern North America treeline, stretching from Alaska to the Northwest Territories, plus a site located in northern Quebec Province. This reconstruction exhibits a pronounced warming from a very cold period during the mid-1800s to a maximum around 1940. By comparison, the two other high-northern-latitude sites discussed here—the indices from northern Scandinavia and from the northern Urals (Figure 7)—do not exhibit quite the same type of variations. The reader should note, however, that a much longer time period is sampled by these other two reconstructions, and that a seasonal (April to August for the Tornetrask index, and June to July for the

TABLE 2 Century Means of Interannual and Interdecadal Variability for 18 Temperature Indices (index sources identified in Table 1)

Century		1	2	3	4	5	6	7	8	9	10	11	12	13	14	15	16	17	18
500s	(ann.)	x	x	x	x	x	x	x	x	x	x	x	.84	x	x	x	x	x	x
	(decad.)	x	x	x	x	x	x	x	x	x	x	x	.20	x	x	x	x	x	x
600s	(ann.)	x	x	x	x	x	x	x	x	x	x	x	.79	x	x	x	x	x	x
	(decad.)	x	x	x	x	x	x	x	x	x	x	x	.27	x	x	x	x	x	x
700s	(ann.)	x	x	x	x	x	x	x	x	x	x	x	.64	x	x	x	x	x	x
	(decad.)	x	x	x	x	x	x	x	x	x	x	x	.40	x	x	x	x	x	x
800s	(ann.)	x	x	x	x	x	x	x	x	x	x	x	.63	x	x	x	x	x	x
	(decad.)	x	x	x	x	x	x	x	x	x	x	x	.35	x	x	x	x	x	x
900s	(ann.)	x	x	x	x	x	x	x	x	x	x	x	.70	.27	x	x	x	x	.87
	(decad.)	x	x	x	x	x	x	x	x	x	x	x	.31	.18	x	x	x	x	.41
1000s	(ann.)	x	x	x	x	x	x	x	x	x	x	x	.66	1.95	x	x	.53	x	.71
	(decad.)	x	x	x	x	x	x	x	x	x	x	x	.25	.53	x	x	.25	x	.28
1100s	(ann.)	x	x	x	x	x	x	x	x	x	x	x	.69	1.70	x	x	.62	x	.46
	(decad.)	x	x	x	x	x	x	x	x	x	x	x	.43	.56	x	x	.20	x	.14
1200s	(ann.)	x	x	1.36	1.25	x	x	x	x	x	x	x	.69	1.85	x	x	.43	x	.59
	(decad.)	x	x	.53	.85	x	x	x	x	x	x	x	.22	1.02	x	x	.16	x	.29
1300s	(ann.)	.93	x	1.10	1.36	x	x	x	x	x	x	x	.68	2.47	x	x	.42	x	.68
	(decad.)	.39	x	.65	.68	x	x	x	x	x	x	x	.30	.66	x	x	.30	x	.19
1400s	(ann.)	.70	x	1.38	1.27	x	x	x	x	x	x	x	.68	2.02	x	x	.42	x	.65
	(decad.)	.47	x	.68	.33	x	x	x	x	x	x	x	.23	.47	x	x	.18	x	.32
1500s	(ann.)	.78	.59	1.13	1.18	x	x	x	x	x	x	x	.60	1.75	1.86	x	.42	.96	.62
	(decad.)	.85	.69	.65	.67	x	x	x	x	x	x	x	.37	.56	.74	x	.17	.23	.14
1600s	(ann.)	.89	.71	1.27	1.43	.45	.32	.31	.63	.22	x	x	.69	1.63	1.80	1.12	.50	.56	.66
	(decad.)	.50	.64	.59	.47	.30	.41	.21	.42	.25	x	x	.43	.53	.66	.49	.14	.14	.19
1700s	(ann.)	.84	.62	1.05	1.36	.42	.31	.25	.53	.23	1.81	.98	.74	1.85	2.01	1.11	.49	.53	.65
	(decad.)	.96	.27	.87	.54	.21	.26	.14	.20	.34	.77	.25	.43	.40	.70	.55	.22	.18	.28
1800s	(ann.)	.97	.78	1.25	1.21	.52	.35	.29	.54	.21	1.89	1.18	.84	1.64	1.67	1.10	.45	.53	.75
	(decad.)	1.07	.55	.62	.77	.26	.29	.21	.23	.26	.31	.50	.35	.95	.96	.29	.17	.18	.38
1900s	(ann.)	1.51	1.02	1.38	1.67	.51	.33	.29	.53	.27	1.67	1.14	.71	1.78	1.24	1.28	.45	.62	.76
	(decad.)	.52	.49	.72	.44	.30	.33	.23	.23	.27	.53	.35	.27	.68	1.10	.59	.21	.22	.33

TABLE 3 Long-term Means of Interannual and Interdecadal Variability for 18 Temperature Indices (index sources identified in Table 1)

	1	2	3	4	5	6	7	8	9	10	11	12	13	14	15	16	17	18
LTM (ann)	.95	.75	1.24	1.33	.47	.33	.29	.56	.23	1.86	1.09	.71	1.91	1.74	1.15	.46	.66	.68
LTM (dec)	.75	.55	.66	.62	.27	.33	.20	.29	.28	.59	.41	.33	.67	.82	.49	.20	.19	.27
Ratio (%)	79	74	54	46	56	100	69	51	122	32	38	47	35	47	42	44	29	39

northern Urals set) interval rather than annual periods is considered.

The two summer temperature reconstructions for South America (Figure 8) are of widely different lengths. A comparison of the overlap period indicates that the Boninsegna series (see Table 1) exhibits a warming trend over the period of record. By contrast, the longer Villalba series exhibits some low-frequency variations but little if any trend over both the full data period and that of the Boninsegna period, beginning about the mid-fifteenth century.

Because of the longer record available, a variability index (described in the section on instrumental data above) was computed using the decadal averages, instead of the annual values. Our purpose here is to examine interdecadal temperature variability, with records spanning several centuries, noting that climate-sensitive tree-ring records are particularly useful for studying decadal-scale climatic fluctuations (Fritts, 1991). The changes in reconstructed climatic vari-

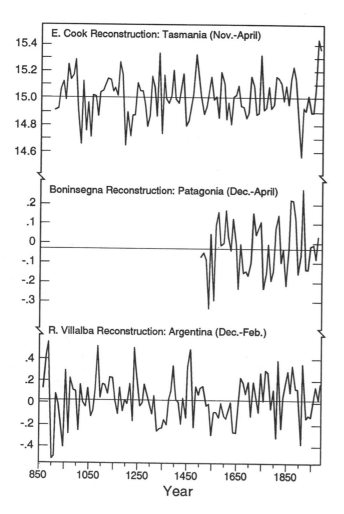

FIGURE 8 Decadal mean values of a reconstruction of summer temperatures for Tasmania, Australia (shown as actual temperatures) and two separate reconstructions for regions in Patagonia, Argentina (shown as anomalies). Units in °C.

ability have been summarized on a century-by-century basis in Table 2. For comparison, we also show in Table 3 the mean values of interannual and interdecadal variability for the full-length record. The variability index was calculated by applying equation (1) to both the annual and decadal-average reconstructed series. We fail to see any consistent trends in interdecadal variability associated with these tree-ring temperature reconstructions (see indices 5-9, 12-13, and 16-18 in Table 2). Interdecadal variability is typically about half the interannual values. This implies that substantial low-frequency variance is present in the paleotemperature record, so that recent high values of reconstructed decadal-mean summer temperature may yet represent an oscillation within the range of natural variability.

Oxygen Isotopes

The oxygen-isotope record has not been directly calibrated against temperature, as has been done for the tree-ring record. However, it is well established that the oxygen-isotope ratio ($\delta^{18}O$) is dependent primarily on the temperature of formation of the precipitation, with increasingly negative $\delta^{18}O$ ratios associated with decreasing temperature. The problems lie in the interpretation of a local ice-core record, not only with respect to local temperature variations, but even in terms of the larger-scale temperature patterns. Several factors control the oxygen-isotope composition of the snow that falls on a given ice body, and temperature is only one of them. The sensitivity of these records to annual temperature variations has been discussed in a number of papers, including the source articles referred to in Table 1. Whether or not these oxygen-isotope records accurately represent a "local" temperature record, we consider them to be sensitive indicators of prevailing climatic conditions within a suitably broad source region. Since our stated purpose is to compare recent changes in a suite of climate-sensitive paleoindicators, we have included them in our comparisons.

Figures 9 and 10 illustrate the changes in decadal averages of the oxygen-isotope ratio of glacier ice cores for different locations in northern Canada and Greenland, and for two low-latitude, high-elevation glaciers (the Quelccaya ice cap in the Peruvian Andes and the Dunde ice cap in the Tibetan Plateau region of China (see Table 1 for sources)). The polar ice $\delta^{18}O$ record, which ends before 1970, shows relatively less warming (trend toward smaller negative values of $\delta^{18}O$) than do the tropical ice cores. Furthermore, the warmest decades in the tropical record occur in the most recent time. The relation of this tropical $\delta^{18}O$ signal to air-temperature changes in the general location of these records may be partially evaluated with independent observations. It is known that significant melting took place at the Quelccaya site during the 1980s (Thompson, personal communication). Increases in tropospheric

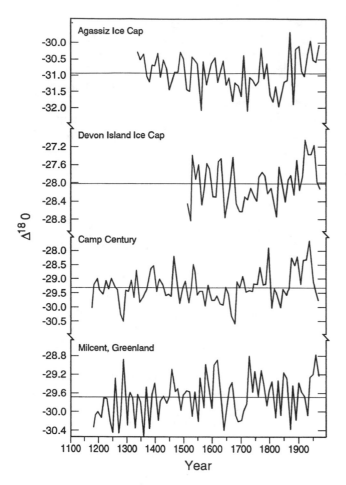

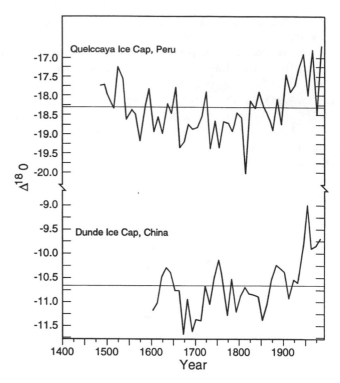

FIGURE 10   The same as Figure 9, but for two tropical ice cores (Quelccaya ice cap, Peru, and Dunde ice cap, China).

FIGURE 9   Decadal-mean values of $\delta^{18}O$ ratios from different polar-region ice cores. (See text for details.)

temperatures and water vapor obtained with tropical radiosondes for the past couple of decades have been documented by Flohn et al. (1992), although the drift toward less negative $\delta^{18}O$ ratios appears to have begun earlier in the ice-core record than in the instrumental climate record.

Changes in interdecadal variability are summarized in Table 2 (see indices 1-4 and 14-15). There is a suggestion that decade-to-decade changes in $\delta^{18}O$ ratios have increased in amplitude at most of these sites. In particular, the calculated interdecadal $\delta^{18}O$ variability at Quelccaya has increased sharply in the last 100 years. This increase is evident even though the greater number of samples in the upper part of the record permitted better averages to be created for the recent decades than for earlier periods. A similar signal is apparent on the Dunde ice cap as well. These changes, however, may reflect a more localized temperature signal, changes in moisture sources for snowfall on the ice caps, or some other unknown cause unrelated to climate. It should also be noted that these two sites are located at very high elevations, and thus differ from most of the other

paleoclimate indices, which are derived from locations at much lower elevations.

Central England Temperature

Decadal means of winter (December-to-January) and summer (June-to-August) mean temperature for central England (Manley, 1974, and subsequent updates) from 1700 to present are shown in Figure 11. The recent decades do not appear to be exceptional in comparison with the total record. The variability curves (not shown) indicate little overall change, with a tendency for the summer and winter seasons to have opposite changes in interdecadal variability (when one increases, the other tends to decrease). Jones and Bradley (1992, their Figure 13.1) have compared the central England temperature record with 12 other long-period station records in Eurasia and North America. Although there are obvious differences among the various climatic series, the twentieth century does not stand out as being a period of higher decadal-scale temperature variability. Most records (although not all) do exhibit the general warming trend of the past two centuries that is associated with the rebound from the Little Ice Age (see Bradley and Jones, 1992).

SUMMARY AND CONCLUSIONS

The purpose of this study was to evaluate whether the climate in recent decades was "changing," both with regard

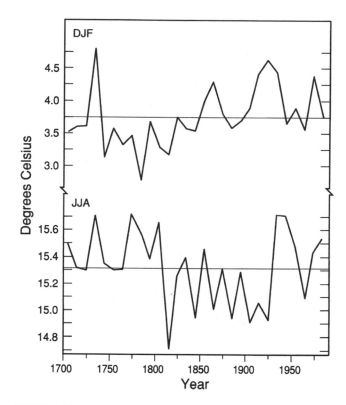

FIGURE 11   Decadal-mean values of central England winter and summer temperature (in °C).

to its mean value as well as with respect to changes in the variance on different time scales. Our findings are mostly mixed, in that the answer depends on the spatial scales selected for analysis.

For the period encompassing the last 100 years, changes in surface air temperature have been greatest and most positive in the period 1970 to 1990. Interannual variability at the largest spatial scales has also increased, although with seasonal differences. By and large, however, increases in variability at these spatial scales are not unique. When one considers smaller regions (in particular different portions of the contiguous United States) the picture changes, so that

the 1980s do not necessarily show up as being a unique period in the instrumental climate history of the country.

In order to have a longer perspective on climatic variations, we utilized data from several paleoclimatic reconstructions of growing-season temperature based on tree-ring records and on $\delta^{18}O$ ratios extracted from different ice-core records. Data from the latter were also taken as a measure of the prevailing air temperature. Annual values were averaged by calendar decades in order to focus on lower-frequency climatic variability. With the exception of the tropical ice cores, the recent decades were not generally unique, in terms of either the average values or increased variability. We note that tropical sea surface temperatures have been generally above the long-term mean since the mid-1970s. It seems plausible that this recent warmth would be reflected in low (less depleted) $\delta^{18}O$ ratios on the Quelccaya and Dunde ice caps, although more positive $\delta^{18}O$ ratios were already evident by the 1950s. It should be emphasized again that the exact relationship between actual past climate and these two high-elevation tropical indices is not fully known. Certainly, the period encompassed by the Little Ice Age is reflected in relatively low $\delta^{18}O$ values, particularly in the Quelccaya record. The development of additional independent ice-core sequences from the Cordillera Blanca in Peru and from extreme western China by L. G. Thompson and colleagues should help in the interpretation of the nature of the climate signals present in these records (e.g., Thompson et al., 1995).

Do we have an answer to the question posed by the title of this study, namely, whether the climate of the twentieth century is different from that of previous centuries? Obviously, the response is not an unequivocal "yes" or "no." As in most instances when one deals with climatic time series, the answer is "It depends." There appear to be differences in regional responses, and they are evident in both the proxy and the instrumental climate records. Insofar as the United States is concerned, the most unusual decade of the last 100 years may have been the 1930s, although the recent period is probably a close second. We can only say that, for most areas of the coterminous United States, the climate of the most recent decades cannot be considered unique, even in the context of the last century.

Commentary on the Paper of Diaz and Bradley

EUGENE M. RASMUSSON
University of Maryland

When Drs. Diaz and Bradley present a paper, it always provides an enormous amount of information. The central question, in terms of a synthesis of the wealth of information provided in this paper, is the question posed by the title: "How different is the climate of the twentieth century"— more specifically the past few decades—"from that of earlier centuries?" In addressing this question the authors examined temperature, decade-to-decade temperature changes, and changes in interannual and interdecadal variability on various spatial scales.

The problem in any analysis of this type, of course, is that of inadequate data distribution in time and space. Thus, the analysis was largely limited to the land areas of the Northern Hemisphere, where both instrumental and proxy data are most plentiful.

It may be well to note that changes in the spatial distribution of observations from decade to decade may affect both the computed area average and the variability. Figure 10 in my paper, which appears later, illustrates the effects of using different data distributions in synthesizing global averages of SST, where the problem is probably more severe than it is over the land areas.

Figure 1 in this commentary also shows estimates of the variability in SST obtained from two different analysis schemes. Two estimates of variability are obtained. One is from the optimal averaging (OA) analysis technique of Vinnikov et al., and I think Dr. Groisman will tell us about that one. Another is derived from the simple box average that has been used—for example, by the U.K. Meteorological Office—in deriving SST area averages. The curves in Figure 10 of my paper show the low-frequency variations, i.e., periods longer than 30 years. There are some differences, but they are not too great, about 0.15 degrees around 1910 and 1920, with the OA staying a little bit closer to the mean for the entire period. Now, if we remove the low-frequency variations so that only the variations on time scales less than 30 years remain (Figure 1 in this commentary), and look at that difference between the two analysis schemes, we see that the level of higher-frequency variability may also depend on the first guess used. The box method and the OA with the previous month as the first guess are similar, but the OA with climatology as a first guess shows less variability during the earlier decades of the series. Thus, relative to the box method, the OA shows an increasing variability with time. This comparison illustrates that one

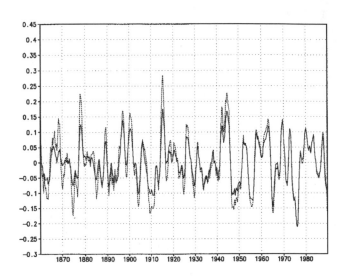

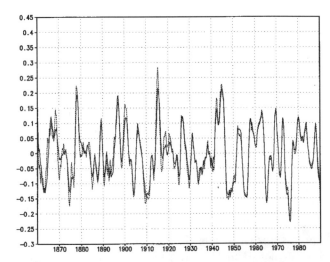

FIGURE 1 Time series of high-frequency variability (periods less than 30 years) of globally averaged sea-surface temperature. Top panel: Dashed curve is obtained by box method (Bottomley et al., 1990). Solid curve is obtained from optimal averaging using climatology as a first guess (Vinnikov et al., 1990). Bottom panel: Curves derived as above except that optimal averaging is obtained by using previous month as the first guess.

must be a bit careful in deriving conclusions about changes in variability from area averages. The estimates of variability may be sensitive to changes in data distribution and the particular analysis scheme used.

Data limitations become extremely severe when one attempts to quantify natural variability on decade-to-century time scales from proxy records. What information can we really derive regarding the patterns of large-scale variability from widely scattered proxy measurements? I think that further effort is required in two areas to maximize and better quantify the information content of the entire climate data base, both instrumental and proxy. First, we need more studies along the lines of data system tests. More specifically, using the more dense data distribution of recent decades, we should determine the loss of information as the distribution of observations is decreased. This will help establish confidence limits for estimates of large-scale averages based on the sparse distribution of observations and proxy series obtained during earlier decades and centuries.

Second, we need to place more emphasis on integration of information from various proxy sources, with the goal of obtaining a better picture of the *pattern* of global variability during earlier centuries. This is admittedly a formidable task when one is dealing with decade-to-century time scales. I do not believe that the concept of an overall integration of information in decade-to-century variations has yet to take root in the proxy community, but I think it should be strongly encouraged as a long-term goal.

Discussion

GROISMAN: Two quick comments, one for Gene and one for Henry. Gene, I just wanted to note that the SST variability does not reflect conditions over the Arctic. There is about a 25 percent difference in amplitude between SST and marine air temperature over the open ocean, the latter being higher, so there may be other reasons for the difference you cite. Henry, the problem with tree rings is that there are very few 1000-year-old trees, so in a time series you can't resolve the same mean variance.

DIAZ: I used only air temperature over land, no SSTs. Also, I started in 1891, when we had fairly good coverage. The changes in data distribution don't have much influence on the large-scale averages. As for the tree rings, we did try to compensate by looking at decadal means and interdecadal variability, and I don't think the middle section has much of a problem.

KEELING: I have a figure that is based on the Jones-Wigley temperature record, with a slightly different smoothing. Some of the early variability and the apparent biennial signal are probably the result of sparse data; beginning in 1951 the record is fairly homogeneous. You can see the 1958, 1961, 1963, 1982 El Niños. I'd like to know whether there is some difference in the quantity of data, or in its processing now that satellite data and ground data are being mixed, that has changed data variability in the last decade.

TRENBERTH: Perhaps when you fit the spline you are taking out more of the variance at the end.

GHIL: Henry, I was puzzled by the apparently much higher interdecadal variability in the second of your first two tree-ring figures.

JONES: It's partly due to the indexing procedure that there is more low-frequency variance in the top curve than the bottom one, though some of it could be real.

KARL: It's important to remember that there are many different ways to classify variability. I suspect that I could show more or less of it just by defining it differently. On this particular figure, I think we need to be aware that one is for a longer season than the other, and the shorter period of a time series will give you higher variability.

DIAZ: Another point is that getting longer time series from proxy records may not improve our understanding of climate change mechanisms.

GROOTES: I want to comment on the Quelccaya ice-core records you used. As you go down from the surface, the seasonal cycle disappears. It reappears about 1880 and continues significant to about 1480. Apparently when there are low values the record is smoothed so little that the seasonal cycle is preserved. As temperature changes it becomes more difficult to calibrate the isotope changes; you may be looking at post-depositional changes rather than the signal itself. Tree rings and ice cores both require particular attention to understanding exactly what you're recording, especially if you are talking about temperature.

MYSAK: For anyone who's interested, I'd like to mention that Brian Luckman at the University of Western Ontario has tree-ring records from the Canadian Rockies for about the last thousand years. He's showing a nice cycle of 150 years or so.

MCGOWAN: Are you comparing tree rings taken from similar altitudes at those different locations? Also, how does the variance of instrumentally measured mountain temperatures compare with that of lowland temperatures?

DIAZ: Most of the data sets are from moderate elevations. As for the high/low elevation temperature variances, the trends are reduced with respect to surface, but they're still comparable.

Natural Climate Variability on Decade-to-Century Time Scales
National Research Council, 1995

Variability of African Rainfall on Interannual and Decadal Time Scales

SHARON E. NICHOLSON[1]

ABSTRACT

This paper examines the rainfall fluctuations that have occurred over the continent of Africa during the past century. Major characteristics include anomaly patterns that are continental in scale, with marked teleconnections between particular sectors of the continent, and quasi-periodic fluctuations with dominant time scales of approximately 2.3, 3.5, and 5 to 6 years. In the Sahelian regions these higher-frequency fluctuations are not apparent; instead, most of the variance is found on time scales of 7 years or longer. Both dry and wet episodes tend to persist for one or two decades at a time; decadal means change by a factor of 2. These characteristics are also apparent in the Sahelian climate record for earlier centuries. Elsewhere on the continent, synchronous changes of rainfall occur, but the "wet" or "dry" decades are not as uniformly wet or dry as in the Sahel. Examples are the 1950s, in which above-average rainfall prevailed over most of the continent, and the 1980s, during which most of the continent was abnormally dry. It is likely that the factors producing higher-frequency interannual variability differ from those producing the decadal-scale fluctuations. Both sea surface temperature and land-surface feedback have been suggested as causes of the unusual persistence in the Sahel, but the latter cannot explain the change of decadal means.

INTRODUCTION

The continent of Africa is primarily tropical or subtropical, so the climatic changes that have taken place have been manifested principally in the rainfall regime. The rainfall fluctuations that occur are extreme; those in the semi-arid regions are probably unmatched anywhere in their magnitude and spatial extent. In the Sahel, for example, decadal averages change by a factor of 2 or more. Most major periods of anomalous rainfall, either individual years or decadal-scale episodes, affect most of the continent.

This paper examines the rainfall fluctuations that have occurred during the twentieth century and compares them with historical episodes. The time scales of the fluctuations are summarized, and the decadal-scale fluctuations described in greater detail. The intra-continental teleconnections, and teleconnections to tropical or global rainfall anomalies, are also considered. Finally, a brief review of causal mechanisms is provided.

Undoubtedly, changes in the thermal regime over Africa have occurred, especially along the continent's poleward

[1]Department of Meteorology, Florida State University, Tallahassee, Florida

extremes and at higher elevations. These are minor compared to rainfall fluctuations; moreover, they have not been systematically studied. Therefore, this paper will be limited to variability of rainfall.

METHODOLOGY

Most of the description of rainfall variability in this paper derives from previous studies made by the author. Analyses are based on a network (Figure 1) of approximately 1400 stations, the statistics of which are fully described in Nicholson et al. (1988). From these, a spatially averaged data set has been derived, utilizing 90 geographical regions (Figure 1) that are homogeneous with respect to the interannual variability of rainfall.

Regional averages are generally expressed as a standardized annual departure from the long-term mean (i.e., the mean over the entire length of record). In select analyses, the rainfall anomalies are also presented as a percentage departure from the long-term mean. About 75 percent of the station records commence before 1925 and continue to 1984 or later. Hence, the length of record is generally at least 60 years.

Regionally averaged rainfall R_j is calculated as

$$R_j = I_j^{-1} \sum_j x_{ij} \qquad (1)$$

where R_j is the regional rainfall departure in the year j, and I_j is the number of stations available in year j. In the above, x_{ij} is a standardized rainfall departure at station i in year j. It is calculated as

$$x_{ij} = \frac{(r_{ij} - \bar{r}_i)}{\sigma_i} \qquad (2)$$

where r_{ij} is the annual total at the station in the year j, $\bar{r}_i$ is the station's long-term annual mean, and σ_i is the variance of annual rainfall at the station.

These regional means are utilized to examine the temporal and spatial structure of rainfall variability over the continent. They also illustrate the magnitude of decadal anomalies. In some analyses 1°-grid averages are utilized instead of the regional averages.

CONTINENTAL-SCALE PATTERNS OF RAINFALL VARIABILITY

Previous studies (e.g., Nicholson, 1986) have utilized a map-classification technique of Lund (1963) to describe the preferred spatial configurations of rainfall anomalies over Africa. This technique is based on linear correlation of map patterns. Most of the rainfall variability over the continent is described by the six "anomaly types" illustrated in Figure 2. Unlike eigenvector techniques, the patterns produced by the Lund method are not orthogonal; hence, these six anomaly types really describe only four basic configurations of anomalies. These include a tendency for above- (or below-) average rainfall over most of the continent (Types 2 and 4) and two patterns (Types 1, 3, 5, and 6) illustrating an opposition between equatorial and subtropical latitudes. Approximately 21 of the 90 years between 1901 and 1990 are represented by a pattern of above-average rainfall in the equatorial regions but below-average rainfall in subtropical latitudes, and 12 by the opposite pattern, which is illustrated by Type 3 in Figure 2. Fourteen of the 70 years are best represented by the Type 2 anomaly pattern, with below-average rainfall throughout most of the continent, and 15 by Type 4, with above-average rainfall in most areas.

These patterns are introduced here for two reasons. First, as will later be seen, they are typical of decadal-scale rainfall fluctuations over Africa. Second, they demonstrate that much of the rainfall variability over the continent can be described using a small number of regional time series. The representative regions, shown in Figure 3, include five latitudinal zones running from the southern margin of the Sahara to the Guinea Coast of the Atlantic, eastern equatorial Africa, and two sectors of southern Africa loosely termed the northern and southern Kalahari.

DECADAL-SCALE FLUCTUATIONS

Fluctuations During Recent Decades

Two major, decadal-scale rainfall anomalies have affected the African continent within the current century:

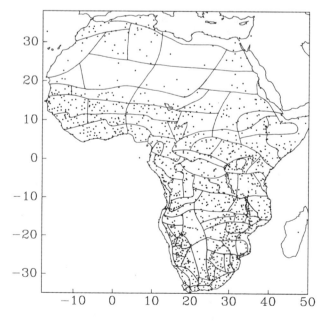

FIGURE 1   Map of rainfall stations in African archive and homogeneous rainfall regions. (From Nicholson et al., 1988.)

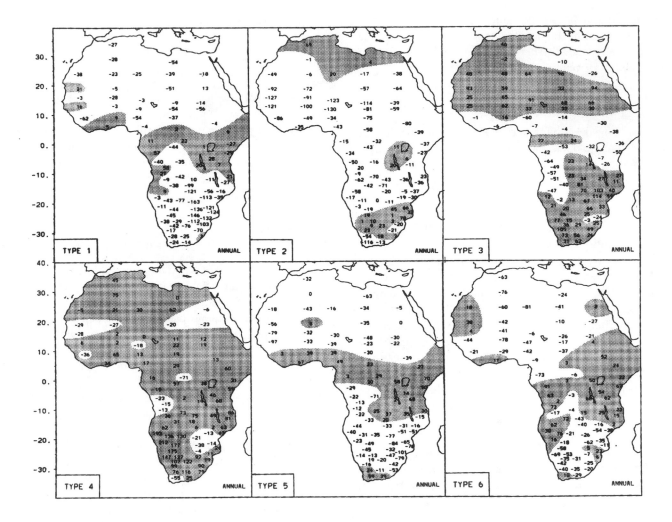

FIGURE 2   The most frequently occurring continental-scale rainfall anomaly types for the twentieth century. (From Nicholson, 1986; reprinted with permission of the American Meteorological Society.) The values represent regionally averaged standardized departures.

TABLE 1   Rainfall Anomalies for the Decade 1980-1989

	1950-1959		1970-1979		1980-1989	
	Depart.* (%)	Stdzd. Depart. (% σ)	Depart. (%)	Stdzd. Depart. (% σ)	Depart. (%)	Stdzd. Depart. (% σ)
Northern Sahel	37	54	−31	−47	−24	−35
Central Sahel	28	71	−22	−55	−31	−82
Southern Sahel	15	62	−13	−53	−20	−85
Soudano-Guinean Zone	3	22	−5	−36	−8	−56
Guinea Coast	3	15	−6	−31	−7	−35
Eastern Africa	−2	−9	0	4	2	15
Northern Kalahari	8	28	6	20	−5	−21
Southern Kalahari	7	23	10	34	−7	−27

*Anomalies are expressed as a percentage departure from the long-term mean and as a standardized departure (ratio of the departure from the mean to the standard departure).

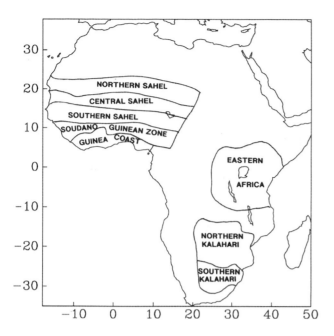

FIGURE 3   Geographical regions described in detail in this study.

relatively wet conditions in the 1950s and relatively arid conditions in the 1980s. The periods of anomalous conditions have not been precisely 10 years in length, nor have they commenced at the onset of each numeric decade (e.g., 1950 or 1960). Nevertheless, these periods and the transition between them are well illustrated from decadal maps and tabulations (Figure 4 and Table 1).

Fluctuations have been particularly large in Sahelian West Africa, where rainfall was about twice as great during the 1950s as during recent years (Table 1). The 1950s (Figure 4a) were also wetter than average over most of the rest of the continent, but abnormally dry in the equatorial regions. The wet episode ended abruptly in the Sahel toward 1960, after which time a change occurred in many other regions.

In the early 1960s, phenomenally wet years occurred in equatorial regions, especially in eastern Africa, while droughts affected much of southern Africa. As an example, Wajir, in northern Kenya with mean annual rainfall of 285 mm, received 612 mm in November 1961; at many Kenyan stations, rainfall that month was five to ten times the long-term mean. The levels of Lake Victoria and other Rift Valley lakes suddenly rose several meters commencing in 1961, abruptly attaining levels unmatched since the end of the nineteenth century. The 1960s as a whole (Figure 4b) was a period of relatively wet conditions throughout the equatorial latitudes and abnormally dry conditions throughout most of the subtropics of both hemispheres. The dry anomaly was particularly strong in the Southern Hemisphere.

The early 1970s were a period of drought over much of

Africa (Nicholson, 1986), but later in the decade a series of tremendously wet years occurred in southern Africa, so that for the decade as a whole, a strong opposition is apparent between the two hemispheres (Figure 4c). This is not a common configuration of anomalies, and had previously occurred in only a few isolated years. A plausible explanation is presented below.

The 1980s (Figure 4d), in strong contrast to the 1950s and unlike other decades, was a period of abnormally dry conditions over most of the continent, including most equatorial regions. The decadal anomalies are largest in the Sahel, on the order of +25 percent in the 1950s and −30 percent in the 1980s (or nearly one standard deviation). In most other areas the departures for the 1950s (Nicholson, 1989) and 1980s (Table 2) are less than 10 percent of the mean.

The continental scale of these anomalies and the abrupt change around 1971 are evident in Figure 5, which shows the linear correlation of the rainfall anomaly pattern of each year since 1901 with anomaly Type 2 in Figure 2. This anomaly type is one of negative departures throughout the continent. Most years from 1950 to 1970 are negatively correlated with it; correlations for 1950-1957, 1961, and 1962 are significant at the 99 percent level. In contrast, nearly every year since 1970 is positively and, in most cases, significantly correlated with the Type 2 pattern.

Teleconnections and Other Characteristics of Rainfall Variability: Comparison of Modern and Historical Periods

The previous section identified several characteristics of African rainfall variability on decadal scales. One is the tendency of anomalous conditions to persist for roughly a decade. Another is the large magnitude of anomalies in the Sahel. A third is the teleconnections within Africa on decadal scales, as illustrated by the anomaly types of Figure 2 and the decadal patterns for the 1950s and 1980s (Figures

TABLE 2   Mean Annual Rainfall (mm) at Select Sahelian Stations for Two Recent Periods

	1950-59	1970-84
Bilma	20	9
Atbara	92	54
Nouakchott	172	51
Khartoum	178	116
Agadez	210	97
Timbuktu	241	147
Nema	381	210
Dakar	609	308
Banjul	1409	791

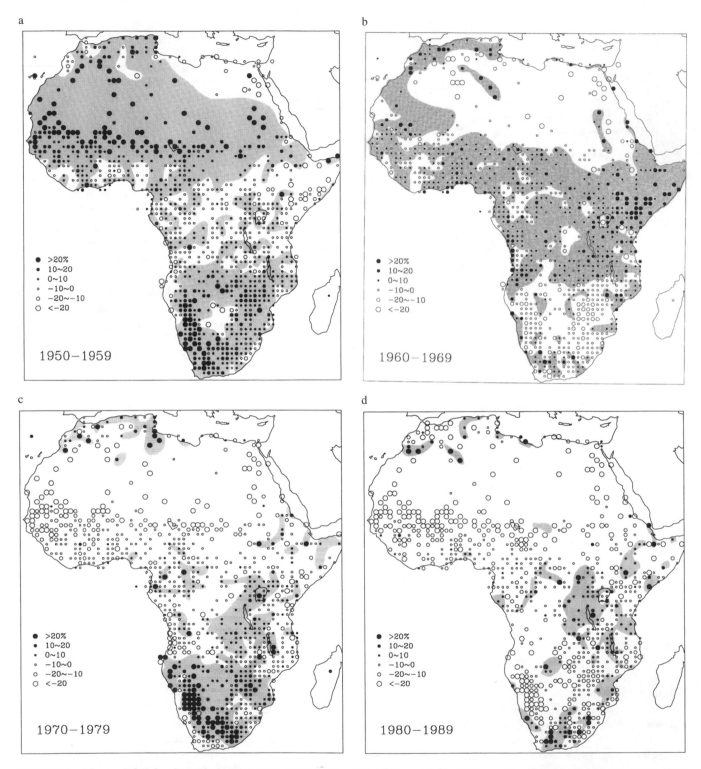

FIGURE 4   Mean rainfall for the periods 1950-1959, 1960-1969, 1970-1979, and 1980-1989 (panels a through d respectively). Rainfall is expressed as a percentage departure from the long-term mean (i.e., the mean for the entire length of record, commencing in 1901 or as soon thereafter as rainfall observations begin). Station data are averaged over 1° squares to facilitate presentation. Shading highlights regions with positive anomalies. (The 1970-1979 map is reprinted from Nicholson, 1994, with permission of Arnold, of the Hodder Headline Group; the 1980-1989 map is reprinted from Nicholson, 1993, with permission of the American Meteorological Society.)

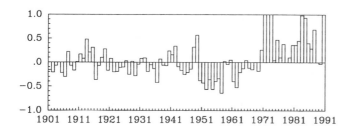

FIGURE 5   Correlation coefficient between the anomaly pattern of each year's rainfall (averaged over the regions shown in Figure 1) and the anomaly pattern corresponding to Type 2 in Figure 2. For the 99 percent confidence level, *r* ranges from about 0.28 to 0.31, depending in part on the number of regions with available data in a given year.

4a and 4d). Overall, the decadal rainfall fluctuations tend to occur quasi-synchronously over the continent. There is a strong tendency for the fluctuations in the subtropical latitudes of both hemispheres to be roughly in phase, as well as a tendency for the fluctuations in the equatorial and subtropical latitudes to be opposed to each other. These same characteristics are evident in the historical past.

Figures 4a and 4d clearly demonstrate that major periods of anomalous rainfall in the Sahel correspond to continental-scale anomalies as well. This is likewise true for historical fluctuations of Sahel rainfall, such as the droughts of the 1820s and 1830s and the 1910s (Figure 6). Both of these periods were times of frequent drought throughout much of the continent. A period of sufficient rainfall from about 1870 to 1895 was part of a continental pattern mirrored by the 1950s pattern (Figure 4a). In the 1950s, lake levels rose dramatically from Chad in the north to Ngami in the south,

and the floods of rivers like the Niger and Nile were consistently high. Near Timbuktu, in the Niger Delta where annual rainfall is now 195 mm, wheat production thrived to such an extent that grain was exported to neighboring regions. In the semi-arid regions of North Africa rainfall probably averaged about 25 to 30 percent above the twentieth-century mean during the period 1870-1895 (Nicholson, 1978).

Interestingly, the historical patterns depicted in Figure 6 were constructed prior to the derivation of the rainfall anomaly types in Figure 2; they were based on proxy data and qualitative indicators. While this origin might render the reconstructions questionable, their reliability is supported by the clear correspondence to the major anomaly types apparent in the modern quantitative rainfall record. They also demonstrate other major characteristics of African rainfall variability established with modern records: the decadal-scale persistence of anomalies in the Sahel, and their continental scale.

In certain cases, global teleconnections to decadal-scale African rainfall fluctuations have also been demonstrated. One example is the change from "wetter" to "drier" conditions around the turn of the century. This change was seen throughout the tropics, with a major shift occurring around 1895 (see, e.g., Kraus, 1955a,b). The anomalously dry 1830s was also a period of globally anomalous climate, often considered to mark the end of the Little Ice Age. The anomalous conditions of the 1950s are also noted in other tropical and subtropical regions such as Central America, India, and Australia. These decadal teleconnections have not, however, been studied systematically.

Fluctuations on Time Scales of Centuries

The climatic fluctuations that have affected Africa on time scales of centuries are difficult to establish because

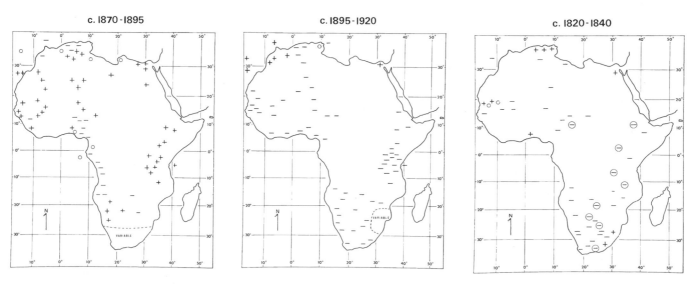

FIGURE 6   African rainfall anomalies for three historical periods. Minus signs denote evidence of drier conditions; plus signs denote evidence of above-average rainfall; zeroes denote near-normal conditions; circled symbols denote regional integrators, such as lakes or rivers. (From Nicholson, 1978; reprinted with permission of Academic Press Ltd.)

there are few long-term climate data sets or even proxy records from which past climatic conditions can be reconstructed. Two notable exceptions are Lake Chad, the fluctuations of which have been well documented through both historical and geological methodologies (Maley, 1976), and the Nile, for which quasi-continuous written records extend back to the seventh century (Toussoun, 1923; Riehl et al., 1979). However, neither record lends itself to completely unambiguous interpretation.

The Lake Chad record suggests two long periods of comparatively wet conditions in the West African Sahel, one in the ninth through fourteenth centuries and another from the sixteenth through nineteenth centuries. The first was synchronous with the Medieval Warm epoch and the second occurred more or less during the Little Ice Age of Europe. The more humid climates in the present-day Sahel, which the Lake Chad record implies, are further supported by numerous historical and geological indicators throughout West Africa. Information for other regions of Africa is more scarce, but the Nile record and those of other East African lakes suggest that the first humid period affected much of eastern equatorial Africa as well. The second appears to have affected that region and much of southern Africa (Nicholson, 1979a, 1981). The decline of the recent wetter conditions took place toward 1800, at a time when the Northern Hemisphere was experiencing the end of the Little Ice Age.

INTERANNUAL VARIABILITY OF RAINFALL

Because of the limited availability of meteorological data for time scales of 10 to 100 years, it may be necessary to evaluate mechanisms of variability on these time scales at least partially through analogy with individual years. For this reason, a description of the interannual variability of African rainfall is relevant here. This section examines the temporal characteristics of rainfall variability in various regions in the domains of both time and frequency. Several of the features exhibited will have to be adequately accounted for by any mechanism that is proposed as explaining variability on 10- to 100-year time scales.

Figures 7 and 8 show the standardized rainfall departure series for select regions of Africa. The most striking feature in any of the series is the extreme persistence of rainfall anomalies in Sahelian West Africa (Figure 7). This persistence was first identified by Bunting et al. (1976), then later by Walker and Rowntree (1977), Nicholson (1979b, 1983), and Lamb (1982). In the Sahel as a whole (a combined series of northern, southern, and central Sahel in Figure 7), regionally averaged rainfall exceeded the long-term mean in every year from 1950 to 1967; it has been below the mean in every year since 1970. Rainfall also tended to be above average throughout most of the period 1927 through 1936 and below average in the 1910s and 1940s. To some

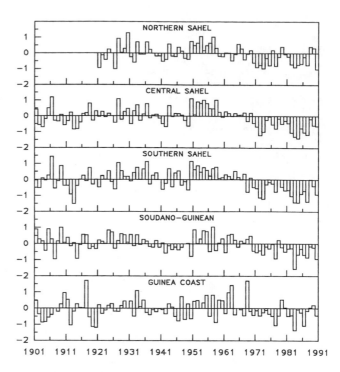

FIGURE 7 Rainfall fluctuations in West and North Africa (1901 to 1990), expressed as a regionally averaged standard departure. Location of the regions is indicated in Figure 2.

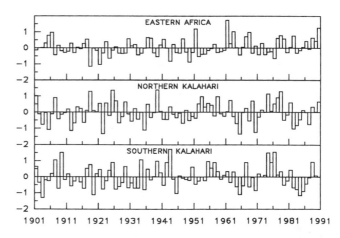

FIGURE 8 Rainfall fluctuations as in Figure 7, but for regions of southern and eastern Africa, as indicated in Figure 2.

degree, this persistence is also apparent in the Soudano-Guinean and Guinean regions further south.

Such long-term persistence is not evident in any of the other regional time series. For example, the longest run of "dry" years is 5 years in the northern Kalahari, 6 years in the southern Kalahari, and only 3 years in eastern Africa (Figure 8). The longest run of "wet" years is 5 years in all three regions. However, longer periods of predominantly wet or dry years have prevailed. These include, for example, the "dry" 1950s in East Africa and the drier periods of 1926 to 1935, the 1960s, and the 1980s in the southern Kalahari.

The contrast between the Sahelo-Saharan region and elsewhere in Africa is underscored by Figure 9. It shows the results of low-pass filtering the rainfall series for the regions in Figure 1 and isolating the percentages of variance of annual rainfall on time scales longer than 7 years. This variance represents a combination of decadal-scale or longer fluctuations and year-to-year persistence of rainfall anomalies. Low-frequency variance is by far the greatest in the Sahelian regions, accounting for as much as 45 to 60 percent of the variance. Elsewhere it is generally between 20 and 35 percent.

Spectral analysis shows that the dominant time scales of rainfall variability throughout the continent are on the order of 2.2 to 2.4, 3.3 to 3.8, and 5 to 6 years (Rodhe and Virji, 1976; Ogallo, 1979; Nicholson and Entekhabi, 1986). These are not the only periodicities that have been detected, but they are the only ones apparent throughout large expanses of the continent and the only ones corresponding to time scales of distinct causal mechanisms such as SST fluctuations or ENSO. The works of Tyson (e.g., 1980), Fleer (1981), and others provide more detail on select regions.

The occurrence of these quasi-periodicities in the regions depicted in Figure 1 were evaluated by Nicholson and Entekhabi (1986). The results are shown in Figure 10. The 5- to 6-year time scale is dominant in equatorial regions; the 3.5-year time scale is dominant in much of southern Africa and in parts of equatorial Africa; the 2.3-year time scale is dominant in the eastern half of equatorial and south-ern Africa. Interestingly, these relatively high-frequency fluctuations (Figure 10) are nearly absent throughout the Sahelo-Saharan region of West Africa, where the spectra are dominated by low-frequency fluctuations (Figure 9). Relatively low-frequency fluctuations have been noted also in parts of South Africa (Tyson, 1980), with time scales of 18 to 22 years being common.

An examination of wet and dry years throughout the continent (e.g., Nicholson et al., 1988) shows other characteristics of interannual variability that must be explained. One is that in many regions interannual variability is associated not with a change in the intensity of the rainy season but by a change in the length of the season or by anomalous rainfall in comparatively dry months. This can be demonstrated for East Africa as well. Throughout most of the region, the main rainy season is April to May, but interannual variability is best correlated with anomalies of the "short rains" of October-November (Figures 11 and 12). The importance of transition-season fluctuations also helps to explain the rainfall's continental-scale coherence over regions (such as the African Northern and Southern Hemispheres) where the rainy seasons are six months out of phase.

Rainfall in East Africa illustrates another interesting characteristic: The factors that produce the climatological-mean

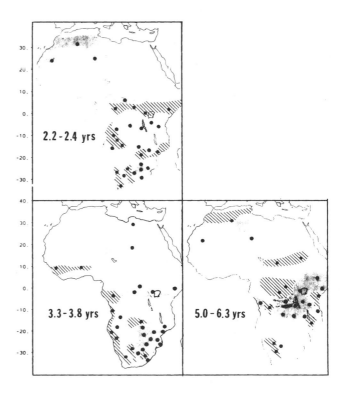

FIGURE 10 Distribution of significant spectral peaks with periods on the order of 2.2-2.4, 3.3-3.8, and 5-6 years. Data are for regions shown in Figure 1, with blanks indicating no significant peak in the regional rainfall. (From Nicholson and Entekhabi, 1986; reprinted with permission of Springer-Verlag.)

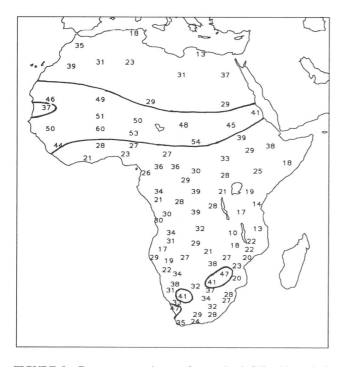

FIGURE 9 Percentage variance of annual rainfall with periods of 7 years or longer for regions depicted in Figure 1 (in circled areas, variance exceeds 40 percent).

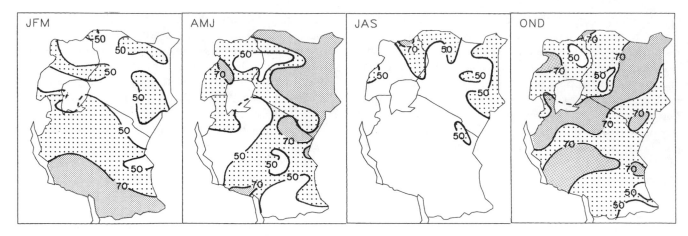

FIGURE 11   Correlation between the seasonal and annual rainfall departure series for East Africa (representing the countries Kenya, Uganda, and Tanzania). Values approximately equal the percentage variance explained by each season: stippling, 50 to 70 percent; shading, greater than 70 percent.

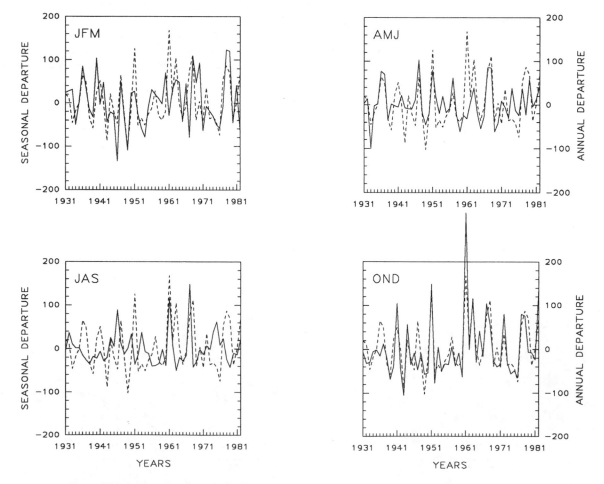

FIGURE 12   Time series of rainfall departures for individual seasons (solid line) compared with the annual rainfall departure series representing East Africa (Kenya, Uganda, and Tanzania). Data are expressed as a percentage standard departure.

patterns are not necessarily those that explain interannual variability. Within this region, mean annual rainfall and its seasonality are highly heterogeneous (Figure 13), being associated with several features of the general atmospheric circulation as well as with regional-scale orographic and lacustrine effects. Nevertheless, the patterns of interannual variability are markedly similar throughout the region (Figure 14), so that there is a remarkable resemblance between the time series for the region as a whole, and for those four climatologically homogeneous subregions. The first eigenvector of annual rainfall shows loadings of the same sign throughout the region; the pattern explains 36 percent of the variability of annual rainfall and 52 percent of the variability during the "short rains" (Nyenzi, 1988).

In summary, there are several notable characteristics of rainfall variability that must be explained by mechanisms for producing this variability. These include, first of all, the continental-scale coherence and the seemingly contradictory contrast between the dominant low-frequency fluctuations in Sahelo-Saharan regions and the high-frequency fluctuations elsewhere, in regions that show strong teleconnections to the Sahel on time scales of 1, 10, and 100 years. The characteristic time scales of 2.3, 3.5, and 5 to 6 years must likewise be explained, as must be the seasonality of the changes and the coherent patterns of interannual variability in climatically heterogeneous regions.

MECHANISMS OF RAINFALL VARIABILITY

Most studies of African rainfall have focused on interannual variability. Numerous authors have demonstrated relationships between SSTs and rainfall in various parts of Africa (e.g., Lamb, 1978; Lough, 1986; Semazzi et al., 1988; Wolter, 1989; Folland et al., 1991; Nicholson and Entekhabi, 1987). For parts of eastern and southern Africa

linkages to ENSO have also been clearly established (by, e.g., Farmer, 1988; Ropelewski and Halpert, 1987; Ogallo, 1987; Harrison, 1983; Lindesay et al., 1986; Nicholson and Entekhabi, 1986, 1987). For Sahel rainfall, a number of other factors, such as changes in the upper-level winds and in the Hadley and Walker circulations, have also been implicated (Kanamitsu and Krishnamurti, 1978; Newell and Kidson, 1984).

Very little attention has been paid to the causes of the lower-frequency variability. The exception is the work by Folland and collaborators (e.g., Folland et al., 1991), which is limited to the Sahel. Virtually no studies have been carried out on the causes of the continental rainfall fluctuations on decadal time scales.

Rowell et al. (1992, 1995) suggest that the twentieth-century trends in Sahel rainfall can be attributed to inter-hemispheric differences in sea surface temperature. The interhemispheric contrast increased in mid-century and decreased sharply as the droughts began in the 1970s and 1980s. This mechanism does not, however, account for the extreme persistence of anomalies in the Sahel, or their lack of persistence in regions such as southern Africa, which otherwise appear to have teleconnections to the Sahel in individual years and on decadal scales.

Land-Surface Feedback in the Sahel

As an example of how this might work, consider a year during which anomalous large-scale circulation triggers drought in the Sahel and in other African regions showing strong teleconnections to the Sahel. When the large-scale forcing returns to normal, so do rainfall conditions in most areas. However, in the Sahel the drought conditions might be reinforced by local, land-surface feedback. In that case they will persist until a year in which a large-scale circula-

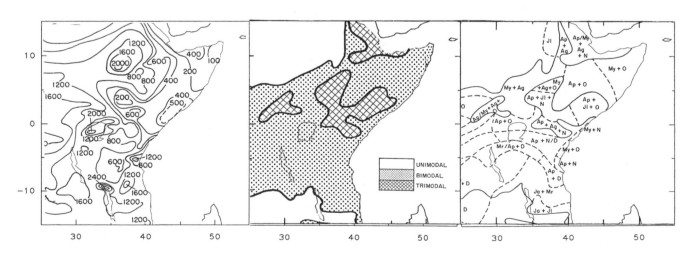

FIGURE 13   Maps of rainfall characteristics over East Africa: mean annual rainfall, number of rainfall maxima in the seasonal cycle, and the months of these maxima. (From Nicholson et al., 1988.)

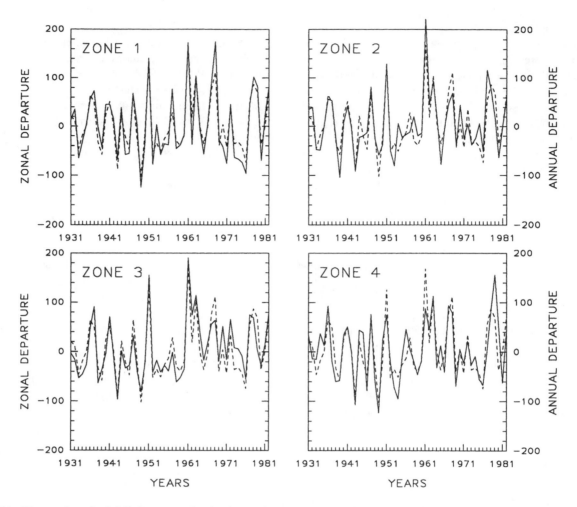

FIGURE 14 Time series of rainfall departures for the four regions shown in the inset map (solid lines) compared to the time series for East Africa as a whole (dashed lines). Values are expressed as a percentage standard departure from the long-term mean.

tion anomaly occurs that is sufficiently great to override that local forcing.

In southern Africa, a region with rainfall teleconnections to the Sahel (see above) but without strong interannual persistence of rainfall anomalies, the interannual variability of surface fluxes is smaller and less spatially coherent. The rain-bearing systems in that region are also less sensitive to surface processes than are those affecting the Sahel.

This might suggest the following hypothetical scenario for explaining the unusual pattern of the 1970s: below-normal rainfall in Sahelo-Saharan regions and above-normal rainfall in corresponding sectors of Southern Hemisphere Africa. The factors producing drought commencing in 1968 were reinforced in the Sahel by land-atmosphere interaction that caused the negative anomalies to persist without interruption throughout the 1970s. In the mid- to late 1970s, some factors in the general atmospheric circulation produced wet conditions in southern Africa and also enhanced rainfall in Sahelo-Saharan regions. In the latter, however, they were

insufficient to override the drought-promoting feedback and resulted in less extreme negative anomalies in years such as 1974 to 1978, but not in rainfall exceeding the long-term mean.

It is interesting to note that when only high-frequency fluctuations are considered, the rainfall fluctuations in the Sahel bear a much stronger similarity to those in the Kalahari (Figure 15) during the past 40 years. The correlation between the resulting "Sahelian" series and the rainfall series for the northern Kalahari is 0.5, just slightly lower than the correlation (0.57) between the series for the northern and southern Kalahari in Figure 8.

The results shown in Figure 15 are consistent with the idea that higher-frequency fluctuations, or yearly departures from a decadal or multi-decadal mean, are forced by similar factors in Sahelian and southern Africa, but that additional causal mechanisms of lower-frequency variability must be sought in the Sahel. A similar contrast in forcing is apparent between various parts of the Sahel and between the early

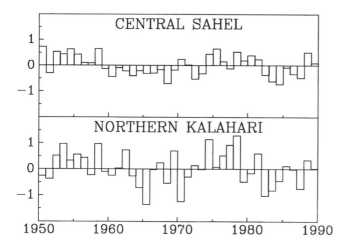

FIGURE 15 Rainfall variations in the Sahel and the Kalahari for the period 1950 to 1990. Rainfall in the Kalahari is a standardized departure from the long-term mean. For the Sahel, standardized departures for 1950 to 1969 are calculated with respect to the mean for that period; those for 1970 to 1990 are calculated with respect to the latter period. For both, the variance is calculated with respect to the entire length of the record. This is roughly equivalent to removing the low-frequency variance from the Sahel series. (From Nicholson, 1993; reprinted with permission of the American Meteorological Society.)

and late rainy-season months. High-frequency forcing is also stronger in June-July than in August-September. This observation is relevant because one hypothesized factor in the feedback, the African Easterly Jet (see below), is operative mainly in the latter period.

The potential feedback mechanisms involve the effect of such factors as soil moisture and dust generation on surface fluxes and atmospheric heating. The dynamic link to the Sahel could be the mid-level African Easterly Jet. The African Easterly Jet is a consequence of the meridional temperature gradient in the region, which in turn is produced by the strong thermal contrast between the dry Sahel-Saharan region and the forests and Atlantic Ocean to the south. It is analogous to the mid-latitude westerly jet stream, in that it provides the instability and energy for the development and maintenance of rain-bearing systems in the Sahel. It has a maximum around 650 or 700 mb, a level where the temperature structure is strongly affected by surface-generated dust. The intensity of the dust, in turn, is highly inversely correlated with rainfall in the Sahel.

The contrast in surface fluxes between wet and dry years

(Lare and Nicholson, 1993) is sufficiently large to produce the observed changes of jet intensity between wet and dry years (Newell and Kidson, 1984). These in turn are sufficiently large to influence the jet's dynamics, and are thus capable of altering such variables as the number, size, and intensity of disturbances in the region.

SUMMARY AND CONCLUSIONS

Numerous studies have firmly established that:

• decadal-scale rainfall fluctuations have been quite marked in Sahelian Africa;
• these fluctuations tend to have a continental spatial scale (although less extreme and persistent than in the Sahel) with distinct teleconnection patterns;
• the characteristics of modern decadal anomalies are also apparent in earlier centuries; and
• the 1950s and, to a lesser extent, 1960s were continentally "wet," while the 1980s and, to a lesser extent, 1970s have been continentally "dry."

Among the major areas in which research is needed are the causes of the decadal-scale fluctuations of African rainfall, particularly the continentally coherent patterns; the role land-surface feedback may play in the decadal-scale anomalies in the Sahel; and the teleconnections between African rainfall anomalies on the decadal scale and global climate anomalies.

Our understanding of decadal-scale fluctuations could be greatly enhanced by comparative analysis of the 1950s and more recent decades. Unfortunately, many of the necessary data are routinely available only for about the last 20 years. However, a great many data, especially upper-air data, do exist for the 1950s and 1960s, but in obscure archives where they are not readily available. Effort should be devoted to creating usable data sets for earlier periods, especially the 1950s.

Our knowledge of African climate fluctuations on the scale of centuries is considerably more vague. In order to establish these regional fluctuations' causal mechanisms and their relationships to global climate fluctuations, they must be more precisely delineated in time and space. This will require the preparation of better proxy data sets. Few attempts have been made to use, for example, tree-ring or varve chronologies. Also, the innovative paleoclimatic techniques that have been developed recently might be appropriate here as well. Creation of these proxy data sets for Africa should be a high priority.

On the Initiation and Persistence of the Sahel Drought

JAGADISH SHUKLA[1]

ABSTRACT

On the basis of analysis of data collected over the Sahel, India, and China, it is conjectured that the occurrence of above-normal rainfall over the Sahel during the decade of the 1950s and below-normal rainfall during the decades of the 1970s and 1980s is a manifestation of natural variations in the planetary-scale coupled ocean-land-atmosphere climate system. In the light of the results of general-circulation model sensitivity experiments, and the fact that the Sahel region is particularly susceptible to changes in land-surface conditions, it is concluded that strong local atmosphere-land interactions over the Sahel region have contributed toward further reduction of rainfall. Both natural climate variability and human activities degrade the land surface in a way that exacerbates the ongoing drought, and therefore there is no guarantee that natural variability can reverse the present trend.

INTRODUCTION

The West African famine during the late 1960s and the 1970s focused the world's attention on the Sahel. The Sahel is a 200- to 500-km wide band across the southern reaches of the Sahara desert. It is ecologically fragile, though it supports nomads, herders, and sedentary farmers; it has a mean rainfall of 300 to 500 mm per year, which is extremely variable. During the past 25 years, rainfall over the Sahel has been significantly lower than the long-term mean. In the recorded meteorological data for the past 100 years, there is no other region on the globe of this size for which spatially and seasonally averaged climatic anomalies have shown such persistence. (See the paper by Nicholson in this section.)

In this paper we shall attempt to address the following questions concerning the Sahel drought:

1. What are the causes of the onset of the drought? In particular, what significant roles did the natural variability of the global climate system and of human activities play in the initiation of the drought?

2. What mechanisms are responsible for the persistence of the Sahel drought?

3. What are the prospects for reversal of the current trend? In particular, what roles might the natural variability of the global climate system and human intervention play in reversing the current trend?

It is of course difficult, if not impossible, to give precise

[1]Center for Ocean-Land-Atmosphere Studies, Institute of Global Environment and Society, Calverton, Maryland

answers to these questions. We will attempt to address them by interpreting past observational data and the results of controlled sensitivity experiments made with general-circulation models (GCMs).

ONSET OF THE SAHEL DROUGHT

Figure 1 shows the 10-year running-mean seasonal rainfall anomalies over Sahel, India, and China. The Chinese rainfall data are available for only a 30-year period (1951-1980); therefore, the Indian and the Sahel rainfall data have also been shown for that period.

We interpret these data to suggest that the occurrence of above-normal rainfall during the decade of the 1950s and the early 1960s, and below-normal rainfall during the late 1960s and the 1970s, is a feature common to the vast areas of African-Asian monsoon, not one confined to the Sahel region. We do not know why rainfall over such a large geographic region, covering two continents, was above normal during the 1950s. We also do not know why the transition from above-normal to below-normal rainfall that took place during the late 1960s occurred over the entire African-Asian monsoon region. However, on the basis of comprehensive GCM sensitivity studies (Folland et al., 1986; Rowell et al., 1992; Xue and Shukla, 1993), it is now generally accepted that the rainfall anomalies, especially over the Sahel, are consistent with, and can be simulated using, the global sea surface temperature (SST) anomalies during these periods.

On the basis of the GCM sensitivity experiments and observations shown in Figure 1, we conjecture that the onset of the Sahel's rainfall deficit in the late 1960s, along with the above-normal rainfall in the 1950s, should be explained as a regional manifestation of the planetary-scale variability of the global coupled ocean-land-atmosphere climate system. Considering the large spatial and temporal extent of these rainfall anomalies, we cannot conceive of any possible mechanism by which local human-induced changes could initiate such major droughts over two continents. We therefore reject the notion, often mentioned in popular literature, that the Sahel drought was initiated by human activities.

We believe that the pertinent question about the Sahel drought is not what initiated the drought (we conjecture it was natural variability) but why the drought has persisted for 25 years.

PERSISTENCE OF THE SAHEL DROUGHT

We further propose that strong interaction between atmospheric and land-surface processes is one of the primary causes of the drought's persistence: A reduction in rainfall (due to natural variability of the climate system) would change the land surface characteristics (increase in albedo and decrease in vegetation), which in turn would cause further reduction in rainfall. The basic mechanisms of atmosphere-land coupling have the potential to perpetuate an initial drought (or an initial excessive-rainfall) regime until they are reversed by an opposite forcing, which could be produced by natural variability.

The extent to which changes in the land-surface properties can influence the local climate depends on the spatial scale of the changes in the land-surface characteristics and the geographic location of the region. We submit that the Sahel region is particularly sensitive to atmosphere-land coupling processes. The Sahel region is geographically unique in that it is adjacent to the largest desert on the earth's surface, and it represents the largest contiguous land surface without extensive mountain terrains. The absence of large-scale topographic forcing in the region is perhaps one of the main reasons for the lack of zonal asymmetry in the rainfall pattern there. As Charney (1975) pointed out in his seminal paper on the dynamics of Sahel drought, climate in the subtropical margins of the deserts is particularly vulnerable to changes in the land-surface properties.

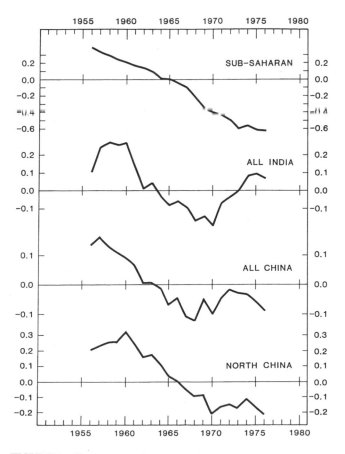

FIGURE 1   Ten-year running mean of normalized rainfall anomalies averaged for all the stations over the sub-Saharan region (10°-25°N, 18°W-25°E), India, and China. North China refers to the area to the north of 35°N.

During the past 20 years there have been a large number of GCM sensitivity studies (Mintz, 1984; Dirmeyer and Shukla, 1993) that show that increases in albedo and decreases in vegetation cover and soil moisture tend to further reduce rainfall over the Sahel. We do not know of any GCM sensitivity study challenging this basic result. We present here the results of a recent GCM sensitivity study (Xue and Shukla, 1993 and an unpublished manuscript) in which the Center for Ocean-Land-Atmosphere Interactions (COLA) model was integrated with two sets of land-surface conditions over the region south of the Sahara desert. In one set of integrations, referred to as the "desertification" experiment, it was assumed that the area approximately between 10° and 20°N and 15° and 40°E was covered by shrubs above bare soil. In the other set of integrations, referred to as the "reforestation" experiment, the area approximately between 15° and 40°E was covered by broadleaf trees above ground cover. The first set of surface conditions (desertification) was considered to represent the exaggerated state of current conditions. The second set of surface conditions (reforestation) represented a hypothetical situation in which vegetation was maintained in the margins of the Sahara Desert. It was an exaggeration of the conditions during the 1950s, when rainfall over the Sahel was above normal, and the latitude of the 200 mm isoline of seasonal rainfall in the western Sahel was at about 18°N. The model was integrated with identical global SST patterns for each change in the land-surface conditions.

Figure 2a shows the difference between the observed rainfall averaged for 1950 plus 1958 and 1983 plus 1984. Figure 2b shows the difference between the desertification and reforestation experiments for seasonal mean rainfall. The similarity between Figures 2a and 2b is remarkable. Both show negative rainfall departures between 10° and 20°N and a positive rainfall departure to the south of the negative departures. The dipole nature of both the observed and the model rainfall anomalies suggests a southward displacement of the mean rainfall pattern.

Since the only changes in these integrations were those in land-surface conditions, it is reasonable to conjecture that if the natural variability of the global climate system were to produce an initial drought, the strong atmosphere-land interaction over the Sahel region could contribute toward the persistence of that drought.

We further propose that since natural changes in the Sahel rainfall have led to large-scale changes in human activities in the region, it is quite likely that the degradation of the land surface was exacerbated by human activities (overgrazing, deforestation, soil desiccation, etc.), thereby producing further reduction in the Sahel rainfall. The exacerbating effects of human activities could have been especially large during the past 30 years because the population density has been much higher than at the beginning of the century.

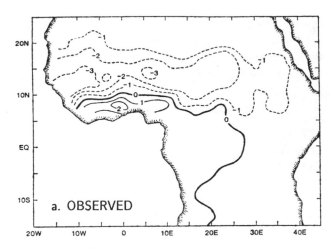

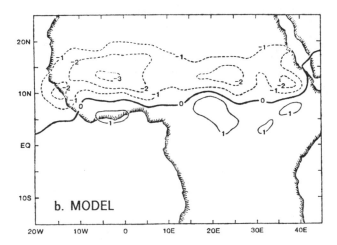

FIGURE 2 (a)   Observed summer (June, July, August, September) rainfall difference, in millimeters per day, for the average of 1983 and 1984 minus the average of 1950 and 1958. (b) Rainfall difference between desertification and reforestation GCM experiments.

The various conjectures put forward in this paper so far have been synthesized in a schematic diagram shown in Figure 3. It is seen that the atmosphere-land interaction has a positive feedback mechanism, so that initial drought caused by the natural variability of the climate system is perpetuated by increases in albedo and decreases in vegetation, soil moisture, and surface roughness. In addition, the human-induced effects also produce an increase in albedo and a decrease in vegetation, soil moisture, and surface roughness. Thus, both the planetary-scale effects and the human-induced effects are mutually reinforcing the dry conditions, which has led to an unprecedentedly long and severe drought over the Sahel.

THE FUTURE

If we accept the premise presented in the beginning of this paper, that the Sahel drought was initiated by natural

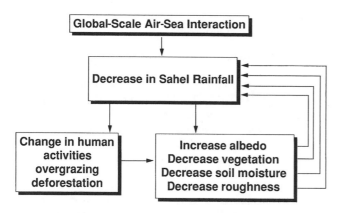

FIGURE 3   Schematic representation of the mechanisms affecting Sahel rainfall. Both the natural variability and the human-induced changes are contributing to the reduction in Sahel rainfall.

variability of the global climate system, it is reasonable to assume that at some unpredictable future time the natural variability (global SST patterns, for example) is likely to produce excessive rainfall over the Sahel and thus reverse the current drought. However, it is unclear whether the changes in the land-surface characteristics can also be reversed. Therefore, even if the global circulation shifts in such a way as to be favorable for reversing the current drought, the global natural changes will have to be stronger than before, and more favorable, to overcome the opposing effects of local atmosphere-land interactions. In fact, it is entirely possible that the Sahel drought will continue indefinitely, because the natural variability effects may never be large enough to counteract the persistent effects of the local land-surface changes.

In a hypothetical GCM sensitivity experiment, Xue and Shukla (unpublished manuscript) have shown that rainfall over the Sahel can be increased by replacing the broadleaf shrubs over ground cover and the shrubs over bare soil by broadleaf trees over ground cover over a large region that includes the Sahel. This suggests that although human effects were not responsible for the initiation of the Sahel drought, a suitable combination of large-scale natural variability and human intervention might contribute toward reversing the current drought. Modification of surface albedo, evapotranspiration, or surface roughness would all be of assistance, though of course it would be extremely difficult to sustain vegetation on a scale sufficiently large to halt the spread of the desert.

Discussion

DESER: Do you think local forcing applies to the string of wet years in the 1950s and 1960s too?

NICHOLSON: Absolutely. What I feel is the critical factor involved in the forcing is the fluxes to the atmosphere, including dust. The land surface forcing should be reversible, or symmetric, for the wet and dry periods.

RIND: How successful have people been in quantifying the albedo change between the wet and dry periods, as far as the land surface is concerned?

NICHOLSON: One study made 10 or 15 years ago took the available satellite data and reconstructed the albedo over about 8 years. The small changes found were not consistent with rainfall fluctuations, and I think basically showed that albedo is not one of the major forcing factors. More recently, Vivian Gornitz searched historical archives for material on land-surface changes, and constructed a beautiful map of West Africa showing changes due to human activities. Again, in terms of absolute values of albedo, it's a very small change, and sometimes in the wrong direction for the Charney mechanism. So I think that not albedo but something coupled to soil moisture, vegetation, and dust will be the major factor.

DIAZ: To what degree do you think that time differences in the frequency of the westward-propagating tropical disturbances affect the variability you're seeing?

NICHOLSON: I think they're key to understanding the region. This may be the one place on the globe where the land-surface forcing per se has a major impact on the overall large-scale dynamics. The jet stream at the mid-levels here, which is critical in disturbance development, is due to nothing other than the temperature gradient across the land surface down to the ocean.

Some of the models done years ago, during the GATE days, showed that if you change the relationship between baroclinic and barotropic instability you can change the characteristics of the disturbances. The difference between the wet 1950s and the dry 1970s can be explained by one or two disturbances a month that produce, say, 150 mm of precipitation a day, where the dry periods have nothing over 50. I think the large-scale circulation is clearly coupled to the land surface there. Incidentally, the dust in that region is right at the level of the African easterly jet, about 650-700 mb, which is another reason I think it is so important.

CANE: I can see that the land-surface processes can affect what's happening, but I'd like to hear what you see as the feedback mechanism for very-long-period changes. How would you cycle in and out, and what would the trigger be?

NICHOLSON: Well, if you assume some large-scale trigger in the general atmospheric circulation, maybe a major change in SST, you could get a continental-scale pattern like the wet 1950s all over Africa. But in other areas the rainfall goes back to normal, whereas in the Sahel the anomalies persist for 10 or even 20 years. There has to be some other smaller-scale forcing that reinforces the existing conditions until it's overridden by some larger trigger in the atmosphere-ocean circulation.

COLE: Ropelewski and Halpert's analysis of global rainfall suggests that the ENSO teleconnection with African rainfall is strongest in the east. Do you see a strong ENSO signal in the variability in the area you've been looking at?

NICHOLSON: There's a small one in the coastal region, but nothing in the central Sahel.

RIND: Shukla, your hypothesis suggests two questions to me. The first is, can it be documented—through satellite observations, for instance—that the vegetation really changed so extremely in those 10° latitude belts? And the second is, did those British experiments in which SST anomalies related so well to the Sahel changes, also produce the drying conditions in China and India between the 1950s and 1980s?

SHUKLA: We have exaggerated the extent of land degradation in our experiments. But there is a very clear shift in the rainfall between the 1950s and the 1980s, and Tucker and his colleagues have shown from the vegetation data that there is a very clear relationship between the two. If you look at the latitude of an isoline, it is definitely going down; the 200-mm isoline has dropped by 2° or 2.5°. So there are changes at the boundary, albeit of less magnitude than what I have chosen for this experiment.

BERGMAN: Sharon, when I was at the Climate Analysis Center, we found amazing discontinuities in precipitation time series in Western Africa that suggested that the station sites for some loca-

tions must have been moved, although there was no record of a change. Did you find this, and were you able to correct for it?

NICHOLSON: There were only a couple of stations in my network that had to be thrown out because of that sort of discontinuity, though we did find differences at the same station because of varied sources. Our approach has been to take aerial averages and hope that the outliers get factored out. But let me make a pitch here for the need to put together an archive of some of the old, forgotten data sources, such as pilot balloon data, to help us understand decadal and longer-scale variability.

RASMUSSON: It is still not clear to me that we are seeing the effect of human intervention. The vegetation seems to move north again in wetter years. Can you see an underlying southward trend?

NICHOLSON: Shukla and I differ somewhat in our views here. I think that so far human intervention has very little to do with it. Many of the papers on desertification are essentially inferring continent-wide desertification from two data points in West Africa. But I don't think that the vegetation's recovery changes any of the ideas about the relation between low-frequency forcing and land surface.

One other thing I'd like to mention with respect to Shukla's argument is that I think one variable has been left out: dust. If I had to put money on any of those land-surface variables, I'd pick dust. It responds to both rainfall and land-surface factors, it has the best memory in the system, and it has shown the most consistent relationship over time with rainfall fluctuations.

SHUKLA: That is certainly a point that ought to be discussed further, if we had time. I'd like to think a little more before I put my money on dust. But let me add just a couple of things. The population pressure has been higher than ever before in the past 20 or 30 years, and that period coincides with the most severe drought in the 100 or so years for which we have reliable instrumental data. Proxy-data evidence of persistent droughts by no means rules out a human role in the current situation. It is also possible that human intervention, such as large-scale agriculture and afforestation, has the potential for reversing the present tendency, whatever its origin.

Continental Snow Cover and Climate Variability

JOHN E. WALSH[1]

ABSTRACT

In terms of its interactions with the atmosphere, snow cover has received the most attention through the albedo-temperature associations that have been detected over daily to monthly time scales. For example, snow cover has been shown to contribute, albeit modestly, to the interannual variability of monthly surface air temperatures over land. On decadal and longer time scales, the more important role of snow cover may be its link with the hydrologic cycle over land. Changes in winter precipitation and/or temperature can, through snow cover, result in substantial temporal and spatial changes of soil moisture and runoff to the oceans. Model results are rather consistent in depicting an alteration of the seasonal cycle of temperature when the timing and amount of spring snow melt are modified. Runoff-induced effects on the ocean stratification in the middle and high latitudes have also been hypothesized, although such arguments need to be substantiated by observational data. Greenhouse-induced changes of high-latitude precipitation, especially snowfall, play significant roles in the scenarios of temperature change projected by global climate models. Data on observed snow cover and temperature variations of recent decades are, in some respects, consistent with the projections of the global models.

INTRODUCTION

Snow cover has long been regarded as an indicator and as a possible agent of climate variability over a range of time scales. Over periods of several days to several weeks, the largest changes in the earth's surface properties result from variations in snow cover on land and on sea ice. Snow cover clearly influences the local values of near-surface atmospheric variables over these time scales. Over time scales of a thousand years or longer, the advance and retreat of the ice sheets depend ultimately on changes in the rates at which snow accumulates and ablates over the continents. However, over the interannual-to-century time scales on which this workshop will focus, the climatic roles of snow cover are poorly understood. Aside from the areal snow coverage of the past 15 to 20 years, the variability of snow cover over decadal time scales is not well documented. Yet model results are sufficiently suggestive of climatic roles of snow cover that one cannot ignore snow cover in projections of climate change over decade-to-century time scales.

---

[1]Department of Atmospheric Sciences, University of Illinois, Urbana

This paper begins with a brief discussion of the physical basis of associations between snow cover and atmospheric variability. Manifestations of these associations in observational data and in model experiments will then be surveyed. While this survey contains several illustrations of shorter-term interactions between snow cover and the atmosphere, the emphasis will be on interactions within the land-atmosphere system over the interannual, decadal, and century time scales. Because sea ice is the focus of another paper in this volume, we limit the discussion to interactions involving snow on land.

PHYSICAL BACKGROUND

A major difficulty in quantifying the climatic role of snow cover is that the distribution of snow is primarily a consequence of the large-scale pattern of atmospheric circulation, which determines the broad features of the distributions of temperature and precipitation over land. Thus, even in the absence of any causal role of snow cover, there can be large statistical correlations between anomalies of snow cover and atmospheric circulation. However, the pattern of atmospheric circulation is itself determined by the distribution of diabatic heating-radiation, conduction and convection of sensible heat, and latent heating. By modifying the exchanges of energy and moisture between the surface and the atmosphere, snow cover alters the distribution of diabatic heating in the atmosphere. For example, the albedo of fresh snow is 0.80 to 0.85 in solar wavelengths, whereas the albedo of bare land or ice-free ocean is typically between 0.05 and 0.30. Snow cover can therefore reduce the solar energy available to the surface by 50 percent or even more, depending on the age and depth of the snow, the vegetative cover, and cloudiness. If this energy reduction is distributed through the lowest 2 km of the atmosphere, it can be equivalent to a cooling of 3°C to 7°C in middle latitudes under clear skies during March (Namias, 1962). Snow cover is also an effective insulator of the underlying surface and an effective radiator of infrared energy. Finally, melting snow represents an effective sink of (latent) heat for the atmosphere and an effective source of moisture for the soil. The subsequent evaporation of this moisture may prolong the tendency for snow to delay the sensible heating of the soil, thereby modifying the phase of the seasonal cycle of surface temperature. The latter hypothesis has provided the basis for several experiments with global climate models (see below).

A striking feature of the present-day distribution of snow cover over land is the virtual absence of seasonal snow in the Southern Hemisphere, where the only large area of land at latitudes in which snow can easily accumulate is the glaciated Antarctic continent. The following discussion of snow cover is therefore limited almost exclusively to the Northern Hemisphere.

SNOW COVER AND ATMOSPHERIC VARIABILITY: SHORT-TERM RELATIONSHIPS

Although the subject of this workshop is climate variability over decade-to-century time scales, we first review relevant studies of snow-atmosphere interactions over shorter time scales. The relevance of these studies stems from the fact that they illustrate interactions that may lead to climate changes over the longer time scales if one component (e.g., snow, atmospheric temperature) is systematically perturbed by some other climate forcing mechanism, either internally or externally. In this sense, the short-term relationships may be regarded as the "building blocks" of long-term change.

Observational Studies

There is little doubt that snow can have substantial impacts on local surface temperature. Analyses of rates at which relatively warm moist air is cooled as it is advected over snow indicate that the loss of heat by conduction to the surface can reduce the surface air temperature by 4°C to 5°C per day (Treidl, 1970). More recently, Petersen and Hoke (1989) showed that the accurate specification of snow cover reduced the error of a regional numerical weather-prediction model's 48-hour forecast of surface temperature (by 8°C to 9°C); the corresponding forecast of precipitation type (rain instead of snow) was also correct over a larger portion of the model domain when the snow cover was accurately prescribed. The radiative impact of the surface albedo enhancement by snow can depress daytime surface temperatures by 5°C to 10°C during spring, as shown by Dewey's (1977) diagnosis of errors in statistical forecasts that ignored snow cover (Figure 1). If this approach is extended to monthly temperature specifications based on upper-air geopotential, specification errors of 5°C to 7°C are found equatorward of the normal snow boundary during months with extremely large positive snow anomalies (Namias, 1985). The errors decrease to 1°C to 3°C when all months over approximately 30 years are included in the statistical sample (Walsh et al., 1985). These impacts of snow are generally larger in the spring, when insolation is stronger. They are, nevertheless, indicative of the changes of mean surface temperatures that could result locally from a systematic advance or retreat of snow cover over decade-to-century time scales, i.e., from a change of the normal position of the snow margin.

It should be noted that the suppression of air temperature by positive anomalies of snow cover is generally confined to the lowest 100 to 200 mb of the atmosphere (Namias, 1985). Because the wintertime troposphere over land areas is characterized by relatively strong static stability even without a suppression of the near-surface temperature, it is unlikely that conditions are favorable for the vertical propagation of the thermal anomalies produced by snow

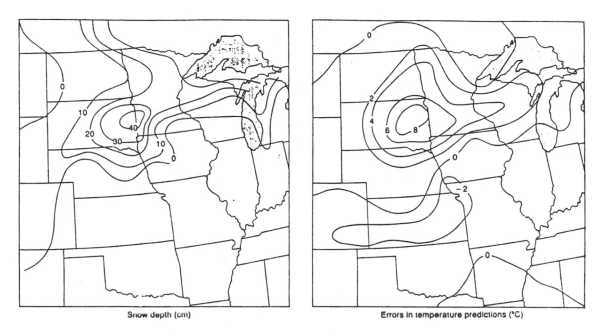

Snow depth (cm)   Errors In temperature predictions (°C)

FIGURE 1   (a) Snow depth (cm) at 1200 UTC on 5 March 1977, and (b) 5-day average of the error (°C) of the daily maximum temperature forecasts for 5-9 March 1977 derived from the National Meteorological Center's Model Output Statistics. (From Dewey, 1977; reprinted with permission of the American Meteorological Society.)

cover. However, Lamb (1955) showed that there is a detectable decrease in the 1000 to 500 mb thickness as a layer of air passes over a large area of snow cover, implying that snow cover can contribute to the maintenance of a trough of cold air, which in turn helps to maintain the snow cover.

Some of the earliest work on the role of snow cover in seasonal atmospheric variability was directed at the Asian monsoon. Blanford (1884), Walker (1910), and others postulated a link between Himalayan winter snow cover and the strength of the Indian summer monsoon by reasoning that extensive snow cover could retard the heating of the Asian landmass. This link has been substantiated in recent years with the aid of satellite-derived measurements of snow cover (Hahn and Shukla, 1976; Dey et al., 1984; Dickson, 1984; Bhanu Kumar, 1988). On the basis of data for the 1967-1980 period, linear correlation coefficients between December-to-March snow extent in the Himalayas and June-to-September monsoon rainfall were found to be approximately -0.6 (Dey et al., 1984; Dickson, 1984).

While the monsoon-snow correlations seem to provide the basis for useful monsoon predictions at ranges of several months, there are several important caveats. First, snow cover over the Himalayan region was not mapped consistently during the first part (1967-1972) of the satellite record (Ropelewski et al., 1984). Second, as shown in Figure 2 (from Bhanu Kumar, 1987), the agreement between the interannual fluctuations of snowfall and monsoon variables has been noticeably poorer since 1980. Inclusion of data from 1981 to 1985 lowered the snow-rainfall correlations

from -0.60 to -0.38 ($r^2 \approx 0.14$); the 95 percent significance threshold for a sample of N = 19 years is approximately 0.45. Third, Indian monsoon rainfall appears to correlate as highly with snow cover over the remainder of Eurasia as it does with Himalayan snow cover (Dickson, 1984). The physical linkage between snowfall and monsoon rainfall may therefore be more complicated than implied by the proposed effect of snow on the timing of the heating of the Himalayan and Tibetan region. For this reason, the modeling studies (e.g., Barnett et al., 1989) described below provide potentially important diagnostic information.

Other observationally derived linkages between snow cover and the atmospheric circulation over seasonal time scales have been addressed by Afanas'eva et al. (1979), who examined the position of the Planetary Upper-air Frontal Zone (PUFZ) over Eurasia during the autumn and spring. Time variations of the PUFZ and the snow boundary correspond closely ($r = 0.78$). However, as noted above, such correlations are at least partially attributable to the fact that the position of major upper-air features (e.g., the jet stream) is a primary determinant of the position of the snow boundary. Interannual variations of snow cover over Eurasia were also examined by Toomig (1981), who found that the annual value of the absorption of solar radiation at Soviet stations is a strong function of the springtime absorption, which in turn depends on the albedo during spring. Toomig also reported modest correlations ($r \approx 0.45$) between the surface albedo of early spring and the surface air temperatures through July at several stations. A similar lag relationship

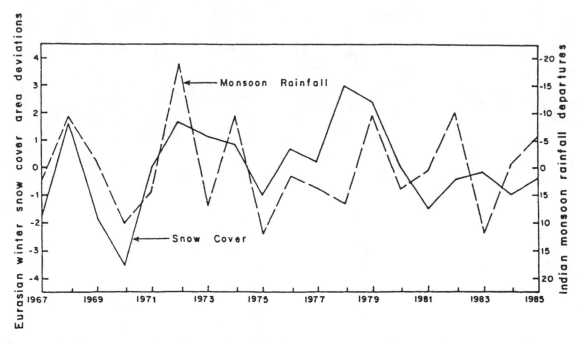

FIGURE 2   Variations of Eurasian winter snow cover and the corresponding summer monsoon rainfall over India. (From Bhanu Kumar, 1988; reprinted with permission of the Journal of the Meteorological Society of Japan.)

was suggested by Lamb (1972), who argued that the exceptionally cold summer of 1968 in northeastern Europe and northwestern Asia was attributable in part to the previous winter's extensive snow cover. Because the snow did not retreat until June and July, the ground remained cold and wet, thereby favoring the persistence of a trough. The attendant cloudiness and frequent cyclones produced above-normal rain and occasional snowfall throughout the summer over the northern Soviet Union. Wahl (1968) suggested that persistent snow may have played a similar role in the unusually cold summers of the 1830s in northern North America.

Feedbacks involving snow cover have been suggested by Namias (1962, 1981), who has argued that extensive snow cover over eastern North America should enhance the coastal baroclinicity and favor strong cyclones along the East Coast. These systems, in turn, reinforce the cold trough and contribute to its persistence over periods of several weeks or longer, as in the winters of 1960-1961 and 1976-1977. Dickson and Namias (1976) have attempted to link extensive snow/extreme cold over eastern North America with anomalies in the winter climatic regime of Greenland and northwestern Europe. According to the authors' hypothesis, the southward displacement and intensification of cyclones on the East Coast of the United States lead to a reduction of cyclone activity over the Iceland-Greenland region. This weakening of the Icelandic low reduces the advection of mild air into the northeastern North Atlantic and northern Europe, thereby increasing the frequency of polar outbreaks over the European land areas. The frequent

recurrence of this abnormal winter regime during the 1960s may have contributed to the relatively extensive European snow cover during that decade. However, the role of snow cover in modifying the atmospheric circulation in such regimes remains speculative because other hypotheses (involving sea surface temperature anomalies, tropical forcing, etc.) can be formulated. Controlled model experiments are needed to explore the physical basis of seasonal anomalies in these specific cases, as has been done for anomalies such as the North American drought of 1988 and the ENSO event of 1982-1983 (e.g., Trenberth and Branstator, 1992; Palmer and Mansfield, 1986).

Iwasaki (1991) has recently reported a tendency for large-scale anomalies of snow cover to persist from December through February, but he found no indication of anomaly persistence from February to March in the longer data record. Iwasaki also reported an apparent lag relationship, in which winters with extensive snow cover over eastern Eurasia tend to be followed by winters with extensive snow cover over North America. Although this lag relationship is statistically significant ($r = 0.52$), at least three counter-examples occurred during the 1980s.

Modeling Studies

As the preceding subsection indicates, various roles have been attributed to snow cover in the evolution and persistence of monthly and seasonal atmospheric anomalies. Most of these roles involve feedbacks, which are notoriously difficult to unravel from observational data. Consequently, model experiments provide a potentially useful vehicle for diagnosing the effects of snow cover. However, the mixed nature of the model-derived conclusions pertaining to the impacts of snow cover will become apparent as the survey proceeds.

In a study of snow-induced effects in non-winter months, Yeh et al. (1983) used a simplified version of the Geophysical Fluid Dynamics Laboratory's general-circulation model (GCM), containing idealized geography and a limited computational domain. The model contained no diurnal cycle, and its cloudiness was prescribed to be zonally uniform and seasonally invariant. The complete removal of the snow cover in mid-March was found to reduce the water available to the soil through snow melt, thus decreasing the soil moisture during the spring and summer in the region of snow removal. The drying of the soil resulted in an increase of surface temperature at high latitudes by 2°C to 8°C for the subsequent 3 to 4 months. The temperature increase extended into the upper troposphere, thereby reducing the meridional temperature gradient and the zonal wind in high latitudes.

A similar conclusion about the snow-hydrology-temperature linkage was obtained from a more realistic GCM by Yasunari et al. (1989). In this experiment, the 5° × 4° version of the Japanese Meteorological Institute GCM was run for 6 months beginning March 1. The experimental runs were identical to the control runs except for the addition of 5 cm (water equivalent) of snow in the snow-covered portion of the 30° to 60°N zone of the Eurasian continent. The results, shown in Figure 3, contained evidence of both (1) an albedo feedback, which suppressed temperatures over lower latitudes (e.g., Tibetan Plateau) by 2°C to 3°C during spring, and (2) a snow-hydrology-temperature linkage, which suppressed temperatures over middle latitudes by 2°C to 3°C during the summer months of June to August. During the summer, the anomalous Eurasian heat sink also appeared to induce a stationary Rossby wave

pattern extending from eastern Asia to northern North America.

The most thorough investigation of spring-summer feedbacks involving snow cover is Barnett et al.'s 1989 study of Eurasian snow impacts on a low-resolution (T21) version of the European Centre for Medium-Range Weather Forecasting's model. In Barnett et al.'s first experiment, snow extent corresponding to observed extremes was prescribed and interactions between snow and the surface hydrology were suppressed in order to isolate the albedo effect. The atmospheric response to the snow anomalies was local and confined primarily to air temperature and upper-air geopotential (but not sea-level pressure). All significant signals vanished when the snow disappeared in the spring, and the albedo effect had no sustained impact on the development of the Asian monsoon.

In Barnett et al.'s second experiment, rates of snowfall over Eurasia were doubled and halved so that the subsequent melt and evaporation could induce changes in the regional hydrology. The two sets of simulations showed statistically significant differences extending through the subsequent two seasons. The results derived from the doubled snowfall were characterized by significantly lower surface and tropospheric temperatures from May through July, higher pressures over Asia and lower pressures over North America, weaker zonal winds over the Arabian Sea, weaker surface convergence over southern Asia, and a weaker monsoon over southeast Asia. The Indian monsoon, however, was not substantially weaker in the "heavy snow" simulations, although this result may be partially attributable to the model resolution. The sea-level pressure signal over Asia and North America is stronger in the model than in the real world; the exchange of mass between the two continents may have been exaggerated in the model because the sea surface temperature (SST) distribution was prescribed climatologically in all the model runs.

In general, the physical mechanisms underlying Barnett et al.'s model response are similar to those of Yeh et al. (1983). The doubling of snowfall by Barnett et al. also produced a general weakening of the wind in the Southern Hemisphere and along the equator. While this response is similar to that which occurs prior to the warm phase of an ENSO event, the model with prescribed SST cannot sustain an ENSO event. In order to address the snow–ENSO link in more detail, Barnett et al. performed additional experiments with a coupled ocean-atmosphere model. The conclusion was that "the snow/monsoon signal has all the characteristics necessary to trigger the Pacific portion of an ENSO event, but the signal is too small by a factor of at least 2. In balance, it appears that snow-induced monsoon perturbations may be one of the (multiple) triggers that can initiate an ENSO cycle" (Barnett et al., 1989, p. 683-684). The elimination of major model biases and the improvement

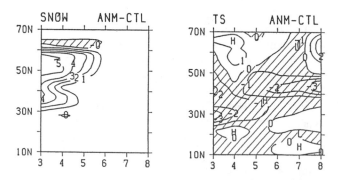

FIGURE 3   Latitude-time (month) sections of anomalies ("Heavy Snow" run minus "Control" run) of snow mass (cm liquid equivalent) and surface air temperature (°C) averaged over the Eurasian continent. (From Yasunari et al., 1989; reprinted with permission of the World Meteorological Organization.)

of the model resolution are clearly high priorities in the context of this large-scale signal involving snow cover.

The studies summarized above suggest that snow cover may play a major climatic role regionally and perhaps even globally. However, the most recently published investigation of snow-climate interactions provides some indications of a negative feedback involving snow and temperature anomalies, at least during late winter and early spring. Cohen and Rind (1991) used the Goddard Institute for Space Studies 9-layer model with $8° \times 10°$ resolution to examine the sensitivities to snow cover during March. The initial (March 1) boundary conditions corresponded to the observationally derived "maximum" and "minimum" snow cover and depth over North America and Eurasia. A key feature of this experiment was that the snow cover was interactive rather than prescribed. In the results of a five-case sample of March simulations, the positive snow anomalies caused only a short-term local decrease in surface temperature. There was no non-local signal, and even the local signal became weak after about 7 days. In the model physics, the reduction in the surface absorption of solar radiation and the increased consumption of latent heat for melting the snow contributed to lower temperatures. However, the remaining terms in the surface energy budget (e.g., long-wave radiation, sensible heat flux) adjusted so that they offset the cooling. Thus a negative feedback limited the impact of the snow anomalies to a slight cooling of about $1°C$—a smaller effect than indicated by the observational studies and by the model experiment of Yasunari et al. (1989). This negative feedback that limits the albedo effect of snow was also found in Barnett et al.'s (1989) first experiment, and the weak signal is not inconsistent with the results of Robock and Tauss (1986), who used a simple, linear, steady-state model. Thus the relatively negligible impact of snow cover in the Cohen and Rind experiment may be due to the fact that their simulations did not extend into the spring season, when the hydrologic role of snow anomalies can become more important. The compatibility of the results from these various model experiments, nevertheless, requires further attention.

SNOW COVER AND ATMOSPHERIC VARIABILITY: DECADE-TO-CENTURY TIME SCALES

The importance of snow cover for climate change over decade-to-century time scales depends on several factors. First, changes in other components of the climate system are likely to alter the large-scale distribution of snow cover. Relatively small shifts in the atmospheric circulation pattern can have major effects on the snow distribution in mountainous areas, e.g., the Rocky Mountains (Changnon et al., 1993). The extent to which snow cover will change in response to a changing climate is not well known; for example, a general warming may increase snow melt and

decrease the fraction of precipitation that falls as snow, but (according to models) it is likely to increase precipitation in high latitudes-where most precipitation falls as snow. In view of these competing effects, it is conceivable that snow extent could decrease while high-latitude snow volume increases. Second, changes of snow cover can trigger a host of potential feedbacks involving air temperature, soil moisture, cloudiness, the phasing of the seasonal cycle, and other variables. The magnitudes and relative importance of these feedbacks are poorly known. The individual feedbacks are notoriously difficult to isolate in observational data. Although various model experiments have addressed individual feedbacks, the isolation of individual feedbacks can be a nontrivial undertaking even in model simulations.

In the section below, we address the first of the two factors listed above by surveying recent analyses of observational data pertaining to snow cover. We then address the issue of snow-related feedbacks in climate change by surveying the recent model experiments that may be most relevant to changes over the decade-to-century time scales.

Observational Studies

Observational studies of historical variations of snowfall, snow depth, and snow water equivalent are confounded by measurement difficulties pertaining to small-scale variations of these variables. Large spatial gradients of all three variables are found over areas containing even moderate topographic features. Moreover, snow gauges are known to "undercatch" snow; the degree of undercatch varies with the type of gauge, and the type of gauge has changed during the period of record at nearly all stations. While satellite data have provided essentially continuous global coverage since the early 1970s, the useful information is generally limited to areal extent. The critically needed mapping of snow depth or water equivalent is not yet achievable over large areas, although algorithms for snow depth in vegetation-sparse areas have been used with some success (Chang et al., 1987).

The most comprehensive analysis of satellite-derived data on snow coverage was made by Robinson and Dewey (1990), who found that the Northern Hemisphere snow cover of recent years is less extensive than that of 10 to 20 years ago. Perhaps coincidentally, the increase of surface temperature over the last few decades is larger over land than over the oceans. The high-latitude land areas, which are generally snow covered during winter, show the strongest warming (Figure 4; see color well); this warming has been strongest in spring and winter. In a recent analysis of North American data, Karl et al. (1993) identified several regions that, because of their high variability of snow cover, have exerted the primary influence on North American snow variations. These regions, which vary seasonally, are ones in which the inverse snow-temperature relationship is strong. Karl et al. also used station data to extend their analysis to the multidecade-to-century time scale. Their findings included a 4 to 5 percent increase of both solid and total precipitation rates over northern Canada during the past four decades, century-scale increases of precipitation over southern Canada and the contiguous United States, and a decrease in the proportion of precipitation falling as snow in southern Canada. These findings are generally consistent with the changes that have been hypothesized to accompany a greenhouse warming. Karl et al. also found that the unprecedented warmth of the 1980s in Alaska was accompanied by a 10 percent increase of annual precipitation in that region. Leathers and Robinson (1993) have examined the winter region (the central United States) in more detail, finding that the concurrent 500 mb height anomalies are generally collocated with the snow/temperature anomalies during December but not during January and February.

The general coincidence of the "marginal snow zone" and the areas of strongest warming over the past several decades (Figure 4) deserves further comment with regard to the possible nature of the forcing. The "land-leading-ocean" feature is characteristic of large-scale forcing such as global warming or the response to major volcanic events. By contrast, natural low-frequency variations will generally manifest themselves in an "ocean-leading-land" pattern because the ocean is the low-frequency source of the forcing on the near-surface atmospheric temperature. The extent to which a large-scale forced temperature response is amplified by the retreat of snow over land is one of the key unknowns in the interpretation of a pattern such as that in Figure 4.

Decadal-scale summaries of snowfall in China have been compiled by Li (1987), who found a decrease of snowfall over China during the 1950s followed by an increase during the 1960s and 1970s; the late 1970s had the largest running-mean values of the 30-year period of record (Figure 5). Li noted a general correspondence between the time series of global mean temperature and mean snow depth in China (as well as an apparent association between years of heavy

snow and ENSO events). Similar compilations have been made for other regions such as the Swiss Alps (Lang and Rohrer, 1987), although the representativeness of such time series is a key issue. From the standpoint of areal coverage, the snowfall records that are potentially most valuable are those of the former Soviet Union. The status of these data is uncertain.

The longest time series of snow cover are generally those coming from single stations (e.g., the Tokyo record dating back to the 1630s (Lamb, 1977)) or the records of annual precipitation derived from polar ice cores (e.g., the Greenland Ice Sheet Project's work in Antarctica and Greenland, as in Alley et al., 1993). The cores represent potentially valuable records of regional snowfall if the spatial representativeness of the point data can be established. The South Pole data, for example, have been used to deduce an increase of snowfall from the 1700s to the early twentieth century (Giovinetto and Schwertfeger, 1966; Lamb, 1977). Analysis of GISP's Greenland cores has only recently begun (Mayewski et al., 1993).

Another type of data for which analyses are in the early phases is the runoff (stream-flow) data for the high-latitude rivers that are fed primarily by snow melt. Mysak et al. (1990) have proposed a mechanism by which an interdecadal (approximately 20-year) cycle results from a feedback loop involving high-latitude precipitation, runoff, arctic sea-ice export to the North Atlantic, and ocean salinity/temperature anomalies. The linkages involving snowfall and runoff have yet to be thoroughly evaluated. In view of the potential implications of this cycle for the global ocean circulation, quantitative analyses of high-latitude snowfall and runoff are being assigned high priority in the upcoming ACSYS (Arctic Climate System) component of the World Climate Research Program. The extension of such analyses to include soil moisture and land surface temperatures also appears to merit high priority in the context of possible greenhouse-induced changes over decadal time scales.

Finally, glaciers are ultimately attributable to continental snowfall over time scales of decades to millennia. The advance and retreat of glaciers have long been regarded as proxy indicators of climate change over these time scales. In the context of global change, the key properties of glaciers are extent, ice volume, and mass balance. Temporal changes in these glacial properties can be complex functions of temperature, precipitation, the seasonality of temperature and precipitation, and topography. Nevertheless, the advance and retreat of glaciers are known to be consistent with century-scale changes of regional climate, e.g., the Little Ice Age.

Field measurements of glaciers are now assembled and reported by the World Glacier Monitoring Service in Zürich. These data, which are summarized by Haberli et al. (1989), include direct measurements of the mass balance of approximately 75 glaciers in the Northern Hemisphere. Statistics on

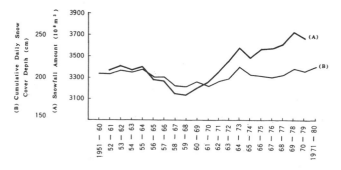

FIGURE 5 Ten-year running mean of horizontally integrated snow depth and amount in China. (From Li, 1987; reprinted with permission of IAHS Press.)

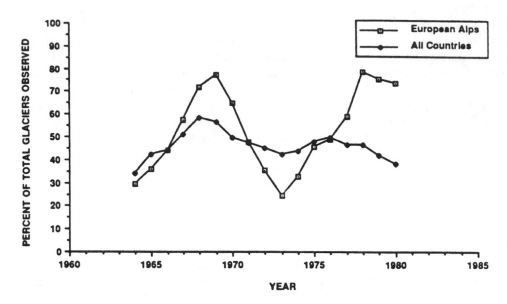

FIGURE 6   Five-year moving averages of percentage of glaciers having positive mass balance. (From Wood, 1988; reprinted with permission of the University of Colorado.)

glacial changes have been compiled for about 500 glaciers worldwide (e.g., Wood, 1988). Figure 6 (from Wood, 1988) shows the trends in glacier mass balance over approximately 20 years, expressed as the percentage of observed glaciers having a positive mass balance. While this percentage for all glaciers is close to 50 percent, the percentage for an individual region such as the European Alps can undergo multi-year excursions well above or below 50 percent.

Modeling Studies

Aside from CO_2-doubling experiments, there have been very few three-dimensional model simulations of the land-atmosphere system over decadal time scales. In this respect, one may argue that less attention has been given to the simulation of land-atmosphere interactions than to the simulation of large-scale ocean variations over decade-to-century time scales. Examples of the latter are the experiments of Bryan (1986), Weaver et al. (1991), and Yang and Neelin (1993). Moreover, most models containing an interactive land surface treat the surface physics quite crudely: the "bucket" method is typically used to handle soil moisture, multi-level soil treatments are generally not included, and runoff often "disappears" from the system.

Most global climate models do include a thermodynamically and hydrologically interactive snow cover. In its assessment of climate models, the IPCC (1990) found that the snow cover simulated by several models was "broadly realistic," leading to the statement that "the simulated snow extent should thus not distort the simulated global radiative feedbacks" (IPCC, 1990, p. 113). Even if one gives the benefit of the doubt to that assertion, the snow cover simu-

lated by all models appears to contain significant errors, particularly over eastern Asia. More comprehensive diagnostic assessments of the simulated snow cover (extent and water equivalent) in the context of both the present climate and projections of climate change are needed.

A more sobering finding concerning model simulations of the feedback between snow and radiation is contained in the IPCC's (1992) update, which cites Cess et al.'s (1991) comparison of the snow feedback (under perpetual April forcing) in 17 global climate models. The temperature sensitivity or feedback parameter, λ, associated with snow cover was found to vary from negative values in some models to a wide range of positive values (Figure 7). The sensitivity parameter in clear-sky regions ranges from values corres-

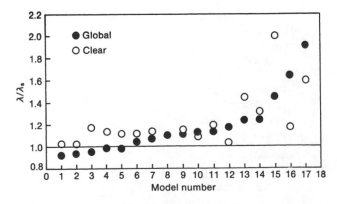

FIGURE 7   Snow feedback parameter for 17 general-circulation models. Solid circles denote global values, open circles denote values for clear-sky regions. (From Cess et al., 1991; reprinted with permission of the American Association for the Advancement of Science.)

ponding to negligible snow feedback to values representing a twofold amplification. Cloudiness complicates the interpretation of the snow feedback even further. The IPCC concludes that "the snow-albedo-temperature feedback processes in models are somewhat more complex" (IPCC, 1992, p. 117) than the conventional interpretation of the positive snow-albedo-temperature enhancement of a perturbation. Since these models also generally lack realistic surface hydrologies, one must view with caution their simulations and/or projections of atmosphere-snow interactions.

The model experiments of Yeh et al. (1983), Barnett et al. (1989), Yasunari et al. (1989), and Cohen and Rind (1991), all of which were summarized above, are potentially relevant to decade-to-century-scale climatic change even though the simulation periods are subdecadal. This relevance stems from the fact that climate changes over decadal time scales may perturb the large-scale snow distribution by amounts comparable to the changes of snow prescribed in the modeling experiments. However, biases introduced by the parameterizations of the snow and surface physics must be addressed before the results of the experiments can be viewed with confidence. A step in this direction has been made by Washington and Meehl (1986), who found that the inclusion of a simple temperature dependence in the parameterization of snow albedo can change substantially the globally averaged surface temperature increase caused by a doubling of CO_2. Ingram et al. (1989) and Covey et al. (1991) also examined the effects of high-latitude surface albedo parameterizations, although their experiments focused on sea-ice albedo. Using an earlier generation of the National Center for Atmospheric Research's model, Williams (1975) found that changes in surface albedo (and SST) influenced a simulated ice-age circulation more than did the orographic changes caused by the glaciation. Further surface-sensitivity experiments with more current atmospheric models and with more realistic treatments of the surface physics and hydrology are needed.

An issue in need of particular attention is the apparent paradox involving recent observational data (on snow and air temperature) and modeling studies of the feedback between snow, soil moisture, and temperature. The modeling studies cited above indicate that a positive anomaly of snow cover tends to depress surface air temperatures through its enhancement of soil moisture for up to several months after the snow melt. The model results seem inconsistent with the observed tendencies toward higher springtime temperatures and increases of snowfall in northern land areas over the past several decades. However, an earlier retreat of snow during the late winter or early spring may be a consequence of the warming, which is most pronounced in the winter and spring seasons, especially if the increases of snow depth have occurred primarily in the northernmost

land areas (as the data suggest they have). The earlier retreat of snow creates the possibility that the albedo effect may offset, or even dominate, the tendency of greater soil moisture to delay the seasonal warming. Since this scenario is especially likely if the upper layers of soil dry rapidly, it is clearly important that models accurately resolve the hydrological and thermal changes in the upper layers of the land surface during the snow-melt period. The potentially high climatic leverage of the snow-melt period should make it a focus of observational and modeling studies, whether the time scales of interest are decadal or century or even longer.

CONCLUSION

Several conclusions emerge from the results surveyed here. The first, which pertains to the strategy for diagnosing the role of snow cover in climate variability, is that both modeling and data analysis are essential and complementary diagnostic tools. Because snow is involved in a variety of interactions within the climate system, and because the distribution of snow is determined largely by other components of the climate system, controlled experimentation with numerical models is a key element of the diagnostic strategy. However, the biases and other limitations of models are such that numerical experiments yield convincing conclusions only when the observational data give some credibility to the model results.

The diagnostic studies of the past few decades appear, at first glance, to have produced a somewhat inconclusive picture of the climatic role of snow cover. However, several scientific conclusions do emerge when the results are viewed in an aggregate sense:

1. Anomalies of snow cover are clearly associated with significant local anomalies of air temperature, at least in the lowest 100 to 200 mb. The duration of these local anomalies generally ranges from several days to several weeks and is often limited by the fluctuations of the snow anomalies themselves over weekly and monthly time scales.

2. Over the past two to three decades, changes of surface air temperature are broadly consistent with changes of snow coverage, especially in winter and spring.

3. There is little or no evidence that the albedo effect of wintertime snow anomalies produces meaningful signals in the large-scale atmospheric circulation.

4. The snow-hydrology-soil moisture feedback appears to be capable of producing a meaningful response in the atmosphere during spring and summer. The scale of this response is at least regional and possibly larger. Regions susceptible to snow-hydrology effects include eastern Asia and the northern portions of Eurasia and North America.

5. The mechanism(s) by which snow influences the large-scale atmosphere are sufficiently complex that rela-

tively sophisticated (e.g., nonlinear) models are required for the diagnosis of snow-atmosphere interactions.

With regard to (4) and (5), the validity of parameterizations involving snow cover in global climate models may be a major constraint on further progress toward an understanding of snow-atmosphere feedbacks. Priority should be given to the parameterization of quantities and processes such as snow surface albedo, the influence of snow on the

boundary layer structure, and the disposition of the liquid water produced by snow melt.

ACKNOWLEDGMENTS

The preparation of this paper was supported by the National Science Foundation through Grants DPP-9214793 and ATM-9319952. We thank Norene McGhiey for typing the manuscript.

Commentary on the Paper of Walsh

DAVID A. ROBINSON
Rutgers University

Rather than commenting on Dr. Walsh's paper, I am going to show you a little about my work on hemispheric snow cover over the last several decades and also on the century time scale I am beginning to put together. This is at Dr. Walsh's suggestion, I hasten to add.

Figure 1 gives you an idea of the annual cycle of snow cover over Northern Hemisphere lands. The area covered by snow ranges from 40 to 50 million km² during the winter over the Eurasian and North American continents to several million km² in summer, primarily on top of the Greenland ice sheet.

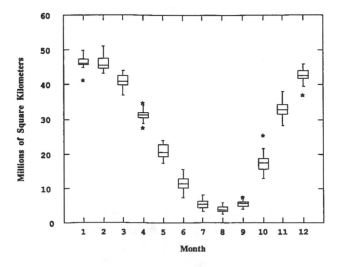

FIGURE 1   Monthly snow cover over Northern Hemisphere lands (including Greenland) between January 1972 and August 1991. The median area of cover is the horizontal line within the 12 monthly boxes, and the interquartile range (ICR) is between the top and bottom of the box. Whiskers show the extreme values between +1 and +1.5 times ICR and between −1 and −1.5 times ICR, and asterisks show values outside these ranges. Values are calculated from NOAA weekly snow charts.

In recent years there has been a dearth of snow cover over Northern Hemisphere lands (cf. Figure 3 in my paper in this volume). It has been true for the last 5 to 6 years over both Eurasia and North America. So it is seen hemisphere-wide, with many record lows set in the last couple of years, particularly 1990. I remind you that this is based on 20 years of observation. Nonetheless, this record, produced weekly by NOAA, happens to be the most consistent long-term satellite-derived data set available for any climatic variable. Charts go back to 1966, but most of us who have looked at the data feel that it is relatively homogeneous and credible only from 1972 on.

Looking at seasonal variations in snow cover over the past 20 years (cf. my Figure 4 in this volume), you see that there has been very little variation over the winter seasons in Eurasia, and even less in North America. This agrees with what Dr. Walsh showed earlier for sea ice. Fall cover has varied a bit more, but spring and summer show the greatest year-to-year variations in cover. In recent years snow cover has decreased in March through June over both North America and Eurasia. These spring decreases in recent years have been most pronounced over northeastern Asia and down the leeward side of the Canadian Rockies into the northern portions of the United States.

So that is what we have seen hemispherically over the past 20 years. Unfortunately, we have no data sets going back longer than that on a hemispheric scale, but there have been some recent efforts to try to extend local and regional data sets back to the turn of the century. For instance, at Denison, Iowa, over the course of this century there appears to be an increase in the number of snow-covered days with ≥1 inch (2.5 cm) of snow on the ground during the fall, winter, and spring.

As for other data sets to examine, we are beginning to sort out daily data from a set containing several hundred Russian stations. In some cases, the records go back to the turn of the century. We have also been examining station data in western China extending back into the 1950s.

Discussion

NORTH: The warming pattern John mentioned, with the leading in the middle of the continental areas and the lagging in the ocean, seems to be a very robust feature. It seems to hold for transient runs for all models, and even toy models have it. I think it's something that needs to be put into the filter when you're trying to do the signal processing I mentioned earlier. You get an increase on the order of 30 percent in the signal-to-noise ratio.

This same land-leading-ocean feature will appear when you have a large-scale forcing. But when it's natural variability at very low frequencies, the ocean tends to lead—for example, with Mt. Pinatubo. It would be interesting to look at the phase relations during the 1940s, when the cooling was so dramatic in the Northern Hemisphere, to see whether the sea ice was leading.

DELWORTH: In the GFDL model for natural variability, for time scales less than 50 years, the continental-region temperatures actually correlate most with the hemispheric means. It's complex, but I think you have to go to longer time scales to get the deep ocean involved.

SHUKLA: I wonder about the apparent paradox of a higher snow amount's being succeeded by warming and reduction in area in the following season.

WALSH: If the snow is disappearing earlier because of the greater springtime warmth, the perturbation of soil moisture due to the extra snow may just be swamped. The additional solar radiation available may be a larger effect than the soil moisture. Also, we don't know that the time scales of the soil perturbations in the model agree with the real world's.

ROBINSON: You might also have different effects deep within the pack and near the periphery.

RASMUSSON: One of the contributors to the recent warming over Eurasia was the exceptionally warm winter of 1982-1983. We were mystified by it, since it was not supposed to be associated with the 1982-1983 El Niño event, but Alan Robock's work on volcanic eruptions suggests that it could have been a response to El Chichon. Also, Dave, I wondered whether you had been able to clear up the snow-cover data problems you were having.

ROBINSON: Yes; I found that two different land masks had been used in the 1970s and 1980s, both of which were wrong. Using GIS techniques, I've come up with a land mask and a procedure that yield consistent and reasonably accurate numbers.

RIND: One caveat: The model run you looked at did not have one additional forcing, the reduction in ozone in the lower stratosphere. That is most effective at polar regions, and would also induce a cooling trend at the surface.

BRYAN: It seems to me that the resemblance to a greenhouse effect that appeared in the slide you showed for the 30-year average may be at least partly an artifact of the averaging—a combination of a very cold event at the beginning of the period and warming over land masses toward the end. I think it could be a little misleading to compare it with the slow warming observed in the greenhouse model experiment.

WALSH: I agree with your comment on the land, but the changes were actually negative over the high-latitude oceans during that period, so I don't think the start can have been unrepresentative.

NORTH: The Hanson-Lebedeff paper of 1988 or so, which was for a longer period leading up to the 1940s, showed the same kind of pattern.

TRENBERTH: I feel obliged to mention that the Southern Hemisphere has snow cover and sea ice too. I know Claire Parkinson has done some work there; can you comment, John?

WALSH: There's essentially no snow there outside the Antarctic continent. The one thing we know from the 20-year satellite record of the South Orkneys is that those data are not representative of the larger-scale changes.

ROBINSON: Ken Dewey and Randy Cerveny are doing a little work with snow over Patagonia and other areas.

JONES: Returning to an earlier point, I wanted to say that the temperature variations at the two Antarctic inland stations and Vostok over the past 35 years are similar to what we've seen around the coast. Since Greenland is a much smaller ice sheet than Antarctica, we may be able to infer that the temperature changes in its interior would also be similar to those measured around the coast.

KUSHNIR: Just a correction to an earlier suggestion: The coldest temperatures in the Atlantic were in the 1970s, not the 1960s, so John may indeed be picking up a trend.

GROISMAN: I'd like to note that it's important that John found the changes in sea ice in summer. Wintertime ice can go back to its old limits, but the new ice is most easily melted off in the spring.

Decadal Variations of Snow Cover

DAVID A. ROBINSON[1]

ABSTRACT

In situ and satellite observations of snow-cover duration show a considerable amount of year-to-year variability on local to hemispheric scales. This variability is often embedded in longer-term fluctuations. In situ data from the central United States indicate multi-decadal fluctuations in the duration of snow over the past century, and hemispheric satellite observations suggest variations lasting for several years or more.

Such analyses have become possible in recent years as a result of the recovery, digitization, and validation of historic in situ observations of snow cover and the availability of two decades of satellite observations of continental snow cover. Efforts are needed to recover additional historic data, to improve the recognition of snow using microwave satellite data, and to create global snow products using all available data sources. With these data available for analysis, knowledge of the spatial and temporal kinematics of snow cover will continue to improve. This will contribute to a better understanding of the role of snow in the climate system and to the utility of snow as an indicator of climate change.

INTRODUCTION

Snow cover is a critical influence on the earth's climatic energy and hydrologic budgets. In many regions it may play an influential role in determining the magnitude of any human-induced climate change, and might be a useful indicator and monitor of such change. To understand better the importance of snow in the climate system, it is essential that accurate information on the temporal and spatial dimensions of snow cover be available. Such data have been obtained through in situ and satellite observations, and recent efforts have begun to locate, assimilate, and validate

these data. Once available, they are being employed in empirical and modeling investigations of snow-cover kinematics and the dynamical aspects of snow within the climate system. An excellent survey of the latter is provided by Walsh (1995) earlier in this section. This paper will concentrate on the distribution of snow cover in space and time over the past two decades and the past century. Much of the discussion will focus on the newly available in situ and satellite data sets. Given the lack of attention paid to secular snow cover data until recently, only limited analyses of these data have taken place. A few examples of these efforts

[1]Department of Geography, Rutgers University, New Brunswick, New Jersey

will be presented. For information on snowfall and its long-term variability across the North American continent, the paper later in this section by Groisman and Easterling (1995) is recommended.

IN SITU OBSERVATIONS

In situ snow-cover data are gathered mainly over land. Only a few short-term studies have measured snow on sea ice or ice sheets (e.g., Hanson, 1980). Most observations on land are made on a once-per-day basis. The general practice is to record the average depth of snow lying on level, open ground that has a natural surface cover. At primary stations, the water equivalent of the snowpack may also be measured. In some regions, snow courses have been established where snow depth, water equivalent, and perhaps other pack properties are measured along prescribed transects across the landscape. Observations are often made only once per month, and the number of courses is extremely limited in North America. More frequent and abundant course data are gathered in the Commonwealth of Independent States, and currently this information is being recovered through a cooperative effort between the U.S. National Snow and Ice Data Center (NSIDC) at the University of Colorado and A. Krenke of the Russian Academy of Sciences.

Current station observations of snow cover are of a sufficient density for climatological study in the lower elevations of the middle latitudes. Elsewhere, while data of a high quality are gathered at a number of locations (Barry, 1983), the spatial and temporal coverage of the information is often inadequate for climate study. There is no hemispheric snow-cover product based entirely on station reports. The U.S. Air Force global snow-depth product depends heavily on surface-based observations as input into a numerical model that creates daily charts with global coverage, but it must rely on extrapolations and climatology in data-sparse regions (Hall, 1986; Armstrong and Hardman, 1991). There have been a number of regional snow-cover products over the years that are based on station data. Of greatest longevity are the Weekly Weather and Crop Bulletin charts, which have been produced since 1935. These, and the daily NOAA charts, are produced for the conterminous United States mainly from first-order station observations. Therefore, neither has a particularly high resolution, and observations may be influenced by urban heat-island effects.

In a number of countries, there are numerous stations with relatively complete records of snow extending back 50 years or more (Barry and Armstrong, 1987). Until recently, most data have remained unverified and disorganized (Robinson, 1989). As a result, few studies have dealt with long-term trends or low-frequency fluctuations of snow over even small regions (e.g., Arakawa, 1957; Manley, 1969; Jackson, 1978; Pfister, 1985). Through the coopera-

tive efforts of a number of scientists and data centers, this situation has begun to be rectified. Examples include the exchange of data through the US/USSR Bi-Lateral Environmental Data Exchange Agreement and between the Lanzhou Institute of Glaciology and Geocryology and both Rutgers University and the NSIDC. These and other data are in the process of being quality controlled, and routines to fill in gaps in snow-cover records are being developed (Hughes and Robinson, 1993; Robinson, 1993a). Clearly, there is a need to continue efforts to identify, assimilate, in some cases digitize, and in all cases validate station and snow-course observations from around the world. These data must also must be accompanied by accurate and complete metadata.

Lengthy in situ records continue to be analyzed for individual stations, and data from networks of stations have begun to be studied on a regional level. For example, marked year-to-year variability in snow-cover duration is recognized over the course of this century at Denison, Iowa (Figure 1). Snow at least 7.5 cm deep has covered the area for as much as 80 percent of the winter, but in a number of years no or only a few days have had a cover this deep. Overall, the duration of winter and spring cover was at a maximum in the 1970s, and fall cover peaked in the 1950s. Other periods of more frequent winter cover include the 1910s and the late 1930s to early 1940s. The 1920s, middle 1940s, and late 1950s to early 1960s were periods with less abundant winter cover. Missing data around 1950 prohibit a direct assessment of winter cover at Denison, but adjacent stations suggest cover was scarce at this time. All three seasons show a greater abundance of snow-cover days in the past 40 years than in the first half of the century.

Efforts are under way to develop gridded snow files for a large portion of the central United States, using data from several hundred stations. Raw and filtered winter records from four of these stations are shown in Figure 2. They are for days with snow cover ≥ 7.5 cm, and all indicate long-term fluctuations on the order of one to several decades. The Nebraska and Kansas stations show maximum durations during the past several decades, with a similar early maximum at Oshkosh, Nebraska, in the 1910s and early 1940s. Late 1920s, early 1950s, and 1970s maxima were observed at Dupree, South Dakota, the latter two ending abruptly shortly thereafter. The North Dakota station had maximum winter snow cover in the 1930s, around 1950, and in the late 1970s. The range in filtered values over the period of record was approximately two weeks in Kansas and Nebraska and seven weeks in South Dakota and North Dakota.

SATELLITE OBSERVATIONS

Snow extent is monitored using data recorded in short-wave (visible and near-infrared) and microwave wave-

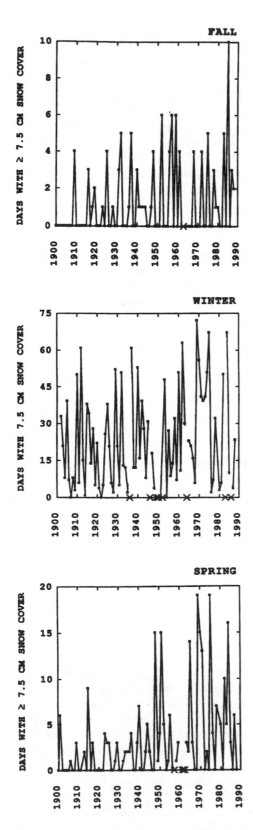

FIGURE 1   Time series of fall (September-November), winter (December-February), and spring (March-May) days with ≥7.5 cm of snow cover at Denison, Iowa. Missing years are marked with an X.

lengths by sensors on board geostationary and polar orbiting satellites. Retrieval techniques, the strengths and limitations of each spectral region for sensing snow, and the snow products derived using short-wave and microwave input are discussed in this section. The secular remote sensing of snow over Northern Hemisphere lands will be the principal focus; only a few efforts have addressed this over Southern Hemisphere lands (Dewey and Heim, 1983) or Arctic sea ice (Robinson et al., 1992).

Short-Wave and Microwave Snow Charting

Short-wave data provide continental coverage of snow extent at a relatively high spatial resolution. Snow is identified by recognizing characteristic textured surface features and brightness. Information on surface albedo and percentage of snow coverage (patchiness) is also gleaned from the data. Shortcomings include (1) the inability to detect snow cover when solar illumination is low or when skies are cloudy, (2) the underestimation of cover where dense forests mask the underlying snow, (3) difficulties in distinguishing snow from clouds in mountainous regions and in uniform, lightly vegetated areas that have a high surface brightness when covered with snow, and (4) the lack of all but the most general information on snow depth (Kukla and Robinson, 1979; Dewey and Heim, 1982).

Microwave radiation emitted by the earth's surface penetrates winter clouds, permitting an unobstructed signal from the surface to reach a satellite. The detection of snow cover from microwave data is possible mainly because of differences in emissivity between snow-covered and snow-free surfaces. Estimates of the spatial extent, as well as of the depth or water equivalent, of the snowpack are derived from equations that employ measurements of radiation sensed by multiple channels in the microwave portion of the spectrum (e.g., Kunzi et al., 1982; McFarland et al., 1987). Estimates of snow cover have been made using microwave data since the launch of the Scanning Multichannel Microwave Radiometer (SMMR) in late 1978. The spatial resolution of the data is approximately several tens of kilometers. Since 1987, close to the time of SMMR failure, the Special Sensor Microwave Imager (SSM/I) has provided data. Both sensors, having nearly the same spectral characteristics, have similar levels of success in monitoring snow extent (cf. the SMMR analyses below).

As with short-wave products, the microwave charting of snow extent is not without its limitations. The resolution of the data makes the detailed recognition of snow cover difficult, particularly where snow is patchy, and it is difficult to identify shallow or wet snow using microwaves. Also, the lack of sufficient ground-truth data on snow volume, wetness, and grain size makes an adequate assessment of the reliability of microwave estimates uncertain. The influence of a forest canopy on microwave emissions in snow-

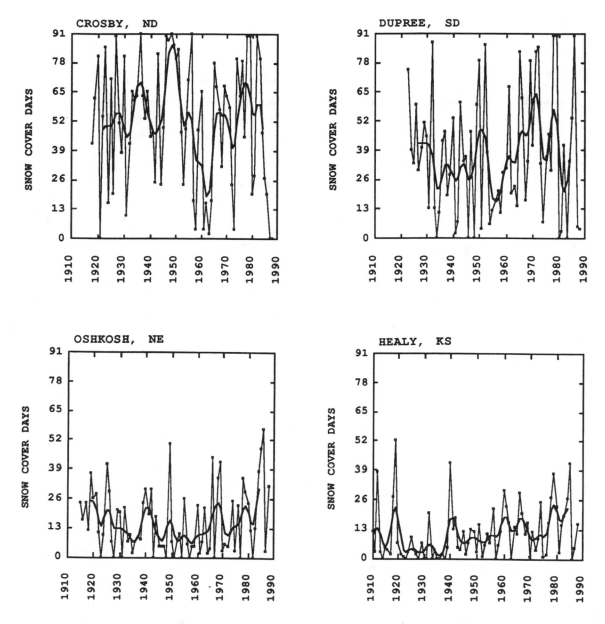

FIGURE 2 Time series of winter days with ≥7.5 cm of snow cover at four stations on the U.S. Great Plains. Also shown smoothed with a nine-point binomial filter, with only those points plotted where nine years are available for filtering (i.e., plotted year ±4 yr).

covered regions, which is currently not well understood, must also be taken into account when estimating snow cover. Because of the region-specific differences in land cover and snowpack properties, no single algorithm can adequately estimate snow cover across Northern Hemisphere lands.

Hemispheric Snow Products: NOAA Weekly Charts

In 1966, NOAA began to map snow cover over Northern Hemisphere lands on a weekly basis (Matson et al., 1986). That effort continues today, and remains the only such hemispheric product. NOAA charts are based on a visual interpretation of photographic copies of short-wave imagery by trained meteorologists. Up to 1972, the sub-point resolution of the meteorological satellites commonly used was around 4 km; since then it has become, and remained, close to 1.0 km. Charts show boundaries on the last day that the surface in a given region was seen. Since May 1982, dates when a region was last observed have been placed on the charts. An examination of these dates shows the charts to be most representative of the fifth day of the week.

It is recognized that in early years the snow extent was underestimated on the NOAA charts, especially during fall. Charting improved considerably in 1972 with the improvement of sensor resolution, and since then charting accuracy has been such that this product is considered suitable for continental-scale climate studies (Kukla and Robinson, 1981). The NOAA charts are most reliable where skies are frequently clear, solar illumination is high, snow cover is reasonably stable or changes slowly, and pronounced local and regional signatures are present owing to the distribution of vegetation, lakes, and rivers.

The NOAA charts are digitized on a weekly basis using the National Meteorological Center Limited-Area Fine Mesh grid. This is an 89 × 89 cell Northern Hemisphere grid, with cell resolution ranging from 16,000 to 42,000 km². If a cell is interpreted as being at least 50 percent snow covered, it is considered to be completely covered; otherwise it is considered to be snow free. Inconsistencies in the designation of a cell as land or water have occurred in the past during the digitization process. This has recently been resolved for the approximately 100 cells in question through the use of digital map files analyzed on a geographic information system (Robinson, 1993b).

A new routine for calculating monthly snow areas from the NOAA data has also recently been developed (Robinson, 1993b). This has eliminated previous inconsistencies resulting from undocumented changes in the methods used by NOAA to calculate the monthly values (Robinson et al., 1991). The new Rutgers Routine calculates weekly areas from the digitized snow files and weights them according to the number of days of a given week that fall in the given month. A chart week is considered to center on the fifth day of the published chart week (cf. above). No weighting has been employed in the NOAA routines.

Hemispheric Snow Products: NASA Microwave Files

Monthly charts of Northern Hemisphere continental snow extent have been produced from SMMR data by NASA scientists (Chang et al., 1990). The only such time series available to date, it covers the interval from November 1978 through August 1987. A single algorithm is used to estimate snow depth on a 0.5° × 0.5° grid. This theoretical algorithm uses the difference in brightness temperatures of 18 and 37 GHz SMMR data to derive a snow-depth/brightness temperature relationship for a uniform snow field. A snow density of 0.3 g/cm³ and a snow-crystal radius of 0.3 mm are assumed; fitting the differences to the linear portion of the 18 and 37 GHz responses permits the derivation of a constant that is applied to the measured differences. This algorithm can be used for snow up to 1 m deep.

The monthly depth estimate for a given grid cell is calculated by averaging depths reported for the five or six pentad charts centered in a given month (SMMR data are gathered every other day and three of these passes are used for each pentad chart; there is a one-day gap between each pentad). If the average is ≥2.5 cm, the cell is considered to be covered with snow for the whole month. This method-

ology biases the snow areas to the high side, especially in those areas and periods where snow cover fluctuates.

Northern Hemisphere Continental Snow Cover: 1972-1992

According to the NOAA snow charts, the extent of snow cover over Northern Hemisphere lands varies from 46.5 million km² in January to 3.9 million km² in August (Table 1); most of August's snow lies on top of the Greenland ice sheet. The past two decades of monthly data are close to normally distributed, and monthly standard deviations range from 0.9 million km² in August to 2.9 million km² in October. The annual mean cover is 25.5 million km² with a standard deviation of 1.1 million km². The snowiest year was 1978, which had a mean of 27.4 million km²; 1990 was the least snowy at 23.2 million km². Eight of the monthly minima for the two decades occurred in 1990.

With only two decades of reliable hemispheric information, it is impossible to identify anything in the way of trends or cycles in the temporal or spatial distribution of snow. What has become recognizable during the period of record is a tendency for multi-year departures in snow cover, in which less pronounced month-to-month and season-to-season departures are embedded. Twelve-month running means illustrate the periods of above-normal cover that occurred in the late 1970s and mid-1980s, and the lower and below-normal extents during the mid-1970s and early 1980s (Figure 3). The most significant lengthy departure during the past 20 years began in the late 1980s and continued into 1992. Of the 65 months between August 1987 and December 1992, only eight had above-normal snow cover. Three of these eight were September, November, and December 1992.

Spring cover has shown pronounced deficits over the past five years in Eurasia and six years in North America; areas in these springs have been at or below lows established before this period (Figure 4). During the same interval, both continents have had low seasonal cover in the fall and summer, although frequently neither continent has been at or approached record low levels. Winter cover has been close to average over the past six years.

The NOAA snow estimates are considered the most accurate figures available. Despite the positive methodologically induced bias of the monthly NASA estimates (cf. the previous section), they range from less than 1 million up to 13 million km² below the NOAA area estimates for those nine years for which both agencies provide estimates. These absolute differences are greatest in the late fall and early winter. In a relative sense, microwave areas are between 80 and 90 percent of short-wave values in winter and spring, 20 to 40 percent of the short-wave estimates in summer, and 40 to 70 percent of short-wave areas in fall. Possible explanations for the significant disparities in the latter two seasons are wet and shallow snows. These are difficult if

TABLE 1   Monthly and Annual Snow Cover (million km²) over Northern Hemisphere Lands, Including Greenland

	Maximum (yr)	Minimum (yr)	Mean*	Median	Standard Deviation
Jan	49.8 (1985)	41.7 (1981)	46.5	46.1	1.8
Feb	51.0 (1978)	43.2 (1990,92)	46.0	45.6	2.0
Mar	44.1 (1985)	37.0 (1990)	41.0	40.8	1.9
Apr	35.3 (1979)	28.2 (1990)	31.3	31.4	1.8
May	24.1 (1974)	17.4 (1990)	20.8	20.6	1.9
Jun	15.6 (1978)	7.3 (1990)	11.6	11.4	2.1
Jul	8.0 (1978)	3.4 (1990)	5.3	5.5	1.2
Aug	5.7 (1978)	2.6 (1988,89,90)	3.9	3.8	0.9
Sep	7.9 (1972)	3.9 (1990)	5.6	5.6	1.1
Oct	26.1 (1976)	13.0 (1988)	17.6	17.5	2.9
Nov	37.9 (1985)	28.3 (1979)	33.0	32.8	2.3
Dec	46.0 (1985)	37.5 (1980)	42.5	42.7	2.3
Annual	27.4 (1978)	23.2 (1990)	25.5	25.4	1.1

*Means are for the period January 1972 through May 1992; extremes are for January 1972 through December 1992.

not impossible to monitor using microwaves. Depth may be the more important of these two variables, given the better agreement in spring, although it has been suggested that unfrozen soil beneath the pack is a major contributor to the underestimates during fall (B. Goodison, pers. commun.). The 85-GHz channel on the SSM/I has shown promise in improving the monitoring of shallow (<5 cm) snow cover (Nagler and Rott, 1991).

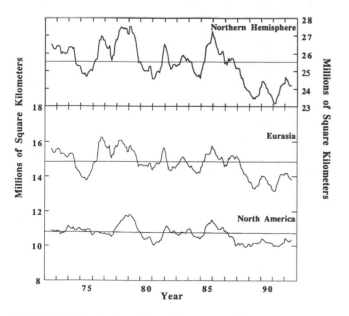

FIGURE 3   Twelve-month running means of snow cover over Northern Hemisphere lands (including Greenland) for the period January 1972 through December 1992. Running means are also shown for Eurasia and North America (including Greenland). Values are plotted on the seventh month of the twelve-month interval.

CONCLUSIONS

Significant strides have been made in recent years in the recovery, digitization, and validation of in situ observations of snow cover and their assimilation into regional networks. The availability of this information, in some cases going back to the turn of the century, complements the more recent global monitoring of snow from satellites. Analyses of both in situ and satellite observations show a considerable amount of year-to-year variability in the duration of snow cover on local to hemispheric scales. However, it is beginning to be recognized that this variability is often embedded in longer-term fluctuations. In situ data from the central United States points to multi-decadal fluctuations in the duration of snow over the past century, and the hemispheric satellite observations suggest variations lasting on the order of several years.

Efforts must continue to ensure the recovery and digitization of all available historic in situ observations, from both stations and snow courses. On the satellite side, the NOAA weekly product must continue to be produced in its present form. To abruptly or, perhaps more seriously, subtly alter the manner in which these charts are produced would severely weaken what is at present the longest and most consistent satellite-derived data set for any surface or atmospheric variable. Coincident with this must be efforts to improve the hemispheric monitoring of snow from microwave data using composited regional algorithms. Ultimately the goal must be to integrate all surface and satellite sources in a series of products reporting snow extent, depth and water equivalent, and regional surface albedo. These data should cover not only Northern Hemisphere lands but also lands in the Southern Hemisphere, ice sheets, and sea ice. Means

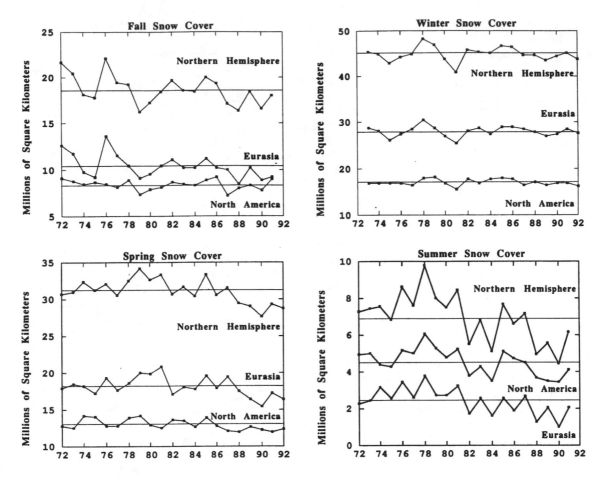

FIGURE 4   Seasonal time series of snow cover over Eurasia and North America (Greenland is excluded).

of integrating satellite and integrative products with earlier in situ observations must also be improved to provide the longest, most consistent records possible.

With adequate data available to analyze, knowledge of the spatial and temporal kinematics of snow cover over recent decades or even the past century will continue to improve. This will contribute to a better understanding of the role of snow in the climate system and to the utility of snow as an indicator of climate change.

ACKNOWLEDGMENTS

Thanks to G. Stevens at NOAA for supplying digital snow data and to A. Chang at NASA for providing microwave data. Thanks also to A. Frei, M. Hughes, and J. Wright at Rutgers for technical assistance. This work is supported by NOAA under grant NA90AA-D-AC518, and by the Geography and Regional Science Program of the National Science Foundation under Grant SES-9011869.

Variability and Trends of Precipitation and Snowfall over North America

PAVEL YA. GROISMAN[1] AND DAVID R. EASTERLING[2]

ABSTRACT

Biases and large-scale inhomogeneities in the time series of measured precipitation and snowfall over the continent are discussed and analyzed. After adjustments and selection of the "best" network, reliable "first guess" estimates of North American snowfall and precipitation have been obtained. Century-long time series of unbiased annual precipitation over the regions to the south of 55°N, and 40-year time series of unbiased area-averaged annual precipitation and snowfall for all of North America, have been developed. Analysis of the trends shows that:

1. During the last hundred years, annual precipitation has increased in southern Canada (south of 55°N) by 13 percent and in the contiguous United States by 4 percent to 5 percent; the main domain of this century-scale precipitation increase, however, is eastern Canada and the adjacent northern regions of the United States.

2. An increase of up to 20 percent increase in annual snowfall and rainfall has occurred during the last four decades in Canada north of 55°N.

INTRODUCTION

During the last century, global surface-air temperature has increased approximately 0.5°C (over Canada it has been about 1°C); the seven warmest years in the instrumental records of global temperature occurred during the last decade (Budyko and Izrael, 1987; Gullett and Skinner, 1992; IPCC, 1990, 1992). With such a large change in the earth's heat balance, we suspect that changes in the water balance of North America may have also occurred. Given the poten-

tial for even greater changes over the next decades, there is an urgent need to obtain reliable results on North American precipitation changes during the last century.

Numerous studies have been made of precipitation over North America and the problems associated with precipitation measurement. The results of these studies can be summarized by saying that no definite century-scale precipitation trends have been found for the United States. (Canadian precipitation cannot be analyzed for the entire century,

---

[1]State Hydrological Institute, St. Petersburg, Russia; current address: Department of Geosciences, University of Massachusetts, Amherst, Massachusetts
[2]NOAA National Climatic Data Center, Asheville, North Carolina

due to lack of data in the northern part of the country.) Fluctuations on decadal time scales are evident, however, which led Klugman (1983) to claim "evidence of climatic change" in the U.S. seasonal precipitation during the 1948-1976 period. Diaz and Quayle (1980) found large-scale changes in contemporary precipitation (1955-1977) as compared to precipitation at the beginning of the century (1895-1920), but these changes were shown to be mainly statistically non-significant. Vining and Griffiths (1985) found an increase over time in the decadal variance of annual precipitation at 10 long-term U.S. stations for the period 1900-1979. Bradley et al. (1987), analyzing the area-averaged percentile of the U.S. annual precipitation, found a precipitation decrease during the period from the 1880s to the 1930s and a general increase thereafter.

There are several reasons for precipitation undercatch by the standard gauges currently being used worldwide (Bogdanova, 1966; WMO, 1991; WWB, 1974). The main factor that reduces the amount of measured precipitation as compared to "ground truth" is wind-induced turbulence over the gauge orifice (Sevruk, 1982). The absence of a wind shield for the gauge (or its poor design) makes the problem even more severe. Field experiments show that an appropriate wind shield (such as the modified Nipher shield used in Canada) reduces wind-related bias in the gauge catch to manageable levels that can then be easily adjusted. All elevated snow gauges in Canada have wind shields, but only about 200 of the approximately 6,000 U.S. gauges that report daily precipitation have a wind shield (Karl et al., 1993a). Furthermore, the Alter shield in use at U.S. stations provides less protection than the Canadian shield (Goodison et al., 1981). As a result, the existing U.S. rain-gauge network measures rainfall with a bias of 3 to 10 percent (Golubev et al., 1992; Sevruk and Hamon, 1984) and snowfall with a bias of up to 50 percent or more (Goodison, 1978; Larkin, 1947; Larson and Peck, 1974). Moreover, in Alaska total biases up to 400 percent in measurements of water equivalent of snowfall have been reported (Black, 1954).

Rain gauges at some of the U.S. primary stations were equipped with the Alter wind shields toward the end of the 1940s. The addition of the wind shield introduced an inhomogeneity into the time-series precipitation data at these stations. Although the stations with Alter shields constitute a small percent of the U.S. stations, they comprise 40 percent of the U.S. National Weather Service stations transmitting meteorological observations over the Global Telecommunication System (GTS) for publication in the *Monthly Climatic Data for the World*.

Canadian methods of making precipitation measurements differ significantly from those used in the United States. This is illustrated by the jump in measured precipitation values often registered by stations on opposite sides of the border between the two countries (WMO, 1979; WWB, 1974). Although the difference in liquid precipitation is small—the Canadian gauge typically measures a few percent more rain than the standard U.S. gauge (Sanderson, 1975)—there is a considerable difference in the solid precipitation measurements (Goodison et al., 1981; Sanderson, 1975). Currently, 85 percent of the Canadian meteorological network uses a 10:1 ratio as a measure of the water equivalence of fresh-fallen snow. Since the 1960s, 15 percent of this primary network has used an elevated snow gauge with a modified Nipher wind shield. This gauge catches solid precipitation effectively even during high-wind conditions (Goodison, 1978; Goodison et al., 1992). Its record differs systematically from the snow-stick measurements continued at other stations (Goodison et al., 1981; Groisman et al., 1993; Karl et al., 1993a). This difference in snow-measurement methods causes both space and scale inhomogeneities, which cannot be avoided by users of such international publications as *World Weather Records* and *Monthly Climatic Data for the World*, since Canada transmits only the data from its primary network over the GTS.

When point precipitation measurements are expanded to area-averaged values in rough terrain, the point values are often less than the "ground-truth area mean", because most meteorological stations are located in valleys, but the vertical gradients in precipitation have been neglected. Figure 1 shows two maps of annual precipitation over North America. The first map (1a), which was constructed during the World Water Balance studies (1974), incorporates estimates of topographic effects on precipitation. However, the second map (1b) simply uses data from 1,900 stations, without any consideration of topographic effects on precipitation. Precipitation in the mountainous West is obviously underestimated on the second map, and even Appalachian-ridge precipitation estimates are noticeably biased.

The area-averaged annual precipitation value for the contiguous United States derived from Figure 1a is about 880 mm, while the value from Figure 1b is 700 mm. When we use data from all 6,000 U.S. cooperative stations, this latter annual total increases to 735 mm (NCDC, 1991) but still remains quite low compared to the value reflecting topography. The same problem has been revealed to exist for Canada by Hare (1980), who showed that annual totals over the country should be increased by 20 percent, on average, to obtain a reliable water balance (i.e., runoff = precipitation − evaporation).

Precipitation measurements are very sensitive to changes in the environment surrounding the gauge, in the gauge type, and in the methods of measurement. A small move of the gauge can result in a twofold change in the values of measured precipitation if the gauge exposure is changed (Karl et al., 1993b; Sevruk, 1982). After spatial averaging, it can be expected that the random errors connected with such shifts will cancel out. However, when systematic changes occur countrywide, they cannot be assumed to cancel out. Several such changes have occurred during the

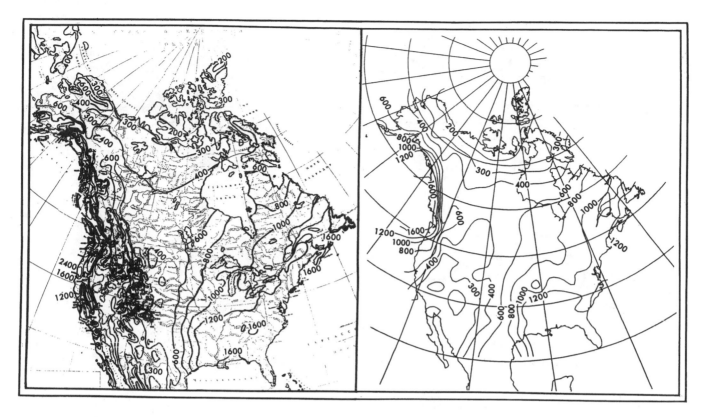

FIGURE 1   Mean annual precipitation over North America: left panel, from WWB (1974); right panel, as estimated using the meteorological network.

last century in the Canadian and the U.S. meteorological primary networks (Figure 2). Area-averaged precipitation time series are especially sensitive to such large-scale inhomogeneities because their noise component has been decreased by averaging. Therefore, when studying trends in these time series, we must be sure that the observed large-scale decadal changes represent true climatic changes rather than artificial changes introduced into the network. Any analysis of precipitation over North America should be preceded by a thorough analysis of the history of observations in the network. This analysis should be followed by selecting a subset of the "best stations" and applying adjustments to the initial data to compensate for all known large-scale inhomogeneities. Our analyses in the following sections give special attention to the selection of the precipitation data and to the adjustments to be made to them.

DATA SELECTION AND PROCESSING

Networks Used

In our studies of precipitation over North America, we selected several networks, each with a special purpose in mind. Figure 3a shows the network that was used to study snowfall. It consists of 223 U.S. first-order stations (includ-

ing Alaska) and 1,107 Canadian stations selected using the requirement that each station have data available for at least 15 years out of the last two decades. The period 1950-1990 was chosen because by then all U.S. first-order stations had been relocated to airports with relatively open sites, and because the networks for northern Canada and Alaska had become dense enough to assure adequate spatial sampling only after World War II. A reliable analysis of changes in snowfall and total precipitation over the entire continent could thus be made only for the last 40 years.

Since the 1960s there have been parallel observations of snowfall (by stick measurements) and its water equivalent (measured by elevated snow gauges) at 335 Canadian stations. These observations were used to evaluate the adjustments made by Groisman (1992) and Karl et al. (1993a) to homogenize time series of solid precipitation in Canada.

Figure 3b shows the network that was used to study century-scale changes of annual precipitation over the contiguous United States and southern Canada. It consists of the same Canadian stations, 45 cooperative Alaskan stations, and 593 "best" cooperative stations selected from the U.S. Historical Climatology Network (Karl et al., 1990). The U.S. stations were selected on the basis of the length of their records (they all began their observations in the nineteenth century) and their having had few station moves or other

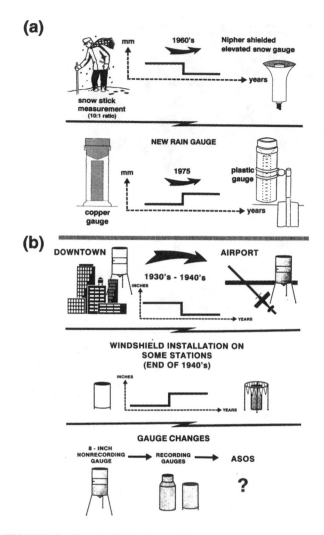

(a)

(b)

FIGURE 2   Sketch of the history of precipitation measurement in North America related to precipitation time-series homogeneity. Major large-scale changes caused inhomogeneities in these time series. (a) Canadian primary network; (b) U.S. primary network.

obvious sources of inhomogeneities. The data are therefore free of the large-scale inhomogeneities shown in Figure 2, but other inhomogeneities may still exist.

The Canadian network, which seems quite large in Figure 3b, declined rapidly in size during the first half of the century (see Table 1). In southern Canada it contained only 190 stations by 1940, and 26 stations in 1900. As a consequence, we selected a special network of Canadian stations that were operating in the 1920s (Figure 3c) to use in to studying Canadian precipitation prior to World War II. The area-averaged precipitation time series based on this network were combined with the corresponding time series developed from the main network shown in Figure 3b.

Data Processing

The precipitation data for the 593 contiguous U.S. stations and the snowfall data were used without adjustments.

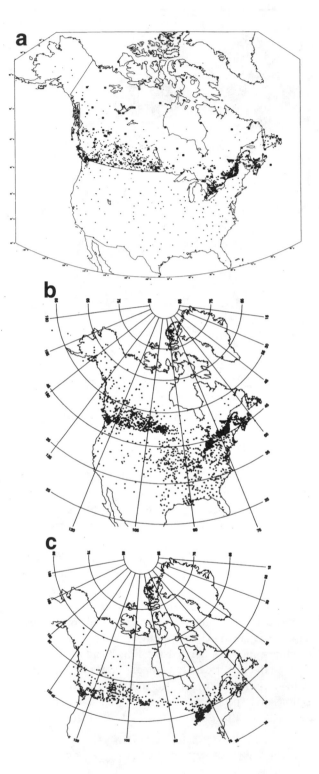

FIGURE 3   Stations used in this paper for analysis of: (a) snowfall (squares depict the primary Canadian stations where the elevated snow gauges were installed in the 1960s); (b) precipitation; (c) accuracy of precipitation changes in southern Canada before World War II.

TABLE 1   Number of Stations with Available Data on Annual Precipitation for Southern Canada

Data Set	Year						Total
	1895	1910	1925	1940	1955	1990	
Main data set used in the study	22	35	119	190	315	765	995
Alternative set of Canadian stations with observations prior to 1925	54	86	323	261	162	24	385
All Canadian stations with observations prior to 1925	70	117	439	384	277	133	523

The Canadian rainfall data were corrected for the gauge change that occurred in the mid-1970s. All rainfall data before 1975 were increased by a factor of 1.025 to approximate the results of Goodison and Louie (1986).

To obtain the water equivalence of snowfall at the Canadian stations, we were able to derive scale adjustments from the set of parallel observations made at the Canadian stations depicted in Figure 3a (full squares). These adjustments reduce snow-stick measurements to match the records of the elevated snow gauge. However, these records still need some corrections for the type (or absence) of wind shield (Goodison, 1978). Larger corrections are required for the U.S. rain gauges (Larson and Peck, 1974; Goodison and Metcalfe, 1989; Peck, 1991) because these gauges do not have wind shields. We hope to be able to use such corrections in the future, but for the moment we can only conclude that the discontinuity across the border of the two countries in measured water equivalent of snowfall is caused by significantly biased measurements in the U.S. data. Therefore, area-averaging of annual precipitation over the regions encompassing both Canada and the United States was done separately for each country. Only snow-stick measurements were used to represent the Canadian solid precipitation. Each area-mean time series was adjusted to its unbiased precipitation value (using the method described by Groisman et al. (1991), which relies on the WWB (1974) estimates of long-term annual precipitation shown in Figure 1a); then time series from the two parts of the region were combined. Furthermore, we prefer to consider the southern Canadian regions separately, due to the instability of their data coverage.

As noted earlier, Table 1 shows the rather variable number of stations providing annual-precipitation data for southern Canada (south of 56°N) for the two non-interlaced networks shown in Figures 3b and 3c. Figure 4 shows time series of annual snowfall and total precipitation area-averaged over Canada south to 55°N, based on these two independent data sets. There is little difference between them during the twentieth century, which confirms the accuracy of the area averaging and the representativeness of the network used. The only noticeable difference is in the 1890s,

when precipitation data were available for only 20 to 30 stations over the main network. Hence, we combined these two networks to develop a century-long area-averaged time series of southern Canadian precipitation. The time series for the period 1891-1960 were developed from the network of stations shown in Figure 3c, together with those stations from the main Canadian network that were operating prior to 1925. These time series were combined with the time series for the period 1921-1990 that was based on the data of the 1,107 stations.

Snowfall Scale Correction

In developing the data set used to construct unbiased area-averaged snowfall time series (Figure 5), we made one

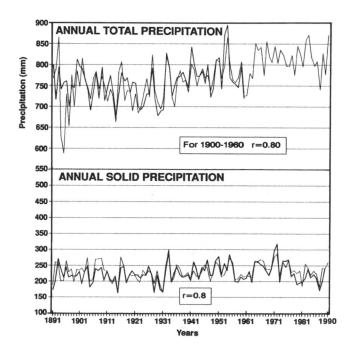

FIGURE 4   Annual snowfall and total precipitation over southern Canada (south of 55°C) estimated by the network depicted in Figures 3b (dashed line) and 3c (solid line).

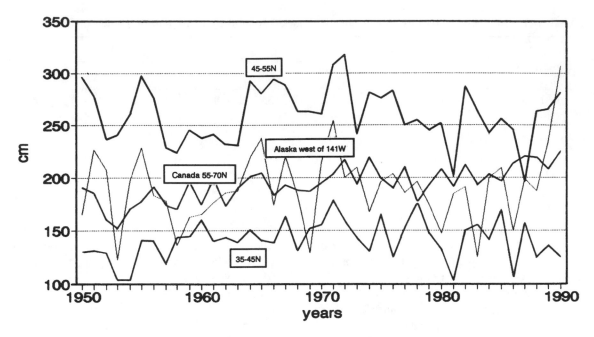

FIGURE 5   Annual snowfall over North America during the last 40 years. The applied scale-correction factors for network elevation are: for Alaska—1.23; for Canada, Zone 55-70°N—1.14, Zone 45-55°N -1.19, and Zone 35-45°N—1.47.

important scale correction that was not performed in the work of Karl et al. (1993a). Bogdanova (1977) showed that in mountains the mean long-term ratio of solid to total annual precipitation rises at a mean rate of 27 percent per 1000 m. We used this to adjust the snowfall time series for network elevation. Without this adjustment, the biases in the regionally averaged snowfall values would be very large. Table 2 provides the mean elevation of the network depicted in Figure 3b, together with true mean surface elevation for large regions of the United States and Canada (the mean elevation of the networks depicted in Figures 3a and 3b scarcely differ). From this table it is clear that with our network it would be possible to estimate occurrence of

traces of snowfall only in the valleys and on the plains of western United States, and completely miss its mountainous portion. So our snowfall adjustment for each region was evaluated in the following way:

1. The mean long-term ratio of solid to total annual precipitation ($rat1$) was estimated for a plane with the elevation of the network depicted in Figure 3a.

2. This ratio was increased using a coefficient of 27 percent km^{-1} to obtain its value ($rat2$) for a plane with the elevation corresponding to the true surface elevation of the region.

3. The value of the estimated area-averaged snowfall was then multiplied by an adjustment scale coefficient equal to $rat2$ divided by $rat1$.

These coefficients were used for every graph in Figure 5. They are, however, conservative estimates of the adjustment needed if indeed the total precipitation area-averaged over the whole region depends on true surface elevation rather than on the elevation of the network used (Karl et al., 1993a).

In the following section, area-averaging by the polygon method (Kagan, 1979; Thiessen, 1911) and scale correction (Groisman et al., 1991) were used in to develop regional, unbiased century-scale annual precipitation time series for the contiguous United States and southern Canada, together with 40-year time series of snowfall and total precipitation over all of North America. So that missing data would not affect the results, we averaged the anomalies from the reference period 1971-1990 and then reconstructed the area-averaged precipitation, adding the corresponding area-

TABLE 2   Mean Elevations of Meteorological Stations in the United States (Historical Climatology Network) and Canada Versus Real Surface Elevations*

	Elevation, m	
Region	Network	Surface
Western United States**	790	945
Eastern United States	145	235
Canada	305	490
Alaska west of 141°W	100	510

*Real surface elevations were estimated from NGDC (1988) data with 5-degree resolution

**"Western" and "Eastern" here refer to west or east of a line drawn from New Orleans to Detroit

averaged value of the mean precipitation for the same reference period. (While processing the data of the Canadian network that was operating in the 1920s, we used the reference period 1921-1940.) These time series provide our first approximation of contemporary changes of North American precipitation. The first experiments using individual gauge corrections (Groisman, 1991) showed that this approximation does not change substantially on a continental scale after the intra-annual variability in gauge errors has been taken into account.

CURRENT CHANGES IN PRECIPITATION AND SNOWFALL

Snowfall Changes During the Last Four Decades

Figure 5 shows the zonally averaged snowfall over Alaska, Canada north of 55°N, and the zones 45°N to 55°N and 35°N to 45°N for the last four decades. The basic statistics of these time series appear in Table 3.

When the spatial correlation function is known, the accuracy of area averaging over these regions using the networks depicted in Figure 3 can be estimated. The method of estimation, which was developed by R.L. Kagan (1979), can be applied to every isotropic two-dimensional field. In North America we used this method only for northern Canada. Since even today the meteorological network in northern

Canada is relatively sparse, we had to estimate the representativeness of the network beforehand in order to analyze the changes over this area.

The spatial correlation function for northern Canadian snowfall can be approximated by the relationship $r(R) = Ce^{-R/R_0}$, where R is the distance; $R_0 = 710$ km, the radius of correlation; and $C = 0.84$. The value of $(1 - C)/C$, which equals 0.19, is a measure of the error of measurements combined with microscale snowfall variability near the stations (Kagan, 1979). For the network selected, the mean-square error of spatial averaging of annual snowfall over northern Canada was estimated to be 17 percent of the theoretical variance of the area-averaged snowfall time series in the first half of 1950s; during the period 1956-1990 the error was about 10 percent. The number and spatial distribution of snowfall-measuring stations for southern Canada and the United States during the last four decades were adequate, so there was no need to verify the representativeness of the area-mean values.

The accuracy of area averaging of the data in northern Canada is high enough to confirm the strong positive linear trend in annual snowfall revealed in the zone (19 percent over four decades). This trend, in conjunction with the trend in rainfall data (not shown), confirms the systematic increase in annual totals of precipitation during the entire period of instrumental observations.

Snowfall in the zone 45°N to 55°N is closely connected to temperature changes on continental and hemispheric

TABLE 3 Statistical Characteristics of Time Series (1950-1990) of Annual Snowfall and Total Precipitation Zonally Averaged over North America (see Figures 5 and 6)

	Annual Snowfall		
	Mean and Standard Deviation, in cm		
Region	Mean	Std. Deviation	Linear Trend ± its Standard Error, in % per Decade
Alaska	190	35	2.7 ± 2.5
Canada, zone 55-70°N	195	17	5.1 ± 0.8
Canada/U.S., zone 45-55°N	260	27	−0.7 ± 1.4
U.S./Canada, zone 35-45°N	140	18	1.8 ± 1.7

	Total Precipitation		
	Mean and Standard Deviation, in mm		
Region	Mean	Std. Deviation	Linear Trend ± its Standard Error, in % per Decade
Alaska	510	59	2.7 ± 1.5
Canada, zone 55-70°N	495	33	4.2 ± 0.6
Canada/U.S., zone 45-55°N	940	41	0.7 ± 0.6
Canada/U.S., zone 35-45°N	905	66	1.9 ± 0.9
United States, zone 28-35°N	1175	137	3.4 ± 1.5

scales; the correlation between this time series and the annual mean surface air temperature over the Northern Hemisphere is about −0.7 (Groisman et al., 1993). Using this relationship, and taking into account the intrinsic coherence of annual snowfall totals in southern Canada (which encompasses the main part of the 45°N to 55°N zone) with regional maximum temperature (Groisman, 1992), we could expect a future decrease in snowfall for this zone if the global warming projected by IPCC (1990) actually occurs.

We realize that the large adjustment coefficient for zonal snowfall in the 35°N to 45°N zone (1.47), while providing a more realistic mean value, cannot be considered a satisfactory solution to the problem of snowfall changes over this zone. Unfortunately for the current concerns about potential climatic change, we still do not have a reliable source of data on the snowfall over the main part of the United States.

Annual Precipitation Changes During the Last Four Decades

Figure 6 shows the unbiased annual precipitation totals for the last four decades, area-averaged over the same four zones used in Figure 5 plus the southern part of the United States (a zone from 28°N to 35°N and to the east of 105°W). The basic statistics of these time series appear in Table 3. It can be seen from this table that annual precipitation has increased during these last decades over the entire continent, but especially significantly over the southeastern United States (14 percent per 40 years) and northern Canada (17 percent per 40 years).

Century-Long Precipitation Changes

Table 4 (adapted from Groisman, 1992) provides the same characteristics as in Table 3 for the century-scale precipitation changes over southern Canada and the contiguous United States. This table confirms the main conclusions derived from Table 3, but shows that, as with the U.S. precipitation, it is unwise to discuss the statistical significance of precipitation trends on the basis of the 41-year record alone (on the century scale the linear trend is not statistically significant).

Given the major terms of the water-balance equation, we can assume that, in general, the variability of annual precipitation drives the variability of annual streamflow (Linsley et al., 1958). Streamflow will thus have approximately the same large-scale features and areas of coherent change as the precipitation used to estimate it. The streamflow data are independently collected and do not exhibit the errors and data problems mentioned in previous sections (they have their own problems and errors, but do not suffer from the rain-gauge problem; being natural area averages, they give representative estimates of changes in the hydrological cycle of the regions under consideration. Therefore, in our study of the spatial correlation of the precipitation field over North America, we divided the area into several homogeneous regions on the basis of the results of a principal component analysis of streamflow data for the United States made by Lins (1985). In his analysis, Lins used a dense network of small watersheds not disturbed by anthropogenic activity. He evaluated five principal components,

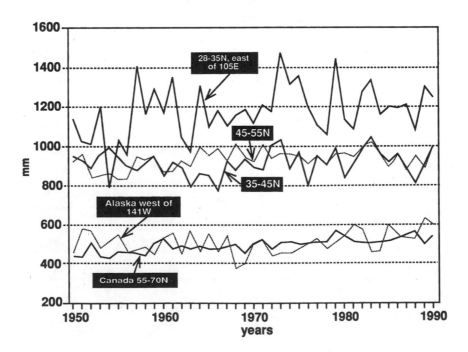

FIGURE 6 Annual precipitation over North America during the last 40 years.

TABLE 4 Statistical Characteristics of the Century-long Time Series of Annual Precipitation, Area-Averaged Over Canada South of 55°N and the Contiguous United States

	Mean and Standard Deviation, in mm		Linear Trend ± its Standard Error, in % per 100 yrs
Region	Mean	Standard Deviation	
Southern Canada	925	56	13.3 ± 1.7
Contiguous United States	870	66	4.5 ± 2.6

each with a distinctive regional pattern: south-central United States, northwestern United States, southwestern United States, north-central United States, and northeastern United States. The generalized map of these regions expanded on the Canadian territory up to 55°N is shown in Figure 7.

In order to identify the independent components of the precipitation pattern, we applied the spatial averaging to the regions depicted in Figure 7. Time series of adjusted annual precipitation area-averaged over these five regions are shown in Figure 8; their correlation matrix appears in Table 5, and their main statistical features in Table 6. From Table 5 we can see that the time series from the different regions are not closely correlated (the north-central region correlated best with other regions, since it has a common

border with each of them). Hence, we can analyze and describe each time series separately.

The debiasing procedure used (Groisman et al., 1991) does not affect trends, but it does change the scale of the time series. For Canada and eastern United States, the corresponding scale factors are not large; they vary from 1.05 (south-central region) to 1.25 (northeastern region). However, in the western United States they become larger (1.5 for the southwestern region and 1.7 for the northwestern United States). This shows that even annual totals are not well measured by the existing meteorological network.

An increase of more than 10 percent per century in annual precipitation over southern Canada (Table 4) is reflected by a corresponding precipitation increase over the northeastern

TABLE 5 Correlation Matrix of the Area-Averaged Century-long Time Series of Annual Precipitation Depicted in Figure 8

U.S. Region*	N-West	N-Central	N-East	S-West	S-Central
Northwestern	1.00	0.44	0.26	0.23	0.16
North-central		1.00	0.29	0.28	0.38
Northeastern			1.00	0.08	0.11
Southwestern				1.00	0.11
South-central					1.00

*See Figure 7 for a depiction of these regions

TABLE 6 Statistical Characteristics of the Area-Averaged Century-long Time Series of Unbiased Annual Precipitation Depicted in Figure 8

	Mean and Standard Deviation, in mm		Linear Trend ± its Standard Error, in % per 100 yrs
Region*	Mean	Standard Deviation	
Northwestern	825	70	5.5 ± 2.9
North-central	845	65	12.1 ± 2.4
Northeastern	1265	90	11.3 ± 2.2
Southwestern	515	90	0.5 ± 6.1
South-central	1115	110	5.4 ± 3.4
All Five Regions Together	925	55	7.6 ± 1.9

*See Figure 7 for a depiction of these regions

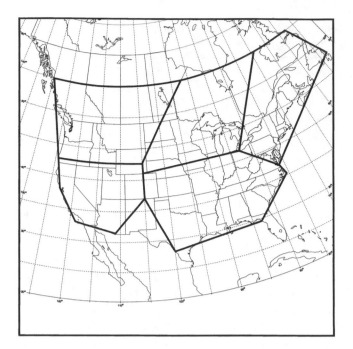

FIGURE 7 Regions over North America (south of 55°N) with spatially homogeneous patterns of streamflow (based on principal-component analysis by Lins (1985)) extended to southern Canada.

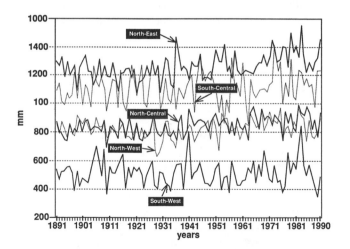

FIGURE 8 Century-long annual precipitation changes over the United States and southern Canada, area-averaged over the five regions depicted in Figure 7.

and north-central regions. Linear trends account for up to 20 percent of the variance of annual precipitation over each of these two regions. However, the trends of 5 to 6 percent per century documented for northwestern and south-central regions are not statistically significant at the 0.05 level, and there have been no systematic changes over the southwestern region during the last 100 years. Therefore, we can conclude that during the last century a large-scale increase has occurred over all of North America, but that the major source of this increase was changes in eastern Canada and

adjacent regions of the United States (the northeastern quarter of the country).

CONCLUSIONS

Standard rain gauges installed at the U.S. and Canadian meteorological networks undercatch "true" precipitation, especially in its solid form. The biases induced by this undercatch have altered during the last century, due to changes in instrumentation and measurement practices. This introduces large-scale inhomogeneities in the time series of the precipitation field over the continent. Climatological studies that do not take into account these inhomogeneities will give misleading results. The low-elevation locations of the existing meteorological network necessitate further research to determine reliable estimates of snowfall changes over the United States. Problems also exist in estimation of total precipitation in the mountainous west of the continent. The scale corrections, which transform area-averaged precipitation data into the unbiased values of "ground true" regional precipitation, increase precipitation estimates in that region by up to 70 percent. These uncertainties in the data restricted our analysis to spatially averaged values of annual precipitation and snowfall over large regions.

After adjustments and selection of the "best" network, reliable "first guess" estimates of North American snowfall and precipitation were obtained. Century-long time series of unbiased annual precipitation over the regions to the south of 55°N, and 40-year time series of unbiased area-averaged annual precipitation and snowfall for all of North America, were developed. Analysis of the trends shows that:

1. During the last hundred years, annual precipitation has increased in southern Canada (south of 55°N) by 13 percent, and in the contiguous United States by 4 to 5 percent.

2. An increase of up to 20 percent in annual snowfall and rainfall has occurred during the last four decades in Canada north of 55°N.

3. In the zone 45°N to 55°N, century-scale total precipitation has increased, while a redistribution between the solid and liquid forms has occurred coherently with variation of the zonal maximum temperature (and global surface-air temperature).

4. The main domain of the century-scale precipitation increase over North America to the south of 55°N (to the north of this latitude the data for the first half of the century are absent) is eastern Canada and adjacent northern regions of the United States.

ACKNOWLEDGMENTS

This paper was completed while the first author worked as a National Research Council senior research associate at the U.S. National Climatic Data Center. Support for David R. Easterling was provided under Department of Energy (DOE) Interagency Agreement DE-A105-90ER60952 and NOAA's Climate and Global Change Program.

Commentary on the Paper of Groisman and Easterling

DAVID A. ROBINSON
Rutgers University

I think Dr. Groisman's presentation has been an excellent example of the advantages and pitfalls of dealing with observational data and of putting data together on a variety of temporal and spatial scales. And I can assure you that the paper contains some tremendous information that, unfortunately, Dr. Groisman had time only to gloss over. I would like to break the study down into a couple of segments to foster further discussion.

The first section I shall simply entitle "obtaining numbers". This includes the problems Dr. Groisman presented that relate to measurement techniques, whether for obtaining liquid rainfall or, more important, getting adequate snowfall measurements. Next, there are the problems in developing an observational network that is homogeneous through time—which, as Dr. Groisman found, was impossible to do for the entire North American continent. He could look only at the southern reaches of Canada and over the United States for the entire century. When it came to looking into the Arctic, he could go back only about 40 years.

Third is the paper's serious criticism, which may have passed people by too quickly, that it is difficult to believe anything that comes out of the primary stations in the United States. The figure that shows the changes at these stations (relocation, gauge changes, wind shields) really substantiates that. So we essentially have different observational networks that vary through time.

Last, there is the whole idea of metadata. We must have enough information on the history of a station or network that we can go back and reconstruct a homogeneous, standardized time series over decade-and century-long periods. While North American metadata is certainly better than African, there are gaps and lapses in the station histories. For precipitation, the impact of station moves is less significant than for temperature. But the gauges that measure precipitation, the wind shields, the 8-inch gauge versus the tipping bucket, and so forth are as important as any problems in measuring temperature in the United States.

It occurs to me to wonder how Dr. Groisman would rank these concerns in terms of importance for obtaining homogeneous precipitation records. To me, the most significant problem is perhaps the switch to the tipping bucket with the introduction of automated surface observing systems at primary U.S. observing stations over the next few years. Some time down the road when we are looking for decadal variations in precipitation, we may have to examine the metadata for the 1990s very carefully to come up with a homogeneous time series.

Another question I have is how deeply inhomogeneities interfere with obtaining true signals. You have to do so much with the data to be able to compare it at different times at an individual station, or among stations or countries. How deeply does this limit our ability to recognize climate change? Tom Karl et al. stated in their 1991 *Science* paper that it would take several decades to recognize a change in precipitation over central North America that might be an anthropogenic signal. I wonder how much these inhomogeneities in the different gauges and the different national recording procedures would extend or contribute to that lag, particularly when it comes to a noisy record like precipitation.

When it comes to point-to-area transformations, the question is where to draw the line. Dr. Groisman has done a lot of extrapolation, particularly in the western reaches of North America, in order to take into account elevational factors or orographic factors for precipitation. In some cases factors as great as a 70 percent increase were applied to the precipitation results of station networks. At some point one must decide whether to stay with point data, or at least data from a similar elevational area, or to stretch and extrapolate the data in order to obtain a more spatially complete and, in terms of absolute values, a more accurate record. Dr. Groisman did make it clear that the extrapolation does not interfere with any type of trend analysis; my question relates only to the absolute values of precipitation and snowfall generated.

Now for evaluating the numbers. Let us begin with the climatologies of snowfall and precipitation. It was interesting that at a given point in time when there was a larger mean, the standard deviations increased. When it comes to climate change, it is important to know whether we see a change in the mean or a change in the variability of climate.

Next Dr. Groisman looked at time series: the 40-year variety, the 100-year variety, precipitation, snowfall, Alaska, Canada, and the lower 48. There are so many interesting things in these time series—for instance, the major increase in snowfall in the Alaskan time series in the 1980s. Is

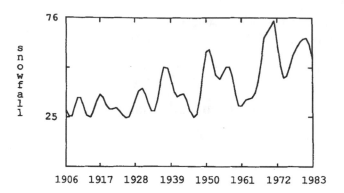

FIGURE 1   Smoothed time series of winter (December to February) snowfall in cm at Napoleon, North Dakota. Data from 1902 through 1987 have been smoothed with a nine-point binomial filter; the only points plotted are those where 9 years are available for filtering (i.e., plotted year ± 4 yr).

this associated with what has been happening in the North Pacific? We know that the North Pacific was rather cool in the 1980s compared to most of the rest of the globe.

I would like to show a few time series of snowfall. The stations are from a network of about 1,000 stations—about 300 of which go back to the turn of the century—over the United States, which I have assembled and quality-controlled in cooperation with NCDC. Figure 1 looks at Napoleon, North Dakota for the past century. An important point here is that these are stick measurements of snowfall and snow depth. Thus, they have no biases due to a change of gauge. We have not yet applied filters and looked for decadal variations. But you can see the suggestion of an increase in snowfall at Napoleon during the century.

Figure 2 is a record of New Brunswick, New Jersey snow going back to 1860. This is not total snowfall, but the number of 5-inch-plus and 10-inch-plus snowfall events. You can see quite a bit of variation across time. I do not see anything in the way of trends or cycles. This small study was inspired by the fact that until March 18, 1992, it was the third least snowy winter in New Brunswick in the last 135 years. Then it snowed 10 inches in 4 days, shooting the near-record to pieces.

It is interesting to note from this figure that the 1973-1974 to 1990-1991 period was the longest interval in which not a single year missed having a 5-inch-plus snow event. The winter of 1991-1992 ended that string. Another item of interest is that in the past 135 years, no year has had more than five 5-inch-plus snow events in central New Jersey.

Another way of examining this time series is to sum the number of 5-inch (12.7 cm) snow events by decade, as in Table 1. Here you can see that the 1980s had the most 5-inch-plus snowfall events of any decade on record. I have shown this chart to people throughout New Jersey, and no one ever believes me. However, the subject of one's

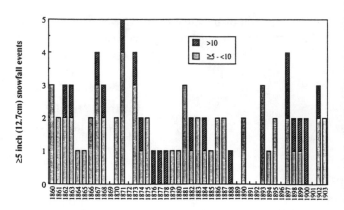

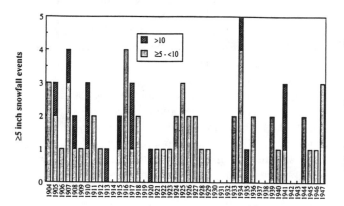

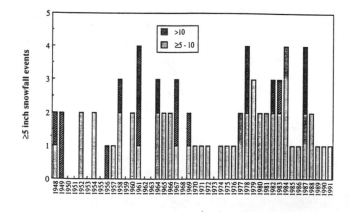

FIGURE 2   Number of ≥5-inch to <10-inch (≥12.7-cm to <25.4-cm) and ≥10-inch snowfall events per winter (October through April) from 1859/1860 to 1990/1991 at New Brunswick, New Jersey. (A snowfall event may be longer than one day, but may not be interrupted by a day without snowfall; no such events were observed for winter 1991/1992.)

perception of climate change is best left for discussion at another time.

Finally, Dr. Groisman talked about associations between snowfall, precipitation, and other climatic variables, and noticed that from 45°N to 55°N there was an inverse rela-

TABLE 1   Snowfall Events* in New Brunswick, N.J. by Decade, 1860–1989

Decade	Number of Snowfall Events*		
	≥5″ to <10″	≥10″	≥5″
1860s**	18	4	22
1870s	13	5	18
1880s	13	3	16
1890s	12	4	16
1900s	15	6	21
1910s	12	6	18
1920s	14	1	15
1930s	10	2	12
1940s	10	5	15
1950s	7	2	9
1960s	11	7	18
1970s	13	2	15
1980s	19	4	23
Total	167	51	218

*Snowfall event may be longer than 1 day, but may not be interrupted by a day without snowfall.
**Decade of the 1860s = winters 1859-60–1868-69.

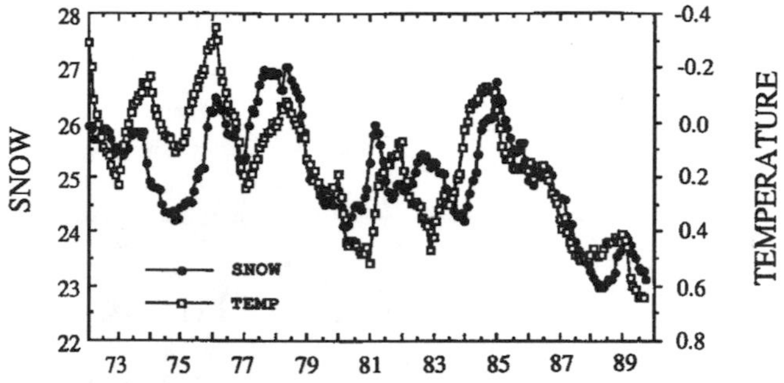

tionship between temperature and snowfall. When examining Northern Hemisphere snow and temperature together, as shown in Figure 3, we have found that both are 12-month running means. The snow, which is from the NOAA

FIGURE 3   Twelve-month running means of snow cover, in millions of km², and surface air temperature, in degrees Celsius, over Northern Hemisphere lands for January 1973 through August 1990. Temperatures are expressed as departures from the reference period, 1951-1970. Temperature data from P.D. Jones. From Robinson et al. (1991).

hemispheric snow charts, is a weekly product that we have averaged into monthly values. The temperature is Phil Jones's data set. Notice the striking relationship between hemispheric temperature—and that is from pole to equator—and Northern Hemisphere snow cover. We are beginning to examine snow and temperature by region and by season. So we see the same general association at this point, at least in a hand-waving sense, as Dr. Groisman found with the snowfall.

Discussion

RIND: You mentioned the relationship between precipitation over North America and global mean temperature. I wanted to add that the warming patterns for the 1930s and the 1980s were very different, just as you said the precipitation patterns were. The 1930s had very large warming at high latitudes with little effect at low latitudes, like an ocean heat-transport change pattern. The 1980s had very uniform warming as a function of latitude, more like what the models are projecting for CO_2 and trace-gas forcing. They may be good examples of two different types of processes.

WALSH: That increase of snowfall in northern Canada is interesting in relation to the borehole measurements mentioned yesterday.

If snowfall increases, it tends to insulate the ground during the coldest part of the year, which can complicate interpretation of borehole measurements. Also, perhaps we should take a closer look at how subsurface temperatures relate to snowfall to see what the implications might be of a continued increase in snowfall in permafrost areas.

TRENBERTH: The summer and winter correlations between precipitation and temperature tend to be very different.

GROISMAN: That's true. We simply wanted to create annual precipitation time series we could relate to some of the early data.

Asymmetric Trends of Daily Maximum and Minimum Temperature: Empirical Evidence and Possible Causes

THOMAS R. KARL[1], PHILIP D. JONES[2], RICHARD W. KNIGHT[1],
GEORGE KUKLA[3], NEIL PLUMMER[4], VYACHESLAV RAZUVAYEV[5],
KEVIN P. GALLO[1], JANETTE A. LINDESAY[6], AND ROBERT J. CHARLSON[7]

ABSTRACT

An examination of monthly maximum and minimum temperatures indicates that the rise of minimum temperature (0.84°C) has occurred at a rate three times that of rise of maximum temperature (0.28°C) over the period 1951 to 1990. This conclusion is based on data from over 50 percent (10 percent) of the Northern (Southern) Hemisphere land mass, accounting for 37 percent of the global land mass. The decrease of the diurnal temperature range is approximately equal to the increase of mean temperature. The asymmetry is detectable in all seasons in most of the regions studied.

The decrease in the daily temperature range is at least partially related to an observed increase in cloudiness. However, an empirical analysis of nearly 50,000 daily observations across 14 stations throughout the United States indicates that the daily temperature range is also sensitive to relative humidity, global radiation, surface stability, thermal advection, and snow cover; earlier work has shown that the temperature range is also sensitive to surface boundary conditions such as soil moisture. Locally, the diurnal range can also respond to increasing urbanization, irrigation, and desertification, but this effect appears to be small in our analysis.

Both general circulation models (GCMs) and radiative-convective models (RCMs) indicate that a decrease in the daily temperature range is likely with increases of CO_2. The magnitude

[1]NOAA National Climatic Data Center, Asheville, North Carolina

[2]Climatic Research Unit, School of Environmental Sciences, University of East Anglia, Norwich, Norfolk, United Kingdom

[3]Lamont-Doherty Earth Observatory, Columbia University, Palisades, New York

[4]Bureau of Meteorology, Melbourne, Australia

[5]Research Institute of Hydrometeorological Information, Obninsk, Russia

[6]Climate Research Group, University of Witswatersrand, South Africa

[7]Department of Atmospheric Sciences and Institute for Environmental Studies, University of Washington, Seattle, Washington

of the range decrease simulated by GCMs, however, is only 10 percent of the magnitude of the simulated mean average temperature increase, whereas the observational record shows nearly equal changes of the daily temperature range and the mean temperature. The clear-sky daytime forcing by sulfate and carbonaceous aerosols, and other indirect effects such as increased cloudiness and albedo, operate in a sense that may reduce the daily temperature range; however, the decadal changes of this geographic and seasonally varying forcing are not well known. At present we do not have an adequate explanation for the observed decrease in the temperature range, or for the lack of daytime warming, yet these changes are clearly important, both scientifically and practically.

BACKGROUND

The mean monthly maximum and minimum temperature is derived from an average of the daily maximum and minimum temperatures. The monthly mean diurnal temperature range (DTR) is defined as the difference between the mean monthly maximum and minimum temperatures. The dearth of appropriate data bases that include information on the daily or mean monthly maximum and minimum temperature has previously impeded our ability to investigate changes in these quantities. The problem has historical roots. It arises because the climatological data that are made available internationally by national meteorological and/or climate data centers are usually derived from the monthly climate summaries (CLIMAT messages) on the Global Telecommunications System (GTS), which do not include information on the maximum or minimum temperatures. The GTS is the principal means by which near-real-time, in situ, global-climate data are exchanged. Moreover, the problem has been exacerbated because the World Meteorological Organization's retrospective data collection projects such as *World Weather Records* and *Monthly Climatic Data of the World* have always been limited to mean monthly temperatures. This has forced climatologists interested in maximum and minimum temperatures either to develop historical data bases on a country-by-country basis (Karl et al., 1991a) or to try to work with the hourly GTS synoptic observations. The former is a slow, painstaking process, and the latter has been limited by poor data quality and metadata (information about the data) and records of short duration (Shea et al., 1992).

The first indication that there might be important large-scale characteristics related to changes of the mean daily maximum and minimum temperatures was reported by Karl et al. (1984a). Their analysis indicated that the DTR was decreasing at a statistically significant rate at many rural stations across North America. Because of data accessibility problems, subsequent empirical analyses continued to focus on data from North America over the next several years (Karl and Quinn, 1986; Plantico et al., 1990). By 1990, however, a United States/People's Republic of China (PRC) bilateral agreement organized by the U.S. Department of Energy and the PRC Academy of Sciences provided the opportunity to analyze maximum and minimum tempera-

tures from the PRC. Also about this time, the Intergovernmental Panel on Climate Change (IPCC) made arrangements with the Australian National Climate Centre to analyze maximum and minimum temperature data from southeastern Australia. The IPCC (1990) reported a significant decrease in the DTR from both of these regions. Meanwhile, work from another data exchange agreement, a bilateral between the United States and the former Soviet Union, came to fruition as a data set of mostly rural maximum and minimum temperatures was developed for the former Soviet Union. Karl et al. (1991a) reported on the widespread decrease of the DTR over the former Soviet Union, China (not including Tibet), and the contiguous United States, as was highlighted by the IPCC (1992).

Additional data from other countries and updates to previous analyses have now been analyzed. Data from the eastern half of Australia, Canada, Alaska, Sudan, Japan, some Pacific island stations, South Africa, and a few other long-term stations in Europe have now been included. Figure 1 shows the area of the globe that has now been analyzed for differential changes of the maximum and minimum temperature. This area now covers over 50 percent of the Northern and 10 percent of the Southern Hemisphere land masses, but still only about 37 percent of the global land mass.

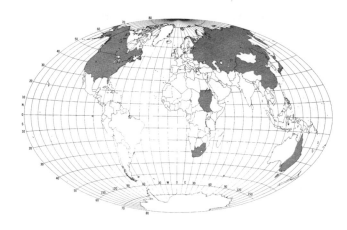

FIGURE 1 Areas of the globe (indicated by shading) that have been analyzed in this paper for changes of mean maximum and minimum temperature.

OBSERVED CHANGES OF MEAN MAXIMUM, MINIMUM, AND DIURNAL RANGE

Spatial and Seasonal Patterns of Contemporary Trends

Daily maximum and minimum temperatures from over 2,000 stations were available for analysis in the countries shaded in Figure 1 during the period 1951 to 1990 (except for Sudan and the former Soviet Union, which had data through 1987 and 1989, respectively). Selected subsets of these data were averaged within various regions of each country. The definition of each region represents a compromise between climatic homogeneity and ensuring an adequate number of stations within its boundaries to limit sampling error. The base period for calculating departures from the average included the years 1951 to 1990 (or slightly fewer years in some countries, as mentioned before). The regions are delineated in Figure 2 where, as in other large-scale studies of the change of the mean annual temperature (Jones et al., 1986a,b; Jones, 1988), the average of the trends of the mean annual maxima and minima reveals a general rise of temperature. A decrease of the minimum temperature within any region is uncommon; it is somewhat more frequent for the maximum temperature, as seen over the United States and China. The differential rate of warming between the maximum and minimum temperatures is apparent, with only a few regions reflecting an increase of the DTR. These weak exceptions occur in central Canada and southeasternmost Australia.

There are some seasonal variations of the rates of decreasing DTR, but they vary from country to country (Table 1). In Japan the decrease is not evident during summer, and it is not as strong during this time over China. In the United States the decrease is weak during the spring but quite strong during the autumn. Alaska and Canada have strong decreases during the spring, with somewhat lesser decreases during the autumn. Over the former Soviet Union the decrease in the DTR is significant throughout the year, but somewhat weaker during the winter. Over Sudan, the rate of the DTR decrease is strong in all seasons except during the summer rainy season, where rains have been very sparse over the past few decades. Over South Africa, decreases in the DTR are greatest in the Southern Hemisphere spring, and DTR actually increases during autumn. In the eastern half of Australia the decrease of the DTR is apparent throughout the year, but is weakest during the Southern Hemisphere summer.

When considered collectively, 60 percent of the trends in Table 1 reflect statistically significant decreases of the DTR. A test for a change point in the trend (Solow, 1987) indicates that for most seasons and areas there is insufficient evidence to suggest a statistically significant change point in the rate of the decrease. The Table 1 temperature trends can be area-weighted to reflect the overall rate of DTR

decrease. Table 2 shows the decrease both north and south of the equator, but without any pronounced seasonal cycle in the Northern Hemisphere. The area of available data in the Southern Hemisphere is too small to permit any general statements about trends in that portion of the globe, but the decrease in the Northern Hemisphere is quite apparent. The rate of the decrease in the DTR ($-1.4°C/100$ years) is comparable to the increase of the mean temperature ($1.3°C/100$ years).

For all areas combined (Figure 3), a noticeable difference between the rates of warming of the minimum and the maximum temperature began in the 1960s. The minimum temperature has continued to warm relative to the maximum through the 1980s. The time series ends in 1989, the year after the major North American drought, because data from the former Soviet Union were not available past 1989. The

TABLE 2   Trends of Temperature (°C/100 years) for Annual and Three-Month Mean Maximum Temperature, Minimum Temperature, and Diurnal Temperature Range for the Areas Denoted in Figure 1 (less Pakistan, northern Finland, and Denmark)

Northern Hemisphere 1951-1990 (50% of land area)			
Seasons	Max	Min	DTR
D-J-F	1.3	2.9	-1.5
M-A-M	2.0	3.2	-1.3
J-J-A	-0.3	0.8	-1.1
S-O-N	-0.4	1.3	-1.7
Annual	0.5	2.0	-1.4

Southern Hemisphere 1951-1990 (10% of land area)			
Seasons	Max	Min	DTR
D-J-F	1.6	2.2	-0.6
M-A-M	1.7	2.5	-0.8
J-J-A	1.0	1.3	-0.4
S-O-N	0.8	2.1	-1.3
Annual	1.3	2.0	-0.8

Globe 1951-1990 (37% of land area)			
Seasons	Max	Min	DTR
D-J-F	1.3	2.9	-1.6
M-A-M	1.9	3.1	-1.2
J-J-A	-0.2	0.8	-1.1
S-O-N	-0.3	1.4	-1.7
Annual	0.7	2.1	-1.4

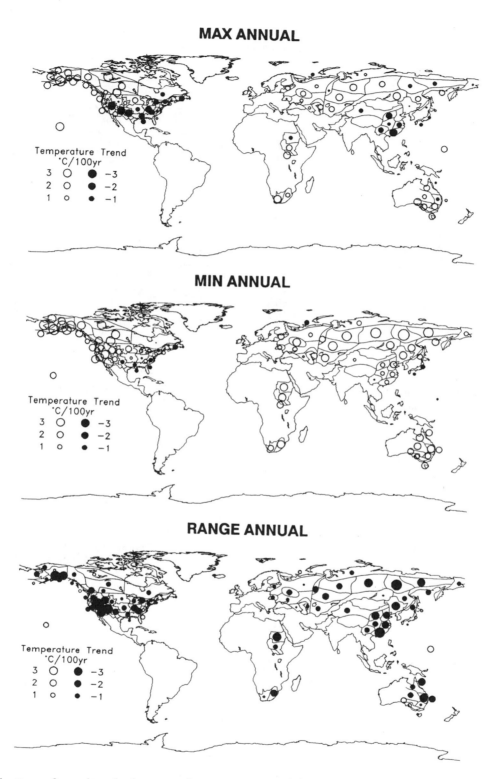

FIGURE 2 Spatial patterns of annual trends of mean maximum temperature, minimum temperature, and diurnal temperature range (mostly 1951-1990) in °C per 100 years. Diameter of circles is proportional to the trend; solid (open) circles represent negative (positive) trends. Circles pertain to regions within each country except for the island stations, e.g., the South Pacific and Hawaii.

TABLE 1   Trends of temperature (°C/100 years) for Annual and Three-Month Mean Maximum Temperature, Minimum Temperature, and Diurnal Temperature Range, Based on a Weighted Average of the Region (by country) in Figure 1.

Alaska 1951-1990 (39 stations)

Seasons	Max	Min	DTR
D-J-F	6.0	*8.8*	*−2.8*
M-A-M	3.1	*6.3*	*−3.2*
J-J-A	0.9	*2.4**	−1.5
S-O-N	−1.4	0.4	*−1.9**
Annual	*2.1*	*4.5*	*−2.4*

Japan 1951-1990 (66 stations)

Seasons	Max	Min	DTR
D-J-F	−0.5	−0.2	−0.2
M-A-M	−0.4	−0.7	0.3
J-J-A	0.5	0.0	0.4
S-O-N	−0.3	−0.5	0.2
Annual	−0.2*	−0.4	0.2

Canada 1951-1990 (227 stations)

Seasons	Max	Min	DTR
D-J-F	1.8	2.1	−0.2
M-A-M	*3.7*	*3.8*	−0.1
J-J-A	0.5	*1.4*	*−0.9*
S-O-N	−2.2	−1.2	*−1.0*
Annual	0.9	1.5	−0.6*

People's Republic of China 1951-1988 (44 stations)

Seasons	Max	Min	DTR
D-J-F	0.5	*3.5*	*−3.0*
M-A-M	−0.8	*1.4*	*−2.2*
J-J-A	*−1.8*	−0.8*	*−1.0*
S-O-N	−0.6	1.0	*−1.6*
Annual	−0.7*	*1.3*	*−2.0**

United States (contiguous) 1951-1990 (494 stations)

Seasons	Max	Min	DTR
D-J-F	−2.3	−0.7	*−1.5*
M-A-M	*2.3*	*2.5*	−0.2
J-J-A	−0.3*	*1.0**	*−1.4*
S-O-N	−1.7	1.3	*−3.0*
Annual	−0.6*	*1.0*	*−1.5*

South Africa 1951-1991 (12 stations)

Seasons	Max	Min	DTR
D-J-F	0.8	*2.0*	−1.2
M-A-M	*2.2*	*1.7*	0.5
J-J-A	1.3	*1.3*	0.0
S-O-N	−0.7	*1.8*	*−2.4*
Annual	0.9	*1.7**	−0.8*

Sudan 1951-1987 (15 stations)

Seasons	Max	Min	DTR
D-J-F	−1.2	*2.7*	*−3.9**
M-A-M	0.4	*3.3*	−2.8
J-J-A	*2.8*	*2.1*	0.7
S-O-N	1.4	*2.5*	*−1.1*
Annual	*0.9*	*2.7*	*−1.7*

Eastern Australia 1951-1991 (44 stations)

Seasons	Max	Min	DTR
D-J-F	*1.8*	*2.3*	−0.4
M-A-M	*1.6*	*2.8*	−1.2
J-J-A	0.8	*1.4*	−0.5
S-O-N	1.3	*2.2*	−0.9
Annual	*1.4*	*2.2*	−0.7

USSR (former) 1951-1990 (165 stations)

Seasons	Max	Min	DTR
D-J-F	2.8	4.2	*−1.3*
M-A-M	2.5	*3.8*	*−1.2*
J-J-A	−0.4*	0.9*	*−1.3*
S-O-N	0.6	2.2	*−1.6*
Annual	1.4	*2.8*	*−1.4*

Key: Trends significant at the .01 level (two-tailed t-test) are bold and italicized
       Trends significant at the .05 level are italicized
       Trends with significant change points are denoted with an asterisk

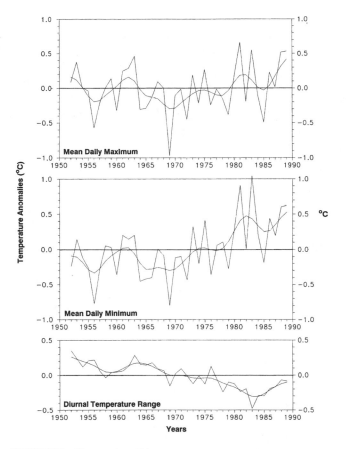

FIGURE 3 Time series of the temperature anomalies of the annual mean maximum temperature, minimum temperature, and diurnal temperature range for 37 percent of the global land mass (areas shaded in Figure 1). Smooth curve was obtained with a nine-point binomial filter with "padded" ends.

end of the time series is significantly affected by the drought. Furthermore, variance of the time series is significantly affected when such large regions drop out of the analysis, which is why the series ends prematurely. Nonetheless, Figure 3 reflects a gradual decrease of the DTR through much of the past several decades.

Longer-Term Variations

Unfortunately, the availability of maximum and minimum temperature data covering the globe is currently limited prior to 1951. In the United States a network of approximately 500 high-quality stations has remained intact since the turn of the century, and in the former Soviet Union a fixed network of 224 stations (165 if only rural stations are used) is available back to the 1930s. The time series from these two countries shown in Figure 4 reflect significant decadal variations in the DTR, which are particularly evident during the dry 1930s and the early 1950s in the United States. The general decrease of the DTR did not

begin in the United States until the late 1950s, and the DTR decreased rather dramatically in the mid-to-late 1970s over the former Soviet Union as part of substantial increases in the minimum temperature. The decrease of the DTR in these two countries is a phenomenon of recent decades. Data are also available further back in time for smaller areas and countries, notably Japan, eastern Australia, and South Africa. Figure 5 indicates that the decrease in the DTR in eastern Australia has occurred rather gradually since the decline in the late 1940s. In South Africa the decrease is predominantly due to the sharp decline in the early 1950s.

A very long record of maximum and minimum temperatures was available from the Klementinum Observatory in Prague, Czech Republic, and another from a benchmark station from northern Finland. Figure 6 portrays a remarkable *increase* of the DTR at the Klementinum Observatory from the early twentieth to the mid-twentieth century, with a substantial decrease since about 1950. The increase coincides with the increase of global mean temperature since the turn of the century, and the decrease occurs when the

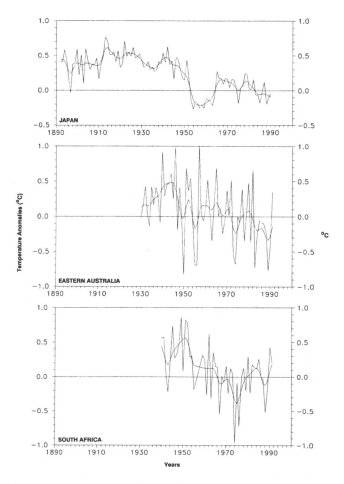

FIGURE 5 Time series of the variations of the diurnal temperature range for Japan, eastern Australia, and South Africa. Smooth curve was obtained with a nine-point binomial filter.

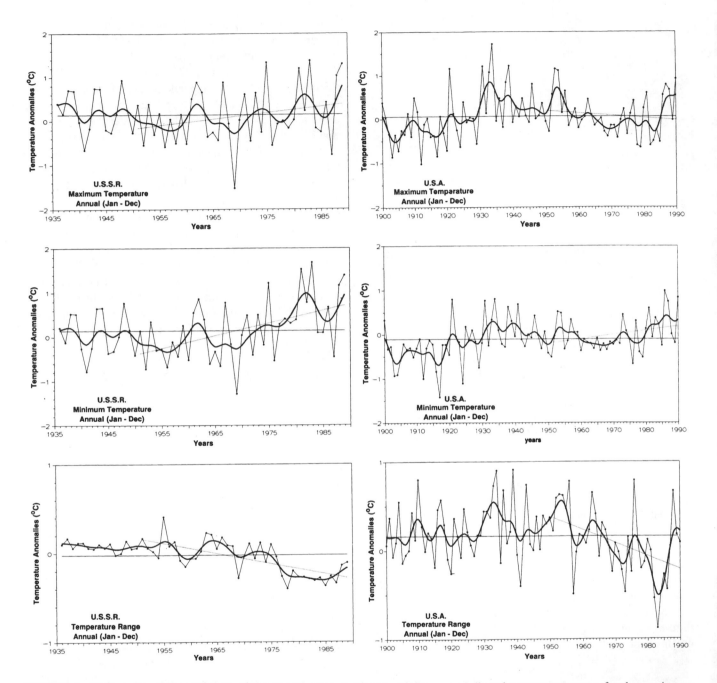

FIGURE 4 Time series of the variations of the annual mean maximum, minimum, and diurnal temperature range for the contiguous United States and the former Soviet Union. Smooth curve was obtained with a nine-point binomial filter. Trends since 1951 are depicted by the dashed line.

mean temperature reflects little overall change. In the first half of the nineteenth century the DTR averages about 0.5°C lower than in the latter part of the century. The DTR at Sodankylä, Finland, also displays a gradual decrease since 1950, but contrary to the Klementinum Observatory's pattern, the decrease is evident back to the turn of the century. The mean temperature at Sodankylä reflects little or no change. The high-frequency variability of the DTR at both

stations is less than that of their respective mean temperatures, but the converse is true for low-frequency variations.

If the data from the Klementinum Observatory truly reflect the regional change of the DTR in central Europe, then the recent decrease of the DTR in this area is less persistent and less substantial than the increase prior to 1950. In light of the variations at the Klementinum Observatory, what makes the results from Figure 2 so remarkable

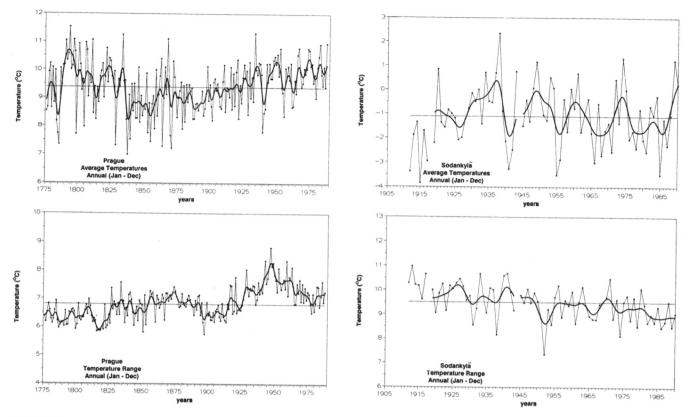

FIGURE 6   Time series of the variations of the annual mean temperature and the annual diurnal temperature range from two long-term stations: the Prague Klementinum-Observatory, Czechoslovakia, and Sodankylä, Finland, which has been designated as a climate reference station. Smooth curve was obtained with a nine-point binomial filter.

is the fact that so many areas share an overall decrease of the DTR.

Recently, other investigators have also compiled information on the change of the DTR over other portions of the globe. Frich (1992) provides evidence to indicate that there has been a general decrease over Denmark since about 1950, based on an analysis of several long-term stations, most located in rural areas. Bücher and Dessens (1991) analyzed a long record of maximum and minimum temperatures from the Pic du Midi de Bigorre Observatory in the Pyrenees at a height of over 2,800 m. Their analysis revealed a significant decrease of the DTR since the late nineteenth century, but an inhomogeneity in the record prevented a continuation of the analysis beyond 1970. Kruss et al. (1992) reported on changes of maximum and minimum temperature between the two 30-year periods 1931 to 1960 and 1961 to 1990 over Pakistan. Despite considerable missing data, they managed to obtain at least 20 years of data from each of the two periods for 35 stations across Pakistan. Their analysis revealed a mix of decreasing and increasing changes of the DTR. In our analysis of Pakistani data we could only manage to identify five stations with adequate data to analyze year-to-year changes, and these were all located in the northern half of the country. These stations also depicted a decrease of the DTR.

Relation to Variations of the Seasonal and Annual Extremes

For a variety of practical considerations, it is important to know whether the decrease in the mean DTR translates to a decrease in the extreme temperature range. Karl et al. (1991a) provide evidence to suggest that, indeed, over the United States and the former Soviet Union (the only areas for which they had access to daily data) there was often a significant and substantial decrease in the seasonal and annual temperature extremes similar to the decrease in the seasonal and annual mean DTR. This similarity is also reflected in the time series of monthly extreme maximum and minimum temperature over Sudan (Jones, 1992).

Data Quality

The reliability of the data used to calculate the changes of the DTR is a critical question. The data presented and discussed here have been subjected to various degrees of quality assurance. The degree to which precautionary measures have been taken to minimize data inhomogeneities varies considerably from country to country. In the United States, a fixed network of stations in the Historical Climatology Network is used (Karl et al., 1990), which largely consists of rural stations that have been adjusted when necessary for random station relocations, changes in instrument heights, systematic changes in observing times (Karl and Quinn, 1986; Karl et al., 1986), the systematic change in instruments during the mid- and late 1980s (Quayle et al., 1991), and increases in urbanization (Karl et al., 1988). The potential warm bias of the maximum introduced by the H083 series of thermometer (Gall et al., 1992) is not a factor in this network, since the H083 instrument was not used in the rural cooperative network.

In the former Soviet Union, the fixed network of 165 stations consists of rural stations (1990 populations less than 10,000 and local surroundings free of urban development). The data from the former Soviet Union have not been adjusted for any random or systematic inhomogeneities. Station histories, however, indicate that there have not been systematic changes in that network's operation over the course of the past 50 years.

In Canada, the results reported here are derived from a set of 227 rural stations (those in areas with population less than 10,000). These data were selected from a network of 373 principal stations, but a large number of urban areas and stations that relocated to airports were eliminated from the analysis.

In Alaska, a network of 39 stations was used. It included most stations that had been operating in that state since the early 1950s, with the exception of the stations in the major cities of Juneau, Fairbanks, and Anchorage. Once again no attempt was made to adjust for station relocations, and the stations consist of a mix of instrument types with some changes at specific sites.

Station histories from China do not reflect any systematic changes in instrumentation, instrument heights, instrument shelters, or observing procedures relative to the maximum and minimum temperature. Our analysis is based on a subset of the more than 150 stations available to us. No attempt was made to correct for random station relocations in the fixed network of 44 stations we finally selected from the larger network. The potential impact of urbanization precluded the use of many stations; we eliminated all stations that were in or near cities with populations over 160,000.

All stations in Australia have recently undergone a thorough homogeneity analysis (Plummer et al., 1995), but the results were not available for this analysis. Instead, stations were selected on the bases of length of record and distance from major areas of urbanization. All of the Australian stations used are in small towns or rural areas, many in post office "back yards."

Fewer than half of the 154 stations available from Japan

were used in this analysis. As with China, many stations were eliminated because of their proximity to major urban areas. An inspection of the station histories reveals a number of network "improvements" related to the automation of the temperature measurements in recent years. A full assessment of the homogeneity of the data awaits a detailed analysis. The station networks from Sudan and South Africa include some stations from urban areas, but countrywide decreases of the DTR are not overwhelmed by these stations. Incomplete information was available regarding systematic changes in instrumentation at these locations over the past several decades, but the data were inspected for station relocations and adjustments made when necessary on the basis of temperature differences from neighboring stations. In total, four stations in South Africa were adjusted using the procedures outlined by Jones et al. (1986a).

INFLUENCES ON DIURNAL TEMPERATURE RANGE

Local Effects

As more data become available from a variety of countries, it becomes difficult to dismiss the general decrease of the DTR over the past several decades as an artifact of data inhomogeneities. Observing networks are managed differently in each country. If local effects are significantly influencing the DTR, then at least three possible sources of change need to be explored: urbanization, irrigation, and desertification. Evidence to support or refute the impact of these human-induced local and regional effects are discussed in the subsections below.

Urban Heat Islands

It is well known that the urban heat island often tends to manifest itself most strongly during the nighttime hours (Landsberg, 1981). In mid-latitude North American cities the urban-rural temperature difference usually peaks shortly after sunset, then slowly decreases until shortly after sunrise, when it rapidly decreases and, for some cities, actually vanishes by midday. In many cities increases in urbanization would differentially warm the minimum relative to the maximum temperature. A number of precautions have been taken to minimize this effect.

In the contiguous United States, the corrections for urban development recommended by Karl et al. (1988) have been applied to the data, and any residual heat-island effect in this analysis should not be an issue. In Canada, only stations with population less than 10,000 were used in the analysis, and the average population of the cities in the proximity of the observing stations was slightly over 1,000. If the Canadian stations behave similarly to stations in the United States, the effect on the diurnal temperature range may be exaggerated by about 0.1°C. An effect of similar magnitude may exist for Alaska. In the former Soviet Union, the population limit for inclusion of any station into the network was 10,000; in addition, however, no station could be within 1 km of any multi-story urban development. If the impact of the effect of urbanization on the DTR in the former Soviet Union is anything like that in the United States, the residual urban heat island effect on the DTR should be at least an order of magnitude smaller than the observed decrease of the DTR (nearly 1.5°C/100 yr).

In China and Japan, a number of tests were conducted to identify the impact of urbanization on the DTR. Three networks of stations were categorized on the basis of population. For China the categories included stations in proximity to cities with populations over 1 million, under 160,000, and in between these two threshold values. Categories in Japan were based on threshold values of 500,000 and 50,000. In the three population categories, proceeding from high to low, China had 23, 42, and 44 stations while Japan had 17, 71, and 66 stations. Figure 7 shows that the decrease of the DTR actually is more marked in China for stations in the lowest population category than for stations in the medium category, while the trend of the average temperature continues to decrease. This unexpected finding suggests that urban-

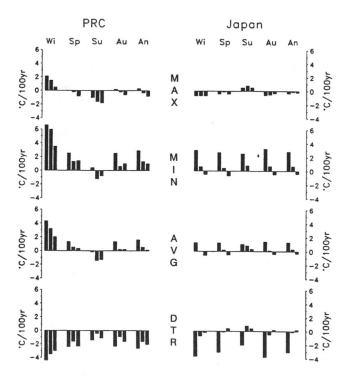

FIGURE 7 Temperature trends calculated for the People's Republic of China and Japan. Abbreviations: Wi, winter; Sp, spring; Su, summer; Au, autumn; MAX, maximum; MIN, minimum; AVG, average; and DTR, diurnal temperature range. The three bars in each of the seasonal and annual-average groupings reflect stations near cities in the high, medium, and low categories (left to right) as defined in the text.

ization effects in China are dissimilar to those in the United States, since they seem not to have affected the maximum and minimum temperature trends in Chinese cities of 500,000 or less. In Japan, however, the impact of urbanization on the DTR is evident even in the lowest population category (less than 50,000), and is even more apparent for the average temperature. On the basis of these analyses, it seems that the impacts of urbanization in China (Wang et al., 1990) are unlikely to significantly affect the trends reported in Tables 1 and 2.

A previous paper by Jones et al. (1990) investigated the impact of increasing urbanization in the land data base used by the IPCC (1990, 1992) to calculate changes of global temperature. The conclusion from the work of Jones et al. (1990) was that any residual urban bias in the land-based average temperature records was about 0.05°C during the twentieth century. A comparison of the average temperature trends derived from the stations used in Tables 1 and 2 with the stations used by Jones (1988) revealed differences in trends from country to country, but virtually identical trends of temperature (within 0.02°C/100 yr) were found when all areas depicted in Figure 1 were considered. This similarity suggests that the degrees of urban-induced bias in these two data sets are of comparable magnitude over the past 40 years, despite the use of substantially different station networks.

Irrigation

It can be argued that increases in irrigation may account for the decrease in the DTR. The evaporation associated with soil moisture would convert sensible to latent heat and thus significantly reduce daytime temperature. In order to test this hypothesis, the correlation coefficient (both the Pearson product-moment and the Spearman rank) was calculated using the values of the trends of the DTR and the change in land area under irrigation from 1950 to 1987 (U.S. Department of Commerce, 1950, 1988) for each of the regions in the United States delineated in Figure 2. No relationship was found between the change in the DTR and the increase of irrigated lands, and in fact many of the largest decreases of the DTR were associated with areas with the smallest increases of irrigation. Considering that the United States has had significant decreases of the DTR over the past several decades, and has had relatively large increases of irrigation relative to other countries over the past 40 years, it seems unlikely that increases of irrigation can be regarded as a serious explanation for the decreases of the DTR.

Desertification

The converse of the theoretical effects of irrigation would result from increased desertification. This might arise from poor land practices such as over-grazing or deforestation. Given the mid- and high-latitude bias of the results reported here, it seems unlikely that desertification would have a significant impact on the results reported in Tables 1 and 2. Moreover, this effect would tend to make the reported decreases underestimates, especially during the warm season, as desertification would increase the maximum and decrease the minimum.

Climatic Effects

Since it seems unlikely that any of the human-induced local effects can provide a satisfactory answer to the widespread decrease of the DTR, a number of climatic variables that differentially affect the maximum and minimum temperature were analyzed to discern which of them most strongly affect the DTR. Over 50,000 days of climatic observations were selected from the stations listed in Table 3 during periods of consistent measurement procedures for the variables defined in Table 4.

The selection of variables to be studied was based on a priori information. For example, increases of the DTR over land have previously been related to snow cover ablation, in simulations by the U.K. Meteorological Office's general circulation model (GCM) with doubled CO_2 (Cao et al., 1992). It is well known that the ability of the surface boundary layer to absorb, radiate, transform, and mix sensible heat differentially affects the maximum and minimum temperature. Relative humidity and cloudiness are two important climate variables that influence these surface-layer properties. In this analysis cloud-related information was contained in two climatic variables, the sky cover (in tenths) and an index of the ceiling height. The ceiling height (CIG)

TABLE 3   Stations and Years Used in Identifying the Sensitivity of the Diurnal Temperature Range to Various Climatic Variables

Stations	Years
Sacramento, CA	1961-69
Tallahassee, FL	1962-70
Indianapolis, IN	1966-74
Worcester, MA	1961-69
Bismarck, ND	1973-81
Scotts Bluff, NE	1971-79
Reno, NV	1961-69
Oklahoma City, OK	1975-83
Pittsburgh, PA	1961-69
Columbia, SC	1971-79
San Antonio, TX	1973-81
Seattle/Takoma, WA	1971-79
Spokane, WA	1966-75
Green Bay, WI	1971-79

TABLE 4   Definitions and Abbreviations of the Climatic Variables Used to Test the Sensitivity of the Diurnal Temperature Range

Variables	Abbreviations
Diurnal temperature range (daily max-daily min)	DTR
Snow cover (binary, if snow depth $\geq$ 2.54 cm)	SNOW
Mean relative humidity (0600 LST* & 1500 LST)	RH
Mean wind speed (0600 LST & 1500 LST)	WS
Mean sky cover (0600 LST & 1500 LST)	SKY
Mean ceiling (0600 LST & 1500 LST)	CIG
Total daily top-of-the-atmosphere solar radiation	TRAD
Day-to-day temperature differences ($|TMP_0 - TMP_{-1}| + |TMP_0 - TMP_{+1}|$)	ΔTMP

*local standard time

was broken down into seven categories. The cloud ceiling is defined as height above ground of the lowest cloud layer that covers 50 percent or more of the sky. The wind speed is an effective measure of the degree of mixing within the surface boundary layer, since it affects and interacts with the frequency and/or intensity of inversions and superadiabatic lapse rates.

In addition, the DTR is affected by the seasonal and latitudinal changes of incoming solar radiation as well as by the magnitude of day-to-day temperature differences. TRAD (top-of-the-atmosphere solar radiation) can also be regarded as a surrogate for the temperature, especially when it is used in conjunction with the other variables listed in Table 4. Karl et al. (1986) demonstrate the impact of the interdiurnal temperature difference on the maximum and minimum temperature. Large interdiurnal temperature differences lead to a large temperature range, even in the absence of a diurnal temperature cycle. These day-to-day changes of temperature are largely controlled by the thermal advection associated with synoptic-scale cyclones and anticyclones.

All of the variables in Table 4 were used in a multiple-regression analysis. Variables were regressed against the square root of the DTR, as opposed to the actual DTR, because the DTR is bounded by zero. Without the transformation, non-normal residuals result in multiple linear-regression analyses, making it more difficult to interpret the results. Figure 8a indicates that the partial-correlation coefficients of each variable with respect to the DTR are often significantly different from the simple linear correlation coefficients, making it difficult to speculate on the effect of changes in any one variable without knowing the changes in (or assuming constancy of) the other variables. Given the huge sample size, very low correlations have high statistical significance (even considering the day-to-day persistence of the DTR). On a local basis, a generalized

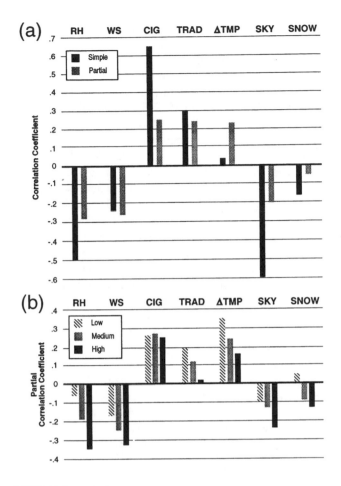

FIGURE 8 Relationship between various climate variables and the diurnal temperature range. (a) Simple and partial correlation coefficients (removing the effects of all other variables) between each variable (defined in Table 4) and the diurnal temperature range. (b) Partial correlation coefficients of each variable for cases partitioned by the total daily solar radiation at the top of the atmosphere: ≤ 170 W m^{-2} (low), ≥ 271 W m^{-2} (high), and remainder (medium).

multiple linear-regression model (one model for all stations and all days) based on the seven climatic variables in Figure 8a explains about 55 percent (53 percent without the square-root transformation) of the daily variance of the DTR. Although the explained variance is substantial, it is apparent that other factors may also need to be considered in order to explain the variations of the DTR (e.g., better representations of the atmospheric stability, external forcing factors, more precise techniques for calculating the mean quantities used in the analysis) and their relationships are not adequately expressed by a linear equation.

On a variable-by-variable basis, the signs of all the partial correlations make qualitative physical sense. It is interesting to note that the partial correlation of ΔTMP with DTR is greater than that of the simple correlation coefficient (Figure 8a), because the correlation between TRAD and ΔTMP

mask the importance of ΔTMP in influencing DTR. The decrease of the partial correlation relative to the simple correlation for the variables RH, CIG, and SKY is to be expected, because changes among these variables are all related to each other. The decrease in the partial correlation of SNOW is particularly noteworthy, especially since Cao et al. (1992) attribute the ablation of snow cover in their model to an increase in the DTR. The empirical results in Figure 8a suggest that SNOW is only weakly related to DTR (7.5 percent of the cases, nearly 4,000, had snow cover), especially by comparison with the other variables.

In order to investigate the linearity, or lack thereof, of the relationships implicit in Figure 8a, the data were partitioned by TRAD. Figure 8b provides strong evidence to suggest that the relationships change with the amount of TRAD. In particular, the partial correlation coefficients of RH, WS, and SKY become stronger as TRAD (and thus temperature) increases. This probably has more to do with a reduction of the maximum temperature than an increase of the minimum. During daylight hours high values of the RH, WS, and SKY are indicative of higher albedos, higher potential evapotranspiration, higher atmospheric water-vapor absorption of incoming radiation, and larger-than-normal mechanical mixing. These factors act to reduce the maximum temperature that would otherwise result from high-intensity TRAD, which would be manifested as sensible heat within the surface boundary layer. The non-linearity of RH as TRAD increases is substantially greater than that of WS (Figure 8b). As the TRAD reaching the surface increases, the temperature increases, and comparatively more of the TRAD can be used for evaporation than for raising surface temperature, as would be anticipated by integrating the Clausius-Clapeyron equation to obtain saturation vapor pressure as a function of temperature.

The reduction of the partial correlation of TRAD with the DTR when the sample is partitioned with respect to TRAD (Figure 8b) relates to the balance between long nights and short days. During the late autumn and early winter in the northern half of the United States (areas which include the lowest partition of TRAD), a moderate increase in the TRAD (by inter-seasonal and latitudinal variations) generally results in a higher DTR. Contrarily, in the warm half of the year, TRAD is more than ample, so the relation between TRAD and DTR is near zero. In fact, a further selection for very high values of TRAD (short nights) leads to small negative partial correlations between TRAD and the DTR.

The reduction of the partial correlation of ΔTMP as TRAD increases is related to the decrease in intensity of the day-to-day changes of temperature during the warm season. Rossby waves and extra-tropical cyclones have lower amplitude, speed, and intensity during the warm season.

A change in sign of the partial correlation coefficient of

SNOW with the DTR (Figure 8b) suggests that the length of night relative to the TRAD is a significant factor in the impact of changes of snow cover on the DTR. In the northern United States, around the winter solstice, TRAD is relatively low. Snow on the ground at this time of year is important because of its excellent insulating properties (it reduces heat flow from the soil), which help lower the nighttime minimum. During the daytime the TRAD is already low, so the amount of solar radiation reflected by the snow cover is no longer so important. As a result, snow cover at this time of year at these latitudes leads to an increase in the DTR. This is not the case as the season progresses or the latitude decreases, as is reflected by the negative partial correlations (lower values of DTR with snow cover) associated with the highest category of TRAD, where there were still nearly 1,000 cases of snow cover. The data suggest that snow-cover ablation will not necessarily lead to an increase of the DTR.

From the above analysis it is apparent that there are many factors, often intricately related, that affect the DTR. Many of these variables are very much related to a greenhouse effect, not necessarily one anthropogenically induced. Overall, however, the two variables related to changes in cloudiness, sky cover, and ceiling height explain the greatest portion of the variance of the DTR. Changes in cloudiness should be one of the first considerations in searching for an explanation of the observed decrease of the DTR (Figure 2). Indeed, when large continental scales are considered, the relationship between cloud amount (or sky cover) with the DTR is quite impressive (Figure 9). Plantico et al. (1990) have already demonstrated that the decrease in the DTR

over the United States is strongly linked to an observed increase in daytime and nighttime cloud cover and to a lowering of cloud ceilings.

Is there a general increase in cloud cover over much of the globe? Empirical evidence by Henderson-Sellers (1986, 1989, 1992) and Jones and Henderson-Sellers (1992) suggests that this may be the case over Canada, the United States, Europe, the Indian subcontinent, and Australia. Analyses of cloud cover changes over China from a network of 58 stations are inconclusive, but there is evidence for a decrease in the sunshine (a 2 to 3 percent decrease in sunshine from the 1950s to the 1980s). The quality of the cloud data (and perhaps the sunshine data) is questionable because the correlation between monthly anomalies of sunshine and cloudiness at many sites is not high. Analyses of changes in cloudiness over the former Soviet Union by Kaiser and Razuvayev (1995) reveal a general increase of cloud cover (about 5 percent) over the period 1936 to 1986, with cirrus increasing in frequency by nearly 20 percent. They also found considerable interannual and interstation variability, so the quality of these data could also be called into question. Nonetheless, the trend over the former Soviet Union is consistent with a decrease in the DTR. Frich (1992) and Bücher and Dessens (1991) also found that decreases in the DTR in Denmark and at the Pic du Midi de Bigorre Observatory occurred concurrently with an increase in cloudiness. An analysis of changes in cloud cover over Japan, using many of the same stations selected for the analysis of the maximum and minimum temperatures, indicates that cloud cover may have increased on an annual basis by nearly 1 percent since 1951, but there is no apparent response in the DTR. Changes in sunshine in Japan are not altogether consistent with the increase in cloud cover, perhaps because a new instrument was introduced into the network in 1986.

LARGE-SCALE ANTHROPOGENIC EFFECTS

Greenhouse Gases

Interest in the possible change of the DTR with increasing anthropogenic greenhouse gases has prompted several modeling groups to publish information from their models regarding the change in projected DTR when CO_2 is doubled. Table 5 summarizes the results of these models. For the GCM experiments, the magnitude of the decrease in DTR is small relative to the overall increase of mean global temperature. Moreover, these experiments reflect a level of CO_2 increase well in excess of present-day values, so even smaller changes would be expected in the observed-temperature record. Outside of the Rind et al. (1989) study, only Cao et al. (1992) focus specifically on the changes of the DTR over land areas. Cao et al. show that their model can reasonably simulate the range of the present-day diurnal

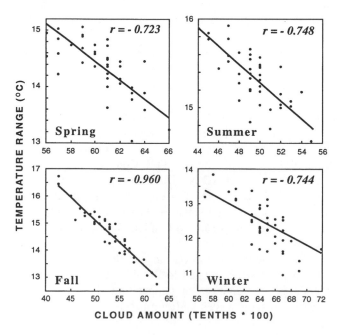

FIGURE 9 Seasonal relationships between U.S. area-average cloud cover and the diurnal temperature range.

TABLE 5 Summary of Modeling Results Showing the Relationship Between Doubled CO_2 Concentrations and Changes in the Diurnal Temperature Range and the Maximum and Minimum Temperatures

		General-Circulation Models				
		Resolution				
Model	Author	Horizontal	Vertical	Ocean	ΔT_{eq}	ΔDTR_{eq}
CCC	Boer (pers. comm.)	T32	10	Mixed layer	3.5	-0.28
GISS	Rind et al. (1989)	8° × 10°	9	Mixed layer	4.2	-0.7 (summer, USA) -0.1 (annual, USA)
UKMO	Cao et al. (1992)	8° × 10°	11	Mixed layer	5.2	-0.17
UKMO	Cao et al. (1992)	5° × 7.5°	11	Mixed layer	6.3	-0.26

Radiative Convective Model (Cao et al., 1992)			
Type	ΔMaximum T_{eq}	ΔMinimum T_{eq}	ΔDTR_{eq}
Fixed absolute humidity	N/A	N/A	-0.05
No surface turbulence	2.5	2.9	-0.4
No evaporation	2.3	2.9	-0.6
Full surface exchange	1.5	2.2	-0.7

Key: ΔT_{eq} is the equilibrium global temperature change (°C) for doubled CO_2 concentrations
ΔDTR_{eq} is the equilibrium global change of the DTR (°C) for doubled CO_2 concentrations
CCC is the Canadian Climate Centre
GISS is the Goddard Institute for Space Studies
UKMO is the U.K. Meteorological Office

temperature cycle, although the modeled diurnal range in mid-latitudes is generally less than the observed.

Cao et al. (1992) also conducted a number of experiments with a one-dimensional radiative-convective model (RCM) to show that the decrease in the diurnal temperature range with doubled CO_2 in that model is due primarily to a water-vapor feedback. A reduction of only 0.05°C in the DTR was observed when the absolute humidity was held constant. Table 5 indicates that increased sensible-heat exchange and evaporation are also important factors leading to a reduction in the DTR in RCM simulations with enhanced CO_2.

Interestingly, the ratio of the DTR decrease relative to the increase of the mean temperature is closer in the RCM to the observed ratio over the past several decades than it is in the GCM (Table 5). The RCM omits the positive feedbacks to the DTR from reductions in cloud and surface albedo included in the GCM simulations (Cao et al., 1992). These feedbacks tend to increase the DTR because of reduced atmospheric (cloud cover) and surface (snow cover) albedo. Other GCMs simulate both increases and decreases in cloudiness from global warming—decreases in much of the troposphere, but increases in the high troposphere, low stratosphere, and near the surface in high latitudes (Schlesinger and Mitchell, 1985). A tendency for a general increase in cloud cover over land (which now seems likely from

the observational evidence) could help explain the large discrepancy between the observed data and the model projections of the ratios of the decrease in the DTR range relative to the mean temperature increase. (This explanation assumes, of course, that the recent warming is induced by increases in anthropogenic greenhouse gases. On the other hand, it leaves questions regarding the cause of the apparent change in cloudiness and how it has affected the mean temperature.)

The ability of present-day GCMs to adequately simulate changes in the DTR resulting from enhanced CO_2 is also affected by surface parameterizations of continental-scale evaporation. As Milly (1992) points out, present-day GCMs can overestimate the surface evaporation because of their failure to properly account for the cooling that occurs with the evaporation. Milly raises concerns about the veracity of the results from studies of soil-moisture changes induced by an increase of greenhouse gases. Accurate projections of the changes in the surface boundary-layer DTR with increases of anthropogenic greenhouse gases will be strongly dependent on adequate simulation of these processes.

Given the dependency of the DTR on surface-layer processes, interactions with the land surface, and cloudiness (which are all areas of significant uncertainties within present-day GCMs), it may not yet be possible to adequately project changes of the DTR with enhanced concentrations of greenhouse gases.

Tropospheric Aerosols

It has recently been shown that increases in sulfate aerosols over and near industrial regions can have a significant impact on the earth's surface temperature (Charlson et al., 1992). Charlson et al. (1991, 1992) provide evidence to indicate that the anthropogenic increase of sulfate (and carbonaceous) aerosol is of sufficient magnitude to compete regionally with present-day anthropogenic greenhouse forcings. This forcing, which is confined primarily to the Northern Hemisphere, is a combination of direct aerosol forcing (especially over land) and indirect aerosol forcing leading to increases in cloud albedo (especially over the marine environment). Charlson et al. (1992) conclude that at present not enough is known to allow the effects of sulfate aerosols on the lifetime of clouds to be estimated. Increased aerosols (cloud condensation nuclei) could lead to smaller droplet sizes, causing a decrease in the fallout rate, which in turn could lead to an increase in cloud cover. Charlson et al. (1991) show a pattern of the geographic regions where direct aerosol forcing should be greatest. It is difficult to identify a direct relation between the pattern and magnitude of the decrease in the DTR (Figure 2) and the anthropogenic radiative forcing (Charlson et al., 1991). For example, on an annual basis over the United States the clear-sky daytime forcing varies from essentially zero (over several western states) to about -4 W m^{-2} (over southeastern states). It also varies with season by roughly a factor of 2 to 3, being greatest during summer.

Recently Penner et al. (1992) have argued that atmospheric aerosols from biomass burning also act to increase the planetary albedo, both directly by clear-sky planetary albedo increases and indirectly through increases in cloud albedo. Since biomass burning is most extensive in subtropical and tropical areas, this effect may be directly relevant to only a small portion of the data analyzed here.

Two tropospheric-aerosol forcings that tend to increase the clear-sky albedo have been identified. Has either of these forcings acted to reduce the maximum temperature and thereby the DTR? In the United States and northern (and perhaps eastern) Europe, where we detected a significant decrease in the DTR, there has actually been a net decrease in sulfur emissions over the past several decades, which would appear to eliminate sulfate aerosols as a cause of the DTR in these areas. However, there are reasons why such a conclusion may be premature. First, there is considerable geographic, seasonal, and secular variation of anthropogenic sulfur emissions. For example, Figure 10 shows the emissions of SO$_2$ gas (precursor of the sulfate aerosol) doubling over the southeastern United States from 1950 to ca. 1980,

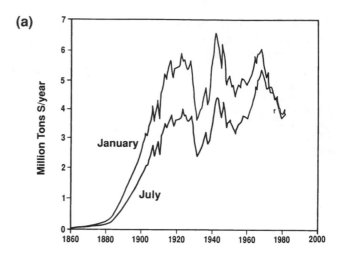

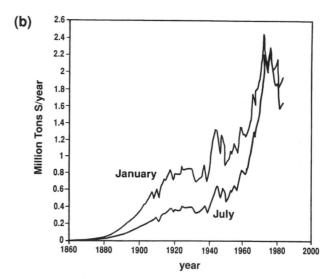

FIGURE 10   Time-dependent emissions of SO$_2$ gas for (a) the northeastern United States and (b) the southeastern United States.

and then decreasing by 25 percent since then, while emissions have remained nearly constant in the northeastern United States. Also, the direct climate forcing due to tropospheric aerosol loading is influenced by (1) the absence of cloud cover, (2) the emission rate, and (3) the residence times of sulfate aerosol in the atmosphere, which are known to be short (about a week, according to Slinn (1983)). Second, in the United States at least, the effective heights, or stack heights, of the sulfur emissions have increased as a consequence of rural electrification in the southeast and, later, of the U.S. Clean Air Act. This increase in stack height may have had an important effect on the lifetime of SO$_2$ and the sulfate aerosols (because of their relatively short lifetimes) and, consequently, on the aerosol concentration. Third, the indirect effects of sulfate aerosols on the DTR must also be considered, e.g., more cloud cover or a changing distribution of cloud characteristics.

CONCLUSIONS

Strong evidence exists for a widespread decrease in the DTR over the past several decades in many portions of the globe. The climatic factors that can affect the DTR are many, but indications are that cloud cover, including low clouds, has increased in many areas where DTR has decreased. These increases in cloud cover could be indirectly related to the observed global warming and increases of greenhouse gases (or perhaps even to the indirect effects of increases in aerosols), could be simply a manifestation of natural climate variability, or could reflect a combination of all three.

Obtaining a robust answer regarding the cause(s) of the decrease in the DTR will require efforts in several areas. First, an organized global effort is needed to develop relevant and homogeneous time series of maximum and minimum temperature, along with information on changes of climatic variables that influence the DTR, such as cloudiness, stability, humidity, thermal advection, and snow cover. Second, improvements in the boundary-layer physics and treatment of clouds within existing GCMs is critically important. Third, the treatment of both anthropic tropospheric aerosols and greenhouse gases must be realistically incorporated into GCMs with a diurnal cycle. Fourth, measurements need to be made to help clarify the role of aerosols. These include (1) data on diurnal, seasonal, and geographic changes in solar radiation, in areas influenced by aerosol forcing and areas free of it; (2) simultaneous data on key aerosol properties (mass-scattering efficiency, wavelength dependence, backscatter fraction/asymmetry factor), concentration, and composition; (3) data related to changes in all the relevant cloud parameters and their coupling to aerosol changes (if any); and (4) improved chemical models to allow more accurate calculation of the sulfate aerosol effects and to provide an estimate of the time course of this aerosol forcing. Finally, imaginative climate-change detection studies that link the observed climate variations to model projections will be required to convincingly support any relationship between anthropogenic changes and the DTR.

It will be difficult to satisfactorily explain the observed changes of the mean temperature until an adequate explanation for the observed decrease in the DTR can be determined. Moreover, the practical implications of projected temperature changes and whether they are likely to continue will be even more difficult to assess.

ACKNOWLEDGMENTS

This work was supported by a U.S. Department of Energy/National Oceanic and Atmospheric Administration (NOAA) interagency agreement and NOAA's Climate and Global Change Program.

We thank the following scientists for providing us with additional data: Reino Heino for the Sodankylä data, and Takehiko Mikami for the Japanese data.

Discussion

GOODRICH: Tom, were you able to use the ISCCP data sets to test your hypothesis that the decrease in the annual range of cloudiness might be associated with temperature?

KARL: I think there might be problems with both the longevity and the quality of those data, and the solar radiation data network is not as comprehensive as one might wish. We really need to get cloud and temperature experts together to see whether relating the two makes sense.

CHARLSON: To determine how cloud cover affects heat balance you would really need to know cloud liquid-water content and mean droplet size, as well as cloud area. All three govern the system albedo. The ISCCP data don't have those, and I'm not sure our current remote-sensing systems have that capability either. If we found a correlation, we wouldn't know whether it was caused by cloud microphysics or dynamics.

SHUKLA: Tom, one of your most dramatic differences was that between the rural and urban areas of Japan. Would it be naive to ask whether radiative forcing associated with urban pollution could be involved?

CHARLSON: Urban haze around most cities actually extends for hundreds of kilometers. Also, neither the soot aerosols nor the light-absorbing gases in smog cause any great amount of heating.

KARL: Various studies have shown that maximum temperature is not much affected by urbanization. But it is a significant point to keep in mind, especially given how it dominates the record in Japan.

KEELING: Many of you have probably heard of the Arctic and Antarctic borehole data that suggest that the heat flux from the interior of the earth is not steady state. Is it possible that these borehole temperatures are showing the effects of nighttime radiative warming?

CHARLSON: The next set of NCAR model calculations will be simulating diurnal temperature range, and a model in Hamburg is doing that too. It will be interesting to hear their results.

RIND: Our GCM looked at diurnal temperature change for doubled CO_2. We found that for a 4° warming we got about a 10 percent change, or 1°C. That was primarily during the summer, so we did not get the one-to-one correspondence you see in the data. I also wanted to mention that the ISCCP data set can't get optical thicknesses for nighttime clouds, since they can't measure the albedos.

KARL: I should perhaps add that while most of the CO_2-doubling models show a decrease in diurnal temperature range of maybe 10 percent of the equilibrium temperature, the simple radiative convective model of Cao et al. shows fairly large decreases. I really think that there's some question whether a GCM can be expected to reproduce the diurnal cycle, given the type of land-surface physics involved.

Decadal Oscillations in Global Temperature and Atmospheric Carbon Dioxide

CHARLES D. KEELING AND TIMOTHY P. WHORF[1]

ABSTRACT

Since 1958, global air temperature has been found to correlate with the concentration of atmospheric CO_2 on interannual time scales, from biennial to decadal. The decadal variations also correlate approximately with the 11-year sunspot cycle. Although CO_2 data are lacking before 1958, global temperature data are available back to the 1850s. No consistent solar correlation is evident in this longer record. Instead, the record shows a decadal pattern that can be expressed by two oscillations having periods of approximately 9 and 10 years. These oscillations show reinforcement in the 1880s and 1970s and interference in the 1920s. Thus, decadal oscillations appear to have been relatively prominent in the latter half of the nineteenth and twentieth centuries, while being weak or absent from about 1905 to 1940. In the absence of an identified mechanism that would explain this pattern, the results must be treated as tentative.

INTRODUCTION

Several years ago we detected a possibly cyclic variation in the concentration of atmospheric CO_2 (Keeling et al., 1989). The period of variation, of the order of 10 years, appeared to be synchronous with a decadal oscillation in air temperature when both records were band-passed to remove high- and low-frequency fluctuations and the influence of fossil-fuel combustion on the CO_2 concentration. As shown in the upper panel of Figure 1, the adjusted 32-year CO_2 record shows maximum concentrations near 1961, 1971, and 1981, the same years in which decadal maxima in air temperature occur. The maxima in CO_2, especially the latter two, nearly coincide with maxima in the number of sunspots (lower panel) as though both CO_2 and temperature were influenced by changes in solar irradiance. The investigation was not extended back to early times, because of a lack of CO_2 data.

Globally averaged surface air temperature anomalies have been determined back to 1854, however (Jones and Wigley, 1991). To extend our analysis, we have accordingly computed the difference between two smoothing splines fit to this temperature record, where the differing stiffnesses of the splines result in band-passed data similar to those shown in Figure 1. The results, plotted in Figure 2 together with the record of sunspots, suggest that the positive correlation of global temperature with sunspot numbers is transi-

---

[1]Scripps Institution of Oceanography, University of California San Diego, La Jolla, California

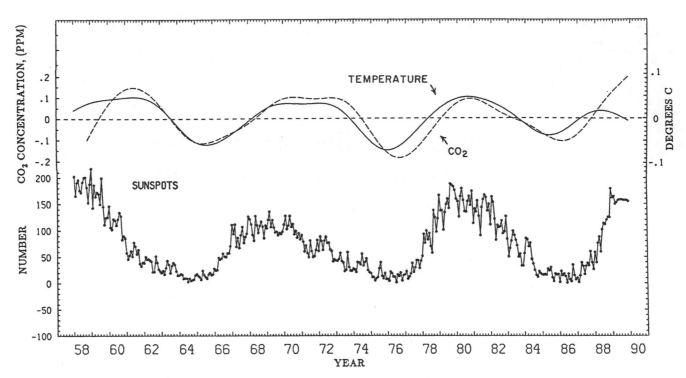

FIGURE 1 Decadal variability of anomalies in global averaged atmospheric CO_2 concentration (in parts per million by volume (ppm)) and in global surface temperature (in °C) compared with decadal variations in sunspot numbers, modified from Keeling et al. (1989, Figure 66). *Upper curves*: Approximately 10-year oscillations in atmospheric CO_2 concentration (dashed curve) and global temperature (solid curve), calculated as differences between splines of different stiffness. For CO_2 the stiffer spline has a standard error σ of 0.245 ppm with respect to monthly averaged data (1958-1989); the less stiff spline has an error of 0.200 ppm. For temperature over the same time period the stiffer spline has a σ of 0.088°C. *Lower curve*: Sunspot numbers (National Geophysical Data Center, 1989), shown as line segments connecting monthly averages. The CO_2 record in 1968 and 1969 was modified from that reported by Keeling et al. (1989) as a result of a reassessment of the original calibrating data.

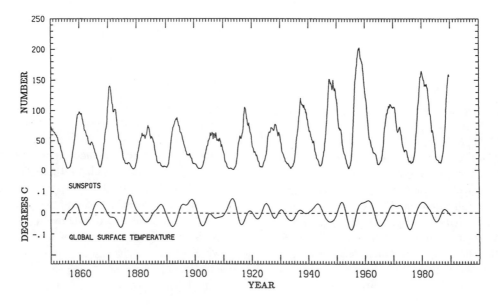

FIGURE 2 Comparison of a 12-month running mean of sunspot numbers (upper curve) with a decadal band-pass filter of global surface temperature in °C (lower curve). The latter is the same as that shown below in panel 2 of Figure 4.

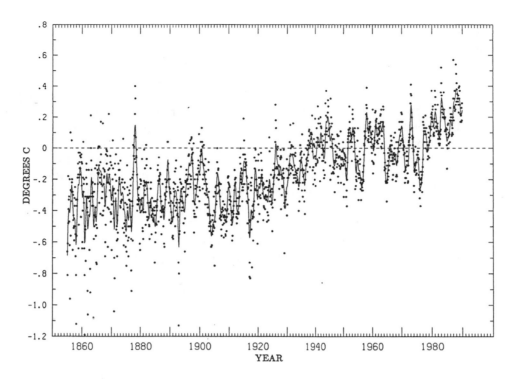

FIGURE 3   Global surface temperature anomaly from 1855 through 1989, in °C (Jones et al., 1986c, and Jones, private communication). Monthly data are shown as dots. The solid curve is a spline fit through this monthly data with a σ of 0.110°C.

tory. The two decadal-scale peaks immediately prior to 1961 (in 1943 and 1952) came well after the sunspot maxima of 1938 and 1948; a series of decadal oscillations appeared before 1905, but the peak temperatures occurred near times of minimum sunspots, rather than near maxima. From about 1905 to 1940, instead of decadal oscillations, the temperature record shows oscillations with periods of only about 6 years.

Although these 6-year and decadal variations may reflect internal oscillations of the coupled atmospheric-oceanic circulation system (James and James, 1989), or simply a noisy record that falsely appears to be periodic from time to time (Gribbin, 1979), the present evidence does not, in our view, conclusively rule out the possibility of a persistent decadal periodicity in temperature. It is our purpose here to describe this possible decadal signal to help decide whether its recent appearance, where it correlates with atmospheric CO_2, is short-lived, or whether instead it might reflect an oscillatory feature of the long-term temperature record, present in spite of the lack of evidence for a simple mechanism (such as solar forcing) that could produce it.

ANALYSIS OF OSCILLATIONS IN AIR TEMPERATURE

To investigate decadal-scale variations in temperature we have made use of a compilation of monthly averaged Northern and Southern Hemisphere temperature anomalies

both over land and in surface sea water, originally summarized by Jones et al. (1986c) but later supplemented by additional data for the years 1854-1860 and 1985-1989, inclusive. The globally averaged data that we used directly in our computations are listed in Appendix A below.

The large scatter in monthly averaged global temperature data (the dots plotted in Figure 3) is not a strong encouragement to look for cyclic phenomena. Nevertheless, if the data are fit to a flexible smoothing spline (Reinsch, 1967) to suppress the high-frequency scatter (the solid curve in Figure 3), persistent periods of warmer and cooler conditions show in the record. The average spacing of warming and cooling events is between 3 and 4 years, depending on how many of the minor fluctuations are considered significant. A large number of the warmer periods occurred close to the times of El Niño events. It seems likely that the latter reflect phenomena that originate mainly in the Indo-Pacific region of the tropics (Quinn et al., 1987; Quinn and Neal, 1992) as a result of interactions of the atmosphere and oceans that arise from imbalances in the coupled system unrelated to any possible external forcing. If only the stronger warm events are counted, the recurrence interval becomes about 9 years (Quinn and Neal, 1992), so that El Niño events cannot be ignored in considering the possible causes of decadal variations in temperature. However, if they are so related, one would not expect them to be closely periodic.

To obtain a closer look at possible decadal-scale fluctuations in temperature, we have fit the individual data points of Figure 3 to three additional smoothing splines of successively greater stiffness (Figure 4, panel 1), chosen to produce increasing degrees of low-pass filtering of the monthly data (Enting, 1987). The choices were subjective, but are not critical to the outcome, because distinct decadal patterns can be produced over a considerable range of stiffnesses. These three spline fits were run on both monthly and yearly averaged data, and the resulting curves showed essentially no difference. The stiffest spline (the curve labeled "1") shows only a tendency for global air temperature to rise irregularly since 1855. The two looser splines (curves "2" and "3"), whose difference is shown in panel 2, are nearly identical to the spline curves whose difference produced the approximately decadal band-pass of temperature that is shown in Figure 1. The difference between curves "1" and "2" (Figure 4, panel 3) approximates a band-pass centered

near 20 years, while the difference between curves "1" and "3" (Figure 4, panel 4) approximates a broader band-pass showing temperature fluctuations on both decadal and bidecadal time scales.

To examine further the spectral character of the global temperature record of Jones and co-workers, we have computed a series of spectra of their record by the maximum entropy method (MEM) (Press et al., 1989). This method of spectral analysis includes both the Thomson multitaper method and the fast fourier transform (FFT) method. Its results are particularly suitable for trying to resolve closely spaced frequencies (Press et al., 1989; Berger et al., 1990). We afterwards established the amplitudes and phases of the identified spectral peaks by least-squares fits of sinusoidal functions, since the maximum entropy method, as discussed by Sonett (1983), is not quantitatively reliable with respect to amplitudes and cannot establish phase relationships.

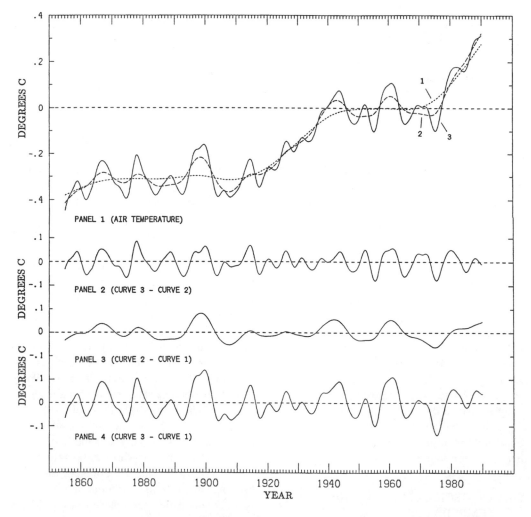

FIGURE 4   Global temperature anomaly from 1855 through 1989, based on annual averages of the data in Figure 3. (Use of monthly data showed negligible difference at these decadal frequencies.) *Panel 1*: Three superimposed spline fits. The dotted curve "1" is a very stiff spline (σ equal to 0.106°C); the dashed curve "2" and solid curve "3" are looser spline curves (σ's of 0.095°C and 0.075°C). *Panel 2*: Curve "3" minus curve "2". *Panel 3*: Curve "2" minus curve "1". *Panel 4*: Curve "3" minus curve "1".

Specifically, we first carried out an MEM analysis of the original monthly "raw" data (shown as dots in Figure 3), with only a linear trend subtracted. The data set consisted of 1620 points. Spectral estimates were obtained for order M (number of poles; see Press et al. (1989)) equal to 750. The periods of the 24 most prominent spectral peaks found are listed in the first column of Table 1 for frequencies greater than 0.6 cycles per year. Since the use of such a large number of poles invites criticism that we have exceeded the limits of the MEM method, we carried out a second analysis in which we first removed the lowest and highest frequencies from the record by subtracting the very stiff curve "1" shown in Figure 4 from the flexible spline shown in Figure 3. The resulting detrended and slightly smoothed temperature record is shown in Figure 5, panel 1. It can be seen by comparison with Figure 3 that almost all of the oscillatory character in the original record with periods between 1 and 30 years is retained in panel 1 of Figure 5. We then calculated the MEM spectrum on the basis of this slightly smoothed record with data points spaced three months apart, beginning with January 1855. The data set consisted of 540 points. Given the degree of smoothing of this record, more closely spaced data would not have contributed any significant additional information, as perhaps they could have done if the raw data had been used. As shown in the second column of Table 1, the power spectrum of the slightly smoothed record, with M set equal to only 250, gives almost the same frequencies for peaks as the spectrum based on the raw data with M set equal to 750.

A check was also made by creating a subset of points, reducing the raw data by a factor of 3 (selecting values for every third month only, beginning with the first). With a linear trend removed as before, we obtained nearly the same spectrum with M set equal to 275 as in the previous two cases.

We then computed the spectrum using the slightly smoothed data spaced three months apart, but with M reduced from 250 to 150. Figure 6 shows this spectrum (indicated by a dashed line) has fewer and less sharp peaks than that computed with M equal to 250 (solid line). In particular, in the decadal-frequency region a single peak with a period of 9.7 years replaces a pair of peaks with periods of 9.3 and 10.2 years. This single peak, having a frequency of approximately the mean of the two peaks in the prior case, shows relative stability in frequency despite the change in M value.

As demonstrated by a synthetic example of a sample of points from the sum of two sinusoids (Press et al., 1989), the MEM method at low pole numbers does not resolve closely spaced oscillations, whereas when MEM is used with high pole numbers, spectral lines may split where there is no firm basis for expecting multiple peaks. These are typically exaggerated in regions of the spectrum having low power spectral density. In our particular example of global temperature, both spectra in Figure 6 indicate high power in the decadal region, whether that power is represented by a single oscillation or a pair. We chose to accept the spectrum showing a pair of peaks, because the spline fit to the temperature data in the decadal band (shown in Figure 4, panel 2) also indicated that the decadal signal vanished in the middle of the record, and this is better represented by a pair of decadal oscillations.

Apart from the changes in frequency caused by the merger of spectral lines as the M value decreased, we did not find noticeable changes in the position of the spectral peaks as we varied our method of analysis. We found, however, that the peaks shifted somewhat in frequency if we used only limited portions of either version of the record—for example, by deleting the data before 1875 that exhibit more high-frequency oscillations than the subsequent record. Nevertheless, for peaks with periods of 6 years or greater, all of the spectra examined were similar to the two versions summarized in Table 1.

TABLE 1　MEM Spectra of Monthly Global Surface Temperature, 1855-1989

Raw Data Spectrum* (years)	Smoothed (years)	Amplitude for 24-Harmonic Fit** (°C)	Non-parametric Multitaper (years)
41.3	31.40	0.017	
22.10	21.69	0.039	20.5
15.28	15.59	0.038	15.5
10.07	12.15	0.017	
9.18	10.22	0.036	
7.509	9.29	0.037	9.1
6.643	7.610	0.014	
6.042	6.670	0.026	
5.218	6.043	0.037	6.1
4.740	5.227	0.040	5.2
4.382	4.758	0.041	4.8
4.128	4.436	0.023	
3.756	4.170	0.026	4.0
3.546	3.746	0.022	3.8
3.283	3.548	0.034	3.5
3.122	3.290	0.025	
2.867	3.127	0.022	
2.716	2.866	0.027	
2.318	2.727	0.009	
2.241	2.320	0.023	
1.998	2.242	0.017	
1.885	1.997	0.020	
1.796	1.884	0.014	
	1.797	0.017	

*From Ghil and Vautard (1991, Figure 2). Their plot also shows an unlabeled, strong peak near 10.2 years.

**From a fit of sine and cosine functions to the smoothed data. The amplitudes are peak to trough.

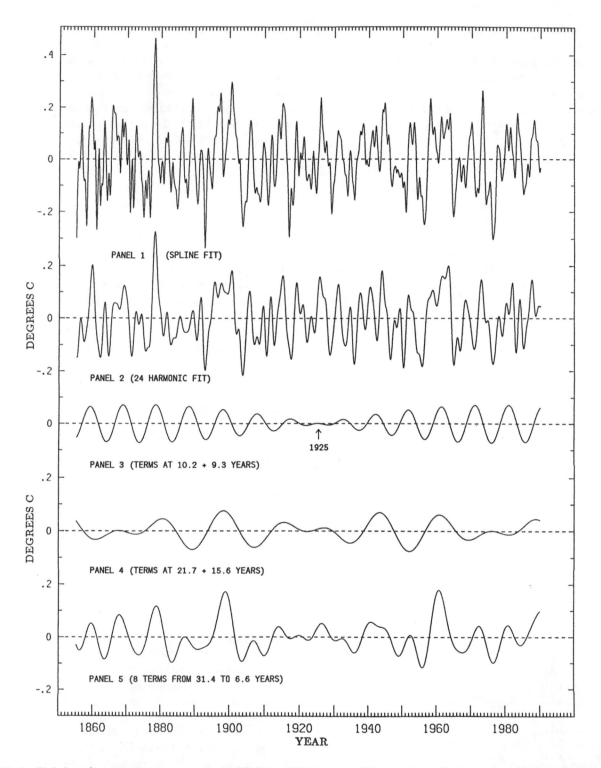

FIGURE 5 Global surface temperature anomaly in °C from 1855 through 1989, together with spectral oscillations derived from it determined from the maximum entropy spectrum. *Panel 1*: Smoothed and detrended temperature anomaly obtained by subtracting from the smooth curve of Figure 3 a spline fit to the yearly averaged anomaly data with a σ of 0.106°C, as in curve 1 of Figure 4. *Panel 2*: Sum of 24 spectral oscillations derived from a fit to the curve in panel 1, which are consistent with the spectral peaks as listed in Table 1, column 2. *Panel 3*: Sum of decadal oscillations. *Panel 4*: Sum of bidecadal oscillations. *Panel 5*: Sum of eight oscillations with periods of 6.67 calendar years or greater (no exclusions). The source of the temperature data is the same as in Figure 3. The time of maximum interference of the decadal oscillations in panel 3 is indicated.

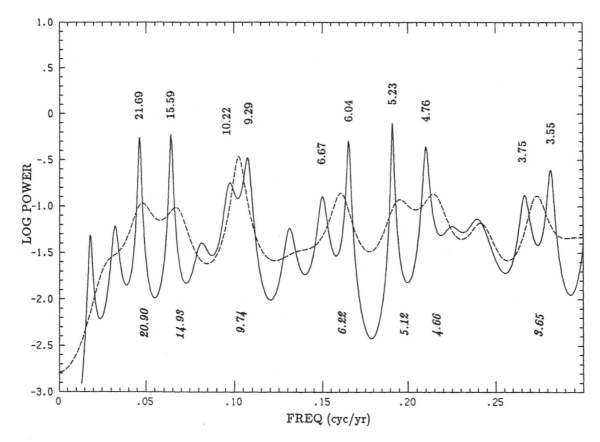

FIGURE 6   Two maximum-entropy spectra of the global surface temperature anomaly, expressed as power (logarithmic scale) versus frequency, in cycles per year. The temperature record used to determine the spectra was expressed as the difference between the flexible spline plotted in Figure 3 and the very stiff spline from curve "1" of Figure 4, panel 1. The solid curve indicates a spectrum with the number of poles, M, equal to 250; the dashed curve shows a spectrum with M equal to 150. The periods of the prominent peaks of the former, in years, are shown in standard characters, those of the latter by figures in italics.

Another study of nearly the same temperature record by Ghil and Vautard (1991) used a combined singular spectrum and multi-taper analysis that provides an F-test for the significance of the peaks. Their study confirms the significance, at the 97 percent confidence level, of the peaks that we found near 6, 9, 15, and 21 years, and showed evidence of several of the other peaks listed in Table 1.

Accepting the peak periods revealed by the second spectrum of Table 1 for further study, we have fit the slightly smoothed temperature record of Figure 5, panel 1, using a non-linear least-squares method (Bevington, 1969, subroutine CURFIT) to a series of sine and cosine functions having the periods indicated for peaks in the MEM spectrum, omitting only very weak peaks and any peaks with periods less than 1.7 years.

The reconstructed time series obtained by summing these 24 computed sinusoidal spectral oscillations reproduces the main peaks and troughs of the original record (Figure 5, panel 2). If individual warming and cold events are scrutinized, admittedly the fit is far from perfect. Allowing, however, that the lowest period of oscillation in the recon-

struction is 1.8 years, one cannot expect a wholly consistent representation of times and amplitudes of short-term interannual fluctuations such as those related to El Niño events. As we verified by carrying out a further reconstruction using 48 harmonics that included oscillations with periods as low as 1.0 years, the correctness of the phasing and amplitudes for short-term interannual fluctuations can be improved by including higher-frequency oscillations in the fit to the temperature data. The amplitudes of the lower-frequency oscillations, which are those of importance to the present study, were only negligibly altered by the addition of higher frequencies in the fit.

The amplitudes of the sinusoidal functions are listed in Table 1. Of the 24 spectral oscillations in the reconstruction, only the two noted earlier, having periods of 9.3 and 10.2 years, occur near to the decadal time scale. Summed, these produce oscillations with an average period of 9.7 years and a maximum amplitude, peak to peak, of 0.15°C. They beat with a recurrence period of 102 years (half of the period of the long-term oscillation produced by the difference of the two shorter-period oscillations) and exhibit interference near 1925 (Figure 5, panel 3). The pattern of interference and reinforcement is distinct because the amplitudes of the separate oscillations, as found by the least-squares fits to the sinusoidal functions, are nearly equal.

Two additional strong spectral oscillations are found in the temperature spectrum at approximately the bidecadal time scale, with periods of 15.6 and 21.7 years. Summed, they have a maximum amplitude of 0.14°C and beat with a period of 55 years (Figure 5, panel 4), which is close to half the beat period of the decadal harmonics. They also exhibit interference in the 1920s, at almost the same time as the decadal oscillations. Furthermore, their summed oscillations, with an average period of 18.1 years, essentially impose variable amplitudes on the decadal oscillations. The addition of weaker harmonics with periods of 6.7, 7.6, 12.2, and 31 years to form a broad low-pass filter of the detrended temperature record does not cancel out this beat pattern (Figure 5, panel 5). The decadal oscillations are still present, with the strongest temperature maxima where the bidecadal oscillations enhance the decadal signal. Between 1905 and 1940, however, only oscillations with periods close to 6

years appear, as was the case for the spline-fit differences of panels 2 and 4 of Figure 4.

As shown in Figure 7, the sinusoidal oscillations of Figure 5, which are derived from spectral analysis, show essentially the same patterns as those previously derived from spline fits. The decadal and bidecadal oscillations are equally prominent, whether they are obtained by spectral analysis or by the use of smoothing splines. Thus, the amplitudes and phases obtained by spectral analysis do not appear to be falsified by the restricted number of degrees of freedom of the spectral compositing.

We conclude that the decadal oscillations that have appeared in the temperature record since the late 1950s may be characteristic of the longer-term record as well. If so, the bidecadal oscillations seen in Figures 5 and 7 may also be characteristic of the longer record, especially since they appear to be an amplitude modulation of the decadal oscillations. Both pairs of oscillations are intermittent, however, being essentially absent for about 35 years near the middle of the temperature record.

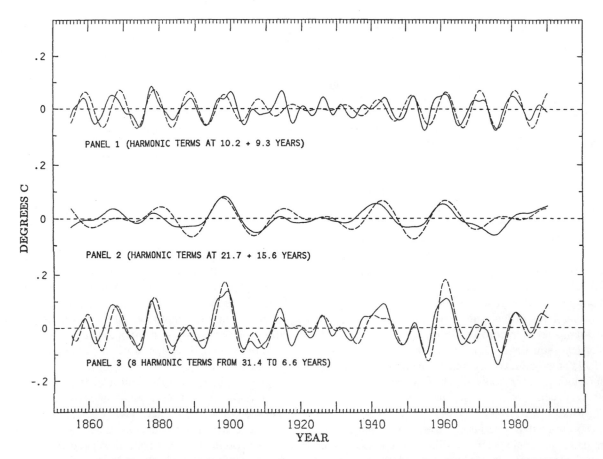

FIGURE 7   Spectral oscillations of the global spectrum of global surface temperature anomaly (dashed curves), compared with the same temperature record that has been band-pass filtered using spline fits (solid curves). *Panel 1*: Sum of decadal oscillations, as in Figure 5, panel 3, and curve "3" minus curve "2" of Figure 4. *Panel 2*: Sum of bidecadal oscillations, as in Figure 5, panel 4, and curve "2" minus curve "1" of Figure 4. *Panel 3*: Sum of eight oscillations, as in Figure 5, panel 5, and curve "3" minus curve "1" of Figure 4.

DISCUSSION

In our quest to understand decadal oscillations that appear in the record of atmospheric carbon dioxide measurements, we have been led to investigate seemingly similar oscillations in global air temperature. Employing a much longer temperature record than that available for atmospheric CO_2, we have found that decadal variations in global air temperature are characteristic of the past 135 years, except for an interval near the middle of the record. This pattern can be reproduced by assuming the existence of two oscillations with slightly different frequencies that beat on the time scale of about 100 years.

Is it possible that these decadal oscillations are caused by a cyclic process? Perceptions of cyclic behavior in the climatic record have in the past involved so many exceptions and inconsistencies that the subject does not enjoy good repute. Compelling evidence that cyclic behavior is actually caused by an identifiable forcing agent such as solar irradiance is difficult to find.

Decadal oscillations have not been emphasized in previous discussions of global temperature, but the existence of bidecadal oscillations has been alleged by Newell et al. (1989) on the basis of marine air-temperature data. An inspection of their Figure 2 indicates that the signal that they found is similar in phase and amplitude to that shown in our Figure 4, panel 4. There is no indication in their analysis of a weakened signal near 1925, however, like that found by our analysis of global air temperature. We are not aware of other finds of either decadal or bidecadal oscillations.

Of possible importance in deciding the significance of the oscillations in temperature that we report here is the indication that the decadal signal appears to be replaced by shorter oscillations, with periods of close to 6 years, from about 1905 to 1940. Although the spectral evidence for this, as shown in the low-passed record of Figure 5, panel 5, is based only on spectral peaks with periods greater than 6.5 years, the strong 6.0-year peak in the MEM spectrum produces an oscillation (not shown) with almost exactly the same phase. Thus a 6.0-year oscillation appears to be a significant contributor to the overall pattern of variability in the global temperature record, although not as prominent as the decadal signal.

The 6.0-year oscillation, and two other adjacent oscillations with periods of 4.8 and 5.2 years (see Figure 6), could well be associated with the low-frequency component of the El Niño/Southern Oscillation (ENSO) phenomenon, as pointed out by Ghil and Vautard (1991) in their spectral analysis of temperature. When these three oscillations are included in a low-frequency band-pass of temperature, warm periods in this band-passed record include most of the strong and many of the moderate El Niño events.

The 4.8- and 6.0-year oscillations, when summed, show an interference pattern with a beat period of 22.4 years. The timing of these beats shows a possible relationship to the 21.7-year spectral oscillation, inasmuch as the warm phases of this bidecadal oscillation tend to occur near the times of reinforcement of the two higher-frequency oscillations. If these higher-frequency oscillations are related to global aspects of the El Niño phenomenon, is it possible that the bidecadal oscillations that we have been discussing are also related? In the absence of a mechanism to explain any of these oscillations, and not even clear proof that they exist, it is not possible to resolve this question, but the topic appears to us to deserve further investigation.

Does the record of atmospheric CO_2 provide any insight into possible causes of the interannual variations in global temperature? At first glance one would not expect so, because interannual variations in CO_2 are presumably a consequence of climatic factors that do not depend on the earth's carbon cycle. Nevertheless, two aspects of the CO_2 record are worth noting.

First, atmospheric CO_2 anomalies show a striking relation to global temperature anomalies with respect to El Niño events (Keeling et al., 1989). Second, on all interannual time scales that can be investigated in the 32-year CO_2 record up to 1990, CO_2 variations tend to lag temperature variations by approximately a half-year (Kuo et al., 1990). The existence of a lag of that length on the decadal time scale is evident in the plot of Figure 1.

This lag, as shown by Keeling et al. (1989), cannot easily be explained as an oceanic phenomenon, because temperature-induced exchange of CO_2 gas at the air-sea interface on a decadal time scale should result in a lag in atmospheric CO_2 variation of about 1.5 years, whereas a change in upwelling of cold water should produce atmospheric CO_2 variations of the opposite phase with respect to temperature, i.e., a lag of about 5 years. On the time scale of El Niño events, Keeling et al. suggest that atmospheric CO_2 responds to global temperature, because vegetation on land tends to release CO_2 during anomalously warm periods. With respect to this terrestrial response, a short lag in CO_2 is to be expected.

Perhaps variations caused by terrestrial vegetation and driven by climatic change also produce decadal variations in CO_2. Indeed, a decadal signal in CO_2 might be simply a modulation in the magnitude or frequency of variations on El Niño time scales. If this should be the case, a nearly constant phase lag in CO_2 with respect to temperature would be expected over a broad range of interannual frequencies, as observed.

In support of such an hypothesis are several papers' assertions of possibly non-random, long-term behavior in El Niño events. Their focus has been mainly on the decadal time scale (see, for example, Hanson et al., 1989; Michaelsen, 1989; and Enfield and Cid, 1990) but there is a hint of a decadal connection in the study of Barnett (1989), as interpreted by Enfield and Cid. To these speculations may be added our suggestion, above, of a possible association of spectral temperature oscillations on El Niño time scales with a bidecadal oscillation.

In conclusion, we suggest that the atmospheric CO_2 record may be of some use in the study of decadal variability in climate. Conclusive evidence of decadal periodicities in climatic parameters will probably be established, however, only if and when a mechanism is found that is physically reasonable and successfully predicts changes in climate.

ADDENDUM

In the oral presentation of this paper, we presented an hypothesis that forcing of ocean temperatures by oceanic tidal dissipation might explain the observed decadal oscillation in global temperature and its disappearance in the early part of this century. We were attracted to the coincidence of the times of prominent decadal temperature signal with the times when the strongest equilibrium oceanic tidal forcing was produced by the gravitational forces of the sun and moon. At these times, within 10 years of 1881 and 1974, episodes of unusually strong tidal forcing occurred at times of the strongest tides of the year, because the sun and moon tended then to be both in unusually close alignment with the Earth and at nearly closest approach. In contrast, in the 1920s, when the decadal signal was weakest, only weaker tidal forcing occurred at times of the strongest tides. Because

strong tidal forcing occurs at intervals of 9 years when the alignments of the sun, moon, and earth are nearly optimal for strong tidal forcing, and otherwise at 3- and 6-year intervals, the decadal signal itself and the 6-year signal in the middle of the record period also correspond to the periodicities of tidal forcing.

It was not possible to prepare an appropriate presentation of the tidal discussion for this volume. An article describing the tidal hypothesis is in preparation.

ACKNOWLEDGMENTS

We are grateful to scientists who gave generously of their time to discuss the subjects of solar phenomena and climatic variation with us in the course of preparation of this article. We specifically thank Philip Jones and Thomas Wigley, Michael Ghil, James Hansen, Harry van Loon and Gerry Meehl, David Parker, and Henry Diaz and Thomas Karl. We are also grateful for discussions with Tim Barnett, Robert Bacastow, Daniel Cayan, and Hugh Hudson, colleagues of ours at the University of California at San Diego. We further thank Drs. Jones and Hansen for supplying us with their temperature data sets, and David Parker for supplying us with extensive global and regional sea surface temperature data in graphic form. Computer time was provided by the San Diego Supercomputer Center. Financial support came from the National Science Foundation via Grant ATM-91-21986 and from the U.S. Department of Energy via Grant FG03-90ER-60940.

APPENDIX A
TABULATION OF TEMPERATURE DATA

Table A1 presents the global-average anomalies of air temperature used in our study. The anomalies are listed for each calendar month, followed in the last column by the annual average. They were obtained by averaging Northern and Southern Hemisphere anomalies. The hemispheric data for 1855 through 1989, expressed both over land and in surface sea water, were supplied to us by P.D. Jones (personal communication of April 1990). These data were supplied to us as monthly averages, whereas for 1987 through 1989 only annual averages were available with respect to temperatures in surface sea water. For these final three years of our data set we estimated global averages for each month from the monthly anomalies supplied to us for temperatures over land.

For the Northern Hemisphere we multiplied the monthly land values for each given year by the ratio of the annual global mean for that year to the mean for land. For the Southern Hemisphere we assumed that no intra-annual variations existed over the oceans, and that temperatures over land contributed 30% to the hemispheric averages. For the full data set we accepted the premise of Jones et al. (1986c) that sea surface and marine air temperatures follow each other closely on interannual time scales, so that combined land and surface sea-water-temperature data portray global-average interannual variations in surface air temperature.

Recently P.D. Jones (personal communication of April 1993) kindly supplied us with updated monthly data for all years of our study. We have repeated our calculations with these data and find only very small changes in spectral amplitudes and frequencies, and thus in the time plots shown in this article.

TABLE A1 Global Surface Temperature Anomalies Used in Our Analysis, in Units of .01°C

Year	Jan	Feb	Mar	Apr	May	Jun	Jul	Aug	Sep	Oct	Nov	Dec	Annual Average
1855	−66	−81	−61	−18	−48	−39	−33	−31	−47	−20	−62	−96	−50
1856	10	−27	−56	−34	−28	5	−22	−20	−28	−28	−65	−30	−27
1857	−46	−43	−51	−50	−59	−32	−37	−25	−50	−55	−69	−33	−46
1858	−81	−112	−56	−37	−33	−19	−19	−10	−17	−3	−64	−49	−42
1859	−19	−9	−13	−3	−5	−13	−17	−6	−44	−4	−9	−24	−14
1860	−10	−64	−60	−40	−19	4	−28	−23	−35	−20	−64	−81	−37
1861	−119	−50	−31	−52	−72	−19	−10	−14	−23	−38	−31	−41	−41
1862	−91	−106	−32	−32	−14	−31	−43	−65	−34	−31	−77	−92	−54
1863	21	0	−8	−14	−23	−31	−37	−29	−22	−28	−31	−37	−20
1864	−108	−58	−28	−50	−43	−23	−13	−38	−44	−56	−49	−73	−49
1865	−11	−74	−51	−14	−16	−23	0	−17	1	−16	−11	−35	−23
1866	1	−22	−19	−9	−42	17	−4	−23	−17	−32	−32	−43	−19
1867	−28	16	−43	−13	−40	−23	−19	−18	−14	−15	−37	−69	−25
1868	−60	−49	−4	−31	−1	−10	−2	−5	−22	−38	−63	−14	−25
1869	−22	22	−39	−17	−13	−23	−27	−5	−9	−41	−47	−42	−22
1870	−11	−65	−36	−30	−20	5	−13	−27	−34	−45	−26	−104	−34
1871	−70	−83	−10	−17	−29	−34	−4	−21	−52	−50	−59	−69	−41
1872	−43	−58	−54	−27	−10	−14	−14	−20	−19	−14	−23	−37	−28
1873	−8	−38	−34	−35	−34	−33	−22	−20	−33	−20	−51	−25	−29
1874	6	−39	−62	−54	−48	−26	−22	−50	−35	−46	−63	−52	−41
1875	−56	−75	−66	−36	−11	−25	−38	−30	−59	−50	−69	−57	−48
1876	−33	−27	−33	−29	−50	−34	−20	−24	−44	−58	−78	−91	−43
1877	−34	−19	−34	−37	−59	−17	2	−15	−5	−5	1	8	−17
1878	−7	22	40	32	−9	7	−6	−5	−6	−1	−17	−40	1
1879	−36	−33	−26	−36	−30	−32	−34	−25	−33	−33	−61	−65	−37
1880	−20	−41	−23	−21	−36	−40	−36	−7	−32	−51	−53	−39	−33
1881	−58	−41	−24	−21	−1	−27	−22	−13	−30	−41	−49	−22	−29
1882	2	−11	−7	−26	−44	−46	−35	−22	−22	−45	−44	−49	−29
1883	−50	−58	−35	−38	−37	−13	−26	−22	−43	−55	−44	−32	−38
1884	−40	−42	−39	−46	−45	−40	−47	−34	−38	−40	−73	−43	−44
1885	−60	−54	−34	−48	−37	−49	−38	−40	−37	−33	−14	−11	−38
1886	−21	−47	−23	−6	−18	−21	−20	−16	−17	−31	−40	−34	−25
1887	−44	−55	−34	−41	−38	−39	−23	−37	−40	−50	−41	−39	−40
1888	−61	−63	−38	−30	−29	−20	−33	−31	−25	−25	−22	−28	−34
1889	−21	−14	4	4	4	−10	−20	−20	−35	−25	−46	−20	−16
1890	−26	−28	−37	−29	−48	−34	−41	−47	−45	−45	−57	−35	−39
1891	−56	−58	−44	−26	−22	−28	−37	−31	−15	−23	−50	−13	−34
1892	−40	−8	−35	−38	−35	−33	−43	−42	−25	−27	−44	−73	−37
1893	−113	−80	−35	−55	−53	−38	−20	−33	−22	−12	−34	−23	−43
1894	−40	−31	−26	−34	−38	−35	−35	−25	−48	−40	−28	−28	−34
1895	−54	−66	−45	−21	−22	−22	−31	−17	−14	−23	−21	−20	−30
1896	−30	−20	−23	−26	−5	−5	−4	3	−17	−11	−23	5	−13
1897	−25	−18	−20	6	7	−4	−1	−8	−10	−3	−34	−34	−12
1898	3	−40	−54	−44	−42	−20	−36	−26	−24	−44	−30	−17	−31
1899	−17	−38	−25	−17	−20	−31	−13	−8	−11	−15	9	−34	−19
1900	−25	−17	−27	−9	−5	4	−2	−1	−4	13	−9	4	−6
1901	1	−26	−20	−10	−17	2	−17	−11	−23	−17	−29	−36	−17
1902	−7	−20	−15	−30	−27	−23	−26	−36	−30	−42	−38	−43	−28
1903	−14	−9	−28	−50	−42	−48	−53	−47	−43	−47	−50	−63	−41
1904	−63	−54	−62	−55	−46	−46	−43	−53	−49	−50	−31	−40	−49
1905	−51	−75	−41	−47	−28	−25	−29	−36	−24	−34	−16	−9	−34
1906	0	−29	−24	−14	−13	−23	−28	−22	−23	−27	−45	−31	−23
1907	−43	−44	−23	−46	−52	−56	−41	−39	−39	−38	−49	−34	−42
1908	−40	−44	−57	−48	−44	−36	−43	−46	−39	−55	−56	−44	−46

TABLE A1   Global Surface Temperature Anomalies Used in Our Analysis, in Units of .01°C, *continued*

Year	Jan	Feb	Mar	Apr	May	Jun	Jul	Aug	Sep	Oct	Nov	Dec	Annual Average
1909	−40	−47	−53	−44	−40	−28	−39	−19	−18	−22	−18	−42	−34
1910	−23	−33	−24	−25	−36	−31	−31	−34	−34	−40	−46	−49	−34
1911	−41	−59	−45	−56	−41	−44	−40	−42	−37	−34	−27	−14	−40
1912	−24	−20	−23	−19	−22	−19	−34	−44	−42	−48	−37	−32	−30
1913	−45	−42	−42	−28	−41	−42	−32	−28	−31	−37	−6	−10	−32
1914	0	−20	−18	−22	−13	−14	−23	−27	−34	−16	−25	−27	−20
1915	−4	19	−23	5	−14	−23	−4	−12	−7	−14	−10	−9	−8
1916	−11	−18	−34	−19	−29	−48	−35	−26	−25	−23	−33	−54	−30
1917	−46	−82	−73	−47	−83	−46	−22	−27	−24	−52	−19	−76	−50
1918	−41	−45	−37	−44	−59	−36	−44	−40	−34	−6	−13	−26	−36
1919	0	−1	−27	−8	−20	−20	−27	−24	−27	−30	−61	−47	−24
1920	−8	−36	−4	−19	−14	−31	−42	−23	−19	−30	−32	−32	−24
1921	−3	−21	−8	−17	−14	−19	−23	−29	−24	−22	−46	−23	−21
1922	−37	−38	−26	−31	−32	−39	−25	−25	−30	−33	−31	−25	−31
1923	−21	−47	−31	−41	−31	−27	−43	−46	−25	−24	7	−2	−28
1924	−34	−28	−31	−27	−28	−17	−30	−36	−34	−26	−35	−53	−31
1925	−38	−28	−14	−19	−23	−19	−18	−10	−17	−33	−13	4	−19
1926	28	18	13	−12	−13	−12	−19	−1	−13	1	−5	−17	−3
1927	−20	−11	−20	−16	−26	−17	−12	−6	−7	−1	−5	−39	−15
1928	−6	−17	−30	−24	−18	−25	−14	−24	−16	−13	−1	−19	−17
1929	−35	−67	−37	−38	−37	−31	−33	−28	−29	−13	−7	−43	−33
1930	−36	−19	−19	−14	−19	−14	−14	−11	−17	−10	4	−5	−15
1931	15	−16	0	−11	−13	2	1	−4	−12	−4	−14	1	−4
1932	16	−22	−17	−4	−19	−11	−17	−17	−5	−3	−15	−19	−11
1933	−21	−22	−21	−16	−14	−14	−20	−14	−22	−13	−26	−40	−20
1934	−22	−22	−34	−22	−8	2	−4	−3	−14	−9	−2	−6	−12
1935	−27	9	−13	−22	−31	−14	−18	−17	−14	−4	−34	−22	−17
1936	−27	−42	−16	−20	−11	−11	0	−7	−11	−10	−6	−1	−13
1937	−15	2	−8	−4	−4	5	0	11	12	8	2	−12	0
1938	1	3	10	9	6	9	10	3	9	21	−2	−14	5
1939	2	−12	−23	−12	3	6	2	1	−20	−16	−5	19	−5
1940	−24	−10	−11	−10	4	4	13	−7	7	−9	−8	27	−2
1941	4	16	2	19	1	−6	2	1	−21	16	4	20	5
1942	14	9	14	13	4	−4	−8	0	8	2	−1	−8	4
1943	−14	5	−19	−2	0	−17	9	−9	4	29	11	21	2
1944	37	13	9	4	11	20	20	23	31	27	8	12	17
1945	12	13	16	13	−11	−20	1	32	14	20	1	−16	6
1946	16	9	3	6	−15	−12	−12	−16	6	−13	−8	−28	−5
1947	−9	−8	−2	2	−8	−4	−6	−4	−12	1	9	−10	−5
1948	8	−18	−22	−6	8	3	−11	−5	−5	−1	−13	−12	−6
1949	11	−22	−10	−6	−2	−11	−6	−7	−1	−2	4	−11	−5
1950	−31	−23	−3	−7	−10	−4	−5	−8	−6	−13	−31	−14	−13
1951	−27	−37	−20	−9	0	4	6	8	12	10	8	8	−3
1952	27	17	4	13	10	5	9	9	8	1	−8	4	9
1953	9	22	25	19	11	9	−2	11	8	10	4	13	12
1954	−26	−7	−7	−24	−18	−20	−21	−9	−5	−8	5	−16	−13
1955	13	−8	−33	−19	−16	−17	−19	−2	−10	−7	−22	−26	−14
1956	−19	−33	−20	−23	−27	−25	−20	−22	−27	−15	−20	−21	−23
1957	−11	−4	3	2	11	8	6	11	7	7	22	28	8
1958	39	26	19	19	13	5	12	6	4	8	12	16	15
1959	14	13	17	14	4	7	10	1	9	−3	−8	6	7
1960	11	25	−16	−9	−10	2	1	3	7	5	−4	23	3
1961	14	24	18	17	13	19	4	9	5	5	8	3	12
1962	13	16	12	9	6	14	10	11	16	21	22	17	14

TABLE A1 Global Surface Temperature Anomalies Used in Our Analysis, in Units of .01°C, *continued*

Year	Jan	Feb	Mar	Apr	May	Jun	Jul	Aug	Sep	Oct	Nov	Dec	Annual Average
1963	1	25	0	8	11	3	17	17	17	27	25	14	14
1964	14	−3	−14	−12	−9	−13	−13	−21	−22	−22	−21	−34	−14
1965	−11	−22	−16	−20	−8	−3	−18	−14	−5	−6	−11	−1	−11
1966	−4	1	2	−4	−8	8	5	−2	2	−8	−6	−20	−3
1967	−10	−17	−4	−3	7	−10	−5	0	−7	10	−4	−4	−4
1968	−21	−18	4	−17	−18	−9	−8	−5	0	4	4	−2	−7
1969	−2	−2	12	17	12	2	12	8	3	5	17	23	9
1970	13	22	6	17	5	4	0	−1	2	−4	1	−15	4
1971	−3	−22	−19	−14	−12	−19	−5	−8	−2	−8	−5	−16	−11
1972	−22	−21	−11	2	4	5	7	9	2	7	12	29	2
1973	35	41	34	23	20	13	8	6	2	7	−2	−1	16
1974	−24	−28	−14	−15	−10	−11	1	−9	−17	−14	−9	−16	−14
1975	−4	−4	−4	−4	−2	−6	−6	−17	−7	−19	−26	−32	−11
1976	−20	−27	−37	−15	−23	−20	−19	−22	−17	−28	−15	−12	−22
1977	−7	17	20	15	11	13	4	4	9	11	21	0	9
1978	13	9	13	4	−1	−2	2	−7	−1	−5	10	2	3
1979	7	−6	14	4	5	13	8	14	12	20	22	37	13
1980	23	19	17	25	27	14	9	8	10	8	19	13	16
1981	37	28	33	25	17	14	11	14	13	14	16	28	21
1982	6	9	3	10	11	2	3	6	11	9	12	36	10
1983	52	42	34	27	23	25	20	25	25	22	36	20	29
1984	25	19	19	11	20	12	11	12	13	9	2	−13	12
1985	12	0	15	7	16	3	10	8	9	9	4	18	10
1986	26	18	22	20	18	22	8	9	10	13	9	20	17
1987	32	57	18	32	27	26	37	24	37	25	28	54	33
1988	48	30	42	38	35	32	32	30	34	28	23	39	34
1989	32	40	37	28	22	17	22	21	17	26	18	29	26

Discussion

CICERONE: I think you're touching on something very funda-mental: the possibility that there may be an external source of climate variability.

TRENBERTH: It seems to me that the key question is what physical mechanism is operating here. Have you any thoughts on that, Dave?

KEELING: Well, there was a paper by John Loder and Christopher Garrett in JGR in 1978 that suggested that an 18.6-year cycle in SST can be seen on the east and west coasts of North America, of opposite phase. They found that the amount of energy involved in the tidal currents on the continental shelves might be sufficient to cause temperature changes. Peter Holloway and his colleagues have done some nice studies off the northwest coast of Australia. They find that the breaking of the internal waves eventually causes turbulence that results in so much mixing that you get an almost linear decrease in temperature from the sea surface downward. I am inclined to think that the tides are causing some sort of turbu-lence that affects the upper ocean.

DOUGLAS: The amplitude of the 18.6-year tide is less than a centimeter. It's very difficult to detect in the tide-gauge records. Where does all the energy go?

KEELING: I think the semi-diurnal and diurnal tides would have to carry most of the energy dissipation. What strikes me is that if you look at the correlation of cooling with the tides themselves, you see that the greatest amount of cooling tends to occur in conjunction with a strong tide.

GHIL: I find it interesting that, using an entirely independent technique, Dave is getting results that are nearly identical to those that I and Robert Vautard published in *Nature* in March 1991. We had peaks at similar locations, including a 15-to-16-year one.

KEELING: The 15-year one seems to me to be the best candidate for some sort of tidal connection because it is not explained by any kind of solar activity.

Observed Variability of the Southern Hemisphere Atmospheric Circulation

DAVID J. KAROLY[1]

ABSTRACT

The dominant modes of interannual variation of the Southern Hemisphere (SH) circulation are described, using results from several studies of grid-point analyses of the SH troposphere. The two leading modes are primarily zonally symmetric, representing out-of-phase variations of height between middle and high latitudes in one case (called the high-latitude mode) and between the tropics and middle latitudes in the other (called the low-latitude mode). Wave-train patterns of anomalies are described also. These include a wavenumber-three mode in winter with large amplitude at high latitudes, a wavenumber-four mode in middle latitudes in summer, a meridional wave-train structure originating over Australia, and a monopole or a meridional wave train over the Pacific Ocean in winter. The two most important processes leading to interannual variations of the SH circulation appear to be the Southern Oscillation and the interaction between waves and mean flow in the SH storm track, which leads to the High Latitude mode.

Since upper-air data and grid-point analyses are available for only a short period, surface data must be used to describe decadal variability of the SH circulation. The barotropic vertical structure of low-frequency variations in the SH means that surface pressure variations are representative of variations throughout the troposphere. Relatively little attention has been paid to decadal variability in the SH. It appears from indices of the two dominant modes of interannual variations, determined using the surface data, that the modes show substantial decadal variability over the last 90 years. Few observational data exist that can be used to directly indicate climate variability on century time scales in the SH. The only possible observational evidence for this comes from proxy data, such as tree rings.

[1]Cooperative Research Centre for Southern Hemisphere Meteorology, Monash University, Clayton, Victoria, Australia

INTRODUCTION

There has been relatively less study of the variability of the Southern Hemisphere (SH) atmospheric circulation than of that of the Northern Hemisphere (NH). There are a number of reasons for this; the lower population and greater ocean areas in the SH mean that observational records are sparser and there have been fewer people to analyze them. In this brief and incomplete review, I will concentrate on studies of the variability of the SH tropospheric circulation, starting at interannual time scales and moving onto longer time scales. This progression is justified by the available observational data and the number of relevant studies, as outlined below.

Detailed and regular observations of the troposphere over much of the SH did not start until the International Geophysical Year in 1957-1958, when the number of SH upper-air sounding stations increased markedly. Operational subjective analyses of the SH circulation have been prepared since that time by the Australian and South African weather services. The introduction of meteorological satellites in the late 1960s led to the development of the first operational objective numerical-analysis scheme for the SH at the Australian Bureau of Meteorology in 1972. Hence, objective numerical analyses for the SH are available for only the last 20 years, and upper-air sounding data are available for only about 30 years.

Hemispheric numerical analyses have been used to investigate the interannual variability of the SH circulation, particularly over the last decade (Trenberth, 1979; Rogers and van Loon, 1982; Mo and van Loon, 1984; Mo and White, 1985; Kidson, 1988a,b; Karoly, 1989a; and others). In addition, radiosonde station data have also been used for similar investigations by Szeredi and Karoly (1987a,b). The results of these studies of the dominant modes of variability of the SH circulation on interannual time scales are summarized in the next section.

Since these upper-air data are available for a short period only, it is not possible to use them to describe variations on decadal time scales or longer. There are longer records of surface meteorological observations for SH stations, some extending back to before the turn of the century. The equivalent-barotropic vertical structure for the low-frequency tropospheric variations (Szeredi and Karoly, 1987a) permits these surface records to be used to provide some information on decadal variability in the SH. However, there have been no comprehensive studies of decadal time-scale variability for the SH. More recently, Jones (1991, hereafter referred to as J91) has reconstructed from station data a gridded mean sea-level pressure data set for the SH that extends back to 1911. J91 and Allan and Haylock (1993) have used this data set to consider decadal variations of the SH circulation. These and a few other studies of long surface records are described briefly below, although the

others have generally concentrated on variations of El Niño and the Southern Oscillation.

In an attempt to provide some evidence of the range of SH variability on century time scales, some climate reconstructions based on proxy data are discussed below. Although these proxy data are available for very few sites in the SH, they constitute the only source of data on natural climate variability on century time scales for the SH. More extensive descriptions of proxy data for the SH are provided in Chapter 5 of this volume.

INTERANNUAL VARIATIONS FROM UPPER-AIR DATA

Relatively less attention has been paid to low-frequency variations of the SH circulation than to those of the NH, probably because of the shorter time period for which analysis data sets are available. Trenberth (1979), Kidson (1988a), and Shiotani (1990) have used zonal-mean data to show that there are large low-frequency variations of the zonal-mean circulation in the SH involving variations of the zonal wind and associated dipole variations of geopotential height. Mo and White (1985, hereafter referred to as MW) carried out a teleconnection analysis for the SH (similar to that of Wallace and Gutzler (1981) for the NH) using 8 years of monthly mean SH analyses. From their grid-point correlation maps, they identified three modes of variation: a zonally symmetric dipole mode at middle to high latitudes, a zonal wavenumber-three mode in winter, and a continent-ocean contrast in summer. Szeredi and Karoly (1987b, hereafter referred to as SK) used the longer period of upper-air station data available for the SH (up to 30 years) to investigate low-frequency variations. They found two dominant modes of monthly variation in the station data that were primarily zonally symmetric, involving opposing departures of height between middle and high latitudes in one case and between the tropics and middle latitudes in the other. The sparse and irregular station network in the SH meant that only the largest-scale modes could be identified.

More recently, the technique of unrotated and rotated principal-component analysis (PCA) has been applied to grid-point analyses of the SH to clarify the typical modes of low-frequency variations of the SH circulation in a number of different studies. These may be separated into two groups according to the time scale of interest: Periods longer than about 50 days, including interannual variations, have been considered by Rogers and van Loon (1982), Kidson (1988b, hereafter referred to as K88), and Karoly (1989b), while another group concentrated on intraseasonal variability. The results are similar for different studies of the same time-scale fluctuations, and differences can usually be explained in terms of differences in analysis method, such as PCA of the correlation (Karoly, 1989b) or covariance matrix (K88), use of different variables (300 hPa height by

Karoly (1989b), 500 hPa height in most other studies), differences in time filters used, or seasonal stratification. The main data sets used have been the Australian SH analyses, available since 1972 (Karoly, 1989b; Rogers and van Loon, 1982; MW) and the European Centre for Medium-range Weather Forecasts (ECMWF) analyses (K88 and many intraseasonal studies), which are available only for 1979 and later but are likely to be of higher quality than the Australian analyses.

In the following discussions, the typical modes of interannual variations are described first, using results from Karoly (1989b). A rotated PCA has been carried out on monthly anomalies of 300 hPa height using 15 years of Australian analyses (1973-1987). Three different types of horizontal structure have been identified for the low-frequency variations in the SH. The leading modes are approximately zonally symmetric, occur in both summer and winter, and explain about a quarter of the variance. They are associated with opposite departures of height for middle compared with high latitudes and for the tropics compared with middle latitudes. The other stable modes are hemispheric-scale wave-like variations, either zonal wave trains around the hemisphere or regional meridional wave trains, both of which seem to be less important than the symmetric modes. All the modes have an equivalent-barotropic vertical structure in the extratropics, and generally are more horizontally extensive than the typical modes of low-frequency variation in the Northern Hemisphere. The equivalent-barotropic vertical structure of low-frequency variations in the SH was identified using radiosonde data by Szeredi and Karoly (1987a).

Zonally Symmetric Modes

The two leading modes from the rotated PCA of monthly anomalies of 300 hPa height are mainly zonally symmetric, representing opposite height variations over different latitude bands. These modes are found in both summer and winter. One mode represents a north-south out-of-phase variation of height between Antarctica and middle latitudes; it will be referred to hereafter as the high-latitude mode. The horizontal structure of this mode is shown in Figure 1a for winter, using correlations of the time series of PC scores for this mode with monthly SH height anomalies. The proportion of the total monthly variance explained by this mode is about 12 percent. Although it is primarily zonally symmetric, there are substantial asymmetries, with larger correlations and gradients over the Indian Ocean and south of Australia. The middle latitudes show evidence of zonal asymmetry resembling a wavenumber-three pattern. This high-latitude mode has an equivalent-barotropic vertical structure with very similar variations in both 500 hPa height and MSLP.

This mode of low-frequency variation in the SH has

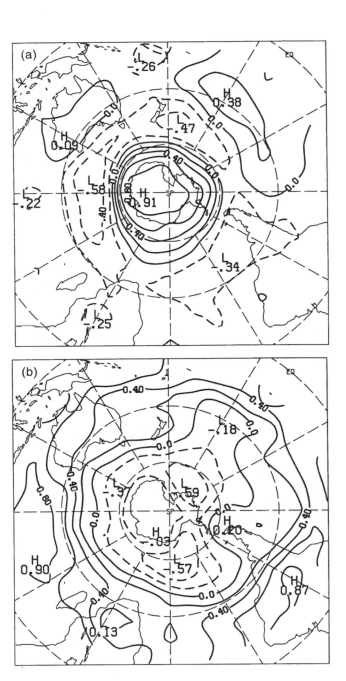

FIGURE 1 Rotated principal-component patterns for monthly 300 hPa height anomalies in winter representing the two leading modes, (a) high-latitude mode and (b) low-latitude mode. Correlations of the time series of the PC scores with monthly anomalies are shown.

been identified by SK using station data; by Rogers and van Loon (1982), MW, and K88 using shorter time periods of analyses; by Shiotani (1990), Kidson (1988a), and Trenberth (1979) using zonal-mean fields; and by Trenberth and Christy (1985) using zonal-mean surface pressure. SK found that it was the second mode of monthly variations of the station data, explaining a smaller fraction of variance than here, whereas K88 found that it was the leading mode of interannual variations of 500 hPa height in the ECMWF analyses. Kidson (1988a) has shown that this mode occurs on all time scales from intraseasonal to interannual.

The other mode found in both seasons represents an out-of-phase north-south variation of height between low and middle latitudes; it is shown in Figure 1b for winter. This mode, hereafter referred to as the low-latitude mode, explains about 13 percent of the variance in winter. Strong departures from zonal symmetry occur in the South American region, with the nodal line running almost parallel to the coast. More intraseasonal variation than interannual variation is found for this mode in winter, but in summer it is mainly associated with interannual variations.

Structures similar to this low-latitude mode have been identified by SK using station data and by Kidson (1988a) using zonal mean wind fields. This mode in the station data explains a larger fraction of variance than here, probably because of the more northerly extent of the analysis region in SK. K88 did not identify this mode, probably because his analysis using the covariance matrix of 500 hPa height variations is restricted to patterns that have large-amplitude anomalies at that height, whereas they are generally small in low latitudes.

MW presented a correlation analysis of the monthly zonal mean height and MSLP fields from their data set (refer to their Figures 1 and 2). They found structures in the zonal-mean fields that are similar to the zonally symmetric modes described here. The two leading modes for the zonal-mean fields, which explain about 70 percent of the variance, have see-saw structures over latitude bands similar to the high-latitude and low-latitude modes shown in Figure 1.

Wave Trains

In addition to the zonally symmetric structures, wavelike patterns were also found in the PCA. They explain less variance and are less stable than the zonally symmetric modes, but are found in one-point correlation maps and composite analyses. In contrast to the NH wavelike modes, the modes found in the SH exert influence over hemispheric scales. They include zonal wavenumber-three patterns and meridional wave trains in winter, and wavenumber-four patterns or continent/ocean contrasts in summer.

Two stable zonal wavenumber-three patterns were found for winter 300 hPa height; they are shown in Figure 2. The centers of action are located at around 60°S for the first version and about 50°S for the second version. The main difference between the two patterns is that one is shifted by about one quarter wavelength relative to the other, so that the extrema of one occur along the nodal lines of the other. The second version shows a stronger relationship (out-of-phase relative to the main centers) with lower latitudes, while the first version shows evidence of a weak

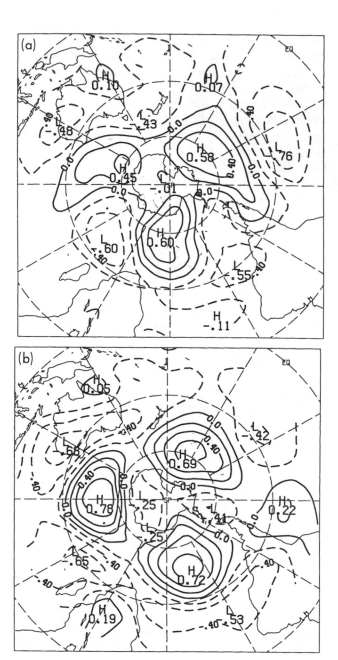

FIGURE 2   Rotated PC patterns of winter 300 hPa height in winter describing the two versions of the wavenumber-three mode, as in Figure 1.

wavenumber-four structure at around 30°S. About 7 percent of the variance is explained by the first version, while the second explains about 5 percent of the variance. Together, wavenumber-three patterns represent about 12 percent of the variance, a total similar to that of each of the zonally symmetric modes.

Using one-point correlation maps, MW identified a zonal wavenumber-three pattern in the SH winter. The two orthogonal versions of the mode found here, with large amplitude at high latitudes, indicate that zonal wavenumber-three anomalies are a common low-frequency pattern in the SH, but that they do not have a strongly preferred phase. Mo and van Loon (1984) describe a change in the amplitude of wavenumber three on long time scales.

As well as the zonal wavenumber-three pattern in winter, two versions of a zonal wavenumber-four pattern were found in summer that are phase-shifted with respect to each other. Together these modes explain about 13 percent of the summer height variance. They both have centers of action around 45°S and show a continent-ocean contrast. MW identified a teleconnection pattern in the SH summer associated with the continent-ocean thermal contrast; this pattern appears to be related to this zonal wavenumber-four pattern found in the upper troposphere.

A meridionally oriented standing-wave pattern resembling a wave train has been found in winter. The pattern for this winter wave train is shown in Figure 3a. Five centers of action can be seen extending from eastern Australia southeastward to high latitudes and equatorward into the southern Atlantic. This mode, which explains about 6 percent of the variance, is not clearly evident in the lower-level variables, where only parts of the wave-train structure are reproduced. Using low-pass-filtered ECMWF anaylses, however, K88 has identified a similar wave-train pattern of height variations.

The third principal component of 300 hPa height in winter explained a relatively large fraction of the variance, but was not stable; it displayed different structures when different analysis methods were used. The pattern for this mode, shown in Figure 3b, essentially represents monopole height variations over the Pacific Ocean, although there is some indication of opposite variations further equatorward and poleward, suggesting a meridional wave train. Although the feature over the central Pacific was relatively stable, the weaker features in this pattern were quite variable.

Mechanisms

Although the above studies have described low-frequency variations in the SH, much less attention has been paid to the mechanisms associated with these observed variations. Karoly (1989a,b) and SK have shown that the interannual variations of the low-latitude mode in summer are are related to the Southern Oscillation and the associated variations of the temperature of the tropical troposphere. In winter, the Southern Oscillation is associated with a pattern of large-amplitude anomalies over the Pacific Ocean, as shown in Figure 3b. Hence the Southern Oscillation is associated with two of the dominant modes of interannual variability in the SH.

The other common interannual variations in the SH are linked to the high-latitude mode. K88, Karoly (1990), and Shiotani (1990) have shown that the zonally symmetric variations of the high-latitude mode are associated with

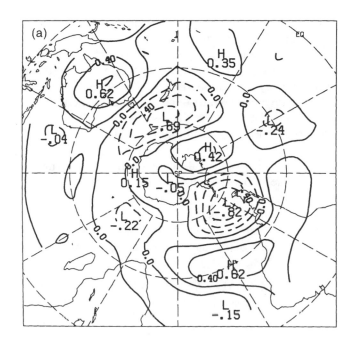

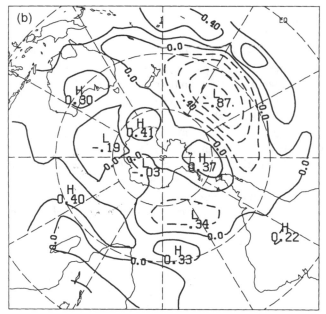

FIGURE 3   Rotated PC patterns for 300 hPa height in winter for the two meridional wave-train modes.

changes in the observed transient-eddy activity and momentum flux in the SH storm track. Trenberth (1984) showed that the large circulation anomalies during the SH winter of 1979, during the Global Weather Experiment, were related to mean flow and eddy variations associated with this high-latitude mode. A general-circulation modeling study of Zwiers (1987) found that the largest interannual variations of the model SH circulation were zonally symmetric opposite variations between high and middle latitudes, very similar to those of the observed high-latitude mode. Simplified atmospheric-modeling studies by James and James (1989) and Robinson (1992) have shown that it is also the dominant mode of low-frequency variations in extended simulations with simplified general-circulation models with zonally uniform boundary conditions. Hence, this zonally symmetric high-latitude mode appears to be a ubiquitous mode of low-frequency variation of the SH, associated with interaction between the mean flow and the transient eddies in the SH storm track.

Little study has been made of the zonal wave-train modes, but it is likely that the quasi-stationary interannual wave-number-three modes are associated with the three SH land masses and continent-ocean thermal contrasts. In addition, the zonal wavenumber-three Rossby wave is almost stationary in the mean SH zonal wind in mid-latitudes.

INTERDECADAL VARIATIONS ASCERTAINED FROM SURFACE DATA

Since upper-air observations are available for the SH for about three decades only, it is not possible to use such data to describe the interdecadal variations of the SH circulation. The only analyses of longer-term variations of the SH troposphere using upper-air station data have concentrated on temperature trends (Angell, 1986; Karoly, 1987). These have shown a small warming trend in the SH lower troposphere over the last three decades.

Some evidence of longer-term changes in the SH circulation is found when the mean sea-level pressure analyses prepared by the South African Weather Bureau for 1951-1958 (Taljaard et al., 1961) are compared with analyses from the Australian Bureau of Meteorology from 1972. Although some differences are due to different analysis methods, some pressure changes appear to be real (Mo and van Loon, 1984; J91).

The equivalent-barotropic vertical structure of low-frequency variations in the SH (Szeredi and Karoly, 1987a) means that longer records of surface observations may be used to describe interdecadal variations. Surface-pressure observations are available for some SH stations for periods extending back to the last century, and sea surface pressure observations over limited SH ship tracks are available for a similar period. It should be noted that these records are often incomplete and are representative of only limited SH regions.

J91 has used station data to reconstruct gridded SH mean sea level pressure (MSLP) fields for most of this century. The quality of this MSLP data set has been assessed by J91 and Barnett and Jones (1992). It is considered to be of reasonable quality over the region 15°S to 60°S from 1951 on, but the reconstructions back to 1911 are useful only near the continents and over the southwestern Pacific Ocean. Jones has used these data to assess changes since 1951 in the strength of the three subtropical anticyclones and the

southern high-latitude westerlies; time series of indices of the strength of the South Pacific anticyclone and the high-latitude westerlies from J91 are shown in Figure 4. They

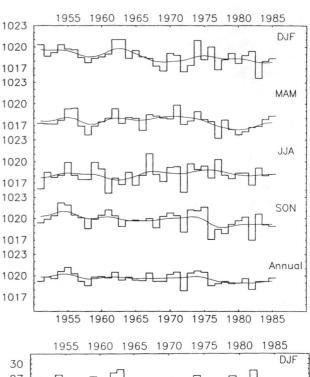

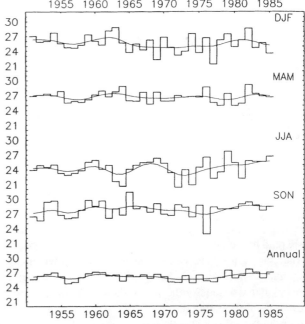

FIGURE 4   Seasonal and annual time series of two indices of the SH circulation: upper panel, South Pacific anticyclone high (average pressure in the region 25°S to 35°S, 80°W to 110°W); lower panel, strength of the high-latitude westerlies (zonal-mean pressure difference 40°S to 60°S). The filtered curve is from a 13-term Gaussian filter designed to suppress variations on time scales less than 10 years. (From Jones, 1991; reprinted with permission of the Royal Meteorological Society.)

show large interannual variations, as well as some fluctuations on decadal time scales. The interannual variations of the South Pacific anticyclone are associated with the Southern Oscillation, and they are larger than those of the anticyclones over the other two SH ocean basins. The index used for the strength of the southern westerlies is also a measure of the high-latitude mode described earlier. It shows pronounced interannual and longer-term variations.

Allan and Haylock (1993) have used the J91 MSLP data set to identify the SH circulation variations associated with the long-term winter rainfall decrease over southwestern Australia. In the Australasian region, they describe variations of MSLP on decadal time scales as well as a longer-time-scale trend, both of which are associated with the regional rainfall variations.

Several indices of the large-scale surface-pressure field in the SH have been used to investigate variations of the SH circulation (Troup, 1965; Trenberth, 1976; Pittock, 1984). The index most frequently used is the Southern Oscillation Index (SOI), the difference between normalized mean sea-level pressure anomalies for Tahiti and those for Darwin. This index represents the magnitude of the pressure oscillation between the eastern Pacific Ocean and Indian Ocean regions associated with the Southern Oscillation. Pittock (1984) defined the Trans-Polar Index (TPI) as the pressure anomaly at Hobart (43°S, 147°E) minus that at Stanley (52°S, 58°W), on the opposite side of the hemisphere. The TPI represents variations of the amplitude of zonal wavenumber-one asymmetries in the pressure field. It probably captures variations in the high-latitude mode to some extent, because of the difference between the latitudes of the two stations. Pittock (1984) has shown that the two dominant modes of common interannual variations of rainfall in Australia, New Zealand, South Africa, and South America are associated with variations of the SOI and the TPI, and that there is little relationship between these two indices. Pittock considered only 30 years of data, 1931 to 1960, and did not directly examine interdecadal variations of these indices and hence of the SH circulation. However, in commenting on several other studies that have considered relatively long time series of the SOI, he noted that there were significant differences in the relationship between the SOI and Australian rainfall between the periods 1930 to 1950 and 1950 to 1980, with a much weaker relationship in the 1930-to-1950 period.

Several recent studies (Hamilton and Garcia, 1986; Elliott and Angell, 1988; Allan et al., 1991; Quinn and Neal, 1992) have examined decadal variability of the Southern Oscillation; it is covered in some detail in Cane et al. (1995), Keeling and Whorf (1995), Rasmusson et al. (1995), and Trenberth and Hurrell (1995), all in this volume. It is sufficient to say here that decadal variations in the Southern Oscillation must be associated with significant decadal variations in the SH circulation.

Time series of the SOI and the TPI since 1896 are shown in Figure 5. There is negligible correlation of the interannual variations between these two time series over the past century, confirming the conclusion of Pittock (1984) that they represent independent modes of variation of the SH circulation. The SOI and TPI show large interannual fluctuations as well as variations on longer time scales. The TPI exhibits decadal variations that are a larger fraction of its interannual variability than does the SOI. The decadal time-scale variations of these two indices must be associated with variations of the SH circulation on these time scales and, almost certainly, with regional climate fluctuations as well.

LONGER-TIME-SCALE VARIATIONS FROM PROXY DATA

It is not possible to use the available surface meteorological observations for the SH to describe variations at century or longer time scales in the SH. The only source of data at these longer time scales is proxy data from tree rings or other sources from a very few sites in the SH. Cook et al. (1991) and Villalba et al. (1989) have used tree rings to reconstruct temperature variations at some SH sites, while Lough and Fritts (1990) have used tree rings to identify

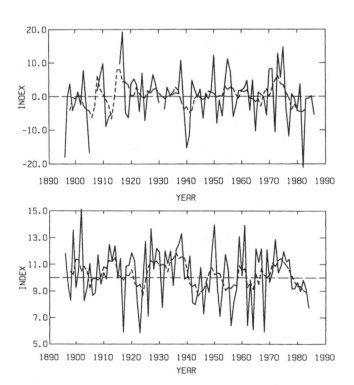

FIGURE 5 Variations of two surface-pressure indices of the SH circulation, with the solid line joining annual values and the dashed line 5-year running averages. Upper panel, Southern Oscillation Index (SOI), normalized Tahiti − Darwin pressure anomalies. Lower panel, Trans-Polar Index (TPI), Hobart-Stanley pressure differences.

SOI fluctuations over several centuries. The temperature reconstructions show marked variability on decadal time scales. The tree-ring temperature reconstruction for Tasmania by Cook et al. (1991) shows a pronounced warming trend over the last few decades, which is consistent with sea surface temperature variations in the SH and in this region over the same period (Folland et al., 1990). The tree-ring series from Tasmania is discussed further in this volume by Cook et al. (1995).

Other proxy evidence of climate variability in the SH has been obtained from coral growth rings (Isdale, 1984), geological lake-level records (Kotwicki and Isdale, 1991), ice cores (Thompson et al., 1984), river-flow records (Wells, 1987; Whetton et al., 1990) and from a variety of sources (Markgraf et al., 1992). In general, however, it is difficult to directly relate these regional proxy data directly to large-scale climate variability in the SH. They provide an indication of the range of climate variability in the source regions for the data. Further work is required to determine the representativeness of some of these regions for the climate of the whole SH.

SUMMARY

The dominant modes of intermonthly and interannual variation of the SH circulation have been described using results from several studies of grid-point analyses of the SH tropospheric circulation. The two leading modes are primarily zonally symmetric, representing out-of-phase variations of height between middle and high latitudes in one case (called the high-latitude mode) and between the tropics and middle latitudes in the other (called the low-latitude mode). These two modes, which explain almost a quarter of the variance of monthly mean 300 hPa height in the SH in both summer and winter, have large interannual variations. The dominance of these zonally symmetric modes by comparison with wavelike or regional patterns in the SH is in contrast to what is typical in the NH. Wave-train patterns of anomalies are also found in the SH circulation.

These include a wavenumber-three mode in winter with large amplitude at high latitudes, a wavenumber-four mode in middle latitudes in summer, a meridional wave-train structure originating over Australia, and a monopole or meridional wave train over the Pacific Ocean in winter. The two most important processes leading to interannual variations of the SH circulation appear to be the Southern Oscillation and the wave, mean-flow interaction in the SH storm track, which leads to the high-latitude mode.

Since upper-air data and grid-point analyses are available for only a very short period in the SH, surface data must be used to describe decadal variability of the SH circulation. Because of the barotropic vertical structure of low-frequency variations in the SH, the surface-pressure variations are representative of variations throughout the troposphere. Relatively little attention has been paid to decadal variability in the SH. It appears from indices of the two dominant modes of interannual variations, determined using surface data, that the modes show substantial decadal variability over the last 90 years.

There are few observational data that can be used directly to indicate climate variability on century time scales in the SH. The only possible observational evidence for this is proxy data, such as tree rings.

ACKNOWLEDGMENTS

This brief review represents my own perspective on studies of observed variability of the SH circulation. It is incomplete, and I apologize for any omissions. I wish to thank Phil Jones for providing Figure 4 and the station data for generating the time series of the TPI in Figure 5, Rob Allan for allowing access to his detailed bibliography on ENSO, and my many colleagues around the world who have provided stimulating discussions on the SH circulation over a number of years. This review was supported in part by a Dedicated Greenhouse Research Grant from the Australian National Greenhouse Advisory Committee. I am grateful to the U.S. National Research Council for the invitation to prepare this brief review.

Commentary on the Paper of Karoly

DAVID E. PARKER
U.K. Meteorological Office

I was particularly interested in Dr. Karoly's video, which raises several questions. I think it would be possible to verify and improve his land-based analyses, because we have large numbers of marine data that we at the Met Office, and Henry Diaz and others at NCAR, are hoping to put together in a blend of our marine data bank and the COADS data base. The COADS data and ours include pressures and winds, so that they can be checked against one another. They should be a good method of filling in this area of the world, which is not very well covered with observations.

Some of you may also have read Neil Ward's work (*J. Climate*, 1992) in which he corrects marine winds. The measured winds appear to have become stronger relative to the geostrophic winds calculated from the pressures. So far, this work has gone back only to 1949, although similar work could be carried out for earlier years. It is very labor-intensive, but one of my recommendations is that the Southern Hemisphere surface data base be improved. An improved surface data base, especially in the Southern Hemisphere, might help us better understand the mechanisms of climate. For example, Section C of the 1992 IPCC Science Supplement shows what happens to estimates of trends in oceanic latent heat fluxes if we apply appropriate corrections to the winds.

Unfortunately, many of the upper-air data from Southern Hemisphere rawinsonde stations are missing, e.g., in Brazil, Antarctica, and southern tropical Africa. I would therefore recommend that everything be done to maintain the network, especially in developing countries and on remote islands. The other problem with the upper-air data is that changes of instrument type and changes of operating practice must be compensated for. To do this, we need adequate station histories. These, as well as the data, are being collated in the Comprehensive Aerological Reference Data Set (CARDS) project under Tom Karl and others at Asheville.

Ultimately, of course, the adjusted data should be put together in a fixed-model-based reanalysis, to yield self-consistent analyses. This is being done by the European Centre for Medium-range Weather Forecasting, and data back to 1958 are being analyzed at NOAA/NMC.

And last, the South Pacific Convergence Zone is an intriguing phenomenon, and vital to our understanding of the Southern Hemisphere climate. So, extra effort should be made to measure and understand it.

Discussion

GHIL: You called your first two EOFs high latitude and low latitude, but what matters is where the nodal line is. Perhaps you have pools of air trapped behind the high-latitude and subtropical jets that happen to have different temperature anomalies.

KAROLY: The interaction between the storm tracks and the main jets is a complicated process that can lead to long-lived low-frequency anomalies. The general symmetry of the anomalies is perhaps more characteristic of the Southern Hemisphere because of the topography. But I agree that it could be described in terms of symmetric air-mass contrast between warmer and colder regions.

GROOTES: Ice-core records for different parts of Antarctica might be able to give you an idea of whether this kind of circulation persisted.

SHUKLA: In the video you showed, almost all the major fluctuations around Australia were coming from the south. Was that a data problem?

KAROLY: It looks that way until the 1950s, but then till the 1970s or 1980s they seem to come more from the north and after that there is no obvious signal. I imagine it's a data problem, but the only way to know is to look at the individual station data. I'm hoping to persuade Phil Jones to do some analysis of this.

TRENBERTH: The dominant mode in the Southern Hemisphere is a double jet structure, with the jets alternating in strength. This mode has been simulated by the Canadian Climate Center and by the Japanese, and in fact turns out to be the dominant global mode. It is entirely derived from natural variability, and seems to have no connection to any external forcing.

KAROLY: It's not clear that internal fluctuations in the atmosphere alone can provide enough longevity to extend it into the decadal range. There might be a link with the Antarctic circumpolar current or the distribution of ocean temperature in the Southern Hemisphere that could provide a longer duration. The Max Planck Hamburg model produces a 50-year episode of this mode, as I remember. It is internal to these sorts of models.

Surface Climate Variations over the North Atlantic Ocean During Winter, 1900-1989

CLARA DESER[1] AND MAURICE L. BLACKMON[2]

ABSTRACT

The low-frequency variability of the surface climate over the North Atlantic during winter is described, using 90 years of weather observations from the Comprehensive Ocean-Atmosphere Data Set. Results are based on empirical orthogonal function analysis of four components of the climate system: sea surface temperature (SST), air temperature, wind, and sea level pressure. An important mode of variability of the wintertime surface climate over the North Atlantic during this century is characterized by a dipole pattern in SSTs and surface air temperatures, with anomalies of one sign east of Newfoundland, and anomalies of the opposite polarity off the southeast coast of the United States. Wind fluctuations occur locally over the regions of large surface-temperature anomalies, with stronger-than-normal winds overlying cooler-than-normal SSTs. This mode exhibits variability on quasi-decadal and biennial time scales. The decadal fluctuations are irregular in length, averaging about 9 years before 1945 and about 12 years afterward. There does not appear to be any difference between the wind-SST relationships on the different time scales.

Another dominant mode of variability is associated with the global surface warming trend during the 1920s and 1930s. The patterns of SST and air temperature change between the periods 1900-1929 and 1939-1968 indicate that the warming was concentrated along the Gulf Stream east of Cape Hatteras. Warming also occurred over the Greenland Sea and the eastern subtropical Atlantic. The warming trend was accompanied by a decrease in the strength of the basin-scale atmospheric circulation (negative phase of the North Atlantic Oscillation). In marked contrast to the dipole pattern, the wind changes occurred downstream of the largest SST anomalies. Hence the gradual surface warming along the Gulf Stream may have been a result of altered ocean currents rather than of local wind forcing.

[1]Cooperative Institute for Research in Environmental Sciences, University of Colorado, Boulder, Colorado
[2]NOAA Climate Monitoring and Diagnostics Laboratory, Boulder, Colorado

INTRODUCTION

Climate variability on decadal and longer time scales is a subject of increasing interest and relevance. Concern over anthropogenic effects on global climate provides a strong impetus to describe and understand the natural modes of variability of the climate system. In this study, we focus on the low-frequency climate fluctuations over the North Atlantic Ocean since the turn of the century.

The North Atlantic is a region of particular importance to the global climate system. The formation of bottom water at high latitudes of the North Atlantic drives a transequatorial thermohaline circulation. Changes in the rate of deep-water production south of Iceland (and hence in the strength of the thermohaline circulation) can have a profound effect on global climate, as can be inferred from paleoclimate data (Broecker et al., 1986; Lehman and Keigwin, 1992) and as is shown by the modeling studies of Rind et al. (1986) and Manabe and Stouffer (1988). Recent ocean modeling experiments indicate that self-sustained oscillations of the thermohaline circulation can occur on decadal and longer time scales (Weaver and Sarachik, 1991).

Low-frequency climate fluctuations in the North Atlantic since World War II have been investigated in several studies. Levitus (1989) used subsurface measurements of temperature and salinity to document a major shift in the ocean circulation in the North Atlantic between the late 1950s and the early 1970s. Knox et al. (1988) showed that the Northern Hemisphere atmospheric circulation experienced an abrupt transition during the early 1960s (see also Flohn, 1986; Shabbar et al., 1988). One of the best-documented examples of low-frequency variability at high latitudes of the North Atlantic is the Great Salinity Anomaly, a freshwater mass that was observed to travel around the subpolar gyre from 1968 to 1982 (Dickson et al., 1988). The freshening of the surface waters was sufficient to halt temporarily deep-water formation in the Labrador Sea (Lazier, 1980). Mysak et al. (1990) discuss the relationship between the Great Salinity Anomaly and interdecadal oscillations in the Arctic climate system.

Studies dealing with the longer (100-year) marine records have tended to emphasize globally averaged surface-temperature variations (cf. Paltridge and Woodruff, 1981; Jones and Kelly, 1983; Barnett, 1984; Folland et al., 1984; Jones et al., 1986c). These studies find that the dominant signal in worldwide temperatures is a warming trend from about 1920 to 1940, with the largest amplitudes at high latitudes.

Bjerknes (1959, 1961, 1962, 1964) investigated air-sea interaction in the North Atlantic on time scales ranging from interannual to interdecadal. Using data from 1890 to 1940, Bjerknes provided compelling evidence that interannual fluctuations in SST are largely governed by wind-induced changes in latent and sensible heat fluxes at the sea surface. However, the long warming trend during the first quarter of this century appears to be linked to a change in the ocean circulation rather than to air-sea energy fluxes (Bjerknes, 1959). Similar conclusions regarding interannual and interdecadal SST variations are reached by Kushnir (1994).

The purpose of this study is to describe the variability of the winter climate over the North Atlantic Ocean, using 90 years of surface marine data. The North Atlantic is the only ocean basin with data coverage that is sufficiently dense for a regional study of climate change since 1900. We focus on the winter season because (1) the atmosphere-ocean coupling is most vigorous and (2) the SST variations reflect significant heat-content anomalies in the upper ocean. We are particularly interested in whether the climate system exhibits preferred time scales of variability, and whether the atmosphere-ocean relationships change with spectral frequency. Our results are based on an objective (empirical orthogonal function) analysis of four components of the climate system: SST, air temperature, wind, and sea level pressure. We will show that the surface climate over the North Atlantic exhibits coherent decadal fluctuations that resemble the variations on interannual time scales. The decadal fluctuations appear to be closely related to sea-ice extent in the Labrador Sea. We will also show that the Gulf Stream was apparently involved in the long-term warming trend during the 1920s to 1930s.

The data and methods are described; the two leading modes of surface-temperature variability during the period 1900 to 1989 and their relation to the atmospheric circulation are documented; and the results are discussed.

DATA AND METHODS

Data

The surface wind, sea level pressure, SST, and surface air temperature data used in this study are from the Comprehensive Ocean-Atmosphere Data Set (COADS), an extensive compilation of weather observations from merchant ships (Woodruff et al., 1987). We have used data gridded by 4° latitude/longitude squares for the period 1900 to 1989. The COADS data are uncorrected for changes in instrumentation, observing practice, ship type, etc. Spurious trends resulting from these changes have been reported in surface wind speed and SST over the tropical oceans (Ramage, 1987; Wright, 1988; Cardone et al., 1990). Our approach to the issue of whether the routine ship-based measurements are sufficiently accurate and homogeneous to allow the detection of real climate signals is to demonstrate physical consistency among independent variables: wind patterns are compared to pressure distributions, and SST fields to surface air temperatures.

Methods

The monthly mean data were converted to anomalies by subtracting long-term monthly mean values for the period 1900 to 1989. These monthly anomalies were then averaged into seasonal anomalies (November to March) to form winter-mean departures from normal. The winter-mean value was considered missing if two or more months in a winter were missing. Missing data have not been replaced.

Empirical orthogonal function (EOF) analysis was used to identify objectively the dominant modes of variability in the North Atlantic climate system during the period 1900 to 1989. EOF analysis was performed separately on the SST, air temperature, sea level pressure, and zonal wind fields. EOFs were calculated using only those grid squares with at least 60 years of winter-averaged data. A minimum of 50 grid squares containing data in each winter was required for reconstructing the principal components (time series of the EOFs).

RESULTS

EOF Modes of Surface Temperature Variability

Figure 1 shows the first two EOFs of SST over the North Atlantic, superimposed on the long-term mean SST distribution. The EOFs are based on un-normalized winter-mean (November to March) anomalies during the period 1900 to 1989. The first EOF (Figure 1a; hereafter referred to as E1 SST), which accounts for 45 percent of the variance, has uniform polarity over the entire basin. The largest loadings occur along the Gulf Stream (as indicated by the tightly packed isotherms in the climatology). The time series of E1 SST (Figure 1b) exhibits a sudden transition from below-normal values to above-normal values around World War II. It is well known that the technique for measuring SSTs aboard ships changed during the early 1940s, and that the bucket temperatures used before World War II were about 0.3° lower than the engine intake samples used later (cf. Wright, 1986; Folland et al., 1984). Over the Gulf Stream this difference can reach approximately 0.8°, due to the rapid evaporative cooling from the bucket samples (Bottomley et al., 1990). Thus, one may be tempted to disregard E1 SST; however, air temperatures exhibit a similar EOF pattern with a more realistic time series (see below).

The second EOF (Figure 1c; hereafter E2 SST), which

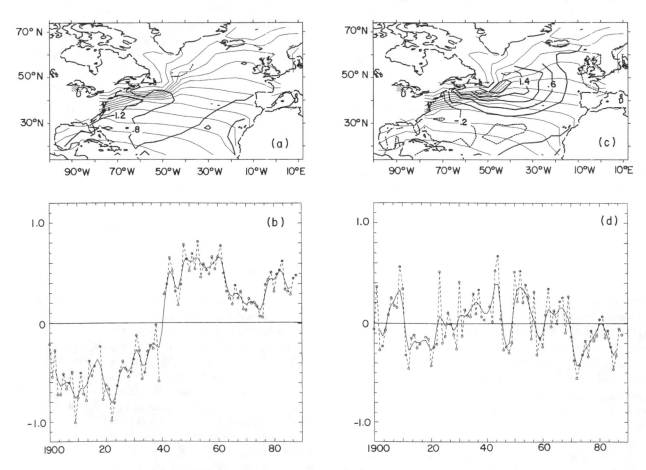

FIGURE 1   (a) EOF 1 of North Atlantic SST anomalies based on un-normalized winter (November-March) means, 1900 to 1989 (bold contours). This mode accounts for 45 percent of the variance. Also shown is the climatological SST distribution (thin contours). (b) Time series of EOF 1 (dashed curve) and smoothed with a 5-point binomial filter (solid curve). The circles denote the winter anomaly, plotted in the year in which November occurs. (c) As in (a) but for EOF 2. This mode accounts for 12 percent of the variance. (d) As in (b) but for EOF 2.

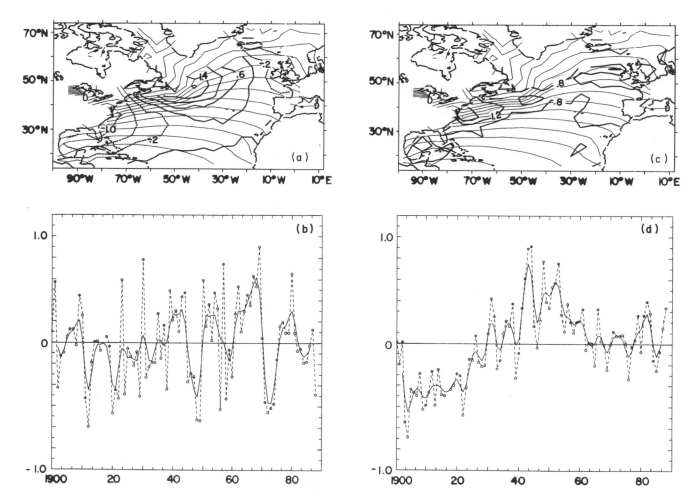

FIGURE 2   (a) EOF 1 of North Atlantic surface air temperature anomalies based on un-normalized winter (November–March) means, 1900 to 1989 (bold contours). This mode accounts for 21 percent of the variance. Also shown is the climatological air temperature distribution (thin contours). (b) Time series of EOF 1 (dashed curve) and smoothed with a 5-point binomial filter (solid curve). (c) As in (a) but for EOF 2. This mode accounts for 17 percent of the variance. (d) As in (b) but for EOF 2.

accounts for 12 percent of the variance, exhibits a center of action east of Newfoundland at the boundary between the subtropical and subpolar ocean gyres. A weaker center of opposite polarity is located off the southeastern United States. The time series of E2 SST (Figure 1d) exhibits quasi-biennial and quasi-decadal fluctuations, as well as longer-term trends.

The first two EOFs of air temperature, shown in Figure 2, are similar to those of SST, but the order of the modes is reversed. E1 of air temperature (Figure 2a), which accounts for 21 percent of the variance, exhibits centers of action off Newfoundland and the southeastern United States; the correlation between the time series of E1 of air temperature and E2 of SST is 0.76.[3] E2 of air temperature (Figure

2c), which accounts for 17 percent of the variance, exhibits maximum amplitude along the Gulf Stream and near western Europe. The time series of E2 (Figure 2d) is dominated by a long warming trend from the 1920s to the 1940s, followed by a cooling trend during the 1950s and 1960s. These trends follow those of global air temperature (Folland et al., 1984). Shorter-period fluctuations are also apparent.

The power spectra of E2 of SST and E1 of air temperature, based on the lag-correlation method, are shown in Figure 3. Both time series exhibit spectral peaks at approximately 10 to 15 years and approximately 2 to 2.5 years, consistent with our visual impression of the time series. E2 of SST also exhibits enhanced power at the lowest spectral estimate. The decadal and biennial peaks in E1 air temperature and the biennial peak in E2 SST are statistically significant at the 95 percent level (a priori) above the red-noise background spectra. The power spectra of E2 of air tempera-

---

[3]All of the correlation coefficients cited in this study are statistically significant at the 99 percent level or higher, based on a one-tailed Student's t-test and taking into account the autocorrelation in the time series according to Leith (1973).

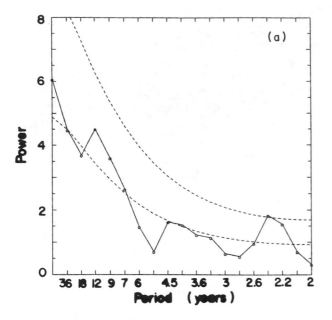

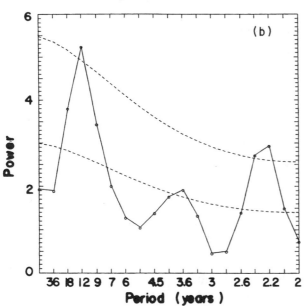

FIGURE 3   (a) Power spectrum of EOF 2 of SST. The thin dashed lines represent the background red-noise spectrum and its 95 percent confidence limit. (b) As in (a) but for EOF 1 of air temperature.

ture and E1 of SST (not shown) are red, reflecting the dominance of the long trends before and after World War II.

The similarity of the leading EOFs of the independently measured fields of surface air temperature and SST lends credence to the patterns.

We have examined the sensitivity of the leading EOF patterns of air temperature to temporal filtering. In view of the power spectrum of E1 of air temperature, we pre-filtered the data to emphasize (a) the biennial time scale and (b) the decadal time scale, and then computed EOFs. The first

EOF of each of the sets of filtered data (not shown) is nearly identical to the first EOF based on unfiltered data. For the biennial time scale, the first EOF accounts for 26 percent of the variance; on the decadal time scale, it accounts for 29 percent. In view of the low-frequency behavior of E2 of air temperature, we computed EOFs based on low-pass filtered data (half-power point at 5 years). The first EOF of the low-pass filtered data (not shown), which resembles the second EOF of the unfiltered data, accounts for 29 percent of the variance of the low-pass filtered data. The correlation coefficient between the time series of E2 (unfiltered) and E1 (low-pass) is 0.88. Thus, although the first and second EOFs of unfiltered air temperature are not well separated according to the criterion of North et al. (1982)—they explain 21 percent and 17 percent of the variance, respectively—the sensitivity experiments described above suggest that they are distinct modes of variability.

RELATION OF SURFACE TEMPERATURE VARIABILITY TO THE SURFACE ATMOSPHERIC CIRCULATION

Dipole Pattern

How are the dominant modes of variability in surface temperature related to anomalies in the surface atmospheric circulation? Figure 4a shows the patterns of SST and surface wind anomalies regressed on the time series of E2 of SST (a similar picture is obtained if E1 air temperature is used in place of E2 SST). Note that the polarity of E2 SST is opposite from that shown in Figure 1c and that the SST pattern has been smoothed in space with a 3-point binomial filter. The relationship between the surface wind and SST anomalies is primarily local: stronger-than-normal westerlies and trade winds are coincident with cooler-than-normal temperatures, and southerly wind anomalies are associated with warmer-than-normal temperatures. The negative SST anomalies east of Newfoundland are located slightly upstream of the largest wind anomalies where the air-sea temperature differences are largest. The local nature of the wind/temperature relationships suggests that changes in the air-sea fluxes and wind-induced vertical mixing processes contribute to the formation of the SST anomalies. Ocean modeling experiments by Luksch et al. (1990) and Alexander (1990) confirm this interpretation. A somewhat surprising result is that the wind/temperature relationships are similar on decadal and biennial time scales (not shown).

Is the wind pattern shown in Figure 4a a preferred mode of variability in the atmosphere? Figure 4b shows the second combined EOF of sea level pressure and zonal wind, based on winter means for 1900 to 1989. This mode accounts for 15 percent of the variance in the combined fields. The meridional wind pattern was obtained by regressing the meridional wind anomalies on the time series of the com-

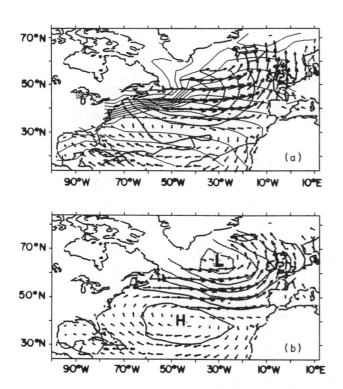

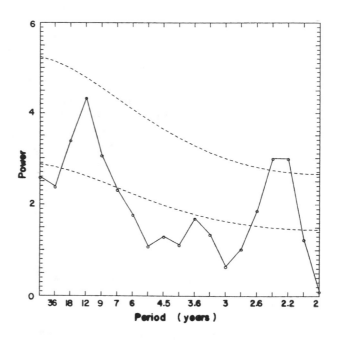

FIGURE 5 Power spectra as in Figure 3a, but for EOF 2 of the combined fields of sea level pressure and zonal wind.

FIGURE 4 (a) SST (bold solid and dashed contours) and wind anomalies (arrows) regressed on the time series of EOF 2 SST. Also shown is the climatological SST distribution (thin contours). Note that EOF 2 of SST here has polarity opposite to that shown in Figure 1c; it has been smoothed in space with a 3-point binomial filter. (b) Combined EOF 2 of sea level pressure and zonal wind, based on winter means of 1900 to 1989. The meridional wind field was obtained by regressing meridional wind anomalies on the time series of the combined EOF. This mode explains 15 percent of the variance in the combined pressure and zonal wind fields.

bined EOF. The wind and pressure distributions in Figure 4b are consistent with the geostrophic relation, lending credence to the patterns. The wind patterns in Figures 4a and 4b are similar, indicating that E2 SST is coupled to a dominant mode of variability in the atmosphere. The power spectrum of the second combined EOF of pressure and zonal wind, shown in Figure 5, exhibits peaks at biennial and decadal time scales, similar to those found for E1 air temperature and E2 SST. The circulation mode represented by the combined EOF corresponds to the West Atlantic pattern as defined by Wallace and Gutzler (1981), although the northern center of action in the sea level pressure field is centered about 5° of longitude east of the corresponding feature in the West Atlantic pattern.

Time series of the various surface parameters to the east of Newfoundland (the center of actions of E1 air temperature and E2 SST) are presented in Figure 6. These records have been low-pass filtered with a 3-point binomial filter. Air temperature and SST were averaged over the region bounded by [52°-40°N, 50°-30°W]; zonal wind was aver-

aged over the region within [48°-38°N, 52°-22°W] and the geostrophic zonal wind was calculated from sea level pressure differences between [40°-32°N, 52°-22°W] and [52°-44°N, 52°-22°W]. Note that the air and sea temperatures and the winds and pressures were measured independently, so the comparisons provide a means of assessing data quality. All four parameters exhibit prominent decadal variability in the area east of Newfoundland; the amplitude of the decadal surface temperature (wind) fluctuations is about 1.5 °C (1.5 m s^{-1}). The correlation between the air temperature and SST time series is 0.88, and that between the measured and geostrophic wind series, 0.80. The high correlations attest to the reality of the decadal signal. The squared coherence between the SST and zonal wind time series (not shown) exceeds the 99 percent significance level at the decadal period, with a phase lag of 180° between the two parameters.

It is apparent from the time series in Figure 6 that the time scale of the quasi-decadal fluctuations was longer after about 1940 than it was before then. This is borne out by calculations of the power spectra for the pre- and post-1944 periods. During the period 1900 to 1944 the decadal spectral peak is centered at 9 years, whereas during 1945 to 1988 it is centered on 12 years (not shown). Thus, the decadal fluctuations are by no means regular.

Gulf Stream Pattern

The dominant features of the time series of E1 of SST and E2 of air temperature are the cold period of 1900 to 1929 and the warm period of 1939 to 1968 (Figures 1b and 2d). These cold and warm epochs may also be identified

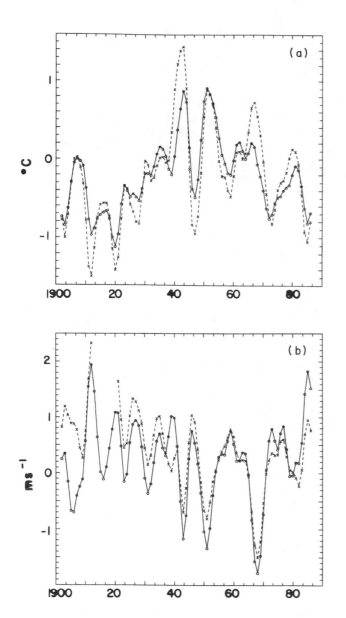

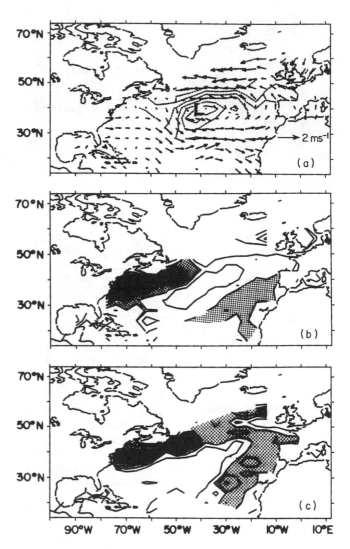

FIGURE 6   Time series of winter anomalies east of Newfoundland of (a) SST (solid) and surface air temperature (dashed) and (b) zonal wind (solid) and geostrophic zonal wind calculated from sea level pressure differences (dashed). All curves are smoothed with a 5-point binomial filter. See text for definition of indices.

FIGURE 7   Difference between the periods 1939 to 1968 and 1900 to 1929 of winter (a) sea level pressure and surface wind, (b) SST, and (c) surface air temperature. In (a) the contour interval is 0.5 mb, with negative contours dashed. The lowest pressure anomaly is −3mb. Wind scale is indicated in lower right. In (b) light shading indicates values between 0.8°C and 1.0°C; heavy shading indicates values greater than 1.0°C. Contour interval is 0.2°C. In (c) light shading indicates values between 0.6°C and 0.8°C; heavy shading indicates values greater than 0.8°C. Contour interval is 0.2°C.

on the basis of globally averaged temperatures (cf. Folland et al., 1984) and North Atlantic basin-averaged temperatures (cf. Bottomley et al., 1990). The surface temperature, wind, and sea level pressure differences between the warm and the cold period are shown in Figure 7. A minimum of 20 winters was required for the period average in each grid square. The surface wind and sea level pressure fields (Figure 7a) are dynamically consistent, with negative pressure anomalies along 40°N and easterly (westerly) wind anomalies to the north (south). The maximum sea level pressure difference is approximately 3 mb, and the largest wind-

speed anomaly is around 1.5 m s⁻¹. This circulation pattern, which resembles the negative phase of the North Atlantic Oscillation (cf. Barnston and Livezey, 1987), represents an overall weakening of the basin-wide wind system.[4] Similar

---

[4]The North Atlantic Oscillation is the dominant mode of variability of the surface atmospheric circulation, accounting for 33 percent of the variance of the winter sea level pressure field during 1900 to 1989 (not shown).

results were obtained by Parker and Folland (1988) from historical sea level pressure analyses. The SST (Figure 7b) and surface air temperature (Figure 7c) anomaly patterns are grossly similar: Both exhibit maximum positive anomalies (about 1°) along the Gulf Stream east of Cape Hatteras, and positive anomalies of smaller amplitude in the eastern ocean. There is some indication of a poleward amplification of the air-temperature anomalies, but the data are sparse north of about 55°. (Stations on Greenland, Iceland, and Scandinavia experienced a warming of approximately 2° during the 1920s and 1930s (Jones and Kelly, 1984).) Although the SST (and to a lesser extent the air-temperature) records suffer from discontinuities around World War II (cf. Jones et al., 1986c), the similarity between the air and sea temperature changes lends credence to the patterns in Figure 7. The "corrected" SST data set of Bottomley et al. (1990) yields a similar pattern, although the warming along the Gulf Stream does not extend as far eastward as in the COADS (not shown).[5]

The wind/temperature relationships are different from those associated with the dipole pattern discussed above. While there is some evidence of wind anomalies overlying SST anomalies in the eastern portion of the basin, the largest surface-temperature anomalies occur in the west, upstream of the atmospheric-circulation anomalies. Whether the temperature anomalies along the Gulf Stream are connected with changes in the strength or position of the current is unknown. However, it is worth noting that a northward shift of the Gulf Stream of only 50 km would be sufficient to produce the observed warming. An alternative explanation for the warming along the Gulf Stream is a reduction in frequency or intensity of cold air outbreaks. This would occur if the temperatures over eastern North America were elevated or the offshore winds were reduced.

To examine the robustness of the patterns and to test whether the patterns are artifacts of data discontinuities around World War II, we computed the linear least-squares trends during the period 1917 to 1939. The data were first smoothed in time with a 3-point binomial filter to enhance the low-frequency signals. The surface wind and sea level pressure fields (Figure 8a) exhibit a cyclonic circulation trend in the central North Atlantic, similar to the results from the epoch analysis. Over the western North Atlantic, the trend analysis shows easterly wind anomalies, which are consistent with the pressure field, whereas the epoch analysis shows weak westerly anomalies. The trends in SST (Figure 8b) and air temperature (Figure 8c) are similar to the results from epoch analysis; the largest warming occurs

[5]Geographically and seasonally varying corrections are incorporated in the Bottomley et al. (1990) SST data set. The corrections are based on a simple model of air-sea heat transfer from an uninsulated bucket and are designed to minimize the difference in the annual cycle of SST before and after 1941.

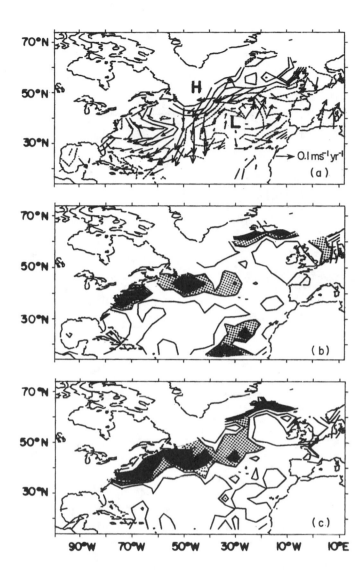

FIGURE 8   Linear least-squares trends during 1917 to 1939 of winter (a) sea level pressure and surface wind, (b) SST, and (c) surface air temperature. In (a) contour interval is 0.05 mb per year, with negative contours dashed. The wind scale is given in the lower right. In (b) and (c) the contour interval is 0.02°C yr^{-1}. Light shading indicates values between 0.04°C yr^{-1} and 0.06°C yr^{-1}; heavy shading indicates values greater than 0.06°C yr^{-1}.

along the Gulf Stream, over the eastern subtropical Atlantic, and near Iceland. The warming along the Gulf Stream is not as continuous, and the warming in the eastern Atlantic not as pronounced, in the trend analysis as in the epoch analysis. As in the epoch analysis, the largest SST trends along the Gulf Stream occur remotely from the wind trends.

The time series of air temperature along the Gulf Stream [44°-36°N, 76°-42°W] and sea level pressure in the region bounded by [46°-34°N, 48°-20°W] and at Ponta Delgada, Azores [40°N, 28°W] are shown in Figure 9. Surface warming along the Gulf Stream occurred during the period 1920 to 1950 (amplitude about 2.5°C). Note that the temperatures

128

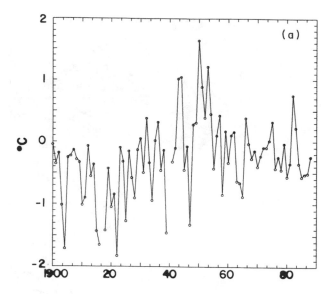

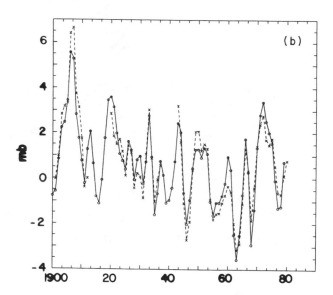

FIGURE 9 Time series of winter anomalies of (a) surface air temperature along the Gulf Stream [44°-36°N, 76°-42°W], and (b) sea level pressure in the central North Atlantic [46°-34°N, 48°-20°W] and at Ponta Delgada, Azores [40°N, 28°W]. The curves in (b) are smoothed with a 3-point binomial filter. The solid (dotted) curve denotes pressures from Ponta Delgada (COADS).

since the mid-1950s have been relatively constant. The COADS and Ponta Delgada sea level pressure time series show remarkable agreement: Both indicate that the warming trend was accompanied by a decrease in the strength of the Subtropical High of approximately 4 mb. The pressure time series give the impression that the downward trend in the atmospheric circulation began about 1905 and lasted until 1962 or so, although the steepest continuous decline occurred during 1920 to 1940.

Bjerknes (1959) found a similar pattern of warming along

the Gulf Stream during the first quarter of this century (he used differences between 1890 to 1897 and 1925 to 1932 to depict the warming trend). The warming was accompanied by a strengthening of the subtropical anticyclone, in contrast to the results of this study. We have attempted to reproduce the results of Bjerknes (1959) by using data from the COADS. While a warming along the Gulf Stream is hinted at in the COADS, it is much weaker than that depicted in Bjerknes (0.2°C versus 2°C; not shown). The pressure and wind patterns derived from the COADS broadly support Bjerknes's results.

DISCUSSION

One important mode of variability of the wintertime surface climate over the North Atlantic during this century is characterized by a dipole pattern in SSTs and surface air temperatures, with anomalies of one sign east of Newfoundland, and anomalies of the opposite polarity off the southeast coast of the United States. This pattern has been noted in the post-World War II data by Wallace et al. (1990), Cayan (1992b), and Kushnir (1994). Wind fluctuations occur locally over the regions of large surface-temperature anomalies, with stronger-than-normal winds overlying cooler-than-normal SSTs. The atmospheric-circulation anomalies resemble the West Atlantic pattern. This mode (both for surface temperatures and for surface winds) exhibits variability on quasi-biennial and -decadal time scales. The decadal fluctuations are irregular in length, averaging about 9 years before 1945 and around 12 years afterward. There does not appear to be any difference between the wind-SST relationships on the different time scales.

The local nature of the atmosphere-ocean relationships exhibited by the dipole pattern suggests that surface wind anomalies contribute to the formation of SST anomalies by altering the fluxes of latent and sensible heat at the ocean surface and the strength of vertical mixing in the upper ocean. Support for this interpretation is given by the ocean modeling studies of Alexander (1990) and Luksch et al. (1989).

Further research is needed to understand the origin of the quasi-decadal cycles in the North Atlantic ocean-atmosphere system. One possibility is that 90 years is not long enough to establish the significance of decadal cycles, and those that have occurred may be random in the sense that they are due to internal atmospheric processes. The decadal SST variations could then be interpreted as simply a reflection of low-frequency modulation of the high-frequency (synoptic and monthly) wind forcing. It should be noted that long (100-year) integrations of atmospheric GCMs, when subject to fixed SST boundary conditions, exhibit prominent low-frequency (decadal and longer) fluctuations (James and James, 1980; Feldstein, 1992).

A more inviting possibility is that the quasi-decadal time

scale is a property of the coupled ocean-atmosphere system. We note that decadal and longer time scale SST variations along the boundary between the sub-polar and sub-tropical ocean gyres are obtained in ocean GCMs as a result of self-sustained oscillations of the thermohaline circulation (cf. Weaver and Sarachik, 1991). In addition, atmospheric GCM experiments indicate that the local atmospheric response to an SST anomaly off Newfoundland is such that weak (strong) winds coincide with high (low) SSTs, reinforcing the original SST anomaly (Palmer and Sun, 1985; Lau and Nath, 1990). The implication of these studies is that observed mid-latitude climate anomalies may result from a positive feedback between the atmosphere and the ocean.

A third possibility is that the decadal atmospheric fluctuations are a response to decadal SST variations outside the North Atlantic. This scenario is difficult to test due to the lack of long marine records in the other ocean basins. Our preliminary results indicate that for the three most recent decadal oscillations, SSTs in the North Pacific are not coherent with those in the Atlantic.

One intriguing link we have found is that between sea ice in the Davis Strait-Labrador Sea region and quasi-decadal fluctuations. Figure 10a shows the time series of winter sea-ice concentration anomalies in the Davis Strait/Labrador Sea region from Agnew (1991), based on data from Walsh and Johnson (1979). The circles denote the winter (December-February) anomaly, plotted in the year in which January occurs. The solid curve shows the data smoothed with a 3-point binomial filter. Decadal variability is evident in the sea-ice record, with peaks occurring in the winters of 1957/1958, 1971/1972, and 1983/1984. Figure 10b shows the sea-ice record superimposed on the inverted time series of the second EOF of winter (November to March) SST (see Figure 1). Both the sea-ice and the SST time series have been detrended by subtracting least-squares parabolas based on the period 1950 to 1988, and smoothed with a 3-point binomial filter. It may be seen that the maxima in sea-ice concentration precede the minima in SST by one to two years. The correlation between the two time series is -0.26 with no lag, -0.62 when sea ice leads SST by 1 year, -0.76 when sea ice leads by 2 years, and -0.62 when sea ice leads by 3 years. The strong lag correlations indicate that, on the decadal time scale, winters of heavy sea ice in the Labrador Sea precede winters of colder-than-normal SSTs east of Newfoundland. It is plausible that the sea-ice anomalies in the Labrador Sea are advected southeastward, resulting in colder-than-normal SSTs east of Newfoundland the following (or second) year. Thus, the quasi-decadal cycle in SSTs east of Newfoundland may result in part from low-frequency Arctic sea-ice variations. Mysak and Manak (1989) and Mysak (1991) have also discussed the possible link between sea ice and SST anomalies in the northern North Atlantic, and Mysak et al. (1990) have postulated the existence of an interdecadal Arctic climate cycle involving

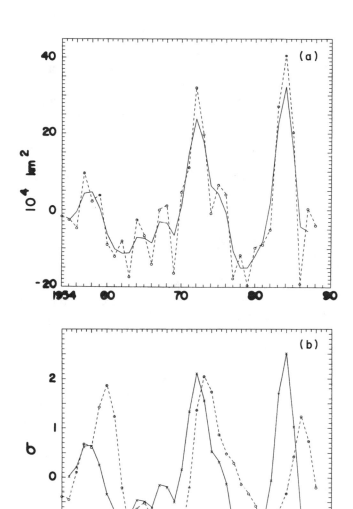

FIGURE 10   (a) Time series of winter sea-ice-area anomalies (10^4 km^2) in the Davis Strait/Labrador Sea region from Agnew (1991), based on data from Walsh and Johnson (1979). The circles denote the winter (December to February) anomaly, plotted in the year in which January occurs. The solid curve shows the data smoothed with a 3-point binomial filter. (b) Sea-ice anomalies from (a) (solid curve) superimposed on the time series of EOF 2 of North Atlantic winter (November to March) SST (dashed curve). Note that the SST time series has been inverted. Both the sea-ice and SST time series have been smoothed with a 3-point binomial filter and detrended by subtracting a least-squares parabola based on the period 1950 to 1988.

runoff, sea ice, and SSTs. Further analysis is needed to elucidate the role of Arctic sea ice in decadal climate variability over the North Atlantic.

Another dominant mode of variability of the wintertime

surface climate over the North Atlantic during this century is associated with the global surface-warming trend during the 1920s and 1930s. The patterns of SST and air temperature change from 1900 to 1929 and from 1939 to 1968 (or, equivalently, the trends between 1917 and 1939) indicate that the warming was concentrated along the Gulf Stream east of Cape Hatteras. Warming also occurred over the Greenland Sea and the eastern subtropical Atlantic. The warming trend was accompanied by a decrease in the strength of the basin-scale atmospheric circulation (negative phase of the North Atlantic Oscillation). In marked contrast to the dipole EOF pattern, the wind changes occurred downstream of the largest SST anomalies. Hence the gradual surface warming along the Gulf Stream may have been a result of altered ocean currents rather than of local wind forcing.

A similar idea has been put forth to explain the cooling in the North Atlantic from the 1950s to the early 1970s (Levitus, 1989; Greatbach et al., 1991; Kushnir, 1994). The cooling trend was accompanied by a decrease in the transport of the Gulf Stream, as diagnosed from subsurface data by Greatbach et al. (1991). The decrease in Gulf Stream transport was in turn traced to bottom-pressure torque effects rather than wind effects (Greatbach et al., 1991).[6]

---

[6]It should be noted that the spatial pattern of cooling during the 1950s to 1970s was somewhat different from the pattern of warming during the 1920s and 1930s (see Kushnir, 1994).

In a recent coupled atmosphere-ocean GCM experiment, Manabe and Stouffer (1988) showed that an intensified oceanic thermohaline circulation is associated with surface warming at high latitudes of the North Atlantic, and with an intensification and poleward shift of the Gulf Stream. The model's atmospheric response to an intensified thermohaline circulation is a weakening of the basin-wide wind system; this is consistent with the reduced baroclinicity. It is tempting to speculate that the climatic trends observed over the North Atlantic during the 1920s and 1930s were due to an intensification of the thermohaline circulation. However, we note that SSTs in the South Atlantic also warmed during this period (cf. Bottomley et al., 1990), contrary to the cooling observed in Manabe and Stouffer's experiment. Further modeling studies are needed to elucidate the role of the ocean circulation in low-frequency climate variability.

ACKNOWLEDGMENTS

We thank Ludmilla Matrosova for her help with the data processing. We are indebted to Harry van Loon for stimulating discussions and to Michael Alexander for his thoughtful review of the manuscript. James Hurrell kindly provided the Bottomley et al. (1990) data set. This work was supported by a grant from NOAA's Climate and Global Change Program.

Commentary on the Paper of Deser and Blackmon

LAWRENCE A. MYSAK
McGill University

Dr. Deser presented a very nice study. I enjoyed it, but I do not have a lot of comments at this stage. Some postwar data sets that have been analyzed show two to three cycles of interdecadal variability in the empirical orthogonal functions, for example. In Dr. Deser's analyses we have now seen decadal-scale variability over a 90-year record, which I for one really appreciate.

I think it is important to emphasize that only the winter data was analyzed. Dr. Deser and I discussed earlier whether this decadal variability shows up in other seasons, such as summer, and thus is a year-round signal. I think the answer is that it is there in the summer, but it is not as strong. It is definitely much more intense in winter. I should perhaps mention here that winter is defined as being from November to March, a five-month period.

The two time scales that I found particularly fascinating in the paper were the decadal and the biennial. While I will not discuss the biennial any further here, I just want to point out some recent work of one of our Ph.D. students at McGill, J. Wang, who also noticed a quasi-biennial cycle in the ice cover in the Labrador Sea and the Baffin Bay region. In addition, he found quite a strong correlation between anomalies in sea-ice extent and the North Atlantic oscillation (which is defined as a winter index).

The decadal time scale Dr. Deser discusses clearly falls between two familiar time scales: the North Atlantic oscillation, which has a period of 7 to 8 years, and the interdecadal cycles in the Arctic, which I shall discuss in my own presentation. Also, many recent modeling studies of the thermohaline circulation in the North Atlantic have shown the existence of internal oscillations with decadal to interdecadal periods.

I should like to list some relevant references, which I believe were not mentioned in the paper. Also, I might remark that many of these papers are by various people at this workshop. One piece of work is the master's thesis done in 1987 by Gordon Fleming at the Naval Postgraduate School. He was supervised by John Walsh, who is here. One of the interesting calculations in the thesis is the cross-correlation analysis of the COADS SST in the Labrador-Greenland-Barents Sea region and the position of the ice edge. Fleming found the two to be inversely correlated, not unlike Dr. Deser's findings for the Davis Strait. However, I believe Fleming was able to show that in some regions the SST actually leads the ice edge and therefore could be used as a predictor.

A recent paper by me and Manak (*Atmosphere-Ocean*, 1989) studying 32 years of sea-ice concentration data clearly showed the existence of a decadal oscillation of the sea-ice cover in the Labrador Sea and in the Barents Sea. The signal in the former region is the same as that shown in Dr. Deser's Figure 10a. In the Greenland Sea, variations in the sea-ice extent seemed to have a much longer time scale, because there the sea ice was dominated by a big ocean event during the 1960s and early 1970s that we have now associated with the Great Salinity Anomaly.

In that paper Manak and I suggested something that might set up an oscillation of 10 to 12 years, namely, the circuit time of the subpolar gyre in the northern North Atlantic. While 12 to 14 years was the circuit time for the GSA, there was nothing to reinforce it within the subpolar gyre region. Thus we abandoned the idea that there might be self-sustained oscillations of this period in this region.

Other papers by Mysak et al. (*Climate Dynamics*, 1990) and Marsden et al. (*JGR*, 1991) showed that salinity and sea-ice extent are highly correlated in the Greenland and Labrador seas, and that there is advection of both ice and salinity anomalies from the Greenland Sea into the Labrador Sea, with an advection time of a few years.

In terms of modeling studies, I first want to mention M. Ikeda (*Atmosphere-Ocean*, 1990), who developed an ocean model for the decadal oscillations in the Barents-Greenland seas. He emphasized that there were feedbacks between the atmosphere and the ocean in the Barents Sea. In 1991, Weaver and Sarachik published in *Atmosphere-Ocean* the first model evidence of a decadal oscillation in the thermoha-

line circulation in the North Atlantic. The physical location of the oscillation is somewhat further south than the Greenland-Labrador Sea, because it is centered between the subtropical and subpolar gyres. In more recent modeling studies with an improved geometry, Weaver and his co-workers have found that the decadal oscillation now extends to 20 years.

There has also been the more recent modeling work by Delworth and his colleagues at GFDL, who find a somewhat longer-time-scale fluctuation in the thermohaline circulation in a coupled atmosphere-ocean model. The period is 30 to 50 years, and the oscillation is found to have a dipole-like structure in the subsurface waters in the central North Atlantic. As for recent data studies looking at decade-to-century time scales, a paper in *Climatic Change* (Stocker and Mysak, 1992) presents a spectrum of the Koch ice index. (The Koch ice index is the number of weeks per year that ice affects the coast of Iceland.) A spectrum of 360 years of this index shows peaks at 90, 27, and 14 years.

Finally, a well-known time series of SST anomalies in the North Atlantic (45°N to 55°N) is presented by Bryan and Stouffer in *J. Marine Systems*. This time series also shows interdecadal variability in the northwest Atlantic. In addition, in the Centre for Climate and Global Change Research's Report No. 91-1 (1991), I suggested that heavy ice conditions in the Greenland Sea lead cold SST anomalies in the northwest Atlantic (i.e., east of Newfoundland) by 4 or 5 years. Similarly, warm SST anomalies there were preceded by generally light ice conditions in the Greenland Sea during the previous 5 years.

Let me now refer you to two figures, Figures 10 and 12a in Stocker and Mysak. The first is a spectrum of the Koch ice index as defined above. During heavy ice conditions in the Greenland Sea, you have a high index; during mild conditions, a low index. This spectrum clearly shows a significant peak at 27 years, which indicates an interdecadal climate cycle in the northern North Atlantic. The second, a spectrum of the famous Central England temperature, shows peaks at around 15 and 7 years and an interdecadal peak at 24 years.

Discussion

NORTH: I wonder whether EOFs are the best way of describing the data. True, they are objective, but can you tell from analyzing them whether the Gulf Stream's temperature is changing or whether it is just shifting location?

DESER: If you look at the anomalies during the cold 30 years, or one of those decades, the warming first appears near Cape Hatteras. Then you see anomalies along the Gulf Stream east of Cape Hatteras, then east of Newfoundland, then in higher latitudes. A northward shift of about 50 km would be sufficient to produce the observed warming, and seems to me to be the simplest explanation.

MCWILLIAMS: The Gulf Stream warming pattern in your last figure is much broader than 50 km. While a mechanistic interpretation is tempting, an advective supply of heat from the Gulf Stream would be related to wind driving, whereas instead the forcing of the anti-cyclonic gyre is actually weakening.

GHIL: I just wanted to comment on the causes of the decadal variability. It seems to me that certain phase differences between fluctuation or oscillation are clear indications of coupling. A coupled atmosphere-ocean phenomenon seems the most plausible to me.

CAYAN: Have you looked at SST minus air temperature together with your Gulf Stream pattern? I wonder whether part of the reason for the change could be a change in the air mass distribution over the Gulf Stream. In extratropical regions the sensible flux is driven a lot by the temperature lapse rate, and to some extent by the wind.

DICKSON: I should like to show a slide, my Figure 14, that suggests two things. The first is that temperature and salinity show similar patterns, and I think were both involved, meaning that both the circulation and the ecosystem might have had large changes. The second is that the century-long trend covered a much larger area of the North Atlantic than we've been talking about— right up to the ice in the north, in fact.

Seasonal-to-Interannual Fluctuations in Surface Temperature over the Pacific: Effects of Monthly Winds and Heat Fluxes

DANIEL R. CAYAN[1], ARTHUR J. MILLER[1], TIM P. BARNETT[1],
NICHOLAS E. GRAHAM[1], JACK N. RITCHIE[1], AND JOSEF M. OBERHUBER[2]

ABSTRACT

Monthly heat fluxes and wind stresses are used to force the Oberhuber isopycnic ocean general-circulation (OPYC) model of the Pacific basin over a two-decade period from 1970 to 1988. The surface forcings are constructed from COADS marine observations via bulk formulae. Monthly anomalies of the fluxes and stresses are superimposed upon model climatological means of these variables, which were saved from a long spin-up. Two aspects of this work are highlighted, both aimed at a better understanding of the atmosphere-ocean variability and exchanges and at diagnosing the performance of the OPYC model in simulating monthly to decadal-scale variability. The first is the evaluation of the data used to force the model ocean, along with its relationship to other observed data. The second is the diagnosis of the processes revealed in the model that are associated with sea surface temperature (SST) variability, including their seasonal and geographic structure.

Although both random and systematic errors arise from the marine data and the bulk formulations, large signals in the air-sea fluxes are nonetheless consistent with the large-scale atmospheric circulation anomalies over the Pacific. This signal is large in a composite prepared from months with similar circulation modes. Also, latent and sensible heat-flux anomaly patterns correspond well to those of SST anomaly tendencies. Considering short-period variations, SST anomaly tendencies have typical magnitudes of $0.3°C$ mo^{-1}. These are associated with monthly mean flux anomalies having typical magnitudes of 50 W m^{-2} and are consistent with observed mixed-layer depths. Decadal anomalies have much smaller magnitudes, perhaps reduced by two orders of magnitude, and it is here that the signal-to-noise problem is more severe. The forcing terms are generally products of variables, so realistic means and fluctuations of these variables are crucial for a successful simulation.

The 19-year simulation of the Pacific basin by the monthly marine data-forced OPYC model displays good skill in reproducing SST variability. These results represent the first

[1]Scripps Institution of Oceanography, University of California, San Diego, La Jolla, California
[2]Max-Planck-Institut für Meteorologie, Hamburg, Germany

hindcast of which we are aware that uses both observed total heat-flux and wind-stress anomalies as forcing for such a long time interval. There is close agreement between the model SSTs and those observed in many regions of the Pacific, including the tropics and the northern extratropics. Besides performing credibly on the monthly time scale, the model captures the essence of low-frequency variability over the North Pacific, including aspects of a marked basin-wide change that occurred in 1976-1977. In the model's detailed heat budget, the anomalous air-sea heat fluxes, entrainment, and to a lesser extent horizontal advection, force thermal-anomaly changes in the mixed layer. Each of these components was apparently involved in the 1976-1977 decadal SST shift.

INTRODUCTION

Our purpose is to describe monthly to decadal variations in surface variability over the Pacific basin. The results presented combine a historical set of monthly marine observations and a two-decade numerical simulation (Miller et al., 1994a,b) that uses the Oberhuber (1993) isopycnic ocean general-circulation model.

The first part of the paper is an analysis of observed monthly surface marine atmospheric variability in relation to sea surface temperature (SST) over the Pacific Ocean. This observational evidence provides background to the second section of the paper, in which monthly wind stresses and heat fluxes are used to force an ocean general-circulation model (OGCM). In the observational section, we concentrate on the variability of the latent and sensible heat fluxes, since they provide a good example of the characteristics and problems involved in bulk formulations using the marine data. The fluxes exhibit monthly variability with large-scale organization, and their effect on the ocean can be detected in the monthly fluctuations of the SST anomaly field.

The second part of the paper reports on applying the marine data parameterizations to a simulation over two decades (1970-1988) of the Pacific Ocean basin using the OPYC OGCM. This run is forced by monthly mean wind stress and fluxes derived from surface marine observations, which are introduced in the first section. In assessing this run, we compare the model SST anomalies with observations, particularly their low-frequency variability. Further insight is provided by the model upper-ocean heat budget, which is not well described by observations.

Aside from diagnosing mechanisms causing monthly SST variability, one motivation for this simulation was to better understand a strong regime-like change in SST that occurred in the North Pacific basin during the mid-1970s (Douglas et al., 1982; Nitta and Yamada, 1989; Trenberth, 1990; Miller et al., 1994a,b; Graham, 1994; Trenberth and Hurrell, 1995, in this volume). Although thorough observational documentation of decadal-scale variability is lacking, glimpses from recent historical episodes suggest that this "gray area" (Karl, 1988) of the variability spectrum contains important climate effects. The actual shift was identified in fall and winter of 1976-1977 when SSTs in the central North Pacific cooled markedly and SSTs along the west coast of North America warmed (Venrick et al., 1987; Trenberth, 1990; Ebbesmeyer et al., 1991). In the atmosphere, the shift involved a basin-scale deepening of the wintertime Aleutian Low System, and appears to have been at least partially instigated by forcing from the tropical Pacific (Graham, 1994; Trenberth, 1995, in this volume), although an alternative theory involves sea-air feedback and advection of thermal anomalies by the North Pacific subtropical gyre (Latif and Barnett, 1994). Accompanying changes in many other physical and biological variables in the North Pacific basin and around its margin were also noted (Venrick et al., 1987; Ebbesmeyer et al., 1991).

The bulk formulae parameterized heat fluxes, calculated from standard marine surface meteorological observations, are the only means of estimating a multi-year time history of the heat exchange over broad regions of the oceans. Similarly, SST has been measured routinely by merchant ships for several decades, and, in the absence of a comprehensive temperature-versus-depth set in the upper ocean, SST is used to infer the variability of the upper-ocean heat content. Although the ocean heat-content structure can be complex, during winter in the extratropics the upper ocean is quite well mixed and SST is a good indicator (White and Walker, 1974).

As is brought out in the section below, there is a rich variability, superimposed upon the climatological mean, that involves both atmospheric and oceanic fluctuations. Three tests of the surface heat budget are discussed here. The first two, which study the data directly, are to relate monthly anomalies of the latent and sensible fluxes to the anomalous atmospheric circulation and to anomalous tendencies of the SST field. The third is to incorporate the monthly bulk-formula wind stress and heat flux as forcing of an extended OGCM run and to compare the simulated versus the observed SST fields. These results represent the first hindcast of which we are aware that uses both observed total-heat-flux and wind-stress anomalies as forcing for such a long time interval.

OBSERVATIONAL EVIDENCE FOR ANOMALOUS AIR-SEA HEAT EXCHANGE

Data

The primary data employed in the observational section of this study, as well as in the model forcing, are gridded monthly mean ship-observed marine surface data for 1946-1986 from the Comprehensive Ocean-Atmosphere Data Set (COADS) (Slutz et al., 1985; Woodruff et al., 1987). From this set, several variables are employed, including the scalar wind or wind speed w, west-to-east and south-to-north horizontal surface wind components u,v, sea level pressure (SLP), specific humidity q_a, air temperature T_a, and sea surface temperature SST. Derived variables include the products $\{w\Delta T\}$, $\{w\Delta q\}$, $\{wu\}$, and $\{wv\}$, where the monthly averages of products of the individual observations are indicated by $\{\ \}$. These products are involved in parameterizing the sensible and latent heat flux, and the horizontal wind-stress components, respectively.

To reduce errors in the data fields and condense the number of data in the analyses, the COADS 2°-gridded "Monthly Summaries Trimmed" data were averaged onto 5° squares, centered at 5° latitude-longitude intersections. This averaging was applied to the anomalies of the 2° squares, rather than their values, to avoid spurious 5° anomalies caused by poor sampling within the domain. Long-term means employed were calculated over 1950 to 1979. Details are provided in four papers by Cayan (1990; 1991, hereafter designated DC1; 1992a, designated DC2; and 1992b, designated DC3).

Bulk Formula Parameterizations

Latent and sensible heat exchange at the ocean surface is an important mechanism for venting ocean heat absorbed from solar radiation to the atmosphere. Globally, approximately 60 percent of the solar radiation absorbed at the earth's surface is released by latent and sensible heating, primarily from the ocean (Sellers, 1965). Gill and Niiler (1973) invoked scaling arguments in the governing equations to infer that, over large scales (≥ 1000 km meridionally and ≥ 3000 km zonally), anomalous open-ocean surface temperature changes are dominated by changes in heat fluxes through the surface, rather than by advective influences. Since they depend on both oceanic and atmospheric conditions, the latent and sensible fluxes vary strongly over space and time. In the extratropics, latent and sensible fluxes are greatest in fall and winter when the near-surface vertical gradients of humidity and temperature are largest and wind speeds are highest (Esbensen and Kushnir, 1981; Isemer and Hasse, 1985, 1987).

The bulk aerodynamic formulae used for the latent and sensible heat fluxes are

$$Q_l = \rho\, L\, C_E\, \{w\Delta q\} \qquad (1)$$

and

$$Q_s = \rho\, c_p\, C_H\, \{w\Delta T\}\ , \qquad (2)$$

where Q_l is the latent flux, Q_s is the sensible flux, L is the latent heat of evaporation of water, and c_p is the specific

heat of air at constant pressure. Again, w is the wind speed, Δq is sea surface saturation minus air specific humidity, and ΔT is SST minus air temperature. C_E and C_H are the transfer coefficients for latent heat and sensible heat, respectively, and are taken to be the modified Bunker coefficients given by Isemer and Hasse (1987). Unlike their mean values (Blanc, 1985), the anomalies of the fluxes are not particularly sensitive to the choice of exchange coefficients. A comparison of monthly anomalies obtained with these exchange coefficients to those from two other schemes yielded strong agreement (DC2).

The sense of the latent and sensible fluxes adopted throughout the paper is *positive* for heat *entering* the ocean from the atmosphere.

Errors in the Latent and Sensible Fluxes

Because of uncertainties in the bulk formulae and because the marine data have inaccuracies and are non-uniformly measured, the parameterized flux estimates contain errors. Although it is not possible to determine precisely the spatial and temporal characteristics of errors in the bulk-formulae-derived heat fluxes, estimates of their magnitudes have been attempted. In many regions the aggregate of the errors in the fluxes considerably exceeds the 10 W m^{-2} tolerance that has been targeted for climate studies (Taylor, 1984). Nevertheless, it will be shown that important climate variability (monthly to decadal time scales) is still contained in the flux estimates. A review of individual monthly maps of the fluxes finds that several have anomalies with magnitudes of 50 W m^{-2} or greater, a "signal" which is well beyond reasonable error estimates. From our observational and modeling experience, the character of this error, as much as its magnitude, is crucial. For example, a 10 W m^{-2} error is probably acceptable if it has a high-frequency, random variability, but it is intolerable if it has low-frequency (decadal-scale) changes.

Both random and systematic errors in marine observations and errors in the bulk formulae contaminate the monthly mean fluxes (Weare, 1989; Taylor, 1984; DC1). Random measurement errors and errors from incomplete sampling of the natural weather fluctuations are reduced by averaging several observations together. Because the mean of the fluxes was removed and the resulting anomalies are generally only a fraction of the mean, the non-time-varying biases are reduced considerably. The largest random errors in the flux estimates appear to be caused by sampling variability, and to a lesser extent by observation errors (DC1). This error was estimated by a Monte Carlo sampling exercise (DC1) where the mix of observations entering particular flux estimates within well-sampled regions was repeatedly re-sampled. The uncertainty of the monthly mean for well-sampled 5° squares (50 or more independent samples per month) is reduced to approximately 40 W m^{-2} or less for the latent flux and to 20 W m^{-2} or less for the sensible flux. The density of observations is relatively high over the extratropics of the North Pacific and North Atlantic, with many 5° squares having 100 or more observations per month. However, the density is marginal-to-poor over the tropics and in the Southern Hemisphere, with fewer than 20 observations per month at many 5° squares. In these sparsely sampled regions, random errors in the monthly flux estimates are likely to be substantial. Consequently, in forcing the OGCM in the region between 20°N and 20°S and data-poor regions south of 20°S, we have resorted to a Newtonian damping scheme for the net flux anomalies.

Systematic errors are caused by biases in the bulk formulae and in the observations, and are not automatically reduced by averaging. Time-varying biases in the fundamental observations caused by changes in instrumental practices are probably a significant source of error. These errors are difficult to quantify, but the marine data have "global" (over the available COADS-covered region) trends in w and ΔT, which translate to changes of approximately 0.8 m s^{-1} and -0.2°C over 1950 to 1986. Because these seem to be at least partially caused by instrument and calibration changes over the period, the fluxes were adjusted to remove the effects of the global average linear trends. This adjustment reduced the linear change in the latent and sensible fluxes over 1950 to 1986 by about 20 W m^{-2} and 5 W m^{-2}, respectively (DC1 and DC2). This adjustment has only minor influence on the high-frequency variability, but it does significantly affect the multi-year numerical model run, because the effects are cumulative. More discussion of the systematic errors in the marine data is provided below.

Anomalous Variability of the Heat-Flux Components

The dependence of the flux anomalies on the mean conditions and the anomalies of the fundamental surface variables was explored in DC1. For the monthly latent flux anomaly

$$Q'_l \sim \overline{w}\Delta q' + w'\overline{\Delta q} + w'\Delta q' , \qquad (3)$$

where an overmark indicates the long-term monthly mean and a prime indicates the monthly anomaly. For monthly means, most of the variation comes from w and Δq, not from the exchange coefficients (DC1). The last term on the right-hand side of (3) does not contribute strongly to latent flux anomalies, because w and Δq monthly anomalies are not locally well correlated over most of the oceans (DC1). A similar relationship involving ΔT instead of Δq describes the monthly sensible flux anomalies.

Equation (3) shows that the joint behavior of mean values and anomalies of wind speed w, the ocean-surface saturation humidity-air humidity difference Δq, and the sea-air temperature difference ΔT must be included to determine the flux anomalies. For atmosphere-ocean models, this interplay between the mean and anomaly components is a challenge.

It is not enough to simulate anomalies of w, Δq, and ΔT; their mean fields must also be properly represented.

There are strong seasonal and geographical modulations of the flux anomalies. In the extratropics the variance is greatest in fall and winter when w, Δq, and ΔT have largest mean values and strongest variability. The largest contributions to the extratropical variance of Q_l anomalies involve $\overline{w}\Delta q'$. Anomalies of Q_s become important north of about 35°N, where their variance is dominated by the term involving $\overline{w}\Delta T'$. In the extratropics, the variance of each of these components is maximum along the western side of the basin, because strong ocean boundary currents and large air-mass differences between the upwind land mass and the sea cause great contrasts. Latent flux anomalies in the tropics often involve greater contributions from $w'\overline{\Delta q}$ than do those in mid-latitudes, although the eastern tropical Pacific has relatively large contributions from $\overline{w}\Delta q'$.

The Δq and ΔT monthly anomalies are correlated (cooler air is usually drier), especially in the extratropics, so the latent and sensible anomalies usually reinforce each other, yielding relatively large net heating anomalies. In the extratropics, flux anomalies are linked to the local wind direction, but in the tropics they are usually not (DC2). North of about 15°N, the largest positive anomalies are associated with northerly to northwesterly winds; they probably result from meridional advection of atmospheric humidity and temperature, and also from changes in the strength of the westerly winds. In the tropics, there is little relationship between wind direction and the flux anomalies, since horizontal gradients of humidity and temperature are weak and the wind direction is relatively steady.

How does the variability of the anomalous latent and sensible fluxes compare with that of the radiative fluxes? The balance of the heat-flux anomaly terms is distinctly different from that of the mean terms. Combined anomalies of Q_l and Q_s often exceed 50 W m^{-2} over regions several hundred kilometers in extent (DC1, DC2, and DC3). These estimates indicate that monthly anomalies of Q_l and Q_s are usually larger than those of the radiative components. Comparing the variances of Q_l and Q_s with those of bulk-formulae-estimated net solar radiative flux Q_{SW} and infrared flux Q_{IR}, DC1 found that: (1) Q_l and Q_s anomalies dominate the monthly anomalous surface heat budget during winter in the extratropics north of 30°N, (2) Q_l anomalies dominate from about 15°N to about 30°N, and (3) Q_l and Q_{SW} anomalies are about equally large from 15°N to 15°S.

Flux Anomalies and Atmospheric Circulation

Q_l and Q_s anomalies have regional-to-basin-scale coherence both the North Atlantic and North Pacific. The first four empirical orthogonal functions (EOFs) of the sum of the Q_l and Q_s anomalies (Q'_{l+s}) account for about half of the total variance of this field in each basin (DC1). That these patterns relate to the anomalous atmospheric circulation is evidence for a realistic flux-anomaly signal. The dominant atmospheric-circulation anomaly modes in the North Atlantic and North Pacific, represented as EOFs of the sea level pressure, produce systematic flux-anomaly patterns (DC2). The first SLP EOF in the North Pacific features a large pressure anomaly in the central North Pacific, and is associated with the well-known "Pacific-North American" (PNA) pattern. Three correlation fields linking the PNA to surface variables are shown in Figure 1: the correlations between winter-month PNA time amplitudes and gridded fields of w (wind speed), Δq, and Q_{l+s} anomalies. This set of maps, and others in DC2, show that the anomaly fields of w, Δq, and latent-plus-sensible flux are remarkably consistent with major SLP anomaly patterns over the North Pacific and North Atlantic. Negative SLP anomalies favor positive w anomalies to their south and negative w anomalies to their north, associated with shifts in storm tracks and changes in the mean wind field. Negative SLP anomalies favor positive Δq (and ΔT) to their west, and negative Δq to their east, probably because of meridional advection of air temperature and humidity. The flux anomalies are a hybrid of these patterns, with enhanced sea-to-air fluxes to the southwest and diminished fluxes to the east of negative SLP anomalies.

For the strong Aleutian low phase of the PNA, the anomalous wind-speed field is dominated by a single zonally oriented high-wind patch to the south of the Low in the central North Pacific, with out-of-phase tendencies to the north and the south. Δq and ΔT anomalies are arranged in pockets to the southwest (positive anomalies) and to the east (negative anomalies) of the low. Farther afield, a positive anomaly center appears in the Gulf of Mexico, reflecting the well-known downstream teleconnection of deep troughs and cold dry air outbreaks leading to positive ΔT and Δq anomalies over the southeast United States. The pattern has a broad center of positive anomalies to the southwest of the Aleutian Low, and a region of negative anomalies east of the Low along the West Coast. In the section of this paper that reports on the ocean model, it is shown that low-frequency changes in the PNA between the 1960s and 1980s were associated with marked variations in heat flux and wind stress, which produced striking variability in the upper-ocean thermal structure.

Flux Anomalies and SST

Does the heat flux drive the SST, or does the SST drive the heat flux? The relationship between SST and the climatological mean of total latent and sensible heat flux appears to support two different points of view. On one hand, the fluxes may be *driven* by the SST: High ocean-to-atmosphere fluxes are found in the tropics where there is warm water and high saturation-vapor pressure, and low fluxes occur

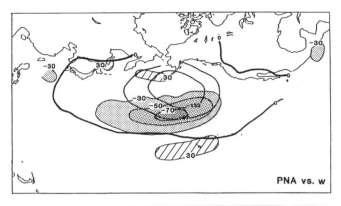

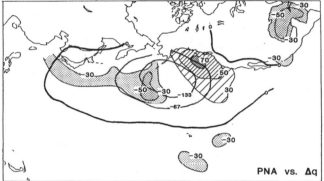

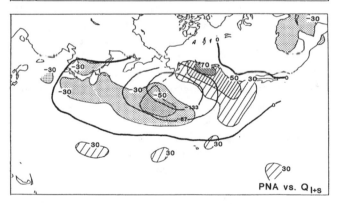

FIGURE 1 The correlation coefficients ($\times$ 100) between time series of the amplitudes of Pacific-North American EOF and our all-grid-point time series of w' (above), $\Delta q'$ (middle), and Q'_{l+s} (below) in the North Pacific for the winter months, 1950-1986. Solid contours show PNA pattern. Significant correlation regions are superimposed upon the EOF patterns; shading indicates regions where the magnitude of the correlation is ≥ 0.3. Significant positive/negative correlations are shown by stippling, hatching in upper two frames; and by hatching, stippling in lower frame.

in high latitudes and along the eastern side of the ocean basins where surface temperatures are cool. Concerning the anomaly relationships, there is evidence for the SST-forcing-flux mechanism in the warm-season extratropics and in the tropics (Cayan, 1990; Liu and Gautier, 1990).

On the other hand, in the extratropics the fluxes appear

to *force* the temperature (Gill and Niiler, 1973; DC3): SST is coolest in late winter following the highest ocean-to-atmosphere fluxes. For the flux-forcing-SST case to be valid, the SST anomaly must play a negligible role in determining the flux, so that the flux is driven by atmospheric conditions—i.e., w, T_a, and q variations dominate the variations of SST. Indirectly, this scenario has been investigated by relating SST anomalies to the large-scale atmospheric circulation (Namias, 1972; Davis, 1976; Wallace et al., 1990). While all of these studies established a definite connection between the monthly or seasonal mean circulation and the SST anomalies, they could only infer the role of anomalous heat flux in forcing SST anomalies. When atmospheric fields are related to contemporaneous monthly SST anomalies, it is difficult to determine whether SST is forcing the atmosphere, both fields are being forced by some other agent, or the atmosphere is forcing the SST.

In testing the phase relationship between anomalies of the estimated flux and SST, the dominant linkage exhibited in the extratropics is that the latent and sensible flux anomalies force the SST anomalies, rather than SST anomalies' forcing the flux anomalies (DC3). Historical time series of flux data provide the basis for directly testing the forcing. Correlations between bulk formulae fluxes and anomalous temperature variations have been documented over the North Pacific by Clark et al. (1974) and Frankignoul and Reynolds (1983), and in the western tropical Pacific by Meyers et al. (1986). As was reported in DC3, the latent and sensible heat-flux anomalies proved to strongly affect monthly changes in SST anomalies over a large portion of the world oceans, in particular the North Atlantic and the North Pacific. A simple thermodynamic model was adopted to relate the flux anomalies to the *tendency* (time rate of change) of the SST anomalies. More discussion of the full mixed-layer thermodynamic equation is given below, and a comprehensive treatment is provided by Frankignoul (1985). Note that the temperature equation predicts the *tendency* of the SST anomaly $\frac{\partial SST'}{\partial t}$, not the anomaly itself; the tendency is represented here by its finite-difference forms, $\frac{\Delta SST'}{\Delta t}$. The flux parameterizations do not contain knowledge of the SST tendency, so the relationship of fluxes to SST-tendency anomalies is an independent test of the influence of the fluxes.

Having a higher-frequency character than the anomaly itself, $\frac{\Delta SST'}{\Delta t}$ has nearly as many independent samples of $\frac{\Delta SST'}{\Delta t}$ as there are months; four decades of records contain approximately 120 independent December-January-February samples.

Spatial Distribution of Flux versus SST Tendency Anomalies

The geographical distribution of the correlations between Q'_{l+s} and $\frac{\Delta SST'}{\Delta t}$ observations for winter months is examined in Figure 2. The 0.3 level is used as a threshold of statistical significance. Meaningful correlations on this map are almost everywhere positive, as they were for zonal averages of the two fields (DC3). Strongest correlations (≥ 0.5) are found mostly between about 25°N and 40°N within the anticyclonic subtropical gyres of the two oceans. For the North Pacific, strongest correlations are east of 180° and extend to the California Current. Though significant in many locations, correlations are weaker in the western North Pacific and from the tropics to 30°S. In the central North Atlantic, the flux and $\frac{\Delta SST'}{\Delta t}$ anomalies have strong correlations in the central subtropical gyre as well as in the high-variance western North Atlantic region. Seasonally, most of the regions have strongest correlations in winter.

Near the equator between 5°N and 5°S in all three basins, Q'_{l+s} and $\frac{\Delta SST'}{\Delta t}$ are not well correlated. Correlations between zonal average flux and SST anomalies (DC3; not shown) indicate that in the tropics, the flux and SST anomalies (not the tendencies) tend to be negatively correlated. This suggests that the flux is driven by SST, presumably because equatorial SST anomalies are governed more by internal ocean processes than by the air-sea heat exchange.

$\frac{\Delta SST'}{\Delta t}$ versus Fluxes during Strong Atmospheric-Circulation Modes

The organization of the latent-plus-sensible flux anomalies (Q'_{l+s}) by the anomalous atmospheric circulation can be used to test the consistency of large-scale links between the flux and SST tendency anomalies $\frac{\Delta SST'}{\Delta t}$. If the SST tendency anomalies are caused by the fluxes, they should have corresponding patterns. In DC3, the organization by the circulation was exploited by using the dominant SLP EOF modes as an index to compare flux-anomaly and corresponding SST-anomaly tendency patterns.

To identify strong winter-circulation months, extreme positive and negative EOF amplitudes were chosen for each SLP EOF (DC2 and DC3). Using this criteria, several (10-30) months of each of the extreme EOF modes (positive and negative amplitudes) were selected. Composites of Q'_{l+s} and $\frac{\Delta SST'}{\Delta t}$ were formed by averaging the fields during the respective extreme months. For brevity, the composites were expressed as the difference between averages of positive (strong) and negative (weak) phase months of the SLP EOFs.

In both northern oceans, the patterns of $\frac{\Delta SST'}{\Delta t}$ are well aligned with the flux anomaly patterns. In the North Pacific, composite $\frac{\Delta SST'}{\Delta t}$ differences and Q'_{l+s} differences corresponding to positive-minus-negative extremes of the PNA pattern are shown in Figure 3. Remarkably, the flux and $\frac{\Delta SST'}{\Delta t}$ signatures of the PNA are marked out more than halfway around the hemisphere. Major centers of $\frac{\Delta SST'}{\Delta t}$ are closely matched to those of Q'_{l+s}, which confirms that

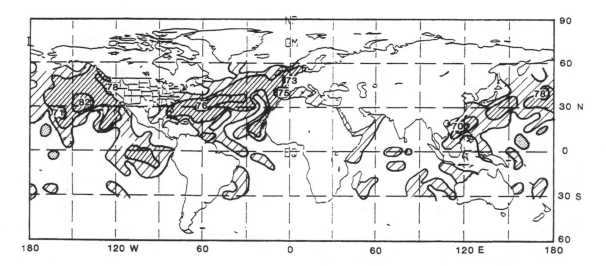

FIGURE 2 Correlation coefficients ($\times$ 100), mapped for global ocean Q'_{l+s} vs. $\frac{\Delta SST'}{\Delta t}$ at each grid point, for the winter months, 1946-1986. Contours at 0.3, 0.5, and 0.7. Light and heavy shadings indicate correlations ≥ 0.3 and ≥ 0.5.

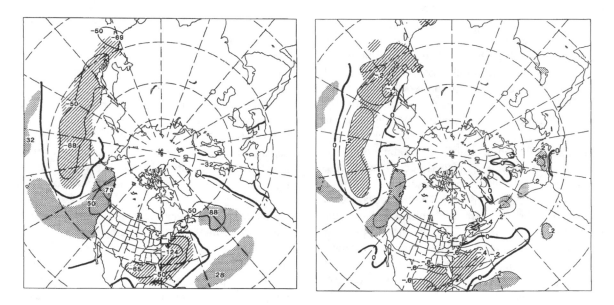

FIGURE 3 Difference of composites of Q'_{l+s} (W m^{-2}; left) and $\frac{\Delta SST'}{\Delta t}$ (°C mo^{-1}; right) associated with positive (deep Aleutian Low phase) vs. negative extremes of EOF 1 of North Pacific SLP. Shading indicates regions where the difference exceeds the 95 percent confidence limit using a two-tailed t-test.

they are causally linked. At the centers, $\frac{\Delta SST'}{\Delta t}$ differences exhibit magnitudes of 0.2°C mo^{-1} to 0.8°C mo^{-1}. Corresponding flux differences range from about 30 to 120 W m^{-2}. Over the North Pacific, the PNA incorporates a swath of positive Q'_{l+s} differences across mid-latitudes (centered at 40°N) from Asia to 140°W, and an arc of negative Q'_{l+s} differences extending from the subtropics at 20°N, 160°E to the eastern North Pacific border, from California to the Gulf of Alaska. Downstream, the strong downstream arm of the PNA produces negative Q'_{l+s} differences in the Gulf of Mexico and the eastern seaboard of the United States,

and positive anomalies in the subtropical North Atlantic (20°N, 50°W) and east of Newfoundland. Throughout the North Pacific and North Atlantic, the PNA-associated $\frac{\Delta SST'}{\Delta t}$ differences mirror the flux differences, having centers co-located with the flux differences. Magnitudes of the corresponding flux and SST anomaly tendencies are consistent with typical observed mixed-layer depths, approximately 100 to 250 m in winter and 10 to 40 m in summer. Similar seasonal behavior and the relation between the flux and SST tendency are replicated in the OGCM, as presented in the following section.

OCEAN MODEL

In this section, we report on an extended run of the OPYC OGCM, which was forced by the wind stress and fluxes derived from the surface marine observations (see Miller et al., 1994a,b). The OPYC model, developed by Oberhuber (1993), consists of eight isopycnal interior ocean layers fully coupled to a surface bulk mixed-layer model, the latter also including a sea-ice model. The interior layers (fixed potential density) and the mixed layer (variable density) have prognostic thicknesses that vary as a function of space and time; the interior layers may have zero thickness, and the mixed layer has a minimum thickness of 5 m. The interior potential densities are specified in such a way as to yield increased vertical resolution (thinner layers) in the thermocline and less in the deep ocean (thick layers). The low horizontal resolution perforce disallows mesoscale eddy variability, which is probably unimportant in the evolution of large-scale SST anomalies. The ocean model solves the full primitive equations for mass, velocity, temperature, and salt for each layer in spherical geometry with a realistic equation of state. The domain is the Pacific Ocean from 70°S to 65°N and 120°E to 60°W. The grid resolution is 77 by 67 points. Resolution is enhanced near the equator and near the eastern and western boundaries. Open-ocean resolution in the middle latitudes is 4°, which is suitable for examining the large-scale variability. The conditions on horizontal solid boundaries are no-slip for velocity and thermally insulating for temperature. The model's Antarctic Circumpolar Current, however, has periodic boundary conditions and unrealistically connects to itself from 60°W to 120°E (half the global circumference). The surface boundary conditions for interior flow are determined by the bulk mixed-layer model, which is forced by the atmosphere. Frictional drag acts between each layer but most strongly along the bottom boundary, which has realistic topography. Horizontal Laplacian friction and diffusion with variable coefficients are also included.

A full discussion of the dynamics is given by Oberhuber (1993), who used the model in Atlantic Ocean modeling studies, and by Miller et al. (1992, designated MOGB below), who used an earlier version of this model for tropical Pacific Ocean circulation studies. Besides the differences in geometry and the forcing functions to be described, the present model differs from MOGB in the following ways: A horizontal finite-differencing scheme based on Bleck and Boudra (1981) that conserves both enstrophy and potential vorticity is implemented in this version of the model. In addition, for a more realistic mixed-layer depth (MLD) in the middle latitudes, the mean turbulent kinetic energy (TKE) input into the mixed-layer equation for entrainment velocity has been altered to provide for month-to-month variability, and is lower than that of the MOGB model. The net result is to reduce the MLD in the middle and high

latitudes of the North Pacific relative to that of MOGB, with the present values being more typical of those observed.

Concerning the OPYC ocean general-circulation model, it is important to point out some potential defects that may obscure our interpretations (Miller et al., 1994a). First, horizontal advection may be underestimated because the model SST climatology is too weak, especially in the Northwest Pacific. Second, the Kuroshio is not well resolved, so the model cannot generate strong western boundary currents. This will impair its ability to advect SST anomalies to nearby open ocean regions. Third, a warm bias in the model winter SST field (approximately 4°C too warm) in the northern North Pacific results in an overly stable stratification that diminishes the effects of entrainment. Last, salinity variations were not included, and these may be important, particularly in high latitudes. Although the OPYC OGCM's rather low resolution has excluded the effects of oceanic mesoscale variability on SST anomaly generation, we expect that such anomalies would have spatial scales too small to provoke important reactions in the large-scale atmospheric fields (Klein and Hua, 1988; Halliwell and Cornillon, 1989; Miller, 1992).

Forcing Functions and Model Runs

Since no ocean model is perfectly realistic, any model forced by observed total heat fluxes (without any feedback) will establish an oceanic temperature climatology that will depart from the real observations. To circumvent this problem, we forced the model with observed *anomalies* of heat fluxes superimposed upon the model climatology rather than with the total observed flux fields. This scheme preserves the model ocean climatology, as follows. We first establish the model climatology by forcing with observed monthly long-term mean wind stresses, TKE input, and total heat fluxes derived from bulk formulae using monthly long-term mean atmospheric observations combined with model ocean temperature. After the oceanic system has reached an acceptably equilibrated state (gauged by the absence of drift in mean SST and MLD), monthly means of pertinent fields are saved for use in the later experiments.

Using the same strategy as that just discussed, the oceanic salinity field is stabilized to climatological average observations. However, the equilibrium salinity is very sensitive to the evaporation-minus-precipitation ($E - P$) field, which is not available at present from observations. Hence, we use Newtonian relaxation to the observed annual mean surface salinity field compiled by Levitus (1982) rather than attempting to specify $E - P$.

Spin-up

During a 35-year-long spin-up period, the model was forced by monthly long-term mean fields of atmospheric wind stress, interpolated from analyses made by the European Centre for Medium-range Weather Forecasting (ECMWF), and total surface heat fluxes (latent, sensible long-wave radiation, and insolation), derived from bulk formulae using model SST and observed atmospheric fields of air temperature (from ECMWF analyses), humidity (COADS), wind speed (ECMWF), and cloudiness (COADS) as inputs. These long-term mean fields, described in Oberhuber (1988), are distinct from the flux and wind-stress anomaly fields, which are formed from individual monthly COADS means as discussed below. Salinity in the surface mixed layer is Newtonially relaxed to Levitus's annual mean observed salinity field so that no evaporation or precipitation fields are needed in the continuity equation.

During the spin-up period, year-to-year changes in SST and mixed-layer depth were monitored to help determine whether the run had developed a reasonable seasonally varying state. Over most of the North Pacific, the SST drift was only a few hundredths of a degree from year 34 to year 35. After the spin-up period was complete, we stored the monthly mean fields of total surface heat flux ($\overline{Q}$) and SST ($\overline{T_s}$) from the final year of spin-up. The field $\overline{Q}$ was used as the mean part of the forcing for all the subsequent runs discussed below. To verify that $\overline{Q}$ was sufficient for mean forcing and to extinguish initial drifts or adjustments in SST, we ran the model for an additional 5 years with $\overline{Q}$ as the surface heat-flux forcing. Since there was little change in mean SST in the North Pacific, the 1970-1988 simulation, outlined below, was commenced from the end of that 5-year period.

Monthly Forcing Fields

Our interdecadal forcing experiments were forced by monthly-mean-plus-anomaly fields of wind stress, total heat flux, and TKE. The wind-stress field is composed of the mean ECMWF analyses, $\overline{\tau}$, plus monthly mean anomalies, τ', derived from the COADS observations. In the extratropics, the COADS wind-stress anomalies are monthly means of the products wu and wv from individual observations. Drag coefficients were taken from Isemer and Hasse (1987); like the heat-flux exchange coefficients, they are weakly dependent upon wind speed and ΔT. However, since the COADS observations are sparse in the low latitudes, we used anomalies of Florida State University (FSU) wind-stress analyses (Goldenberg and O'Brien, 1981) in the $\pm 20°$ latitude zone. In an overlap region, of approximately $10°$ at $20°N$ and $20°S$, the COADS and FSU flux anomaly fields were smoothly merged.

The TKE variations from month to month were estimated using the observed monthly mean wind speed. We invoked the Weibull parameterization of Pavia and O'Brien (1986) for relating the monthly mean of the wind speed cubed, $\{w^3\}$, to monthly mean wind speed, $\{w\}$, of COADS. Choosing a representative value of this parameterization from the North Pacific regions shown by Pavia and O'Brien yields $\{w^3\} \approx 1.9\{w\}^3$. Comparison of 10 years (1970-1979) of $\{w\}^3$ with $\{w^3\}$ estimated from individual COADS observations exhibits fair agreement: $\{w^3\}/\{w\}^3$ ranged from 1.3 to 2.5 over most $2°$ squares of the North Pacific. After computing TKE from the monthly mean cubic wind speed and removing the long-term monthly mean, we added the resulting anomalies to the mean ECMWF TKE fields that had been extracted from the model spin-up period.

For the total heat-flux anomalies, we adopted a similar strategy. We used Q' as determined from the COADS observations (after DC1 and the discussion above) poleward of $20°$ latitude. As described in the Observational Evidence section above, the COADS latent and sensible heat fluxes were formed from monthly averages of products of individual observations. Latent and sensible heat exchange coefficients taken from Isemer and Hasse (1987) are weakly dependent on wind speed and ΔT. In the tropics, ship weather reports are sparse, so we could not easily apply the COADS flux anomalies. However, since there is some evidence that the role of heat fluxes in the tropics is to damp SST anomalies (Liu and Gautier, 1990; Cayan, 1990), we used a Newtonian damping scheme for $20°S$ to $20°N$ (in this zone the bulk-formulae flux anomalies are excluded). The Newtonian scheme was formulated as $Q' = \alpha T'$, where T' is the model SST anomaly. The spatially variable parameter α was determined by analyzing the total heat-flux output of the ECHAM T42 model (Roeckner et al., 1992) run that was forced from 1970-1985 with observed SST anomalies. The parameter field α varies geographically over the domain; it was determined by regressing the observed T' with the ECHAM model's Q' output. Typical values of α vary from -10 W m^{-2} °C^{-1} in the eastern tropical Pacific to near -40 W m^{-2} °C^{-1} in the western Pacific.

A point to be made is that the heat-flux parameterizations do depend on the SST, but they do not "build in" the observed SST anomaly variability in the resultant simulation. Actually, the latent and sensible flux formulations depend on the wind speed and the difference $SST - T_a$, which is not well correlated to the SST anomaly. Also, in the governing thermodynamic equation (see equation 4 in the next section), it is the temperature-anomaly tendency, not the anomaly itself, that responds to the flux forcing. Thus, in the simulation the model temperature is free to drift (possibly away from the observed SST state), governed only by the observed heat flux and wind forcings.

Adjustment of the Marine Weather Data

One of the biggest problems with using marine observations is that they contain artificial long-period fluctuations (Cayan, 1990, 1992a; Michaud and Lin, 1992), which cause spurious low-frequency behavior in the fluxes. Specifically, the marine data contain substantial trends in the most important fundamental variables: wind speed, ΔT, and possibly Δq, even where these data are averaged over the largest spatial scales (DC1). The wind speed trends have been noted by several previous authors (Ramage, 1987; Cardone et al., 1990; Posmentier et al., 1989) and appear to be instrumental artifacts instead of natural variability. Although we were unable to identify a simple instrumental bias responsible for negative trends in ΔT and Δq, we did not trust wholesale decreases in oceanic ΔT and Δq, so we also filtered the ultra-low/large-scale variability from these.

We made two attempts to remove time-varying changes in the fundamental variables involved in the flux calculations. As described in DC1 and above, our first attempt calculated the area-averaged linear trend of w and ΔT, where the area considered was the entire COADS date coverage over the global ocean. Then, the equivalent latent and sensible flux changes were computed for each month of each $5°$ grid point, by adding the linear change in w and ΔT to their respective long-term monthly means, inserting these into the bulk formulae, and computing the fluxes over 1950-1988. By subtracting the long-term monthly mean value from each month of the 1950-1988 time history, a "correction" to the latent and sensible fluxes was obtained to reduce the apparent time-varying bias. This version of the fluxes still left apparently spurious low-frequency changes. Further inspection indicated three problems. First, the time-varying bias for ΔT, which generally decreases over time, was *not* linear, but contained low-frequency undulations over the period since 1950. Second, inspection of Δq, which had not been corrected, revealed similar low-frequency fluctuations. (The wind-speed trend, on the other hand, was quite nearly a linear increase of about 1 m s^{-1} over 1950-1988.) Third, changes in the net outgoing terrestrial flux from ΔT trends were not removed in the original correction. (This was because the initial flux-anomaly study focused on variations in the latent and sensible heat components, not the net flux.)

In our second attempt to adjust the flux (Miller et al., 1994a), we more accurately removed the basin-scale low-frequency variation in (generally decreasing) ΔT, by including a similar adjustment for variation (also generally decreasing) in Δq, and retaining the adjustment for w increases. The flux "correction" was calculated in the same manner as before, by perturbing the long-term mean data entering the bulk formulae with the time-varying adjustments. However, in this case, the adjustments to ΔT, Δq, and w were determined from a basin-scale EOF analysis, rather than from a linear trend. The EOF analysis was performed on smoothed versions (13-month running mean filter applied twice) of three COADS data fields encompassing the Pacific basin from $30°S$ to $60°N$ and $130°E$ to $75°W$. For each of the three smoothed variables, the spatial pattern of the first EOF exhibited one sign over virtually the entire field. The time coefficients for each of the first EOFs con-

tained the time variation that was portrayed by the linear trend (w increasing, ΔT and Δq decreasing). The first EOFs of the smoothed ΔT, Δq, and w fields accounted for 55 percent, 43 percent, and 73 percent of their variability, respectively. The projection of the first EOF back onto the smoothed data constituted the time-varying adjustments that were then applied in computing the flux corrections.

These adjustments were applied to the respective bulk formulae for the latent, sensible, and net terrestrial (infrared) heat fluxes. Since the first EOF of the smoothed wind speed was very nearly linear and showed relatively little spatial variation, we did not readjust the COADS wind-stress anomalies, which were corrected using the original COADS global-average linear trend of 0.8 m s^{-1} over 1950-1988. These second versions of the corrected fluxes were employed in the "R-run" simulation experiments discussed below.

Mechanisms of SST Variability

To help determine the physical processes forcing SST anomalies, we examine the time variability of the anomalies of the terms in the surface mixed-layer heat budget, along with MLD anomalies and SST anomalies. The surface heat budget is governed by

$$\frac{\partial T_s}{\partial t} = \frac{Q}{\rho c_p H} - \boldsymbol{u} \cdot \nabla T_s - \frac{w_e}{H}(T_s - T_o) + \kappa \nabla^2 T_s \quad (4)$$

Here the SST (T_s) tendency is balanced by heat-flux input, horizontal advection, vertical mixing/entrainment, and diffusion. Q is the net surface heat flux, $\boldsymbol{u}$ is the horizontal velocity, w_e is the entrainment velocity, T_o is the temperature of the ocean just underlying the mixed layer (whose depth is H), and κ is the horizontal diffusivity. Monthly means of these quantities are computed during integration, and the climatological monthly means are computed and subtracted from the stored fields to obtain the anomalies.

The mean and standard deviations of the terms contributing to the heat budget (Equation 4) of the mixed layer are illustrated in Figure 4 for the North Pacific basin, the California Coast, and the central North Pacific. In addition to the heat-flux terms, the mean tendency of SST, the mean and standard deviation of SST, and the mean and standard deviation of MLD are shown for each region. All three regions are heated in spring and summer (April through September) and cool in fall, winter, and early spring (October-March), as indicated by $\frac{\Delta SST'}{\Delta t}$. In the climatological mean, the shallowest and deepest MLDs occur in the central North Pacific where the wind forcing undergoes a large annual cycle. This heating and cooling is mirrored by the shoaling and deepening of the mixed layer, which ranges between about 30 m and 150 m. SST and MLD exhibit anomalous year-to-year variability that is a significant fraction of the long-term mean, with typical standard deviations

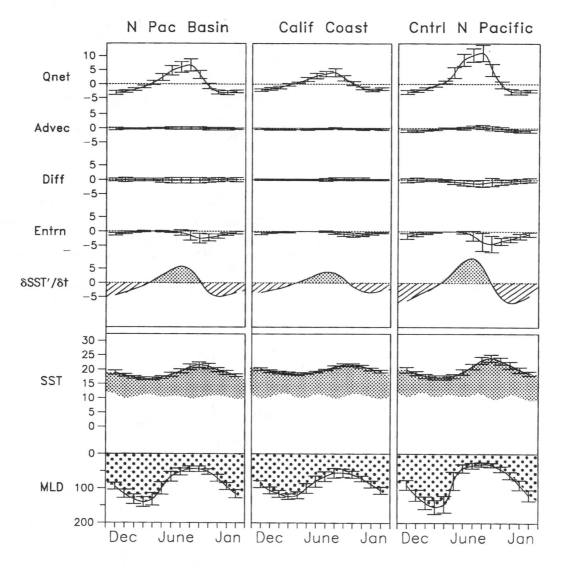

FIGURE 4 Long-term mean and standard deviation of model heat budget terms $\frac{\Delta SST'}{\Delta t}$, SST (°C), and mixed layer depth (m) for the North Pacific basin, the California Coast, and the central North Pacific regions. Units are 0.1°C mo^{-1} for heat-budget terms and $\frac{\Delta SST'}{\Delta t}$.

of 1°C and 20 m, respectively. Of the forcing terms in the heat balance, the heat flux has the largest mean and standard deviation (which is in line with previous studies, e.g., Frankignoul, 1985; Haney, 1985; Luksch and von Storch, 1992), followed by the entrainment. The heat flux is in phase with the climatological mean SST tendency, swinging from positive in spring-summer to negative during fall-winter.

Entrainment acts only in one direction: to deepen the mixed layer and to cool the temperature. Entrainment exhibits its maximum effect in fall. Presumably this is because the combination of strong winds and a sharp gradient at the bottom of the mixed layer has its greatest potential in fall. Advection, although a small component, acts in phase with the net heat flux in its mean behavior. Diffusion, which acts to dissipate temperature extremes, is out of phase with

the SST and about 90° phase-shifted (but usually of opposite sign) from the heating terms.

The spatial pattern of the primary linkages between heat-budget components and $\frac{\Delta SST'}{\Delta t}$ are shown by the maps of the local correlations over the Pacific basin in Figures 5a through 5c, which appear in the color well. Note that $\frac{\Delta SST'}{\Delta t}$ is approximated here by the centered difference between the SST means for the months before and after the month considered. The correlations, calculated separately for heat flux, entrainment, and advection, were obtained for the set of all months of 1970 through 1988. Correlations between heat flux and $\frac{\Delta SST'}{\Delta t}$ (Figure 5a) are relatively high (> 0.6) at nearly all grid points north of 20°N. There is a strong similarity of the model correlations to those derived from observations (Figure 2), with the strongest correlations (exceeding 0.8) in the lower middle latitudes of the eastern North Pacific and weakest correlations to the northwest, east of Kamchatka. The heat-flux correlations are small in the subtropics, but this is where we prescribed the fluxes to be in phase with the SST via the Newtonian damping scheme. In the tropical Pacific, the correlations are negative and moderately strong, indicating that ocean dynamics processes play a significant role in the model's SST anomaly variability. (Note, however, that the tropical heat flux vs. $\frac{\Delta SST'}{\Delta t}$ relationship may be artificial because the fluxes are prescribed using the Newtonian damping scheme.) Interestingly, in data-available regions of the Southern Hemisphere, the correlations again are relatively strong and positive, implying that heat flux is a substantial component of the model's anomalous temperature budget. For entrainment vs. $\frac{\Delta SST'}{\Delta t}$ (Figure 5c), fairly high (≈ 0.5) correlations occur at scattered locations poleward of about $10°$, and right along the equator. Concerning advection (Fig-

ure 5b), greatest correlations (> 0.5) with $\frac{\Delta SST'}{\Delta t}$ are found in the subtropics and tropics, with strongest centers developed south of the equator between $5°S$ and $25°S$ along South America and between $15°S$ and $15°N$ north of Australia in the western tropical Pacific. In the North Pacific extratropics, the correlations are generally not very strong, but there is a mild positive correlation region (≈ 0.4) south of the Aleutian Islands, extending into the Gulf of Alaska.

SST Response: Model vs. Observations

As a first look at the skill of the simulation, we examine the time series of area averages of model and observed SST anomalies (Figure 6). Both model and observed anomalies are defined with respect to their monthly mean climatology for the 1970-1988 time interval. We focus on two key regions plus a North Pacific basin average. One key region is the California Coastal region, bounded by $135°$ and

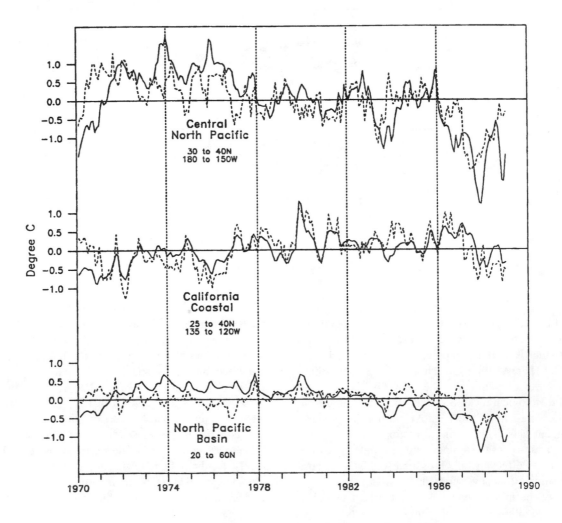

FIGURE 6 Regional averages of simulated (solid) and observed (dashed) SST anomalies. Correlation coefficients between model and observed are 0.44, 0.71, and 0.67 for the basin-wide, coastal, and central North Pacific regions, respectively.

120°W and by 25° and 45°N, where substantial warming occurred during the 1976-1977 climate shift. The other is in the central North Pacific region, where cooling occurred during the shift, and is bounded by 180° and 150°W and by 30° and 40°N. The basin average was taken in a region corresponding to 130° E to 110°W and 20° to 60°N.

The amplitudes of simulated SST anomalies are slightly lower than the observed anomalies. However, short-term variations in SST, particularly for the coastal region, are remarkably similar considering the uncertainty in the observed heat-flux, TKE, and wind-stress forcing fields. The correlation coefficients between model and observed anomalies are 0.44, 0.71, and 0.67 for the basin-wide, coastal, and mid-Pacific regions, respectively.

Like the observations, the model SST contains variability on monthly to decadal time scales. Reasonable agreement appears in the two key regions, where both major breaks and low-frequency trends are similar. In particular, there is a striking agreement in the longer time-scale variability, with cooling in the central North Pacific and warming along the Coastal region (Miller et al., 1994a). On the other hand, over the entire North Pacific, the model exhibits a long-term variation in SST, which is not evident in the observations (a residual effect of the long-term variation of heat flux, as described earlier).

The shorter-period model SST variability in Figure 6 has many features in common with the observed SST variability, particularly for the Coastal region. Correlation coefficients between the model and observed SST time series in the central North Pacific and California Coastal regions are 0.67 and 0.71, respectively. In interpreting these, note that the SST contains variability on all time scales, and the correlation is biased toward representing those with the greatest variance. The 1976-1977 warming observed in the Coastal region is clearly evident in the model's response, as is the 1976-1977 cooling of the Mid-Pacific region.

Finally, the spatial pattern of local skill appears in Figure 7 (see color well), which shows the correlation between modeled and observed SST anomalies (all 228 months, 1970-1988). Confirming the validity and impact of the heat flux and wind stress calculated with the bulk formulae, correlations are relatively high (and positive) northward of 20°N, as well as over a broad swath along the tropical Pacific from 165°E to South America. Many grid points exhibit correlations in excess of 0.7. Interestingly, the highest correlations are found in the same area as the largest correlations betweenthe observed latent and sensible fluxes and $\frac{\Delta SST'}{\Delta t}$ in Figure 2. Skill is also poor in areas along the western and northern boundaries and just north of the equator near Central America. Although there is a small region of fairly high skill at 40° to 50°S between 120° and 90°W, the model generally performs poorly in the South Pacific extratropics, where data is scarce. Along the immediate

tropical strip (5°N to 5°S), the high skill presumably arises primarily from wind-forced ocean dynamics instead of sea-air fluxes. The model skill is low in the region 10° to 20°N of the North Pacific, where the flux was specified using a Newtonian damping instead of from COADS observations. This suggests that the idealized Newtonian flux scheme is not realistic, and supports the validity of the bulk-formulae fluxes.

For better understanding of the forcing of the anomalous heat budget, $\frac{\Delta SST'}{\Delta t}$ was correlated with monthly anomalies of each of the terms on the right-hand side of Equation 4, with the sum of these four terms, and with the MLD anomaly. Results are shown in Figure 8 for the averages over the North Pacific basin, the California Coastal region, and the central North Pacific region. The relationships in the three regions are quite similar. Consistent with the maps shown in Figure 5, the strongest influence on the mixed-layer temperature anomaly is the net heat-flux anomaly, with correlations usually exceeding 0.7. The next most important term is entrainment, which is important when the mixed layer begins to deepen, and the heat accumulated in summer is vented to the atmosphere. Correlations of $\frac{\Delta SST'}{\Delta t}$ and entrainment are greatest in fall and early winter, some exceeding 0.8. The correlations with entrainment are lowest in spring when mixed-layer depth is decreasing, and remain low until fall when it begins to deepen again. Advection is not as strongly correlated with $\frac{\Delta SST'}{\Delta t}$ as the heat flux and the entrainment are, but advection does exhibit correlations exceeding 0.5 in winter for the central North Pacific and

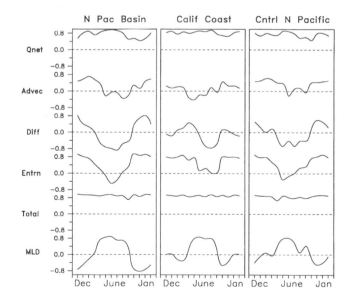

FIGURE 8 Correlation between $\frac{\Delta SST'}{\Delta t}$ and: each of the anomalous model heat-budget forcing terms (right-hand side of Equation 4); their total; and anomalous MLD for the North Pacific basin, California Coastal, and central North Pacific regions. Stratified by month, from 1970-1988.

North Pacific basin regions during winter. Model advection is generally not an important influence on SST anomalies in the Coastal region. There is an interesting seasonal cycle in the link between MLD and $\frac{\Delta SST'}{\Delta t}$, which has its strongest expression for the North Pacific basin average. In fall-early winter (October-January), the two are out of phase: The SST anomaly tends to decrease ($\frac{\Delta SST'}{\Delta t}$ is negative) when the mixed layer is anomalously deep, and vice versa. This implies that when large heat losses and strong wind mixing produce decreasing winter SST anomalies, they also produce a deeper MLD. In summer, the sign of the relationship reverses: The SST anomaly tends to increase when the mixed layer is anomalously deep. This suggests that in summer, when MLD is shallow, the mechanisms that extract heat from the mixed layers produce decreasing SST anomalies when the layer is thinner and the gradient below the mixed layer is stronger (and vice versa).

Analogous correlations (not shown) were also computed for the Niño 4, Niño 3, and Niño 2 regions along the equator from the international date line to South America. The largest contributions to model SST-anomaly variability in these equatorial regions are entrainment, advection, and diffusion. Strong correlations linking equatorial Pacific SST anomalies to advection and entrainment are indicated by the maps in Figures 5b and 5c. In the tropics, SST anomaly tendencies are generally in phase with MLD anomalies (warming coincides with deeper mixed layers), but not strongly so.

The 1976-1977 SST Shift

To evaluate the model's performance in simulating the spatial pattern of the 1976-1977 shift, the wintertime SST difference field for the six years after the shift (December 1976-February 1977 through December 1981-February 1982) minus the six years before the shift (December 1970-February 1971 through December 1975-February 1976) was calculated, following Graham (1991). Figure 9 (see color well) shows these fields, with the observed SST differences for comparison. The model captures the principal extratropical observations, namely, a warming in the Coastal region and a cooling in the Mid-Pacific region. In the tropics and subtropics, observations contain a swath of warm water throughout the equatorial region and across the southeastern North Pacific. The model captures only a vestige of this warming, which appears mostly in the western half of the tropics. This discrepancy may relate to the use of Newtonian damping rather than observed fluxes in the ±20° zone. The warming in the equatorial Pacific points to possible dynamic effects due to wind-stress anomalies alone. Mechanisms for the SST shift are suggested by the comparable-difference maps of the winter fields of SLP, net heat flux, and pseudo-rate of kinetic energy transfer by the wind shown in Figures

10a, 10b, and 10c. These maps clearly show a deepened winter Aleutian Low during the post-1976 period (see also Venrick et al., 1987; Trenberth, 1990; Trenberth and Hurrell, 1995, in this volume). The deepened low sustained stronger winds and greater storm activity across the central North Pacific, and generally warmer, moister air masses along the West Coast. This pattern represents a tendency for strong positive PNA patterns in the post-1976 period. These conditions resulted in a large region of increased heat loss in the western and central North Pacific and decreased heat loss in the eastern North Pacific. Increased wind mixing apparently occurred in the central North Pacific, particularly between 30° and 40°N along the southern fringe of the anomalous low-pressure center. Both the heat-flux and wind-mixing patterns are consistent with the SST difference field, shown in Figure 9. Also, the heat-flux difference map is quite similar to the latent-and-sensible heat-flux anomaly maps associated with months having large PNA EOF amplitudes, which appear in Figures 1 and 3.

Inspection of the seasonal time series of the heat-budget components at the individual regions helps to understand

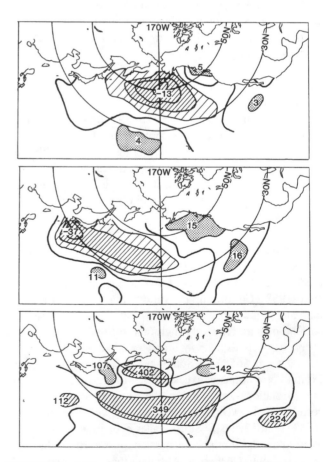

FIGURE 10   Difference fields of observed winter SLP (mb; top), heat flux (W m^{-2}; middle), and {w^3}, the pseudo-total kinetic energy (m^3 s^{-3}; bottom), for the winters of 1976-77 through 1981-82 minus those of 1970-71 through 1975-76.

the evolution of variability features such as the 1976-1977 shift (Miller et al., 1994a,b). In the Coastal region, heat-flux input tends to be about four times larger than either horizontal advection or entrainment effects. In the mid-Pacific region, in contrast, which is nearer the more variable winds of the storm track, heat-flux input is only twice as large as horizontal advection, and typically the same size as entrainment effects. Diffusion tends to be slightly weaker than horizontal advection, but inspection of the time series shows that diffusion simply acts in opposition to the cumulative effects of the other heating term.

During the six-month period preceding the 1976-1977 climate shift, a long period of warming via heat-flux input occurred in the Coastal region (not shown). Although this period was not particularly strong, it was persistent compared to most of the rest of the time intervals (Miller et al., 1994a). What makes this period particularly striking, however, is that the strongest occurrence (in the 1970-1988 time interval) of warming by horizontal advection took place during fall 1976 and the subsequent winter. Together, these two effects resulted in a mixed layer 10 to 15 m shallower and more than half a degree of warming in the surface temperature. In spite of a strong loss of heat during the late winter of 1977, the system managed to remain in a shallow mixed-layer, warm-SST state through the winter of 1988. Throughout 1977, several months of strong heat-flux cooling began to cool the upper ocean toward the mean, but strong heating in summer 1979 thinned the mixed layer and warmed the SST to such a degree that the mixed-layer warmth persisted into winter 1980.

In the Mid-Pacific region, the 1976-1977 climate shift is likewise instigated by the effects of an anomalously long and strong period of cooling by horizontal advection combined with sizable cooling by heat flux (Figure 11). During the six-year period preceding the shift, anomalously weak entrainment effects (partially due to lower TKE input) serve to maintain the system in a shallow mixed-layer, warm-SST state. In the years following the shift, the effects of stronger entrainment are confined to the summer months.

It therefore appears that the 1976-1977 shift in both the Coastal and Mid-Pacific regions was caused by an unusual atmospheric state that took hold several months before the 1976-1977 winter, and by conditions in the winter atmospheric circulation that persisted for several winters thereafter. This atmospheric circulation produced large-scale shifts in ocean-current advection that acted in concert with large-scale heat-transfer processes to significantly alter the upper-ocean thermal structure, and thus its stratification. The Mid-Pacific region subsequently remained in that perturbed state through the maintenance effects of reduced TKE input.

SUMMARY AND DISCUSSION

Two experimental approaches indicate that bulk parameterizations yield fairly realistic estimates of anomalous sea-

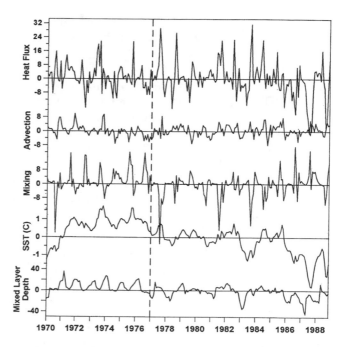

FIGURE 11 Monthly anomaly time series (1970-1988) of simulated surface heat-budget components (net heat-flux term, horizontal advection term, and entrainment term), along with simulated mixed-layer depth (m) and SST (°C), for the central North Pacific. Dashed vertical line marks January 1977, the approximate time at which the large North Pacific SST shift occurred.

air fluxes on seasonal-to-interannual scales over the tropical and northern extratropical oceans. Monthly parameterizations of wind stress and heat flux derived from COADS historical summaries provide reasonable coverage over much of the Pacific basin back to the 1960s. The validity of these derived variables was demonstrated in two ways: (1) by the consistency of the spatial and temporal variability of heat-flux anomalies with independently observed data on the atmospheric circulation and SST tendency; and (2) by the success in simulating two decades of Pacific Ocean SST anomalies with a model driven by the parameterized wind and heat-flux forcings. The bulk parameterizations appear to be adequate for useful diagnoses of several unobserved thermal and dynamic processes (advection, vertical mixing, etc.) that operate in the upper ocean on seasonal-to-decadal time scales.

The model's air-sea heat fluxes were shown to be consistent with two sets of independently observed data. First, the latent and sensible flux anomalies are spatially organized by the atmospheric circulation into systematic patterns about the major cyclonic and anticyclonic features. This organization occurs because the fluxes are determined by wind speed and by air-mass temperature and humidity, which have characteristic structures relating to the general circulation. The flux anomaly/circulation patterns that appear in the North Pacific are confirmed by very similar patterns found in the North Atlantic (not shown, but see DC1 and DC2). Second, the modeled latent and sensible heat-flux anomaly patterns correspond well to those of SST anomaly tendencies, with regions of increased sea-to-air heat exchange overlying regions of decreasing SST, and vice versa. Monthly SST anomaly tendencies have typical magnitudes of $0.3°C$ mo^{-1}, and correspond to monthly mean flux anomalies having typical magnitudes of 50 W m^{-2}. Together, the magnitudes of the heat flux and SST tendency are consistent with observed mixed-layer depths.

We also tested the ability of monthly heat fluxes and winds to force a Pacific Ocean general-circulation model over an extended period (1970-1988). The forcings were monthly means of the bulk-formula parameterizations; in this experiment, we superimposed the monthly *anomalies* of the fluxes and wind parameterizations upon the climatological forcing that was used to spin up the model to its "stable" annual cycle. In general, the simulation agreed well with observed SST where weather reports (wind, temperature, and humidity) were densely sampled in the extratropical region north of 20°N, but compared quite poorly where they were sparsely sampled. Model SST-anomaly performance was best in the eastern North Pacific between 20°N and 40°N, and along the equatorial strip. It performed poorly between 5°N and 20°N, in the western tropical Pacific, and in most of the southern Pacific. Miller et al. (1994a) showed that the model was remarkably successful in simulating the first three empirical modes of the winter North Pacific SST anomaly.

In the model, anomalous heat fluxes and wind-driven entrainment provide the primary extratropical forcing of the upper-ocean thermal field. The model clearly shows that in the extratropics, the mixed layer deepens with winter anomalous cooling, and shoals with anomalous warming. SST in the region from 5°N to 20°N was poorly simulated,

probably because of poor representation of the heat-flux anomalies by a Newtonian scheme in the ±20° strip. Near the equator, however, where SST anomalies reflect ocean dynamic response to wind forcing, the model anomalies closely resembled those observed.

At lower temporal frequencies, the model captures some important elements of the interannual-to-decadal variability, including aspects of the marked Pacific-basin "shift" that occurred in the winter of 1976-77. Multiple causes appear to be at work in creating the shift, including horizontal advection, anomalous heat fluxes, and wind stirring. This result suggests that a major portion of the shift was forced locally (albeit organized over a large scale) by the atmosphere, rather than by an internal adjustment of the ocean's thermal field. The latter mechanism deserves closer attention, since it is a prime candidate for driving the North Pacific decadal variability exhibited in recent coupled ocean/atmosphere model results of Latif and Barnett (1994).

These results suggest that ocean models forced by parameterized monthly surface observations are capable of generating important aspects of the anomalous SST variability, on monthly to multi-year time scales. Evidently the lack of synoptic weather variability that characterizes real winds and fluxes is not fatal to the model results. Note, however, that synoptic effects are aggregated into the monthly averages of the fluxes and wind stress, as the nonlinear products within the bulk formulae are taken on an observation-by-observation basis. A more sensitive issue in the simulation of decadal and longer ocean histories is the treatment of the lowest frequencies of the marine data. In the model run conducted here, we removed the largest-spatial-scale, low-frequency (trend-like) behaviors in the wind speed, ΔT, and Δq, since these seemed likely to have arisen from instrumental artifacts. This issue is problematic, though, and longer model runs will be needed to further test these assumptions.

Commentary on the Paper of Cayan et al.

STEPHEN E. ZEBIAK
Lamont-Doherty Earth Observatory

I think Dr. Cayan has given us the opportunity to consider a lot of fairly serious questions, particularly in relation to decadal time scales and ocean modeling. I have noted a few questions that came to my mind. Some of them are really quite general questions, but perhaps they can focus our discussion.

My first two points really deal with the whole issue of using surface heat fluxes for ocean modeling purposes. Dr. Cayan talked about the issue of biases, and I think that

when we come to consider decadal time scales, this is an overwhelmingly big problem. The technique that he used was to take components of the flux fields, subject them to EOF analyses, and to look at the first mode and essentially take it out.

I wonder whether we can come to some agreement as to whether that is the best way to proceed. For instance, some of the other talks, such as that of Dr. Wallace yesterday, viewed some of the first empirical orthogonal factors

(EOFs) as the real signal for decadal variability. Dr. Cayan's paper uses the same technique to remove a signal with similar time scales. The products of these terms, as he pointed out, are perhaps a little questionable.

In view of this and of the fact that, as Dr. Cayan noted, the ocean accumulates the errors on long time scales, does it really make sense to try to force ocean models with heat fluxes estimated this way? We do have alternatives. It may be that the traditional place to break the atmosphere and the ocean—the interface—is actually the worst place to make the break. Perhaps it makes more sense, when running an ocean model, to include some sort of planetary boundary-layer model. In that case, the break is made at the top of the boundary layer, which also has problems, but the strength of the coupling at that interface may be less sensitive than that at the ocean surface. Similarly, atmospheric models could be run with ocean mixed layers rather than breaking at the interface, which has often been done.

From Dr. Cayan's results and many others, there is a lot of evidence that the atmosphere forces the ocean on seasonal-to-interannual time scales in middle latitudes, and for the most part it goes the other way around in the tropics. But can we really say that those same relations hold at lower frequency? Our intuition tells us that as we go to a lower and lower frequency, the ocean may be able to contribute more in terms of its dynamics. And in view of the problems with estimating fluxes and so forth, I wonder whether we can answer that question just by looking at surface variables. Perhaps we have to have more information about what goes on beneath the surface of the ocean.

Dr. Cayan's paper points out that the ocean model uses a technique of relaxing the salinity field to the Levitus climatology. It seems to me that the technique introduces a process that is really unphysical, as well as a time scale that is not in the real problem. The alternative is to do other kinds of flux corrections; they also have problems, of course, which I shall deal with later.

Take the issue of climatologies, for example. While Dr. Cayan's paper does not actually show the climatology, I believe it mentions that some of the model results differed considerably from the observations, at least in terms of sea surface temperature.

So how do we deal with mismatches in climatology and things like that that really plague us? One alternative is to improve the model; we would all like to do that, but it is not easy and it takes a long time. The second approach is to ignore them. That has been done, but it is probably not the best choice either. Other methods include relaxation and flux-correction techniques. The flux correction taken to its extreme is essentially making an anomaly model. Of the options, I would propose that until such time as we can improve the models to get the right climatology, thinking in terms of an anomaly model might be the most fruitful approach.

My last point concerns the fact that in his simulations Dr. Cayan got very different-looking EOF structures of simulated SST when he looked at just winter seasons and when he looked at all months, whereas in the observations, the EOF structures looked quite similar. I should like to know what is going on in the model in the other seasons that is different from nature.

Discussion

CAYAN: The time variability I removed is, I think, a wave-like, dynamic phenomenon, not the sort Mike Wallace mentioned. As to whether it makes sense to force ocean models with estimated fluxes, it seems to me that it is accounting for a useful portion of the variability. But I think this model data set can yield some revealing analysis, and Art Miller and I plan to redesign our experiments to incorporate some of what we've been talking about.

LINDZEN: It's always seemed to me that the value of an EOF analysis could be gauged by how many EOFs you need to account for a given amount of variance compared with some objective criterion such as spherical harmonics. If it's, say, one-twentieth as many, EOFs are clearly meaningful.

My other question is on Steve Zebiak's comment on what drives what in which latitudes. Why do you think that the ocean would play a more important role if you extrapolated to longer time scales?

ZEBIAK: I don't think we can completely exclude the possibility that things like gyre circulation could affect SST on those time scales.

BRYAN: I think the ocean's effect is least in the very short and very long time scales, since you reach climatic equilibrium. Its effect should be greatest on the intermediate scales, which of course is what we're concerned with at this workshop.

SARACHIK: Well, it seems to me that in our models the SST determines the mean climate, and so would have the largest effect on long time scales.

LINDZEN: If you take the model to T equals infinity, I think SST would become independent of the ocean.

ROOTH: I want to comment further on Steve's second point, which I think raises the question of what model tests like this are for. The success of an experiment in a simple situation may say that the flux data are reasonably good, but it doesn't say much about the mechanics of the responses for more complicated processes.

SHUKLA: Dan, I'd like to know whether you see the potential for a good mixed-layer model, with some dynamics, that would give us some idea of SST evolution on seasonal time scales.

CAYAN: To do that you'd need to be able to predict the fluxes on fairly high-frequency time scales, not to mention a number of other processes. These models are handicapped, just like the statistical models, by our inability to predict the fluxes.

Atmospheric Observations Reference List

Afanas'eva, V.B., N.P. Esakova, and R.V. Klimentova. 1979. Relation of the planetary upper-air frontal zone to the position of the snow limit during fall and spring. Meteor. Gidrol. 9:110-112.

Agnew, T. 1991. Simultaneous winter sea ice and atmospheric circulation anomaly patterns. Preprints of the Fifth AMS Conference on Climate Variations, Denver, Colo. American Meteorological Society, Boston, Mass., pp. 362-365.

Alexander, M.A. 1990. Simulation of the response of the North Pacific Ocean to the anomalous atmospheric circulation associated with El Niño. Climate Dynamics 5:53-65.

Allan, R.J., and M.R. Haylock. 1993. Circulation features associated with the winter rainfall decrease over southwestern Australia. J. Climate 6:1356-1367.

Allan, R.J., N. Nicholls, P.D. Jones, and I.J. Butterworth. 1991. A further extension of the Tahiti-Darwin SOI, early ENSO events and Darwin pressure. J. Climate 4:743-749.

Alley, R.B., D.A. Meese, C.A. Shuman, A.J. Gow, K.C. Taylor, P.M. Grootes, J.W.C. White, M. Ram, E.D. Waddington, P.A. Mayewski, and G.A. Zielinski. 1993. Abrupt accumulation increase at the Younger Dryas termination in the GISP2 ice core. Nature 362:527-529.

Angell, J.K. 1986. Annual and seasonal global temperature changes in the troposphere and low stratosphere, 1960-85. Mon. Weather Rev. 114:1922-1930.

Arakawa, H. 1957. Climatic change as revealed by the data from the Far East. Weather 12:46-51.

Armstrong, R.L., and M. Hardman. 1991. Monitoring global snow cover. Proceedings of the IEEE International Geoscience and Remote Sensing Symposium (IGARSS '91). IGARSS Volume 4, Institute of Electrical and Electronics Engineers, New York, pp. 1947-1949.

Barnett, T.P. 1984. Long-term trends in surface temperature over the oceans. Mon. Weather Rev. 112:303-312.

Barnett, T.P. 1986. Detection of changes in the global tropospheric temperature field induced by greenhouse gases. J. Geophys. Res. 91:6659-6667.

Barnett, T.P. 1989. A solar-ocean relation: Fact or fiction? Geophys. Res. Lett. 16:803-806.

Barnett, T.P., and P.D. Jones. 1992. Intercomparison of two different Southern Hemisphere sea level pressure data sets. J. Climate 5:92-99.

Barnett, T.P., and M.E. Schlesinger. 1987. Detecting changes in global climate induced by greenhouse gases. J. Geophys. Res. 92:14772-14780.

Barnett, T.P., L. Dümenil, U. Schlese, E. Roeckner, and M. Latif. 1989. The effect of Eurasian snow cover on regional and global climate variations. J. Atmos. Sci. 46:661-685.

Barnett, T.P., M.E. Schlesinger, and X. Jiang. 1991. On greenhouse gas signal detection strategies. In Greenhouse-Gas-Induced Climatic Change: A critical appraisal of simulations and observations. M.E. Schlesinger (ed.). Elsevier, Amsterdam, pp. 537-558.

Barnston, A.G., and R.E. Livezey. 1987. Classification, seasonality, and persistence of low-frequency atmospheric circulation patterns. Mon. Weather Rev. 115:1083-1126.

Barry, R.G. 1983. Arctic Ocean ice and climate: Perspectives on a century of polar research. Ann. Assoc. Am. Geogr. 73:485-501.

Barry, R.G., and R.L. Armstrong. 1987. Snow cover data management: The role of WDC-A for glaciology. Hydrol. Sci. J. 32:281-295.

Berger, A., J.L. Mélice, and I. van der Mersch. 1990. Evolutive spectral analysis of sunspot data over the past 300 years. Phil. Trans. Roy. Soc. London A 330:529-541.

Bevington, P.R. 1969. Data Reduction and Error Analysis for the Physical Sciences. McGraw-Hill, New York, 336 pp.

Bhanu Kumar, O.S.R.U. 1987. Seasonal variation of Eurasian snow cover and its impact on the Indian summer monsoon. In Large Scale Effects of Seasonal Snow Cover. B.E. Goodison, R.G. Barry, and J. Dozier (eds.). IAHS Press, Wallingford, U.K., pp. 51-60.

Bhanu Kumar, O.S.R.U. 1988. Interaction between Eurasian winter snow cover and location of the ridge at the 500 hPa level along 75°E. J. Meteorol. Soc. Japan 66:509-514.

Bjerknes, J. 1959. The recent warming of the North Atlantic. In the Atmosphere and the Sea in Motion, the Rossby Memorial Volume. B. Bolin (ed.). Rockefeller Institute Press in association with Oxford University Press, New York, pp. 65-73.

Bjerknes, J. 1961. Climatic change as an ocean-atmosphere problem. In Proceedings of the Rome Symposium, organized by UNESCO and the World Meteorological Organization, 1961. UNESCO, New York, pp. 297-321.

Bjerknes, J. 1962. Synoptic survey of the interaction of the sea and atmosphere in the North Atlantic. Vilhelm Bjerknes Centenary volume, Geofys. Publikas. XXIV, No. 3, pp. 115-145.

Bjerknes, J. 1964. Atlantic air-sea interaction. Adv. Geophys. 10:1-82.

Black, R.F. 1954. Precipitation at Barrow, Alaska, greater than recorded. AGU Trans. 35(2):203-206.

Blanc, T.V. 1985. Variation of bulk-derived surface flux, stability, and roughness results due to the use of different transfer coefficient schemes. J. Phys. Oceanogr. 15:650-669.

Blanford, H.F. 1884. On the connection of the Himalayan snowfall with dry winds and seasons of drought in India. Proc. Royal Soc. London 37:3-32.

Bleck, R., and D.B. Boudra. 1981. Initial testing of a numerical ocean circulation model using a hybrid coordinate (quasi-isopycnic) vertical coordinate. J. Phys. Oceanogr. 11:755-770.

Bloomfield, P. 1992. Trends in global temperature. Climatic Change 21:1-16.

Bogdanova, E.G. 1966. Investigation of precipitation measurement losses due to the wind. Trans. Main Geophys. Observ. (Leningrad) 195:40-62 (in Russian).

Bogdanova, E.G. 1977. Annual totals and seasonal variation of solid, mixed, and liquid precipitation in mountains of South America. Mat. Glyatsiol. Issled. Khron. Obsuzhd. 30:130-135 (in Russian).

Bolin, B., B.R. Döös, J. Jaeger, and R.A. Warrick (eds.). 1986. The Greenhouse Effect, Climatic Change and Ecosystems: A Synthesis of Present Knowledge. John Wiley and Sons, New York, 541 pp.

Boninsegna, J. A. 1992. South American dendroclimatological records. In Climate Since A.D. 1500. R.S. Bradley and P.D. Jones (eds.). Routledge, London, pp. 446-462.

Bottomley, M., C.K. Folland, J. Hsiung, R.E. Newell, and D.E. Parker. 1990. Global Ocean Surface Temperature Atlas (GOSTA). Joint Project of the Meteorological Office, Bracknell, U.K., and the Massachusetts Institute of Technology. HMSO, London, 20+iv pp. and 313 plates.

Bradley, R.S., and P.D. Jones (eds.). 1992. Climate Since A.D. 1500. Routledge, London, 680 pp.

Bradley, R.S., H.F. Diaz, J.K. Eischeid, P.D. Jones, P.M. Kelly, and C.M. Goodess. 1987. Precipitation fluctuations over northern hemisphere land areas since the mid-19th century. Science 237:171-275.

Briffa, K.R., P.D. Jones, T.S. Bartholin, D. Eckstein, F.H. Schweingruber, W. Karlén, P. Zetterberg, and M. Eronen. 1992a. Fennoscandian summers from A.D. 500: Temperature changes on short and long time scales. Climate Dynamics 7:111-119.

Briffa, K.R., P.D. Jones, and F. Schweingruber. 1992b. Tree-ring density reconstructions of summer temperature patterns across western North America since 1600. J. Climate 5:735-754.

Broecker, W.S., D.M. Peteet, and D. Rind. 1986. Does the ocean-atmosphere system have more than one stable mode of operation? Nature 315:21-26.

Bryan, F. 1986. High-latitude salinity effects and interhemispheric thermohaline circulations. Nature 323:301-304.

Bryan, K., and R. Stouffer. 1991. A note on Bjerknes' hypothesis for North Atlantic variability. J. Mar. Systs. 1:229-241.

Bücher, A., and J. Dessens. 1991. Secular trend of surface temperature at an elevated observatory in the Pyrenees. J. Climate 4:859-868.

Budyko, M.I., and Yu.A. Izrael (eds.). 1987. Anthropogenic climatic changes. Gidrometeoizdat, Leningrad, 406 pp.

Bunting, A.H., M.D. Dennett, J. Elston, and J.R. Milford. 1976. Rainfall trends in the West African Sahel. Quart. J. Roy. Meteorol. Soc. 102:59-64.

Cane, M.A., S.E. Zebiak, and Y. Xue. 1995. Model studies of the long-term behavior of ENSO. In Natural Climate Variability on Decade-to-Century Time Scales. D.G. Martinson, K. Bryan, M. Ghil, M.M. Hall, T.R. Karl, E.S. Sarachik, S. Sorooshian, and L.D. Talley (eds.). National Academy Press, Washington, D.C.

Cao, H.X., J.F.B. Mitchell, and J.R. Lavery. 1992. Simulated diurnal range and variability of surface temperature in a global climate model for present and doubled CO_2 climates. J. Climate 5:920-923.

Cardone, V.J., J.G. Greenwood, and M.A. Cane. 1990. On trends in historical marine wind data. J. Climate 3:113-127.

Cayan, D.R. 1990. Variability of Latent and Sensible Heat Fluxes over the Oceans. Ph.D. dissertation, University of California, San Diego, 220 pp.

Cayan, D.R. 1991. Variability of latent and sensible heat fluxes estimated using bulk formulae. Atmos.-Ocean 30:1-42.

Cayan, D.R. 1992a. Latent and sensible heat flux anomalies over the northern oceans: The connection to monthly atmospheric forcing. J. Climate 5:354-369.

Cayan, D.R. 1992b. Latent and sensible heat flux anomalies over the northern oceans: Driving the sea surface temperature. J. Phys. Oceanogr. 22:859-881.

Cayan, D.R., A.J. Miller, T.P. Barnett, N.P. Graham, J.N. Ritchie, and J.M. Oberhuber. 1995. Two decades of a Pacific Ocean GCM simulation driven with monthly surface marine observations. In Natural Climate Variability on Decade-to-Century Time Scales. D.G. Martinson, K. Bryan, M. Ghil, M.M. Hall, T.R. Karl, E.S. Sarachik, S. Sorooshian, and L.D. Talley (eds.). National Academy Press, Washington, D.C.

Cess, R.D., G.L. Potter, J.P. Blanchet, G.J. Boer, R. Colman, and 21 others. 1991. Intercomparison and interpretation of snow-climate feedback processes in seventeen atmospheric general circulation models. Science 253:888-892.

Chang, A.T.C., J.L. Foster, and D.K. Hall. 1987. Nimbus-7 SMMR derived global snow cover parameters. Ann. Glaciol. 9:39-45.

Chang, A.T.C., J.L. Foster, and D.K. Hall. 1990. Satellite sensor estimates of northern hemisphere snow volume. Int. J. Remote Sens. 11:167-171.

Chang, F.-C., and J.M. Wallace. 1987. Meteorological conditions during heat waves and droughts in the United States Great Plains. Mon. Weather Rev. 115:1253-1269.

Changnon, D., T.B. McKee, and N.J. Doesken. 1993. Annual snowpack patterns across the Rockies: Long-term trends and associated 500-mb synoptic patterns. Mon. Weather Rev. 121:633-647.

Changnon, S.A., Jr. 1987. Climate fluctuations and record-high levels of Lake Michigan. Bull. Amer. Meteorol. Soc. 68:1394-1402.

Chapman, W.L., and J.E. Walsh. 1993. Recent variations of sea ice and air temperature in high latitudes. Bull. Amer. Meteorol. Soc. 74:33-47.

Charlson, R.J., J. Langner, H. Rodhe, C.B. Leovy, and S.G. Warren. 1991. Perturbation of the Northern Hemisphere radiative balance by backscattering from anthropogenic sulfate aerosols. Tellus 43AB:152-163.

Charlson, R.J., S.E. Schwartz, J.M. Hales, R.D. Cess, J.A. Coakley, Jr., J.E. Hansen, and D.J. Hofmann. 1992. Climate forcing by anthropogenic aerosols. Science 255:423-430.

Charney, J.G. 1975. Dynamics of deserts and drought in the Sahel. Quart. J. Roy. Meteorol. Soc. 101:193-202.

Clark, N.E., L. Eber, R.M. Laurs, J.A. Renner, and J.F.T. Saur. 1974. Heat exchange between ocean and atmosphere in the eastern North Pacific for 1961-71. NOAA Technical Report NMFS SSRF-682, Department of Commerce, Seattle, Washington, 108 pp.

Cohen, J., and D. Rind. 1991. The effect of snow cover on the climate. J. Climate 4:689-706.

Cook, E.R., and L.A. Kairiukstis (eds.). 1990. Methods of Dendrochronology: Applications in the Environmental Sciences. Kluwer Academic Publishers, Dordrecht, 394 pp.

Cook, E., T. Bird, M. Peterson, M. Barbetti, B. Buckley, R. D'Arrigo, R. Francey, and P. Tans. 1991. Climatic change in Tasmania inferred from a 1089-year tree-ring chronology of Huon pine. Science 253:1266-1268.

Cook, E.R., T. Bird, M. Peterson, M. Barbetti, B. Buckley, and R. Francey. 1992. The Little Ice Age in Tasmanian tree rings. In Proceedings of the International Symposium on the Little Ice Age. T. Mikami (ed.). Tokyo Metropolitan University, pp. 11-17.

Cook, E.R., B.M. Buckley, and R.D. D'Arrigo. 1995. Interdecadal temperature oscillations in the Southern Hemisphere: Evidence from Tasmanian tree rings since 300 B.C. In Natural Climate Variability on Decade-to-Century Time Scales. D.G. Martinson, K. Bryan, M. Ghil, M.M. Hall, T.R. Karl, E.S. Sarachik, S. Sorooshian, and L.D. Talley (eds.). National Academy Press, Washington, D.C.

Court, A. 1967-68. Climatic Normals as Predictions: Parts I-V. U.S. Air Force Cambridge Research Laboratories scientific reports. NTIS AD-657 358; AD-672 268; AD-687 137; AD-687 138; AD-688 163.

Covey, C., K.E. Taylor, and R.E. Dickinson. 1991. Upper limit for sea-ice albedo feedback contribution to global warming. J. Geophys. Res. 96:9169-9174.

D'Arrigo, R.D., and G.C. Jacoby, Jr. 1992. Dendroclimatic evidence from northern North America. In Climate Since A.D. 1500. R.S. Bradley and P.D. Jones (eds.). Routledge, London, pp. 296-311.

Davis, R. 1976. Predictability of sea surface temperature and sea level pressure anomalies over the North Pacific Ocean. J. Phys. Oceanogr. 6:249-266.

Deser, C., and M.L. Blackmon. 1993. Surface climate variations over the North Atlantic Ocean during winter: 1900-1989. J. Climate 6:1743-1753.

Deser, C., and M.L. Blackmon. 1995. Surface climate variations over the North Atlantic Ocean during winter: 1900-1989. In Natural Climate Variability on Decade-to-Century Time Scales. D.G. Martinson, K. Bryan, M. Ghil, M.M. Hall, T.R. Karl, E.S. Sarachik, S. Sorooshian, and L.D. Talley (eds.). National Academy Press, Washington, D.C.

Dewey, K.F. 1977. Daily maximum and minimum temperature forecasts and the influence of snow cover. Mon. Weather Rev. 105:1594-1597.

Dewey, K.F., and R. Heim, Jr. 1982. A digital archive of Northern Hemisphere snow cover, November 1966 through December 1980. Bull. Am. Meteorol. Soc. 63:1132-1141.

Dewey, K.F., and R. Heim, Jr. 1983. Satellite observations of variations in Southern Hemisphere snow cover. NOAA Technical Report NESDIS 1, Washington, D.C., 20 pp.

Dey, B., O.S.R.U. Bhanu Kumar, and S.N. Kathuria. 1984. Himalayan snow cover and the Indian summer monsoon activity. Final Report, NSF Grant ATM-8109177, Howard Univ., Washington, D.C., 30 pp.

Diaz, H.F. 1983. Some aspects of major dry and wet periods in the contiguous United States, 1895-1981. J. Clim. Appl. Meteorol. 22:3-16.

Diaz, H.F., and R.S. Bradley. 1995. Documenting natural climatic variations: How different is the climate of the twentieth century from that of previous centuries? In Natural Climate Variability on Decade-to-Century Time Scales. D.G. Martinson, K. Bryan, M. Ghil, M.M. Hall, T.R. Karl, E.S. Sarachik, S. Sorooshian, and L.D. Talley (eds.). National Academy Press, Washington, D.C.

Diaz, H.F., and R.G. Quayle. 1980. The climate of the United States since 1895: Spatial and temporal changes. Mon. Weather Rev. 108:249-266.

Dickson, R.R. 1984. Eurasian snow cover versus Indian monsoon rainfall—An extension of the Hahn-Shukla results. J. Clim. Appl. Meteorol. 23:171-173.

Dickson, R.R., and J. Namias. 1976. North American influence on the circulation and climate of the North Atlantic sector. Mon. Weather Rev. 104:1255-1265.

Dickson, R.R., J. Meincke, S.A. Malmberg, and A.J. Lee. 1988. The "Great Salinity Anomaly" in the northern North Atlantic 1968-1982. Prog. Oceanogr. 20:103-151.

Dirmeyer, P., and J. Shukla. 1993. Observational and modeling studies of the influence of soil moisture anomalies on atmospheric circulation (Review). In Prediction of Interannual Climate Variations. J. Shukla (ed.). Springer-Verlag, New York, pp. 1-24.

Douglas, A.V., D.R. Cayan, and J. Namias. 1982. Large-scale changes in North Pacific and North American weather patterns in recent decades. Mon. Weather Rev. 110:1851-1862.

Dutton, E.G., R.S. Stone, D.W. Nelson, and B.G. Mendonca. 1991. Recent Interannual Variations in Solar Radiation, Cloudiness, and Surface Temperature at the South Pole. J. Climate 4:848-858.

Ebbesmeyer, C.C., D.R. Cayan, D.R. McLain, F.H. Nichols, D.H. Peterson, and K.T. Redmond. 1991. 1976 step in the Pacific climate: Forty environmental changes between 1968-75 and 1977-1984. In Proceedings of the Seventh Annual Pacific Climate (PACLIM) Workshop (Asilomar, April 1990). J.L. Betancourt and V.L. Tharp (eds.). Interagency Ecological Studies Program Technical Report 26, California Department of Water Resources, Sacramento, Calif., pp. 115-126.

Elliott, W.P., and J.K. Angell. 1988. Evidence for changes in Southern Oscillation relationships during the last 100 years. J. Climate 1:729-737.

Elliott, W.P., M.E. Smith, and J.K. Angell. 1991. Monitoring tropospheric water vapor changes using radiosonde data. In Greenhouse-Gas-Induced Climatic Change: A Critical Appraisal of Simulations and Observations. M.E. Schlesinger (ed.). Elsevier, Amsterdam, pp. 311-328.

Enfield, D.B. and L. Cid. 1990. Statistical analysis of El Niño/Southern Oscillation over the last 500 years. In Toga Notes, Volume 7, Nova University, Dania, Florida, pp. 1-4.

Enting, I.G. 1987. On the use of smoothing splines to filter CO_2 data. J. Geophys. Res. 92:10977-10984.

Esbensen, S.K., and Y. Kushnir. 1981. The Heat Budget of the Global Ocean: An atlas based on estimates from surface marine observations. Report No. 29, Climatic Research Institute and Department of Atmospheric Sciences, Oregon State University, Corvallis, Oregon, 219 pp.

Farmer, G. 1988. Seasonal forecasting of the Kenya coast short rains, 1901-1984. J. Climatol. 8:489-497.

Fisher, D.A., and R.M. Koerner. 1983. Ice core study: A climatic link between the past, present and future. In Climatic Change in Canada 3. C.R. Harington (ed.). Syllogeus No. 49, pp. 50-69.

Fleer, H. 1981. Large-Scale Tropical Rainfall Anomalies. Bonner Meteorologische Abhandlungen, No. 26, 114 pp.

Flohn, H. 1986. Singular events and catastrophes now and in climatic history. Naturwissenschaften 73:1851-1862.

Flohn, H., A. Kapala, H.R. Knoche, and H. Machel. 1990. Recent changes of the tropical water and energy budget and of midlatitude circulations. Climate Dynamics 4:237-252.

Flohn, H., H.R. Kapala, H.R. Knoche, and H. Mächel. 1992. Water vapour as an amplifier of the greenhouse effect: New aspects. Meteorol. Zeitschrift, N.F. 1: 122-138.

Folland, C.K., D.E. Parker, and F.E. Kates. 1984. Worldwide marine temperature fluctuations 1856-1981. Nature 310:670-673.

Folland, C.K., T.N. Palmer, and D.E. Parker. 1986. Sahel rainfall and worldwide sea temperatures. Nature 320:602-607.

Folland, C.K., T.R. Karl, and K.Ya. Vinnikov. 1990. Observed climate variations and change. In Climate Change: The IPCC Scientific Assessment. J.T. Houghton, G.J. Jenkins, and J.J. Ephraums (eds.). Prepared for the Intergovernmental Panel on Climate Change by Working Group I. WMO/UNEP, Cambridge University Press, pp. 195-238.

Folland, C.K., J. Owen, M.N. Ward, and A. Colman. 1991. Prediction of seasonal rainfall in the Sahel region using empirical and dynamical methods. J. Forecast. 10:21-56.

Frankignoul, C., 1985. Sea surface temperature anomalies, planetary waves, and air-sea feedback in the middle latitudes. Rev. Geophys. 23:357-390.

Frankignoul, C., and R.W. Reynolds. 1983. Testing a dynamical model for midlatitude sea surface temperature anomalies. J. Phys. Oceanogr. 13:1131-1145.

Frich, P. 1992. Cloudiness and diurnal temperature range. In Proceedings of the Fifth International Meeting on Statistical Climatology. Environment Canada, Toronto, pp. 91-94.

Friis-Christensen, E., and K. Lassen. 1991. Length of the solar cycle activity closely associated with climate. Science 254:698-700.

Fritts, H.C. 1976. Tree Rings and Climate. Academic Press, London, 567 pp.

Fritts, H.C. 1991. Reconstructing Large-Scale Climatic Patterns from Tree-Ring Data. University of Arizona Press, Tucson, 286 pp.

Fritts, H.C., G.R. Lofgren, and G.A. Gordon. 1979. Variations in climate since 1602 as reconstructed from tree rings. Quat. Res. 12:18-46.

Gall, R., K. Young, R. Schotland, and J. Schmitz. 1992. The recent maximum temperature anomalies in Tucson: Are they real or an instrumental problem? J. Climate 5:657-665.

Ghil, M., and R. Vautard. 1991. Interdecadal oscillations and the warming trend in global temperature time series. Nature 350:324-327.

Gill, A.E., and P.P. Niiler. 1973. The theory of the seasonal variability in the ocean. Deep-Sea Res. 20:141-177.

Giovinetto, M.B., and W. Schwertfeger. 1966. Analysis of a 200-year snow accumulation series from the South Pole. Archiv. Met. Geophys. Biokl. A15:227-250.

Goldenberg, S.B., and J.J. O'Brien. 1981. Time and space variability of tropical Pacific wind stress. J. Phys. Oceanogr. 11:1190-1207.

Golubev, V.V., P.Ya. Groisman, and R.G. Quayle. 1992. An evaluation of the U.S. standard 8-inch nonrecording rain gauge at the Valdai polygon, USSR. J. Atmos. Ocean. Technol. 49:624-629.

Goodison, B.E. 1978. Accuracy of Canadian snow gauge measurements. J. Appl. Meteor. 17:1542-1548.

Goodison, B.E., and P.Y.T. Louie. 1986. Canadian methods for precipitation measurement and correction. In Proceedings of the International Workshop on Correction of Precipitation Measurements. Instruments and Observing Methods Report No. 25, WMO/TD No. 104, World Meteorological Organization, Geneva, pp. 141-145.

Goodison, B.E., and J.R. Metcalfe. 1989. Canadian participation in the WMO Solid Precipitation Measurement Intercomparison: Preliminary results. In Proceedings of the WMO/IAHS/ETH International Workshop on Precipitation Measurement (St. Moritz). Swiss Federal Institute of Technology, Zürich, pp. 121-125.

Goodison, B.E., H.L. Ferguson, and G.A. McKay. 1981. Comparison of point snowfall measurement techniques. In Handbook on Snow: Principles, Processes, Management and Use. D.M. Gray and D.M. Male (eds.). Pergamon Press, Oxford, pp. 200-210.

Goodison, B.E., V.S. Golubev, T. Gunter, and B. Sevruk. 1992. Preliminary results of the WMO Solid Precipitation Measurement Intercomparison. In Proceedings of WMO Technical Conference on Instruments and Methods of Observation (Vienna).

Instruments and Observing Methods Report No. 49, World Meteorological Organization, Geneva, pp. 161-165.

Gordon, A.L., S.E. Zebiak, and K. Bryan. 1992. Climate variability and the Atlantic Ocean. EOS 73:161-165.

Graham, N.E. 1994. Decadal-scale climate variability in the tropical and North Pacific during the 1970s and 1980s: Observations and model results. Climate Dynamics 10:135-162.

Graybill, D.A., and S.G. Shiyatov. 1992. Dendroclimatic evidence from the northern Soviet Union. In Climate Since A.D. 1500. R.S. Bradley and P.D. Jones (eds). Routledge, London, pp. 393-414.

Greatbach, R.J., A.F. Fanning, and A.D. Goulding. 1991. A diagnosis of interpentadal circulation changes in the North Atlantic. J. Geophys. Res. 96:22009-22023.

Gribbin, J. 1979. The search for cycles. In Climatic Change. J. Gribbin (ed.). Cambridge University Press, Cambridge, pp. 139-149.

Groisman, P.Ya. 1991. Unbiased estimates of precipitation change in the Northern Hemisphere extratropics. In Proceedings of the Fifth Conference on Climate Variations (Denver). American Meteorological Society, Boston, pp. 42-45.

Groisman, P.Ya. 1992. Studying the North American precipitation changes during the last 100 years. In Proceedings of the Fifth International Meeting on Statistical Climatology (Toronto). American Meteorological Society, Boston, pp. 75-79.

Groisman, P.Ya., and D.R. Easterling. 1995. Variability and trends of precipitation and snowfall over North America. In Natural Climate Variability on Decade-to-Century Time Scales. D.G. Martinson, K. Bryan, M. Ghil, M.M. Hall, T.R. Karl, E.S. Sarachik, S. Sorooshian, and L.D. Talley (eds.). National Academy Press, Washington, D.C.

Groisman, P.Ya., V.V. Koknaeva, T.A. Belokrylova, and T.R. Karl. 1991. Overcoming biases of precipitation measurement: A history of the USSR experience. Bull. Amer. Meteorol. Soc. 72:1725-1733.

Groisman, P.Ya., R.W. Knight, T.R. Karl, and R.R. Heim, Jr. 1993. Inferences of the North American snowfall and snow cover with recent global temperature changes. In Glaciological Data Report GD-25 (Snow Watch 1992), World Data Center A for Glaciology, Boulder, Colo., pp. 44-51.

Gullett, D.W., and W.R. Skinner. 1992. The State of Canada's Climate: Temperature Change in Canada 1895-1991. SOE Report No. 92-2, Atmospheric Environment Service, Environment Canada, Ottawa, 36 pp.

Haberli, W., P. Maller, P. Allan, and H. Bijsch. 1989. Glacier changes following the Little Ice Age—A survey of the international data base and its perspectives. In Glacier Fluctuations and Climatic Change. J. Oerlemans (ed.). Kluwer Academic Publishers, Dordrecht, 77-101.

Hahn, D.G., and J. Shukla. 1976. An apparent relationship between snow cover and Indian monsoon rainfall. J. Atmos. Sci. 33:2461-2462.

Hall, S.J. 1986. Air Force Global Weather Central Snow Analysis Model. Technical Note 86/001, USAF Airforce Global Weather Central, Offut Air Force Base, Neb., 48 pp.

Halliwell, G.R., and P. Cornillon. 1989. Large scale SST anomalies associated with subtropical fronts in the western North Atlantic during FASINEX. J. Mar. Res. 47:757-775.

Hamilton, K., and R.R. Garcia. 1986. El Niño-Southern Oscillation events and their associated midlatitude teleconnections 1531-1841. Bull. Am. Meteorol. Soc. 67:1354-1361.

Hammer, C.U., H.B. Clausen, W. Dansgaard, N. Gundestrup, S.J. Johnsen, and N. Reeh. 1978. Dating of Greenland ice cores by flow models, isotopes, volcanic debris and continental dust. J. Glaciol. 20:3-26.

Haney, R.L. 1985. Midlatitude sea surface temperature anomalies: A numerical hindcast. J. Phys. Oceanogr. 15:787-799.

Hansen, J.E., and A.A. Lacis. 1990. Sun and dust versus greenhouse gases: An assessment of their relative roles in global climate change. Nature 346:713-718.

Hanson, A.M. 1980. The snow cover of sea ice during the Arctic Ice Dynamics Joint Experiment, 1975-1976. Arct. Alp. Res. 12:215-226.

Hanson, K., G.W. Brier, and G.A. Maul. 1989. Evidence of significant nonrandom behavior in the recurrence of strong El Niño between 1525 and 1988. Geophys. Res. Lett. 16:1181-1184.

Hare, F.K. 1980. Long-term annual surface heat and water balances over Canada and the United States south of 60°N: Reconciliation of precipitation, run-off and temperature fields. Atmos.-Ocean 18(2):127-153.

Harrison, M.S.J. 1983. The southern oscillation, zonal equatorial circulation cells and South African rainfall. In Preprints of the First International Conference on Southern Hemisphere Meteorology. American Meteorological Society, Boston, pp. 302-305.

Hartmann, D.L., M.E. Ockert-Bell, and M.L. Michelsen. 1992. The effect of cloud type on earth's energy balance: Global analysis. J. Climate 5:1157-1171.

Henderson-Sellers, A. 1986. Cloud changes in a warmer Europe. Climatic Change 8:25-52.

Henderson-Sellers, A. 1989. North American total cloud amount variations this century. Glob. Planet. Change 1:175-194.

Henderson-Sellers, A. 1990. Review of our current information about cloudiness changes this century. In Observed Climate Variations and Change: Contributions in Support of Section 7 of the 1990 IPCC Scientific Assessment. D.E. Parker (ed.). Intergovernmental Panel on Climate Change, Geneva, pp. XI.1-XI.12.

Henderson-Sellers, A. 1992. Continental cloudiness changes this century. Geogr. J. 27(3):255-262.

Holloway, P.E. 1991. On the dissipation of internal tides. In Tidal Hydrodynamics. B.B. Parker (ed.). John Wiley and Sons, New York, pp. 449-468.

Hughes, M.G., and D.A. Robinson. 1993. Creating temporally complete snow cover records using a new method for modeling snow depth changes. In Glaciological Data Report GD-25 (Snow Watch 1992), World Data Center A for Glaciology, Boulder, Colo., pp. 150-163.

Hughes, M.K., P.M. Kelly, J.R. Pilcher, and V.C. LaMarche, Jr. (eds.). 1982. Climate from Tree Rings. Cambridge University Press, Cambridge, 223 pp.

Hurst, H.E. 1957. A suggested statistical model of some time series which occur in nature. Nature 180:494.

ICSU/UNEP/UNESCO/WMO. 1993. Draft Plan for Global Climate Observing System. World Meteorological Organization, Geneva, Switzerland.

Ikeda, M. 1990. Decadal oscillations of the air-ice-ocean system in the Northern Hemisphere. Atmos.-Ocean 28:106-139.

Ingram, W.J., C.A. Wilson, and J.F.B. Mitchell. 1989. Modelling climate change: An assessment of sea ice and surface albedo feedbacks. J. Geophys. Res. 94:8609-8622.

IPCC. 1990. Climate Change: The IPCC Scientific Assessment. J.T. Houghton, G.J. Jenkins, and J.J. Ephraums (eds.). Prepared for the Intergovernmental Panel on Climate Change by Working Group I. WMO/UNEP, Cambridge University Press, 365 pp.

IPCC. 1992. Climate Change 1992: The Supplementary Report to the IPCC Scientific Assessment. J.T. Houghton, B.A. Callander, and S.K. Varney (eds.). Prepared for the Intergovernmental Panel on Climate Change by Working Group I. WMO/UNEP, Cambridge University Press, 200 pp.

Isdale, P. 1984. Fluorescent bands in massive corals record centuries of coastal rainfall. Nature 310:578-579.

Isemer, H.-J., and L. Hasse. 1985. The Bunker Climate Atlas of the North Atlantic Ocean. Volume 1: Observations. Springer-Verlag, Berlin, 218 pp.

Isemer, H.-J., and L. Hasse. 1987. The Bunker Climate Atlas of the North Atlantic Ocean. Volume 2: Air-Sea Interactions. Springer-Verlag, Berlin, 252 pp.

Iwasaki, T. 1991. Year-to-year variation of snow cover area in the Northern Hemisphere. J. Meteorol. Soc. Japan 69:209-217.

Jackson, M.C. 1978. Snow cover in Great Britain. Weather 33:298-309.

James, I.N., and P.M. James. 1989. Ultra-low-frequency variability in a simple atmospheric circulation model. Nature 342:53-55.

Johnsen, S.J., W. Dansgaard, H.B. Clausen, and C.C. Langway, Jr. 1972. Oxygen isotope profiles through the Antarctic and Greenland ice sheets. Nature 235:429-434.

Jones, P.A., and A. Henderson-Sellers. 1992. Historical records of cloudiness and sunshine in Australia. J. Climate 5:260-267.

Jones, P.D. 1988. Hemispheric surface air temperature variations: Recent trends and an update to 1987. J. Climate 1:654-660.

Jones, P.D. 1991. Southern Hemisphere sea level pressure data: An analysis and reconstructions back to 1951 and 1911. Int. J. Climatol. 11:585-607.

Jones, P.D. 1992. Maximum and minimum temperature trends over Sudan, 1950-1982. In Proceedings of the International Temperature Workshop. D.E. Parker (ed.). Climate Research Technical Note No. 30, Hadley Centre, UKMO, Bracknell, Berkshire, pp. 19-22.

Jones, P.D., and R.S. Bradley. 1992. Climatic variations in the longest instrumental records. In Climate Since A.D. 1500. R.S. Bradley and P.D. Jones (eds.). Routledge, London, pp. 246-248.

Jones, P.D., and K.R. Briffa. 1995. Decade-to-century-scale variability of regional and hemispheric-scale temperatures. In Natural Climate Variability on Decade-to-Century Time Scales. D.G. Martinson, K. Bryan, M. Ghil, M.M. Hall, T.R. Karl, E.S. Sarachik, S. Sorooshian, and L.D. Talley (eds.). National Academy Press, Washington, D.C.

Jones, P.D., and P.M. Kelly. 1983. The spatial and temporal characteristics of Northern Hemisphere surface air temperature variations. J. Climatol. 3:243-252.

Jones, P.D., and T.M.L. Wigley. 1991. Global and hemispheric anomalies. In Trends '91: A Compendium of Data on Global Change. T.A. Boden, R.J. Sepanski, and F.W. Stoss (eds.).

U.S. Department of Energy, Oak Ridge National Laboratory, Environmental Sciences Division Publication No. 3746, pp. 512-517.

Jones, P.D., S.C.B. Raper, R.S. Bradley, H.F. Diaz, P.M. Kelly, and T.M.L. Wigley. 1986a. Northern Hemisphere surface air temperature variations, 1851-1984. J. Clim. Appl. Meteorol. 25:161-179.

Jones, P.D., S.C.B. Raper, R.S. Bradley, H.F. Diaz, P.M. Kelly, and T.M.L. Wigley. 1986b. Southern Hemisphere surface air temperature variations, 1851-1984. J. Clim. Appl. Meteorol. 25:1213-1230.

Jones, P.D., T.M.L. Wigley, and P.B. Wright. 1986c. Global temperature variations between 1861 and 1984. Nature 332:430-434.

Jones, P.D., P.Ya. Groisman, M. Coughlan, N. Plummer, W.-C. Wang, and T.R. Karl. 1990. Assessment of urbanization effects in time series of surface air temperature over land. Nature 347:169-172.

Kagan, R.L. 1979. Averaging of Meteorological Fields. Gidrometeoizdat, Leningrad, 212 pp. (in Russian).

Kaiser, D.P., and V.N. Razuvayev. 1995. Cloud cover and type over the former USSR, 1936-83: Trends derived from the RIHMI-WDC 223-station 6- and 3-hourly meteorological data base. In Proceedings of the Sixth International Meeting on Statistical Climatology. University College, Galway, Ireland, pp. 419-422.

Kalnay, E., and R. Jenne. 1991. Summary of the NMC/NCAR Reanalysis Workshop of April 1991. Bull. Amer. Meteorol. Soc. 72:1897-1904.

Kanamitsu, M., and T.N. Krishnamurti. 1978. Northern summer tropical circulations during drought and normal rainfall months. Mon. Weather Rev. 106:331-347.

Karl, T.R. 1988. Multi-year fluctuations of temperature and precipitation: The gray area of climate change. Climate Change 12:179-197.

Karl, T.R., and P.M. Steurer. 1990. Increased cloudiness in the United States during the first half of the twentieth century: Fact or fiction? Geophys. Res. Lett. 17:1925-1928.

Karl, T.R., and C.N. Williams, Jr. 1986. An approach to adjusting climatological time series for discontinuous inhomogeneities. J. Clim. Appl. Meteorol. 26:1744-1763.

Karl, T.R., G. Kukla, and J. Gavin. 1984a. Decreasing diurnal temperature range in the United States and Canada from 1941-1980. J. Clim. Appl. Meteorol. 23:1489-1504.

Karl, T.R., R.E. Livezey, and E.S. Epstein. 1984b. Recent unusual mean winter temperatures across the contiguous United States. Bull. Am. Meteorol. Soc. 65:1302-1309.

Karl, T.R., C.N. Williams Jr., and P.J. Young. 1986. A model to estimate the time of observation bias associated with mean monthly maximum, minimum and mean temperatures for the United States. J. Clim. Appl. Meteorol. 25:145-160.

Karl, T.R., H.F. Diaz, and G. Kukla. 1988. Urbanization: Its detection and effect in the United States climate record. J. Climate 1:1099-1123.

Karl, T.R., J.D. Tarpley, R.G. Quayle, H.F. Diaz, D.A. Robinson, and R.S. Bradley. 1989. The recent climate record: What it can and cannot tell us. Rev. Geophys. 27:405-430.

Karl, T.R., C.N. Williams, Jr., F.T. Quinlan, and T.A. Boden. 1990. United States Historical Climatology Network (HCN)

Serial Temperature and Precipitation Data. Report ORNL/ CDIAC-30, NDP-019/R1, Carbon Dioxide Information Analysis Center, Oak Ridge National Laboratory, Oak Ridge, Tenn., 83 pp. plus appendices.

Karl, T.R., G. Kukla, V.N. Razuvayev, M.J. Changery, R.G. Quayle, R.R. Heim, Jr., D.R. Easterling, and C.B. Fu. 1991a. Global warming: Evidence for asymmetric diurnal temperature change. Geophys. Res. Lett. 18:2253-2256.

Karl, T.R., R.R. Heim, Jr., and R.G. Quayle. 1991b. The greenhouse effect in central North America: If not now, when? Science 251:1058-1061.

Karl, T.R., P.Ya. Groisman, R.R. Heim, Jr., and R.W. Knight. 1993a. Recent variations of snow cover and snowfall in North America and their relation to precipitation and temperature variations. J. Climate 6:1327-1344.

Karl, T.R., R.G. Quayle, and P.Ya. Groisman. 1993b. Detecting climate variations and change: New challenges for observing and data management systems. J. Climate 6:1481-1494.

Karl, T.R., P.D. Jones, R.W. Knight, G. Kukla, N. Plummer, V. Razuvayev, K.P. Gallo, J.A. Lindesay, and R.J. Charlson. 1995. Asymmetric trends of daily maximum and minimum temperature: Empirical evidence and possible causes. In Natural Climate Variability on Decade-to-Century Time Scales. D.G. Martinson, K. Bryan, M. Ghil, M.M. Hall, T.R. Karl, E.S. Sarachik, S. Sorooshian, and L.D. Talley (eds.). National Academy Press, Washington, D.C.

Karoly, D.J. 1987. Southern hemisphere temperature trends: A possible greenhouse gas effect? Geophys. Res. Lett. 14:1139-1141.

Karoly, D.J. 1989a. Southern Hemisphere circulation features associated with El Niño-Southern Oscillation events. J. Climate 2:1239-1252.

Karoly, D.J. 1989b. Low frequency variations of the southern hemisphere circulation. In Extended Abstracts, Third International Conference on Southern Hemisphere Meteorology and Oceanography. American Meteorological Society, Boston, pp. 105-109.

Karoly, D.J. 1990. The role of transient eddies in low-frequency zonal variations of the Southern Hemisphere circulation. Tellus 42A:41-50.

Karoly, D.J. 1995. Observed variability of the Southern Hemisphere atmospheric circulation. In Natural Climate Variability on Decade-to-Century Time Scales. D.G. Martinson, K. Bryan, M. Ghil, M.M. Hall, T.R. Karl, E.S. Sarachik, S. Sorooshian, and L.D. Talley (eds.). National Academy Press, Washington, D.C.

Kay, P.A., and H.F. Diaz (eds.). 1985. Problems and Prospects for Predicting Great Salt Lake Levels. Center for Public Affairs and Administration, University of Utah, 309 pp.

Keeling, C.D., and T.P. Whorf. 1995. Decadal oscillations in global temperature and atmospheric carbon dioxide. In Natural Climate Variability on Decade-to-Century Time Scales. D.G. Martinson, K. Bryan, M. Ghil, M.M. Hall, T.R. Karl, E.S. Sarachik, S. Sorooshian, and L.D. Talley (eds.). National Academy Press, Washington, D.C.

Keeling, C.D., R.B. Bacastow, A.F. Carter, S.C. Piper, T.P. Whorf, M. Heimann, W.G. Mook, and H. Roeloffzen. 1989. A three-dimensional model of atmospheric CO_2 transport based on observed winds: 1. Analysis of observational data. In Aspects of Climate Variability in the Pacific and the Western Americas. D.H. Peterson (ed.). Geophysical Monograph 55, American Geophysical Union, Washington, D.C., pp. 165-236.

Kelly, P.M., and T.M.L. Wigley. 1992. Solar cycle length, greenhouse forcing, and global climate. Nature 360:328-330.

Kidson, J.W. 1988a. Indices of the Southern Hemisphere zonal wind. J. Climate 1:183-194.

Kidson, J.W. 1988b. Interannual variations in the Southern Hemisphere circulation. J. Climate 1:1177-1198.

Klein, P., and B.L. Hua. 1988. Mesoscale heterogeneity of the wind-driven mixed layer: Influence of a quasigeostrophic flow. J. Mar. Res. 46:495-525.

Klein, S.A., and D.L. Hartmann. 1993. Spurious changes in the ISCCP dataset. Geophys. Res. Lett. 20:455-458.

Klugman, M.R. 1983. Evidence of climate change in United States seasonal precipitation data, 1948-1976. J. Clim. Appl. Meteorol. 22:1367-1376.

Knox, L.J., K. Higuchi, A. Shabbar, and N.E. Sargent. 1988. Secular variation of northern hemisphere 50 kPa geopotential height. J. Climate 1:500-511.

Koerner, R.M. 1977. Devon Island ice cap: Core stratigraphy and paleoclimate. Science 196:15-18.

Kotwicki, V., and P. Isdale. 1991. Hydrology of Lake Frome: El Niño link. Paleogeogr. Paleoclimatol. Paleoecol. 84:87-98.

Kraus, E.B. 1955a. Secular changes of tropical rainfall regimes. Quart. J. Roy. Meteor. Soc. 81:198-210.

Kraus, E.B. 1955b. Secular changes of east coast rainfall regimes. Quart. J. Roy. Meteor. Soc. 81:430-439.

Kruss, P.O., K.A.Y. Khan, F.M.Q. Malik, M. Muslehuddin, and A. Majid. 1992. Cooling over monsoonal Pakistan. In Fifth International Meeting on Statistical Climatology. Environment Canada, Toronto, p. 27.

Kukla, G., and D.A. Robinson. 1979. Accuracy of snow and ice monitoring. In Glaciological Data Report GD-5 (Workshop on Snow Cover and Sea Ice Data), World Data Center A for Glaciology, Boulder, Colo., pp. 91-97.

Kunzi, K.F., S. Patil, and H. Rott. 1982. Snow-cover parameters retrieved from Nimbus-7 scanning multichannel microwave radiometer (SMMR) data. I.E.E.E. Trans. Geosci. Rem. Sen., GE-20, pp. 452-467.

Kuo, C., C. Lindberg, and D. Thomson. 1990. Coherence established between atmospheric carbon dioxide and global temperature. Nature 343:709-714.

Kushnir, Y. 1994. Interdecadal variations in North Atlantic sea surface temperature and associated atmospheric conditions. J. Climate 7:141-157.

Lamb, H.H. 1955. Two-way relationship between snow or ice limit and 1000-500 mb thickness in the overlying atmosphere. Quart. J. Roy. Meteor. Soc. 8:172-189.

Lamb, H.H. 1972. Climate: Present, Past and Future. Volume 1: Fundamentals and Climate Now. Methuen, London, 613 pp.

Lamb, H.H. 1977. Climate: Present, Past and Future. Volume 2: Climatic History and the Future. Methuen, London, 835 pp.

Lamb, P.J. 1978. Large-scale tropical Atlantic surface circulation patterns associated with sub-Saharan weather anomalies: 1967 and 1968. Tellus 30:240-251.

Lamb, P.J. 1982. Persistence of sub-Saharan drought. Nature 299:46-47.

Landsberg, H.E. 1981. The Urban Climate. Academic Press, New York, 285 pp.

Lang, H., and M. Rohrer. 1987. Temporal and spatial variations of the snow cover in the Swiss Alps. In Large Scale Effects of Seasonal Snow Cover. B.E. Goodison, R.G. Barry, and J. Dozier (eds.). IAHS Press, Wallingford, U.K., pp. 79-92.

Lare, A.R., and S.E. Nicholson. 1993. Contrasting conditions of surface water balance in wet years and dry years as a possible land-surface-atmosphere feedback mechanism in the West African Sahel. J. Climate 7:653-668.

Larkin, H.H., Jr. 1947. A comparison of the Alter and Nipher shields for precipitation gages. Bull. Amer. Meteorol. Soc. 28:200-201.

Larson, L.W., and E.L. Peck. 1974. Accuracy of precipitation measurements for hydrologic modeling. Water Resour. Res. 10:857-863.

Latif, M., and T.P. Barnett. 1994. Causes of decadal climate variability over the North Pacific and North America. Science 266:634-637.

Lau, N.C., and M.J. Nath. 1990. A general circulation model study of the atmospheric response to extratropical SST anomalies observed in 1950-79. J. Climate 3:965-989.

Lazier, J.R.N. 1988. Temperature and salinity changes in the deep Labrador Sea 1962-1986. Deep-Sea Res. 35:1247-1253.

Leathers, D.J., and D.A. Robinson. 1993. The association between extremes in North American snow cover extent and United States temperatures. J. Climate 6:1345-1355.

Lehman, S.J., and L.D. Keigwin. 1992. Sudden changes in the North Atlantic circulation during the last deglaciation. Nature 356:757-746.

Leith, C.E. 1973. The standard error of time-average estimates of climatic means. J. Appl. Meteorol. 12:1066-1069.

Levitus, S. 1982. Climatological Atlas of the World Ocean. NOAA Professional Paper 13, U.S. Government Printing Office, Washington, D.C., 173 pp.

Levitus, S. 1989. Interpentadal variability of temperature and salinity at intermediate depths of the North Atlantic Ocean, 1970-74 versus 1955-59. J. Geophys. Res. 94:6091-6131.

Li, P. 1987. Seasonal snow resources and their fluctuations in China. In Large Scale Effects of Seasonal Snow Cover. B.E. Goodison, R.G. Barry, and J. Dozier (eds.). IAHS Press, Wallingford, U.K., pp. 93-104.

Lindesay, J.A., M.S.J. Harrison, and M.P. Haffner. 1986. The Southern Oscillation and South African rainfall. S. African J. Science 82:196-198.

Lins, H.F. 1985. Streamflow variability in the United States: 1931-1978. J. Clim. Appl. Meteorol. 24:463-471.

Linsley, R.K., Jr., M.H. Kohler, and J.L.H. Paulhus. 1958. Hydrology for Engineers. McGraw-Hill, N.Y., 340 pp.

Liu, W.T., and C. Gautier. 1990. Thermal forcing on the tropical Pacific from satellite data. J. Geophys. Res. 95(C8):13209-13217.

Loder, J.W., and C. Garrett. 1978. The 18.6-year cycle of sea surface temperature in shallow seas due to variations in tidal mixing. J. Geophys. Res. 83:1967-1970.

London, J., S.G. Warren, and C.J. Hahm. 1991. Thirty-year trend of observed greenhouse clouds over the tropical oceans. Adv. Space Res. 11(3):45-49.

Lorenz, E.N. 1986. The index cycle is alive and well. Namias Symposium, Library of Congress No. 86-50752, pp. 188-196.

Lough, J.M. 1986. Tropical Atlantic sea-surface temperatures and rainfall variations in sub-Saharan Africa. Mon. Weather Rev. 114:561-570.

Lough, J.M., and H.C. Fritts. 1990. Historical aspects of El Niño/Southern Oscillation—Information from tree rings. In Global Ecological Consequences of the 1982-83 El Niño-Southern Oscillation. P.W. Glynn (ed.). Oceanography Series 52, Elsevier, Amsterdam, pp. 285-321.

Luksch, U., and H. von Storch. 1992. Modeling the low-frequency sea surface temperature variability in the North Pacific. J. Climate 5(9):893-906.

Luksch, U., H. von Storch, and E. Maier-Reimer. 1990. Modeling North Pacific SST anomalies as a response to anomalous atmospheric forcing. J. Mar. Syst. 1:51-60.

Lund, I.A. 1963. Map-pattern classification by statistical methods. J. Appl. Meteorol. 2:56-65.

Maley, J. 1976. Les variations du lac Tchad depuis un millénaire: Conséquences paléoclimatiques. Palaeoecol. Africa 9:44-47.

Manabe, S., and R.J. Stouffer. 1988. Two stable equilibria of a coupled ocean-atmosphere model. J. Climate 1:841-866.

Mandelbrot, B.B., and Wallis, J.R. 1969. Some long-run properties of geophysical records. Water Resour. Res. 5:321-340.

Manley, G. 1969. Snowfall in Britain over the past 300 years. Weather 24:428-37.

Manley, G. 1974. Central England temperatures: Monthly means 1659-1972. Quart. J. Roy. Meteorol. Soc. 100:389-405.

Marsden, R.F., L.A. Mysak, and R.A. Myers. 1991. Evidence for stability enhancement of sea ice in the Greenland and Labrador Seas. J. Geophys. Res. (Oceans) 96: 4783-4789.

Markgraf, V., J.R. Dodson, A.P. Kershaw, M.S. McGlone, and N. Nicholls. 1992. Evolution of late Pleistocene and Holocene climates in the circum-South Pacific land areas. Climate Dynamics 6:193-211.

Matson, M., C.F. Ropelewski, and M.S. Varnadore. 1986. An Atlas of Satellite-Derived Northern Hemispheric Snow Cover Frequency. National Oceanic and Atmospheric Administration, Washington, D.C., 75 pp.

Mayewski, P.A., G. Holdsworth, M.J. Spencer, S. Whitlow, M. Twickler, M.C. Morrison, K.K. Ferland, and L.D. Meeker. 1993. Ice core sulfate from three northern hemisphere sites: Source and temperature forcing implications. Atmos. Environ. 27A:2915-2919.

McFarland, G.D. Wilke, and P.W. Harder II. 1987. Nimbus 7 SMMR investigation of snowpack properties in the northern Great Plains for the winter of 1978-1979. I.E.E.E. Trans. Geosci. Rem. Sen. GE-25:35-46.

McGuffie, K., and A. Henderson-Sellers. 1989. Is Canadian cloudiness increasing? Atmos. Ocean 26:608-633.

Meyers, G., J.R. Donguy, and R.K. Reed. 1986. Evaporative cooling of the western equatorial Pacific Ocean by anomalous winds. Nature 323(6088):523-526.

Michaels, P.J., and D.E. Stooksbury. 1992. Global warming: A reduced threat? Bull. Amer. Meteorol. Soc. 73:1563-1577.

Michaelsen, J. 1989. Long-period fluctuations in El Niño amplitude and frequency reconstructed from tree-rings. In Aspects of Climate Variability in the Pacific and the Western Americas. D.H.

Peterson (ed.). Geophysical Monograph 55, American Geophysical Union, Washington, D.C., pp. 69-74.

Michaud, R., and C.A. Lin. 1992. Monthly summaries of merchant ship surface marine observations and implication for climate variability studies. Climate Dynamics 7:45-55.

Miller, A.J. 1992. Large-scale ocean-atmosphere interactions in a simplified coupled model of the midlatitude wintertime circulation. J. Atmos. Sci. 49:273-286.

Miller, A.J., J.M. Oberhuber, N.E. Graham, T.P. Barnett. 1992. Tropical Pacific Ocean response to observed winds in a layered general circulation model. J. Geophys. Res. 97:7317-7340.

Miller, A.J., D.R. Cayan, T.P. Barnett, N.E. Graham, and J.M. Oberhuber. 1994a. Interdecadal variability of the Pacific Ocean: Model response to observed heat flux and wind stress anomalies. Climate Dynamics 9:287-302.

Miller, A.J., D.R. Cayan, T.P. Barnett, N.E. Graham, and J.M. Oberhuber. 1994b. The 1976-77 climate shift of the Pacific Ocean. Oceanography 7:21-26.

Milly, P.C.D. 1992. Potential evaporation and soil moisture in general circulation models. J. Climate 5:209-226.

Mintz, Y. 1984. The sensitivity of numerically simulated climates to land surface boundary conditions. In The Global Climate. J. Houghton (ed.). Cambridge University Press, pp. 79-105.

Mitchell, J.M., Jr. 1976. An overview of climate variability and its causal mechanisms. Quat. Res. 6:481-493.

Mo, K.C., and H. van Loon. 1984. Some aspects of the interannual variation of mean monthly sea level pressure on the Southern Hemisphere. J. Geophys. Res. 89D:9541-9546.

Mo, K.C., and G.H. White. 1985. Teleconnections in the Southern Hemisphere. Mon. Weather Rev. 113:22-37.

Mysak, L.A. 1991. Current and Future Trends in Arctic Climate Research: Can Changes in the Arctic Sea Ice be Used as an Early Indicator of Global Warming? Centre for Climate and Global Change Research Report No. 91-1, McGill University, Montreal.

Mysak, L.A., and D.K. Manak. 1989. Arctic sea-ice extent and anomalies, 1953-1984. Atmos.-Ocean 27:376-405.

Mysak, L.A., D.K. Manak, and R.F. Marsden. 1990. Sea-ice anomalies observed in the Greenland and Labrador Seas during 1901-1984 and their relation to an interdecadal Arctic climate cycle. Climate Dynamics 5:111-133.

Nagler, T., and H. Rott. 1991. Intercomparison of snow mapping algorithms over Europe using SSM/I data. Report to the SSM/I Products Working Team, World Data Center A for Glaciology, Boulder, Colo., 12 pp.

Namias, J. 1962. Influence of abnormal heat sources and sinks on atmospheric behavior. Proceedings, Symposium on Numerical Weather Prediction. Meteorological Society of Japan, Tokyo, pp. 615-627.

Namias, J. 1966. Nature and possible causes of the northeastern United States drought during 1962-65. Mon. Weather Rev. 94:543-554.

Namias, J. 1972. Large-scale and long-term fluctuations in some atmospheric and oceanic variables. Nobel Symposium 20:27-48.

Namias, J. 1981. Snow covers in climate and long-range forecasting. Glaciological Data GD-11 (Snow Watch 1980), World Data Center A for Glaciology, Boulder, Colo., pp. 13-26.

Namias, J. 1985. Some empirical evidence for the influence of snow cover on temperature and precipitation. Mon. Weather Rev. 113:1542-1553.

National Climatic Data Center (NCDC). 1991. Climate Variations Bulletin, Historical Climatology Series 4-7, NOAA NCDC, Asheville, N.C., 35 pp.

National Geophysical Data Center (NGDC). 1988. Data Announcement 88-MGG-02, NOAA, Boulder, Colo. 1 p.

National Geophysical Data Center (NGDC). 1989. Solar Indices Bulletin. NOAA NGDC, Solar-Terrestrial Physics Division, Boulder, Colo.

National Research Council (NRC). 1982. Carbon Dioxide and Climate: A Second Assessment. National Academy Press, Washington, D.C., 72 pp.

Newell, N.E., R.E. Newell, J. Hsiung, and W. Zhongxiang. 1989. Global marine temperature variation and the solar magnetic cycle. Geophys. Res. Lett. 16:311-314.

Newell, R.E., and J.W. Kidson. 1984. African mean wind changes in Sahelian wet and dry periods. J. Climatol. 4:1-7.

Nicholson, S.E. 1978. Climatic variations in the Sahel and other African regions during the past five centuries. J. Arid Environ. 1:3-24.

Nicholson, S.E. 1979a. The methodology of historical climate reconstruction and its application to Africa. J. African Hist. 20:3-24.

Nicholson, S.E. 1979b. Revised rainfall series for the West African subtropics. Mon. Weather Rev. 107:620-623.

Nicholson, S.E. 1981. The historical climatology of Africa. In Climate and History. T.M.L. Wigley, J.J. Ingram, and G. Farmer (eds.). Cambridge University Press, pp. 249-270.

Nicholson, S.E. 1983. Sub-Saharan rainfall in the years 1976-1980: Evidence of continued drought. Mon. Weather Rev. 111:1646-1654.

Nicholson, S.E. 1986. The spatial coherence of African rainfall anomalies: Interhemispheric teleconnections. J. Clim. Appl. Meteorol. 25:1365-1381.

Nicholson, S.E. 1989. African drought: Characteristics, causal theories, and global teleconnections. In Understanding Climate Change. A. Berger, R.E. Dickinson, and J.W. Kidson (eds.). American Geophysical Union, Washington, D.C., pp. 79-100.

Nicholson, S.E. 1993. An overview of African rainfall fluctuations of the last decade. J. Climate 6:1463-1466.

Nicholson, S.E. 1994. Recent rainfall fluctuations in Africa and their relationships to past conditions over the continent. Holcene 4:121-131.

Nicholson, S.E. 1995. Variability of African rainfall on interannual and decadal time scales. In Natural Climate Variability on Decade-to-Century Time Scales. D.G. Martinson, K. Bryan, M. Ghil, M.M. Hall, T.R. Karl, E.S. Sarachik, S. Sorooshian, and L.D. Talley (eds.). National Academy Press, Washington, D.C.

Nicholson, S.E., and D. Entekhabi. 1986. The quasi-periodic behavior of rainfall variability in Africa and its relationship to the Southern Oscillation. Arch. Meteor. Geophys. Bioklimatol. A34:311-348.

Nicholson, S.E., and D. Entekhabi. 1987. Rainfall variability in equatorial and Southern Africa: Relationships with sea-surface temperatures along the southwestern coast of Africa. J. Clim. Appl. Meteorol. 26:561-578.

Nicholson, S.E., J. Kim, and J. Hoopingarner. 1988. Atlas of

African Rainfall and Its Interannual Variability. Florida State University, Tallahassee, 252 pp.

Nitta, T., and S. Yamada. 1989. Recent warming of tropical sea surface temperature and its relationship to the Northern Hemisphere circulation. J. Meteorol. Soc. Japan 67:375-383.

North, G.R., T.L. Bell, R.F. Calahan, and F.J. Moeng. 1982. Sampling errors in the estimation of empirical orthogonal functions. Mon. Weather Rev. 110:699-706.

Nyenzi, B.S. 1988. Mechanisms of East African Rainfall Variability. M.S. Thesis, Department of Meteorology, Florida State University, Tallahassee, 184 pp.

Oberhuber, J.M. 1988. An Atlas Based on the COADS Data Set: The budgets of heat, buoyancy, and turbulent kinetic energy at the surface of the global ocean. Report No. 15, Max-Planck-Institut für Meteorologie, Hamburg.

Oberhuber, J.M. 1993. Simulation of the Atlantic circulation with a coupled sea ice-mixed layer-isopycnal general circulation model. Part I: Model description; Part II: Model experiment. J. Phys. Oceanogr. 22:808-845.

Ogallo, L. 1979. Rainfall variability in Africa. Mon. Weather Rev. 107:1133-1139.

Ogallo, L. 1987. Relationships between seasonal rainfall in East Africa and the Southern Oscillation. J. Climatol. 7:1-13.

Palmer, T.N., and D.A. Mansfield. 1986. A study of wintertime circulation anomalies during past El Niño events using a high resolution general circulation model. II: Variability of the seasonal mean response. Quart. J. Roy. Meteorol. Soc. 112:639-660.

Palmer, T.N., and Z. Sun. 1985. A modelling and observational study of the relationship between sea surface temperature in the north-west Atlantic and the atmospheric general circulation. Quart. J. Roy. Meteor. Soc. 111:947-975.

Paltridge, G., and S. Woodruff. 1981. Changes in global surface temperature from 1880 to 1977 derived from historical records of sea surface temperature. Mon. Weather Rev. 109:2427-2434.

Parungo, F., J.F. Boatman, H. Sievering, S. Wilkinson, and B.B. Hicks. 1994. Trends in global marine cloudiness and anthropogenic sulfur. J. Climate 7:434-440.

Pavia, E.G., and J.J. O'Brien. 1986. Waybill statistics of wind speed over the ocean. J. Climate Appl. Meteorol. 25:1324-1332.

Peck, E.L. 1991. Hydrometeorological Data Collection Design and Analysis for the Lake Ontario Drain Basin: Final Report, Phase 1. Coordinating Committee on Great Lakes Basin Hydraulic and Hydrometeorologic Data, Hydrometeorology and Modeling Subcommittee, International Joint Commission, Washington, D.C.

Penner, J.E., R.E. Dickinson, and C.A. O'Neill. 1992. Effects of aerosol from biomass burning on the global radiation budget. Science 256:1432-1433.

Petersen, R.A., and J.E. Hoke. 1989. The effect of snow cover on the Regional Analysis and Forecast System (RAFS) low-level forecasts. Wea. Forecast. 4:253-257.

Pfister, C. 1985. Snow cover, snow lines and glaciers in Central Europe since the 16th century. In The Climatic Scene. George Allen and Unwin, Winchester, Mass., pp. 154-174.

Pittock, A.B. 1984. On the reality, stability and usefulness of Southern Hemisphere teleconnections. Aust. Meteorol. Mag. 32:75-82.

Plantico, M.S., T.R. Karl, G. Kukla, and J. Gavin. 1990. Is recent climate change across the United States related to rising levels of anthropogenic greenhouse gases? J. Geophys. Res. 95:16617-16637.

Plummer, N., Z. Lin, and S. Torok. 1995. Trends in the diurnal temperature range over Australia since 1951. Atmos. Res. 237:79-86.

Posmentier, E.S., M.A. Cane, and S.E. Zebiak. 1989. Tropical Pacific trends since 1960. J. Climate 2:731-736.

Press, W.H., B.P. Flannery, S.A. Teukolsky, and W.T. Vetterling. 1989. Numerical Recipes (FORTRAN Version). Cambridge University Press, New York, 702 pp.

Quayle, R.G., D.R. Easterling, T.R. Karl, and P.M. Hughes. 1991. Effects of recent thermometer changes in the cooperative station network. Bull. Amer. Meteorol. Soc. 72:1718-1723.

Quinn, W.H., and V.T. Neal. 1992. The historic record of El Niño events. In Climate Since A.D. 1500. R.A. Bradley and P.D. Jones (eds.). Routledge Press, London, pp. 623-648.

Quinn, W.H., V.T. Neal, and S.E.A. de Mayolo. 1987. El Niño occurrences over the past four and a half centuries. J. Geophys. Res. 92:14449-14461.

Ramage, C.S. 1987. Secular change in reported surface wind speed over the ocean. J. Climate Appl. Meteorol. 26:525-529.

Rasmusson, E., X. Wang, and C.F. Ropelewski. 1995. Secular variability of the ENSO cycle. In Natural Climate Variability on Decade-to-Century Time Scales. D.G. Martinson, K. Bryan, M. Ghil, M.M. Hall, T.R. Karl, E.S. Sarachik, S. Sorooshian, and L.D. Talley (eds.). National Academy Press, Washington, D.C.

Reinsch, C.H. 1967. Smoothing by spline functions. Num. Math. 10:177-183.

Riehl, H., M. el-Bakary, and J. Meitin. 1979. Nile river discharge. Mon. Weather Rev. 107:1546-1553.

Rind, D., D. Peteet, W. Broecker, A. McIntyre, and W. Ruddiman. 1986. The impact of cold North Atlantic sea surface temperatures on climate: Implications for the younger Dryas cooling (11-10K). Climate Dynamics 1:3-33.

Rind, D., R. Goldberg, and R. Ruedy. 1989. Change in climate variability in the 21st century. Climate Change 14:5-37.

Robinson, D.A. 1989. Evaluation of the collection, archiving and publication of daily snow data in the United States. Phys. Geog. 10:120-130.

Robinson, D.A. 1993a. Historical daily climatic data for the United States. In Preprints of the Eighth Conference on Applied Climate (Anaheim). American Meteorological Society, Boston, Mass., pp. 264-269.

Robinson, D.A. 1993b. Monitoring Northern Hemisphere snow cover. In Glaciological Data Report GD-25 (Snow Watch 1992), World Data Center A for Meteorology, Boulder, Colo., pp. 1-25.

Robinson, D.A. 1995. Decadal variations of snow cover. In Natural Climate Variability on Decade-to-Century Time Scales. D.G. Martinson, K. Bryan, M. Ghil, M.M. Hall, T.R. Karl, E.S. Sarachik, S. Sorooshian, and L.D. Talley (eds.). National Academy Press, Washington, D.C.

Robinson, D.A., and K.F. Dewey. 1990. Recent secular variations in the extent of Northern Hemispheric snow cover. Geophys. Res. Lett. 17:1557-1560.

Robinson, D.A., F.T. Keimig, and K.F. Dewey. 1991. Recent variations in Northern Hemisphere snow cover. In Proceedings of the 15th Annual Climate Diagnostics Workshop. NOAA National Weather Service, Washington, D.C., pp. 219-224.

Robinson, D.A., M.C. Serreze, R.G. Barry, G. Scharfen, and G. Kukla. 1992. Interannual variability of snow melt and surface albedo in the Arctic basin. J. Climate 5:1109-1119.

Robinson, W.A. 1992. The dynamics of the zonal index in a simple model of the atmosphere. Tellus 43A:295-305.

Robock, A., and J.W. Tauss. 1986. Effects of snow cover and tropical forcing on mid-latitude monthly mean circulation. Glaciological Data GD-18 (Snow Watch 1985), World Data Center A for Glaciology, Boulder, Colo., pp. 207-214.

Rodhe, H., and H. Virji. 1976. Trends and periodicities in East African rainfall data. Mon. Weather Rev. 104:307-315.

Roeckner, E., K. Arpe, L. Bengtsson, S. Brinkop, L. Domenil, M. Esch, E. Kirk, F. Lunkeit, M. Ponater, B. Rockel, R. Sausen, U. Schlese, S. Schubert, and M. Windelband. 1992. Simulation of the Present-Day Climate with the ECHAM Model: Impact of model physics and resolution. Report No. 93, Max-Planck-Institut für Meteorologie, Hamburg.

Rogers, J.C., and H. van Loon. 1982. Spatial variability of sea level pressure and 500 mb height anomalies over the Southern Hemisphere. Mon. Weather Rev. 110:1375-1392.

Ropelewski, C.F., and M.S. Halpert. 1987. Global and regional scale precipitation and temperature patterns associated with El Niño/Southern Oscillation. Mon. Weather Rev. 115:1606-1626.

Ropelewski, C.F, A. Robock, and M. Matson. 1984. Comments on "An apparent relationship between Eurasian spring snow cover and the advance period of the Indian summer monsoon." J. Clim. Appl. Meteorol. 23:341-342.

Rossow, W.B., and B. Cairns. 1995. Monitoring changes of clouds. Climatic Change 31:305-347.

Rowell, D.P., C.K. Folland, K. Maskell, J.A. Owen, and M.N. Ward. 1992. Modelling the influence of global sea surface temperature on the variability and predictability of seasonal Sahel rainfall. Geophys. Res. Lett. 19:905-908.

Rowell, D.P., C.K. Folland, K. Maskell, and M.N. Ward. 1995. Variability of summer rainfall over tropical north Africa (1906-1992): Observations and modelling. Quart. J. Roy. Meteor. Soc. 121:669-704.

Saltzman, B. (ed.). 1983. Theory of Climate. Volume 25, Advances in Geophysics, Academic Press, New York, 505 pp.

Sanderson, M. 1975. A comparison of Canadian and United States standard methods of measuring precipitation. J. Appl. Meteorol. 14:1197-1199.

Schlesinger. M.E. 1984. Climate model simulations of CO_2-induced climatic change. Adv. Geophys. 26:141-235.

Schlesinger, M.E. (ed.). 1991. Greenhouse-Gas-Induced Climatic Change: A Critical Appraisal of Simulations and Observations. Elsevier, Amsterdam, 615 pp.

Schlesinger, M.E., and J.F.B. Mitchell. 1985. Model projection of the equilibrium climatic response to increased carbon dioxide. In Projecting the Effects of Increasing Carbon Dioxide. M.C. MacCracken and F.M. Luther (eds.). Report DOE/ER-0237, U.S. Department of Energy, pp. 57-80.

Sellers, W.D. 1965. Physical Climatology. University of Chicago Press, 272 pp.

Semazzi, F.H.M., V. Mehta, and Y.C. Sud. 1988. An investigation

of the relationship between sub-Saharan rainfall and global sea surface temperatures. Atmos.-Ocean 26:118-138.

Sevruk, B. 1982. Methods of correction for systematic error in point precipitation measurement for operational use. Operational Hydrology Report No. 21, World Meteorological Organization, Geneva, 91 pp.

Sevruk, B., and W.R. Hamon. 1984. International Comparison of National Precipitation Gauges with a Reference Pit Gauge. Instruments and Observing Methods Report No. 17, World Meteorological Organization, Geneva, 111 pp.

Shea, D., R. Jenne, and C. Ropelewski. 1992. NCAR data sets with daily maximum and minimum temperatures. In Proceedings of the International Temperature Workshop. D.E. Parker (ed.). Climate Research Technical Note No. 30, Hadley Centre, UKMO, Bracknell, Berkshire, pp. 27-29.

Shiotani, M. 1990. Low-frequency variations of the zonal mean state of the Southern Hemisphere troposphere. J. Meteorol. Soc. Japan 68:461-471.

Shukla, J. 1995. The initiation and persistence of the Sahel drought. In Natural Climate Variability on Decade-to-Century Time Scales. D.G. Martinson, K. Bryan, M. Ghil, M.M. Hall, T.R. Karl, E.S. Sarachik, S. Sorooshian, and L.D. Talley (eds.). National Academy Press, Washington, D.C.

Skaggs, R.H. 1975. Drought in the United States, 1931-40. Ann. Assoc. Amer. Geogr. 65:391-402.

Slutz, R.J., S.J. Lubker, J.D. Hiscox, S.D. Woodruff, R.L. Jenne, D.H. Joseph, P.M. Steurer, and J.D. Elms. 1985. Comprehensive Ocean-Atmosphere Data Set: Release I. Climate Research Program, NOAA Environmental Research Laboratories, Boulder, Colo., 268 pp. (NTIS PB86-105723).

Solow, A.R. 1987. Testing for climate change: An application of the two-phase regression mode. J. Clim. Appl. Meteorol. 26:1401-1405.

Sonett, C.P. 1983. Is the sunspot spectrum really so complicated? In Weather and Climate Responses to Solar Variations. B.M. McCormac (ed.). Colorado Associated University Press, Boulder, pp. 607-613.

Spencer, R.N., and J.R. Christy. 1990. Precise monitoring of global temperature trends from satellites. Science 247:1558-1562.

Stocker, T.F., and L.A. Mysak. 1992. Climatic fluctuations on the century time scale: A review of high-resolution proxy data and possible mechanisms. Climatic Change 20:227-250.

Szeredi, I., and D.J. Karoly. 1987a. The vertical structure of monthly fluctuations of the Southern Hemisphere troposphere. Aust. Meteorol. Mag. 35:19-30.

Szeredi, I., and D.J. Karoly. 1987b. The horizontal structure of monthly fluctuations of the Southern Hemisphere troposphere from station data. Aust. Meteorol. Mag. 35:119-129.

Taljaard, J.J., W. Schmitt, and H. van Loon. 1961. Frontal analysis with application to the Southern Hemisphere. Notos 10:25-58.

Taylor, P.K. 1984. The determination of surface fluxes of heat and water by satellite microwave radiometry and in situ measurements. In Large-Scale Oceanographic Experiments and Satellites. C. Gautier and M. Fixe (eds.). Series C: Mathematical and Physical Sciences, Vol. 128, pp. 223-246.

Thiessen, A.H. 1911. Precipitation averages for large areas. Mon. Weather Rev. 39:1082-1084.

Thompson, L.G., E. Mosley-Thompson, and B.M. Arnao. 1984.

El Niño-Southern Oscillation events recorded in the stratigraphy of the tropical Quelccaya ice cap, Peru. Science 226:50-53.

Thompson, L.G., E. Mosley-Thompson, W. Dansgaard, and P.M. Grootes. 1986. The Little Ice Age as recorded in the stratigraphy of the tropical Quelccaya ice cap. Science 234:361-364.

Thompson, L.G., E. Mosley-Thompson, M.E. Davis, J. Bolzan, J. Dai, N. Gundestrup, T. Yao, X. Wu, L. Klein, and Z. Zichu. 1990. Glacial stage ice core records from the subtropical Dunde ice cap, China. Ann. Glaciol. 14:288-297.

Thompson, L.G., E. Mosley-Thompson, M.E. Davis, P.-N. Lin, K.A. Henderson, J. Cole-Dai, J.F. Bolzan, and K.-B. Liu. 1995. Late glacial stage and Holocene tropical ice core records from Huascaran, Peru. Science 269:46-50.

Toomig, K.G. 1981. Correlation of mean annual albedo and short-wave radiation balance with these parameters in early spring. Meteor. Gidrol. 5:48-52.

Toussoun, O. 1923. Mémoire sur l'histoire du Nil. Mém. Inst. Égypte 9:63-213.

Treidl, R.A. 1970. A case study of warm air advection over a melting snow surface. Bound. Layer Meteorol. 1:155-168.

Trenberth, K.E. 1976. Fluctuations and trends in indices of the Southern Hemisphere circulation. Quart. J. Roy. Meteorol. Soc. 102:65-75.

Trenberth, K.E. 1979. Interannual variability of the 500mb zonal mean flow in the Southern Hemisphere. Mon. Weather Rev. 107:1515-1524.

Trenberth, K.E. 1984. Interannual variability of the Southern Hemisphere circulation: Representativeness of the year of the Global Weather Experiment. Mon. Weather Rev. 112:108-123.

Trenberth, K.E. 1990. Recent observed interdecadal climate changes in the Northern Hemisphere. Bull. Amer. Meteorol. Soc. 71:988-993.

Trenberth, K.E. (ed.). 1992. Climate Systems Modeling. Cambridge University Press, 788 pp.

Trenberth, K.E., and G.W. Branstator. 1992. Issues in establishing causes of the 1988 drought over North America. J. Climate 5:159-172.

Trenberth, K.E., and J.R. Christy. 1985. Global fluctuations in the distribution of atmospheric mass. J. Geophys. Res. 90:8042-8052.

Trenberth, K.E., and J.W. Hurrell. 1995. Decadal climate variations in the Pacific. In Natural Climate Variability on Decade-to-Century Time Scales. D.G. Martinson, K. Bryan, M. Ghil, M.M. Hall, T.R. Karl, E.S. Sarachik, S. Sorooshian, and L.D. Talley (eds.). National Academy Press, Washington, D.C.

Troup, A.J. 1965. The Southern Oscillation. Quart. J. Roy. Meteorol. Soc. 91:490-506.

Tyson, P.D. 1980. Temporal and spatial variation of rainfall anomalies in Africa south of latitude 22° during the period of meteorological record. Climate Change 2:363-371.

U.S. Department of Commerce. 1950. Census of Agriculture. Irrigation of Agricultural Levels. Vol. III. (Available through the U.S. Government Printing Office.)

U.S. Department of Commerce. 1988. Farm and Ranch Irrigation Survey. Vol. 3, Part 1. (Available through the U.S. Government Printing Office.)

van Loon, H., and J. Williams. 1978. The association between

mean temperature and interannual variability. Mon. Weather Rev. 106:1012-1017.

Venrick, E.L., J.A. McGowan, D.R. Cayan, and T. Hayward. 1987. Climate and chlorophyll a: Long-term trends in the central North Pacific Ocean. Science 238:70-72.

Villalba, R. 1990. Climatic fluctuations in northern Patagonia during the last 1000 years as inferred from tree-ring records. Quat. Res. 34: 346-360.

Villalba, R., J.A. Boninsegna, and D.R. Cobos. 1989. A tree-ring reconstruction of summer temperature between A.D. 1500 and 1974 in western Argentina. In Extended Abstracts, Third International Conference on Southern Hemisphere Meteorology and Oceanography. American Meteorological Society, Boston, pp. 196-197.

Vining, K.C., and J.F. Griffiths. 1985. Climatic variability at ten stations across the United States. J. Clim. Appl. Meteorol. 24:363-370.

Vinnikov, K.Ya., P.Ya. Groisman, and K.M. Lugina. 1900. Empirical data on contemporary global climate changes (temperature and precipitation). J. Climate 3:662-677.

Wahl, E.W. 1968. A comparison of the climate of the eastern United States during the 1830's with the current normals. Mon. Weather Rev. 96:73-82.

Walker, G.R. 1910. Correlations in seasonal variations of weather. Mem. India Meteorol. Dept. 21:22-45.

Walker, J., and P.R. Rowntree. 1977. The effect of soil moisture on circulation and rainfall in a tropical model. Quart. J. Roy. Meteorol. Soc. 103:29-46.

Wallace, J.M., and D.S. Gutzler. 1981. Teleconnections in the geopotential height field during the Northern Hemisphere winter. Mon. Weather Rev. 109:784-812.

Wallace, J.M., C. Smith, and Q. Jiang. 1990. Spatial patterns of atmosphere/ocean interaction in the northern winter. J. Climate 3:990-998.

Walsh, J.E. 1995. Continental snow cover and climate variability. In Natural Climate Variability on Decade-to-Century Time Scales. D.G. Martinson, K. Bryan, M. Ghil, M.M. Hall, T.R. Karl, E.S. Sarachik, S. Sorooshian, and L.D. Talley (eds.). National Academy Press, Washington, D.C.

Walsh, J.E., and C.M. Johnson. 1979. An analysis of Arctic sea ice fluctuations, 1953-1977. J. Phys. Oceanogr. 9:580-591.

Walsh, J.E., W.H. Jasperson, and B. Ross. 1985. Influences of snow cover and soil moisture on monthly air temperature. Mon. Weather Rev. 113:756-768.

Wang, W.-C., Z. Zeng, and T.R. Karl. 1990. Urban heat islands in China. Geophys. Res. Lett. 17:2377-2389.

Ward, M.N. 1992. Provisionally corrected surface wind data, worldwide ocean-atmosphere surface fields and Sahel rainfall variability. J. Climate 5:454-475.

Washington, W.M., and G.A. Meehl. 1986. General circulation model CO$_2$ sensitivity experiments: Snow-sea ice albedo parameterizations and globally averaged surface air temperature. Climatic Change 8:231-241.

Weare, B.C. 1989. Uncertainties in estimates of surface heat fluxes derived from marine reports over the tropical and subtropical oceans. Tellus 41A:357-370.

Weaver, A.J., and E.S. Sarachik. 1991. Evidence for decadal variability in an ocean general circulation model: An advective mechanism. Atmos.-Ocean 29:197-231.

Weaver, A.J., E.S. Sarachik, and J. Marotzke. 1991. Fresh water flux forcing of decadal and interdecadal oceanic variability. Nature 353:836-838.

Wells, L.E. 1987. An alluvial record of El Niño events from northern coastal Peru. J. Geophys. Res. 92C:14463-14470.

Whetton, P., D. Adamson, and M.A.J. Williams. 1990. Rainfall and river flow variability in Africa, Australia, and East Asia linked to El Niño-Southern Oscillation events. Geol. Soc. Aust. Symp. Proc. 1:21-26.

White, W.B., and A.E. Walker. 1974. Time and depth scales of anomalous subsurface temperature at ocean weather stations P, N, and V in the North Pacific. J. Geophys. Res. 79:4517-4522.

Wigley, T.M.L. 1989. Possible climate change due to SO_2-derived cloud condensation nuclei. Nature 339:365-367.

Wigley, T.M.L. 1991. Could reducing fossil-fuel emissions end global warming? Nature 349:503-505.

Williams, J. 1975. The influence of snow cover on the atmospheric circulation and its role in climatic change: An analysis based on results from the NCAR global circulation model. J. Appl. Meteorol. 14:137-152.

Wolter, K. 1989. Modes of tropical circulation, southern oscillation, and Sahel rainfall anomalies. J. Climate 2:149-172.

Wood, F.B. 1988. Global alpine glacier trends, 1960's to 1980's. Arct. Alp. Res. 20:404-413.

Woodruff, S.D., R.J. Slutz, R.L. Jenne, and P.M. Steurer. 1987. A comprehensive ocean-atmosphere data set. Bull. Amer. Meteorol. Soc. 68:1239-1250.

World Meteorological Organization (WMO). 1979. Climatic Atlas of North and Central America, Vol. 1. WMO/UNESCO, Geneva, 39 pp.

World Meteorological Organization (WMO). 1991. Final Report of the Fifth Session of the International Organizing Committee for the WMO Solid Precipitation Measurement Intercomparison. WMO, Geneva, 19 pp.

World Water Balance and Water Resources of the Earth (WWB). 1974. Gidrometeoizdat, Leningrad, 638 pp. (in Russian; published in English by UNESCO Press in 1978).

World Glacier Monitoring Service. 1993. Fluctuations of Glaciers: 1895-1990 (Vol. VI). IAHS (ICSU)/UNEP/UNESCO, Paris, 321 pp. plus plates and charts.

Wright, P.B. 1986. Problems in the use of ship observations for the study of interdecadal climate change. Mon. Weather Rev. 114:1028-1034.

Wright, P.B. 1988. An atlas based on the COADS data set: Fields of mean wind, cloudiness and humidity at the surface of the global ocean. Report No. 14, Max Planck-Institut für Meteorologie, Hamburg, 70 pp.

Wunsch, C. 1992. Decade-to-century changes in the ocean circulation. Oceanography 5:99-106.

Xue, Y., and J. Shukla. 1993. The influence of the land surface properties on Sahel rainfall. Part I: Desertification. J. Climate 6:2232-2245.

Yang, J., and J.D. Neelin. 1993. Sea-ice interactions with the thermohaline circulation. Geophys. Res. Lett. 20:217-220.

Yasunari, T., A. Kitoh, and T. Tokioka. 1989. The effect of Eurasian snow cover on the summer climate of the Northern Hemisphere: A study by the MRI-GCM. In Research Activities in Atmospheric and Oceanic Modelling. G. Boer (ed.). WMO/TD-No. 332, World Meteorological Organization, Geneva, pp. 7.37-7.38.

Yeh, T.C., R. Wetherald, and S. Manabe. 1983. A model study of the short term climatic and hydrological effects of sudden snow-cover removal. Mon. Weather Rev. 111:1013-1024.

Zwiers, F.W. 1987. A potential predictability study conducted with an atmospheric general circulation model. Mon. Weather Rev. 115:2957-2974.

Atmospheric Modeling

MICHAEL GHIL

A mature (albeit incomplete) understanding of the atmospheric component of the climate system is now available, thanks to the models of varying complexity and detail that have been developed over the last few decades. Because of the dominant role the atmosphere plays in weather variability on time scales of hours to weeks, as well as the practical importance of weather forecasting, observations of atmospheric phenomena began before observations of other climate subsystems, and are still the most plentiful. On the other hand, on the decade-to-century time scale, the atmosphere—with its characteristic time of months to years, at most—does not play as dominant a role in the complicated collective behavior of the above-mentioned family of subsystems. (This topic is discussed further in the introduction to Chapter 4.)

The purpose of reviewing atmospheric modeling in the decade-to-century context is therefore twofold: (1) as a useful in-depth illustration of the broader climate modeling enterprise, since it has the longest history and largest variety of models, as well as the best data sets for validation, and (2) as the source of indispensable building blocks for a hierarchy of coupled climate models. The papers in this section discuss useful aspects of the validating data sets and of the methodology for their analysis, as well as an important subset of the great variety of atmospheric models. This introduction provides a quick review of the breadth of atmospheric models, as well as of validation methodology. Attention is drawn to those aspects that are covered in greater depth and detail by the papers that follow.

THE MODELING HIERARCHY

Decade-to-century-scale variability, natural and anthropogenic, is but one of the problems of climate dynamics. Various problems, in terms of temporal and spatial scales, have been addressed with a variety of models. These models span the entire spectrum of detail and complexity, from zero-dimensional models that lump the entire atmosphere into one single variable—global temperature—to very highly resolved, three-dimensional general-circulation models (GCMs). Intermediate models can have one or two spatial dimensions.

Each type of model has its strengths and weaknesses; none is perfectly reliable, nor can it address all climate problems on decade-to-century or any other time scales. To obtain, therefore, a reliable estimate of climate sensitivity to external forcing—natural or anthropogenic—two approaches must be employed: (1) the systematic use of a complete hierarchy of climate models, with simpler models being tailored to the in-depth study of specific climate processes or phenomena, while more complex models are geared mainly to simulating a wide range of observed details (Schneider and Dickinson, 1974); and (2) the validation of the models in this hierarchy against each other and against existing data sets.

The simplest zero-dimensional (0-D) model is applied in this section by Wigley and Raper (1995) and by Lindzen (1995) to decade-to-century-scale problems. It equates the rate of change of the climate system's heat content to the balance of incoming-minus-outgoing radiation. The heat content equals the heat capacity times the average temperature of the system. Since the heat capacity of water is much higher than that of air, the rate of change of the heat content depends crucially on the volume of water that is included in the system, along with the total volume of the atmosphere: The former can extend from the oceanic mixed layer, of a few tens of meters, to the entire depth of the ocean, of several kilometers. Accordingly, the atmospheric temperature—assumed to be in equilibrium with the appropriate volume of water—will change more or less rapidly for a given change in the radiation balance.

The radiation balance is often expressed, in the context of decade-to-century variability or of paleoclimate studies, as the sum of a feedback term, dependent on global average temperature, and of a prescribed source term. The feedback term, in its simplest form, is linear, i.e., proportional to temperature. In this version of a 0-D model, equilibrium temperature equals the source term divided by the constant of proportionality, or gain. The change in temperature is thus directly proportional to the prescribed change in radiation balance and inversely proportional to the feedback gain. In the context of decade-to-century-scale climate variability, the externally prescribed change in radiation balance is due to an increase in greenhouse-gas concentration, which leads to a more positive balance at the surface, or to an increase in the concentration of aerosols of natural or anthropogenic origin (e.g., volcanoes or industry respectively), which leads to a more negative surface balance; insolation changes are admittedly smaller, on this time scale, and can contribute in either direction.

The characteristic time needed for temperature to reach a new equilibrium when the source term has been changed is inversely proportional to the heat capacity. This time can be as short as a few years or as long as millennia, depending on the amount of water assumed to be in equilibrium with the surface temperature. The importance of these two model constants, gain and heat capacity, in validating such 0-D linear models is discussed by Lindzen (1995).

Somewhat more complex than the 0-D atmospheric models are the one-dimensional (1-D) models, which fall into two main categories: energy-balance models (EBMs) and radiative-convective models (RCMs). In EBMs, the spatial coordinate treated explicitly is latitude; the rate of change of the heat content of each latitude belt is equal to the incoming radiation minus the outgoing, as in the 0-D models discussed first, plus a redistribution term between belts. These models were introduced, independently of each other, by Budyko (1969) and Sellers (1969) to study the expected cooling of climate as a result of increased aerosol loading. Their main feature was the nonlinear dependence of planetary reflectivity, or albedo, on surface temperature, which determined the presence or absence of ice at a given latitude. As a result of this nonlinear ice-albedo feedback, tempera-

ture was shown to drop precipitously as the insolation received at the surface decreased below a certain threshold.

Ghil (1976), Held and Suarez (1974), and North (1975), using slightly different formulations of EBMs, showed that this precipitous drop was due to the presence of multiple equilibria and the ensuing hysteresis. The existence of multiple model equilibria and their dependence on insolation changes was confirmed by Wetherald and Manabe (1975) with a somewhat simplified GCM, illustrating for the first time the power of the hierarchic modeling concept. With their increased geographic detail, 1-D models have been used to address such interesting decade-to-century-scale problems as polar amplification of climate perturbations—i.e., the fact that surface-air temperature changes are larger at high than at low latitudes. Polar amplification, which has been noted in a number of GCM studies, can be easily understood in the 1-D EBM context (Ghil, 1976).

In RCMs (Ramanathan and Coakley, 1978), the resolved dimension is height, and the main mechanism under study is the interaction of radiation with clouds. A number of papers by Cess (1976), Manabe and Wetherald (1967), Schneider (1972), and others considered the effects of relative humidity, partial cloud cover, cloud height, and other factors on the vertical distribution of heat flux and temperature. RCMs are often used for the off-line testing of radiative and cloudiness formulations, or parameterizations, of sub-grid-scale thermodynamic processes in GCMs. The performance of GCMs is also examined by intercomparison with RCM results, as well as with 0-D and 1-D EBMs.

The second model type, intermediate between the simplest, 0-D models and the most detailed ones, the GCMs, is given by the two-dimensional (2-D) models; these are of three kinds, according to the phenomena they concentrate on and the spatial coordinates they resolve. Zonally symmetric, meridional-plane models essentially combine and extend the main features of 1-D EBMs and RCMs. Considerable attention was given to the development of such models in the 1960s and early 1970s by B. Saltzman and colleagues (e.g., Saltzman and Vernekar, 1971; see also Saltzman's 1978 review). Zonally symmetrized versions of GCMs have also been found useful in the study of intraseasonal and interannual variability (Goswami and Shukla, 1984). Two-dimensional meridional models are widely applied to coupled dynamics-chemistry problems in the stratosphere (Andrews et al., 1987; Kaye, 1987) that arise in connection with ozone depletion in high latitudes.

Another kind of 2-D model extends EBMs to greater geographic detail, resolving both latitude and longitude. In fact, one subset of these, the thermodynamic models of J. Adem and collaborators (e.g., Adem, 1964), preceded the 1-D EBMs introduced by Budyko (1969) and Sellers (1969) that were so thoroughly analyzed in the mid-1970s. The results of the former—like those of the 2-D meridional models in the previous paragraph—attracted less attention than the results of the 1-D EBMs because of the lack of interesting nonlinear feedbacks, such as the ice-albedo feedback. Various applications of 2-D EBMs to paleoclimate studies are reviewed by Crowley and North (1991). In this section North and Kim (1995) consider a randomly forced 2-D EBM in the context of decade-to-century-scale variability.

Finally, there are 2-D models of barotropic planetary flow as well, resolved in latitude and longitude. These focus on the dynamics that is not contained in 2-D EBMs. Highly truncated barotropic models (Charney and DeVore, 1979) have provided insight—similar to that of the 1-D EBMs for atmospheric thermodynamics—into the possibility of multiple equilibria in atmospheric dynamics. More highly resolved versions of such models have shown a rich variety of flow regimes (Legras and Ghil, 1985) and of intraseasonal oscillations (Ghil et al., 1991). Intermediate between these 2-D models and GCMs are baroclinic flow models with two to five layers and limited sub-grid-scale parameterizations (Reinhold and Pierrehumbert, 1982; Simmons and Hoskins, 1979). Such models can play an important role—complementary to high-resolution GCMs—in determining the change in frequency and severity of persistent anomalies in the atmospheric circulation, given changes in mean temperature or in surface insolation.

The application of three-dimensional (3-D) GCMs to decade-to-century-scale and paleoclimate variability is discussed in this section by Rind and Overpeck (1995). A general introduction is provided by Washington and Parkinson (1986). GCMs are computer models based on the equations governing the 3-D dynamics and thermodynamics of the atmosphere. They provide today a fairly detailed and reliable simulation of the present climate. The ultimate goal of GCM modeling is to achieve such a simulation by deriving all the terms in the model equations from first principles of atmospheric dynamics, physics, and chemistry. In practice, given the incomplete state of our knowledge in these fields, and the limited spatial resolution permitted by any computational device, GCMs contain many semi-empirical terms; their empirical parameters are calibrated ("tuned"), explicitly or implicitly, to the present climate. Hence, the adequate simulation of the latter does not guarantee that a GCM's climate sensitivity will be an equally good approximation of the natural climate's sensitivity.

The major conclusion of this brief review is that modeling and predicting natural climate variability on decade-to-century time scales will require the use of the full range of atmospheric models available. GCM results on the steady-state and transient response of climate to greenhouse-gas and aerosol loading are routinely interpreted in terms of the simple linear 0-D EBM presented at the beginning of this section (Hansen et al., 1985). The intermediate models—1-D and 2-D—described here capture better the climate system's essential inhomogeneities and non-linearities (Ghil

and Childress, 1987). They should therefore play a more important role in understanding and complementing GCM results than has been the case heretofore.

VALIDATION METHODOLOGY

In some sense, model validation can never be complete, since we are making models to describe or predict behavior that we cannot observe directly in nature; that is a primary goal of modeling. On the other hand, confidence in a model's predictions increases as additional aspects of its behavior come to agree with those observations we have. Validation methodology is thus essential to our confidence in model performance.

A model can be validated against other models, whose behavior either is better understood or has been previously tested against observations, or directly against observations. To be more precise, a physical model is usually tested against some statistical model of reality that is itself fitted to the raw observations. The role of a statistical model is more that of a go-between in comparing a dynamical model with reality than that of a competitor. Ideally, the same statistical model should fit equally well both the data and the solution of the dynamical, physical, or—more generally—understanding-based model. For the purpose of illustration in this brief review, the discussion that follows emphasizes validation in the time and frequency domain. Validation in the space and wave-number domain is at least as important for decade-to-century-scale variability. Questions that pertain to changes in the position and intensity of mid-latitude storm tracks or of tropical cyclones, changes in the location and extent of droughts or floodings, or changes in other persistent anomalies in the workings of the climate system cannot be reliably addressed without proper validation of the models used in the latter domains. The main tools of such validation are mass-, energy-, and momentum-flux diagnostics, along with the examination of the spatial patterns of the meteorological fields of primary interest, such as surface air temperature and precipitation.

North and Kim (1995) review in this section the classical estimation and prediction theory of random processes due to Kolmogorov (1941) and Wiener (1956), and its application to climate signals by Barnett (1986) and Bell (1986). Considerable attention has been given recently in the geosciences in general and in climate signal analysis in particular to two relatively novel methodologies: the multi-taper method and singular spectrum analysis.

The multi-taper method (MTM) introduced by Thompson (1982) uses an optimal set of tapers, based on Slepian's (1978) work, rather than the single, empirical taper used in classical spectral filters (Jenkins and Watts, 1968). It permits high spectral resolution, as well as confidence intervals that are based on coherence; these intervals are therefore independent of spectral amplitude, unlike those provided by classical Blackman-Tukey filters. MTM can thus detect small-amplitude organized variability with measurable statistical confidence. Kuo et al. (1990) applied MTM to demonstrate coherence between the instrumental temperature and carbon dioxide (CO_2) records.

Yiou et al. (1991) applied evolutive spectral analysis based on MTM to the study of high-frequency paleoclimate variability in the Vostok ice core. In this context, it is important to notice that partial CO_2 pressure in the air bubbles trapped in the core (Barnola et al., 1987) rises at the same time as the temperature recorded isotopically in the ice itself—for example, from the last glacial maximum into the Holocene, about 20,000 years ago. But temperature drops thousands of years before the CO_2 level does, as happened with the shift from the previous climate optimum into the last glaciation, about 120,000 years ago. This indicates that, at least on paleoclimate time scales, temperature does not respond to CO_2 changes alone, but may be influenced by other factors or may interact with CO_2 in a nonlinear fashion that distinguishes between phase relations during warming and cooling episodes. On decade-to-century time scales, the still-incomplete atmospheric observational evidence is only marginally useful at present in diagnosing phase relationships among various internal and external variables (Karl et al., 1995, in this volume).

Singular spectrum analysis (SSA) was developed by Broomhead and King (1986) and Vautard and Ghil (1989). It uses data-adaptive basis functions that are the time analogs of empirical orthogonal functions (EOFs) in space (Preisendorfer, 1988; North and Kim, 1995). Pairs of such temporal EOFs can efficiently describe anharmonic, nonlinear oscillations. Using SSA on instrumental records of global surface-air temperature (Jones et al., 1986; IPCC, 1990), Ghil and Vautard (1991) detected organized climate variability with peaks near 10, 15, and 25 years. The first two peaks were confirmed by considering the spatial coherence of the same temperature data set using SSA (Allen and Smith, 1994) and MTM (Mann and Park, 1993) respectively.

A 31-year peak was also found in 2,000-year-long tree-ring records from Tasmania by Cook et al. (1995, in this volume), and a 27-year peak in the Koch index of sea-ice extent around Iceland by Stocker and Mysak (1992). The longest continuous instrumental record of local surface air temperature is the central England record (Manley, 1974; Parker et al., 1991). Figure 1 displays the power spectrum of this 335-year-long record (Plaut et al., 1995). Similar interdecadal peaks were also found, using entirely different spectral methods, by Keeling and Whorf (1995, in this volume) in global temperature records. One plausible explanation for such internal variability on interdecadal time scales lies in changes of the ocean's thermohaline circulation (Weaver et al., 1991; Quon and Ghil, 1992).

Wallace (1995, in this section) separates carefully the interdecadal variability apparent in temperature records

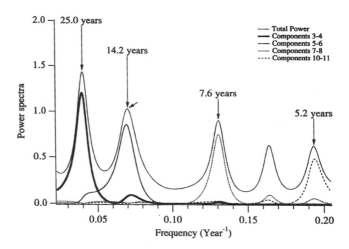

FIGURE 1   Stack spectrum of the central England temperature record, using SSA combined with a low-order, robust, maximum-entropy method. Each line shows the power spectrum of a particular oscillation, isolated as a pair of temporal EOFs, essentially removing the broad-band variability in which the peaks are embedded (see text for details). (From Plaut et al., 1995; reprinted with permission of the American Association for the Advancement of Science.)

from higher-frequency variability due to the El Niño/Southern Oscillation phenomenon and volcanic eruptions. He emphasizes the need for using satellite data in considering spatial details of natural and anthropogenic variability, especially for temperatures aloft.

To derive full benefit from sophisticated methods of signal detection and analysis, the same methods have to be applied to the existing data and to model simulations. This emphasizes again the need for intercomparing detailed, but short, GCM simulations with the longer, but less detailed, time series generated by simple and intermediate models, as well as with the data themselves. These comparisons need to be carried out in both the spatial and the temporal domains.

A rich panoply of atmospheric models has been developed over the last 40 years, from zero-dimensional EBMs to fully three-dimensional GCMs, passing through one- and two-dimensional EBMs, RCMs, and dynamic flow models. These models have been widely used to study anthropogenic climate change due to increases in aerosol loading or greenhouse-gas concentrations. Their use for the study of internal climate variability on the time scales of interest here has only begun.

Understanding climate response to external forcing, natural or anthropogenic, requires that one understand slow shifts in equilibrium behavior, and then—more important—slow shifts in internal variability. Even for the former, passive response, considerable uncertainty exists at present as to amplitude and timing. The IPCC estimates the global surface-air temperature increase for a doubling of present CO_2 concentration to be anywhere from 1.5°C to 4.5°C

(IPCC, 1990); for a sudden doubling, decades to a century may elapse before a substantial fraction of this response will occur. The study of internal variability, and of its interaction with the external forcing, will require improving each type of model, and enhancing the intercomparisons between model results. A more intensive use of the hierarchical modeling concept can be guided by the ideas of nonlinear dynamics, which provide a common thread in the analysis of complex behavior for time-dependent models with arbitrary spatial detail.

Given the shortness of instrumental records, model validation on the decade-to-century time scale requires the use of sophisticated methods for their analysis and interpretation. A number of such methods have been applied recently to climate records, with promising results, and are being extended from the purely temporal to the combined space-time domain. These methods need to be refined further and applied systematically to model results, as well as to the existing records. Similar features detected in a consistent manner, by applying the same statistical methods, in models and data will increase our confidence in the former, our understanding of the latter, and hence our ability to predict climate change on the time scales of interest.

Modeling Low-Frequency Climate Variability: The Greenhouse Effect and the Interpretation of Paleoclimatic Data

TOM M.L. WIGLEY[1] AND SARAH C.B. RAPER[2]

ABSTRACT

Detection of the decade-to-century time-scale effects of any external forcing agent on global-mean temperature requires a knowledge of internally generated natural variability on these time scales. Methods for estimating the magnitude of this variability are described. While the observed twentieth-century warming agrees well with model estimates based on greenhouse-gas and aerosol forcing alone, it is shown that changes of similar magnitude could have occurred solely through internally generated variability. Similarly, observed changes in global-mean temperature over the past 10,000 years (in particular, the numerous Little Ice Age events) could have been internally generated. At the very least, if these events were externally forced (e.g., by solar irradiance changes) then internal variability would probably have modified the response markedly from a "pure" response to external forcing, making the interpretation of the paleoclimatic record extremely difficult.

INTRODUCTION

Understanding the past is often said to be the key to predicting the future. The reverse is equally true: Attempts to predict the future can help us to understand the past. These maxims will be illustrated below.

An important but nebulous question raised in the context of the greenhouse effect is, "Have we detected it?" This issue concerns not just anthropogenic greenhouse-gas-induced climate change but past climate changes as well. The primary question in both cases is the identification of an externally forced change (a signal) in the presence of natural variability (background noise). Of course, detection requires more than just identifying a climate change that is consistent with some hypothesis—it requires also the demonstration of a cause-effect relationship and the attribution of some part of the observed changes to the particular cause. The present paper deals mainly with identification, rather than attribution.

We begin by comparing predicted and observed changes in global-mean temperature over the instrumental period, highlighting the role of low-frequency natural variability. We then review methods that may help to elucidate the character of this variability. Finally, we broaden the detection issue to consider the detection of external forcing effects

[1]University Corporation for Atmospheric Research, Boulder, Colorado
[2]Climatic Research Unit, University of East Anglia, Norwich, United Kingdom

in general, and interpret the paleoclimatic record in this general detection context. The key distinction here is between externally forced and internally generated low-frequency natural variability. Separating these factors in past data appears to be a very difficult task.

CLIMATE CHANGE SINCE PRE-INDUSTRIAL TIMES

The most obvious detection-related question is: Do the observed changes in global-mean temperature agree with predictions of changes induced by human activities? To answer this question we need an estimate of the predicted changes (the signal), appropriate observational data, and suitable statistical methods to compare the two—i.e., to decide whether any similarity between them is more than might be expected to occur by chance. The last aspect requires a knowledge of the low-frequency character of natural climate variability.

The anthropogenic global-mean temperature signal to date (1765-1990) is subject to considerable uncertainty. Predictions of this signal depend both on the radiative forcing changes that have occurred and on how temperature responds to such changes (which, in turn, is determined largely by the climate sensitivity). Both factors introduce uncertainties. The imposed forcing, while undeniably large, is uncertain by a factor of two (1.06 to 2.06 W m^{-2}, global-mean). Following Wigley and Raper (1992), the total forcing may be broken down into the effects of CO_2 (1.50 W m^{-2}), CH_4 (0.56 W m^{-2}), N_2O (0.10 W m^{-2}), halocarbons (0.02 to 0.28 W m^{-2}, depending on whether one includes the possible effect of halocarbon-induced stratospheric ozone changes), and SO_2-derived sulphate aerosols (-0.38 to -1.13 W m^{-2}).

Most of the forcing uncertainty derives from the aerosol component. The lower limit for the total global forcing, about 1 Wm^{-2}, corresponds to the high aerosol forcing case. While the magnitude of the aerosol forcing is highly uncertain, its upper bound is probably constrained by observational evidence of hemispheric-scale temperature changes. Because most of the forcing is in the Northern Hemisphere, a global-mean aerosol forcing in excess of about -1 W m^{-2} would imply such a large interhemispheric forcing contrast[3] that the effects would likely be obvious in the observed temperature data (Wigley, 1989).

The response to this man-made forcing, i.e., the global-mean temperature signal, is subject to still further uncertain-

---

[3]The 1.06 W m^{-2} lower limit for total forcing arises when the global-mean aerosol forcing is -1.13 W m^{-2}. It corresponds to forcings of 1.96 W m^{-2} for the Southern Hemisphere mean and only 0.16 W m^{-2} for the Northern Hemisphere. The climate implications of such a contrast are uncertain, but simple models suggest that it should lead to a substantial contrast in hemispheric-mean temperature trends, which has not been observed.

ties. These arise mainly from uncertainties in the climate sensitivity and in the damping effect of oceanic thermal inertia. For "best-guess" forcing (1.44 W m^{-2} over 1765-1990), simple model calculations (using an energy-balance model with an upwelling-diffusion ocean) give a warming over 1880-1990 of 0.3 to 0.6°C (corresponding to a climate sensitivity of 1.5 to 4.5°C for $2 \times CO_2$) (Wigley and Raper, 1992). The range of possible values is slightly increased if one accounts for uncertainties in the model parameters that determine ocean mixing, and is considerably larger if one accounts for forcing uncertainties.

The above model-based range for 1880-1990 warming is, in fact, the same as the range that the Intergovernmental Panel on Climatic Change (Folland et al., 1990) gives for the observed warming over this period. What is the significance of this consistency? From paleoclimatic data, we know that the climate system has a high level of natural variability. A greenhouse-warming skeptic could claim, therefore, that the observed warming was merely (or largely) natural variability, that the consistency was a fluke, and that the model-predicted warming was too high. Alternatively, greenhouse-warming extremists might claim that the model prediction was too low, and that natural variability had actually offset a much larger anthropogenic signal. Clearly, a crucial issue is to unravel the effects of man from natural variability on the time scale of 50 years or more. Unfortunately, our quantitative knowledge of natural variability is at least as poor as our knowledge of the enhanced greenhouse effect. Furthermore, as we will show below, the magnitudes of any externally forced temperature signal and of low-frequency internal variability are both dependent on the climate sensitivity, making their separation even more difficult.

QUANTIFYING LOW-FREQUENCY VARIABILITY

How do we obtain information about natural 50-year time scale fluctuations in *annual global-mean* temperature? There are two possible approaches: using observational data, or using models.

Using Observational Data

If observational data are to be used directly, then one must resort to paleoclimatic data because of the limited duration of the available instrumental record. Paleoclimatic information, however, is invariably local or regional rather than global, and the data are usually seasonally specific and subject to uncertainties because of the complex links between proxy data and temperature (see, e.g., Bradley and Jones, 1992a). It is therefore very difficult to estimate changes in annual global-mean temperature prior to 1860.

In spite of the data uncertainties, there is strong evidence that substantial 50-year time-scale changes in global-mean temperature have occurred over the past 10,000 years (i.e., the period over which we have the best data and during which the planet's albedo-related boundary conditions have probably varied little) (Röthlisberger, 1986; Grove, 1988). Since this evidence is indirect and based on limited geographical coverage, we can only offer speculative estimates of the magnitude of these annual global-mean temperature changes. Most probably, the fluctuations have been within a band 0.5°C wide (Wigley and Kelly, 1990). Dating and data uncertainties do not allow us to compare the recent warming with similar warmings in the past in any useful quantitative way, so we cannot say whether the twentieth-century warming is occurring at an unprecedented rate.

Using Models

An alternative approach is to try to estimate 50-year time-scale variability using either a statistical or a deterministic model. In both cases, information from the observed, instrumental record of global-mean temperature changes over the past 130 years is used to estimate the spectrum of natural variability over the full range of frequencies.

In the statistical approach, the data are fitted to a standard-time series model of some sort (such as an nth-order autoregressive process or some more complex model), and the fitted model is used to extend the spectrum to lower frequencies (e.g., Bloomfield and Nychka, 1992; Kheshgi and White, 1993).

A deterministic approach has been used by Wigley and Raper (1990, 1991). Here, a simple climate model is employed to extend the spectrum of observed variability: The method used is to force the model stochastically with white-noise forcing, the magnitude of which is chosen to match the observed high-frequency variability (1- to 10-year time scale). It should be noted that this approach yields information only on *passive* internal variability, i.e., variability that arises in the absence of ocean circulation changes. Low-frequency variability may also occur, for example, through changes in the ocean's thermohaline circulation, as evidenced by recent coupled ocean/atmosphere general-circulation model (GCM) results (e.g., Washington and Meehl, 1989; Manabe et al., 1991; Cubasch et al., 1992) and by much longer simulations with an ocean GCM (Mikolajewicz and Maier-Reimer, 1990) and with a two-dimensional ocean model (Mysak et al., 1993). It has been speculated that such circulation changes may be coupled to surface temperature changes, leading to feedbacks that would amplify low-frequency climate variability (e.g., Gaffin et al., 1986; Piehler and Bach, 1992; Harvey, 1992). Model simulations of passive internal variability, therefore, may only represent a lower bound to overall internally generated variability.

An example of the output from a simulation of passive

internal variability is given in Figure 1. For a simple linear first-order model (e.g., a one-box ocean model) white-noise forcing as input is transformed to red-noise temperature output, with an enhancement of variability at low frequencies. For more complex models (such as that used here), the output spectrum is still basically reddish in character, in that there is a marked enhancement of the power at low frequencies.

The passive internal variability modeling approach gives results very similar to those of the statistical model of a fractionally integrated white-noise process that is preferred by Bloomfield and Nychka (1992) (see Figure 2). The deterministic modeling approach yields additional physical insights. It shows that high-frequency variability (≤ 10 years) is virtually independent of the climate sensitivity ($\Delta T_{2\times}$), and mainly reflects the magnitude of the forcing and the heat capacity of the upper mixed layer of the ocean (Wigley and Raper, 1991). For similar reasons, the response to short-lived forcing events, such as those of volcanic eruptions, is also independent of $\Delta T_{2\times}$. This is why interannual variability can tell us nothing directly about $\Delta T_{2\times}$. At low frequencies (≥ 30 years), however, the temperature-response spectrum of passive internal variability depends critically on $\Delta T_{2\times}$.

If we had evidence of large natural climate excursions in the past on the 50-year time scale, there would be two

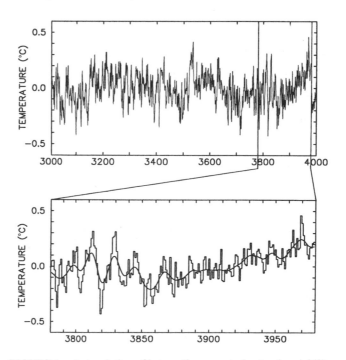

FIGURE 1   A simulation of internally generated natural variability of global-mean temperature (using $\Delta T_{2\times} = 2.5$°C). The upper panel shows the third 1,000 years of a 100,000-year run; the lower panel shows an enlargement of 200 years of this record. (From Wigley and Raper, 1991; reprinted with permission of Elsevier Science Publishers BV.)

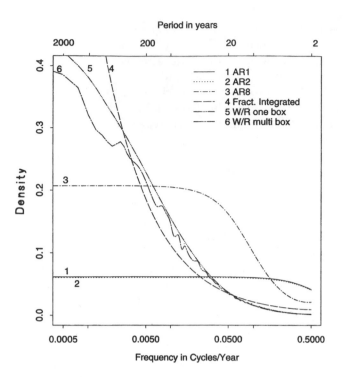

FIGURE 2 Spectral densities for observed and simulated global-mean temperature series. Items 1 to 4 are spectra of the observed data based on different assumptions regarding the underlying statistical structure of the data. Item 5 is a theoretical response spectrum for white-noise forcing derived from a global upwelling-diffusion energy-balance climate model. Item 6 is the spectrum based on data from a more complex form of the same model forced with white noise. (From Bloomfield and Nychka, 1992; reprinted with permission of Kluwer Academic Publishers.)

possible interpretations. First, if the climate sensitivity were low, then passive internal variability and any response to global-mean external forcing would be small, so we would have to invoke either significant changes in the thermohaline circulation or (less likely on these time scales) major changes in the spatial patterns of forcing (as in the Milankovitch effect). Alternatively, if neither of the latter two mechanisms was operating, we would have to infer that the climate sensitivity was relatively large (i.e., $\Delta T_{2\times}$ in the range suggested by GCMs, 1.5°C to 4.5°C), since a large sensitivity is required for both passive internal and global-mean externally forced changes to be substantial.

The Importance of Climate Sensitivity

In the following, we assume that the climate sensitivity lies in the commonly accepted range of 1.5°C to 4.5°C, and consider some of the consequences. The dependence of passive low-frequency variability in global-mean temperature on $\Delta T_{2\times}$ has two important implications. First, as noted earlier, it means that the problem of separating a low-frequency (e.g., greenhouse) signal from low-frequency

natural variability is one that is with us no matter what the value of $\Delta T_{2\times}$; both signal and noise increase as $\Delta T_{2\times}$ increases. On the other hand, the dependence of low-frequency variability on $\Delta T_{2\times}$ gives us the possibility of estimating $\Delta T_{2\times}$ from the spectral character of the past (paleoclimatic) climate record. Unfortunately, as noted above, our quantitative knowledge of this past record is poor, so we cannot use it to derive any reliable information regarding $\Delta T_{2\times}$, at least on the basis of recent (the past 10,000 years) paleoclimatic data. Attempts have been made, however, to use longer-time-scale paleoclimatic data. These point to $\Delta T_{2\times}$ being around 2.5°C, but with a range of uncertainty similar to that deduced from GCMs (see, for example, the recent work by Hoffert and Covey, 1992).

The deterministic modeling approach can also be used to estimate statistical confidence limits on natural century-time-scale trends in global-mean temperature. The results, which depend on $\Delta T_{2\times}$, show that a trend of more than 0.3°C per century is unlikely (95 percent confidence level) on the basis of passive internal variability alone. Since other forms of natural variability exist (due to external forcings at low frequencies, or due to ocean circulation changes), a natural trend of about 0.5°C per century is certainly possible, although such a natural trend is as likely to be negative as positive.

Both external forcings (man-made and/or natural) and internally generated variability could have contributed noticeably to the twentieth-century trend. The relative importance of these factors over this period is uncertain. Nevertheless, we can infer from the very existence of the trend that $\Delta T_{2\times}$ must be far from negligible. Given that there is no evidence of thermohaline circulation changes large enough to cause such a large overall trend (see, e.g., Wigley and Raper, 1987), a substantial climate sensitivity is necessary no matter what the cause of the trend. External forcing can clearly have a marked response only if $\Delta T_{2\times}$ is substantial. For example, if man-made forcing were the major factor, then $\Delta T_{2\times}$ would have to be 2°C or more (Wigley and Raper, 1992). For passive internal variability to contribute substantially to the trend, $\Delta T_{2\times}$ must also be of similar magnitude (Wigley and Raper, 1990). Thus, although we cannot claim to have detected the enhanced greenhouse effect on the basis of the data alone (note the conditional clause), the data do give us valuable insights into the way the climate system works.

INTERPRETING THE PALEOCLIMATIC RECORD

Detecting the enhanced greenhouse effect has a strong parallel in trying to interpret the paleoclimatic record. One of the justifications commonly used for work in paleoclimatology is that the data can provide information on the causes of past climate change. At least for the past 10,000 years, however, identifying causes, or even distinguishing exter-

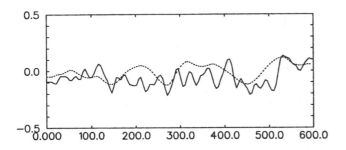

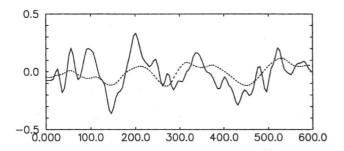

FIGURE 3   Two contrasting simulations of global-mean temperature changes over A.D. 0 to A.D. 600. During this period, the forcing shows noticeable but relatively weak fluctuations (range about 0.5 W m^{-2}). In the upper simulation, the "observed" (signal plus noise) temperature fluctuations (full line) bear no resemblance to the signal (dashed line). Temperatures remain roughly constant over A.D. 0 to A.D. 500 and then rise rapidly by about 0.3°C over about 10 years, but this rise is unrelated to the forcing. In the lower panel, there are major fluctuations of magnitude far exceeding the largest of the fluctuations in the signal. The rapid cooling over A.D. 100 to A.D. 140 is followed by an even larger warming (0.7°C over A.D. 140 to A.D. 200) and a subsequent long-term cooling trend over A.D. 200 to A.D. 420. None of these events is a reflection of the forcing changes.

nally forced as opposed to stochastically forced events, could be difficult. The reason for this is that the potential magnitude of internal variability, as revealed by the modeling work described above, is similar to the range of observed changes over the past 10,000 years. Thus, even if the major climate excursions of the past (such as the Little Ice Age) were externally forced, internal variability may mask these to such an extent as to make their interpretation problematic.

To demonstrate this, we use the last 2,000 years of a hypothesized record of past external (solar) forcing derived from radiocarbon data from precisely dated tree-ring samples (from Stuiver and Braziunas, 1989). The magnitude of this forcing is uncertain, so we choose the scale to give a range of modeled global-mean temperature changes of 0.5°C for $\Delta T_{2\times} = 2.5$°C. (The corresponding forcing range is about 1 W m^{-2}, similar to the magnitude of man-made forcing over the past few centuries.) To simulate internally generated natural variability, we superimpose on this solar forcing record additional white-noise forcing. The magnitude of the white-noise forcing is chosen so that the

high-frequency (≤ 10 years) variability of the modeled temperature changes matches that of the twentieth-century temperature data. This produces a low-frequency internal variability background whose magnitude depends on $\Delta T_{2\times}$.

Even though both the response to external forcing (i.e., the signal) and the magnitude of passive internal variability (i.e., the noise) depend on $\Delta T_{2\times}$, this does not necessarily mean that the signal-to-noise ratio in our simulations is independent of $\Delta T_{2\times}$. If we were to keep the externally forced range of temperature changes at 0.5°C (to do this, we would have to assume larger solar forcing if we used a smaller $\Delta T_{2\times}$), then the signal would stay constant, but the noise would vary with $\Delta T_{2\times}$. If, however, we kept the forcing fixed, then the relative magnitude of signal to noise would be effectively independent of $\Delta T_{2\times}$ (in this case, lower $\Delta T_{2\times}$ would lead to a lower externally forced range and a similarly reduced internal variability range). It is not clear which would be the better approach. We avoid the dilemma by considering results for only a single value of $\Delta T_{2\times}$ (2.5°C); this is all that is required to illustrate the problem of interpreting the past record.

Figures 3 and 4 give selected examples showing the combined externally forced and internally generated variations in annual global-mean temperature compared with the

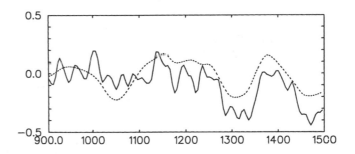

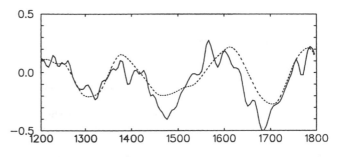

FIGURE 4   In the upper panel, the cooling event evident in the signal around A.D. 1000 to 1100 (dashed line) is entirely absent from the simulated record, while the subsequent two forced cooling events around A.D. 1300 and A.D. 1500 show up quite clearly. In the lower panel, there are three cooling signals (A.D. 1300, 1500, and 1700), all of which show up in the simulated observed record. In the last event, however, the observed cooling leads that in the signal by 20 to 50 years. The strength of the signal in all three events is similar, but the realized magnitude differs markedly.

externally forced component alone. (Note that these are not necessarily typical examples—they represent cases that might be expected to occur about 10 percent of the time). In many instances, there are marked differences between signal and signal-plus-noise. For example, even when the external forcing component is small, warm and cold intervals may occur that are as large as many of the major externally forced episodes; external forcing effects are often masked by internal variability; and noticeable "phase shifts" occur in which the "observed" (signal-plus-noise) changes appear to lead the forcing by a number of decades.

What does this mean for interpretation of the observed paleoclimatic record? For these data, we only have a single manifestation of an ensemble of internally generated possibilities. Since signals and (passive) noise can be of similar magnitude (albeit with relatively low probability), it would be possible to completely misinterpret the observed record even if it were perfect (i.e., a faithful representation of annual, global-mean temperature changes). It should not be surprising, therefore, that attempts to associate cause and effect in the paleoclimatic record have invariably met with only limited success (see, e.g., Wigley and Kelly, 1990). Even with perfect forcing and response data, we would not expect to get highly significant results when comparing theory with observations.

This brings us back to our initial statement. It would indeed be useful to understand the causes of past climate change, but we are unlikely to unravel this puzzle easily. At least for the past 10,000 years, the forcing (as deduced from a noise-free interpretation of observed data) is such that signal and noise may be of similar magnitude. Anthropogenic forcing over the past few centuries has almost certainly been larger in magnitude than any natural forcing that has occurred at the annual, global-mean level. Thus, detection of the enhanced greenhouse effect (given that we are still enhancing it!) is probably an easier task than explaining the past, and efforts to improve future climate predictions are likely to add more to our understanding of the past than vice versa.

Discussion

LINDZEN: I'd like to comment on something common to Jerry North's and Tom Wigley's papers. Both use very simple models, which I myself favor. But because of this simplicity, they can be forced only in a certain way, and they also lack the feedback that is one of the determinants of the delay imposed by the ocean's heat capacity. The latter must be kept in mind in the interpretation of such models' results.

ROOTH: I consider it a serious flaw that these models treat the ocean in such a lumped way as regards delay effects. When you change surface temperature by a couple of degrees, you want to have a proportional change decaying toward deep water.

NORTH: I want to make it clear that the Wigley-Raper model and the one I talked about are exactly the same, except that ours has geography. Now about the delay: Last year we did a little study of the ramp warming with that model, and found that it matters very much when you start the ramp. Two times are very important: One is how long it takes for the model to adjust to the linear ramp, and the other is the delay. The delay is frequently a few decades, in our models as well as the coupled GCMs, but the adjustment time seems to be extremely model dependent. In these simple models it can be 150 years or so, which limits the warming you can have, whereas the GCMs get onto the ramp so quickly it's hard to get slow warming. I don't know the answer to this dilemma.

[UNKNOWN]: One crucial thing we don't know from the data is the natural relaxation time. A shoulder you see on some of these power spectra, between a flat spectrum to the left and a slope to the right, is occurring effectively at 10 years. If this break in the slope is an empirical measurement of a response time, that has important implications.

LINDZEN: We must be careful not to construe model spin-up as a characteristic delay.

MYSAK: I just wanted to agree with Claes about ocean response to SST changes, and to say that I think we have learned a lot about the natural variability of the ocean's thermohaline circulation in the last few years. Modeling work at several places has shown natural oscillations in the range of 100 to 300 years.

GHIL: It seems to me that we have a spectrum of more or less predictable types of climatic behavior that ranges from totally boring to completely surprising. We go from a predictable trend through periodic, quasi-periodic, and deterministically aperiodic behavior all the way to random noise. Natural variability is the sum of all these things, and we need methods that can distinguish among them.

Detection of Forced Climate Signals

GERALD R. NORTH AND KWANG-YUL KIM[1]

ABSTRACT

A method of signal detection used in communication theory can be applied to the problem of detecting forced climate signals. This method calls for the construction of an optimal filter (averaging procedure) which is then applied to a data stream that varies in both space and time such that the natural variability is suppressed to the maximum extent possible. The construction of such filters requires a good knowledge of the waveform of the signal and the frequency-dependent empirical orthogonal functions for the natural variability. Although we cannot yet supply reliable estimates of either of these from data or from coupled general-circulation-model simulations, we present some samples that are based on simple stochastic models. These exercises suggest that the signal due to sunspot forcing, while quite weak, can be considerably enhanced by using optimal filters. Greenhouse warming, modeled with a ramp increase of temperature, presents a large signal-to-noise ratio. Either the observed warming signal is highly significant, or the surface climate has experienced a very rare natural fluctuation over the last century.

INTRODUCTION

To detect a faint signal that is superimposed on random fluctuations, it is helpful to know as much as possible about the characteristics of the fluctuations and the form of the signal. In the detection and attribution of forced climate signals we are faced with this generic problem. But precisely in what form do we need variability statistics? And how can we objectively employ variability statistics in our detection strategy? Can sensible statistical tests be applied using this information? Is there an optimal procedure that will assist us in formulating our questions clearly? Some of the answers lie in a class of techniques known to signal-processing engineers for over half a century, dating to the work of Wiener (e.g., as summarized in his 1949 book) and Kolmogorov (1941). Such techniques were first connected with the climate problem by Hasselmann (1979) and Bell (1982, 1986). Other approaches to the climate-change detection problem have been taken by Barnett and colleagues (e.g., Barnett, 1986, 1991). Several others are summarized in a recent book edited by Schlesinger (1991). Here I summarize the signal-processing approach as detailed by North et al. (1992).

---

[1]Department of Meteorology, Texas A&M University, College Station, Texas

Our goal is to develop a filter (a weighted averaging process) that can be applied to the historical data stream in space and time so that the signal or "forced part" of the stream is optimally enhanced by comparison with the natural variability or "noise." By a filter we mean a linear operation on the data stream such as a weighted integral over the globe and over the past. This space- and time-dependent weighting function essentially constitutes a filter. This function must take place despite an imperfect observing system and an imperfect knowledge of the signal. To pose the problem formally, we seek a filter that minimizes the mean squared error between our estimator for the signal and the true signal. Implicit is the assumption that the data stream $T^{data}(r,t)$ can be written as a sum of signal and noise parts,

$$T^{data}(r,t) = T_S(r,t) + T_N(r,t) \qquad (1)$$

where $T_S(r,t)$ is the forced signal and $T_N(r,t)$ is the natural variability.

This assumption has been shown to be adequate for a general-circulation-model (GCM) simulation of the surface temperature field for an idealized planet (North et al., 1992). We conjecture that it holds for the more general questions of interest, but it is particularly problematic for studies of the world ocean. The mathematical expression for a filtered data stream, $\hat{T}_S(r,t)$, is

$$\hat{T}_S(r,t) = \int_D \int_{-\infty}^{\infty} \Gamma(r,t;r',t') \, T^{data}(r',t')dt' d\Omega' \qquad (2)$$

where D is the spatial domain and $\Gamma(r,t;r',t')$ is the filter kernel.

AN EXAMPLE

Consider a planet that experiences periodic (forced) solar luminosity. Simulations of climate indicate that the resulting spectrum of the global average temperature contains a simple red-noise continuum with a sharp peak at the forcing frequency (North et al., 1992), provided the thermal response amplitude is no more than a few degrees. If we know the imposed forcing and response frequency from theory, we can filter the data stream so that only a narrow band of frequencies around the peak is retained. The longer the time series of data, the narrower can be the filter.

Much more can be done if the temporal profile (waveform) being detected is known precisely. In the case of the oscillating solar luminosity, for example, the phase of the imposed sinusoid is known as well as the frequency. Since the noise is independently distributed between sine and cosine components, we can further select within the frequency band to eliminate phases that are unlike the signal. For example, if the signal is all sine component, we can eliminate the cosine component with a simple projection filter. Since the random-noise component for a stationary process distributes variance equally between the sine and

cosine components, we can reduce the variance of the noise by a further factor of two, or reduce the noise standard deviation by the square root of two. In this way we can hope to increase the signal-to-noise ratio considerably more than can be accomplished with a simple frequency band-pass filter.

In fact, there is even more enhancement potential at our disposal. If the forced response has a characteristic geographical signature at the responding frequency, we can use our filter to select only geographically similar patterns. This procedure can be employed to eliminate any patterns occurring in the natural variability that are dissimilar (orthogonal) to the signal pattern. It follows that very faint signals can be seen if we know enough about them a priori and if enough of the noise is orthogonal to the signal. Our experimentation with stochastic energy-balance models suggests that at least another factor of two in the signal-to-noise ratio can be gained from this extension to pattern recognition.

While the sinusoidal example is interesting, and potentially useful for the important sunspot-response detection problem (it is important mainly because it may tell us something about climate sensitivity at the decadal scale), it is more idealized (being nearly a sharp tone) than the one we face in the greenhouse problem, which is spread over a broad continuum of frequencies.

In contrast to the periodically forced system, the greenhouse problem presents a single ramp-like waveform with a characteristic geographical response pattern (land-surface temperatures tend to lead slightly over ocean-surface temperatures). Hence, the response will be composed of a large number of sinusoidal frequencies in the Fourier sense. Since the pattern of response of the surface-temperature field depends strongly on the forcing frequency, we will find it natural to use frequency-dependent empirical orthogonal functions (fdEOFs) as an expansion basis (see, e.g., North, 1984). We stress that this is not just a convenient choice, but one that arises naturally from the formalism as the optimum for such purposes.

MEAN-SQUARE-ERROR FORMALISM

In this section we formulate the mean square error (MSE) for the random component of the error made with our estimator of the signal for a given realization. Unfortunately, in the climate problem we have only one realization to work with. In this paper we dispense with mathematical detail, referring the reader to the paper by North et al. (1995) for detailed derivations. The error in question is the difference for an individual realization between the estimator of the signal (filtered data stream) and the actual signal, whose shape in space and time must be known a priori. We form the square of this error and average it over an ensemble of realizations of the same waveform embedded in natural variability. The natural variability is uncorrelated from one realization to another. The geographical dependence of the natural fluctuations will be taken to be stationary in time. The mathematical expression for the MSE is

$$\epsilon^2 = \langle (T_S - \hat{T}_S)^2 \rangle \tag{3}$$

After formulating the expression for the MSE, we find that it can be written in terms of a weighted space and time integral over the unknown filter. If we take the variation of the MSE with respect to infinitesimal changes in the unknown filter function in space and time and set the resulting expression equal to zero, we end up with an integral equation for which the filter is the unknown. An important special case is that for which a no-bias constraint is imposed. We would like to make our estimator of the signal be unbiased. For this case, the ensemble average of the filtered data stream is identical to the true signal except that it is multiplied by a constant factor that is realization dependent. The constraint can be incorporated by the use of the familiar method of Lagrange multipliers. The result is again an integral equation that has to be solved for the optimal filter as its unknown,

$$\int \rho(r,t;r',t')\Gamma(r,t;r',t')dr'dt' = -\Lambda T_S(r,t) \tag{4}$$

where $\rho(r,t;r',t')$ is equal to $\langle T_N(r,t)T_N(r',t') \rangle$ and Λ is the Lagrange multiplier.

OPTIMAL-FILTER SOLUTION

The integral equation for the optimal filter is a simple nonhomogeneous linear integral equation, in which the desired filter function occurs under the integral sign, weighted by a kernel. The kernel of the integral equation happens to be the space-time lagged covariance of the field in question (e.g., the surface temperature field). This tells us that the optimal filter is strongly dependent on the structure of the space-time covariance of the natural variability. In addition, the optimal filter depends on the signal itself. Later we will discuss solutions for which there is some uncertainty in the signal (as actually occurs).

In solving a linear integral equation in which the unknown occurs underneath the integral sign, weighted by a kernel that is symmetric in its arguments (e.g., a covariance), it is advantageous to expand the kernel into its eigenfunctions. Since the kernel is symmetric and well behaved, we can assume the eigenfunctions form a complete basis set with positive eigenvalues that satisfy

$$\int \rho(r,t;r',t')\Psi_n(r',t')dr'dt' = \lambda_n\Psi_n(r,t) \tag{5}$$

where $\Psi_n(r,t)$ is the frequency-dependent EOF and λ_n is the eigenvalue. The eigenfunctions of a covariance kernel are known as the empirical orthogonal functions (EOFs). The

eigenvalues are the variance attributable to the corresponding EOF "mode." We can reduce the integral equation to so-called diagonal form by expanding all quantities in terms of the eigenfunctions of the kernel. This is equivalent to inverting a matrix by using a basis in which the matrix is diagonal, which simply requires inverting the elements along the diagonal.

Since the natural variability can be taken to be stationary in time, the temporal part of the problem can be solved using the Fourier (integral) basis. The spatial part of the problem is solved at each frequency by using the EOF basis set for that frequency. The choice of frequency-dependent EOFs completely diagonalizes the space- and time-dependent kernel of the integral equation, rendering the problem invertible. Hence, the problem of constructing an optimal filter amounts to obtaining an adequate knowledge of the fdEOFs and an a priori knowledge of the signal shape in space and time. In principle, each could be obtained from simulations with sufficiently reliable coupled ocean–atmosphere climate models.

The signal waveform in space and time and the data stream are to be expanded into the frequency-dependent EOF basis set. After the insertion of these into the integral equation, we find the following formula for the optimal filter:

$$\Gamma^{opt}(r,t;r',t') = T_S(r,t) \frac{\sum_m \dfrac{T_{Sm}\Psi_m(r',t')}{\lambda_m}}{\sum_n \dfrac{T_{Sn}^2}{\lambda_n}}. \tag{6}$$

PROPERTIES OF THE OPTIMAL FILTER

The unbiased optimal filter for a known non-random signal turns out to be exactly proportional to the known signal waveform. This space- and time-dependent shape factor can be moved outside the integral sign when the data are filtered (see equations (2) and (6)). The filtered data stream then is the imposed signal, as a function of space and time, multiplied by a realization-dependent dimensionless scale factor whose expectation value is unity. The signal-to-noise ratio is the reciprocal of the standard deviation of the scale factor. This standard deviation is a key property of the optimal filter. The square of the signal-to-noise ratio can be written as a sum of terms, each of which represents the contribution from a particular fdEOF mode:

$$(SNR)^2 = \sum_n \frac{T_{Sn}^2}{\lambda_n}. \tag{7}$$

This representation is particularly convenient, since it allows us to see how adding more modes in the expansion affects the performance of the filter. A filter in which the series is truncated is an unbiased suboptimal approximation to the optimal filter.

Each term in the sum for the signal-to-noise ratio squared is the ratio of the square of the signal projected onto mode n divided by the fdEOF eigenvalue corresponding to that mode (7). Hence, each term can be interpreted as the ratio of the square of the signal amplitude to the noise variance for that mode. If the signal is confined to only a few fdEOF modes, the sum will terminate quickly. If it is broad-band, the sum will have many terms, each one of which is positive. A highly desirable situation is one in which the signal projections are large, for modes with small eigenvalues. Clearly, it is important to estimate the rough size of the signal-to-noise calculation before performing any data manipulation or extensive model simulations, since detection of the forced signal may be impossibly difficult even with an optimal filter.

PILOT MODELS FOR CLIMATE SIGNAL DETECTION

Energy-balance models (EBMs) have proven to be able to simulate the geographical distribution of the seasonal cycle of surface temperature (North et al., 1983; Hyde et al., 1989). When EBMs are forced with white noise (in space and time), they have also been shown to reproduce the geographical distribution of the variance of natural fluctuations in frequency bands from two months to a few years (Kim and North, 1991). Even the geographic distributions of spatial correlation characteristics from EBMs are in good agreement with observational data for these same frequency bands. The only new parameter in the noise-forced simulations is the strength of the noise forcing. Even this cancels out in the computation of correlations. Such a forced model enables us to compute the fdEOFs exactly. We dub them frequency-dependent theoretical orthogonal functions (fdTOFs), since they come from a theoretical model rather than from data. When we computed these fdTOFs and compared them to those computed from 40 years of surface-temperature data, they were in reasonable agreement (Kim and North, 1993).

We have extended the noise-forced EBMs (nfEBMs) to include a deep ocean similar to that used in a number of studies by the IPCC (IPCC, 1992; see also Hoffert et al., 1980). The model consists of land and ocean distributed on the globe's surface. The ocean consists of a mixed layer of 75 m depth atop a deep ocean, which transports heat by way of uniform upwelling and uniform thermal diffusion. Such an ocean model develops a thermocline at a depth of about 1 km for a reasonable choice of the two new parameters (Kim et al., 1992). The new coupled model still reproduces a good seasonal cycle and second-moment variance and covariance characteristics. Its lower-frequency behavior

is thought to be somewhat improved over that of the earlier mixed-layer-only version (Kim and North, 1992).

The new coupled (linear) model is capable of producing fdTOFs for the natural-variability and forced-climate signals. The static sensitivity of the model is somewhat low as climate models go: 2°C for a doubling of carbon dioxide and 1.2°C for a 1 percent increase in solar constant. These last quantities do not depend on the ocean, since they are static (for transition from equilibrium to equilibrium) characteristics. The transient behavior of the model is fairly well understood. For a ramp in greenhouse forcing starting in the year 1750, the model produces about a 0.5°C increase over the last century. One important signature or fingerprint of this type of forcing is that the mid-Asia minus mid-Pacific differences over the last century are about 0.13°C. Hence, in this class of models there is a potentially significant land-sea contrast in the warming that can be exploited for pattern recognition. In greenhouse warming, land leads ocean by a potentially detectable amount. If warming is caused by internal variability of the oceans, we would expect ocean to be leading.

Next we consider a few explicit numerical examples in which the signals and noise are computed from nfEBMs (e.g., Kim and North, 1992; Kim et al., 1992).

SUNSPOT EXAMPLE

We have measurements of the solar constant taken from several independent satellites over the last 12 years. There

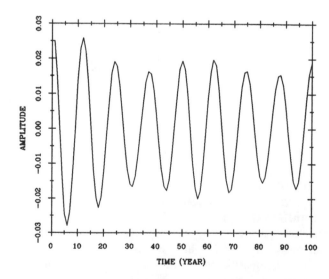

FIGURE 1   Response over the last century of the globally averaged temperature to sunspot-cycle forcing as computed by the Kim-North-Huang (1992) model. The variation of sunspot number is converted to a variation in solar constant using the last solar cycle. The Kim-North-Huang model is an energy-balance model of the surface temperature over land and the mixed-layer temperature over ocean atop an upwelling-diffusion deep ocean.

does appear to be a small oscillation of the solar constant that is consistent with a modulation by the sunspot number. The amplitude of the oscillation appears to be about 0.1 percent. If we use this one cycle as a calibrator and ask what the temporal variation over the last century would have been, we can proceed to compute the climatic response of the earth. Figure 1 shows the curve of the global average temperature as computed by the Kim-North-Huang model. The response is a disturbed 11-year cycle with an amplitude of about 0.02°C, a very faint signal considering that the standard deviation of one-year averages for natural variations is about 10 times larger. This simple exercise suggests a signal-to-noise ratio of about 0.1, implying that the signal is virtually undetectable. On the other hand, if we use more information, which is fortunately at our disposal (at least from the model simulations), we can do much better.

Next we show the results of optimal filtering of the data to detect the surface-temperature response to the sunspot cycle. Figure 2 shows the geographical pattern of response to an 11-year periodic solar forcing in this model. When this pattern is included along with the precise phase and amplitude information represented in Figure 2, the signal-to-noise ratio squared can be plotted as a function of the number of fdTOFs included as shown in Figure 3. A monotonic increase in the signal-to-noise ratio can be seen as more fdTOFs are included. Hence, we have shown that the method is capable of improving signal-to-noise ratio through

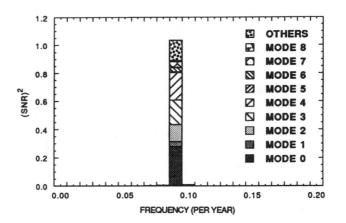

FIGURE 3   Theoretical signal-to-noise ratio squared for the sunspot-forcing problem. The bar graph shows contributions from different frequencies. Within each bar is the partitioning of contributions from different spatial EOFs at that frequency. One sees that the contribution from spatial-pattern recognition is significant.

the inclusion of geographical, frequency, and phase information in the filter. The signal-to-noise value of the order of unity is still short of acceptable for purposes of detection, but the sunspot cycle is used here mainly for illustration (although we think more can be done with the sunspot problem). Significant degradation will occur when we start to use real data instead of simulated data.

GREENHOUSE-WARMING DETECTION

Confirming the existence of greenhouse warming consists of detecting a ramp-like increase of temperature over the last century. Our simulations show that the differences between the 1980 and the 1880 surface temperature fields should look like the map shown in Figure 4, which clearly shows the land-leading effect. If we expand our ramp-like response into sine and cosine harmonics of 100 years and into the fdEOFs at these frequencies, we can estimate the signal-to-noise ratio squared, as we did in the sunspot case. The result is shown in Figure 5. We find that greenhouse-warming detection is not helped as much by the addition of pattern recognition as sunspot-cycle detection, but the signal is large enough to be detected with considerable confidence. Thus, if our assessment of the low-frequency variance in the model is correct, either the greenhouse-warming ramp is real, or a very rare natural fluctuation with a positive slope of 0.5°C per century is occurring. (By a rare natural fluctuation we mean one that is 0.25°C above normal and lasts for a century.)

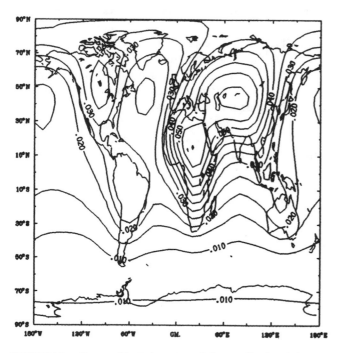

FIGURE 2   Geographical signature of the amplitude of the surface-temperature response for solar-constant oscillations of 0.1 percent at a period of 11 years, as computed in the Kim-North-Huang model.

CONCLUSIONS

Signal-processing methods can significantly enhance our ability to detect forced climate response. It is important to

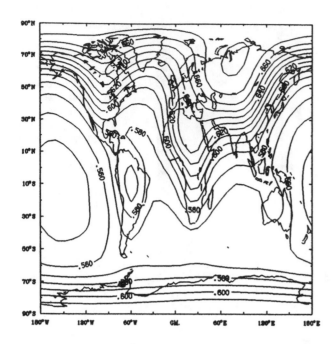

FIGURE 4   Temperature-difference map for 1980 to 1890, as computed by the Kim-North-Huang model.

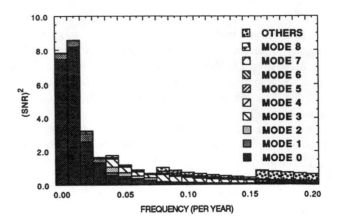

FIGURE 5   The square of the theoretical signal-to-noise ratio, as in Figure 3, for the greenhouse-warming problem. The signal is a (repeating) ramp upward of 100 years' duration.

note that the word "detect" can be misleading. We do not mean that we show that such a signal is present or not present. Our procedure aims to measure the strength of the signal, given its waveform in space and time. We need a good model of the signal beforehand, and we need a good understanding of the lagged covariances so that we can compute the fdEOFs—or more likely the fdTOFs, since long enough strings of data will not be available. It is especially important to have good low-frequency fdTOFs. Once these quantities are in place, we can construct an optimal filter for the data stream. It is possible to relax some of these requirements, such as the need to have perfect knowledge of the signal. We may know only its probability distribution, for example. This kind of statistical information can also be used to construct an optimal filter at the cost of some reduction of the signal-to-noise ratio.

We feel that it will be possible to use the techniques just mentioned to construct filters capable of achieving a signal-to-noise ratio greater than 10 for simulated data. Moving to the inclusion of real data is a large leap, however. First, the actual data are imperfect, coming as they do from a sparse network that is itself time dependent. The effects of variations in the observational network can be included in the filter formalism, and we intend to test some of them in the near future. Second, we lack complete knowledge of the signal. We hope to show the consequences of poor signal knowledge through modeled examples in future papers.

Finally, it must be re-emphasized that our method depends on the assumption that signals (whose waveform is calculable a priori) and stationary random fluctuations can be linearly separated. These components do not depend upon each other in our formulation. It is likely that none of these assumptions holds strictly for the climate system, but it would seem unwise to take a more complicated approach until the proposed one has been demonstrated to fail.

Discussion

GHIL: Aren't you actually trying to distinguish between two kinds of signal, one anthropogenic and one that could be called a deterministic part of natural variability? It seems to me, furthermore, that singular spectrum analysis, which is a data-adaptive version of the Wiener-Kolmogorov filter, does some of the same things rather better. Myles Allen at Oxford has developed a test like the one Dick Lindzen wants, and when he applied it to a long time series for central England he got interdecadal natural-variability peaks.

NORTH: Perhaps I didn't explain carefully enough that I have termed the natural variability "noise" with respect to the anthropogenic "signal". Also, I see no reason to adopt fancy techniques to deal with a stationary time series.

LINDZEN: I think we need to address the physics of these primitive models you and Tom Wigley have used, which have a forcing function that pulses.

MOREL: I recently had the opportunity to see the results obtained by the Hamburg group. They have run four different global-warming experiments at 30-year intervals, with widely differing responses. In fact, their warming curves are so different it is impossible to distinguish between variability and warming.

NORTH: Our empirical orthogonal functions suggest that we have nailed down the eigenvalue spectrum extremely well. Also, when we used the model on the NCAR data set, we found that the pattern of variance was totally dominated by the land-sea contrast. At lower frequencies the El Niño signal begins to appear, which seems to me to be a good indicator of usefulness.

Constraining Possibilities Versus Signal Detection

RICHARD S. LINDZEN[1]

ABSTRACT

Rather than concentrate on the problem of greenhouse signal *detection*, one can exploit the intimate relation of climate sensitivity to length of ocean delay in order to *constrain* possibilities. It is noted that the observed record of surface temperature is "broadly" consistent with equilibrium sensitivity to CO_2 doubling of between 0°C and 4.8°C—provided that the ocean delay for the higher sensitivity is of the order of 160 years. If ocean delay can be independently established as being shorter, then one must reduce the maximum consistent sensitivity. It is suggested that volcanic responses can be used to estimate ocean delay. Preliminary results suggest very short delays.

REMARKS

As noted in the papers in this section by North and Kim (1995) and by Wigley and Raper (1995), simple box-diffusion-upwelling models yield useful insights into the issues of the seasonal and long-term delay in climate response to changes in geophysical forcing quantities that results from the ocean's heat capacity. The point, quite simply, is that the more rapidly heat is mixed downward into the ocean, the greater the ocean's effective heat capacity and the longer the delay. Also evident is the fact that the delay itself is time dependent and is different for each transient problem and for the annual cycle.

The geometry of simple energy-balance box-diffusion-upwelling models is shown in Figure 1. No attempt will be made here to repeat the mathematical development of

such a model. However, certain points are worth noting. The equilibrium response to forcing, ΔS, is $\Delta T_{eqbm} = \Delta S/a$. Also, the intensity of the coupling of the atmospheric climate system to the ocean is given by a. As noted by Hansen et al. (1985), the larger ΔT_{eqbm} is for a given ΔS, the smaller a must be, and with the weaker coupling, the ocean delay will be longer.

Reasonable parameter choices are $h = 75$ m, $k = 1.5$ cm^2 sec^{-1}, and upwelling, which limits downward diffusion to about $H = 400$ m. A common choice for a is 1.55 W m^{-2} deg^{-1} (viz., Sellers, 1969; North and Kim use a somewhat different value, but for our purposes the small differences are not of consequence). Other values involve negative or positive feedbacks in the atmospheric climate system. If we refer to the original choice of a as a_o, then

[1]Center for Meteorology and Physical Oceanography, Massachusetts Institute of Technology, Cambridge, Massachusetts

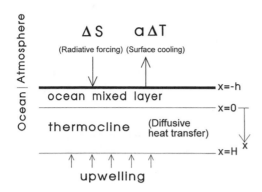

FIGURE 1   Schematic of the geometry of the box-diffusion-upwelling model used in these remarks.

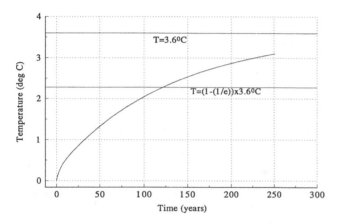

FIGURE 2   Temperature versus time for the impulsive doubling of CO_2 at $t = 0$ in a model with $\Delta T_{2 \times CO_2} = 3.6°C$.

$a = a_o/g$, where g = system gain, and the relation between g and total feedback, f, is $g = 1/(1 - f)$. Negative f's lead to $g < 1$, while $f > 1$ implies an unstable system. If we associate a doubling of CO_2 with a ΔS (at the surface as opposed to the "top of the atmosphere") of 1.8 W m^{-2}, then, in the absence of feedbacks, $\Delta T = 1.2°C$. Models that yield $\Delta T_{2 \times CO_2} = 4°C$ involve a gain of 3.3 (or $f = 0.7$). An idea of the nature of ocean delay can be seen in Figure 2, where we consider the transient response to an impulsive doubling of CO_2 in a system where $\Delta T = 3.6°C$. Clearly, the approach to equilibrium is not simply exponential. However, for simplicity of discussion, we will identify a response with a time scale τ, corresponding to the time it takes ΔT to reach to within $(1 - 1/e)$ of its equilibrium value when the system is impulsively forced. Figure 3 shows how τ varies with g. Over the range of interest, the dependence is almost linear. (As Hansen et al. (1985) note, the relation becomes quadratic as $H \rightarrow \infty$.)

A main point of these remarks is that in dealing with these simple models, it is occasionally useful to consider τ rather than g, and $\Delta T_{2 \times CO_2}$ rather than $\Delta S_{2 \times CO_2}$ as variables to focus on. Doing so will suggest constraints on both τ

and $\Delta T_{2 \times CO2}$, which can be determined from considering responses to both volcanos and increasing greenhouse gases. It must be emphasized that it is just as useful to use data to constrain the likely response to doubled CO_2 as it is to detect the actual response (which might prove small).

Figure 4 shows T vs. t for the IPCC "business as usual" (BAU) emissions scenario (which leads to a quadrupling of "effective" CO_2 by 2100; viz., Houghton et al., 1990). The different curves correspond to various choices of equilibrium $\Delta T_{2 \times CO_2}$. We see (consistent with IPCC results) that expectations for current warming range from 0.15°C for $\Delta T_{2 \times CO_2} = 0.24°C$ to 0.8°C for $\Delta T_{2 \times CO_2} = 4.8°C$. The observed warming over the past century of 0.45°C ± 0.15°C (viz., Figure 5) corresponds to about $\Delta T_{2 \times CO_2} = 1.2°C$,[2] but given the natural variability in T, it is difficult to rule out any of the choices on the basis of the observed global

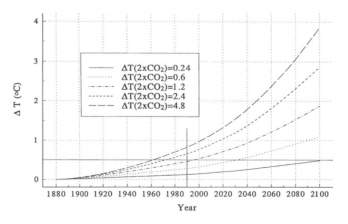

FIGURE 3   Characteristic ocean delay (τ) versus gain (g). See text for discussion.

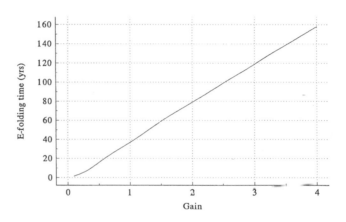

FIGURE 4   Temperature change since 1880 for IPCC "business as usual" emissions scenario and various model sensitivities.

---

[2]Wigley and Raper consider the impact of sulfate cooling; however, according to recent calculations of Kiehl and Briegleb (1993), this has been reduced to 0.3 W m^{-2} and is no longer of such great consequence.

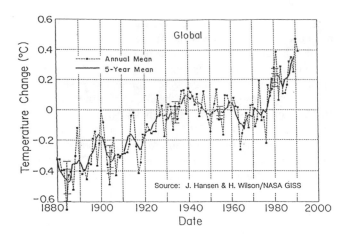

FIGURE 5   Global mean temperature since 1880 (Hansen and Wilson, personal communication).

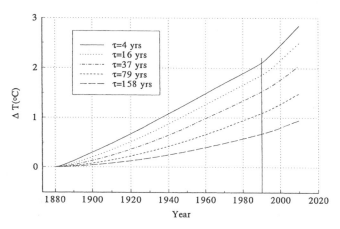

FIGURE 6   Temperature changes since 1880 for IPCC "business as usual" emissions scenario and $\Delta T_{2 \times CO_2} = 4°C$ for various choices of characteristic ocean delay.

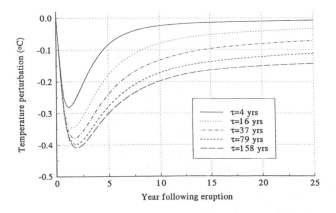

FIGURE 7   Response of global temperature to Krakatoa-type eruption for various choices of characteristic ocean delay.

warming. It is also worth pointing out that none of the choices of $\Delta T_{2 \times CO_2}$ leads to more than 3.5°C of warming by the year 2100—despite the fact that effective CO_2 is assumed to have quadrupled, and hence for $\Delta T_{2 \times CO_2} = 4.8°C$ the equilibrium warming in 2100 would be 9.6°C. This dramatically illustrates the importance of ocean delay.

The question we deal with is whether we can do better than the above in constraining expectations. We wish to suggest that the answer may well be yes. Figure 6 shows T vs. t for $\Delta T_{2 \times CO_2} = 4°C$ and various choices of τ (again using the IPCC BAU scenario). Here we see that for $\Delta T_{2 \times CO_2} = 4°C$ to be compatible with the observed warming, τ must be greater than 100 years. The point is that natural variability is unlikely to cancel out more than about 0.5°C of warming over relatively short periods. There are, indeed, model results that suggest both $\Delta T_{2 \times CO_2} = 4°C$ and short τ's (Manabe et al., 1991), but as we see here, such suggestions are highly problematic.

We finally turn to volcanos. Wigley and Raper (1995) note that atmospheric response to volcanic dust is largely independent of τ because of the short time scales involved. These claims are true only for the first year or two after eruption. For longer times, there are significant dependencies. We represent the forcing from a volcano as $\Delta S = kt$ for $t \leq 3$ months, and $\Delta S = S_0 e^{-t/d}$ for $t > 3$ months, where $S_0 = k \times 3$ months = 9.3 W m^{-2}, and $d = 13$ months. These values are crudely chosen to correspond to Krakatoa (Oliver, 1976). Figure 7 shows the response to a single eruption for various choices of τ (or, equivalently, of gain, using Figure 2). We see that we can expect a cooling of 0.25-0.35°C within a year or two of eruption, regardless of climate gain.[3] However, for large τ (or system gain) long-term behavior is very different. For short τ's ($\tau < 16$ years or $g < 0.5$) the cooling maximizes by $t = 1$ year and decays to relatively undetectable levels within less than 10 years. However, for longer τ's almost half the maximum cooling persists for many years. Also, the maximum cooling seems to occur at $t = 2$ years rather than 1 year. Neither of these features has been much remarked on. What is going on is that the short-term volcanic cooling is disequilibrating the surface temperature vis-à-vis both the atmosphere and the ocean. For long τ's, the atmospheric coupling is weak and the ocean delay plays a larger role. Detecting such differences for a single volcano is likely to be non-trivial. Figure 8, from Hansen et al. (1992), shows predicted responses to Pinatubo in a general-circulation model with different assumed greenhouse-warming scenarios. The upper panel clearly shows that Pinatubo leaves

---

[3]Hansen et al. (1992) claimed that the prediction of cooling following the eruption of Pinatubo constituted an "acid test" of their model. They failed to indicate what aspect of their model was being tested. The prediction certainly did not constitute a test of the predictions of climate response to increasing greenhouse gases.

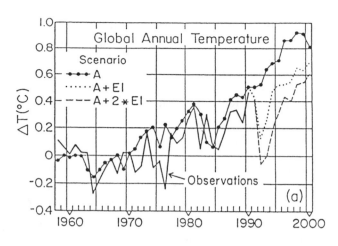

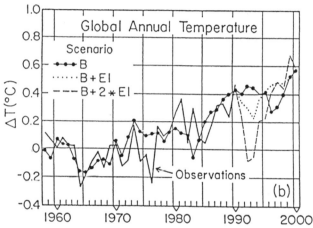

FIGURE 8 Response to Pinatubo for various scenarios calculated by Hansen et al. (1992). See text for details. (From Hansen et al., 1992; reprinted with permission of the American Geophysical Union.)

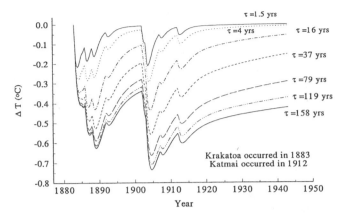

FIGURE 9 Calculated response to all volcanos between Krakatoa and Katmai for various choices of model gain (expressed in terms of characteristic ocean delay).

global temperature cold compared to the expected warming for a long time; however, the lower panel fails to show this persistence in a model where expected greenhouse warming is reduced. A reading of the text of Hansen et al. (1992)

TABLE 1   Ersatz Distribution of Volcanic Eruptions

Year of Eruption	Actual Volcano(s)	Relative Magnitude
1883	Krakatoa	1.000
1885	Falcon Island	0.100
1886	Tarawera, Niafu	0.500
1888	Bandai San, Ritter Island	0.375
1892	Bogoslav (1890), Awu	0.150
1902	Una Una (1898), Mt. Pelée, Soufrière	0.400
1903	Santa Maria, Minami Iwoshima (1904)	0.630
1907	Shtyubelya Sopka	0.150
1912	Taal (1911), Katmai, Sakurashima (1914)	0.185

shows that this scenario also included an additional volcanic eruption in 1995!

There would appear to be some possibility of distinguishing different τ's by considering sequences of volcanos, since larger τ's imply cumulative effects for a sequence of volcanos while smaller τ's imply largely independent responses to each individual volcano within a sequence.

These considerations should be relevant to the period 1883–1912 (Krakatoa to Katmai), when there were a number of major eruptions whose total dust production was about three times that of Krakatoa alone (Oliver, 1976). This period was followed by a relative absence of eruptions until about 1950. We have modeled the volcanism of the period 1883–1912 by an ersatz distribution in which various closely spaced eruptions are combined. This is described in Table 1. Figure 9 shows the response to this distribution of volcanism for various choices of τ. Clearly, for larger τ's (greater than 16 years) each eruption adds cooling to the response of successive volcanos, leading to cumulative, long-lasting cooling following the period 1883–1912, while for smaller τ's (less than 16 years) the response has little cumulative character, although the large number of eruptions leaves the temperature depressed for the period 1883–1912.

On the whole, the temperature record shows no evidence of a cumulative effect of volcanism (viz., Figure 5),[4] suggesting that appropriate τ's are less than 16 years, while Figure

[4]Granted, the warming trend seen after 1920 might be an extension of a similar trend that would have occurred during 1883–1912, had it not been for the cooling due to the volcanos. However, there is no evidence of such a trend before 1883. In addition, the warming ceased after 1940. Finally, had the volcanos been masking a strong trend before 1912, the net warming that would have occurred would have been unprecedented and in excess of anything we could account for by greenhouse considerations—especially given the long ocean delay that would pertain to large greenhouse sensitivity.

6 suggests that $\tau < 100$ years is incompatible with $\Delta T_{2\times CO_2}$ = 4°C.[5] Indeed, reference to Figure 3 suggests that $\tau <$ 16 years corresponds roughly to g 0.5 and $\Delta T_{2\times CO_2} < 0.6$°C.

Certainly, these matters require more careful analysis. However, the present remarks suggest that consideration of a variety of known climate-forcing perturbations as probes of the climate system may allow the use of a simple methodology for constraining our expectations for greenhouse warming. For example, if responses to volcanos imply short

---

[5]Conceivably, support for this conclusion might also emerge if the climate response to volcanos tends to peak at one year rather than two years after eruption (viz., Figure 7).

characteristic ocean delays, then we would also expect small climate responses to increasing CO_2. One will of course want to understand in detail the physics determining climate sensitivity. However, this hardly justifies ignoring the possibility that it may be possible to bypass these details in approaching the issue of sensitivity directly.

ACKNOWLEDGMENTS

I should like to acknowledge the support of the National Science Foundation under Grant 8520354-ATM and of the National Aeronautics and Space Administration under Grant NAGW-525.

Modeling the Possible Causes of Decadal-to-Millennial-Scale Variability[1]

DAVID H. RIND[2] AND JONATHAN T. OVERPECK[3]

ABSTRACT

There are at least five possible causes of decade-to-century-scale climate variability: inherent ("random") variability in the atmosphere (i.e., no external forcing); inherent or forced variability in the atmosphere-ocean system (e.g., North Atlantic Deep Water fluctuations); solar variability (e.g., the Maunder Minimum); variability in volcanic aerosol loading of the atmosphere (e.g., Tambora); and atmospheric-trace gas variability (e.g., CO_2, methane). Modeling experiments conducted for each of these potential mechanisms show that they have somewhat different signatures, which may allow for discrimination between them in the climate record. The necessary magnitudes of forcing required from the different mechanisms to produce the past climate variations also allow us to estimate the likelihood of their affecting future global change. We conclude that none of them can be dismissed, and we should be prepared for the possibility that several have acted in concert to produce the variability seen in the climate record.

INTRODUCTION

The paleoclimate record contains many examples of apparently significant decade-to-century-scale climate variability, including variations in both temperature and moisture balance (Overpeck, 1991; Bradley and Jones, 1992a). The last millennium appears to have featured at least two widespread cool periods lasting a century or more, beginning approximately A.D. 1275 and A.D. 1510, respectively (Bradley, 1990). Shorter cool periods arose in the decades of the 1590s to 1610s, 1690s to 1710s, 1800 to 1810s and 1880s to 1900s, while prolonged warm periods occurred in the 1650s, 1730s, 1820s, 1930s to 1940s, and 1980s (Bradley, 1990). The global synchronicity of these temperature trends, and their geographic distribution, are in many cases not clear, especially for the earlier time periods. Nor are the causative factors understood in most cases. This state of affairs clearly hinders our ability to predict the anthropogenic climate signature, since its magnitude is likely to be similar to that of these past perturbations, at least for the next few decades.

---

[1]A somewhat more detailed version of this paper has appeared as Rind and Overpeck (1994).
[2]Goddard Space Flight Center, Institute for Space Studies, New York, New York
[3]NOAA Paleoclimate Program, National Geophysical Data Center, Boulder, Colorado

There are at least five potential explanations for global-scale decadal to millennial climate variations. Natural, unforced variability, associated with either atmospheric dynamics or interactions within the climate system (without including ocean circulation changes), is one candidate; this potential explanation obviates the need to search for "causative" factors of the observed climate changes, and simply relies on the mechanisms inherent to the system. Extending this concept to include ocean circulation changes broadens the range of possibilities. Fluctuations in the atmosphere-ocean system are observed to occur today and might be a natural characteristic over longer time scales, although they could equally well come into existence as a response to changes in climate forcing. Solar variability exists as a favorite "wild card" possibility, legitimized to some extent by recent observations of solar irradiance changes during the most recent solar cycle. Volcanic aerosol injections appear to cool the climate, and a clustering of such events, or the lack of such events, would likely lead to a change in mean climate. Finally, variations in greenhouse trace gases such as CO_2 and methane have been observed to occur over the last several hundred years, with cumulative climate impact.

Given the lack of quantitative measurements of some of these forcings over time, one approach to estimating the cause(s) of decade-to-century-scale variability is to model each of these perturbations in a generic sense and compare the results to climate reconstructions. The model used for experiments done specifically for this paper is the GISS general circulation model (GCM) (Hansen et al., 1983b), run at $8° \times 10°$ resolution, with sea surface temperatures either prescribed or calculated, as indicated in Table 1. This model has a sensitivity of 4.2°C warming for $2 \times CO_2$ (or an increase of 2 percent in the solar constant) (Hansen et al., 1984), with a high-latitude/low-latitude sensitivity of about a factor of 2. The quantitative calculations given below are a function of this model sensitivity. Calculations from other GCMs are noted where appropriate.

The model results indicate that each of the potential

explanations of climate variability listed above has its own characteristic signature that may provide a basis for discrimination in the climate record. In the subsequent sections we will discuss evidence suggestive of the various forcings, as well as relevant modeling studies depicting the possible nature of the climatic response. We concentrate mostly on temperature, since it is this parameter that is most often recoverable in the paleoclimatic data on these time scales. In the discussion section, we will briefly review what evidence is needed to decide on which causes are most likely.

NATURAL VARIABILITY

Natural variability suggests an unforced oscillation in climate parameters. It can arise primarily from atmospheric dynamics or from atmospheric thermodynamics. In either mode it must alter the net heating of the surface, either locally, if that is all that is involved, or globally. For oscillations of multidecadal scale to be set in motion, portions of the system with large heat capacities or time constants (ocean, cryosphere) must be involved, or interactions setting in motion positive feedbacks (water vapor, cloud cover), which ultimately reverse direction, must arise. The ocean mixed layer by itself has an e-folding time constant of a few years, with an impact that can be lengthened via other feedback processes (Hansen et al., 1985). An example of the "reversing-direction" process, somewhat analogous to the mechanism proposed by Ramanathan and Collins (1991) although operating on a longer time scale, could be as follows: Warmer sea surface temperatures lead to an increase in high-level clouds and thus to a further greenhouse amplification of sea surface temperatures; eventually the clouds become sufficiently optically thick to reduce incoming solar energy, which set in motion a cooling of the surface and additional water-vapor feedbacks. Wigley and Raper (1990) suggested that high-frequency climate forcing, perhaps associated with clouds, could extend to a century-scale oscillation of a few tenths of a degree Celsius via the ocean's thermal inertia.

TABLE 1 GCM Experiments Discussed in the Text

Experiment	Duration	Ocean Temperatures
Unforced variability	100 years	Calculated SSTs, with mixed-layer depth (MLD) averaging 125 m
Reduced North Atlantic SSTs	6 years	Prescribed colder North Atlantic SSTs, 1/5 full ice-age cooling
Reduced solar inputs of -0.25% or -2%	55 years each	Calculated SSTs with MLD of 65 m
Added volcanic aerosols $\tau = 0.15$	55 years	Calculated SSTs with MLD of 65 m

We begin with the possibility of dynamically induced climate changes. Inherent variability in the atmosphere, possibly associated with non-linear dynamics, is thought to be capable of generating intraseasonal variability. Hansen and Sutera (1986) utilized a planetary-wave activity index to resolve two modes of Northern Hemisphere winter 500 mb height distributions, one with low and one with high amplitudes. The differences between the height fields are centered over the Aleutians, north central Canada, and western Europe. If these two modes represent interannual tendencies that might be selectively activated, they could conceivably set in motion climate feedbacks (snow, sea ice, ocean mixed-layer depth) with climate implications and longer time scales. It is not known whether the modes are forced or simply the result of non-linear dynamics. This type of "natural" variability must be considered speculative when applied to the climate domain. Additional decadal variability may be found in meridional modes of the coupled ocean-atmosphere system that propagate from pole to pole over 3- to 20-year time periods (see, e.g., Mehta, 1992). Their relationship to climate variability is also unproven.

Hansen et al. (1988) ran a GCM with a mixed-layer ocean for 100 years (Table 1). The model produced an oscillation of 0.5°C (peak to peak) over that interval without any external forcing (Figure 1, top). As emphasized by Barnett et al. (1992), an inversion and slight time shift of

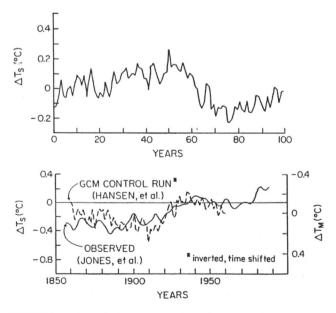

FIGURE 1 Top: Global-mean (unforced) surface air temperature variations in a 100-year control run with an atmospheric GCM and mixed-layer ocean model. (From Hansen et al., 1988; reprinted with permission of the American Geophysical Union.) Bottom: An inversion and time shift of the results shown in the top figure, compared with observed temperature changes over the past 100+ years. (From Barnett et al., 1992; reprinted with permission of the American Geophysical Union.)

this record results in matches to a number of the features in the observed temperature record from 1850 to 1950 (Figure 1, bottom). The implication is that the system is capable of producing such temperature oscillations without the need for causative factors (i.e., forcing).

As analyzed by Barnett et al. (1992), the primary contributions to the temperature oscillation occurred in the tropics (Figure 2); over 90 percent of the variability was located within 30° latitude of the equator. The oscillation was largely the result of a cloud cover/sea surface temperature correlation: Low clouds increase as the ocean cools, further reducing the net heating of the surface. At least part of this correlation was due to the sub-grid-scale temperature parameterization used in the GCM. In contrast to these results, the observed variability over the last century has been dominated by high-latitude contributions (Hansen et al., 1983a). Thus there is good reason to doubt that the variability found in the GISS GCM 100-year run is a convincing indicator of the natural variability in the climate system. The inclusion of additional processes, such as ocean dynamics, variations in cloud optical thickness, and the heat exchange with the ocean below the mixed layer would improve the model's realism (Barnett et al., 1992).

Manabe et al. (1991) produced similar-looking temperature global variations in a 100-year simulation with a fully coupled atmosphere-ocean GCM. The greatest persistence was found in the polar oceans of both hemispheres, due to their weaker thermal damping (reduced surface and radiative heat fluxes) and deep mixing. The relevance of this run to real-world oscillations is limited by the continuously specified input of heat and salinity to the ocean; while this input does not vary from year to year, its very existence will perturb the model's own natural variability, and the necessity for incorporation of these fluxes raises doubts about the model's accuracy.

In yet another unforced 100-year simulation with a coarse-resolution model and a mixed-layer ocean, Houghton et al. (1991) found decadal and longer time-scale variability in sea surface temperatures; in fact, the sea surface temperature at low and middle latitudes had a distinctly red spectrum. The percentage of variance in the low-frequency range (4 to 100 years) showed a small increase with latitude, with high-latitude effects associated with decadal-scale sea-ice variability (as has been noted by Mysak et al. (1990) in the Greenland and Labrador seas). The simulated global surface-air temperature did not display significant multidecadal-scale variability, mainly because little such variability arose in the tropical middle-latitude domain (in contrast to the GISS model results). Cloudiness was not allowed to change in these experiments, which it could in the GISS model.

In summary, the unforced decade-to-century-scale oscillations of global temperature that characterize the natural climate system may well be on the order of several tenths

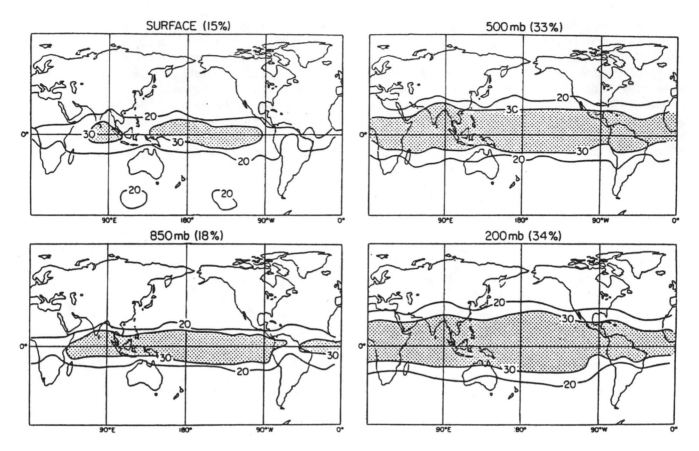

FIGURE 2 Geographical distribution of the first empirical orthogonal function (EOF) of the surface air temperature changes from Figure 1. The percentage of variance represented by this EOF is given for the different levels. (From Barnett et al., 1992; reprinted with permission of the American Geophysical Union.)

of a degree Celsius, although this has not been proven satisfactorily. At the very least, the model results imply that oscillations of this magnitude may be expected to occur without changes in external forcing.

VARIABILITY OF THE OCEAN-ATMOSPHERE SYSTEM

The El Niño/Southern Oscillation (ENSO) phenomenon represents a well-studied example of atmosphere-ocean oscillations. The ability of this type of localized forcing to alter global patterns of temperature and precipitation indicates that it could have an effect on globally averaged climate (Ropelewski and Halpert, 1987), and the El Niño occurrence itself is generally associated with a global mean temperature change of about 0.1 to 0.2°C. The prevalence of El Niños during the past several decades and the lack of pronounced La Niñas undoubtedly have influenced global temperatures (Trenberth, 1990). A key question is whether the frequency of El Niño occurrences fluctuates with time. The results of analysis are equivocal; for example, there are some indications that fewer events took place during

the period from the 1920s to the 1950s (Cooper et al., 1989), while others claim that 1925 to 1932 was a period of unusual activity (Quinn et al., 1987). Cole (1992) has investigated the variance of ENSO activity as reflected in coral records for the past century and found that the dominant frequencies have changed significantly. This observation serves to emphasize that ocean-atmosphere oscillations may themselves be the result of external forcing, a subject we will return to below.

Changes in the ocean heat transport, associated with changes in the wind-driven circulation or the large-scale thermohaline circulation, represent an obvious method of inducing climate variability (see, e.g., Weyl, 1968; Watts, 1985; Rind and Chandler, 1991). The "Great Salinity Anomaly" in the North Atlantic was apparently associated with colder temperatures during the late 1960s and 1970s and reduced deep-water formation (Brewer et al., 1983; Lazier, 1980). The Younger Dryas cooling event (from ca. 11,000 to 10,000 years B.P.) is a much-cited example of North Atlantic Deep Water (NADW) reduction that led to cooling, at least in the region of the North Atlantic (e.g., Broecker et al., 1985). The same type of process may have played a role in the Little Ice Age, judging from the pattern of cold, dry winters over Europe (Pfister, 1992). As in the case of altered El Niños, variations in deep-water production could be a natural characteristic of the system, or they might be forced by other mechanisms; the Younger Dryas event may have been initiated by melting of land ice associated with the general deglaciation (Broecker et al., 1988), while the Great Salinity Anomaly could have been a random perturbation on decadal time scales.

How cold was the Little Ice Age, and how much would ocean heat transports have had to be reduced to produce the observed cooling? Mountain glaciers advanced by some 100 to 200 m (Porter, 1975; Broecker and Denton, 1989), equivalent to about one-fifth of their reduction since the full ice age (Rind and Peteet, 1985). If the cooling was also one-fifth as large, we can reduce the North Atlantic sea surface temperatures by one-fifth their full glacial reduction and use a GCM to estimate the resulting regional and global changes in atmospheric temperature. The temperature perturbation associated with this North Atlantic cooling (Table 1), which is reminiscent of a reduction in NADW formation, is shown in Figure 3. Cooling is basically constrained to downstream locations in Eurasia, and immediately upstream in the vicinity of the eastern United States. Note that if sea surface temperatures in other parts of the ocean had been involved (e.g., the North Pacific) the cooling would have been more widespread. Other aspects of the model simulation also appear realistic: Associated with the colder sea surface temperatures there is higher pressure and lower precipitation over Europe, in good agreement with reconstructions for the Little Ice Age (Pfister, 1992).

The ocean heat-transport reduction necessary to produce this cooling is given in Figure 4. Peak reductions are 25 percent in the North Atlantic, concurrent with increased poleward heat transport in the South Atlantic. For the North Atlantic as a whole, poleward heat transports are reduced by 20 percent; this can be contrasted with estimates of North Atlantic transport reduction during the full ice age of some 70 percent (Miller and Russell, 1989). If the heat transport changes are at all proportional to changes in NADW production, it would imply a Little Ice Age change in NADW production two-sevenths the magnitude of the full ice age reduction, a fraction somewhat similar to that of the advance of the ice lines in the Little Ice Age compared with the advance in the full ice age.

SOLAR VARIABILITY

The hypothesis of solar forcing of weather and climate is at least a century old (Blanford, 1891). The relationships between sunspots and various weather phenomena have been described in numerous publications, some implying statistical significance. The percentage of variance explained increased dramatically when sunspot-cycle effects were partitioned according to the phase of the equatorial zonal wind, the quasi-biennial oscillation (QBO) (van Loon and Labitzke, 1988). While the reality of this effect is still being debated (e.g., Barnston and Livezey, 1989; Baldwin and Dunkerton, 1989), its existence could have climatic implications to the extent that it would heighten the sensitivity of the troposphere to perturbations (i.e., ultraviolet radiation fluctuations) that directly affect the stratosphere.

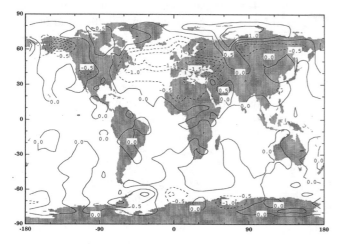

FIGURE 3   Annual average surface-temperature change in the GISS GCM due to colder North Atlantic sea surface temperatures. Temperatures were reduced by one-sixth of the full glacial cooling for the North Atlantic, in accordance with the relative reductions in alpine glaciers for the two time periods, from 30°N to the pole, consistent with the location of observed sea surface temperature changes in the Younger Dryas (Rind et al., 1986). (From Rind and Overpeck, 1994; reprinted with permission of Pergamon Press.)

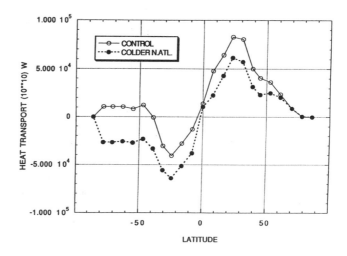

FIGURE 4   Implied Atlantic Ocean poleward heat transports, between longitudes 70°W and 10°E for the current climate control run, and the experiment with reduced North Atlantic sea surface temperatures. Ocean heat transports are calculated from the net energy forcing at the ocean surface, as described in Miller and Russell (1989). (From Rind and Overpeck, 1994; reprinted with permission of Pergamon Press.)

Investigations of solar forcing of climate variability have focused on longer-term periodicities associated with the near-absence of sunspot cycles (e.g., Stuiver et al., 1991). Sunspot records have been deduced from observations, although they are quite uncertain prior to 1800, and from variations in carbon and beryllium isotopes. In periods of high solar activity, the stronger solar wind deflects cosmic rays and decreases the production of cosmogenic isotopes, so solar activity is typically thought to be responsible for much of the natural decade-to-century-scale changes in ^{14}C and ^{10}Be (Stuiver et al., 1991; Stuiver and Braziunas, 1988). Long-term ^{14}C variations can be determined by comparing ^{14}C dates with dates obtained from analyses of tree rings, whereas records of ^{10}Be variations are obtained from ice cores. Although care needs to be taken in interpreting ^{14}C changes, since they are affected by the intensity of the geomagnetic-field polar dipole and carbon-cycle/climate variations, the covariation between ^{14}C and ^{10}Be suggests strong solar control of isotope production for the time scales of concern here (Stuiver et al., 1991).

Spectral analyses of the ^{14}C record indicate apparent fluctuations in periods ranging from decades up to several thousand years, including the 11- and 22-year (Hale) cycles, the 88-year (Gleisberg) cycle, and approximately 200-year and approximately 2500-year cycles (see, e.g., Damon and Linick, 1986; Stuiver et al., 1991). Hints of these cycles have also been identified in the climate record, e.g., the 2500-year cycle in marine, glacier, and polar ice core records (Pestiaux et al., 1987; Denton and Karlen, 1973; Wigley and Kelly, 1990; Dansgaard et al., 1984), and the 88- and 200-year cycles in varved sediments (Halfman and Johnson, 1988; Peterson et al., 1991; Anderson, 1992). Whether these are truly periodic in nature is doubtful, and the relationship of the geologic indicators to climate is itself uncertain, but the similarities in periodicities are indicative of a general coincidence of apparently reduced ^{14}C production and colder temperatures (de Vries, 1958; Eddy, 1976), with the Maunder Minimum/Little Ice Age being the most recent example.

To quantify this connection, it is necessary to know the magnitude of the possible solar-irradiance variations. Recent observations during the last solar cycle (solar cycle 21) indicated close to 0.1 percent peak-to-peak change from solar maximum to solar minimum (e.g., Willson and Hudson, 1991). What these observations imply about past solar-irradiance variations is a more difficult question. Various indicators of insolation changes have been generated. These include ^{10}Be and ^{14}C variations, indicative of solar activity (Beer et al., 1988, 1990; Stuiver et al., 1991); solar-radius variations, thought to be representative of solar luminosity (Ribes et al., 1987); the envelope of the sunspot cycle, perhaps representative of solar irradiance and apparently correlated with decade-to-century-scale sea surface temperature variations (Reid, 1991); and the length of the solar cycle, also seemingly correlated (Friis-Christensen and Las-

sen, 1991). These latter correlations have been used to suggest the possibility of solar dominance of climate variability over the last several centuries. In recent publications, Kelly and Wigley (1992) and Schlesinger and Ramankutty (1992) both found that including solar-irradiance variations related to the length of the solar cycle improved their reconstruction of global temperature changes for the last 150 years. Nevertheless, it is difficult to translate these solar properties into absolute irradiance changes and prove the causal connection.

Another approach to understanding solar variations has been to observe other stars that have a nature similar to the sun's. Decadal-scale stellar-activity variations are apparent (Radick et al., 1990), including the complete absence of "sunspot" conditions (Baliunas and Jastrow, 1990). Accompanying variations in luminosity are as much as 0.4 percent, or somewhat higher than the variation recorded in solar cycle 21. While the relevance of these observations to solar-irradiance variations is uncertain, they do serve to provide some constraint on the magnitude of likely changes.

Lean et al. (1992) estimate that the Maunder Minimum reduction in solar insolation was on the order of 0.25 percent, due to changes in sunspots, faculae, and network radiation. Current GCMs give a climate sensitivity of approximately 4°C for a 2 percent change in solar constant (Hansen et al., 1984); whether the 0.25 percent reduction is sufficient depends on how cool the Little Ice Age was. Estimates range from 0.5°C (Wigley and Kelly, 1990) to 1°C to 1.5°C (Crowley and North, 1991). The nominal value we have used above, about one-fifth the cooling of the full glacial age, implies values of 0.7°C to 1.0°C (e.g., see Rind and Peteet (1985) for estimates of the full ice-age cooling of 3.5°C to 5°C). A GCM simulation with the GISS model, using the 0.25 percent reduction, produced a global annual-average cooling of about 0.45°C, which is probably somewhat smaller than would be required to explain the Little Ice Age if solar variability were acting alone. Of course, if the climate sensitivity were greater or the global temperature change smaller than indicated, the required solar variability would be reduced.

The geographical variation of the temperature change generated by the 0.25 percent reduction (Table 1) is shown in Figure 5. Significant cooling occurs in the tropics, and there is no obvious high-latitude amplification of the climate change. Note that not all regions are cooler than they are today; the forcing is small enough that regional changes in advection patterns can produce local warming. A similar response occurred in the colder-North-Atlantic experiment, which produced advective warming over the Arctic (Figure 3). These results imply too that advection changes may also alter the sources of water vapor sufficiently to confound isotope/temperature reconstructions when the cooling is of such small magnitude (Cole et al., 1992).

The low-latitude cooling led to a reduction in the intensity

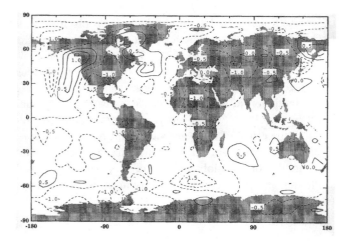

FIGURE 5 Annual average surface-air temperature changes resulting from a flat spectrum decrease in the solar irradiance of 0.25 percent in the GISS GCM. Results are for years 46 to 55 of the simulation. (From Rind and Overpeck, 1994; reprinted with permission of Pergamon Press.)

of the tropical portion of the Hadley circulation by 5 to 10 percent. Because reductions in Hadley-cell intensity are a possible cause of ENSO events, this result is qualitatively consistent with the Enfield and Cid (1991) correlations of strong El Niños and low solar activity. As will be discussed below, volcanic aerosol could also play this role, perhaps more efficiently.

An additional problem is the wavelength distribution of the radiation fluctuations. During the last solar cycle, 20 percent of the observed solar-irradiance variations have been in the ultraviolet range, having wavelengths short of 300 nm, with magnitudes as presented in Figure 6 (Lean, 1987; Cebula et al., 1992). This radiation is all absorbed in the middle and upper atmosphere, above the tropopause. In a

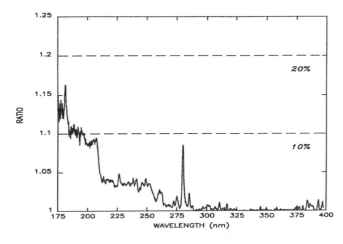

FIGURE 6 Solar ultraviolet irradiance variability as observed in solar cycle 21 (Lean, 1992, personal communication). (From Rind and Overpeck, 1994; reprinted with permission of Pergamon Press.)

GCM experiment with an increase in total solar radiation of 0.5 percent, but with all wavelengths restricted to less than 300 nm, we found that there was no increase in solar heating at levels below the tropopause. Thus, the ability of solar-irradiance variations to replicate observed climate variability will depend on the frequency of the radiation affected, in addition to its magnitude. A change in UV radiation absorbed in the stratosphere might conceivably be able to influence climate if the impact on tropospheric dynamics described by van Loon and Labitzke (1988) proves to be real.

VOLCANIC AEROSOLS

Previous studies (Lamb, 1970; Bray, 1974; Porter, 1981; Grove, 1988) have suggested that the trends in global climate observed during the years A.D. 1000 to 1800 could largely be explained by variations in volcanic aerosol forcing. Porter (1986) related acidity records in the Greenland ice cores to alpine glacial fluctuations in the Northern Hemisphere and found a reasonably good correlation, with increased acidity during periods of glacial advance. It is recognized that high-latitude ice cores may overemphasize mid- and high-latitude volcanic eruptions (Bradley and Jones, 1992b), which should be less effective in altering global climate than low-latitude eruptions. Furthermore, cold periods with low precipitation may have higher acidity concentrations because there has been less dilution from snowfall. Nevertheless, as the recent estimates of a 0.5°C cooling associated with Mt. Pinatubo (Hansen et al., 1992) suggest, volcanic aerosols (primarily sulfuric acid) have the capability to significantly alter the global climate (Rampino and Self, 1984).

What magnitude and frequency of volcanic eruptions are required to produce the cooling estimated for the Little Ice Age? We can use several recent GCM experiments with Mt. Pinatubo (Hansen et al., 1992; Rind et al., 1992; Pollack et al., 1993) to provide estimates. Aerosol radiative forcing depends on geographical distribution, optical depth, size distribution, composition, and altitude; however, as shown by Lacis et al. (1992), the forcing of tropospheric climate is primarily a function of aerosol-column optical depth, assuming a global distribution. In a study of transient volcanic aerosol, a volcanic eruption with a peak global optical thickness of 0.18 (on the order of what has been observed for Mt. Pinatubo) and an *e*-folding time of 12 months (so that in 3.5 years 97 percent of the aerosols will have been removed) produced a peak cooling of 0.5°C in year 2, and a three-year average cooling of 0.3°C (Hansen et al., 1992). In another experiment, with one-half the optical thickness and the same residence time, the cooling was reduced by a factor of 2. If this linearity holds for increased optical thickness values as well, it implies that to produce a global cooling of some 0.5°C to 1.0°C on a three-year average, via a transient volcanic eruption, would require a volcanic-aerosol injection of 2 to 3 times the magnitude of Mt. Pinatubo's. The Tambora eruption of 1815 may have been of this magnitude (Rampino and Self, 1982).

However, if volcanic aerosols were maintained continuously in the atmosphere, i.e., their climate forcing remained relatively constant, then the ocean would have sufficient time to respond to them and to initiate positive feedbacks amplifying the cooling. With an optical thickness of 0.15 (equivalent to a 2 percent decrease in solar constant), the GISS model produces an equilibrium cooling of some 4.7°C after 50 years (Rind et al., 1992; Pollack et al., 1993). To produce a 0.5°C to 1°C cooling would take an optical thickness approximately one-sixth as large, or 0.025. This is some five times larger than the "non-volcanic" background that existed during much of the time from 1900 to 1960, and it could be achieved by the eruption of a Mt. Pinatubo once every eight years, or, equivalently, of an El Chichon every fifth year. Over the last 100 years, we have had two events of the latter magnitude (Mt. Agung and El Chichon), and one of approximately twice this magnitude (Mt. Pinatubo). Thus, volcanic-aerosol-induced cooling of the appropriate magnitude would require a substantial increase in large volcanic injections. (Also, to be effective globally, they would have to originate in low latitudes, although a big high-latitude eruption might have a significant regional or hemispheric impact.)

Is there any way to distinguish global volcanic-aerosol cooling from reduced solar insolation? We have run two experiments with the GISS GCM and GCMAM (the version of the model that extends up to 85 km altitude). In one, the solar constant was reduced by 2 percent; in the other, the volcanic-aerosol optical depth was increased to 0.15, which is equivalent to a 2 percent reduction in solar energy reaching the surface (Table 1). The relevant differences between the results from the two experiments are given in Table 2. A primary difference is that the reduction in solar irradiance cools the lower stratosphere, while volcanic aerosols induce warming in that region, via short-wave and long-wave absorption (Rind et al., 1992). In conjunction with the slightly greater sea surface cooling in the tropics from the aerosol experiment, volcanoes appear to increase the static stability at low and subtropical latitudes more than does a simple solar energy reduction. With increased stability, the Hadley cell is weakened more in the volcano experiment, and the precipitation gradient between the tropics and Northern Hemisphere subtropics is somewhat reduced. (For example, if a record of precipitation for the Little Ice Age were available at these latitudes, such a change in gradient (about 20 percent) might be observable.) Wetter subtropical deserts might occur more easily with volcanic-aerosol forcing than with a solar-irradiance reduction; however, both types of forcing appear to weaken the Hadley circulation somewhat.

In this section we have been concerned with the influence of volcanic aerosols, but tropospheric aerosols may well have increased over the past few centuries, thus cooling the climate (Charlson et al., 1992). The increase in aerosol-cooling effect through time does not allow for any obvious influence on earlier cold oscillations, but if the magnitude of the reduction in solar irradiance is as large as hypothesized in the IPCC (1992) report, aerosols would counter much of the effect of trace-gas variations, discussed below. There is still much uncertainty concerning historical variations in the direct global effect of tropospheric aerosol scattering, and even more uncertainty in aerosols' influence on cloud cover.

TRACE-GAS VARIATIONS

Ice-core evidence of trace-gas concentrations during the last several centuries has indicated that CO_2 levels were at approximately 270 ppm for some time prior to this century (Neftel et al., 1985) (they are currently close to 360 ppm), and that methane has increased from around 0.8 ppmv to its current value of 1.72 ppmv (Rasmussen and Khalil, 1984). Such variations in greenhouse gases undoubtedly have had an effect on the climate system. They complicate the issue of the Little Ice Age cooling because they can add a linear trend to whatever fluctuations might have occurred—that is, the cooling that would be required to reduce current temperatures to Little Ice Age values is increased due to the trace-gas-induced warming over the last several hundred years.

To estimate the influence that these trace-gas variations may have had, two facts need to be known: The equilibrium response of the system, and the response that has occurred so far. The equilibrium response can be calculated with the formulae presented in Hansen et al. (1988). For CO_2, the global-average radiative surface-temperature change due to a change in gaseous concentration X (in ppm) can be calculated as

$$\Delta T_0 = F(X) - F(X_0) \qquad (1)$$

where

$$F(X) = ln\,(1 + 1.2X + 0.005X^2 + 1.4\times 10^{-6}X^3). \qquad (2)$$

A CO_2 reduction from 315 to 270 ppm thus gives a ΔT (radiative) of 0.30°C.

The equilibrium feedback factor to radiative perturbations in GCMs is between 3 and 4 (Hansen et al., 1984). Therefore, in equilibrium the CO_2 change would provide for a cooling of about 1°C.

For methane, the equilibrium temperature change to a change in methane (in ppmv) can be calculated as

TABLE 2   Differences Between Experiment with Volcanic Aerosal Optical Depth Increased to 0.15 and Experiment with Solar Irradiance Reduced by 2 Percent

Latitude (degrees)	T (68 mb)	Surf Temp., Global (°C)	Surf Temp., Land (°C)	Precip., Global (mm/day)	Precip., Land (mm/day)
90	−1	0.8	0	−0.1	0
82	−1	0.3	−0.7	−0.1	−0.1
74	−1	0	−0.5	0	−0.1
67	0	0	0	0	0
59	0	0.1	−0.3	0	0
51	1	−0.4	−0.6	0	−0.1
43	1	−0.1	−0.6	0	0
35	2	−0.3	−0.8	0	0
27	2	−0.5	−0.6	0	0
20	2	−0.6	−0.7	0	−0.1
12	3	−0.6	−0.7	0	−0.1
4	3	−0.7	−0.9	−0.1	−0.4
−4	3	−0.7	−0.9	−0.3	−0.4
−12	3	−0.8	−0.6	−0.3	−0.1
−20	2	−0.8	−0.2	0	−0.2
−27	1	−0.4	−0.5	−0.2	−0.1
−35	1	−0.4	0.1	0	0
−43	1	−0.4	0.7	−0.1	0
−51	1	0.8	0.2	0	0.2
−59	0	1	−0.2	0	0.1
−67	0	0.3	0	0	0
−74	0	−0.1	0.3	0	−0.1
−82	−1	0	0	0	0
−90	−1	0.3	0	0	0
GLOBAL	1.4	−0.34	−0.45	−0.04	−0.08

$$\Delta T_0 = G(X) - G(X_0) \qquad (3)$$

where

$$G(X) \approx \frac{0.394X^{0.666} + 0.16e^{(-1.6X)}}{1 + 0.169X^{0.62}} \qquad (4)$$

A methane reduction from 1.5 ppmv to 0.7 ppmv corresponds to a radiative temperature change of 0.08°C. With the estimated feedback, this gives an equilibrium temperature change of close to 0.3°C.

The observed variation in CO_2 and methane in an equilibrium calculation would thus correspond to a temperature change over the last several hundred years of about 1.3°C, with the currently estimated feedback factors. This alone would be sufficient to account for much of the Little Ice Age cooling, without the additional forcing generated by recently added trace gases (e.g., CFCs). (However, this correspondence also ignores the potential impact of increases in tropospheric aerosols, which, as noted above, may have acted to force the system in the opposite direction, with a magnitude similar to that of the trace-gas effect.) It therefore raises the question of what fraction of this equilibrium response has been observed, i.e., what the response

time of the system is. Hansen et al. (1985) discuss the various estimates for this value, noting that it is a function of both ocean mixing (which determines the effective heat capacity of the system) and climate sensitivity (which is indicative of the magnitude of the feedbacks that must come gradually into play; note that if there are no feedbacks, then all the forcing arises instantaneously from the initial perturbation, reducing the delay in system response).

With an effective vertical diffusion through the base of the mixed layer on the order of 1.5 cm^2 s^{-1}, as implied by transient ocean tracers (Broecker et al., 1980), and a feedback factor of the magnitude noted above, the CO_2-induced warming from 1850 to 1980 amounts to approximately one-third of the equilibrium response (Hansen et al., 1985). For an approximately similar phasing of methane changes, as indicated by the above references, the proportion of equilibrium warming should be similar. Thus we might conclude that only one-third of the equilibrium trace-gas response has been realized, or approximately 0.4°C to 0.5°C. Nevertheless, if the Little Ice Age was 1°C cooler than today, half of this could be a trace-gas effect, with most of the related warming occurring in the twentieth century. Since the trace gases are globally distributed, the climate changes they induce would be similar to those shown for solar radiation reduction (Figure 5), since doubled CO_2 and a 2 percent solar constant increase yield similar patterns of model response (Hansen et al., 1984).

DISCUSSION

The different forcings and natural variability discussed above have in many cases different geographical expressions (see, e.g., Figures 2, 3, and 5, and Table 1). For example, the forcing that is of radiative importance globally, such as reduced insolation or continual volcanic-aerosol injections, has a larger impact at inland locations away from the moderating influence of the ocean. In contrast, changes in ocean heat transport and NADW production have maximimum influence in coastal locations. The former type of forcing would therefore imply maximum Little Ice Age cooling in central Eurasia, decreasing toward the Atlantic coast, whereas the opposite would prevail if NADW-formation changes were acting alone.

It might therefore appear to be possible to distinguish between them through reference to paleoclimatic data for the last several centuries. A joint NSF/NOAA project, Analysis of Rapid and Recent Climate Change (ARRCC), has begun reconstructing the geographical climate distribution for several time periods during the last few hundred years, with this goal in mind. Worldwide distributions of temperature and precipitation changes will be needed for this purpose. Whether the data are sufficiently comprehensive and accurate remains to be seen.

A complementary approach involves reconstructing the potential climate forcing for this time period, whether it be solar irradiance, volcanic aerosols, trace gases, or ocean circulation. As noted above, there are many empirical estimates of solar-irradiance variations, but they need to be quantified on the basis of physical principles. The actual volcanic-aerosol properties are similarly difficult to constrain (Bradley and Jones, 1992b). It may be possible to deduce ocean-circulation changes through biological/geochemical studies in high-resolution sediment and coral cores, but that is not yet certain.

With both the climate response and climate forcing somewhat ambiguous for these time periods, in this paper we have had to rely on current climate models to suggest possible links. The major problem here is that we cannot be sure that the models have the proper climate sensitivity, either globally or locally. Analysis of climate forcing on the time scale of glacial/interglacial epochs suggests that we probably know the global sensitivity to within a factor of 2 (Lorius et al., 1990). The shorter-term response is hard to calibrate, and its latitudinal expression is likewise somewhat uncertain, beyond the relative certainty that there is a greater thermal response at high latitudes, especially in winter.

Given the above uncertainties, it does not appear as though we can rule out any of the proposed mechanisms for forcing climate fluctuations on the decade-to-century time scale, although there is so far little proof that any of them has a magnitude sufficient to strongly influence global temperatures. The radiative forcing due to the observed trace-gas increase is the climate perturbation that can be quantified best; our ability to quantify its historical effect is limited by uncertainty in the climate response time, and the potential offsetting influence of increased tropospheric aerosols. From the perspective of recent observations in the current century, it is perhaps more likely that a sustained insolation reduction of 0.25 percent could have occurred (associated with the Maunder Minimum) than that volcanic eruptions on the order of El Chichon could have arisen every five years. However, we do not know how typical this last century is, nor do we know what portion of the estimated larger insolation variations might have occurred at visible wavelengths capable of penetrating to the troposphere. While the unforced variations that arose in the GISS GCM may not ultimately prove to be physically realistic, this does not imply that unforced fluctuations are not possible, with perhaps some feedback involving ocean temperatures that operate on a longer time scale. The results also indicate that global cooling on the order of 0.5°C to 1°C is not sufficient to guarantee cooler temperatures everywhere, since advective changes, perhaps caused by the climate perturbation, can dominate the local response. Therefore, the possibility of global-scale forcings should not be dismissed if local regions of warming are uncovered.

Finally, it is certainly possible to have several of these effects working together. Reductions in high-latitude temperatures associated with volcanic aerosols or diminished insolation, if they are sufficiently large, might have initiated changes in high-latitude deep-water formation, especially in the North Atlantic; however, such variations could simply arise naturally, in conjunction with random atmospheric forcing. The lower concentrations of CO_2 and methane found prior to the current anthropogenically influenced levels must have affected the climate system, perhaps preconditioning high-latitude regions, or at least acting in concert with other cooling mechanisms. The search for "the cause" of decade-to-century-scale variability may ultimately uncover a system that responds to a variety of forcings and exhibits unforced variability. The climate record of the past several centuries may be difficult to decipher if we explore it with a strictly linear cause/effect approach.

ACKNOWLEDGMENTS

Modeling studies done in conjunction with the ARRCC project have been supported by the National Science Foundation and the National Oceanographic and Atmospheric Administration. Climate modeling at GISS is supported by the NASA Climate Program Office. We thank R. Healy and P. Lonergan for help with model experiments; J. Hansen, R. Bradley, M. Hughes, G. Jacoby, and L. Thompson for discussions or critical comments on the manuscript; and J. Lean for providing estimates of solar variability. J.M. Wallace provided substantial comments in review.

Commentary on the Paper of Rind and Overpeck

JOHN M. WALLACE
University of Washington

Drs. Rind and Overpeck have given us a very interesting and informative summary of a coordinated series of climate sensitivity experiments designed to investigate the potential importance of five sources of climate variability: (1) inherent (random) variability of the climate system, not taking into account variability in the ocean circulation, (2) coupled atmosphere-ocean interactions, including phenomena such as the ENSO cycle and the thermohaline circulation, (3) variability in solar output, (4) forcing due to episodic injections of volcanic aerosols, and (5) increases in greenhouse gases. I was struck by the fact that they had independently chosen nearly the same categories that I had used in interpreting the observational record in my paper. This coincidence serves to illustrate that many of us share a common view as to the nature and probable causes of interdecadal climate variability.

Climate modeling is such a broad topic that no single paper of this length can hope to be comprehensive, but I think that the numerical experiments discussed in their review are well chosen and provide a great deal of food for thought. They may be viewed as a first step toward a quantitative assessment of the relative importance of various proposed mechanisms that contribute to climate variability on the interdecadal time scale.

The authors conclude that none of the five interpretations of interdecadal climate variability that they considered should be categorically discounted at this point. I agree that we need to be open to the possibility that any of these interpretations might ultimately prove to be relevant, but I also agree with the earlier suggestion that we have a duty to try to assess the relative importance of the various climate forcing mechanisms. I believe the modeling evidence reviewed by Drs. Rind and Overpeck does, in fact, suggest that some mechanisms are more important than others.

To start with, I think it can be argued that solar variability should be near the bottom of the list. In order to elicit a significant response to solar variability on the time scale of the sunspot cycle in the experiments that they described, it was necessary to artificially inflate the solar forcing of the troposphere by treating the observed temporal variability in the solar output as if it were independent of wavelength, whereas in reality it involves only ultraviolet radiation, which is absorbed in the upper atmosphere. I would also place aerosols injected by volcanic eruptions near the bottom of the list because, in order to elicit a significant interdecadal signal, it is necessary to invoke changes in aerosol loading on time scales much longer than the typical 1- to 2-year residence times observed in association with recent eruptions.

Another mechanism that I would be inclined to dismiss is decadal-scale changes in the ENSO cycle. I would do so not on the basis of the numerical experiments but on the basis of the observational evidence presented in my paper, which indicates that there is very little temporal correlation between the ENSO signal in the tropics and the observed interdecadal variability and mean temperature of the extratropical regions of the globe. And it is mainly in extratropical latitudes that one finds the large interdecadal variability in paleoclimate records.

That leaves us with greenhouse gases, the final topic discussed in Dr. Rind's paper, and inherent (random) variability of the climate system, including variations in North Atlantic deep-water formation, which might account for the dip in Northern Hemisphere temperatures during the 1960s and 1970s. I suppose that aerosols generated by sources other than volcanic eruptions (e.g., industry) might also be important. I propose this priority list for our discussion.

Discussion

RIND: Well, for solar variability we could look at sunspot cycle length and variations in the sun's radius. During the last solar cycle people actually did try to measure this, but they couldn't relate it to the observed solar radiance variations. As for volcanos, we could claim that the acidity in ice cores indicates a change in eruption frequency. The forcing is not well constrained in this area, so it's very difficult to confirm or discount an apparent correlation.

MCWILLIAMS: In your categorization, you apologized for declaring the ocean variability to be external. How about declaring the greenhouse gases to be external variability too? Aren't they both really part of the internal variability?

RIND: Yes, in the sense that you can change nothing in the model and still get those fluctuations in the global surface air temperature. That's internal to the model rather than to the system, though.

SHUKLA: In a paper several years ago Lorenz showed that with his smallest GCM he could reproduce 100-year records simply as a result of the internal dynamics of the system. On what basis can we distinguish between his model and our complex GCMs, and what criteria can we use to decide how much to believe the results?

RIND: With something like the NSF/NOAA ARRCC program, which has both a modeling and an observational component, it's hard to tell whether the forcing is wrong or the model is poor when the results don't agree with the observations. A complicated 3-D model can at least show geographic distributions you can relate to what happened.

LEHMAN: One thing that seems to be diagnostic of your ocean cooling experiment is the appearance of warming over the Eurasian Arctic. I think that might be in part an artifact of the limitations of the 18-ka climate data set, which you take a fifth of an increment of to apply to your model. There is no resolution of SST variability between the Norwegian Sea and the Eurasian Arctic in the data

set. Looking at deglaciation and the Holocene, we find that if the import of ocean deep water into the Norwegian Sea is cut back by just a small amount the Barents Sea fills it rapidly. The downstream effects of cooling over the northeast Atlantic might be offset by local cooling of the sea surface in the Barents Sea. Now I might be tempted to use this warming as diagnostic, but I wanted to introduce this reservation about the climate data set.

RIND: Have you any data set that will tell you what happened to the Norwegian Sea during the Little Ice Age?

LEHMAN: Yes, we see evidence for a reduction in the inflow of warm Atlantic water. Our SST proxy records are not very sensitive, though, so we can't put a number of degrees on the amount of cooling that occurred.

RIND: Would you relate this to actual changes in North Atlantic Deep-Water production at that time?

LEHMAN: We can find records with resolution sufficient to capture these sorts of changes in the trajectory of surface currents, but when we look for similar records of the changes in abyssal circulation we run into the problem of ocean-sediment deposition rates—4 cm for 1000 years. We have not yet found a suitable location for making such measurements.

LINDZEN: One comment on Jim Hansen's volcano simulation: Remember that was superimposed on his nominal greenhouse warming. I think our results are actually pretty close. Also, a propos of the forcing, the interannual variabilities turn out to be so huge that I wonder whether we shouldn't ask how the system remains as stable as it does, rather than looking for the very small forcings we call external.

RIND: Indeed, if the interannual variability and net forcing are as large in the models, we should try to diagnose why it remains stable.

Natural and Forced Variability in the Climate Record

JOHN M. WALLACE[1]

ABSTRACT

Extended records of surface air temperature at land stations and tropospheric temperature are examined, with emphasis on horizontal structure, seasonality, and detection of specific signals such as El Niño and volcanic eruptions. Interdecadal trends in hemispheric mean temperature show up clearly in data for summertime and the transition seasons, even without any filtering of monthly mean data. In wintertime they tend to be obscured by large local temperature fluctuations associated with regional teleconnection patterns. A distinctive El Niño/Southern Oscillation (ENSO) signature is evident in averages over the tropical belt, where surface- and upper-air temperatures exhibit a remarkably consistent pattern of interannual variability. However, the ENSO cycle is not evident in time series of spatially averaged temperatures in the extratropical regions of the globe. The El Chichon and Pinatubo eruptions show up clearly in globally averaged lower-stratospheric temperatures based on MSU Channel 4 data, but they tend to be obscured by dynamical phenomena in hemispheric averages or averages over particular latitude belts. Short-lived cooling signatures of a few of the major volcanic eruptions are evident in the mean summertime land surface-air temperatures of the Northern Hemisphere extratropics. Eruptions that took place in high northern latitudes during spring and summer produced the strongest signatures. The warming that has been observed in surface-air temperatures in both hemispheres since the late 1970s is not as clearly evident in temperatures aloft. The observed pattern of temperature trends is suggestive of a secular decrease in the mean static stability of the extratropical regions of both hemispheres, perhaps in response to an increase in the optical thickness of the atmosphere.

INTRODUCTION

Much of the empirical work on interdecadal climate variability has been directed toward the production of time series of surface-air temperature and sea surface temperature averaged over the globe and over the Northern and Southern Hemispheres separately. Through an extended effort at data collection and quality control, the reliability of these time series has been substantially improved, and the record length has been extended beyond a century. Gridded analyses of

---

[1]Department of Atmospheric Sciences, University of Washington, Seattle, Washington

these data are now available to the research community (Jones et al., 1985, 1986a).[2] Efforts are under way to extend the climate record back further into the past, by combining the few available direct measurements with a selection of much longer proxy records, as described elsewhere in this volume. These efforts are complemented by statistical studies aimed at determining the minimum spatial sampling requirements for obtaining reliable global and hemispheric means (North and Kim, 1995).

The surface data are supplemented by a continuous record of satellite-based microwave temperature soundings, beginning in 1979, which provide global coverage (Spencer and Christy, 1990, 1992, 1993). Monthly mean gridded data are available for Channels 2 and 4 of the microwave sounding unit (MSU). The former (MSU-2) provide a measure of the mean temperature of the troposphere (surface to 300 hPa) and the latter (MSU-4) the mean temperature of the lower stratosphere centered near the 70 hPa level.

Although spatial patterns have been examined in a few studies (e.g., Jones and Briffa, 1992; Jones and Kelley, 1983; van Loon and Williams, 1976), hemispheric or global mean temperatures have been the focal point for most of the empirical work on interdecadal temperature variations. The emphasis on hemispheric means is justified by the structure of the leading empirical orthogonal function (EOF) of annual mean surface temperature, which is of the same sign throughout most of the Northern Hemisphere (Barnett, 1978). Even more impressive in this respect is the marked similarity between time series of hemispheric means computed separately for land areas, sea surface temperature, and nighttime marine air temperature (Folland et al., 1990) and the time series of Bradley et al. (1987), presented in Figure 1; these series indicate that on the interdecadal time scale, as China goes, so goes the Northern Hemisphere and vice versa. The same is true of other regions of comparable size (see, e.g., Diaz and Bradley, 1995).

The foregoing evidence of spatial consistency notwithstanding, there still remain questions concerning the interpretation of hemispheric temperature trends in surface-air temperature. In this brief review, we will consider the following issues:

(1) Are the unseasonably warm wintertime temperatures over the high-latitude continents during the past decade entirely responsible for the observed warming trend in the Northern Hemisphere, or are they a consequence of a persistent pattern of wintertime circulation anomalies, superimposed on a more fundamental warming trend that is affecting the entire hemisphere during all seasons?

[2]Updated versions are available from the Carbon Dioxide Information Analysis Center (CDIAC), Environmental Sciences Division, Oak Ridge National Laboratory, Oak Ridge, TN 37831, FAX (615) 574-2232.

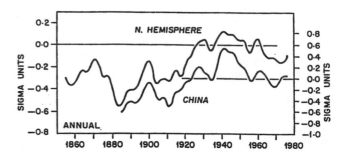

FIGURE 1   Time series of normalized annual mean surface-air temperature averaged over China and the entire Northern Hemisphere. (After Bradley et al., 1987; reprinted with permission of Springer-Verlag.)

(2) Is the zonally averaged temperature signature associated with the El Niño/Southern Oscillation (ENSO) largely a tropical phenomenon, or is it felt globally?

(3) Are the interdecadal trends in surface-air temperature and temperatures aloft virtually identical, or is there evidence of systematic trends in the static stability of the troposphere?

STRUCTURE AND SEASONALITY OF VARIATIONS IN SURFACE AIR TEMPERATURE

The variability of wintertime temperatures over the high-latitude continents (specifically western Canada and Siberia) is substantially larger than anywhere else in the hemisphere, and anomalies in these two regions are influenced by the Pacific-North American (PNA) teleconnection pattern (Gutzler et al., 1988). Wintertime and springtime warming over these regions has been implicated in the recent rise in the mean surface-air temperature during the 1980s (Trenberth, 1990; Folland et al., 1990, 1992; Jones and Briffa, 1992). Indeed, the spatial signature of the PNA pattern is clearly evident in the field of 1,000 to 500 hPa thickness anomalies for the 1980s, shown in Figure 2. Comparison of the wintertime mean and the more bland annual mean fields in Figure 2 suggests that this regional pattern is largely a wintertime phenomenon.

Lest the significance of the high-latitude continents be overstated, it should be noted that trends in the PNA pattern and wintertime temperatures over the high-latitude continents do not always determine the sign of the trend in the mean surface-air temperature over the Northern Hemisphere. A notable example is the period of the late 1950s and early 1960s, when the mean temperature of the Northern Hemisphere extratropics fell slightly, even though wintertime temperatures over western Canada rose substantially in connection with a reversal in the predominant polarity of the PNA pattern (van Loon and Williams, 1976; Wallace and Zhang, 1993).

Perhaps as a result of regional features such as the PNA

TABLE 1   Temporal Correlation Between Area-Weighted, Normalized Seasonal-Mean Surface-Air Temperature over China, and All Northern Hemisphere Land Masses (after Bradley et al., 1987)

	Spring	Summer	Autumn	Winter	Annual
Unsmoothed data	0.66	0.67	0.68	0.54	0.81
Smoothed (as in Figure 1) data	0.92	0.93	0.90	0.68	0.95

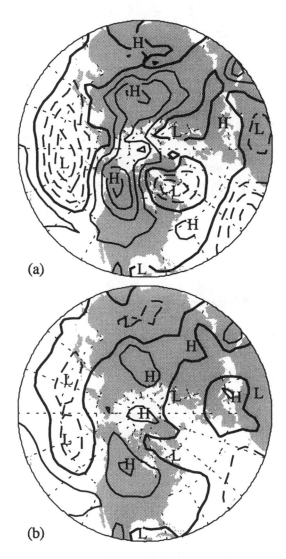

(a)

(b)

FIGURE 2   Difference in thickness of the 1000-500 hPa layer, 1981-1990 minus the period 1951-1980. (a) Wintertime means, (b) annual means. Contour interval 10 m (0.5K). Negative values are dashed. (After Wallace et al., 1993; reprinted with permission of the American Meteorological Society.)

air temperature is dominated by regional-scale features reminiscent of the PNA pattern, in sharp contrast to the monopolar structure of the leading mode of annual mean temperature. And when the annual mean time series shown in Figure 1 are broken down by season (Table 1), strong positive correlations are observed between the China series and the hemispheric mean series for every season except winter.

Since the low-frequency variability during the warm season is less strongly influenced by circulation anomalies, hemispheric mean temperatures observed during the warm season should be of particular interest from the standpoint of detecting thermodynamically induced hemispheric or global trends. Seasonal mean time series of hemispheric mean surface-air temperature presented by Jones and Briffa (1992) are suggestive of a more favorable "signal-to-noise ratio" for the detection of interdecadal variability in the plots for the warm season. This distinction is brought out clearly in Figures 3 and 4, which show time series of monthly mean surface-air temperature anomalies averaged over the extratropical Northern and Southern Hemispheres,[3] respectively, with the months May to October and November to April plotted separately. These figures, and several of the subsequent figures in this paper, are based on the data set compiled by Jones et al. (1985, 1986a) under the sponsorship of the U.S. Department of Energy. The "DOE" identifier on individual curves also refers to these data. Analogous plots (not shown) based on 3-month means (May-through-July and August-through-October versus November-through-January and February-through-April) also exhibit substantially more scatter during the cold seasons.

Hence, it can be concluded that the recent upward trend in surface air temperature averaged over the Northern Hemisphere is a reflection of a genuine, year-round hemispheric warming. The more dramatic rises in wintertime temperatures over western Canada and Siberia are of considerable interest in their own right, but it seems unlikely that they are the fundamental cause of the hemispheric warming, which shows up just as clearly in the warm-season time series as in the cold-season time series.

pattern, local wintertime temperatures generally tend to be less coherent with the mean temperature of the Northern Hemisphere than those observed during other seasons. For example, in the EOF analysis of Barnett (1978) cited in the previous section, the leading mode of wintertime surface-

[3]The abrupt increase in variance of the Southern Hemisphere time series in 1958 coincides with the establishment of stations in Antarctica.

The choice of how to define the hemispheric averages in Figures 3 and 4 is, of course, rather arbitrary. In this particular case it does not make much difference whether one considers the entire hemisphere, the extratropics poleward of 20° latitude, or a smaller polar-cap region. As one moves the outer boundary poleward, the interdecadal trends become somewhat larger, but so does the sampling noise associated with regional and short-term variability, so one's overall impression of the "signal-to-noise ratio" does not change a great deal.

THE SIGNAL OF EL NIÑO/SOUTHERN OSCILLATION IN THE GLOBAL TEMPERATURE FIELD

Walker and Bliss (1932) showed that their Southern Oscillation Index (SOI)[4] modulates surface-air temperature, averaged over the tropics, as well as over certain regions of higher latitudes, such as southwestern Canada. This result was confirmed by the later studies of Newell and Weare (1976), Angell and Korshover (1978), Newell (1979), and Horel and Wallace (1981), which showed that the ENSO signature is evident in time series of mean tropical tropospheric temperature, lagged by about a season or two relative to sea surface temperatures in the equatorial Pacific. This effect is clearly evident in time series of the radiance in Channel 2 of the MSU carried aboard the TIROS satellites (an indicator of the mean temperature of the 1,000 to 300 hPa layer) shown in Figure 5. The upper time series is unsmoothed monthly mean sea surface temperature (SST) averaged over the equatorial Pacific from 160°E to the South American coast and from 5°N to 5°S, which shows clearly the signature of the three most recent warm episodes (1982-1983, 1986-1988, and 1991-1992). The corresponding MSU-2 time series, averaged over the latitude belt from 20°S to 20°N shown just below it, exhibits the same features, some of them lagged by about a season, as noted in the studies cited above. A remarkably similar signature is evident in the mean surface-air temperature of the tropics based on gridded data from land stations, shown in the lowest curve. The much larger amplitude of the SST time series and the phase lag between SST and the other time series are consistent with the notion that the ENSO signature in tropical tropospheric temperature is a forced response to local perturbations in the heat balance at the air-sea interface in the equatorial eastern Pacific. The fact that the ENSO signature is virtually identical in the MSU-2 and land-

<hr>

[4]The SOI is the difference between normalized sea level pressures at Tahiti (11°S, 150°W) and Darwin (12°S, 131°E). It is a measure of the strength of the zonal pressure gradient that drives the trade winds in the equatorial Pacific, which, in turn, govern the volume and coldness of the water brought to the surface by equatorial upwelling.

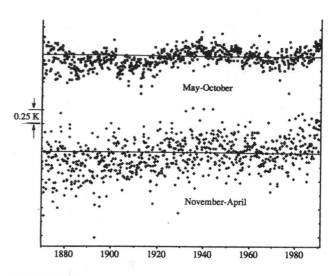

FIGURE 3   Scatterplot of the time series of monthly mean surface-air temperature anomalies, averaged over all land gridpoints in the Northern Hemisphere extratropics (25° to 90°), based on the U.S. Department of Energy data set, plotted separately for the calendar months May to October and November to April, as indicated. One small tick mark on the vertical scale is equivalent to 0.5K.

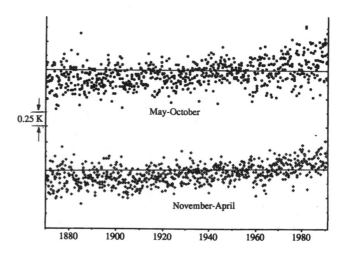

FIGURE 4   Temperature anomaly time-series scatterplot as in Figure 3, but for the Southern Hemisphere extratropics.

surface temperature time series, but slightly larger in the former, suggests that it is transmitted from the equatorial Pacific to the remainder of the tropics via thermally direct circulations in the free atmosphere.

A strong correspondence between equatorial Pacific SST and surface-air temperature averaged over the tropical belt is also evident in the longer time series shown in Figure 6. Because of the slight phase lag between the SST anomalies and the land-surface temperature anomalies, the correlation coefficient between the two time series is only about 0.5; nonetheless, the correspondence between warm and cold phases of the ENSO cycle and tropical tropospheric

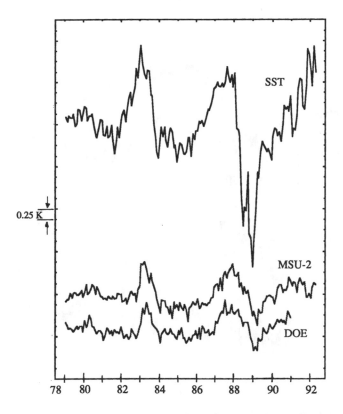

FIGURE 5   Three time series. Upper series: equatorial Pacific sea surface temperature anomalies for the region 160°E to 90°W, 5°N to 5°S, which corresponds to the regions included in Niño 3 and Niño 4 in the Climate Diagnostic Bulletin. Middle series: tropospheric temperature anomalies, as sensed by the Microwave Sounding Unit (MSU-2) carried aboard the TIROS N satellites, averaged over the tropical belt (20°N to 20°S). Lower series: surface-air temperature over all land grid points within the tropical belt (20°N to 20°S). The temperature scale is the same for all three curves: one small tick mark is equivalent to 0.25K. MSU data courtesy of Roy Spencer, NASA Marshall Space Flight Center, Huntsville, Ala.. (After Yulaeva and Wallace, 1994; reprinted with permission of the American Meteorological Society.)

temperature anomalies is unmistakable. A correlation coefficient of about 0.8 can be obtained by performing a transformation on the SST time series, using it as the input for a stochastically forced climate model based on the formulation of Hasselmann (1976) (see also Frankignoul, 1985, Eq. 20). This formulation contains only two free parameters. One is analogous to a characteristic damping time of the forced climate system, which can be determined from the Stefan-Boltzmann Law, under the assumption that tropospheric temperature anomalies are radiatively damped. The other is analogous to a heat capacity (in this case, that of the troposphere and perhaps some part of the oceanic mixed layer in the passive regions of the tropics). This coefficient can be determined empirically to optimize the agreement between the output of the model and the tropospheric time

series. The output of the model, shown as the middle curve in Figure 6, captures most of the features in the land-surface temperature time series immediately above it. For further details, see Yulaeva and Wallace (1994).

Vestiges of the same warm and cold episodes are evident in global mean MSU-2 time series shown in Figure 7, but the ENSO signal is diluted by about a factor of 3 by being combined with unrelated features in the extratropical time series, which are also shown in the figure. Time series of 850 to 300 hPa thickness based on zonal averages of longer records of rawinsonde observations for selected latitude belts also give the impression that the ENSO signal is largely confined to the tropical belt (Angell, 1988). It is evident from Figure 7 that the interannual variability of tropical tropospheric temperatures in association with the ENSO cycle is much larger than that of extratropical tropospheric temperatures. One obtains similar impressions from the results of EOF analysis of cosine-weighted, zonally averaged temperature anomalies derived from MSU-2 instruments. The leading mode is largely confined to the tropics and symmetric about the equator (Yulaeva and Wallace, 1994).

Figure 8 shows longer records of land surface-air temperature for the same latitude belts. As in the previous figure, the temperature scales for all four curves are the same, but in this case a 3-month running mean filter has been applied to suppress the high-frequency variability in the extratropi-

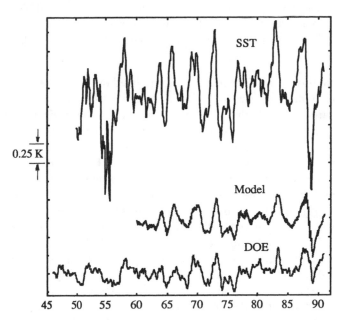

FIGURE 6   Time series of (upper) equatorial Pacific SST, as in Figure 5, (lower) surface-air temperature over all land gridpoints within the tropical belt (20°N to 20°S), and (middle) simulated tropical tropospheric response to the SST time series, based on a simple thermodynamic model as described in Yulaeva and Wallace (1994). One small tick mark on the vertical scale is equivalent to 0.5K.

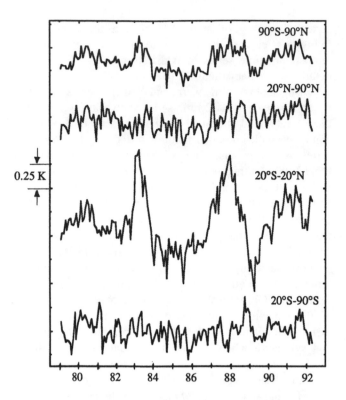

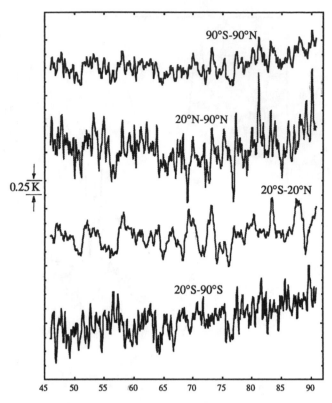

FIGURE 7   Tropospheric temperatures based on the MSU-2 data averaged over the entire globe, the extratropical Northern Hemisphere (poleward of 20°N), the tropical belt 20°S to 20°N, and the extratropical Southern Hemisphere (poleward of 20°S). The temperature scale is the same for all four curves: one small tick mark is equivalent to 0.25K. (After Yulaeva and Wallace, 1994; reprinted with permission of the American Meteorological Society.)

FIGURE 8   Land-surface temperatures based on the DOE data set averaged over the entire globe, the extratropical Northern Hemisphere (poleward of 20°N), the tropical belt 20°S to 20°N, and the extratropical Southern Hemisphere (poleward of 20°S). The temperature scale is the same for all four curves: one small tick mark is equivalent to 0.25K.

cal time series. The relationships pointed out in the previous paragraph appear to be valid during earlier decades. In particular, note the weak coupling between the tropical and extratropical time series on the interannual time scale.

It has been suggested that the prevalence of the warm phase of the ENSO cycle during the late 1970s and 1980s has contributed to the global warming observed during this period (e.g., see Trenberth, 1990; Trenberth and Hurrell, 1995). Inspection of Figure 8 suggests that some of the warming observed within the tropical belt from the late 1970s onward could be attributed to the absence of a well-defined cold phase of the ENSO cycle between 1976 and 1987. On the other hand, this feature could equally well be interpreted as a gradual upward temperature trend that is independent of the ENSO variability. The latter interpretation is supported by the fact that equatorial Pacific SST (Figure 6) was substantially colder during 1988 than at any time since the early 1950s, yet tropical land-surface temperatures (Figure 8), although cool relative to other periods during the 1980s, were not as cool as those observed on a number of occasions during the 1960s and early 1970s.

The time series of extratropical Northern and Southern

Hemisphere surface-air temperature in Figure 8 are not particularly well correlated with one another. The only common element is the upward trend during the past decade. Since the ENSO signal is common to the northern and southern tropics, it contributes to the correlation between the two hemispheres when the spatial average is extended all the way to the equator, as in many published time series (e.g., Spencer and Christy, 1992, Figure 12). It is evident from our figure that when the tropical belt is excluded in defining the hemispheric means, the correlation between the hemispheric time series drops dramatically.

Hence, we can conclude that the interannual variations in zonally averaged temperature associated with the ENSO cycle are largely restricted to the tropics, even though the ENSO signal is apparent in global mean temperature.

TRENDS IN SURFACE-AIR TEMPERATURE VERSUS TEMPERATURES ALOFT

Low-frequency variability in temperatures aloft over the extratropical Northern Hemisphere exhibits the same type of structural characteristics and seasonality as variability in

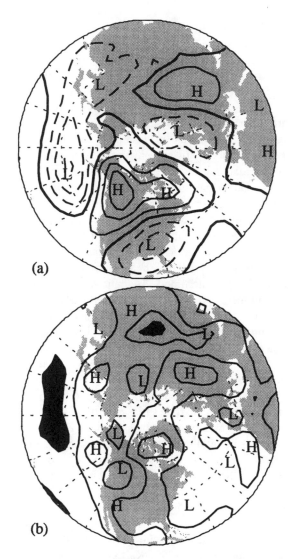

(a)

(b)

FIGURE 9 Leading EOFs of (a) wintertime (December to February) and (b) summertime (June to August) mean hemispheric (poleward of 20°N) 1000-500 hPa thickness field based on NMC analyses for the period 1946 to 1989, scaled such that they represent the thickness perturbations corresponding to an amplitude of unity for the corresponding expansion coefficient contour interval 0.2. Negative values are dashed; regions of negative values are shaded in the lower panel. The numbers at the upper right are the percentages of the hemispherically integrated variance explained by that mode. Courtesy of Yuan Zhang, Department of Atmospheric Sciences, University of Washington.

temperatures at the earth's surface (Wallace et al., 1993). The wintertime variability of the 1,000 to 500 hPa thickness field, a measure of the mean temperature of the lower troposphere, is dominated by a PNA-like pattern (Figure 9, upper panel). The summertime variability is dominated by a relatively featureless "background field" (Figure 9, lower panel) whose expansion coefficient is highly correlated with the time series of hemispheric mean thickness. Even though

the wintertime mode exhibits higher amplitude, the leading EOF of annual mean thickness closely resembles the leading summertime pattern, which explains a higher percentage of the hemispherically integrated variance than its wintertime counterpart, and is evidently present year round. For fluctuations with periods of 5 years or longer, variations in hemispheric mean thickness account for half the hemispherically integrated variance of the local thickness time series.

There is some evidence of systematic differences in the long-term trends in surface-air temperature versus temperatures aloft. Figure 10 compares hemispherically averaged temperature time series based on (a) surface-air temperature over land, (b) mean tropospheric temperature as sensed by Channel 2 of the MSU, and (c) mean virtual temperature of the 1,000 to 500 hPa layer based on the operational analyses of the National Meteorological Center (NMC), all for the area poleward of 20°N. All three time series have been smoothed by taking 3-month running means, and the DOE climatology has been adjusted to eliminate the annual march in the 1979 to 1990 period of record. Neither of the upper-air time series exhibits as pronounced a warming trend as that evident in the time series of surface-air temperature, nor do the hemispheric mean time series of Angell (1988) or Oort and Liu (1993), which are based on radiosonde data. The corresponding MSU-2 time series of

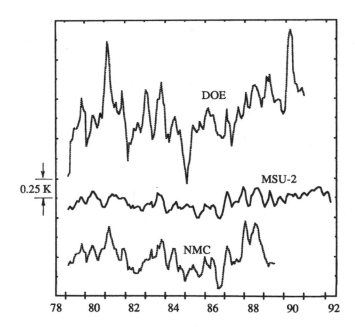

FIGURE 10 Time series of 3-month running means of (upper) surface air temperature over land, based on the DOE data set, (middle) tropospheric temperature based on MSU-2, and (lower) mean virtual temperature of the 1000-500 hPa layer based on the NMC operational analyses, all averaged over the Northern Hemisphere poleward of 20°N. The climatology for the DOE data is based on the 1979-1990 reference period. The temperature scale is the same for all three curves: one small tick mark is equivalent to 0.25K.

tropospheric temperature in the extratropical Southern Hemisphere also fails to reproduce the pronounced warming that is evident in the surface-air temperature record, as shown in Figure 11.

Figure 12 shows a comparison of the same surface-air temperature and NMC upper-air temperature time series for the extratropical Northern Hemisphere, based on a longer period of record. Other discrepancies are evident around 1955 and 1963, when the mean 1,000 to 500 hPa temperature in the NMC analyses dropped substantially, while the surface temperatures cooled by a lesser amount. Lambert (1990) has suggested that both these features are reflections of changes in analysis procedures, while Shabbar et al.

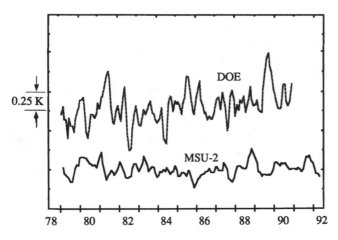

FIGURE 11   Series as in the two top curves of Figure 10, but for the Southern Hemisphere poleward of 20°S. The temperature scale is the same for both curves: one small tick mark is equivalent to 0.1K.

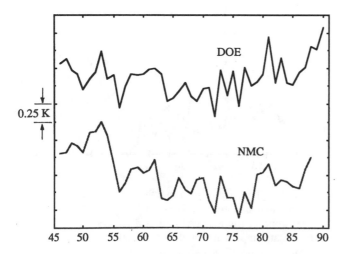

FIGURE 12   Series as in the top and bottom curves in Figure 10, but of annual means based on the calendar year. The temperature scale is the same for both curves: one small tick mark is equivalent to 0.25K.

(1990) have linked the second to a change in climate regimes.

There are many reasons to be suspicious of the sharp downward temperature trend during the late 1950s and early 1960s indicated by the NMC analyses. Changes in radiosonde sensors, reporting times, and analysis procedures that occurred during this period could have introduced spurious trends or discontinuities in the record. Yet in view of the marked differences between temperature trends in the MSU-2 and DOE data sets during the 1980s, it would be imprudent to dismiss the possibility that the cooling aloft might have been somewhat more pronounced than that at the earth's surface. A tendency toward stronger cooling aloft than at the surface would be indicative of a long-term decrease in the stratification of the lower troposphere.

A contributing factor may be the shift in the predominant polarity of the wintertime PNA pattern between the early 1950s and the 1980s. This change was marked by cooling of several degrees aloft over the North Pacific and warming by a comparable amount at the earth's surface over western Canada and much of Siberia (Gutzler et al., 1988; Wallace et al., 1993). Since this effect was observed only during the cold season over limited regions of the hemisphere, it seems unlikely that it could be the main reason for the difference between the time series of surface-air temperature and temperatures aloft over the Northern Hemisphere. Of course, it also would not account for the difference between the recent temperature trends in the MSU-2 and DOE time series in the Southern Hemisphere, which is in the same sense.

A downward trend in static stability is also suggested by the marked difference between the long-term trends in daily maximum and daily minimum temperatures since the 1940s (Plantico et al., 1990; Karl et al., 1995). Minimum temperature exhibits a pronounced upward trend relative to maximum temperature, resulting in a decrease of several percent in the mean diurnal temperature range during this period of record. The decrease is observed during all seasons and over most geographical regions. Higher daily minimum surface-air temperatures could be an indication of weaker nighttime inversions, which would contribute to a decrease in the mean static stability of the lower troposphere.

LOWER STRATOSPHERIC TEMPERATURE

Figure 13 shows the time series of lower stratospheric temperatures based on Channel 4 of the MSU, whose level of unit optical depth is near 70 hPa. The global curve is the same as the one published by Spencer and Christy (1993). Consistent with earlier studies based on radiosonde data (Newell, 1970; Angell and Korshover, 1978), it shows distinct signatures of the two major volcanic eruptions during this period (El Chichon in April 1982 and Pinatubo in June 1991) whose plumes reached stratospheric levels. The

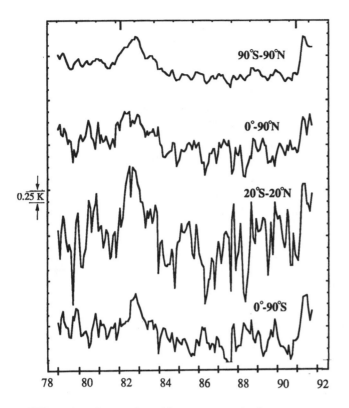

FIGURE 13 Time series of lower stratospheric temperature anomalies, relative to the 1979-1991 reference period, as sensed by the stratospheric channel of the microwave sounding unit (MSU4). Curves show the average over the entire globe, the average over the Northern Hemisphere (0° to 90°N), the average over the tropics (20°N to 20°S), and the average over the Southern Hemisphere (0° to 90°S). The times of the El Chichon and Pinatubo eruptions are indicated near the top of the figure. One small tick mark on the vertical scale is equivalent to 0.25K. Data courtesy of Roy Spencer, NASA Marshall Space Flight Center, Huntsville, Ala. (After Yulaeva et al., 1994; reprinted with permission of the American Meteorological Society.)

volcanic signature shows up much more clearly in the globally averaged time series than in the time series for the tropical belt or for individual hemispheres. Over a wide range of frequencies, much of the large, dynamically induced temperature variability involves compensating warming and cooling in different latitude belts. For example, the correlation between month-to-month temperature changes in the tropics (20°N to 20°S) and extratropics is −0.79 (−0.86 when the 3 months subsequent to the Pinatubo eruption are excluded). The dynamically induced temperature perturbations are almost entirely filtered out by the global averaging, leaving only the smaller radiatively induced perturbations (Yulaeva et al., 1994).

During the period of record 1979 through 1992, the globally averaged MSU-2 and MSU-4 time series are virtually uncorrelated. In the presence of the large ENSO signal in tropospheric temperatures, the cooling signatures associ-

ated with the El Chichon and Pinatubo eruptions do not stand out clearly in Figures 5 and 7. As pointed out by Bradley (1988) and Mass and Portman (1989), the Katmai eruption in the Aleutians in June 1912 was followed by a pronounced dip in Northern Hemisphere monthly mean temperatures during the warm season (Figure 3, upper plot). There was an analogous, although somewhat weaker, dip following the eruption of Ksudach (also in the Aleutians) three years earlier. There is no clear signature of the ENSO cycle in the global-mean stratospheric time series, although Randel and Cobb (1994) and Yulaeva and Wallace (1994) have shown evidence of a geographically localized signature.

DISCUSSION

The ENSO and volcanic signatures are reflections of natural variability in the earth system: The former is generated by processes within the more limited atmosphere-ocean climate system, and the latter is externally forced. The remaining variability in the climate record is predominantly interdecadal in time scale. It is not clear how much of it is due to natural causes and how much is a consequence of anthropogenic influences. Depending on whether aerosols or greenhouse gases predominate, and the strength and polarity of various feedbacks, anthropogenic influences might serve either to warm or to cool the troposphere. Hence, the absence of a pronounced upward trend in tropospheric temperatures as sensed by the MSU-2 instrument during the past decade does not necessarily preclude the possibility that a detectable anthropogenic signal exists. A gradual increase in the optical thickness of the atmosphere is, in fact, suggested by the secular decrease in static stability observed over the course of the past 40 years in the Northern Hemisphere extratropics and over at least the past decade in the Southern Hemisphere extratropics, coupled with the marked increase in nighttime temperatures relative to daytime temperatures at many land stations. Increases in greenhouse gases and aerosols could both be contributing in the same sense to these trends. Such a signature would be consistent with the results of many general-circulation models' simulations of the response to increasing CO_2, which show the largest temperature rises at the earth's surface at higher latitudes, in regions subject to frequent low-level temperature inversions.

The sharp rise in Northern Hemisphere surface-air temperatures from 1920 to 1940 (Figures 3 and 4) coincided with a northward shift of the Gulfstream (Bjerknes, 1959; Deser and Blackmon, 1995). The subsequent dip during the 1960s and 1970s may be related to a reduction in the salinity of the surface waters of the subpolar North Atlantic that was observed during this period (Dickson et al., 1988; Delworth et al., 1995; Dickson, 1995; Mysak, 1995), which might have tended to suppress the thermohaline circulation, thereby reducing the poleward heat transport. It has also been suggested that increasing concentrations of sulfate aerosols over the Northern Hemisphere might have been responsible for the cooling during that period (Charlson et al., 1991, 1992). Hence, there is no shortage of mechanisms that may have contributed to the observed interdecadal variability in the climate record. The problem is to quantify the arguments, through the use of climate models, so that the dominant mechanisms can be identified (Battisti, 1995; Rind and Overpeck, 1995).

The results presented in the foregoing sections serve to illustrate the need for a flexible, pragmatic approach toward the smoothing of monthly mean data, and stratifying of data by season and by latitude belt. For example, in displaying tropospheric data, a three-way partitioning into means for the tropics and Northern and Southern Hemisphere extra-tropics appears to be more informative than the more commonly used two-way partitioning by hemisphere, because it isolates the tropical ENSO signal. For stratospheric data it may be best not to do any partitioning at all, since global averaging tends to filter out spurious, dynamically induced temperature fluctuations.

ACKNOWLEDGMENTS

I would like to thank Yuan Zhang and Elena Yulaeva for providing material from their graduate research projects, and Clifford Watson for assistance with the preparation of the figures. Clifford Mass offered a number of helpful comments. The work was supported by the National Science Foundation under Grant 9215512.

Commentary on the Paper of Wallace

HENRY F. DIAZ
NOAA/Environmental Research Laboratories

Before we begin a discussion, I just want to encapsulate some of the primary conclusions or at least principal points Dr. Wallace discussed. First, I think that the ENSO is manifested primarily in tropical temperatures. When you look at global indices or hemispheric indices, what you are seeing is essentially the folding of the tropical signal into the overall record. A very important point that we should consider in the context of the previous talks and the slides that Professor Lindzen showed is that the volcanic temperature signal on the hemispheric temperature record is relatively short-lived. It seems to me that we are talking about thermal relaxation on the order of 3 to 5 years from the time when a major volcanic eruption takes place, through the drop in the temperatures, to the recovery. In that sense, I think that we are not talking about decades but less than a decade.

The volcanic effect also says something about the sensitivity of climate to greenhouse perturbations. There is something interesting about the difference between the Northern and Southern Hemispheres' temperature responses. Even in the 1979-to-present record, the atmospheric response is quite rapid, and there are clear differences between the two hemispheres. We should try to discover whether they are due primarily to the fact that the Southern Hemisphere reflects the ocean damping effects, and whether the Antarctic ice cap has something to do with the difference. Also, the lower stratosphere has a much more spatially homogenous response than the troposphere, and perhaps long-term monitoring of lower stratospheric temperatures would turn out to be a useful tool. Large temperature changes in the lower stratosphere have been projected, and I think we should consider how the observations and the predictions are coming together.

Dr. Wallace mentioned that interdecadal temperature signals are different for surface and upper air indices. This is suggested by data for the last decade, although the record is not really long enough that we can say so with certainty. The correlation between, say, Jim Angell's 63-station radiosonde network and temperatures at the surface is sufficiently high to explain about 50 percent of the variance. For comparison, the correlation between that radiosonde network and the MSU data is about 0.8 or higher.

Discussion

TRENBERTH: Let me comment a moment on the trend aspect of our analysis of the MSU data. Our paper comparing it with the IPCC data set for surface temperature will be in the September issue of the *Journal of Climate*, by the way. Over North America, the MSU Channel 2 data has about 0.2°C per decade less trend than the surface record. My best guess is that this results from the way in which all the differences between the satellites getting the TOVS series of data are taken out.

WALLACE: That indeed is a point to be taken seriously, but I don't think we should dismiss the possibility that these differences might prove to be real. Another five years of data should settle the matter.

TRENBERTH: Yes, that's true; there are real physical differences. Over North America the correlation is better than 0.9, whereas over the rest of the world it's not that good.

SHUKLA: I had the impression from various Oort papers that a large part of the vertically integrated three-dimensional temperature for the Northern Hemisphere is the ENSO signal.

WALLACE: Yes, I think his results are consistent with what I have indicated—that the ENSO signal shows up clearly in the globally averaged temperature time series, but not in the time series for extratropical latitudes.

KAROLY: I've actually performed a correlation of SST with upper-air temperatures, stratified by latitude and elevation. It's very strong up to 25° north and south, being dominated by direct tropical SST, but between 25° and the poles you get structures that vary in sign in different latitude bands, and the correlation is negligible.

RASMUSSON: What about the differences between land and sea surface temperatures, though both seem to have been increasing in the last couple of decades?

WALLACE: The low-frequency interdecadal trends exhibit the same signature in land-based and marine data sets, as can be seen from the IPCC report's figures. But I suspect that those two data sets show quite different scatter around the trend lines for individual months or seasons, year by year. I think that if you found a point in the time series for land that was way below the trend line, the corresponding one for the marine series would be way above the trend line.

RASMUSSON: Different parts of the ocean certainly have different temporal responses; the tropical temperatures have ramped up very quickly since the mid-1970s.

LINDZEN: I think you made a crucial point, Mike, about the difference between detection and interpretation. For example, data reflecting crustal movement languished for years before the concept of plate tectonics gave them meaning. Also, when we talk about detection, we should be careful not to equate detection of CO_2 increase with detection of climate change. Some of the measures of CO_2 change have to do with radiative flux, which varies strongly with the time of day. Also, one thing I would very much like to see done is further work comparing measurements made at surface level with those made at 500 millibars, particularly as regards lapse-rate variability. Some of Oort's recent work suggests that surface temperature changes are amplified at 500 mb globally, while some of the MSU data indicate that the integral changes over a certain region may be reduced.

KEELING: As I remember, Oort found a substantial signal, about 6 months delayed from the tropical signal, at 200 mb, which he used as a proxy for integrated temperature.

KAROLY: There is a very strong 200-mb signal, but if you average around the hemisphere the latitudinally averaged signal in the extratropics is weak.

GHIL: I'd like to make two remarks. The first is about the spatial patterns. It's quite clear that signals such as the so-called ENSO interannual signal have associated spatial patterns. Since this signal is relatively homogeneous throughout the tropics, which is a large part of the globe, it will dominate any global average. Also, spatial patterns of various signs outside the tropics are going to cancel each other out if you do zonal averages.

My second remark relates to Dick Lindzen's point about the volcanic signal and its two-part decay. Mike showed us clusters of colder months that disappeared after a year or two, that being the fast-decay part. I wonder whether interdecadal variability might not be masking the slow-decay part.

RIND: It would be interesting to clarify the nature of the system's long response time. A volcanic eruption is such a quick signal that you might observe a much shorter response than a system with so many feedbacks would otherwise exhibit; the longer responses could be drowned in the rest of the variability.

TRENBERTH: I'd like to pick up on one of Michael Ghil's points. The Pacific is not in fact coherent with the ENSO signal; the western Pacific tends to have a sign opposite to the eastern Pacific's. The tropics are not necessarily a coherent whole either. However, the ENSO signal is so large in the eastern Pacific that it is twice as high as the standard deviation and four times or more the variance.

WALLACE: There's more on this issue in my paper, but for the purposes of this discussion, I think the tropics can be divided into the 'active' eastern Pacific, where SST is directly influenced by equatorial upwelling, and the 'passive' remainder. I think the ENSO signal is also present, with smaller amplitude and lagged by about a season, in the passive region. Furthermore, I believe that stations in that region, which includes nearly all the land stations in the DOE data set, all exhibit very similar ENSO-related time variability.

We have performed some experiments with a thermodynamic model of the type used by Klaus Hasselmann. With this simple model you can transform an SST time series that is representative of the active region into a time series that looks a lot like the DOE time series for the tropical land stations. So I believe the relationship between the interannual temperature variabilities in different parts of the tropics is much simpler than those for other variables like precipitation and cloudiness.

GROISMAN: I wanted to say that I completely agree that there

has to be more than one decay time, and neither the initial decay time nor a sustained signal is an acceptable approximation of the long decay times. And second, don't forget how much global temperature variance differs by season. And sometimes just a change in the Y-axis scale will make a trend look good.

WALLACE: It's only the shorter-term variability—the "noise"—that's larger in winter; the interdecadal component appears to be the same for both. That's why the scatter plots for the summer months look better.

Atmospheric Modeling Reference List

Adem, J. 1964. On the physical basis for the numerical prediction of monthly and seasonal temperatures in the troposphere-ocean-continent system. Mon. Weather Rev. 92:91-104.

Allen, M.R., and L.A. Smith. 1994. Investigating the origins and significance of low-frequency climate modes of variability. Geophys. Res. Lett. 21:883-886.

Anderson, R.Y. 1992. Possible connection between surface winds, solar activity and the Earth's magnetic field. Nature 358:51-53.

Andrews, D.G., C. Leovy, and J.R. Holton. 1987. Middle Atmosphere Dynamics. Academic Press, Orlando, Florida, 489 pp.

Angell, J.K. 1988. Variations and trends in tropospheric and stratospheric global temperature, 1958-87. J. Climate 1:1296-1313.

Angell, J.K., and J. Korshover. 1978. Estimate of global temperature variations in the 100-30 mb layer between 1958 and 1977. Mon. Weather Rev. 106:1422-1432.

Baldwin, M.P., and T.J. Dunkerton. 1989. Observations and statistical simulations of a proposed solar cycle/QBO/weather relationship. Geophys. Res. Lett. 16:863-866.

Baliunas, S., and R. Jastrow. 1990. Evidence for long-term brightness changes of solar-type stars. Nature 348:520-523.

Barnett, T.P. 1978. Estimating variability of surface temperature in the northern hemisphere. Mon. Weather Rev. 106:1353-1367.

Barnett, T.P. 1986. Detection of changes in the global tropospheric temperature field induced by greenhouse gases. J. Geophys. Res. 91:6659-6667.

Barnett, T. 1991. An attempt to detect the greenhouse-gas signal in a transient GCM simulation. In Greenhouse-Gas-Induced Climatic Change: A Critical Appraisal of Simulations and Observations. M.E. Schlesinger (ed.). Elsevier, Amsterdam, pp. 559-568.

Barnett, T.P., A.D. Del Genio, and R. Ruedy. 1992. Unforced decadal fluctuations in a coupled model of the atmosphere and ocean mixed layer. J. Geophys. Res. 97:7341-7354.

Barnola, J.M., D. Raynaud, Y.N. Korotkevitch, and C. Lorius. 1987. Vostok ice-core provides 160,000-year record of atmospheric CO_2. Nature 329:408-414.

Barnston, A.G., and R.E. Livezey. 1989. A closer look at the effect of the 11-year solar cycle and the quasi-biennial oscillation in the Northern Hemisphere 700 mb height and extra tropical North American surface temperature. J. Climate 2:1295-1313.

Battisti, D.S. 1995. Decade-to-century time-scale variability in the coupled atmosphere-ocean system: Modeling issues. In Natural Climate Variability on Decade-to-Century Time Scales. D.G. Martinson, K. Bryan, M. Ghil, M.M. Hall, T.R. Karl, E.S. Sarachik, S. Sorooshian, and L.D. Talley (eds.). National Academy Press, Washington, D.C.

Beer, J., U. Siegenthaler, G. Bonani, R.C. Finkel, H. Oeschger, M. Suter, and W. Wolfli. 1988. Information on past solar activity and geomagnetism from ^{10}Be in the Camp Century ice core. Nature 332:675-679.

Bell, T.L. 1982. Optimal weighting of data to detect climatic change: Application to the carbon dioxide problem. J. Geophys. Res. 87:11161-11170.

Bell, T.L. 1986. Theory of optimal weighting of data to detect climatic change. J. Atmos. Sci. 43:1694-1710.

Bjerknes, J. 1959. The recent warming of the North Atlantic. In Atmosphere and Sea in Motion, the Rossby Memorial Volume. B. Bolin (ed.). Rockefeller Inst. Press in association with Oxford Press, New York, pp. 65-73.

Blanford, H.F. 1891. The paradox of the sun-spot cycle in meteorology. Nature 43:583-587.

Bloomfield, P., and D. Nychka. 1992. Climate spectra and detecting climatic change. Climatic Change 22:275-288.

Bradley, R.S. 1988. The explosive volcanic eruption signal in northern hemisphere continental temperature records. Climatic Change 12:221-243.

Bradley, R.S. 1990. Pre-instrumental climate: How has climate varied during the past 500 years? In Greenhouse Gas-Induced Climatic Change: A Critical Appraisal of Simulations and Observations. M. Schlesinger (ed.). Elsevier, Amsterdam, pp. 391-412.

Bradley, R.S. 1992. Instrumental records of past global change: Lessons for the analysis of noninstrumental data. In Global Changes of the Past. R.S. Bradley (ed.). UCAR/Office of Interdisciplinary Earth Studies, Boulder, Colo., pp. 103-116.

Bradley, R.S., and P.D. Jones (eds.). 1992a. Climate Since A.D. 1500. Routledge, New York, 679 pp.

Bradley, R.S., and P.D. Jones. 1992b. Records of explosive volcanic eruptions over the last 500 years. In Climate Since A.D. 1500. R.S. Bradley and P.D. Jones (eds.). Routledge, New York, pp. 606-622.

Bradley, R.S., H.F. Diaz, P.D. Jones, and P.M. Kelly. 1987. Secular fluctuations in temperature over northern hemisphere land areas and mainland China since the mid-nineteenth century. In The Climate of China and Global Climate. Ye D., Fu C., Chao J., and M. Yoshino (eds.). China Press and Springer-Verlag, pp. 77-87.

Bray, J.R. 1974. Glacial advance relative to volcanic activity since A.D. 1500. Nature 248:42-43.

Brewer, P.G., W.S. Broecker, W.J. Jenkins, P.B. Rhines, C.G. Rooth, J.H. Swift, T. Takahashi, and R.T. Williams. 1983. A climatic freshening of the deep Atlantic (north of 50°N) over the past 20 years. Science 222:1237-1239.

Broecker, W.S., and G.H. Denton. 1989. The role of ocean-atmosphere reorganizations in glacial cycles. Geochim. Cosmochim. Acta 53:24652501.

Broecker, W.S., T.H. Peng, and R. Engh. 1980. Modeling the carbon system. Radiocarbon 22:565-598.

Broecker, W.S., D.M. Peteet, and D. Rind. 1985. Does the ocean-atmosphere system have more than one stable mode of operation? Nature 315:21-26.

Broecker, W.S., M. Andree, W. Wolflie, H. Oeschger, G. Bonani, J. Kennett, and D. Peteet. 1988. The chronology of the last deglaciation: Implications to the cause of the Younger Dryas event. Paleoceanography 3:1-19.

Broomhead, D.S., and G.P. King. 1986. Extracting qualitative dynamics from experimental data. Physica D 20:217-236.

Budyko, M.I. 1969. The effect of solar radiation variations on the climate of the Earth. Tellus 21:611-619.

Cebula, R.P., M.T. DeLand, and B.M. Schlesinger. 1992. Estimates of solar variability using the solar backscatter ultraviolet (SBLTV) 2 Mg 11 Index from the NOAA 9 satellite. J. Geophys. Res. 97:11613-11620.

Cess, R.D. 1976. Climate change: An appraisal of atmospheric feedback mechanisms employing zonal climatology. J. Atmos. Sci. 33:1831-1843.

Charlson, R.J., J. Lagner, H. Rodhe, C.B. Leovy, and S.G. Warren. 1991. Perturbations of the northern hemisphere radiative balance by backscattering from anthropogenic sulfate aerosols. Tellus 43AB:152-163.

Charlson, R.J., S.E. Schwartz, J.M. Hales, R.D. Cess, J.A. Coakley Jr., J.E. Hansen, and D.J. Hofmann. 1992. Climate forcing by anthropogenic aerosols. Science 255:423-430.

Charney, J.G., and J.G. DeVore. 1979. Multiple flow equilibria in the atmosphere and blocking. J. Atmos. Sci. 36:1205-1216.

Cole, J.E. 1992. The spectrum of recent variability in the southern oscillation: Results from a Tarawa atoll coral. Ph.D. Thesis, Columbia University, 302 pp.

Cole, J.E., D. Rind, and R. Fairbanks. 1992. Isotopic responses to climate variability in the GISS GCM: Interannual vs. glacial/interglacial mechanisms. EOS, spring supplement (abstract), p. 151.

Cook, E.R., B.M. Buckley, and R.D. D'Arrigo. 1995. Interdecadal temperature oscillations in the Southern Hemisphere: Evidence from Tasmanian tree rings since 300 B.C. In Natural Climate Variability on Decade-to-Century Time Scales. D.G. Martinson, K. Bryan, M. Ghil, M.M. Hall, T.R. Karl, E.S. Sarachik, S. Sorooshian, and L.D. Talley (eds.). National Academy Press, Washington, D.C.

Cooper, N.S., K.D.B. Whysall, and G.R. Bigg. 1989. Recent decadal climate variations in the tropical Pacific. J. Climate 9:221-242.

Crowley, T.J., and G. North. 1991. Paleoclimatology. Oxford University Press, New York, 339 pp.

Cubasch, U., K. Hasselmann, H. Höck, E. Maier-Reimer, U. Mikolajewicz, B.D. Santer, and R. Sausen. 1992. Time-dependent greenhouse warming computations with a coupled ocean-atmosphere model. Climate Dynamics 8:55-69.

Damon, P.E., and T.W. Linick. 1986. Geomagnetic-heliomagnetic modulation of atmospheric radiocarbon production. Radiocarbon 28:266-278.

Dansgaard, W., S.J. Johnson, H.B. Clausen, D. Dahl-Jensen, N. Gundestirup, C.H. Hammer, and H. Oeschger. 1984. North Atlantic oscillations revealed by deep Greenland ice cores. Geophys. Monogr. 29:288-298.

Delworth, T., S. Manabe, and R.J. Stouffer. 1995. North Atlantic interdecadal variability in a coupled model. In Natural Climate Variability on Decade-to-Century Time Scales. D.G. Martinson, K. Bryan, M. Ghil, M.M. Hall, T.R. Karl, E.S. Sarachik, S. Sorooshian, and L.D. Talley (eds.). National Academy Press, Washington, D.C.

Denton, G.H., and W. Karlen. 1973. Holocene climatic variations—Their pattern and possible causes. Quat. Res. 3:155-205.

Deser, C., and M.L. Blackmon. 1995. Surface climate variations over the North Atlantic Ocean during winter: 1900 to 1989. In Natural Climate Variability on Decade-to-Century Time Scales. D.G. Martinson, K. Bryan, M. Ghil, M.M. Hall, T.R. Karl, E.S. Sarachik, S. Sorooshian, and L.D. Talley (eds.). National Academy Press, Washington, D.C.

de Vries, H. 1958. Variations in concentration of radiocarbon with time and location on Earth. Proc. Koninkl. Nederl. Adak. Wet. Ser B. 61:94-102.

Diaz, H.F., and R.S. Bradley. 1995. Documenting natural climatic variations: How different is the climate of the twentieth century from that of previous centuries? In Natural Climate Variability on Decade-to-Century Time Scales. D.G. Martinson, K. Bryan, M. Ghil, M.M. Hall, T.R. Karl, E.S. Sarachik, S. Sorooshian, and L.D. Talley (eds.). National Academy Press, Washington, D.C.

Dickson, R.R. 1995. The local, regional and global significance of exchanges through the Denmark Strait and Irminger Sea. In Natural Climate Variability on Decade-to-Century Time Scales. D.G. Martinson, K. Bryan, M. Ghil, M.M. Hall, T.R. Karl, E.S. Sarachik, S. Sorooshian, and L.D. Talley (eds.). National Academy Press, Washington, D.C.

Dickson, R.R., J. Meincke, S.A. Malmberg, and A.J. Lee. 1988. The "great salinity anomaly" in the northern Atlantic 1968-82. Prog. Oceanogr. 20:103-151.

Eddy, J.A. 1976. The Maunder Minimum. Science 192:1189-1202.

Enfield, D.B., and S. Luis Cid. 1991. Low-frequency changes in El Niño/Southern Oscillation. J. Climate 4:1137-1146.

Folland, C.K., T.R. Karl, and K.Ya. Vinnikov. 1990. Observed climate variations and change. In Climate Change: The IPCC Scientific Assessment. J.T. Houghton, G.J. Jenkins, and J.J. Ephraums (eds.). Prepared for the Intergovernmental Panel on Climate Change by Working Group I. Cambridge University Press, Cambridge, pp. 195-238.

Folland, C.K., T.R. Karl, N. Nicholls, B.S. Nyenzi, D.E. Parker, and K.Ya. Vinnikov. 1992. Observed climate variability and change. In Climate Change 1992: The Supplementary Report to the IPCC Scientific Assessment. J.T. Houghton, B.A. Callander, and S.K. Varney (eds.). Prepared for the Intergovernmental Panel on Climate Change by Working Group I. Cambridge University Press, Cambridge, pp. 135-170.

Frankignoul, C. 1985. Sea-surface temperature anomalies, planetary waves and air-sea feedbacks in the middle latitudes. Rev. Geophys. 23:357-390.

Friis-Christensen, E., and K. Lassen. 1991. Length of the solar cycle: An indicator of solar activity closely associated with climate. Science 254:698-700.

Gaffin, S.R., M.I. Hoffert, and T. Volk. 1986. Nonlinear coupling between surface temperature and ocean upwelling as an agent in historical climate variations. J. Geophys. Res. 91:3944-3950.

Ghil, M. 1976. Climate stability for a Sellers-type model. J. Atmos. Sci. 33:3-20.

Ghil, M., and S. Childress. 1987. Topics in Geophysical Fluid Dynamics: Atmospheric Dynamics, Dynamo Theory and Climate Dynamics. Springer-Verlag, New York, N.Y., 485 pp.

Ghil, M., and R. Vautard. 1991. Interdecadal oscillations and the warming trend in global temperature time series. Nature 350:324-327.

Ghil, M., M. Kimoto, and J.D. Neelin. 1991. Nonlinear dynamics and predictability in the atmospheric sciences. Rev. Geophys., Supplement (U.S. National Report to the IUGG 1987-1990), pp. 46-55.

Goswami, B.N., and J. Shukla. 1984. Quasi-periodic oscillations in a symmetric general circulation model. J. Atmos. Sci. 41:20-37.

Grove, J.M. 1988. The Little Ice Age. Methuen, London, 498 pp.

Gutzler, D.S., D. Richard, D.A. Salstein, and J.P. Peixoto. 1988. Patterns of interannual variability in the northern hemisphere wintertime 850 mb temperature field. J. Climate 1:949-964.

Halfman, J.D., and T.C. Johnson. 1988. High-resolution record of cyclic climatic change during the past 4 ka from Lake Turkana, Kenya. Geology 16:496-500.

Hansen, A.R., and A. Sutera. 1986. On the probability density distribution of planetary-scale atmospheric wave amplitude. J. Atmos. Sci. 43:

Hansen, J., D. Johnson, A. Lacis, S. Lebedeff, P. Lee, D. Rind, and G. Russell. 1983a. Climatic effects of atmospheric carbon dioxide—A response. Science 220:874-875.

Hansen, J.E., G. Russell, D. Rind, P. Stone, A. Lacis, S. Lebedeff, R. Ruedy, and L. Travis. 1983b. Efficient three-dimensional global models for climate studies: Models I and II. Mon. Weather Rev. 111:609-662.

Hansen, J., A. Lacis, D. Rind, G. Russell, P. Stone, I. Fung, R. Ruedy, and J. Lerner. 1984. Climate sensitivity: Analysis of

feedback mechanisms. In Climate Processes and Climate Sensitivity. J.E. Hansen and T. Takahashi (eds.). Geophys. Monogr. Ser. 29:130-163.

Hansen, J., G. Russell, A. Lacis, I. Fung, D. Rind, and P. Stone. 1985. Climate response times: Dependence on climate sensitivity and ocean mixing. Science 229:857-859.

Hansen, J., I. Fung, A. Lacis, D. Rind, S. Lebedeff, R. Ruedy, and G. Russell. 1988. Global climate changes as forecast by Goddard Institute for Space Studies three-dimensional model. J. Geophys. Res. 93:9341-9364.

Hansen, J., A. Lacis, R. Ruedy, and M. Sato. 1992. Potential climate impact of Mount Pinatubo eruption. Geophys. Res. Lett. 19:215-218.

Harvey, L.D.D. 1992. A two-dimensional ocean model for long-term climatic simulations: Stability and coupling to atmospheric and sea ice models. J. Geophys. Res. 97:9435-9453.

Hasselmann, K. 1976. Stochastic climate models. I. Theory. Tellus 28:473-485.

Hasselmann, K. 1979. On the signal-to-noise problem in atmospheric response studies. In Meteorology over the Tropical Oceans. D.B. Shaw (ed.). Royal Meteorological Society, Bracknell, Berkshire, England, pp. 251-259.

Held, I.M., and M.J. Suarez. 1974. Simple albedo feedback models of the icecaps. Tellus 36:613-628.

Hoffert, M.I., and C. Covey. 1992. Deriving global climate sensitivity from paleoclimate reconstructions. Nature 360:573-576.

Hoffert, M.I., A.J. Callegari, and C.-T. Hsieh. 1980. The role of deep sea heat storage in the secular response to climatic forcing. J. Geophys. Res. 85:6667-6679.

Horel, J.D., and J.M. Wallace. 1981. Planetary-scale atmospheric phenomena associated with the Southern Oscillation. Mon. Weather Rev. 109:813-829.

Houghton, D.D., R.G. Gallimore, and L.M. Keller. 1991. Stability and variability in a coupled ocean-atmosphere climate model: Results of 100-year simulations. J. Climate 4:557-577.

Hyde, W., J.T. Crowley, K.-Y. Kim, and G.R. North. 1989. Comparison of GCM and energy balance model simluations of seasonal temperature changes over the past 18000 years. J. Climate 2:864-887.

IPCC. 1990. Climate Change: The IPCC Scientific Assessment. J.T. Houghton, G.J. Jenkins, and J.J. Ephraums (eds.). Prepared for the Intergovernmental Panel on Climate Change by Working Group I. WMO/UNEP, Cambridge University Press, 365 pp.

IPCC. 1992. Climate Change 1992: The Supplementary Report to the IPCC Scientific Assessment. J.T. Houghton, B.A. Callander, and S.K. Varney (eds.). Prepared for the Intergovernmental Panel on Climate Change by Working Group I. WMO/UNEP, Cambridge University Press, 200 pp.

Jenkins, G.M., and D.G. Watts. 1968. Spectral Analysis and its Applications. Holden-Day, San Francisco, 525 pp.

Jones, P.D., and K.R. Briffa. 1992. Global surface air temperature variations over the twentieth century. Part I. Spatial, temporal, and seasonal details. The Holocene 2:165-179.

Jones, P.D., and P.M. Kelly. 1983. The spatial and temporal characteristics of northern hemisphere surface air temperature variations. J. Climatol. 3:243-252.

Jones, P.D., S.C.B. Raper, B.D. Santer, B.S.G. Cherry, C.M. Goodess, P.M. Kelly, T.M.L. Wigley, R.S. Bradley, and H.F.

Diaz. 1985. A Grid Point Surface Air Temperature Data Set for the Northern Hemisphere. Technical Report TR022, U.S. Department of Energy, Carbon Dioxide Research Division, 251 pp.

Jones, P.D., S.C.B. Raper, B.S.G. Cherry, C.M. Goodess, and T.M.L. Wigley. 1986a. A Grid Point Surface Air Temperature Data Set for the Southern Hemisphere. Technical Report TR027, U.S. Department of Energy, Carbon Dioxide Research Division, 73 pp.

Jones, P.D., T.M.L. Wigley, and P.B. Wright. 1986b. Global temperature variations between 1861 and 1984. Nature 322:430-434.

Jones, P.D., P.M. Kelly, G.B. Goodess, and T.R. Karl. 1989. The effect of urban warming on the northern hemisphere temperature average. J. Climate 2:285-290.

Karl, T.R., P.D. Jones, R.W. Knight, G. Kukla, N. Plummer, V. Razuvayev, K.P. Gallo, J.A. Lindesay, and R.J. Charlson. 1995. Asymmetric trends of daily maximum and minimum temperature: Empirical evidence and possible causes. In Natural Climate Variability on Decade-to-Century Time Scales. D.G. Martinson, K. Bryan, M. Ghil, M.M. Hall, T.R. Karl, E.S. Sarachik, S. Sorooshian, and L.D. Talley (eds.). National Academy Press, Washington, D.C.

Kaye, J. 1987. Analysis of the effect of zonal averaging on reaction rate calculations in two-dimensional atmospheric models. J. Geophys. Res. 92:11965-11970.

Keeling, C.D., and T.P. Whorf. 1995. Decadal oscillations in global temperature and atmospheric carbon dioxide. In Natural Climate Variability on Decade-to-Century Time Scales. D.G. Martinson, K. Bryan, M. Ghil, M.M. Hall, T.R. Karl, E.S. Sarachik, S. Sorooshian, and L.D. Talley (eds.). National Academy Press, Washington, D.C.

Kelly, P.M., and T.M.L. Wigley. 1992. Solar cycle length, greenhouse forcing and global climate. Nature 360:328-330.

Kheshgi, H.S., and B.S. White. 1993. Does recent global warming suggest an enhanced greenhouse effect? Climatic Change 23:121-139.

Kiehl, J.P., and B.P. Briegleb. 1993. The relative roles of sulfate aerosols and greenhouse gases in climate forcing. Science 260:311-314.

Kim, K.-Y., and G.R. North. 1991. Surface temperature fluctuations in a stochastic climate model. J. Geophys. Res. 96:18573-18580.

Kim, K.-Y, and G.R. North. 1992. Seasonal cycle and second-moment statistics of a simple coupled climate system. J. Geophys. Res. 97:20437-20448.

Kim, K.-Y., and G.R. North. 1993. EOF analysis of surface temperature field in a stochastic climate model. J. Climate 6:1681-1690.

Kim, K.-Y., G.R. North, and J. Huang. 1992. On the transient response of a simple coupled climate system. J. Geophys. Res. 97:10069-10081.

Kolmogorov, A. 1941. Interpolation and extrapolation of stationary time series. Bull. Acad. Sci. USSR, Ser. Math. 5:3-14. (in Russian)

Kuo, C., C. Lindberg, and D.J. Thompson. 1990. Coherence established between atmospheric carbon dioxide and global temperature. Nature 343:709.

Lacis, A., J. Hansen, and M. Sato. 1992. Climate forcing by stratospheric aerosols. Geophys. Res. Lett. 19:1607-1610.

Lamb, H.H. 1970. Volcanic dust in the atmosphere; with a chronology and assessment of its meteorological significance. Philos. Trans. Roy. Soc. London A226:425-533.

Lambert, S.J. 1990. Discontinuities in the long-term Northern Hemisphere 500-millibar dataset. J. Climate 3:1479-1484.

Lazier, J.R.N. 1980. Oceanographic conditions at ocean weather ship Bravo, 1944-1974. Atmos.-Ocean 18:227-238.

Lean, J. 1987. Solar ultraviolet irradiance variations: A review. J. Geophys. Res. 92:839-868.

Lean, J., A. Skumanich, and O. White. 1992. Estimating the sun's radiative output during the Maunder Minimum. Geophys. Res. Lett. 19:1591-1594.

Legras, B., and M. Ghil. 1985. Persistent anomalies, blocking and variations in atmospheric predictability. J. Atmos. Sci. 42:433-471.

Lindzen, R.S. 1995. Constraining possibilities versus signal detection. In Natural Climate Variability on Decade-to-Century Time Scales. D.G. Martinson, K. Bryan, M. Ghil, M.M. Hall, T.R. Karl, E.S. Sarachik, S. Sorooshian, and L.D. Talley (eds.). National Academy Press, Washington, D.C.

Lorius, C., J. Jouzel, D. Raynaud, J. Hansen, and H. Le Treut. 1990. The ice-core record: Climate sensitivity and future greenhouse. Nature 347:139-145.

Manabe, S., and R.T. Wetherald. 1967. Thermal equilibrium of the atmosphere with a given distribution of relative humidity. J. Atmos. Sci. 24:241-259.

Manabe, S., and R.T. Wetherald. 1975. The effects of doubling the CO_2 concentration on the climate of a general circulation model. J. Atmos. Sci. 32:3-15.

Manabe, S., K. Bryan, and M.J. Spelman. 1990. Transient response of a global ocean-atmosphere model to a doubling of atmospheric carbon dioxide. J. Phys. Oceanogr. 20:722-749.

Manabe, S., R.J. Stouffer, M.J. Spelman, and K. Bryan. 1991. Transient responses of a coupled ocean-atmosphere model to gradual changes of atmospheric CO_2. Part I: Annual mean response. J. Climate 4:785-818.

Manley, G.Q. 1974. Central England temperatures: Monthly means 1659 to 1973. Quart. J. Roy. Meteorol. Soc. 100:389-405.

Mann, M.E. and J. Park. 1993. Spatial correlations of interdecadal variation in global surface temperatures. Geophys. Res. Lett. 20:1055-1058.

Mass, C.F., and D.A. Portman. 1989. Major volcanic eruptions and climate: A critical evaluation. J. Climate 2:566-593.

McDermott, D., and E.S. Sarachik. 1995. Thermohaline circulations and variability in a two-hemisphere sector model of the Atlantic. In Natural Climate Variability on Decade-to-Century Time Scales. D.G. Martinson, K. Bryan, M. Ghil, M.M. Hall, T.R. Karl, E.S. Sarachik, S. Sorooshian, and L.D. Talley (eds.). National Academy Press, Washington, D.C.

Mehta, V.M. 1992. Meridionally propagating interannual-to-interdecadal variability in a linear ocean-atmosphere model. J. Climate 5:330-342.

Mikolajewicz, U., and E. Maier-Reimer. 1990. Internal secular variability in an ocean general circulation model. Climate Dynamics 4:145-156.

Miller, J., and G. Russell. 1989. Ocean heat transport during the last glacial maximum. Paleoceanography 4:141-155.

Mysak, L.A. 1995. Decadal-scale variability in ice cover and climate in the Arctic Ocean and Greenland Sea. In Natural Climate Variability on Decade-to-Century Time Scales. D.G. Martinson, K. Bryan, M. Ghil, M.M. Hall, T.R. Karl, E.S. Sarachik, S. Sorooshian, and L.D. Talley (eds.). National Academy Press, Washington, D.C.

Mysak, L.A., D.K. Manak, and R.F. Marsden. 1990. Sea ice anomalies observed in the Greenland and Labrador seas during 1901-1984 and their relation to an interdecadal Arctic climate cycle. Climate Dynamics 5:111-133.

Mysak, L.A., T.F. Stocker, and F. Huang. 1993. Century-scale variability in a randomly forced, two-dimensional thermohaline ocean circulation model. Climate Dynamics 8:103-116.

Neftel, A., E. Moor, H. Oeschger, and B. Stauffer. 1985. Evidence from polar ice cores for the increase in atmospheric CO_2 in the past two centuries. Nature 315:45-47.

Newell, R.E. 1970. Stratospheric temperature change from the Mt. Agung volcanic eruption of 1963. J. Atmos. Sci. 27:977-978.

Newell, R.E. 1979. Climate of the ocean. Am. Sci. 67:405-416.

Newell, R.E., and B.C. Weare. 1976. Ocean temperatures and large-scale atmospheric variations. Nature 262:40-41.

North, G.R. 1975. Analytical solution to a simple climate model with diffusive heat transport. J. Atmos. Sci. 32:1301-1307.

North, G.R. 1984. Empirical orthogonal functions and normal modes. J. Atmos. Sci. 41:879-887.

North, G.R., and K-Y. Kim. 1995. Detection of forced climate signals. In Natural Climate Variability on Decade-to-Century Time Scales. D.G. Martinson, K. Bryan, M. Ghil, M.M. Hall, T.R. Karl, E.S. Sarachik, S. Sorooshian, and L.D. Talley (eds.). National Academy Press, Washington, D.C.

North, G.R., J. Mengel, and D.F. Short. 1983. Simple energy balance model resolving the seasons and the continents: Application to the astronomical theory of the ice ages. J. Geophys. Res. 88:6576-6586.

North, G.R., K.-Y. Yip, R. Leung, and R. Chervin. 1992. Forced and free variations of the surface temperature field in a GCM. J. Climate 5:227-239.

North, G.R., K.-Y. Kim, S.S.P. Shen, and J.W. Hardin. 1995. Detection of forced climate signals. Part I: Theory. J. Climate 8:401-408.

Oliver, R.C. 1976. On the response of hemispheric mean temperature to stratospheric dust: An empirical approach. J. Appl. Meteorol. 15:933-950.

Oort, A.H., and H. Liu. 1993. Upper air temperature trends over the globe, 1958-1989. J. Climate 6:292-307.

Overpeck, J.T. 1991. Century-to-millennium scale climatic variability during the late Quaternary. In Global Changes of the Past. R. Bradley (ed.). UCAR/Office for Interdisciplinary Earth Studies, Boulder, Colo., pp. 139-173.

Parker, D.E., T.P. Legg, and C.K. Folland. 1991. A New Daily Central England Temperature Series, 1772-1991. Tech. Note CRTN11, Hadley Centre, Meteorological Office, Bracknell, Berks., U.K.

Pestiaux, P., J.C. Duplessy, and A. Berger. 1987. Paleoclimatic variability at frequencies ranging from 10^{-4} cycle per year to 10^{-3} cycle per year—Evidence for nonlinear behavior of the climate system. In Climate History, Periodicity, and Predictability. M.R. Rampino, J.E. Sanders, W.S. Newman, and L.K. Konigsson (eds.). Van Nostrand Reinhold, New York, pp. 285-299.

Peterson, L.C., J.T. Overpeck, N. Yipp, and J. Imbrie. 1991. A high-resolution Late-Quaternary upwelling record from the anoxic Cariaco Basin, Venezuela. Paleoceanography 6:99-119.

Pfister, C. 1992. Monthly temperature and precipitation in central Europe 1525-1979: Quantifying documentary evidence on weather and its effects. In Climate Since A.D. 1500. R.S. Bradley and P.D. Jones (eds.). Routledge, New York, pp. 118-142.

Piehler, H., and W. Bach. 1992. The potential role of an active deep ocean for climatic change. J. Geophys. Res. 97:15507-15512.

Plantico, M.S., T.R. Karl, G. Kukla, and J. Gavin. 1990. Is recent climate change across the United States related to rising levels of anthropogenic greenhouse gases? J. Geophys. Res. 95:16617-16637.

Plaut, G., M. Ghil, and R. Vautard. 1995. Interannual and interdecadal variability in 335 years of Central England temperature. Science 268:710-713.

Pollack, J.B., D. Rind, A. Lacis, and J. Hansen. 1993. GCM simulations of volcanic aerosol forcing. I: Climate changes induced by steady state perturbations. J. Climate 6:1719-1742.

Porter, S.C. 1975. Equilibrium-line altitudes of late Quaternary glaciers in the Southern Alps, New Zealand. Quat. Res. 5:27-47.

Porter, S.C. 1981. Recent glacier variations and volcanic eruptions. Nature 291:139-142.

Porter, S.C. 1986. Pattern and forcing of Northern Hemisphere glacier variations during the last millennium. Quat. Res. 26:27-48.

Preisendorfer, R.W. 1988. Principal Component Analysis in Meteorology and Oceanography. C.D. Mobley (ed.). Elsevier, New York, 425 pp.

Quinn, W.H., V.T. Neal, and S.E.A. de Mayolo. 1987. El Niño occurrences over the past four and a half centuries. J. Geophys. Res. 92:14449-14461.

Quon, C., and M. Ghil. 1992. Multiple equilibria in thermosolutal convection due to salt-flux boundary conditions. J. Fluid Mech. 245:449-483.

Radick, R.R., G.W. Lockwood, and S.L. Baliunas. 1990. Stellar activity and brightness variations: A glimpse at the Sun's history. Science 247:39-44.

Ramanathan, V., and J.A. Coakley. 1978. Climate modeling through radiative-convective models. Rev. Geophys. Space Phys. 16:465-489.

Ramanathan, V., and W. Collins. 1991. Thermodynamic regulation of ocean warming by cirrus clouds deduced from the 1987 El Niño. Nature 351:27-32.

Rampino, M., and S. Self. 1982. Historic eruptions of Tambora (1815), Krakatau (1883), and Agung (1963): Their stratospheric aerosols and climatic impact. Quat. Res. 18:127-143.

Rampino, M., and S. Self. 1984. Sulfur-rich volcanic eruptions and stratospheric aerosols. Nature 310:677-679.

Randel, W.J., and J.B. Cobb. 1994. Coherent variations of monthly mean total ozone and lower stratospheric temperature. J. Geophys. Res. 99D:5433-5447.

Rasmussen, R.A., and M.A.K. Khalil. 1984. Atmospheric methane in the recent and ancient atmospheres: Concentrations, trends, and interhemispheric gradient. J. Geophys. Res. 89:11599-11605.

Reid, G.C. 1991. Solar total irradiance variations and the global sea surface temperature record. J. Geophys. Res. 96:2835-2844.

Reinhold, B.B., and R.T. Pierrehumbert. 1982. Dynamics of weather regimes: Quasi-stationary waves and blocking. Mon. Weather Rev. 110:1105-1145.

Ribes, E., J.C. Ribes, and R. Barthalot. 1987. Evidence for a larger Sun with a slower rotation during the seventeenth century. Nature 326:52-55.

Rind, D., and M. Chandler. 1991. Increased ocean heat transports and warmer climate. J. Geophys. Res. 96:7437-7461.

Rind, D., and J. Overpeck. 1994. Hypothesized causes of decade-to-century-scale climate variability: Climate model results. Quat. Sci. Rev. 12:357-374.

Rind, D.H., and J.T. Overpeck. 1995. Modeling the possible causes of decadal-to-millennial-scale variability. In Natural Climate Variability on Decade-to-Century Time Scales. D.G. Martinson, K. Bryan, M. Ghil, M.M. Hall, T.R. Karl, E.S. Sarachik, S. Sorooshian, and L.D. Talley (eds.). National Academy Press, Washington, D.C.

Rind, D., and D. Peteet. 1985. Terrestrial conditions at the last glacial maximum and CLIMAP sea surface temperature estimates: Are they consistent? Quat. Res. 24:1-22.

Rind, D., D. Peteet, W.S. Broecker, A. McIntyre, and W. Ruddiman. 1986. The impact of cold North Atlantic sea surface temperatures on climate: Implications for the Younger Dryas cooling (11-10K). Climate Dynamics 1:3-33.

Rind, D., N.K. Balachandran, and R. Suozzo. 1992. Climate change and the middle atmosphere. Part 11: The impact of volcanic aerosols. J. Climate 5:189-208.

Ropelewski, C.F., and M.S. Halpert. 1987. Global and regional scale precipitation patterns associated with the El Niño/Southern Oscillation. Mon. Weather Rev. 115:1606-1626.

Röthlisberger, S.C. 1986. 10,000 Jahre Gletschergeschichte der Erde. Verlag Sauerländer, Aärau, Switzerland, 416 pp.

Saltzman, B. 1978. A survey of statistical-dynamical models of the terrestrial climate. Adv. Geophys. 20:183-304.

Saltzman, B. 1983. Climatic systems analysis. Adv. Geophys. 25:173-233.

Saltzman, B., and A.D. Vernekar. 1971. An equilibrium solution for the axially symmetric component of the earth's macroclimate. J. Geophys. Res. 76:1498-1524.

Schlesinger, M. (ed.). 1991. Greenhouse-Gas Induced Climate Change: A Critical Appraisal of Simluations and Observations. Elsevier, Amsterdam, 615 pp.

Schlesinger, M.E., and N. Ramankutty. 1992. Implications for global warming of intercycle solar irradiance variations. Nature 330:330-333.

Schneider, S.H. 1972. Cloudiness as a global climatic feedback mechanism: The effects on the radiation balance and surface temperature variations in cloudiness. J. Atmos. Sci. 29:1413-1422.

Schneider, S.H., and R.E. Dickinson. 1974. Climate modeling. Rev. Geophys. Space Phys. 12:447-493.

Sellers, W.D. 1969. A global climatic model based on the energy balance of the earth-atmosphere system. J. Appl. Meteorol. 8:392-400.

Shabbar, A., K. Higuchi, and J.L. Knox. 1990. Regional analysis of northern hemisphere 50 kPa geopotential heights from 1946 to 1985. J. Climate 3:543-557.

Simmons, A.J., and G.J. Hoskins. 1979. The downstream and upstream development of unstable baroclinic waves. J. Atmos. Sci. 36:1239-1254.

Slepian, S. 1978. Prolate spheroidal wave functions, Fourier analysis and uncertainty. V: The discrete case. Bell Syst. Tech. J. 1371-1430.

Spencer, R.W., and J.R. Christy. 1990. Precise monitoring of global temperature trends from satellites. Science 247:1558-1562.

Spencer, R.W., and J.R. Christy. 1992. Precision and radiosonde validation of satellite gridpoint temperature anomalies. Part II. A tropospheric retrieval and trends during 1979-90. J. Climate 5:858-866.

Spencer, R.W., and J.R. Christy. 1993. Precision lower stratospheric monitoring with the MSU: Technique, validation and results 1979-1991. J. Climate 6:1194-1204.

Stocker, T.F., and L.A. Mysak, 1992. Climatic fluctuations on the century time scale: A review of high-resolution proxy data and possible mechanisms. Climatic Change 20:227-250.

Stuiver, M., and T.F. Braziunas. 1988. The solar component of the atmospheric ^{14}C record. In Secular Solar and Geomagnetic Variations in the Last 10,000 Years. F.R. Stephenson and A.S. Wolfendale (eds.). Kluwer Academic Publishers, Dordrecht, pp. 245-266.

Stuiver, M., and T.F. Braziunas. 1989. Atmospheric ^{14}C and century-scale solar oscillations. Nature 338:405-408.

Stuiver, M., T.F. Braziunas, B. Becker, and B. Kromer. 1991. Climatic, solar, oceanic, and geomagnetic influences on late-Glacial and Holocene atmospheric $^{14}C/^{12}C$ change. Quat. Res. 35:1-24.

Thompson, D.J. 1982. Spectrum estimation and harmonic analysis. Proc. IEEE 70:1055-1096.

Trenberth, K.E. 1990. Recent observed interdecadal climate changes in the northern hemisphere. Bull. Amer. Meteorol. Soc. 71:988-993.

Trenberth, K.E., J.R. Christy, and J.W. Hurrell. 1992. Monitoring global monthly mean surface temperatures. J. Climate 5:1405-1423.

Trenberth, K.E., and J.W. Hurrell. 1995. Decadal climate variations in the Pacific. In Natural Climate Variability on Decade-to-Century Time Scales. D.G. Martinson, K. Bryan, M. Ghil, M.M. Hall, T.R. Karl, E.S. Sarachik, S. Sorooshian, and L.D. Talley (eds.). National Academy Press, Washington, D.C.

van Loon, H., and K. Labitzke. 1988. Association between the 11-year solar cycle, the QBO, and the atmosphere. Part II: Surface and 700 mb in the Northern Hemisphere in winter. J. Climate 1:905-920.

van Loon, H., and J. Williams. 1976. The connection between the trends of mean temperature and circulation at the surface: Part I: Winter. Mon. Weather Rev. 104:365-380.

Vautard, R., and M. Ghil. 1989. Singular spectrum analysis in nonlinear dynamics, with applications to paleoclimatic time series. Physica D 35:395-424.

Vautard, R., P. Yiou, and M. Ghil. 1992. Singular spectrum analysis: A tool kit for short, noisy, chaotic signals. Physica D 58:95-126.

Walker, G.T., and E.W. Bliss. 1932. World Weather V. Mem. Roy. Meteorol. Soc. 4:53-84.

Wallace, J.M. 1992. Effect of deep convection on the regulation of tropical sea-surface temperature. Nature 357:230-231.

Wallace, J.M. 1995. Natural and forced variability in the climate record. In Natural Climate Variability on Decade-to-Century Time Scales. D.G. Martinson, K. Bryan, M. Ghil, M.M. Hall, T.R. Karl, E.S. Sarachik, S. Sorooshian, and L.D. Talley (eds.). National Academy Press, Washington, D.C.

Wallace, J.M., Y. Zhang, and K.-H. Lau. 1993. Structure and seasonality of interannual and interdecadal variability of the geopotential height and temperature fields in the northern hemisphere troposphere. J. Climate 6:2063-2082.

Washington, W.M., and G.A. Meehl. 1989. Climate sensitivity due to increased CO_2: Experiments with a coupled atmosphere and ocean general circulation model. Climate Dynamics 4:1-38.

Washington, W.M., and C.L. Parkinson. 1986. An Introduction to Three-Dimensional Climate Modeling. Oxford Univ. Press, Oxford/New York, 422 pp.

Watts, R.G. 1985. Global climate variations due to fluctuations in the rate of deep water formation. J. Geophys. Res. 90:8067-8070.

Weaver, A.J., E.S. Sarachik, and J. Marotzke. 1991. Nature 353:836-838.

Wetherald, R.T., and S. Manabe. 1975. The effect of changing the solar constant on the climate of a general circulation model. J. Atmos. Sci. 32:2044-2059.

Weyl, P.K. 1968. The roles of ocean in climate change. A theory of the ice ages. Meteor. Monogr. 8:37-62.

Wiener, N. 1949. Time Series. MIT Press, Cambridge, 163 pp.

Wigley, T.M.L. 1989. Possible climatic change due to SO_2-derived cloud condensation nuclei. Nature 339:365-367.

Wigley, T.M.L., and P.M. Kelly. 1990. Holocene climatic change: [14]C wiggles and variations in solar irradiance. Philos. Trans. Roy. Soc. London A330:547-560.

Wigley. T.M.L., and S.C.B. Raper. 1987. Thermal expansion of sea water associated with global warming. Nature 330:127-131.

Wigley, T.M.L., and S.C.B. Raper. 1990. Natural variability of the climate system and detection of the greenhouse effect. Nature 344:324-327.

Wigley, T.M.L., and S.C.B. Raper. 1991. Internally generated natural variability of the climate system and detection of the greenhouse effect. In Greenhouse-Gas-Induced Climatic Change: A Critical Appraisal of Simulations and Observations. M.E. Schlesinger (ed.). Elsevier Science Publishers, Amsterdam, pp. 471-482.

Wigley, T.M.L., and S.C.B. Raper. 1992. Implications for climate and sea level of revised IPCC emissions scenarios. Nature 357:293-300.

Wigley, T.M.L., and S.C.B. Raper. 1995. Modeling low-frequency climate variability: The greenhouse effect and the interpretation of paleoclimatic data. In Natural Climate Variability on Decade-to-Century Time Scales. D.G. Martinson, K. Bryan, M. Ghil, M.M. Hall, T.R. Karl, E.S. Sarachik, S. Sorooshian, and L.D. Talley (eds.). National Academy Press, Washington, D.C.

Willson, R.C., and H.S. Hudson. 1991. The sun's luminosity over a complete solar cycle. Nature 351:42-44.

Yiou, P., C. Genthon, M. Ghil, J. Jouzel, H. Le Treut, J.M. Barnola, C. Lorius, and Y.N. Korotkevitch. 1991. High-frequency paleo-variability in climate and CO_2 levels from Vostok ice-core records. J. Geophys. Res. 96:20365-20378.

Yulaeva, E., and J.M. Wallace. 1994. The signature of ENSO in global temperature fields derived from the microwave sounding unit. J. Climate 7:1719-1736.

Yulaeva, E., J.R. Holton, and J.M. Wallace. 1994. On the cause of the annual cycle in tropical lower stratospheric temperatures. J. Atmos. Sci. 51:169-174.

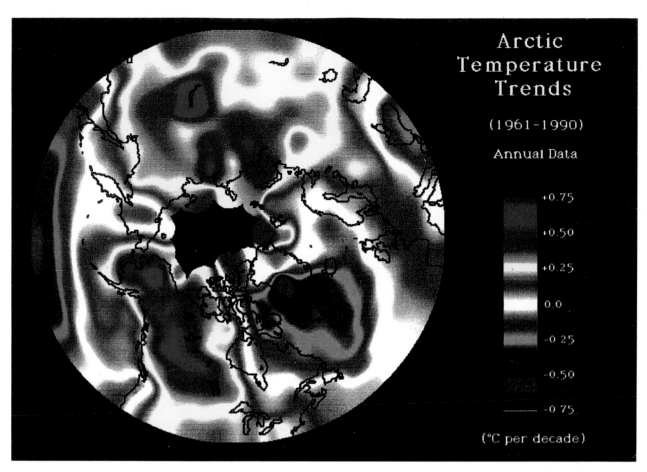

WALSH, FIGURE 4   Trends of temperature (°C per decade) computed using linear regression onto annual temperatures for 1961-1980. A Cressman analysis with a 400 km radius of influence has been applied to the computed trends. (From Chapman and Walsh, 1992; reprinted with permission of the American Meteorological Society.)

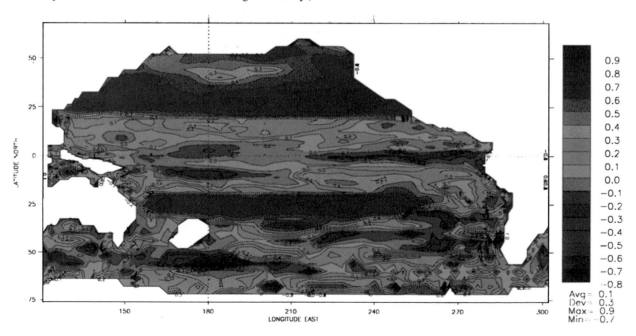

CAYAN ET AL., FIGURE 5a   Map of local correlations between $\frac{\Delta SST'}{\Delta t}$ and net heat-flux term on the right-hand side of Equation 4. Note that in the tropical strip ($\pm 20°$ latitude), a Newtonian heat-flux parameterization is invoked and observed heat-flux anomalies are excluded.

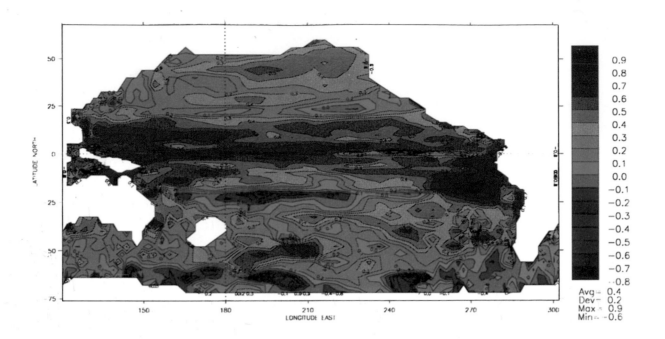

CAYAN ET AL., FIGURE 5b Map of local correlations between $\frac{\Delta SST'}{\Delta t}$ and the horizontal temperature advection term on the right-hand side of Equation 4.

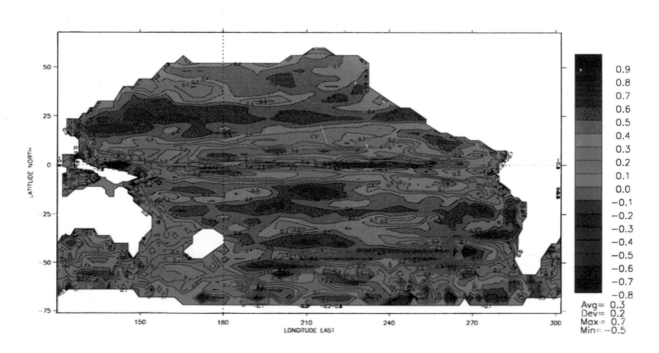

CAYAN ET AL., FIGURE 5c Map of local correlations between $\frac{\Delta SST'}{\Delta t}$ and the entrainment term on the right-hand side of Equation 4.

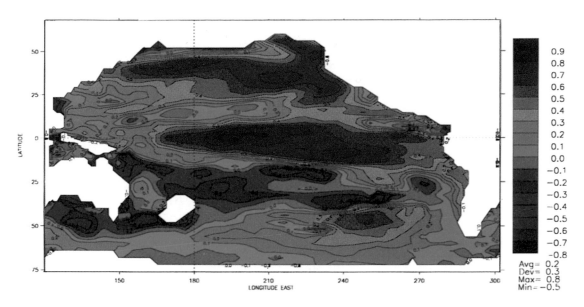

CAYAN ET AL., FIGURE 7   Map of local correlation coefficients between model and observed SST anomalies for all months, 1970-1988. Note that in the tropical strip (± 20° latitude), a Newtonian heat-flux parameterization is invoked and observed heat-flux anomalies are excluded.

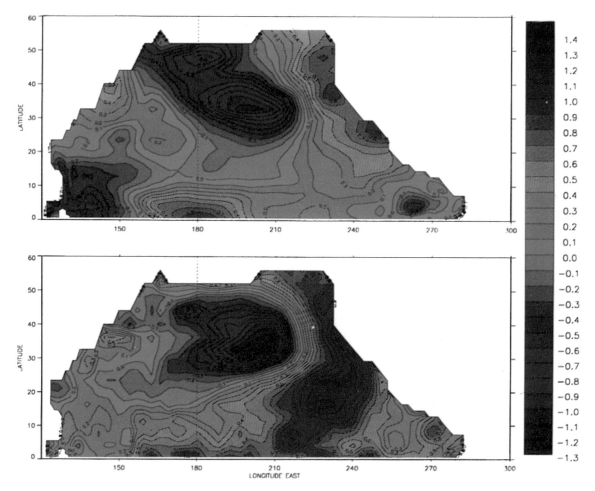

CAYAN ET AL., FIGURE 9   Difference fields of SST (°C) for the winters of 1976-77 through 1981-82 minus those of 1970-71 through 1975-76, for the model (top) and observations (bottom).

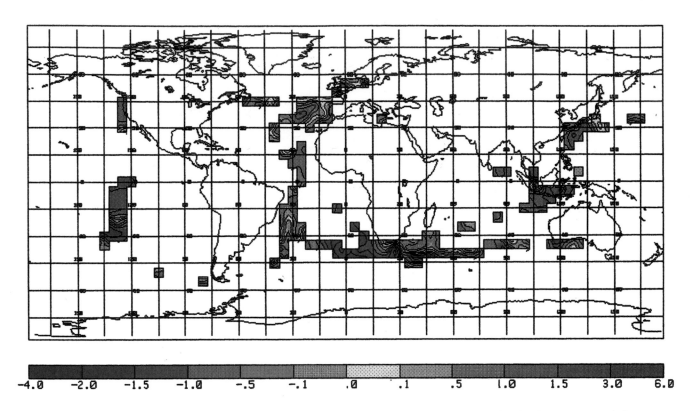

PARKER ET AL., FIGURE 1a   MOHSST5 anomalies (°C) (w.r.t. 1951-1980) without filled values, January 1878.

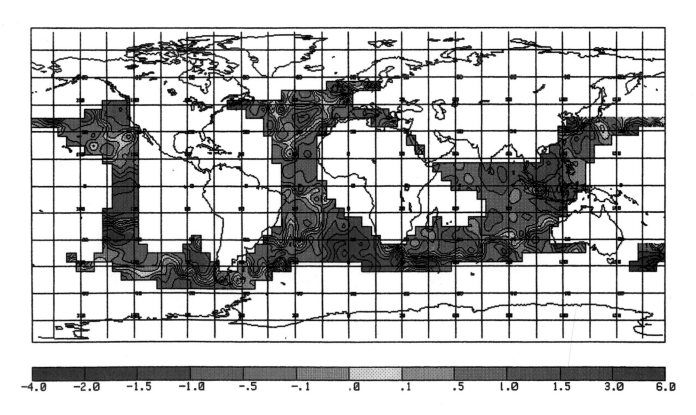

PARKER ET AL., FIGURE 1b   MOHSST5 anomalies (°C) with filled values, January 1878.

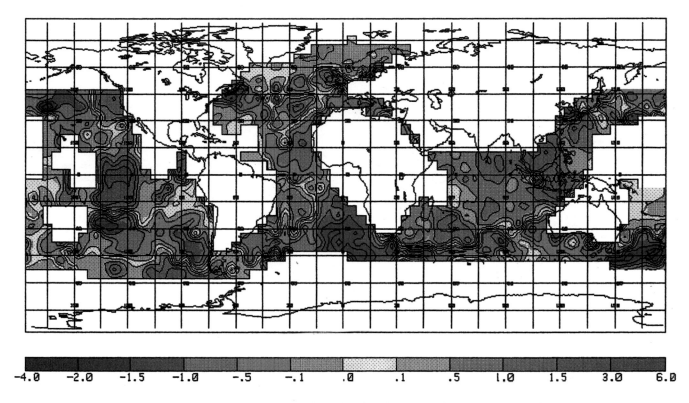

PARKER ET AL., FIGURE 1c   Enhanced MOHSST5 anomalies (°C), January 1878.

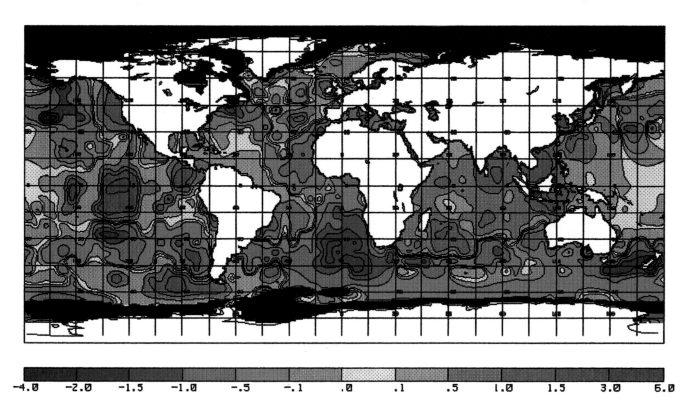

PARKER ET AL., FIGURE 1d   GISST 1.0 anomalies (°C), January 1878. Sea ice is shaded black.

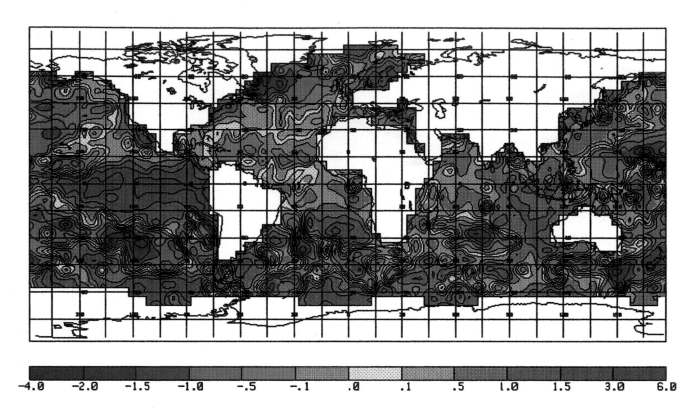

PARKER ET AL., FIGURE 3a   MOHSST5 anomalies (°C) with filled values, January 1983.

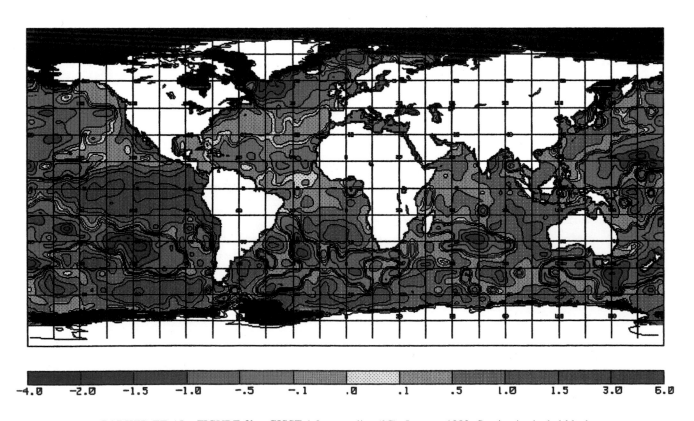

PARKER ET AL., FIGURE 3b   GISST 1.0 anomalies (°C), January 1983. Sea ice is shaded black.

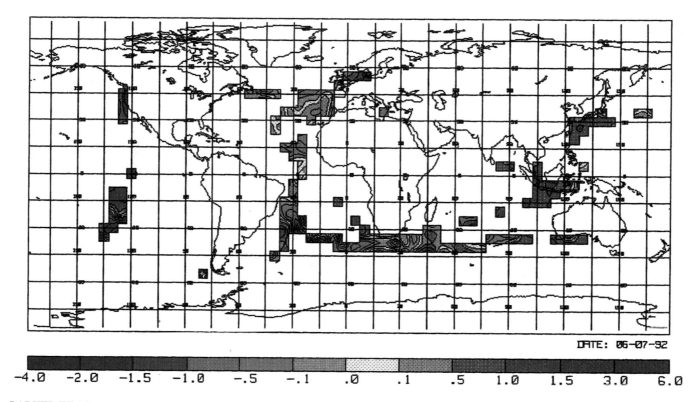

PARKER ET AL., FIGURE 4a   MOHSST5 anomalies (°C) without filled values, January 1983, with coverage limited to that of January 1878.

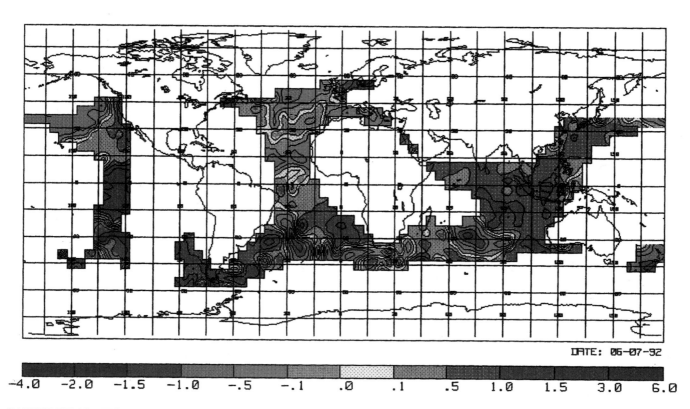

PARKER ET AL., FIGURE 4b   MOHSST5 anomalies (°C) with filled values, January 1983, with coverage of input data limited to that of 1877-1878.

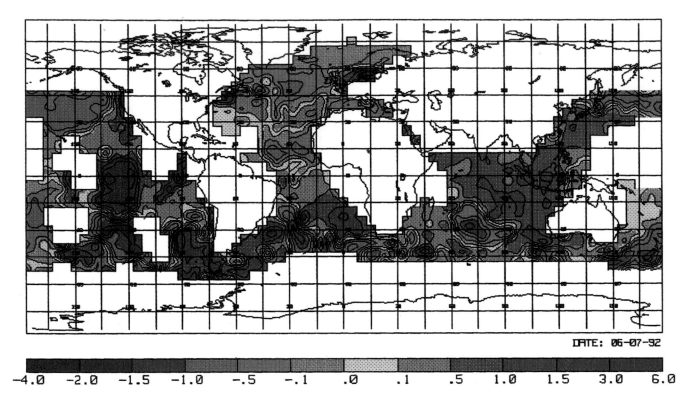

DATE: 06-07-92

−4.0 −2.0 −1.5 −1.0 −.5 −.1 .0 .1 .5 1.0 1.5 3.0 6.0

PARKER ET AL., FIGURE 4c   Enhanced MOHSST5 anomalies (°C), January 1983, with coverage of input data limited to that of 1877-1878.

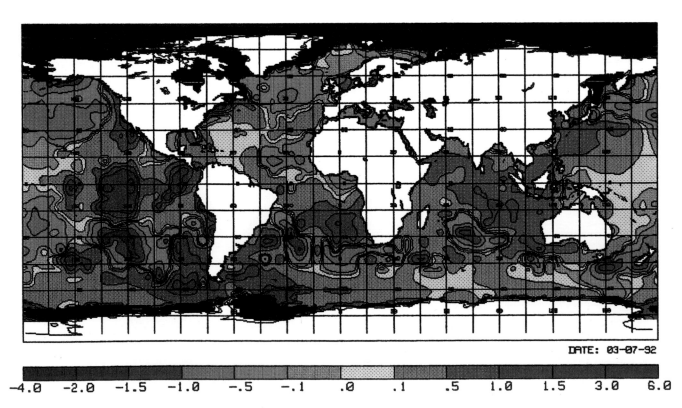

DATE: 03-07-92

−4.0 −2.0 −1.5 −1.0 −.5 −.1 .0 .1 .5 1.0 1.5 3.0 6.0

PARKER ET AL., FIGURE 4d   Globally complete SST anomalies (°C) and sea ice, January 1983, with coverage of input SST data limited to that of 1877-1878. Sea ice is shaded black.

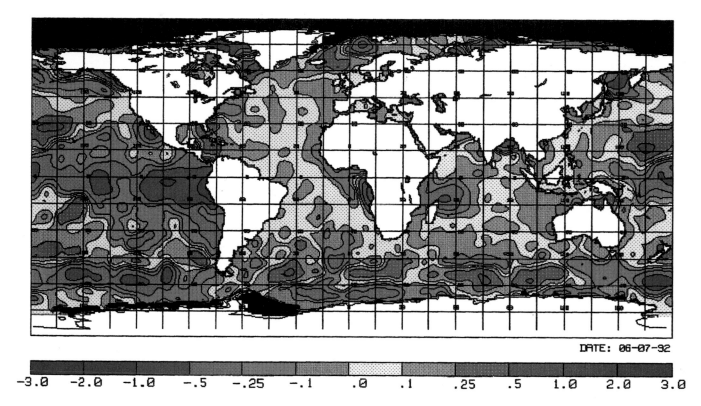

DATE: 06-07-92

-3.0 -2.0 -1.0 -.5 -.25 -.1 .0 .1 .25 .5 1.0 2.0 3.0

PARKER ET AL., FIGURE 5a   Bias (°C) of the reduced-sampling analysis of 1982-1983.

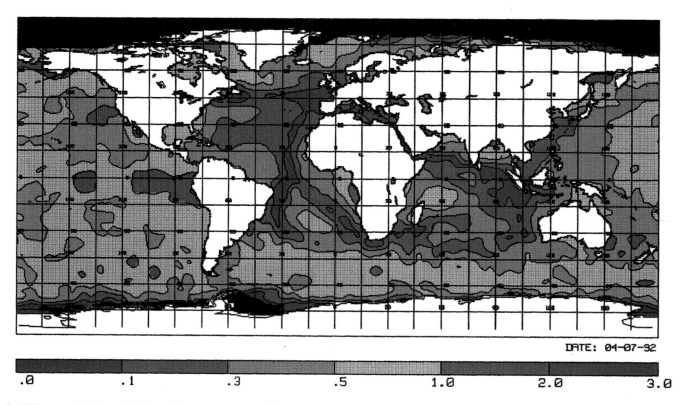

DATE: 04-07-92

.0 .1 .3 .5 1.0 2.0 3.0

PARKER ET AL., FIGURE 5b   Root-mean-square differences (°C) between GISST 1.0 and the reduced-sampling analysis, 1982-1983.

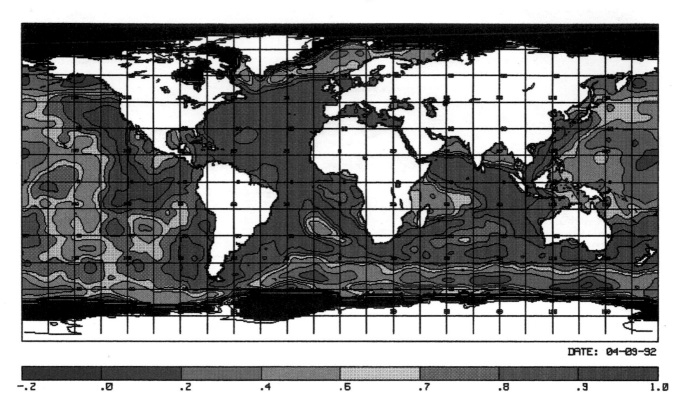

PARKER ET AL., FIGURE 6   Correlations between GISST 1.0 and the reduced-sampling analysis, 1981-1990.

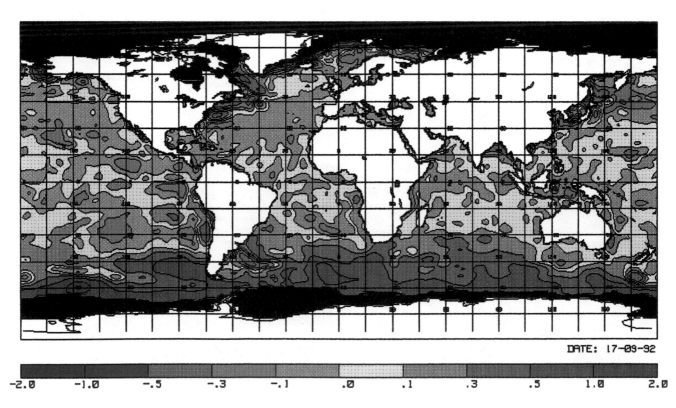

PARKER ET AL., FIGURE 7   Bias (°C), GISST 1.1 minus Reynolds and Marsico (1993) analysis, 1982-1991.

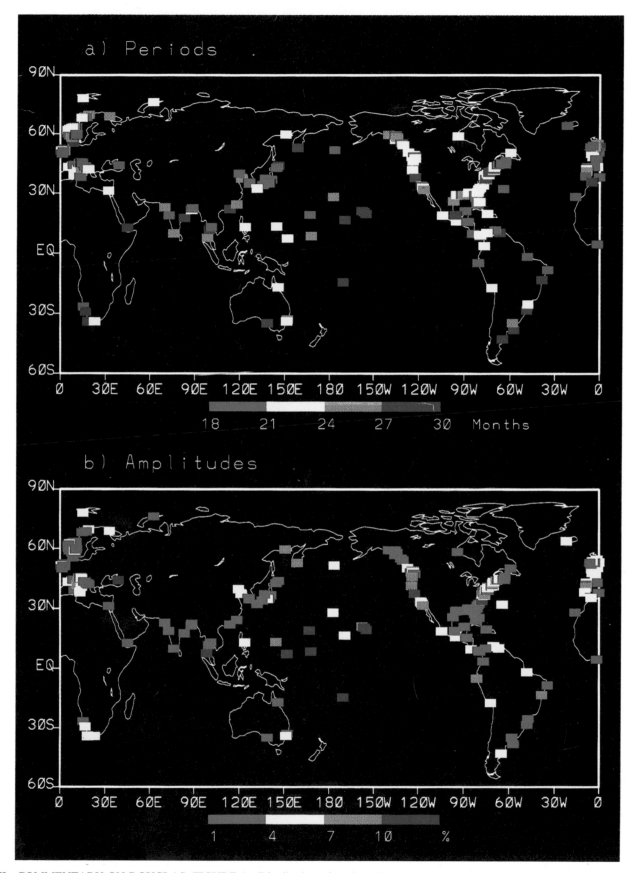

GHIL, COMMENTARY ON DOUGLAS, FIGURE 1   Distribution of stations that show an oscillatory feature with periods from 18 to 30 months. (a) distribution of period; (b) distribution of the amplitudes of the quasi-biennial oscillation, as a percentage of the total variance at each station (courtesy of Y. Sezginer-Unal).

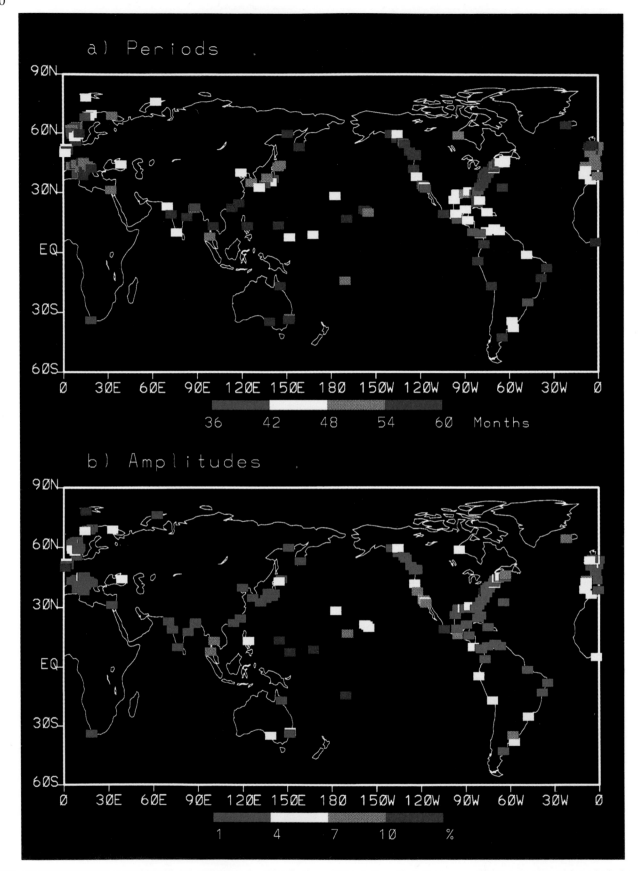

GHIL, COMMENTARY ON DOUGLAS, FIGURE 2   Distribution of the stations that show an oscillatory feature with periods from 36 months to 60 months. (a) distribution of the period; (b) distribution of the amplitudes of the low-frequency ENSO signal, as a percentage of the total variance at each station (courtesy of Y. Sezginer-Unal).

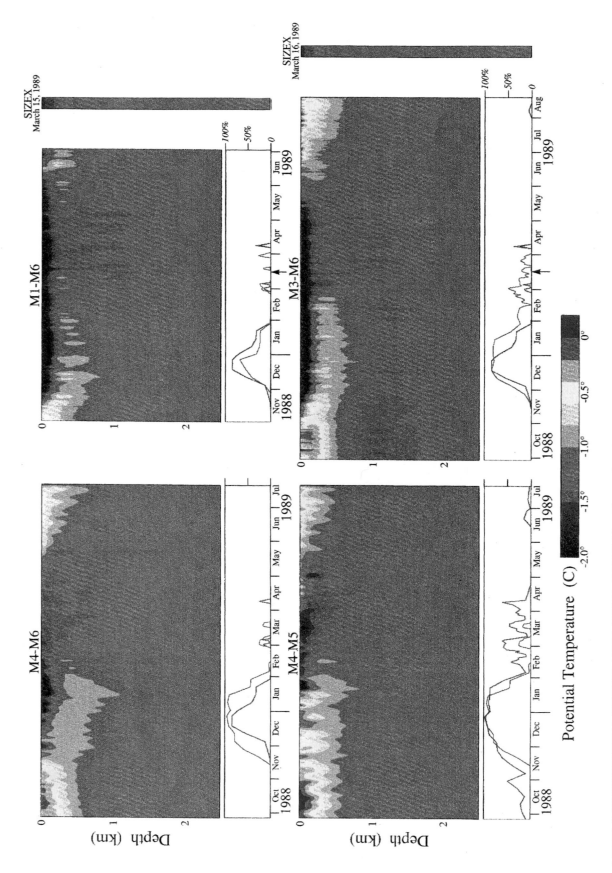

MUNK, DISCUSSION FOLLOWING LAZIER PAPER, FIGURE 1, Time series of range-averaged potential temperature profiles along four paths of the tomographic array. Total ice concentration in percent (from SSMM/I data) at the path endpoints are plotted below the images. Range-averaged profiles, computed from Seasonal Ice Zone EXperiment (SIZEX) data obtained on 15-16 March 1989 between moorings 1 and 6 and moorings 3 and 6 are given for comparison. (From SIZEX group, 1989; reprinted with permission of the Nauseu Remote Sensing Center.)

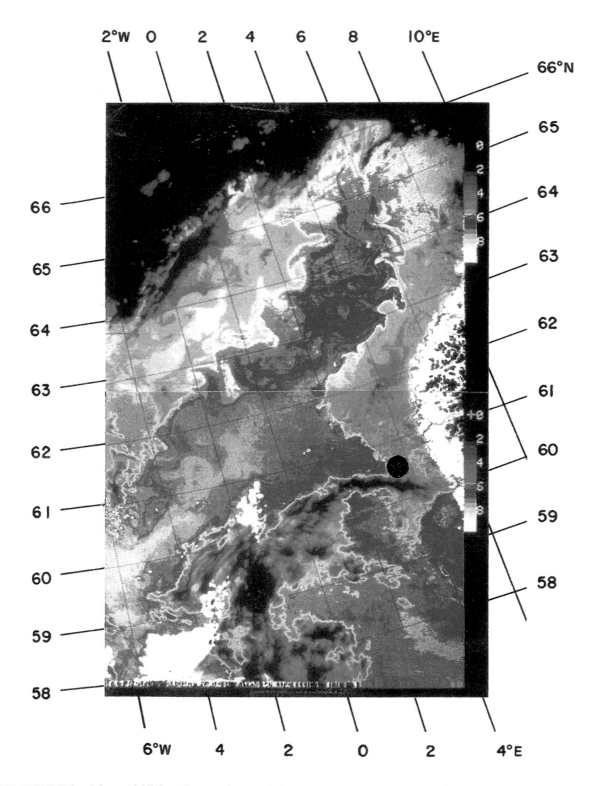

LEHMAN, FIGURE 8   Advanced VHR radiometer image of the eastern Norwegian Sea, showing the entry of warm Atlantic waters (warm colors) between the Faeroe and Shetland islands and the mixing of Atlantic waters into the North Sea. The site of Troll 3.1 is marked (●). The dark, northeast-arching swath near the core site reflects cloud cover. (From Johannessen, 1986; reprinted with permission of Springer-Verlag.)

3

THE OCEAN

Introduction

Climate studies are motivated by the curiosity we all have about the weather, and our desire to predict it in order to make economic projections. The general public tends to focus on local weather—and thus generally on weather over land rather than over the ocean—as well as on fairly short-scale climate variations, several years rather than decades to millennia. For these reasons it is not always appreciated outside the scientific community that the ocean is an essential component of the coupled climate system. Understanding and modeling the ocean and its coupling to the atmosphere, land, and biosphere is vital. In the realm of fisheries, climate variations in the ocean itself can be seen to have an economic impact; recognizing this, coastal states have supported oceanography.

Approaches to studying the ocean's role in climate can be divided into two types: understanding and modeling the ocean as part of the fully coupled climate system, and observing, quantifying, and modeling the dynamics of the ocean itself. Clearly there is great overlap, and a full grasp of the latter is necessary for progress in the former. The ocean's most obvious and direct importance to land-based climate variations lies in the fact that it sets the surface temperature that forces the atmosphere over three-quarters of the planet; distribution of sea ice is important also, since it affects the planetary albedo and the amount of ocean/atmosphere heat exchange. Predicting the surface temperature of the ocean and the extent of sea ice is not a simple exercise, however: It involves atmospheric forcing, lateral circulation, and vertical overturning. The last is affected by the salinity distribution, and salinity depends on factors that are similar to those influencing the ocean temperature.

Because of its great thermal inertia relative to that of the atmosphere, the ocean has a significant effect on climate. The two most commonly mentioned climate phenomena in which the ocean's role is important are El Niño, which is purely natural variability, and global warming, which is partly anthropogenic. Much progress has been made in observing and modeling both the oceanic and the atmospheric components of El Niño; other obvious climate effects are also strongly tied to the ocean, such as the higher temperatures in northern Europe relative to those of northeastern Canada. The papers in this volume discuss a decadal climate oscillation in the North Atlantic involving both the atmosphere and ocean, and a longer-period oscillation in the North Pacific, neither of which is firmly tied to El Niño. Our understanding of the ocean's role in much longer-scale variations, such as glaciations, has also improved greatly through the use of proxy records and coupled ocean-atmosphere modeling, both of which are represented in this volume. Thus our vocabulary of climate variations, even if limited to those we can quantify today, is already much broader than those that have drawn the most public attention.

In order to dissect the relative roles of the ocean and atmosphere in climate, it is necessary to both observe and model. Modeling is particularly important because of the relative lack of long-term observations. The merchant marine data set is the most comprehensive in time and space, since ship observations have been reported and archived for many years. This data set is limited to the sea surface, however, and is good only in regions of high merchant marine activity. The time series of water-column ocean observations reported in this volume are representative of what is available globally; at best they are repeated measurements of temperature and salinity through the water column, collected at regular intervals and at the same location over decades. Such time series are very restricted spatially; they are usually localized to the coastal waters of a state or country that depends on fisheries. The time series that are more widely spatially distributed, such as those from the defunct weather ships, tend to be single points separated by a long distance from the next station. The upper water column has been better measured since the 1970s using ships of opportunity, but only for temperature and not salinity. (Relatively little from this extensive data set is reported in this volume.) Proxy records from sediment cores relate primarily to much longer time scales than the decade-to-century ones that are the focus of this volume, and consist of very few data points. Thus we are operating observationally from an extremely limited data set. In the last few years, attention has been turned to extracting as much information as possible from the data we do have, and progress has been remarkable under the circumstances. Many fine examples of this work are presented in this chapter.

Ocean modeling relevant to climate has progressed greatly in the last decade, due to the growth of computing power, the availability of a community ocean model, and an increased focus on these problems by a small community of modelers. Enormous advances have been made in understanding the El Niño problem in the tropical Pacific. Several modeling-related papers in this volume present new insights into the role of the lateral and overturning modes of ocean circulation in producing climate oscillations.

The current public concern about global warming and other issues related to climate and its prediction has resulted in international plans for greatly enhanced ocean observations in space and time. Major efforts include the placement of monitoring equipment in the tropical Pacific for El Niño prediction, a global ocean-circulation observational experiment designed to enhance our knowledge of the circulation as it exists today and provide better data for modeling, and establishment of global monitoring. Global monitoring should be focused on time series of ocean properties that affect climate and/or reflect climate change; these include temperature and velocity, and, as we are recognizing more and more, salinity. For this monitoring to be effective, measurements must be made in the upper ocean worldwide; at locations likely to be indicative of climate, they should be made throughout the relevant portion of the water column. Perhaps even more important than location, however, will be some assurance that a time series will be maintained indefinitely. If these two conditions can be met, monitoring should be begun as soon as possible. It is therefore critical that as much knowledge as possible be synthesized from currently available data and models, to ensure that the ocean observing systems will be both efficient and comprehensive. The papers in this chapter, which reflect the increasing attention being paid to climate issues by the oceanographic research community, represent a large advance in our knowledge of ocean variability on decade-to-century time scales.

Ocean Observations

MELINDA M. HALL

INTRODUCTION

Natural variability in the ocean has periods ranging from seconds to millennia. Those phenomena that have the most obvious impact on human affairs tend to recur periodically as well: daily, such as the tides; sporadically, such as storm surges or tsunamis; and seasonally, such as the simple warming of coastal waters in summer. Most of these events are predictable to varying degrees. It is now recognized that occurrences of the El Niño-Southern Oscillation (ENSO), a phenomenon of global scale that has tremendous socio-economic consequences, are quasi-periodic (over a term of several years) and are therefore within the realm of predictability as well. Our understanding of these examples of natural variability, and hence our ability to predict them, are derived from our past experience with them—in other words, repeated observations of the same event—as well as from theoretical models based on ocean physics. Identifying the effects of anthropogenically induced changes in the ocean is a subtle problem, for there are few precedents against which models can be tested. But a prerequisite for prediction in any case is a knowledge of the natural variability inherent in the system, and an understanding of the physics that drives that variability.

The study of natural variability at periods of decades to centuries presents particular challenges. A primary difficulty derives from the fact that the observed variability that we surmise to be associated with climate change is generally smaller in magnitude than variability due to other causes, and is sometimes at the limits of instrumental accuracy. Long time series are therefore required to deconvolve its signal from the much more energetic influences of seasonal and other types of variability. Long in situ records are inherently difficult to obtain, however, due to the hostile nature of the very environment we are trying to observe. Indeed, because oceanographic data will never be quite complete enough to "solve" the problem, there is a natural interdependency between the observations and modeling efforts. Models can provide globally complete fields, but data will always be required for their initialization, calibration, and validation.

On the other hand, regarding the observational effort, it is important to note that oceanographic variability tied to atmospheric forcing may be much stronger in isolated areas. For example, it is now recognized that the production of deep water in the northern North Atlantic is intimately related to the global climate, and thus changes in its production are either the result of, or harbingers of, more widely spread climate changes. Although it is almost impossible to directly measure the amounts of water convectively overturned each year, much qualitative and some quantitative information regarding production in previous years can be inferred from an examination of the variability of water properties at locations downstream from the source waters, in the deep western boundary current that carries these waters to the midocean. Swift (1995) clearly outlines these arguments; focusing particularly on the deep-water formation in the northern North Atlantic, he provides a good introduction to how one documents, studies, and interprets decadal changes.

Before returning to these ideas in more detail, the reader might find a brief history of ocean observations to be useful.

HISTORY OF OCEAN OBSERVATIONS

Quantitatively useful ocean observations date back only to about the turn of the century, which brought several important advances to the field of oceanography, and might be said to mark the start of a "modern" era. Around this time, empirical formulas were developed relating salinity, chlorinity, and density. These allowed precise salinity measurements to be made, since samples could be titrated to determine clorinity. Coincidentally, although the mathematics governing fluid dynamics had been studied for centuries, general physical theories of ocean circulation also developed with great rapidity beginning around the turn of the century. Many of these advances can be attributed to Scandinavian researchers (for a more detailed history, see the Introduction in Sverdrup et al., 1942).

The oceanographic expedition of the German research vessel *Meteor*, in 1925-1927, was led by Georg Wüst. It is notable for (at least) two contributions: First, Wüst's careful attention to accuracy and detail rendered the *Meteor* data useful as a baseline for comparison with later measurements of temperature, salinity, and dissolved oxygen. Second, Wüst conceived of and popularized the "core" method for determining the circulation of water masses. This method is based on the assumption that water parcels acquire their physical characteristics when they are in contact with the atmosphere at the sea surface, and that they retain these characteristics as they sink and flow into the ocean. Thus, Wüst concluded, the large-scale circulation in the ocean is reflected in the patterns of the temperature, salinity, and oxygen distributions. This concept is of fundamental importance to observations of deep-water production and circulation, particularly in recent decades when we have been able to measure many chemical constituents of anthropogenic origin (Schlosser and Smethie, 1995). By the middle of this century, temperature was being determined accurately to within about $\pm 0.02°C$ and salinity to within about 0.02 permil. The next baseline for observational oceanography was the International Geophysical Year, carried out in the mid- to late 1950s. This coordinated series of expeditions sought to map the physical properties of the entire Atlantic Ocean on a somewhat regular grid, and the resulting data provide the second "snapshot," three decades after Wüst's work, of the North Atlantic temperature and salinity structure. (Fuglister's 1960 atlas presents these data.)

Clearly, because of the limited accuracy of most measurements before 1900, there exist relatively few examples of long time series of measurements useful to the study of climate change. Roughly century-long global or regional records derived from operational measurements of such quantities as sea-surface temperature, sea-ice cover and extent, and sea-level measurements have been accessible to observers much longer than quantitatively useful deep ocean measurements. Decades-long time series of deep-ocean properties do exist, but are generally either mid-ocean and very isolated, or extensive spatially but limited to coastal waters. Fisheries provide strong economic motivation for such programs as the 40-year time series from the CalCOFI hydrographic cruises off the coast of California, or some of the repeated hydrographic data sets maintained for years off the coast of Japan. For several decades, a number of mid-ocean stations were occupied regularly by the ocean weather ships, for the purpose of providing marine weather forecasts.

Although long time series collected explicitly for climate studies or other research purposes are virtually nonexistent, an outstanding exception is the time series of temperature and salinity from the Panulirus station, located in deep water just off the coast of Bermuda, which already has contributed to studies looking at long-term variability in properties of the North Atlantic. Finally, there exists a vast archive of expendable bathythermograph (XBT) data collected from merchant ships, which is global in extent but has remarkably dense coverage in the North Pacific, where the NORPAX program is in its third decade. Although the XBT measures temperature as a function of depth to only 400 or 750 m (sometimes 1500 m), prediction of decade-to-century-scale ocean variability will require emphasis on such upper-ocean monitoring.

Parker et al. (1995) describe how useful baseline data sets can be constructed from historical records to yield a more comprehensive record of, in this case, monthly sea-ice and sea surface temperature fields dating back to January 1871. Since the resulting time series may exceed a century in length, it can be used for forcing and testing numerical models designed to examine variability at decade-to-century time scales. Mysak et al. (1990), analyzing sea-ice concentration and ice-limit data collected over almost 90 years, have found decadal-scale fluctuations in sea-ice extents, and have related them to other processes in the Arctic in a "negative feedback loop." Mysak (1995) reviews the evidence for such self-sustained climatic oscillations, presents more recent evidence strengthening these conclusions, and suggests links between the Arctic cycle and interdecadal variability at lower latitudes. That some of these observed interdecadal fluctuations are regular implies that they may in fact be predictable. Douglas (1995) reaches somewhat more negative conclusions in analyzing two sets of tide-gauge records (80 years and 141 years long): He finds no statistical evidence for acceleration of global sea-level rise, which is predicted to accompany global warming. A major difficulty is that the interdecadal signal overwhelms any longer-term trend. However, an understanding of the physics involved in the sea-level rise would allow this interdecadal signal to be removed from tide-gauge records, and would reduce the record length required for detecting acceleration of the rise.

These analyses illustrate both the possibilities and the difficulties involved in dealing with data collected by operational measurements. These should be borne in mind in planning future observing systems, for it is clearly on such operational measurements that we will depend if we are to acquire sufficient data coverage to monitor climate variability.

INTO THE PRESENT

The explosion of the electronics industry and the advent of the Space Age with its technological advances have had obvious implications for oceanographic observations. The development of electronic CTD (conductivity-temperature-depth) instruments now allows virtually continuous vertical sampling of the water column, whereas individual bottle samples are typically spaced 10-25 m apart in shallow waters and may be separated by several hundred meters at depth. Temperature is regularly measured to millidegree precision with an accuracy of $\pm 0.002°C$, and salinity is measured to a precision of 0.001 permil, with typical accuracies on the order of $\pm 0.002\%$. These accuracies are capable of revealing local and regional changes of water-mass properties over time even at depth, where the magnitude of the variability is usually $< 0.01°C$ and 0.02 permil (see, for example, Levitus et al. (1995)). Besides obtaining the necessary accuracy of measurements, it is essential to establish time series of velocity, temperature, and salinity. These goals are being accomplished with the use of new technologies and improvements to existing instruments: continuation, expansion, and extension of XBT collection to high-resolution, deeper sampling; implementation of global arrays of surface drifters and subsurface floats; acoustic tomography; and satellite measurements. In addition, expendable CTDs (XCTDs) are becoming a viable though still expensive means of increasing coverage of salinity as well as temperature observations in thermocline waters.

Both surface drifters and subsurface floats have been in use for decades as a means of measuring absolute water velocities, which cannot be determined from hydrographic data alone. However, recent design improvements have increased their usefulness, as well as their lifetimes. Surface drifters, for example, are now drogued properly and designed for minimal windage to sample surface velocities accurately. Subsurface floats can be programmed to follow an isopycnal surface rather than a constant-pressure surface, and to change their depth periodically to provide vertical sampling; they can also be equipped with temperature and conductivity sensors for sampling hydrographic properties. Floats either transmit their data to acoustic transceivers moored on the ocean bottom, which must then be retrieved, or they surface periodically to telemeter their position and other stored data to a satellite, which transmits the information to a shore-based lab, allowing near-real-time data analysis. Floats can live up to four years, are easily deployed, and are generally considered "expendable."

Acoustic tomography takes advantage of the changing speed of sound in seawater, due to changes in density. At mid-depths a "channeling" effect allows transmitted acoustic signals to travel thousands of kilometers with little attenuation. (Note the summary of Dr. Munk's speech in this section.) Moreover, since density is a strong function of temperature, the measured travel time of an acoustic signal between two transceivers is related to the heat content of the water between them, suggesting the use of large arrays of acoustic transceivers as a potential tool for monitoring long-term changes of heat content at transoceanic scales.

Another technological advance of the past two decades is the development of remote sensing capabilities, that is, observations of the sea surface from instruments mounted on satellites in orbit around the earth. Different frequency bands are exploited to image different aspects of the ocean's surface. For example, infrared (IR) imagery can be used to deduce and map sea surface temperature, but because its ability is limited by the extent of cloud cover over the ocean, it is a more useful tool in subtropical and tropical latitudes than near the poles. On the other hand, microwaves penetrate the cloud cover, and several satellite-borne instruments are based on this frequency band. Radar altimeters can be used to determine the absolute distance between satellite and sea surface; scatterometers yield information on wind speed and direction over the sea surface; and synthetic aperture radar (SAR) can be used to map or image a wide variety of dynamical features at the sea surface and in the upper ocean. Clearly, satellites offer the potential of global coverage in space and more or less continuous temporal coverage of the ocean's surface. However, for future monitoring capability, it is essential that consistency be maintained: Sequential satellite missions must provide continuity in time, and they must sample in overlapping frequency bands —something past measurements have not.

The past several decades also have led to the almost routine sampling of a host of other physical and chemical properties, including concentrations of helium and tritium, halocarbons ("freons"), and radiocarbon, which exist in trace quantities in the ocean. Some occur naturally, and some have been anthropogenically produced; in some cases the anthropogenically induced signal overwhelms an existing natural signal. All of these quantities act as "tracers" of water masses in the Wüstian sense: Once a water parcel has acquired its characteristic value of a tracer from contact with the atmosphere, it retains that value as it sinks and participates in the ocean circulation. Wüst's core method for tracing deep-water flow is thus appropriate, with one fundamentally important difference: Unlike temperature and salinity, these trace substances carry time information. Bomb tests of the early 1960s and the ever-growing use of halocarbons in industry since the 1930s are among the sources for these tracers. It is fairly well known at what rate over time they have been injected into the atmosphere, and/or at what rate they decay or are destroyed. Schlosser and Smethie (1995) describe the nature and measurement of these "transient tracers." (Because of the particularly sparse nature of the observations in space and time, they emphasize the need to apply a model for interpreting the data most of the time.) They demonstrate the utility of tracers for studying decadal-scale variability by presenting two specific examples, and suggest that transient tracers, with their unique time-history information, be employed as part of an ongoing climate monitoring system.

TOWARD THE FUTURE

We noted earlier that deep convection occurring in the marginal seas surrounding the North Atlantic provides the sources for waters found in the North Atlantic deep western boundary current. Besides the tracer-based studies presented by Schlosser and Smethie, evidence for interdecadal variability has been documented in temperature and salinity records for other areas of the North Atlantic. Examples are presented by Swift, Lazier, and Dickson in this section. Swift (1995) discusses the freshening in recent decades of both deep and upper waters of the northern North Atlantic, and speculates that it is related to long-term shifts in the wind-driven ocean transport. Lazier (1995) argues that although in a broad sense LSW is characterized by a relative salinity minimum coincident with a relative stratification minimum at depths of 1000 to 2000 m in the ocean, its properties cannot be tracked properly over time by plotting temperature or salinity in the traditional way, on a constant-density surface, since LSW is not necessarily formed at a constant density year after year.

The work by Dickson (1995), who describes interdecadal variability of physical exchanges and transfers in the Irminger Sea and through the Denmark Strait, is a good example of the impact of geographically isolated variability on local, regional, and global scales. Locally, the impact is socio-economic, affecting nearby cod fisheries. Regionally, the variability is tied in with the Great Salinity Anomaly (Dickson et al., 1988) through anomalous ice and fresh-water production and export from the Arctic. Finally, the deep water formed in the Irminger Sea contributes to the total North Atlantic Deep Water production, which in turn is part of the global thermohaline circulation.

It should be pointed out that interdecadal variability is not limited to marginal seas and boundary currents, although such examples are prominent. Levitus et al. (1995) document interdecadal changes in the temperature and salinity fields in the interior of both the subpolar and subtropical

gyres of the North Atlantic, by applying appropriate averaging techniques to the vast but irregular (in space, time, and quality) historical data base of the North Atlantic.

In addition to developing our ability to observe interdecadal changes in pivotal areas likely to be associated with more widespread climate change, we would of course like to be able to predict future changes. This will require continued effort in coupled ocean-atmosphere model development, using historical data for testing and validation purposes, as well as acquisition of real-time data as input for prediction of future climate states. The oceanic community recognizes these issues and has begun to address them in recent years with several programs of broad scope. Among these is the Tropical Ocean and Global Atmosphere (TOGA) Program which was initiated to increase understanding of ENSO events, but has contributed as well to our data base in the Pacific, especially the equatorial Pacific. Recently, TOGA has successfully made the transition from a scientific investigation to an operational monitoring system, maintaining an extensive upper-ocean network in all three oceans, with greatest concentration in the tropical Pacific. Though geographically limited, it might be regarded as a prototypical model for more extensive future systems. At high latitudes, increased effort is now being applied to understanding the complex interactions of the atmosphere/ocean/sea-ice system, as we have come to realize its significant role in determining the thermohaline circulation. This effort includes both more observational work than historically has been possible, and intensive modeling studies by a variety of individuals.

Two programs under way that are more attuned to longer periods of variability are the Atlantic Climate Change Program (ACCP) and the World Ocean Circulation Experiment (WOCE). The first of these seeks to determine the nature of interactions between the meridional circulation of the Atlantic Ocean, sea surface temperature and salinity, and the global atmosphere. Attaining this goal, it is noted, will require documentation of the general characteristics of decadal/century modes of Atlantic variability for model validation. WOCE, which is internationally coordinated and funded, has as its primary scientific objective "to understand the general circulation of the global ocean well enough to be able to model its present state and predict its evolution in relation to long-term changes in the atmosphere" (U.S. WOCE Office, 1989). WOCE includes both observational and modeling components, and also addresses data management issues. The Global Ocean Observing System (GOOS) comprises the operational extensions of programs such as GOOS and ACCP; its design will rely to a large extent on the scientific background provided by the research experiments. This international project, with its large amount of support, will be the vehicle for collecting much long-term data useful for modeling climate prediction. Numerous other programs that are under way or in the planning stages seek to understand the myriad other processes contributing to climate.

The collection of papers presented in this section demonstrates that the tools exist to observe decade-to-century time scales of variability in the ocean, although there are clearly gaps in our understanding of underlying physical processes.

The remaining challenge is to determine how to distribute necessarily limited resources among these different tools, in order to create a viable operational observing network that can be maintained well into the future. This challenge calls for close cooperation between observationalists and modelers, oceanographers and atmospheric scientists, and the academic and political communities.

Marine Surface Data for Analysis of Climatic Fluctuations on Interannual-to-Century Time Scales

DAVID E. PARKER, CHRIS K. FOLLAND, ALISON C. BEVAN, M. NEIL WARD, MICHAEL JACKSON, AND KATHY MASKELL[1]

ABSTRACT

The sea surface temperature (SST) data base of the Bottomley et al. (1990) Global Ocean Surface Temperature Atlas has recently been augmented with COADS data, better corrections for uninsulated and semi-insulated buckets have been applied, and other improvements have been made. A spatially and temporally interpolated version of the data set has been blended with historical sea-ice data and the 1951 to 1980 Bottomley et al. climatology to give monthly "globally complete" fields since 1871. One version of the data set includes satellite SST data from 1982 onward. To help explore more accurately the relationships between atmospheric circulation, surface climatic parameters, and SST, refined adjustments to marine wind data to compensate for progressive changes in observing practices have been derived using pressure data back to 1949. The improved winds, along with other atmospheric data, can also be used in the verification of numerical model simulations of the atmosphere forced with the new SST and sea-ice data set.

INTRODUCTION

The improvement of the data base of marine meteorological observations is a crucial prerequisite for most studies of climatic variation. Although "frozen-grid" experiments (e.g., Bottomley et al., 1990; Folland et al., 1990; Folland and Parker, 1992) suggest that multidecadal hemispheric and global mean sea surface temperature (SST) anomalies have been estimated with some reliability since at least the late nineteenth century, confidence in these estimates will be improved by any increase in coverage of, and improved analysis of, the data. Better analyses on shorter time scales are certainly needed. Moreover, regional or ocean-basin-scale studies will greatly benefit from improvements to the data. Most important, simulations of recent climate using numerical models require globally complete and, as far as possible, unbiased SST fields as input and, in addition, require reliable and reasonably complete coverage of mean-sea-level pressure and surface winds for use in verification. Satisfactory data bases of these latter parameters are currently lacking, except for short periods of modern surface-pressure data.

---

[1]Hadley Centre for Climate Prediction and Research, Meteorological Office, Bracknell, Berkshire, England, U.K.

In this paper we present a new, "globally complete" monthly analysis of SST and sea ice, known as the Global Sea Ice and Sea Surface Temperature (GISST) data set. GISST 1.0 and 1.1 have already been created; plans for future versions are indicated. We also describe recent improvements in the analysis of trends in marine surface winds.

OUTLINE OF CREATION OF GISST 1.0

GISST 1.0 is a monthly data set that extends from January 1871 to December 1990. We created it in the following stages.

1. The monthly 5° latitude × longitude Meteorological Office Historical Sea Surface Temperature Data Set (MOHSST4, Bottomley et al., 1990) was augmented with 2° latitude × longitude data from the Comprehensive Ocean-Atmosphere Data Set (COADS, Woodruff et al., 1987) after these had been averaged into 5° boxes and subjected to rudimentary extreme-value quality control. The resulting data set is known as MOHSST5. The results were converted to anomalies from the Bottomley et al. (1990) 1951 to 1980 climatology.

2. Improved corrections to compensate for the use of uninsulated and semi-insulated buckets (Folland, 1991; Folland and Parker, 1995) were applied to the data up to 1941.

3. Missing and extreme monthly 5° latitude × longitude area SST anomalies were replaced by the mean of four or more spatially adjacent anomalies if available, or, in their absence, by the mean anomaly of the two adjacent months at the same location if available.

4. The coverage was further enhanced by replacing missing values with weighted SST anomalies from up to 5 months either side of the target month. Weights decreased with elapsed time before or after the target month.

5. Using the Bottomley et al. (1990) globally complete high-resolution 1951 to 1980 climatology, the fields of 5° latitude × longitude SST anomalies output by step (4) were converted to 1° latitude × longitude SST values.

6. Sea-ice extent information from a wide variety of sources was added.

7. SSTs were assigned in a special way to data-void 1° latitude × longitude areas adjacent to ice edges and to any data-void open-water areas that were climatologically ice covered.

8. SSTs were extended into the remaining missing areas using the Laplacian of the 1951 to 1980 climatology (Bottomley et al., 1990; Reynolds, 1988).

9. The resulting 1° latitude × longitude SST analysis was smoothed to retain anomaly variations with about 5° resolution.

TECHNIQUES AND QUALITY CONTROLS USED FOR GISST

The heading numbers below refer to the step numbers in the section above.

1. *MOHSST5*

a. Any values less than −1.8°C were set to −1.8°C. Values for the Caspian Sea were omitted because they, and the Bottomley et al. (1990) climatology there, appear to be unreliable for unknown reasons.

b. The addition of the COADS data resulted in an improvement in seasonal coverage, relative to MOHSST4, approaching 20 percent of the global ocean between the 1870s and World War I, with large improvements in the eastern Pacific. See Figure C3(a) of Folland et al. (1992).

2. *Bucket Corrections*

The thermodynamic theory is given by Folland (1991) and Folland and Parker (1995). The semi-empirical technique used for derivation of the corrections is outlined by Bottomley et al. (1990) and is presented in full by Folland and Parker (1995), whose major differences from Bottomley et al. (1990) include a more rigorous formulation of the heat exchanges affecting wooden buckets and revised estimates of the historical variations of the types of buckets used. In particular, newly uncovered evidence led Folland and Parker to assume, despite considerable uncertainty, that 80 percent of buckets were wooden in 1856 with a linear transition to all-canvas or other uninsulated types in 1920. This brought their estimate of this factor into better agreement with that of Jones et al. (1991), although their estimates of the actual corrections for wooden buckets did not agree because they made different assumptions about the heat transfers involved. The bucket corrections used in the present paper follow Folland et al. (1992) in assuming 100 percent wooden buckets in 1856 and a slightly different specification for these buckets from that used by Folland and Parker (1995) but the resulting corrections generally only differ by a few hundredths of a degree Celsius.

3. *Filling and Quality Control*

a. Any anomalies exceeding 7°C in magnitude were recorded as missing. This slack criterion allowed anomalies in major El Niño events to be accepted. In future versions of GISST, a geographically varying threshold is to be used.

b. For each 5° latitude × longitude box, the average anomaly for the eight surrounding 5° latitude × longitude boxes was calculated, provided at least four had data. This average was then substituted in the box if the existing anomaly was missing or differed from it by more than 2.25°C. This criterion was chosen empirically following careful tests on monthly 5° latitude × longitude fields taken from a range of years since 1860 and covering the entire annual cycle (Colman, 1992).

c. Next, for each box, the average anomaly for the previous and the subsequent months was calculated, if both were available. This average was then substituted in the box if the existing anomaly was still missing or differed from it by more than 2.25°C.

d. Processes b and c were carried out three times altogether. The substitution of missing data greatly augmented the global coverage in data-sparse years while maintaining spatial coherence (compare Figures 1a and 1b in the color well). The effects in recent years were greatest along the boundaries between well-sampled areas and major data voids, e.g., in the Southern Ocean.

4. *Further Enhancement*

a. Where data were still missing for a 5° latitude × longitude box, a search was made up to 5 months backward and forward to find the nearest anomalies. If both anomalies were available for months -1 and $+1$, their average was substituted for the missing value. Otherwise, any available anomaly a_n observed n months before (n negative) or after (n positive) the target month was multiplied by a reduction factor $0.6^{|n|}$. The search was continued with increasing $|n|$ until the sum of the reduction factors used ($\Sigma \delta_n 0.6^{|n|}$ where $\delta_n = 0$ for missing data, 1 for available data) reached 0.6; note that both anomalies were used when available from equidistant months. The average of the reduced or "muted" anomalies ($p^{-1}\Sigma a_n \delta_n 0.6^{|n|}$), where p is the number of anomalies used, was substituted for the missing value. The empirically chosen reduction factors are consistent with the global annual average of the monthly lag correlations presented in Bottomley et al. (1990), but no geographical or seasonal variation has been allowed.

b. A further spatial quality control was carried out. This was designed to reduce any grid-scale incoherence introduced by (a) above, especially where anomalies were rapidly changing in time or were much larger than the newly introduced muted anomalies. The procedure corresponded to item (b) in the section on Filling and Quality Control, but as few as two neighboring anomalies were used, and no missing boxes were substituted. If only a single neighboring anomaly was available, the mean of it and the anomaly being checked was used in the same way. Isolated 5° anomalies exceeding ± 2.25°C were reduced to ± 2.25°C.

The step-by-step effects of the "filling" and enhancement stages on sparse data can be seen by comparing Figures 1a, 1b, and 1c for January 1878. For recent years with far more data, the effects were much smaller.

5. *Conversion from 5° to 1° Resolution and to Absolute SST Values*

This step was an essential preparation for the incorporation of sea-ice fields as well as for the Laplacian interpolation (see below), in which it was necessary to preserve climatological gradients of SST. The 5° resolution monthly anomalies output after the "further enhancement" described in the previous subsection were added to the Bottomley et al. (1990) globally complete 1° resolution monthly climatological SST for 1951 to 1980. This climatology was assigned to 1° boxes in 5° areas without anomalies.

6. *Sea Ice*

The sources of sea-ice data are listed in Table 1. The NOAA analyses from 1973 onward are largely satellite based (Ropelewski, 1990). Note that published manuscript climatologies were used for earlier times, so that the same calendar-monthly ice cover was used in successive years, as opposed to the use of observed, interannually changing ice cover for more recent times, i.e., 1953 onward for the Arctic and

TABLE 1   Sources of Sea-ice Data

a)	ARCTIC	
	Up to 1943	German 1919-1943 climatology (Deutsches Hydrographisches Institut, 1950)
	1944-1952	Interpolation to recent climatology (1953-1982)
	1953-1972	Observed data provided by J. Walsh (Walsh, 1978)
	1973 onward	Observed data provided by NOAA (Walsh, 1991)
b)	ANTARCTIC	
	Up to 1939	German 1929-1939 climatology (Deutsches Hydrographisches Institut, 1950)
	1940-1946	Interpolation to Russian 1947-1962 climatology
	1947-1962	Russian 1947-1962 climatology (Tolstikov, 1966)
	1963-1972	Interpolation to recent climatology (1973-1982)
	1973 onward	Observed data provided by NOAA (Walsh, 1991)

1973 onward for the Antarctic. We fully recognize the very uncertain nature of the earlier climatologies but, in the absence of evidence to the contrary, consider that they provide a better estimate than modern data. The data were interpolated to 1° latitude × longitude resolution. Each oceanic box was designated either "ice" or "water."

7. *Assignment of SSTs near Sea Ice*

We chose to produce a globally complete temperature field by Laplacian interpolation, preserving the second derivative of the climatology. To do this, we needed to specify conditions along an external boundary completely enclosing the region to be filled (Figures 2a-f). Reynolds and Marsico (1993) used the observed ice limits as their boundary, setting ice points to −1.8°C. However, this method rarely yields the expected positive anomalies when ice cover is less extensive than its climatological normal. This is because −1.8°C is also assigned to ice in the climatology, so that the ice edge is given a zero anomaly of SST. Then, since the Laplacian of the SST sea-ice climatology is completely preserved, the computed SST between the retreated ice edge and the climatological ice edge deviates from −1.8°C only to the extent that the nearest observed SST implies an anomalous gradient of SST. An observed negative SST anomaly thus even yields SSTs below −1.8°C in the anomalous open-water area (dotted line in Figure 2e). These cold biases also influence the SST analysis between the climatological ice edge and the nearest observations of SST; see also section C3.1.2.2 of Folland et al. (1992). Furthermore, if ice is anomalously advanced toward warmer water, we could expect local SST gradients to be enhanced; thus, an assumption of −1.8°C at the ice edge and approximate climatological SST gradients seaward, as implied by the Laplacian interpolation, may also yield negative temperature biases (dotted line in Figure 2f). This is particularly to be expected when there are strong gradients very close to the ice edge at or below the 1° latitude × longitude resolution of our analysis, a documented occurrence (Muench et al., 1985).

We therefore used specially computed values of SST near ice edges in data-void regions. In a particular year and month any of the following four types of situations might occur (Figure 2a):

a. Ice areas that are also climatologically ice covered. These areas were set to −1.8°C.

b. Ice areas that are climatologically open water. These areas were also set to −1.8°C.

c. Open water areas that are climatologically ice covered. Anomalous water areas of 1° were set to the mean of the Bottomley et al. (1990) climatological

SSTs for the relevant calendar month for all 1° latitude × longitude areas adjacent to the climatological ice edge in a 19° latitude × longitude area centered on the target area. The representative temperature for target box A in Figure 2b, for example, would be set to the mean of the climatological SSTs in those boxes marked with a cross.

d. Boundary areas. Boundary areas were defined as 1° open-water areas adjacent to (i.e., sharing a side with) any area in the above three categories. Ice areas and anomalous water areas were thus separated from the open sea by a boundary area 1° latitude or longitude wide. (See Figure 2b.) The SST in a boundary area was calculated as the climatological value plus one-half the temperature anomaly in an adjacent anomalous ice or anomalous water square. (See Figures 2c and 2d.) If more than one adjacent area had anomalous conditions, preference was given to the areas lying to the north and south, e.g., case C in Figure 2d. If none of the adjacent areas had anomalous

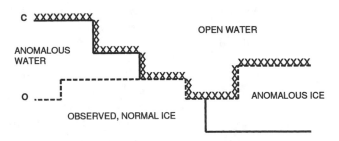

FIGURE 2a   Calculation of a representative temperature for anomalous water areas.

KEY: C   Climatological ice edge
       O   Observed ice edge
       X   Boundary areas (see text).

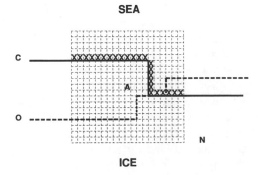

FIGURE 2b   Designation of areas near ice edge.

KEY: C   Climatological ice edge
       O   Observed ice edge
       A   Target area
       N   19° latitude x longitude area centered on target area
       X   Climatological SSTs for these boxes are used to calculate representative temperature for area A.

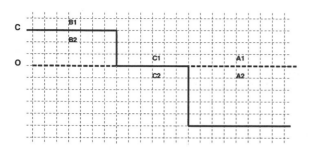

FIGURE 2c   Calculation of boundary temperatures.

KEY: C   Climatological ice edge
     O   Observed ice edge
     Al etc. 1° boxes: see footnotes

FOOTNOTES

A.   If T_{A2} = Temperature in 1° box A2 = $-1.8°C$ for observed ice, and N_{A1}, N_{A2} = Normals in 1° boxes A1, A2, then T_{A1} = Temperature in 1° box A1 = $N_{A1} + 0.5 (T_{A2} - N_{A2})$ = $N_{A1} + 0.5 (-1.8 - N_{A2})$.

B. T_{B2} = Representative temperature assigned to 1° box B2 (see text and Figure 2b)

$T_{B1} = N_{B1} + 0.5 (T_{B2} - N_{B2})$
$= N_{B1} + 0.5 (T_{B2} + 1.8)$ because $N_{B2} = -1.8°C$ for climatological ice.

C. $T_{C1} = N_{C1}$ because the ice edge is at its normal position.

-0.8	0.0	0.0	0.5	1.0	1.0
-1.5	-1.3	-1.2	-0.6	0.0	0.0
-1.8	-1.8	-1.8	-1.3	-0.7	0.0
-1.8	-1.8	-1.8	-1.4	-0.8	-1.0
-1.8	-1.8	-1.8	-1.8	-1.8	-1.8

1° BOX NORMALS (°C)

L	L	L	L	L	L
B	B	B	L	L	L
W	W	W	B	B	B
I	I	I	I	I	I
I	I	I	I	I	I

DESIGNATION OF AREAS

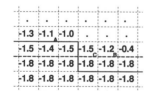

-1.3	-1.1	-1.0			
-1.5	-1.4	-1.5	-1.5	-1.2	-0.4
-1.8	-1.8	-1.8	-1.8	-1.8	-1.8
-1.8	-1.8	-1.8	-1.8	-1.8	-1.8

ASSIGNED TEMPERATURES (°C)

FIGURE 2d   Sample calculations of SSTs near ice edges.

KEY TO AREA DESIGNATIONS:

C   Climatological ice edge
O   Observed ice edge
I   Observed ice areas set to $-1.8°C$
W   Anomalous open-water areas: SSTs calculated as in text
B   Boundary SSTs: see text and Figure 2c
L   Areas left blank for filling in the Laplacian stage.

FOOTNOTES

A. $-1.3 + 0.5 (-1.4 - (-1.8)) = -1.1°C$
B. $-0.7 + 0.5 (-1.8 - (-0.8)) = -1.2°C$
C. $-1.3 + 0.5 (-1.8 - (-1.4)) = -1.5°C$

using ice anomaly to southward, not to westward

conditions, the climatological value alone was used. If an observed SST was found in the boundary area, this was used in preference to the computed value. The SSTs in the boundary areas were then used as the external boundary conditions in the Laplacian interpolation process (Figures 2e and 2f). Reducing the ice-edge anomalies by a half had ensured that the values supplied to the Laplacian process were not extreme and had strengthened the temperature gradient at the edge of anomalously frozen areas (Figure 2f).

8. *Laplacian Interpolation*

a. The 1° resolution SSTs were first smoothed 1:2:1 east-west, then north-south, simplifying the smoothing to 1:1 for boxes next to land. Anomalous water areas were omitted from this smoothing, as were boundary 1° areas (Figure 2a) next to sea ice or to missing data. In this process 1° boxes adjacent to boxes still containing climatology would have been affected by the climatology, muting their anomalies. They were therefore regarded as data void, along with the boxes containing pure climatology.

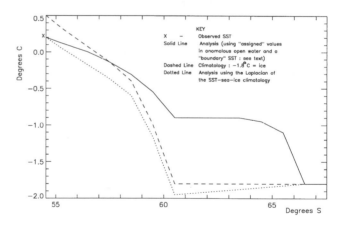

FIGURE 2e   Example of SST analysis near retreated ice edge.

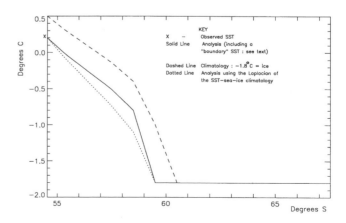

FIGURE 2f   Example of SST analysis near advanced ice edge.

b. Laplacians of the 1951 to 1980 MOHSST4 climatology (Bottomley et al., 1990) were calculated for data-void 1° boxes.

c. SSTs for the data-void 1° boxes were then computed by solving Poisson's equation with the Laplacians as forcing and the values in data-filled boxes as boundary conditions, as has been done by Reynolds (1988) and Bottomley et al. (1990) in constructing their climatology.

9. *Final Smoothing*

The 1° resolution absolute SST data output by the Laplacian interpolation were converted to anomalies, and the anomalies were smoothed 1:2:4:6:4:2:1 east-west, then north-south. Anomalies rather than actuals were smoothed, to avoid smoothing strong SST gradients. After smoothing, the anomalies were converted back to SSTs. SST gradients were thus retained with approximately 1° resolution, whereas the anomalies, which vary much more slowly geographically, were retained on about their original 5° resolution.

Examples of the final result of this process, in the form of anomalies for January 1878 and January 1983, are shown in Figures 1d and 3b (see color well). These fields (and others not shown) suggest that in future versions of GISST, the anomalies should be smoothed with a spatially lower-pass filter. Even in the 1980s, monthly (as opposed to seasonal) in situ SST anomalies are often unreliable on a 5° space scale (Folland et al., 1993).

INITIAL ASSESSMENT OF GISST 1.0

A sampling experiment was carried out to assess the reliability of the analysis technique when used on sparse data. In it, the complete analysis, including the Laplacian stage, was repeated for each month of the El Niño years 1982 to 1983, but with the basic SST data prior to the "filling and quality control" stage omitted from 5° boxes where there were no data in the corresponding months of the El Niño years 1877 to 1878. In this experiment, January 1877 corresponds to January 1982, and December 1878 corresponds to December 1983. Also, data for 1981 and 1984 needed for this experiment were reduced to the coverages available in the corresponding months of 1876 and 1879. Figures 4a, b, c, and d, which should be compared with Figures 3a and b, illustrate the results of the experiment for January 1983. Figure 5a shows the two-year mean difference between the reduced-data analysis and GISST 1.0. (All are in the color well.)

Biases are largest in the eastern tropical Pacific, where the analysis of the reduced data base underestimated the strength of the 1982 to 1983 warm El Niño event. This underestimation exceeded 1°C in some places at the peak of the event in late 1982 to early 1983 (compare Figures 3b and 4d). It is likely therefore that GISST 1.0 itself will have underestimated the strength of the 1877 to 1878 El Niño, owing to the sparser data base then. Figure 5b presents the root-mean-square (rms) difference field between the GISST 1.0 and the reduced-data analyses; it is based on 24 monthly values at each location. The rms differences (which include a bias component) are less than 0.5°C over most of the Atlantic and Indian Ocean north of 40°S, but exceed 0.5°C over much of the Pacific and the Southern Ocean, and are over 1°C in the eastern tropical Pacific, where the reduced-data analyses underestimate the El Niño warmth. Cosine-latitude-weighted summary statistics for the globe and for the east tropical Pacific (20°N to 20°S, east of 170°W) are given in Table 2. The global rms difference (which also includes a bias component) is around 0.5°C before and after the El Niño but reaches 0.7°C in early 1983, when the underestimation in the east tropical Pacific averages nearly 0.7°C (compare also Figures 3b and 4d). However, the global field correlations rise from about 0.55 before the event to about 0.7 during the event and maintain this level, possibly because the El Niño intensifies the overall global anomaly pattern signal. During the El Niño the global bias reaches -0.15°C. For the two-year period as a whole, however, the global bias is only -0.03°C, suggesting, in accord with the "frozen grid" analyses in Bottomley et al. (1990) and Folland et al. (1990), that the sparse coverage available in 1877-1878 does not severely prejudice estimation of annual or multi-annual global mean SST anomalies. Similar, but apparently slightly better, results (not shown) were obtained when a corresponding experiment was made using the El Niño years 1972-1973 with the 1877-1878 coverage. The apparent improvement may have resulted from the slightly reduced coverage of observations in 1972-1973 relative to 1982-1983.

In a further, longer experiment the analysis for 1981 to 1990 was repeated using coverage for 1881 to 1890. Biases (not shown) were within ± 0.1°C in most of the Atlantic and the Indian Ocean, and generally exceeded ± 0.3°C only in the Southern Ocean and parts of the North Pacific. Hemispheric and global average biases were within ± 0.01°C. Root-mean-square differences generally exceeded 0.5°C only in the latter areas and in parts of the tropical Pacific and the far northern Atlantic. Correlations (Figure 6, again in the color well) based on 120 values at each location, exceeded 0.8 over most of the Atlantic, much of the eastern Pacific, and much of the Indian Ocean. Low correlations over the Southern Ocean and the central and northern Pacific emphasize the need to acquire more historical data for these regions.

These sampling experiments only assess the impact of reduced areal coverage of 5° box SST data. They do not provide any measure of the effects of increased scatter in the individual monthly 5° box values resulting from, for

TABLE 2   Summary Statistics for the Reduced-Sampling Test of GISST 1.0 Analysis

| | | Globe | | | Tropical East Pacific | |
| | | rms Difference (°C) | Field Correlation | Bias* (°C) | rms Difference (°C) | Bias* (°C) |
Year	Month					
1982	1	.51	.57	− .02	.55	− .17
1982	2	.52	.57	+ .01	.55	+ .03
1982	3	.55	.55	− .05	.62	− .12
1982	4	.51	.54	− .08	.52	− .17
1982	5	.51	.57	− .06	.57	− .26
1982	6	.51	.64	+ .00	.65	− .36
1982	7	.53	.63	+ .10	.49	− .13
1982	8	.54	.56	+ .06	.60	− .19
1982	9	.55	.67	+ .06	.73	− .16
1982	10	.54	.74	+ .01	.76	− .35
1982	11	.60	.70	− .06	1.09	− .50
1983	12	.66	.67	− .12	1.24	− .59
1983	1	.70	.67	− .14	1.19	− .63
1983	2	.69	.65	− .15	1.11	− .68
1983	3	.67	.65	− .09	.94	− .54
1983	4	.58	.71	− .03	.84	− .21
1983	5	.60	.71	+ .00	.77	− .16
1983	6	.56	.78	− .09	.82	− .36
1983	7	.57	.78	− .07	.88	− .36
1983	8	.52	.76	− .05	.79	− .22
1983	9	.46	.72	− .03	.53	− .12
1983	10	.49	.62	− .03	.58	− .23
1983	11	.43	.72	+ .01	.54	.06
1983	12	.42	.74	+ .02	.49	− .08
WHOLE PERIOD				− .03		− .28

*Bias is "reduced-sampling analysis" minus "GISST"

example, a smaller number of constituent observations. Also, the reduced detail of the earlier sea-ice data is ignored. To take advantage of satellite data, which are expected to improve the analyses from the 1980s onward especially in the Southern Ocean, the experiments will be repeated using GISST 1.1.

The sampling experiments on GISST 1.0 do suggest that the interpolation techniques have not introduced systematic long-term biases or false trends into the analysis. There may, however, be some systematic uncertainties owing to shortcomings in the bucket corrections resulting from the assumptions made about measurement techniques etc. (Folland and Parker, 1995). On a global average, the uncertainty arising from these is likely to be of the order of ± 0.1°C, especially before 1900 (Folland et al., 1992); but in regions where the corrections themselves are large, e.g., the Gulf Stream and Kuroshio in winter, an uncertainty of at least ± 0.25°C may be expected.

A 1951-1980 monthly climatology has been created from GISST 1.0. MOHSST5 is more consistent with this than with the Bottomley et al. (1990) climatology. Relative to the latter, there were persistent anomalies of one sign in

MOHSST5 in substantial parts of the Southern Ocean. These persistent anomalies were reduced or eliminated by referencing MOHSST5 to the GISST climatology. We regard the GISST climatology as the more reliable because the additional data input to GISST will have reduced the influence of the highly interpolated Alexander and Mobley (1976) climatology used by Bottomley et al. (1990) in data-sparse areas.

REVISIONS TO GISST

For GISST 1.1 we have incorporated satellite-based SSTs from R. W. Reynolds (NOAA) from 1982 onward. Because of their extensive coverage, they replaced the filling and enhancement processes, and the assignment stage except for 1° boxes adjacent to observed sea ice and Antarctica. They also almost entirely replaced the Bottomley et al. (1990) climatology in the Laplacian stage, resulting in an improved analysis. The Laplacian of the climatology was first used to make a complete field from the satellite data; then the Laplacian of this field was used to make the in situ data field complete. As expected, GISST 1.1 is warmer in the Southern Ocean than the combined in situ satellite analysis by Reynolds and Marsico (1993), because of our different assignment of SSTs near sea ice. Typical differences are of the order of 0.5°C (see Figure 7 in the color well).

The historical in situ SST data base is to be expanded by incorporating millions of hitherto undigitized observations from the United States, Japan, and possibly the United Kingdom, Russia, and Norway. These data will also be incorporated into COADS, and an optimum data bank will be created by blending COADS and the U.K. Meteorological Office marine data base observation by observation. (See Parker (1992) as well as Komura and Uwai (1992), whose report suggests that there are about 4 million undigitized Japanese marine observations for the period 1890 to 1932.)

In collaboration with J. Walsh (University of Illinois), additional historical sea-ice data have been located and are to be digitized and combined with the Kelly (1979) sea-ice data. It is hoped that these data will be incorporated into later versions of GISST, replacing some of the existing sea-ice data.

Possible major changes to the analysis technique include:

1. Use of characteristic anomaly patterns in the background field for the Laplacian stage. The result may be an improved analysis of El Niño events in data-sparse years.

2. Refinement of the interpolation processes. The use of geographically and seasonally varying lag correlations and of optimum averaging (Gandin, 1963) is being considered.

MARINE WINDS

We are undertaking an extensive study of the dynamical consistency between sea level pressure (SLP) and near-surface wind observations in the COADS dataset. The chief motivation is the suggestion that changes in observational practice on board ships may have introduced a spurious upward trend in reported ship winds (e.g., Ramage, 1987; Cardone et al., 1990). Ward (1992) devised a method for deriving the seasonal mean geostrophic wind from COADS 2° latitude × longitude seasonal mean SLP data. A comparison of the trends in the observed winds and the derived geostrophic winds suggested a spurious upward trend in the magnitude of the observed wind of some 16 percent on average over the period 1949 to 1988. This work is being extended in two major ways:

1. Analysis is being performed on the 10° latitude × longitude (10° box) scale to improve spatial data coverage. The procedure works by first forming a 10° box anomaly data set in the way described in Ward (1992) and then adding the 10° box normal to the 10° box anomaly to get an actual value for the given 10° box.

2. A wind is derived from the 10° box grid of SLP values using a method that performs better in the tropics than does the geostrophic approximation in Ward (1992). The new method assumes a three-way balance of forces between pressure gradient, Coriolis force, and friction. The last is assumed to be linearly related to wind speed (through a constant, called the coefficient of surface resistance) and to directly oppose the wind direction. Such a system is often said to describe "balanced frictional flow," and the horizontal momentum equation can be written (see Gordon and Taylor, 1975):

where

$$-fv = P_x - ku$$
$$fu = P_y - kv$$

$$P_y = -\frac{1}{\rho}\frac{\partial p}{\partial y}, \quad P_x = -\frac{1}{\rho}\frac{\partial p}{\partial x} \qquad (1)$$

and p = surface pressure, ρ = density of air, f = Coriolis parameter, k = coefficient of surface resistance, u,v = zonal, meridional wind, and x,y = zonal, meridional distance.

Such a formulation leads to the following equations for the u and v wind components:

$$u = \frac{kP_x + fP_y}{k^2 + f^2}$$

$$v = \frac{kP_y - fP_x}{k^2 + f^2} \qquad (2)$$

Assumptions have to be made about the value of k. It can be shown that, under the assumptions of this method, k is related to the backing angle (α) of the observed wind from the geostrophic wind:

$$\tan \alpha = \frac{k}{f} \qquad (3)$$

For the studies described here, the value of k used is varied with latitude from about 2.5×10^{-5} at 50° latitude to 1.5×10^{-5} near the equator. These values are broadly consistent with the backing angle of the mean wind in directionally steady regions as calculated using climatological fields, and they are also consistent with the results of previous studies that have used other methods to estimate k (e.g., Brummer et al., 1974, Hastenrath, 1991).

To illustrate a typical result, we show time series of the December to February observed and derived u-wind component for the 10° box centered at 15°N, 45°W (see Figure 8a). The required pressure gradients are estimated as finite differences using the four boxes to the north, south, east, and west. Strictly speaking, the derived wind is applicable to a 20° latitude × 20° longitude region centered at 15°N, 45°W, so the observed zonal wind is calculated as a weighted average of the box centered at 15°N, 45°W (weight = 1.0) and the four surrounding boxes (each with weight = 0.75). The agreement between detrended versions

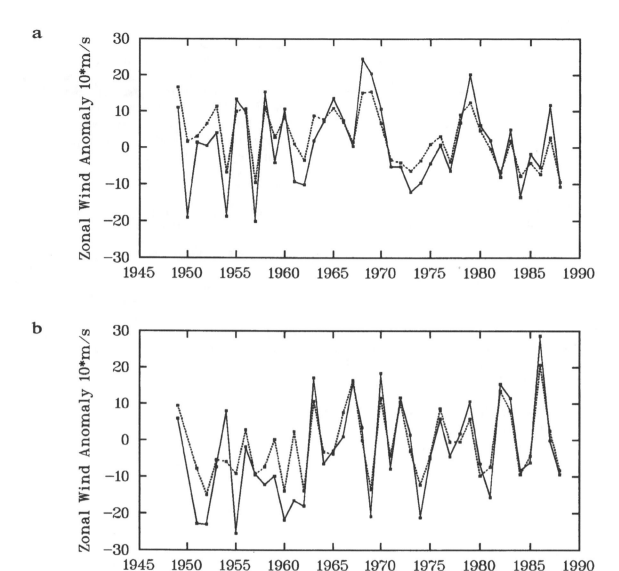

FIGURE 8   Time series of observed zonal wind anomaly (dotted line) and derived zonal wind anomaly (solid line) based on seasonal mean SLP (see Eq. 2 in text). (a) December to February for 10° latitude × 10° longitude box centered at 15°N, 45°W. (b) As for (a), but March to May centered at 55°N, 15°W. Some statistics for the series are given in Table 3.

of the derived and reported wind is excellent (Table 3, column a). This gives high confidence in the quality of the detrended component of the data and in the method used to derive the wind from the SLP. Good agreement in the detrended component of the derived and observed wind is generally found for most 10° boxes, except for some regions adjacent to the equator that show poor agreement. This is probably because Eq. 1 is less applicable close to the equator, and SLP needs to be resolved on a finer spatial scale.

The quality of the data on the longer time scale is questioned because of the difference in the trends of the two series in Figure 8a. The observed wind shows a stronger trend toward increasing wind strength than does the derived

wind (Table 3). Such a result is found in most of the 10° boxes analyzed in each of the four seasons. An average spurious upward trend in the magnitude of the observed wind of about 14 percent over 1949 to 1988 is suggested by the new derived wind series. This is very similar to the value of about 16 percent derived using the different analysis scheme of Ward (1992). However, the better data coverage of the new analysis scheme is leading to the identification of a somewhat different regional pattern of biases from that in Ward (1992), which is being studied further.

One region where Ward's (1992) trend biases were substantially different from the global average was in the North Atlantic near 55°N, where the average spurious wind speed

TABLE 3   Statistics for the Wind Time Series Shown in Figure 8

Wind Series	(a)	(b)
Observed Trend	−0.27	+0.19
Derived Trend	−0.03	+0.40
$r(O,D)$	+0.87	+0.89
$r(O_{dt}, D_{dt})$	+0.94	+0.88

NOTES:

Wind series (a) and (b) are those shown in Figure 8 (a) and (b), respectively.

Trends are in $10 \times$ m s^{-1} yr^{-1}. For series (a), mean zonal wind is negative (i.e., easterly), so negative trend implies strengthening wind. For series (b), mean zonal wind is positive.

$r(O,D)$ is the correlation between observed wind and derived wind.

$r(O_{dt}, D_{dt})$ is the correlation after removing the linear trend from the series.

change over 1949 to 1988 was suggested by the pressure data to be negative. Figure 8(b) illustrates this using the newly derived wind for March to May for the 10° box centered at 55°N, 15°W. The relationship between the observed and the derived winds seems to change around 1962. After about 1962, there is particularly good agreement between the observed and derived winds with no appreciable systematic difference in the two series. However, before about 1962, the year-to-year agreement is less good, and there is an appreciable bias; the derived wind is more easterly than the observed. This bias leads to differing linear trends in the observed and derived winds (Table 3); this difference in trends broadly supports the results of Ward (1992). Similar results are found for other nearby boxes in this and other seasons. The discrepancy before 1962 may be caused by biases in either the SLP or the wind data, or by the method used to derive the wind from the SLP data. These possibilities are to be investigated.

"CLIMATE OF THE TWENTIETH CENTURY" INTEGRATIONS

Atmospheric simulations are being commenced using the climate version of the U.K. Meteorological Office Unified Model forced with GISST 1.1. The simulations will require much-improved surface pressure and marine surface wind data sets for their verification. We anticipate substantial benefits from interactions between data analysis, modeling, and studies of the mechanisms of recent climatic variations.

ACKNOWLEDGMENTS

This work was partially supported by the Commission of the European Communities under Contract EPOC-0003-C(MB), and by the U.K. Department of the Environment under Contract PECD/7/12/37.

Commentary on the Paper of Parker et al.

THOMAS L. DELWORTH
NOAA/Geophysical Fluid Dynamics Laboratory

As someone who is mostly familiar with modeling and the efforts that go into it, I am very appreciative of the tremendous effort that goes into constructing observational data sets. The data sets are of critical importance in assessing the ability of models to simulate climate and its variability. I have a few comments and a couple of examples from modeling work to emphasize the importance of the work that Dr. Parker is doing, particularly with regard to the importance of sea ice for low-frequency climate variability.

Examination of low-frequency variability contained in a multi-century integration of a coupled ocean-atmosphere model reveals that there are very long time scales associated with a number of oceanic variables in the polar latitudes of the Southern Hemisphere. In particular, there is a tremendous amount of low-frequency variability in the ice thickness, which then has a substantial impact on the model atmospheric variability. Sea ice, in some regions, can virtually disappear for times as long as a decade, with the result that surface air temperatures are much higher during this period. When you look at interdecadal model variability, features such as this are quite striking. It is critical to increase our observational data base for quantities such as sea ice in order to be able to assess whether such model features exist in the real climate system. Some of the data sets produced by Dr. Parker should help with that.

A key point in his manuscript is the theme of attempting to obtain more information from the available observations through the use of empirically motivated assumptions about the spatial and temporal scales of variations in the observed data. It is therefore critical to keep in mind when utilizing this data the assumptions that go into it. Two key assumptions are the degree of persistence in time and the spatial structure of the data. In general, this represents a powerful technique for supplementing existing data sets. There is, however, a potential for this technique to underestimate variability in data-scarce regions. This potential bias must be kept in mind.

An additional point Dr. Parker made is the critical importance of recovering and digitizing existing observations from data-poor regions. Allow me to comment on the importance of this with an additional modeling example. In characterizing low-frequency variability, a simple technique is to compute at each grid point the serial correlation of data time series. This is a measure of the persistence in the data and thus the inherent time scales. The results of such computations using annual mean sea surface temperature computed in a coupled model shows some very intriguing features. The longest time scales of sea surface temperature anomalies appear to be associated with higher latitudes where deep-water formation is occurring in the model. There are also very long time scales in the circum-Antarctic region. These are features that we would like to investigate in the observations, but the limitations in observational data sets make that difficult. For example, computing serial correlations of observed annual mean sea surface temperature from MOHSST4 produces a map with large data-void regions. In particular, the very intriguing model features in variability occur in regions with insufficient observations to assess their validity. Therefore, the technique that Dr. Parker is trying to use to extract more information from the observations available is really a critical one.

The techniques of data analysis described in the manuscript check for physical consistency between independent data. In particular, comparisons were made between derived and observed wind fields, and inconsistencies between the two were noted and studied. This is an important technique, and is an appropriate method to assess what features are real and what features are not.

Finally, as Dr. Parker mentioned, these data sets are of great utility in conducting simulations of the climate of the twentieth century. It would be beneficial to have other variables available in such a format. For example, land-surface processes might have a role in climate variability. One can envision a similar data set of time-varying land-surface characteristics over the twentieth century.

Discussion

RASMUSSON: There's much to be said for working only with real data. I hope you will also make available fields that are real data only.

PARKER: Yes, MOHSST5 is available, and we'd like to improve it by digitizing all the data possible.

KUSHNIR: If you have a good approximation to the covariance matrix structure, you can use that with more confidence than Laplacian methods to fill up missing regions.

PARKER: That would be similar to the eigenvector technique, which I think would be a good one here. We just haven't got that far yet. Also, I think getting more real data is important; for example, our pre-1953 sea ice is all from an old climatology, so the sea-ice anomalies are the same from year to year.

TRENBERTH: The use of covariances in either space or time depends very much on the nature of your signal. For instance, when you put in greenhouse warming a global-scale structure is superimposed on your fields, and larger spatial and temporal correlations are implied than you might otherwise get.

PARKER: Indeed. What would really be nice would be to have designators categorizing the quality of the data for each of the months you're presenting—for instance, with the Air Force's cloud-cover analysis, designating which are real data and which climatology.

KARL: David, when you said you tried to make the frequency of observations consistent with that of the nineteenth century, did you actually reduce the number of observations in a grid box to match the earlier ones?

PARKER: No, we had to use the box values, so we were just testing for errors resulting from the interpolation scheme. Selecting observations would have been a major computational exercise. But we are aware of that problem.

KEELING: When you are adding the COADS data to your own data set, what do you do when you have both? Can you identify the data sources and tell which are original observations?

PARKER: For the moment we are inserting the COADS SST only if the box is missing from our data set. We hope that with Scott Woodruff's help we will be able to blend the two sets, removing duplicates, by 1994. Ultimately we'd like to have everything that can be digitized.

KEELING: How hard would it be, then, to average your data by anything other than month?

PARKER: In our scheme the data are in five-day periods, or pentads, so that you could do half-months or three-month periods. We also have data sets of the bucket corrections we used so that anyone who wants to remove them from the data as they stand can do so.

JONES: We have now digitized the spring and summer months of that Danish sea-ice chart series that runs from the turn of the century to about 1960. I believe John Walsh is working on the winter ones.

WALSH: We're aggregating the narrative reports into a data set for winter. But I'd like to add that no one has yet taken advantage of the spatial character of these reports the way David has with SSTs. There is also a set of reports going back to the turn of the century from an Antarctic ice station, and some sort of spatial extrapolation procedure used on both of these might give us some measure of interannual variation in certain limited regions.

GHIL: In addition to data origin identifiers and such, I would really like to have error bars with your data.

PARKER: Well, there's a first approximation already there—the root-mean-square error fields—though they might be a lower limit, considering Tom's comment about the number of observations per grid box.

GROISMAN: I wanted to mention the dangers of using the German and Russian climatology of sea ice for the 1920s to 1940s, since sea ice was retreating considerably during this period. I hope that some day you will be able to use some Russian arctic observation data now stored in the Arctic Institute—not digitized, of course.

LEVITUS: I'm happy to say that we at NOAA are making arrangements with the appropriate officials right now to do just that.

Decadal-Scale Variability of Ice Cover and Climate in the Arctic Ocean and Greenland and Iceland Seas

LAWRENCE A. MYSAK[1]

ABSTRACT

Analyses of 90 years of sea-ice concentration and ice-limit data from the Arctic Ocean and marginal seas reveal the presence of decadal-scale fluctuations in sea-ice extents, especially in the Greenland and Iceland seas. It has recently been proposed (Mysak et al., 1990) that such fluctuations may be part of an interdecadal Arctic climate cycle that can be described in terms of a reversing or negative feedback loop. An important property of this cycle is the suppression, every other decade, of convective overturning in the Iceland Sea for several winters (which occurred, for example, at the time of the "Great Salinity Anomaly" (GSA) in the 1960s, when sea-ice extents were also large). Such anomalies are preceded by increases in runoff and ice production in the western Arctic, which, it is argued, may partly cause such events as the GSA because of ice-anomaly advection from the Arctic into the Greenland Sea and subsequent ice melt in the Iceland Sea to the south. To find further evidence for this Arctic climate cycle, a lagged cross-correlation analysis of regional ice anomalies derived from sea-ice concentration data for the period 1953 to 1988 has been performed by Mysak and Power (1992). They found that both positive and negative ice anomalies in the Beaufort Sea consistently lead those in the Greenland and Iceland seas by 2 to 3 years. Mysak and Power also showed that high Mackenzie River runoffs in the mid-1980s led positive ice and negative salinity anomalies in the Greenland and Iceland seas in the late 1980s, which together resemble another (although weaker) GSA-like event.

A Boolean delay-equation model of this cycle has recently been developed by Darby and Mysak (1993). Among other things, they found that both ice and upper-ocean salinity advections from the western Arctic through to the Greenland and Iceland seas are needed in the model to create the observed structure of the Great Salinity Anomaly in the Iceland Sea.

[1]Centre for Climate and Global Change Research and Department of Atmospheric and Oceanic Sciences, McGill University, Montreal, Quebec

INTRODUCTION

Over the past two decades there have been many studies of the nature and causes of interannual variability of sea-ice cover in the Arctic Ocean and marginal seas. Observational studies have been carried out by Walsh and Johnson (1979), Mysak and Manak (1989), Parkinson (1991), and others, whereas model simulations of atmospherically forced interannual variability of sea-ice extent and concentration have been performed by, for example, Walsh et al. (1985) and Fleming and Semtner (1991). Mysak and Manak (1989) also noted that the low-pass-filtered areal sea-ice anomalies in the Barents and Greenland seas contain well-defined decadal-scale fluctuations, and they hypothesized that some of these might be related to the Great Salinity Anomaly (GSA; see Dickson et al., 1988), a widespread freshening of the surface waters in the subpolar gyre of the North Atlantic during the 1960s and 1970s.

In Mysak et al. (1990), hereafter referred to as M³, the decadal-scale fluctuations of ice cover in the Greenland and Iceland seas and attendant salinity anomalies in the Iceland sea (defined here as the region between Jan Mayen Island (near 71°N, 8°W) and Iceland) were further analyzed for the period 1901 to 1984, using updated Walsh and Johnson (1979) sea-ice concentration and Danish Meteorological Institute ice-limit data (Sear, 1988). In an attempt to formulate a comprehensive theory of the origin, evolution, and decay of these anomalies in the Greenland and Iceland seas, M³ linked a number of hydrological, oceanic, and atmospheric processes in the Arctic in the form of a negative feedback loop (Kellogg, 1983), which could give rise to self-sustained climatic oscillations in the Arctic for periods of 15 to 20 years. Under this scenario, GSA-like events could be regarded as cyclic rather than isolated, and as part and parcel of a sequence of complex air-ice-sea interactions.

In this paper a modified (and simplified) form of the feedback loop proposed by M³ is described, and evidence is presented for the occurrence and possible origin of another GSA-like event during the 1980s, which was predicted by M³. Because of the close relationship between salinity and sea-ice anomalies, we shall hereafter refer to these events as GISAs—great ice and salinity anomalies. Some results of a recent cross-correlation analysis (Mysak and Power, 1992), which show that sea-ice anomalies in the western Arctic consistently lead those in the Greenland Sea by 2 to 3 years, are also presented below. The question of whether GISAs have occurred prior to the 1960s is briefly addressed, and some possible links of GISAs with lower-latitude interdecadal climate fluctuations are described. (Within this context, we regard the 1960s-1970s GSA as also a GISA.)

NEGATIVE FEEDBACK LOOP FOR AN INTERDECADAL ARCTIC CLIMATE CYCLE

Figure 1 shows a six-component feedback loop of an interdecadal Arctic climate cycle that represents a modified

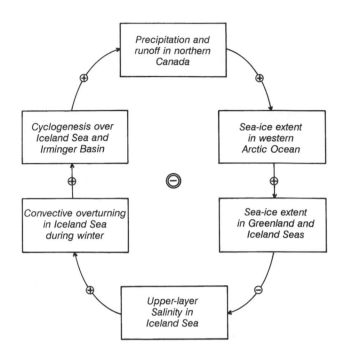

FIGURE 1   Negative feedback loop linking Iceland Sea-Irminger Basin cyclogenesis, northern Canadian river runoff, sea-ice extent, and salinity and convective overturning in the Iceland Sea (from Mysak and Power, 1992). This is a modified (and simplified) version of the ten-component negative feedback loop originally proposed by Mysak et al. (1990) to account for interdecadal Arctic climate oscillations with periods of about 15 to 20 years.

(and simplified) version of the ten-component loop in M³. The plus (minus) appearing between two boxes indicates that an increase in the first would cause an increase (decrease) in the second. Since the number of negative signs around the loop is odd, it represents a reversing or negative feedback loop (Kellogg, 1983). Therefore, in the absence of other strongly damping factors, a perturbation transferred from any one component to the next can theoretically result in a reversal of the sign of the initial perturbation. It has been estimated by M³ that the period of the climate cycle, which is twice the loop circuit time, is about 15 to 20 years.

A key feature of the loop in Figure 1 is that large negative salinity (and positive ice) anomalies in the Iceland Sea are hypothesized to be due, at least in part, to prior large runoffs from North America into the western Arctic (see the box at top of the loop). This concept of a remote forcing for anomalies like the GSA contrasts with earlier hypotheses that suggest that such anomalies are due mainly to local or neighboring atmospheric effects (Dickson et al., 1975; Pollard and Pu, 1985; Serreze et al., 1992; Walsh and Chapman, 1990a; see also the discussion in Mysak and Power, 1991). As noted in M³ (see their Figure 17), during the mid-1960s there were indeed anomalously large runoffs from North America into the Arctic and subarctic, which may have been associated with the climatic jump at that time (see

Figures 20 to 22 in M^3). During the early to mid-1960s, however, the Siberian River runoffs were generally below normal (Cattle, 1985) and hence would not have contributed to a positive sea-ice anomaly in the western Arctic. The above-average North American runoffs led to large sea-ice concentration anomalies in the Beaufort Sea in the western Arctic during the mid-1960s (see Figure 19 in M^3 and also Figure 3c below), which in turn were exported out of the Arctic into the Greenland and Iceland seas via the Beaufort gyre and Transpolar Drift Stream over a 2- to 3-year period (Colony and Thorndike, 1984). The melting of the large sea-ice anomalies over several summers in the Iceland Sea would then have led to the GSA, which peaked in 1968 (see Figure 6 in M^3).

It is important to emphasize that the Beaufort Sea ice anomalies described above are based on ice concentration data only. As W. Chapman has suggested (personal communication), it is conceivable that during the winter and spring months, when the ice is essentially land-locked there, the ice thickness could be increasing due to the lower near-surface salinities brought about by the increased runoffs the previous summer. Such thicker ice would then be exported out of the Arctic (via the Beaufort gyre and transpolar drift stream) into the Greenland Sea, and hence add to the ice and fresh-water anomaly in the latter region. A more complete set of ice-thickness data for the Arctic (and the western Arctic in particular) may thus provide additional support for the negative feedback-loop theory.

Due to the stabilizing influence of the GSA, convective overturning during winter in the Iceland Sea, which normally brings warm Atlantic water to the surface, was suppressed, which in turn reduced the sea surface temperatures (SSTs) there (see Dickson et al., 1988). This reduction in SST, we argue, would tend to reduce cyclogenesis in this region, as well as further south in the Irminger Basin because of the advection of cold SSTs by the East Greenland Current. It is hypothesized that a lowered SST leads to reduced cyclogenesis, precipitation, and runoff over North America, and therefore causes a flip in the sign of the perturbations in the next circuit of the feedback loop. A partial confirmation of the above hypothesis has been obtained by Serreze et al. (1992), who showed that during the peak GSA years (1967-1971) the frequency of anticyclones during winter increased over northern Canada, especially over the Canadian Arctic Archipelago, which implies drier conditions in this region. Nevertheless, at this stage the link between Iceland Sea-Irminger Basin cyclogenesis and northern-Canada precipitation should be regarded with some degree of skepticism until further studies have been carried out. The idea of such a link is consistent with recent teleconnection results of Walsh and Chapman (1990b), however. They showed that winter sea level pressure (SLP) anomalies in the Iceland Sea and Irminger Basin are highly correlated

($0.6 < r < 0.8$) with those at the base point 75°N, 90°W in the CAA (see their Figure 12b).

Despite the above possible weakness, the feedback-loop theory has had two successful applications. First, using a cycle time of about 20 years, M^3 predicted that there should be another GISA in the Greenland and Iceland seas in the late 1980s. This conjecture was verified through an examination of February sea-ice concentration data made by Mysak and Power (1991), and evidence of anomalous runoffs that might have generated this anomaly will be presented below. Second, the feedback loop in Figure 1 has been used by Darby and Mysak (1993) as a guide to developing a Boolean delay-equation (BDE) model of the interdecadal Arctic cycle. The BDE model contains six variables that represent the state of precipitation in northern Canada (1 or 0, corresponding to a high or low state), the state of ice and salinity conditions in the western Arctic and in the Iceland Sea, and the convective state in the Iceland Sea. For a variety of initial conditions, the model successfully simulates a 20-year cycle in the Boolean variables. A particularly novel result is that by allowing for different time scales for ice and salinity advection from the western Arctic through to the Greenland and Iceland seas, an ice anomaly in the Iceland Sea can persist longer than an ice anomaly in the western Arctic, a finding that is in agreement with observations.

THE 1980s GISA

Figure 2b shows that during the winters of 1987 and 1988 a large anomaly in sea-ice extent existed in the Greenland and Iceland seas, since the 0.9 ice-concentration contour extends well beyond the east Greenland coast as compared to its climatological position (Figure 2a). Concurrent with this ice anomaly, which also shows up in the late 1980s as a peak in the areal sea-ice anomaly time series for this region (see Mysak and Power, 1992), was the reduction of convection in the Greenland Sea during winter 1988 (Rudels et al., 1989) and the appearance of low-salinity water there during February and March 1989 (GSP Group, 1990). If the latter fresh-water anomaly advected southward into the Iceland Sea and suppressed convection there, then these features taken together suggest that a moderately sized GISA occurred in the Iceland Sea in the late 1980s. As in the case of the 1960s GSA, the large Greenland-Iceland sea-ice anomalies in the late 1980s appeared to have advected into the Labrador Sea by the early 1990s and thus contributed to extremely heavy sea-ice conditions and cooler air temperatures along the coast of Newfoundland in May 1991 (Globe and Mail, May 31, 1991). (It has also been argued that such positive ice anomalies off Newfoundland could be partly due to anomalous offshore winds that forced coastal ice seaward (J. Elliot, personal communication, 1991).)

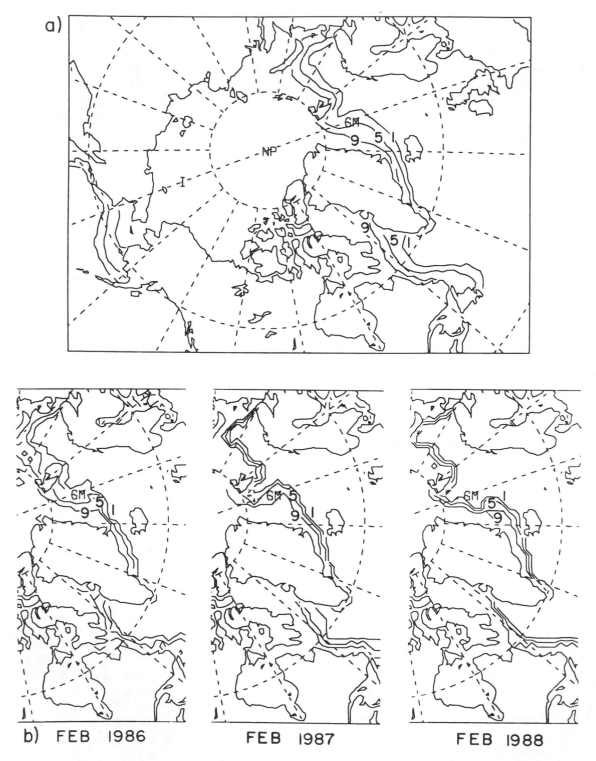

FIGURE 2   Sea-ice extent. (a) February climatology in the Arctic for the period 1953-1988. The ice concentration contours are labeled in tenths (1 indicates 0.1, etc.). In general, the sea-ice extent in the Greenland and Iceland seas is greatest during February (Mysak and Manak, 1989). (b) Monthly mean sea-ice extent in the Greenland and Iceland seas for February 1986, 1987, and 1988. While the 0.9 contour off the east Greenland coast is near its climatological position for February 1986 (see (a)), during 1987 and 1988 this contour extends well out into the Greenland and Iceland seas; in particular, it is much closer to Iceland. (From Mysak and Power, 1991; reprinted with permission of the Canadian Meteorological and Oceanographic Society.)

If the 1980s GISA is an integral part of the interdecadal climate cycle shown in Figure 1, it is conceivable that it too could have been partly generated by large runoffs in the western Arctic during the mid-1980s. Figure 3a shows that during the mid-1980s the Mackenzie River runoff was indeed slightly above average. This led to below-average salinities on the Beaufort shelf to the north of the Mackenzie delta (Figure 3b) and, for the period 1984-1986, to above-average areal sea-ice anomalies (Figure 3c) in the western Arctic subregion B_1 (identified in Figure 4). Heavy ice conditions in the Beaufort Sea can also be seen at this time in the ice atlas of Mysak and Wang (1991) (see also Figure 4b

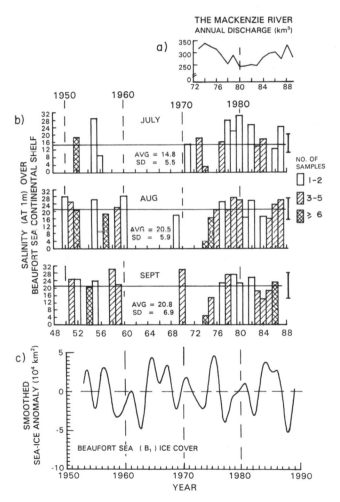

FIGURE 3   (a) Annual Mackenzie River runoff at the Arctic Red River (a city on the Mackenzie River) for 1973-1989 (data courtesy of R. Lawford). (b) Salinity at 1 m depth over the southeastern Beaufort Sea continental shelf (in subregion B_1, shown in Figure 4) in July, August, and September for 1950-1987 (data from Fissel and Melling, 1990). (c) Smoothed areal sea-ice anomalies in the Beaufort Sea subregion B_1 for 1953-1988. (From Mysak and Power, 1992; reprinted with permission of the Canadian Meteorological and Oceanographic Society.) In deriving the time series in (c), the monthly areal sea-ice-extent anomalies were low-pass filtered so as to remove all fluctuations with periods shorter than 30 months.

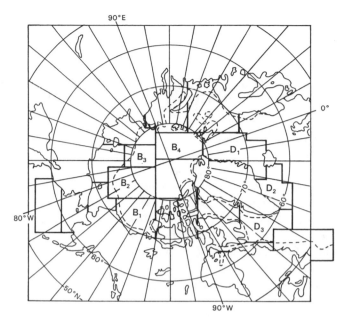

FIGURE 4   Subregions used in the cross-correlation analysis of low-pass-filtered sea-ice anomalies derived from ice-concentration data given on a $1° \times 1°$ (latitude) grid (developed by Walsh and Johnson, 1979) for 1953-1988. (Adapted from Mysak and Power, 1992; reprinted with permission of the Canadian Meteorological and Oceanographic Society.) In this paper cross-correlation results are presented only for the subregions labeled B_1, B_2, B_3, B_4, D_1, D_2, and D_3. The dashed line denotes the 200 m isobath.

in Mysak and Power, 1992). It is also interesting to note the sharp ice-anomaly peak in 1975, evident on Figure 3, which presumably occurred in response to the high runoff and low salinities in 1974; similarly, we note the sequence of low ice anomalies from 1978 to 1982, which correspond to low runoff and high shelf salinities at that time. Figure 3c also indicates the large ice anomalies from 1964 to 1967, which, as was pointed out by M^3, were likely due to the very large runoffs in 1964-1966. (Mysak and Power (1992) present evidence that shows that local winds may also have played a role in producing the Beaufort Sea ice anomalies.)

It has often been suggested by critics of the interdecadal-cycle theory characterized by Figure 1 that runoff from the Siberian rivers, which in total is 4 to 5 times that of the Mackenzie, should contribute substantially to sea-ice anomalies in the Arctic and subsequently in the Greenland and Iceland seas. Here we counter (as does M^3, in more detail) that, because the Eurasian shelf is very wide and shallow, this runoff gets well mixed there with the central Arctic Ocean waters that also flow onto the shelf. Such mixing takes place mainly through tides and eddies in this region; also, convective overturning on the shelf mixes the surface runoff with the deeper shelf water. Fluctuations in the Siberian river flow thus have little direct bearing on the creation of sea-ice anomalies in the central Arctic. A similar conclusion was also reached by Ikeda (1990), who considered effects of runoff on sea ice in the Barents Sea.

CROSS-CORRELATION ANALYSIS OF SEA-ICE ANOMALIES IN THE WESTERN ARCTIC AND GREENLAND SEA

To verify that sea-ice anomalies produced in the Beaufort Sea are advected into the Greenland and Iceland seas in 2 to 3 years' time (as proposed in the feedback loop), we first grouped according to subregion the monthly sea-ice concentration data (Walsh and Johnson, 1979) that had been collected on a square $1° \times 1°$ (latitude) grid for the period 1953-1988. Figure 4 depicts these subregions; B_1 to B_4 are in the Beaufort Sea and Canada Basin, and D_1 to D_3 are in the Greenland, Iceland, Irminger, and Labrador seas, the latter two seas being source regions of North Atlantic Deep Water. Then a monthly time series of the areal sea-ice anomalies was computed for each subregion (i.e., the concentration departure from the long-term mean for each month at a grid point in the subregion was multiplied by the grid area). The procedure is similar to that followed by M^3, who studied the advection of Greenland Sea ice anomalies into the Labrador Sea via five subregions (see Figures 2 and 3 in M^3). The monthly time series for each subregion was next low-pass filtered to eliminate high-frequency components, i.e., those periods shorter than 30 months. (A description of the filtering process can be found in Power and Mysak (1992).) An example of one such low-passed time series is shown in Figure 3c.

Note that the subregions B_1 through B_3 were chosen to follow the Beaufort gyre ice-drift pattern in the deep part of the Arctic Ocean, so that the anomaly advection would not be contaminated by smaller-scale ice motions on the Eurasian shelf. The lagged cross-correlation functions for the low-passed fluctuations in B_1 versus those in the other subregions were computed and examined for asymmetries about zero lag (e.g., as seen in Figure 4 of M^3). The maximum lagged correlation coefficients obtained from this analysis are given in Table 1. To find the 95 percent significance level for the correlations, the number of degrees of freedom (for simultaneously correlated data) was estimated to be $2N/30 = 29$, where $N = 432$ (total number of points of data). Then from Table 13 in Pearson and Hartley (1966), we obtained, using a one-tailed test for normally distributed

data, a 95 percent significance level of $r = 0.3$. For lagged cross-correlations, the estimated number of degrees of freedom decreases with lag; the 95 percent significance level for the correlation coefficient r, which is estimated to be 0.3 for simultaneously correlated data, slowly increases to 0.36 for a lag of 120 months.

The results in Table 1 show that the low-pass-filtered sea-ice anomalies in B_1 lead those in the Greenland Sea (subregion D_1) by just over 2 years, and, moreover, that there appears to be a continuous advection of ice anomalies by the Beaufort gyre, the Transpolar Drift Stream, and the East Greenland Current all the way into the Labrador Sea. (Although the maximum cross-correlations for B_1 versus D_2 and D_3 are not significant at the 95 percent level, the correlations between data from adjacent regions are significant (Mysak and Power, 1992) and support the above conclusion.) Chapman and Walsh (1991) found that (unfiltered) monthly Beaufort sea-ice anomalies also led those in the Greenland Sea (see their Figure 12); however, they did not estimate the time lag involved nor did they carry out an analysis of the sea-ice fluctuations between subregions B_1 and D_1. A detailed discussion of the sea-ice anomalies described here, and also of those in the other subregions shown (but not labeled) in Figure 4, is given in Mysak and Power (1992). The possible role of wind forcing in creating these anomalies is examined by Mysak and Power (1992) as well.

To estimate the average advection speed of the anomalies from the Beaufort Sea into the Greenland Sea and beyond, a cumulative lag plot was constructed (Figure 5). From the best linear fit to the data, we find an average advection speed of about 2000 km/yr, or roughly 5 km per day, which lies well within the range of typical drift speeds for sea ice, namely 1 to 10 km per day (Chapman and Walsh, 1991).

EVIDENCE FOR EARLIER GISAs

In M^3 the ice-limit data of the Danish Meteorological Institute (DMI) for the period 1901-1956, which give the summer ice-edge positions in the Greenland, Iceland, Irminger, and Labrador seas, were analyzed for evidence of earlier GSA-like events. It was noted that in response to large North American runoff increases in 1931-1932 and 1945-1947 (see Figure 17 in M^3), the sea-ice extent in the Greenland and Iceland seas increased noticeably a few years

TABLE 1 Maximum Lagged Correlation Coefficients (r) for Smoothed Areal Sea-Ice Anomalies in Subregion B_1 (see Figure 4) Versus Those in Other Subregions, with the Lag at Each Maximum Correlation

	B_2	B_3	B_4	D_1	D_2	D_3
r_{max} (with B_1 leading)	0.27	0.33	0.53	0.40	0.27	0.21
lag (months) at r_{max}	2	5	28	27	36	39

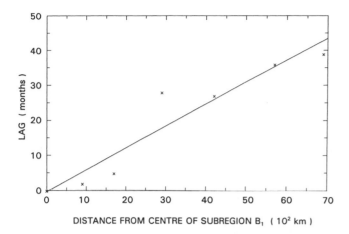

FIGURE 5 Best linear fit to the lag at maximum cross-correlation (taken from Table 1) versus distance from center of subregion B_1 to center of subregion D_3 via the centers of subregions B_2, B_3, B_4, D_1, and D_2 shown in Figure 4. (From Mysak and Power, 1992; reprinted with permission of the Canadian Meteorological and Oceanographic Society.)

later (see Figure 15 in M^3), in agreement with the right-hand side of the feedback loop in Figure 1. Moderate peaks can also be seen in the annual Koch ice index (the number of weeks per year that ice affects the coast of Iceland) at about the same time as the large ice limits, namely, around the early 1930s and late 1940s (Figure 10 in M^3). Mysak and Power (1991) showed that the North American runoff in Figure 17 in M^3 (a 38-year time series) leads the Koch ice index by 3 to 5 years, which provides further evidence of the correlation between runoff and sea ice implicit in the feedback loop. (The above lag correlation has also been recomputed (R. Tyler, personal communication, 1992) using a 50-year runoff record (1918-1967); it was again found that the runoff leads the Koch ice index by about 5 years.)

More recently, the spectrum of the Koch ice index for the period 1600-1970 has been computed by Stocker and Mysak (1992) in their analysis of high-resolution proxy data from the Holocene epoch. As well as a peak at around 90 years, they found significant energy at 27 years (see their Figure 10), which corroborates the finding of such a signal in a previously computed Koch ice-index periodogram using a shorter time series (Figure 25 in M^3). This 27-year peak is also remarkably close to the 24-year peak in the spectrum of the central England temperature record (Stocker and Mysak, 1992) and also to the 20-year oscillation seen in the Greenland ice-core oxygen-isotope records, which extend over a 700-year period in the recent Holocene (Hibler and Johnsen, 1979). In view of the latter results, other ice-core data from glaciers in the Canadian high Arctic should be examined for evidence of interdecadal climate oscillations. From very short ice cores extracted from the Devon ice cap, F. Koerner (personal communication, 1992) found evidence of large accumulations in the mid-1960s,

which is consistent with the sudden precipitation increase that was observed at that time in the Canadian high Arctic (Bradley and England, 1978). Thus the analysis of longer, high-resolution ice cores from the Canadian north could be quite revealing from the viewpoint of Arctic interdecadal variability.

POSSIBLE CONNECTIONS OF GISAs WITH LOWER-LATITUDE INTERDECADAL CLIMATE VARIABILITY

We shall divide the discussion on this issue into three parts:

1. We could suppose (perhaps naively) that the interdecadal Arctic climate cycle described in this paper is essentially self-contained (like some theories of ENSO, which involve air-sea interactions in only the tropical Pacific region) and is thus independent of lower-latitude variabilities. This assumes, for example, that the variability of Pacific inflow through the Bering Strait is unimportant, that exchanges between the Arctic Ocean and the Canadian Arctic Archipelago have little effect, and that atmospheric interdecadal fluctuations in lower latitudes essentially peter out in the Arctic.

2. Even if the above is essentially correct, it is nevertheless conceivable that the Arctic climate cycle could induce interdecadal variability at lower latitudes.

3. Alternately, lower-latitude interdecadal variability in the climate system (and in the ocean in particular) may very well have an influence on or trigger some aspects of Arctic interdecadal variability. We will expand on the second and third possibilities below.

The Arctic as a Trigger for Lower-Latitude Fluctuations

One way in which the interdecadal Arctic cycle could influence lower-latitude climate is through the connection between the Arctic Ocean and the Greenland Sea via Fram Strait. According to Aagaard and Carmack (1989), because the waters in the Greenland Sea are weakly stratified, they are "delicately poised with respect to their ability to sustain convection," and therefore the latter process is sensitive to the amount of fresh water exported from the Arctic. A modest increase in this water supply could severely reduce convection in the Greenland Sea. However, it is not believed (R. Dickson, personal communication, 1992) that this reduction in convection occurred during the time of the GSA. It is also conceivable that alternating periods of heavy and light Greenland-Iceland sea-ice anomalies, which are accompanied by relatively cold and warm upper-ocean tem-

peratures respectively (see Figure 6 in M³), could result
(via advection) in decadal-scale SST anomalies in the southwestern part of the subpolar gyre a few years later, an idea
proposed by Mysak (1991). For example, our Figure 6,
which depicts the SST anomalies averaged over 45°N to
55°N in the Atlantic Ocean, clearly indicates the existence
of alternating cold and warm periods in the northwestern
Atlantic, i.e., at around 30°W. To the left of the time axis
is a sequence of solid and dotted lines that indicate heavy
or light ice conditions, respectively, in the Greenland and
Iceland seas. In several cases, we observe that heavy ice
conditions precede negative SST anomalies in the northwestern Atlantic by approximately 5 years (the advection
time from the Greenland and Iceland seas to this region);
similarly, light ice conditions in the Greenland and Iceland
seas precede positive SST anomalies, also by about 5 years.
While the correspondence between heavy (light) ice conditions and warm (cold) SST anomalies in the northwestern
Atlantic is not completely one-to-one, the relationship is
sufficiently encouraging to warrant further investigation and
increased monitoring of high-latitude climate parameters.

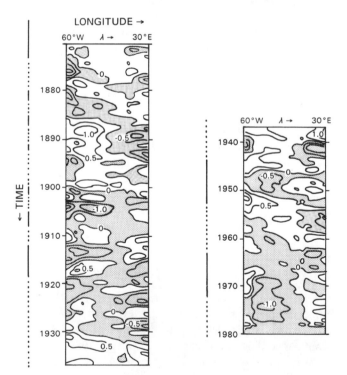

FIGURE 6 Low-pass-filtered and detrended sea surface temperature anomalies averaged over the latitude band 45°N to 55°N as
a function of time and longitude in the North Atlantic and Baltic
Sea. (Adapted from Bryan and Stouffer, 1991; reprinted with permission of Elsevier Scientific Publishers.) Solid and dotted lines
on the time axis indicate, respectively, periods of relatively large
and small (or near-normal) sea-ice extents in subregion D₁ (see
Figure 4) of the Greenland and Iceland seas, as estimated from
the sea-ice data described in Figures 3, 10, and 15 of Mysak et
al. (1990).

A second way in which Arctic sea-ice changes could
induce climate variability at lower latitudes is via the surface
albedo: Anomalously large sea-ice extents increase the
albedo, and vice-versa. However, the change in the area
covered by the sea ice that can be ascribed to natural
decadal-scale variability is only of the order of a few percent;
such variations are therefore unlikely to induce changes in
the polar energy budget large enough to affect the lower
latitudes via atmospheric circulation. The Southern Hemisphere sea-ice cover, however, may show larger changes
on the interdecadal time scale, in which case Arctic albedo
changes and climate need to be discussed together with the
natural variability of the southern polar region.

The possibility that polar (in contrast to just Arctic) and
high-latitude climate fluctuations on decadal time scales may
be related, with the former driving the latter, is suggested by
the spatial structure of the third empirical orthogonal function of SST that has been computed by Folland et al. (1986a).
(For a clear picture of this EOF, which accounts for 6
percent of the SST variance, see Figure 2 in Bryan and
Stouffer (1991).) The largest amplitudes of this EOF, whose
temporal fluctuations have an interdecadal time scale, are
in the northeast Pacific, the northwest Atlantic, and a broad
range over the South Atlantic and South Indian oceans.
(Folland et al. (1986b) have shown that this EOF is closely
linked with Sahel rainfall.) At the same time, we note
that because of the particularly large SST gradients in the
northeast Pacific and northwest Atlantic, it is conceivable
that this EOF's variations may be driven by interdecadal
variability in the Arctic. A recent study by Walsh and
Chapman (1990b) showed that monthly Arctic SLP fluctuations in winter are associated with or teleconnected to SLP
changes in the northern North Atlantic. In a continuation
of this study, using data from all seasons, Power and Mysak
(1992) showed that at low frequencies (periods of 5 years
and longer) this teleconnection pattern persists; in addition,
they found that at low frequencies another teleconnection
pattern exists between centers in the Arctic and the North
Pacific. However, neither of these studies addresses the
question of cause and effect, so this information should
be taken mainly as a starting point for further work on
determining how the polar regions (and the Arctic in particular) drive lower-latitude fluctuations.

Lower-Latitude Forcing of Arctic Fluctuations

The third viewpoint noted earlier, namely, that lower-
latitude interdecadal fluctuations trigger Arctic climate fluctuations on this time scale, perhaps has the most widespread
appeal. There have been many recent observations of interdecadal variability at middle and tropical latitudes in the
climate system (for a list of references, see Weaver et al.,
1991). Many investigators believe that these fluctuations
may be due to internal oscillations in the thermohaline

circulation—an idea that appears to originate with Bjerknes (1964), and one that has also been developed by many ocean modelers (see, e.g., Weaver et al., 1991; Delworth et al., 1993). This meridional overturning circulation is global, and transports heat from low to high latitudes. Thus, changes in this transport would produce high-latitude changes in the atmospheric surface temperatures at locations where the thermohaline circulation "ventilates" (in the North Atlantic and in the Southern Ocean). Such changes in the poleward heat transport in the North Atlantic have recently been observed by Greatbatch and Xu (1992). Furthermore, such changes in the heat transport could have produced the type of SST anomalies seen in Figure 6. In particular, Greatbatch and Xu found increased transports (at 54.5°N) in the 1950s and reduced transports in the 1970s, which correspond to the warm and cold SSTs seen at these times in Figure 6.

The influence of the aforementioned SST anomalies on the atmospheric circulation has been studied by Peng and Mysak (1993), among others. They showed that during the warm 1950s there was a teleconnection pattern of high-low-high sea level pressure centers established across the North Atlantic, Europe, and western Siberia. During the cold 1970s, however, there was a very large low-pressure system extending from the North Atlantic to central Europe. These two circulation patterns in the atmosphere had very different effects on the precipitation and runoff over western Siberia. Peng and Mysak found that during periods of warm SST anomalies there tended to be less precipitation and hence less runoff into the Arctic, whereas during the cold years there was more precipitation and runoff into the Arctic. One important conclusion from this study is therefore that the fresh-water budget in the Arctic is indeed affected by lower-latitude interdecadal fluctuations in the ocean. However, it does not appear that the induced changes are directly related to the interdecadal Arctic climate cycle, because the runoff into the Siberian shelf does not seem to have an immediate effect on sea-ice concentration in the central Arctic.

In a recent study by Higuchi et al. (1991), it was shown that the standing-eddy poleward heat transport in the lower troposphere in the Northern Hemisphere exhibits strong decadal-scale variability. In particular, in years of very large or very weak transports, the ice margin in the Greenland Sea is substantially reduced or expanded accordingly. Thus in this case the atmosphere is a medium that can transmit interdecadal signals into the Arctic region and possibly influence the sea-ice cover.

While many of the above suggestions are rather speculative, we believe it worthwhile to look further at connections between middle- and polar-latitude interdecadal variabilities in the climate system. In particular, we should examine the Southern Ocean and Antarctic regions for interdecadal variability to see whether there are any manifestations (e.g., ice anomalies) similar those found in the Arctic. However, one should bear in mind that if these northern and southern variations are linked by the influence of the thermohaline circulation, there will be long time lags (of the order of a hundred years) between related events in the two polar regions, because the overturning time scale of the thermohaline circulation in the Atlantic basin is a few centuries (see, e.g., Mysak et al., 1993). Also, another important factor that should be kept in mind is the different geographies of the two polar regions: The Arctic Ocean is surrounded by a land mass, whereas the Southern Ocean surrounds the Antarctic continent.

In conclusion, it perhaps is worth making one final point, which concerns both polar regions as well as the lower latitudes. It can be summed up colloquially as follows: "What sea surface temperature is to interannual variability in the tropics, sea surface salinity is to interdecadal variability at high latitudes." This contrast arises because of the very different effects of SST and sea surface salinity anomalies on the atmosphere, a relation well known to ocean and climate modelers (Weaver et al., 1991). Air temperature responds fairly quickly to changes in the sea surface temperature (on time scales of the order of weeks to months), whereas precipitation is not so immediately affected by surface salinity changes. There can be feedbacks to the atmosphere through the hydrological cycle, ocean circulation, and sea-ice formation, drift, and decay, but, as described in this paper, they are on time scales of the order of decades.

ACKNOWLEDGMENTS

The author is grateful for financial support received over the past several years from the Canadian Natural Sciences and Engineering Research Council, the Canadian Atmospheric Environment Service, Fonds FCAR (Québec), and the U.S. Office of Naval Research. It is also a pleasure to thank Ann Cossette for typing the first draft, John Walsh for helpful discussions, Douglas Martinson for his interest in and review of this paper. The comments of Bill Chapman on an earlier draft of this paper are gratefully acknowledged.

Commentary on the Paper of Mysak

DOUGLAS G. MARTINSON
Lamont-Doherty Earth Observatory

I want to start by commending Dr. Mysak for attempting to combine a wide variety of observations and processes into a single feedback model. Having said that, I want to play a "friendly devil's advocate." First, as you point out, there is a decrease in the Siberian river runoff into the Arctic that coincides with an increase in the Mackenzie River runoff. What is the net effect of that sort of competition?

That leads to the next point, the influence of runoff on ice growth. Are you suggesting that the runoff influences the ice growth through its impact on stability? If so, is the Beaufort Sea less stable than the region receiving the Siberian river inflow? Or are you suggesting some other mechanism? Alternatively, might the increased runoff be indicative of some other atmospheric anomaly (e.g., winds, temperature, cloud cover) that more directly affects the ice cover? Also, do you know whether the increased areal distribution of ice actually represents more fresh water? Spreading of the ice in the Greenland Sea by increased winds would also lead to thinner ice, in which case the net volume of fresh water has not necessarily changed.

Now for some general points. You suggest that local convection changes the SST, which in turn influences the cyclogenesis. As Walter Munk asked previously, can the small spatial scales of convection significantly influence SST? In other words, how exactly do you envisage the convection's influencing SST to the extent that it could affect cyclogenesis? Might not the lateral influences be more important? Also, if the cyclogenesis in the Greenland Sea is increased, it is still difficult for me to imagine that it drives increased precipitation so far to the west.

Finally, a comment regarding the shutdown of convection in response to an increase in advected sea ice (fresh water). I just want to point out that in the Weddell Sea today, a large output of glacial ice from the Filchner, Ronne, and Larson ice shelves is streaming ice down the eastern side of the Antarctic peninsula, and that is the region where we have the largest source of Antarctic bottom water. If you track this ice, it reenters the Weddell in the vicinity of the Greenwich meridian where the open ocean convection was observed in the mid-1970s. Therefore, the relationship between advected ice and a convection shutdown must not be as straightforward as it might otherwise seem.

Those are a few interesting lead-in points, and I appreciate the model you have given us to talk about.

Discussion

GROISMAN: As I remember, 70 to 80 percent of the fresh-water input into the Arctic from Eurasia comes from rivers, some of which are highly variable in volume. You would have to take that variability into account in that region.

MYSAK: The tremendous mixing due to tides and even convection on the wide continental shelf in the Arctic Ocean that borders Eurasia seems to minimize Eurasian runoff water's impact on sea ice. In the Beaufort Sea region, where the shelf is very narrow, there seems to be an immediate effect.

MOREL: One must be careful not to interpret an association of anomalies, even if it occurs more than once, as a causal relationship, particularly when it involves precipitation. The models tend to yield different anomalies.

TRENBERTH: If sea ice melts, it creates fresh water that substantially affects salinity. But I believe that observational evidence suggests that you get more ice created because it's cold than because the water is fresher.

MYSAK: For a given air temperature you'll get more ice if the water's fresher. Also, some data I looked at didn't show a very good correlation between cold-air temperature anomalies and sea-ice formation, but then it was limited to spring and summer.

LEHMAN: I'd like to comment on topics from the last several presentations. First, Bob Dickson said that he didn't feel the net overflow in the Greenland/Icelandic/Norwegian Sea area was being affected—in other words, that its strength was invariant on, say, decadal time scales. It seems to me that the topographic barrier acts to filter out variability there on short time scales. I should note, however, that I disagree with him about the short-time-scale stability of the thermohaline circulation.

Now let's consider some consensus views on the relative transports for the two different limbs of the North Atlantic Deep Water: overflow water and the Labrador Sea water (ignoring recirculation). You might have about 6 sverdrups for the former, and 7 to 8 for the latter, which is not insignificant. Together they constitute a potent climate forcing agent.

If the modelers can get the deep-water formation north of the ridges right, we can explore the sensitivity of these limbs to the same kind of salinity forcing. I predict that the Labrador Sea limb will respond on short time scales, and represent an important effect on the thermohaline circulation.

RIND: It has been suggested that at the time the last glacial cycle started there might actually have been a shutoff of the GIN seas component, while the Labrador Sea component continued.

GHIL: If North Atlantic Deep Water is formed by Arctic intermediate water, we need to know what controls the formation of the Arctic intermediate water so we can put it into our models.

LEHMAN: The convection and water formation at both levels seem to be responding to salinity forcing at the surface, but even if the forcing fails to drive convection there's enough water backed up behind the sills to keep overflowing. I know I've simplified by ignoring the wind-driven gyre, but things might be simpler in the Labrador Sea. It would be interesting to apply the same salinity forcing to it as to the Norwegian Sea, and see whether you get the climate vibration you might expect.

GROOTES: This morning almost everyone said that we have some evidence for a 10- or 20-year time scale change, but not a long enough record. I just wanted to mention that my Figure 4 definitely shows a fluctuation on that scale. The Greenland ice sheet serves as a high-resolution monitor for climate variability in the Greenland and North Atlantic area, and I think we ice-core people need to get together with the oceanographers and modelers to put this kind of proxy data to better use.

Long-Term Sea-Level Variation

BRUCE C. DOUGLAS[1]

ABSTRACT

Published values for the long-term, global-mean sea-level rise determined from tide-gauge records range from about 1 to 3 mm per year. I have argued (Douglas, 1991) that the scatter of the estimates appears to have arisen from use of data from gauges located at convergent tectonic plate boundaries or affected by land subsidence or uplift due to isostatic rebound from the last deglaciation, and especially data reflecting the effects of large interdecadal and longer sea-level variations on short ($<$50 years) records. Using only long and nearly complete records from locations where tectonic plates do not collide, and correcting for rebound with the Tushingham and Peltier model (1991), I obtained for 21 stations in nine oceanic regions during 1880-1980 a value of 1.8 ($\pm$0.1) mm/yr for global sea-level rise. Because low-frequency signals in the sea-level record even more seriously affect estimates of acceleration, I examined (Douglas, 1992) the longest tide-gauge records for evidence of a past, globally coherent, nonlinear component of sea-level rise. For the 80-year period 1905 to 1985, 23 essentially complete tide gauge records in 10 geographic groups were found to be suitable for analysis. These yielded an apparent global acceleration of -0.011 ($\pm$0.012) mm/yr^2. A larger, less uniform set of 37 records in the same 10 groups, having an average length of 92 years and covering the 141 years from 1850 to 1991, gave a value of 0.001 ($\pm$0.008) mm/yr^2. Thus there is no evidence for an acceleration in the past 100+ years that could be termed significant, either statistically or by comparison with the value (0.15 to 0.2 mm/yr^2) predicted to accompany global warming. The large interdecadal fluctuations of sea level so severely affect estimates of regional and global sea-level acceleration for time spans of less than about 50 years that future tide-gauge data alone cannot serve as a leading indicator of climate change. If the interdecadal fluctuations of sea level could be understood in terms of their forcing mechanisms and removed from the records, then tide-gauge data would be useful over shorter spans.

[1]National Oceanographic Data Center, National Oceanic and Atmospheric Administration, Washington, D.C.; present address, Department of Geography, University of Maryland, College Park, Maryland

INTRODUCTION

Concern over the consequences of global warming has inspired many examinations of historical tide-gauge records to determine the rate of sea-level rise. Excellent summaries of the sea-level-rise problem are available in the National Research Council's *Sea-Level Change* (NRC, 1990) and the Intergovernmental Panel on Climate Change's report *Climate Change: The IPCC Scientific Assessment* (IPCC, 1990). Interest in sea-level rise is high due to its obvious practical impact, and because of its scientific value as an indicator of global change. In the latter respect, global sea level is similar to earth orientation (Carter et al., 1989). Both give measures that are necessary (but not sufficient) conditions for evaluating model predictions.

Published values for global sea-level rise for the last 50 to 100 years vary from about 1 to 3 mm/yr (Barnett, 1990), with formal uncertainties ranging from 0.15 to 0.90 mm/yr. While there is not much doubt that sea level is rising, the scatter of results makes impossible a meaningful interpretation of the global balance of water in its various forms and locations.

The large disparity of results arises in part (Barnett, 1990) because authors analyze data in different ways, and select and group tide-gauge stations according to different criteria. Emery and Aubrey (1991) have pointed out in addition the importance of vertical crustal movements on estimates of sea-level rise. Also, Pugh (1987) and Munk et al. (1990) noted the importance of decadal variations, and Sturges (1987) further warned of the influence of very-low-frequency variations of sea level on estimates of sea-level rise. To minimize these latter problems, I selected or rejected sea-level time series from tide gauges (Douglas, 1991) according to length and completeness of record, general agreement with other nearby stations over a common time interval, and freedom from obvious tectonic effects. Stations were grouped according to oceanic region. The station groups surviving this selection process formed an extremely consistent set, particularly after correction for the effects of post-glacial rebound (PGR) by the ICE-3G model of Tushingham and Peltier (1991). In fact, the rms difference of sea-level rise data for widely separated oceanic regions from long records exceeding 60 years in length, after correction for PGR, was so good (0.4 mm/yr) that the data could be aggregated without regard to record length. This obviated the need for statistical methods such as empirical orthogonal function (EOF) analysis. The value I obtained (Douglas, 1991) for the global (linear) sea-level rise was 1.8 mm/ yr ±0.1. The small value of the uncertainty reflects the consistency of sufficiently long records.

Woodworth (1990) and Douglas (1992) have further considered the question of an acceleration of global sea level. Using somewhat different approaches, both concluded that there is no evidence of a statistically significant acceleration (i.e., quadratic variation) of global sea level over the past century. However, these authors differed considerably in their views of how quickly a regional or global acceleration of sea level could be detected in future tide-gauge records. Woodworth concluded that a regional acceleration could be confirmed at the 95 percent confidence level within a few decades. Douglas was far more pessimistic, because of the chaotic occurrence of large and durable low-frequency sea-level variations in the records.

THE SEA-LEVEL RECORD

The IPCC report gives for the "business as usual" scenario of global warming an additional sea-level change of 18 cm by the year 2030 and 44 cm by 2070, corresponding to accelerations of about 0.22 mm/yr^2 in the former case and 0.14 mm/yr^2 for the latter (IPCC, 1990). Verification of these or similar predictions at an early date is necessary in order to establish confidence in climate models and forecasts.

Determinations of sea-level acceleration suffer from the same problems as determinations of linear sea-level-rise value, with an important exception, viz., steady vertical crustal movements. The most pervasive of these, post-glacial rebound, is linear over the time span for which tide-gauge data are available (Peltier and Tushingham, 1989) and is of no consequence in computation of the acceleration in a record. Any other temporally linear vertical crustal movement, such as that associated with colliding tectonic plates at places with long earthquake-recurrence times, is similarly not a factor. As a result, many more tide-gauge records are usable for an analysis of apparent global sea-level acceleration than for determining the linear trend.

Although data from more gauges are available for analysis of apparent sea-level acceleration than for the case of the linear rise, this is of little significance. The reason is that at low frequency, spatial coherence of sea level is very great. Thus a region such as the Baltic, which contains a score of long records, is effectively a single measurement system for examining decade-to-century-scale and longer sea-level variations. Figure 1 illustrates this phenomenon. The data are monthly mean values of sea level from the Permanent Service for Mean Sea Level (PSMSL) (Spencer and Woodworth, 1991) that have been detrended over the common time period 1895 to 1965 and smoothed by a Gaussian filter with full width at half maximum of four years. Five typical sea-level records, in ascending order from the southernmost part of the Baltic northward to Stockholm and then well eastward into the Gulf of Finland, are displayed. These low-pass-filtered records are striking for their spatial and temporal coherence and the long relative

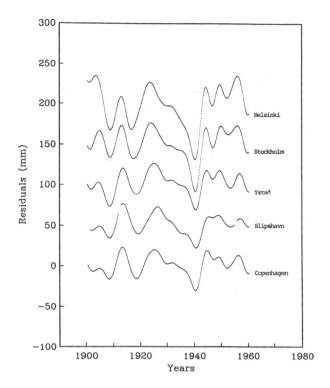

FIGURE 1   Detrended and low-pass-filtered sea-level records in the Baltic. Note the extraordinary coherence at low frequency.

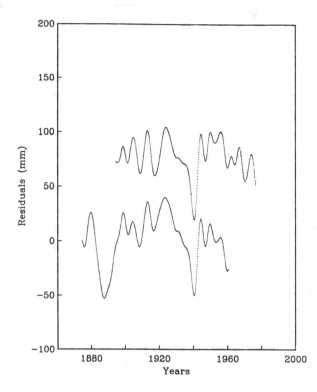

FIGURE 2   Low-pass-filtered and detrended records for Stockholm (top) and Nedre Sodertalje (bottom).

decline from about 1925 to 1940. This low-frequency "noise" corrupts a determination of acceleration from these records, and is clearly not independent from one site to another. Such large and persistent oscillatory sea-level events, present in most tide-gauge time series, radically affect the value of apparent sea-level acceleration derived from records even many decades in length. Figure 2 shows that even a near-100-year record may not be long enough to obtain a representative sample of low-frequency sea-level variation at a site. It presents the filtered and detrended sea-level records for Nedre Sodertalje and Stockholm, sites only about 25 km apart that have records nearly the same length, at just over 90 years. The difference between them is that the Nedre Sodertalje series covers an earlier period, before the Stockholm record began, but one that exhibited a large, persistent sea-level variation comparable in amplitude to the event of 1925 to 1940.

Such large and enduring sea-level variations are not unusual in tide-gauge records. Figure 3 shows the filtered and detrended record for San Francisco, the longest continuous record available for the United States. Clearly visible are interdecadal fluctuations as well as larger, lower-frequency oscillations. These signals obviously corrupt determinations of rise and acceleration. Figure 4 shows the results for the linear rise at San Francisco computed from the original unfiltered monthly mean values for a sliding 30-year win-

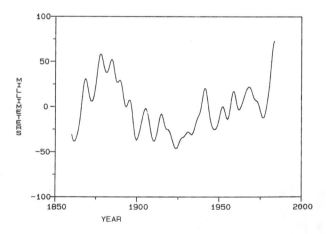

FIGURE 3   Low-pass-filtered and detrended sea levels for San Francisco. Note the existence of very large low-frequency signals in the record.

dow. The 30-year linear rise value at San Francisco varies from $+5$ to -2 mm/yr over the record. Figure 5 presents the acceleration estimates (in mm/yr^2) for the same sliding 30-year window. These values are, of course, even more variable than the linear term values, ranging from $+0.5$ to -0.5 mm/yr^2. Clearly, determining a meaningful global acceleration value requires using a well-distributed set of oceanic regions that lack interregional coherence.

REGIONAL AND GLOBAL SEA-LEVEL ACCELERATION

In the Douglas (1991) paper on global sea-level rise, the condition was enforced that all records used had to be about 80 percent or more complete because of the presence of low-frequency sea-level variations. From the comments above, it is clear that the apparent acceleration determined for a site can also be affected by a data gap, so the same condition needs to be applied for an analysis of acceleration.

As we have seen, record length is a critical parameter in the selection of data sets. Figure 6 displays quantitatively how record length affects the apparent acceleration estimate by displaying it as a function of data span for PSMSL records more than 10 years long. The scatter of accelerations for record lengths less than about 40 to 50 years is very

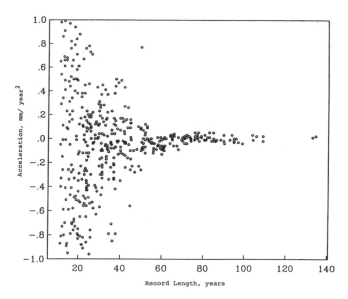

FIGURE 6   Apparent accelerations of sea level for records more than 10 years in length. Low-frequency variations of sea level heavily corrupt the computation of an acceleration parameter for records less than about 50 years in length.

great; the reason for the striking increase in consistency beginning with records over 50 years long is unknown. Of course, the scatter of shorter records in Figure 6 is due to absorption of interdecadal variability into the estimate of acceleration. The figure clearly shows the site dependence of apparent sea-level acceleration, even for records that are many decades in length.

Although the disparity of values in Figure 6 is discouraging, it is worth pointing out that the number of values with short records available is large, since many new gauges have been installed in recent decades. The oceanic regions with records greater than 75 years in length that were selected for this paper do not represent the possibilities for the recent past or near future. But as was seen in the example of the Baltic, a new tide gauge is really significant for the sea-level trend and acceleration problem only if it is located in an area not previously sampled. Even allowing for some argument as to what constitutes an oceanic region, it is unlikely that the number of regions chosen for this paper (10) could be doubled. Since a doubling of independent samples reduces the error of the mean by only 40 percent, the importance of understanding and eliminating interdecadal and longer low-frequency variations of sea level is underscored.

Table 1 presents the regional oceanic groups of stations used in Douglas (1992) for an analysis of apparent sea-level acceleration during the period approximately 1850 to 1990. The largest group is the one covering the North Sea near Esbjerg, extending to the Baltic entrance, and thence to the Gulf of Finland. Four of the groups contain only one

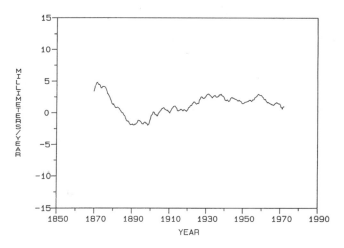

FIGURE 4   Sea-level rise in mm/yr for San Francisco, for a sliding 30-year window of the original monthly mean values.

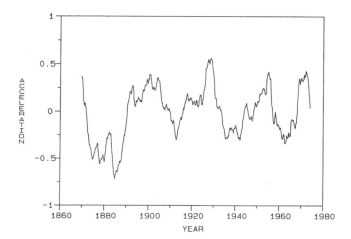

FIGURE 5   Apparent sea-level acceleration in mm/yr^2 for San Francisco, for a sliding 30-year window. The amplitude of the low-frequency excursions is about twice the value anticipated to result from future global warming.

TABLE 1   Stations with Records Exceeding 75 Years During 1850-1991

Oceanic Group (No.) and City	Start	End	Span	Acceleration* (mm/yr²)
(1) Goteborg	1887	1969	82	0.01
Varberg	1887	1982	95	−0.03
Ystad	1887	1982	95	−0.03
Kungholmsfort	1887	1982	95	0.00
Landsort	1887	1982	95	−0.02
Nedre Sodertalje	1869	1966	97	−0.03
Stockholm	1889	1982	93	−0.01
Bjorn	1892	1970	85	0.00
Ratan	1892	1982	90	−0.02
Oulu/Uleaborg	1889	1988	99	−0.04
Vaasa/Vasa	1883	1988	104	−0.01
Lyokki	1858	1937	79	0.02
Lypyrtti	1858	1937	79	0.01
Helsinki	1879	1988	109	0.02
København	1889	1970	81	−0.01
Fredericia	1889	1970	80	−0.01
Aarhus	1888	1969	81	−0.01
Esbjerg	1889	1970	81	−0.01
(2) Aberdeen II	1862	1966	104	0.03
North Shields	1895	1991	96	0.00
Newlyn	1915	1991	76	−0.01
Brest	1807	1991	184	0.01
Cascais	1882	1987	106	0.02
(3) Marseille	1885	1988	103	−0.02
Genova	1884	1989	105	0.00
Trieste	1905	1991	86	−0.02
(4) Bombay	1878	1987	109	−0.02
(5) Tonoura	1894	1984	90	−0.04
(6) Sydney	1897	1991	94	0.03
Auckland	1904	1989	85	−0.02
(7) Honolulu	1905	1989	84	−0.02
(8) Seattle	1899	1989	90	0.03
San Francisco	1854	1989	134	0.02
San Diego	1906	1989	83	0.01
(9) Buenos Aires	1905	1988	83	0.05
(10) Baltimore	1902	1989	86	−0.01
New York	1856	1989	133	0.01

*Accelerations are for entire records

station apiece. The average record length of the stations in Table 1 is 92 years.

Acceleration values for each tide-gauge record are shown in the last column of the table. There is some scatter, but the majority of them are far lower in absolute magnitude than the anticipated future acceleration (up to 0.2 mm/yr²) of sea-level rise. It is also worth noting that the 37 values are evenly distributed as to sign. Table 1 suggests that the apparent acceleration of sea level in the last century has been very small; even the sign is not obvious. The real acceleration, if any, in existing long records is too small compared to low-frequency undulations in the data to permit precise determination.

The apparent acceleration of sea level for each of the groups in Table 1 was next estimated according to the following scheme. For any group, a unique trend for each station and a single acceleration value satisfying all stations were estimated simultaneously. Thus records of disparate length in a group are correctly weighted, and records that reach very far back influence the calculated value of acceleration for the group. The group accelerations are given in Table 2. Note that they too are evenly divided as to sign.

The standard deviations shown in Table 2 for the group-acceleration values are based on an estimate of the noise level of the data. They have been adjusted to give an a posteriori variance of unit weight of 1. The larger groups

TABLE 2 Apparent Acceleration Values for the Groups Shown in Table 1

Group Number	Acceleration (mm/yr^2)	Standard Deviation*
(1)	−0.010	.002
(2)	0.011	.003
(3)	−0.008	.005
(4)	−0.015	.006
(5)	−0.039	.010
(6)	0.015	.007
(7)	−0.016	.012
(8)	0.022	.003
(9)	0.046	.012
(10)	0.006	.003

Mean Acceleration = 0.001 mm/yr^2 (±0.008)

*The error of the mean was computed from the residuals about the mean, not from the errors of the individual estimates.

give apparently more precise results only because of the huge number of values in them. Therefore, when the mean acceleration of the groups shown in Table 2 was calculated, each group estimate was given equal weight. These ten groups give a more meaningful estimate of global sea-level acceleration and its uncertainty than do the simply aggregated data, because their individual estimates constitute a more statistically independent sample. The 1-sigma error of the mean shown in Table 2 is based on the residuals about that mean, and for this case of 10 samples the 95 percent confidence interval is 2.26 times larger, or nearly 0.02 mm/yr^2. Thus it can be concluded that at the 95 percent confidence level, the mean acceleration over roughly the past century, across all groups, is of the order of 10 percent of the acceleration estimated for the early years of most global-warming scenarios.

It is possible to find a single, long (1905 to 1985) period in which each of the 10 groups in Table 1 has at least one usable member. The acceleration estimates for the period 1905 to 1985 are shown in the last column of Table 3. Both the mean value and the uncertainty are larger than in the case of the longer records, and once again it can only be concluded that the deviation from a linear rise is small compared to the acceleration predicted to accompany global warming.

Figure 6 suggests that a solution with records at least 50 years long should give a small uncertainty for acceleration. This is verified by computing an estimate for the apparent mean acceleration of the 10 groups in Table 3 from 1935 to 1985. The result is 0.003 (±0.03) mm/yr^2.

An issue as interesting as the apparent past acceleration of sea level is whether sea level can be used as a "leading indicator" of climate change, either regionally or globally, in a reasonable length of time, say a few decades. Since

TABLE 3 Stations with Nearly Complete Records from 1905-1985*

Oceanic Group (No.) and City	Acceleration (mm/yr^2)	Group Acceleration
(1) Varberg	−0.028	
Ystad	−0.017	
Kungholmsfort	0.031	
Landsort	0.009	
Stockholm	−0.001	
Ratan	−0.019	
Oulu/Uleaborg	−0.006	
Vaasa/Vasa	−0.008	
Helsinki	0.034	0.000
(2) North Shields	−0.027	
Cascais	−0.021	−0.024
(3) Trieste	−0.011	−0.011
(4) Bombay	−0.084	−0.084
(5) Tonoura	−0.064	−0.064
(6) Sydney	0.047	
Auckland	−0.009	0.019
(7) Honolulu	−0.013	−0.013
(8) Seattle	0.044	
San Francisco	0.029	
San Diego	0.019	0.031
(9) Buenos Aires	0.041	0.041
(10) Baltimore	−0.011	
New York	−0.015	−0.013

Mean Acceleration of Groups = −0.011 mm/yr^2 (±0.012)

*The formal uncertainty of each record is about the same, around 0.015 mm/yr^2; the uncertainty was calculated from the scatter about the mean.

an acceleration of 0.2 mm/yr^2 amounts to only 1 cm in 10 years and 4 cm in 20, the outlook for using tide-gauge measurements by themselves is clearly not good. These values are small compared to the interannual-to-interdecadal variations of sea level, casting doubt on the likelihood of obtaining a useful index of sea-level acceleration relatively quickly, say in a few decades. A simple test can be made by computing independent 20-year global solutions from the existing tide-gauge record. The result of subdividing

TABLE 4 Apparent Accelerations and Errors for the 80-year Period 1905-1985 Subdivided into Four 20-year Groups

Interval	Acceleration (mm/yr^2)	Standard Deviation
1905-25	−0.219	0.18
1925-45	−0.129	0.20
1945-65	−0.162	0.15
1965-85	+0.265	0.20

the set of records constituting the 1905 to 1985 data set into four 20-year subsets and determining the global acceleration parameter for each is shown in Table 4. Note that the standard deviations are an order of magnitude larger than the value for the entire 80-year period. What we see is what Figure 6 indicates: The 1-sigma uncertainty of apparent global sea-level acceleration determined from 20-year records is comparable to the acceleration of sea level anticipated by climate models.

CONCLUSIONS

Consideration of a global set of tide-gauge records made since 1850 indicates that the apparent acceleration of global sea level in that period has been small, much less than the acceleration predicted to accompany future greenhouse warming. However, confidently determining the future value of global sea-level acceleration from tide-gauge data alone would appear to require 50 years or more. But this requirement is too pessimistic for a final conclusion about using sea level as an indicator of global warming. The difficulty is caused by the large interannual-to-interdecadal variations of sea level. If these can be measured or modeled and removed from the data, acceleration will be revealed more clearly. This has been accomplished for one site, Bermuda, for a few decades. Roemmich (1990) shows that the interdecadal variations of sea level, which are on the order of 10 cm there, are explained by the density changes above 2000 m depth to within the accuracy of the measurements. It seems clear that a relatively small number of island, and properly selected coastal, tide gauges that are distributed in locations around the globe at which the effects of meteoro-

logical forcing and the ocean's response are thoroughly understood, could give a reliable estimate of sea-level acceleration in a much shorter time than the current tide-gauge network.

Finally, satellite altimetry offers a new method for observing low-frequency variations of sea level. Even though the accuracy of point values of sea-level variation obtained from satellite data is lower than that of data from tide gauges, the global coverage given by satellites has the potential of yielding very precise values for global sea-level changes. Cheney et al. (1991) have obtained agreement within 3 to 4 cm (rms) with individual island tide gauges for Geosat monthly mean values of sea level. Miller and Cheney (1990) have shown that much better regional results are possible. They obtained agreement within 9 mm (rms) between the average of 14 tropical Pacific gauges and the average of corresponding Geosat sea-level series. More recently, Nerem (1995) has obtained from TOPEX/POSEIDON data millimeter-level precision for 10-day estimates of global mean sea level. This confirms the potential of satellite altimeter data for monitoring sea level. A coordinated effort consisting of tide-gauge measurements, altimetric satellites, meteorological data, water-column-density observations, and modeling should make it possible to observe an acceleration of sea level soon enough for it to be of value in evaluating climate models and forecasts.

ACKNOWLEDGMENT

I wish to thank Ed Herbrechtsmeir for making the computations for this paper. This investigation was carried out under the auspices of the NOAA Climate and Global Change Program, directed by the NOAA Office of Global Programs.

Commentary on the Paper of Douglas

MICHAEL GHIL
UCLA

I should like to mention first that Dr. Douglas has a recent paper in the *Journal of Geophysical Research*, brief and very much to the point; it emphasizes even more than his talk did that the issue of trends and accelerations is tightly coupled with the issue of diagnosing and maybe understanding the interannual and interdecadal variability. I hope it will reassure Dr. Lehman that there is indeed some work going on in that area.

À propos of Dr. Douglas's work, I should like to show you a plot of a few selected records out of the 800 or so tide gauges available over the world. It reflects work I have been doing jointly with a student of mine, Yurdanur

Sezginer (Sezginer and Ghil, 1992). We selected those sea-level records that had no big gaps and were long enough to permit the study of interannual-to-interdecadal variability. We did spectral analyses first for each record, and then for the entire data set.

A quasi-biennial oscillation (QBO), which is defined in this case as having a period somewhere between 18 and 30 months, can be seen in these tide gauges; in Figure 1a (see Ghil, commentary on Douglas, in the color well) the darker points correspond to longer periods. More interesting is that you can see that these oscillations are distributed over the entire globe, but tend to be stronger in the central Pacific and along the west coast of the Americas (Figure 1b).

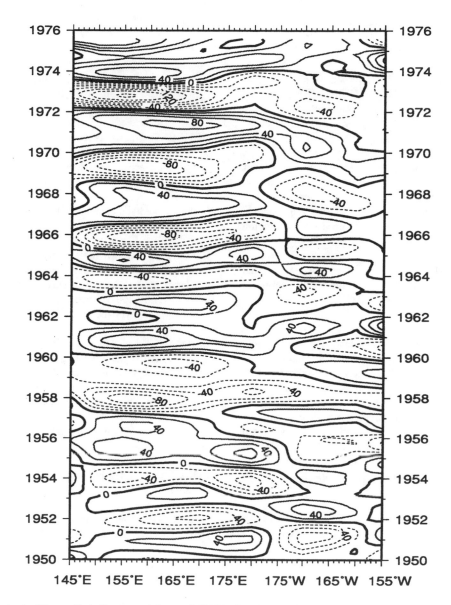

FIGURE 3 Time/longitude (Hovmöller) diagram of the total ENSO signal in sea-level height between approximately 145°E and 155°W. The diagram is based on spatial interpolation onto a regular grid of the SSA-filtered time series at eight stations slightly north of the equator (see text for details); latitude differences between stations are neglected. Contour interval is 20 mm (courtesy of Y. Sezginer-Unal).

Following the distinction made by Gene Rasmusson (Rasmusson et al., 1991) in the wind data, we also looked at the low-frequency mode (LFM) associated with ENSO (Figure 2, again in the color well). In that case, the amplitudes are even more tightly connected to the tropical Pacific and the west coast of the Americas. While a QBO-type oscillation does seem to be more global, this LFM is really concentrated in the tropical Pacific.

As for the spatial patterns, we took a group of eight stations just a little bit north of the equator in the western tropical Pacific and did a Hovmöller diagram (Figure 3). We wanted to look at the eastward-propagating signal in the quasi-biennial component of the zonal winds Dr. Rasmusson

mentioned. We actually combined the two components, QBO and LFM. In this plot, which goes from 145°E to 155°W and includes data from 1950 to 1976, you can see that there is not so much a propagating as a standing signal, and actually there seems to be something of a nodal line about 175°W.

So I think that we are getting on with this business of trying to define the spatial modes of interannual and interdecadal variability, which obviously is more interesting than just looking at spectral peaks. And perhaps what we should be noticing is that the peaks are in the same places, not in different places, for various records.

Discussion

BRYAN: Do you suppose that the basic reason we don't see sea-level changes corresponding to the temperature changes in the climatic record is that the main signal, which reflects recovery from the Little Ice Age, is so large as to mask any other trends?

DOUGLAS: Well, the current rate of rise is close to 2 mm/yr, half of which is unaccounted for by glaciers and upper-layer warming. It's true that it may reflect a short-term readjustment, since we know archeologically that we haven't had a 4-m rise in the last 2000 years. But I think the only thing we can conclude from the tide-gauge record is that there hasn't been any acceleration in the last 100 or 150 years.

GHIL: Some of the high-latitude records suggest that the internal variability in sea level substantially exceeds the warming trend of about 20 mm in 10 years.

LEVITUS: The interannual sea-level changes at Bermuda are highly correlated with the steric changes from the Panulirus station. We've found large-scale, basically mechanical displacements of the whole pycnocline in the subtropical gyre—but no simple explanation for them.

RASMUSSON: One of the great advantages we have in studying the ENSO cycle is that we know where to index it to get good results. Some of the things you were showing, Bruce, suggested that you could pick out index stations too. I think it's very important in maximizing results, and also for setting priorities on which stations should be maintained.

MUNK: I'd like to point out that a lot can be learned from seasonal variations in sea level—a 1-year period—which seems not to have been done adequately. To a first order, the seasonal variation is steric; you get a significant agreement between sea level and volumetric changes. As far as I know, no one has attempted to account for the observed seasonal variation in terms of the seasonal variation in wind-stress curl, which I think would be worthwhile.

My other point is that tide-gauge data will not really be useful until it is corrected for the structural variability. It's unacceptable to have the geologic noise be about the same magnitude as the oceanic noise. The technology for getting accuracy of a few millimeters in the movement of the ground to which the tide gauge is attached is available. Also, it seems to me that you'd learn more from a few gauges from which the crustal movement could be removed than from many observation points subject to unknown geologic variability.

DOUGLAS: I agree completely. In fact, my original proposal called for a dozen or so super-stations, where we could do density and geodetic measurements, but unfortunately they were too expensive to get funded.

KARL: Could you elaborate a little on the acceleration you didn't find and how that relates to global warming?

DOUGLAS: Well, we have nearly 150 years of good records with reasonable distribution, and overall they don't show any global acceleration of sea-level rise. That says that the 25 percent increase in CO_2 over the last 100 years has not caused anything that would accelerate sea-level rise—yet.

LEHMAN: I'd just like to point out that there is no consensus on the mass-balance status of the Antarctic ice sheet. Recent Greenland ice-sheet data seem to support Zwaly's controversial estimates, which suggest that it might actually have a positive mass balance. There are large uncertainties about the effect of ice volume on sea level; if it's offsetting a steric or thermally induced rise, it could be an important indicator that the ocean is absorbing greenhouse heat.

DOUGLAS: We really need a satellite-mounted laser to monitor the Arctic and Antarctic.

DESER: The large accelerations in sea level at Charleston and Bermuda between about 1920 and 1940 correspond to a time when there was warming along the Gulf Stream. I suspect that any differences between them might reflect changes in transport in the Gulf Stream.

GHIL: Ingemar Holmström has performed the data reduction you suggest, applying empirical orthogonal functions. He found very little independent information in the Baltic tide gauges; the first EOF contains some 90 percent of the variance.

I think we could sum all this up by saying that tide gauges may have relatively little to say about global warming, but they have a lot to say about air-sea interaction on interdecadal time scales.

Acoustic Thermometry of Ocean Climate[1]

WALTER H. MUNK[2]

A number of the workshop participants have alluded to the good spatial coherence of climate fluctuations, present and past, measured at separated sites. A system of climate observations should take into account the fact that variability on decadal and century time scales is generally associated with large spatial scales. In the ocean, this means gyre and basin scales of the order of say, 10 megameters (10,000 km).

I wish to discuss briefly a program of measuring climate-related fluctuations in ocean temperatures. The speed of sound is a good temperature indicator, increasing by 4 m/s per degree centigrade. Travel time between separated sites A and B yields a spatial average of temperature along the transmission path between A and B. The separation can be as great as 10 megameters, a distance well matched to climate scales. In most cases such a spatial average is preferable to traditional point measurements.

The ocean is a very good propagator of sound. It has exceptional acoustic properties because of the existence of the "sound-fixing and ranging" (SOFAR) channel, a classical wave guide, typically at a depth of 1000 m. The SOFAR channel owes its existence to a minimum in sound speed, with sound increasing upward from the SOFAR axis with increasing temperature, and increasing downward from the axis with increasing pressure. Sound is trapped in the wave guide, and the attenuation at the lossy top and bottom boundaries is largely avoided.

In January 1991 we carried out a feasibility test to determine whether 10 megameter ranges were achievable. We would like to be able to measure changes of 5 millidegrees Kelvin per year. This number derives from various model studies of greenhouse warming, from temperature changes inferred from a sea-level rise of 2 mm per year, and from a few direct observational time series. An increase of 5 millidegrees per year translates to a decrease in travel time by 0.2 s per year. To measure this we need a time precision of better than 0.05 s per year. This in turn demands that we use stable and coded acoustic signals from electrically driven (non-explosive) sources.

Our 1991 feasibility test was aimed at determining whether such electric sources were of sufficient intensity to be heard at 10-megameter ranges, and whether the codes could maintain sufficient fidelity to be read at such distances. Both questions have been affirmatively answered. In fact, usable signals were read at 17-megameter ranges, almost halfway around the globe.

We are now aiming at developing a global acoustic network that can resolve variations on a gyre and basin scale. The array will be designed to monitor ambient variability on decadal and century scales, as well as to detect a possible greenhouse-induced ocean warming.

---

[1]Summary of remarks made at dinner on 24 September 1992
[2]Scripps Institution of Oceanography, University of California San Diego, La Jolla, California

Transient Tracers as a Tool to Study Variability of Ocean Circulation

PETER SCHLOSSER[1,2] AND WILLIAM M. SMETHIE, JR.[1]

ABSTRACT

During the past decades, a variety of anthropogenic trace substances have been delivered to the ocean surface at relatively well-known rates. These transient tracers are used to study deep-water formation processes and rates, and to estimate mean residence times of waters found in the thermocline. Long-term observations of transient tracers indicate that they are useful parameters for studying ocean variability. On the basis of a review of the main elements of the tracer methodology and tracer time series from the Greenland/Norwegian seas and the Deep Western Boundary Current along the east coast of North America we discuss the potential of transient tracers as a tool for studying variability of ocean circulation.

INTRODUCTION

Prediction of the potential impact of greenhouse gases on our present climate requires understanding of the major elements of the climate system, including the ocean. Of particular concern is the understanding of the natural variability of the climate system, which has to be subtracted from any observed climatic trend before its significance can be evaluated.

There are numerous tools that can be used to study variability in the ocean, such as long-term observations of the classical parameters, temperature and salinity, or of currents. A more recently developed tool is the measurement of trace substances of natural or anthropogenic origin that carry time information. The time information can be derived from radioactive decay (radioactive clock) or from time-dependent delivery of the tracers to the surface waters from which they spread into the deep basins (dye tracers). In principle, both steady-state tracers and transient tracers can be used to detect variability in the ocean. A basic requirement for the application of the tracer method to detection of ocean variability is that multiple observations be made over a period of time of a certain tracer in a specific water mass or at a fixed location. Unfortunately, few such tracer time series exist, and the existing observations cover a rather short time span.

However, the available data clearly demonstrate the potential of tracers for studies of ocean variability (e.g., Rhein, 1991; Schlosser et al., 1991a). In this contribution we review the principles of the transient-tracer methodology and discuss its potential for studies of ocean variability on the basis of two selected examples.

---

[1]Lamont-Doherty Earth Observatory of Columbia University, Palisades, New York
[2]Department of Geological Sciences, Columbia University, New York City

TRANSIENT TRACERS

This section reviews the properties of the transient tracers that are currently most widely used in oceanographic studies. We do not attempt to provide a complete assessment of all potential transient tracers, but concentrate on those which in our opinion hold the greatest promise for future regional and global studies.

Tritium/³He

Tritium (^{3}H) is produced naturally in the upper atmosphere by interaction of cosmic rays with oxygen and nitrogen. It is oxidized to HTO and takes part in the hydrological cycle. The production rate of 0.5 ± 0.3 atoms $cm^{-2}sec^{-1}$ (Craig and Lal, 1961) leads to natural tritium concentrations in continental precipitation of about 5 TU (see, e.g., Roether, 1967; 1 TU or tritium unit means a tritium-to-hydrogen ratio of 10^{-18}). Natural tritium concentrations of ocean surface waters have been estimated to be close to 0.2 TU (see, e.g., Dreisigacker and Roether, 1978). Such relatively low tritium concentrations can be well measured with state-of-the-art technology (see below). However, the natural tritium signal was masked more or less completely by delivery of bomb tritium to the atmosphere, mainly during the surface nuclear-weapons tests in the early 1960s. The bomb tritium levels exceeded the natural tritium concentrations in continental precipitation by about 2 to 3 orders of magnitude (Weiss et al., 1978). The tritium concentrations of Northern Hemisphere surface waters were raised from about 0.2 to about 17 TU (Dreisigacker and Roether, 1978).

Tritium decays to ^{3}He with a half-life of 12.43 years (Unterweger et al., 1980). Tritiogenic ^{3}He elevates the ^{3}He/^{4}He ratio of the waters in the ocean by up to roughly 50 percent (typical values are between 0 percent near the surface and about 10 to 20 percent in the lower thermocline). For much of the ocean, the tritiogenic ^{3}He component can be separated from the other helium components found in ocean waters (mainly atmospheric and mantle helium; see, e.g., Roether, 1989, and Schlosser, 1992). In these cases an apparent tritium/^{3}He age can be calculated. (For details, see the section on tracer ratios below.)

Tritium was traditionally measured radiometrically, using either scintillation counters or proportional counters. The precision and detection limits achieved by radiometrical measurement are about ± 2.5 to ± 5 percent and 0.05 to 0.08 TU, respectively (Östlund and Brescher, 1982; Weiss et al., 1976). These values have been improved significantly by mass spectrometric measurement of tritium using the ^{3}He ingrowth method (see, e.g., Clarke et al., 1976; Bayer et al., 1989; Jenkins et al., 1991). State-of-the-art mass spectrometric systems achieve precision of ± 1 to ± 2 percent and detection limits below 0.005 TU.

^{3}He is measured mass spectrometrically. Typical precision of the ^{3}He/^{4}He ratio measurement is about ± 0.2 percent, and precision of the ^{4}He measurement is of the order of ± 0.5 to ± 1 percent (Bayer et al., 1989; Jenkins et al., 1991).

Halocarbons

During the past decade several halocarbons have come into use as tracers of oceanographic processes. These substances include CFC-11, CFC-12, CFC-113, and carbon tetrachloride. They all have strong anthropogenic sources and are chemically stable in the troposphere. However, they decompose in the stratosphere, and their decomposition products cause ozone destruction. They are also greenhouse gases, and thus a considerable effort has been made to determine and to monitor their atmospheric concentration as a function of time. The well-known atmospheric time histories of these substances allow a boundary condition for their input to the oceans to be accurately determined, which is a fundamental requirement for obtaining quantitative information about ocean processes.

CFC-11 (CCl_3F) and CFC-12 (CCl_2F_2) are by far the most widely measured halocarbons in the ocean. CFC-11 has been employed extensively as a propellant in aerosol spray cans, and also in many manufacturing processes, particularly the expansion of plastic foams. CFC-12 is the primary coolant used in refrigerators, and is used in air conditioners as well. The manufacture of CFC-12 began in the mid-1930s and that of CFC-11 in the early 1940s; both substances came into wide use in the 1950s. The atmospheric lifetimes for CFC-11 and CFC-12 have been estimated to be 74 and 111 years, respectively (Cunnold et al., 1986).

CFC-11 and CFC-12 are measured by gas chromatography with an electron-capture detector. Maximum concentrations in high-latitude surface waters are greater than 6 pmol kg^{-1} for CFC-11 and 3 pmol kg^{-1} for CFC-12. Concentrations as low as 0.01 pmol kg^{-1} can be detected. Measurements can be made at sea at a rate of 60 to 70 samples per day. The precision of the measurement, as reported by several different laboratories, is the larger of 0.01 pmol kg^{-1} or ± 1 percent. An intercalibration cruise conducted in 1989 showed that this level of agreement was approached for CFC-11 for six different laboratories, but for CFC-12 the level of agreement was the larger of 0.02 pmol kg^{-1} or ± 2.5 to ± 3 percent (Wallace, 1992). The SIO 1986 calibration scale, which is widely used for oceanographic measurements, is estimated to be accurate to ± 1.3 percent for CFC-11 and ± 0.5 percent for CFC-12 (Bullister and Weiss, 1988).

CFC-113 (CCl_2FCClF_2) has been used during the last couple of decades as a cleaning solvent in industry, particularly in the electronics industry. Its atmospheric concentration has been increasing much more rapidly than that of

CFC-11 or CFC-12 during this time, and Khalil and Rasmussen (1986) have estimated a 13 percent annual rate of increase from measurements made at Point Barrow, Alaska from 1983 to 1985. Like those of CFC-11 and CFC-12, the atmospheric lifetime of CFC-113 is long; it has been estimated to be between 136 and 195 years (Golombek and Prinn, 1989). Its potential as a tracer in oceanography was demonstrated by Wisegarver and Gammon (1988) from measurements in the North Pacific subarctic gyre, but few measurements have been reported since then.

Like CFC-11 and CFC-12, CFC-113 is measured by electron-capture gas chromatography. It has proven to be much more difficult to measure than CFC-11 or CFC-12. One reason for this is severe contamination problems. Much of the equipment used for the measurement, such as pressure regulators and valves, is cleaned with CFC-113 when it is manufactured. This causes problems not only with the analysis but also with preparation of standards. Another major problem has been interference by naturally occurring halocarbons. These problems are now being overcome, but there are not sufficient data to determine the precision and accuracy to which CFC-113 can be measured. Theoretically, it should be possible to make the measurement to the same accuracy and precision as have been achieved for CFC-11 and CFC-12.

Carbon tetrachloride (CCl_4) is another man-made halocarbon that has been produced in significant quantities since the early 1900s. Widely employed for dry cleaning and as an industrial solvent, it is currently used as feedstock in the manufacture of CFC-11 and CFC-12. The most recent estimate of its atmospheric lifetime is 40 years, and its current rate of increase in the atmosphere is 1.3 percent per year (Simmonds et al., 1988). The input of carbon tetrachloride to the ocean extends back in time 3 to 4 decades earlier than the CFC-11 and CFC-12 input, and thus it has penetrated the subsurface waters of the ocean farther than CFC-11 or CFC-12.

CCl_4 can also be measured by electron-capture gas chromatography, and this procedure is routinely used for air samples. Procedures for routine analysis of seawater samples are currently under development (e.g., Krysell and Wallace, 1988; Wallace et al., 1992).

$^{85}Krypton$

^{85}Kr is a radioactive noble gas with a half-life of 10.76 years. It is formed when uranium and plutonium undergo fission. Although there is a small natural abundance, anthropogenic sources are over 6 orders of magnitude greater than natural sources (Rozanski, 1979). The major anthropogenic source is fission in nuclear reactors used for energy and plutonium production; there was also some production during the atmospheric nuclear-weapons testing in the 1950s and early 1960s. In reactors, most of the ^{85}Kr is trapped in the fuel rods and released to the atmosphere when the fuel rods are reprocessed. This is done almost exclusively in the Northern Hemisphere, and accounts for the approximately 20 percent difference in atmospheric concentrations between the Northern and Southern Hemispheres.

^{85}Kr is measured by counting the β-particles emitted during radioactive decay. Approximately 200 liters of seawater are required for each analysis, and samples must be counted for about one week. Thus, only a few hundred samples have been measured. The accuracy of the measurement is about ±5 percent for near-surface water and ±10 percent for deeper water with concentrations less than 25 percent of surface values (Smethie and Mathieu, 1986).

Bomb ^{14}C

Natural ^{14}C is produced mainly in the stratosphere by the $^{14}N(n,p)^{14}C$ reaction. It enters the ocean as $^{14}CO_2$ via gas exchange between the ocean and the atmosphere. Natural ^{14}C concentrations in ocean surface waters range from about −50 to −150 permil (in the Δ notation; for details of the notation, see Stuiver and Pollach, 1977). During the surface tests of nuclear weapons, mainly in the late 1950s and early 1960s, a significant amount of bomb ^{14}C was added to the natural inventory, increasing the atmospheric ^{14}C concentrations by roughly 90 percent. Observation of the invasion of bomb ^{14}C into the ocean can be used to study ocean variability. The behavior of ^{14}C differs from that of CFCs or ^{85}Kr, because the exchange time of ^{14}C with the atmosphere is modified by the buffering effect of carbonate and bicarbonate ions dissolved in seawater. Typical times for equilibration of a 100 m thick water layer are of the order of a decade for ^{14}C, while they are only of the order of several weeks for CFCs or ^{85}Kr. ^{14}C also yields information on the average CO_2 invasion rate into ocean surface water. Until recently, ^{14}C was measured radiometrically using low-level proportional counters (e.g., by Schoch and Münnich, 1981). This technique allows very precise measurements (1σ error: ±2 to ±4 permil). However, it requires collection of about 200 liters of water per sample and CO_2 extraction at sea, which limits the currently available number of ^{14}C data. The relatively new AMS (accelerator mass spectrometry) technique requires less than 0.5 liters of water per sample. Routine procedures for CO_2 extraction from the water and preparation of graphite targets have been developed (e.g., Schlosser et al., 1987; Kromer et al., 1987). The best precision currently reached for AMS measurements of oceanic water samples is about ±5 to ±10 permil (see, e.g., Kromer et al., 1987), but this precision is sufficient for studies of the penetration of bomb ^{14}C into the ocean.

BOUNDARY CONDITIONS

Tritium/^3He

Tritium is delivered to the ocean by water-vapor exchange, precipitation, and river runoff (see, e.g., Weiss and Roether, 1980). Using a hydrological model and tritium data in precipitation observed at numerous stations of the WMO/IAEA network, Weiss and Roether estimated the delivery of tritium to the oceans as a function of time and space. Using a complex atmospheric circulation model, Koster et al. (1989) derived tritium fluxes to the ocean similar to those estimated by Weiss and Roether (1980). Dreisigacker and Roether (1978) converted the flux boundary condition given by Weiss and Roether (1980) into a concentration boundary condition for the North Atlantic (between 20 and 60°N). This "Dreisigacker/Roether function" (Figure 1) was used either directly or in modified form in many tracer studies, among them Sarmiento (1983), Thiele et al. (1986), Smethie et al. (1986), Thiele and Sarmiento (1990), Heinze et al. (1990), and Schlosser et al. (1991a). Recently, Doney et al. (1992) reevaluated the IAEA tritium data using a factor analysis. They derived tritium concentrations in precipitation as a function of time and space with an estimated error of ±3 to ±10 percent. These data can be used to derive a relatively accurate flux boundary condition for tritium.

Tritiogenic ^{3}He is exchanged between the atmosphere and the ocean on a time scale of the order of 1 week to 1 month. Since the atmospheric helium concentration (5.24 ppm; Glueckauf and Paneth, 1945) as well as the ^{3}He/^{4}He ratio (1.384 $\times$ 10^{-6}; Clarke et al., 1976) are constant for practical purposes, the ^{3}He boundary condition for ocean surface water in equilibrium with the atmosphere is well defined. However, in areas of deep convection or in ice-covered regions, significant ^{3}He excesses above solubility equilibrium can be observed (see, e.g., Fuchs et al., 1987; Schlosser et al., 1990).

Halocarbons

The first accurate measurements of CFC-11 and CFC-12 in the atmosphere were made in the mid-1970s. Since 1978, high-quality continuous measurements of both substances have been made at a number of different locations around the earth (see, e.g., Cunnold et al., 1986). McCarthy et al. (1977) have shown that the amount of CFC-11 and CFC-12 released to the atmosphere can be determined from the industrial production data, and atmospheric concentrations prior to the mid-1970s can be estimated from the amount of CFCs released. The cumulative amount of CFC-11 and CFC-12 in the atmosphere can be calculated as a function of time from this release data and the atmospheric lifetimes of CFC-11 and CFC-12. The atmospheric lifetimes have been estimated to be 57 to 105 years with a best estimate of 111 years for CFC-12 (Cunnold et al., 1986). The CFC inventory as a function of time can be calculated from

$$I_i = I_{i-1} - \lambda \frac{I_i + I_{i-1}}{2} + R_{i-1} \qquad (1)$$

where I is the atmospheric CFC inventory in January of a given year, R is the amount of CFC released during a given year, λ is the reciprocal of the atmospheric lifetime of CFCs, and i denotes the year. Inventories are converted to concentrations by normalizing them to concentrations measured after the mid-1970s.

Smethie et al. (1988) used the procedure described above to estimate the CFC-11 and CFC-12 atmospheric concentrations as a function of time. The Chemical Manufacturers Association (1983) global release data were used; they are thought to be the most reliable up to 1976 because they contain Russian data for the 1968-to-1975 period. The normalization was done using 1976 data based on the SIO 1986 concentration scale. (Normalizations could have been done using data from later years, but the inventories are less accurate in later years because Russian data are not available after 1975.) The calculation was made using the extreme values for the atmospheric lifetimes, but this had only a small effect on estimated concentrations. The maximum difference in concentrations was 2.3 percent in 1950 decreasing to 0.3 percent in the mid-1970s for CFC-11, and 5 percent decreasing to 0.4 percent for CFC-12.

CFC-11 and CFC-12 concentrations for the Northern Hemisphere, based on the best estimate of the atmospheric lifetimes prior to 1976 and measurements after 1976, are shown in Figure 2. The production and use of CFCs occur mainly in the Northern Hemisphere, and concentrations of both CFC-11 and CFC-12 are 7 to 8 percent higher in the Northern Hemisphere than in the Southern Hemisphere (Cunnold et al., 1986), with a sharp decrease in concentration across the intertropical convergence zone. CFC-11 and CFC-12 are well mixed in both hemispheres, although concentrations above coastal waters can be a few percent higher than elsewhere in the atmosphere remote from land, particularly near the northeastern United States and Europe (Prather et al., 1987). The annual increase in CFC-11 and CFC-12 atmospheric concentrations is between 4 and 5 percent per

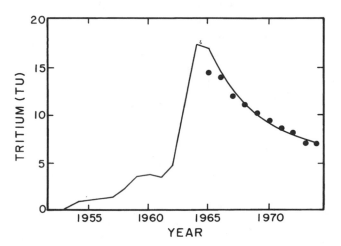

FIGURE 1   Tritium concentration in North Atlantic surface water as a function of time. (From Dreisigacker and Roether, 1978; reprinted with permission of Elsevier Science Publishers.)

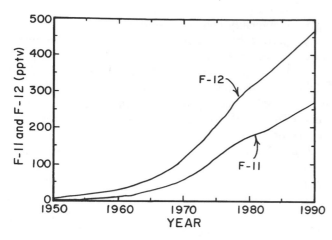

FIGURE 2   Atmospheric CFC-11 and CFC-12 concentration as a function of time for the Northern Hemisphere.

year today, but exceeded 20 percent per year in the past (Figure 3).

CFC-11 and CFC-12 enter the ocean from the atmosphere by gas exchange with an average equilibration time of the order of 1 month (Broecker and Peng, 1984). Therefore, most of the surface ocean is in equilibrium with the atmosphere, and the surface ocean concentration can be calculated from the atmospheric concentration and the CFC-11 and CFC-12 solubility. These solubilities are well known from laboratory measurements (Warner and Weiss, 1985); they are strongly dependent on temperature (Figure 4). In regions of the ocean where deep convection occurs, gas exchange may not be rapid enough to equilibrate the convecting waters, and low CFC concentrations have been observed in such regions. Bullister and Weiss (1983) observed CFC-11 and CFC-12 concentrations to be at 77 percent of satura-

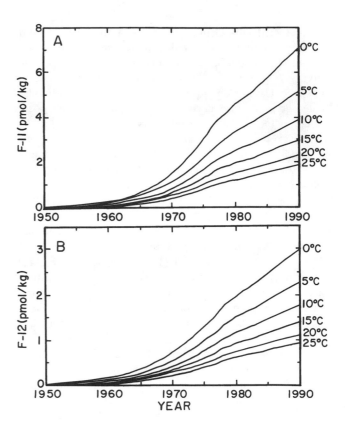

FIGURE 4   CFC-12 (a) and CFC-11 (b) concentrations of Northern Hemisphere surface water in equilibrium with the atmosphere for different temperatures.

tion in the Greenland Sea, and Wallace and Lazier (1988) observed recently formed Labrador Sea Water to be at 65 percent saturation. Since deep convection is an important process in the formation of many subsurface water masses, it is important to understand the extent to which these regions reach equilibrium with the atmosphere.

Atmospheric measurements of CFC-113 have been reported in the literature by two groups, the Oregon Graduate Center and the Climate Monitoring and Diagnostics Laboratory of NOAA. These results are reported on different calibration scales, and Golombek and Prinn (1989) have indicated that the concentrations reported on the Oregon Graduate Center scale may be 30 percent too low. The atmospheric concentration of CFC-113 can be estimated from industrial release data, as were those of CFC-11 and CFC-12. Such an estimate is presented in Figure 5. This curve was normalized to the concentration of CFC-113 at 35° over the western Pacific Ocean in 1987 measured by the NOAA Climate Monitoring and Diagnostics Laboratory (Thompson et al., 1990). Measurements made by the Oregon Graduate Center that have been increased by 30 percent are also plotted. The curve estimated from the industrial release data fits the observations well.

CFC-113 enters the ocean from the atmosphere by gas

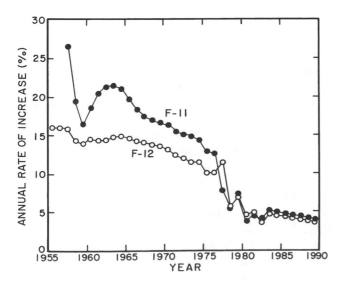

FIGURE 3   Annual rate of increase of the atmospheric CFC-11 and CFC-12 concentrations for the Northern Hemisphere.

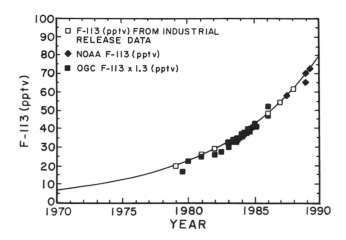

FIGURE 5 Northern Hemisphere atmospheric CFC-113 concentration as a function of time. The solid line is an exponential fit to the CFC-113 concentrations estimated from industrial-release data normalized to NOAA observations.

exchange, and as with CFC-11 and CFC-12, the surface ocean concentration should be in equilibrium with the atmosphere except in regions of active deep convection. Laboratory measurements of the solubility of CFC-113 are still in progress (Bu and Warner, 1995), but Wisegarver and Gammon (1988) estimated it from surface ocean observations where saturation levels of CFC-11 and CFC-12 were measured. It lies between the solubilities of CFC-11 and CFC-12.

The concentration of carbon tetrachloride in the atmosphere has been continuously measured since 1978, and its concentration prior to this time can be determined from industrial release data, as was done for CFC-11 and CFC-12. Its concentration in the Northern Hemisphere is 6 to 7 percent higher than in the Southern Hemisphere (Simmonds et al., 1988).

Carbon tetrachloride, CCl_4, enters the ocean by gas exchange, as discussed above for other halocarbons. However, it is not as chemically stable in seawater as these other halocarbons. It slowly hydrolyzes, and Jeffers and Wolfe (1989) have estimated half-lives of 40 years at 25°C, 468 years at 10°C, and 2790 years at 0°C. This limits its usefulness in warm thermocline waters, though not in the cold deep water that comprises most of the ocean.

The atmospheric concentrations of the halocarbons discussed above are all well known as a function of time since the mid-1970s. Prior to this time, concentrations must be estimated from industrial production data. This is not so important for CFC-113, since it has been in use in significant quantities only since the early 1970s. For CFC-11 and CFC-12 the industrial production data are of high quality; they are thought to include 85 percent of global production. Thus, it is difficult to imagine that concentrations estimated from these data would be more than 5 to 10 percent in

error. The production records for CCl_4 are not as accurate as those for CFC-11 and CFC-12, and concentrations estimated from these data may have a larger error.

$^{85}Krypton$

The ^{85}Kr atmospheric concentration as a function of time is known from measurements taken at various locations over the earth since the 1950s (Rozanski, 1979). In the remote atmosphere there is a smooth increase with time, but there can be considerable variability in air masses that have been exposed to nuclear fuel reprocessing plants (Weiss et al., 1986). The concentrations in the Northern and Southern Hemispheres from a global atmospheric model fitted to observations (Rath, 1988) are shown in Figure 6. ^{85}Kr enters the ocean by gas exchange with a roughly 30-day equilibration time, like the halocarbons discussed above.

Bomb ^{14}C

The $\Delta^{14}C$ values of atmospheric CO_2 as a function of time are fairly well known (Figure 7). The histories of $\Delta^{14}C$ in near-surface waters can be reconstructed from ^{14}C measurements of growth-ring-dated corals (see Figure 8; examples appear in Druffel and Linick (1978) and Nozaki et al. (1978)). Using the known atmospheric ^{14}C concentration, the CO_2 exchange rate, and the surface water ^{14}C concentration, we can estimate the expected ^{14}C inventories for any station in the ocean (see, e.g., Broecker et al., 1985a). Comparison of calculated and observed inventories provides information on water mass transport. Observations made at different times yield information on variability of oceanic transport.

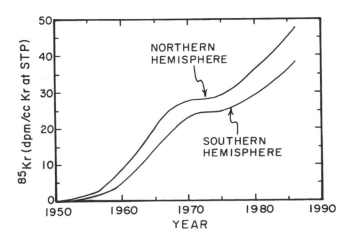

FIGURE 6 Atmospheric ^{85}Kr concentration as a function of time from Rath's (1988) 2-D model fitted to observations.

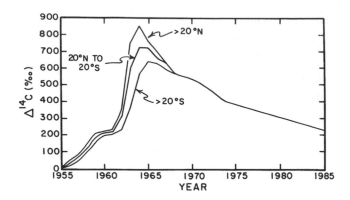

FIGURE 7 $\Delta^{14}C$ in atmospheric CO_2 as a function of time. (From Broecker et al., 1985a; reprinted with permission of the American Geophysical Union.)

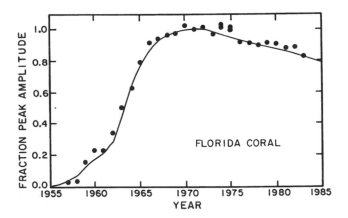

FIGURE 8   Shape of $\Delta^{14}C$ as a function of time for ocean surface water, based on ^{14}C measurements of growth-ring-dated Floridian corals made by Druffel and Linick (1978). (From Broecker et al., 1985a; reprinted with permission of the American Geophysical Union.)

TRACER RATIOS

The presence of any of the substances discussed in this paper indicates that the water has interacted to some extent with the atmosphere within the past several decades. Since the measured concentration is a function of the initial concentration in the source water, which can be determined from the atmospheric time history, and the amount of mixing that has occurred since the water parcel left the source region, a model is needed to place further constraints on the age of the water.

Various tracers have different input functions, and thus tracer ratios vary in a unique way as a function of time. These ratios can be used as a water-mass age indicator. If the ratios are not altered by mixing, which is the case when a water parcel mixes with tracer-free water, they give a relatively accurate estimate of the time since the water parcel became isolated from the surface. Mixing of waters

with different values of a specific tracer ratio can affect the age derived from this ratio in the same way as for absolute tracer concentrations. When a transient tracer first begins entering the ocean, the assumption of mixing with tracer-free water as a water parcel flows through the ocean is valid. However, with time the adjacent water begins to accumulate tracer by mixing, and this water, now tagged with tracer, mixes with newly formed subsurface water. Pickart et al. (1989) have shown for the Deep Western Boundary Current (DWBC) of the North Atlantic that this causes the CFC-11/CFC-12 ratio to be underestimated, and thus the age to be overestimated. This will be the case for any tracer/tracer ratio in which both tracers increase with time without radioactive decay. Thus, ages estimated by such a methodology are upper limits, and the inferred current speeds are lower limits.

Thiele and Sarmiento (1990), simulating the tracer ratio development in a Stommel-type gyre (Stommel, 1948), estimated the effect of mixing on ages derived from tritium/^{3}He and CFC-11/CFC-12 ratios. They found that for a wide range of parameters their simulated tritium/^{3}He ages were in very good agreement (about 10 percent) with a simulated true ventilation age. Deviations of simulated CFC-11/CFC-12 ratio ages from their true ventilation age were significantly larger, particularly for lower ages, because the CFC-11/CFC-12 ratio has been constant since the mid-1970s. Overall, the Thiele and Sarmiento simulations as well as the work done by Jenkins (1987) on tritium/^{3}He show that certain tracer-ratio ages are very close to real ventilation ages in typical oceanographic settings.

Several tracer ratios that have potential for determining water-mass ages or mean residence times are discussed below. A summary of these ratios is presented in Table 1.

Tritium/^{3}He Age

Parallel measurement of the radioactive mother/daughter nuclides tritium and ^{3}He allows us to calculate the tritium/^{3}He age of a water mass (see, e.g., Jenkins and Clarke, 1976). The tritium/^{3}He age has to be considered an apparent age. It reflects the true age of a water parcel only in cases where eddy diffusion is negligible compared to advection. Its behavior is non-linear with respect to mixing (see, e.g., Jenkins, 1974; Schlosser, 1992). Addition of water free of tritium and tritiogenic ^{3}He does not affect the tritium/^{3}He age of this water mass. This means that if young water is mixed into an old water body with no tritium and tritiogenic ^{3}He, the apparent tritium/^{3}He age of the mixture is the same as that of the young water component. (For details of the tritium/^{3}He dating method, see, e.g., Schlosser (1992).) However, simple model simulations show that the apparent tritium/^{3}He age of a water parcel is close to the mean residence time of this water if the age is below about 10 to 13 years (Figure 9; Wallace et al., 1992).

TABLE 1   Summary of Tracer Ratios, Together with Approximate Age Ranges and Age Resolutions

Tracer Ratio	Approximate Age Range	Age Resolution
Tritium/^{3}He age	mid-1960s to present	$\pm \approx 3$ months (present) to several years (1963)
CFC-11/CFC-12 ratio	mid-1950s to mid-1970s	± 1 to ± 2 years
CFC-113/CFC-11 ratio	1980 to present	$\approx \pm 1$ year
CCl$_4$/CFC-11 ratio	mid-1950s to present	$\approx \pm 1$ to several years
Tritium/^{85}Kr ratio	mid-1960s to present	± 1 to ± 4 years
Tritium/CFC-11 ratio	mid-1970s to present	± 1 to ± 2 years

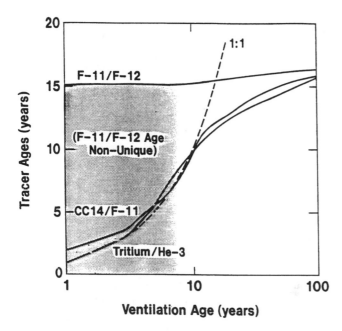

FIGURE 9   Apparent tracer ages obtained by a single mixed-box model versus ventilation age (mean residence time) of this box. The calculations were performed for the period between 1930 and 1987 (1987 values are plotted). The apparent tritium/^{3}He ages (and CCl$_4$/CFC-11 ages) are close to the mean residence time of the mixed box for mean residence times below about 10 to 13 years. (From Wallace et al., 1992; reprinted with permission of Pergamon Press.)

The apparent tritium/^{3}He age of a water parcel is independent of its original tritium concentration. It therefore is a much better tool for estimation of mean residence times than absolute tritium concentrations alone. For waters with mean residence times below about a decade, the tritium/^{3}He age is close to being a steady-state parameter. In addition, the time resolution of the tritium/^{3}He age is very good (of the order of several months). The combination of these properties makes the tritium/^{3}He age an ideal tool for studies of thermocline ventilation (see, e.g., Jenkins, 1987).

CFC-11/CFC-12 Ratio

From Figure 2 it can be seen that the atmospheric concentrations of CFC-11 and CFC-12 increased at different rates

until the mid-1970s. CFC-11 increased at a faster rate, so the CFC-11/CFC-12 ratio increased with time for different seawater temperatures, as shown in Figure 10. (The CFC-11/CFC-12 ratio in seawater increases at the same rate as that of the atmosphere, but the absolute values are different due to the different solubilities of CFC-11 and CFC-12 in seawater.) The CFC-11/CFC-12 ratio can be used to estimate ages of water parcels that left the surface between the mid-1950s and mid-1970s (see, e.g., Weiss et al., 1985; Smethie, 1993) with a resolution of ± 1 to ± 2 years. As discussed previously, these apparent ages are upper limits because of the effect of mixing. The ratio has been more or less constant since the mid-1970s, and for the past 15 years it is not useful for dating except to indicate that a water parcel was formed in the mid-1970s or later.

CFC-113/CFC-11 Ratio

CFC-113 has great potential as a tracer because of its rapid increase (13 percent per year) during the past 15 years. It ideally complements CFC-11 and CFC-12, whose ratio has been constant since the mid-1970s. A plot of the CFC-113/CFC-11 ratio for the atmosphere (Figure 11) indicates that the ratio is increasing at a rate of about 8 percent per year. The ratio in water will increase at the same rate, but

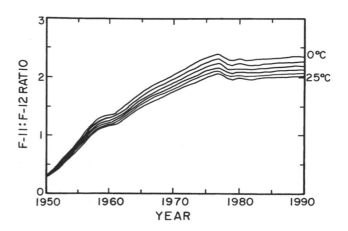

FIGURE 10   CFC-11/CFC-12 ratios as a function of time and temperature of Northern Hemisphere surface water.

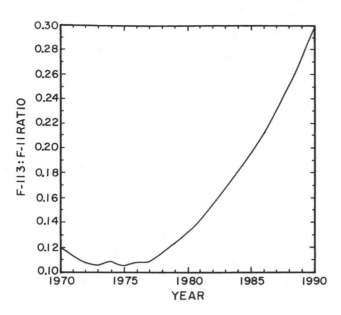

FIGURE 11   CFC-113/CFC-11 ratio as a function of time for the Northern Hemisphere.

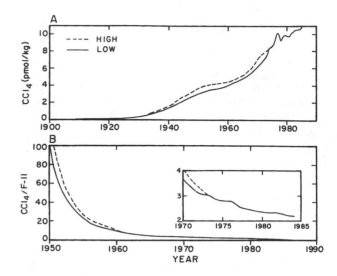

FIGURE 12   (a) Equilibrium surface water concentration of CCl_4 in the Northern Hemisphere. (b) CCl_4/CFC-11 ratio for Northern Hemisphere surface ocean water in equilibrium with the atmosphere. (From Krysell and Wallace, 1988; reprinted with permission of the American Association for the Advancement of Science.)

will be different due to the differences in the solubilities of CFC-11 and CFC-113. If both CFC-11 and CFC-113 are measured to ±2 percent, an age resolution of about 6 months can be achieved. However, the CFC-113 concentration in seawater is roughly an order of magnitude less than the CFC-11 concentration; for water that has been isolated from the surface for several years, the CFC-113 concentration will be less than 0.3 to 0.4 pmol kg^{-1}. A more realistic measurement error for this concentration range is ±4 to ±5 percent, resulting in an age resolution at present of about 1 year.

CCl_4/CFC-11 Ratio

Although carbon tetrachloride has been entering the ocean for a longer period of time than CFC-11, its rate of increase has generally been less than the rate of increase of CFC-11, and thus the CCl_4/CFC-11 ratio has been continually decreasing with time (see Figure 12, from Krysell and Wallace, 1988). From 1950 to 1975, this ratio provides another estimate of water-mass age in addition to that provided by the CFC-11/CFC-12 ratio. Since 1975, the CFC-11 concentration has been increasing at 4 to 5 percent per year, and CCl_4 at 1.3 percent per year. Therefore, the ratio is decreasing at about 3 percent per year, and if both CFC-11 and CCl_4 are measured to ±1 percent, an age resolution of ±1 year can be achieved. An age resolution of ±2 years is more realistic.

Tritium/[85]Krypton Ratio

Since the early 1960s tritium input to the ocean from the bomb spike has been continually decreasing, and [85]Kr

input has been continuously increasing. Therefore, the tritium/[85]Kr ratio decreases strongly with time. Both substances are radioactive, but decay at nearly the same rate (the half-life of tritium is 12.43 years, only slightly larger than 10.76 years for [85]Kr). The small difference in decay rate causes the ratio to change by about 1 percent per year, but this change can be calculated and a correction applied. Smethie and Swift (1989) determined the time history of the tritium/[85]Kr ratio for Denmark Strait Overflow Water (DSOW) (Figure 13); from the measured ratio just downstream of Denmark Strait, they estimated that pure DSOW resided behind the Denmark Strait sill for no longer than a year. They also used the measured ratio in Gibbs Fracture Zone Water at the same location to estimate that about

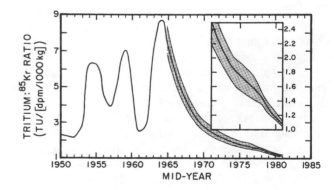

FIGURE 13   Tritium/[85]Kr ratio of Iceland Sea surface water as a function of time (corrected for radioactive decay). (From Smethie and Swift, 1989; reprinted with permission of the American Geophysical Union.)

7.5 years was required for it to be transported from its source region. The age resolution for the ^{3}He/^{85}Kr ratio is ±1 to ±4 years.

CFC-11/Tritium Ratio

In the Southern Hemisphere, tritium/^{3}He dating is severely limited, due to the low tritium concentration in surface waters and a relatively high mantle-^{3}He component. In addition, the CFC-11/CFC-12 ratio is fairly insensitive for the time after about 1975. This situation limits the palette of available tracer ratios for dating of young waters in the Southern Hemisphere. However, recent measurements of tritium and CFC-11 in the Weddell Sea indicate that the CFC-11/tritium ratio might be a useful parameter for tracer-ratio dating in this region (Schlosser et al., 1991b). The ratio must be corrected for radioactive decay of tritium. The decay-corrected CFC-11/tritium ratio increases as a function of time (Figure 14) at a rate of roughly 10 percent per year, yielding a time resolution of about ±1 to ±2 years.

EVALUATION OF TRANSIENT TRACER DATA

Since concentrations of transient tracers are continually increasing in the ocean, a time series of a single tracer usually does not reveal ocean variability directly. To determine whether there is variability, a model must be applied to the data. Depending on the hydrographic situation, the models can range from simple box models to complex general circulation models. If the model parameterizes water-mass formation (including exchange with the atmo-

sphere), circulation, and mixing in a reasonable way and can fit the time series with a set of constant (with respect to time) parameters, then little or no variability is suggested. If one or more of the parameters used in the model (for example, exchange rates between the individual water masses) must be changed as a function of time to fit the data, then there is variability. In special cases qualitative information can be derived from a single tracer time series. For example, if the concentration of a tracer with increasing surface concentrations increases in the deep water for a certain period of time and then suddenly stays at a constant level, there is a strong indication that the deep-water formation rate has changed.

Time series of tracer/tracer ratios might be used in special cases to determine whether there is variability without a model. If there is no variability in formation, circulation, or mixing (mixing has to be weak), water-mass ages estimated from the ratios should not change with time. Changes in the age would indicate variability in the formation rate of a specific water mass, its exchange rate with other water masses, or its transfer time from the formation region to the location of observation.

Straightforward application of tracer-ratio dating might work best on isopycnals in the thermocline (Thiele and Sarmiento, 1990) or in advectively dominated boundary currents (Smethie, 1993; see below).

EXAMPLES OF TRACER TIME SERIES

Atlantic Deep Western Boundary Current

CFC-11 and CFC-12 observations in the deep core of the DWBC in the Atlantic between 32° and 44°N were taken in 1983, 1986, and 1990 (Figure 15). Four sections across the DWBC were obtained in 1983, three sections in 1986, and five sections in 1990, with some overlap between the three surveys. In all of these sections there were two high-CFC-concentration cores adjacent to the western boundary: an upper core with a potential temperature of about 4.5°C at about 850 m depth, and a deep core with a potential temperature of about 2°C at about 3500 m depth (Figure 16). The upper core appears to be formed by winter-time convection in the southern Labrador Sea region (Pickart, 1992a), but is too warm to be classical Labrador Sea Water (Pickart, 1992a; Fine and Molinari, 1988). The lower core consists of the waters that flow over the Denmark Strait sill and the Scotland-Iceland Ridge (Smethie, 1993). The source waters for both cores interact extensively with the atmosphere and thus become tagged with high levels of CFCs. The high CFC concentrations within the cores indicate that the water in these cores formed more recently than adjacent deep water. This is also apparent in the CFC-11/CFC-12 ratios, which are higher in the cores than in the adjacent water (Figure 16).

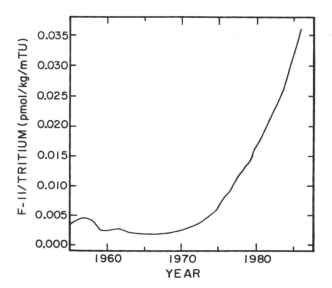

FIGURE 14   CFC-11/tritium ratio for Weddell Sea surface water (corrected for tritium decay).

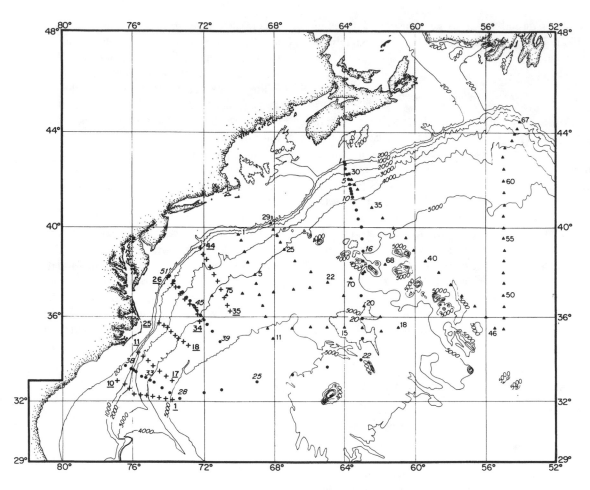

FIGURE 15   Station locations for the 1983, 1986, and 1990 surveys of the Deep Western Boundary Current.

These observations can be used in two ways to examine temporal variability in the DWBC: (1) Ages of the cores can be estimated from the CFC-11/CFC-12 ratio and compared for the different sections, as discussed in the previous chapter. (2) The CFC concentration can be used to identify the most recently formed component of the DWBC, as Pickart (1992b) has done using oxygen, and the temperature and salinity characteristics of this component can be compared at different locations and at different times at the same location along the western boundary current. This has been done for the deep CFC maximum for the 1983, 1986, and 1990 surveys.

For all three surveys the CFC-11 and CFC-12 concentrations in the deep core of the DWBC decrease in the downstream direction, and increase with time (Figure 17). These patterns are as expected, since the CFC-11 and CFC-12 concentrations have been continually increasing with time in the source waters. Ages of the deep core water estimated from the CFC-11/CFC-12 ratio increase in the downstream direction; this is also as expected, since younger water should be found closer to the source region. These ages increase in a linear fashion with distance from the source

region, indicating that the apparent mean velocity of the current does not vary with distance from the source region (Figure 17). This apparent mean velocity is about 1.2 cm sec^{-1}. It should be kept in mind that mixing causes the CFC-11/CFC-12 ratio age to be overestimated and the mean velocity of the current to be underestimated (Pickart et al., 1989). The important implication of these observations for temporal variability in the DWBC is that the estimated CFC-11/CFC-12 ratio ages from the three surveys lie along the same trend line when plotted versus distance from the source region (Figure 17). The estimated ages have an error of ±1 to ±1.5 years, which propagates to a ±5 to ±10 percent error in apparent mean velocity, and there is roughly ±10 percent variability in the calculated apparent mean velocity. This means that the age of the deep core between 32° and 44°N, as indicated by the CFC-11/CFC-12 ratio, did not change by more than about 10 percent between 1983 and 1990, which implies that the combination of water-mass formation rate, transport, and mixing with adjacent water has been constant to within about ±10 percent over this time period.

As mentioned above, the maximum CFC concentration

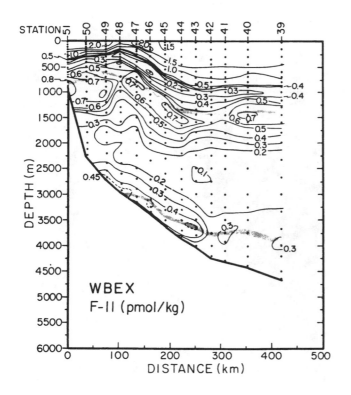

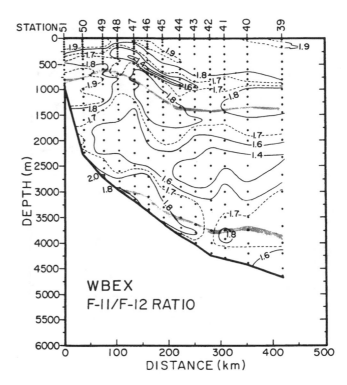

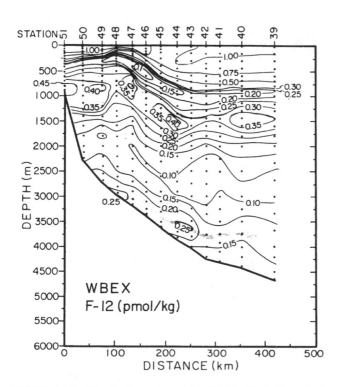

FIGURE 16 Vertical sections of CFC-11, CFC-12, and the CFC-11/CFC-12 ratio across the Deep Western Boundary Current at 36°N taken in 1986. (From Smethie, 1993; reprinted with permission of Pergamon Press Ltd.)

in the DWBC can be used to identify its most recently formed component. In this way the three surveys discussed above can be used to examine variability of temperature and salinity in the DWBC with time and distance along its flow path. Variability along the flow path also yields information on temporal variability because the DWBC can be thought of as a sequence of consecutive vintages of source water. A plot of temperature, salinity, and density versus distance from the source region shows some variability in space and time. The maximum differences between potential temperature, salinity, and density along the flow path are 0.17°C, 0.019 psu, and 0.026, respectively (Figure 18). The maximum differences between observations at the same location are 0.05°C, 0.007 psu, and 0.008. The combination of these findings and the observation that the water mass ages do not vary significantly with time suggests that small changes in the temperature/salinity characteristics of the source water can occur without changing the rate of formation, transport, and mixing of the deep core of the DWBC.

Some of the variability in the potential temperature and salinity of the DWBC core may be caused by not sampling the absolute CFC maximum due to vertical (100 to 150 m) and horizontal (20 to 30 km) sample spacing. Therefore, closer spacing should be used in future studies.

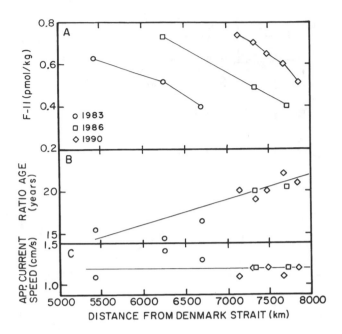

FIGURE 17 CFC-11 concentration (pmol kg⁻¹), CFC-11/CFC-12 ratio age, and apparent mean current velocity calculated from the ratio age versus distance from Denmark Strait. Surveys were conducted in 1983, 1986, and 1990.

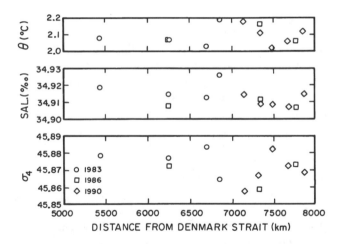

FIGURE 18 Potential temperature, salinity, and σ_4 versus distance from Denmark Strait for surveys conducted in 1983, 1986, and 1990.

European Polar Seas

Tracer data (mainly tritium and ³He, with some CFC-11/CFC-12 data) were collected between 1972 and 1988 at stations located in the Greenland and Norwegian seas (Figure 19). A small data set was also collected in the Eurasian Basin of the Arctic Ocean in 1984 and 1987 (Figure 19). The tracer concentrations in the deep waters (depth > 1500 m) of the Greenland and Norwegian seas are relatively homogeneous. Therefore, the concentrations in Greenland

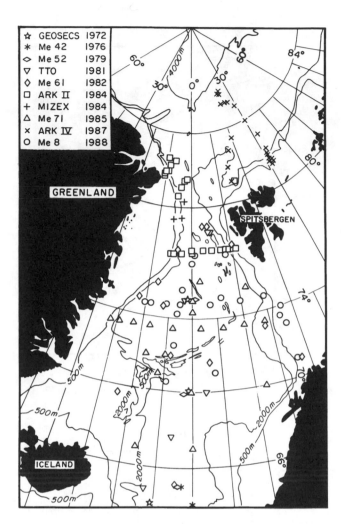

FIGURE 19 Tracer stations in the European polar seas occupied between 1972 and 1988. (From Schlosser et al., 1991; reprinted with permission of the American Association for the Advancement of Science.) Tracer data from these stations are used to construct the time series discussed in the text and displayed in Figure 20.

Sea Deep Water (GSDW) and Norwegian Sea Deep Water (NSDW) can be averaged for each survey. The time series of the averaged tritium and CFC-11 data of GSDW (Schlosser et al., 1991; Rhein, 1991; see Figure 20) show that the tritium concentration follows closely the curve expected from radioactive decay of tritium between 1980 and 1988, while the CFC-11 concentrations remained constant between 1982 and 1989. In addition, the tritium/³He age of GSDW increased more or less linearly between 1980 and 1988. These results suggest that almost no deep-water formation took place in the Greenland Sea in the period between 1980 and 1989.

A change in deep-water production was suggested as well by observations of temperature and salinity (GSP, 1990; Clarke et al., 1990). However, the salinity signals are small (of the order of 0.005 psu), and the T/S signal is definitely

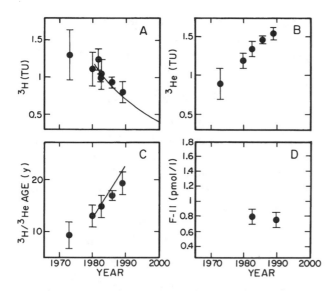

FIGURE 20   Temporal evolution of (A) the average tritium, (B) ³He, (C) tritium/³He age, and (D) CFC-11 values of Greenland Sea Deep Water (GSDW; depth > 1500 m). (From Schlosser et al., 1991; reprinted with permission of the American Association for the Advancement of Science.)

too small to be evaluated quantitatively. In contrast, the transient tracer signals are relatively large.

For quantitative evaluation of the transient tracer time series, a model is needed that simulates the basic exchange processes in the study area. In the case of the GSDW, we have to simulate the deep water formation and the coupling of the deep Greenland Sea to the Norwegian Sea and the Eurasian Basin of the Arctic Ocean. For this purpose, we modified a simple box model developed by Heinze et al. (1990) (Figure 21). The model allows deep-water formation

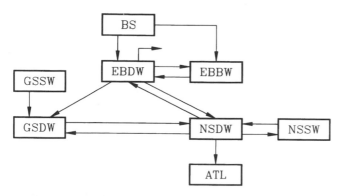

FIGURE 21   Box model used to simulate the GSDW tracer concentrations as a function of time. (NSDW: Norwegian Sea Deep Water; GSDW: Greenland Sea Deep Water; EBDW: Eurasian Basin Deep Water; EBBW: Eurasian Basin Bottom Water; BS: Barents Sea Water; GSSW: Greenland Sea Surface Water; NSSW: Norwegian Sea Surface Water; Atl: Overflow into Atlantic. (From Schlosser et al., 1991; reprinted with permission of the American Association for the Advancement of Science.)

in the main basins (Greenland Sea, Norwegian Sea, and Eurasian Basin). Exchange of deep water between the basins is implemented on the basis of hydrographic studies (Swift et al., 1983; Aagaard et al., 1985; Swift and Koltermann, 1988; Smethie et al., 1988). The model is then tuned to simulate a variety of steady-state and transient parameters in the deep waters (temperature, salinity, tritium, ³He, CFC-11, ⁸⁵Kr, and ³⁹Ar). The parameters used to tune the model are the deep water formation rates and the exchange rates of deep water between the individual basins.

The GSDW tracer time series can be reproduced by the model only if the formation rate of GSDW is variable (Figure 22; Schlosser et al., 1991a). The model requires a reduction of the GSDW formation rate by about 80 percent starting in 1980 (±2 years). With such an assumption the model fits the observations perfectly. Simulations of the tracer concentrations expected for the period between 1990 and 2000 for two scenarios with reduced deep-water formation rate (Figure 22, curve 2) and high deep-water formation rate (Figure 22, curve 3) show the sensitivity of the transient tracer approach for this particular region. The currently ongoing monitoring of tritium/³He and CFC-11/CFC-12 in the Greenland Sea should be a very good indicator of changes in the GSDW formation rate.

CONCLUSIONS

Existing data indicate that transient tracers are potentially valuable tools for studies of ocean variability on time scales

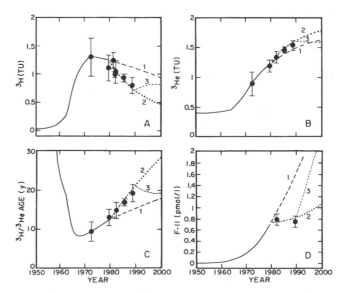

FIGURE 22   Comparison of observed and simulated tracer concentrations of GSDW for three scenarios: (1) constant renewal rate of GSDW; (2) GSDW renewal rate reduced by 80 percent starting in 1980; (3) GSDW formation rate reduced by 80 percent starting in 1980, followed by increase to full deep-water formation rate starting in 1990. A-D are as in Figure 20. (From Schlosser et al., 1991; reprinted with permission of the American Association for the Advancement of Science.)

of several years to several decades. Successful application requires careful selection of the best set of parameters for each case study. Typically, time series of a single tracer do not directly provide information on ocean variability. Therefore, models capable of parameterizing water-mass formation, including exchange of tracers with the atmosphere, ocean circulation, and mixing along flow paths, have to be applied to isolate the individual processes that affect the tracer concentration as a function of time. Tracer ratios might yield information on ocean variability more directly. Evaluation of the full potential of tracers as tools to study ocean variability clearly requires more and longer time series. Such time series could be obtained as part of the planned Global Ocean Observing System.

Long-term observations of transient tracers appear to be most useful for studies of deep- and bottom-water formation, as well as for investigation of the thermocline ventilation.

ACKNOWLEDGMENTS

We are grateful to Patty Catanzaro for drafting the figures. This work was supported by the National Science Foundation under contracts OCE 91-11404 (PS), OCE 89-17801 (WMS), and OCE 90-19690 (WMS), and by the National Oceanic and Atmospheric Administration under contract NA16RC0512-01 (PS). LDEO Contribution No. 5295.

Commentary on the Paper of Schlosser and Smethie

JAMES H. SWIFT

Scripps Institution of Oceanography

The transient tracers, as Dr. Schlosser showed, provide real-world data on the time and space scales of ocean variability. In many cases, these are the only measurable information on these time scales, so I consider them extremely valuable for the conceptual understanding required for climate-type studies. (Like Dr. Schlosser, I am careful not to call them "climate studies," because that is not what we are working with here.)

The observational network for providing the data for this type of study has many shortfalls, which is probably due to a shortage of resources. But I also have questions for Dr. Schlosser and the other workshop participants about whether we also need to fine-tune our notions of where we should carry out these studies.

My first question is whether there is a mismatch between the physical oceanography programs that provide water to people for the tracer studies, and the ideal data sets, you might say, required from the tracer studies. The freon, radiocarbon, and helium data are basically dependent on cruises that are, to a very great extent, planned and carried out by people in other fields. I wonder what sort of impact that has had on our ability to use the tracers.

Second, I notice that the many different groups carrying out tracer measurements are all under resource pressures. Is there a definite commitment from the investigators, in either the long or the short term, to getting the data from different sources together into a common set that we can use for global studies? We have a regional study in the western boundary undercurrent, a Greenland Sea study, and Antarctic studies. I should very much like to see their results being pooled into a single data base as the physical oceanography data are beginning to be.

Discussion

SCHLOSSER: The tracer groups are indeed working on establishing a common data set, but we still have some way to go before we can realize it. As for the tracer studies, I think we're getting closer to having the right platforms in the right places; in the process studies we're approaching an ideal tracer strategy. Of course, labor and resources continue to be limitations.

LEHMAN: You mentioned that recent observations suggest

that ventilation has not recurred since the downturn in the 1980s. It was speculated that the GSA had quelled convection, but if ventilation hasn't recurred since the GSA's surface dissipation it would weaken any steric link between the two.

SCHLOSSER: We just don't know whether the GSA was involved, or how long it should take to restore the original mode. We're dealing mainly with the water below 1500 m, but we also don't see a strong signal in the surface layer where we might expect it. We're just going to have to look at these complex dynamics more carefully.

LEHMAN: I noticed too that your data set indicates no diminution of the strength of the overflow or western boundary current. Before the shutoff of deep ventilation, the formation rate was about one-half a sverdrup, which means it would take 30 years to fill the basin. The overflow is controlled by the damming effect of the sills, so at 10 years after the shutoff we should now be one-third of the way through a cycle that should affect the overflow.

SCHLOSSER: But the overflow seems to be regulated to a very large extent by intermediate-depth convection, as Bob Dickson has pointed out. We may be seeing a phenomenon that is largely decoupled from the global circulation, but can give us insight into regional dynamics.

MYSAK: Isn't it in the Icelandic Sea rather than the Greenland Sea that the GSA caused the shutdown?

SWIFT: In my opinion, the overflows haven't changed; it's only the type of water that has changed. The current continues with whatever's at hand. The Arctic Ocean also outflows through the Denmark Strait, so you get high-salinity pulses from its intermediate waters as well.

RIND: Can you tell from the tracer data how much of the vertical mixing is associated with small-scale convection and how much with larger-scale overturning?

SCHLOSSER: We can only look at rates at present, not processes. We'd need a more elaborate model to do that.

TALLEY: Could you say something about the relative importance of sea-ice formation in open-ocean convection or in ventilating the Greenland, Norwegian, or Icelandic Seas?

SWIFT: In the conceptual model of Rudels and Quadfasel, which has the cyclic overturn, convection proceeds until the surface layer freezes. Then the brine released from further ice formation penetrates the warmer, saltier water underneath, bringing up heat, melting the ice, and starting another cycle. Deep-water formation appears to occur in the regions where the surface salinity and the heat and salt underneath are balanced so that this cycle can continue. The ice-formation process has to be parameterized as well.

WEAVER: I don't quite understand this concept of a western boundary current with constant magnitude but changing temperature and salinity. Wouldn't there be feedbacks that would force it to vary?

SWIFT: Well, that question is related to something I'd wanted to mention earlier. The models I've seen here seem to favor deep-ocean convection from the surface—the classic large-scale overturn. This focuses attention on "chimneys", which are relatively small-scale instabilities. But in the ocean the processes that prevail will be those that produce the densest water the fastest. For example, where there are continental shelves or ice shelves at high latitudes the local density can be greatly enhanced, partly because cooling is concentrated on a smaller total volume and dense products can accumulate, and you get the spreading Peter mentioned. This process may account for half the "deep convection" in the Arctic mediterranean seas. The point is that the environmental sensitivities of the continental-shelf processes are different from that of open ocean convection, which could affect the sensitivities used in modeling deep-water formation.

I guess that doesn't really answer your question, but I think we need to keep it in mind.

A Few Notes on a Recent Deep-Water Freshening[1]

JAMES H. SWIFT[2]

ABSTRACT

In 1981 much of the deep-water column of the northern North Atlantic was approximately 0.02 fresher than previous measurements had shown. These deep waters are closely tied to the sea surface in nearby water-mass formation regions, and the salinity change probably reflects the penetration into the ocean interior of the "great salinity anomaly" in the surface waters described by Dickson et al. (1988). The surface-layer freshening in the 1970s propagated into the deep water, because during winter the downward penetration of the cooled water depends on its density and not on its temperature or salinity alone. The net result of a freshening of the surface water is to produce colder, less saline water for a given density. This produced the changes seen in the 1981 data.

Until the 1980s there was virtually no modern-day evidence of any large-scale alteration in deep-ocean salinity. Oceanographers had not been accumulating high-precision deep salinity data long; nonetheless, the concept of a "steady-state ocean" was broadened by tradition to include the deep temperature and salinity fields. Small variations in deep-water salinity were observed in the western North Atlantic, particularly on the often-run Woods Hole-to-Bermuda section, but these were near the noise level and did not receive wide attention.

The GEOSECS (Geochemical Ocean Sections) program provided sparse, large-scale coverage of the North Atlantic in 1972. While the deep GEOSECS temperatures and salinities were similar to those from past expeditions, the GEOSECS data did show new features, notably a deep tongue of tritium-enriched water extending southward from the Denmark Strait. The presence of the deep tritium signal was a signal to oceanographers, showing that the bottom layers in the northwest Atlantic clearly responded on a time scale of 10 years or less to the injection of tritium into the atmosphere by thermonuclear testing.

Indeed, in general the best location to examine the deep ocean for evidence of recent property changes is the northern North Atlantic: It is close to the deep-water mass formation regions, and has been surveyed by expeditions 10 to 20 years apart.

[1]This article is adapted from an article by the author in *Glaciers, Ice Sheets, and Sea Level: Effect of a CO₂-Induced Climatic Change*, DOE/ER/60235-01, National Research Council, Washington, D.C., 1985, pp. 129-138.

[2]Scripps Institution of Oceanography, University of California at San Diego, La Jolla, California

The apparent sensitivity of the deep northwest Atlantic on 10-year (or less) time scales added immediate interest to the data gathered in 1981—nine years after GEOSECS— by the Transient Tracers in the Ocean North Atlantic Study (TTO/NAS). A total of 250 high-quality hydrographic-geochemical stations were occupied north of 17°N, a region covered by only 43 stations during GEOSECS. The TTO track provided some basin-scale sections, thus giving adequate coverage and facilitating comparison with other data from different years. Early reports in the summer and fall of 1981 from TTO/NAS indicated detectable changes in deep-ocean properties in the northern North Atlantic during the nine years following GEOSECS. Brewer et al. (1983) outlined the essentials of the new observations: a shift toward colder, fresher deep water in the northern North Atlantic and a larger freshening in the upper layers there.

The northern North Atlantic plays a significant role in the oceans' contribution to climate response, because it can be thought of as the primary deep ventilator of the world ocean. As Reid (1981) and others have shown, recently ventilated, or oxygenated, deep water spreads from the North Atlantic into the other oceans. Part of these high-oxygen waters derives from dense outflows to the North Atlantic from the Greenland, Iceland, and Norwegian seas, and part from processes within or over the deep northern basins of the Atlantic.

We can examine the northern sources by following a long section around the northern North Atlantic (Figure 1), following portions of the western boundary undercurrent. Note that the section and the undercurrent cut through a fracture zone in the mid-ocean ridge. This section passes near the overflows and the deep-water formation zone in the Labrador Sea. The sections in Figure 2 start in the Iceland-Scotland Overflow on the right, go through the

Charlie-Gibbs Fracture Zone and up to the northern Irminger and Labrador seas in the middle, then around the Grand Banks on the left. (There is a great deal of information in these sections—here I will discuss only a few of the features.) The middle section shows salinity versus potential density (σ_2, referred to 2000 db), not depth, and only for the denser portions of the water column. Because the Labrador Sea data are from winter cruises, when the density of the surface waters is higher, the upper line in that area follows the potential density at the surface. The bottom contour is bottom density, and it shows the contributions of dense water from the two main overflows.

There are differences in the salinities and final densities of the overflows. The Iceland-Scotland Overflow mixes with much warmer, saltier, and less dense water and forms a relatively high-salinity deep layer, underneath lower-salinity Labrador Sea Water. The Denmark Strait Overflow is a little fresher to begin with and mixes mostly with the layers immediately above, which are not so different. Thus the Denmark Strait Overflow retains its relatively low salinity and high density, and forms the bottom water in the northwest Atlantic.

The bottom section shows dissolved oxygen versus σ_2, along the same path. The three northern North Atlantic Deep Water sources are shown by the sources of relatively well-oxygenated water in the Labrador Sea, the Denmark Strait Overflow, and the Iceland-Scotland Overflow. The relatively well-oxygenated deep layer spreading south around the Grand Banks shows the propagation of the younger, overflow-derived deep waters southward via the western boundary undercurrent.

One of the results of this data analysis was an improved understanding that these deep-water components are closely tied to the winter sea surface and so are potentially sensitive to environmental changes. But during the 1957-1972 period emphasized in Figure 2, no such changes could be discerned in the deep temperature and salinity fields.

In the 1970s however, the surface layers of the northern North Atlantic freshened, by about 0.1 psu. European oceanographers call this "the great 1970s salinity anomaly" (cf., e.g., Dickson et al., 1988). The section in Figure 2 was well enough reoccupied in 1981 to permit comparison of the deep-water salinities before and after the surface-water freshening.

By 1981 deep-water salinities had changed throughout the northern North Atlantic. Figure 3 shows the salinity-versus-potential-density section from Figure 2, and the same section in 1981, contoured for the same ranges; it is clear that the 1981 salinities were lower than their earlier counterparts. The deep freshening was about 0.02 psu, and, since densities in each layer are about the same, we can see there was also an overall cooling, which in this case was about 0.15°C.

Figure 4 shows this deep temperature-salinity shift at

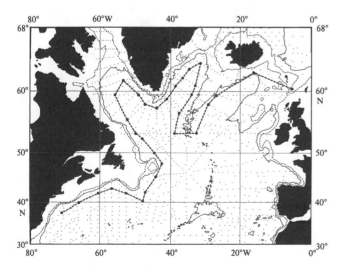

FIGURE 1   Map showing path of the section shown in Figures 2 and 3. The 500 m and 2000 m isobaths are shown. (From Swift, 1984b; reprinted with permission of Pergamon Press, Ltd.)

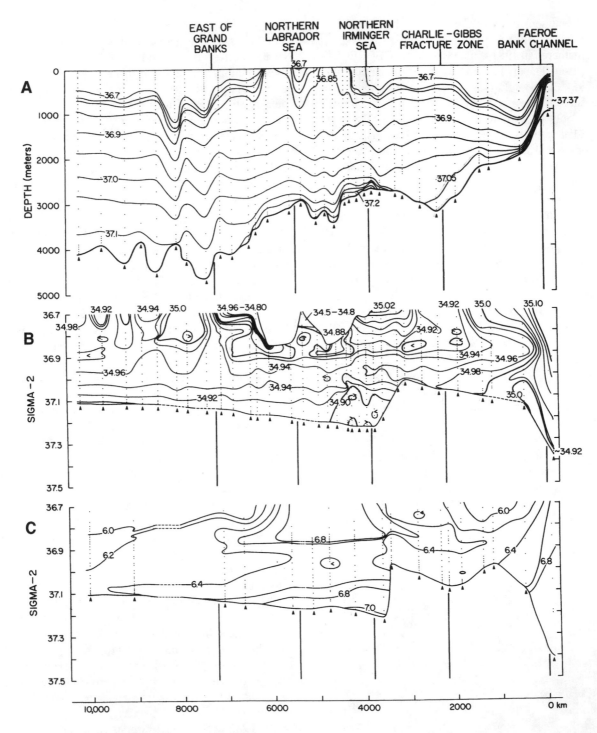

FIGURE 2   Vertical section from the Faeroe-Shetland Channel to the eastern U.S. continental margin of (A) σ_2 versus depth, (B) salinity (psu) versus σ_2, and (C) dissolved oxygen (ml/l) versus σ_2 (potential density anomaly referenced to 2000 decibars). Sections (A) and (B) are drawn from a composite of historical data (*Chain*, 1960, 1972; *Crawford*, 1960, 1965; *Erika Dan*, 1962; *Atlantis*, 1964; *Hudson*, 1967; *Knorr*, 1972, 1976) along the long, winding path shown in Figure 1. The only post-1972 data in (A) and (B) are from a single *Knorr* 1976 station in the Faeroe-Shetland Channel. Section (C) is drawn from April-October 1981 TTO/NAS data along a nearly identical path (see Swift, 1984a). Section (A) is contoured only for $\sigma_2 > 36.7$, and (B) and (C) show properties only where $\sigma_2 \geq 36.7$. The contour intervals are (A) 0.05 kg/m³, (B) 0.02 psu (for S $\geq$ 34.8 psu), and (C) 0.2 ml/l. (From Swift, 1984b; reprinted with permission of Pergamon Press, Ltd.)

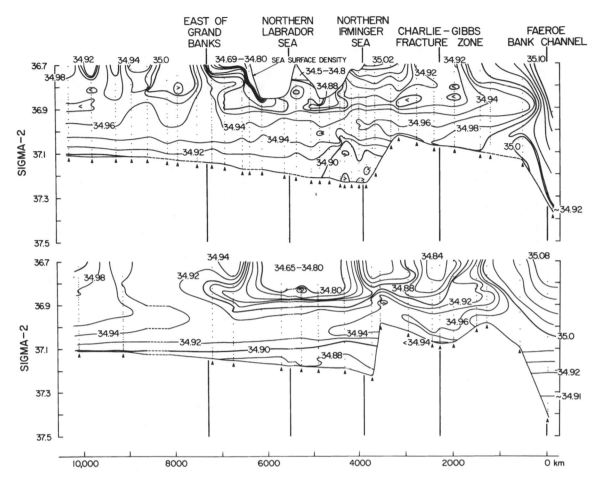

FIGURE 3   Sections of salinity (psu) versus σ_2 on a long, winding path from the Faeroe Bank Channel to the waters off the U.S. east coast (location shown in Figure 1). The upper section is from Figure 2, and the lower section is drawn from April-October 1981 TTO/NAS data. The only post-1972 data in the upper section are from a single *Knorr* (1976) station in the Faeroe-Shetland Channel. (From Swift, 1984a; reprinted with permission of the American Geophysical Union.)

one location, in this case northeast of the Grand Banks. Four deep stations from 1960 to 1972 showed virtually the same potential temperature-salinity correlation, at least for deep waters colder than about 3°C. But the 1981 station at this location shows noticeably colder, fresher water—the changes are four standard deviations above the noise level. These changes occurred in virtually every deep-water layer in the North Atlantic north of 50°N.

The surface-layer freshening in the 1970s propagated into the deep water because in winter, in the water-mass formation areas, the downward penetration of the cooled water depends on its density and not on its temperature or salinity alone. The net result of a freshening of the surface water is to produce colder, less saline water for a given density. This produced the changes seen in the 1981 data.

The origin of the low salinities is an interesting puzzle. Figure 5 shows surface salinities before the salinity shift. The formation of dense waters in the Labrador and Greenland seas owes much to the salty water carried north at this

and deeper levels. But much fresher water is plentiful along the boundaries. At first, I supposed that some shift in forcing in the 1970s allowed the fresh boundary waters to penetrate seaward into the deep gyres where the deep waters are formed. But we can see in Figure 5 that even in the pre-anomaly period the fresh waters do not just flow south out of the northern North Atlantic. Instead their salinities are increased by vertical mixing and by lateral mixing, with much of the fresh water incorporated into the gyres by the time they reach the Grand Banks. If during the 1970s the fresh waters had simply spread away from the boundaries in the north and been mixed into the deeper layers there, then some areas of the far northeastern Atlantic farther "downstream" from the fresh source might be expected to increase in salinity, lacking the accustomed supply of fresh water from the west. But since all the upper waters north of 50°N freshened, perhaps there was some net increase in fresh-water supply, or decrease in salty water supply, after 1972.

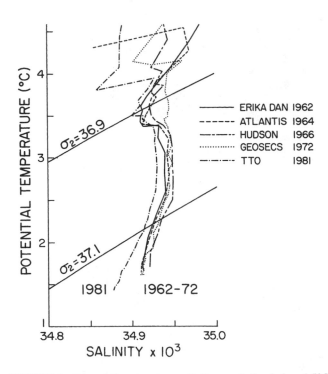

FIGURE 4 Potential temperature-salinity correlation below 4.5°C for four stations from 1962-1972 and from TTO/NAS (1981 data), all from a location northeast of the Grand Banks extending to about 3500 m. (From Swift, 1984b; reprinted with permission of Pergamon Press, Ltd.)

There may have been long-term changes in the atmospheric circulation coincident with this freshening. For an example of possible changes, consider Figure 6, which shows the difference between average wind-driven transport in the period 1973-1981 and that from 1955-1972. The changes are an average of about 30 percent of the mean at each grid point, and so may be significant. The negative values indicate a greater southward interior surface transport in recent years. Since the greatest negative values occur over the North Atlantic Drift, I wonder whether the high salts there might have been swept southward, partially shutting down the supply of salty water to the northeast Atlantic, and hence to the Labrador and Greenland, Iceland, and Norwegian seas. But Dickson et al. (1988) show that the freshening in the Faeroe-Shetland Channel took place late in an ordered sequence of events that began eight years earlier in the East Greenland Current, suggesting that it was a pulse-like increase in the volume of the fresh-water source that was at least in part responsible for the overall freshening.

We should remember that although these deep-water salinity changes (e.g., those in Figure 3) were the largest observed (at the time), they were still small in magnitude (only 20 parts per million). It may be that in the sinking zones only a small shift of the balance between precipitation and evaporation, or something similarly small, is at work here. In either event, this does illustrate the possible ties to large-scale atmospheric forcing of the deep thermohaline circulation, making the salinity shift, as we trace its propagation, one more tool with which to study the responsible processes.

FIGURE 5 Salinity (psu) at the sea surface. (From Reid, 1979; reprinted with permission of Pergamon Press, Ltd.)

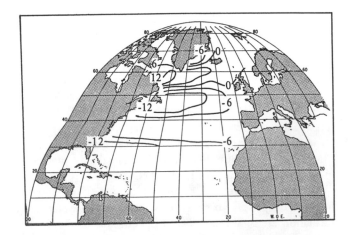

FIGURE 6 Change in integrated Sverdrup transport, average from 1973-1981 minus the average from 1955-1972. The transports were calculated from multi-year average sea level pressures on a 5° grid. Data from the Scripps Institution of Oceanography Climate Research Group. Owing to a minor (but non-recoverable) computational error, the values shown are not in sverdrups, but in nontraditional units. The average changes were 30 percent of the 1955-1972 base-period values.

The Salinity Decrease in the Labrador Sea over the Past Thirty Years

JOHN R.N. LAZIER[1]

ABSTRACT

Temperature and salinity data obtained in the Labrador Sea in 1966 and 1992 show that the average temperature and salinity of the 3500 db water column have decreased by 0.46°C and 0.059, respectively. The change in salinity is a maximum of approximately 0.09 at 2000 db in the Labrador Sea Water (LSW), an intermediate water mass that is renewed via deep convection in severe winters. The salinity difference diminishes to 0.05 in the Northwest Atlantic Bottom Water (NWABW) near the bottom and to 0.03 in the Northeast Atlantic Deep Water (NEADW), which lies between the LSW and the NWABW. An interrupted time series of salinity between 1962 and 1992 shows that anomalously low salinity occurred in the upper layer (0 to 250 db) between 1967 and 1972 and between 1978 and 1985. The salinity decrease in the LSW (1000 to 1500 db) occurs during two periods of rapid decline, 1972 to 1976 and 1988 to 1990, following the periods of low-salinity surface water. This suggests that deep convection is limited during the periods of low surface salinity and that, when convection resumes, the low-salinity surface water is mixed down to intermediate depths and serves to decrease the salinity of the intermediate LSW. Below the Labrador Sea Water the water column is strongly stratified, and the changes in salinity are due to changes in the properties of the incoming NWABW and NEADW rather than convection.

INTRODUCTION

The circulation in the Labrador Sea, first described by Smith et al. (1937), is dominated by a slow cyclonic gyre bordered by three strong currents (Figure 1). To the east is the West Greenland Current over the continental shelf and slope of southwest Greenland, to the west is the Labrador Current over the Labrador shelf and slope, and to the south is the North Atlantic Current flowing eastward at about 51°N. The West Greenland Current is a continuation of the East Greenland Current flowing south out of the Arctic along the eastern side of Greenland. It passes through the Denmark Strait and changes into the West Greenland Current after it rounds Cape Farewell at the southern tip of Greenland. On the inshore side of the West Greenland Current the water is cold (<0°C) and low in salinity (<34) (Grant, 1968). On the offshore side of this arctic water is a warm (4°C) saline (>34.95) band of water, which comes

---

[1]Bedford Institute of Oceanography, Dartmouth, Nova Scotia

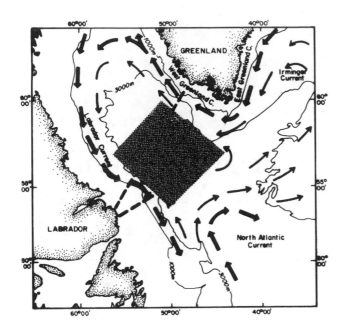

FIGURE 1   A map showing the general circulation of the Labrador Sea. The WOCE AR7/W line across the Labrador Sea and the CTD line used in the 1980s Labrador Current study are shown as dashed lines. The data used in this analysis were selected from the shaded region.

from the Irminger Current, a northern branch of the North Atlantic Current flowing north through the Irminger Sea toward Iceland. A significant portion of the West Greenland Current flows north into Baffin Bay, but most of the flow is thought to turn westward to the south of Davis Strait to become an offshore component of the Labrador Current.

The Labrador Current is a continuation of the Baffin Island Current flowing out of Baffin Bay through Davis Strait. It continues south over the continental shelves and slopes of Labrador and Newfoundland to the southern tip of the Grand Banks of Newfoundland. Like the West Greenland Current, the Labrador Current transports cold (<0°C) low-salinity water (<34) over the shelf and warm (>4°C) saline (>34.90) water over the slope.

The North Atlantic Current begins near the southern end of the Grand Banks of Newfoundland and brings subtropical water (T > 12°C, S > 35.5) north through the Newfoundland Basin into the southern Labrador Sea. Around 51°N, 44°W the current turns sharply east in what Worthington (1976) called the Northwest Corner. Although the subtropical water does not continue directly into the central Labrador Sea, the flow does appear to have some influence as far as 55°N, 50°W, where a weak northward flow turns east (Lazier and Wright, 1993).

The temperature field associated with the West Greenland and Labrador Currents is illustrated in the 1990 temperature

section across the Labrador Sea shown in Figure 2. Over the continental shelves of both Greenland and Labrador is the cold (<0°C) arctic water. And over the continental slopes is the warmer Irminger water (>4°C) extending about 50 km out into the basin at 300 to 400 m depth.

In the central portion of the Labrador Sea (Figures 2 and 3) lies the nearly homogeneous Labrador Sea Water (LSW), which is renewed by deep convection in severe winters. From approximately 500 to 2000 db the temperature range is only 0.2°C, from 2.8 to 3.0°C, and the salinity range is only 0.02, from 34.82 to 34.84. The density as measured by the potential density anomaly ($\sigma\theta$) referenced to 1500 db, $\sigma_{1.5}$, varies over the same pressure interval by 0.02 kg m^{-3}, from 34.65 to 34.67 kg m^{-3}. This is an average

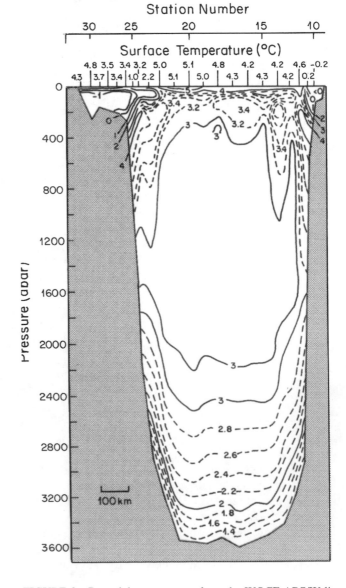

FIGURE 2   Potential temperature along the WOCE AR7/W line in July 1990.

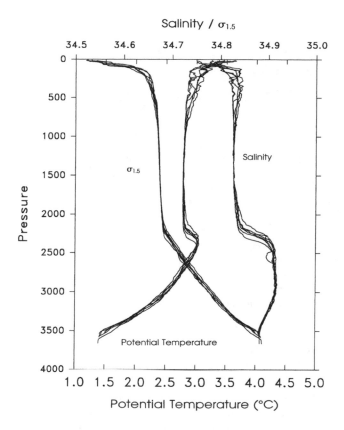

FIGURE 3 Profiles of potential temperature (θ), salinity, and the potential density anomaly referenced to 1500 db ($\sigma_{1.5}$, in kg m^{-3}) versus pressure (in db) at four stations in the middle of the AR7/W line in July 1990.

vertical density gradient of 1.33×10^{-5} kg m^{-4}. (The renewal of the LSW is discussed by Lazier (1973), Clarke and Gascard (1983), Gascard and Clarke (1983), and Clarke and Coote (1988), while its circulation and distribution are considered by Talley and McCartney (1982).)

Below the LSW (Figure 3) the vertical density gradient is about an order of magnitude greater than in the LSW. The value of $\sigma_{1.5}$ varies from 34.67 kg m^{-3} at 2100 db to 34.90 kg m^{-3} at 3500 db, for an average vertical gradient of 1.6×10^{-4} kg m^{-4}. Also, the variations of temperature and salinity below the LSW are much greater than those in it. In this deep stratified layer two water masses are usually identified: the Northeast Atlantic Deep Water (NEADW) and the Northwest Atlantic Bottom Water (NWABW). The NEADW is characterized by a salinity maximum S ≈ 34.92 between 2300 and 3000 db and potential temperatures of 2°C and 3°C. According to Swift (1984b), this water mass includes a component of the dense overflow from the Norwegian Sea through the Faeroe Bank Channel. This is modified by mixing with the water on the downstream side of the sill and flows into the western basin of the North Atlantic through the Charlie-Gibbs fracture zone. It then flows around both the Irminger and Labrador

seas and is further modified through isopycnal mixing with the water flowing over the Denmark Strait.

The NWABW lies at the bottom of the water column and is identified by the decrease in salinity below the NEADW. Swift et al. (1980) conclude that this water is formed in winter at the sea surface north of Iceland. As it is the densest water in the region, it is not substantially modified through isopycnal mixing as it flows from the Denmark Strait to the Labrador Sea.

The purpose of this paper is to describe the decrease in salinity that has occurred throughout the water column over the past 30 years. The data base (Figure 4) begins with the observations from the *Erika Dan* cruise (Worthington and Wright, 1970). This is followed by the data collected at Ocean Weather Ship (OWS) Bravo throughout the years 1964 to 1973 inclusive and from stations on the 1966 cruise by *C.S.S. Hudson* (Lazier, 1973; Grant, 1968). The remaining data were obtained through the Transient Tracers of the Ocean (TTO) program's cruise to the Labrador Sea in 1981 and from various cruises made under the auspices of the Bedford Institute of Oceanography (BIO). The BIO efforts include 1976 and 1978 winter studies of the deep convection as well as the 1984 to 1988 data collected as part of a 10-year study of the Labrador Current (Lazier and Wright, 1993). The 1990 and 1992 data are from the World Ocean Circulation Experiment (WOCE) repeat hydrographic line AR7/W across the Labrador Sea collected by BIO. The positions of the conductivity-temperature depth (CTD) lines used for the Labrador Current study and the WOCE program are shown in Figure 1.

In the following section, data collected in June 1992 are compared with data obtained in 1966 in order to illustrate the gross changes in the water column. This comparison is followed by a presentation of time series of salinity from various layers. These demonstrate how the decrease in salinity in the LSW is determined through convection, which

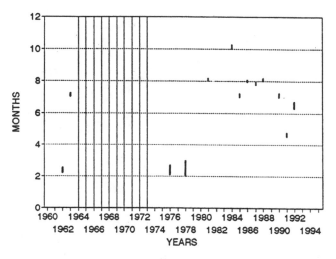

FIGURE 4 Periods of data collection in the central Labrador Sea.

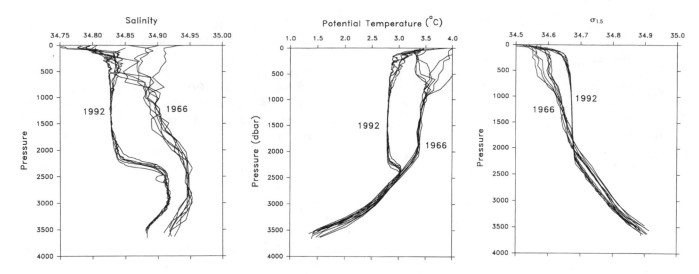

FIGURE 5   Temperature and salinity profiles obtained at six stations in the central Labrador Sea in March 1966 and at six stations in June 1992. Left, profiles of salinity; center, profiles of potential temperature; right, profiles of $\sigma_{1.5}$.

mixes low-salinity surface water down to intermediate depths, while the decreases in the NEADW and NWABW are controlled more by the salinity of the inflowing water. In a later section the changes in the temperature and density of the core of the LSW between 1962 and 1992 are presented.

1966 COMPARED TO 1992

Data obtained from six of the stations in the central Labrador Sea in March 1966 by *C.S.S. Hudson* are compared in Figures 5 and 6 with those obtained in June 1992 from six stations in roughly the same region. All these diagrams

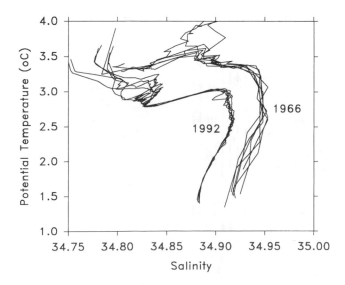

FIGURE 6   Potential temperature versus salinity for the stations plotted in Figure 5.

show that a remarkable change has occurred in the region over the 26-year interval. The average salinity over the whole water column has dropped from 34.92 to 34.861. This is equivalent to a loss of about 200 kg of salt per m^2 of surface area or the addition of about 6 m of fresh water per m^2. The decrease in salinity is not constant throughout the water column but ranges from around 0.03 in the NEADW in the pressure interval between 2500 to 3000 db to around 0.04 in the NWABW at 3500 db. The greatest decrease is approximately 0.09 at 2000 db in the LSW.

The average temperature through the water column decreased by 0.46°C, from 3.144°C in 1966 to 2.686°C in 1992. This represents a total heat loss of 6.7×10^6 kJ m^{-2}, equivalent to a continuous loss over the 26 years of 8 W m^{-2}. Most of the heat loss occurs in the LSW between 200 and 2300 db. In the NEADW and the NWABW the temperatures are only slightly less at a given pressure than they were in 1992.

The most significant difference between the deep temperature distributions of the two years is the erosion of the upper part of the NEADW by convection in the upper 2000 db. The profiles from 1966, for instance, suggest that the base of the LSW occurs around 2000 db at a temperature of 3.4°C, while in 1992 the base is around 2300 db at a temperature of 2.8°C. What were the upper layers of the NEADW in 1966 appear to have been incorporated into the LSW by 1992. Thus the water mass at a given density surface may change over time. This is discussed more fully later in the paper. The temperature maximum in the 1992 profiles at about 2400 db is another feature generated by the greater cooling in the LSW relative to the NEADW.

The profiles of $\sigma_{1.5}$ in Figure 5 indicate that the average density of the LSW is slightly greater in 1992 than it was in 1966, but that the vertical density gradient is less. The strength of the gradient is thought to reflect the intensity of convection during the preceding winter. Thus the higher gradient in 1966 is evidence of a year when convection did not seem to be active (Lazier, 1973), while the lower gradient in 1992 is indicative of recent convection. The higher density in 1992 must reflect a higher-than-normal heat loss to the atmosphere in the years between 1966 and 1992, since the lowering of salinity over these years would tend to decrease the density. The heat loss has been sufficient to overcome the increased stratification due to the lower salinity, and cause convection to 2000 m or more.

The final comparison between 1966 and 1992 is shown by the temperature-versus-salinity curves in Figure 6. Here again the drop in salinity in the three main water masses is obvious, as is the fact that the difference in salinity is greater in the LSW than in the two deeper water masses.

In the following analysis, all the available data are used to construct time series of salinity in three layers in order to examine variations in time.

TIME SERIES OF SALINITY CHANGES

In Figure 7 the average salinities of three layers are plotted for all the years of available data. For each year,

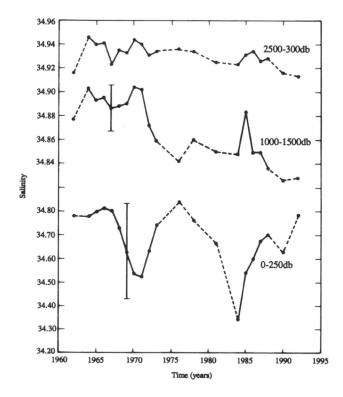

FIGURE 7 Average salinity in three layers between 1962 and 1992. The salinity scales for the 1000-to-1500 db and the 2500-to-3000 db curves are expanded five times relative to that for the 0-to-250 db curve. The vertical lines represent the maximum standard deviations associated with the data from OWS Bravo.

the data obtained within the shaded region in Figure 1 have been averaged together. In the years when data were collected at OWS Bravo, observations were collected in all months, so the averaged values approximate annual averages. The annual number of observations in the Bravo data varies a great deal, as does the number at each pressure level. In 1964, for example, 66 observations are recorded in the surface layers, but in 1969 (the peak year) there are 450 observations, and in 1973 there are 248. In the deep water the number of observations is more constant through the years. At 3000 db in 1966, 19 observations were obtained; in 1969 there are 11; and in 1973 there are 21. The standard deviations associated with the averages range from 0.08 to 0.2 in the near-surface layer, and from 0.01 to 0.02 in the deep layers.

In the years when Bravo data are not available, the values in Figure 7 have been obtained from one or two cruises. In these cases the averages are not annual averages, and do include any seasonal variation. Previous analyses of the Bravo data (Lazier, 1980) indicate that there are no significant seasonal signals in the intermediate or deep waters, so seasonal aliasing of the interannual signal will be a problem only in the 0-to-250 db curve. Since data were obtained at Bravo for 10 years, it should be possible to define the annual signal for the 0-to-250 db layer and remove it from the data collected in the non-Bravo years. This was not done, however, because the salinity in the upper layer at Bravo changed greatly over the 10 years of observation as a result of the arrival of the low-salinity pulse associated with the Great Salinity Anomaly. This large shift in the middle of the time series would distort the seasonal signal and make it unrepresentative of either the Bravo years or any of the other years.

The first curve plotted in Figure 7 represents the salinity in the surface layer from 0 to 250 db. The second curve is the average salinity from 1000 to 1500 db; it illustrates the changes in the LSW. The third curve shows the changes in the NADW from 2500 to 3000 db.

The surface layer is dominated by two large salinity decreases, in 1967 to 1971 and 1978 to 1985. The first of these is contained in the observations from OWS Bravo; it was discussed by Lazier (1980), who found that the lower salinity occurred during a period of mild winters. These higher winter temperatures, in combination with the increased stratification due to the higher vertical salinity gradient, limited the convection in winter to the upper few hundred meters. In 1972, however, a severe winter caused a renewal of convection to intermediate depths. The 1967-1972 low-salinity anomaly, which was also observed at other locations around the sub-polar gyre of the North Atlantic Ocean, became known as "the Great Salinity Anomaly." Dickson et al. (1988) examined these observations, and were able to trace the path of the anomaly around the gyre;

they determined its speed of advance to be approximately 0.03 m s^{-1}.

The second period of low salinity (1978-1985) coincides with one observed by Myers et al. (1989) over the West Greenland shelf at Fylla Bank. In the middle of the Labrador Sea, however, there are only a few stations available, and the minimum is poorly sampled in both time and space. The fact that the salinity of the minimum appears to be less than that of the one in the 1967-1971 period is probably due to this poor sampling, which has allowed part of the seasonal cycle to be included.

In the intermediate layer, between 1000 and 1500 db, there is not a large change in salinity during the periods of low surface values, in 1967 to 1971 and 1978 to 1985. The most significant changes in this layer occur after the period of low salinity at the surface, when convection mixes the low-salinity water down to intermediate depths. For example, in 1972 and 1973, when the atmospheric conditions became more severe and convection returned, the low-salinity upper layer became mixed down into the intermediate layer. This is shown by the decrease in salinity from about 34.90 to 34.85 in the intermediate layer in 1972 and 1973, and the compensating increase in the upper layer from 34.53 to 34.84. A simple calculation of the salinity resulting from mixing the top 250 db at a salinity of 34.53 with the 250-to-1500 db layer at 34.90 yields a 1500 db layer with a salinity of 34.84. This agrees with the observed value, within 0.01, and suggests that salt is conserved through the vertical mixing.

During the 1978-to-1985 salinity minimum there was again no significant drop in the intermediate layer's salinity. As in the earlier case, this is presumably because convection was limited to shallower depths. The salinity increase in 1985 seems unreasonably large, as no comparable increase occurred in either of the other layers. It is possible that the CTD salinity calibrations are incorrect at intermediate pressures but correct at higher pressures, but various checks have not yet found this to be so.

The second significant drop in the salinity of the LSW occurs between 1988 and 1990, when it went from 34.85 to less than 34.83. This decrease is similar to that of 1972 to 1976, in that it occurs after the minimum in the surface layer, and a restoration of convection in the late 1980s is the assumed cause. In contrast to the changes in 1972 to 1973, when the salinity increase in the upper layer occurred at the same time as the salinity decrease in the LSW layer, there appears to be a delay of about 3 years between the time when the salinity in the upper layer increased (1985-1987) and the 1988-1990 decrease in the LSW layer. There are two factors that may have caused this lag. The first is the fact that the salinity of the upper layer is badly aliased by the seasonal signal, so the actual decrease in salinity of the upper layer that is due to mixing with the intermediate layer is impossible to detect. The second factor is that

mixing between the upper and intermediate layer is not the only way in which the salinity of the intermediate layer can change. Since that layer is fed partly by a component of the Denmark Strait overflow, its salinity will increase if the salinity of the Denmark Strait overflow increases as described below.

The third curve in Figure 7 shows the salinity changes in the 2500-to-3000 db layer that represents the NEADW. Since this water is in the more strongly stratified region of the water column, vertical mixing is small, and there should not be any connection between the salinity changes and events in the upper layer. For example, this layer does not exhibit decreases in salinity in 1972 to 1976 and 1988 to 1990 similar to those in the LSW. But there is definitely a decrease over the years, especially from the maximum of 34.94 in the mid-1960s to the present-day value of 34.915. The curves for the deep and intermediate layers also appear to be correlated during the periods when convection is not changing the LSW properties. This is especially true during the 1960s, when the two curves show increases and decreases of about the same magnitude at about the same time. A possible explanation of this correlation may be the influence of the water overflowing the Denmark Strait.

According to Mann (1969), the overflow water, which had a salinity of 34.93 in March 1967 and has a temperature that ranges between 0.5°C and −0.5°C, mixes with the warmer overlying water. This mixed water is not dense enough to reach the bottom, but it is of the same density range as the NEADW and the LSW. Thus, when these water masses flow south along the eastern slope of Greenland, the water on all the density surfaces is affected by either the dense overflow water at the bottom or the less dense mixed water that mixes with the LSW and the NEADW. When there is no convection in the central Labrador Sea, the changes in salinity in the intermediate and deep layers may therefore be regulated by the salinity of the overflow water and its mixtures. When convection is active, the salinity changes in the LSW are partly due to the convection and partly due to the changing properties of the mixed water formed south of the Denmark Strait.

TEMPERATURE AND DENSITY CHANGES IN THE LABRADOR SEA WATER

From the curves in Figures 3, 5, and 6, it is possible to define the properties of the LSW. For instance, the values of the temperature or salinity minima at intermediate depth can be recorded. But the water mass covers such a large pressure interval that the value of the temperature or salinity minimum and the pressure at which it occurs are often ambiguous. A more accurate way of defining the water properties is to plot temperature or salinity against density. Examples of temperature-versus-density curves are given in Figure 8. These curves are from the 1966 and 1992 stations plotted in Figure 5. In addition to illustrating the warmest and coldest years, the curves present conditions when a definite minimum exists (1992) and when only an inflection point occurs (1966) in the LSW. The LSW can therefore be defined in 1992 by the temperature minimum of 2.8°C at a $\sigma_{1.5}$ of 34.675 kg m^{-3}. In 1966 the water mass is defined by the inflection point in the curve at a temperature of 3.4°C at a $\sigma_{1.5}$ of 34.655 kg m^{-3}.

By plotting all cruises in a similar way, a list of $\sigma_{1.5}$ versus minimum temperature values was obtained for the LSW. These are plotted against each other in Figure 9. As can be seen, from the beginning of the record in 1962, the temperature of the LSW increased from slightly greater than 3.2°C to 3.6°C in 1970 to 1971. At the same time the $\sigma_{1.5}$ decreased from about 34.65 to 34.63 kg m^{-3}. These changes occurred during the period when deep convection was not taking place. The LSW was isolated from the upper layer, and the properties changed only through horizontal and vertical diffusion. In 1972 convection began again and caused the core temperature of the LSW to drop to 3.0°C at a density of $\sigma_{1.5} = 34.65$ kg m^{-3}.

Again in the 1980s, during the period of low surface salinities, convection was limited in depth, and the LSW became slightly warmer and less dense. This phase culminated in 1987, which is circled by a dashed line in the figure to indicate the indefinite nature of the determination. Convection appears to have begun again in 1988 and continued in 1990 and 1992.

The average $\sigma_{1.5}$, determined from the 18 values posted in Figure 9, is 34.650 kg m^{-3}, with a standard deviation of 0.013 kg m^{-3}. Thus all the values lie within one standard deviation of the mean, except for the years 1970, 1971, and

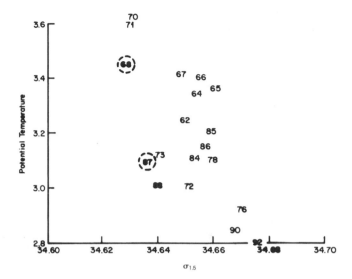

FIGURE 9   The value of potential temperature at the LSW minimum plotted against the $\sigma_{1.5}$ value where the minimum occurs. The numbers are the last two digits of the years of observations. When the values are indistinct, as in 1968 and 1987, the year is circled with a dashed line.

1968 on the low-density side and 1990 and 1992 on the high-density side. Without these five exceptional years, the density of the LSW is constant to within one standard deviation of the mean, while the temperature and salinity of the LSW have decreased by about 0.5°C and about 0.06. These figures support the suggestion of Clarke and Gascard (1983) that the density of the LSW remains constant over time because density changes due to temperature and salinity changes are equal and opposite.

The changes in density, although small through most years, are significant; in the 1960s a $\sigma_{1.5}$ of 34.675 kg m^{-3} represented the lower limit of the NADW, but in 1992 it was at the core of the LSW. It is clear that such a sigma surface cannot be used to monitor changes in the properties of either the LSW or the NEADW.

SUMMARY AND CONCLUSIONS

The changes in temperature and salinity at all depths in the Labrador Sea over the past 30 years have been demonstrated through the presentation of temperature and salinity data from the central portion of the sea. The changes between 1966 (during the most saline conditions) and 1992 (during the least saline) were presented in vertical profiles and θ-S curves. These showed that over the whole water column in 1992 there was approximately 200 kg m^{-2} less salt than in 1966 and about 6.7 × 10^6 kJ m^{-2} less heat. The diagrams also demonstrated that the change in salinity was not the same at all depths, but ranged from around 0.03 in the NEADW to about 0.04 in the NWABW and about 0.09 in the LSW. The changes in salinity with time

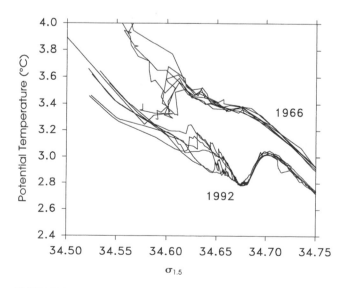

FIGURE 8   Potential temperature versus $\sigma_{1.5}$ curves for the same stations plotted in Figure 5.

were demonstrated using plots of average salinity in layers representing the near-surface conditions (0 to 250 db), the LSW (1000 to 1500 db), and the NEADW (2500 to 3000 db). The record in the upper layer was dominated by two periods of anomalously low salinity. The one in 1967 to 1971 is identified with the Great Salinity Anomaly, while the one in 1978 to 1985 is smaller and less well known. The anomalously fresh water was subsequently mixed down to intermediate depths by convection, which caused the salinity of the LSW to decrease from 34.90 to 34.85 in 1972 to 1973 and from 34.85 to 34.83 in 1988 to 1992. The salinity decreases in the NEADW and NWABW were shown to differ from that in the intermediate LSW in that no influence of the deep convection was found. It was

suggested that the conditions in the NEADW and NWABW are determined by isopycnal mixing between the mixed water created south of the Denmark Strait and the NEADW flowing into the Labrador Basin through the Charlie-Gibbs fracture zone in the mid-Atlantic Ridge.

The properties of the LSW were defined by the mid-depth temperature minimum or inflection in temperature versus $\sigma_{1.5}$ curves. A plot of the temperature at the temperature minimum versus the corresponding $\sigma_{1.5}$ demonstrated that the $\sigma_{1.5}$ of the LSW has varied from a low of 34.63 kg m^{-3} in 1970 to 1971 to a high of 34.675 kg m^{-3} in 1992. Because of this change in density over the years, it has been suggested that the properties of the water mass could not be accurately tracked through time by plotting temperature or salinity on a density surface.

Commentary on the Paper of Lazier

LYNNE D. TALLEY
Scripps Institution of Oceanography

I am greatly impressed by all the data that have been collected over the years. People need to understand how much individual effort goes into making sure that data continue to be collected, and Dr. Lazier is particularly to be commended for making sure that we have kept measurements going in the Labrador Sea. The Labrador Sea is one of the most important places in the world ocean for looking at the overall cycles of ventilation and overturn, and ultimately for understanding what determines SST.

Let me summarize some of the major points in Dr. Lazier's paper. The most important is that the Labrador Sea is a very significant site for deep convection. Labrador Sea Water affects a very big part of the water column in the northern North Atlantic; Figure 1 shows where it is found at mid-depth. Two water masses are in opposition—Labrador Sea Water, which is relatively fresh, and Mediterranean Water, which is salty. The salty Mediterranean Water is what makes the North Atlantic the saltiest of all oceans, and hence the likeliest for deep convection. The Labrador Sea Water provides an important fresher balance to the Mediterranean Water.

Why do we care about this? This may sound a little too simplistic, but on the very longest time scales and global spatial scales, what determines where deep water is formed, basin by basin, is which ocean is saltier and which ocean is fresher. Saltier surface water can be driven to higher density with cooling, so the salinity is what will determine the location of deep-water formation. The North Atlantic is currently relatively warm and salty—salty being the word to underline—and the deep water is formed there. In con-

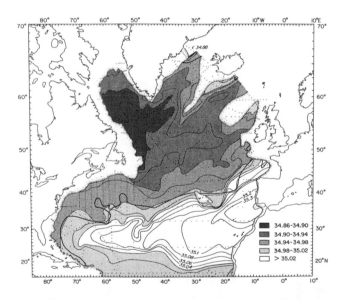

FIGURE 1   Salinity at the core of the Labrador Sea Water as defined by its potential vorticity minimum and at a potential density anomaly of 34.72 relative to 1500 dbar south of the heavy curve. The depth of the surface is approximately 1500 meters. (From Talley and McCartney, 1982; reprinted with the permission of the American Meteorological Society.)

trast, the North Pacific is a fresher ocean, and it receives its relatively salty deep water from the North Atlantic and Antarctic. Its surface layer is very strongly stratified or capped off in the subarctic gyre due to salinity.

The important consequence of this for modeling and for understanding the effect on the atmosphere is that the surface layer can be colder in capped oceans than in overturning oceans. Simplistically, relatively fresh surface water can be cooled to freezing without overturning. So on very long time scales, the saltier ocean has the greater influence, and that probably comes down to the one that has the evaporative Mediterranean.

For deep-water formation in the Atlantic, the depth to which the overturn extends—and probably the surface temperature as a result—depends on the history of the overturn. Thus, if you want to understand the surface temperature or the temperature of the upper 500 meters, you have to be able to follow the history of the overturn.

I have one question I would like to pose to both Dr. Lazier and Dr. Dixon: Is there a possibility of capping off the North Atlantic completely in the current regime? You see the anomaly coming through. Could that anomaly be large enough or sustainable enough that North Atlantic conditions would be significantly changed?

It is clear to the oceanographic community that salinity is really important; it controls ventilation and overturning. Sea ice, evaporation, precipitation—all of which are things that we do not measure very well—are therefore important as well. If you have a global observing system without salinity measurements, you will be looking only at the result of what is going on; you will not be getting at the mechanism.

Obviously, salinity is not the whole answer; you need to know what the atmosphere is doing, and the forcing and all the interactions. But without salinity, you are stuck because the salinity controls overturn. The TOGA people are finding that to be true in the tropical Pacific. If you are thinking about where to monitor in a global observing system, you need to follow the history of the overturns, identify where overturns generally appear, and monitor them all the way through the depth of overturn. In the North Atlantic, you have to care about Labrador Sea Water down to where it shows up, and about overflow water—in short, the whole water column in the northern North Atlantic. In the North Pacific, I agree with Claes Rooth that you do not need to worry about deep water, but you might want to measure to 1500 meters. In each of the ocean basins some sort of argument can be made for where to monitor and, if you are looking at the water-mass structure, how deep to go.

We know very little about time scales, since we have an incredible dearth of time series. We are fortunate to have had the ocean weather ships in the past, as well as efforts such as John Lazier's to continue making measurements in the Labrador Sea. In the Pacific, there are equivalent data sets only in the California Current and around Japan.

Dr. Lazier's paper has demonstrated nicely the effect of two consecutive salinity anomalies. In both the convection was capped, with big property changes down below. A trend toward decreasing salinity and temperature can also be seen.

I had several questions in relation to the data shown. I was interested in the little blips of higher salinities during the capping. What is the source of the high salinity at depth during capping?

Second, I wondered whether there are enough data to tell whether there are 10- to 15-year cycles, or whether the overall decrease is part of a longer cycle. The earlier data are rather sporadic, so it may not be appropriate to use them to look at trends. And how are the two cycles related to the cycles that Syd Levitus showed?

And my last question is, how cold would you have had to make the water during those capping events in order to break through, and what kind of heat flux would it require?

Discussion

LAZIER: Capping the low-salinity water in the late 1970s was coincident with mild weather, and the same was true in the early 1980s. The succeeding severe weather resulted in mixing. Incidentally, two recent records indicate that the very-low-salinity Labrador Sea water is showing up in the Irminger Sea.

DICKSON: I think the real question was how you cap convection to the point that you interrupt North Atlantic Deep-Water production in some way. To start with, I'd say that we should be using the word "entrainment" rather than "convection" there. It's what's coming over the sill and doubling itself you worry about, not the deep-water convection where it got started.

Now if we compare the Denmark Strait overflow with the Mediterranean outflow, the latter starts off denser, comes up a steeper slope, and gets up to a faster speed. It therefore entrains a large volume of unrelated, much less dense water—it multiplies itself by three—and comes down the slope with lots of turbulence. It ends up coming across at mid-depth, while the colder, more homogeneous Denmark Strait water goes all the way to the bottom.

For the Greenland Sea, it would be easier to effect model changes by altering what is entrained rather than the density of the top 400 meters of the Icelandic Sea. And that could indeed change the production of deep water.

SWIFT: I wonder whether Walter Munk would comment on the scales of convection.

MUNK: Yes, I have a slide [see Munk, discussion after Lazier paper, in the color well] from a Greenland Sea experiment of

Peter Worcester and some others—their paper's in press—that might answer a question asked earlier. It's a tomographic representation of one full year, summer to summer, in the Greenland sea along four different sections. The data were taken every four hours, so it has unparalleled time resolution. You can see that you have very sudden—within about a day—penetration of cold water down to 1000 meters. Presumably this is a local effect, taking place at one spot along the 300-km transmission path. On the average, the upper water was lighter than the lower water when this occurred, so it was not a question of a massive overturning convection but of penetrating, chimney-like downward motions that formed some deep water.

LEHMAN: Obviously salt and entrainment are very important in determining some of the mechanisms and manifestations of the formation and composition of deep water. I think the question is how much water and heat content go north as a result of these processes we're monitoring. I'm beginning to agree with Wally Broecker that processes north of the Faeroe Ridge could almost be viewed as a single common-batch process. The key to some of the recent variability may lie in the Labrador Sea, where the water masses that are in the area of entrainment downstream are modified. We might also be seeing a real change in the flux of the formation of upper North Atlantic Deep Water. Climatologists really need to keep in mind that we may be able to explain the climate record in terms of how much heat these processes draw north with them.

BRYAN: We certainly need to remember that as we weigh the alternatives of instrumenting the whole Atlantic to look at the northward flow of the Norwegian current, or concentrating on the western boundary where we have currents moving southward.

LINDZEN: In the tropics the mean configuration of temperature and humidity is generally believed to be stable with respect to convection. Convection occurs due to local breakdowns in the mean, and cannot be anticipated on the basis of the mean itself. Therefore, when you look at mixtures and transports, the absence of gross instability is not necessarily germane to convection.

ROOTH: Also, in high latitudes there is a special factor that contributes to the significance of salinity anomalies: colder water is more compressible. When you cool water to the density at which convection can set in, a larger amount of potential energy is available, which enhances the intensity and depth of penetration of convection events. The correlation between the intensity of deep ventilation and the nature of the near-surface salinity anomalies is thus very important there.

The Local, Regional, and Global Significance of Exchanges through the Denmark Strait and Irminger Sea

ROBERT R. DICKSON[1]

ABSTRACT

The physical exchanges and transfers that have been observed to take place in the Irminger Sea and through the Denmark Strait are of local, regional, and global significance. The factors that control these processes, and their effects, are described according to these three distinct scales. The principal local effect is suggested to be a century-long modulation of the effectiveness of egg and larval drift between Iceland and Greenland, due to long-period wind-induced changes in the strength of the Irminger/West Greenland current system, with major implications for the cod fishery at both ends of the drift path. The exchanges of greatest regional importance are those associated with the approximately 14-year drift of the Great Salinity Anomaly around the Subpolar Gyre; here the key components are the anomalous production and export of ice and fresh water from the high Arctic, the northerlies that brought it south in a swollen East Greenland Current, and the local processes that preserved the fresh-water layer north of Iceland prior to its export through the Denmark Strait. The transfers of global importance are the dense outflows of Arctic Intermediate Water and upper Polar Deep Water south through the deeper layers of the Denmark Strait, which contribute significantly to North Atlantic Deep Water production and thus drive the global thermohaline circulation. Some interactions between these three scales of processes are described, and some suggestions are made for improving the match between observation and simulation.

INTRODUCTION

This paper has two main aims: first, to describe the large-amplitude fluctuations that have characterized the physical environment of the Greenland-Iceland region at time scales of decades to those of a century or more; and second, to begin (but by no means complete) a description of their causes, controls, and effects over space scales that range from the local through the regional to the global and to provide some partial glimpses on how these scales may interrelate. With limited space available, this account will necessarily be brief; fortunately an extensive body of literature already exists to provide the details.

[1]Fisheries Laboratory, Ministry of Agriculture, Fisheries, and Food, Lowestoft, Suffolk, U.K.

THE LOCAL SCALE: THE PHYSICAL MEDIATION OF BIOLOGICAL EXCHANGES BETWEEN ICELAND AND GREENLAND

During the twentieth century, the high latitudes of the Atlantic sector have undergone a considerable wave of warming and salinification that included the Iceland-Greenland region but was by no means confined to it. This change is illustrated in Figure 1 with reference to three of the longest series; the sea surface temperature anomalies for the West Greenland Banks from 1876-1974 (Figure 1a; Smed's data, in a succession of contributions to Annales Biologiques), the salinity of the upper North Atlantic Central Water layer passing northward through the Faeroe-Shetland Channel in

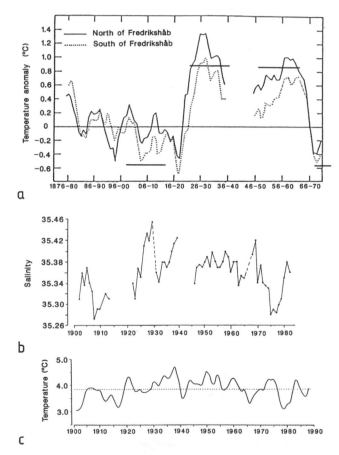

a

b

c

FIGURE 1   Increased warming and salinification during the current century. (a) Surface temperature anomalies for West Greenland, 1876-1974 (Smed's data, from Buch and Hansen, 1988; reprinted with permission). (b) the salinity of the North Atlantic Water in the Faeroe-Shetland Channel, 1902-82 (from Dooley et al., 1984; reprinted with permission of the International Council for the Exploration of the Sea); and (c) 3-year running averages of yearly temperature along the Kola Section of the Barents Sea, 1900-90 (from Loeng, 1991; reprinted with permission of the Norsk Polarinstitutt), plotted to a common time base. The horizontal bars in Figure 1a refer to the warm and cold periods for which wind fields are compared in Figure 3.

1902-1982 (Figure 1b), and the 0-to-200 m mean temperature along the Kola meridian (33°30'E) in the Russian sector of the Barents Sea, 1900-1990 (Figure 1c). Many other temperature series—some from the open ocean (see Kushnir, 1994)—could have been adduced to make this point and, although long time series of salinity are less commonly available, we have sufficient glimpses of the salt record around the northern North Atlantic to suggest that the salinification was similarly pervasive, similarly protracted, and similarly extreme.

Thus as salinities rose to unprecedented values in the Faeroe-Shetland Channel in the late 1920s and 1930s (Tait, 1957; Dooley et al., 1984; see Figure 1b), extreme values were also being encountered further around the subpolar gyre in the U.S. Ice Patrol Standard Section running southwest from Cape Farewell. Some were so extreme as to provoke skepticism; as Harvey (1962) reminds us, "apart from 1933 when the salinity observations were thought to be in error, the highest salinity recorded by the US Coast Guard in any of their 22 observations along this section between 1928 and 1959 was a value of 35.07% observed in both August 1931 and July 1934."

In all cases, the period of peak warming and salinification occurred between the early 1930s and the late 1960s, and in a series of classic papers by Saemundsson (1934), Stephen (1938), Jensen (1939), Taning (1943, 1949), and Fridriksson (1949), we are left in no doubt that this wave of warming and salinification was accompanied by radical northward shifts in the distributional boundaries for a wide range of marine species.

In other words, both the circulation and the ecosystem of the subpolar gyre appear to have undergone a slow evolution over large space and time scales during the present century.

In socio-economic terms, the impact of this widespread change reached a focus in the Iceland-Greenland sector, amplified through an apparent change in the effectiveness of egg and larval drift in the Irminger/West Greenland current system. The body of evidence is circumstantial but appears to be self-consistent. In Figure 2 below, Buch and Hansen (1988) indicate that the warming on the West Greenland Banks was accompanied by an explosive development of the West Greenland cod fishery (solid line), which rose from a negligible tonnage in the early 1920s to over 300,000 tons/year in the 1950s and 1960s and a maximum of more than 450,000 tons/year before abruptly declining once again between the late 1960s and the present as cooler conditions returned. Such "cod periods" have been documented in the past at West Greenland, in the 1820s and 1840s (Hansen, 1949), but this change represented a return of cod to West Greenland after an absence of at least 50 to 70 years (Buch and Hansen, 1988).

The hypothesis is that this change was both initiated and maintained by an increased exchange of cod larvae from

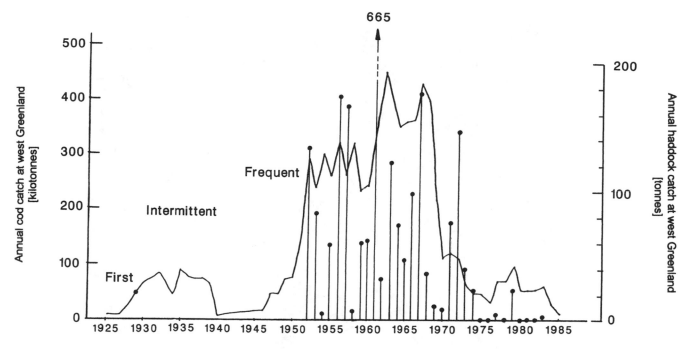

FIGURE 2   The catch of cod at West Greenland, 1925-85 (solid line; from Buch and Hansen, 1988; reprinted with permission) and the corresponding catch of haddock, 1952-83 (dots; data from Hovgård and Messtorff, 1987).

the spawning grounds off southwest Iceland during the warm decades, compared with the cold periods that preceded and succeeded them. Several items of evidence sustain this theory.

1. *The Use of Haddock as a Tracer of Larval Exchange.* Unlike cod, haddock do not spawn successfully at West Greenland, so if adult haddock are caught there they must have drifted there as larvae. (The closest known spawning site for haddock lies upstream near the cod spawning grounds off southwest Iceland.) When we compare the (small) international catch of haddock at West Greenland with the (large) international catch of cod (Figure 2), we find essentially similar trends of change in both species since 1952 when annual haddock catch statistics first became available (listed in Hovgard and Messtorff, 1987). There is even a qualitative similarity before that, with the first specimen of haddock reported from Greenland waters in 1929, infrequent occurrences in the 1930s, and frequent catches in 1945 (see Hovgard and Messtorff). Furthermore, in the years in which young fish surveys cover much of the drift path, the year (1984) with highest recorded abundance of 0-group cod off East Greenland (Vilhjalmsson and Magnusson, 1984) immediately preceded a very high abundance of 1-group cod at West Greenland (ICES, 1986). Thus while it is possible that the first cod colonizing the West Greenland Banks established a self-sustaining stock there, purely through the amelioration of the marine climate, the parallelism between cod and haddock catches along the western banks seems to argue that recruitment to the West Greenland

cod stock was fed to a significant extent by a time-varying larval drift from Iceland.

2. *The Analysis of Changes in the Windfield.* Kushnir (1994 and personal communication) has analyzed the windfield changes that are associated with both short-term interannual variation of Atlantic SST and its long-term interdecadal variability. With Bjerknes (1964) he concludes that while the short-term variations in SST are governed by local forcing (windspeed and heat exchange), the interdecadal changes are governed by the non-local dynamics of the large-scale circulation. Figure 3a and b are truncated extracts from Kushnir's North Atlantic study, showing, for the area north of 40°N only, the change in windfield during winter (December to April) that occurred during the warming of the 1920s and the cooling of the late 1960s. The sense of the change has been made comparable by subtracting the 15-year cold-year average from the 15-year warm-year average in each case. Thus Figure 3a shows the change in the mean winter wind field from the cold period of 1900-1914 to the warm years of 1925-1939, while Figure 3b shows the equivalent change from the more recent cold episode of 1970-1984 to its antecedent warm period, 1950-1964. (These four periods are indicated by horizontal bars on Figure 1a.)

Since the Icelandic cod spawn off the Reykjanes Peninsula from March to May (Jonsson, 1982), these changes in the windfield overlap the season of egg liberation and early larval drift. And in each case the sense of the change from cold to warm periods has been one of increased easterly

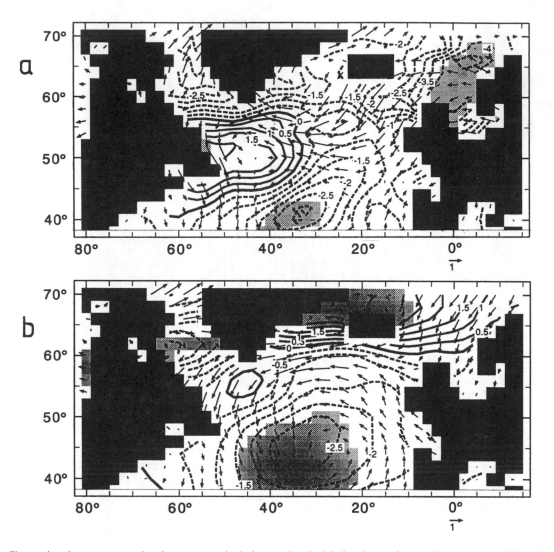

FIGURE 3   Change in winter-mean sea level pressure and winds associated with the change from cold to warm conditions in the northern North Atlantic. (a) Changes from 1900-14 (cold) to 1925-39 (warm); (b) changes from 1970-84 (cold) to the antecedent warm period, 1950-64. (Adapted from Kushnir, 1994; reprinted with permission of the American Meteorological Society.) These periods are indicated by horizontal bars in Figure 1a.

airflow overlying the westward-flowing Irminger current. While the simulation of the actual effect on the Irminger current awaits further refinement of the Comprehensive Ocean-Atmosphere Data Set (COADS) (Kushnir, pers. comm.), this result is at least qualitatively consistent with the hypothesis of increased larval exchange between Iceland and Greenland in the warm years of this century via the Irminger/West Greenland current system.

3. *The Return Migration of Adults to Iceland.* Economically, the effects of such changes are not confined to the effect on the cod catch at West Greenland. It is now becoming apparent that cod year classes that survive the drift to West Greenland as larvae may later return as adults to Iceland. These are not trivial events! Schopka (1991) shows that the return of the 1945 year class represented an unpre-

dicted increment of over 700,000 tons of 8-year-old fish, worth more than £1 billion at today's prices (Figure 4).

The year-classes that Schopka lists as including adult immigrants to Iceland—1936, 1937, 1938, 1942, 1945, 1950, 1953, 1961, 1962, 1963, 1973, and 1984—are mostly those of the warm epoch, reinforcing the idea of more frequent emigration as larvae at that time, but even the infrequent occurrences of more recent years are of importance both economically and in terms of understanding which management measures were appropriate for the local Icelandic stock.

While certain components of the problem remain unresolved, the body of available evidence suggests that the local effect of a changing physical exchange between Iceland and

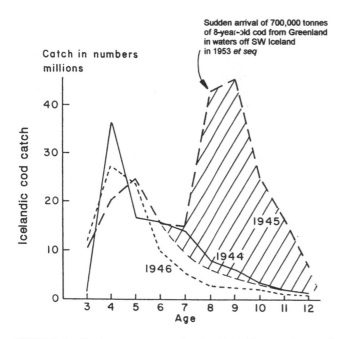

FIGURE 4   Catch-at-age curves for the 1944-46 year classes of cod at Iceland (from Schopka, 1991; reprinted with permission of the Northwest Atlantic Fisheries Organisation), but with the likely contribution of returning adults from Greenland shown hatched.

Greenland is an economic or socio-economic one, acting to augment recruitment to the West Greenland cod population via larval drift from Iceland during years and decades when the wind field in winter and spring acted to strengthen the Irminger/West Greenland current system. A proportion of these fish may return as adults to Iceland.

Physically mediated exchanges of a similar type may also occur farther downstream along the same current system. As Templeman (1981) points out: "Cod occasionally migrate from the Newfoundland area to West Greenland and vice versa (Templeman, 1974, 1979). Also in some years, especially 1957, there was considerable drift of cod larvae from Greenland toward the coasts of Baffin Island and Labrador (Hansen, 1958; Hermann et al., 1965)." (See also Templeman, 1965, and Danke, 1967.)

THE REGIONAL SCALE: THE "GREAT SALINITY ANOMALY" AND ITS LOCAL AND REMOTE CONTROLS

The widespread freshening of the upper 500 to 800 m of the northern North Atlantic, which has come to be known as the "Great Salinity Anomaly" (GSA), represents one of the most persistent and extreme variations in global ocean climate yet observed anywhere in this century. Dickson et al. (1988) describe it as largely an advective event, originating in the accumulation and preservation of an anomalously fresh near-surface layer in the seas north of Iceland in the middle to late 1960s, which was transferred through the

Denmark Strait to the open North Atlantic in the late 1960s and was subsequently traceable around the subpolar gyre for over 14 years until its return to the Greenland Sea in 1981-1982. It is the first direct measure we have of the strength of the gyre circulation (about 3 cm s^{-1} on average). Figure 5 dates the passage of the GSA from its point of origin, through the Denmark Strait and around the Northern Gyre, while Figure 6 presents best-guess estimates of the salt deficit associated with the GSA at a series of points around this circuit.

The extreme nature of this event can be judged from the facts that as it passed south along the Labrador coast in 1971-1973 it represented a total salt deficit of some 72 × 10^9 tons; that in the eastern Atlantic, the observed freshening of about 0.1 psu to 500 m depth was the equivalent of adding an "extra" 1.4 m of fresh water at the ocean surface (Pollard and Pu, 1985); that as it passed through the Faeroe-Shetland Channel in 1976 or so, the salinities in the North Atlantic and Arctic Intermediate Water masses were at their lowest since records began in 1902 (see Figure 1b); and that as it passed by northern Norway and west Spitsbergen in 1978-1979, the Norwegian Atlantic Current, by any conventional definition, ostensibly contained no Atlantic Water (i.e., water of salinity of 35 or more).

The chronology of this great event (Dickson et al., 1988) and its biological effects (e.g., Blindheim and Skjoldal,

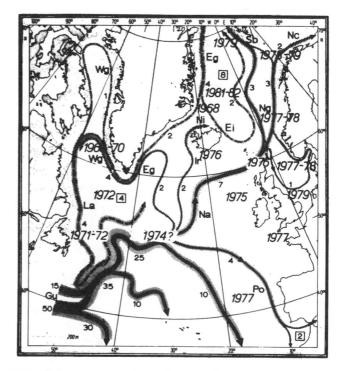

FIGURE 5   Transport scheme for the 0-1000 m layer of the northern North Atlantic, with dates of the GSA salinity minimum superimposed. (From Dickson et al., 1988; reprinted with permission of Pergamon Press.)

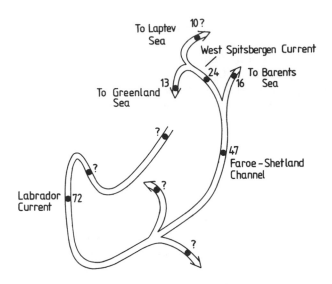

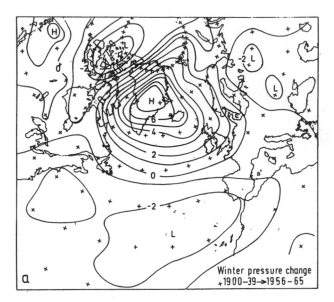

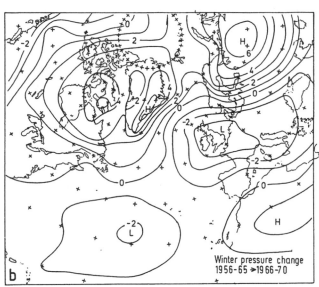

FIGURE 6   Estimates of salt deficit ($\times 10^9$ tonnes) during the passage of the Great Salinity Anomaly around the northern North Atlantic. (From Dickson et al., 1988; reprinted with permission of Pergamon Press.)

1993) have already been thoroughly discussed and need not be repeated here. In the present context, the more relevant points for discussion are, first, to identify the near- and far-field controls that promoted and maintained such an extreme and protracted change and, second, to assess in the light of these controls whether we are dealing with a unique, occasional, or regular (even cyclical) event.

Various factors seem to have contributed to the amplitude and persistence of the GSA. These are conveniently discussed under four headings: (1) factors supporting the build-up of anomalously high pressure at Greenland during winters of the 1960s; (2) factors affecting the upstream supply of fresh water; (3) local factors preserving the fresh-water layer north of Iceland, and (4) factors contributing to the persistence of the GSA downstream.

Factors Supporting the Build-up of High Pressure at Greenland During Winters of the 1960s

Dickson et al. (1988) ascribe the GSA primarily to the build-up of high pressure at Greenland from the mid-1950s to the late 1960s, in every season of the year but especially during winter (Figure 7). This change directed an abnormally strong northerly airflow over the Norwegian and Greenland seas, and as these northerlies increased to maximum strength in the late 1960s, the hydrographic character of the cold East Greenland and East Icelandic currents was altered. The East Greenland Current swelled in volume and became cooler and fresher as an increasing proportion of polar water was brought south from the Arctic Ocean to seas north of Iceland; the East Icelandic Current, which had been an ice-free Arctic Current during 1948-1963, became

FIGURE 7   Mean change of winter sea level pressure (mb) between (a) 1900-39 and 1956-65 and (b) 1956-65 and 1966-70. (From Dickson et al., 1975; reprinted with permission of Macmillan Magazines Ltd.)

a polar current in 1965-1971, transporting and preserving drift ice.

Thus, it is the anomalous ridge at Greenland that is held responsible for bringing an increased supply of ice and fresh water southward through the Greenland and Iceland Seas to the points of outflow into the North Atlantic, and thus for initiating the GSA (while, incidentally, terminating the warm epoch in Greenland waters, which formed the subject of the preceding section). After reaching a peak (5-year mean) anomaly of more than +12 mb in winters 1966-1970, the ridge rapidly collapsed in the 1970s (recorded by Dickson et al., 1975), but by then the GSA was already a distant signal en route around the subpolar gyre.

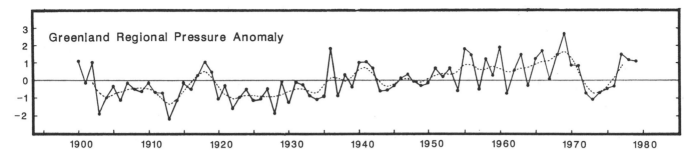

FIGURE 8   Normalized winter-mean pressure anomaly for the region 60-70°N, 30-65°W from 1900-1979. (From Rogers, 1984; reprinted with permission of the American Meteorological Society.)

The two most relevant questions in the present context are how unusual the Greenland Ridge of the 1960s was, and what factors might have contributed to its development. The first of these is addressed in Figures 8 and 9, which provide both a long-term and a hemispheric perspective. Figure 8 shows that the ridging that developed over Greenland in the late 1960s was unprecedented in a record of some 75 years' duration and, more remarkably, that it formed the end point of a steadily increasing trend in winter pressure there, which has been under way almost since the turn of the century. (The subsequent collapse of the Ridge in the 1970s is the event reported by Dickson et al. (1975).)

Even more remarkable is the spatial isolation of this change. In illustrating the difference in winter mean SLP between 1904-1925 and 1955-1971 (i.e., across the two endpoints of the above trend), Rogers shows (Figure 9 above) that the principal pressure change is confined to a restricted cell centered over Greenland, paired with a lesser

change of opposite sign near the Azores. Not surprisingly, then, the North Atlantic Oscillation (NAO = Azores − Iceland pressure difference) was at its century-long minimum also during the late 1960s (Figure 10).

Part of the reason for the shorter-term intensification of ridging at Greenland and decrease in the NAO between the mid-1940s and the late 1960s may be the processes that promote winter storm development in the zone of maximal land-sea temperature contrast along the U.S. eastern seaboard half a wavelength upstream. The reason for supposing such a connection lies in the notion that a steady strengthening of pressure at Greenland may more properly be regarded as a weakening of the statistical Iceland low. (A weakening Iceland low in twentieth-century Januaries has been noted by van Loon and Madden (1983).)

Dickson and Namias (1976) offer some support to this link by showing that a decadal change from a regime of warm winters (1948-1957) over the southeastern United States to a regime of cold winters there (1958-1969; Figure 11) was accompanied by a steepening of the thermal gradient at the coast, a sharp coastwise increase in the mean winter cyclone frequency (Figure 12) and a significant (approximately 350 nmi) southwestward retraction of the zone of maximum storm frequency as storms matured faster offshore. Thus, it remains a possibility that the record intensification of the Greenland High at this time (the 1960s) might in part be a reflection of a southwestward withdrawal of the statistical Iceland low.

The Upstream Supply of Ice and Fresh Water

The total salt deficit of the GSA as it passed the Labrador coast (72 × 10^9 tons—see Figure 6) is equivalent to a fresh-water excess of 2000 km³. As Aagaard and Carmack (1989) point out, this quantity of fresh water is too great to have its source in the Iceland Sea itself, but as the equivalent of about one-half of the annual fresh-water transport of the East Greenland Current where it enters the Greenland Sea from the Arctic Ocean, it can be accounted for by only a moderate perturbation of that outflow (e.g., a 2-year period of a fresh-water flux 25 percent above normal) and repre-

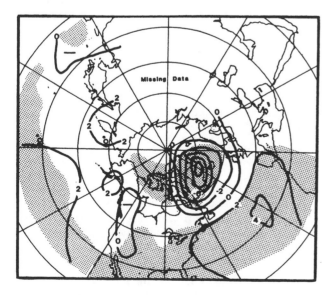

FIGURE 9   Difference in winter mean SLP between 1904-25 and 1955-71. Stippled areas are significant at the 95% level. (From Rogers, 1984; reprinted with permission of the American Meteorological Society.)

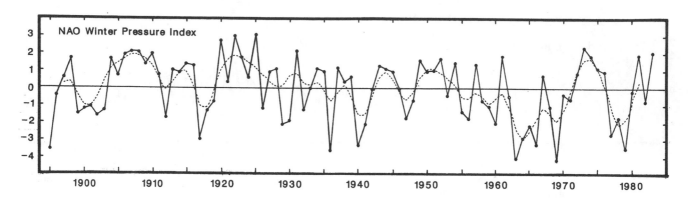

FIGURE 10   Winter index of the North Atlantic Oscillation (1895-1983) based on the mean normalized pressure difference between Ponta Delgada, Azores and Akureyri, Iceland. (From Rogers, 1984; reprinted with permission of the American Meteorological Society.)

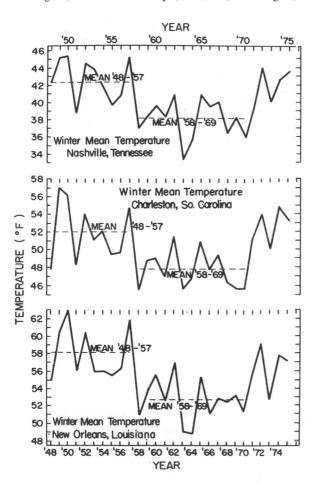

FIGURE 11   Winter mean air temperatures at Nashville, Charleston, and New Orleans, from 1947-48 to 1974-75. (From Dickson and Namias, 1976; reprinted with permission of the American Meteorological Society.)

sents an insignificant drain on the 100,000 km³ or so of the Arctic fresh-water reservoir. Thus, the fresh-water supply to the GSA is likely to have its origins upstream, in the processes of ice production and supply within the Arctic Ocean.

By comparing the timing of sea-ice variations in seven subregions that encircle the polar ocean over a 32-year period, Mysak and Manak (1989) were able to confirm a significant out-of-phase relationship between the ice-concentration anomalies of the Beaufort and Chukchi seas and the Greenland Sea. Later, Manak and Mysak (1989) and Mysak and Power (1991) extended and added detail to this sequence of relationships by showing, first, that fluctuations of the Mackenzie River discharge promote sea-ice anomalies of the same sign in the Beaufort and Chukchi seas about one year later and, second, that the export of large build-ups of sea-ice from the western Arctic via the Beaufort Gyre and the Transpolar Drift Stream would contribute to anomalies of ice extent and (from ice melt) fresh-water supply in the Greenland Sea some 3-4 years after that.

Thus, annual runoff from the North American continent into the western Arctic precedes Koch's sea-ice severity index for the coast of Iceland by around 5 years (Figures 13a, b, c), and Mysak and Power (1991) conclude that the record freshets to the Arctic Basin in 1964-1966 (2 or more standard deviations from the mean; see Figure 13a) were significant precursors and suppliers of fresh water to the GSA, as it developed in the Iceland and Greenland seas during the middle to late 1960s.

Local Preservation of the Fresh-water Layer in the Iceland Sea

The record extent of sea ice that developed in the Greenland and Iceland seas during the late 1960s (maximum in 1968) was partly the result of local as well as remote forcing, and was in fact a "telltale" of a local process that was fundamental to the preservation of an accumulating anomaly as large as the GSA: namely, by the mid-1960s the proportion of Polar Water in the East Greenland Current was so great that surface salinities north of Iceland decreased below the critical value of 34.7 (Figure 14). Below this value the surface layers will not reach a sufficiently high density,

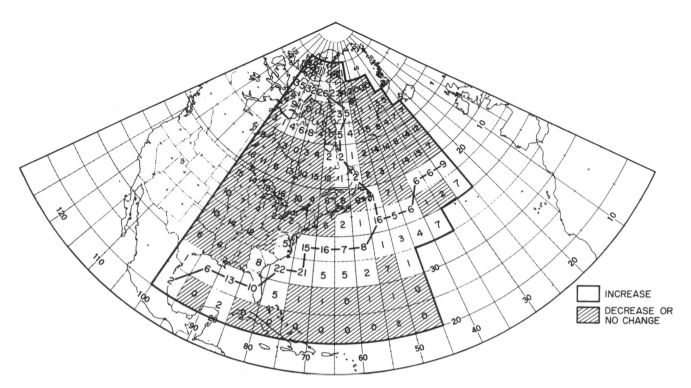

FIGURE 12   Change in the total number of cyclones between the "warm SE" and "cold SE" groups of winter months in the southeastern USA. (From Dickson and Namias, 1976; reprinted with permission of the American Meteorological Society.)

even at the freezing point, to mix with the underlying, more saline water (Figure 15); instead, ice will tend to form (Malmberg, 1969). Thus, not only was more fresh water transported south toward Iceland during the 1960s (from whatever source) under the stress of an anomalous northerly airflow, but in the years of peak polar influence (1965-1971), the extremely cool, fresh character of the upper 200-300 m was preserved since it was not mixed out by convective overturn (Dickson et al., 1988).

Downstream Preservation of the Fresh-water Layer

Once the GSA has passed through the Denmark Strait and is undergoing its long circuit of the subpolar gyre, we are forced to be even more conjectural about the forces that might have maintained its integrity. One suggestion that is perhaps better based than most comes from Clarke (1984); it posits an interaction between the GSA and local environmental conditions as it passed south along the Labrador coast (see Figure 9 of Dickson et al., 1988). From the late 1960s until 1971-1972, and possibly for unconnected reasons, there was progressively less deep winter convection at OWS Bravo in the central Labrador Sea (Lazier, 1980). As a result, said Lazier, the surface layer at Bravo freshened and the offshore salinity gradient weakened, so that the offshore fresh-water flux by eddy-diffusive processes would have weakened also. For this reason, Clarke suggests that

the GSA fresh-water signal would have been maintained rather than mixed offshore during its long circuit of the Labrador shelf and slope. Although convection at Bravo and the offshore salinity gradient were rapidly restored following the severe winter of 1972, the GSA signal had largely passed.

Periodicity

Our present patchy understanding of the GSA event suggests that the Arctic Ocean provided the fresh water, the northerlies of the Greenland High brought it south, and conditions both north of Iceland and downstream prevented it from being mixed out (vertically and laterally). However, we have only an inconclusive answer to its likely periodicity. If the strength of the winter high-pressure cell at Greenland in the 1960s was a critical factor, as Dickson et al. (1988) claim, then that circumstance was a secular event—the product of a slow evolution over most of this century. If the maxima in Koch's sea-ice index indicate the same process of accumulation of fresh water and suppression of convection north of Iceland as we describe in Figure 15, then there may be grounds for belief that an earlier, if lesser, salinity minimum in 1910-1914 (see, for example, Figure 1b) reflected a previous iteration of the GSA set of processes (Dickson et al., 1984). However, numerous authors have pointed to the scope for multiannual or decadal periodicity

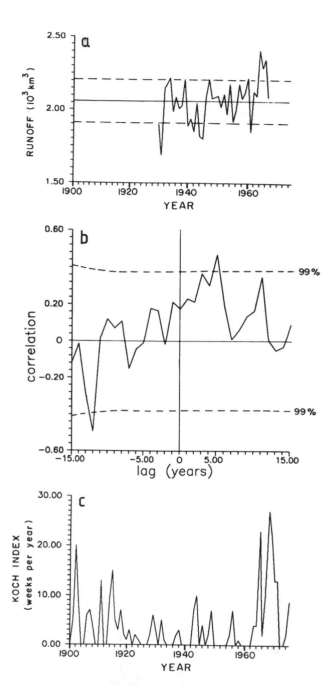

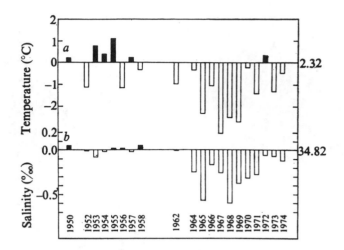

FIGURE 14   Anomalies of (a) temperature and (b) salinity in June at 25 m depth in a study area between Iceland and Jan Mayen (67-69°N, 11-15°W). The long-term means, 1950-58, are also shown. (From Malmberg, 1973; reprinted with permission.)

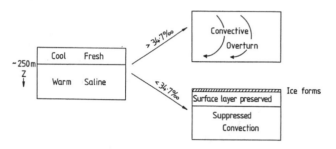

FIGURE 15   Schematic diagram illustrating the suppression of convection north of Iceland as upper-ocean salinities decrease below 34.7. (From Dickson et al., 1988; reprinted with permission of Pergamon Press.)

FIGURE 13   (a) Annual runoff from the North American continent into the Arctic Basin, 1930-67, with mean (solid line) and ± 1 standard deviation (dashed line) indicated; (b) lagged cross-correlation between runoff (above) and Koch Index of ice severity (below) at the Icelandic coast, 1915-75; (c) the Koch sea-ice severity index for Iceland. (All from Mysak and Power, 1991; reprinted with permission of the Canadian Meteorological and Oceanographic Society.)

that is inherent in such processes. The approximately 14-year circulation period of the subpolar gyre is plainly one such control, but the decadal regimes in southeastern U.S. air temperature shown in Figure 11, the 6.7-year spectral maximum in the North Atlantic oscillation index (Rogers,

1984), the multi-year advective time lag of ice drift from the Beaufort Sea to Greenland and Labrador (Mysak and Power, 1991; Mysak and Manak, 1989) plainly have the potential to lend a degree of periodicity to the process. Conceptual models for these periodic anomalies—some of great complexity (see Figure 3 of Mysak and Power, 1991)—have been proposed (see also Darby and Mysak, 1993; Ikeda, 1990). Under this rationale the Great Salinity Anomaly was merely a more obvious, high-amplitude itera-tion of a regular event.

THE GLOBAL SCALE: PRODUCTION OF NORTH ATLANTIC DEEP WATER AND THE THERMOHALINE CIRCULATION

To demonstrate a global role for Denmark Strait exchanges, there can be few more eloquent illustrations than Figure 16 (from Charles and Fairbanks, 1992), in which the $\delta^{13}C$ history in benthic forams from the sea-floor sediments of the Southern Ocean is used to indicate the changing content of the North Atlantic Deep Water (NADW) in the abyssal circulation over the past 6,000 to 18,000 years. Since Circumpolar Deep Water consists of both NADW with a high $\delta^{13}C$ content and Indo-Pacific Deep Water with low $\delta^{13}C$, the increase in $\delta^{13}C$ content at the end of the Ice Age is thought to reflect the rapid resumption of full NADW production at that time.

Whether or not we believe that NADW production was stopped or merely much reduced during the last glaciation, it is plain that the glacial-interglacial signal was reflected in the amount of NADW in circulation. Since the dense overflows across the Greenland-Scotland Ridge (plus consequent entrainment) constitute the largest two of the four main sources of NADW (see Figure 17, taken from Swift

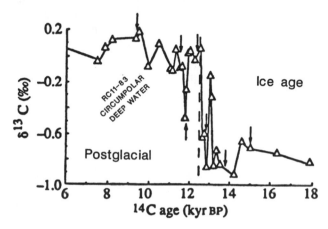

FIGURE 16 $\delta^{13}C$ record of benthic forams in Southern Ocean core, indicating the changing production of North Atlantic Deep Water, 6,000-18,000 BP. (From Charles and Fairbanks, 1992; reprinted with permission of Pergamon Press.)

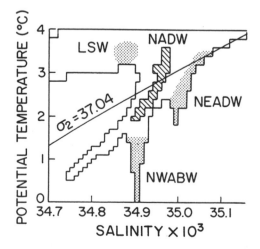

FIGURE 17   Potential temperature vs. salinity diagram for North Atlantic deep waters colder than 4°C, with the characteristics of NADW shown by hatching. (From Swift, 1984b; reprinted with permission of Pergamon Press.)

(1984b)), it is equally plain that a time-varying exchange between the sub-Arctic seas and the deep open Atlantic has had a global importance in modulating the thermohaline circulation (THC) on 100- to 1000-year time scales.

The less obvious question concerns the stability or instability of the thermohaline circulation on decade-to-century time scales, which form the focus of this symposium. Such shorter-term changes have often been invoked, most frequently to provide the feedback mechanism required in "recurrent GSA" models. Typically these conceptualizations include:

1. The suppression of deep convection in the Greenland and Iceland seas (e.g., Mysak and Power, 1991). As Aagaard and Carmack (1989) point out, the main convective gyres in these seas are rather delicately poised with respect to their ability to sustain convection, requiring only a modest redistribution of a small part (about 150 km³) of the vast Arctic fresh-water reservoir (approaching 100,000 km³) to achieve shutdown.

2. The effect of suppressed deep convection on the strength of the thermohaline circulation. This process has been suggested at many scales up to and including the "haline catastrophe theory" of Broecker et al. (1985b), in which runoff from the rapid melting of continental ice sheets provokes a radical renewed suppression of NADW production during deglaciation.

3. Compensatory changes in the influx of heat and salt to the Norwegian Sea due to changes in the strength of the thermohaline outflow to the North Atlantic. A direct relationship was demonstrated, for example, in the simulation by Manabe and Stouffer (1988).

From these and similar arguments, the thermohaline circulation has indeed been associated with the decadal scale of climate change—for example, by Weaver et al. (1991) and, most recently, by Yang and Neelin (1993 and personal communication). In the latter model, a recurrent 13.5-year periodicity is obtained using a GSA feedback loop local to the North Atlantic in which brine rejection from heavy ice production stimulates the thermohaline circulation and increases the poleward heat flux in compensation, thus melting ice, suppressing convection, and weakening the thermohaline circulation once again.

A number of objections to the simplifications of such models remain:

• Without realistic bathymetry (the Yang/Neelin model has a constant depth of 4000 m from 70°S to 70°N, for example), the strength of the thermohaline circulation really is just a function of the effectiveness of deep convection. In reality, the presence of a Greenland-Scotland Ridge with a sill depth in the Denmark Strait of only 600 m drives home the point that it is intermediate water production that drives the global overturning cell, and this automatically means that the convection-THC link is much more tenuous. As Aagaard and Carmack (1989) point out, we can envisage a system "in which mid-depth convection (which is the main source of the Denmark Strait overflow) occurs, albeit involving waters of reduced salinity, while the deeper convection, which renews the densest waters in the system, is shut down." Also, although the bulk of the outflow through the Denmark Strait has been shown by Swift et al. (1980) to consist of locally formed Arctic Intermediate Water, the 2.5 Sv of that local AIW production is augmented with about 0.5 Sv of upper Polar Deep Water whose origins lie north of Fram Strait in the Arctic Ocean (Rudels and Quadfasel, 1991). Thus, even if properly characterized, the convection of the Greenland and Iceland seas is not the sole factor involved.

• The T-S characteristics of the Greenland Sea Deep Water did not appear to feel the influence of the widespread surface freshening of the late 1960s and early 1970s. The time series of Figure 18 are interpreted by Meincke et al. (1992) as showing conditions favorable to the renewal of Greenland Sea Deep Water at that time. This seems important if GSA freshening events are to be held responsible for modulating the thermohaline circulation with their recurrent signal. (However, Schlosser et al. (1991a) report a cessation of deep-water formation from 1980 onward, coincident with the return of the GSA signal to the Greenland Sea.)

• While the hydrographic characteristics of the Denmark Strait outflow have been observed to change (Brewer et al., 1983; Lazier, 1988), we have no direct observational evidence that the overflow transport actually changes on the decadal time scales that current THC models predict. Our direct current measurements are of only 4 years'

duration and are too short by themselves to permit comment on the question of decadal change in overflow transport (Dickson and Brown, 1994), even though they do appear to emphasize the steadiness rather than the variability of the outflow (Figure 19). A slightly longer perspective is provided by the downstream behavior of tracers, which appears to confirm this impression of steadiness in the overflowing stream. As Schlosser and Smethie point out (1995, in this section), the "age of the deep core between 32°N and 44°N did not change by more than 10% between 1983 and 1990, which implies that the combination of water-mass formation rate, transport, and mixing with adjacent water has been constant to within about ± 10 percent over this time period."

If these points seem unduly critical of current models, they are not intended to be so. There is no doubt that our ability to trace, understand, and simulate these large-amplitude events at high latitudes will open up an important avenue to improved climate prediction. Even if they turn out to be less than cyclic in character, the slow shifts in the ocean-atmosphere system that these events set in motion will still be of use in forecasting effects, and these effects in turn will still be of sufficient significance to make any improvement in prediction worthwhile. The above are merely suggestions for points that future models will have to address—in other words, reasons for model improvement rather than model abandonment. Recent simulations by Weaver (personal communication) appear to support the idea of an invariant THC on decadal time scales.

CONCLUSIONS

1. Although the evidence is circumstantial, there does appear to have been a physically mediated, century-long variation in the exchange of heat and larvae between Iceland and Greenland via the Irminger Current, with major socio-economic impact.

2. Although it contains periodic elements, the Great Salinity Anomaly is neither a periodic nor a frequently recurring phenomenon. One or possibly two such events are thought to have occurred during this century, and the event we observed in 1968-1982 was driven by a secular change in the winter pressure field at Greenland.

3. Although its proxies clearly reflect the glacial–interglacial signal, the global thermohaline circulation has not yet been demonstrated to vary at decadal time scales. Simulations of the process are not yet adequately realistic in topography, in the role of deep convection (as opposed to Arctic Intermediate Water formation), in the processes that might cap deep convection, and in the periodicity of these processes.

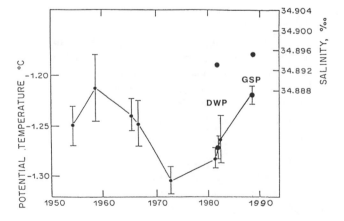

FIGURE 18   Time series of average potential temperature below 2000 m in the Greenland Basin. (From Meincke et al., 1992; reprinted with permission of the International Council for the Exploration of the Sea.)

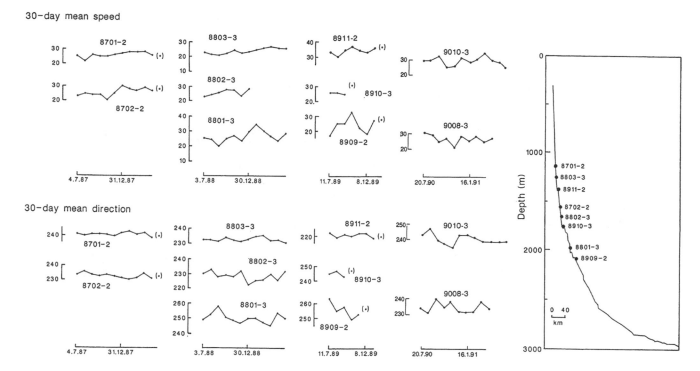

FIGURE 19 Thirty-day means of near-bottom current speed and direction in the Denmark Strait Overflow, taken from successive deployments of instruments set at similar depths on the East Greenland Slope. The "87", "88", and "89" instruments are from the Angmagssalik Array. The "90" instruments are from equivalent depths on the TTO Array, 160 km upstream. (From Dickson and Brown, 1994; reprinted with permission of the American Geophysical Union.)

Discussion

TALLEY: Is it possible that the Great Salinity Anomaly, singular though it was, could have circled around the basin a few times before dying away?

DICKSON: Our rather poor records of salinity were at their worst just at that point, so we didn't observe that. It's certainly possible—it has a subpolar gyre to go 'round in, and remember Meincke's long studies of the flow across the mid-Atlantic ridge and what gets turned back—but would it make any difference once it's been diluted?

GHIL: Where did you get the periodicities you showed, like the 7-year North Atlantic oscillation?

DICKSON: Well, there were quite a few. The 6.7-year peak you mention was from Rogers. Jerry Namias and I have long talked about another, which is the relationship between conditions at the U.S. eastern seaboard and the high pressures in Greenland. We used air and sea temperature and calculated the seaboard baroclinic gradients for cold and warm groups of years. Then we looked at the storm frequency for those periods, and found an increase along a line up the coast where the gradient was heightened, and a corresponding periodic retraction of the Iceland low.

SARACHIK: You've made the case for the decoupling of convection and genesis and overflow. Would you care to speculate about what can turn off the thermohaline circulation and the production of deep water in the Atlantic?

DICKSON: Knut Aagaard's papers show that it's very delicately poised; the fresh-water flow into the Greenland Sea can stop convection dead. It would take 31 years of the same trend to make a difference at the depth of the Denmark Strait Sill, but during the Ice Age there could have been something quite dramatic. My feeling is that if something so light were formed it would set up a sort of shallow thermohaline circulation. We don't really know yet how intermediate and deep-water formation mesh.

KUSHNIR: It seems possible that very cold temperatures in the central North Atlantic trigger certain atmospheric conditions that pull ice and fresh water from the Arctic Sea.

MCGOWAN: Bob, was the increased cod population a matter of improved survival or different spatial distribution?

DICKSON: Well, at first people thought it was increased survival because of the improved climate on the West Greenland Banks. But that's not true of haddock, and the haddock catch mirrors the change in cod, so we now think it's a time-varying conveyor belt that relocates larvae.

Natural Climate Variability on Decade-to-Century Time Scales
National Research Council, 1995

Observational Evidence of Decadal-Scale Variability of the North Atlantic Ocean

SYDNEY LEVITUS[1], JOHN I. ANTONOV[1], ZHOU ZENGXI[2], HARRY DOOLEY[3], VLADIMIR TSERESCHENKOV[4], KONSTANTIN SELEMENOV[4], AND ANTHONY F. MICHAELS[5]

ABSTRACT

Temperature and salinity time series 27 to 45 years in length are presented for two locations in the North Atlantic. These data are from Ocean Weather Station "C" (52.75°N, 35.5°W), located in the subarctic gyre, and Station "S" (32.16°N, 64.5°W), located in the subtropical gyre. Decadal-scale variability is evident in the deep ocean as well as the upper ocean at these locations. At Station "S", annual mean temperature at 1750 m depth increased by 0.3°C from 1960 to 1990, with salinity also increasing. At OWS "C" annual mean temperature and salinity in the 1,000-to-1,500 m layer increased on the order of tenths of a degree centigrade and 0.02-0.04 psu respectively between 1966 and 1974. From 1975 to 1985 both these parameters decreased by somewhat larger amounts than the earlier increase. At a minimum, these changes indicate that a redistribution of heat and salt have occurred in the North Atlantic Ocean on decadal time scales.

INTRODUCTION

At the beginning of the twentieth century support for research in the field of physical oceanography was justified by the belief that variability of ocean environmental quantities such as temperature and salinity might be responsible for the interannual variability of fisheries. The International Council for the Exploration of the Sea (ICES) was estab-lished at this time (1902) for the purpose of coordinating such studies on an international scale. The governments of the various north European countries that founded ICES recognized the requirement to undertake such studies multi-nationally. Many of the founders of the field of physical oceanography, such as V. W. Ekman and M. Knudsen, were scientific members of this organization.

One of the earliest descriptions of interannual variability

[1]NOAA National Oceanographic Data Center, Washington, D.C.
[2]Department of Meteorology, University of Maryland, College Park, Maryland
[3]International Council for Exploration of the Sea, Copenhagen, Denmark
[4]State Oceanographic Institute, Moscow, Russia
[5]Bermuda Biological Station for Research, Inc., Ferry Reach, Bermuda

was provided by Helland-Hansen and Nansen (1909), who observed that the flow of North Atlantic water into the Norwegian Sea exhibited interannual variability. Because of a lack of appropriate data, due in part to the expense of oceanographic measurement programs and the disruptions caused by the two world wars, studies describing temporal variability of open ocean conditions during the twentieth century are relatively few. The importance of determining the role of the ocean as part of the earth's climate system has promoted greater involvement of international and intergovernmental scientific organizations in examining the issue of temporal variability of the world ocean.

The World Climate Research Program published results (WCRP, 1982) from a meeting convened to discuss existing long-term time series of oceanographic parameters, particularly temperature and salinity but also biological and chemical parameters. The Intergovernmental Oceanographic Commission is publishing analyses (IOC, 1983, 1984, 1986, 1988) of time series of oceanographic parameters "to demonstrate the importance and usefulness of time series data to the understanding of oceanic and atmospheric processes." Levitus (1989a) briefly reviews some of these studies.

We have analyzed temperature and salinity time-series data at two locations in the North Atlantic Ocean—Ocean Weather Station "C" (OWS "C"), located in the subarctic gyre, and Station "S", located in the subtropical gyre. We present our results in the following sections. In order to place these time series in oceanographic context, which will assist in interpretation, we present Figures 1-3. These show the climatological annual mean distribution of temperature and salinity at depths of 100 m, 1500 m, and 1750 m, respectively, for the North Atlantic Ocean; the locations of OWS "C" and Station "S" are marked by the letters C and S. These climatological distributions represent the output from objective analyses of climatological means over 1° squares of all temperature and salinity data in the files of the National Oceanographic Data Center (NODC), Washington, D.C. as of December, 1988. The analysis techniques used are the same as those described by Levitus (1982). OWS "C" was located in the vicinity of the Ocean Polar Front (OPF), an upper-ocean feature associated with the North Atlantic Current. In the deeper layers OWS "C" occupies a location affected by the properties of deep water formed in the Labrador Sea (Talley and McCartney, 1982). Station "S" is located near Bermuda just south of the Gulf Stream recirculation.

VARIABILITY AT OCEAN WEATHER STATION "C"

Variability of Upper Ocean Thermal Structure at OWS "C"

Ocean Weather Station "C" was located in the North Atlantic at approximately 52.75°N, 35.5°W. For the period

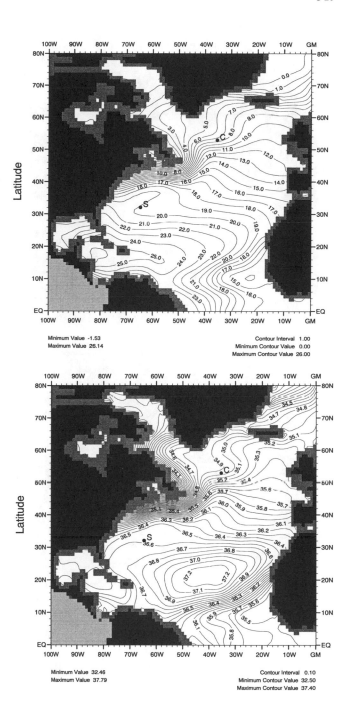

FIGURE 1 Climatological distribution of annual mean temperature (upper panel, in °C) and annual mean salinity (lower panel, in psu) at a depth of 100 m for the North Atlantic Ocean. "C" indicates the location of Ocean Weather Station "C", and "S" the location of Station "S".

1946-1964 OWS "C" was occupied by ships of the U.S. Coast Guard. During that period, only temperature profiles made using mechanical bathythermographs (MBTs) were taken. From 1964 through 1974, hydrographic casts (reversing thermometers to measure temperature and seawater samples gathered for salinity determinations) were made, as

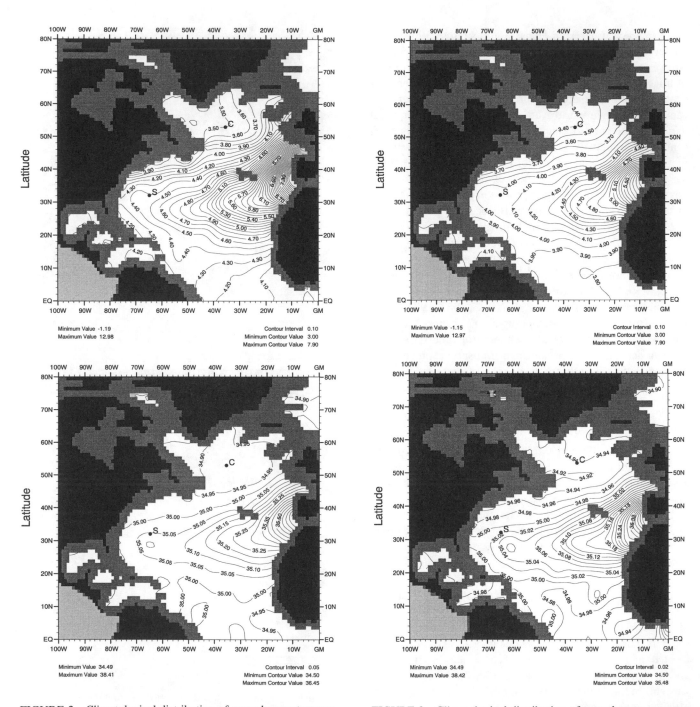

FIGURE 2   Climatological distribution of annual mean temperature (upper panel, in °C) and annual mean salinity (lower panel, in psu) at a depth of 1500 m for the North Atlantic Ocean. "C" indicates the location of Ocean Weather Station "C", and "S" the location of Station "S".

FIGURE 3   Climatological distribution of annual mean temperature (upper panel, in °C) and annual mean salinity (lower panel, in psu) at a depth of 1750 m for the North Atlantic Ocean. "C" indicates the location of Ocean Weather Station "C", and "S" the location of Station "S".

well as MBT casts in some years. Hannon (1979) described some results of analyses of these hydrographic data. From 1975 through 1990, research ships from the Former Soviet Union (FSU) occupied this ocean weather station. They took Nansen casts and conductivity-temperature-depth (CTD) profiles, in addition to some MBT profiles. Some

of the FSU data for OWS "C" were available only in the form of daily means. Therefore, we have averaged all data to form daily means before forming other averages so as to use all data in a consistent manner.

Figure 4a displays the time series of annual mean temperature at a 100 m depth at OWS "C" for the period 1947-

1989. Individual monthly means are computed as the average of all daily mean profiles in each month. Each annual mean is computed as the average of the available monthly means for each year. All years plotted in Figure 4a had at least one monthly mean for each of the four seasons. The vertical bars centered at each annual mean represent ± 1 standard error of the monthly means in each year about the annual mean.

There are two prominent features of this series. The first is the approximately decadal-scale oscillation of about 2°C magnitude, with peaks occurring in 1956, 1965, and 1979. The years for which we have both station data and MBT data indicate good agreement between these two different sets of measurements, and give us confidence in the accuracy and representativeness of annual mean MBT data that have been averaged as we have done. Although the accuracy of any individual MBT measurement is an order of magnitude less than that of a reversing thermometer measurement (0.2°C versus 0.02°C), the accuracy of monthly and annual mean values is increased by averaging many observations. Figure 4a clearly demonstrates that the signal-to-noise ratio

is small enough that averaged MBT data are adequate to define the observed signal (2°C peak-to-trough difference in annual mean temperature).

The monthly mean data used in Figure 4a was derived from both MBT and station daily means. The MBT data was available from 1946 through 1981. During the 1947-1954 period and again in 1960-1967, measurements were taken quite frequently, sometimes daily. During the 1970s relatively few observations were made, although a couple of times the annual frequency again reached 150 or higher (as it had been during 1955-1959). The station data, which covers 1964 through 1990, reflects reasonably frequent observations (again, about 150 per year) for the period 1967-1973, and then near-daily readings for 1976-1985. Both the MBT data and the station observations were scattered fairly evenly throughout the year.

Figure 4b shows the time series of annual mean sea surface temperature (SST) for OWS "C". The same features observed at 100 m occur at the sea surface. To provide further evidence of the existence of both the downward trend and the decadal-scale oscillation, Figure 5 presents

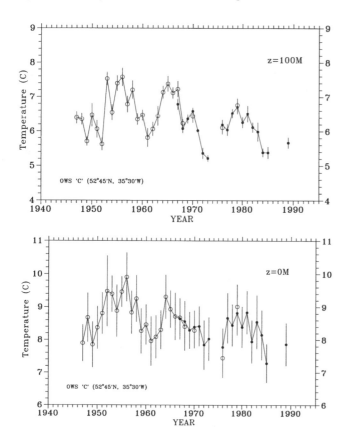

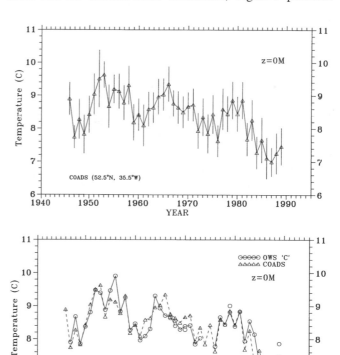

FIGURE 4 Time series of annual mean temperature (°C) from OWS "C" data. The vertical bars centered at each annual mean data point represent ± 1 standard error of the monthly means for that year about the annual mean for that year. Upper panel, at a depth of 100 m; lower panel, at the sea surface.

FIGURE 5 Time series of annual mean temperature (°C) at 52.5°N, 35.5°W at the sea surface. The vertical bars centered at each annual mean data point represent ± 1 standard deviation of the monthly means for that year about the annual mean for that year. Upper panel, COADS data; lower panel, COADS data and OWS "C" time series data set.

the time series of annual mean temperature at 52.5°N, 35.5°W taken from objective analyses of SST observations from the Comprehensive Ocean Atmosphere Data Set (COADS) of surface marine observations (Woodruff et al., 1987). The COADS data have been gridded and analyzed using the same procedures described by Levitus (1982) as part of a joint project with A. Da Silva of the University of Wisconsin. (We emphasize that the OWS "C" time series are averages of the data taken by ships at the OWS "C" location.) Figure 5a shows the annual mean temperature at this grid point as computed from the 12 monthly fields of objectively analyzed SST data for each year. Figure 5b shows the annual mean time series for the OWS "C" surface data and the COADS data time series plotted together to facilitate comparison. Some differences exist between these two data sets, but the comparison indicates that both data sets display the same negative trend as well as the same decadal-scale oscillation. The year of maximum temperature of the decadal-scale oscillation in each series may differ by 1 to 3 years in each figure. We attribute this to differences in data density and differences in the way these annual means have been computed. The decreasing SST of the North Atlantic Ocean over this time period has been described in previous studies (e.g., Folland et al., 1984).

To further describe the temporal variability at this location, Figure 6 presents the time-depth temperature field at OWS "C" over the 0-125 m depth range. This figure makes it very clear that both the negative trend and decadal-scale oscillations noted in Figures 4 and 5 extend coherently over the upper water column. There is no phase propagation with depth of the interannual variations.

Beginning in the mid-1960s, very-low-salinity water was observed at various locations along the periphery of the subarctic gyre of the North Atlantic. Dickson et al. (1988) have suggested that this low-salinity (and relatively cold) water advected out of the Arctic Ocean into the subarctic gyre of the North Atlantic and circulated around this gyre.

They identified changes in salinity that occurred at OWS "C" in the 1970s as being associated with the "Great Salinity Anomaly" (GSA). Levitus (1989b) also presented evidence for this. Dooley et al. (1984), however, suggested that the changes at OWS "C" were associated with horizontal movement of the OPF. The fact that several oscillations have occurred in the temperature data at OWS "C" suggests that the OPF might be shifting its axis. It does not seem likely to us that the cold, fresh waters of the GSA have traveled around the subarctic more than once. Further examination of the historical data will be required to test these different hypotheses. Perhaps both phenomena are responsible to some degree for the decadal-scale variability at OWS "C". Quite possibly the GSA was a phenomenon independent of the process responsible for the oscillations we describe in Figures 4 to 6.

To place the variability described by Figure 6 in perspective, Figure 7 presents the climatological annual cycle of OWS "C" upper-ocean temperature data. At a 100 m depth the annual range of temperature is approximately 1.8°C. A clear propagation of phase with depth is observed.

Variability of Deep-Ocean Temperature and Salinity at OWS "C"

To document the variability of deep ocean conditions at OWS "C," Figure 8 presents time series of temperature and salinity for the 1964-1990 period. Again, all data for the 1964-1973 period are U.S. Coast Guard observations (Hannon, 1979), and the 1975-1990 data are Russian observations. While only a few observations were available for the

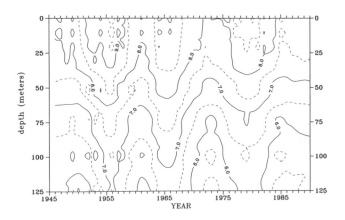

FIGURE 6   Annual mean temperature (°C) as a function of depth and time, based on the OWS "C" time series data.

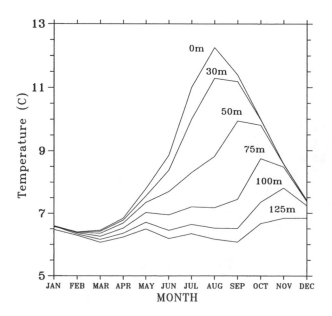

FIGURE 7   Climatological annual cycle of temperature (°C) based on the OWS "C" data series.

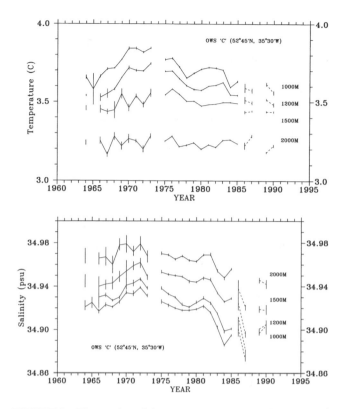

FIGURE 8   Time series of deep ocean temperature (upper panel, in °C) and deep ocean salinity (lower panel, in psu) at OWS "C" location.

periods 1964-1975 and 1986-1990, readings were taken almost daily from mid-1975 through 1985.

Figure 8 shows that annual mean temperature increased by 0.25°C between 1964 and 1971 at a depth of 1000 m, which is consistent with the results of Levitus (1989c) and Antonov (1990). Temperature then decreased a total of 0.20°C by 1984 at this same depth. The magnitude of these temperature changes decreased with increasing depth. Comparison of the two panels of Figure 8 reveals that, in general, changes in salinity were positively correlated with temperature changes at these depths. During the 1965-1972 period, the salinity data at the 1000 m level exhibit a positive trend that led to an increase of about 0.02 psu in the annual mean. After 1972 the trend reversed sign, and salinity decreased by 0.05 psu by 1984. There is some evidence of a further decrease in salinity during the post-1985 period. The amount of data available in this period is limited, however, so we choose not to emphasize this portion of the record. Data do exist for this period, and these series will be updated in a future work. One important feature to note is that the salinity at a depth of 2000 m shows a decrease beginning in 1975, but there is no corresponding change in temperature at this depth.

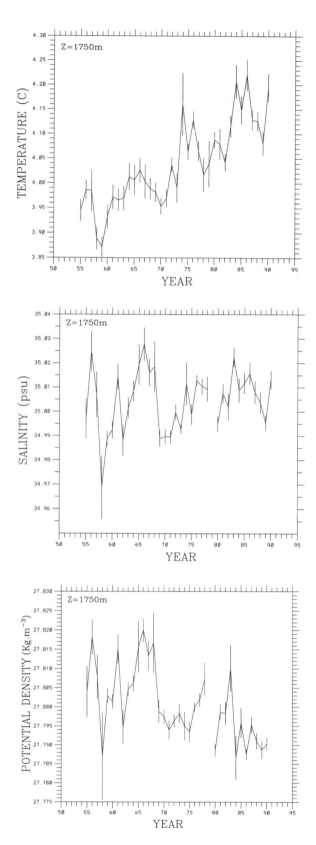

FIGURE 9   Time series at 1750 m depth at Station "S". Upper panel, temperature (°C); middle panel, salinity (PSU); lower panel, potential density (kg m^{-3}).

VARIABILITY OF DEEP-WATER TEMPERATURE AND SALINITY AT THE STATION "S" LOCATION

Ocean Weather Station "S", located off Bermuda (32.16°N, 64.5°W), was established in 1954 through the efforts of Henry Stommel (WHOI and BBSR, 1988). Beginning in 1954, hydrographic profiles have been made on a monthly or semimonthly basis. This time series is one of the longest in existence for an oceanographic quantity, particularly for the deep ocean.

The work of Roemmich (1985) presents some of the most recent analyses of the data from this time series. Roemmich found evidence of decadal-scale trends in the deep ocean temperature off Bermuda. He smoothed the annual mean values of his time series with a 5-year running mean, however, and presented no evidence of decadal-scale oscillations or interannual variability.

We have used the data from the OWS "S" series, as well as other hydrographic data taken within a one-half degree radius of this location, and present here the results of an analysis of the resulting combined data series at 1750 m depth. Measurements at OWS "S" were made fairly regularly for the entire 1954-1990 period, mostly about 4 per month, with occasional periods of heavier sampling.

Figure 9 shows the time series of annual mean temperature, salinity, and potential density at a depth of 1750 m for the 1955-1990 period. As in our earlier figures, the vertical bars centered at each annual mean data value represent the range of ± 1 standard deviation of the monthly means for each year about the annual mean. The dominant feature in the temperature series is the positive trend during the 1959-1990 period. This trend resulted in a net warming of about 0.3°C over a 30-year period. Superimposed on this trend are decadal-scale oscillations with a range of about 0.10°C to 0.15°C. To better describe these oscillations, we have detrended the 1959-1990 portion of the time series. The warmer portions of these oscillations are centered about the years 1965, 1975, and 1985. Figure 9b shows the salinity series at 1750 m. The minimum salinity of this series occurred in 1958, one year earlier than the temperature minimum. There is a small positive linear trend in salinity. In addition, the decadal-scale oscillations of temperature have counter-parts in the salinity record. A positive correlation is observed between the two detrended parameters during the 1960s and 1970s, but it is less strong in the 1980s.

CONCLUSION

The time-series results presented in this paper indicate that statistically significant decadal-scale changes of temperature and salinity have occurred in both the subarctic and the subtropical gyres of the North Atlantic Ocean. These time-series results further characterize the variability of the North Atlantic described in a series of scientific papers published over the last 12 years. These papers include, but are certainly not limited to, the works of Lazier (1980, 1988, 1995), Brewer et al. (1983), Dickson et al. (1988), Dooley et al. (1984), Mysak (1995), Roemmich (1985), Roemmich and Wunsch (1984), Swift (1984a, 1985), Taylor (1983), Levitus (1989a,b,c; 1990), Antonov (1990), and Antonov and Groisman (1988).

Several statements can be made on the basis of the results of this study and the work of others mentioned above:

1. At a minimum, oceanographers have succeeded in documenting the large-scale redistributions of heat and salt in the North Atlantic Ocean that occurred in the 1946-1990 period.

2. The decadal-scale variability of the North Atlantic Ocean we have described may represent the ocean component of decadal-scale interactions between the various parts of the earth's ocean-atmospheric-cryosphere climate system. The existence of such variability provides the observational basis for future diagnostic and modeling studies.

3. Any program for monitoring the world ocean in order to determine its role in the earth's climate system must include an examination of these changes in the deep ocean.

ACKNOWLEDGMENTS

Funding for the work described in this paper was provided by the NOAA Climate and Global Change Program and the NOAA Earth Sciences Data and Information Management Program. This paper is BBSR Contribution No. 1349.

Commentary on the Paper of Levitus et al.

JOHN R.N. LAZIER
Bedford Institute of Oceanography

Dr. Levitus and his co-authors suggest that the low-frequency temperature variations in the upper layers at Ocean Weather Ship (OWS) Charlie may be related to the lower-than-normal salinity that appeared in the subpolar gyre between 1968 and 1978. To explore this possibility I wish to examine some features of this phenomenon, which has come to be known as the Great Salinity Anomaly (Dickson et al., 1988).

In the latter half of the 1960s the salinity in the Iceland Sea became anomalously fresh (Dickson et al., 1975). Subsequently, water of lower-than-normal salinity was observed at various stations around the subpolar gyre. Taken together, these observations led Dickson et al. (1988) to propose that all these signals were due to a single pulse of low-salinity water moving around the subpolar gyre. They demonstrated that the low-salinity water moved out of the Iceland Sea into the North Atlantic Ocean via the East Greenland Current, and thence to the Labrador Sea (1972), the Labrador Current (1971-1972), OWS Charlie (1974), the Faeroe-Shetland Channel (1976), the Barents Sea (1978), and the Greenland Sea (1981-1982). The average rate of advance over the 14-year circuit was approximately 0.03 m s⁻¹. The total salt deficit was estimated to be 72×10^{12} kg along the Labrador shelf, but by the time the anomaly moved north through the Faeroe-Shetland Channel into the Barents and Greenland Seas it was reduced to about 47×10^{12} kg.

The decrease in salinity caused by the passage of the anomaly is illustrated in Figure 1. Here the time series of salinity averaged over the upper 800 m at OWS Charlie is presented, along with eight similar values I calculated for September of the years 1966 to 1973 at OWS Bravo. The latter have been displaced along the time axis; the OWS Bravo values were actually obtained three years earlier than the year indicated.

The salinity variation through the years is clearly similar in magnitude at the two weather ships, but the minimum occurs at OWS Bravo about three years before it occurs at OWS Charlie, which supports the conclusion of Dickson et al. (1988) that the low-salinity water could have moved from OWS Bravo to Charlie. Farther east, the salinity variation through these years is similar to that at the weather ships. This is illustrated by the five-year running mean of the surface salinity in Rockall Channel shown in Figure 2. According to Dickson et al. (1988), the decrease in the Rockall Channel of about 0.08 between the late 1960s and 1976 occurs about 9 months after the decrease at OWS Charlie, which fits well with their advection hypothesis.

It is unfortunate that Levitus et al. were unable to present the salinity data for OWS Charlie to compare with the temperature data, since the Great Salinity Anomaly is noted for a change in salinity rather than a change in temperature. The temperature records in their Figures 4, 5, and 6 do exhibit minima in the mid-1970s at the same time that the salinity is minimum in my Figure 2; however, the temperature data also indicate minima of similar magnitude in the early 1960s and mid-1980s that do not appear to be connected to the Great Salinity Anomaly or any other large-scale salinity minimum.

Another possibility raised by Levitus is that the low salinity signal at OWS Charlie could be due to long-period horizontal displacements of the water masses, which would bring water of lower salinity into the region of OWS Charlie.

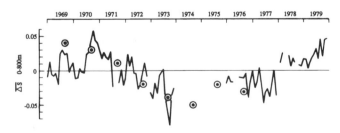

FIGURE 1   The 0-800 m average salinity at OWS Charlie (solid line), adapted from Dickson et al. (1988), and at OWS Bravo (circles) for September of the years 1966 to 1976. The data for OWS Bravo have been plotted 3 years late to match the variation at OWS Charlie.

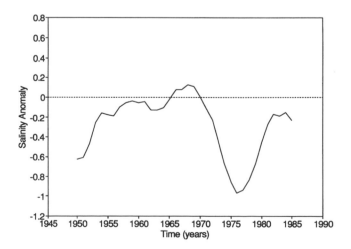

FIGURE 2   Five-year running mean of the winter sea surface salinity anomaly in Rockall Channel, adapted from Ellett and Edelsten (1983) and Ellett (pers. comm.).

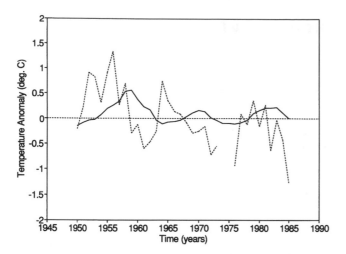

FIGURE 3   The annual average sea surface temperature anomaly (dotted line), adapted from Levitus's Figure 5, and the 5-year running mean of the winter sea surface temperature anomaly in Rockall Channel (solid line), adapted from Ellett and Edelsten (1983) and Ellett (pers. comm.).

If this is the mechanism for the variation at OWS Charlie it should also be the mechanism for the variations at OWS Bravo, Rockall Channel, and the other places where the anomaly signal has been observed. At 50°N 51.5°W, OWS Bravo is near the center of the cyclonic gyre that dominates the circulation of the Labrador Sea. There are no horizontal gradients that could be moved to lower the salinity of the top 800 m of the water column without drastically altering the dynamics of the sea, and no such alteration has been observed. This is also true in the Rockall Channel, so the interpretation of Dickson et al. (1988), an advecting anomaly, fits more comfortably with the observations.

To look further at the relationship between OWS Charlie and the Rockall Channel I have prepared Figure 3, which shows the sea-surface temperature variations at these two points. The data for OWS Charlie is from Dr. Levitus's Figure 5, and the surface temperature anomaly (for winter) in the Rockall Channel is from Ellett and Edelsten (1983) and Ellett (pers. comm.). The amplitudes of the variations at OWS Charlie are larger than at Rockall; this is in part because they are based on annual averages, while those at Rockall are 5-year running means. The interesting feature is that in each record there is a variation of about 10 years' period that occurs at Rockall roughly 5 years later than at OWS Charlie, whereas Dickson et al. (1988) demonstrated that the Great Salinity Anomaly moved from OWS Charlie to Rockall Channel in 9 months. If we assume that the temperature variations at OWS Charlie are connected to the Great Salinity Anomaly, we are then faced with the problem of explaining why the variations due to salinity move to Rockall Channel five times faster than those due to temperature.

Discussion

GOODRICH: Was there a demonstrable air-temperature signal associated with the presumably colder, fresher water of the Great Salinity Anomaly?

LEVITUS: I'm not sure. But there certainly was for this decadal-scale oscillation and for the decreasing trend.

GROISMAN: If I had only two stations on land, I wouldn't dare claim that I was seeing interdecadal variability in surface temperature. If your two ocean stations are near strong temperature fronts, how can you be sure from those observations that long-term changes aren't the result of a 200-km shift east or west in such a front?

LEVITUS: We're mostly trying to document the variability, and we have confirmation from other data sets for that. There may be some contribution from front changes in the upper ocean, but I really don't think that they can explain the horizontal shifts in the deep ocean that we're seeing at OWS "C".

MOREL: I have two questions. First, can we associate these demonstrated changes in the ocean that are on decadal or longer time scales with actual changes in climate, such as atmospheric circulation? Ocean variability would not qualify as climate research otherwise. And second, could you suggest a practical sampling strategy for a global climate—not just ocean—observing system?

LEVITUS: I think that what's most important in connection with both your questions is bringing together the huge amount of data that already exists. We at NOAA are digitizing data from the U.S., Europe, and Russia. We are seeing a decadal-scale oscillation in the COADS, subsurface, and SAT data, but it will take several years before we can evaluate all this historical data and design a better global observing system.

One of the important things we've already learned is that we need long-term commitments if we are to look at long-time-scale problems. As we saw at the Panulirus station, you can have 10 years with no changes and suddenly see very large changes. The assurance of continuing financial support is as essential as good design for an observing system.

MYSAK: In response to Pierre Morel's first question, I wanted to point out that Rosanne D'Arrigo's reconstructions of tree rings from Labrador, Newfoundland, and Nova Scotia very clearly reflect cooling at the time of the GSA.

DESER: And I'd like to emphasize that the decadal SST variability is clearly reflected in the atmosphere as high as the 50-mb level. It's influencing storm tracks, and in fact you can see it in rainfall over western Ireland.

BRYAN: This question of why we should measure deep-ocean temperature when the atmosphere can respond only to SST is a long-standing one. I think there are two reasons. The first is that the deep-sea signals are signatures of past events that may be important for unraveling both the mechanism and the magnitude of the exchanges of heat between the ocean and atmosphere. The second is that unless the temperature changes we see are perfectly compensated for by salinity, the thermohaline circulation is responding to the pressure fields within the ocean itself, and those circulation changes do have a large effect on SST in areas of strong currents.

GHIL: Pierre, I really think we can safely assume that the ocean is part of the climate system. Also, it seems to me that suggesting that all the ocean is good for is giving you SSTs is like saying that all the atmosphere is useful for is heat fluxes.

MOREL: Many phenomena in the ocean, like the tides, are not part of climate, any more than rainbows in the atmosphere are.

GOOS, as currently defined, covers everything that can be measured in the ocean, which is why I was asking about a sampling strategy. I feel that a cause-and-effect relationship between decadal variability in the ocean and decadal variability in the atmosphere needs to be shown before we can consider the former a climate phenomenon.

MARTINSON: That's odd; I would have said that the atmospheric moisture reflected in rainbows was part of climate.

ROOTH: Getting serious, now, the temperature anomalies on a 10-year time scale at OWS "C" mean that you can have fairly substantial changes in moist static energy at the surface. Being an old cloud-physics man, I think that we need to look at that, not just dry energy, as we try to feed ocean temperature back into climate. As for location of observation stations, we do need to understand the processes in the deep in the Atlantic, because of the deep-water formation and the convective effects, which involve large temperature differences. The deep circulation in the Pacific appears to be much less important to understanding what goes on at the surface on decadal time scales.

LEHMAN: It seems to me that I've been hearing some of the first evidence connecting ocean surface variability with real changes in the interior. I think modelers should really be homing in on this evidence of surface-to-deep-ocean coupling that directly implicates thermohaline circulation in the variability of the surface.

Ocean Observations
Reference List

Aagaard, K., and E.C. Carmack. 1989. On the role of sea ice and other fresh water in the Arctic circulation. J. Geophys. Res. 94:14485-14498.

Aagaard, K., J.H. Swift, and E.C. Carmack. 1985. Thermohaline circulation in the Arctic Mediterranean seas. J. Geophys. Res. 90:4833-4846.

Alexander, R.C., and R.L. Mobley. 1976. Monthly average sea-surface temperatures and ice-pack limits on a 1° global grid. Mon. Weather Rev. 104:143-148.

Antonov, J. 1990. Recent climatic changes of the vertical thermal structure of the North Atlantic Ocean and the North Pacific Ocean. Meteor. Gidrol. 4:78-87 (in Russian).

Antonov, J., and P. Groisman. 1988. Sea water temperature change below the seasonal thermocline in the North Atlantic. Meteor. Gidrol. 3:57-63 (in Russian).

Barnett, T.P. 1990. Recent changes in sea level: A summary. In Sea-Level Change. National Research Council, Studies in Geophysics, National Academy Press, Washington, D.C., pp. 37-51.

Bayer, R., P. Schlosser, G. Bönisch, H. Rupp, F. Zaucker, and G. Zimmek. 1989. Performance and blank components of a mass spectrometric system for routine measurement of helium isotopes and tritium by the ^{3}He ingrowth method. Sitzungsber. Heidelb. Akad. Wissensch., Math. Naturwiss. Kl., Jahrgang 1989, 5, Springer-Verlag, pp. 241-279.

Bjerknes, J. 1964. Atlantic air-sea interaction. Adv. Geophys. 10:1-82.

Blindheim, J., and H.R. Skjoldal. 1993. Effect of climatic change on the biomass yield of the Barents Sea, Norwegian Sea and West Greenland large marine ecosystems. In Large Marine Ecosystems. K. Sherman, M. A. Lewis, and B. D. Gold (eds.). American Association for the Advancement of Science, Washington, D.C., pp. 185-198.

Bottomley, M., C.K. Folland, J. Hsiung, R.E. Newell, and D.E. Parker. 1990. Global Ocean Surface Temperature Atlas (GOSTA). Joint Project of the Meteorological Office, Bracknell, U.K., and the Massachusetts Institute of Technology. HMSO, London, 20iv pp. and 313 plates.

Bradley, R.S., and J. England. 1978. Recent climatic fluctuations of the Canadian high Arctic and their significance for glaciology. Arct. Alp. Res. 10:715-731.

Brewer, P.G., W. S. Broecker, W.J. Jenkins, P.B. Rhines, C.G. Rooth, J.H. Swift, T. Takahashi, and R.T. Williams. 1983. A climatic freshening of the deep North Atlantic (north of 50°N), over the past 20 years. Science 222:1237-1239.

Broecker, W.S., and T.-H. Peng. 1984. Gas exchange rates between air and sea. Tellus 26:21-35.

Broecker, W.S., T.-H. Peng, G. Östlund, and M. Stuiver. 1985a. The distribution of bomb radiocarbon in the ocean. J. Geophys. Res. 90:6953-6970.

Broecker, W.S., D.M. Peteet, and D. Rind. 1985b. Does the ocean-atmosphere system have more than one stable mode of operation? Nature 315:21-26.

Brummer, B., E. Augstein, and H. Riehl. 1974. On the low-level wind structure in the Atlantic trade. Quart. J. Roy. Meteorol. Soc. 100:109-121.

Bryan, K., and R. Stouffer. 1991. A note on Bjerknes' hypothesis for North Atlantic variability. J. Mar. Sys. 1:229-241.

Bu, X. and M.J.W. Warner. 1995. Solubility of Chlorofluorocarbon 113 in water and seawater. Deep-Sea Res. 42:1151-1161.

Buch, E., and H.H. Hansen. 1988. Climate and cod fishery at West Greenland. In Long-Term Changes in Marine Fish Populations. T. Wyatt and M.G. Larraneta (eds.). Instituto de Investigaciones Marinas, Vigo, Spain, pp. 345-364.

Bullister, J.L., and R.F. Weiss. 1983. Anthropogenic chlorofluoromethanes in the Greenland and Norwegian Seas. Science 221:265-268.

Bullister, J.L., and R.F. Weiss. 1988. Determination of CCl$_3$F and CCl$_2$F$_2$ in seawater and air. Deep-Sea Res. 35:839-853.

Cardone, V.J., J.G. Greenwood, and M.A. Cane. 1990. On trends in historical marine wind data, J. Climate 3:113-127.

Carter, W.E., D.G. Aubrey, T. Baker, C. Boucher, C. LeProvost, D. Pugh, W. Peltier, M. Zumberge, R. Rapp, B. Schutz, K. Emery, and D. Enfield. 1989. Geodetic Fixing of Tide Gauge Bench Marks. Technical Report WHOI-89-31, Woods Hole Oceanographic Institution.

Cattle, H. 1985. Diverting Soviet rivers: Some possible repercussions for the Arctic Ocean. Polar Res. 22:485-498.

Chapman, W.L., and J.E. Walsh. 1991. Long-range prediction of regional sea ice anomalies in the Arctic. Wea. Forecast. 6:271-288.

Charles, C.D., and R.G. Fairbanks. 1992. Evidence from Southern Ocean sediments for the effect of North Atlantic deep-water flux on climate. Nature 355:416-419.

Chemical Manufacturers Association, Fluorocarbon Program Panel. 1983. World production and release of chlorofluorocarbons 11 and 12 through 1981. FPP 83-F, Washington, D.C.

Cheney, R.E., W.J. Emery, B.J. Haines, and F. Wentz. 1991. Recent improvements in Geosat altimeter data. EOS, Trans. AGU 72:577-580.

Clarke, R.A. 1984. Changes in the western North Atlantic during the decade beginning in 1965. C.M. 1984/Gen:17, International Council for the Exploration of the Sea, Copenhagen, 7 pp. (mimeo).

Clarke, R.A., and A.R. Coote. 1988. The formation of Labrador Sea Water. Part III: The evolution of oxygen and nutrient concentration. J. Phys. Oceanogr. 18:469-480.

Clarke, R.A., and J.C. Gascard. 1983. The formation of Labrador Sea Water. Part I: Large-scale processes. J. Phys. Oceanogr. 13:1764-1778.

Clarke, R.A., J.H. Swift, J.L. Reid, and K.P. Koltermann. 1990. The formation of Greenland Sea Deep Water: Double diffusion or deep convection? Deep-Sea Res. 37:1385-1424.

Clarke, W.B., W.J. Jenkins, and Z. Top. 1976. Determination of tritium by mass spectrometric measurement of ^{3}He. Int. J. Appl. Rad. Isotopes 27:515-522.

Colman, A.W. 1992. Development of worldwide marine data eigenvectors since 1985. CRTN 28, 26 pp. (Available from the National Meteorological Library, London Road, Bracknell, Berks, RG12 2SZ, U.K.)

Colony, R., and A.S. Thorndike. 1984. An estimate of the mean field of Arctic sea ice motion. J. Geophys. Res. 89:10623-10629.

Craig, H., and D. Lal. 1961. The production of natural tritium. Tellus 13:85-105.

Cunnold, D.M., R.G. Prinn, R.A. Rasmussen, P.G. Simmonds, F.N. Alyea, C.A. Cardelino, A.J. Crawford, P.J. Fraser, and R.D. Rosen. 1986. Atmospheric lifetime and annual release estimates for $CFCl_3$ and CF_2Cl_2 from 5 years of ALE data. J. Geophys. Res. 91:10797-10817.

Danke, L. 1967. Erster bericht über kabeljaumarkierung 1964 bis 1966 bei Labrador. Fischerie Forsch. 5(1):87-91.

Darby, M.S., and L.A. Mysak. 1993. A Boolean delay equation model of an interdecadal Arctic climate cycle. Climate Dynamics 8:241-246.

Delworth, T.L., S. Manabe, and R.J. Stouffer. 1993. Interdecadal variations of the thermohaline circulation in a coupled ocean-atmosphere model. J. Climate 6:1993-2011.

Deser, C., and M.L. Blackmon. 1993. Surface climate variations over the North Atlantic Ocean during winter: 1900-1989. J. Climate 6:1743-1753.

Deutsches Hydrographisches Institut. 1950. Atlas of ice conditions in the North Atlantic Ocean and general charts of the ice conditions of the North and South Polar regions. Deutsches Hydrog. Inst. Publ. No. 2335, Hamburg, 24 pp. and 34 colored charts (in German).

Dickson, R.R. 1995. The local, regional, and global significance of exchanges through the Denmark Strait and Irminger Sea. In Natural Climate Variability on Decade-to-Century Time Scales. D.G. Martinson, K. Bryan, M. Ghil, M.M. Hall, T.R. Karl, E.S. Sarachik, S. Sorooshian, and L.D. Talley (eds.). National Academy Press, Washington, D.C.

Dickson, R.R., and J. Brown. 1994. The production of North Atlantic Deep Water: Sources, rates and pathways. J. Geophys. Res. 99(C6):12319-12341.

Dickson, R.R., and J. Namias. 1976. North American influences on the circulation and climate of the North Atlantic sector. Mon. Weather Rev. 104:1255-1265.

Dickson, R.R., H.H. Lamb, S.-A. Malmberg, and J.M. Colebrook. 1975. Climatic reversal in the northern North Atlantic. Nature 256:479-482.

Dickson, R.R., S.-A. Malmberg, S.R. Jones, and A.J. Lee. 1984. An investigation of the earlier great salinity anomaly of 1910-14 in waters west of the British Isles. C.M. 1984/Gen:4, International Council for the Exploration of the Sea, Copenhagen, 15 pp. + 15 figs. (mimeo).

Dickson, R.R., J. Meincke, S.-A. Malmberg, and A.J. Lee. 1988. The "Great Salinity Anomaly" in the northern North Atlantic, 1968-1982. Progr. Oceanogr. 20:103-151.

Doney, S.C., D.M. Glover, and W.J. Jenkins. 1992. A model function of the global bomb tritium distribution in precipitation, 1960-1986. J. Geophys. Res. 97:5481-5492.

Dooley, H.D., J.H.A. Martin, and D.J. Ellett. 1984. Abnormal hydrographic conditions in the northeast Atlantic during the 1970s. Rapp. P.-v. Rëun. Cons. Int. Explor. Mer 185:179-187.

Douglas, B.C. 1991. Global sea level rise. J. Geophys. Res. 96:6981-6992.

Douglas, B.C. 1992. Global sea level acceleration. J. Geophys. Res. 97:12699-12706.

Douglas, B.C. 1995. Long-term sea-level variation. In Natural Climate Variability on Decade-to-Century Time Scales. D.G. Martinson, K. Bryan, M. Ghil, M.M. Hall, T.R. Karl, E.S. Sarachik, S. Sorooshian, and L.D. Talley (eds.). National Academy Press, Washington, D.C.

Dreisigacker, E., and W. Roether. 1978. Tritium and Sr-90 in North Atlantic surface water. Earth Planet. Sci. Lett. 38:310-312.

Druffel, E.M., and T.W. Linick. 1978. Radiocarbon in annual coral rings of Florida. Geophys. Res. Lett. 5:913-916.

Ellett, D.J., and D.J. Edelsten. 1983. Hydrographic conditions in the central Rockall Channel during 1980. Ann. Biol. 37:63-66.

Emery, K.O., and D.G. Aubrey. 1991. Sea Levels, Land Levels, and Tide Gauges. Springer-Verlag, New York.

Fine, R., and R.L. Molinari. 1988. A continuous deep western boundary current between Abaco (26.5°N) and Barbados (13°N). Deep-Sea Res. 35:1441-1450.

Fissel, D.B., and H. Melling. 1990. Interannual variability of the oceanographic conditions in the southeastern Beaufort Sea. Can. Contr. Rep. Hydro. Ocean Sci. No. 35, 102 pp. (plus 6 microfiches).

Fleming, G.H., and A.J. Semtner, Jr. 1991. A numerical study of interannual ocean forcing on Arctic sea ice. J. Geophys. Res. 96:4589-4603.

Folland, C.K. 1991. Sea temperature bucket models used to correct historical sea surface temperature data in the Meteorological Office. CRTN 14, 29 pp. (Available from the National Meteorological Library, London Road, Bracknell, Berks, RG12 2SZ, U.K.)

Folland, C.K., and D.E. Parker. 1992. The instrumental record of surface temperature: How good is it and what can it tell us about climate change and variability? In Proceedings, Fifth International Meeting on Statistical Climatology and Twelfth American Meteorological Society Conference on Probability and Statistics in the Atmospheric Sciences (Toronto). American Meteorological Society, Boston, Mass., pp. J1-J6.

Folland, C.K., and D.E. Parker. 1995. Correction of instrumental biases in historical sea surface temperature data. Quart. J. Roy. Meteorol. Soc. 121:319-367.

Folland, C.K., D.E. Parker, and F.E. Kates. 1984. Worldwide marine temperature fluctuations 1856-1981. Nature 310:670-673.

Folland, C.K., D.E. Parker, M.N. Ward, and A.W. Coleman. 1986a. Sahel rainfall, northern hemisphere circulation anomalies and the worldwide sea surface temperature changes. Memorandum 7a, Long Range Forecasting Climate Research, Meteorological Office, Bracknell, U.K. 49 pp.

Folland, C.K., T.N. Palmer, and D.E. Parker. 1986b. Sahel rainfall and worldwide sea temperatures. Nature 320:602-607.

Folland, C.K., T.R. Karl, and K.Ya. Vinnikov. 1990. Observed climate variations and change. In Climate Change: The IPCC Scientific Assessment. J.T. Houghton, G.J. Jenkins, and J.J. Ephraums (eds.). Prepared for the Intergovernmental Panel on Climate Change by Working Group I. WMO/UNEP, Cambridge University Press, pp. 195-238.

Folland, C.K., T.R. Karl, N. Nicholls, B.S. Nyenzi, D.E. Parker, and K.Ya. Vinnikov. 1992. Observed climate variability and change. In Climate Change 1992: The Supplementary Report to the IPCC Scientific Assessment. J.T. Houghton, B.A. Callander, and S.K. Varney (eds.). Prepared for the Intergovernmental Panel on Climate Change by Working Group I. WMO/UNEP, Cambridge University Press, pp. 135-170.

Folland, C.K., R.W. Reynolds, M. Gordon, and D.E. Parker. 1993. A study of six operational sea surface temperature analyses. J. Climate 6:96-113.

Fridriksson, A. 1949. Boreo-tended changes in the marine vertebrate fauna of Iceland during the last 25 years. Rapp. P.-v. Rëun. Cons. Int. Explor. Mer 125:30-35.

Fuchs, G., W. Roether, and P. Schlosser. 1987. Excess ³He in the ocean surface layer. J. Geophys. Res. 92:6559-6568.

Fuglister, F.C. 1960. Atlantic Ocean atlas of temperature and salinity profiles and data from the International Geophysical Year of 1957-1958. Woods Hole Oceanographic Institution Atlas, Series 1, 209 pp.

Gandin, L.S. 1963. Objective Analysis of Meteorological Fields.

Gidrometeor. Izdat., Leningrad. (Translated from Russian by the Israeli Program for Scientific Translations, Jerusalem, 1966, 242 pp.)

Gascard, J.C., and R.A. Clarke. 1983. The formation of Labrador Sea Water. Part II: Mesoscale and smaller-scale processes. J. Phys. Oceanogr. 13:1779-1797.

Glueckauf, E., and F.A. Paneth. 1945. The helium content of atmospheric air. Proc. Roy. Soc. Lond. A 185:89-98.

Golombek, A., and R.G. Prinn. 1989. Global three-dimensional model calculations of the budgets and present day atmospheric lifetimes of CF₂ClCF₂Cl₂ (CFC-113) and CHClF₂. Geophys. Res. Lett. 16:1153-1156.

Gordon, A.H., and R.C. Taylor. 1975. Computations of surface layer air parcel trajectories, and weather, in the oceanic tropics. International Indian Ocean Expedition, Meteorological Monographs No. 7. University Press of Hawaii, Honolulu, 112 pp.

Grant, A.B. 1968. Atlas of Oceanographic Sections. Rept. AOL 68-5, Bedford Institute of Oceanography, 80 pp. Unpublished manuscript.

Greatbatch, R.J., and J. Xu. 1992. On the transport of volume and heat through sections across the North Atlantic: Climatology and the pentads 1955-1959, 1970-1974. J. Geophys. Res. 98:10125-10143.

Greenland Sea Project (GSP) Group. 1990. Greenland Sea Project: A venture toward improved understanding of the oceans' role in climate. EOS 71:750-751, 754-755.

Hannon, L.J. 1979. North Atlantic Ocean Station Charlie Terminal Report 1964-1973. Oceanographic Report No. CG 373-79, U.S. Coast Guard Oceanographic Unit, Washington, D.C.

Hansen, P.M. 1949. Studies on the biology of the cod in Greenland waters. Rapp. P.-v. Rëun. Cons. Int. Explor. Mer 123:1-83.

Hansen, P.M. 1958. Danish research report, 1957. ICNAF Annual Proc. 8:27-36.

Harvey, J.G. 1962. Hydrographic conditions in Greenland waters during August 1960. Ann. Biol. Copenh. 17:14-17.

Hastenrath, S. 1991. Climate dynamics of the tropics. Kluwer Academic Press, Dordrecht, 488 pp.

Heinze, C., P. Schlosser, K.P. Koltermann, and J. Meincke. 1990. A tracer study of the deep water renewal in the European polar seas. Deep-Sea Res. 37:1425-1453.

Helland-Hansen, B., and F. Nansen. 1909. The Norwegian Sea. Report on Norwegian Fishery and Marine Investigations, Vol. II(2), Kristiana, Bergen.

Hermann, F., P.M. Hansen, and S.A. Horsted. 1965. The effect of temperatures and currents on the distribution and survival of cod larvae at West Greenland. ICNAF Spec. Pub. 6:389-395.

Hibler, W.D. III, and S.J. Johnsen. 1979. The 20-year cycle in Greenland ice core records. Nature 280:481-483.

Higuchi, K., C.A. Lin, A. Shabbar, and J.L. Knox. 1991. Interannual variability of the January tropospheric meridional eddy sensible heat transport in the northern latitudes. J. Meteor. Soc. Japan 69:459-472.

Hovgård, H., and J. Messtorff. 1987. Is the west Greenland cod mainly recruited from Icelandic waters? An analysis based on the use of juvenile haddock as an indicator of larval drift. SCR Doc. 87/31, Northwest Atlantic Fisheries Organisation, Dartmouth, Nova Scotia, 18 pp. (mimeo).

ICES. 1986. Report of the working group on cod stocks off East Greenland. C.M. 1986/Assess:11, International Council for the Exploration of the Sea, Copenhagen, 52 pp. (mimeo).

Ikeda, M. 1990. Decadal oscillations of the air-ice-ocean system in the Northern Hemisphere. Atmos.-Ocean 28:106-139.

IOC. 1983. Time Series of Ocean Measurements, Volume 1. 1983 Intl. Oceanogr. Comm. Tech. Ser. 24, UNESCO, Paris.

IOC. 1984. Time Series of Ocean Measurements, Volume 2. 1984 Intl. Oceanogr. Comm. Tech. Ser. 24, UNESCO, Paris.

IOC. 1986. Time Series of Ocean Measurements, Volume 3. 1986 Intl. Oceanogr. Comm. Tech. Ser. 31, UNESCO, Paris.

IOC. 1988. Time Series of Ocean Measurements, Volume 4. 1988 Intl. Oceanogr. Comm. Tech. Ser. 33, UNESCO, Paris.

IPCC. 1990. Climate Change: The IPCC Scientific Assessment. J.T. Houghton, G.J. Jenkins, and J.J. Ephraums (eds.). Prepared for the Intergovernmental Panel on Climate Change by Working Group 1. WMO/UNEP, Cambridge University Press, 365 pp.

Jeffers, P.M., and N.L. Wolfe. 1989. Hydrolysis of carbon tetrachloride. Science 246:1638-1639.

Jenkins, W. 1974. Helium isotope and rare gas oceanology. Ph.D. thesis, McMaster University, Hamilton, Ontario, Canada, 171 pp.

Jenkins, W.J. 1987. ^{3}H and ^{3}He in the Beta Triangle: Observation of gyre ventilation and oxygen utilization rates. J. Phys. Oceanogr. 17:763-783.

Jenkins, W.J., and W.B. Clarke. 1976. The distribution of ^{3}He in the western Atlantic Ocean. Deep-Sea Res. 23:481-494.

Jenkins, W.J., D.E. Lott, M.W. Davis, S.P. Birdwhistell, and M.O. Matthewson. 1991. Measuring helium isotopes and tritium in seawater samples. In WOCE Operations Manual, WOCE Hydrographic Programme Report WHPO 91-1 (WOCE Report No. 68/91), Woods Hole, Mass., 21 pp.

Jensen, A.S. 1939. Concerning a change of climate during recent decades in the Arctic and Subarctic regions, from Greenland in the west to Eurasia in the east, and contemporary biological and geophysical changes. Det. Konigl. Danske Videnskabernes Selskab., Biologiske Meddelelser 14 (8), 75 pp.

Jones, P.D., T.M.L. Wigley, and G. Farmer. 1991. Marine and land temperature data sets: A comparison and a look at recent trends. In Greenhouse-Gas-Induced Climatic Change: A Critical Appraisal of Simulations and Observations. M.E. Schlesinger (ed.). Elsevier, Amsterdam, pp. 153-172.

Jonsson, E. 1982. A survey of spawning and reproduction of the Icelandic cod. Rit. Fiskideildar 6:1-45.

Kellogg, W.W. 1983. Feedback mechanisms in the climate system affecting future levels of carbon dioxide. J. Geophys. Res. 88:1263-1269.

Kelly, P.M. 1979. An Arctic sea ice data set, 1901-1956. In Glaciological Data Report GD-5, World Data Center-A for Glaciology, Boulder, Colo., pp. 101-106.

Khalil, M.A.K., and R.A. Rasmussen. 1986. Trichlorotrifluoromethane (F-113): Trends at Pt. Barrow, Alaska. In Geophysical Monitoring for Climate Change, No. 13, Summary Report 1984. E.C. Nickerson (ed.). NOAA Environmental Research Laboratory, Boulder, Colo.

Komura, K., and T. Uwai. 1992. The collection of historical ships' data in Kobe Marine Observatory. Bull. Kobe Mar. Obs. Japan 211:19-29.

Koster, R.D., W.S. Broecker, J. Jouzel, R.J. Suozzo, G.L. Russell, D. Rind, and J.W.C. White. 1989. The global geochemistry of bomb-produced tritium: General circulation model compared to available observations and traditional interpretations. J. Geophys. Res. 94:18305-18326.

Kromer, B., C. Pfleiderer, P. Schlosser, I. Levin, K.O. Münnich, G. Bonani, M. Suter, and W. Wölfli. 1987. AMS ^{14}C measurement of small volume oceanic water samples: Experimental procedure and comparison with low-level counting technique. Nucl. Instr. Methods Physics Res. B29:302-305.

Krysell, M., and D.W.R. Wallace. 1988. Arctic Ocean ventilation studied using carbon tetrachloride and other anthropogenic halocarbon tracers. Science 242:746-749.

Kushnir, Y. 1994. Interdecadal variations in North Atlantic sea surface temperature and associated atmospheric conditions. J. Climate 7:141-157.

Lazier, J.R.N. 1973. The renewal of Labrador Sea Water. Deep-Sea Res. 20:341-353.

Lazier, J.R.N. 1980. Oceanographic conditions at Ocean Weather Ship BRAVO, 1964-74. Atmos.-Ocean 18:227-238.

Lazier, J.R.N. 1988. Temperature and salinity changes in the deep Labrador Sea, 1962-86. Deep-Sea Res. 35:1247-1253.

Lazier, J.R.N. 1995. The salinity decrease in the Labrador Sea over the past thirty years. In Natural Climate Variability on Decade-to-Century Time Scales. D.G. Martinson, K. Bryan, M. Ghil, M.M. Hall, T.R. Karl, E.S. Sarachik, S. Sorooshian, and L.D. Talley (eds.). National Academy Press, Washington, D.C.

Lazier, J.R.N., and D.G. Wright. 1993. Annual velocity variations in the Labrador Current. J. Phys. Oceanogr. 23:659-678.

Levitus, S. 1982. Climatological Atlas of the World Ocean. NOAA Prof. Paper 13. U.S. Government Printing Office, Washington, D.C., 178 pp.

Levitus, S. 1989a. Interpentadal variability of temperature and salinity at intermediate depths of the North Atlantic Ocean, 1970-74 versus 1955-59. J. Geophys. Res. (Oceans) 94:6091-6131.

Levitus, S. 1989b. Interpentadal variability of salinity in the upper 150 m of the North Atlantic Ocean, 1970-74 versus 1955-59. J. Geophys. Res. (Oceans) 94:9679-9685.

Levitus, S. 1989c. Interpentadal variability of temperature and salinity in the deep North Atlantic Ocean, 1970-74 versus 1955-59. J. Geophys. Res. (Oceans) 94:16125-16131.

Levitus, S. 1990. Interpentadal variability of steric sea level and geopotential thickness of the North Atlantic Ocean, 1970-74 versus 1955-59. J. Geophys. Res. (Oceans) 95:5233-5238.

Levitus, S., J.I. Antonov, Zhou Z., H. Dooley, V. Tsereschenkov, K. Selemenov, and A.F. Michaels. 1995. Observational evidence of decadal-scale variability of the North Atlantic Ocean. In Natural Climate Variability on Decade-to-Century Time Scales. D.G. Martinson, K. Bryan, M. Ghil, M.M. Hall, T.R. Karl, E.S. Sarachik, S. Sorooshian, and L.D. Talley (eds.). National Academy Press, Washington, D.C.

Loeng, H. 1991. Features of the physical oceanographic conditions of the Barents Sea. In Proceedings of the Pro Mare Symposium on Polar Marine Ecology, Trondheim, 12-16 May 1990. E. Sakshaug, C.C.E. Hopkins, and N.A. Oritsland (eds.). Polar Res. 10(1):5-18.

Malmberg, S.-A. 1969. Hydrographic changes in the waters between Iceland and Jan Mayen in the last decade. Jokull 19:30-43.

Malmberg, S.-A. 1973. Astand sjavar milli Islands og Jan Mayen, 1950-72. Aegir 66:146-148.

Manabe, S., and R. Stouffer. 1988. Two stable equilibria of a coupled ocean-atmosphere model. J. Climate 1:841-866.

Manak, D.K., and L.A. Mysak. 1989. On the relationship between Arctic sea ice anomalies and fluctuations in northern Canadian air temperature and river discharge. Atmos.-Ocean 27:682-691.

Mann, C.R. 1969. Temperature and salinity characteristics of the Denmark Strait overflow. Fuglister 60th Anniversary Volume, Deep-Sea Res. 16 (Suppl.):125-137.

McCarthy, R.L., F.A. Barrer, and J.P. Gessen. 1977. Production and release of CCl_3F and CCl_2F_2 (fluorocarbons 11 and 12) through 1975. Atmos. Environ. 11:491-497.

Meincke, J., S. Jonsson, and J.H. Swift. 1992. Variability of convective conditions in the Greenland Sea. ICES Mar. Sci. Symp. 195:32-39.

Miller, L., and R.E. Cheney. 1990. Large scale meridional transport in the tropical Pacific Ocean during the 1986-1987 El Niño from Geosat. J. Geophys. Res. 95:17905-17919.

Muench, R.D., J.L. Newton, and R.L. Rice. 1985. Temperature and salinity observations in the Bering Sea winter Marginal Ice Zone. May 1985 MIZEX Bulletin, U.S. Army Cold Regions Research and Engineering Laboratory (CRREL) Special Report 85-6, pp. 13-30.

Munk, W., R. Revelle, P. Worcester, and M. Zumberge. 1990. Strategy for future measurements of very-low-frequency sea-level change. In Sea-Level Change. National Research Council, Studies in Geophysics, National Academy Press, Washington, D.C., pp. 221-227.

Myers, R.A., J. Helbig, and D. Holland. 1989. Seasonal and interannual variability of the Labrador Current and West Greenland Current. C.M. 1989/C:16, International Commission for the Exploration of the Sea, 18 pp. (mimeo).

Mysak, L.A. 1991. Current and Future Trends in Arctic Climate Research: Can Changes in the Arctic Sea Ice be Used as an Early Indicator of Global Warming? Centre for Climate and Global Change Research Report No. 91-1. McGill University, Montreal, 35 pp.

Mysak, L.A. 1995. Decadal-scale variability of ice cover and climate in the Arctic Ocean and Greenland and Iceland Seas. In Natural Climate Variability on Decade-to-Century Time Scales. D.G. Martinson, K. Bryan, M. Ghil, M.M. Hall, T.R. Karl, E.S. Sarachik, S. Sorooshian, and L.D. Talley (eds.). National Academy Press, Washington, D.C.

Mysak, L.A., and D.K. Manak. 1989. Arctic sea-ice extent and anomalies, 1953-1984. Atmos.-Ocean 27:376-405.

Mysak, L.A., and S.B. Power. 1991. Greenland Sea ice and salinity anomalies and interdecadal climate variability. Climatol. Bull. 25:81-91.

Mysak, L.A., and S.B. Power. 1992. Sea-ice anomalies in the Arctic Ocean and Greenland-Iceland Sea during 1953-1988 and their relation to an interdecadal climate cycle. Climatol. Bull. 26:147-176.

Mysak, L.A., and J. Wang. 1991. Climatic atlas of seasonal and annual Arctic sea-level pressures, SLP anomalies and sea-ice concentrations, 1953-1988. Centre for Climate and Global Change Research Report No. 91-4, McGill University, Montreal, 194 pp.

Mysak, L.A., D.K. Manak, and R.F. Marsden. 1990. Sea-ice anomalies observed in the Greenland and Labrador Seas during 1901-1984 and their relation to an interdecadal Arctic climate cycle. Climate Dynamics 5:111-133.

Mysak, L.A., T.F. Stocker, and F. Huang. 1993. Century-scale variability in a randomly forced, two-dimensional thermohaline ocean circulation model. Climate Dynamics 8:103-116.

National Research Council (NRC). 1990. Sea-Level Change. Studies in Geophysics, National Academy Press, Washington, D.C., 234 pp.

Nerem, R.S. 1995. Global mean sea level variation from TOPEX/POSEIDON altimeter data. Science 268:708-710.

Nitta, T., and S. Yamada. 1989. Recent warming of tropical sea surface temperature and its relationship to the Northern Hemisphere circulation. J. Meteorol. Soc. Japan 67:375-383.

Nozaki, Y., D.M. Rye, K.K. Turekian, and R.E. Dodge. 1978. A 200-year record of carbon-13 and carbon-14 variations in a Bermuda coral. Geophys. Res. Lett. 5:825-828.

Östlund, H.G., and R. Brescher. 1982. GEOSECS Tritium. Tritium Laboratory Data Report No. 12, Rosenstiel School of Marine and Atmospheric Science, University of Miami, 295 pp.

Parker, D.E. 1992. Blending of COADS and U.K. Meteorological Office marine data sets. In Proceedings of the International COADS Workshop. H.F. Diaz, K. Woller, and S.D. Woodruff (eds.). NOAA/ERL Special Publication, Boulder, Colo., pp. 61-72.

Parker, D.E., C.K. Folland, A.C. Bevan, M.N. Ward, M. Jackson, and K. Maskell. 1995. Marine surface data for analysis of climatic fluctuations on interannual-to-century time scales. In Natural Climate Variability on Decade-to-Century Time Scales. D.G. Martinson, K. Bryan, M. Ghil, M.M. Hall, T.R. Karl, E.S. Sarachik, S. Sorooshian, and L.D. Talley (eds.). National Academy Press, Washington, D.C.

Parkinson, C.L. 1991. Interannual variability of the spatial distribution of sea ice in the north polar region. J. Geophys. Res. 96:4791-4801.

Pearson, A.S., and H.O. Hartley. 1966. Biometrika Tables for Statisticians, Vol 1. Cambridge University Press, 270 pp.

Peltier, W.R., and A.M. Tushingham. 1989. Global sea level rise and the greenhouse effect: Might they be connected? Science 244:806-810.

Peng, S., and L.A. Mysak. 1993. A teleconnection study of interannual sea surface temperature fluctuations in the northern North Atlantic and the precipitation and runoff over western Siberia. J. Climate 6:876-885.

Pickart, R.S. 1992a. Water mass components of the North Atlantic deep western boundary current. Deep-Sea Res. 39:1553-1572.

Pickart, R.S. 1992b. Space-time variability of the Deep Western Boundary Current oxygen core. J. Phys. Oceanogr. 22:1047-1061.

Pickart, R.S., N.G. Hogg, and W.M. Smethie, Jr. 1989. Determining the strength of the Deep Western Boundary Current using the chlorofluoromethane ratio. J. Phys. Oceanogr. 19:940-951.

Pollard, R.T., and S. Pu. 1985. Structure and circulation of the upper Atlantic Ocean northeast of the Azores. Prog. Oceanogr. 14:443-462.

Power, S.B., and L.A. Mysak. 1992. On the interannual variability of Arctic sea-level pressure and sea ice. Atmos.-Ocean 30:551-577.

Prather, M., M. McElroy, S. Wofsy, G. Russell, and D. Rind. 1987. Chemistry of the global troposphere: Fluorocarbons as tracers of air motion. J. Geophys. Res. 92:6579-6613.

Pugh, D.T. 1987. Tides, Surges, and Mean Sea Level. John Wiley and Sons, New York, 472 pp.

Qiu, B., and T.M. Joyce. 1992. Interannual variability in the mid- and low-latitude western Pacific. J. Phys. Oceanogr. 22:1062-1079.

Ramage, C.S. 1987. Secular change in reported surface wind speeds over the ocean. J. Climate Appl. Meteorol. 26:525-528.

Rath, H.K. 1988. Simulation der globalen ^{85}Kr und ^{14}CO$_2$ Verteilung mit Hilfe eines zeitabhängigen, zweidimensionalen Modells der Atmosphäre. Ph.D. Dissertation, University of Heidelberg.

Reid, J.L. 1979. On the contribution of the Mediterranean Sea outflow to the Norwegian-Greenland Sea. Deep-Sea Res. 26:1199-1223.

Reid, J.L. 1981. On the mid-depth circulation of the World Ocean. In Evolution of Physical Oceanography. B.A. Warren and C. Wunsch (eds.). MIT Press, Cambridge, Mass., pp. 70-111.

Reynolds, R.W. 1988. A real-time global sea surface temperature analysis. J. Climate 1:75-86.

Reynolds, R.W., and D.C. Marsico. 1993. An improved real-time global sea-surface temperature analysis. J. Climate 6:114-119.

Rhein, M. 1991. Ventilation rates of the Greenland and Norwegian seas derived from distributions of the chlorofluoromethanes F-11 and F-12. Deep-Sea Res. 38:485-503.

Roemmich, D. 1985. Sea level and the variability of the ocean. In Glaciers, Ice Sheets, and Sea Level: Effect of a CO$_2$-Induced Climatic Change. DOE/ER/G0235-1, Department of Energy, Washington, D.C., pp. 104-115.

Roemmich, D. 1990. Sea level and the thermal variability of the ocean. In Sea-Level Change. National Research Council, Studies in Geophysics, National Academy Press, Washington, D.C., pp. 208-217.

Roemmich, D., and C. Wunsch. 1984. Apparent changes in the climatic state of the deep North Atlantic. Nature 307:447-450.

Roether, W. 1967. Estimating the tritium input to groundwater from wine samples: Groundwater and direct run-off contribution to Central Europe surface waters. In Isotopes in Hydrology (Proc. Symp. Vienna, 1966). IAEA, Vienna, pp. 73-91.

Roether, W. 1989. On oceanic boundary conditions for tritium, on tritiugenic ^{3}He, and on the tritium/^{3}He age concept. In Oceanic Circulation Models: Combining Data and Dynamics. D.L.T. Anderson and J. Willebrand (eds.). Kluwer Academic Publishers, Dordrecht, pp. 377-407.

Rogers, J.C. 1984. The association between the North Atlantic oscillation and the southern oscillation in the northern hemisphere. Mon. Weather Rev. 112:1999-2015.

Ropelewski, C.F. 1990. A discussion of sea ice data sets. In Observed Climate Variations and Change: Contributions in Support of Section 7 of the 1990 IPCC Scientific Assessment. D. Parker (ed.). Intergovernmental Panel on Climate Change, Geneva, pp. XX.1-XX.6.

Rozanski, K. 1979. Krypton-85 in the atmosphere 1950-1977, a data review. Environ. Int. 2:139-143.

Rudels, B., and D. Quadfasel. 1991. The Arctic Ocean component in the Greenland-Scotland overflow. C.M. 1991/C:30, International Council for the Exploration of the Sea, Copenhagen, 10 pp. + 14 figs. (mimeo).

Rudels, B., D. Quadfasel, H. Friedrich, and M.-N. Hossaias. 1989. Greenland Sea convection in the winter of 1987/1988. J. Geophys. Res. 94:3223-3227.

Saemundsson, B. 1934. Probable influence of changes of temperature on the marine fauna of Iceland. Rapp. P.-v. Réun. Cons. Int. Explor. Mer 86:1-6.

Sarmiento, J.L. 1983. A simulation of bomb tritium entry into the Atlantic Ocean. J. Phys. Oceanogr. 13:1924-1939.

Schlosser, P. 1992. Tritium/^{3}He dating of waters in natural systems. In Isotopes of Noble Gases as Tracers in Environmental Studies. IAEA, Vienna, pp. 123-145.

Schlosser, P., and W.M. Smethie, Jr. 1995. Transient tracers as a tool to study variability of ocean circulation. In Natural Climate Variability on Decade-to-Century Time Scales. D.G. Martinson, K. Bryan, M. Ghil, M.M. Hall, T.R. Karl, E.S. Sarachik, S. Sorooshian, and L.D. Talley (eds.). National Academy Press, Washington, D.C.

Schlosser, P., C. Pfleiderer, B. Kromer, I. Levin, K.O. Münnich, G. Bonani, M. Suter, and W. Wölfli. 1987. Measurement of small volume oceanic ^{14}C samples by accelerator mass spectrometry. Radiocarbon 29:347-352.

Schlosser, P., G. Bönisch, B. Kromer, K.O. Münnich, and K.P. Koltermann. 1990. Ventilation rates of the waters in the Nansen Basin of the Arctic Ocean derived from a multitracer approach. J. Geophys. Res. 95:3265-3272.

Schlosser, P., G. Bönisch, M. Rhein, and R. Bayer. 1991a. Reduction of deepwater formation in the Greenland Sea during the 1980s: Evidence from tracer data. Science 251:1054-1056.

Schlosser, P., J.L. Bullister, and R. Bayer. 1991b. Studies of deep water formation and circulation in the Weddell Sea using natural and anthropogenic tracers. Mar. Chem. 35:97-122.

Schoch, H., and K.O. Münnich. 1981. Routine performance of a new multi-counter system for high-precision ^{14}C dating. In Methods of Low-Level Counting and Spectrometry. IAEA, Vienna, pp. 361-370.

Schopka, S.A. 1991. The Greenland cod at Iceland, 1941-1990 and its impact on assessment. SCR Doc. 91/102, Northwest Atlantic Fisheries Organisation, Dartmouth, Nova Scotia, 7 pp. (mimeo).

Sear, C.B. 1988. An index of sea ice variations in the Nordic Seas. J. Climatol. 8:339-355.

Serreze, M.C., J.A. Maslanik, R.G. Barry, and T.L. Demaria. 1992. Winter atmospheric circulation in the Arctic Basin and possible relationships to the Great Salinity Anomaly in the northern North Atlantic. Geophys. Res. Lett. 19:293-296.

Simmonds, P.G., D.M. Cunnold, F.N. Alyea, C.A. Cardelino, A.J. Crawford, R.G. Prinn, P.J. Fraser, R.A. Rasmussen, and R.D. Rosen. 1988. Carbon tetrachloride lifetimes and emissions determined from daily global measurements during 1978-1985. J. Atmos. Chem. 7:35-58.

SIZEX group. 1989. SIZEX Experiment Report. Technical Report 23, Nansen Remote Sensing Center, Bergen, Norway.

Smethie, W.M., Jr. 1993. Tracing the thermohaline circulation in the western North Atlantic using chlorofluorocarbons. Progr. Oceanogr. 31:51-99.

Smethie, W.M., Jr., and G. Mathieu. 1986. Measurements of krypton-85 in the ocean. Mar. Chem. 18:17-33.

Smethie, W.M., Jr., and J.H. Swift. 1989. The tritium:krypton-85 age of Denmark Strait Overflow Water and Gibbs Fracture Zone Water just south of Denmark Strait. J. Geophys. Res. 94:8265-8275.

Smethie, W.M., Jr., H.G. Östlund, and H.H. Loosli. 1986. Ventilation of the deep Greenland and Norwegian seas: Evidence from krypton-85, tritium, carbon-14, and argon-39. Deep-Sea Res. 33:675-703.

Smethie, W.M., Jr., D.W. Chipman, J.H. Swift, and K.P. Koltermann. 1988. Chlorofluoromethanes in the Arctic Mediterranean seas: Evidence for formation of bottom water in the Eurasian Basin and deep water exchange through Fram Strait. Deep-Sea Res. 35:347-369.

Smith, E.H., F.M. Soule, and O. Mosby. 1937. The Marion and General Green expeditions to Davis Strait and Labrador Sea. Bull. U.S. Coast Guard 19:1-259.

Spencer, N.E., and P.L. Woodworth. 1991. 1991 Data Holdings of the Permanent Service for Mean Sea Level. Permanent Service for Mean Sea Level, Bidston, Birkenhead.

Stephen, A.C. 1938. Temperature and the incidence of certain species in western European waters. J. Animal Ecol. 7:125.

Stocker, T.F., and L.A. Mysak. 1992. Climatic fluctuations on the century time scale: A review of high-resolution proxy data and possible mechanisms. Climatic Change 20:227-250.

Stommel, H. 1948. The westward intensification of wind-driven ocean currents. EOS, Trans. AGU 29:202-206.

Stuiver, M., and H.A. Polach. 1977. Reporting of ^{14}C data. Radiocarbon 19:355-363.

Sturges, W.E. 1987. Large-scale coherence of sea level at very low frequencies. J. Phys. Oceanogr. 71:2084-2094.

Sturges, W., and B.G. Hong. 1995. Wind forcing of the Atlantic thermocline along 32°N at low frequencies. J. Phys. Oceanogr. 25:1706-1715.

Sverdrup, H.U., M.W. Johnson, and R.H. Fleming. 1942. The Oceans: Their Physics, Chemistry, and General Biology. Prentice-Hall, Englewood Cliffs, N.J., 1087 pp.

Swift, J.H. 1984a. A recent θ-S shift in the deep water of the northern North Atlantic. In Climate Processes and Climate Sensitivity. J.E. Hansen and T. Takahashi (eds.). Geophysical Monograph 29 (Ewing Volume 5), American Geophysical Union, Washington, D.C., pp. 39-47.

Swift, J.H. 1984b. The circulation of the Denmark Strait and Iceland-Scotland overflow waters in the North Atlantic. Deep-Sea Res. 31:1339-1355.

Swift, J.H. 1985. A few comments on a recent deep-water freshening. In Glaciers, Ice Sheets, and Sea Level: Effect of a CO$_2$-Induced Climatic Change. DOE/ER/G0235-1, Department of Energy, Washington, D.C., pp. 104-115.

Swift, J.H. 1995. A few notes on a recent deep-water freshening. In Natural Climate Variability on Decade-to-Century Time Scales. D.G. Martinson, K. Bryan, M. Ghil, M.M. Hall, T.R. Karl, E.S. Sarachik, S. Sorooshian, and L.D. Talley (eds.). National Academy Press, Washington, D.C.

Swift, J.H., and K.P. Koltermann. 1988. The origin of Norwegian Sea Deep Water. J. Geophys. Res. 93:3563-3569.

Swift, J.H., K. Aagaard, and S.-A. Malmberg. 1980. The contribu-

tion of the Denmark Strait overflow to the deep North Atlantic. Deep-Sea Res. 27:29-42.

Swift, J.H., T. Takahashi, and H.D. Livingston. 1983. The contributions of the Greenland and Barents seas to the deep water of the Arctic Ocean. J. Geophys. Res. 88:5981-5986.

Tait, J.B. 1957. The hydrography of the Faroe-Shetland Channel, 1927-52. Marine Res. 2:309.

Talley, L.D., and M.S. McCartney. 1982. Distribution and circulation of Labrador Sea water. J. Phys. Oceanogr. 12:1189-1205.

Taning, A.V. 1943. Remarks on the influence of the higher temperatures in the northern waters during recent years on distribution and growth of some marine animals. Cons. Int. Explor. Mer, Ann. Biol. 1:76.

Taning, A.V. 1949. On changes in the marine fauna of the northwestern Atlantic area, with special reference to Greenland. Rapp. P.-v. Réun. Cons. Int. Explor. Mer 125:26-29.

Taylor, A.H. 1983. Fluctuations in the surface temperature and surface salinity of the north-east Atlantic at frequencies of one cycle per year and below. J. Climatol. 3:253-269.

Templeman, W. 1965. Relation of periods of successful year classes of haddock on the Grand Bank to periods of success of year classes for cod, haddock and herring in areas to the north and east. ICNAF Spec. Publ. 6:523-533.

Templeman, W. 1974. Migrations and intermingling of Atlantic cod (Gadus morhua) stocks of the Newfoundland area. J. Fish Res. Bd. Can. 31:1073-1092.

Templeman, W. 1979. Migrations and intermingling of stocks of Atlantic cod, Gadus morhua, of the Newfoundland and adjacent areas from tagging in 1962-66. ICNAF Res. Bull. 14:5-50.

Templeman, W. 1981. Vertebral numbers in Atlantic cod, Gadus morhua, of the Newfoundland and adjacent areas, 1947-71, and their use for delineating cod stocks. J. Northw. Atl. Fish Sci. 2:21-45.

Thiele, G., and J.L. Sarmiento. 1990. Tracer dating and ocean ventilation. J. Geophys. Res. 95:9377-9391.

Thiele, G., W. Roether, P. Schlosser, R. Kuntz, G. Siedler, and L. Stramma. 1986. Baroclinic flow and transient-tracer fields in the Canary-Cape Verde Basin. J. Phys. Oceanogr. 16:814-826.

Thompson, T.M., J.W. Elkins, J.H. Butter, B.B. Hall, K.B. Egan, C.M. Brunsen, J. Sczechowski, and T.H. Swanson. 1990. Annual report of the Nitrous Oxide and Halocarbons Group. In Climate Monitoring and Diagnostic Laboratory Report No. 18 (Summary Report 1990). W.D. Komhyr (ed.). NOAA CMDC, Boulder, Colo., pp. 64-72.

Tolstikov, E.I. (ed.). 1966. Atlas of the Antarctic, Vol. 1. Glav. Uprav. Geod. Kart. MG SSSR, Moscow, 23 pp. + 225 pp. of color maps (in Russian).

Trenberth, K.E., and J.W. Hurrell. 1994. Decadal atmosphere-ocean variations in the Pacific. Climate Dynamics 9:303-319.

Tushingham, A.M., and W.R. Peltier. 1991. ICE-3G: A new global model of late Pleistocene deglaciation based upon geophysical predictions of post-glacial relative sea level change. J. Geophys. Res. 96:4497-4523.

U.S. WOCE Office. 1989. U.S. WOCE Implementation Plan (Implementation Report No. 1). College Station, Texas, 176 pp.

Unterweger, M.P., B.M. Coursey, F.J. Schima, and W.B. Mann. 1980. Preparation and calibration of the 1978 National Bureau of Standards tritiated-water standards. Int. J. Appl. Radiat. Isotopes 31:611-614.

van Loon, H., and R.A. Madden. 1983. Interannual variations of mean monthly sea-level pressure in January. J. Climate Appl. Meteor. 22:687-692.

Vilhjalmsson, H., and J.V. Magnusson. 1984. Report on the 0-group fish survey in Icelandic and East Greenland waters. C.M.1984/H:66, International Council for the Exploration of the Sea, Copenhagen, 26 pp. (mimeo).

Wallace, D.W.R. 1992. WOCE chlorofluorocarbon intercomparison cruise report. WOCE Hydrographic Programme Report WHPO 92-1 (WOCE Report 83/92), Woods Hole, Mass., 31 pp.

Wallace, D.W.R., and J.R.N. Lazier. 1988. Anthropogenic chlorofluoromethanes in newly formed Labrador Sea Water. Nature 332:61-63.

Wallace, D.W.R., P. Schlosser, M. Krysell, and G. Bönisch. 1992. Halocarbon ratio and tritium/^{3}He dating of water masses in the Nansen Basin, Arctic Ocean. Deep-Sea Res. 39:5435-5458.

Walsh, J.E. 1978. A data set on Northern Hemisphere sea ice extent, 1953-76. In Glaciological Data Report GD-2, World Data Center-A for Glaciology, Boulder, Colo., pp. 49-51.

Walsh, J.E. 1991. Operational satellites and the global monitoring of snow and ice. Glob. Planet. Change 4(1-3):219-224.

Walsh, J.E., and W.L. Chapman. 1990a. Arctic contribution to upper-ocean variability in the North Atlantic. J. Climate 3:1462-1473.

Walsh, J.E., and W.L. Chapman. 1990b. Short-term climatic variability of the Arctic. J. Climate 3:237-250.

Walsh, J.E., and C.M. Johnson. 1979. An analysis of Arctic sea ice fluctuations. J. Phys. Oceanogr. 9:580-591.

Walsh, J.E., W.D. Hibler III, and B. Ross. 1985. Numerical simulation of northern hemisphere sea ice variability, 1951-1980. J. Geophys. Res. 90:4847-4865.

Ward, M.N. 1992. Provisionally corrected surface wind data, worldwide ocean-atmosphere surface fields and Sahelian rainfall variability. J. Climate 5:454-475.

Warner, M.J., and R.F. Weiss. 1985. Solubilities of chlorofluorocarbons 11 and 12 in water and seawater. Deep-Sea Res. 32:1485-1497.

Weaver, A.J., E.S. Sarachik, and J. Marotzke. 1991. Freshwater flux forcing of decadal and interdecadal oceanic variability. Nature 353:836-838.

Weiss, W., and W. Roether. 1980. The rates of tritium input to the world oceans. Earth Planet. Sci. Lett. 49:435-446.

Weiss, W., W. Roether, and G. Bader. 1976. Determination of blanks in low-level tritium measurement. Int. J. Appl. Radiat. Isotopes 27:217-225.

Weiss, W., J. Bullacher, and W. Roether. 1978. Evidence of pulsed discharges of tritium from nuclear energy installations in Central European precipitation. In Behaviour of Tritium in the Environment (Proc. Symp. San Francisco). IAEA, Vienna, pp. 17-30.

Weiss, R.F., J.L. Bullister, R.H. Gammon, and M.J. Warner. 1985. Chlorofluoromethanes in the Upper North Atlantic Deep Water. Nature 314:608-610.

Weiss, W., H. Stockburger, H. Sartorius, K. Rozanski, M. del Milagro Perez Garcia, and H.G. Östlund. 1986. Mesoscale transport of krypton-85 originating from European sources. Nucl. Instr. Methods Phys. Res. B17:571-574.

Wisegarver, D.P., and R.H. Gammon. 1988. A new transient tracer: Measured vertical distribution of CCl_2FCClF_2 (F-113) in the North Pacific subarctic gyre. Geophys. Res. Lett. 15:188-191.

Woodruff, S.D., R.J. Slutz, R.L. Jenne, and P.M. Steurer. 1987. A comprehensive ocean-atmosphere data set. Bull. Amer. Meteorol. Soc. 68:1239-1250.

Woods Hole Oceanographic Institute (WHOI) and Bermuda Biological Station for Research (BBSR). 1988. Station "S" off Bermuda: Physical Measurements, 1954-1984. Library of Congress Catalogue No. 88-51552, 189 pp.

Woodworth, P.L. 1990. A search for accelerations in records of European mean sea level. Intl. J. Climatol. 10:129-143.

Worcester, P.F., J.F. Lynch, W.M.L. Morawitz, R. Pawlowicz, P.J. Sutton, B.D. Cornuelle, O.M. Johannessen, W.H. Munk, W.B. Owens, R. Shuchman, and R.C. Spindel. 1993. Evolution of the large-scale temperature field in the Greenland Sea during 1988-89 from tomographic measurements. Geophys. Res. Lett. 20:2211-2214.

World Climate Research Programme (WCRP). 1982. Time Series of Ocean Measurements. World Climate Rept. 21, World Meteorological Organization, Geneva, Switzerland, 400 pp.

Worthington, L.V. 1976. On the North Atlantic Circulation. No. 6, The Johns Hopkins Oceanographic Studies, The Johns Hopkins University Press, 110 pp.

Worthington, L.V., and W.R. Wright. 1970. North Atlantic Ocean Atlas of Potential Temperature and Salinity in the Deep Water Including Temperature, Salinity and Oxygen Profiles from the Erika Dan Cruise of 1962. Woods Hole Oceanographic Institution, Woods Hole, Mass., 24 pp. + 58 plates.

Yang, J., and J.D. Neelin. 1993. Sea-ice interaction with the thermohaline circulation. Geophys. Res. Lett. 20:217-220.

Ocean Modeling

KIRK BRYAN

Ocean models are becoming an increasingly useful tool for studying the ocean circulation. They are expected to play a major role in the analysis of the new data now being collected by the international World Ocean Circulation Experiment (WOCE) and the observing systems motivated by the Tropical Ocean and Global Atmosphere (TOGA) Program, which are slated for continuation under the Global Ocean-Atmosphere-Land System (GOALS) program. Ocean models are also essential for the assimilation of information to be collected by other observing programs such as the Global Climate Observing System (GCOS), the Global Ocean Observing System (GOOS), and those planned by the World Climate Research Programme (WCRP). While the use of models to provide a more detailed picture of the ocean circulation is an interesting and important subject in its own right, this volume focuses on climate variation on decade-to-century time scales. Due to the great thermal inertia of the ocean (roughly three orders of magnitude greater than that of the atmosphere), the oceans play a primary role in longer-term climate variations. The papers on ocean modeling that appear in this volume report on the development of models specifically designed for studying ocean climate variability, and on studies aimed at showing what mechanisms could cause climatic variations much longer than the El Niño time scale yet less than the near-millennium time scale of the deep abyssal waters of the ocean.

There will always be a debate as to the appropriate level of complexity of an ocean climate model. We now have very simple models in which the description of the ocean is reduced to a few degrees of freedom, and very complex

models that can simulate the interaction of processes on scales of a hundred kilometers with the ocean circulation on a planetary scale. As the papers in this volume illustrate, no single level of complexity is uniquely appropriate. Different types of models provide their own special insights into the role of the ocean in climate. In the final analysis, we must work toward developing a well-designed hierarchy of models of increasing complexity.

Perhaps the simplest conceptual model that illustrates the role of the ocean's thermal inertia is Hasselmann's (1976) model of the response of the ocean to stochastic fluctuations of heating by the atmosphere, including radiative feedback. This view of air-sea interaction has been reviewed recently by Wunsch (1992). In a seminal study, Stommel (1961) explored another aspect of air-sea interaction. He pointed out that the atmospheric feedback to the flux of heat and the net flux of water at the ocean surface are not the same. The flux of heat causes a change of sea surface temperature (SST), while the net flux of water due to the difference between precipitation and evaporation changes the sea surface salinity. While the atmosphere responds directly to anomalies in SST, there is no corresponding response to sea surface salinity. At the same time, the salinity of seawater can have a very important effect on the density and pressure fields in the ocean, particularly in high latitudes where the surface temperature is low. In a very simple, two-box model of the thermocline circulation, Stommel showed how this asymmetry in temperature and salinity boundary conditions could lead to multiple equilibrium states for the thermohaline circulation, and hence to multiple equilibrium states of the earth's climate. Many studies of ocean climate, including several contributions to this volume, are based on mixed boundary conditions for temperature and salinity, which is essentially the idea behind Stommel's 1961 model.

Recently a revisionist view has been presented in studies by Zhang et al. (1993), Greatbatch and Zhang (1995), Kleeman and Power (1995), Rahmstorf and Willebrand (1995), Cai and Godfrey (1995), and Marotzke and Stone (1995). They emphasize that the mixed boundary conditions used in many previous studies may have greatly overestimated the strength of atmospheric feedback in response to sea surface temperature. The degree of asymmetry between the ocean boundary conditions for temperature and salinity has thus been greatly exaggerated, and as a consequence the time-dependent behavior of the models must be interpreted with a great deal of care. This difficulty in constructing appropriate boundary conditions for an ocean model arises from the attempt to approximate complex air-sea interaction in a very simple way. It will eventually be solved by using more realistic fully coupled models of the ocean and atmosphere.

Stommel's model is revisited in a review by Bryan and Hansen (1995). Combining Stommel's model with Hasselmann's concept of stochastic forcing, Bryan and Hansen construct a highly simplified model of climate variability in which a water mass corresponding to the northern North Atlantic interacts with the remainder of the ocean to the south and the atmosphere above. The time-dependent behavior of this model provides insight into the way in which the thermohaline circulation in the present climate regime acts to stabilize climate variations with a period longer than a few decades. At present, the poleward density gradient in the North Atlantic is dominated by differences in temperature rather than in salinity. The model is simple enough to allow exploration of a full range of boundary conditions in which temperature feedback ranges from very strong to very weak. For realistic forcing, the model becomes unstable only when the north-south gradients of salinity become far more important than they are at present, or if stochastic forcing of the salinity field becomes much larger. Such conditions may have prevailed during the close of the last ice age.

As soon as we attempt to extend the ideas taken from the simple conceptual or "toy" models to more complete models based on the actual equations of fluid motion that govern the ocean circulation, the adequacy and reliability of present models come into question. McWilliams (1995) reviews these topics, and offers specific suggestions for the improvement of existing models. Since stratification in the ocean is relatively weak compared to that in the atmosphere, many of the significant scales of motion for heat exchange are an order of magnitude smaller in the ocean than in the atmosphere. At the same time, the very great heat capacity

of the oceans means that the times required to reach climatic equilibrium are much longer there than in the atmosphere. The ocean is thus a classic example of a "stiff" system. McWilliams's approach to this difficulty is to make maximum use of models of relatively low resolution in which the effects of mesoscale eddies are represented in closure schemes. The parameters in the closure schemes are fitted to observations, if available, or to higher-resolution models.

Weaver (1995) does an excellent job of summarizing many studies of ocean models in which air-sea interaction is approximated by the mixed boundary conditions alluded to in connection with Stommel's 1961 model. Through careful examination of a wide range of numerical experiments, Weaver has been able to separate results that are robust from those that may be artifacts introduced by specific experimental methods.

Studies by McDermott and Sarachik (1995) and Barnett et al. (1995) are based on mixed boundary conditions similar to those used in the studies summarized by Weaver (1995). In addition, Barnett et al. explore the effects of stochastic fresh-water fluxes. While the recent studies mentioned earlier indicate that the atmospheric feedback to SST may be greatly exaggerated in these numerical experiments, the results will still be valuable in interpreting future climate models with more realistic simulation of air-sea interaction.

As pointed out by Barnett et al. (1995), the ideal model for studying low-frequency climate variability is a fully coupled ocean-atmosphere model, since at present we simply do not know how to design a simple stochastic model that can replicate with fidelity the known feedbacks in heating, evaporation-minus-precipitation, and wind stress that take place in nature. The paper by Delworth et al. (1995) in the coupled-model section of this volume illustrates this approach to modeling decade-to-century-scale climate variability. As pointed out previously, the ocean represents what in theoretical mechanics would be termed a "stiff" system, with a wide range of important time and space scales. Coupling an ocean model to an atmospheric model produces an even stiffer system, due to the relatively short time scales in the atmosphere. In consequence, there is an even greater computational requirement for numerical experiments with coupled models.

Delworth et al. (1995) have integrated a coupled model for the equivalent of many centuries. The most interesting feature of their simulation is a 40- to 50-year climatic variability centered in the North Atlantic that is closely associated with changes in the intensity of the thermohaline circulation. When Bjerknes (1964) studied the historical record of SST in the North Atlantic, he was particularly struck by the return to warmer conditions in the period 1925-1935 from the much colder conditions that had existed in the North Atlantic in the earlier part of the century. He hypothesized that this warming was related to a change in the overturning circulation of the North Atlantic, and that the same phenomenon could not take place in the Pacific because the thermohaline circulation there had such a different character. Subsequent to the publication of Bjerknes's study in 1964, much colder conditions returned to the North Atlantic, so we now have good oceanographic and meteorological observations for those conditions.

This simulation of multi-decadal climate variability involving the thermohaline circulation appears to be the most realistic modeling result for this type of air-sea interaction obtained so far, because the coupling of detailed ocean and atmosphere models removes the very significant difficulties in mixed-boundary-condition models that tended to overestimate atmospheric feedback. It must be kept in mind, however, that all models are still crude representations of the real coupled system, and much future work needs to be done in understanding the details of large-scale air-sea interaction. In summing up the status of modeling the ocean's role in multi-decadal climate variability, we see that two somewhat different mechanisms have been put forward. In both cases the thermohaline circulation acts to restore the system to its equilibrium state and set the time scale. In one mechanism the high-latitude ocean is freshened at the surface. This freshening causes an unstable response that drives the system from equilibrium. This mechanism is examined by Weaver (1995). Although this mechanism has been demonstrated only in models with mixed boundary conditions, which may exaggerate atmospheric feedback, it may still have validity.

The other mechanism consists of stochastic forcing that drives the ocean away from its equilibrium state. This mechanism is illustrated in the simple model of Bryan and Hansen (1995). In the coupled models, both mechanisms may play a role, and it may be difficult to distinguish between them. Interest in air-sea interaction on El Niño time scales has stimulated the development of a hierarchy of ocean models for the tropical oceans. Predictions of the interannual variability in the tropical Pacific using these models have been remarkably successful, and are gradually being adapted as operational models for monitoring and prediction. The papers presented in this volume show that a promising start has been made in developing a new class of ocean-circulation models that are addressed to the study of higher-latitude processes. Through a study of the communication between the ocean's surface and the deep ocean, which takes place via the thermohaline circulation, we are beginning to get some insights on the role of the high-latitude ocean as the long-term "memory" of the climate system. Modeling the polar and sub-polar oceans is in many ways much more difficult than modeling the equatorial oceans. It involves many long-standing unanswered questions in oceanography. The papers in this volume demonstrate, however, that an enthusiastic and concerted effort is being made to develop the ocean models needed to study high-latitude air-sea interaction, which appears to be the key to understanding decade-to-century-scale climate change.

Sub-Grid-Scale Parameterizations in Oceanic General-Circulation Models

JAMES C. MCWILLIAMS[1]

ABSTRACT

For studies of global and regional climate variability on time scales from a season to hundreds of years, oceanic numerical models must represent through parameterization the important processes that occur on scales smaller than can be resolved on the model grid. These sub-grid-scale processes usually include mesoscale eddies, internal waves, double diffusion, tides, coastal waves, planetary boundary-layer turbulence, interior mixing by shear and convective instabilities, and aspects of atmospheric forcing or topography or marginal seas below the resolution threshold. Furthermore, for completeness as an oceanic component of a climate system model, these physical oceanographic elements must be augmented by parameterizations of both sea ice and oceanic biogeochemical processes. This paper reviews present practices for these different parameterization elements and describes some ideas for implementing them better. Particular emphases are given to nearly isopycnal tracer transports by mesoscale eddies and to vertical transports both inside and outside the planetary boundary layer.

INTRODUCTION

The ocean participates in the dynamics of climate variability on all time scales. It provides an enormous storage reservoir for heat, water, nutrients, CO_2, and other chemical quantities relevant to the radiatively active trace gases. It absorbs and releases these properties in exchanges with the atmosphere with a broad range of time delays and geographical redistributions resulting from internal transport processes by currents. These currents range in scale from systematic motions across ocean basins (i.e., 10^4 km) to eddies that mix fluid within a centimeter.

As yet, I believe we have seen only a glimmer of the multivariate, multidimensional character of large-scale climate variability on time scales of decades and centuries. Furthermore, it seems infeasible by any current technologies to measure the oceanic storage and transport in any comprehensive fashion. Even if this could be done, it would provide no firm predictive basis for anticipating future climate

[1]National Center for Atmospheric Research, Boulder, Colorado, and Department of Atmospheric Sciences and Institute of Geophysics and Planetary Physics, University of California, Los Angeles, California

changes. This can only be done through skillful modeling of the relevant oceanic behaviors. Yet modeling has its own feasibility limits; only scales of motion above a certain threshold can be included in the discretized partial differential equation system that comprises an ocean model. Current computational technology sets this threshold at about 100 km in horizontal scale, if the goal is to span the globe and investigate the nearly equilibrium-state variability; there are analogous limitations in the vertical and time scales. Only on a human time scale of a decade or more might we expect this threshold to be reduced by a physically significant degree (i.e., by a factor of ten—with a resulting increase of about 10^3 in computation—to match the requirements for accurate mesoscale-eddy resolution).

However, no ocean model makes good dynamical sense if restricted to merely the larger scales of motion. For example, the Reynolds number at the threshold scale is about 10^{10}, so the necessary dissipation for equilibration must occur on very much smaller scales. These missing motions must be parameterized so as to provide their essential effects in the dynamical balances of the motions resolved in the model; we refer to them generically as sub-grid-scale (or SGS) motions. In the context of an oceanic climate model, the SGS motions include mesoscale eddies, internal waves, double diffusion, tides, coastal waves, planetary boundary layer (or PBL) turbulence forced by surface fluxes of momentum and buoyancy, interior turbulence resulting from shear instability or convection, and any effects of atmospheric forcing or topography or marginal seas that lie below the resolution threshold. In addition, for a complete oceanic component of a climate system model, parameterizations are needed for both sea ice and oceanic biogeochemical processes; the presence of sea ice strongly suppresses the surface heat and moisture fluxes, and oceanic biochemical processes are important in the global budgets of the radiatively active gases.

I advance the following proposition which, although it cannot be proven in general, is a summary of considerable experience with oceanic model solutions and a reasonable, albeit conservative, expectation for future solutions:

> Except for short-term transients, large-scale oceanic model solutions will be sensitively dependent upon the SGS forms employed.

Illustrations of this are the sensitivity of the mean meridional overturning circulation to the value of the vertical diffusivity for tracers, and the occurrence of substantial shifts in this circulation due to small changes in boundary or initial conditions (Bryan, 1986b,c). Thus, skillful modeling of the oceans' roles in climatic variability requires skillful parameterizations. As we shall see below, I believe the achievement of the latter lies largely in the future, although not necessarily a distant future. Even substantial future decrements in

the computational threshold scale will leave many processes still SGS. The parameterization problem is here to stay.

This paper summarizes present practices for the different parameterization components in oceanic climate models and describes some ideas for implementing them better. Its scope is intended to be broad, but the depth of discussion is highly uneven among topics, with emphases given to those areas in which I am currently working. It is intended only as a guide to other, more primary publications in which more complete information is given.

GENERAL PRINCIPLES

We can identify two approaches to parameterization that are distinct in principle but may be blurred in practice:

1. Hypothesize the desired effect by the missing motions and the functional form leading to it, subject to consistency constraints (e.g., conservation integrals) that minimize undesirable side effects. The seemingly inescapable unknown coefficients in these forms are then chosen either to obtain some desired magnitude for the resulting solutions or to match the gross magnitude of observed eddy fluxes.
2. Determine and calibrate a parameterization using fields of eddy fluxes, from solutions or observations, that fully resolve the motions being parameterized.

The distinction between these approaches is the relatively greater roles of scientific imagination in the former and of established facts about the dynamical interactions across the SGS threshold in the latter. Thus, there is a natural succession from the former to the latter. In the examples below, we will see that we are on the verge of realizing the second approach for mesoscale eddies and PBL turbulence. For other processes, though, we are still largely confined to the first approach. In general, however, I believe that accurate, well-resolved numerical solutions of SGS processes can provide the bases for testing (and inspiring) better parameterizations. We should expect this to be a long path, which must be walked from both ends: Small-scale processes must be examined (i.e., modeled) to see how they act on larger-scale distributions, and large-scale models must be run with various parameterizations to see which aspects their solutions are sensitive to.

In considering the effect of fluctuating or small-scale motions on the transport of mean or large-scale property distributions, a traditional view is that of diffusion or mixing. This is often referred to as the Fickian hypothesis; viz., SGS fluxes act simply to mix the mean field for some property s, with down-gradient fluxes proportional to the product of the mean gradient and a scalar eddy diffusivity μ,

$$\overline{u's'} = -\mu\nabla\overline{s}. \qquad (1)$$

The overbar denotes an average over the symmetry coordinates, such as time or, in some examples below, the zonal direction, or else simply a low-pass filter in space and time; the prime denotes a deviation field. In its simplest form—the one common in the SGS parameterizations currently in ocean models—μ is assumed to be a constant, which might occur if there were spatial homogeneity and isotropy, a significant scale gap, and weak dynamical coupling between the mean and deviation fields. However, in flows with significant inhomogeneity or non-stationarity and often little scale separation, as is usual in the ocean and atmosphere, this assumption is incorrect. So μ can be expected to depend on the space and time coordinates, in part through its dependence on the mean flow structures. In general the eddy-flux and mean-field-gradient vectors will not be parallel, so that the relation (1) will not be satisfied for any scalar field μ; thus, consistent extensions of (1) may be required, and two examples of this are given below for mesoscale eddies and the PBL. One possible extension, to make it a second-rank tensor, is not generally attractive because it is so strongly underdetermined from its defining relation; however, a diagonal tensor form can be a useful framework for representing anisotropy. Nevertheless, the Fickian relation and its modest extensions can be viewed as a definition of the diffusivity, which then can be used as a language for discussing SGS influences. It does have the advantage of acting, both locally and integrally, to limit and deplete the mean fields (for μ > 0), as seems usually to be true for eddy motions.

Two additional cautions should be stated about the Fickian formalism. Some flow structures need to be viewed as a whole, such as the PBL where parcel trajectories traverse the full layer depth on short time scales. Other processes, particularly those associated with wave propagation through inhomogeneous media, can be strongly nonlocal; this is expressed in the well-known Eliassen-Palm or "non-acceleration" theorems (see, e.g., Andrews and McIntyre, 1976) indicating that eddy fluxes can cause a net forcing of the mean fields only where eddy dissipation occurs (e.g., near critical surfaces). In principle, both of these behaviors can be represented in $\mu(\mathbf{x},t)$ or $\mu[\bar{s}]$ forms, though the latter behavior presents the greater challenge in this regard. Fortunately, most of the important SGS motions in the ocean may be more turbulent than wave-like, so the effects of non-locality may not be too severe.

MESOSCALE EDDIES

Mesoscale eddies typically have horizontal scales near the first internal deformation radius (tens of km), vertical scales ranging from the pycnocline depth to the full ocean depth (ones of km), and time scales of weeks and months. They play several essential roles in the general circulation.

1. Mesoscale eddies cause an enstrophy cascade from the large-scale circulation to small-scale dissipation, as is well known in two-dimensional and geostrophic turbulence

(Batchelor, 1969; Charney, 1971; McWilliams, 1984, 1989). In the presence of a large-scale circulation, this provides substantial lateral stresses, regulates the interiorward separation of western boundary currents, establishes the horizontal boundary conditions on velocity and stress, and effects an exchange of energy between the mean and eddies by means of horizontal Reynolds stress work. This latter can be expressed as

$$\overline{KE} \rightarrow KE' = -\left\langle \overline{u'v'} \frac{\partial \overline{u}}{\partial y} \right\rangle \sim \left\langle \nu_h \left(\frac{\partial \overline{u}}{\partial y}\right)^2 \right\rangle, \quad (2)$$

where we have assumed zonal symmetry and a zonal mean flow for simplicity. *KE* is kinetic energy, the overbar and prime denote mean and eddy components, and the angle brackets denote a volume average. The approximate magnitude of the associated horizontal eddy viscosity is $\nu_h \sim 10^4$ m^2 s^{-1}, and the expected energy conversion is a significant fraction of the work done by the surface wind stress.

2. Mesoscale eddies effect the downward penetration of the wind-driven circulation and control its baroclinic structure. The downward penetration is as if there were a vertical stress of order $\nu_v V/H \sim 10^4$ m^2 s^{-1} (i.e., the order of the surface wind stress), which requires $\nu_v \sim 1$ m^2 s^{-1} for $V \sim 0.1$ m s^{-1} and $H \sim 1$ km characteristic of the gyre-scale circulations; yet it is clear from observations that vertical Reynolds stresses, $\overline{u'w'}$ and $\overline{v'w'}$, are many orders smaller than this requirement (see, e.g., Gregg, 1987). The mesoscale process that effects this penetration is isopycnal form stress: we define an internal stress

$$\tau_i = \overline{p' \frac{\partial \eta'}{\partial x}} = \frac{f}{N^2} \overline{v'b'}, \quad (3)$$

where p is the pressure and η is the elevation of an isopycnal surface. Thus, τ_i is an integrated form stress over an isopycnal surface; its vertical divergence then is a horizontal acceleration (here in the zonal direction). The second expression in (3) arises from the small-amplitude form of the definition of η ($\eta = -b/N^2$, where b is the buoyancy field $-g\rho/\rho_0$ and N^2 is the vertical gradient of b) and geostrophic balance ($fv' = \partial p'/\partial x$). It also is associated with an exchange of energy between the mean and the eddies through buoyancy work,

$$\overline{PE} \rightarrow PE' = -\langle \overline{wb} \rangle = -\left\langle \frac{\overline{v'b'}}{N^2(z)} \frac{\partial \overline{b}}{\partial y} \right\rangle, \quad (4)$$

that is also of the same order as the wind work. Here *PE* is potential energy, and we have made use of the quasigeostrophic approximation and again assumed zonal symmetry. It is easy to see that substitution of a Fickian form for buoyancy flux implies energy losses from the mean fields.

3. Mesoscale eddies stir and make inevitable the mixing of both passive and dynamically active tracers, primarily along isopycnal surfaces (Iselin, 1939; Montgomery, 1940).

Dispersal statistics of neutrally buoyant floats in the ocean indicate that there is an effective diffusivity $\kappa_i \sim 10^4$ m^2 s^{-1}, but again with appreciable geographical variability (McWilliams et al., 1983). κ_i can also be related to isopycnal form stress (see below). There is the interesting possibility that these transports may often occur primarily in long-lived coherent vortices, and thus be highly non-local in that little "mixing" need occur over the long distances separating the generation and destruction sites (McWilliams, 1985).

At present the common practice in eddyless ocean models is to achieve (1) with a constant ν_h and to partially achieve (2) and (3) with a constant, and usually somewhat smaller, horizontal eddy diffusivity κ_h. The latter is formally equivalent to a κ_i at leading order in Rossby number ($Ro = V/fL$); however, it is clear that substantial differences in the tracer distributions result from κ_h versus κ_i, since the former implies substantial and excessive diapycnal fluxes across sloping isopycnal surfaces.

We have developed a novel parameterization scheme for the latter of these two roles, isopycnal form stress and isopycnal tracer diffusion (Gent and McWilliams, 1990; McWilliams and Gent, 1994). (The first role, horizontal momentum diffusion, seems less problematic in some ways, but it also needs to be extended beyond the present common practice of spatially constant viscosity.) The parameterization is chosen to have the following properties, which we collectively dub "quasi-adiabatic"; their rationale is that we want the quasi-adiabatic eddyless solutions to correspond to the large-scale components of an adiabatic, eddy-resolving solution.

1. All domain-averaged moments of potential density ρ are conserved, as is the volume between any two isopycnals.

2. With insulating boundary conditions, the domain average of any advectively conserved tracer S is conserved between any two isopycnals, and higher moments of S decrease in time if S has any gradients on the isopycnal surfaces.

3. The equation for a tracer S is satisfied identically by ρ.

Its mathematical form is as follows. In isopycnal coordinates, the equation for the thickness gradient $\partial h / \partial \rho$ is a combination of mass conservation (continuity) and the internal energy equation, viz.,

$$\frac{\partial^2 h}{\partial t \partial \rho} + \nabla_\rho \cdot \left(\frac{\partial h}{\partial \rho} \boldsymbol{u} \right) + \nabla_\rho \cdot \boldsymbol{F} = 0 , \qquad (5)$$

where $\boldsymbol{F}$ is the isopycnal thickness flux by eddies that is the quantity to be parameterized. (Overbars are implied for all the fields in this section, and all vectors are only horizontal.) Equivalently, in either an isopycnal or physical-height coordinate frame, the associated internal-energy equation is

$$\frac{D\rho}{Dt} = Q , \qquad (6)$$

where Q is determined by

$$\frac{\partial h}{\partial \rho} Q = \int^\rho \nabla_\rho \cdot \boldsymbol{F} \, d\rho . \qquad (7)$$

For any advectively conserved tracer S, the equation is

$$\left(\frac{\partial}{\partial t} + \boldsymbol{u} \cdot \nabla_\rho \right) S + \left(\frac{\partial h}{\partial \rho} \right)^{-1} \boldsymbol{F} \cdot \nabla_\rho S = D[S] , \qquad (8a)$$

where D represents diffusion on an isopycnal surface with diffusivity κ_i

$$D[S] \equiv \left(\frac{\partial h}{\partial \rho} \right)^{-1} \nabla_\rho \cdot \left(\kappa_i \frac{\partial h}{\partial \rho} \nabla_\rho S \right) . \qquad (8b)$$

A particularly simple (and almost Fickian) choice for $\boldsymbol{F}$ and Q is

$$\boldsymbol{F} = -\frac{\partial}{\partial \rho} (\kappa_i \nabla_\rho h)$$

$$\frac{\partial h}{\partial \rho} Q = -\nabla_\rho \cdot (\kappa_i \nabla_\rho h)$$

$$Q = \nabla_z \cdot (\kappa_i \nabla_z \rho) - \frac{\partial}{\partial z} \left(\kappa_i \nabla_z \rho \cdot \frac{\nabla_z \rho}{\rho_z} \right) , \qquad (9)$$

where the final expression is in physical-height coordinates. For $\kappa_i > 0$, any coordinate or resolved-flow-field dependence that respects the insulating boundary conditions and still satisfies the quasi-adiabatic constraints is acceptable. Furthermore, κ_i generally provides a down-gradient vertical flux of momentum (as an isopycnal form stress) and a loss of resolved-flow potential energy (as a baroclinic instability). We note that this form with $D \neq 0$ and $\boldsymbol{F} = Q = 0$ has previously been proposed (Redi, 1982) and implemented (Cox, 1987), but it has been found to be unsatisfactory in eddyless solutions because it cannot accomplish the necessary second role (above) for mesoscale transports, since $\boldsymbol{F} \cdot \nabla_\rho \rho = D[\rho] = 0$.

This parameterization is an appropriate component of a balanced or primitive equation model of the general circulation, since it respects the finite-Ro deformations of isopycnal surfaces.

We have obtained eddyless solutions to equations (5) through (9) in the linear balance equations for a linear, single-component equation of state, $\rho \propto T$, and a steady surface-wind stress that symmetrically forces a double gyre pattern. In these solutions the basin is a 4000 $\times$ 5000 $\times$ 5 km rectangular volume, and the gravest deformation radius is 45 km. Several results from these solutions are shown in Figures 1 through 4. A qualitatively sensible steady circulation results from the new parameterization with a spatially constant κ_i in combination with an also constant ν_h. It exhibits the broad pattern of the wind driving only in the upper ocean, but has cells of enhanced recirculation near the western boundary at all depths (Figure 1). The isopycnal mixing induces mean meridional-circulation cells (Figure 2) and an energy cycle in which eddy generation is mimicked as diffusive dissipation (Figure 3; note that these parameterized

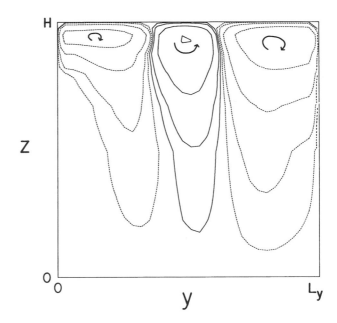

FIGURE 2   The mean meridional stream function $\xi(y,z)$ for the zonally averaged circulation for the same solution as in Figure 1. It is defined such that

$$\overline{v}^x = \frac{\partial \xi}{\partial z}, \quad \overline{w}^x = -\frac{\partial \xi}{\partial x}.$$

The contours have a logarithmic distribution in unspecific non-dimensional units.

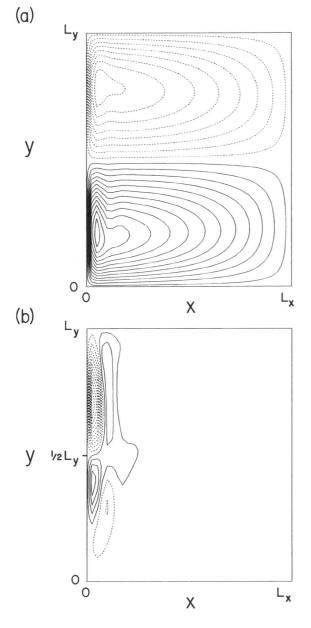

(a)

(b)

FIGURE 1   Horizontal stream function $\psi(x,y)$ at (a) upper-pycnocline and (b) deep-vertical levels in steady state for the wind-driven circulation with an isopycnal mixing parameterization in a linear balance-equation model. The wind pattern is symmetric about the middle latitude ($y = L_y/2$), with an eastward extremum there and equal westward extrema at $y = 0$ and $y = L_y$. The contour intervals are 0.1 and 0.005, respectively, in non-dimensional units.

eddy work terms are comparable to the mean wind work). However, these solutions are sensitive to spatial dependence in κ_i, in ways that can influence, for example, the separation structure of the mean Gulf Stream (Figure 4). A more complete report is in McWilliams and Gent (1994).

Obviously there is a need to go beyond constant diffusivities if the eddyless solutions are to have the same large-scale circulations as those in equivalently forced eddy-resolving solutions. We currently are in the process of fitting plausible functional forms for ν_h and κ_i to the eddy-flux patterns of eddy-resolving solutions for the wind-driven circulation. In particular we wish to examine the hypothesis that ν_h and κ_i are often linked so as to make the down-gradient flux of potential vorticity the common situation (see, e.g., McWilliams and Chow, 1981; Marshall, 1981; Rhines and Young, 1982).

In Figures 5 through 8 are some preliminary results from this study. Notice that here the time-mean streamfunction has about the same peak amplitude near the western boundary in the middle of the gyres as in the analogous eddyless solution (Figure 1). Thus, the parameterization of equations (5) through (9) with a spatially constant diffusivity is working as it should, in some overall sense, in effecting the downward penetration of the wind-driven circulation; however, clearly it is not correct in matching the detailed patterns

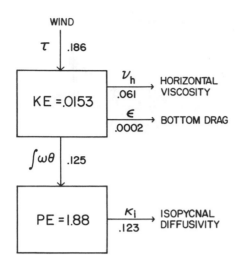

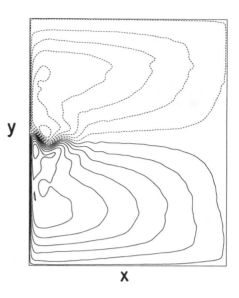

FIGURE 3   The energy budget in steady state for the same solution as in Figure 1, also in unspecified non-dimensional units.

of circulation. When we examine the patterns of the mean fields and their associated eddy fluxes (Figures 6 and 7), we see that both are structurally fairly smooth and simple. However, this does not necessarily imply that they will satisfy a Fickian relationship with equally simple patterns of eddy diffusivity. Figure 8 shows some examples of "minimal" eddy diffusivities, which are defined, e.g., for the horizontal diffusion of vorticity, to be

$$\nu_h^{(\zeta)} = -\frac{\overline{u'\zeta'} \cdot \nabla\overline{\zeta}}{(\nabla\overline{\zeta})^2} . \tag{10}$$

The sense in which this is minimal is that it is a solution of the more general relation

$$\nu_h^{(\zeta)}\nabla\overline{\zeta} = -\overline{u'\zeta'} + e_z \times \nabla E^{(\zeta)} , \tag{11}$$

for which the scalar field $E^{(\zeta)}$ contributes nothing to the eddy fluxes parallel to $\nabla\overline{\zeta}$. Thus, (11) is an extension of the strictly Fickian relation (1). Note that any choice for $E^{(\zeta)}$ provides an equivalent effect on the large-scale circulation when the divergence of the left side is taken, and any determination of ν simply from knowledge of the mean field and eddy fluxes is inherently non-unique. However, rather than simply determining the eddy diffusivity as a spatial function that satisfies (11) for a particular solution, our goal should be to fit eddy diffusivities with dynamically plausible functional dependences on the large-scale flow fields, with the hope that these functional relations may turn out to be more general. In doing so, we can simultaneously avail ourselves of the arbitrariness of E in order to discover the best functional forms. Thus, in Figure 8 one can be struck with the structural complexity of the minimal eddy diffusivities, including their extensive regions of negative values (which are likely to be computationally ill

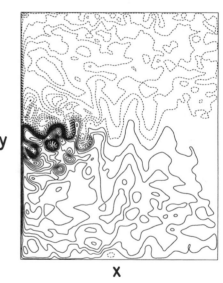

FIGURE 5   Time-mean and instantaneous $\psi(x,y)$ in the upper pycnocline in an eddy-resolving, balance-equation variant of the problem defined in Figure 1. The contour interval is 0.2 in the same non-dimensional units.

behaved in a diffusion operator). Once the functional fitting has been made, however, the final diffusivity patterns may be quite different from those in Figure 8.

PLANETARY BOUNDARY LAYERS

The planetary boundary layers are usually the sites of the most intense three-dimensional turbulence in the oceans. Their spatial scales are bounded from above by the boundary layer depth (i.e., 10 to 10^3 m). Typical adjustment times for the PBL range from an eddy turnover time ($\sim 10^2$ to 10^3 s) to an inertial time ($\sim 1/f = 10^4$ s); both are short compared to the climatic times of interest.

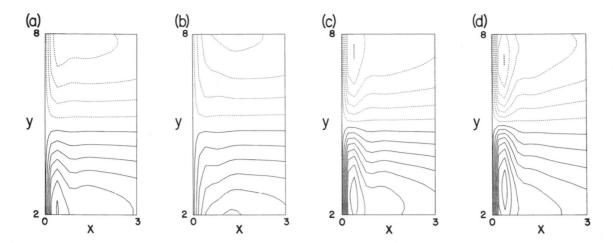

FIGURE 4 Upper-layer $\psi(x,y)$ in the region of western-boundary current separation (i.e., $0 \leq x \leq 1500$ km and $1000 \leq y \leq 4000$ km) for different spatial/functional dependences of κ_i. Panel (a) is for the same solution as in Figure 1, which has a spatially uniform diffusivity; (b) has a linear dependence on the local kinetic energy density KE; (c) has the diffusivity vanishing in the western boundary current; and (d) has the combined features of (b) and (c). The contour interval is 0.2 in non-dimensional units.

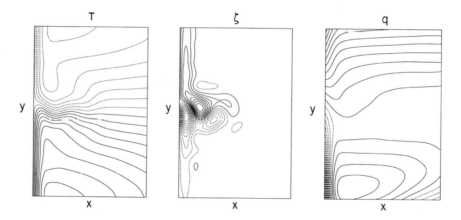

FIGURE 6 Time-mean and horizontally filtered temperature, vorticity, and potential-vorticity fields in arbitrary units in the middle pycnocline for the same solution as in Figure 5. Only a west-central 2000 km by 3000 km portion of the domain is displayed.

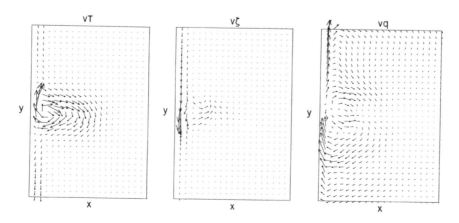

FIGURE 7 Filtered horizontal eddy fluxes of temperature, vorticity, and potential vorticity, in the same format as in Figure 6. Flux vectors have been deleted at the western boundary.

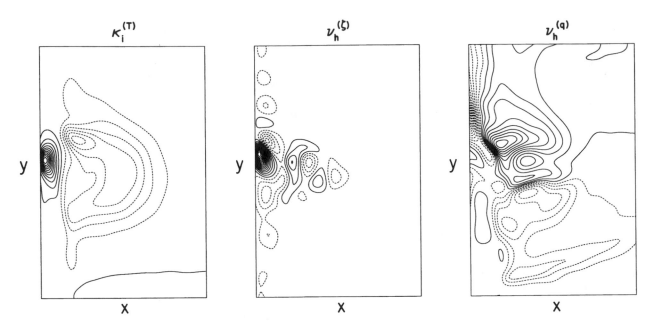

FIGURE 8   "Minimal" eddy diffusivities based on the quantities in Figures 6 and 7. Units are again arbitrary.

The PBL at the upper surface of the ocean provides an important reservoir for material properties being exchanged with the atmosphere, with a sequestration time of not more than a year. It also serves as a conduit transmitting atmospheric surface forcing to the interior, both for momentum (i.e., the Ekman velocity at the interior edge of the PBL) and material properties (often referred to as ventilation or subduction), which can lead to very much longer sequestration times. The transmissivity is much more efficient and inexorable for momentum than for the material properties; nevertheless, ventilation provides the principal sources and sinks for the interior, primarily isopycnal, tracer transports by mesoscale eddies (see above). The PBL at the bottom of the ocean provides a drag force on the adjacent currents, and, where there is an intersection with an isopycnal surface, as is particularly likely for a sloping bottom, it mixes material properties with locally diapycnal fluxes due to the three-dimensional motions in the PBL (Armi, 1978; Garrett, 1991; McCready and Rhines, 1991).

Present common practice in large-scale ocean models is to use linear Ekman relations at the top and bottom to transmit the boundary stresses to the interior motions, although in some instances a quadratic bottom-drag law is substituted, heuristically rationalized by analogy with non-rotating wall-bounded shear layers. Often the material-property transports are effected simply by extending the interior vertical diffusivities (see below) to the bounding surface. Insofar as these diffusivities are usually made large under conditions of static instability (or infinite, in an instantaneous adjustment to neutral stability), to some extent this procedure does mimic the PBL in at least this regime of penetrative vertical convection. Typically the vertical grid

resolution near the upper boundary is taken to be so coarse (i.e., 50-100 m) that very little of the near-surface variability (e.g., the seasonal cycle) can be represented.

Nevertheless, it is something of a scientific oddity that, although many different types of PBL parameterizations have been proposed by meteorologists and oceanographers, they have seldom been used in atmospheric and oceanic general circulation models. If there is a conventional wisdom for what ought to be used in ocean models, probably it would be either a mixed-layer model (as in Garwood, 1977; Gaspar, 1988) or a second-order moment-closure model (as in Mellor and Yamada, 1982).

We have recently begun using yet a different type of PBL model that assumes neither that the mean profiles are vertically well-mixed—which they often are not—nor that the turbulent transports are wholly locally determined at any given vertical level—which clearly is incorrect, for example, in convective situations where boundary-generated plumes often traverse the layer. Also, in contrast to many of the previous, regime-specific proposals, we believe our parameterization to be valid in a broad range of conditions for the surface-buoyancy and -momentum fluxes and the mean profiles. Its essential basis is the hypothesis that the transport coefficients have a universal shape, but a vertical scale and amplitude that are particular to the local surface fluxes and mean profiles (O'Brien, 1970; Troen and Mahrt, 1986; Large et al., 1994).

We express the vertical turbulent fluxes of momentum, buoyancy, and passive scalars in the oceanic boundary layer in terms of a vertical diffusivity profile $K_x(d)$, where x is any one of the above properties and d is the distance from the ocean surface. In certain circumstances a counter-gradient profile $\gamma_x(d)$ is also required. This term arises from an analysis of the turbulent temperature-variance budget (Deardorff, 1972; Holtslag and Moeng, 1991) and from observations in the atmosphere that show non-zero buoyancy fluxes where the local buoyancy gradient is zero or even weakly of the opposite sign (i.e., a local deviation from the Fickian form (1) is required to avoid large negative μ values). Throughout the boundary layer, $h \geq d \geq 0$, the fluxes are given by

$$\overline{w'x'(d)} = - K_x(d) \left[\frac{\partial \overline{x}}{\partial z} - \gamma_x(d) \right]. \qquad (12)$$

The diffusivity profile is formulated to conform to known and desirable properties of boundary-layer turbulence. In the surface layer, $d < \epsilon h$ (with $\epsilon \approx 0.1$), the semi-empirical results of Monin-Obukhov similarity theory should apply (Lumley and Panofsky, 1964). An important aspect of this is that the dimensionless flux profiles are universal functions, $\phi_x(\zeta)$, of the stability coordinate $\zeta = d/L$, where L is the Monin-Obukhov length scale,

$$L = \frac{(u^*)^3}{kB_0}, \qquad (13)$$

k is the von Karman constant, B_0 is the surface buoyancy flux into the ocean, and $(u^*)^2$ is the kinematic surface wind stress. This leads to logarithmic mean-property profiles near the boundary; the associated fluxes are about 20 percent of their surface values at $d = \epsilon h$ and approach their surface values linearly as $d \to 0$. Since there is no turbulent transfer either across the surface, or at $d = h$ if the interior mixing is zero, the diffusivities must satisfy $K_x(0) = K_x(h) = 0$. The universal shape is assumed to be cubic in d, which has had rough empirical confirmation; thus,

$$K_x(d) = w_x h \left(\frac{d}{h} \right) \left(1 - \frac{d}{h} \right)^2. \qquad (14)$$

The turbulent velocity scale w_x is formulated for consistency with similarity theory as a simple combination of the surface forcing velocities u^* and w^* (which is equal to $(-B_0 h)^{1/3}$ if B_0 is less than 0).

The counter-gradient term γ_x is non-zero only for the buoyancy field and for passive scalars in unstable forcing conditions (i.e., $B_0 < 0$); thus, we restrict attention to $x = s$ (for scalar). As suggested by Deardorff (1972), it has been successfully parameterized as

$$\gamma_s = C_s \frac{\overline{w'x'_0}}{w_s h} \qquad (15)$$

with $C_s = 6.2$.

The thickness of the oceanic planetary boundary layer h is determined as the smallest value of d at which a bulk

Richardson number Ri_b across the boundary layer achieves a certain critical value Ri_c. Thus it primarily depends on the buoyancy profile $b(d)$ and the velocity profile $\mathbf{v}(d)$, but there is also a (usually weak) dependence on the turbulent intensity:

$$Ri_b(d) = \frac{d[\overline{b}(0) - \overline{b}(d)]}{|\overline{v}(0) - \overline{v}(d)|^2 + v_t^2(d)} = Ri_c. \qquad (16)$$

The turbulent velocity v_t combines with the mean velocity difference in the denominator of (14); its primary role is to establish the correct entrainment flux at the interior edge of the PBL when the mean velocity is weak or absent (i.e., primarily in the case of free convection). There is strong empirical evidence that this entrainment buoyancy flux $\overline{w'b'}_e$ is a fixed fraction, β_T, of the surface buoyancy flux $B_0 = -\overline{w'b'}_0$, viz., $\beta_T = 0.20$ (see, e.g., Tennekes, 1973; this result is sometimes known as Turner's rule). With the parameterization forms above, it is achieved for

$$v_t^2(d) = \sqrt{\frac{\beta_T}{c_t \epsilon}} \left(\frac{w_s d\, N(d)}{k Ri_c} \right), \qquad (17)$$

where $c_t = 5.8$ is a constant that arises from the empirically determined stability function ϕ_s. Thus, from (16) and (17), h is larger, and hence the entrainment layer is thicker, where $N(h)$ is smaller.

We have used this model, together with a new parameterization of internal vertical mixing described below that joins smoothly onto the PBL mixing at $d = h$, to calculate a variety of solutions, on time scales from hours to decades. Among these is one coupled to a one-dimensional radiative-convective atmosphere for the Ocean Weather Station Papa site in the North Pacific Ocean (Figures 9 through 11). The quality of its annual-cycle simulation is very good by the standards of previous upper-ocean PBL models (Martin, 1985), and we have found that an important part of this success is the inclusion of a diurnal cycle in insolation and an idealized, periodic storm cycle in the surface winds. To assess the consequences for climate modeling of accurately representing the oceanic annual cycle, we can compare Figure 11 with another coupled solution whose "ocean" is no more than a constant-depth mixed layer; it shows differences in tropospheric temperatures of up to 5 K in the annual cycle.

We are now in the process of extending and calibrating this new parameterization using large-eddy simulation solutions for the PBL under a variety of stress and buoyancy forcing conditions (e.g., Wyngaard and Moeng, 1993; McWilliams et al., 1993). We are particularly concerned with representing the asymmetry of scalar diffusivities forced by either surface or entrainment fluxes under convective conditions (Wyngaard and Brost, 1984). This effect is absent from the present parameterization forms, but it may be quite important for the diverse oceanic combinations of

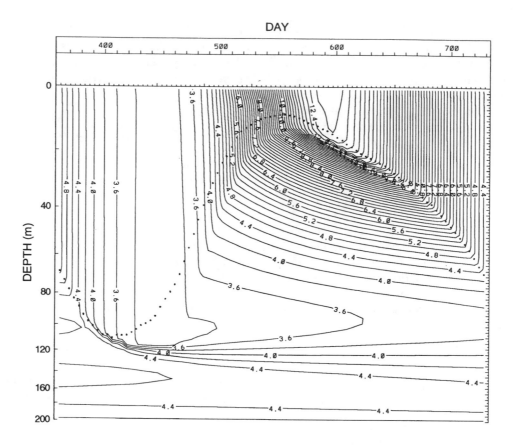

FIGURE 9 Four-day averages of oceanic $T(z,t)$ in the upper North Pacific Ocean for a one-dimensional model using the PBL and interior vertical-mixing parameterizations of Large et al. (1994) and the radiative-convective atmosphere of Briegleb (1991), together with a specified surface wind consisting of the mean annual cycle (Trenberth et al., 1990) and an idealized "storm" cycle with a period of 4.1 days; the wind enters only in the bulk formulae for surface fluxes. Shown here is the second year of a two-year integration. The contour interval is 0.2 K. The dots indicate the base of the PBL, $z = -h(t)$.

temperature and salinity stratification and surface forcing that occur.

INTERIOR VERTICAL MIXING

Under most circumstances, the vertical—more precisely, diapycnal—transports in the oceanic interior by small-scale motions are quite small. The associated eddy diffusivities are perhaps $\nu_v \sim 10^{-4}$ m^2 s^{-1} and $\kappa_v \sim 10^{-5}$ m^2 s^{-1}, and it is uncertain whether such small fluxes are important for climate variability on any except the longest time scales (millennia). However, under some circumstances these fluxes are locally very much larger and thus clearly of significance to the large-scale distributions of material properties.

The relevant small-scale processes are (1) shear and buoyancy (i.e., Kelvin-Helmholtz and convective) instabilities, which are regulated to some degree by the local, or gradient, Richardson number,

$$Ri_g = \frac{N^2}{(\overline{\nu}_z)^2}, \qquad (18)$$

(2) double diffusive overturning motions, which are regulated by the ratio of temperature and salinity contributions to the density gradient,

$$R_\rho = \frac{\alpha \overline{T}_z}{\beta \overline{S}_z} \qquad (19)$$

and (3) internal wave breaking, which is regulated by properties such as wave steepness or proximity to a critical surface that are not expressible directly in terms of the large-scale wave environment. For this latter reason, we prefer to distinguish (1) and (3) in principle, although some proposals have been made that (3) should also depend on Ri_g (e.g., Gargett and Holloway, 1984).

There is clear evidence from observations (see, e.g., Peters et al., 1988) that ν_v and κ_v are greatly enhanced for small and negative values of Ri_g. The extant measurements, however, have failed to collapse onto a universal curve, perhaps due to uncertainties in the determination of both the turbulent fluxes and the mean profiles. As yet this process has not been well modeled.

Conventional practice for parameterization of vertical

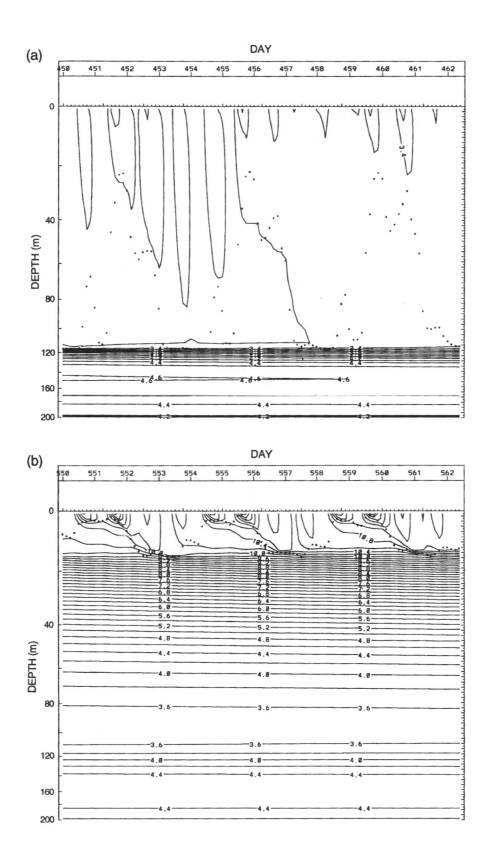

FIGURE 10   Instantaneous oceanic $T(z,t)$ from the same solution as in Figure 9 for 12-day intervals in late winter (with contour interval of 0.1 K) and middle summer (with a contour interval of 0.2 K).

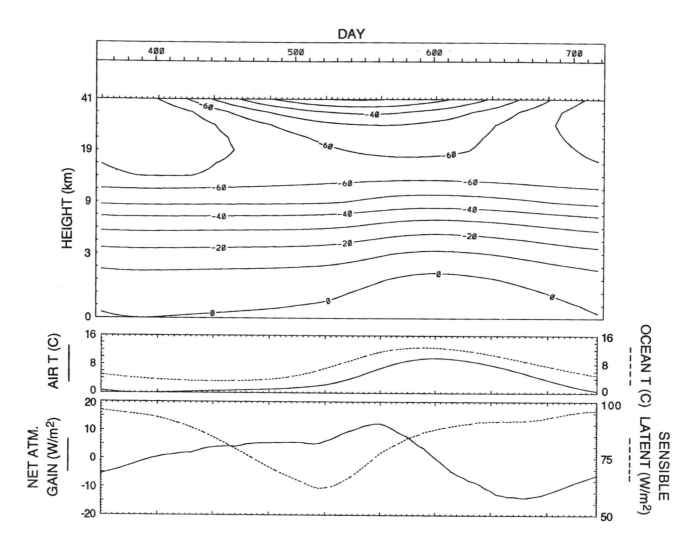

FIGURE 11 For the same solution as in Figure 9: atmospheric $T(z,t)$ with contour interval 20.0 K; surface $T(t)$ for the air (solid) and ocean (dashed); net atmospheric heat gain (solid) and net oceanic heat loss due to surface sensible and latent heat fluxes (dashed).

mixing is to use constant diffusivities with a large local enhancement in κ_v (and sometimes μ_v) where N^2 is very small or negative, sometimes even one as extreme as an instantaneous convective adjustment. However, there has been increasingly widespread use of $\nu_v, \kappa_v[Ri_g]$ functions, usually of the form proposed by Pacanowski and Philander (1981), which has the qualitatively correct monotonic dependence on Ri_g, but is not particularly close to the observed values.

We have developed a variant formulation that explicitly includes all the processes enumerated above and matches the observations more closely, in spite of their uncertainties (Large et al., 1994). This parameterization is used in the solutions shown in Figures 9 through 11; we believe it contributes substantially to their quality, particularly during rapid oscillations in the PBL depth (as in Figure 10, e.g.) where the sub-PBL region often has smallish Ri_g values.

SGS TOPOGRAPHY AND MARGINAL ZONES

The necessary roles played by dynamical processes operating near bottom topography and in the coastal regions and marginal seas include the following:

1. Topographic Form Stress. Form stress can arise due to non-uniform pressure forces on the variable topography at the lower boundary in the ocean, just as it does on an airplane wing (or, as above, on a deformable isopycnal surface). The zonal form stress is defined, assuming geostrophy, by

$$\tau_b = p \overline{\frac{\partial H}{\partial x}} = -\overline{fvH} , \qquad (20)$$

where $H(x,y)$ is the variation of the bottom elevation about its mean. This stress enters into a mean zonal-momentum balance at the lowest level in an ocean model.

It seems clear that topographic stresses often are of significance to the large-scale flow. This has been shown in idealized contexts by Bretherton and Haidvogel (1976), Holloway (1978, 1987), and Treguier (1989). It seems particularly relevant for the Antarctic Circumpolar Current (McWilliams et al., 1978; Treguier and McWilliams, 1990; Wolff et al., 1991), and it also is probably so for the continental slopes (Holloway, 1992). Unfortunately, it currently is not included as a SGS process in most large-scale ocean models.

2. Material Property Mixing. The boundary regions of the ocean, including the continental shelves and marginal seas, have especially high biological productivity, anomalous chemical sources because of river runoff and human pollution, and strong local mixing because of both strong tidal flows and shallow depths that permit the boundary layer to encompass the whole water column. These processes are not commonly part of the SGS parameterizations in large-scale ocean models.

3. Rapid Communication in the Boundary Wave Guide. The sides of ocean basins can support a rich variety of coastal waves (LeBlond and Mysak, 1978). In the simplest geometry, with vertical sides, these waves are Kelvin waves. They have an off-shore scale of the deformation radius (tens of km) and travel at the rapid speed of short gravity waves. They play an important dynamical role in communicating changes in the boundary-pressure distribution along the coastline and establishing along-shore currents of deformation-radius width. Since this scale lies in the SGS range for large-scale ocean models, there is a question of how to incorporate these effects adequately. Milliff and McWilliams (1994) have shown that the outcomes of these communication events by waves can at least sometimes be accurately represented by simple, integral consistency constraints on the large-scale fields.

SEA ICE

A sea-ice parameterization is a necessary element of an oceanic climate model. In its simplest, most often used form, it is merely a thermodynamic model for a temperature profile within a layer of ice of a certain thickness. The parameterization allows for storage of heat and water and alteration of the air-sea fluxes. Yet it is also important to model both sea ice's concentration, since the fraction of open water (or leads) makes an enormous difference to the air/sea heat and water fluxes, and its horizontal movements.

Modeling the concentration and movements requires the inclusion of mechanical dynamics as well.

In our own modeling studies, we are following the formulation of Hibler (1979), in part as it has been extended by Lemke et al. (1990) and by Flato and Hibler (1990).

BIOGEOCHEMICAL PROCESSES

Oceanic biochemistry is a necessary element of any oceanic climate model that addresses either the oceanic distributions of nutrients, oxygen, and so on, or the global cycle for CO_2, et al. I hesitate to declare a common current practice for ocean models, but guidance can be found in the recent study by Sarmiento et al. (1993).

Scott Doney, David Glover, and Raymond Najjar are developing a simple ecosystem model for the upper ocean that is based on the flow of nitrogen among organic and inorganic constituents. A solution for the annual cycle in the Sargasso Sea near Bermuda is shown in Figure 12. There are some attractive features: the timing of the spring phytoplankton bloom that follows the wintertime renewal of nutrients by deep PBL mixing and ends with their depletion in the well-mixed layer; the summertime subsurface maximum of phytoplankton that live on the border between nutrients in the seasonal pycnocline and penetrating solar radiation; and the somewhat deeper subsurface maximum in nutrients associated with the remineralization of sinking particles below the solar penetration.

ACKNOWLEDGMENTS

It seems increasingly clear to me that any serious climate modeling that goes beyond the initial conception of new possibilities requires the cooperation of many scientists. The scope and tasks are so large that pooled knowledge and labor seem nearly essential. In this spirit, I would like to thank my current partners in the work touched on in this paper: Bruce Briegleb, Gokhan Danabasoglu, Scott Doney, Peter Gent, Jeff Kiehl, Bill Large, Chin-Hoh Moeng, Jan Morzel, Ralph Milliff, Nancy Norton, Breck Owens, Mike Spall, Peter Sullivan, and John Wyngaard. In addition, I thank Kirk Bryan for his comments both during the workshop and in a review of this manuscript. The work is sponsored, under various contracts, by the National Science Foundation and the National Oceanic and Atmospheric Administration.

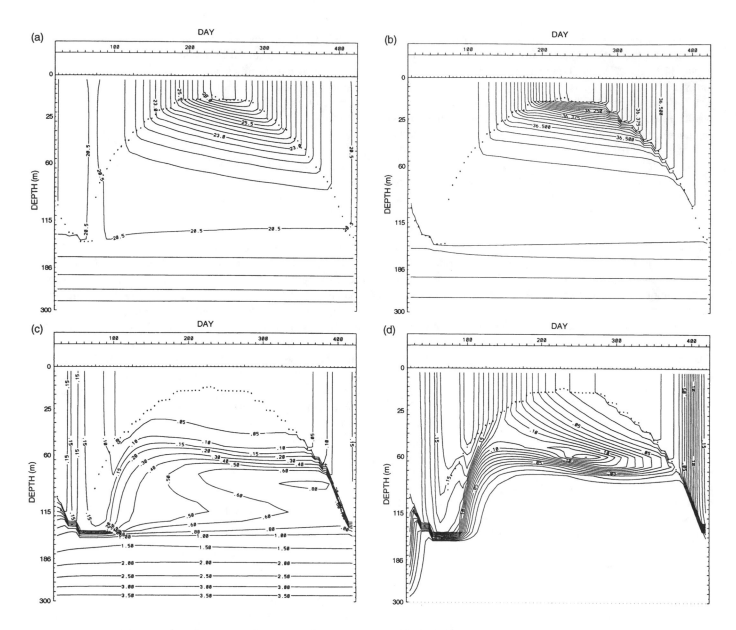

FIGURE 12   Depth and time dependence for (a) temperature [°C], (b) salinity [‰], (c) nutrients [mmol/m³ of equivalent nitrogen], and (d) phytoplankton [mmol/m³ of equivalent nitrogen] in a Sargasso Sea, annual-cycle solution with the physical dynamics of Large et al. (1994) and the ecosystem dynamics of Doney and Najjar. Again, the dots indicate the base of the PBL, where $z = -h(t)$.

Commentary on the Paper of McWilliams

KIRK BRYAN
NOAA/Geophysical Fluid Dynamics Laboratory

In reading this paper, I felt that it is very good news indeed that our field has reached the point at which somebody with Dr. McWilliams's qualifications in geophysical fluid dynamics is taking climate seriously enough to devote his full attention to improving ocean models for climate.

I have a couple of questions: First, does it make sense to design a very general ocean climate model rather than one for each problem? Second, can we represent these climatically important features such as overflows and deep western boundary currents—and this has come up in conversations I have had here with Drs. Dickson and Lazier—without going down to eddy-resolving resolution? Even if we are not greatly interested in mesoscale eddies, the boundary currents require very high resolution. I am a little more optimistic than Dr. McWilliams; I think that with massively parallel computers we will be able to resolve longer time scales and shorter spatial scales somewhat sooner than his prediction of 10 or 20 years. Admittedly, trying to run global calculations to something like equilibrium is a very stiff requirement.

Let me just comment on some of the topics that this paper addresses. One item is the parameterization of thickness mixing. I feel that this is a very important contribution. Another is lateral boundaries. I am a little skeptical as to whether a 100-km resolution is going to be adequate. Third, the approach to the planetary boundary layer seems like a good one.

One other point that Claes Rooth brought up earlier is that convection is not mentioned in the paper. Convection is often pointed out as a major weakness of these models. Is that taken care of by thickness mixing and mixing along isopycnals?

Discussion

MCWILLIAMS: The rationale for focusing on the upper ocean, at least for a certain range of time scales as we develop and try out new parameterizations, is the following. At present we are confronted with very serious climate drifts that appear at early times in most coupled calculations, as indicated by the large-amplitude flux corrections that are commonly used to control the equilibrium state in such models. These drifts seem to be associated with near-interface processes, in that the air-sea fluxes of momentum and latent and sensible heat are controlled by the planetary boundary layers. The atmospheric boundary layer interacts strongly with the stratus and cumulus clouds that in turn strongly regulate the precipitation. Since full-depth ocean models are very expensive to use and have ill-defined quasi-equilibrium states except on very long time scales, I think it's useful to give an upper-ocean domain a climatological lower boundary condition in order to be able to efficiently calculate evolution on shorter time scales, from months to decades. The resulting model should be capable of calculating reasonably accurately the variability of large-scale material properties, like T and S, in the upper ocean. In this domain we could use fairly high vertical-grid resolution, make calculations reasonably cheaply, look at coupled solutions with the atmosphere and sea ice, and try to sort out some of the coupled dynamics, including the sources of spurious climate drift.

We are taking the GFDL model as our starting point, and putting it in a form in which we can put the sigma coordinate at the bottom so that it can be a full ocean model. As we look at shorter variability time scales we don't expect that to be our primary mode. While we're trying to deal with complex general circulation questions, we certainly don't want to spend our energies on a proliferation of models. But I do see an interim utility in trying to isolate upper-ocean interactions with climate.

MYSAK: Away from boundaries we can get away with an upper-ocean model. But if interannual and even decadal variabilities are perturbations in the seasonal cycle, for which topography is very important for many of the flows, then such models will not describe real climate variability at high latitudes.

MCWILLIAMS: My sense is that except for essentially barotropic currents, which are of limited importance here, there isn't much penetration to depth. I agree that in the northern high latitudes topographic complexity is a real problem, but the full ocean models can't deal with that either. I don't see it as a prime obstacle to progress in, say, global coupling issues.

CANE: I'd like to comment on the need for one grand model versus many smaller ones. I find that a complicated model tends to distract you from things that might require attention, such as the mixed layer. I don't think there's a way of doing it all.

Second, I'm not sure that eddy resolution is the most important issue. For instance, topography might be important for the deep-circulation modes. But at present, with a 100-km grid, we can't get the width of the Gulf Stream right. That means we can't get its speed and transport right either, which we need for some of the air-sea interaction problems. There may be other approaches, aside from a full high-resolution model, we should be looking at.

MCWILLIAMS: Surface material-property distributions near the Gulf Stream tend to be a pretty broad envelope, since there are both advection by the narrow stream and buffeting by eddies. For climate you need to get the current's transport right, but not necessarily its narrowness and speed maximum.

CANE: I think you'd need to get the advective transport of warmer waters about right in order to get the fluxes right. I think it could be done without introducing tremendous resolution everywhere.

MCWILLIAMS: It's a burden on the sub-grid-scale parameterization. Progress will be made in both resolution and parameterizations, but I think we need to make what you might call the mesoscale calculations before we look at the details of, say, topography. Someone needs to examine topographic parameterization. In the end, of course, it might have to be done with local high-resolution grids for particularly complex regions.

BERGMAN: Is it possible to use variable grids to resolve features like the western boundary currents, yet still have a model that is not excessively complex overall?

BRYAN: Non-uniform grids are used extensively in engineering calculations. But you must then use a very implicit type of calculation, or the time step will be limited by the smallest grid size.

BERGMAN: Are semi-lagrangian approaches being tried for ocean GCMs as well as atmospheric GCMs?

MCWILLIAMS: Yes, some people are pursuing this idea; semi-lagrangian schemes that are sort of shape-preserving keep you from being embarrassed by things like negative concentrations. I myself feel that we can develop multi-grid, fully implicit problem-solvers that can be run at Courant-Friedrichs-Levy numbers of 10 or so, which is a considerable potential economy over semi-lagrangian schemes.

DICKSON: It seems to me that you need a comprehensive observing system to give you a long-period look at the ocean against which you could match a comprehensive climate model's results.

MCWILLIAMS: I quite agree. But please invite the modelers to join the design process!

A Stochastic Model of North Atlantic Climate Variability on Decade-to-Century Time Scales

KIRK BRYAN[1] AND FRANK C. HANSEN[2]

ABSTRACT

A conceptual model of North Atlantic climate variability is based on a simple two-box representation of the thermohaline circulation of the ocean. The model is linearized about a basic state, which corresponds approximately to the present ocean climate of the North Atlantic. Stochastic forcing, which represents the random effects of atmospheric cyclones and anticyclones passing over the ocean surface, drives the model away from its equilibrium state. The model transforms this stochastic forcing with equal power at all frequencies into a red-noise response in ocean temperature and salinity. At frequencies less than the thermohaline circulation's time scale, the solution is an equilibrium response and the amplitude of model ocean climate becomes independent of frequency.

Damping of salinity variations in the model is due to the thermohaline circulation. Ocean temperature variations are damped by both the thermohaline circulation and interaction with the atmosphere at the ocean surface.

The model illustrates how air-sea interaction involving the thermohaline circulation could produce a continuous spectrum without peaks. Stochastic forcing amplitudes corresponding to the climate of the last few thousand years produce a nearly linear response of the model. Large perturbations of the hydrological cycle typical of the close of the last ice age produce a chaotic response.

INTRODUCTION

The instrumental climate record now extends over a century. Although the record contains many gaps in both space and time, it is still possible to find very-large-scale climate variations that extend over several decades (Folland et al., 1986). These fluctuations are quite distinct from the climate variations related to the El Niño phenomenon. The El Niño fluctuations have a time scale of three to four years and tend to have their greatest amplitude in the equatorial Pacific. The El Niño climate variations are associated with changes in tropical atmospheric convection, and cause changes in air temperature extending up to the base of the

[1]NOAA Geophysical Fluid Dynamics Laboratory, Princeton University, Princeton, New Jersey
[2]Atmosphere and Ocean Sciences Program, Princeton University, Princeton, New Jersey

stratosphere. On the other hand, the very-low-frequency climate variations tend to be amplified in polar latitudes and have their greatest amplitude near the earth's surface.

The most unambiguous evidence for low-frequency climate variations with a period of several decades comes from surface temperature records over land. Figure 1 shows the historical temperature record averaged by latitude belts (Hansen and Lebedeff, 1987). Several features stand out in Figure 1. First, there is a clear upward trend in temperature over the past century, which is greatest in the higher latitudes of the Northern Hemisphere. Second, very-low-frequency

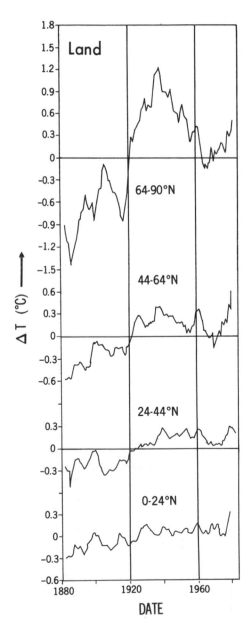

FIGURE 1 Zonally averaged surface temperatures over land, compiled by Hansen and Lebedeff (1987). Note the low-frequency variability superimposed on the upward trend over the past century.

variations are superimposed on that trend. The most obvious features are pronounced minima at the beginning of the century and in the late 1960s and early 1970s. In between these two relatively cold events there was a period of relatively rapid warming in the 1920s. The most recent part of the record is not shown in Figure 1, but measurements indicate that Northern Hemisphere temperatures are rising rapidly in the 1980s, as they did in the 1920s (IPCC, 1990).

Bjerknes (1964) studied the pre-World War II climatic record, particularly the sea surface temperature archives of the British Meteorological Office. He was impressed by the relatively rapid rise of Northern Hemisphere temperature in the 1920s, which followed an anomalously cold period at the beginning of the century. In his analysis of the sea surface temperature and the surface atmospheric pressure fields at midlatitudes he found a distinct difference between the fluctuations that had a time scale of seasons and years and the climatic fluctuations, which were on decadal time scales. Bjerknes concluded that the atmosphere played the dominant role in those fluctuations with a yearly time scale, while the ocean played the dominant role in the climatic fluctuations. He was aware of the heat-balance calculations carried out by Sverdrup (1957) and realized that the North Atlantic is much more important than the North Pacific in the poleward transport of heat in higher latitudes. He reasoned that fluctuations in the North Atlantic poleward heat transport could be the cause of decadal-scale Northern Hemisphere climate variations. Bjerknes' hypothesis received relatively little attention in 1964. Since so little was known about ocean circulation at the time, his ideas probably seemed to be difficult to check in any way. On the other hand, his explanation of the onset of the Southern Oscillation (Bjerknes, 1969) in terms of the Walker circulation was widely accepted within a few years after publication.

A few years previous to the publication of Bjerknes' study of the Atlantic, Stommel (1961) developed a simple model of the thermohaline circulation. He pointed out that air-sea interaction has a very different effect on the temperature field of the ocean than on the salinity field. The radiation balance of the atmosphere directly responds to the sea surface temperature field, while a similar feedback cannot exist for salinity. When this asymmetry was taken into account, Stommel demonstrated, his simple model of the thermohaline circulation could take on two stable solutions for the same external boundary conditions. In one solution the thermohaline solution was dominated by the thermal component of the density gradient, and in the other solution the salinity component of the density gradient was most important. In recent years it has been possible to test the concepts of Stommel's two-box thermohaline model in more complete two- and three-dimensional models (see Weaver and Hughes, 1992, for a review). Through these recent numerical studies the fundamental importance of Stommel's 1961 paper has become more widely recognized.

Hasselmann (1976) demonstrated in a simple way how the heat-storage capacity of the upper ocean acts to integrate random heat impulses from the atmosphere. This integrating property of the upper ocean greatly amplifies the response of sea surface temperature to low-frequency heating inputs. In Hasselmann's stochastic model the excursions of sea surface temperature about equilibrium are taken to be the result of a random-walk process, which is limited in amplitude by atmospheric feedback. In Hasselmann's model, air-sea interaction is purely local within the ocean. However, this elegantly simple model runs into difficulties when an attempt is made to generalize it to include salinity as well as temperature. For a random-walk process on the temperature-salinity plane driven by stochastic forcing there is no physically plausible mechanism to limit extreme anomalies of vertically averaged salinity, as there is for temperature. This difficulty is removed only by allowing for a nonlocal process that permits exchange with other parts of the world ocean. A stable regime in Stommel's two-box model of the thermohaline circulation driven by stochastic forcing represents the simplest extension of Hasselmann's ocean climate model, which can include salinity as well as temperature.

The motivation for attempting to construct a simple toy model of this kind is the success of much more complex numerical models that illustrate oscillations of thermohaline circulation. Examples are studies by Weaver and Sarachik (1991b), Mikolajewicz and Maier-Reimer (1990), and Delworth et al. (1995, in this volume). The calculations by Mikolajewicz and Maier-Reimer were carried out for a numerical model of the world ocean. Surface temperature is damped toward observed values, but stochastic forcing is used to simulate observed variations of the net water flux at the surface. It was found that stable oscillations of the Atlantic thermohaline circulation took place with a maximum amplitude at a period of several centuries. Delworth et al. studied a fully coupled ocean-atmosphere model. Their results illustrate climate fluctuations consistent with Bjerknes' (1964) hypothesis for Atlantic climate variability. Variations in the strength of the thermohaline circulation are correlated with decadal-scale sea surface temperature anomalies that seem to be quite realistic when compared with analyses of the Comprehensive Ocean-Atmosphere Data Set (COADS) by Kushnir (1994). The aim of the present study is to construct a simpler framework for understanding the important results of these physically complete but highly complex models.

HASSELMANN'S MODEL

Since Hasselmann's (1976) stochastic model of climate is an important point of departure for the present study, we will review it briefly. Consider a reservoir of upper ocean water as shown in Figure 2a. Let T' be the departure of the temperature from its climatological average. The governing equation is

$$\frac{d}{dt} T' = -\frac{\lambda}{D} T' + Q , \qquad (1)$$

where D is the depth of the reservoir, and Q is the average temperature change in the reservoir due to heating, which is associated with the random fluctuations of cyclones and anticyclones at the ocean surface. Temperature fluctuations are damped by the term, $-\lambda T'/D$, on the right-hand side of (1). This negative feedback term represents the combined effects of long-wave radiation, evaporative cooling, and sensible heating, all of which are closely related to ocean surface temperature. Let

$$Q(t) = \sum_{\omega} Q_{\omega} e^{i\omega t} . \qquad (2)$$

Hasselmann (1976) assumes that the effect of "weather" over the ocean can be represented as white noise, where $|Q_{\omega}|$ is thus uniform at all frequencies. The power spectrum of the solution of (1) may be written as

$$|T_{\omega}|^2 = \frac{|Q_{\omega}|^2}{\omega^2 + \lambda^2/D^2} . \qquad (3)$$

The spectrum given by (3) has two very different regimes. In the high-frequency regime, where $\omega \gg \lambda/D$, "white noise" forcing gives rise to a "red noise" response. This can be understood physically as the effect of the memory of the ocean. Hasselmann (1976) describes it as a random-walk process in one dimension. Excursions from the origin become longer and longer over greater and greater time scales.

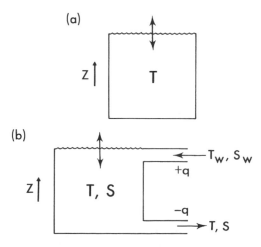

FIGURE 2 (a) A schematic diagram of Hasselmann's (1976) stochastic climate model. (b) A generalized model, such as that of Stommel (1961), that includes salinity and communication to other regions of the ocean.

The negative-feedback term provides a restraint to the random-walk process at very low frequencies, $\omega \ll \lambda/D$. At this point, $|T_\omega|$ becomes simply proportional to $|Q_\omega|$, which has a uniform value at all frequencies. In this second regime, forcing and negative feedback are in exact balance. We will designate this low-frequency part of the spectrum as an equilibrium response regime.

THE EFFECTS OF SALINITY

Suppose we consider a more general case, in which the same reservoir model has departures of both temperature and salinity from equilibrium. Random variations of excess evaporation over precipitation will be associated with the same weather events responsible for stochastic heating in the original Hasselmann model. Instead of a one-dimensional random-walk process, we can think of a two-dimensional random walk in the T-S plane. A true random walk is physically unlikely, because heating and evaporation minus precipitation tend to be negatively correlated.

There is one insurmountable difficulty to a purely local model of this type. No feedback mechanism can be invoked that will damp excursions in salinity in the same way that temperature excursions are damped by interaction with the atmosphere. A random walk process on the T-S plane would thus be limited in the T-direction, but could attain extreme states of very high or very low salinity. The Great Salt Lake and the Great Lakes are examples of isolated water bodies with extremely high and low salinity values, respectively. Thus we are forced to abandon the elegantly simple model of Hasselmann and include nonlocal effects. Figure 2b shows how a one-box model can be generalized to include communication with the world ocean, thus avoiding an infrared catastrophe in salinity.

STOMMEL'S MODEL

Fortunately, the equations governing the nonlocal model shown in Figure 2b are almost the same as those for Stommel's (1961) classical two-box model of the thermohaline circulation. Stommel's model has recently been revisited by Huang et al. (1992). More complex versions of the model have also been investigated by Marotzke (1990) and Birchfield et al. (1990). The most important result of Stommel's (1961) original study was the demonstration that the thermohaline circulation can have multiple equilibrium states in response to only small changes in surface forcing. This fundamental idea has been the basis for many recent studies with much more elaborate models (Bryan, 1986b; Manabe and Stouffer, 1988).

Let V and D be the volume and depth of the reservoir, which represents the subarctic gyre of the North Atlantic. Let T and S be the average temperature and salinity of the

subarctic gyre, while T_ω and S_ω are the fixed temperature and salinity of the rest of the World Ocean. Let

$$\delta T^* = T_\omega - T^* \quad , \tag{4}$$

where T^* is a fixed reference temperature for the subarctic box. The equations for T and S are then

$$\frac{d}{dt}(T_\omega - T) = -(|q|/V)(T_\omega - T) - \lambda(T_\omega - T - \delta T^*)/D \tag{5}$$

$$\frac{d}{dt}(S_\omega - S) = -(|q|/V)(S_\omega - S) + \delta(E - P)S_0/D \quad . \tag{6}$$

Here q is the exchange of water between the subarctic gyre and the remainder of the ocean, as indicated in Figure 2. The effect of the first term on the right-hand side of both (5) and (6) is to diminish the contrast between the world ocean and the subarctic box. A key time scale of the problem is given by V/q. For the present circulation of the North Atlantic, this overturning time scale is about 25 years. The second term on the right-hand side of (5) is the heating term, which forces the temperature of the subarctic box to depart from T_ω, the temperature of the world ocean. Air-sea interaction forces T toward the reference value of $T_\omega - \delta T^*$ on a damping time scale given by D/λ. The second term on the right-hand side of (6) represents the effect of north-south contrasts in evaporation and precipitation that force the salinity of the subarctic box to depart from the salinity of the world ocean. Note that (5) includes a feedback due to air-sea interaction, while (6) has no corresponding term. In our toy model, the forcing terms on the right-hand side of (5) and (6) will have both a steady component and a stochastic component.

Stommel (1961) specified that the transport, q, of the thermohaline circulation should be proportional to the density difference between the reservoirs, such that

$$q = \mu \left[\alpha(T_\omega - T) - \beta(S_\omega - S)\right] \tag{7}$$

where α and β are expansion coefficients for temperature and salinity, respectively, and μ is a constant of proportionality relating transport to the density difference. For the North Atlantic thermohaline circulation, an appropriate value of μ would be approximately 10 sverdrups (10^7 m^3 s^{-1}) per sigma unit of density (kg m^{-3}) (Roemmich and Wunsch, 1985).

To gain some insight on the sensitivity of the model to specified parameters, a rescaling is useful. Let $\overline{\delta T^*}$ be the time-averaged value of δT^*; then

$$\hat{T} = \frac{(T_\omega - T)}{\overline{\delta T^*}} \tag{8}$$

$$\hat{S} = \frac{\beta(S_\omega - S)}{\alpha \overline{\delta T^*}} \tag{9}$$

$$\hat{t} = \frac{\mu \alpha \overline{\delta T^*} t}{V} \tag{10}$$

$$\hat{Q} = \frac{\delta T^*}{\overline{\delta T}} \qquad (11)$$

$$\hat{E} = \frac{\beta \delta(E - P)S_0 V}{D_\mu (\alpha \overline{\delta T^*})^2} \qquad (12)$$

$$\eta = \frac{\lambda V}{\mu \alpha D \overline{\delta T^*}} \qquad . \qquad (13)$$

Huang et al. (1992) chose to scale the Stommel 2×1 box model by the damping time scale, D/λ. We have chosen to use the circulation time scale, V/q. This introduces a significant simplification, which we will point out later. It also has the advantage that the circulation time scale can be estimated from measurements of overturning at 24°N (Roemmich and Wunsch, 1985), while the damping time scale due to air-sea interaction is more difficult to estimate. Temperature is scaled by the difference between the reference temperatures of the subarctic box and the world ocean. The density difference due to salinity is scaled by the density difference associated with $\overline{\delta T^*}$. $\hat{Q}$ is defined so that it is unity plus a stochastic component that averages to zero with respect to time. $\hat{E}$ is scaled by the circulation time and a factor to remove the salinity dimension. η is the circulation time scale divided by the atmospheric thermal damping time scale. An appropriate value for this parameter is discussed below. Substituting equations (7)-(13) into (4)-(6), we obtain

$$\partial_t \hat{T} = -(|\hat{T} - \hat{S}| + \eta)\hat{T} + \eta \hat{Q} \qquad (14)$$

$$\partial_t \hat{S} = -|\hat{T} - \hat{S}|\hat{S} + \hat{E} \qquad . \qquad (15)$$

Steady-state solutions of a model very much like (14) and (15) were originally explored by Stommel (1961). He found a maximum of three possible steady-state solutions. In one of these solutions, $\hat{T}$ is greater than $\hat{S}$, indicating that the density gradient between the boxes is dominated by temperature. Stommel identified this solution with present climate. Another stable solution existed for $\hat{S}$ greater than $\hat{T}$, which corresponds to the case of a reversed and weak thermohaline solution dominated by salinity. These solutions are also discussed in detail by Huang et al. (1992), who also explore more elaborate models that include several degrees of freedom in the vertical direction. Eq. (15) is the non-diffusive form of a model described by Spall (1992), although, unlike our model, Spall's includes a mixed layer.

In (14) and (15), the derivatives on the left-hand side are taken with respect to nondimensional time. For convenience, the steady-state values of $\hat{T}$ and $\hat{S}$ will henceforth be referred to as T and S.

STEADY SOLUTIONS CORRESPONDING TO NORTH ATLANTIC CLIMATE

The motivation for this study is the question of whether small oscillations of the thermohaline circulation could be responsible for decadal climate variations over the North Atlantic, as was proposed by Bjerknes (1964). For this reason, we wish to make a detailed study of a linearized version of Stommel's model in the vicinity of a steady-state solution that approximates the present circulation of the North Atlantic. It is important for this purpose to examine the physical interpretation and plausible range of the model parameters. Table 1, in which V corresponds to the volume of the Atlantic from the surface to the bottom between 52° and 75°N, presents their values. We see from Table 1 that the basic time scale is about a few decades for a thermohaline circulation of 13.3×10^6 m^3 s^{-1} and a volume corresponding to the subarctic gyre of the North Atlantic. A plausible damping time scale is more difficult to estimate. Using a value of λ given by Haney (1971), we get a damping time scale of only a decade. Using a smaller value of λ more appropriate for a large geographical area and a depth on the order of the basin depth (Schopf, 1985), we obtain a damping time scale of nearly a century. Corresponding values of η range from 0.25 to 2.5. The depth over which feedback to the air is relevant for fluctuations on a decade-to-century time scale is the depth of the basin. The fact that the time scales of thermal damping and of the thermohaline circulation are nearly the same is a key feature of the physics of our model.

If we consider the steady-state version of (14) and (15), the curves corresponding to (14) are those shown as solid lines in Figure 3. The curves corresponding to (15) are those shown as dashed lines. The manifold of solid curves corresponds to different values of the damping coefficient η. The different dashed lines correspond to different values of the salinity forcing parameter, E. Solutions correspond to intersection points of these two curves. Note that at most two intersection points exist for T > S, which corresponds to the present state of the North Atlantic, where the thermal component of the density gradient dominates the salinity component.

The thermal equation (14) gives a family of nearly straight lines for different values of η. As η increases, the lines become more nearly horizontal and are asymptotic to T = 1 as η goes to infinity. On the other hand, the salinity

TABLE 1   Plausible Values of Model Parameters

Parameter	Symbol	Value
Depth	D	4×10^3 m
Volume	V	1×10^{16} m^3
Transport	$\mu \alpha \overline{\delta T^*}$	1.33×10^7 m^3 s^{-1}
Feedback Parameter	λ	$0.133 - 1.33 \times 10^{-5}$ m s^{-1}
Circulation Time Scale	$V/(\mu \alpha \overline{\delta T^*})$	7.5×10^8 sec ($\approx$24 yr)
Damping Time Scale	D/λ	$0.3 - 3.0 \times 10^9$ sec
Damping Coefficient	$\lambda V/(\mu \alpha \overline{\delta T^*}D)$	$0.25 - 2.5$

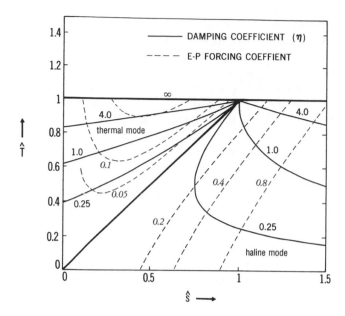

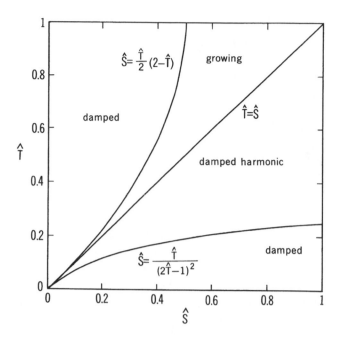

FIGURE 3   Steady-state solutions of the model are given by the intersection points of the solid lines (bold-faced numbers), which correspond to the different values of the damping parameter, and the dashed lines (italic numbers), which correspond to different values of E, the nondimensional evaporation-minus-precipitation forcing. The diagonal line separates the thermal regime, in which temperature differences dominate the poleward density gradients, from the haline, in which salinity dominates.

FIGURE 4   A map of the T-S plane showing the results of a linear-stability analysis of steady-state solutions of the model. Unstable solutions are only found in the region T(2-T)/2 < S < T. In the remaining regions, solutions are either simply damped or harmonically damped.

equation (15) gives a family of curves that penetrate further below T = 1 for weaker haline forcing.

STOCHASTIC FORCING OF THE THERMALLY DOMINATED REGIME

In this section we will consider the results obtained by stochastic forcing of a version of the Stommel model with no restoring of the salinity field. The stability of the Stommel two-box model has been discussed by previous authors (Huang et al., 1992; Marotzke, 1990; Stommel, 1961; Walin, 1985). The results of the linear analysis of our model are shown in Figure 4. Damped harmonic solutions exist for $T < S < T/(2T - 1)^2$. Unstable real roots exist in the region bounded by $S = T(2 - T)/2$ and $T = S$. The remaining areas have simple damped solutions. In the North Atlantic, geologic evidence suggests that a thermally dominated thermohaline regime has existed since the close of the last ice age, 10,000 years ago. Therefore, it appears that the stable regime in which $S < T(2 - T)/2$ in our model corresponds to the present climate of the North Atlantic. The remainder of this paper is concerned with the analysis of the response to forced oscillation about the stable equilibrium in the thermally dominated regime of the model.

To check our calculations we calculated the linear response analytically and compared it to the results of direct numerical integration of the full nonlinear model with very small stochastic forcing. It is appealing to think of the effect of air-sea fluxes associated with cyclones and anticyclones passing over the ocean as causing a random walk in vertically integrated water-mass properties. Results from the GFDL coupled model show that heating and net evaporation-minus-precipitation at the ocean surface are negatively correlated. The results of Delworth et al. (1995) show that surface fluxes tend to heat and simultaneously freshen the ocean surface or, conversely, cool the ocean surface and make it more saline. Rather than making heating and evaporation-minus-precipitation independent random variables, we made them proportional to one another, with opposite sign, in our stochastic model.

Analytic spectra for the three cases are shown in Figure 5. Figure 5a corresponds to the case in which $\eta = 2$ and the steady component of $E = 0.1$. Q' is 10 times larger than E' and of opposite sign, causing the temperature fluctuations to be larger than the salinity fluctuations at all frequencies. At high frequencies, both spectra have a slope of ω^{-2}. Damping becomes important for salinity only at frequencies less than 0.1 cycles/unit time, which corresponds to a period of 240 years. Figure 5b shows a case that is the same as that of Figure 5a, except that $\eta Q'$ is now twice $-E'$. $|T^2|$ is exactly four times longer than $|S^2|$ at high frequencies, but a crossover point is reached at a period between 50 and 100 years. The effect of removing the thermohaline coupling term is shown in Figure 5c. This case corresponds to the

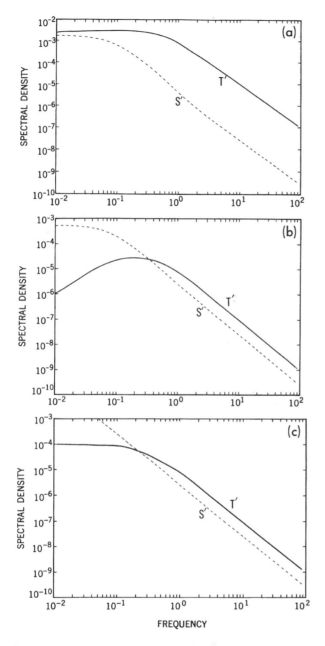

FIGURE 5 The spectral density of density variations due to temperature (solid line) and salinity (dashed line) as a function of frequency in cycles per unit of nondimensional time. Eta, the thermal damping parameter, is equal to 2, and the steady state component of E equals 0.1. (a) Random heating is 10 times the salinity forcing. (b) The same as (a), but the heating is equal to the salinity forcing. (c) The same as (b), but the thermohaline circulation is eliminated, corresponding to the Hasselmann (1976) model.

Hasselmann (1976) model. Temperature is damped by interaction with the atmosphere, but salinity increases without bounds as frequency decreases. This is the infrared catastrophe for salinity noted earlier.

The phase relations predicted by the model are shown

in Table 2 for periods corresponding to 48, 96, and 240 years, assuming that one unit of nondimensional time is equivalent to 24 years. Amplitude and phase relationships are for $E = 0.1$ and $\eta = 2.0$, and different ratios of E' to Q'. θ_T is the phase of T with respect to the thermohaline circulation. θ_{-S} is the phase of $-S$ with respect to the thermohaline circulation. (A) corresponds to a clockwise orbit on the T-S plane, while (B) and (C) correspond to counterclockwise orbits. Of particular interest are the phase relationships between the density anomalies, which are a function of temperature and salinity, in both the subarctic gyre and the thermohaline circulation. Delworth et al. (1995) find that in their three-dimensional coupled model, density anomalies due to temperature lead the thermohaline circulation, and density anomalies due to salinity tend to lag slightly. Since $\hat{T}$ is proportional to $T_\omega - T$, it is also proportional to the thermal density component in the polar box. On the other hand, the haline density component is proportional to $-S$.

The first entry in the table corresponds to a positive correlation between E' and Q' perturbations, which have equal magnitude. The model shows that the response in temperature and salinity are nearly equal, and the $-S$ component leads the thermohaline circulation, while the T component lags behind. Delworth et al. (1995) find just the opposite phase relationship: Thermal density perturbations tend to lead the thermohaline circulation. Their calculations also show that perturbations of heating and E − P are negatively correlated over the subarctic gyre. It thus seems more appropriate that $E' = -Q'$.

Entry (B) in Table 1 shows this case, and we see that the phase of the thermal-density component now leads the thermohaline circulation, and the phase of the haline component lags behind in reasonable agreement with the more detailed model of Delworth et al. (1995). Case (C) is similar to (B), except that heating perturbations are reduced in

TABLE 2 Model-Predicted Phase Relations

Frequency (cycles/ unit time)	0.10	0.25	0.50
Period (years)	240	96	48
(A) $E' = Q'$			
$|T|/|S|$	0.7	1.0	1.5
θ_T	−188°	−69°	−38°
θ_{-S}	36°	76°	112°
(B) $E' = -Q'$			
$|T|/|S|$	0.3	0.8	1.3
θ_T	−58°	−37°	−21°
θ_{-S}	−17°	−28°	−28°
(C) $E' = -2Q'$			
$|T|/|S|$	0.2	0.4	0.7
θ_T	96°	57°	33°
θ_{-S}	−11°	−20°	−21°

amplitude by a factor of 2. As a result, the negative amplitude of $|T|$ is smaller relative to $|S|$, but the phase relationships are qualitatively the same.

CONCLUSIONS

The aim of this paper is to develop a very simple model of Atlantic climate variability on decadal and longer time scales by combining elements of existing models of Stommel (1961) and Hasselmann (1976). In this model random forcing at the ocean surface drives the water mass distribution of the subarctic gyre of the North Atlantic away from its climatologically balanced and stable steady state. A random forcing by unstable cyclones and anticyclones passing over the ocean would produce a uniform forcing of both high and low frequencies. Feedback through air-sea interaction is assumed to be purely negative, restoring temperature to its equilibrium state, but with no corresponding feedback for salinity.

As shown by Hasselmann (1976), the storage capacity of the ocean greatly amplifies the response of the ocean to forcing at very low frequencies. Extremely large responses at very low frequencies are tempered by thermal damping and by the effects of the thermohaline circulation that responds to density gradients that build up between the subarctic gyre and the ocean at lower latitudes. The combined effect of these two mechanisms is to flatten the response spectra for salinity and temperature at low frequencies. Details of the forcing and thermal damping determine the exact amplitude and phase of the temperature and salinity fluctuations.

Although the response of the simple toy model of this study is devoid of the sharp spectral peaks that have been found in more complex models, we can reproduce the gross features of the results of those complex models. The response of the model to white-noise forcing is simply red noise at frequencies less than the basic time scale of the thermohaline circulation, and an equilibrium response at frequencies much less than that of the thermohaline circulation. The response of the model to small perturbations is always stable in the thermally dominated regime, corresponding to present-day climate. Since there is no possibility of resolving the upper thermocline, instabilities that turn on or shut off convection are excluded. The thermohaline circulation always acts to restore the ocean climate to its equilibrium state.

The simplicity of the model allows some interesting insights on the phase relationships between the thermohaline circulation and temperature and salinity fluctuations. If fluctuations of heating and $E - P$ are negatively correlated, as suggested by the coupled-model results of Delworth et al. (1995), the toy model predicts that density perturbations due to temperature in the subarctic box will lead the thermohaline circulation, while density perturbations corresponding to salinity will lag behind. This result corresponds to the phase relationships found in the more detailed three-dimensional model of Delworth et al. (1995). On the other hand, very different phase relationships are found in the toy model when heating and $E - P$ are taken to be positively correlated.

The realistic phase behavior of the model may be described in terms of counterclockwise orbits in the $T - S$ plane, caused by the fact that temperature is damped much more than salinity due to the air-sea interaction. Counterclockwise orbits are required by a negative correlation between stochastic heating and stochastic evaporation-minus-precipitation.

It is assumed that forcing is sufficiently small that the system oscillates in an essentially linear fashion about its equilibrium state. This would seem to be appropriate for the climate that has existed in the North Atlantic since the last ice age, and is consistent with the simulations of Delworth et al. (1995). At the close of the last ice age the hydrological cycle of the North Atlantic was strongly perturbed by the melting of large ice sheets. In that case much larger excursions from equilibrium would be expected, and it is not clear whether useful insights could be obtained with such a simple model as we have described in this study.

ACKNOWLEDGMENTS

The authors would like to thank Tom Delworth, Syukuro Manabe, and Edward Sarachik for generously sharing their ideas and results with us. This research was supported in part by funding from the Atlantic Climate Change Program of the NOAA Office of Climate and Global Change.

Discussion

ROOTH: I think that was a very nice introduction to the whole business of thermohaline influences. I'd just like to make a couple of cautionary comments. First, we need to be very careful about extending our experience with simple models to more complex cases. The strong constraints on the climate system mean that it's easy to come up with simple, first-order explanations, but you can't quantitatively improve on those by adding a little complexity here and another there. You have to add a lot of stuff.

Another thing is that you need to keep a number of limitations in mind. For instance, Kirk's model has a well-mixed domain, when actually features like the stratification within the sub-Arctic gyre are very important. You automatically exclude the possibility of high-latitude haloclines. Then there's the question of whether you can treat the world ocean as an infinite, unresponsive basin. With this model we are simply seeing a random-walk process that creates a variability in the intensity of the effect of the overturning on the two competing influences. This loop has a damping effect on the anomalies. Also, only temperature is damped by the surface layer.

BRYAN: I should perhaps mention that originally, when I was thinking about Atlantic climate variability, I viewed Hasselmann's model as what you might call a default model: worth checking observations against to see whether going to a more complex model would be justified.

TALLEY: Going back to your observations: When you say "salinity-dominated regime," do you mean one that can't be overturned no matter how cold you make it? Isn't there a situation in the North Pacific like that right now that we could compare with the Atlantic?

BRYAN: I have thought a little bit about the application of this stochastic model to the North Pacific. As I understand it, though, Bryden's measurements suggest that the poleward transport there is accomplished by horizontal gyres rather than the thermohaline circulation.

MYSAK: Have you thought of complicating the model by allowing for stratification in each layer?

BRYAN: It's very tempting, but the whole virtue of the mechanism we're suggesting lies in the elegant simplicity of Stommel's model.

LEHMAN: I think if we view this box as a representation of the Atlantic, it would not be unreasonable to compare its implications with the records in the Greenland ice sheet. We now have four cores that show a strongly bimodal behavior that could be likened to the thermal and saline modes in your model. The thermal mode would be the Holocene, and the saline would correspond to events recorded in the glacier that are characterized by high-frequency, high-amplitude changes in temperature or O_2. These periods, which last 2000 to 4000 years, are marked by very sudden changes of 5° to 10°C within 50 to 100 years.

BRYAN: Actually, this simple model is not very applicable to the saline-dominated ice-age mode. When salinity dominates the density distribution, you must have a model that includes vertical stratification.

TALLEY: Isn't there an implicit salinity feedback in that model?

BRYAN: The model doesn't really include any instabilities. It can't handle the Great Salinity Anomaly, for instance. I think the simplest model that could do that might be a 2×20 or 2×30 box model, which would permit you to have a rather small vertical diffusivity. Tom Stocker has gotten amazingly good results mimicking 3-D models with 2-D models, but I think even his model could be simplified by having two degrees of freedom in the north-south direction and an infinite number in the vertical.

CESSI: Does the fact that there are two equilibria in the regime you were showing explain the oscillation?

BRYAN: The ice-cap evidence suggests that over the past 2000 or 4000 years there has not been a climate 'flip' like what you find during the ice ages.

MARTINSON: Kirk, you mentioned that your model did strange things because it had temperature feedback but no salinity feedback. There's a salinity feedback through the sea-ice field in high-latitude regions that affects the thermohaline circulation; as the ice starts to form and you lose the heat you start to salinate the water. Local process models are actually of critical importance in understanding sea-ice distribution, as well as upper-ocean stability and stratification around the Antarctic region. I realize you can't put that process in, since your model has no stratification, but that salt-temperature coupling might be something to look at further.

MYSAK: The same thing is true for the salinity feedback into the atmosphere. It's also a way of getting in the longer-time-scale effects—decadal and longer—in a natural way. Just as with El Niño, rapid atmosphere-ocean interactions have longer-time-scale effects.

BRYAN: There's no doubt that ice formation could be important, but it redistributes the salinity only within the column. There's no horizontal transport, so you still have the runaway effect I mentioned, which would allow the salinity of an isolated reservoir to go to extreme values.

ROOTH: You may be talking about only 5 percent of the model's temperature range, but you're applying it at a critical point. Some years ago, a student of mine named Bill Peterson looked at the relative penetration of two competing convective plumes in a container as a function of their relative buoyancy-source strengths. It turns out that if the basin is sufficiently deep you get extremely high sensitivity to small anomalies. If the two plumes are nearly in balance and you kick the buoyancy source slightly, the weaker one will terminate in a relatively shallow range while the stronger will go all the way to the bottom. You can generalize this to stochastic perturbations and get pronounced bistability. This toy-model explanation also has something to say about why you found a more vigorous circulation when there was an asymmetry in the forcing.

GHIL: We've been through this exercise before with energy-balance models, trying to force them from one equilibrium into another. It turns out that for the levels of stochastic forcing available it takes just short of the age of the universe to kick them over. I think it might be time to consider going from Hasselmann's 1976 first-order stochastic differential equation to a second-order one like Steve Koonin's—that is, from a passive, stable response to stochastic forcing (Hasselmann) to oscillatory response (Koonin). We might even get an answer to Tony Socci's question about why a small forcing may have more effect than a large one, which we can't address with models lacking internal variability, like Hasselmann's.

BRYAN: Well, I think that great amplification of a very small forcing at low frequencies is simply due to the fact that the ocean has most of the heat capacity of the climate system. It would be interesting to know whether that atmosphere always tends to damp climate back to equilibrium. The research on fluxes that Dan Cayan and others are doing could be used to investigate whether Hasselmann's idea makes sense.

Decadal-to-Millennial Internal Oceanic Variability in Coarse-Resolution Ocean General-Circulation Models

ANDREW J. WEAVER[1]

ABSTRACT

The ocean's thermohaline circulation, driven by fluxes of fresh water and heat through the ocean's surface, is an important mechanism for the transport of heat from low to high latitudes. Changes in the intensity of the thermohaline circulation, and hence its poleward heat transport, would have a significant effect on global climate.

Here, the results of a number of experiments conducted using a coarse-resolution ocean general-circulation model (OGCM) in idealized basins are reviewed. They illustrate the importance of fresh-water flux, thermal, and wind forcing in exciting decadal-to-millennial variability of the thermohaline circulation. A brief discussion of the shortcomings of these models and some suggestions for future research are also presented.

Recent experiments for a coarse OGCM simulation of the North Atlantic are described as well. This model is driven by annual mean Hellerman and Rosenstein winds; Levitus sea-surface restoring temperature; and Schmitt, Bogden, and Dorman fresh-water flux fields (mixed boundary conditions). Various parameterizations of Arctic fresh-water export into the North Atlantic are included to examine the internal variability properties of the North Atlantic thermohaline circulation.

It is found that internal variability with about a 20-year period develops under steady forcing, provided there is a sufficiently weak Arctic fresh-water flux through the Canadian archipelago into the Labrador Sea. The variability is robust over a range of parameterizations of Arctic fresh-water export. Over an oscillation, large variations occur in the deep-water formation rate, especially in the Labrador Sea, and hence in the poleward transport of heat. The importance of topography is also addressed, and a detailed physical discussion of the mechanism and time scale for the oscillation is presented.

[1]School of Earth and Ocean Sciences, University of Victoria, Victoria, British Columbia

INTRODUCTION

There has been a good deal of scientific, economic, and even political interest in studying potential changes in our global climate and their influences on our environment. The search for an understanding of climate change, both past and present, has led directly to the ocean and in particular to the oceans' thermohaline circulation. The ocean, with its large thermal stability and its potential to store both anthropogenic and natural greenhouse gases, also serves as an important regulator of climate. It is the buffer that moderates temperature fluctuations during the course of a day, from season to season and even from year to year. One only has to compare the maritime climate of Victoria, British Columbia (48°25'N, 123°22'W), which has average temperatures of 4°C in January and 16°C in July, with the continental climate of Winnipeg, Manitoba (49°54'N, 97°14'W), which has average temperatures of −18°C in January and 20°C in July, to see the moderating effect of the ocean. The ocean also acts as a large-scale conveyor that transports heat from low to high latitudes, thereby reducing latitudinal gradients of temperature. Much of the oceanic heat transport is thought to be associated with the thermohaline circulation. In the North Atlantic, intense heat loss to the overlying atmosphere causes deep water to be formed in the Greenland, Iceland, and Norwegian seas. These sinking regions are fed by warm, saline waters brought by the thermohaline circulation from lower latitudes. No such deep sinking exists in the Pacific. Again, if one compares the climates of Bodö, Norway (67°17'N, 14°25'E), which has an average January temperature of −2°C and an average July temperature of 14°C, to that of Nome, Alaska (64°30'N, 147°52'W), which has an average January temperature of −15°C and an average July temperature of 10°C (both being at similar latitudes and on the western flanks of continental land masses), one sees the impact of this oceanic poleward heat transport.

The thermohaline circulation is driven by the flux of buoyancy through the ocean surface. This buoyancy flux can be broken down into two competing components—heat and fresh-water fluxes. High-latitude cooling and low-latitude heating tend to drive a poleward surface flow, high-latitude sinking, and a deep equatorward return flow, whereas high-latitude excess precipitation over evaporation and low-latitude excess evaporation over precipitation (except in a relatively narrow belt at the Intertropical Convergence Zone) tend to brake this thermally driven overturning. The existence of multiple equilibria and the stability and variability properties of the thermohaline circulation depend fundamentally on the competing properties of temperature (T) and salinity (S) in the net surface-buoyancy forcing of the ocean, and in particular on the fundamental difference in the coupling of T and S between the ocean and the atmosphere. Variations in the stability or variability properties of the ocean's thermohaline circulation, and

hence its associated poleward transport of heat, would have significant impact on both local and global climate.

The introduction of a new generation of fast supercomputers and workstations has allowed researchers to undertake long-time integrations of coarse-resolution ocean general-circulation models (OGCMs). These integrations have revealed numerous intriguing results pertaining to the stability and variability of the ocean's thermohaline circulation. In particular, during integrations of uncoupled (ocean-only) GCMs, self-sustained variability of the models' thermohaline circulation has been found on time scales ranging from decades to millennia. The purpose of this paper is to review some of these recent OGCM studies in order to illustrate the types of spontaneous thermohaline variability that may arise. Furthermore, some new experiments are discussed in which a coarse-resolution North Atlantic model is driven by annual mean Levitus (1982) restoring temperatures, the annual mean Schmitt et al. (1989) North Atlantic fresh-water flux field, and the annual mean Hellerman and Rosenstein (1983) wind-stress field.

The structure of this paper is as follows. First, some observations of century-to-millennial time scale variability in the air-sea ice climate system are discussed. A few observations of climate variability on the shorter (decadal to interdecadal) time scales are also briefly summarized. The nature and time scales of the internal, self-sustained variability of the thermohaline circulation found in coarse resolution OGCMs are then reviewed. Here, particular attention is focused on the relative importance of fresh-water flux, thermal, and wind forcing in driving the variability. The results of some recent experiments conducted in an idealized coarse-resolution North Atlantic basin driven by realistic forcing fields are then described. Finally, a summary and discussion are presented.

OBSERVATIONS OF CENTURY-TO-MILLENNIAL CLIMATE VARIABILITY

On the basis of ice-core records for the last glacial period, Oeschger et al. (1984) suggested that the climate system had two quasi-stable modes of operation between which the system oscillated in the transition between glacial and postglacial times. Broecker et al. (1985) further postulated that the two modes described by Oeschger et al. (1984) were characterized by the presence or absence of significant North Atlantic Deep Water (NADW) formation. There is indeed much evidence that during glaciations deep ocean temperatures were colder (Labeyrie et al., 1987) and that more Antarctic Bottom Water (AABW) flowed into the North Atlantic (Duplessy et al., 1988), while NADW formation was substantially reduced (Boyle and Keigwin, 1987), all of which tend to support the idea that the present-day thermohaline circulation is not unique. Furthermore, Ruddiman and McIntyre (1977) found that the surface waters off

Britain were of the order of 7°C cooler during glacial times than today, which, as noted by Broecker (1989), would be consistent with the absence of the Atlantic thermohaline conveyor and associated poleward heat transport.

Recently much attention has been given to trying to understand the climate oscillations that have taken place between the last glacial period and the present interglacial period. One example of such an oscillation is the Younger Dryas cold event, which took place between about 11,000 and 10,000 years before the present (BP). Keigwin et al. (1991) found that during the Younger Dryas (and at three other times since the last glaciation, about 14,500, 13,500, and 12,000 years BP) NADW production was substantially reduced or even eliminated, which tends to support the hypothesis of Broecker et al. (1985) regarding the existence of more than one quasi-stable mode of operation of the thermohaline circulation. The transitions into and out of these mini-glaciations are thought to have been very rapid. For example, Dansgaard et al. (1989) suggest that the Younger Dryas event ended abruptly within a 20- to 50-year period leading to the present interglacial period. Broecker et al. (1990) have proposed that during glacial times, when the northern end of the Atlantic Ocean is surrounded by ice sheets, a stable mode of operation of the conveyor belt for NADW is not possible. They propose the following millennial-time-scale oscillation: When the NADW conveyor is shut down and there are growing ice sheets, there is little oceanic salt export from the Atlantic to the other world basins. If a net evaporation over the North Atlantic is assumed, the salinity continues to increase. When a critical salinity is reached, deep convection and subsequently the conveyor turn on, transporting and releasing heat to the North Atlantic and thereby melting the ice sheets. The flux of fresh water into the North Atlantic from the melting ice sheets eventually shuts off the conveyor, and the process begins anew.

Variability in the earth's climatic system on the century time scale is also evident in oxygen-isotope and other proxy records such as those of Dansgaard et al. (1970) at Cape Century in northwest Greenland (see Stocker and Mysak, 1992, for more details). Over the last 10,000 years (since the Younger Dryas event) Dansgaard et al. found that the dominant climatic variability exhibited an energy peak at the 350-year period. This variability may well be linked to fluctuations of the thermohaline circulation over its overturning time scale (Mikolajewicz and Maier-Reimer, 1990).

OBSERVATIONS OF DECADAL-TO-INTERDECADAL CLIMATE VARIABILITY

On shorter time scales the air-sea-ice climate system also exhibits decadal-to-interdecadal variability. For example, signals of decadal-to-interdecadal time scales are exhibited by global surface-air temperatures (Ghil and Vautard, 1991),

sea surface temperature (SST) anomalies (Loder and Garrett, 1978), West African rainfall and the landfall of intense hurricanes on the U.S. coast (Gray, 1990), properties of NADW formation (Dickson et al., 1988; Lazier, 1980; Roemmich and Wunsch, 1984; Schlosser et al., 1991), temperature and salinity characteristics and circulation of the North Atlantic (Greatbatch et al., 1991; Levitus 1989a,b,c; Levitus, 1990), Arctic sea-ice extent (Mysak and Manak, 1989; Mysak et al., 1990), runoff from the Eurasian land mass (Cattle, 1985; Ikeda, 1990), and global sea-level pressure (Krishnamurti et al., 1986). While many of these studies have been restricted to relatively short time series, the Greenland ice-core data of Hibler and Johnsen (1979) clearly show a 20-year oscillation in the North Atlantic for oxygen isotope records spanning the years 1244 to 1971.

The source of this variability may once more be linked to internal fluctuations of the thermohaline circulation. Indeed, this hypothesis was originally put forward by Bjerknes (1964) in his attempt to explain decadal-to-interdecadal changes in long time series of SST in the subpolar North Atlantic (see Bryan and Stouffer, 1991, for a more complete discussion of Bjerknes' paper).

VARIABILITY OF THE THERMOHALINE CIRCULATION IN COARSE-RESOLUTION GENERAL-CIRCULATION MODELS

Over the past few years it has become evident that the thermohaline circulation may not be static, and that it may indeed undergo natural, internal variability on the decadal-to-millennial time scale. Numerous OGCM simulations have found such natural variability under steady, imposed surface-boundary conditions (strictly speaking, although the wind and fresh-water flux are time-invariant, the surface heat flux may vary with time if the sea surface temperature changes, as only the restoring temperature is time-invariant).

Variability of the thermohaline circulation found in coarse-resolution OGCMs can be roughly classified according to fundamental time scale: diffusive, overturning, and horizontal advection. This classification is used below in a brief review of recent modeling efforts aimed at understanding the variability properties of the thermohaline circulation in OGCMs.

Mixed Boundary Conditions

The heat and fresh-water flux coupling between the ocean and the atmosphere occur on different time scales and involve different physical processes. The lag of SST behind the seasonal cycle of insolation, which is on the order of 6 weeks (Bretherton, 1982), is conventionally parameterized in ocean models as a response to changing atmospheric conditions. The dependence of long-wave emission, sensible heating, and atmospheric humidity (and hence latent heat fluxes) on temperature allows the use of a simple, linear, Newtonian-damping boundary condition. The upper layer of the ocean (or the reservoir representing it in a box model or a more complicated model), is restored to an appropriate reference temperature on a fast time scale, 1 to 2 months (Haney, 1971). The boundary condition therefore takes the form of a variable flux (in watts per square meter; positive Q_T means heat out of the ocean),

$$Q_T = \frac{\rho_0 C_p \Delta z_1}{\tau_R}[T_1(\lambda,\phi) - T_a(\lambda,\phi)] , \qquad (1)$$

where $T_1(\lambda,\phi)$ is the upper-ocean box (with thickness Δz_1) temperature at longitude λ and latitude ϕ. $T_a(\lambda,\phi)$ is the atmospheric reference temperature, C_p is the specific heat at constant pressure (approximately 4000 J kg^{-1} °C^{-1}), ρ_0 is a reference density (approximately 1000 kg m^{-3}), and τ_R is a restoring time scale (Haney, 1971).

In ocean models it is appropriate to represent fresh-water fluxes at the ocean surface (due to evaporation, precipitation, river runoff, or ice formation) as a surface boundary condition on salinity. However, evaporation is mainly a function of the air-sea temperature difference, while the distribution of precipitation depends on complicated small- and large-scale atmospheric processes. A Newtonian boundary condition on salinity (units are g salt m^{-2} s^{-1}) as shown in (2),

$$Q_s = \frac{\rho_0 \Delta z_1}{\tau_R}[S_1(\lambda,\phi) - S_a(\lambda,\phi)] , \qquad (2)$$

then cannot be justified physically; it implies a definite time scale (τ_R) for the removal of salinity anomalies, which is not observed. Furthermore, (2) implies that the amount of precipitation or evaporation at any given place depends on the local sea surface salinity $S_1(\lambda,\phi)$, which is clearly incorrect. To resolve this problem in uncoupled ocean models, the imposition of either specified salinity fluxes Q_S or a salinity flux that depends weakly on the atmosphere-ocean temperature difference is preferred. The salinity fluxes Q_S may then be converted to implied fresh-water fluxes (P − E, in m yr^{-1}) by

$$P - E = - \frac{cQ_s}{\rho_0 S_0}, \qquad (3)$$

where S_0 is a constant reference salinity (about 34.7 psu) and c = 3.16×10^7 is the number of seconds in a year. A constant reference salinity is used in (3) instead of the local salinity $S_1(\lambda,\phi)$ so that when (3) is integrated over the surface of the ocean, zero net P − E corresponds to zero net Q_S.

Surface boundary conditions that involve a Newtonian restoring condition on temperature and a specified flux on salinity are termed mixed boundary conditions. While these boundary conditions are admittedly crude, they do reflect the different nature of the observed sea surface salinity (SSS) and SST coupling between the ocean and the atmosphere. A

discussion of some further enhancements to these boundary conditions, which might be employed in future OGCM studies, is presented later. In the uncoupled models discussed below these boundary conditions will usually be used.

Due to the lack of open-ocean observations of surface wind speed, mixing ratios of air above the sea surface (needed to determine evaporation through bulk formulae), and precipitation, it is common to obtain a surface fresh-water flux for use in uncoupled ocean models by spinning up a model to equilibrium under restoring boundary conditions on both T and S and then diagnosing the salt flux at the steady state. That is, (1) and (2) are used in the initial spin-up, and then at steady state the right-hand side of (2) is diagnosed at each grid box to yield a two-dimensional salt-flux field. (This field can then be converted to an implied fresh-water flux using (3).) The rationale for this approach is that by spinning up the model using some specified climatological surface restoring fields, one obtains an equilibrium in which the surface fields of T and S are climatologically correct. The diagnosed P − E field is then that field which, in theory, should yield the climatological SSS field. Furthermore, the equilibrium under restoring boundary conditions is also an equilibrium under the diagnosed mixed boundary conditions. Paradoxically, however, if the model simulates the SSS field exactly under restoring boundary conditions, P − E goes to zero.

Diffusive Time-Scale Variability

Marotzke (1989) spun up a single-hemisphere OGCM under restoring boundary conditions on temperature and salinity with no wind forcing. Switching to the diagnosed mixed boundary conditions and adding a small fresh perturbation to the high-latitude salinity budget at equilibrium, precipitated a polar halocline catastrophe. Several thousand years later the system evolved into a quasi-steady state with weak equatorial downwelling (a weak inverse circulation). This state was not stable, since low-latitude diffusion acted to make the deep waters warm and saline while horizontal diffusion acted to homogenize these waters laterally. Eventually, at high latitudes the deep waters became sufficiently warm that the water column became statically unstable and rapid convection set in. As in his zonally averaged model (Marotzke et al., 1988), the result was a flush in which a violent overturning (up to 200 Sv) occurred, whereby the ocean lost in a few decades all the heat it had taken thousands of years to store. At the end of the flush the system continued to oscillate for a few decades until the circulation once more collapsed. In the presence of wind forcing, Marotzke (1990) found that no flush existed. The inevitability of the occurrence of a flush under a purely buoyancy-forced, diffusive regime was illustrated in an analytic model developed by Wright and Stocker (1991).

Weaver and Sarachik (1991a) undertook experiments of similar design to those of Marotzke (1989, 1990). Contrary to the findings of Marotzke (1989, 1990), they observed the occurrence of flushes even when wind forcing was included (Figure 1a). Once more, in the collapsed state (Figure 2a) low-latitude diffusion and the subsequent horizontal homogenization of these waters tended to warm the deep waters (Figure 1b) until static instability was detected at high latitudes. The result was the onset of a violent flush (Figure 2b) that, as in Marotzke (1989, 1990), released all the heat stored over hundreds of years in a matter of a few decades. In two-hemisphere experiments Weaver and Sarachik (1991a) found that flushes still occurred, although they were slightly weaker and of one-cell (pole-to-pole) structure.

The apparent discrepancy between the work of Marotzke (1989, 1990) and Weaver and Sarachik (1991a) regarding

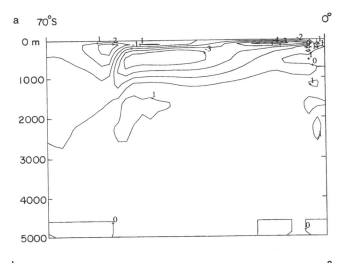

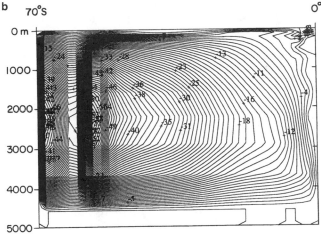

FIGURE 2   (a) Collapsed thermohaline state corresponding to year 1971 (refer to Figure 1) immediately before the flush shown in (b) at year 2067. All contours are in Sv; 1 Sv ≡ 10^6 m^3 s^{-1}. (From Weaver and Sarachik, 1991a; reprinted with permission of the American Meteorological Society.)

the occurrence of flushes was resolved by Weaver et al. (1993). They showed that the existence of flushes is linked to both the importance of fresh-water flux relative to thermal forcing and the strength of the wind forcing compared to the high-latitude freshening. The latter balance was also investigated in detail in Marotzke (1990). Comparing the equilibria obtained under restoring boundary conditions with and without wind, he found that at middle and high latitudes the thermohaline circulation provided the dominant transport mechanism for the meridional fluxes of heat and salt. Moreover, the strength of the meridional overturning was little influenced by the wind field, except for the Ekman transport in the top layer and its return flow, which takes place in the 200 meters below the top layer. Because the stratification is nearly homogeneous in near-surface layers at high latitudes, the Ekman cells (i.e., Ekman transport plus return flow) contribute very little to the meridional

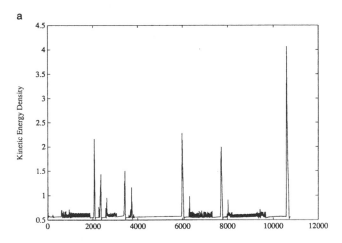

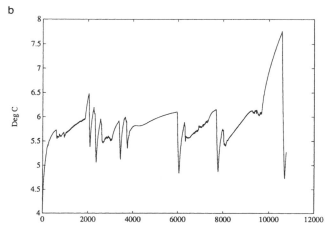

FIGURE 1   (a) Kinetic energy density (10^{-1} kg m^{-1} s^{-2}); (b) basin mean temperature (°C) throughout one of the single-hemisphere integrations of Weaver and Sarachik (1991a). The sharp peaks in (a) and (b) represent the occurrence of flushes, whereas the more rapid oscillations indicate decadal variability. (From Weaver and Sarachik, 1991a; reprinted with permission of the American Meteorological Society.)

transports. The situation changes, however, after the polar halocline catastrophe has occurred in, for example, the Northern Hemisphere: The surface layer is very fresh, compared to the layers below, but the northward transport of more saline water compensates for the southward Ekman transport of very fresh water, resulting in a net northward transport of salt. Moreover, the northward salt transport due to the horizontal subtropical gyre increases substantially during the collapsed phase of the thermohaline overturning.

When the thermohaline circulation has collapsed, the wind-driven northward transport of salt amounts to about half the value of the total transport for the spun-up steady state. Thus, the high-latitude surface freshening is counteracted by the wind-driven salt transport, which in the case of Marotzke (1989, 1990) was strong enough to make the high-latitude surface waters sufficiently saline again that deep convection resumed and the thermohaline circulation reestablished itself. Strong surface freshening (as in Weaver and Sarachik, 1991a) cannot be compensated for by the wind-driven salt transport, and the thermohaline circulation remains in the collapsed state until a flush sets in.

Weaver et al. (1993) further examined the robustness of these flushes in the presence of a stochastic term added to the imposed surface fresh-water flux field. In particular, they showed that as the magnitude of the stochastic term increased, the frequency of the flushing events increased, while their intensity decreased. When there was no stochastic forcing, deep-ocean temperatures warmed up to about 9°C in their model before static instability appeared at high latitudes, inducing convection and a flush. With the inclusion of a stochastic term in the fresh-water forcing field, flushes tended to occur earlier, before the ocean had warmed as much, and even earlier still as the magnitude of the stochastic forcing was increased. With increasing magnitude of the stochastic fresh-water flux forcing, there is an increasing probability that an evaporation anomaly will occur that is sufficiently large to induce convection and hence the onset of a flush. The basin mean temperature will not have warmed as much, so the ocean will lose less heat during the (thus weaker) flushing event. A similar result regarding the frequency and intensity of flushes was found when a seasonal cycle was imposed on the fresh-water flux field (Myers and Weaver, 1992).

Century-Time-Scale Overturning Variability

The time scale between the aforementioned flushes is diffusive and hence long (hundreds to thousands of years). A second fundamental period for variability occurs on the overturning time scale (Mikolajewicz and Maier-Reimer, 1990; Weaver et al., 1993; Winton and Sarachik, 1993). Mikolajewicz and Maier-Reimer (1990), in an uncoupled global ocean model that was driven by mixed surface boundary conditions and wind stress and to which a stochastic

freshwater-flux forcing term had been added, found internal variability with a dominant period of 320 years (their overturning time scale). This variability was manifested in salinity anomalies, which they traced around the overturning gyre of the Atlantic Ocean.

Winton and Sarachik (1993), in a planetary geostrophic ocean model (where the velocity field is exactly geostrophic), obtained similar 350-year variability; they interpreted the oscillation as a manifestation of a large-scale Howard-Malkus loop oscillation (as described by Welander, 1986). They argued that the presence of a positive salinity anomaly in the low-latitude surface regions would tend to slow the meridional overturning slightly, since thermal effects tend to accelerate the thermohaline circulation and haline effects to brake it. The weakened thermohaline circulation would then be more affected by the specified flux on salinity, which would act to intensify the positive anomaly at low latitudes and induce a negative salt anomaly at high latitudes. When the low-latitude salinity anomaly reached the high latitudes, convection and an intensified thermohaline circulation would ensue. The whole process would begin anew when the saline anomaly resurfaced at low latitudes. Thus their oscillation had a rapid phase, which was associated with the saline anomaly's being at low latitudes, and a slow phase in which the salinity anomaly was at high latitudes or in the deep ocean. This oscillation may well be linked to the 350-year period variability found in the Cape Century ice-core records of Dansgaard et al. (1970), discussed earlier.

Weaver et al. (1993) also found variability of overturning time scale in their runs, which were conducted in a thermally dominant regime. That is, if the surface fresh-water flux forcing field was sufficiently weak that it played only a minor role in driving the thermohaline circulation, the only thermohaline variability that existed under a stochastically forced regime was on the overturning time scale. This variability is analogous to the loop oscillation of Winton and Sarachik (1993).

Decadal and Interdecadal Time-Scale Variability

In many of the long time integrations of Weaver and Sarachik (1991a,b) under mixed boundary conditions (see, e.g., Figure 1a), self-sustained internal variability on the decadal-to-interdecadal time scale was found. This variability was linked to the turning on and shutting off of high-latitude convection and the subsequent generation and removal of east-west steric height gradients that caused the thermohaline circulation to intensify and weaken on a decadal time scale (Weaver and Sarachik, 1991b). They showed that this variability was associated with the propagation to the eastern boundary of warm, saline anomalies, generated in a localized region of net evaporation in the mid-ocean, between the subpolar and subtropical gyres. The separated western boundary current provided the source of warm, saline water required to initiate the anomaly development. Advection set the oscillation time scale, which was given by the length of time it took a particle to be advected from the mid-ocean region, between the subpolar and subtropical gyres, to the eastern boundary and then, as subsurface flow, toward the polar boundary. In more complicated geometry, they argued, one would expect this time scale to be slightly longer since the advective paths would no longer be along straight lines.

Figure 3, from Weaver and Sarachik (1991b), illustrates the meridional overturning stream function throughout the course of one particular oscillation in a single, Southern Hemisphere basin. During the oscillation the poleward heat transport changed by as much as a factor of 3 at certain latitudes. For example, at 26°S, during the most intense stage of the oscillation (Figure 3d) 0.29 petawatts (1 petawatt = 10^{15} W) of heat was being transported poleward, whereas during the weakest phase (Figure 3h) this was reduced to only 0.11 petawatts. The changes in heat transport corresponded directly to changes in the heat lost to the overlying atmosphere at high latitudes, since the ocean stored little heat during the oscillation. If such internal variability were to exist in the real ocean, it would evidently have a profound effect on global climate.

Weaver et al. (1991) attempted to understand why Marotzke (1989, 1990, 1991) and Marotzke and Willebrand (1991) almost never saw such spontaneous internal decadal oceanic variability (although one of the two-basin experiments described in Marotzke (1990) indicated the presence of decadal-to-interdecadal-scale oscillations in the region of the Antarctic Circumpolar Current), whereas Weaver and Sarachik (1991a,b) almost always found such variability. Through a systematic analysis of the forcing fields used in these works, they concluded that the presence of decadal-to-interdecadal variability was linked to an area of negative P − E at middle to high latitudes, and fresh-water gain further north. The meridional gradients in the fresh-water flux forcing field also had to be sufficiently strong that the system was in a haline-dominant regime. Figure 4a illustrates the basin-averaged surface heat flux over the course of the integration of one of the single-hemisphere experiments of Weaver et al. (1991), while Figure 4b shows the power spectral density of this curve over the first 8,214 years of integration. What is readily evident from this figure is that the dominant variability is in the decadal band, with associated basin-averaged surface heat-flux anomalies of between + 6 and − 10 W m^{-2} over one oscillation. Although the integration shown in Figure 4a eventually reached a steady state, Weaver et al. (1993) showed that the inclusion of a stochastic component in the fresh-water flux forcing field continually excited the decadal variability, with the result that no equilibrium was ever reached.

Weaver et al. (1991) suggested that when the thermoha-

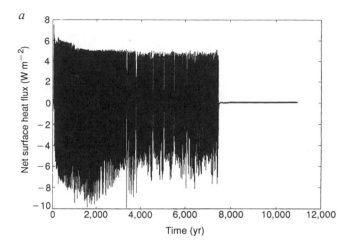

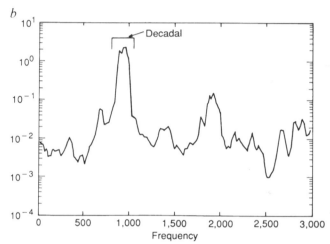

FIGURE 4   (a) Net basin averaged surface heat flux in W m^{-2} throughout the haline-dominated integration of Weaver et al. (1991). (b) Power spectral density (using a 256-point fast Fourier transform) of the surface heat-flux curve shown in (a) for the first 8,214 years of integration. The x-axis in (b) corresponds to the number of cycles over the 8,214 years of integration. (From Weaver et al., 1991; reprinted with permission of Macmillan Magazines, Ltd.)

line circulation was weak, it slowly passed through the region with negative P − E, and hence the surface waters became more saline. A warm, saline surface anomaly then developed through convection, and this anomaly was advected to the eastern boundary by the mean flow, from which it was convected to the deeper ocean (as in Weaver and Sarachik, 1991b). This led to the subsequent generation of the reverse cell seen in Figures 3a and 3b and Figures 3i and 3j, which in turn caused the thermohaline circulation to intensify. The intensified thermohaline circulation passed rapidly through the evaporative region; hence, the surface waters did not become as saline. Deep water then formed at high latitudes until high-latitude freshening dominated and the thermohaline circulation slowed down. This whole process repeated itself, with the time scale of the oscillation

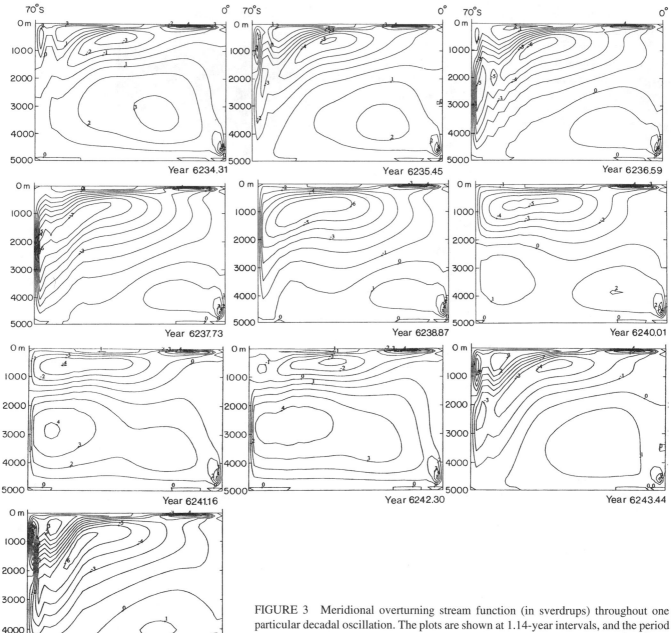

FIGURE 3   Meridional overturning stream function (in sverdrups) throughout one particular decadal oscillation. The plots are shown at 1.14-year intervals, and the period of the oscillation is 8.4 years. Positive contours indicate a clockwise transport. (From Weaver and Sarachik, 1991b; reprinted with permission of the Canadian Oceanographic and Meteorological Society.)

again determined by the time needed for salinity and temperature anomalies, formed in the local evaporative region, to be advected to the northern boundary.

The aforementioned OGCM simulations used time-invariant surface forcing fields (recall that the actual heat flux is time-varying but the restoring temperatures are fixed). In reality the oceanic surface forcing is not steady; apart from the seasonal cycle, the atmosphere and ocean are continually interacting with each other on small space and time scales. Weaver et al. (1993) considered these latter interactions as stochastic perturbations of the background forcing fields. They showed that the decadal internal variability still persisted when a stochastic component was added to the fresh-water forcing. Furthermore, Myers and Weaver (1992) showed that seasonally varying the surface forcing did not substantially alter the results.

The internal decadal variability discussed above is not limited to single-hemisphere, idealized ocean models. As shown by Weaver and Sarachik (1991a), in two hemispheres under symmetric (with respect to the equator) forcing, deca-

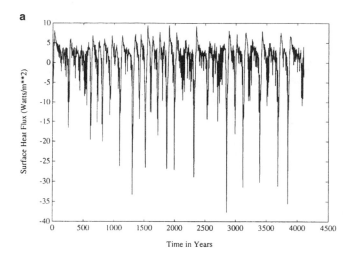

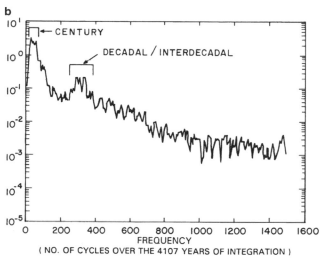

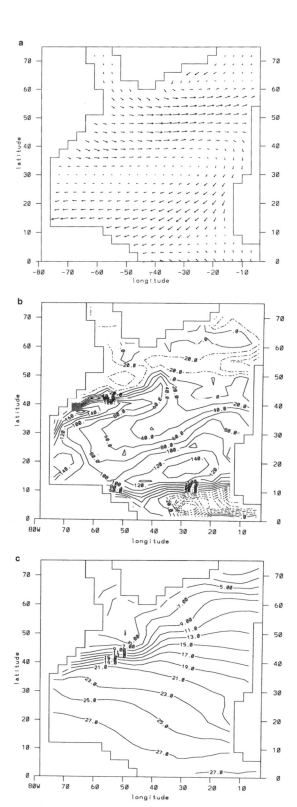

FIGURE 5   (a) Net basin averaged surface heat flux in W m^{-2} as a function of time for the stochastically forced experiment of Weaver et al. (1993). (b) Power spectral density (using a 512-point fast Fourier transform) of the surface heat-flux curve shown in (a) for the first 4,107 years of integration. (From Weaver et al., 1993; reprinted with permission of the American Meteorological Society.)

dal variability could occur in either hemisphere, essentially independent of the other hemisphere. Furthermore, in some of the two-basin results of Hughes and Weaver (1994), decadal-to-interdecadal variability was observed in either the North Atlantic or southern oceans, depending on the form of the fresh-water flux forcing field.

Combined Decade-to-Century Thermohaline Variability

As a final example of the type of variability that can exist in idealized OGCMs of the thermohaline circulation, the results of one of the experiments of Weaver et al. (1993) are presented. In this single-hemisphere experiment a stochastic term was added to the mean fresh-water flux forcing field. In the absence of stochastic forcing, the mean

FIGURE 6   Forcing fields used to drive the 3° × 3° North Atlantic OGCM. (a) Hellerman and Rosenstein (1983) annual mean surface wind stress in dynes cm^{-2}; (b) annual mean fresh-water flux field of Schmitt et al. (1989) in cm/yr; (c) Levitus (1982) annual mean sea surface temperature in °C. The maximum vector shown in (a) corresponds to 1.5 dynes cm^{-2}.

fresh-water flux forcing field led to the occurrence of quasi-periodic intense flushing events on which was superimposed very weak decadal-to-interdecadal variability. Weaver et al. set the standard deviation of the fresh-water flux to 80 mm per month, which is equal to the globally averaged annual mean precipitation (Baumgartner and Reichel, 1975).

Figure 5a illustrates the net basin-averaged surface heat flux throughout the 4,107 years of integration, while Figure 5b is the power spectral density of the curve shown in Figure 5a. What is readily evident is that variability on numerous time scales is present. Strong decadal-to-interdecadal variability is superimposed on flushes and loop oscillations, yielding a time series that looks chaotic. In a two-hemisphere version of this experiment, a similar pattern occurred.

It is tempting to speculate that the thermohaline circulation of the real ocean might behave in a manner similar to this stochastic forcing experiment. That is, transitions would be occurring between states with relatively strong and relatively weak thermohaline circulations; superimposed on this one might expect background decadal variability with moderate flushing events. This suggests the possibility that a major source of decade-to-century climatic variability resides in the intrinsic dynamics of the ocean's thermohaline circulation. Moreover, the chaotic behavior observed in Figure 5a would make the ocean circulation essentially unpredictable on climatic time scales.

INTERDECADAL VARIABILITY IN A COARSE-RESOLUTION MODEL OF THE NORTH ATLANTIC

The OGCMs discussed in the previous section were largely restricted to idealized flat-bottomed basins forced by zonally averaged restoring temperatures, idealized winds (with no meridional component), and P − E fluxes diagnosed from equilibria obtained under restoring boundary conditions. A natural question that arises is: Does this variability carry over to more realistic GCMs that incorporate irregular coastlines, topography, and more realistic surface forcing fields? The purpose of this section is to present some recent simulations of the North Atlantic thermohaline circulation driven by realistic forcing fields.

Description of the Numerical Model

The model used is the Cox (1984) version of the Bryan-Cox OGCM as described by Weaver et al. (1994). The resolution of the model is $3° \times 3°$, with 20 vertical levels ranging in thickness from 50 m at the surface to 300 m at the bottom of the basin. In experiments 1 to 5 the basin is assumed to be flat-bottomed with a depth of 4,020 m. In experiments T1 to T3 bottom topography is incorporated into the model.

The model was driven at the surface by the Hellerman and Rosenstein (1983) annual mean North Atlantic wind-stress field (Figure 6a), Schmitt et al. (1989) annual mean fresh-water fluxes (Figure 6b), and by restoring the SST to the Levitus (1982) annual mean climatology (Figure 6c) with a 50-day time scale. The P − E field was converted to a salt-flux field using equation (3), with S_0 chosen to be equal to the basin-averaged salinity (35.12 psu) of the North Atlantic Ocean as calculated from the data of Levitus (1982).

Over the top 210 m of the northern boundary either 0.0, 0.1, or 0.2 Sv of Arctic fresh-water flux was incorporated. The residual of the net evaporation from the Schmitt et al. (1989) data less the fresh-water flux through the northern boundary was added as an additional fresh-water source throughout the top 210 m at the equatorial boundary. This was done in order to keep the basin mean salinity constant throughout all integrations.

Table 1 lists the distinguishing characteristics of all eight

TABLE 1   Characteristics of Eight Experiments with a Coarse-Resolution Model of the North Atlantic*

Experiment Number	Arctic Flux in Sv	Where Arctic Flux was Applied	Topography	Internal Variability?	Period of Variability
1	0.0	N/A	flat	yes	22 years
2	0.1	Labrador + GIN** Seas	flat	no	N/A
3	0.2	Labrador + GIN Seas	flat	no	N/A
4	0.1	GIN Seas	flat	yes	17 years
5	0.1	East Greenland Current	flat	yes	17 years
T1	0.1	East Greenland Current	full	no	N/A
T2	0.1	East Greenland Current	dredge of Labrador Sea	yes	17 years
T3	0.1	East Greenland Current	flat + full Labrador Sea	no	N/A

*Details of these experiments in interdecadal variability can be found in Weaver et al. (1994).
**GIN stands for Greenland, Iceland, and Norwegian

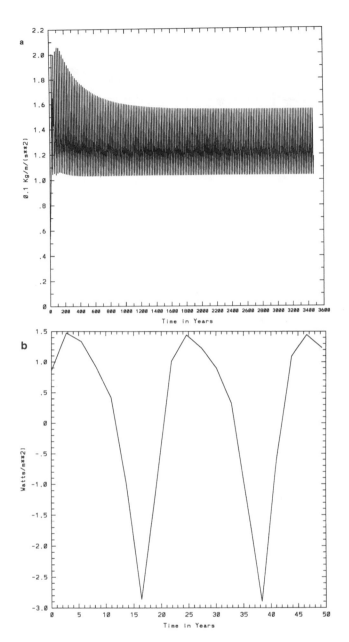

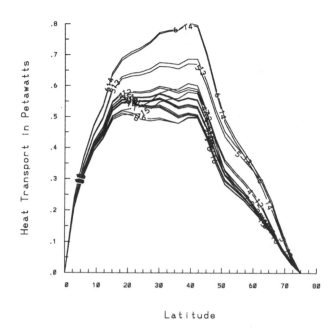

FIGURE 8   Poleward heat transport in petawatts (1 PW $\equiv$ 10^{15} W) at 2.7-year intervals throughout the 49 years shown in Figure 7b. The number n on each curve corresponds to $2.7 \times n$ years along the x-axis of Figure 7b.

FIGURE 7   (a) Kinetic energy density (10^{-1} kg m^{-1} s^{-2}) throughout the entire integration of experiment 1; (b) basin mean surface heat flux (in W m^{-2}) during the last 49 years of integration of experiment 1. Notice that in (b) two complete 22-year period oscillations are resolved. Positive contours indicate that the ocean is gaining heat.

experiments; a more detailed discussion of the model used may be found in Weaver et al. (1994).

Interdecadal Variability of the North Atlantic Thermohaline Circulation

In this and the next subsection, attention is focused on experiment 1, in which there was no parameterization of

Arctic fresh-water input into the North Atlantic (see Table 1). Figure 7a shows the kinetic-energy density of the model ocean throughout the 3,472 years of integration. What is readily evident from this time series is that the system very rapidly entered a limit cycle, the period of which is about 22 years (Figure 7b).

Over the course of the oscillation the poleward heat transport varied from a maximum of about 0.8 petawatts at 39°N to a minimum of 0.5 petawatts (Figure 8). This change was associated with the shutting off and turning off of deep-water formation at the northern boundary of the Labrador Sea.

Figures 9 and 10 show the meridional overturning in the whole North Atlantic and in the Labrador Sea region, respectively. Plots are shown at the weak phase (Figures 9a and 10a) and strong phase (Figures 9b and 10b) of the oscillation. In the weak phase about 16 Sv of deep water is forming, about 8 Sv of it in the Labrador Sea near 63°N. Gradients in the stream function, and hence the vertical and meridional velocity, are relatively weak at this phase in the oscillation; the energy of the system (Figure 7a) and the poleward heat transport (Figure 8, lines 8 and 16) are therefore at a minimum. The ocean basin is also beginning to slowly take up the heat that it lost in the rapid phase of the oscillation (Figure 7b). As the thermohaline circulation in the Labrador Sea intensifies, so does the poleward heat transport (Figure 8), until deep-water formation is suddenly triggered at the northern boundary of the Labrador Sea (Figures 9b and 10b). At this stage the thermohaline circula-

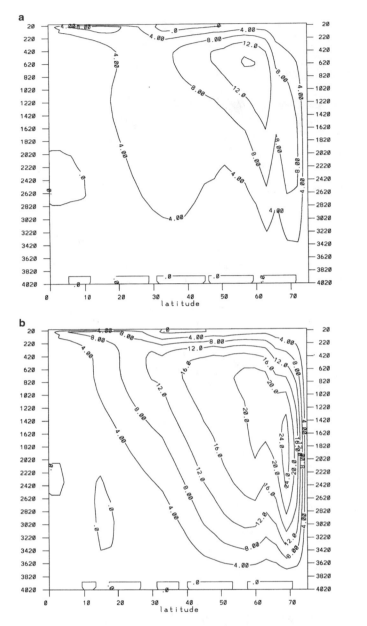

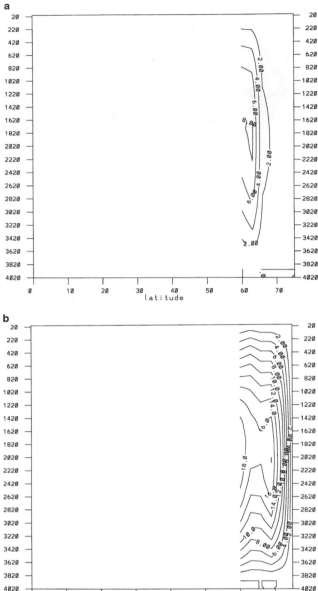

FIGURE 9   Meridional overturning stream function (Sv) for the whole North Atlantic basin throughout one particular interdecadal oscillation. The two plots correspond to the (a) weak phase and (b) strong phase of the oscillation. The year in the title of each figure corresponds to the x-axis of Figure 7b.

FIGURE 10   Meridional overturning stream function (Sv) for only the Labrador Sea region of the model domain throughout one particular interdecadal oscillation. The two plots correspond to the (a) weak phase and (b) strong phase of the oscillation. The year in the title of each figure corresponds to the x-axis of Figure 7b.

tion is most intense (Figure 7a), as is its associated poleward heat transport (Figure 8, lines 6 and 14). Furthermore, the heat that the ocean had gained over the previous 13 years is rapidly lost to the overlying atmosphere (Figure 7b).

Mechanism and Time Scale of the Variability

The discussion of the physical mechanism and time scales involved in the oscillation begins from the weakest phase

of the oscillation. The mechanism for the oscillation is quite different from that found by Weaver and Sarachik (1991b) and Weaver et al. (1991).

At the weak phase of the oscillation there is thermally driven deep convection all across the Labrador Sea (Figure 11a); hence, there is a very weak east-west pressure gradient. Consequently, the meridional geostrophic flow is weak at very high latitudes in the Labrador Sea, so there is no overturning there (Figure 10a). There is still a strong north-

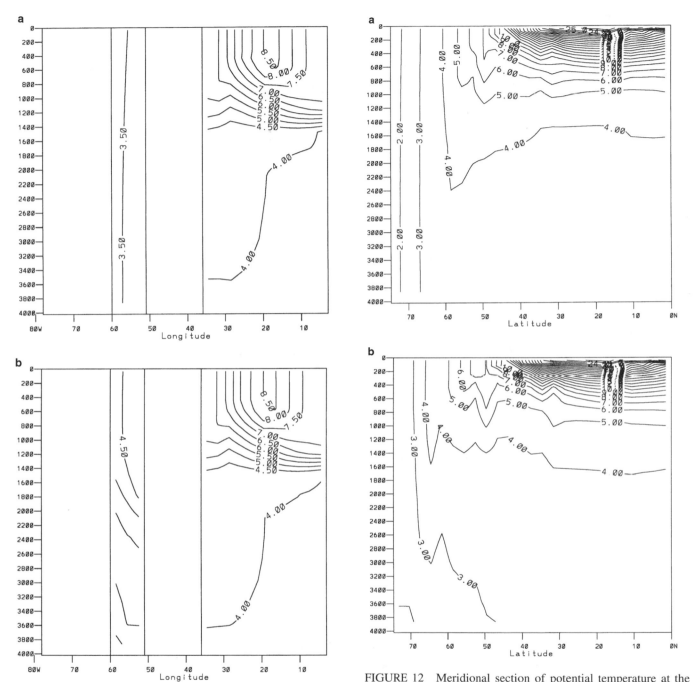

FIGURE 11   Horizontal section of potential temperature at 64.5°N at the (a) weak phase and (b) strong phase of the oscillation. The year in the title of each figure corresponds to the x-axis of Figure 7b.

FIGURE 12   Meridional section of potential temperature at the first grid point (1.5°) off the western boundary of the model domain (see Figure 6). The figures correspond to the (a) weak phase and (b) strong phase of the oscillation. The year in the title of each figure corresponds to the x-axis of Figure 7b.

south pressure gradient due to the surface restoring boundary condition on temperature (Figure 12a). Associated with this north-south pressure gradient is a zonal flow (Figure 13a) that converges at the western boundary of Greenland (near 66°N), causing sinking there and a return deep zonal flow below (Figure 14a).

The meridional velocity field converges on the boundary

at the tip of Greenland, so at this stage the main component Labrador Sea overturning is formed between 60° and 63°N. Notice that during the previous strong phase of the oscillation, warm saline waters are brought into the Labrador Sea. The surface boundary conditions remove SST anomalies faster than they remove SSS anomalies; hence, the net surface density gradient during the weak phase still increases

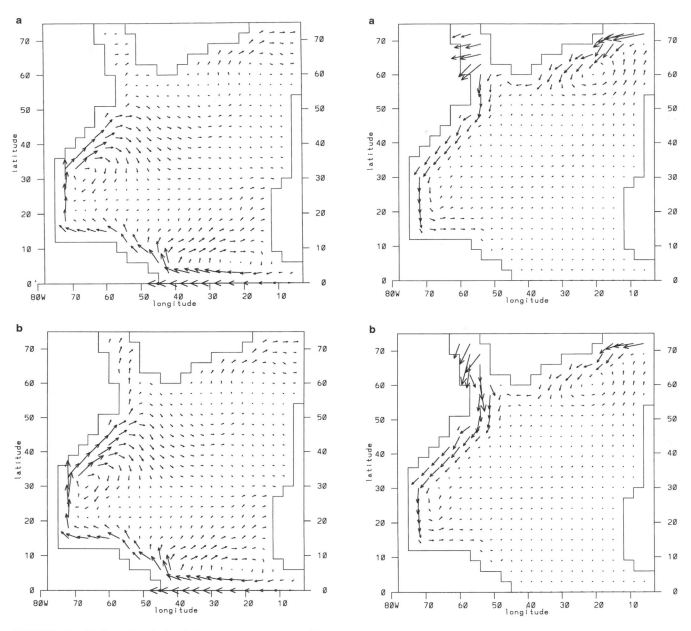

FIGURE 13 Horizontal-velocity field at the top model level at the (a) weak phase and (b) strong phase of the oscillation. The year in the title of each figure corresponds to the x-axis of Figure 7b.

FIGURE 14 Horizontal-velocity field at a depth of 2970 m at the (a) weak phase and (b) strong phase of the oscillation. The year in the title of each figure corresponds to the x-axis of Figure 7b.

as one progresses northward into the Labrador Sea, because temperature gradients dominate over salinity gradients. Since the SST and SSS are nearly zonally uniform, and convection penetrates to the bottom of the basin, there is only a very weak east-west pressure gradient in this weak phase. Deep water therefore does not form at the northern boundary. Overturning in the Greenland-Iceland-Norwegian seas now dominates the total overturning at high latitudes (Figure 9a).

The surface restoring boundary condition on temperature is given by equation (1) with $T_l(\lambda,\phi)$ replaced by the Levitus

(1982) annual mean SST, $T_L(\lambda,\phi)$ (Figure 6c), together with a restoring time scale of $\tau_R = 50$ days. During the intermediate phase of the oscillation, the warm saline waters in the high-latitude region of the Labrador Sea are being cooled toward the Levitus climatological temperature. Since convection extends to the bottom of the basin in the Labrador Sea on a rapid time scale ($H^2/A_{TV} \approx 1/2$ year for average basin depth $H = 4020$ m and vertical diffusivity $A_{TV} = 10^4$ cm^2 s^{-1}) condition (1) is equivalent to restoring the 4020 m deep basin to the surface value with an e-folding time scale of 11 years (assuming $T_1 - T_L$ is constant). Thus during this 11-year period, the entire northern region of the Labrador Sea is being cooled from top to bottom.

As the high-latitude region of the Labrador Sea is cooled, the meridional surface pressure gradient increases, since temperature gradients dominate over salinity gradients as in the weak phase. This increasing meridional pressure gradient implies an increasing zonal overturning, which, through divergence (and hence upwelling) at the western boundary and convergence (and hence downwelling) at the eastern boundary, leads to an enhanced zonal pressure gradient and hence an enhanced meridional overturning.

The time scale for the set-up of the zonal pressure gradient is linked to the zonal overturning time scale in the Labrador Sea, which, assuming a 10 cm s^{-1} surface current, 1 cm s^{-1} deep return flow, and vertical velocities of the order of 0.01 cm s^{-1} (from figures not shown), is about 4 years.

As the meridional overturning begins to increase slightly at high latitudes, the cold and now fresh water starts to move slowly equatorward in a deep western boundary current. Once this cold water has traveled a few degrees southward it suddenly shuts off convection from below at the western boundary (e.g., Figure 11b), because the cold fresh deep water is denser than the overlying warm saline water. The zonal pressure gradient is then further enhanced; hence, the overturning strengthens even further. The time scale for this southward advection of this cold fresh water is approximately 1 year, assuming an advective speed of 2 cm s^{-1} over a distance of 600 km.

Once convection is shut off in the west, the overturning starts to increase rapidly, and the cold fresh high-latitude water mass is advected equatorward (in a deep western boundary current) and replaced by a warm salty water mass that originated in the surface waters at lower latitudes (Figure 12b). This is now the strong phase of the oscillation when the system is most energetic (Figure 7a), the heat loss is greatest (Figure 7b), and the poleward heat transport is the strongest (Figure 8, lines 6 and 14). Deep water is now forming at the northern boundary of the Labrador Sea (Figures 10b and 14b), and the flow in the Labrador Sea is strongest and oriented meridionally.

The time scale associated with the removal of the cold fresh high-latitude water mass and its replacement by the warm saline mass is linked to the overturning time scale in the Labrador Sea region. Assuming a 10 cm s^{-1} surface flow (Figure 13b), a 3 cm s^{-1} deep flow (Figure 14b) over a distance of 25° (that is, from the latitude where the western boundary current separates from the coast, 50°N (Figure 13b), to the northern boundary, 75°N), and a vertical velocity of 0.01 cm s^{-1}, this time scale is about 5 years.

Since the deep basin of the Labrador Sea is now filled with warm saline water, convection from the surface to the bottom of the basin is induced everywhere through cooling to the Levitus SST field. Hence, the zonal pressure gradient

that drives the enhanced thermohaline circulation is reduced, and the whole process begins again.

The total time scale of the oscillation according to the discussion above is therefore:

11 years — Time scale for the cooling of Labrador Sea

4 years — Time scale for Labrador Sea zonal overturning and subsequent set-up of meridional pressure gradient

1 year — Time scale for the equatorward advection of cold fresh water in the deep western boundary current

5 years — Labrador overturning time scale for the replacement of cold fresh-water mass by warm saline-water mass

21 years — Approximate total time scale

This agrees fairly well with the period of 22 years found in the model solutions.

It should be noted that the time period required for the cooling of the Labrador Sea does not scale linearly with τ_R. For example, in an additional run that was performed with $\tau_R = 100$ days instead of 50 days, the time scale of variability remained nearly the same (now 21 years). This might be expected, since $T_1 - T_L$ in the region of the Labrador Sea (Equation 1 with $T_a = T_L$) was, on average, about twice as large as in the integration with $\tau_R = 50$ days. As a result, $(T_1 - T_L)/\tau_R$ and hence the heat flux out of the Labrador Sea were similar in both integrations. The remainder of the mechanism was identical to that discussed above, although deep ocean temperatures were warmer in the second case.

Role of Arctic Fresh-water Forcing and Topography

In experiments 2 through 5 a flux boundary condition was applied to the top 210 m of the northern boundary in order to parameterize a transport of fresh water from the Arctic. Table 1 summarizes the location and magnitude of the flux that was applied. In experiments 2 and 3, in which a fresh-water source was added at the northern boundary of the Labrador Sea, the internal variability discussed in the last subsection was suppressed. The external freshening was so strong as to cap convection in the Labrador Sea, so the aforementioned mechanism for the variability could not occur.

If the 0.1 Sv was not spread all along the northern boundary, but concentrated only in the Greenland-Iceland-Norwegian seas region (experiment 4) or only in the two grid boxes next to the Greenland coast in the eastern North Atlantic (experiment 5), the variability still occurred, although on a slightly shorter time scale (17 years). In the case when Arctic fresh water was put directly into the East Greenland Current region, the SSS field (not shown) did a fairly reasonable job of reproducing the climatological Levitus SSS field, although it is clear that runoff from rivers, such as the Amazon and the St. Lawrence, and outflow from marginal seas, such as the Mediterranean, should be incorporated if one expects the Schmitt et al. (1989) P − E fields to drive a more realistic ocean climatology.

The role of bottom topography in the stability and variability of the thermohaline circulation also has not been treated adequately. Most of the uncoupled OGCM simulations described above assumed a flat ocean bottom. Moore and Reason (1993) suggest that some of the internal variability of the thermohaline circulation might be sensitive to the inclusion of bottom topography, although it is not clear how well this topography is resolved in their 12-level model. In experiments T1 to T3 bottom topography was incorporated into the model; the results are summarized in Table 1.

In experiment T1 full topography was used in the North Atlantic model. This topography had the effect of making the Labrador Sea everywhere less than 840 m deep. The net result was that the variability was suppressed. In experiment T2, the Labrador Sea region was dredged to be flat (4,020 m deep), with full topography elsewhere in the domain. In this case the variability reappeared with the same time scale and dynamics as in experiments 4 and 5. In the final experiment, experiment T3, the topography in the Labrador Sea region was left alone, and the rest of the ocean (with the exception of continental shelf regions) was dredged to be 4,020 m deep. The internal variability was once more suppressed.

The conclusions that one can draw from these experiments are limited. The Labrador Sea region is very poorly resolved (being only three tracer points wide); hence, the topography used was not representative of the region. Due to smoothing of the topography (continental shelf regions with deep ocean), the Labrador Sea ended up being far too shallow. Hence, deep convection, strong east-west pressure gradients, and a fully developed thermohaline circulation could not occur. Much finer horizontal resolution must be used before a more complete understanding of the role of topography can be obtained.

SUMMARY

From the foregoing discussion it is clear that in OGCMs forced by mixed boundary conditions, internal variability of the thermohaline circulation can exist on the decadal-to-millennial time scale. The most important component of the system—in fact, the one that ultimately determines the model's response—is the strength of the fresh-water flux forcing versus the thermal forcing. Of secondary importance is the strength of the fresh-water flux forcing versus the wind-driven salt transport. If the fresh-water flux forcing is weak, no internal variability exists. If the regime is haline dominated and high-latitude freshening dominates over the wind-driven northward salt transport, a polar halocline catas-

trophe and subsequent flushing events may set in. As shown by Weaver et al. (1993), for several P − E fields that differ well within the range of observational uncertainties, the thermohaline circulation in an OGCM responds in a completely different way. This has important implications for the interpretation and modeling of both paleo- and present climate and climate change. If the results of the idealized OGCMs hold for comprehensive climate models as well, a relatively small error in the fresh-water exchange between ocean and atmosphere may produce dramatic changes in the thermohaline circulation, with a significant impact on the global climate.

The models discussed above exhibit serious shortcomings in their formulation of the atmospheric coupling. For example, there is no feedback between the SST and the hydrological cycle; a warm SST anomaly should cause enhanced evaporation, which in turn would be likely to change the overall P − E pattern. If the thermohaline circulation is vigorous, heat transport is large and high-latitude SST should rise. The relative influence of the fresh-water flux forcing would thereby increase by comparison with the thermal forcing (which actually is given by the temperature contrast between high and low latitudes), which would tend to destabilize the state of vigorous meridional overturning. On the other hand, increased high-latitude SST leads to increased evaporation, and it is conceivable that the total atmospheric water-vapor transport from low to high latitudes would be reduced. This would reduce the high-latitude freshening and thus stabilize the thermohaline circulation. More modeling studies are needed to investigate the effect of a more realistic formulation of boundary conditions on an ocean-only model. In particular, one might expect that a realistic atmosphere would damp some of the oscillatory behavior discussed herein (e.g., Zhang et al., 1993). Nevertheless, a robust result of the above models is that spontaneous variability on a decadal or millennial time scale is possible in an ocean model, for a realistic set of parameters.

All of the long integrations that were discussed in this section were done using non-eddy-resolving ocean models without an ice component. It is not clear whether the spontaneous variability found in coarse-resolution models will carry over to more realistic eddy-resolving OGCM simulations. This is particularly worrying for the interdecadal variability found in the North Atlantic simulations. It was evident that the Labrador Sea was the most dynamically important region of the model, yet the resolution in the Labrador Sea was very coarse. Furthermore, feedbacks associated with a prognostic ice component coupled to an ocean model may act to stabilize the variability found in the OGCMs.

It is interesting to point out recent results from fully coupled atmosphere-ocean-ice GCM climate simulations being conducted at NOAA's Geophysical Fluid Dynamics Laboratory. The thermohaline circulation of the North

Atlantic Ocean (and the high-latitude air-sea-ice system in general) was found to undergo variability on the 30- to 50-year time scale in these long-term coupled integrations (Delworth et al., 1993). This result tends to lend credibility to the internal variability found in uncoupled OGCMs. It is not clear, however, to what extent the variability is preconditioned by the use of the flux-correction term in the coupled model. If the flux-correction term is large in magnitude and of a form that contains much high-latitude structure (as is the case in the flux-correction term used by Manabe and Stouffer, 1988—see their Figure A1), one might expect the oceanic variability to be determined by this structure, with the atmosphere providing a stochastic forcing that excites the variability. The high-latitude variability in the atmo-sphere and ice components of the climate system would then be forced by the internal thermohaline dynamics. This speculation follows from Weaver et al. (1993), in which it was shown that the internal variability in uncoupled ocean models was linked to the structure of the mean fresh-water flux field.

ACKNOWLEDGMENTS

The author acknowledges support in the form of NSERC, AES, DFO, and FCAR operating grants. I am also grateful to P. Myers and S. Aura for their assistance with some of the North Atlantic experiments, and to J. Marotzke for providing the Schmitt et al. (1989) $P - E$ fields.

Commentary on the Paper of Weaver

EDWARD S. SARACHIK
University of Washington

I should like to comment first on the heat transport in Dr. Weaver's model. I remember that when Drs. Bryan and Lewis made their first global circulation model the heat transport was quite sluggish—less than a petawatt globally, I think. And in Dr. Weaver's Atlantic we have something of the order of 0.8 petawatt, basically because more than one heat source is sending cold water southward. So the branches of the thermohaline circulation really do seem to matter. There is sort of a general rule: The more sinking regions you accurately simulate, the higher the heat transport will be.

Maybe I can structure this discussion by raising some general questions. As you can tell, Dr. Weaver and I think alike, so he has mentioned some of them already. But I should like to reemphasize that the reason we do all of this is that the thermohaline circulation really is very poorly observed. We have some time information from the paleo indicators, like what Keigwin and Lehman do—where they can see whether there is inflow into the Nordic seas 10,000 years ago, for example. There is sediment and tracer information where you can actually follow freons, ^{14}C, and various other things around loops. But there is no measure, except very indirect ones, of vertical velocity. I feel that these sorts of models are the only way of getting real intuition about the kinds of circulations that are possible: in particular, thermohaline circulations. As we build up intuition about the types of circulations and types of variability possible, we develop a language to describe thermohaline circulations, and this language is then applicable to more realistic simulations. In fact, I think this progression could be seen in the papers we heard earlier today.

It seems to me that some general statements can be made. In these forced ocean models the variability increases when you switch to mixed boundary conditions; that is clear. No restoring-boundary-condition model has ever showed any instability, to my knowledge. As you increase the freedom to shut down and turn on the deep-water sources, you seem to get more variability. Just today, we have seen examples that Dr. Weaver showed of the Labrador deep source, and I suspect that if you put enough fresh water into the south by some sort of melting process, you would find variability in the deep-water source in the Antarctic also.

The basic question, of course—one that involves the relevance of these sorts of forced ocean models to climate or to reality—is the coupling to the atmosphere. One way of phrasing it is, "Do these oscillations continue to exist if you couple the ocean to an atmosphere?" Now you have to be very careful, because when you do couple it to a model atmosphere, you simply are not going to get the fluxes and the circulations that we get by using observed winds and observed salinities. It is a major problem of the atmospheric models that they do not rain in the right place, and it is a major problem of coupled atmosphere-ocean models in general that you have to make flux corrections in order to get the salinity of the upper ocean correct.

So the issue of how you answer the question is difficult. There will be coupled atmosphere-ocean models that will have variability or no variability, but they will not be coupled atmosphere-ocean models that have the same sort of forcing that you use to get the variability in the forced ocean models. Maybe we should do it the other way: Take the forcing from the coupled models and see whether the forced ocean models have the same sort of variability, then answer the question of whether the atmosphere is stabilizing or destabilizing.

All the models we have seen are coarse-resolution models, not out of any deep feeling about what resolution to use but out of economic necessity and practicality. Dr. Barnett indicated the amount of data you get when you do global ocean models, although his was relatively coarse. It is just that these coarse-resolution models are manageable; they can actually be done on work stations, as Dr. Weaver's group and my group do.

It is hard to say whether the sensitivities of these models will be the same as those of higher-resolution models, because sensitivity tests simply have not been done for coarse-resolution models. But my own presentation clearly indicated the sensitivity to forcing. There will also be sensitivities to geography. The coarse-resolution, simplified sector models have used zonally symmetric forcing. When you use zonally asymmetric forcing, as Dr. Weaver did, you do

get sinking in the Labrador Sea. Other factors are topography and ice. Since we are talking about fresh-water forcing, any process that contributes to fresh-water creation is important. It certainly seems that to do realistic simulations we will have to get all of those fresh-water processes correct.

Let me add some thoughts about what is going to happen in the future. Coupled atmosphere-ocean models will ultimately be the standard for looking at variability in the ocean. But it seems to me that they will not become so until the flux corrections are no longer needed. In other words, as the models get better, I expect the variability to become more believable.

The whole issue of eddies that Dr. McWilliams raised is one that I do not know how to deal with, because running eddy-resolving models is so complicated and time consuming. I do not expect to see comparable results from eddy-resolving models for another 10 or 20 years. At the moment we simply do not know how important eddies are for thermohaline variability.

Discussion

WEAVER: You'll find more topography results in my paper; I didn't have time to present them all. Topography is certainly important; when I put full topography into the entire ocean basin, it kills the variability. Dredge the Labrador Sea, and it comes back. But it's really a problem with the Labrador Sea, which is only three or four grid points wide in places; you end up smudging a 100-m shelf into a 3000-m ocean floor.

TALLEY: To get the overturning right on a global model, I think you need to consider salt sources, such as the Mediterranean. There's nothing similar in the Pacific.

GOODRICH: I wonder whether Peter Schlosser would care to comment on whether the areas of long-term convective shutdown he has observed resemble any of the mechanisms we've heard described in these modeling papers.

SCHLOSSER: It's really an 80 percent or so reduction of the deep-water formation rate in the Greenland Sea that began around 1980 and still persists, as far as we know. We don't know what drives this phenomenon. Among the speculations are a salinity change in the surface layers related to the return of the GSA to the Greenland Sea, and greater mixing of fresh water from the East Greenland Current into the center. When we tried a similar assessment for the upper water column we saw hardly any trend over the period with the drastic deep-water change.

GHIL: I don't think that very high levels of stochastic forcing are compatible with true atmospheric forcing, and I don't find activating millennial flushes more often to be an appealing physical explanation of interdecadal variability.

Going to a more general question, what strategy can we use to validate some of these results, considering the rather poor set of observations we have?

WEAVER: On your first topic, I should perhaps explain that my decadal variability is not linked to the flushes; they occur in a separate, dominant period. It's in the century-to-millennial band that the flushes can be excited. On your second, I'd have to say that as a modeler I don't feel competent to design a field-observation comparison program.

DICKSON: Let me amplify Peter's point a bit. Although deep convection in the Greenland Sea does seem to have slowed down since 1980 when the GSA came back, there was an even greater fresh-water anomaly passing south along the east of Greenland in the 1960s as the GSA formed. But at that time the deep convection in the Greenland Sea was in full swing. It's difficult to understand how when it was undiluted it couldn't cap convection there, while now it's back in a diluted form it's offered as one explanation of capping convection in the Greenland Sea.

What concerns me is that people are starting to look at what they perceive to be recurrences of the GSA as the cause of change in the thermohaline circulation —which they in turn perceive to be driven by variations in deep-water formation in the Greenland Sea—and they're getting decadal fluctuations out of it as a result. I have a lot of problems with that; the main one is topographic: deep convection in the Greenland Sea simply doesn't get out, and there just isn't any convection to the bottom of the Labrador Sea at present.

Ed was right when he said that the problem is one of modeling intermediate-water formation, which is very difficult, as Aagaard and Carmack noted. But if you truly want realism, you need to put in intermediate-water formation, the Greenland-Scotland Ridge, and the lack of convection to the bottom of the Labrador Sea.

SWIFT: As Bob mentioned, what affects the overflows may be more intermediate-water formation than deep-water. I'm not at all sure that there ever was deep convection in the Iceland Sea, at least in the present-day climate. My impression is that the GSA caused a freshening of the waters there involved in intermediate-depth convection, which meant that the Denmark Strait overflow freshened. Meanwhile, the fresh pulse rounded the tip of Greenland and mixed into the central Labrador Sea Water. The Iceland-Scotland Overflow Water probably took longer to respond to the fresh pulse, because a good deal of warm, salty Atlantic Water is resident at the sills and mixes with the outflow. The deep freshening signal is the sum of these components, each with a slightly different time of freshening.

MYSAK: Bob, is it true that convection was suppressed in the region of the Icelandic Sea, but not in the Greenland Sea, by the GSA?

DICKSON: Yes, it is. There is a tendency to call that area a gyre, but actually they are horses of a different feather. The Greenland, Icelandic, and Norwegian Seas are quite distinct, almost isolated. When the GSA caused salinity to drop below 34.7 in the area north of Iceland, suppressing deep convection, the peak convection in the Greenland Sea was still going like a train. I think Aagaard and Swift are right in saying that the Arctic intermediate water from the Icelandic Sea rarely moves over the convective center of the Greenland Sea.

ROOTH: I'd like to switch horses here, and ask Andrew a question. It's been proposed that for salt accumulation in subtropical gyres there is a critical temperature-salinity vertical-gradient ratio that allows 'salt fingers' (double diffusion) to preferentially transport salt down to intermediate- or deep-water levels. Do you know whether you are exceeding those salt-flux situations? And do you know whether anyone has used that as a constraint?

WEAVER: I must admit I've never checked that myself; I don't know whether anyone's looked into it.

CANE: I'd like to raise the issue of coarse resolution. It seems to me that it forces you to use—at least with the current numerical procedures—very high diffusivity characteristics, which will affect circulation stability. It's hard to say what that would mean when you can't get to a parameter range that resembles the real ocean.

Coarse resolution will generally suppress eddies, though Kirk's heat-transport results suggest that the resulting noise reduction might be a good thing. But then eddies are tied to mean circulation in ways that noise is not.

What worries me the most, though, is the necessity for simplifying the topography. Things move so slowly in the ocean when they don't have some slopes to rub up against.

The other thing I'd like to comment on is the question of coupling to the atmosphere. It seems to me that we might make one small step forward by replacing the ocean with a swamp-mixed layer and putting an atmospheric boundary layer on top of the ocean. It's difficult to calculate changes in the wind that way, or to do anything very convincing about precipitation changes, but you can get variations in heat flux and evaporation that way.

WEAVER: A Ph.D. student at the University of Victoria is taking a sector model, putting a simple thermodynamic ice model and a simple energy-balance model on top of it, and specifying some ad hoc hydrological cycle within it, as a first step toward the coupling you'd like to see. I agree with you entirely about the eddies and topography, but most of us just haven't the computing resources needed.

MCWILLIAMS: Let me add to the list of model-credibility questions the plausibility of the flushing modes. I think it might be worth looking at using some non-hydrostatic and, if you will, more defensible parameterizations of small-scale vertical fluxes.

Thermohaline Circulations and Variability in a Two-Hemisphere Sector Model of the Atlantic

DAVID MCDERMOTT AND EDWARD S. SARACHIK[1]

ABSTRACT

A coarse-resolution, simplified-geometry (sector) model of the Atlantic is described. Important features of this model include a Northern Hemisphere sinking region, a Southern Hemisphere sinking region, and a re-entrant channel to simulate the effects of the Antarctic Circumpolar Current. The circulation of the model ocean is described and diagnosed under restoring boundary conditions for both temperature and salinity. The model proves capable of simulating the grosser aspects of the deep-water Atlantic circulation. The variability of the thermohaline circulation under a switch to mixed boundary conditions (restoring on temperature and flux condition for salinity) is described.

INTRODUCTION

The thermohaline circulation (THC) of the Atlantic has a sinking branch, mostly in the Nordic seas and partly in the Labrador Sea, that puts cold water beneath the warm surface water. It also has a southern sinking branch, mostly in the Weddell Sea but partly in the Ross Sea. Cold, dense waters from these sinking regions spread out through the world ocean and are generally responsible for the maintenance of the stratification of the world oceans against the effects of heat diffusion from the surface. The export of cold water from high latitudes and its replacement by warmer surface waters implies a net heat transport toward the sinking regions: The THC is generally recognized as an important determinant of the mean climatic state of the earth's surface and overlying atmosphere.

The water from the Antarctic sinking region is extremely cold and relatively fresh, so when it sinks it is more compressible and therefore denser than the more saline deep water formed in the North Atlantic. The deep water formed in the Antarctic sinks to the bottom and becomes Antarctic Bottom Water (AABW, see Figure 1). The sinking from the Nordic seas becomes North Atlantic Deep Water (NADW), which overlies the AABW. Sector models with only one deep-water source cannot reproduce the interleaving of water masses seen in the deep ocean. Intermediate waters invade the ocean through wintertime convection and through the strong saline outflow of the Mediterranean. Intermediate water is much harder to simulate in a numerical model than deep water; this paper will concentrate on the deeper waters.

The water masses in the ocean can be identified by standard hydrographic measurements and by following the inputs of transient tracers at the surface through specialized

[1]Department of Atmospheric Sciences, University of Washington, Seattle, Washington

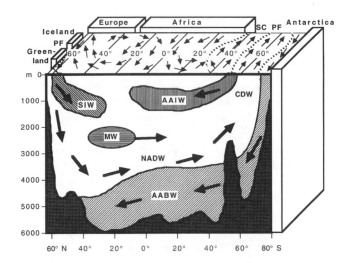

FIGURE 1 Schematic of water-mass distribution in the Atlantic (updated from Dietrich and Ulrich, 1968). AABW: Antarctic Bottom Water; NADW: North Atlantic Deep Water; CDW: Circumpolar Deep Water; SIW: Sub-Arctic Intermediate Water; AAIW: Antarctic Intermediate Water; MW: Mediterranean Water; PF: Polar Front; SC: Subtropical Convergence.

measurements; both give information about the broad outlines of the thermohaline circulation. The thermohaline circulation is characterized by extremely small velocities (especially vertical velocity), so direct measurement of the THC is impossible. Analytic and numerical models have therefore played a major role in our approach to understanding the THC. In particular, simple three-dimensional coarse-resolution sector models of the dynamics of the THC forced by steady mixed-boundary conditions (pioneered by F. Bryan, 1986a, 1986b) have begun to contribute a great deal to our understanding of the mean and variable THC.

Sector models contain the rudimentary dynamics needed to model the THC and its variability in a simplified context that allows many numerical experiments to be performed. On the other hand, the applicability of these simplified-model results to the real ocean will always be in question until they have been confirmed by experiments using more complex models with higher resolution, realistic geography and topography, and coupling to the cryosphere. Even then, the applicability of more complex forced ocean models to climate will be in doubt unless realistic, interactive coupling to the atmosphere can take place as well. (Fully coupled global atmosphere-ocean models are the ultimate tool for understanding the climate, but at present they are so expensive and complicated that they can be run only at a very small number of institutions.)

Coarse-resolution sector models have been used to investigate decadal and longer-term variability under steady mixed boundary conditions (see Weaver (1995), in this section, for a review and references). In sector models with two hemispheres, equatorially symmetric restoring bound-ary conditions on both temperature and salinity bring about a symmetric circulation with a sinking motion in the high latitudes of each hemisphere and a rising one at the equator. Upon a switch to mixed boundary conditions (restoring on temperature but with fluxes for fresh water) asymmetric pole-to-pole circulations arise, with sinking in one hemisphere and rising in the other (F. Bryan, 1986b; Weaver and Sarachik, 1991a). With mixed boundary conditions (Weaver and Sarachik, 1991a), a transient chaotic state develops (in which symmetric circulations sometimes, though rarely, occur) that ultimately leads to a single asymmetric pole-to-pole steady-state circulation.

This paper presents a simplified model for investigating a more realistic geometry—one with a circumpolar current in one (the southern) hemisphere—in a sector context. Following a brief description of the model, the steady-state thermohaline circulations under restoring boundary conditions for both temperature and salinity are discussed, and then the variability of the model's physical quantities with a switch to mixed boundary conditions.

THE MODEL

The model used in the experiments described below is the Modular Ocean Model (MOM), a modular form of the Cox (1984) version of the Bryan-Cox Ocean General Circulation Model. The latter is a full primitive-equation model (in spherical coordinates) described by Bryan (1979) and Bryan and Lewis (1979). The MOM employs the finite-difference scheme described by Bryan (1969) and Cox (1984), and the polynomial approximation to the UNESCO equation of state described by Bryan and Cox (1972). We use a coarse horizontal resolution of 4° in latitude by 3.75° in longitude. The basin extends from 72°S to 72°N and is bound by meridians separated by 56.25° of longitude. A single re-entrant channel extends longitudinally 3.75° and latitudinally from 64°S to 48°S, and from the surface to a sill at approximately 2500 m depth. The basin depth ramps down from the sill depth to its full depth of 5000 m in the three grid spaces surrounding the channel opening, both to the north and south and to the east and west, and is uniformly 5000 m elsewhere. Vertical resolution is in 20 levels, so that numerical problems involving spurious circulations are avoided (Weaver and Sarachik, 1991a).

Cyclic boundary conditions are applied to the meridional boundaries at the slot, creating the re-entrant channel. Coarse-resolution models present difficulties in representing straits and narrow and silled channels. The B-grid used in the Cox-Bryan model requires a channel width of at least two tracer grid points to define one velocity point within the channel. As discussed in Toggweiler et al. (1989), the horizontal viscosity required to define boundary currents prevents a realistic flow from occurring in a channel that contains only one velocity point. We therefore create a re-entrant channel that includes three velocity grid points and four tracer grid points.

The lateral walls and bottom all have insulating boundary conditions. A no-slip condition is applied at the walls, and the bottom is assumed to be impermeable. There is no bottom friction. The model uses the rigid-lid approximation at the surface in order to filter out high-frequency external gravity waves. The imposed surface wind stress, which is based on Hellerman and Rosenstein (1983), has only a zonal component. The zonal wind stress is shown in Figure 2b. The wind stress curl is anticyclonic between 15° and 45° latitude in each hemisphere and cyclonic elsewhere.

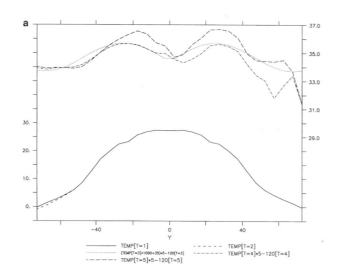

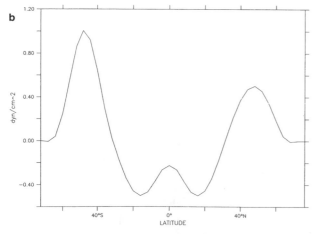

FIGURE 2   (a) The restoring temperature (lower curves) and salinity (upper curves) used in Experiments 1 and 2. The dashed temperature curve shows the slightly colder southern temperatures used in Experiment 2. The smooth (dotted) salinity curve, the restoring salinity used for Experiments 1 and 2, is a rough compromise between the dashed salinity curve (zonally averaged Atlantic curve from Levitus, 1982) and the dot-dashed salinity curve (zonally averaged world ocean salinity from Levitus, 1982). (b) The zonal wind stress applied in all experiments, in dynes/cm².

The temperature is restored to prescribed values with a linear damping coefficient, with a time constant of 50 days. The reference temperature profile for the restoring boundary condition in Experiment 1 was obtained by averaging the world ocean values for Northern and Southern Hemisphere sea surface temperatures from Levitus (1982). In Experiment 2, the values for the Southern Ocean south of 50°S are lowered slightly in order to investigate the sensitivity of the circulation to the temperature of the sinking water. Both temperature profiles are shown in Figure 2a. In Experiments 1 and 2 the surface salinity is similarly restored to the idealized reference salinity values shown in Figure 2a.

To evaluate the salinity flux, Experiment 3 (mixed boundary conditions) used the equilibrium state of Experiment 1 but allowed it to run for an additional 50 years after the initial 7000-year integration utilized in Experiments 1 and 2. The surface salinity flux is averaged for that time period, and this average value is used to restart the model from the equilibrium state of Experiment 1. (This salinity flux is shown in Figure 9a.) The acceleration techniques of Bryan (1984) are used to speed the convergence of the model. The baroclinic velocity and barotropic vorticity equations are integrated with a time step of 2400 seconds, and the tracer equations are integrated with time steps of 4 days. The vertical eddy viscosity and vertical heat and salt diffusivity are set to 1.0 cm²/s everywhere. Horizontal eddy viscosity is set to 6×10^9 cm²/s, and horizontal tracer diffusivity is set to 1×10^7 cm²/s. The method of complete convection, the asymptotic limit of the diffusive convection parameterization, is used for all experiments. It has been described by Yin and Sarachik (1994).

THE STEADY THERMOHALINE CIRCULATION

Experiment 1: Symmetric Restoring Temperatures

Figure 3a shows the zonally averaged stream function for Experiment 1 under restoring boundary conditions on both temperature and salinity. Experiment 1 shows three distinct thermohaline cells: an Arctic cell 16 Sv in strength with sinking at the northern wall, an Antarctic cell over 6 Sv in strength with sinking at the southern wall, and a cell 6 Sv in strength with sinking on the north side of the re-entrant current and southward surface flow all the way from the equator. Surface-confined cells on either side of the equator, locally driven by the equatorial Ekman divergence caused by easterly trade winds, are also evident.

Note that the surface stream function between 30°N and 40°N is oriented northward when the winds at those latitudes are westerly, implying southward Ekman flow near the surface. Examination of the surface currents (Figure 4a) indicates that the southward branch of the zonally averaged stream function in Figure 3 is dominated by a southward boundary current, while the interior flow is northward in

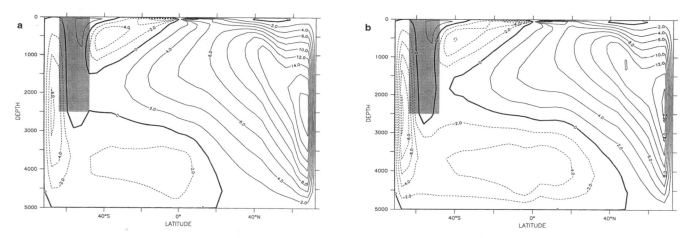

FIGURE 3   Zonally averaged streamfunction in sverdrups. (a) Experiment 1 with symmetric restoring conditions. (b) Experiment 2 with restoring to slightly colder southern temperatures. Clockwise circulations shown solid, anticlockwise shown dashed. Contour interval is 2 Sv. Shaded area indicates location of re-entrant channel.

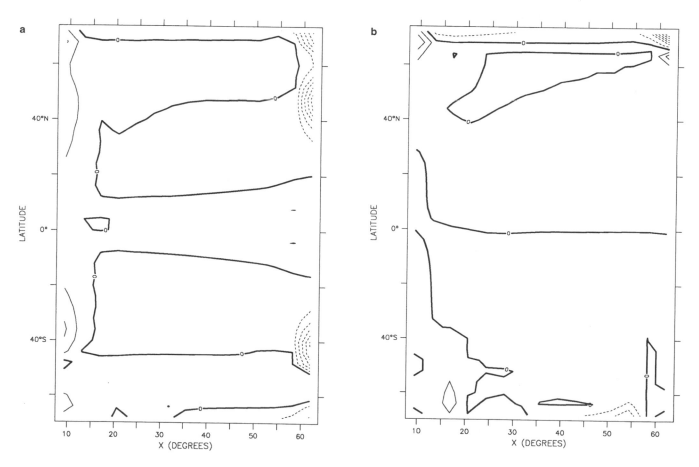

FIGURE 5   Vertical velocities from Experiment 1 (symmetric restoring temperature): (a) at 150 m, (b) near the bottom at 4800 m. Contour interval is 0.0005 cm/s.

the westerlies and southward in the easterlies, as expected. Similar considerations hold for a limited region of the surface stream function near 30°S.

Also disguised by the zonally averaged stream function shown in Figure 3a are the details of the mean sinking motion. Near the surface, there are three regions of sinking in the basin (Figure 5a). There is sinking along the entire northern boundary, joined to sinking along a considerable portion of the eastern boundary north of about 20°N. There is sinking in the southeast corner of the basin poleward of the slot. Finally, there is a sinking region on the eastern boundary in the Southern Hemisphere north of the slot: This is the sinking that corresponds to the shallow counterclockwise meridional cell in Figure 3a. Figure 5b shows that the only flow that reaches the bottom is the sinking along the northern wall, which is joined to intense sinking in the northeast corner of the basin, and a small region of sinking near the southeast corner.

Examination of the currents clarifies the movement of the water from those sinking regions in which water does not reach the bottom. Figure 4c shows the currents at 2000 m. The sinking water along the northern part of the eastern boundary feeds a westward current flowing along 50°N, which then becomes part of the deep southward-flowing western-boundary deep current. This broad current (about 11° of longitude) extends from about 1000 m down to the bottom while north of the equator, and from 1000 m down to only 4000 m from the equator to its termination at 20°S; it lies directly below the equally broad (in this coarse-resolution model) northward-flowing western-boundary current. Below 4000 m, the southward-flowing current turns eastward at the equator, leaving southward flow only above 4000 m. This remaining southward flow turns eastward at about 20°S and flows back across the basin to the eastern boundary.

The other region of shallow sinking (centered at 40°S just north of the slot on the eastern boundary) extends only to 1000 m. It feeds a westward-flowing current (Figure 4b), which ultimately turns northward above the southward-flowing western-boundary current.

The water that does reach the bottom along the entire northern boundary flows southward and westward upon reaching the bottom (Figure 4d). The negative vertical velocity, which goes to zero at the bottom, is accompanied by anticyclonic flow; the zero vertical velocity, which increases to positive values in the interior near-bottom upwelling regions, is accompanied (through the Sverdrup relation $fw_z = v$) by cyclonic flow, in particular northward flow toward the northern sinking regions. The southern sinking water in the southeast corner of the basin reaches the bottom and flows northwestward. The near-bottom vertical velocity is downward and, again by the Sverdrup relation, implies northward motion.

The combination of the northward-flowing warm western-boundary current and the deeper southward-flowing cold western-boundary countercurrent produces a strong northward heat flux (Figure 6). This flux is consistent with the input of heat through the surface in low latitudes and the strong removal of heat at higher northern latitudes (Figure 9b). The vertically and zonally integrated meridional oceanic heat flux is northward north of 8°S; it has a peak value of 0.61 PW.

The eastward flow through the slot (the analog of the Antarctic Circumpolar Current) is relatively intense. The eastward flow, which extends below the sill to about 3000 m, is caused by westward-flowing deeper water's hitting the sill, upwelling, and returning eastward.

Barely evident in Figure 3a is the so called "Deacon Cell," a cell with rising motion at lower depths to the south of the re-entrant current, northward surface currents (consistent with westerly winds), and sinking still further north. This cell, which appears in more realistic ocean models that have geography and topography (Toggweiler et al., 1989), is responsible for bringing old ^{14}C to the surface.

Experiment 2: Colder Southern Restoring Temperature

Figure 3b shows the zonally averaged stream function for Experiment 2, which has slightly colder restoring temperatures in the south (with 0.6°C maximum difference at the southern boundary). Compared to the results of Experiment 1, the Antarctic Cell has markedly increased in strength, and the apparent penetration of Antarctic Bottom Water has moved considerably northward. The northern thermohaline cell has weakened by 2 sverdrups.

Examination of the vertical velocities near the surface (Figure 7a) and near the bottom (Figure 7b) shows that in

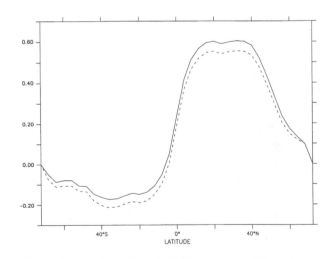

FIGURE 6   Zonally and vertically integrated meridional heat flux for Experiment 1 (solid line, symmetric restoring temperature) and Experiment 2 (dashed line, slightly colder southern temperature). Units are petawatts.

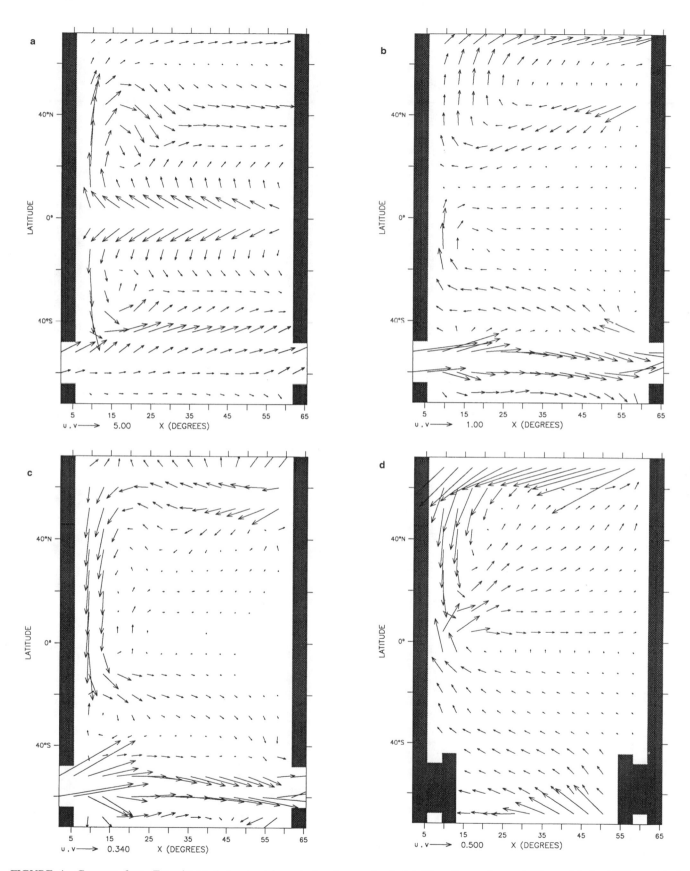

FIGURE 4 Currents from Experiment 1 (symmetric restoring temperature). (a) Near the surface, at 25 m. (b) At 1000 m. (c) At 2000 m. (d) Near the bottom, at 4800 m. Length of vector indicates velocity in cm.

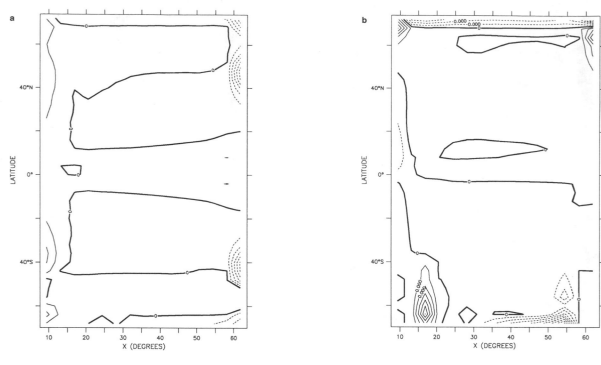

FIGURE 7 Vertical velocities from Experiment 2 (slightly colder southern restoring temperature): (a) at 150 m, (b) near the bottom at 4800 m. Contour interval is 0.0002 cm/s.

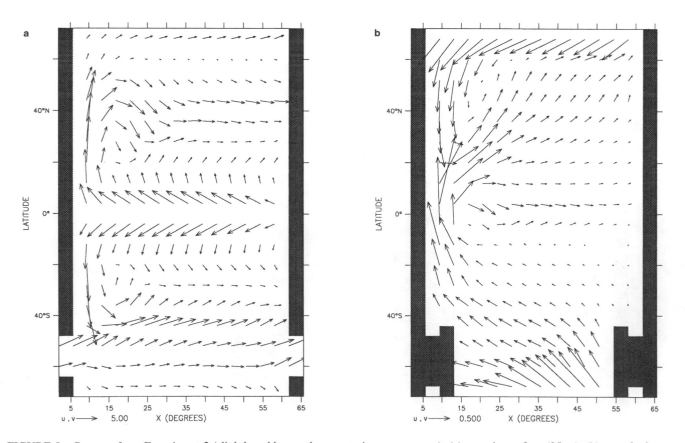

FIGURE 8 Currents from Experiment 2 (slightly colder southern restoring temperature): (a) near the surface (25 m), (b) near the bottom at 4800 m.

Experiment 2 the vertical velocities near the surface are similar to, but slightly stronger than, those in Experiment 1, and that the sinking velocities become much stronger with increasing depth. Although the sinking along the entire northern wall is a good deal stronger in Experiment 2 than in Experiment 1, the zonally averaged sinking is weaker (causing the weakening of the overturning stream function) because of the small region of upwelling in the northwest corner. As can be seen from Figure 7b, all the sinking is stronger in Experiment 2.

The stronger sinking at depth leads to stronger near-bottom currents. In particular, the water leaving the sinking region in the southeast corner of the basin flows northward and westward (Figure 8b), as it does in Experiment 1 (compare Figure 4d), but now creates a noticeable upwelling region against the ramp connected to the slot sill. It also flows further northward along the western boundary in the bottom thousand meters. This more northward penetration of bottom water shows up in the zonally averaged stream function as a more northward penetration of the southern overturning cell (Figure 3b compared to Figure 3a).

The more northward penetration of colder bottom water emanating from the sinking region in the southeast corner

for Experiment 2 implies a more southward (or less northward) total heat flux. Figure 6 shows that this is indeed the case.

THERMOHALINE VARIABILITY

When the boundary conditions at the end of Experiment 1 are switched to mixed boundary conditions (i.e., conditions that restore the symmetric temperature profile and the diagnosed salinity fluxes shown in Figure 9a; we called this Experiment 3), strong haloclines form in both the Northern and the Southern Hemispheres, shutting off deep sinking in both regions. Intermediate water is formed north of the gap in the Southern Hemisphere, which moderates the diffusive warming of the deep ocean. As the deep ocean warms, the circulation exhibits a series of small-amplitude oscillations about 1400 years after the switch to mixed boundary conditions (see Figure 10a). The variability begins as fairly regular cycles with periods of just over four years, and the period is seen to decrease. After 450 years of oscillations the period is reduced to just over 3 years, and then for approximately 200 years the variability becomes increasingly irregular. The circulation then shifts to the steady state

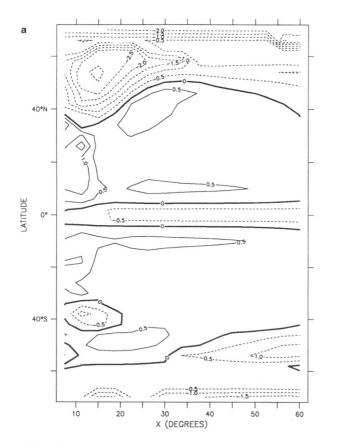
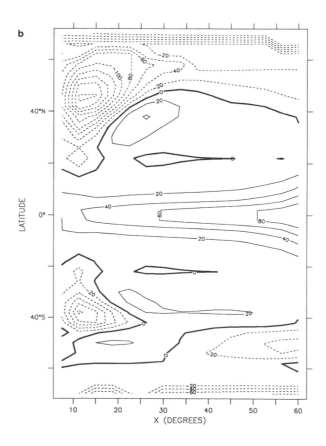

FIGURE 9   (a) The diagnosed salinity flux (converted to fresh-water flux in m/yr) from the equilibrium state of Experiment 1. Contour interval is 0.5 m/yr. (b) The heat flux through the surface for the equilibrium state of Experiment 1 in w/m². Contour interval is 20 W/m².

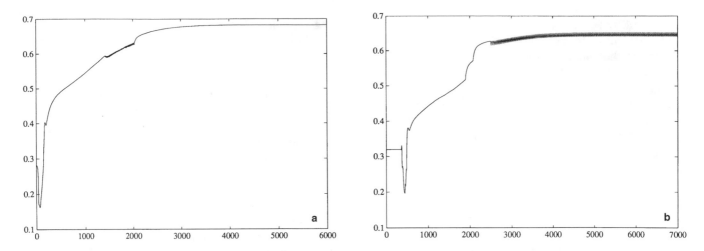

FIGURE 10   The kinetic energy per unit volume averaged over the entire ocean volume (units of 0.1 kg m^{-1} s^{-2}) after the switch to mixed boundary conditions from the steady state of (a) Experiment 1 (symmetric restoring boundary conditions) and (b) Experiment 2 (cold south restoring boundary conditions).

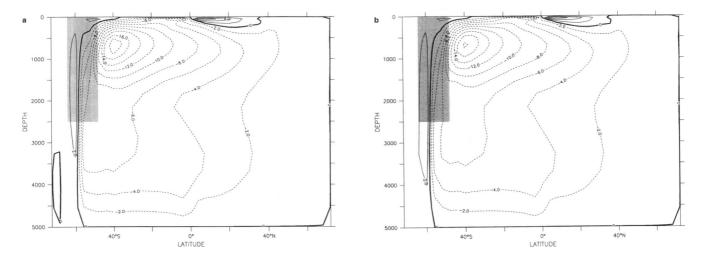

FIGURE 11   Zonally averaged steady-state stream function (in sverdrups). (a) Stream function corresponding to the steady state reached in Figure 10a. (b) Stream function corresponding to a time average over 30 years in Figure 10b. Contour interval 2 Sv. Shaded area indicates location of re-entrant channel.

shown in Figure 11. The high-latitude freshening is so large that sinking no longer occurs at high latitudes, and all the deep sinking occurs in the latitude of the slot in the Southern Hemisphere. Figure 11a may be considered the analog of the pole-to-pole circulation in the two-hemisphere symmetric case of Weaver and Sarachik (1991a), except that here the restoring salinity flux is itself asymmetric.

A closer look at the variability shows it to be most strongly evident in changes in convection in the western side of the basin at 58°S. Figure 12 shows contours of salinity at 58°S for four consecutive years. Throughout this period there is convection to the north of this latitude to about 2000 m, and a strong halocline capping the latitude

band to the south. All deep convection is occurring at 58°S. In Figure 12a, the southern halocline extends to 58°S in the western part of the basin. Fresh surface water is advected northward to produce this cap, but convection continues at depth. One year later, in Figure 12b, we see the capped region has moved slightly eastward. The eastern edge of the capped region advances no further than this; as can be seen in Figure 12c, however, convection advances in the west, shrinking the size of the cap. At the end of the cycle, in Figure 12d, convection occurs across the basin at 58°S. Advection of fresh surface water in the west produces a capping of convection there, and the cycle continues.

When the salinity flux of Experiment 2 is diagnosed, we

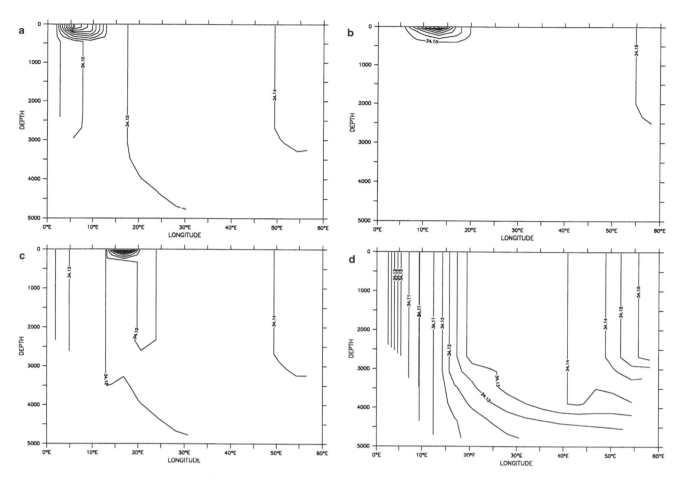

FIGURE 12   Vertical-longitudinal sections of salinity at 58°S in Experiment 3 for four consecutive years during the period of variability. Contour interval in (a) is 0.02 ppt, in (b) is 0.05 ppt, in (c) is 0.02 ppt, and in (d) is 0.01 ppt.

find the fluxes shown in Figure 13. Note that the fluxes are very similar to the fluxes diagnosed from Experiment 1 except near the southern boundary, where Experiment 2 convection is very much stronger (compare Figure 9). When the steady-state Experiment 2 is then switched to mixed boundary conditions (Experiment 4), the circulation undergoes the evolution shown in Figure 10b. The initial stages of the circulation are similar to those in Experiment 3, with the formation of strong haloclines in both northern and southern regions, and the initiation of intermediate-water formation at approximately the latitude of the gap. As this water formation strengthens and extends to become deep-water formation, the circulation develops a very regular 30-year oscillation. The overturning stream function averaged over one 30-year period is shown in Figure 11b.

The variability in Experiment 4 is also most strongly evident in changes in convection in the western side of the basin at 58°S. The extension of the southern halocline to 58°S latitude is shown in Figure 14. The strength of the stratification in this region varies over the course of the 30-year cycle, but convection from the surface breaks out only

at the column located at 5.6°E and 58°S. As can be seen in Figure 15b, the resulting variability in overturning stream function is fairly small.

The surface forcing at this location comes from cooling and evaporation, both of which tend to destabilize the column. Advection brings cold fresh water into the upper levels from the south. Figure 15a shows how the meridional velocity to the south of the variably convecting region changes through one cycle. The advection from the south decreases until convection from the surface begins (year 11) and then increases until the convection is shut off (year 19). The column upstream of this location (to the west) is convecting down to the sill depth. The water being advected from the west is vertically homogenous in temperature and salinity, and is both warmer and saltier than the water found in the column at 5.6°E and 58°S. This zonal velocity (not shown) into the location of convective variability decreases while convection occurs, and increases while convection is not occurring. While this oscillation has only a small effect on the large-scale overturning, local effects are significant. Figure 16 shows the surface heat flux in the region of convective variability over one 30-year cycle.

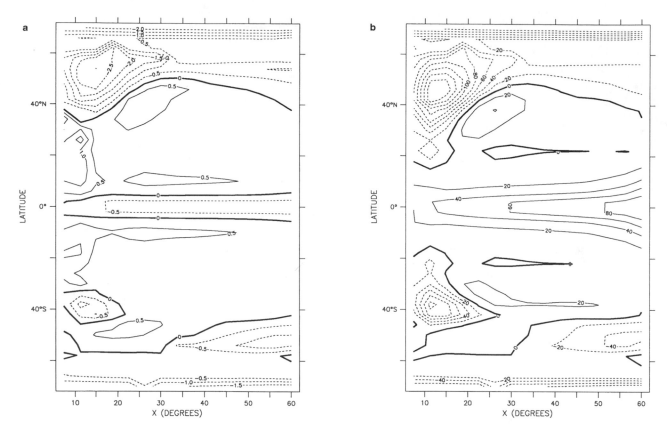

FIGURE 13   (a) The diagnosed salinity flux (converted to fresh-water flux in m/yr) for the equilibrium state of Experiment 2. Contour interval is 0.5 m/yr. (b) The heat flux through the surface for the equilibrium state of Experiment 2 in w/m². Contour interval is 20 W/m².

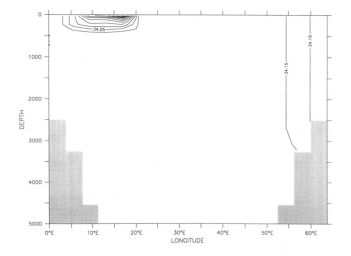

FIGURE 14   Vertical-longitudinal sections at 58°S of salinity in Experiment 4 at a time when cap extends to 5.6°E. Contour interval is 0.05 ppt.

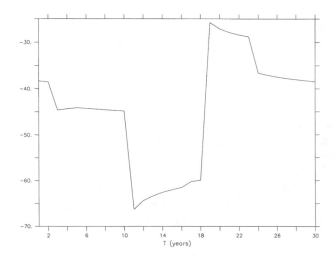

FIGURE 16   Surface heat flux for one cycle in region of convective variability, in W/m².

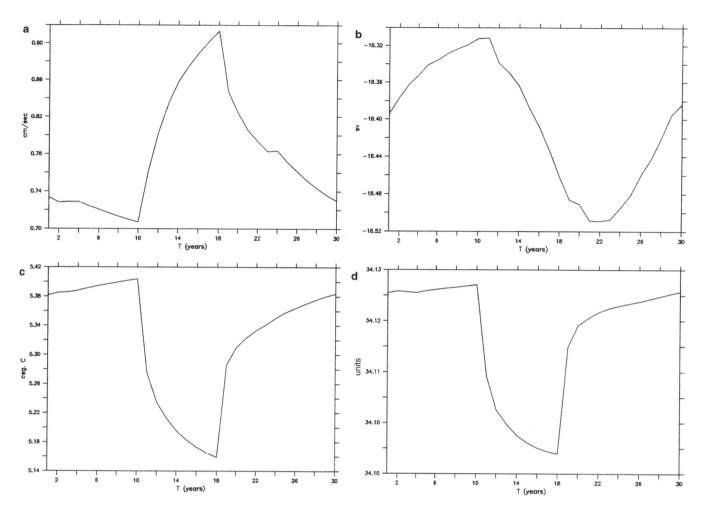

FIGURE 15 (a) Meridional velocity averaged over the top 112 m at 5.6°E, 62°S. (b) Minimum value of zonally averaged stream function (in sverdrups) during 30-year cycle (see Figure 11b). (c) Average temperature below 1500 m in region of convective variability. (d) Average salinity below 1500 m in region of convective variability, in ppt.

CONCLUSION

We have constructed an idealized model of the Atlantic basin and described the steady circulations that resulted from forcing under symmetric restoring boundary conditions (Experiment 1). A small reduction in restoring temperature in the Southern Hemisphere (Experiment 2) led to a rather significant change in the intensity of the vertical circulations. Upon a switch to mixed boundary conditions, the state produced by Experiment 1 progressed to a final steady state with only transient variability, while the state resulting from Experiment 2 settled into a regular oscillation with a period of approximately 30 years.

It has become clear that several different kinds of variability are possible under constant mixed boundary conditions. Each region of deep sinking in the ocean can be capped by fresh water and can therefore be variable. Weaver (1995, in this section) shows that variability arising from the variability of the sinking region in the Labrador Sea involves processes different from variability arising from the variability that involves the sinking region in the Nordic Seas. Similarly, this paper shows variability involving the sinking region poleward of the Antarctic Circumpolar Current.

We conclude with some questions and qualifications of the above work.

The simple type of coarse-resolution sector model presented here has a number of features that resemble the real ocean, and therefore proves useful in isolating these processes for detailed study. In particular, we saw that a small change in the restoring temperature led to rather large changes in the thermohaline circulation, even at high northern latitudes, and to even larger changes in behavior under a switch to mixed boundary conditions. While we cannot yet explain this sensitivity, we note that such unexpected sensitivities, if shown to be realistic in more complex models, may have important climatological consequences.

The basic question that arises from this and all other

ocean-only models is what the effect of the atmosphere is on the variability. Coupling the atmosphere to the ocean in tropical models tends to induce instabilities that were not present in models of either the atmosphere or the ocean. Whether the atmosphere is stabilizing or destabilizing for the variability exhibited in ocean-only models must await detailed simulations with coupled atmosphere-ocean models.

ACKNOWLEDGMENTS

The authors are indebted to Michael Winton and Fenglin Yin for very helpful discussions, to Michael Ghil for a detailed reading of the first draft of this paper, and to Ed Harrison and Steve Hankin of NOAA/PMEL for the use of FERRET, a diagnostics and graphics program that was used extensively in this research. This work was supported as part of the Atlantic Climate Change Program by a grant from the NOAA/Office of Global Programs to the University of Washington Experimental Climate Forecast Center. David McDermott was also supported as a Graduate Fellow for Global Change by the Department of Energy, and part of this research was performed using the resources located at the Advanced Computing Laboratory of Los Alamos National Laboratory.

Commentary on the Paper of McDermott and Sarachik

MICHAEL GHIL
University of California, Los Angeles

I very much enjoyed this paper. However, it needs to be put into the broader context of the other papers on decade-to-century variability in the ocean's thermohaline circulation (THC) presented at this workshop. There is considerable activity going on in this area, and we will get together a small Climate System Modeling Program (CSMP) workshop at UCLA in October 1993 to clarify some of the issues further.

My view of the scientific method is that we proceed from detection of phenomena via understanding to simulation, and eventually on to prediction. There are some short-cuts, such as statistical predictions that try to circumvent understanding and simulation, but we have full confidence only when all the stages of this progression have been visited.

Now, this gradual approach to the entire cognitive process also implies that there should be a gradual approach to one particular step, which is the understanding of the models. So I would just like to refer you to a couple of relevant figures that are halfway between Dr. Bryan's 'toy' model and Dr. Sarachik's complex one (Figures 9, 14, and 16 of Quon and Ghil, 1992), and talk about the issue of multiple equilibria in such simple models, as touched upon by Claes Rooth.

It has been observed in a number of THC models of different complexity that, under symmetric forcing with respect to the equator, you can have either two-cell symmetric circulation or one-cell pole-to-pole circulation. My illustrations show that essentially you can have all the intermediate steps. In other words, you can go from the symmetric circulation to the completely antisymmetric one gradually. Actually, this is a more realistic approach to the Atlantic's THC, as presented in McDermott and Sarachik's

Figure 1 (see also Figure 1 of Ghil et al., 1987). There is North Atlantic Deep Water (NADW) reaching into the high southern latitudes, but there is Antarctic Bottom Water (AABW) spreading under it past the equator. So while the Quon and Ghil (1992) model is only two-dimensional, it manages to capture this asymmetry (see Figure 9 there), with one cell (NADW) dominating the other (AABW) without suppressing the latter entirely.

The linear stability analysis for this highly resolved, albeit two-dimensional, model yields a regime diagram (Figure 14 of Quon and Ghil). This diagram demonstrates that increasing either the imposed pole-to-equator temperature gradient at the surface or the salinity flux across the boundary will produce a gradual transition between two-cell symmetric circulations and completely antisymmetric circulations. A gradual increase of the measure of asymmetry is thus possible. Mathematically, this gradual increase is captured by a pitchfork bifurcation (Figure 16 in Quon and Ghil).

Now, Dr. Sarachik started with symmetric forcing and then also imposed slightly asymmetric forcing. Actually, when these mixed boundary conditions were enforced, in one case the response of his model was to switch to another steady state; in the other case, it went to oscillatory behavior.

To conclude, I believe that oscillations shown by some of these ocean THC models or coupled atmosphere-ocean GCMs have something to do with the spectral peaks detected in global and local temperature series (e.g., Ghil and Vautard, 1991; also Cook et al., 1995, and Keeling and Whorf, 1995, both in this volume). We need to identify certain oscillations in the models with what we think we observe, since confronting models with observations is the name of the game in the physical sciences.

Discussion

DESER: What is the mechanism for beginning a new cycle of the eastward-moving salinity anomalies?

SARACHIK: The Antarctic Circumpolar Current is a re-entrant current in our model. But of course it's hard to say how any oscillation begins; there's no particular "start" spot in the cycle.

PHILANDER: You mentioned that you were concerned about the consequences of a 0.6°C perturbation. Could it reflect a large change in the westerlies, via Ekman drift?

SARACHIK: Perhaps. What most concerns me is the sensitivity these sorts of model seem to exhibit. I'm wondering whether that's the result of something in the idealized configuration, the re-entrant current, or what.

CANE: Eli Tziperman and colleagues have done some work recently that relates to the time constants you put into the restoring terms for temperature and salinity and whether you go to a flux condition. In some sense, I think one of the characteristics of most of these ocean-model studies with time-constant forcing is that the atmosphere is perhaps not active enough. There are various ways of calculating the sensitivity of the heat flux to a change in surface temperature; except for the truly coupled GCMs, I think that the models tend to hold the atmosphere more fixed than is appropriate for its heat capacity, which makes the time shorter. If that were loosened up, and the time lengthened, we'd get somewhat different answers for all these cases—the points at which you go from stable to unstable behavior and the like.

Low-Frequency Ocean Variability Induced by Stochastic Forcing of Various Colors

TIM P. BARNETT,[1] MING CHU,[2] ROBERT W. WILDE,[3] AND
UWE MIKOLAJEWICZ[4]

ABSTRACT

A primitive-equation global ocean general-circulation model with realistic geography has been forced with a variety of fresh-water flux models. The models' response appears as a number of fundamental eigenmodes. If the high-latitude forcing is of the order of 1 to 2 mm/day, then the dominant response has a period of roughly 300 years with significant variability in all the world's oceans. This response, partially documented by Mikolajewicz and Maier-Reimer (1990, 1991), occurs for forcing that is red or white in space and white in frequency. The transport of the Circumpolar Current fluctuates in response to this mode by a factor of two. In the Atlantic, the mode has a meridional circulation that extends over the whole basin/water column and a horizontal circulation that is clockwise. The most energetic portion of the Pacific circulation is more closely confined to the surface and more confused, although the associated attractor is fairly regular.

INTRODUCTION

Increasing attention has been focused on the role of the ocean in forcing climate variations with time scales from decades to centuries. Among the various components of the global climate system, only the ocean seems at this stage of understanding to have the requisite thermal inertia. The atmosphere appears to have little memory beyond a few weeks, due to its small density and thermal capacity. Major changes in climate are associated with shifts in the planetary ice sheets, but the time scale of these changes is of the order of 1000 years or more, apparently too long for them to be a prime mechanism in forcing decade-to-century variations in other climate system components. Finally, changes in external forcing (e.g., solar radiation) and in some internal forcing (e.g., volcanoes) either have yet to be established or are currently thought to be too weak to affect other than local climate changes. The oceans thus remain as the most likely source of climate variability on time scales of 10 to 100 years or more.

[1]Scripps Institution of Oceanography, La Jolla, California
[2]California Institute of Technology, Pasadena, California
[3]Department of Physics, Harvard University, Cambridge, Massachusetts
[4]Max Planck-Institut für Meteorologie, Hamburg, Germany

Recent work by a number of climate modelers has suggested that modest changes in atmospheric forcing can lead to radical changes in oceanic circulation that could have a dramatic impact on the global heat balance. The pioneering study of Stommel (1961) was apparently the first to suggest that the ocean's thermohaline circulation could exist in different, but stable, states. A spate of recent work with more complex models has confirmed that early work (see the comprehensive review of Weaver et al., 1993) and revealed many of the properties of the state transitions, as well as the sensitivity of the model results to subtle changes in the specification of the forcing—in these cases the flux of fresh water into and out of the oceans.

Most of the work to date has been done with simplified ocean models or steady-state fluxes of fresh water. The important work of Mikolajewicz and Maier-Reimer (1990, 1991) added a stochastic forcing term in the fresh-water flux and also used a full three-dimensional ocean general-circulation model (OGCM). Their result showed a quasi-periodic fluctuation in the thermohaline circulation; it had a time scale of the order of 300 years, and was most apparent in the Atlantic Ocean. Stocker and Mysak (1992) find significant spectral peaks in paleoclimate data with about the same period. Recently, Weaver et al. (1993) and Mysak et al. (1993) repeated that experiment (with a more simplified OGCM and rather arbitrarily defined flux fields) and described the relative importance of the stochastic term to the general model behavior.

The current paper expands on these earlier studies by examining the response of a realistic OGCM to different types of stochastic forcing. The model response to purely climatological forcing is compared to the additional response induced by stochastic forcing that is (1) white in both space and time, (2) red in space and white in time, and (3) red in both space and time, with the degree of redness determined by the coupling coefficients between the sea-surface temperature (SST) and SST gradients and the fresh-water flux field. This represents a coupled ocean-atmosphere model of a type that does not seem to have been previously explored.

DESCRIPTION OF MODELS AND EXPERIMENTAL DESIGN

This section describes the ocean and stochastic atmospheric models used in this study. It also describes briefly how the coupled ocean-atmosphere model with feedback was constructed.

Ocean General-Circulation Model

The OGCM used here is identical to that used by Mikolajewicz and Maier-Reimer (1990, 1991; MMR hereafter), which is described in detail by Maier-Reimer et al. (1993).

It is a linear, primitive-equation model with a horizontal resolution of approximately 3.5° (an E-type grid) and 11 levels in the vertical. Unlike most of the modeling studies discussed above, this OGCM uses realistic geometry and ocean bathymetry. This is an important characteristic if one wishes to infer that the model results have relevance to the real world. The numerics are handled in such a way as to allow a time step of 30 days, thereby making extended integration feasible even on a workstation. At the surface (only) the heat-balance equation has a seasonally varying Newtonian damping to observed surface-air temperature climatology, while the salt balance uses a seasonally varying fresh-water flux to give realistic surface salinity fields (mixed boundary conditions). Seasonally varying wind stress (Hellerman and Rosenstein, 1983) is also used to force the model.

Again unlike most of the modeling work discussed above, the OGCM includes a simplified thermodynamic sea-ice model wherein the heat flux through the ice is proportional to the temperature difference between the underlying water and air temperature, and inversely proportional to the thickness of the ice. If one is interested in forcing an OGCM with fresh-water flux, then some representation of sea ice seems mandatory for a realistic simulation. In addition, the full UNESCO equation of state for sea water (see UNESCO, 1981) is used by the model to help properly represent the large-scale density changes associated with mixing. More details on the model and its construction can be found in the references above. These references also show that the model does a reasonable job of reproducing the major features of the ocean's general circulation as well as the distribution of temperature and salinity in the deep ocean.

Atmospheric Models

The basis for constructing several of the atmospheric models is a 10-year integration of the Hamburg climate model (ECHAM3, T42 resolution; see Roeckner et al., 1992) forced by anomalous global SSTs. This integration was conducted as part of the Atmospheric Model Intercomparison Project (AMIP). The results, made available to us courtesy of L. Bengtsson, give monthly, globally gridded anomalies of fresh-water flux from the model and the observed SST and SST gradients that produced them. The distribution of the long-term mean and standard deviations of the fresh-water flux (E − P) are shown on Figure 1. In another study, we show that the flux and atmospheric moisture fields produced by this model compare well with observations of these same quantities where such comparisons are possible (Pierce et al. (unpublished manuscript, 1994)). The following atmospheric "models" were derived from this basic data set.

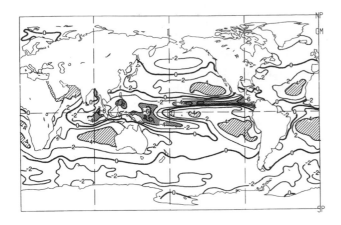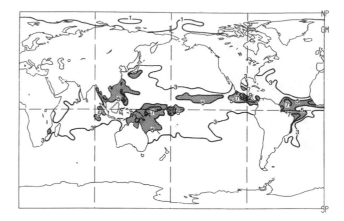

FIGURE 1   Mean (left) and standard deviation (right) of (E − P) from ECHAM3 T42 AGCM AMIP run.

White-White (WW) Atmospheric Model

At a given ocean-model grid point the anomalous fresh-water flux for each time step was obtained from a set of normally distributed random numbers with zero mean and standard deviation of s mm/day. Values of s in the range 1 to 3 mm/day are approximately equal to the amplitude of the seasonal cycle of fresh-water flux that forces the OGCM. These values are also approximately equal to rms values obtained from the AMIP run, and to an independent estimate obtained by Roads et al. (1992) from the National Meteorological Center (NMC) analysis. The same procedure was used at each individual ocean-model grid point, so that the anomalous fresh-water flux that drove the OGCM had no spatial correlation in its covariance field and was uncorrelated in both space and time, i.e., white in both wavenumber and frequency space. We refer to this as the white-white or WW model.

Red-White (RW) Atmospheric Model

The AMIP fresh-water flux anomaly field was represented as a series of empirical orthogonal functions (EOFs). The first 15 of these spatial functions $e_n(x)$ and their associated eigenvalues l_n were retained. At each time step of the OGCM, an anomalous global fresh-water flux field was constructed as the linear combination of products of the $e_n(x)$ and n-random number with zero mean and standard deviation l_n. This model had the spatial structure of the fresh-water flux field from the T42, which was highly spatially correlated (red in wavenumber space) while being random in time (white in frequency space). Note that each EOF mode carried the same variance, l_n, as in the AMIP run. This type of stochastic atmospheric model is conceptually similar to but quantitatively different from that employed by MMR, and we refer to it here as the RW model.

Red-Red (RR) Atmospheric Model with Feedback

The AMIP data were used to develop a regression model relating the SST and SST gradients to the anomalous fresh-water flux. The model used the 15 leading EOFs of the AMIP fresh-water flux and SST fields. The regression model was nearly global in nature, covering all ocean points where sea ice never occurred. In most regions it captured 80 to 90 percent of the variance in the original AMIP fresh-water flux data set (only 15 EOFs).[5] However, the interesting fact was that the regression model captured a minimum of 96 percent of the variance associated with the first 15 EOFs. This result in turn suggests the highly linear relation between the SST and SST gradients and fresh-water flux. The basic approach to constructing such an atmospheric model can be found in Barnett et al. (1993).

Incorporation of the SST and SST gradient into the model obviously allows for a feedback between the ocean and pseudo-atmosphere and represents a type of coupled model not previously attempted on a global scale (although such a coupled model has produced good simulations of El Niño events (Barnett et al., 1993)). The relatively slow changes in SST ensure that the atmospheric model will be red in frequency space. The large-scale spatial correlations in both SST and fresh-water flux produce a wavenumber dependence that is also red, so this model was called the red-red or RR model.

Experimental Setup

The basic model was run for 4000 years of simulated time forced only by the seasonal cycle. The state of the ocean at that time was taken as the initial condition for all subsequent runs. Four different basic simulations were

---

[5]The main model skill (and the AMIP signal) lay between 30°N to 30°S. It captured only 20 to 50 percent in the highest latitudes, and this turned out to be an important shortcoming.

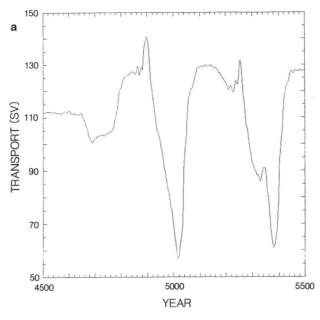

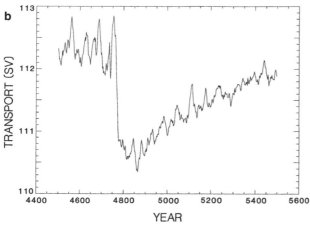

FIGURE 2   Transport (sverdrups) through the Drake Passage for OGCM forced by white-white noise (a) and red-white (b) noise.

carried out: The first was a continuation of the climatology run for an additional 1500 years. Results from this run have the designator C. Each of the three stochastic, anomalous fresh-water flux models was mated to the OGCM and run for 1500 years. The first 500 years of each integration were discarded, leaving the remaining 1000 years for analysis. The designators for these runs are WW, RW, or RR depending on which atmospheric model was used. All three anomaly runs included the same seasonal cycle, fresh-water flux, wind stress, and pseudo-heat flux forcing used in the climatology run.

RESULTS

This section briefly summarizes some of the more interesting results of the four experiments described above. Since this research effort is in the very early stages, these results are presented in descriptive form.

Antarctic Circumpolar Current Response

The model-simulated transport through the Drake Passage is shown in Figure 2 for several of the experiments. The C run shows nearly constant transport close to that observed, with annual variations of less than 1 percent (no illustration shown). The WW run, on the other hand, for $s = 2$ mm/day shows large variability, wherein the transport can change by 50 percent (Figure 2a). The time scale for this fluctuation is of the order of 300 years; it seems to be the mode of variation previously found by MMR (see below). By contrast, the RW run (Figure 2b) shows a step-like jump but otherwise low interdecadal variability. The feedback or RR run (not shown) is much like the RW run but with almost no high-frequency variability.

We deduced from the above results that the spatial correlation in the fresh-water forcing was not particularly important to the model response. Similarly, we concluded that it is the magnitude of the forcing in the high latitudes (only) that really affects the model behavior. We confirmed these conclusions by rerunning the RW case, but with the magnitude of the forcing increased by a factor of 5 for latitudes above 40°, where the RW atmospheric model was deficient in energy. This increased forcing brought the magnitude of the flux up to the order of 1 mm/day and corrected for the model's low variability in that region. The resulting transport through the Drake Passage (Figure 3) now resembles that found in the WW run with regard to both magnitude and time scale. In addition, similar experiments showed that it was the E − P flux south of 40°S that generated the MMR mode, while the flux in the northern ocean was of little significance other than local.

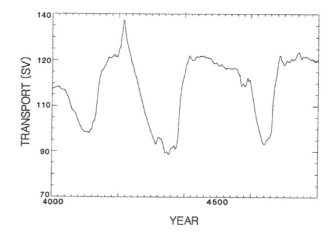

FIGURE 3   Same as red-white transport in Figure 2b, but run with variability above 40° latitude increased by a factor of 5.

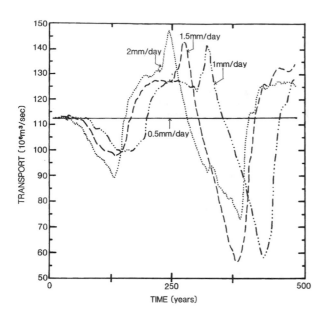

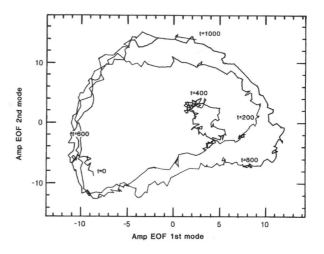

FIGURE 6   As in Figure 5, but for the Pacific section. The trajectory of the variability (the attractor) is shown in the phase space of the first two principal components.

FIGURE 4   Drake Passage transport for white-white forcing with different levels of fresh-water flux variability. Atmospheric models and observation/analyses suggest that values of 1 to 2 mm/day are realistic in the higher-latitude regions.

The sensitivity of the Antarctic Circumpolar Current (ACC) transport to the magnitude of the forcing in the WW run was studied by repeating the experiment for values of s between 0.5 and 2.0 mm/day (Figure 4). The large-scale mode is excited rather uniformly by values above 1.0 mm/day, but not by values of 0.5 mm/day or less (which resemble the control run). Note the similar shape of the responses in the range 1 to 2 mm, as well as a magnitude that is essentially independent of s. As the magnitude of the forcing is reduced, the mode takes longer to become excited. Apparently a threshold value between 0.5 to 1.0 mm is required to trigger the mode, and this fact suggests a highly nonlinear generation mechanism.

Meridional Structure

Salinity and temperature data from the WW run[6] were saved along key meridional sections in both the Pacific and the Atlantic and subjected separately to EOF analysis. The principal components (PCs) for modes 1 and 2 for both oceans are shown in Figure 5 in standard format and in Figure 6 in two-dimensional phase space. It is clear that the roughly 300-year oscillation described above extends into the high latitudes of both major oceans. In the Atlantic, the PCs are in quadrature, and this means that the salinity anomalies propagate. The sense of the motion revealed by the EOFs (Figure 7) is that more (or less) saline water moves northward from Antarctica in the near-surface waters to the central North Atlantic where it sinks and returns at depth to the Southern Ocean. While this is happening, an

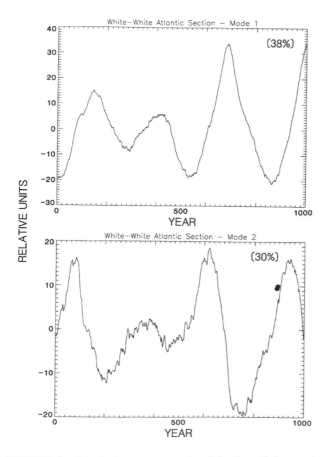

FIGURE 5   Principal components for Atlantic salinity section EOFs: white-white forcing.

[6]Given the similarity in the responses of the OGCM to different atmospheric forcings, we concentrate in the rest of the paper on the results of the WW run.

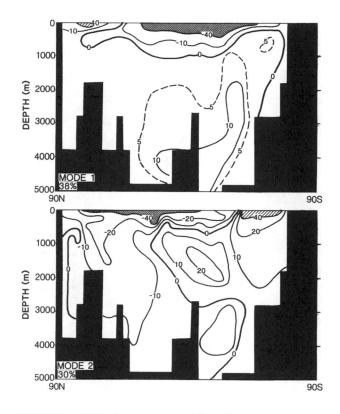

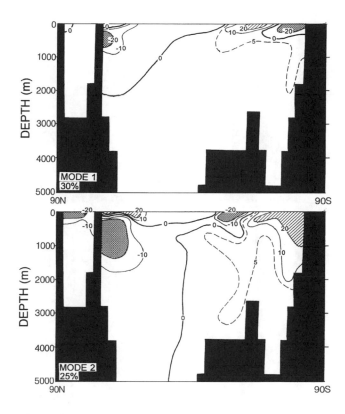

FIGURE 7   EOFs that accompany Figure 5.

FIGURE 8   EOFs that accompany Figure 6.

anomaly of opposite sign is following the same trajectory, but 180° out of phase spatially. This is just the signal described by MMR, but it is produced here by totally white forcing as opposed to the RW forcing used in their experiment. As we saw above, the signal was also produced by our version of the RW forcing provided the magnitude of the high southern- latitude forcing was large enough.

The form of the signal in the Pacific (Figure 8) is somewhat like that found in the Atlantic. The two PCs form a rather simple attractor (Figure 6) and are in quadrature again, suggesting a propagating signal in the salinity field. The time scale is roughly 300 years, as found above. A major difference between the two oceans is that the signal does not penetrate in strength to the same depth it was found to in the Atlantic. The signal appears to propagate from one end of the ocean to the other, but is closely confined to the upper layers of the water column and has more spatial structure than in the Atlantic.

The simple correlation between the ACC transport fluctuations (Figure 2) and the variability of the salinity along the Atlantic-Pacific sections is shown in Figure 9. In the Atlantic, the plus-minus signature of the correlation field is indicative of the propagating signal discussed above. Note that major reductions in the strength of the ACC accompany reduced salinity over most of the South Atlantic to depths of the order of 1000 m. This is countered by increased salinity in the North Atlantic and in the deeper parts of the

entire Atlantic. The signature of the ACC variability in the Pacific is identical to that in the Atlantic, but only in the North Pacific. Most of the Pacific, even to the greatest depths, is more or less in phase with the ACC signal. However, the absolute values of the correlation in the Pacific are less than those found in the Atlantic.

Horizontal Structures

The horizontal structure of the low-frequency MMR signal is investigated in Figure 10, which shows the first and second EOFs of the Atlantic salinity anomaly at a depth of 700 m from the WW run. The associated PCs (not shown) suggest motion such that EOF1 leads EOF2. The first EOF shows the North and South Atlantic having anomalies of opposite sign. But note the negative anomaly extending into the Gulf of Mexico region. The second EOF shows the positive anomaly that occupied the North Atlantic has rotated and moved clockwise such that it is now off Africa. The negative anomaly has also moved clockwise so that it now covers most of the North Atlantic. These clockwise motions suggest that the anomalies are being influenced mainly by advection. Much the same type of anomaly motion is seen at 2000 m also.

The simple correlation between the ACC index and the salinity anomalies at 700 m and 2000 m is given in Figure 11. The entire Indian, North Pacific, and North Atlantic

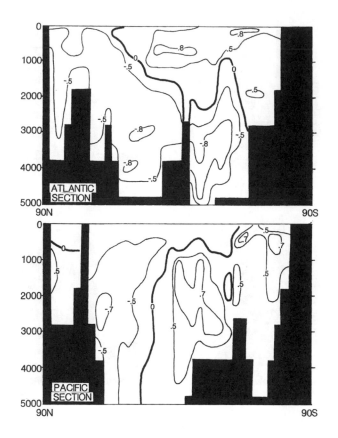

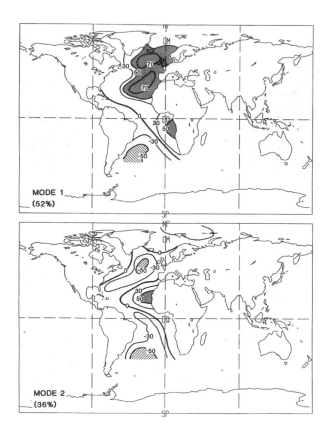

FIGURE 9   Correlation between Drake Passage transport and salinity variability along the meridional sections for the white-white forcing experiment.

FIGURE 10   Leading EOFs of Atlantic salinity field at 700 m from the white-white run.

oceans are in antiphase with the ACC at 700 m. Much the same pattern holds in these regions at 2000 m. The South Pacific and South Atlantic vary in phase with the ACC at both depths, but an unexpectedly strong signal (positive) is apparent in the western Equatorial Pacific. At 700 m the strongest signals are in the Indian Ocean, but at 2000 m this distinction is held by the South Pacific and the North Atlantic. It is obvious that the ocean mode of variation discussed here is truly global in extent.

Levels of Variability

The rms variability of the salinity field from the WW and RR runs is shown in Figure 12 for depths of 75 m. The RR run produces far more variance in the salinity field, especially in the deep ocean (e.g., 2000 m, not shown) where rms values of 0.06 to 0.08 psu are common in higher latitudes. The spatial distribution of the variability between the two runs clearly shows the importance of large-scale air-sea interactions in forcing the model. If the RR run is at all realistic, then the coupling between the two media needs to be taken into account in studies of interdecadal variability. It is interesting that the spatial structure of the

modal response was similar between all the runs, even if the levels of variance were not. This suggests the response is a leading "eigenmode" of the global ocean model that is easily excited by a wide range of forcing.

SUMMARY

A reasonably sophisticated, realistic OGCM has been forced with annual cycles of wind stress, temperature, and fresh-water flux, and also with anomalies of fresh-water flux. The latter anomaly model simulations range from forcing that is white in both space and time to a model that is red in both domains and also incorporates feedback between the fresh-water flux and local SST.

The results of these simulations show a richly structured response. One prominent mode is that discovered by Mikolajewicz and Maier-Reimer (1990, 1991). We found that mode is driven principally by anomalous fresh-water flux in the higher latitudes of the Southern Hemisphere. The spatial structure and details of the forcing are not as important to the mode as the magnitude of the forcing. Monthly anomalies greater than 1 mm/day will excite the mode, while values below 0.5 mm/day do not. Atmospheric-model results and NMC analyses suggest that realistic values of anomalous fresh-water flux in the high latitudes are of the order

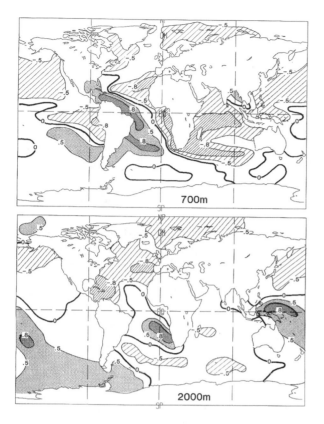

FIGURE 11 Correlation between Drake Passage transport and salinity variations at 700 m and 2000 m for the 1000 year white-white run.

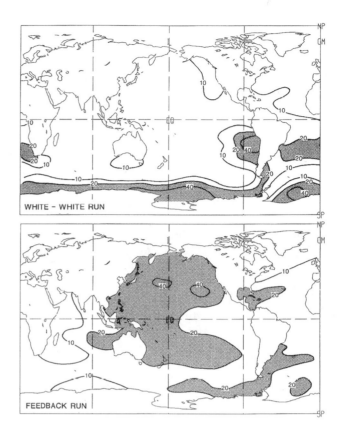

FIGURE 12 Standard deviation of salinity (psu) at 75 m from two noise-forced runs.

of 1 to 2 mm/day, so the forcing required for this mode in the OGCM may not be unrealistic.

The principal mode referred to above has expressions in all major oceans, and is also associated with changes in the transport of the ACC by a factor of two. The mode in the Atlantic is associated with a meridional overturning of the entire ocean and a horizontal circulation that is clockwise. The mode in the Pacific does not penetrate in strength as deeply as in the Atlantic, and it appears to have a more difficult time extending across the equator.

The levels of interdecadal variability produced by the model for virtually all of the atmospheric forcing models we used was surprisingly large in view of the coarse OGCM resolution. If these levels of variance were found in the real-world ocean, they would represent an important level of change.

ACKNOWLEDGMENTS

This work was supported by the Department of Energy through its Computer Hardware, Advanced Mathematics, and Model Physics (CHAMMP) effort under contract DE-FG03-91-ER61215, and by the National Science Foundation under Grant NSF ATM88-14571-03. The Deutsche Forschungsgemeinschaft and the Sonderforschungbereich 318 provided support for U. Mikolajewicz. M. Schultz helped with the drafting and T. Tubbs carried through many of the key computations. R. Wilde was supported under the NSF program for undergraduate research (REU).

Commentary on the Paper of Barnett et al.

KEVIN E. TRENBERTH
National Center for Atmospheric Research

Dr. Barnett has given us a very "colorful" talk dealing with an ocean GCM that has very simple forcing. The nature of the forcing, as he mentioned, is critical in terms of its magnitude in particular, and its special character to a lesser degree. The first thing that I thought of—probably a minor point—with regard to the fresh-water forcing is the aspect dealing with runoff from rivers into the ocean. In fact, the precipitation exceeds evaporation over land in general. Is that in the model? If you just model E − P over the ocean, there will be imbalances, because there E exceeds P. What does that do to the overall fresh-water balance?

Dr. Barnett used three kinds of forcings. The main one he talked about was the white/white, which seems somewhat unrealistic. He used a global standard deviation value of 2 mm/day. That number may be reasonable globally, but should be much larger than that in the tropics and about half that in high latitudes. But large numbers at high latitudes were one of the critical things in getting this oscillation.

My other concern is that I consider the model to be unnaturally constrained. Any time you use an idealized atmosphere, the nature of the fluxes into the model ocean and the ability of the atmosphere to feed back and adjust in various ways are limited. For instance, can the climate system adjust to compensate for the fresh-water flux? The parameterization derived here for the red/red case, especially in the tropics, does indicate that the fresh-water flux can be altered substantially by changes in sea temperatures. Certainly other aspects of land-ocean differences introduce similar complexities. Therefore, the relevance of something like this to the real world is, I think, a very open question.

Dr. Barnett did not talk about mechanisms, although maybe other atmospheric modelers will. Perhaps, because the model integrations are made over wide areas, the spatial structure does not matter much; if the area mean is the main thing that counts, the result is a red spectrum whatever you do. Wide-area integration also implies that there is a random-walk process that will result in perturbations in the fresh-water flux. The perturbations will affect or even shut down the thermohaline circulation, which will alter the heat balance because of the heat transports involved, which in turn will cause changes in temperatures.

This process is reflected in the temperature variations the model exhibits at high latitudes. Dr. Barnett asks, "Is this natural variability something that is going to confound us when we look at the greenhouse effect?" But ultimately, shutting down the thermohaline circulation will change the temperatures enough that they will probably cause the thermohaline circulation to jerk back into action at some point and advect fresh water around. Might this be part of the mechanism that results in an oscillation? I do not know. I think the bottom line is the question of these models' relevance to the real world.

Discussion

BARNETT: That last question does need to be kept in mind, but for a full-ocean GCM our model does quite a good job of reproducing the main features of the global ocean.

TRENBERTH: Isn't the forcing at high latitudes much higher than in the AMIP run?

BARNETT: Not much. We picked 2 mm/day because it's a fairly good global average, and I think it's fairly realistic. I believe some other models using a smaller figure still get the oscillation.

SARACHIK: Why did you choose to show the transport of the Antarctic circumpolar current, and what was the structure of the changes in it?

BARNETT: It's a simple diagnostic for the system. If you have no signal there, you won't have much anywhere.

BRYAN: I'd like to respond to Kevin's comment about runoff. For the North Atlantic, I believe that runoff is dominant at very high latitudes simply because the coastline is so extensive by comparison with the ocean area. If you also take into account the tremendous Arctic fresh-water discharge, the total runoff is much greater than the local net precipitation.

BARNETT: This runoff effect has been included in a couple of models for the Hamburg greenhouse runs, and it's my impression that it didn't make much difference. If it is large, though, it should be fairly simple to put into these kinds of models.

TRENBERTH: Runoff might affect the nature of feedbacks, and clearly on time scales that cover the melting of major ice caps it would become critical.

WEAVER: It seems to me that you do have a sort of parameterization of the runoff, since you obtain your mean fresh-water flux from diagnosing a spun-up state that you got from observed salinities.

MYSAK: Have you done any sensitivity studies for your model's parameters?

BARNETT: No, though others have. It's tuned to today's climate. I don't think it's particularly diffusive, for instance.

MYSAK: One of the things we were interested in doing with the very simple two-dimensional thermohaline model was the Maier-Reimer experiment. We looked at its sensitivity to diffusivities, for example, to see just how robust the 200-to-300-year oscillation was. We found that it was fairly robust over a fairly wide range, and very robust for a certain set of parameters. For the extreme of a very diffusive ocean, the 200-year oscillation was damped out, and the system alternated between sinking in the north and sinking in the south.

LEHMAN: I have two questions. What is the sensitivity of this model to fresh-water forcing? Convection in the model Maier-Reimer used collapsed with a .02-sverdrup increase to the northern part of the North Atlantic. Second, do you know why the model is more sensitive to Southern Hemisphere forcing than Northern?

TRENBERTH: It might simply be the respective sizes of the oceans. If you're doing random forcing that has no spatial structure, you need a large area to integrate over to get a decent-sized signal.

BARNETT: The odd thing is that you get about the same answer when you use white noise as when you use something with nice big spatial scales.

YOUNG: Have you thought about computing the gain by calculating the linear eigenmodes?

BARNETT: I believe we've done that numerically by forcing it with white noise. The transfer function of the model is essentially the empirical orthogonal function. But we've barely begun to look at it.

MARTINSON: The fluctuations that come out of these models are very interesting. Of course it's hard to equate them with reality when some of the regions appear to be so sensitive that you could spit off a ship and cause the ice cover to disappear. I think in most cases there is a whole set of self-regulating feedbacks that prevent the system from overturning like that. I'd like to get some of us together to look at the fine-scale local processes in terms of your larger-scale results and see whether indeed your model is representative in what you might call a budget sense.

BARNETT: I'd be glad to join you. We did manage to take sea ice into account and not destroy it too badly—though our answers weren't very different from those of models without any—but we really don't know how well the integral properties represented by this model resemble reality.

MYSAK: My feeling is that on the 200-to-300-year time scale the sea ice is probably not going to influence the results.

TRENBERTH: Ah, but having ice in the model allows for many other feedbacks to the atmosphere that make the system even more complex.

Ocean Modeling
Reference List

Andrews, D.G., and M.E. McIntyre. 1976. Planetary waves in horizontal and vertical shear: The generalized Eliassen-Palm relation and the mean zonal acceleration. J. Atmos. Sci. 33:2031-2048.

Armi, L. 1978. Some evidence for boundary mixing in the deep ocean. J. Geophys. Res. 83:1971-1979.

Barnett, T.P., N. Graham, M. Latif, S. Pazan, and W. White. 1993. ENSO and ENSO-related predictability. Part 1: Prediction of equatorial Pacific sea surface temperature with a hybrid coupled ocean-atmosphere model. J. Climate 6:1545-1566.

Barnett, T.P., M. Chu, R. Wilde, U. Mikolajewicz. 1995. Low-frequency ocean variability induced by stochastic forcing of various colors. In Natural Climate Variability on Decade-to-Century Time Scales. D.G. Martinson, K. Bryan, M. Ghil, M.M. Hall, T.R. Karl, E.S. Sarachik, S. Sorooshian, and L.D. Talley, eds. National Academy Press, Washington, D.C.

Batchelor, G. 1969. Computation of the energy spectrum in homogeneous, two-dimensional turbulence. Phys. Fluids 12:233-238.

Baumgartner, A., and E. Reichel. 1975. The World Water Balance. Elsevier, New York.

Birchfield, G.E., H. Wang, and M. Wyant. 1990. A bimodal climate response controlled by water vapor transport in a coupled ocean-atmosphere box model. Paleoceanography 5:383-395.

Bjerknes, J. 1964. Atlantic air-sea interaction. Adv. Geophys. 10:1-82.

Bjerknes, J. 1969. Atmospheric telecommunications from the equatorial Pacific. Mon. Weather Rev. 97:163-172.

Boyle, E.A., and L.D. Keigwin. 1987. North Atlantic thermohaline circulation during the past 20,000 years linked to high-latitude surface temperature. Nature 330:35-40.

Bretherton, F.P. 1982. Ocean climate modeling. Prog. Oceanogr. 11:93-129.

Bretherton, F., and D. Haidvogel. 1976. Two-dimensional turbulence above topography. J. Fluid Mech. 78:129-154.

Briegleb, B.P. 1991. Description of CCM2 Radiative/Convective Model. Climate Modeling Section Note No. 4, National Center for Atmospheric Research, Boulder, Colo., 11 pp.

Broecker, W.S. 1989. The salinity contrast between the Atlantic and Pacific Oceans during glacial time. Paleoceanography 4:207-212.

Broecker, W.S., D.M. Peteet, and D. Rind. 1985. Does the ocean-atmosphere system have more than one stable mode of operation? Nature 315:21-26.

Broecker, W.S., G. Bond, and M. Klas. 1990. A salt oscillator in the glacial Atlantic? 1. The concept. Paleoceanography 5:469-477.

Bryan, F. 1986a. Maintenance and Variability of the Thermohaline Circulation. Ph.D. dissertation, Princeton University, 254 pp.

Bryan, F. 1986b. High latitude salinity effects and interhemispheric thermohaline circulations. Nature 323:301-304.

Bryan, F. 1986c. Parameter sensitivity of primitive equation ocean general circulation models. J. Phys. Oceanogr. 17:970-985.

Bryan, K. 1969. A numerical method for the study of the circulation of the World Ocean. J. Comput. Phys. 3:347-376.

Bryan, K. 1979. Models of the world ocean. Dyn. Atmos. Oceans 3:327-338.

Bryan, K. 1984. Accelerating the convergence to equilibrium of ocean climate models. J. Phys. Oceanogr. 14:666-673.

Bryan, K., and M.D. Cox. 1972. An approximate equation of state for numerical models of ocean circulation. J. Phys. Oceanogr. 2:510-517.

Bryan, K., and F.C. Hansen. 1995. A toy model of North Atlantic climate variability on a decade-to-century time scale. In Natural Climate Variability on Decade-to-Century Time Scales. D.G. Martinson, K. Bryan, M. Ghil, M.M. Hall, T.R. Karl, E.S. Sarachik, S. Sorooshian, and L.D. Talley, eds. National Academy Press, Washington, D.C.

Bryan, K., and J.L. Lewis. 1979. A water mass model of the world ocean. J. Geophys. Res. 84:2503-2517.

Bryan, K., and R. Stouffer. 1991. A note on Bjerknes' hypothesis for North Atlantic variability. J. Mar. Systs. 1:229-241.

Cai, W., and S.J. Godfrey. 1995. Surface heat flux parameterizations and the variability of thermohaline circulation. J. Geophys. Res. 100:10679-10692.

Cattle, H.. 1985. Diverting Soviet rivers: Some possible repercussions for the Arctic Ocean. Polar Record 22:485-498.

Charney, J. 1971. Geostrophic turbulence. J. Atmos. Sci. 28:1087-1095.

Cook, E.R., B.M. Buckley, and R.D. D'Arrigo. 1995. Interdecadal temperature oscillations in the Southern Hemisphere: Evidence from Tasmanian tree rings since 300 B.C. In Natural Climate Variability on Decade-to-Century Time Scales. D.G. Martinson, K. Bryan, M. Ghil, M.M. Hall, T.R. Karl, E.S. Sarachik, S. Sorooshian, and L.D. Talley (eds.). National Academy Press, Washington, D.C.

Cox, M. 1987. Isopycnal diffusion in a z-coordinate ocean model. Ocean Model. 74:1-5.

Cox, M.D. 1984. A Primitive Equation, Three Dimensional Model of the Ocean. GFDL Ocean Group Technical Report No. 1, NOAA, 143 pp.

Dansgaard, W., S.J. Johnsen, H.B. Clausen, and C.C. Langway. 1970. Climatic record revealed by the Camp Century ice core. In The Late Cenozoic Glacial Ages. K.K. Turekian (ed.). Yale University Press, pp. 37-56.

Dansgaard, W., J.W.C. White, and S.J. Johnsen. 1989. The abrupt termination of the Younger Dryas climate event. Nature 339:532-534.

Deardorff, J.W. 1972. Theoretical expression for the counter-gradient vertical heat flux. J. Geophys. Res. 77:5900-5904.

Delworth, T., S. Manabe, and R.J. Stouffer. 1993. Interdecadal variability of the thermohaline circulation in a coupled ocean-atmosphere model. J. Climate 6:1993-2011.

Delworth, T., S. Manabe, and R. Stouffer. 1995. North Atlantic interdecadal variability in a coupled model. In Natural Climate Variability on Decade-to-Century Time Scales. D.G. Martinson, K. Bryan, M. Ghil, M.M. Hall, T.R. Karl, E.S. Sarachik, S. Sorooshian, and L.D. Talley (eds.). National Academy Press, Washington, D.C.

Dickson, R.R., J. Meincke, S.-A. Malmberg, and A.J. Lee. 1988. The Great Salinity Anomaly in the northern North Atlantic 1968-1982. Prog. Oceanogr. 20:103-151.

Dietrich, G., and J. Ulrich. 1968. Atlas zur Ozeanographie. Bibliogr. Institut, Mannheim, 79 pp.

Duplessy, J.C., N.J. Shackleton, R.G. Fairbanks, L. Labeyrie, D. Oppo, and N. Kallel. 1988. Deepwater source variations during the last climate cycle and their impact on the global deepwater circulation. Paleoceanography 3:343-360.

Flato, G.M., and W.D. Hibler. 1990. On a simple sea-ice dynamics model for climate studies. Ann. Glaciol. 14:72-77.

Folland, C.K., D.E. Parker, M.N. Ward, and A.W. Colman. 1986. Sahel rainfall, northern hemisphere circulation anomalies and worldwide temperature changes. Long Range Forecasting and Climate Research, Memorandum #7a. Meteorological Office, Bracknell, U.K., 49 pp.

Gargett, A., and G. Holloway. 1984. Dissipation and diffusion by internal wave breaking. J. Mar. Res. 42:15-27.

Garrett, C. 1991. Marginal mixing theories. Atmos.-Ocean 29:313-339.

Garwood, R.W. 1977. An oceanic mixed-layer model capable of simulating cyclic states. J. Phys. Oceanogr. 7:446-468.

Gaspar, P. 1988. Modeling the seasonal cycle of the upper ocean. J. Phys. Oceanogr. 18:161-180.

Gent, P.R., and J.C. McWilliams. 1990. Isopycnal mixing in ocean models. J. Phys. Oceanogr. 20:150-155.

Ghil, M., and R. Vautard. 1991. Interdecadal oscillations and the warming trend in global temperature time series. Nature 350:324-327.

Ghil, M., A. Mullhaupt, and P. Pestiaux. 1987. Deep water formation and Quaternary glaciations. Climate Dynamics 2:1-10.

Gray, W.M. 1990. Strong association between West African rainfall and U.S. landfall of intense hurricanes. Science 249:1251-1256.

Greatbatch, R.J., and S. Zhang. 1995. An interdecadal oscillation in an idealized ocean basin forced by constant heat flux. J. Climate 8:81-91.

Greatbatch, R.J., A.F. Fanning, A.D. Goulding, and S. Levitus. 1991. A diagnosis of interpentadal circulation changes in the North Atlantic. J. Geophys. Res. 96:22009-22023.

Gregg, M. 1987. Dyapycnal mixing in the thermocline: A review. J. Geophys. Res. 92:5249-5286.

Haney, R.L. 1971. Surface thermal boundary condition for ocean circulation models. J. Phys. Oceanogr. 1:241-248.

Hansen, J., and S. Lebedeff. 1987. Global trends of measured surface air temperature. J. Geophys. Res. 92:13345-13374.

Hasselmann, K. 1976. Stochastic climate models. Part I. Theory. Tellus 28:289-305.

Hellerman, S., and M. Rosenstein. 1983. Normal monthly wind stress over the world ocean with error estimates. J. Phys. Oceanogr. 13:1093-1104.

Hibler, W.D. 1979. A dynamic thermodynamic sea ice model. J. Phys. Oceanogr. 9:815-846.

Hibler, W.D., and S.J. Johnsen. 1979. The 20-year cycle in Greenland ice core records. Nature 280:481-483.

Holloway, G. 1978. A spectral theory of nonlinear barotropic motion above irregular topography. J. Phys. Oceanogr. 8:414-427.

Holloway, G. 1987. Systematic forcing of large-scale geostrophic flows by eddy-topography interaction. J. Fluid Mech. 184:463-476.

Holloway, G. 1992. Representing topographic stress for large-scale ocean models. J. Phys. Oceanogr. 22:1033-1046.

Holtslag, A.A.M., and C.-H. Moeng. 1991. Eddy diffusivity and countergradient transport in the convective atmospheric boundary layer. J. Atmos. Sci. 48:1690-1698.

Huang, R.X., J.R. Luyten, and H.M. Stommel. 1992. Multiple equilibrium states in a combined thermal and saline circulation. J. Phys. Oceanogr. 22:231-246.

Hughes, T.M.C., and A.J. Weaver. 1994. Multiple equilibria of an asymmetric two-basin ocean model. J. Phys. Oceanogr. 24:619-637.

Ikeda, M. 1990. Decadal oscillations of the air-ice-ocean system in the northern hemisphere. Atmos.-Ocean 28:106-139.

IPCC. 1990. Climate Change: The IPCC Scientific Assessment. J.T. Houghton, G.J. Jenkins, and J.J. Ephraums (eds.). Prepared for the Intergovernmental Panel on Climate Change by Working Group I. WMO/UNEP, Cambridge Univ. Press, 365 pp.

Iselin, C. 1939. The influence of vertical and lateral turbulence on the characteristics of the waters at mid-depths. Eos, Trans. AGU 20:414-417.

Keeling, C.D., and T.P. Whorf. 1995. Decadal oscillations in global temperature and atmospheric carbon dioxide. In Natural Climate Variability on Decade-to-Century Time Scales. D.G. Martinson, K. Bryan, M. Ghil, M.M. Hall, T.R. Karl, E.S. Sarachik, S. Sorooshian, and L.D. Talley (eds.). National Academy Press, Washington, D.C.

Keigwin, L.A., G.A. Jones, and S.J. Lehman. 1991. Deglacial meltwater discharge, North Atlantic deep circulation and abrupt climate change. J. Geophys. Res. 96:16811-16826.

Kleeman, R., and S.B. Power. 1995. A simple atmospheric model of surface heat flux for use in ocean modeling studies. J. Phys. Oceanogr. 25:92-105.

Krishnamurti, T.N., S.-H. Chu, and W. Iglesias. 1986. On the sea level pressure of the southern oscillation. Arch. Meteor. Geophys. Bioklimatol., Series A, 34:385-425.

Kushnir, Y. 1994. Interdecadal variations in North Atlantic sea surface temperature and associated atmospheric conditions. J. Climate 7:141-157.

Labeyrie, L.D., J.C. Duplessy, and P.L. Blanc. 1987. Variations in mode of formation and temperature of oceanic deep waters over the past 125,000 years. Nature 327:477-482.

Large, W.G., J.C. McWilliams, and S.C. Doney. 1994. Oceanic vertical mixing: A review and a model with a non-local K-profile boundary layer parameterization. Rev. Geophys. 32:363-403.

Lazier, J.R.N. 1980. Oceanographic conditions at Ocean Weather Ship BRAVO, 1944-1974. Atmos.-Ocean 18:227-238.

LeBlond, P.H., and L.A. Mysak. 1978. Waves in the Ocean. Elsevier, Amsterdam, 602 pp.

Lemke, P., W.B. Owens, and W.D. Hibler. 1990. A coupled sea ice-mixed layer-pycnocline model for the Weddell Sea. J. Geophys. Res. 95:9513-9525.

Levitus, S. 1982. Climatological Atlas of the World Ocean. NOAA Professional Paper 13, U.S. Department of Commerce: National Oceanic and Atmospheric Administration, Washington, D.C., 173 pp. + 17 microfiches.

Levitus, S. 1989a. Interpentadal variability of temperature and salinity at intermediate depths of the North Atlantic Ocean, 1970-74 versus 1955-59. J. Geophys. Res. 94:6091-6131.

Levitus, S. 1989b. Interpentadal variability of salinity in the upper 150 m of the North Atlantic Ocean, 1970-74 versus 1955-59. J. Geophys. Res. 94:9679-9685.

Levitus, S. 1989c. Interpentadal variability of temperature and salinity in the deep North Atlantic, 1970-74 versus 1955-59. J. Geophys. Res. 94:16125-16131.

Levitus, S. 1990. Interpentadal variability of steric sea level and geopotential thickness of the North Atlantic Ocean, 1970-74 versus 1955-59. J. Geophys. Res. 95:5233-5238.

Loder, J.W., and C. Garrett. 1978. The 18.6-year cycle of sea surface temperature in shallow seas due to variations in tidal mixing. J. Geophys. Res. 83:1967-1970.

Lumley, J.A., and H.A. Panofsky. 1964. The Structure of Atmospheric Turbulence. John Wiley and Sons, New York City, 239 pp.

Maier-Reimer, E., U. Mikolajewicz, and K. Hasselmann. 1993. Mean circulation of the Hamburg LSG OGCM and its sensitivity to the thermohaline surface forcing. J. Phys. Oceanogr. 23:731-757.

Manabe, S., and R.J. Stouffer. 1988. Two stable equilibria of a coupled ocean-atmosphere model. J. Climate 1:841-866.

Marotzke, J. 1989. Instabilities and multiple steady states of the thermohaline circulation. In Oceanic Circulation Models: Combining Data and Dynamics. D.L.T. Anderson and J. Willebrand (eds.). NATO ASI series, Kluwer, Dordrecht, pp. 501-511.

Marotzke, J. 1990. Instabilities and multiple equilibria of the thermohaline circulation. Ph.D. Dissertation, Ber. Institut für Meereskunde, Kiel, 126 pp.

Marotzke, J. 1991. Influence of convective adjustment on the stability of the thermohaline circulation. J. Phys. Oceanogr. 21:903-907.

Marotzke, J., and P.H. Stone. 1995. Atmospheric transports, the thermohaline circulation, and flux adjustments in a simple coupled model. J. Phys. Oceanogr. 25:1350-1364.

Marotzke, J., and J. Willebrand. 1991. Multiple equilibria of the global thermohaline circulation. J. Phys. Oceanogr. 21:1372-1385.

Marotzke, J., P. Welander, and J. Willebrand. 1988. Instability and multiple steady states in a meridional-plane model of the thermohaline circulation. Tellus 40A:162-172.

Marshall, J. 1981. On the parameterization of geostrophic eddies in the ocean. J. Phys. Oceanogr. 11:257-271.

Martin, P.J. 1985. Simulation of the mixed layer at OWS November and Papa with several models. J. Geophys. Res. 90:903-916.

McCready, P., and P. Rhines. 1991. Buoyant inhibition of Ekman transport on a slope and its effect on stratified spin-up. J. Fluid Mech. 223:631-661.

McDermott, D., and E.S. Sarachik. 1995. Thermohaline circulations and variability in a two-hemisphere sector model of the Atlantic. In Natural Climate Variability on Decade-to-Century Time Scales. D.G. Martinson, K. Bryan, M. Ghil, M.M. Hall, T.R. Karl, E.S. Sarachik, S. Sorooshian, and L.D. Talley, eds. National Academy Press, Washington, D.C.

McWilliams, J. 1984. The emergence of isolated coherent vortices in turbulent flow. J. Fluid Mech. 146:21-43.

McWilliams, J. 1985. Sub-mesoscale coherent vortices in the ocean. Rev. Geophys. 23:165-182.

McWilliams, J. 1989. Statistical properties of decaying geostrophic turbulence. J. Fluid Mech. 198:199-230.

McWilliams, J.C. 1995. Sub-grid-scale parameterization in ocean general-circulation models. In Natural Climate Variability on Decade-to-Century Time Scales. D.G. Martinson, K. Bryan, M. Ghil, M.M. Hall, T.R. Karl, E.S. Sarachik, S. Sorooshian, and L.D. Talley, eds. National Academy Press, Washington, D.C.

McWilliams, J.C., and J.H.S. Chow. 1981. Equilibrium geostrophic turbulence: A reference solution in a β-plane channel. J. Phys. Oceanogr. 11:921-949.

McWilliams, J.C, and P.R. Gent. 1994. The wind-driven ocean circulation with an isopycnal-thickness mixing parameterization. J. Phys. Oceanogr. 24:46-65.

McWilliams, J.C., W.R. Holland, and J.H.S. Chow. 1978. A description of numerical Antarctic circumpolar currents. Dyn. Atmos. Oceans 2:213-291.

McWilliams, J., and 20 co-authors. 1983. The local dynamics of eddies in the western North Atlantic. In Eddies in Marine Science. A. Robinson (ed.). Springer-Verlag, Berlin, pp. 92-113.

McWilliams, J.C., P.C. Gallacher, C.-H. Moeng, and J.C. Wyngaard. 1993. Modeling the oceanic planetary boundary layer. In Large Eddy Simulation of Complex Engineering and Geophysical Flows. B. Galperin and S. Orszag (eds.). Lecture Notes in Engineering, Cambridge University Press, pp. 441-454.

Mellor, G.L., and T. Yamada. 1982. Development of a turbulence closure model for geophysical fluid dynamics. Rev. Geophys. Space Phys. 20:851-875.

Mikolajewicz, U., and E. Maier-Reimer. 1990. Internal secular variability in an ocean general circulation model. Climate Dynamics 4:145-156.

Mikolajewicz, U., and E. Maier-Reimer. 1991. One example of a natural mode of the ocean circulation in a stochastically forced ocean general circulation model. In Strategies for Future Climate Research. M. Latif (ed.). Available from Max Planck-Institut für Meteorologie, Hamburg, Germany, pp. 287-318.

Milliff, R.A., and J.C. McWilliams. 1994. The evolution of boundary pressure in enclosed ocean basins. J. Phys. Oceanogr. 24:1317-1338.

Montgomery, R. 1940. The present evidence on the importance of lateral mixing processes in the ocean. Bull. Am. Meteorol. Soc. 21:87-94.

Moore, A.M., and C.J.C. Reason. 1993. The response of a global ocean general circulation model to climatological surface boundary conditions for temperature and salinity. J. Phys. Oceanogr. 23:300-328.

Myers, P.G., and A.J. Weaver. 1992. Low-frequency internal oceanic variability under seasonal forcing. J. Geophys. Res. 97:9541-9563.

Mysak, L.A., and D.K. Manak. 1989. Arctic sea-ice extent and anomalies, 1953-1984. Atmos.-Ocean 27:376-405.

Mysak, L.A., D.K. Manak, and R.F. Marsden. 1990. Sea-ice anomalies in the Greenland and Labrador Seas during 1901-1984 and their relation to an interdecadal Arctic climate cycle. Climate Dynamics 5:111-133.

Mysak, L.A., T.F. Stocker, and F. Huang. 1993. Century-scale variability in a randomly forced, two-dimensional thermohaline ocean circulation model. Climate Dynamics 8:103-116.

O'Brien, J.J. 1970. A note on the vertical structure of the eddy exchange coefficient in the planetary boundary layer. J. Atmos. Sci. 27:1213-1215.

Oeschger, H., J. Beer, U., Siegenthaler, and B. Stauffer. 1984. Late-glacial climate history from ice cores. In Climate Processes and Climate Sensitivity. J.E. Hansen and T. Takahashi (eds.). Geophysical Monographs 29, Maurice Ewing Volume 5, American Geophysical Union, Washington, D.C., pp. 299-306.

Pacanowski, R., and G. Philander. 1981. Parameterization of vertical mixing in numerical models of tropical oceans. J. Phys. Oceanogr. 11:1443-1451.

Peters, H., M. Gregg, and J. Toole. 1988. On the parameterization of equatorial turbulence. J. Geophys. Res. 93:1199-1218.

Pierce, D.W., T.P. Barnett, and U. Mikolajewicz. Unpublished manuscript, 1994. On the competing roles of heat and fresh water flux in forcing thermohaline oscillations.

Quon, C., and M. Ghil. 1992. Multiple equilibria in thermosolutal convection due to salt-flux boundary conditions. J. Fluid Mech. 235:449-483.

Rahmstorf, S., and J. Willebrand. 1995. The role of temperature feedback in stabilizing the thermohaline circulation. J. Phys. Oceanogr. 25:787-805.

Redi, H. 1982. Oceanic isopycnal mixing by coordinate rotation. J. Phys. Oceanogr. 12:1154-1158.

Rhines, P.B., and W.R. Young. 1982. Homogenization of potential vorticity in planetary gyres. J. Fluid Mech. 122:347-367.

Roads, J.O., S.-C. Chen, J. Kao, D. Langley, and G. Glatzmaier. 1992. Global aspects of the Los Alamos general circulation model hydrologic cycle. J. Geophys. Res. 97(D9):10051-10068.

Roeckner, E., K. Arpe, L. Bengtsson, S. Brinkop, L. Dumenil, M. Esch, E. Kirk, F. Lunkeit, M. Ponater, B. Rockel, R. Sausen, U. Schlese, S. Schubert, and M. Windelband. 1992. Simulation of the present-day climate with the ECHAM model: Impact of model physics and resolution. Rept. No. 93, Max Planck-Institut für Meteorologie, Hamburg.

Roemmich, D., and C. Wunsch. 1984. Apparent changes in the climatic state of the deep North Atlantic Ocean. Nature 307:447-450.

Roemmich, D., and C. Wunsch. 1985. Two transatlantic sections: Meridional circulation and heat flux in the subtropical North Atlantic Ocean. Deep-Sea Res. 32:619-664.

Ruddiman, W., and A. McIntyre. 1977. Late Quaternary surface ocean kinematics and climate change in the high-latitude North Atlantic. J. Geophys. Res. 82:3877-3887.

Sarmiento, J.L., R.D. Slater, M.J.R. Fasham, H.W. Ducklow, J.R. Toggweiler, and G.T. Evans. 1993. A seasonal three-dimensional ecosystem model of nitrogen cycling in the North Atlantic euphotic zone. Glob. Biogeochem. Cycl. 7:415-450.

Schlosser, P., G. Bönisch, M. Rhein, and R. Bayer. 1991. Reduction of deepwater formation in the Greenland Sea during the 1980s: Evidence from tracer data. Science 251:1054-1056.

Schmitt, R.W., P.S. Bogden, and C.E. Dorman. 1989. Evaporation minus precipitation and density fluxes for the North Atlantic. J. Phys. Oceanogr. 19:1208-1221.

Schopf, P.S. 1985. Modeling tropical sea-surface temperature: Implications of various atmospheric responses. In Coupled Ocean-Atmosphere Models. J. Nihoul (ed.). Elsevier Oceanography Series No. 40, Amsterdam, pp. 727-734.

Spall, M.A. 1992. Variability of sea surface salinity in stochastically forced systems. Climate Dynamics 8:151-160.

Stocker, T.F. and L.A. Mysak. 1992. Climatic fluctuations on the century time scale: A review of high-resolution proxy data and possible mechanisms. Climate Change 20:227-250.

Stommel, H. 1961. Thermohaline convection with two stable regimes of flow. Tellus 13:224-230.

Sverdrup, H.U. 1957. Oceanography. In Handbuch der Physik. Springer-Verlag, Berlin, 48:630-638.

Tennekes, H. 1973. A model for the dynamics of the inversion above a convective boundary layer. J. Atmos. Sci. 30:558-567.

Toggweiler, J.R., and B. Samuels. 1992. Is the magnitude of the deep outflow from the Atlantic Ocean actually governed by the Southern Hemisphere winds? In The Global Carbon Cycle. M. Heimann (ed.). NATO ASI Series, Springer-Verlag, Berlin, pp. 303-331.

Toggweiler, J.R., K. Dixon, and K. Bryan. 1989. Simulations of radiocarbon in a coarse-resolution world ocean model. I: Steady state, pre-bomb distributions. J. Geophys. Res. 94:8217-8242.

Treguier, A. 1989. Topographically generated steady currents in barotropic turbulence. Geophys. Astrophys. Fluid Dyn. 47:43-68.

Treguier, A., and J. McWilliams. 1990. Topographic influences on wind-driven, stratified flow in a β-plane channel: An idealized model for the Antarctic Circumpolar Current models. J. Phys. Oceanogr. 20:321-343.

Trenberth, K.E., W.G. Large, and J.G. Olson. 1990. The mean annual cycle in global ocean wind stress. J. Phys. Oceanogr. 20:1742-1760.

Troen, I.B., and L. Mahrt. 1986. A simple model of the atmospheric boundary layer: Sensitivity to surface evaporation. Bound.-Layer Meteorol. 37:129-148.

UNESCO. 1981. Tenth report of the Joint Panel on Oceanographic Tables and Standards. UNESCO Technical Paper on Marine Science No. 36, UNESCO, Paris.

Walin, G. 1985. The thermohaline circulation and control of ice ages. Paleogr. Paleoclim. Paleoecol. 50:323-332.

Weaver, A.J. 1995. Decadal-to-millennial internal oceanic variability in coarse-resolution ocean general-circulation models. In Natural Climate Variability on Decade-to-Century Time Scales. D.G. Martinson, K. Bryan, M. Ghil, M.M. Hall, T.R. Karl, E.S. Sarachik, S. Sorooshian, and L.D. Talley, eds. National Academy Press, Washington, D.C.

Weaver, A.J., and T.M.C. Hughes. 1992. Stability and variability of the thermohaline circulation and its link to climate. In Trends in Physical Oceanography. No. 1, Research Trends Series, Council of Scientific Research Integration, Trivandrum, India, pp. 15-70.

Weaver, A.J., and E.S. Sarachik. 1990. On the importance of vertical resolution in certain ocean general circulation models. J. Phys. Oceanogr. 20:600-609.

Weaver, A.J., and E.S. Sarachik. 1991a. The role of mixed boundary conditions in numerical models of the ocean's climate. J. Phys. Oceanogr. 21:1470-1493.

Weaver, A.J., and E.S. Sarachik. 1991b. Evidence for decadal variability in an ocean general circulation model: An advective mechanism. Atmos.-Ocean 29:197-231.

Weaver, A.J., E.S. Sarachik, and J. Marotzke. 1991. Internal low frequency variability of the ocean's thermohaline circulation. Nature 353:836-838.

Weaver, A.J., J. Marotzke, E. Sarachik, and P. Cummins. 1993. Stability and variability of the thermohaline circulation. J. Phys. Oceanogr. 23:39-60.

Weaver, A.J., S.M. Aura, and P.G. Myers. 1994. Interdecadal variability in an idealized model of the North Atlantic. J. Geophys. Res. 99:12423-12441.

Welander, P. 1986. Thermohaline effects in the ocean circulation and related simple models. In Large-Scale Transport Processes in Oceans and Atmosphere. D.L.T. Anderson and J. Willebrand (eds.). NATO ASI series, Reidel, Dordrecht, pp. 163-200.

Winton, M., and E.S. Sarachik. 1993. Thermohaline oscillations induced by strong steady salinity forcing of ocean general circulation models. J. Phys. Oceanogr. 23:1389-1410.

Wolff, J.-O., E. Maier-Reimer, and D.J. Olbers. 1991. Wind-driven flow over topography in a zonal β-plane channel: A quasigeostrophic model of the Antarctic Circumpolar Current. J. Phys. Oceanogr. 17:236-264.

Wright, D.G., and T.F. Stocker. 1991. A zonally averaged ocean model for the thermohaline circulation. Part 1: Model development and flow dynamics. J. Phys. Oceanogr. 21:1713-1724.

Wunsch, C. 1992. Decade-to-century changes in the ocean circulation. Oceanography 5(2):99-106.

Wyngaard, J.C., and R.A. Brost. 1984. Top-down and bottom-up diffusion of a scalar in the convective boundary layer. J. Atmos. Sci. 41:102-112.

Wyngaard, J.C., and C.-H. Moeng. 1993. Large-eddy simulation in geophysical turbulence parameterization. In Large Eddy Simulation of Complex Engineering and Geophysical Flows. B. Galperin and S. Orszag (eds.). Lecture Notes in Engineering, Cambridge University Press, pp. 349-366.

Yin, F.L., and E.S. Sarachik. 1994. An efficient convective adjustment scheme for ocean general circulation models. J. Phys. Oceanogr. 24:1425-1430.

Zhang, S., R.J. Greatbatch, and C.A. Lin. 1993. A reexamination of the polar halocline catastrophe and implications for coupled ocean-atmosphere modeling. J. Phys. Oceanogr. 23:287-299.

4

COUPLED SYSTEMS

Coupled Systems: An Essay

EDWARD S. SARACHIK

Coupled atmosphere-ocean-land-cryosphere models are basic tools in the study of climate and its variability. Since the atmosphere is sensitive to changes in lower-boundary conditions on long enough time scales, we must simulate the time evolution of these conditions in order to ensure the consistent simulation of the atmosphere. The time scales of atmospheric sensitivity depend on the geographic region of interest: The tropical atmosphere responds to sea surface temperature (SST) variability on monthly and longer time scales, while it has not been shown that the mid-latitude atmosphere responds significantly to SST unless the anomaly lasts for several years. Furthermore, variations in mid-latitude soil moisture seem to affect the distribution of precipitation over the continents seasonally. Variations of snow cover and sea ice have also been implicated in atmospheric variability beyond the seasonal time scale.

In turn, the evolution of the lower boundary conditions is partly determined by atmospheric processes, so coupled models become essential for simulating the mutually consistent evolution of the interacting systems. It is safe to say that if we are interested in decade-to-century-scale climate variability, the global atmosphere must be coupled to the global ocean, to the global land surface, and to global snow and ice.

While this realization has been with us since the beginning of climate modeling, progress in coupled modeling over the past decade has been fitful and hard won. The basic problem has been one of resources: A 100-year run of a coupled model consisting of a global atmosphere of modest resolution, with land processes parameterized, cou-pled to a global coarse-resolution ocean, with sea ice, uses a major part of a dedicated supercomputer. Increasing the resolution by just a factor of two increases the computer demands by an order of magnitude. If we are to understand and simulate climate variability on decade-to-century time scales, model runs of thousands of years are required. Up to this time, fully coupled models of satisfactory (but never sufficient) resolution have been run only at major institutions having access to large amounts of supercomputer time.

As computers become more capable, resource problems are ameliorated and the real problems of physical climate simulation come to the fore. The fundamental problem has been that the coupling of a reasonably well-understood atmospheric model to a reasonably well-understood oceanic model has produced a coupled model that is not only *not* well understood but also exhibits unexpected and unac-counted-for properties. The sensitivities of the two models to errors in each other, which are not apparent when each model is run in decoupled mode, seem to produce unusual sensitivity in the coupled model (Ma et al., 1994). It has become increasingly clear that a coupled model is a unique beast, with properties distinct from those of the component models. Coupled modeling therefore requires a quite differ-ent set of outlooks and approaches from those needed for modeling the component systems.

The coupled-model papers that appear in this chapter can best be put into perspective by recounting a bit of the history of coupled climate modeling, by pointing out where we now stand with respect to coupled modeling (and the data needed to support such modeling), and by suggesting some future directions and problems likely to be addressed over the next few years.

HISTORY

Only a decade after the first numerical general-circulation model (GCM) of the atmosphere had been constructed (Phillips, 1956), the first attempt at a coupled general-circulation model (CGCM) was made in a remarkably prescient series of three papers (Manabe, 1969a,b; Bryan, 1969) published as a single issue of the *Monthly Weather Review*. The model was geographically simplified (it consisted of a sector of the globe, bounded by meridians, covering only a third of the zonal extent of the globe, a bit more than half the sector was covered by land), and the solar driving was without annual variation, but it contained most of the physics now recognized as important for the climate problem. Water vapor and its changes of phase were computed explicitly; a rudimentary land hydrology model was included (the "bucket" model) that allowed for land evaporation and runoff; snow and land ice were parameterized; and radiative transfer for visible and infrared radiation was explicitly calculated using the specified clouds. The only major specification was cover from three types of clouds (low, middle, and high), as a function of latitude for use in the radiative transfer calculations. Rainfall and snowfall were explicitly calculated.

The ocean had five levels in the vertical; computed salinity explicitly; used an equation of state for density as a function of the calculated salinity, temperature, and pressure; and included a parameterization for sea ice. Coupling at the surface was accomplished by fluxes of heat and momentum through the surface into the ocean and by sensible and latent heat transfer into the atmosphere from the surface. SST was determined interactively by thermodynamic processes in both the ocean and the atmosphere.

The coupled model could be run for only 100 years of ocean model time, due to computational limitations, but at the end of this time it had reached a quasi-equilibrium in which only the deeper parts of the ocean were still changing. The resulting distribution of surface temperature, while not directly comparable to observations, looked quite reasonable, with the ocean heat transport warming higher latitudes and cooling the tropics. The modeled atmosphere developed eddies and had a wind and thermal structure similar to that observed, while the ocean developed a thermocline and had a density and current structure similar to that observed. Systematic problems were found in the lack of an intertropical convergence zone over the ocean, in a too deep and diffuse thermocline, and in a lack of sufficient meridional heat transport by the ocean circulation. No significant decadal variability was seen in the coupled model.

All succeeding CGCMs followed the basic themes set out in the original Manabe-Bryan papers (Figure 1, from Manabe (1969b), is still the best summary of CGCMs and continues to be widely used). In subsequent years, geography and topography have become more realistic, resolution has improved (but is still severely limited), clouds are now explicitly calculated instead of prescribed, radiation schemes have become more sophisticated and now include aerosols, land-surface parameterizations are more complete (they now describe vegetative types and evapo-transpiration), and ocean models now include more detailed bottom topography and more sophisticated mixing parameterizations. Many organizations other than GFDL are now running longer-term global coupled models, including groups at NCAR, NASA, DOE, and a few universities.

A major spur to a quite different type of coupled modeling came with the investigation of the ENSO phenomenon in the equatorial Pacific. A simplified coupled model, developed by Zebiak and Cane (1987), specified the annual cycle in both the atmosphere and the single-layer ocean (with embedded surface layer) and calculated the anomalies departing from this annual cycle. The model was successful not only in simulating the equatorial aspects of ENSO in and over the tropical Pacific but also at predicting aspects of ENSO a year or so in advance (see Cane, 1991). Only the upper portion of the equatorial ocean was modeled, since only the part above the thermocline is needed to simulate short-term variability (i.e., months to a year or two). Resolution near the equator was enhanced to fully resolve uniquely equatorial processes, especially equatorial waves and upwelling.

More complicated CGCMs without an annual cycle have also been successful in modeling ENSO variability (e.g., Philander et al., 1992). At this point, models with an annual cycle in solar forcing have had some success (e.g., Nagai et al., 1993; Latif et al., 1993) but still have difficulties in simulating the annual response as well as the full range and amplitude of interannual variability. (A recent review is

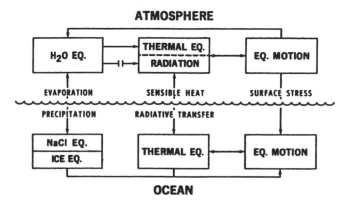

FIGURE 1 Box diagram of coupled-model structure. (From Manabe, 1969b; reprinted with permission of the American Meteorological Society.)

given in Neelin et al. (1992), and a full intercomparison of the climatology of this type of coupled model in Mechoso et al. (1995).) These models tend to have the oceans active only in selected regions; elsewhere, the ocean is relaxed to climatology, while the atmosphere has global extent.

PROBLEMS WITH GLOBAL CGCMS

Coupled models used to simulate ENSO (such as Philander et al., 1992) generally have only the tropical ocean active and use enhanced resolution in the equatorial area. Because these models are designed for interannual studies and are not configured for longer-term variability, they generally lack the mechanisms that maintain the upper ocean's thermal structure, especially the thermohaline circulation (THC), and so gradually diffuse away their thermocline. These models are therefore limited in the length of time over which they can be usefully run.

These types of CGCMs succeed in reproducing some aspects of the time-averaged climate, the annual cycle, and the interannual variability. Unrealistic features persist, however, especially off the western coast of South America, where simulated SSTs tend to be too high. Such problems could be remedied by adjusting the surface fluxes to make the climatology move closer to observed values, but attempts are under way to avoid such "flux corrections" by including or improving parameterizations of the necessary processes—for example, the stratus clouds off the coast of Peru that keep SSTs low.

Global CGCMs used for longer-term studies (e.g., for the response to anthropogenic increases in the greenhouse gases) must correctly simulate the basic climatology of the observed climate system, i.e., they must correctly simulate both the mean state and the annual cycle. It would be most desirable to correctly simulate the climatology without the need for flux corrections. A recent CGCM of Manabe and Stouffer (1988)—a model similar to the one used in the paper of Delworth et al. (1995) in this chapter, but with a sun lacking annual variation—illustrates the difficulties involved in achieving this goal. The model produces a deficit of salt (more properly, an excess of fresh water) at high latitudes and prevents the deep sinking of ocean parcels. As a result, the THC does not exist, and the heat and salt transports into high latitudes are reduced. The higher latitudes are thus too fresh and too cold, and sea ice extends too far south, conditions that guarantee that the THC cannot get started. If the high-latitude ocean is artificially salted by a constantly imposed saline flux correction and the THC is on, the circulation is helped to stay on by its own delivery of salt to high latitudes. On the other hand, the artificially imposed salt flux is not, by itself, adequate to start the THC; a steady state exists with imposed salt flux but no THC.

The lessons from the Manabe-Stouffer model are that there can be two climate states (one with and one without a THC) in the presence of identical external forcings, and that it is difficult to achieve a good simulation of the surface salinity field unless both the atmospheric and oceanic processes that control salinity are correctly modeled. Salinity at the surface of the ocean is changed by the difference between evaporation and precipitation, by runoff from land, by freezing or melting of sea ice, by advection and subsequent melting of icebergs, by advection, convergence, and divergence of salinity by ocean currents, and by mixing of salinity downward into the ocean. All of these processes, some quite poorly measured and understood, must be modeled correctly to ensure a proper high-latitude salinity budget and hence a correct THC.

The global annual cycle is relatively well documented in the instrumental record. Since there is no guarantee that a CGCM will respond correctly to the imposition of the annually varying external solar forcing, the modeled annual cycle provides a major test of CGCMs. But if a CGCM responds only annually to an annually varying sun, it would miss the variability on all other time scales that comprise the climatology: The annual cycle is the long time average over all the variability present, and variability other than annual may contribute to the observed annual cycle. Since at present flux corrections are still needed to prevent climate drift of CGCMs, the global annual cycle in these models cannot be considered to have been independently modeled.

VARIABILITY

Since the ENSO cycle is so important a signal in the tropical atmospheres and oceans, it is not surprising that inter-decadal modulations of this signal are also important. Rasmusson et al. (1995, in this chapter) present evidence that the ENSO is modulated by long (century-scale) variability, although they make the point that the records are not nearly long enough (or good enough) to characterize this variability more precisely. Cane et al. (1995, also in this chapter) speculate that such regimes of enhanced or suppressed ENSO variability are internally generated, at least in the Zebiak-Cane model, and show that longer data records might be capable of resolving the precise mechanisms. Trenberth and Hurrell (1995, in this chapter) hypothesize that North Pacific decadal variability, which is connected with the Pacific North American pattern, can be linked to this decadal variability of the ENSO phenomenon.

At this time no CGCM that has a resolution near the equator high enough to simulate the processes known to be important for ENSO, has produced (or can produce) long enough simulations to tell us whether such regimes exist. In order to be able to run these models for long periods of time, equatorial high resolution in the ocean must be combined with the accurate simulation of the THC; satisfying both these conditions in the ocean component of a CGCM is beyond the capabilities of the present generation of super-computers.

It used to be thought that the response time of the THC was so long that decadal and longer variability would occur simply as a result of the accumulated response of higher-frequency forcings from the atmosphere. Recent results (see Weaver, 1995, in this volume) have indicated that there exist purely oceanic mechanisms for the generation of inter-decadal variability, namely, the internal dynamics of the THC itself in response to *steady* fresh-water forcing from the atmosphere.

The question arises as to whether or not such variability would be present in a fully consistent CGCM, i.e., whether or not the atmosphere would increase or damp the THC variability that would exist in an ocean-only model. The paper by Delworth et al. (1995) in this chapter describes a fully coupled model, driven by an annually varying sun, that has high-latitude salt-flux corrections and heat-flux corrections, both varying as a function of the time of year. The results indicate that inter-decadal or longer variability survives the coupling. The mechanism for the SST and surface salinity involves the variability of the oceanic THC but is modified and complicated by the feedbacks inherent in a fully coupled model.

FUTURE PROBLEMS

Coupled models are gradually coming into their own. They have been quite successful at enabling us to understand and predict a wide range of phenomena, from the ENSO cycle to the responses to the anthropogenic increase of the radiatively active gases, especially carbon dioxide (see, e.g., Manabe et al., 1991, 1992). As computer resources become more available, coupled models are being run at more and more institutions. While it is still true that higher-resolution CGCMs can be run only on supercomputers, the advent of workstation computing has now made it possible for individual investigators to begin to run similar coarse-resolution coupled models.

The success of coupled models depends on the ability of the component models to realistically simulate key climate processes. It is therefore true, and always will be, that progress in atmospheric and oceanic modeling will prompt progress in coupled models. It is also true, however, that coupled models require special attention to processes not ordinarily emphasized in decoupled modeling. For example, atmospheric models must now consider the details of boundary-layer processes near the ocean surface in order to successfully simulate the fluxes of heat and momentum in response to a given SST. Similarly, mixed-layer processes in the ocean must be able to successfully simulate the SST for specified fluxes of heat and momentum from the atmosphere. The distribution of atmospheric rainfall, runoff from land, and sea ice growth and advection become impor-

tant processes in guaranteeing the success of THC simulation. Because each of the component models is very sensitive to small errors in the other, process simulation that would be acceptable in a decoupled model can lead to unacceptable results in a coupled context. Successful simulation of one component in response to forcing in the other is a necessary, but by no means sufficient, condition for the success of the coupling.

We see that successful coupling demands improvements in the component models; indeed, the future of coupled models will depend on such improvements. Flux corrections can provide a temporary fix for model problems, but totally believable coupled models require simulating accurate climatologies without the need for flux corrections. Simulating decadal and longer variabilities also requires ocean models that correctly maintain the upper ocean's thermal structure; in practice this means that the THC must be correctly modeled. In order to confirm that natural variability has been successfully simulated, longer and more accurate data sets must be available. In this regard, paleoclimate indicators become especially valuable, and other proxy data sets (tree rings, corals, sediment cores, ice cores, etc.) become essential to mapping the domains of variability in which to test the coupled models. In this chapter Battisti (1995) points out not only the unique role that proxy data play *in* modeling but also the equally unique role that understanding and exploration *through* modeling play when data are so sparse and difficult to come by.

The ultimate test of understanding and simulating climate variability is the ability to predict that variability. We now know that certain aspects of seasonal-to-interannual variability, especially aspects of ENSO, are predictable, but no one knows whether decadal variability is deterministically predictable—and, if it is, which data are needed as initial conditions. Even if longer-term variability is not predictable, the ability of models to successfully simulate the spectrum of climate variability is a necessary prerequisite to our full understanding of the natural climate system.

CONCLUSION

As we have seen, the basic problem in coupled modeling is the correct simulation of the climatology, particularly the annual cycle. Since climate variability and the annual cycle contribute to each other, this goal is something of a moving target. We have to simulate the annual cycle correctly in order to simulate variability, but we have to simulate the variability correctly in order to simulate the annual cycle. We may therefore expect progress to be rather slow.

Progress in coupled modeling will be accelerated by:

- Improvements in understanding and modeling the atmosphere, ocean, land, and cryosphere separately. The primary physical processes needing better parameterization are clouds and water vapor in the atmosphere, mixing in the ocean, evapotranspiration and small-scale hydrology in the land, and sea-ice extent and growth in the cryosphere.

• Advances in understanding the nature of coupling, especially the general question of the sensitivities of each system to small errors in the other.

• General increases in available resources and computational infrastructure, allowing a wider community to gain access to coupled modeling.

• Improvement and extension of time series of physical

quantities in the atmosphere, ocean, land, and cryosphere. This can be achieved by reanalysis of existing model data (e.g., daily weather analyses), data archaeology, improvements in existing observing networks and data handling, new techniques of paleoclimate analysis . . . or by instituting new measuring networks and waiting till the time series is long enough.

While this last technique may seem the least efficient, future generations will appreciate our efforts as much as we would be grateful to previous generations, had they been prescient enough to do the same for us.

Decade-to-Century Time-Scale Variability in the Coupled Atmosphere-Ocean System: Modeling Issues

DAVID S. BATTISTI[1]

ABSTRACT

A primary limitation in the study of the decade-to-century (hereafter termed "intermediate") time-scale variability in climate is that the instrumental records for all climate variables either do not exist or are too short to detect these phenomena with any measure of statistical confidence. Hence, the methodologies and strategies of the research activities related to the intermediate time scales of climate variability will be distinctly different from those related to interannual variability. While climate models are currently used mainly to simulate phenomena that are already well observed, the models themselves will frequently be the instruments with which scientists identify intermediate-scale climate phenomena. Ascertaining the veracity of a climate model must therefore be a primary activity in the study of intermediate-scale climate variability. Proxy data will play an important role in documenting climate variability on intermediate time scales and in evaluating the climate variability simulated by the models.

   In this paper I argue that it is extremely important that the models used for intermediate-scale climate studies be validated a priori by assessing how accurately many well-documented "target" phenomena are represented in each model. Specific target phenomena are suggested, such as the seasonal and diurnal cycles of the state variables that are well-documented in nature, and the fluxes of energy and mass at the media interfaces. A quantitative comparison should be made of the processes that are responsible for these cycles in the model and those that are observed. In addition, valuable information on the veracity of the models will result from an assessment of how well the model's performance matches the well-documented interannual variability in the climate system.

[1]Department of Atmospheric Sciences, University of Washington, Seattle, Washington

INTRODUCTION

An increasing number of earth scientists have become interested in understanding natural variability in the climate system on century and especially on decadal time scales. This paper pertains to the climate variability within the atmosphere and ocean system on these time scales, which will be referred to as the "intermediate" time scales.[2] The potentially critical role of the land hydrology and cryosphere for the intermediate-time-scale fluctuations in climate is discussed elsewhere (see the Atmospheric Observations section of Chapter 2). The variability in the atmosphere-ocean system on the intermediate time scales is poorly documented at present by comparison with both shorter (interannual) and longer (e.g., 10^3 to 10^6 years) time-scale oscillations. Information on the longer-scale variability is abundant because the transitions between glacial and interglacial conditions are large enough changes in the global climate system to be clearly defined in the global geological record— e.g., in the stratigraphy of the sediment and ice deposits, and in the radio-isotopic composition and distribution of the embedded flora and fauna. These proxy data have been instrumental in the studies of long-term climate variability because an accurate measure of the absolute elapsed time is available from the decay of the isotopes of the ubiquitous carbon.

Compared with the sub-millennial vacillations in climate, the variability of the climate system on the intermediate time scales is thought to be rather small in amplitude. Until recently, there was little direct evidence for either local or global inhomogeneous climate variations on these time scales. The instrumental records of the climate variables prior to the turn of the century are, in isolation, inadequate for documenting the decade-to-century-scale climate variability; they exist for only a limited number of state variables, and the data for these variables are largely confined to very near the earth's surface (usually within 10 m) and are sparsely distributed. Historians and scientists have used phenological data to infer variability on interannual time scales. Phenological data are, by definition, available only in the regions of human habitation and are susceptible to the vagaries of human perception. In his seminal contribution to the study of European history, Braudel (1949) combined the evidence from literary references with a variety of phenological data and concluded that the entire Mediterranean area experienced abnormally cold and wet winters in the

early seventeenth century—the Little Ice Age. (Braudel's data included the time of year of river and lake floods, the years of significant frost damage to olive trees, and the times and yields of various agricultural harvests.)

A much more rigorous historical portrait of the climate variability on intermediate time scales is provided by proxy data, although they are limited in usefulness because their interpretation is not straightforward. Proxy data relevant to intermediate-scale climate variability include $\delta^{18}O$ concentration in glacial ice, lake varves, loesses, pollen data, and coral and tree-ring data (see, e.g., Bradley, 1991). The proxy and phenological data have been used together to construct a rather detailed and comprehensive assessment of the change in climate since the Little Ice Age; a brief summary is presented in Crowley and North (1991).

In the early 1960s Stommel and Bjerknes published studies that had a benign impact on the community for more than two decades, and now provide focal points for scientists working on the intermediate-scale climate variability. Stommel (1961) hypothesized that the general thermohaline circulation of the oceans might have multiple stable regimes, and demonstrated this hypothesis using a two-box analog model for an ocean basin. Stommel's hypothesis is supported by the recent coupled atmosphere/global-ocean general-circulation model studies of Manabe and Stouffer (1988), who achieved two statistically steady states for the meridional circulation in the Atlantic Ocean—with and without a vigorous thermohaline circulation. Bryan (1986) found that the transition between strong and weak mean meridional circulations in the abyssal ocean could happen in less than a century in a sector ocean model. Recently, Levitus (1989a,b) demonstrated that there was a change in the deep North Atlantic hydrographic structure from the 1950s to the 1970s, and supposed this change to be associated with deep convection (Figure 1). Together, these studies and related analytical, modeling, and observational studies have elevated the deep ocean circulation into the arena of potential mechanisms for—or indicators of—climate variability on the intermediate time scales.

In 1964, Bjerknes examined the contemporaneous fields of sea surface temperature (SST) and sea-level pressure (SLP) from the North Atlantic Ocean and noted differences in the 5-year mean climatologies of 1920-1924 and 1930-1934 (these pentads were chosen on the basis of the Azores-minus-Iceland SLP index). Bjerknes argued that these differences reflected a local climate change resulting from an interaction between the surface gyre circulation of the North Atlantic and the overlying atmosphere. He also implied that there were quantitative changes in the oceanic equator-to-pole heat transport. In retrospect, it can be seen that Bjerknes's method for inferring decadal changes in the pentads' means was inappropriate; the interannual variability of the coupled atmosphere-ocean system in the North Atlantic is large (see, e.g., Wallace et al., 1992; Kushnir, 1994), so a

---

[2]The customary definition of climate is used in this paper: the aggregate statistical moments of the appropriate state variables over a prescribed period of time. Thus, changes in the climate on the intermediate time scales allow changes in the variability within the climate system on intermediate and all shorter time scales, including changes in the diurnal and annual cycles, and changes in the amplitude or pattern of the interannual climate phenomena, such as El Niño/Southern Oscillation.

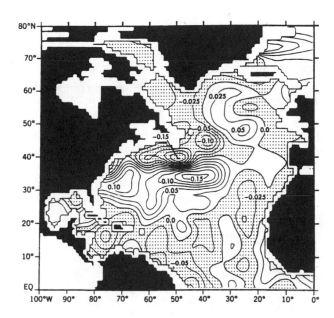

FIGURE 1 The difference in density (in kg m^{-3}) at 500 m depth between the two pentads, 1970-74 and 1955-59. Regions with negative values are stippled and denote higher density during the 1955-59 pentad. (From Levitus, 1989a; reprinted with permission of the American Geophysical Union.)

long-term change inferred from differences in 5-year means is extremely unreliable. Nonetheless, Bjerknes's hypothesis that ocean dynamics play an important role in interdecadal variability is supported by recent studies of the decadal variability of the North Atlantic atmosphere-ocean system (e.g., Kushnir, 1992; Pan and Oort, 1983).

Nearly all the numerical and theoretical research activities on intermediate-scale climate variability have taken place during the last half-decade, for a variety of reasons. First, there is a growing demand that scientists assess and predict the anthropogenic impact on climate. Essential (but insufficient) to accomplish this goal are an accurate statement of the present climate from observations of the state variables, a rigorous program that results in multiple independent forecasts of the anthropogenically forced change in climate, and a comprehensive inventory of the intermediate variability in the present natural climate system so one can plan an efficient monitoring strategy to confidently assess the accuracy of the forecast climate change from the future observed climate.

There is another reason why the interest in the intermediate time scale variability of the climate system has sharply peaked during the past 5 to 10 years that is external to the greenhouse warming problem. The TOGA and EPOCS programs of the 1980s resulted in the documentation, simulation, and skillful model prediction of the El Niño/Southern Oscillation (ENSO) phenomenon. Through these extraordinarily successful programs scientists have explicitly demon-

strated for the first time that rich variability in the climate system can result solely from the interaction between the oceans and the atmosphere. More important, this period of research marked the advent of a new era. With a few notable exceptions, for the first time the atmospheric scientists began to focus on the sub-monthly circulation anomalies in the troposphere, and the oceanographers began to abandon the default assumption of a world ocean in a quasi-steady state.[3]

The research activities of the last decade also created a modest population of scientists that are actively performing basic research on both oceanic and atmospheric circulation, and on the response of a climate system composed of atmosphere coupled with the global oceans. As a result, there are now numerous studies that document coordinated interannual variability in the atmosphere-ocean system, and many studies wherein isolated phenomena have been simulated and the essential physics documented. Thus, the extraordinary interest of the scientific community in identifying and analyzing the variability in the full climate system on decade-to-century time scales through modeling and observational studies can be attributed to both the research focus on the interannual variability of the coupled atmosphere-ocean climate system and the practical problems that have arisen in the detection of an anthropogenically forced climate change.

In this paper, I discuss the constraints inherent in assessing both the actual variability of the climate system on the intermediate time scales and the physical and dynamical processes that are likely to be responsible for this variability. The methods for validating the "modes" of intermediate-scale climate variability that are produced by numerical and analytical models necessarily represent a change from the traditional modus operandi. A good example of the unique blend of modeling and observational research and monitoring efforts that is required to assess intermediate-scale variability is found in the charter for the Atlantic Climate Change Program (ACCP) of the National Oceanic and Atmospheric Administration (NOAA). I describe the ACCP and briefly review the evidence for a directly observed variability in the atmosphere/ocean/sea-ice system that has recently become a focus of the ACCP. The implications for modeling and modeling strategies are discussed, and a summary is presented at the end of the paper.

THE LIMITATIONS OF THE HISTORICAL INSTRUMENTAL RECORD

The primary limitation on the study of climate variations on the intermediate time scale is that the instrumental record

---

[3]In *The Evolution of Physical Oceanography: Scientific Surveys in Honor of Henry Stommel*, Wunsch writes: "Until very recently, the ocean was treated as though it had unchanging climate with no large-scale temporal variability." There are no references to Stommel's 1961 paper in the entire volume, which was published in 1981!

for all climate variables is too short to permit the detection of these phenomena. Prior to the 1950s, the only maritime instrumental records useful for these studies are those for SST and air temperature. Although SST data are available along the major global shipping routes from about 1900 (see, Pan and Oort, 1990), spatial and temporal coverage for these variables is adequate only across the North Atlantic (Figure 2). For the continental areas, potentially useful data are available for more of the climate variables as far back as the mid-1800s. These data, however, are uneven in their spatial distribution. Prior to World War II there are essen-

tially no instrument-based data for the state of the atmosphere above the surface. For the ocean, subsurface data are limited to infrequent and isolated transects through the ocean, usually across the North Atlantic. Thus, prior to the 1950s the instrumental data records exist for only a few key variables in isolated regions, and provide only a blurred glimpse at climate variability. The data are insufficient for deducing the attending atmospheric and oceanic circulations and heat transport, and the energy exchange between the media.

The post-World War II instrumental data base for the

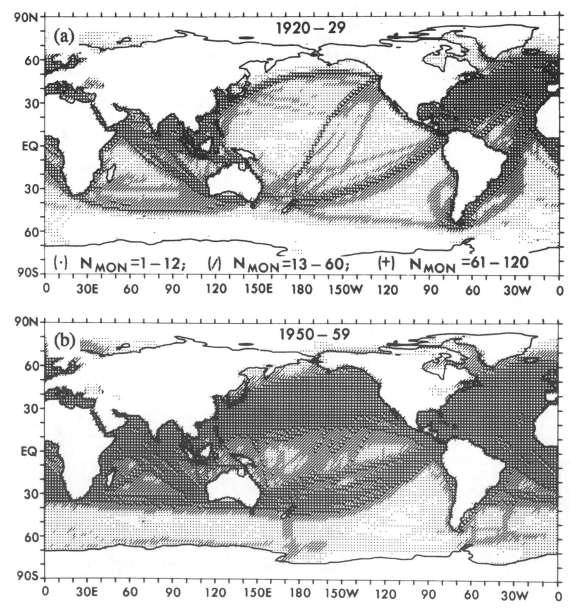

FIGURE 2   The climatological data base for sea-surface temperature. Shown in (a) is the number of months in which there exists at least one observation in a 2° latitude by a 2° longitude area during the decade 1920-29. A dot indicates 1-12 months (out of 120) with data, a slash 13-60 months, and a plus sign 61-120 months. Panel (b): as in (a), but for 1950-59. (From Pan & Oort, 1990; reprinted with permission of Springer-Verlag.)

state of the atmosphere is rather complete[4] over the Northern Hemisphere, especially over the continents, but it is not suitable for studying decadal climate variability in the Southern Hemisphere. A very uneven spatial and temporal record of the hydrographic and current structure of the world oceans is available starting after World War II, but it is unlikely that the data coverage is sufficient permit deduction of a posteriori decadal variations in the ocean climate.

Fundamental to understanding the coupled atmosphere-ocean system is a knowledge of how energy and constituents are exchanged between the two media. On this score, even the recent instrumental record is clearly deficient. For example, there is large uncertainty as to the annual cycle of the turbulent exchange of heat and momentum alone; estimates of the variability in these energy exchanges on the intermediate time scales would be premature.

In summary, it is not possible to discern the past variability of the climate system on the century scale from the instrumental data sources. Recent decadal variations in the atmosphere-ocean system can perhaps be deduced with some confidence from the instrumental data base, although incompletely. The ample spatial voids in the data set (especially in the subsurface oceans) introduce some uncertainty, into determining whether the variability is locally confined or is of global extent. Because of the severe limitations of the existing instrumental record, proxy data sets will play an important role in documenting the intermediate-scale climate variability and, perhaps, in evaluating simulated climate variability (discussed below).

THE TRADITIONAL MODUS OPERANDI AND THE ATLANTIC CLIMATE CHANGE PROGRAM

The Traditional Modus Operandi

The history of atmospheric sciences and oceanography is replete with examples of community-wide intensive research activities that are focused on the precise documentation and analysis of specific, observed phenomena that represent perturbations from a mean (state) that are statistically significant, e.g., the midlatitude cyclone, the Quasibiennial Oscillation (QBO), and the Gulf Stream. Major programs have also commonly had as a centerpiece a well-observed phenomenon. Examples include GATE (easterly waves, mesoscale tropical convection), POLYMODE (long-lived coherent eddies in the ocean thermocline), ERICA (rapidly deepening cyclones), CLIMAP (reconstruction of the climate of the last ice age), and the upcoming EPOCS effort (the annual cycle and the boundary layer circulation in the Pacific).[5]

The Tropical Ocean and Global Atmosphere (TOGA) program provides a good illustration of the traditional research strategies, and the profound effects the limited data base will have on the modus operandi in research on the variability of climate on intermediate time scales. The ENSO phenomenon was the centerpiece for TOGA, although midlatitude phenomena were also documented and modeled in this program. It is important to recall that prior to TOGA the ENSO was already a reasonably well-documented phenomenon, being a large-scale, large-amplitude perturbation in the atmosphere-ocean system. The emphasis of the TOGA program was on providing an understanding of how and why this climate anomaly was manifested and assessing the predictability of the phenomena. In contrast, for intermediate-scale climate variability the target phenomena are smaller in amplitude, are derived from only a few realizations, and are not completely defined by the historical data.

The methodologies and strategies of the research activities related to the intermediate-scale climate variability will be distinctly different from those related to interannual variability for two additional reasons: (1) the inherent limitations of the data base of directly observed climate state variables (discussed above), and (2) the constraints imposed by limited computational resources coupled with the uncertainty as to the veracity of the simulated phenomena because of the parameterization of the small-scale processes and the (still) poorly understood physics. The science plan, priorities, and ongoing activities of the ACCP and of the nascent Global Ocean-Atmosphere-Land System program (GOALS) duly reflect these constraints.

The Atlantic Climate Change Program

The ACCP was formally initiated after a workshop held at the Lamont-Doherty Earth Observatory of Columbia University in July 1989. The goals of this program are as follows:

• To determine the seasonal-to-decadal and multidecadal variability in the climate system due to interactions between the Atlantic Ocean, sea ice, and the global atmosphere using observed data, proxy data, and numerical models.

• To develop and utilize coupled ocean-atmosphere models to examine seasonal-to-decadal climate variability in and around the Atlantic Basin, and to determine the predictability of the Atlantic climate system on seasonal-to-decadal time scales.

• To observe, describe, and model the space-time variability of the large-scale circulation of the Atlantic Ocean and determine its relation to the variability of sea ice and sea surface temperature and salinity in the Atlantic Ocean on seasonal, decadal, and multidecadal time scales.

• To provide the necessary scientific background to design an observing system of the large-scale Atlantic Ocean circulation pattern, and develop a suitable Atlantic

---

[4]Certain state variables are better measured than others. Water vapor, which may be a central component in decadal climate variations, is only crudely measured above the middle troposphere.

[5]Much of the remaining activity can be categorized as process-oriented studies or studies that relate to weather prediction.

Ocean model in which the appropriate data can be assimilated to help define the mechanisms responsible for the fluctuations in Atlantic Ocean circulation.[6]

The ACCP is an interdisciplinary program that involves atmospheric scientists, oceanographers, and paleoclimatologists. Currently the program shows an appropriate balance between modeling efforts and analysis of the historical proxy and instrumental data. Furthermore, observational programs have been launched to help determine the link between intermediate-time-scale SST anomalies and variability in the thermohaline circulation by means of long-term monitoring of the deep Western Boundary Current. Observational studies that are ongoing or anticipated include tracer inventories, hydrologic monitoring of Fram Strait, and monitoring of the upper ocean thermal structure through the Atlantic Voluntary Observing Ships Special Observing Project and surface-drifter deployments.

Two different patterns of variability in the North Atlantic atmosphere-ocean system have been identified in the historical data. Deser and Blackmon (1991; 1995, in this volume), using the historical observational data of sea-surface temperature, sea-level pressure, and the zonal wind over the Atlantic Basin, have demonstrated that there is a complex wintertime atmosphere-ocean interaction in the North Atlantic with a preferred time scale of about 10 years. The anatomy of this decadal variability appears to be somewhat different from that of the interannual climate anomaly in the North Atlantic documented in Wallace et al. (1992).

Using the same surface fields and focusing on the changes in the Atlantic atmosphere-ocean system over multidecadal periods during the last century, Kushnir (1992) found the relationship between changes in the SST and changes in the overlying surface atmospheric circulation significantly different from that associated with higher-frequency variability in the Atlantic climate system. In this multidecadal transition, the atmospheric circulation changes are consistent with a local quasigeostrophic response to the changes in the ocean SST (see Figure 3). The hydrographic record for the same period indicates that the SST changes are echoed in the deep ocean, where concomitant changes in salinity are found (Lazier, 1988; Levitus, 1989a,b; see also Figure 1). Thus, as is not true for the higher-frequency "modes," evidence suggests that changes in the thermohaline circulation and sea-ice export from the Arctic Ocean are associated with this multidecadal transition.

In order to identify the climate variability in and around the Atlantic basin on decade-to-century and longer time scales, ACCP-sponsored studies have been begun that will determine the extent to which the instrumental records used in the studies mentioned earlier can be augmented (e.g.,

---

[6]An extended presentation of the Atlantic Climate Change Program Science Plan can be found in NOAA (1992a).

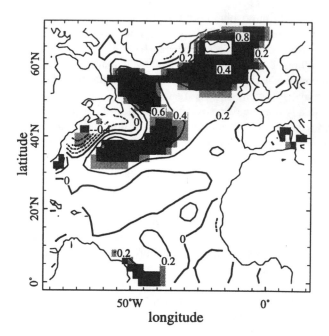

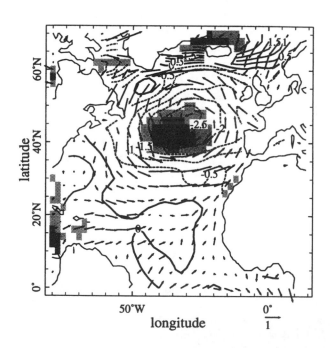

FIGURE 3   Decadal time-scale climate anomalies in the wintertime (December-April) in the North Atlantic. Panel (a): the difference between the average wintertime SSTs between 1950 and 1964 and between 1970 and 1984. The former period (1950-1964) was characterized by warmer-than-normal water in the winter in the North Atlantic, the latter (1970-1984) by anomalously cold water. The contour interval is 0.2°C. Panel (b): as in (a), except for the attendant atmospheric surface variables: sea-level pressure (contour interval 0.5 mb) and vector winds (a 1.0-m/s vector is found below the panel). (From Kushnir, 1994; reprinted with permission of the American Meteorological Society.)

surface salinity and hydrographic data). Nonetheless, the state of the (Atlantic) climate system, as realized from direct measurements of the climate state variables, will remain insufficient to permit assessment of the natural climate variability on decadal and longer time scales, especially away from the earth's surface. Thus, within the ACCP there are also ongoing studies to identify sources of proxy data that will define and constrain the variability in the Atlantic climate system. These studies include an analysis of dendroclimatic and ice-core data surrounding the North Atlantic as well as an analysis of deep-sea, high-sedimentation-rate cores. Encouraging preliminary results from these studies were presented by D'Arrigo et al. (1992), Bond et al. (1992), and Keigwin and Boyle (1992) at the 1992 ACCP Principal Investigators' meeting.

The usefulness of a single proxy data source for inferring the intermediate-time-scale variability in climate or a climate state variable is often limited for two reasons. First, common proxy data are not usually available for the entire globe (e.g., coral is found only in the tropics). Thus, a proxy source, in isolation, cannot be used to differentiate local and global climate variations. Second, climatic implications from a single proxy source can be ambiguous. For example, the cambial activity of trees depends on both seasonal temperatures and precipitation; for many species of trees it is difficult to determine uniquely the relationship between these two climatic variables and the cambial activity. Many proxy climate indicators must be used to ensure consistency and remove any ambiguity in the historical climate record assembled from the proxy data, and to better assess whether the climate variability is regionally confined or global in extent. The utility of this effort is illustrated by the remarkable progress made on the climate of the last glacial maximum (CLIMAP, 1981). In this regard, the IGBP Past Global Change project (PAGES) is timely (IGBP, 1992; see also Bradley, 1991).

IDENTIFYING POTENTIAL CLIMATE VARIABILITY ON DECADE-TO-CENTURY TIME SCALES FROM NUMERICAL MODELS: STRATEGIES AND LIMITATIONS

Numerical models will be the primary tools by which the mechanisms responsible for decade-to-century-scale climate variability are identified. More important, these same models will frequently be the instruments that scientists use to identify target intermediate-time-scale climate phenomena, because instrumental data exist only for the last century and, as mentioned above, for only a limited domain of the climate system. For the same reasons it will often be difficult to determine whether simulated phenomena are ever realized in nature. Thus, our confidence that a simulated phenomenon could occur will result only from an a priori assessment of

how accurately the model simulates many well-documented phenomena.

A model to be used for studying natural variability in the climate system on the intermediate time scales must include complete and interactive modules for the four central media: the atmosphere, global oceans, the global terrestrial marine biosphere, and sea ice. An important aspect of this system for the intermediate-time-scale climate studies is an accurate and complete representation of the hydrologic cycle (OCP, 1989). Climate variability models are the result of marrying the individual models for the four media, each of which has first been tested in isolation by prescribing the appropriate boundary conditions for the state of the adjacent media. (When appropriate, the flux of energy is prescribed at the boundaries.) The tests of the uncoupled models do not guarantee a good coupled climate model, but are practical first steps.

Finally, full general-circulation models (GCMs) for the atmosphere and global oceans are required to study the intermediate-time-scale climate variability. There is no evidence from the observational data that the intermediate-time-scale climate variability is regionally confined—for example, to within a hemisphere, an ocean basin, or to the near-surface ocean—as the interannual climate variations seem to be.

Prerequisite Constraints for the Uncoupled Modules

The foremost test of the uncoupled component modules is the ability of the model to reproduce the seasonal cycle, and thus the annual mean state, when forced by the imposed boundary conditions. The annual cycle is an test case for validating uncoupled component modules, because in most cases the annual cycle is well known for many of the climate state variables; in some cases the product moments are also reasonably well known (e.g., the meridional heat transport in the atmosphere). For the coupled atmosphere and land-surface modules, the diurnal cycle provides a second excellent test. The diurnal cycle of near-surface fields of air temperature, moisture, clouds, and wind are well-observed quantities over land. An accurate simulation of these cycles throughout the year provides a rigorous test of the impact of the combined boundary-layer physics and surface-flux parameterizations that act to maintain the simulated climate.

The general-circulation models for the atmosphere (AGCMs) are routinely validated by comparing the simulated climatology with that observed for variables or features that include the jet structure, variation of the height of selected geopotential surfaces in the troposphere, the zonal mean distribution of temperature, zonal wind, and SLP. It is of crucial importance for climate variability, however, that the models accurately simulate the observed annual cycle of all the boundary-layer fluxes: momentum, sensible, convective, latent, and radiative fluxes at the surface, and outgoing longwave flux at the top of the atmosphere.

The sea-ice models should be required to provide an adequate simulation of the seasonal cycle of surface temperature, sea-ice thickness, and ice advection and production—all quantities that are qualitatively known from observations (e.g., Walsh et al., 1985). The ocean models must be able to simulate the climatological annual mean circulation and hydrographic structure. In the tropics there is a significant and well-documented annual cycle in the upper-level currents and hydrography that provides additional constraints on the model. Especially important is the accurate simulation of the global SST. Until recently the SST was constrained in many ocean modeling studies by a rapid forced relaxation to a prescribed climatology.

A relaxation to a climatologically observed hydrography has also been used in and below the thermocline in many numerical studies of ocean general circulation (Hibler and Bryan, 1987; Semtner and Chervin, 1988); in some instances maintenance of the thermocline is avoided by limiting the length of integrations, so equilibrium is never achieved. This practice has been useful for short-term simulations of the upper-ocean circulation away from convective regions (e.g., Philander et al., 1987) but is clearly inappropriate for the study of climate and intermediate-scale climate fluctuations. The limited observational data indicate that the entire ocean domain can be involved in climate variations on these time scales (e.g., Levitus, 1989a), although it is not clear, for instance, whether the deep ocean is necessarily an active player or just a barometer for the changing surface processes via the resultant convection.

Independent tests of the component models should also include constraints that can de deduced from certain tracers that are well observed and whose annual cycle and mean distributions are understood. For the atmosphere, these include the ^{85}Kr, Freons, and CO_2, which have already been used to independently validate the mean cross-equatorial exchange rate of one AGCM (Tans et al., 1989). For the ocean, the bomb-produced tracer ^{14}C can be utilized to assess the accuracy of the simulated Atlantic Ocean circulation and the concomitant mixing processes (Toggweiler et al., 1989). Further insights into the oceanic mixing processes and convection, both of which are parameterized in ocean models, may be obtained by comparing the Lagrangian transport in models with observed distributions of other tracers, including the chlorofluorocarbons.

Finally, there are well-observed phenomena in the atmosphere and ocean that occur on interannual time scales that can be used to help validate the models. For the atmosphere models these include the QBO and the Southern Oscillation, and for the Pacific Ocean models, the El Niño.

There is a tremendous amount of work to be done on documenting the impact of the parameterization of unresolved physics on the simulated large-scale, low-frequency climate. Recent studies have demonstrated that different parameterizations for convection in both atmosphere and ocean GCMs can result in qualitatively different mean circulations and climate variability; an example is discussed below. Similarly, the recent advances in understanding the dynamics and thermodynamics of individual clouds must be extended to yield quantitative descriptions and parameterizations of the energy and mass transport by an ensemble of clouds on the scale of an AGCM grid.

Recent Results: Overturning the Rocks

In this section I will use three examples that I believe presage the results that will be achieved during the 1990s in modeling intermediate-time-scale climate variations.

Example 1. Over the last decade, studies have been published on the response of the wintertime Northern Hemisphere atmosphere to the principal mode of the observed interannual SST anomaly in the North Pacific Ocean (e.g., Pitcher et al., 1988). In these studies, which utilized AGCMs, investigators prescribed a perpetual January insolation to ensure statistically significant results with limited computational resources. The model's response to the prescribed SST anomaly in each of these experiments was contrary to that observed (e.g., Wallace and Jiang, 1987). The polarity of the model's geopotential anomaly was *independent* of the polarity of the forcing anomaly. Recently, Lau and Nath (1990; hereafter LN) examined the response of the GFDL AGCM to the observed 1950-1979 SST and an annual cycle in insolation.[7] Upon isolating the circulation anomalies associated with the North Pacific SST anomalies, LN found that the model anomalies were indeed consistent with those observed (i.e., a quasi-linear relationship between anomalies in SST and geopotential). Kushnir and Lau (1992) used the same model as LN and repeated the perpetual-January experiments of Pitcher et al. (1988). The results of this study were consistent with those of the earlier perpetual-January experiments, and contrary to LN and the observational data. Thus, the discrepancies between the physics of the observed atmospheric response to SST anomalies and that found in the simulations of Pitcher et al. and of Kushnir and Lau can be explained by the prescribed unrealistic (perpetual-January) forcing. The moral here is that compromises in the experimental plan that are made because of computational constraints may lead to fallacious conclusions.

Example 2. James and James (1989) employed a primitive-equation model of the atmosphere (T21, five layers), prescribing the annual cycle as the only long-term forcing; slow variations in SST or insolation were not permitted in their experiment. In this model, the variance in the largest-scale structures in the circulation was on a decadal or longer

---

[7]The reader is cautioned there are significant differences between the SST forcing used by LN and that used in the earlier perpetual-January studies. Also, a different AGCM is employed.

time scale (Figure 4). Here the moral seems to be that variability on the intermediate time scale may be internal to one or more of the principal climate media, and coincidental to any variations in either the external forcing or in the surrounding media.

Example 3. A similar cautionary note is found in the studies of Weaver and Sarachik (1991a,b; hereafter WS). WS employed a stand-alone ocean-sector model to examine the variability in the thermohaline circulation (THC). They found that under steady mixed surface-boundary conditions (i.e., linear relaxation to a prescribed surface temperature, and a flux boundary condition on surface salinity) the meridional circulation was not steady; rather, three "climate states" were realized: a collapsed THC, a vigorous THC, and a stage with highly energetic decadal oscillations in the circulation. The decadal oscillations result from the interaction between the convection in the polar regions and the buoyancy advection in the subtropical and polar-gyre circulations (Figure 5). These oscillations are accompanied by up to a threefold change in the poleward transport of heat. Throughout the oscillation, the deep ocean essentially acts as a reservoir, and thus works to maintain the long-term mean thermocline. Eventually, the THC collapses, and the decadal oscillations cease. The WS studies clearly indicate the potential for rich and unexpected intermediate-scale variability that is internal to the ocean.

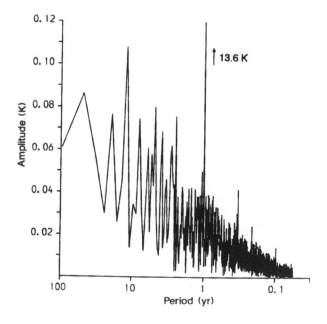

FIGURE 4 Spectrum of mean pole/pole temperature difference from a 96-year integration of an atmospheric primitive-equation model. The only long-term forcing is a prescribed annual cycle in the radiative relaxation temperature; no energy exchange is allowed at the surface of the top of the atmosphere. Note the spectral peak at 12 years. (From James and James, 1989; reprinted with permission of Macmillan Magazines Ltd.)

These and additional results presented in WS also serve to illustrate the state of affairs in ocean climate modeling. Here, I will only point out that the stability of the climate states found in WS depended on the type of convective scheme used, the type of boundary condition set for fresh water, whether synchronous or asynchronous time stepping was used, and the horizontal resolution.

Together, these three examples illustrate the importance of extensive and meticulous examination of all the uncoupled models that will be used to study the intermediate-time-scale variations in climate. Such an examination should document the following prior to constructing a full climate model for use in studying intermediate-scale climate variability: the ability of all models to simulate multiple observed phenomena, the sensitivity of the solutions to the parameterized physics, and the sensitivity of the component models to the chosen numerical methods of solution and to the shortcuts that are motivated by computational constraints (e.g., perpetual-January insolation, asynchronous time stepping).

Validating the Performance of the Coupled Models

The numerical models that will be used to study intermediate-time-scale variability in climate will necessarily allow interactions between the atmosphere and global oceans and include modules for the land-surface hydrology and cryosphere. The climate system models will be validated by comparing the simulated modes of variability with those that are indicated by a careful analysis and synthesis of the multiple, contemporaneous proxy indicators for the global climate state. Stream One of PAGES is a very ambitious program to reconstruct the global climate since 2000 years BP using the available proxy data sources collectively. The temporal resolution of this climate record is expected to be decadal or better, so the reconstruction will provide the modeling community with "target" phenomena. The data from this program, which is not yet under way, will not be available for quite some time. It is reasonable to assume, however, that the climate models least prone to yielding non-physical solutions on the intermediate time scales are the models that can accurately simulate many of the directly observed and well-documented higher-frequency phenomena in the climate system. Thus, a prerequisite for a model for intermediate-time-scale climate studies is accurate reproduction of many of the already well-observed target phenomena and features.

The same target phenomena that are appropriate for testing the stand-alone component models are also appropriate for testing the climate system model. For example, the system model should adequately simulate the observed diurnal cycle in the near-surface atmosphere over land. It also should accurately simulate the annual cycle in each medium, especially for the state variables adjacent to the media inter-

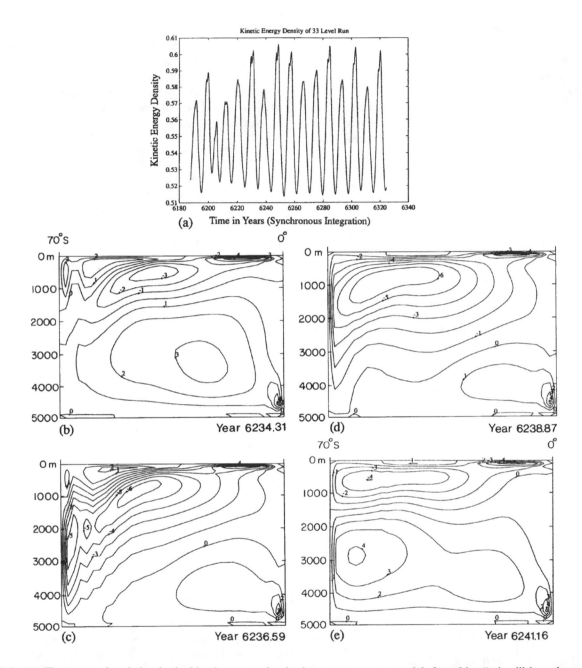

FIGURE 5   (a) The temporal variation in the kinetic energy density in an ocean-sector model, forced by "mixed" boundary conditions. The regular oscillations at 8.6 years are associated with cyclic changes in the meridional overturning circulation; the four characteristic stages are displayed in panels (b) through (e). Positive contours denote clockwise transport. (From Weaver and Sarachik, 1991b; reprinted with permission of the Canadian Meteorological and Oceanographic Society.)

faces and for the energy and mass exchange between the media. Wherever possible, both annual and diurnal cycles should be validated against those observed.

Important new constraints are available to the climate system model on the annual and especially the interannual time scales. Subtle discrepancies between the observed and simulated interfacial fields and fluxes formed with the stand-alone atmosphere and ocean models are often amplified in

the coupling process. This amplification leads to a coupled climate state that is qualitatively different from either of the states achieved by the two uncoupled models in isolation. These "setbacks," however, are fruitful because they often indicate a physical deficiency in a component model that, when corrected, yields a more accurate climate model. The unpublished studies by Gordon (1989), Latif et al. (1988), and Mechoso et al. (1991) provide excellent examples of how the coupling of atmosphere and ocean models can be used to identify the subtle but key processes involved in the maintenance of the present mean climate.

The observations also clearly demonstrate that certain coordinated atmosphere-ocean modes on the interannual time scale exist only because of the interaction between these media. The most robust example is ENSO, although there is increasing evidence for one or more western Northern Hemisphere coupled atmosphere-ocean modes. The absence of these unforced natural "modes" or phenomena in a climate system model can be helpful in isolating serious deficiencies in the model physics or in the formulae that govern the exchange of energy and mass between the media.

CONCLUSIONS AND DISCUSSION

Within the scientific community interest is growing in identifying and analyzing the variability in the full climate system on decade-to-century, or intermediate, time scales through modeling and observational studies. The resulting increased research activity can be attributed both to the advances in research on the interannual variability of the coupled atmosphere-ocean climate system and to the practical problems that have arisen in the detection of an anthropogenically forced climate change. The primary limitation on the study of climate variation on the intermediate time scale is that the instrumental records for many climate variables either do not exist or are too short to detect a phenomenon with any statistical measure of confidence. Proxy data for climate, therefore, will play an important role in documenting the past climate variability on the intermediate time scales and, most likely, in evaluating the simulated climate variability. This approach represents a significant change from the traditional modus operandi, wherein the models are used mainly to simulate phenomena that are already well observed. The goals and research activities supported by the ACCP illustrates these points.

Numerical models will be the primary tools by which the mechanisms responsible for decade-to-century-scale climate variability are identified and, most important, will frequently be the instruments scientists use to identify the intermediate climate phenomena. Thus, it is extremely important that the veracity of climate system models be evaluated a priori by assessing how accurately they simulate many well-documented phenomena. These "target" phenomena include the seasonal and diurnal cycles of the state variables that are well documented in nature, the fluxes of energy and mass at the media interfaces. Also necessary is a quantitative comparison of the processes that are responsible for these cycles in the model with those that are observed. Other prerequisites for a climate system model should include its ability to simulate the robust natural interannual variability that is observed within the individual media (e.g., the QBO) and in the coupled atmosphere-ocean system.

Where are we now? At present, no climate *system* models include a satisfactory treatment of the cryosphere and land hydrology for the intermediate-scale climate problem. In fact, there are only two climate models that have the potential to resolve the quantitative three-dimensional circulation of both the atmosphere and the global oceans: the Hamburg model (Cubash et al., 1991, unpublished manuscript) and the GFDL model (Manabe et al., 1992). Both of these models require flux corrections to achieve a mean climatology that is qualitatively consistent with that observed.[8] To date these and other models have been used primarily to simulate the climate of the recent ice age and to estimate the potential for large climate changes due to an increase in the atmospheric CO_2 concentration (IPCC, 1990; Delworth et al., 1995, in this chapter). These climate perturbations are extremely large in both amplitude and in spatial scale by comparison with the climate variations that are anticipated—and have thus far been observed—on the intermediate time scale. In addition, the agents responsible for the glacial cycles are unlikely to be relevant to the intermediate climate problem. Thus, the success of the climate system models in reproducing geologic time-scale variations will be of little use in assessing the veracity of the intermediate-scale variability that is produced by the same model. For these and many other reasons it is unclear whether the decade-to-century-scale variability produced by any of the existing climate system models has any bearing on the variability in the true climate system.

The research activities over the last 10 years have led to incredible progress in the modeling of climate, especially on the geologic time scales. On the intermediate time scale, significant progress has also been made. Important recent results achieved using stand-alone models of the individual media and the GFDL complete climate-system model will provide a great deal of guidance in establishing the scientific priorities for the next decade. These priorities include improvements in the parameterization schemes for certain small-scale processes, and the identification of key physics through extensive model-sensitivity experiments. The results of studies that address these and other central issues will be instrumental in the design of reliable models of decade-to-century-scale variability in the climate system.

ACKNOWLEDGMENTS

This work was supported by the National Science Foundation (Grant ATM 8822980) and the National Oceanic and Atmospheric Administration's Office of Global Programs and EPOCS. This is contribution number 217 to the Joint Institute for the Study of the Atmosphere and Oceans (JISAO).

---

[8]The flux corrections for the Hamburg model have not been published. The water flux correction in the GFDL model is comparable to the observed mean flux.

Commentary on the Paper of Battisti

GERALD R. NORTH
Texas A&M University

Dr. Battisti's presentation made a broad sweep over many of the issues that we have talked about this week. He has thus given me the opportunity to touch on a few of these in relation to my own paper.

I tend to approach the question of long-time-scale change in an engineering fashion: I wonder whether we can tell from all the different records that we have whether there really is any anthropogenic forcing, and whether we are seeing any response to it. I recognize that there are many other reasons for studying these time scales, but this is certainly an important one. I find it particularly interesting that as we try to make some kind of statistical test of what we have seen in the last century, it is the frequencies just bordering on the decadal range from below for which we really need answers.

I should like to talk a little about two Hasselmann-type models, which could also be called default models. The first one is noise forcing on a mixed-layer slab model with geography; the other one is a deep-ocean extension of that, but still very crude: It distorts the spectrum somewhat at low frequencies. One of the things I noticed as we went through the week—Ed Sarachik and Bob Dickson have already mentioned it—is the failure of these models to generate intermediate water very well. That has a very important impact right in the frequency band $(100 \text{ yr})^{-1}$ to $(10 \text{ yr})^{-1}$. Also, one of the odd things about our deep-ocean model is that it cannot get information down to the thermocline area very well. The giant models adjust so quickly onto the ramp curve that I suspect they might be getting information down there too fast.

All this makes me wonder whether the kinds of things that Jim McWilliams brought up earlier might be fairly serious—for instance, when we construct our bottom topography out of stair steps. Going out in Reynolds number on that lovely bifurcation diagram that he showed is equivalent to going to higher and higher resolution. In our present models we have gone past only a couple of bifurcations, and it is hard to tell whether we are seeing fictitious oscillations in our model world that might go away if we were to go just a little bit further. I do not mean to criticize the people who are working on these models; I think they are doing what needs to be done. Consider, for instance, some of the things that Dr. Rooth and Dr. Barnett showed. I cannot help suspecting that if we were to look at the Mikolajewicz/Maier-Reimer model, we might see a peak at 300 years (recall the film loop that we viewed earlier this week), and in the Delworth et al. work we might see a peak in the 50-year range. These would, of course, have a large impact on our ability to pick out a signal or to infer how rare such an excursion might be in, say, the last century. Clearly, we have a lot to do before we can unravel these problems, so I again raise Dr. McWilliams's issue.

I should also like to mention once more something that Dick Lindzen pointed out on the first day, since I think it is important but easy to miss. Up at the top of the ocean there is a "valve" that allows the radiation to go out and be absorbed. It amounts to an effective cooling coefficient, with all the feedbacks and everything that must go into it, and thus controls the spectrum and basically the sensitivity of the atmospheric model. If we have an otherwise good atmospheric model that does not do the clouds right, its sensitivity may be wrong by a factor of 2 or even 3, and that error will actually control the spectrum in the low-frequency range. It will have a rather important influence on fluctuations, even at the decadal level.

I hope that Dr. Battisti will open the discussion with some ideas about where we should be going in modeling, and particularly how fine the scales must be to get at some of the questions that we need to answer.

Discussion

BATTISTI: Explicitly examining the mechanisms and processes that yield variations in climate would take more computational power than I for one can fathom, even if we looked at only a subset of the system. I think what we should try for is determining the sensitivity of the simulated climate variability to the parameterization schemes for the unresolved physics. That sensitivity is likely to be a strong function of model resolution.

SARACHIK: You know, it seems to me that we should start concentrating on the signal rather than the noise. Some of the variability we've been talking about is undoubtedly scientifically interesting, but it's relatively local. It seems to be bounded by something on the order of one-third or one-half of a degree of temperature. My inclination is to look at longer time scales, something like 1000 years, and a signal big enough to be called climate change.

REIFSNYDER: I'd like to take philosophical issue with your statement that models are used to test hypotheses. I'd say that models are simply functional expressions of hypotheses, and any meaningful testing must involve independent data sets. Testing against data used in the specification of the model, or against general phenomena, won't tell you anything about predictive power.

BATTISTI: Well, that's true. The solution to a model is really the solution to a set of equations. But what worries me is the possibility that the results of complex, numerical GCMs will be used to build hypotheses of how model phenomena come about, without adequate attention to whether those phenomena appear in the observational or proxy data bases. I'd rather see models used to test hypotheses about the mechanisms responsible for observed phenomena. That is, we should be testing our understanding of a phenomenon, not defining one.

LINDZEN: I think that our interest has traditionally been signal detection—greenhouse warming, for instance. That implicit search for something dramatic worries me. I think we should be equally concerned with the constraints on detectable phenomena provided by the data themselves.

LEVITUS: It seems to me essential that we understand decadal-scale variability better before we can go on to longer time scales or develop better models. First we have to be able to parameterize the processes better and to understand the system on shorter time scales.

WEAVER: The people who complain that this or that process hasn't been included in a model seem not to understand that you can't possibly do a systematic analysis by putting everything in at once.

RIND: A propos of some comments by both Dave and Ed, I'd like to mention again that the NSF/NOAA ARRCC—Analysis of Rapid and Recent Climate Change—program is currently working on reconstructing the big climate changes over the past 1000 years, which includes two cold periods and one warm. We want to put together a worldwide picture of how the climate then compares with today's, and ultimately see whether models can reproduce it.

MARTINSON: Jerry, I just wanted to add that I'm glad you emphasized that it's almost impossible to evaluate the role of sea ice in long-term changes when a model has *prescribed* flux corrections, since in reality so much of that flux is driven by sea ice itself.

North Atlantic Interdecadal Variability in a Coupled Model[1]

THOMAS L. DELWORTH, SYUKURO MANABE, AND RONALD J. STOUFFER[2]

ABSTRACT

A fully coupled ocean-atmosphere model is shown to have irregular oscillations of the thermohaline circulation in the North Atlantic Ocean with a time scale of approximately 40 to 50 years. The fluctuations appear to be driven by density anomalies in the sinking region of the thermohaline circulation combined with much smaller density anomalies of opposite sign in the broad, rising region. Anomalies of sea surface temperature associated with this oscillation induce surface air temperature anomalies over the northern North Atlantic, the Arctic, and northwestern Europe. The spatial pattern of sea surface temperature anomalies bears an encouraging resemblance to a pattern of observed interdecadal variability in the North Atlantic.

INTRODUCTION

Substantial variability in the North Atlantic Ocean occurs on time scales of decades; this variability is thought to be associated with fluctuations in the large-scale meridional overturning (Gordon et al., 1992). This meridional overturning consists of cold, saline water sinking at high latitudes of the North Atlantic and flowing equatorward at depth, upwelling throughout the world oceans, and returning as a northward flow of warm, salty water in the upper layers of the Atlantic ocean. This overturning largely determines the oceanic component of the northward transport of heat from the tropics to higher latitudes and is thus essential for the maintenance of climate in the North Atlantic region (Manabe and Stouffer, 1988). Fluctuations in the intensity of this overturning have the potential to affect climate (Bjerknes, 1964) in both the ocean and the atmosphere. Interdecadal fluctuations in this overturning are thus of potentially critical importance, not only for their direct impact on the climate of the North Atlantic and European regions but also because of the implications of such variability for the detection or modification of anthropogenic climate change. The presence of interdecadal variability in the coupled ocean-atmosphere system makes the detection of anthropogenic climate change more difficult.

The incompleteness of the observational record makes it difficult to study variations in this overturning and the associated processes solely on the basis of the observational data. As a complementary approach, the output from a 200-

---

[1] A more complete discussion of these results may be found in Delworth et al. (1993).
[2] NOAA Geophysical Fluid Dynamics Laboratory, Princeton University, Princeton, New Jersey

year integration of a coupled ocean-atmosphere model is used to study the stability and variability of this meridional overturning (hereafter referred to as the thermohaline circulation, or THC). The output from this model, developed at NOAA's Geophysical Fluid Dynamics Laboratory, forms the basis for the analyses presented here. Several previous modeling studies (Mikolajewicz and Maier-Reimer, 1990; Weaver and Sarachik, 1991b; Weaver et al., 1991; Winton and Sarachik, 1993) have used ocean-only models to study the variability and stability of the thermohaline circulation, in contrast to the fully coupled ocean-atmosphere model employed in the present investigation. Since the mechanisms governing the fluxes of heat and fresh water across the ocean surface differ substantially between the two types of models, the models might be expected to have different time scales for variations in the THC. While the quantitative results presented here will depend on the details of the model formulation and the physical parameterizations employed, it is hoped that the physical processes identified in this model study are robust and are responsible for interdecadal variability in the North Atlantic. The present study demonstrates that substantial variability can occur in the coupled ocean-atmosphere system on time scales of decades in the absence of external forcings such as changing concentrations of greenhouse gases.

MODEL AND EXPERIMENTAL DESIGN

The oceanic component of the coupled model solves the primitive equations of motion with an approximate horizontal resolution of 3.75° longitude by 4.5° latitude, and 12 unevenly spaced vertical levels. A simple thermodynamic balance is used to predict the presence and thickness of sea-ice, which is advected by surface currents as long as the sea ice thickness is less than 4 m. The atmospheric model solves the primitive equations using the spectral transform technique, and has an approximate horizontal resolution of 7.5° longitude by 4.5° latitude. The atmospheric model, which has 9 unevenly spaced levels in the vertical, incorporates seasonally varying insolation and terrestrial radiation. Cloud cover is predicted whenever the relative humidity exceeds a critical value (99%). At continental surfaces, the heat and water budgets and their interaction with the atmosphere are computed.

The coupled model is global in domain, incorporating realistic geography smoothed to the model resolution. The atmosphere and ocean interact through the surface fluxes of heat, fresh water, and momentum. The surface fluxes of heat and fresh water into the ocean are adjusted at each grid point in order to obtain a more realistic simulation of climate. The surface flux adjustments are derived from preliminary integrations of the separate oceanic and atmospheric models. The derived flux adjustments do not vary from year to year and are not dependent on the anomalies of

surface temperature and salinity. Thus, they do not explicitly affect the strength of the feedback processes that reduce the anomalies of temperature and salinity at the ocean surface. The model and the flux adjustments are described in more detail in Manabe et al. (1991).

Starting from an initial condition in quasi-equilibrium, the coupled model is time-integrated over a period of 200 years. The mean rate of change of global mean sea surface temperature for the model over this period is very small (0.01°C per century).

RESULTS

Comparison with Observed Variability

An assessment of the model's ability to simulate climate variability is indicated in Figure 1 by comparing a spectrum of model generated sea surface temperature (SST) to a spectrum of observed SST (Frankignoul and Hasselmann, 1977). The observed SSTs are from ocean weather ship India in the North Atlantic (59°N, 19°W), while the model SSTs are from the model grid point closest to the location of ocean weather ship India. Although the model tends to underestimate the total variance (proportional to the area beneath the spectrum), the overall agreement between the two spectra is good.[3] One should note, however, that the present model does not resolve mesoscale eddies, which may contribute significantly to the variability of sea surface temperature at seasonal to annual time scales.

Meridional Overturning Index and Variability

The stream function describing the meridional circulation in the Atlantic basin (200-year mean) is shown in Figure 2. The circulation follows the direction of the arrows; its magnitude is proportional to the gradient of the contours. Water flows northward in the upper layers, sinks at high latitudes, and flows southward at depth. The region in which most of the sinking occurs in the model North Atlantic is principally confined to the latitudinal belt from 52°N to 72°N; hereafter, the term "sinking region" will refer to the

---

[3]In subsequent figures and the text, it is demonstrated that the intensity of the meridional overturning in the model North Atlantic has enhanced variance at a time scale of 40 to 50 years, but the spectrum of model SST in Figure 1 is not characterized by enhanced variance in this frequency band. As shown in Figure 4a, there is a spatial pattern of SST anomalies associated with these overturning variations. The point chosen for the SST spectrum in Figure 1 is not located in this region, and thus does not have enhanced variance at the 40-to-50-year time scale. Spectra of model SSTs from the region of maximum SST change shown in Figure 4a are characterized by enhanced variance at the 40-to-50-year time scale.

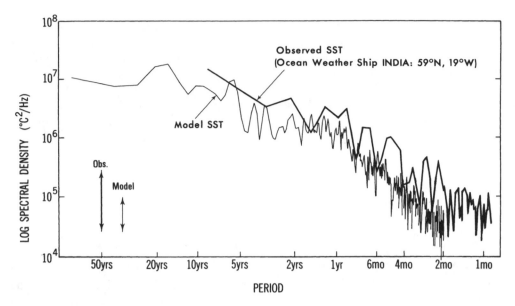

FIGURE 1   Heavy, solid line: spectrum of sea surface temperature from observed data (ocean weather ship India, 59°N, 19°W). Light, solid line: spectrum of model sea surface temperature from grid point closest to 59°N, 19°W. The arrows in the lower left corner denote 95 percent confidence intervals about the spectra. (From Delworth et al., 1993; reprinted with permission of the American Meteorological Society.)

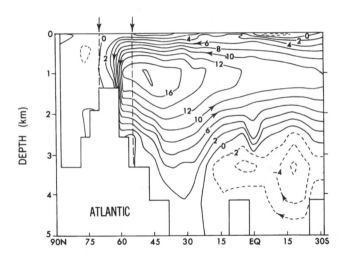

FIGURE 2   Stream function of the meridional overturning in the Atlantic Basin. Flow follows the direction of the arrows. Units are sverdrups (10^6 m^3/s). The dashed lines denote the sinking region as defined in the text. (From Delworth et al., 1993; reprinted with permission of the American Meteorological Society.)

region defined by this latitudinal belt and extending from the western to the eastern boundaries of the Atlantic basin.

In order to concisely describe variations in this meridional overturning, one needs a measure of the intensity of the thermohaline circulation in the North Atlantic. This intensity is defined each year as the maximum stream-function value representing the annual mean meridional circulation. A time series of this index, which represents the fluctuations in the intensity of the annual mean THC,

is shown in Figure 3a. Visual inspection of this figure reveals substantial variability on interdecadal time scales. The spectrum of this time series, shown in Figure 3b, demonstrates enhanced variance in the 40-to-50-year time scale. Analyses of an extension of this integration (to 1000 years in length) suggest that these irregular oscillations are present in the period beyond 200 years, but are characterized by a somewhat longer time scale. In addition, the spectrum from the longer THC time series contains a more clearly defined broad spectral peak.

It should be noted that the index, by definition, primarily describes variations in the North Atlantic section of the THC. Subsequent analyses show that the climatic variations in the model associated with variations in this index are not hemispheric in nature, but are better characterized as regional to the North Atlantic and Arctic.

Sea Surface Temperature, Salinity, and Surface Air Temperature Variations

The spatial pattern of the changes of model sea surface temperature associated with fluctuations in the intensity of the thermohaline circulation is shown in Figure 4a. These differences are computed by subtracting the mean of four decades having anomalously small values of the THC index from the mean of four decades with anomalously large values of the THC. The pattern of SST change bears an encouraging resemblance to a pattern of observed interdecadal SST variation computed by Kushnir (1994) and shown in Figure 4b. The field in Figure 4b was computed by

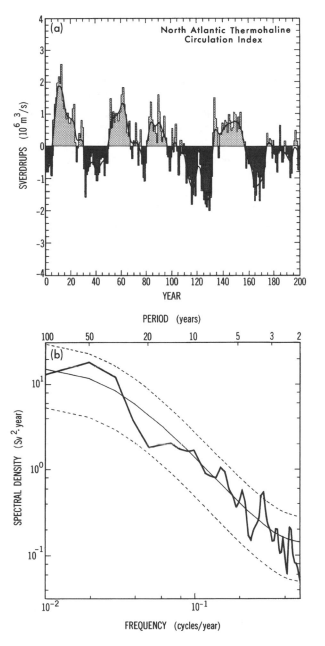

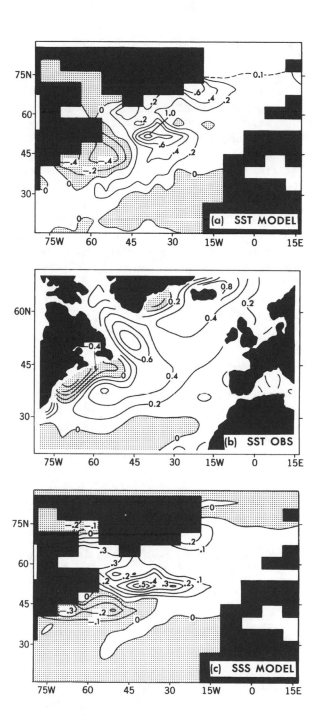

FIGURE 3 (a) Time series of the annual mean intensity of the index of the meridional overturning in the North Atlantic. Units are sverdrups. (The long-term mean of 18.28 sverdrups has been removed from the time series prior to plotting.) Heavy, solid line is a smoothed time series computed by applying a 13-point binomial filter to the annual mean data (approximately a 10-year low-pass filter). (b) Heavy, solid line denotes spectrum of the thermohaline circulation index time series shown in (a). Thin, solid line denotes the least-squares best fit of a theoretical red-noise spectrum to the spectrum of the THC. Dashed lines denote 95 percent confidence limits about the red noise spectrum. Note: The spectrum was computed by taking the Fourier transform of the autocovariance function, using a maximum of 50 lags and a Tukey window (Chatfield, 1989, Chapter 7). (From Delworth et al., 1993; reprinted with permission of the American Meteorological Society.)

FIGURE 4 (a) Differences in annual mean model sea surface temperature between four decades with anomalously large thermohaline circulation index values and four decades with anomalously small thermohaline circulation index values. Units are °C. Values less than zero are stippled. (b) Difference in observed sea surface temperature between the periods 1950-1964 (warm period) and 1970-1984 (cold period). Units are °C. Values less than zero are stippled. (Adapted from Kushnir, 1994; reprinted with permission of the American Meteorological Society.) (c) Differences in annual mean model sea surface salinity, computed in the same manner as for Figure 4a. (From Delworth et al., 1993; reprinted with permission of the American Meteorological Society.)

subtracting the 1970-1984 mean (cold period) from the 1950-1964 mean (warm period). These 15-year periods were chosen on the basis of analyses of areal average SSTs in the North Atlantic (see Kushnir (1994) for details). There is a resemblance between the model and observational results, in terms of both the spatial pattern and the magnitude. It is tempting to speculate that the observed changes in SST may have arisen because of changes in the intensity of the THC in the North Atlantic. The pattern of model sea surface salinity (SSS) changes associated with fluctuations in the intensity of the thermohaline circulation, shown in Figure 4c, is somewhat similar to the pattern of model SST changes. Additional analyses have shown that the SST and SSS patterns in Figure 4 can be interpreted as the result of anomalous advection of the climatological mean SST and SSS distributions by the anomalous currents associated with variations in the THC. It should be noted that the pattern of model ocean temperature and salinity changes at 295 m depth (not shown) also resembles the observational results of Levitus (1989a), in which differences in ocean temperature and salinity were computed for two 5-year periods comprising 1955-1959 and 1970-1974.

The changes in surface air temperature associated with fluctuations in the intensity of the thermohaline circulation, computed in the same manner as for SST and SSS, are shown in Figure 5. The geographical extent of the surface air temperature changes is not restricted to the North Atlantic, but encompasses large regions of northern Europe and the Arctic. (Note that these surface air temperature results are from the winter season only.) Over the Arctic Ocean and its vicinity, surface air temperature changes are particularly large in winter. This seasonal dependence in the high latitudes results partially from changes in sea-ice thickness associated with changes in the intensity of the thermohaline circulation. When the THC is anomalously strong, the enhanced transport of heat into high latitudes results in a reduction of mean sea-ice cover (not shown), thereby permitting an enhanced heat flux from the ocean to the atmosphere during the winter months when the mean surface air temperatures are substantially lower than that of the ocean at high latitudes. During the summer, the mean surface air temperature over the Arctic Ocean is equal to or greater than the ocean temperature at high latitudes, and there is little or no heat flux from the model ocean to the atmosphere. However, the enhanced surface absorption of summer insolation due to the reduced coverage of sea ice also contributes to the reduction of sea ice thickness and the increase of surface air temperature in winter.

The above results highlight the potential importance to climate of such fluctuations in the thermohaline circulation over the North Atlantic, Europe, and the Arctic. These variations arise in the absence of any external forcing at this time scale.

Ocean Density Variations

It is desirable to examine in greater detail the spatial and temporal relationships between the irregular oscillations in the intensity of the thermohaline circulation and variations in the three-dimensional fields of salinity, temperature, and density. These relations are quantified by computing, at each grid point, linear regressions between the time series of temperature, salinity, and density versus the time series of the thermohaline circulation. The regressions were computed as:

$$y(t) = a\,x(t+\tau) + b \qquad (1)$$

where $y(t)$ can be salinity, temperature, or density at each grid point, t is time, a is the slope of the regression line, $x(t+\tau)$ is the time series of the THC, τ is the lag, and b is the intercept of the regression line. These regressions were computed at various lags in order to provide a three-dimensional picture of the evolution of the oceanic state as the thermohaline circulation fluctuates. Prior to the regression analyses, all time series were first detrended and filtered in order to effectively remove fluctuations with time scales less than approximately 10 years. The analyses were also performed without filtering; they yielded phase relationships similar to those described below. The slope of the regression line (referred to hereafter as the regression coefficient) esti-

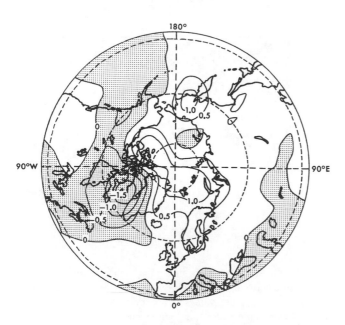

FIGURE 5 Differences in surface air temperature computed in a manner similar to that used for Figure 4a, except that only data from the months of December, January, and February was used. Values greater than 0.5 are densely stippled, values less than −0.5 are lightly stippled. A polar projection is used. Actual coastlines are used in the construction of this map in order to facilitate orientation. (From Delworth et al., 1993; reprinted with permission of the American Meteorological Society.)

mates the change in temperature, salinity, or density for a unit change in the intensity of the THC.

In order to provide a large-scale measure of conditions in the sinking region, the regression coefficients between density (ρ') and the thermohaline circulation were averaged vertically and horizontally over the sinking region. As can be seen in Figure 6, the density time series leads the thermohaline circulation time series by a few years, suggesting that fluctuations in density in the sinking region induce the fluctuations in the THC. In addition, the density term is

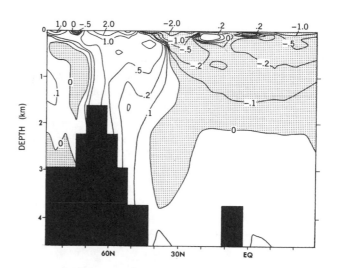

FIGURE 7   Latitude by depth cross-section in the Atlantic basin of the zonal mean of the regression coefficients between the time series of density and the thermohaline circulation. The regression coefficients are computed at lag -3 years, indicating that the density values at year $(n - 3)$ are correlated with values of the THC index at year n. This lag corresponds to the phase with the maximum vertically averaged density perturbation in the sinking region. Units are $(g/cm^3) \times 10^{-5}/Sv$. Values less than zero are stippled. This cross-section can be interpreted as the anomaly density structure three years prior to a maximum in the thermohaline circulation. (From Delworth et al., 1993; reprinted with permission of the American Meteorological Society.)

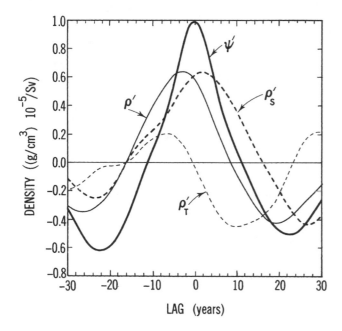

FIGURE 6   At each grid point at each model level, linear regression coefficients were computed between the time series of annual mean density, temperature, and salinity at that grid point versus the time series of the thermohaline circulation. The regression coefficients were computed using various lags in order to obtain an estimate of the temporal evolution of the three-dimensional fields of density, temperature, and salinity with respect to variations in the thermohaline circulation. These regression coefficients were then averaged vertically and horizontally over the sinking region of the North Atlantic (defined in text) and plotted as a function of lag in years. For example, the regression coefficients plotted at lag -5 ($+5$) years indicate conditions 5 years prior (subsequent) to a maximum in the thermohaline circulation. The heavy, solid line (ψ') denotes the regression coefficients of the THC index with itself (thus representing a "typical" fluctuation). The thin, solid line (ρ') represents the regression coefficients between density and the THC index. The dashed line (ρ'_S) denotes the regression coefficients for the density changes attributable solely to changes in salinity versus the THC index, while the dotted line (ρ'_T) represents the regression coefficients for the density changes attributable solely to changes in temperature versus the THC index. Note that ρ'_T is inversely related to temperature, whereas ρ'_S is proportional to salinity. The units for ρ', ρ'_T, and ρ'_S are $(g/cm^3) \times 10^{-5}/Sv$. (From Delworth et al., 1993; reprinted with permission of the American Meteorological Society.)

decomposed into individual contributions from temperature and salinity anomalies, and the regression coefficients between these two terms and the THC are also plotted. This figure indicates that density fluctuations attributable to salinity anomalies (ρ'_S) are almost in phase with, but slightly behind, the thermohaline circulation, whereas density fluctuations attributable to temperature anomalies (ρ'_T) lead the thermohaline circulation by approximately $90°$ in phase. The sum of the effects of temperature and salinity anomalies results in a density time series that leads the thermohaline circulation by several years, as indicated in this figure. The phase differences between ρ'_S and ρ'_T shown in Figure 6 are important to the quasi-oscillatory nature of the fluctuations.

Temperature-induced density anomalies (ρ'_T) serve as a negative feedback on the fluctuations of the THC. The largest negative ρ'_T values (corresponding to the warmest vertically averaged temperatures) occur after the maximum in the THC, thereby reducing the mean density in the sinking region, and thus weakening the intensity of the THC. In contrast, the largest positive values of ρ'_T occur before the maximum in the THC, thereby increasing the mean density in the sinking region and enhancing the intensity of the THC. Thus, temperature-induced density anomalies play and important role in these fluctuations.

To examine the latitude-depth distribution of the density

anomalies, the zonal means of the linear regression coefficients between density and the thermohaline circulation are shown in Figure 7. The lag chosen (≈ 3 years) corresponds to the time when the vertically averaged density anomaly in the sinking region is close to its maximum. This figure shows that, in the sinking region, positive density anomalies extend very deep, indicating the downward penetration of positive salinity anomalies. Note also the smaller negative density anomalies to the south of the sinking region. The distribution of the density anomalies described above forms a solenoidal field that intensifies the THC. As previously shown in Figure 6, the temporal variations of the vertically averaged density field in the sinking region precede the variations in the thermohaline circulation's intensity, implying that the former drives the latter.

A map of the regression coefficients between the time series of the near-surface density anomalies and the THC is shown in Figure 8. The largest density anomalies occur in the vicinity of the Labrador Sea and south of Greenland, with smaller values in the Greenland and Norwegian seas. Thus, the fluctuations of the model THC at this time scale appear to be related primarily to convection in the Labrador Sea and in the region south of Greenland. Variations in the near-surface currents associated with fluctuations in the THC support this conclusion. Shown in Figure 9 are the regression coefficients between surface currents and the THC. At each grid point, regression coefficients are computed between the meridional component of the surface currents and the THC, as well as between the zonal component of the currents and the THC. These regression coefficients are then plotted in vector form in the figure, so that the length of an arrow in the zonal (meridional) direction indicates the magnitude of the coefficient. Note that the current changes associated with fluctuations in the THC are

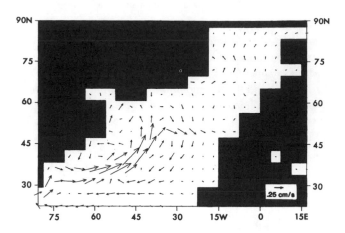

FIGURE 9   Regression coefficients between surface currents in the ocean and the intensity of the thermohaline circulation. Regressions were computed with zero lag. At each grid point, regression coefficients were computed between the zonal component of the current and the THC, as well as between the meridional component of the current and the THC. These regression coefficients were then plotted at each grid point in vector form, in such a way that the length of the arrow in the zonal (meridional) direction denotes the magnitude of the regression coefficients between the zonal (meridional) component of the current and the THC. The vector flow field shown in this figure may be interpreted as the change in surface currents associated with a unit increase in the intensity of the thermohaline circulation. Scaling and units are indicated by the arrow in the lower right-hand corner. (After Delworth et al., 1993; reprinted with permission of the American Meteorological Society.)

primarily confined to the western North Atlantic, scarcely affecting the Greenland and Norwegian seas. These changes in the current structure are not limited to the surface, but are found throughout the top 1 km.

The fluctuations of temperature and salinity in the sinking region, which determine the density variations and thus the intensity of the THC, are themselves generated by variations in the currents associated with fluctuations in the intensity of the THC. Budget analyses of the vertically averaged water column in the sinking region have demonstrated that the most important factors governing the heat and salt budgets for the sinking region appear to be the horizontal transports of heat and salt into the sinking region from that portion of the Atlantic that is south of the sinking region (for the heat budget, the surface flux is also important). These budget analyses have demonstrated that in addition to transport of heat and salt by the zonal mean meridional overturning, there are substantial transports of heat and salt into the sinking region by the horizontal gyre component of the flows. (Here, the horizontal gyre component of the flow is defined at each grid point as the departure of the meridional flow from the zonal mean meridional flow.) The phase of the gyre transport appears to precede the phase of the transport by the zonal mean meridional flow, suggesting

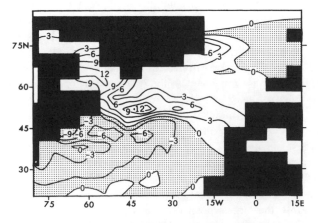

FIGURE 8   Regression coefficients of surface density versus the time series of the intensity of the thermohaline circulation. Regressions were computed with zero lag. Values less than zero are stippled. Units are $(g/cm^3) \times 10^{-5}/Sv$. (From Delworth et al., 1993; reprinted with permission of the American Meteorological Society.)

an important role for variations in the gyre transports in the model. Thus, these variations in the transport by the horizontal gyre component of the flow are an integral component of the THC fluctuations.

SUMMARY AND DISCUSSION

It has been shown that an irregular oscillation of the thermohaline circulation in the North Atlantic ocean appears in a coupled ocean-atmosphere model. This oscillation, which has a time scale of approximately 40 to 50 years, appears to be driven by density anomalies in the sinking region of the North Atlantic ocean combined with much smaller density anomalies of opposite sign in the broad, rising region. These density variations are in turn induced by fluctuations of the thermohaline circulation itself. Although it has not been confirmed here, one can speculate that the oscillation is triggered by nearly random surface-buoyancy forcing of heat and water fluxes.

Sea surface temperature fluctuations associated with the variations in the thermohaline circulation have a spatial pattern that bears an encouraging resemblance to a pattern of observed interdecadal variability (Kushnir, 1994). The SST anomalies induce surface air temperature anomalies of substantial magnitude over the North Atlantic, northern Europe, and the Arctic. The interdecadal time scale of this variability points out the potential difficulty of distinguishing natural variability of the coupled climate system from anthropogenic climate change, which would be expected to occur on similar time scales. Bjerknes (1964) suggested that low-frequency changes in SST over the northern North Atlantic are associated with variations in the intensity of the thermohaline circulation. The coupled-model simulation illustrates a physically plausible mechanism for such a complex low-frequency air-sea interaction. All that can be concluded is that the model variability is broadly consistent with what little evidence is available. Future studies of both the instrumental record and proxy paleoclimatic data have the potential for producing new evidence to further constrain models of North Atlantic climatic variability.

Commentary on the Paper of Delworth et al.

STEPHEN E. ZEBIAK
Lamont-Doherty Earth Observatory

I read Dr. Delworth's paper with great interest. I think it is the only result of its kind that we have: a coupled GCM that is producing decadal-scale oscillations that are sufficiently like what we observe to warrant careful study and scrutiny.

The paper mostly presents diagnostics of the oscillation. It left me with my appetite whetted but feeling somewhat disappointed. I am sure that I speak for many of us in saying that we are very anxious for further diagnostics and attempts to understand what is going on. From what was presented, I still do not understand what the essential mechanisms of this oscillation are. For instance, Dr. Delworth indicated differences in phase between the temperature and salinity components in the density field, but I am not sure from what I saw today why we would expect those phase differences.

Let me note a few questions. The first is the obligatory question about flux correction. Given the choice between not making flux corrections and thus having a totally unphysical climatology, and making flux corrections in order to have something more realistic-looking to start with, I think the flux correction is the better of the two. But with a phenomenon like this you can look to see whether the structure and the amplitude of the flux corrections that you are putting into the model give you confidence that the model processes, as reflected in your means, are not likely to yield results that will not be physical or realizable in the real world.

Next, I was struck by the fact that the variance of the coupled model seems to be lower than that in nature for almost all frequencies across the entire spectrum. I am curious to hear why that should be true.

I also noted with interest that on the figure showing the spectrum of the thermohaline circulation the only peak outside the 95 percent confidence interval derived from a simple red-noise spectrum occurred at four years. Dr. Delworth's method of analysis precludes looking at that so far, but I wonder whether there is something there of interest.

The paper showed a broad-band frequency spectrum in the first 200 years, but it seemed to be favoring 40- to 50-year periods. When the integration was carried further, it looked as though it might be shifting. I wonder whether it would be robust if the model were run for, say, 5000 years. Would it disappear or turn into something else? From my own experience, I know that with ENSO models that are chaotic, it is possible to get a certain behavior for quite a long time that then turns into something else. I should be interested in hearing any additional results from longer running that might bear on this topic.

One very important question is the nature of the atmosphere-ocean interaction. For the oscillations you get, in what ways are the atmosphere and the ocean coupled? We obviously have coupling from fresh-water fluxes, from heat fluxes, and from momentum fluxes, and it would be of great interest to figure out whether, say, the wind-driven part of the ocean circulation has anything to do with what happens, or whether the primary coupling takes place through fresh-water fluxes and heat fluxes alone.

The reductionist approach is one way to attack this. For instance, if the atmospheric model is run with just an ocean mixed layer, do you get a mode that has anything in common with this? Similarly, the ocean could be run with just the mean wind field and precipitation fields prescribed from the atmospheric model, so the only coupling would happen through the atmospheric boundary-layer fluxes. Experiments like this might help determine what aspects of the coupling among the many in the full coupled model are primary in this particular oscillation.

I have already alluded to the differences in the temperature and salinity parts of the density fluctuations. As the paper mentions, one hypothesis is that the fluctuations are driven by noise. Another possibility is that the coupling between the atmosphere and ocean is sufficient to generate those fluxes without noise.

Discussion

DELWORTH: Many of the issues you've raised are fundamental and not easily soluble. With regard to the flux corrections, you should look at the spatial pattern, see where they're largest, and make some assessment of their importance. They can indeed be said to affect low-frequency variability, since without them there would be no overturning at all, but we can only speculate how much.

Let me show you again the spectrum of the thermohaline index I defined over the first 200 years, my Figure 3. If you compute a spectrum based on 600 or 800 years, the diminution of variance at the very-low-frequency end becomes clearer, but still doesn't pass the 95 percent limit. I tend to view the spectrum as characterized by an area of broad-band enhanced variance; I wouldn't say it's periodic. I also don't feel that the peak at four years has any meaning.

You asked whether these fluctuations are robust as the integration continues. On an index for years 200 to 400 you can still see the oscillation, but on a somewhat longer time scale. You see similar fluctuations on the others, all the way out to 1000 years. It's an open question whether this variability would be maintained in an even longer integration.

The analyses I've done don't suggest any role for the wind-driven circulation in these fluctuations. There is a coupling in terms of the sensible and latent heat fluxes that does appear to play some role; these fluxes are affected by mean wind speed and by large-scale changes of surface pressure and wind. I should point out that Yochanan Kushnir has done similar analyses with COADS pressure data, and the model results don't correspond well with observations, so there is certainly room for experimentation. Incidentally, the temperature and salinity differences between the phases is very important, because temperature provides a strong negative feedback on the density in the sinking regions, tending to stabilize these fluctuations.

MOREL: Disregarding the gyre for the time being, would you characterize this change in the overturning as essentially internal to the Atlantic Basin, or as a global signature?

DELWORTH: The integration is global, and the various analyses were computed globally, but the variations are pretty much confined to the North Atlantic on this time scale.

LEVITUS: Have you tried to trace any of these temperature/salinity anomalies throughout the model?

DELWORTH: I see it more as changing intensity of the circulation operating on the mean gradients to produce these anomalies than as an anomalous water mass propagating through the model.

DIAZ: Have you seen anything indicating an out-of-phase sea surface temperature relationship between the North and South Atlantic?

DELWORTH: I haven't looked for that. Defining the index the way I do steers the analyses toward the North Atlantic.

SARACHIK: In the simpler sector models of the forced ocean we find two types of variability, very-long-term and the advective shorter-period variability you show. Your flux correction basically stabilizes the long-term thermohaline circulation, I think, confining you to the latter type.

DELWORTH: Well, my time series suggest that there may also be a lower-frequency mode, though it's hard to see in a 1000-year integration.

ROOTH: Manabe and Stouffer's two-mode experiment showed that there are two thermohaline circulation modes in the model even with the flux corrections in place.

CANE: There's been a lot of interest in the oscillations, but actually your response looks like a red-noise process. You're getting all kinds of durations for these changes, which may in fact be correct. Also, when you were discussing the effects on the atmosphere you ended up talking about sea ice. Can you amplify a little on what role it might have in the ocean circulation as well as its effect on the atmosphere?

DELWORTH: I was expecting sea ice to play a strong role by changing sea surface salinity, but I didn't find it in the model. I did see movement of sea ice out of the Arctic into the convective regions, but it was on a higher frequency—5- to 10-year time scales—than these thermohaline fluctuations. In fact, the sea ice is virtually in phase with the fluctuations. In this model, and on this time scale, the ice appears to be reacting to changes in circulation rather than shaping them.

DICKSON: The top of your north wall is at about 1600 or 1800 meters. If you raised that to 600 meters, which would reflect reality, would you be able to get a deep enough overturning circulation in the model?

BRYAN: This is a persistent problem with these models. If we had the wall that high, we would get no overflow at all. But when we bring it down we get too much entrainment over the wall. Given their narrowness, it may be impossible to resolve the currents in a model like this.

Natural Climate Variability on Decade-to-Century Time Scales
National Research Council, 1995

Model Studies of the Long-Term Behavior of ENSO

MARK A. CANE, STEPHEN E. ZEBIAK, AND YAN XUE[1]

ABSTRACT

The ENSO cycle is irregular. As a result, time series of ENSO indicators exhibit variability from decade to decade.

There are at least two plausible suggestions in the literature for this low-frequency variability. ENSO could be a low-order dynamical system with chaotic dynamics, or it could be that there is no aperiodicity inherent in ENSO but that the ENSO system is jiggled by the external noise generated by the rest of the climate system. A third possibility is that the low-frequency behavior of the ENSO cycle is driven by variations in some other part of the climate system. Although possible in principle, thus far this last possibility has little support from the data record or from model simulations.

We consider these possibilities within the framework of models that have produced long-term aperiodic simulations of ENSO-like variations. We argue that in the models, at least, the ENSO cycle results from low-order chaotic dynamics. Model results are used to explore the possibility of unequivocally distinguishing chaotic dynamics from truly random behavior, which is difficult to do in the far more limited records of observational data. We conclude that the 100 years of instrumental data are probably insufficient, but despite added difficulties in interpretation, the much longer time series from proxy records may suffice.

INTRODUCTION

It is a well-known and well-documented fact that the El Niño/Southern Oscillation (ENSO) cycle is irregular. The paper in this chapter by Rasmusson et al. (1995, henceforth RWR), enhances this documentation. There is variation in the amplitude of the extreme phases of the cycle—warm and cold events—and in the interval between events. There is variation in the evolution of events: no two cycles are exactly alike, and occasionally the Southern Oscillation signals in the west Pacific assert their independence from the El Niño sea surface temperature (SST) signatures in the east. Furthermore, it is well documented that ENSO has global concomitants and global consequences (RWR; Ropelewski and Halpert, 1987, 1989; Glantz et al., 1991). These, too, vary in time. Indian monsoon rainfall, for example, shows some variability at the approximately 4-year

---

[1]Lamont-Doherty Earth Observatory of Columbia University, Palisades, New York

time scale characteristic of the ENSO cycle, but also varies on a variety of other time scales.

What is the reason for the irregularities in the global ENSO cycle? Generically, possible causes may be classified into forcings external to the ENSO cycle and nonlinear chaotic dynamics internal to ENSO. Such a division implicitly assumes that "ENSO physics" is a subset that may be disentangled from the larger system. We return to this point below.

If the forcing is outside of the physical ocean-atmosphere system, then there is no difficulty in accepting it as "external." Two examples that have been suggested in the literature are volcanic eruptions (Handler, 1990) and solar activity (Anderson, 1989). These two seem to have some currency. There have been others, including underwater tectonic activity and the influence of biological productivity on the penetration of sunlight. Although the case for volcanic eruptions has been pressed forcefully, a careful reconsideration leaves it without evidentiary support (Nicholls, 1988). The case for solar activity is quite tentative and will not be considered further here.

Forcings within the physical climate system but external to ENSO may be divided into "base state," the terminology used by RWR, and "noise." By "base state" forcings we mean changes in the large-scale, low-frequency, global background state in which the ENSO phenomena are embedded. RWR detect no coherent relationship between variations in ENSO and in the base-state parameters they consider. By "noise" we mean factors external to the ENSO process with time scales that are short compared with the dominant time scales of ENSO. Their effects on ENSO are quasi-random. For example, Schopf and Suarez (1988) suggest that incursions of synoptic-scale systems from higher latitudes into the tropical Pacific are responsible for the aperiodicity of El Niño (cf. Battisti, 1988).

An alternate hypothesis is that the aperiodicity is a consequence of nonlinear chaotic dynamics internal to the ENSO system of interactions. The concept of dynamical chaos is by now well known, the most celebrated example being the three-variable system of Lorenz (1963). ENSO participates in the global climate system, which has an enormous number of variables, so this hypothesis requires that a relatively low-order ENSO subsystem be distinct from the system as a whole. This is possible, even reasonable, because it is consistent with the generally successful current paradigm that accounts for the principal characteristics of ENSO (see below).

In this paper we speculate on these three possible causes of the variability of the ENSO cycle: variations in the base state, in the noise, and in the internal dynamics. Our principal tool in this investigation is a numerical model (Zebiak and Cane, 1987; henceforth ZC) that has been used to predict the ENSO cycle a year or more ahead and to simulate its long-term behavior. Ultimately, a theory for the causes

of ENSO variability must be verified against real-world data. However, as the discussion of RWR vividly illustrates, the shortness of the instrumental record and the ambiguities of the proxy record make this problematical. Part of our purpose here is to use the model to explore how long or what sort of record might suffice to distinguish among alternatives.

The next section very briefly reviews the model and sketches a case for its suitability for the purpose at hand. We also sketch the theory for ENSO that is needed for what follows. The three sections that follow consider variability due to base state changes, noise, and chaotic dynamics. A discussion section concludes the paper.

THEORY AND MODEL

The ZC model is one of several that have led to a converging view of ENSO as an internal oscillation of the tropical Pacific ocean-atmosphere system (Zebiak and Cane, 1985; ZC; Schopf and Suarez, 1988; Hirst, 1988; Battisti, 1988; see Cane (1991) for a recent summary account). Although this view continues to be refined and elaborated (Chao and Philander, 1993; Jin and Neelin, 1992; Latif ct al., 1993), it is fair to say that we now have a widely accepted paradigm for the generation and maintenance of the ENSO cycle.

The core of the paradigm is the two-way ocean-atmosphere interaction hypothesized by Bjerknes (1969). Briefly, the thermal contrast between the cold eastern equatorial Pacific and the western Pacific warm pool drives the easterly trades at the surface. The easterlies create the contrast by driving the upwelling and advection that make the eastern Pacific cold. Thus there is a positive feedback in this coupling. The same feedback in reverse gives the El Niño state: if the east warms, the trades relax, the east warms further, the relaxation of the trades increases, and so on.

What was lacking in Bjerknes' scenario was a mechanism for the cycle, the transition from one state to the other. The answer lies in equatorial ocean dynamics. When the east warms anomalously, the layer of warm water above the thermocline thickens. This excess must be compensated for by a deficit of warm water elsewhere in the basin. Equatorial dynamics dictates that it will be to the west, off the equator, in the form of equatorial Rossby wave packets. The Rossby waves propagate west until they encounter the boundary. There their mass deficit signal is reflected eastward in equatorial Kelvin waves. This signal thins the warm layer in the east and eventually brings the warm event to an end and initiates a cold phase. The phase of cold waters in the east is necessarily accompanied by a warming in the west, which will eventually work its way around to initiate a new warm phase. And so the cycle continues ad infinitum.

Certain features of this "delayed oscillator" paradigm are important to our discussion. The essential interactions are of large scale and low frequency; the number of degrees of freedom in the ENSO cycle could well be quite small. Only the tropical Pacific is involved; the rest of the globe is assigned no role. Of course, this does not preclude ENSO influences on the entire climate system.

The ZC model, which we rely on in the present study, was built with the Bjerknes hypothesis in mind. It is a model for the interactions of the atmosphere and ocean in the tropical Pacific. Although simplified in many respects, it produces aperiodic interannual oscillations that have many features in common with observed oscillations. The same model has been used in largely successful forecasts a year and more in advance of ENSO (Cane et al., 1986; Barnett et al., 1988; Cane, 1991).

Both the atmospheric and oceanic components of the model describe perturbations about the mean climatological state, with the monthly climatology specified from observations (Rasmusson and Carpenter, 1982). The atmospheric dynamics take the form of the steady, linear shallow-water equations on an equatorial beta-plane, with the circulation forced by a heating anomaly that depends partly on local heating associated with SST anomalies and partly on the low-level moisture convergence. The latter effect is nonlinear because latent heating occurs only when the total wind field is convergent, and this is a function of the background mean convergence as well as the calculated anomalous convergence.

The ocean model is set up in a rectangular basin extending from 124°E to 80°W, and 29°N to 29°S. The dynamics begin with the shallow-water equations, but are modified by the addition of a shallow frictional layer of a constant 50-meter depth to allow for the intensification of wind-driven currents near the surface. The thermodynamics describe the evolution of temperature anomalies in the model surface layer. All terms, including nonlinear advection components, are retained. Finally, temperature anomalies below the base of the surface layer are parameterized in terms of vertical displacements of the model thermocline (identified with movements of the layer interface).

The next few figures, from Zebiak and Cane (1991), are intended to allow the interannual variability of the model to be compared with that of the observations, thereby (we hope) establishing its credibility for the purpose at hand. Figure 1 shows a time series of NINO3 (90°W to 150°W, 5°N to 5°S) averaged SST anomalies from a 1024-year run of the model. Inspection of this time series reveals a wide range of behavior: there are well-marked periods of large-amplitude, rather regular oscillations and other periods of more chaotic, smaller-amplitude variability. Also of note are the period of nearly 20 years with negligible anomalies (near year 500), and periods of similar length with anomalies of only one sign. The other indices of Figure 1 describe these properties quantitatively and were devised as follows. The 1024-year (monthly) time series was divided into 51

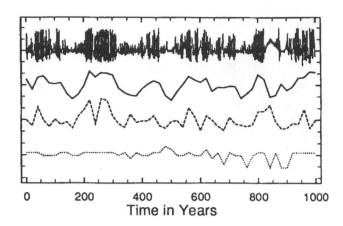

FIGURE 1   Time series of the NINO3-averaged SST anomaly from a 1024-year coupled-model simulation (top). Below are the three indices representing variance (heavy solid line), fraction of variance (dashed line), and dominant frequency (dotted line) for successive 24-year segments. (See text for definition of indices.) The heavy solid line indicates the overall intensity of ENSO activity for each 25-year period throughout the simulation. The dashed and dotted curves measure respectively the degree of regularity and the preferred period of El Niño cycles. Two regime types are evident: one in which strong and regular ENSO cycles of approximately 4-year periodicity persists, and another with weaker, more erratic behavior.

segments of 24 years' length (overlapping each other by 4 years). Within each segment, the mean, standard deviation, and power spectrum were computed. (A Welch window was applied prior to FFT calculation.) The spectral band with largest power was determined, as well as the fraction of total (interannual) variance contained in that band. Shown are the standard deviation, dominant frequency, and fraction-of-variance indices. The latter is a measure of the degree of regularity; nearly periodic oscillations result in large variance accompanying the dominant frequency, while chaotic variations result in variance spread among many bands. Two regimes are apparent: one with large variance, high regularity, and a dominant 4-year period; the other with lower variance, low regularity, and mixed periods. Thus, the indices provide a quantitative measure of the behavior patterns conspicuous in the original time series.

The distributions of the indices (among the 51 realizations) are shown in Figure 2. The standard deviation and fraction-of-variance distributions are notably broad, reflecting the distinct regimes. The favored frequency band centers on 4 years, but adjacent bands are also represented (3 to 6 years). For purposes of comparison, these same indices were computed from the single 18-year realization of observed SST anomalies between 1970 and 1987 (NOAA Climate Analysis Center analysis) and are indicated on the distribution plots. The observed values fall well within the range of the model-derived distributions, showing that, at least

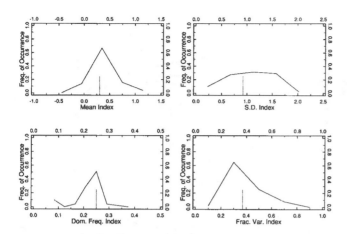

FIGURE 2   Distributions of mean sea-surface temperature (°C), standard deviation, dominant frequency (years^{-1}), and fraction-of-variance indices from the 1024-year simulation. The values of each index computed from observed sea-surface temperature between 1970 and 1987 are indicated by vertical line segments.

for this period, the temporal characteristics of observed interannual variability are consistent with those of the model.

To examine the spatial structures, we computed empirical orthogonal functions (EOFs) of model fields on the basis of monthly values from the 1024-year simulation. Figure 3 shows the first four EOFs of SST anomalies. The first EOF, with 83 percent of the total variance, is clearly the mature El Niño signal. The others represent various phases of the life cycle of model warm and cold episodes. Corresponding EOFs derived from observed SST anomalies (1970 to 1987) are shown in Figure 4. Although there is correspondence between the patterns, two biases of the model are apparent: one underestimation of variability in the South American coastal region, and another in the vicinity of the dateline at the equator. Also, whereas the first four EOFs of the model fields contain 93 percent of the total variance, only 68 percent of the variance is accounted for in the observed fields. This partly reflects the absence of non-ENSO signals (that is, "noise") in the model fields. In addition, the model anomaly patterns are more consistent from event to event than in nature, as evidenced by the fact that the second EOF variance is a smaller fraction of the first for the model than in reality.

EFFECTS OF VARIATIONS IN BASE-STATE PARAMETERS

Because ZC is a perturbation model, the model framework allows a clean separation between ENSO and the base state. We may then consider the effect on model ENSO characteristics of changes in the model base state. Five parameters defining this state were chosen for a systematic study. They reflect the oceanic equivalent depth (Par 1),

the sharpness (Par 2) and the amplitude (Par 3) of the mean thermocline, the strength of atmospheric heating associated with SST anomalies (Par 4), and the atmospheric friction (Par 5). A detailed treatment of these parameters can be found in ZC. For this study, each of the five parameters was allowed to assume each of three values—the standard one, a 5 percent decrease, and a 5 percent increase—in all possible permutations. This amounts to 3^5 or 243 simulations, each of which was run for 100 years (starting from the same initial conditions). The purpose of this study was to determine whether rather modest parameter changes could significantly change the model behavior, and if so, in what sense. Because of the large number of simulations, it is possible to assign statistical significance to the results, avoiding the uncertainties inherent in comparisons of individual realizations. After each 100-year simulation had been broken into four segments, the segments' characteristics were evaluated using the four indices defined above. To examine the individual effects of each parameter, the total set of results was grouped into three subsets corresponding to the three values of that parameter. Each of the subsets then contained 324 realizations. The cumulative distributions of each of the four indices were computed for each of the three subsets and then compared (Figure 5). In this form, the significance of differences in the distributions can be assessed using the Kolmogorov-Smirnov statistic (Conover, 1980). The 99 percent confidence interval derived in this manner is displayed in each distribution plot; there is only a 1 percent probability that two distributions (of size 324) differing by more than this amount represent the same process.

All five of the parameters produce significant changes in at least one of the indices, and the net effect of each parameter is unique. In the case of Par 1, increasing values produce smaller mean and standard deviation, and a tendency toward higher frequencies. Increasing Par 2 results only in decreasing the standard deviation on the high end (that is, decreasing the amplitude of the largest warm/cold events). Increasing Par 3 acts to increase the mean (marginally) and the standard deviation, while giving a more dominant 4-year period. For Par 4, the result of an increase is a larger mean and standard deviation, and a smaller fraction of variance (more irregular). Finally, increasing Par 5 leads to much reduced mean and standard deviation, a greater likelihood of dominant periods less than 4 years, and more regular oscillations. All of these effects are consistent and identifiable amid the background of natural variability, showing a genuine sensitivity to externally determined factors making up the background climate state. However, in most cases it would have been impossible to identify systematic changes without a large sample to work with: 324 realizations were used, whereas 100 years of data would yield only 5. Another consequence is that since many of the parameter settings yield distributions not significantly

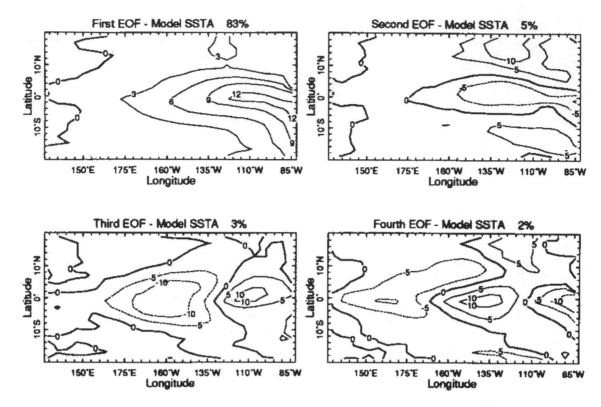

FIGURE 3   First four EOFs of the model SST anomalies, calculated from monthly values of the 1024-year simulation.

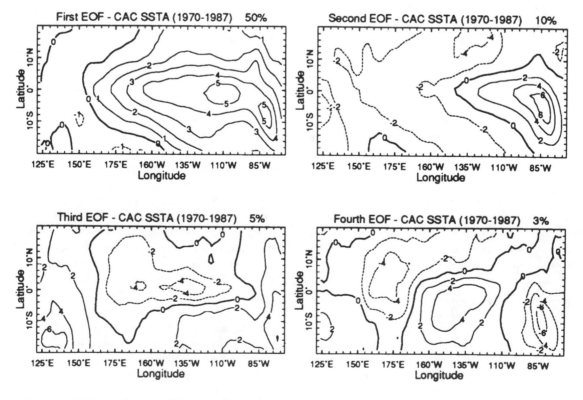

FIGURE 4   First four EOFs of observed SST anomalies, calculated from monthly values between the years 1970 and 1987, inclusive (Climate Analysis Center analysis).

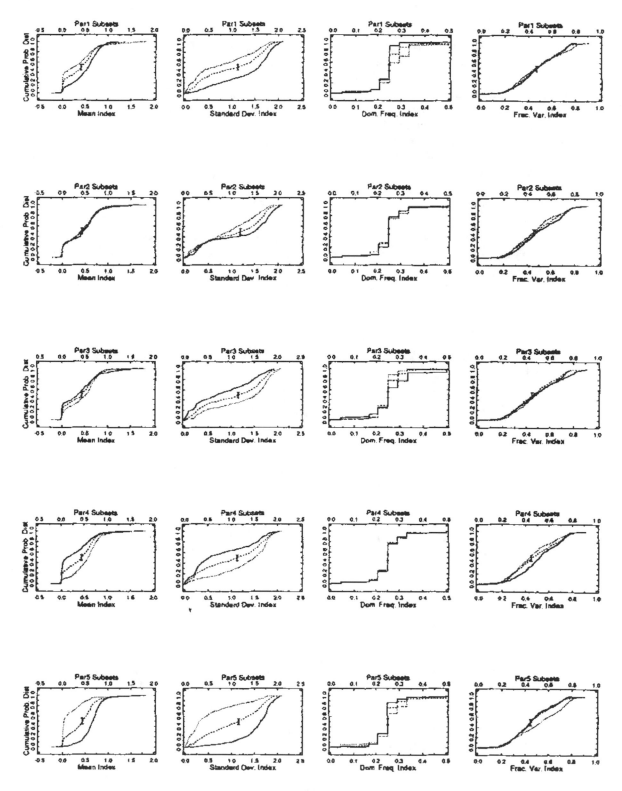

FIGURE 5 Cumulative distributions of mean SST, standard deviation, dominant frequency, and fraction-of-variance indices from the set of 243 simulations with variations in five model parameters. Across each row, the total set of results in each plot has been stratified according to the three distinct values of the indicated parameter. The heavy solid line represents the distribution for the subset with a 5 percent reduction of the parameter, the dashed line represents the subset with the standard value, and the dotted line represents the subset with a 5 percent increase. Also indicated in each plot (vertical bar) is the separation interval beyond which any pair of distributions may be considered distinct at the 99 percent level.

different from the order-of-100 years of instrumental data, a unique best-model fit to an observations model (i.e., a best set of parameter values) cannot be determined.

Figure 6 illustrates the issue. A number of 256-year time series with different parameter sets are shown. The variability within each series is more impressive than the differences between series. An obvious implication is that the effects of base-state changes of this magnitude are not detectable against the inherent natural variability of ENSO. Assuming that the model response to such changes is realistic, such base-state changes (e.g., due to greenhouse effects) would cause real changes in the climate system. However, we could not show them to be significant until they had persisted through tens of generations. Perhaps, then, it is fortunate that we are unaware of any compelling evidence for such changes in the recent past (cf. RWR). Any variations in such features as mean atmospheric temperature or decadally averaged tropical Pacific SST are more likely to be consequences than causes of ENSO variations.

INFLUENCE OF NOISE

Zebiak (1989) added to the ZC model a random wind forcing of a form intended to simulate observed intraseasonal variability in the tropical Pacific. That study demonstrated that this added noise did not alter the prediction skill of the model, the broader implication being that the intraseasonal variations have little effect on the predictability of ENSO at lead times up to two years. More relevant to present interests, the spectra presented there (his Figure 5) were virtually identical at periods longer than a year, indicating that the added noise made no difference in overall model behavior. Of course, this does not tell us the reasons for the ZC model's aperiodicity.

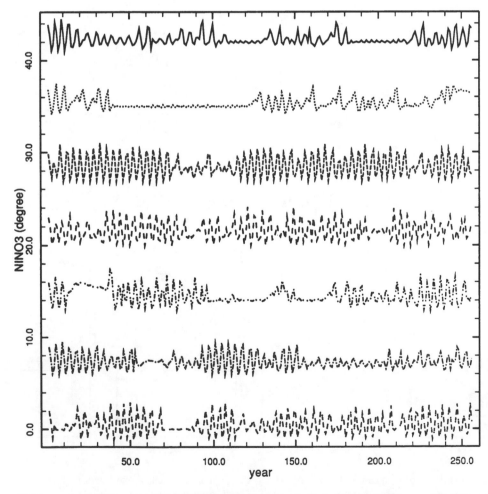

FIGURE 6 A set of 256-year-long time series of the NINO3 SST anomaly. Each is from a model run with different parameter settings; each began with the same initial conditions. Using a code where the nth digit is 0, 2, or 1 according to whether Par n is decreased by 5 percent, increased by 5 percent, or left at the standard value used in ZC, from top to bottom the cases are 02011, 02022, 11111, 12010, 12021, 12111, 12121. For example, the third curve is the standard case, while for the one above it Par 1 and Par 3 were decreased 5 percent while Par 2, Par 4, and Par 5 were increased 5 percent.

The time series of NINO3 from the model runs (Figures 1 and 6) appear to have characteristics associated with chaotic nonlinear dynamics: They are aperiodic and exhibit broad-band variability, with identifiable regimes persisting from a few years to more than a century. Nonetheless, it is not altogether straightforward to establish rigorously that these time series result from nonlinear dynamics.

Since the model is relatively simple, one might try an appeal on formal grounds: The model is nonlinear, and most of its variance is in an ENSO mode (e.g., Figure 3), so what other explanation could there be? However, not all of its variance is ENSO; there is other variability at smaller time and space scales. It may be argued that this acts as noise, preventing the model from settling into a periodic orbit. Battisti (1988) found his version of the ZC model to be periodic unless noise was added. Schopf and Suarez (1988) showed that in their model the irregular behavior disappeared when the atmospheric component of their model was replaced by a linear one. They attribute the difference to extratropical synoptic variability, which is effectively a source of noise for the ENSO dynamics.

However, the change in atmospheric model makes the interpretation ambiguous. Inevitably, it must alter the parameters influencing the ENSO physics, and it is well established that even small parameter changes may render an aperiodic model perfectly periodic (Jin and Neelin, 1992, and references therein). Such a change in model behavior can be induced, for example, by changes in any model parameters affecting the strength of the coupling between ocean and atmosphere (e.g., ZC; Münnich et al., 1991).

We will take the following approach to investigate the possibility that the model's irregular behavior is caused by noise. A long monthly time series (256 years) from a run of the fully nonlinear ZC model is best fit by a linear first-order autoregressive (AR) model:

$$u(t+1) = Au(t) + e(t) \qquad (1)$$

where u is a vector of model state space variables, t is time in months, and e is the residual in the fit, which we model as white in time; i.e., $e = Br$, where r is a vector of independent random numbers uncorrelated in time. Our objective is to determine whether the behavior of the original nonlinear coupled model can be distinguished from that of this linear model driven by white noise.

The procedure for construction of the AR model is a variant of that described by Blumenthal (1991) and Blumenthal et al. (1992). Grid-point data from a long run of the ZC model are saved monthly. The original multivariate model grid-point time series are then transformed into a set of multivariate EOFs and associated time series (PCs). A truncated set of these principal components (PCs) constitutes the AR model state space $u(t)$. In the results reported here, we retain the first 20 components, which capture over 95 percent of the variance in the model run. (Over 95 percent

of the variance in the key model variables of surface wind, SST, and thermocline displacement is retained.) We note that this set can reconstruct the NINO3 index, an area average, with negligible error.

Given the time series $u(t)$, the transition matrix A is found in the usual way as the best fit (in the least-squares sense) to the AR model equation (1). In recognition of the importance of the seasonal cycle, we elaborate the usual procedure by calculating a different A for each of the 12 monthly transitions (cf. Blumenthal, 1991). Once the A's have been determined, the 12 monthly values for B are calculated from the covariance of the residual time series $e(t)$.

Figure 7 demonstrates the ability of the AR model to reproduce the behavior of the fully nonlinear coupled model in the short run—for a few years, at least. Clearly, this "best fit" model is a good fit indeed. We next want to see whether the long-term—interdecadal and beyond—behavior of the nonlinear ZC model can be distinguished from that of the AR model. If so, it would strongly suggest that ZC's long-term behavior is not well described as the response to a noise forcing of an essentially linear system. It is more likely a consequence of nonlinearities internal to the ENSO cycle itself.

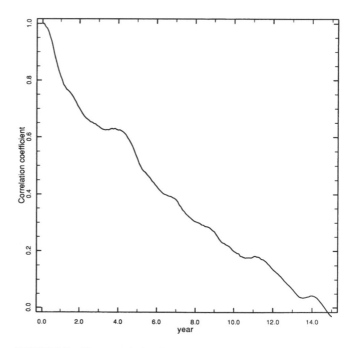

FIGURE 7   The correlation between the NINO3 index from the ZC model run and that obtained from the unforced linear AR model $u(t+1) = Au(t)$ of Eq (1). The correlation is taken over the ensemble of AR model runs initialized from each of the monthly states in a 256-year ZC model run. Since the unforced AR model runs decay, the correlation must ultimately go to zero. That the correlation remains high for a few years demonstrates the ability of the AR model to reproduce the behavior of the fully nonlinear coupled model in the short run.

A NINO3 index time series typical of runs of the AR model is shown in Figure 8. The format is the same as in Figure 1. The similarities are more compelling than the differences, although there appears to be a discernible tendency for the regimes in the ZC run to persist longer. A comparison of power spectra of the NINO3 index from the ZC run and the AR run (Figures 9 and 10) adds to the

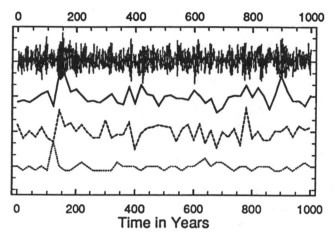

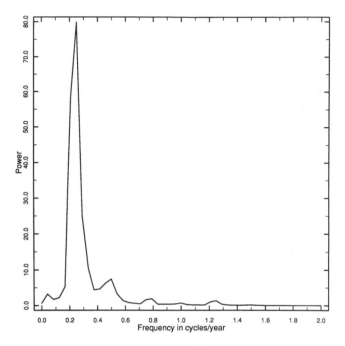

FIGURE 8   Time series of NINO3 SST anomaly from a 1024-year AR model run (top). The three indices below represent variance (heavy solid line), fraction of variance (dashed line), and dominant frequency (dotted line) for successive 24-year segments. Compare to Figure 1.

FIGURE 10   Power spectrum of the NINO3 SST anomaly from a typical 1024-year run of the AR model. See Figure 9.

impression that the two are quite similar. The AR model reproduces all the peaks at periods of a year and longer. (This was true of all the AR runs we made with different initial seeds for the random number generator, although, as should be expected, the runs differ slightly in the amplitudes of the peaks.) Figure 11 shows for the AR model the same distributions of regime indicators as were presented for the ZC model in Figure 2. The distribution of means is quite similar, as is the distribution of dominant frequencies, although there are somewhat more occurrences of longer

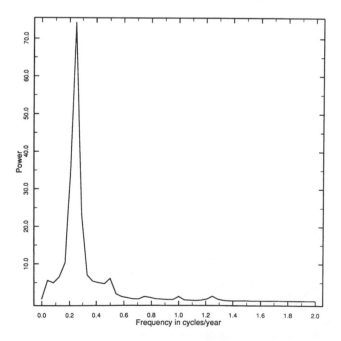

FIGURE 9   Power spectrum of the NINO3 SST anomaly from a run of the ZC nonlinear model. As described in the text, the 1024-year (monthly) time series was divided into 24-year-long segments, overlapping each other by 4 years, and a Welch window was applied. The spectrum shown is the average over all segments.

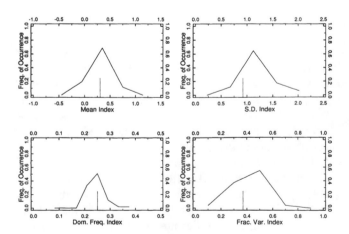

FIGURE 11   Distributions of mean SST, standard deviation, dominant frequency, and fraction-of-variance indices from a 194,560-year run of the AR model. (Compare Figure 2.) This run is a long enough sample to approximate closely the true distributions of the AR process (viz. the error bars on Figure 12).

periods for the ZC model. The AR model shows fewer instances of high or low standard deviations, that is, fewer 24-year periods either without events or dominated by strong ENSO events. Also, as the fractional variance index shows, the AR model tends not to have a single frequency dominating a 24-year period. While the ZC model clearly has periods with quite regular and quite large events as well as periods of rather little ENSO activity, the AR model exhibits a more uniform mix of such behavior within 24-year segments. The noise-driven linear model is less regime-like.

Applying the Kolmogorov-Smirnov test to these distributions shows this difference to be statistically significant, as is evident in Figure 12. In particular, the 1024-year (51-sample) ZC model run can be distinguished at the 98 percent level from the AR process by its differing standard-deviation and fractional-variance indices. The mean and the dominant frequency distributions are indistinguishable from those of an AR process. Note that if it were based on a run of only 100 years (5 samples), the ZC distributions would be indistinguishable from those of an AR process at the 98 percent level—or even at the 90 percent level. With 500 years (25 samples) the standard deviation distribution is just significant at the 90 percent level; even at this level the others are not distinguishable from the AR process.

Again, we are able to reach this conclusion because of the sufficiently large samples available to us with millennium-long simulations. With a 100-year record the distributions would have only 5 points, making it virtually impossible to establish statistical significance. Prospects for using long time series from paleo proxy data will be considered in the discussion section.

INTERNAL DYNAMICS

The analysis above builds a case for attributing the aperiodicity in the ZC model to nonlinear dynamics intrinsic to the ENSO cycle. Under the assumption of linear dynamics, it was shown to be unlikely that the regime-like behavior could be driven by noise. The possibility remains that nonlinearities are essential and the noise is somehow important.

The ZC model could be doctored to eliminate all non-ENSO variability (e.g., by increasing frictional effects), but then any change in ENSO behavior is likely to have more to do with the alterations in model parameters than with the absence of noise. We take a different approach by showing that a still simpler model, one containing nothing but ENSO physics, can be aperiodic.

The model of Münnich et al. (1991, MCZ hereafter) is

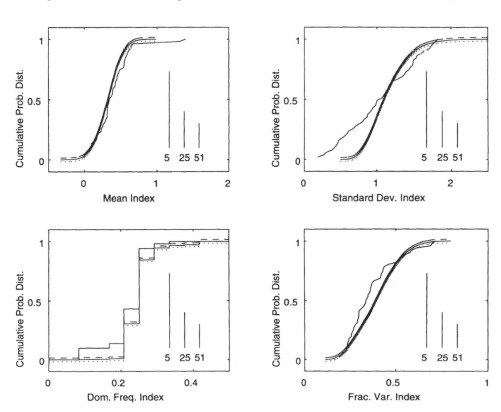

FIGURE 12 Cumulative probability distributions of mean SST, standard deviation, dominant frequency, and fraction of variance from a 1024-year run of the ZC model with standard parameters (solid curves), and the same distributions from the very long run (194,560 years) of the AR model. The 98 percent confidence limits on the AR model are given by the dashed and dotted lines. The vertical lines are the 98 percent confidence limits for 5, 25, and 51 samples, corresponding to runs of 100, 500, and 1024 years.

one of a number of conceptual models intended to illustrate and explore the Bjerknes *cum* delayed oscillator paradigm (Battisti and Hirst, 1989; Schopf and Suarez, 1988; for a recent elaboration and synthesis see Jin and Neelin, 1992). Cane et al. (1990) give a derivation and heuristic justification (also see Jin and Neelin, 1992). This will not be repeated here; instead we will stipulate that the arguments might fail to convince the determined skeptic that the MCZ model embodies the essence of the ZC model. We offer it as an analog, a demonstration that a prototype delayed-oscillator model is capable of aperiodic behavior.

The model equation is the solution for wind-forced motions described by the shallow-water equations on an equatorial beta plane. The wind forcing is purely zonal, is centered longitudinally in the model ocean basin, and has a fixed spatial form. The wind amplitude, A, depends solely on the thermocline height anomaly at the eastern end of the equator, h. The model may be reduced to the form of an iterated map (Eq. (6) of MCZ), which we write as

$$h(t) = \sum_n \alpha_n A(t-4n+3) + \sum_n \beta_n h(t-4n) \qquad (2)$$

The only model variables are A and h. Time t is nondimensionalized by half the time it takes for an equatorial Kelvin wave to cross the basin. The α's and β's are numbers; in addition to n, they depend only on the meridional scale of the wind and the magnitude of the friction. In principle, the sums are from 1 to infinity, but convergence is rapid, so in practice they are cut off at $N = 10$.

In its primary nonlinear form, the model is closed by specifying a function $A(h)$ for the dependence of wind amplitude on thermocline displacement. The function chosen (Eq. (9) and Figure 2 of MCZ), a linear piece inserted smoothly between two hyperbolic-tangent segments, is fashioned after the shape of the tropical thermocline.

The slope of $A(h)$ at $h = 0$ is a measure of the strength of the coupling between atmosphere and ocean. Figure 13 illustrates the change of behavior as this coupling strength κ is increased. For very small κ the model decays. As it increases, there is a bifurcation resulting in regular oscillations with a period of around 2 years. For still larger κ there is a period doubling to a period of about 4 years. At $\kappa = 2.16$ the period triples, and for $\kappa = 2.33$ aperiodic behavior appears.

As discussed in MCZ, with this simplest version of the model it is a somewhat delicate matter to coax out aperiodic behavior. Either of two elaborations in the direction of greater realism makes this behavior quite robust. One is to add an annual cycle to κ, taking crude account of effects of the mean annual cycle in such factors as SST and wind speed. The other is to include in A the asymmetry between the warm (upper) and cold (lower) regions of the thermocline. Figure 14 is an example of the behavior obtained when both factors are included. The behavior is markedly

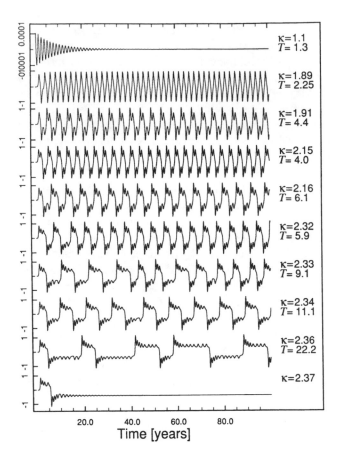

FIGURE 13   Time series of thermocline displacement $h(t)$ as calculated from the iterated map (2) for various values of the coupling strength κ. The function $A(h)$ is symmetric about the center of the thermocline and does not include an annual cycle. (See text for details.) T, the average period in years, was determined from the output. (From Münnich et al., 1991; reprinted with permission of the American Meteorological Society.)

irregular, although with a dominant period of around 4 years and noticeable power at 2 years. These characteristics occur for a wide range of parameter settings (although Figure 14 is one of the better specimens).

In this section we have recounted a model built as a paradigm for ENSO, stripped of any physics able to generate natural "noise" external to the ENSO cycle. In common with the real world and a number of numerical models, including ZC, it generates aperiodic behavior with a dominant period of about 4 years and some power at quasi-biennial periods. It adds to the argument that the most likely cause of the regime-like behavior of ENSO is nonlinear dynamics intrinsic to the ENSO cycle.

DISCUSSION

Is the natural variability of ENSO in the real world due to changes in the base state, forcing of the ENSO system

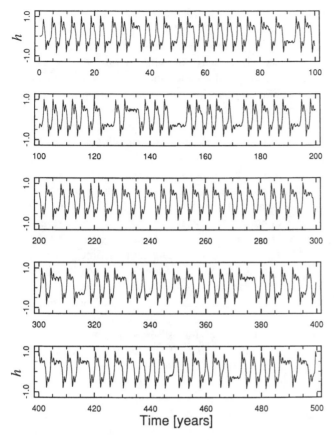

FIGURE 14 Time series of thermocline displacement $h(t)$ as calculated from the iterated map (2) for an asymmetric thermocline shape, and an annual cycle modulating the strength of the coupling between ocean and atmosphere. (From Münnich et al., 1991; reprinted with permission of the American Meteorological Society.)

by climatic noise, or dynamics internal to the ENSO cycle itself? Can this hypothesis be verified for nature's ENSO?

The existence of models and theories for ENSO provides a valuable foundation for investigating its long-term variability. In this paper we have explored the possible causes for the irregularity of the ENSO cycle in a model, that of Zebiak and Cane (1987). The model is clearly sensitive to relatively small changes in its base state: Long model runs allowed us to make statistically reliable statements as to the effects of various parameter changes. However, the intrinsic variability within a run made with fixed parameters is quite often more marked than the difference between runs with different parameters (viz., Figure 6). Discerning systematic parameter effects typically requires far longer time series than the 100 years of instrumental records. We see no way to attribute any of the variability of the past 100 years to variations in the base state (cf. RWR).

It is also unlikely that we will be able to detect reliably the effect of greenhouse warming on ENSO in the near future. However, experience with models will allow the inference that changes in the base state bias the climate

system so that certain ENSO characteristics (e.g., a more regular cycle) become more likely (cf. Zebiak and Cane, 1991).

We examined the possibility that natural noise, such as intrusions of mid-latitude synoptic systems, could force aperiodicity in the ENSO cycle. We concluded that this is a viable mechanism, but were able to show that the behavior of the ZC model could be distinguished from the most realistic noise-forced linear auto-regressive model we could construct. Again, achieving confidence in this conclusion required a model time series much longer than 100 years.

We were thus left with the presumption that the aperiodicity of the ZC model is due to chaotic nonlinear dynamics intrinsic to the model's ENSO cycle. This was bolstered by a consideration of a noise-free conceptual model (Münnich et al., 1991) of the Bjerknes-delayed oscillator paradigm. In common with the ZC model, it produced records dominated by an approximately 4-year cycle, but with power at quasi-biennial periods as well as interdecadal regime-like behavior.

According to Occam's Razor, these results recommend the following as a working hypothesis: The principal cause of all the variability associated with the ENSO cycle is low-order chaotic dynamics of ocean-atmosphere interactions within the tropical Pacific. "All the variability" includes the basic quasi-quadrennial cycle, quasi-biennial variability (Rasmusson et al., 1990; Barnett, 1991; Ropelewski et al., 1992), and interdecadal variability (RWR). It includes all ENSO-related variability outside the tropical Pacific, such as that in the Indian monsoon rainfall record. This working hypothesis is not (yet) contradicted by any known facts. It is the minimum hypothesis needed to account for the observed global variability at all time scales associated with ENSO.

As noted early and often, the shortness of instrumental records makes it difficult—more likely impossible—to establish statistically significant answers. Certainly the sort of statistics we have examined in the model context would not allow it.

Our model results suggest that paleoclimatic proxy time series can be long enough to provide statistically significant answers. However, proxy data bring their own difficulties, stemming from the uncertainties in exactly what is being measured. Characteristics of a tree-ring series, or ice cores, or corals surely relate to large-scale climatic conditions, but it is equally certain that to some extent they also reflect local conditions. Moreover, if it is not clear precisely what combination of climatic variables they do measure, then the relationship between the proxy measurement and the output of a model will be problematic.

The issue may be rephrased to bring it closer to the present study. In testing whether a model is a suitable explanation for some observational data, the loss of predictability precludes reliance on anything as straightforward as a correlation. Whether it is due to noise or to intrinsic nonlinear dynamics, we know a priori that the model will not closely reproduce long time behavior. Instead, we must look to see whether it reproduces more general characteristics, including statistical behavior. The underlying rationale for the relevant statistical tests is to check whether some measure obtained from two different samples could have been generated by the same random process. For example, we tested to see whether the distribution of standard deviations obtained from the nonlinear ZC model (Figure 2) might have been generated by a first-order auto-regressive process.

Now it is perfectly clear before we start that an ice-cap accumulation record, for example, is not generated by the same process as eastern equatorial Pacific SSTs. Both NINO3 SST and the Quelccaya ice cap are influenced by ENSO (Thompson et al., 1984) but are otherwise determined by quite different processes. It is not obviously possible to extract some statistic from records of the two that will reflect only the ENSO process, effectively "filtering out" everything else.

We calculated the same distributions as in Figure 2 for a 500-year record of accumulation in the Quelccaya ice cap (Thompson et al., 1984). As anticipated, even the most cursory inspection showed that the distributions were quite different from the NINO3 model results of Figure 2; accumulation is generated by a different process, and we failed to extract any usable characterization of the long-term behavior of ENSO.

Using an idea of Y. Zhao's, Zhao, Cane, and Zebiak have tried a different statistic. First, the time series are band-passed with a narrow filter centered at 4 years to reduce the non-ENSO "noise". Complex demodulation is then used to define an instantaneous amplitude and frequency. While for most of the time the frequency is almost constant (corresponding to periods near 4 years), it occasionally exhibits strong changes ($>3.5\ \sigma$). Figures 15 and 16 illustrate time series for the ice cap accumulation and for a ZC model run, respectively. The intervals between these transition times are termed "regime durations," and the distributions of these regime durations are taken as characterizing the underlying process. By testing for differences in the distributions, it is found that for a few parameter settings (e.g., the case 02011 shown in the top curve of Figure 6) the model series is not significantly different from

the ice-cap accumulation record, while for others (e.g., the standard setting of ZC), it is. Perhaps the ice-cap proxy data can be used to distinguish among different candidate models, its shortcomings notwithstanding.

The statistic used to reach this conclusion, regime duration, is unfamiliar and indirect, requiring a multistep process to extract it from the time series. We advance it tentatively, a piece of evidence asking to be bolstered by further investigations and additional tests. We imagine that many readers find it unpersuasive. Be that as it may, concern with a particular test should not obscure the deeper issue.

There is now great interest in studying natural variability on time scales of decades and longer. Since the climate system contains both nonlinear chaotic subsystems and sources of what may be viewed as noise, even an impeccable model cannot reproduce a long natural time series. Since detailed predictions decades ahead are not possible, we must resort to statistical characteristics to check models and theories of long-term natural variability. The instrumental record can suggest many hypotheses, but is too short to allow rigorous tests. Time series long enough to contain many realizations of decadal variability can be found only in paleoclimatic proxy records. Because proxy records typically represent a nonlinear convolution of many climatic processes, both large-scale and local, novel methods are required to extract usable information. It will make the task easier to have available a number of simultaneous observational records suggesting by a clear hypothesis. And such records are essential; no matter how well a model reproduces short-term variations, only a test against long records can justify confidence in its long-term behavior.

ACKNOWLEDGMENTS

Valuable discussions with Zhao Yuechen and Yochanan Kushnir are gratefully acknowledged. Thanks to Virginia DiBlasi-Morris for her help in preparing the manuscript. This work was supported by grant NA16-RC-0432-01 from the National Oceanic and Atmospheric Administration's Office of Global Programs and grant ATM-89-21804 from the National Science Foundation. Lamont-Doherty Earth Observatory Contribution No. 4986.

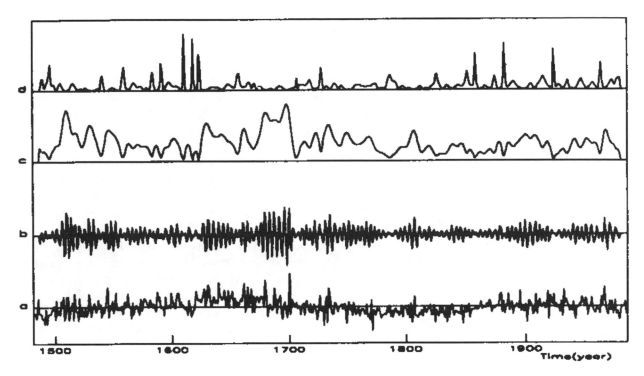

FIGURE 15 (a) Accumulation in the Quelccaya ice cap (courtesy of L. Thompson). (b) Band-passed accumulation record, filter centered at 4 years. Instantaneous (c) amplitude and (d) frequency, as obtained by complex demodulation; see text. (From an unpublished manuscript by Zhao, Cane, and Zebiak.)

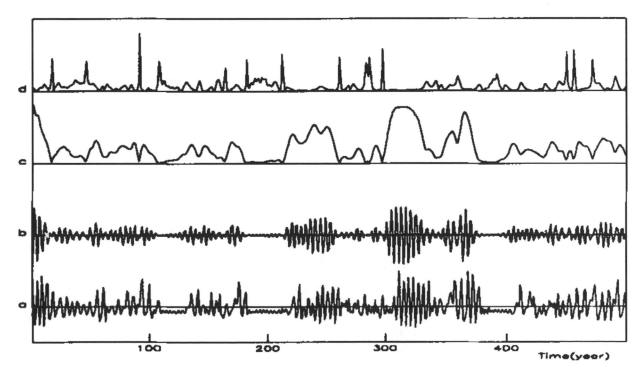

FIGURE 16 NINO3 SST anomaly from ZC model run 02011; see Figure 6. To be compared with the ice-cap accumulation record of Figure 15. (a) unfiltered record, (b) band-passed record, (c) instantaneous amplitude, and (d) instantaneous frequency. See Figure 15 and text. (From an unpublished manuscript by Zhao, Cane, and Zebiak.)

Commentary on the Paper of Cane et al.

S. GEORGE H. PHILANDER
Princeton University

Let me summarize. Julia Cole gave us an excellent talk and presented us with persuasive evidence that there is decadal variability in ENSO. Tomorrow morning Gene Rasmusson, using a completely different data set, will confirm those facts. Dr. Cane addresses the causes of the decadal variability of ENSO described by Drs. Cole and Rasmusson.

I thought his was a very clear presentation, and there is an excellent paper that accompanies it. I confess that I did not pay much attention to all the statistics in the paper; it seems to confirm the adage that statistics serves us in the same way that a lamppost serves a drunk: primarily for orientation rather than illumination. Those of you who are disoriented as regards the use of short-term records should by all means go through the paper. As regards illumination, I feel that the reason why ENSO is variable is still open to discussion. Dr. Cane considered three mechanisms: chaos, noise, and base-state modulation. He voted for chaos, invoking William of Occam.

The only substantive criticism I have of this paper concerns the discussion of a typical time series from the model. (See Figures 1 and 6 of Cane, Zebiak, and Xue above.) In these records there are actually two time scales, so that there are two issues to be addressed. Dr. Cane lumps them together. First, there are decades when the model behaves in a quasi-periodic manner, and there are other decades when nothing seems to happen, so one question is: Why is the time series non-stationary? A separate question is the following: When the model has quasi-periodic behavior, what causes the broad-band spectrum? In other words, why is the oscillation not perfectly periodic?

From simulations with other coupled ocean-atmosphere models we know that simulations of the Southern Oscillation can reproduce perfectly periodic oscillations unless "noise" is introduced. But if noise causes the aperiodicity, and we cannot predict noise, then why is it that we can predict El Niño? Tools (models) to do so exist, but I do not believe that we understand why El Niño is predictable.

A separate issue is the "non-stationarity" of the time series. Why is El Niño more prominent during some decades than others? Tropical oceanographers take a great deal for granted before they build their models. They assume that there is a thermocline and that its depth is given so that certain phase speeds for waves are also assumed. If they were to increase the depth of the thermocline and change the phase speeds and so on, they would be in a different regime altogether, possibly one without El Niño.

It seems to me that such changes probably involve the global thermohaline circulation, which determines the depth of the tropical thermocline. So it is quite possible that the tropical and extratropical oceanic circulations cannot be isolated from each other if the decadal variability in Dr. Cole's records is to be explained.

In summary, it seems to me that Dr. Cane's interesting study leaves some crucial questions unanswered. To attribute the variability in the model to chaos is a form of poetry. Even if it were chaos, surely there would still be something physically different in the system at different times. Exactly what changed the system from the one state to the other?

Discussion

CANE: I've just been thinking about the issue of statistics. It seems to me that the maturity of a subject is inversely related to the number of spectra shown when it's discussed. But we may have to fall back on statistics to make judgments about the models or mechanisms proposed as explanations for irregular behavior and long-term variability. How do we distinguish between internal chaotic dynamics and changes in background conditions like circulation? I can't really rule out changes in the base state, but it is incumbent on those who offer them as explanations to identify the particular change and a mechanism by which it operates. I have generated these series with models to demonstrate that in fact chaos could account for the results. William of Occam's principle is particularly appealing to those of us who don't like to multiply hypotheses.

SARACHIK: When you fitted the noise model to the nonlinear model it gave you a good fit. Why are you so sure noise models won't work?

CANE: The noise model is interesting; we fit a month-to-month transition and it can mimic the spectral peaks out through 4 years. But it seems to be incapable of giving you the lower-frequency variability—longer than the 4 years—exhibited by the other model.

LEVITUS: Does your rather sweeping generalization that the El Niño phenomenon is contained within the Pacific mean that we don't need to monitor the Pacific to understand heat flux and the role of the ocean as part of the earth's climate system?

CANE: No more than it means you need to monitor only the tropical Pacific. There are other modes of variability, of which at least a few have some sort of internal dynamic that gives them coherent characteristics over the time scales of interest here.

MYSAK: Ten or 12 years ago we did a statistical analysis correlating sea level, SST, and salinity in the northeast Pacific with fish population, weights, and so on. The dominant spectral peaks that emerged were about 5 or 6 years apart—not El Niño, not decadal. Have you any suggestion what they might be connected with?

PHILANDER: I think we're back to the lamppost: orientation only, remember.

MCGOWAN: I don't remember whether I pointed out on my time series of the California Current that zoöplankton showed a huge El Niño response in 1958-59 and 1982-83, though not to any of the other El Niños.

RIND: It seems to me that the various possible causes other than El Niño should leave evidence elsewhere that could be looked for. We should be combining our information on El Niño variations with observations taken elsewhere to look for correlations.

RASMUSSON: As you get more remote from what I'd call the core region, you may get responses only to the larger events.

TRENBERTH: I have the impression that your conclusions pertain mainly to a model that is still very idealized. Also, changes have been observed in the base state in the real world — for instance, the shift in the location of the South Pacific convergence zone over this century.

CANE: I don't see that base-state changes contradict my main thrust—namely, that everything seems to be caused in these tropical regions—and I'd like to see evidence that it's otherwise before we abandon this minimum hypothesis. As for the spectra, I admit that our model leaves out a lot of climate-system processes, but by adding noise we can generate broad-band spectra that will match Julie Cole's.

KARL: Did you limit yourself to a first-order autoregressive process? If so, why?

CANE: Yes, I fitted that because I used a complete state/space representation that captures all the variance in the model; it's a multivariant first-order process that mimics the equations we started from.

KARL: Was there any correlation in the noise left over?

KAROLY: If you used some sort of maximum-entropy spectral estimation you'd get something like a high-order AR process. Mark, did you find that modifying your nonlinear model parameters to fit the ice-core records made any difference to the model's predictability characteristics on the 2-year time scale?

CANE: These runs started from the same initial conditions. They all go through a negative phase, in about the same way, and then start to diverge. For a small slice of the present they seem to be equally good, yet some of them have quite different behaviors for longer times. That's important to keep in mind if the plan is to develop a good model of the present so it can be used for exploring variability on all time scales.

TALLEY: George, when you said El Niño is predictable, what sort of time period did you have in mind?

PHILANDER: Over longer periods, such as months, it's predictable.

ROOTH: When you ran the model with different base states did you keep the base state consistent for each run? I should think it would be reasonably feasible to put in a strictly harmonic variation, which would make it easier to detect any response. The effects of external signals might exist even if the spectra can't always be recognized as distinct.

CANE: We didn't try to vary the base-state parameters. Maybe we could have excited some resonance, but we really had to limit the number of questions to keep it manageable.

ROOTH: I was thinking not so much about resonance as about any process with a nonlinear trigger phase. It would be sensitive to the superposition of two biases during that critical period, so that a background variable might influence the transitional effects quite substantially.

Secular Variability of the ENSO Cycle

EUGENE M. RASMUSSON[1], XUELIANG WANG[2],
AND CHESTER F. ROPELEWSKI[3]

ABSTRACT

Secular changes can occur in both multi-decadal climate means and multi-decadal measures of variability. We have examined the secular variability of the ENSO cycle as revealed by commonly used indices, i.e., sea level pressure and sea surface temperature from the low-latitude core region of the oscillation. We view the low-frequency variations (periods longer than approximately 30 years) as a varying base state upon which the ENSO cycle is super-imposed.

The following are the major findings of the analyses:

- ENSO-cycle variance for 31-year periods has changed by a factor of two or more during the past century. The cycle was quite pronounced late in the nineteenth century, was relatively weak from 1920 to 1950, and has increased in intensity since then.
- The century-scale variation in equatorial sea surface temperature was broadly similar to that in globally averaged sea surface temperature.
- No obvious relationship could be detected between variations in the base-state parameters we analyzed (equatorial sea surface temperature and Pacific-Indian Ocean sector sea level pressure) and variations in the intensity of the ENSO cycle.
- Regional statistics, such as those derived from the Quinn et al. (1987) compilation of strong and very strong El Niño events in Peru, cannot be considered a reliable index of basin-scale ENSO-cycle variability.
- The century-scale variations in ENSO-cycle intensity broadly correspond to changes in all-India monsoon-season rainfall variability, to the modulation of the intensity of drought episodes over the U.S. Great Plains during the twentieth century, and, less clearly, to the century-scale variation in Sahel rainfall.

[1]Department of Meteorology, University of Maryland, College Park, Maryland
[2]RDS Corporation, Greenbelt, Maryland
[3]NWS/NOAA Climate Analysis Center, Camp Springs, Maryland

INTRODUCTION

Secular changes can occur in both multi-decadal climatic means and multi-decadal measures of variability. Changes in the mean indicate changes in the "base-state climate," while changes in variability are more closely related to changes in the frequency and/or intensity of droughts, heat waves, frost occurrence, and other features of year-to-year variability. Changes in variability may have a greater socio-economic impact than changes in the mean—as, for example, the ENSO does.

Three basic questions come to mind in considering the secular variability of the El Niño/Southern Oscillation (ENSO) cycle:

1. What do we know about secular variability in the character of the ENSO cycle (amplitude, frequency, character of regional responses, etc.) on time scales of a few decades to a century?

2. Can the "observed" secular variations be related to changes in base-state climate parameters? On what time scales?

3. What are the climate processes associated with secular variability of the ENSO cycle?

Continuous time series spanning at least several hundred years, from enough points to resolve both temporal and spatial changes in the character of the cycle, are required to fully describe multi-decadal to century-scale ENSO variability. Since only a few instrument time series approach or exceed a century in length, this requirement is not remotely satisfied. Proxy reconstructions and historical records represent promising indirect sources of information for extending the instrument record, but few solid conclusions can yet be drawn from such proxy sources.

Returning to the first of the three questions posed earlier in this section, a partial picture of the general nature of ENSO-cycle variability during the past 100 to 150 years can be distilled from existing instrumental data. The primary objective of this paper is to provide such a description. Regarding the second question (relationship of secular variability to changes in base-state parameters), the observations do not provide the information required to quantify important base-state parameters such as stratification, depth of the thermocline, or strength of atmosphere/ocean coupling (Cane et al., 1995). Thus we simply compared secular changes in variability to secular changes in parameters that could be quantified by the observational data, i.e., equatorial sea surface temperature (SST) and sea level pressure (SLP) at selected stations in the Pacific-Indian Ocean sector of the tropics. The third question (processes involved) also cannot be effectively addressed with this limited data base. For insight into this question obtained from model simulations, see Cane et al. (1995) in this section.

Salient characteristics of the ENSO cycle are first described as background. The nature of the instrumental data base and proxy sources of information are then briefly reviewed. Evidence of secular variability derived from instrumental records is summarized; both the published results and the results from new analyses by the authors are reviewed. The new results and their implications are discussed, and conclusions are summarized.

ENSO CYCLE

After years of analysis of global correlation patterns in surface temperature, pressure, and precipitation, Sir Gilbert Walker (Walker, 1924) identified three large atmospheric oscillations: two "Northern Oscillations", centered in the North Atlantic and the North Pacific, and a larger, more pervasive "Southern Oscillation" (SO), whose centers of action are in the tropics. In a subsequent paper (Walker and Bliss, 1932), he characterized the SO as follows: "When pressure is high in the Pacific Ocean it tends to be low in the Indian Ocean from Africa to Australia; these conditions are associated with low temperatures in both these areas, and rainfall varies in the opposite direction to pressure. Conditions are related differently in winter and summer, and it is therefore necessary to examine separately the seasons of December to February and June to August."

Almost a half century later, Bjerknes (1969) described the link between above-normal SST and enhanced convection in the central equatorial Pacific. He further related the SST variations to Walker's SO, thus forging a link between the SO and SST anomalies, both positive and negative, over the entire span of the equatorial Pacific, including the El Niño phenomenon of the eastern equatorial Pacific. This system of coupled ocean/atmosphere interactions, and the associated global pattern of atmospheric teleconnections, has come to be known as the El Niño/Southern Oscillation cycle.

The meaning attached to phenomenological names such as the SO and El Niño has evolved over the years as meteorologists and oceanographers have attempted to adapt earlier descriptive terms to an evolving understanding of the nature and scope of these phenomena. This has sometimes moved full circle, first away from and then back to an earlier use of a term. The result has been a high degree of semantic anarchy in the literature (Aceituno, 1992) and a need to clearly define how terms are to be used.

The ENSO cycle is the dominant quasi-regular mode of global climate variability. Over an imprecisely defined "core region" that broadly corresponds to the area of the SO surface-pressure dipole referred to in the previously quoted Walker-Bliss definition, the SO is reflected as a broad-band (2-6 year) maximum in the variance spectra of pressure, wind, air temperature, SST, and precipitation (Rasmusson et al., 1990; Ropelewski et al., 1992). A discussion of our present understanding of the processes which give rise to the ENSO cycle, as well as the atmospheric processes associated with remote teleconnections, are beyond the scope of this paper. Suffice it to say that the current consensus view is that the ENSO cycle is an internal oscillation of the tropical ocean-atmosphere system (Cane et al., 1995, in this section).

In this paper, El Niño is considered to be an unusual warming of the normally cool surface waters along the west coast of South America, primarily affecting the extreme eastern equatorial Pacific and the coastal regions of southern Ecuador and northern Peru. It is therefore viewed as one of the regional responses to the basin-scale ENSO-cycle rhythm described by Bjerknes (1969). The coastal El Niño warm episode typically runs its course in about a year (Rasmusson and Carpenter, 1982).

The phase and amplitude of the ENSO cycle is usually indexed by sea level pressure variations at stations representative of the SO pressure dipole or by SST anomalies in the central and eastern equatorial Pacific. In characterizing and indexing the ENSO cycle it is important to give due consideration to the seasonality of the anomaly fields, as emphasized by Walker and Bliss (1932). For example, basin-scale anomaly patterns in the low latitudes typically reach maximum amplitude during the last half of the year, and often change sign during the March-May period. Thus 12-month averages are more definitive if they span the May-April "ENSO year" rather than the calendar year (Wright et al., 1988).

DATA CONSIDERATIONS

Instrument Data

During the past four decades, the distribution of surface meteorological observations has been sufficient to permit the documentation of ENSO-cycle variations over the core region of the oscillation in considerable detail. Prior to that time, the data distribution, or at least that represented by the observations which have found their way into the readily accessible archives, is far less satisfactory. Little in the way of surface meteorological data from merchant ships is available during the two world wars. This deficiency is superimposed on a progressively poorer sampling in the ENSO core region from both merchant ships and land stations as one moves backwards into the late nineteenth century. Little reliable information can be distilled from the surface marine data prior to 1875, and no meteorological station time series are available from the Pacific prior to this date. In summary, a good deal of information on the variability of the ENSO cycle over the past 100-120 years can be obtained from instrument records. The record can be extended backward for a few more decades, albeit much

less securely, using surface pressure data from Djakarta and a few Indian stations.

Historical Information

The behavior of the ENSO cycle prior to the last half of the nineteenth century can be deduced only from indirect evidence contained in historical records and paleo-climate reconstructions. The most notable historical data are the descriptive records from Peru that provide anecdotal evidence of El Niño-related rainfall anomalies, e.g., descriptions of severe or prolonged desert flooding, massive crop failures, disease, or insect infestations. Using this material and information from ships' logs, Quinn et al. (1987) compiled a chronology of El Niño occurrences since 1525.

In evaluating this chronology, one must remember that the regional El Niño anomalies, while related to ENSO-cycle swings, also exhibit a significant degree of independent behavior (Deser and Wallace, 1987), particularly in their intensity. The Quinn et al. El Niño chronology should therefore be viewed as related to, but not necessarily a good index of, the ENSO cycle itself. This caveat will be further dealt with in the Discussion section below.

Paleoclimate Reconstructions

Paleo-ENSO research has developed rapidly during the past decade and is now being pursued on a broad front. There are a large number of candidates for annual/seasonal-resolution proxy reconstructions. An overview of this subject can be found in Enfield (1992). High-resolution proxy variables are discussed by Baumgartner et al. (1989). A volume edited by Diaz and Markgraf (1992) presents a collection of paleo-ENSO papers. Ortlieb and Macharé (1992) have edited a volume of extended abstracts from the 1992 Paleo-ENSO International Symposium held in Lima, Peru. We discuss briefly three of the proxy candidates which may in due time yield important results: corals, tree rings, and ice cores.

The most logical candidates for ENSO proxy reconstructions are the ENSO-sensitive natural processes of the core region. Particularly promising are proxy records derived from the paleochemistry of coral reefs (Shen et al., 1992; Fairbanks et al., 1992). A variety of chemical tracers have been developed in corals from both sides of the equatorial Pacific Ocean for the purpose of establishing the variability of surface ocean parameters (SST, changes in insolation, upwelling, vertical mixing, precipitation, river discharge) on annual to century time scales. The paper by Cole et al. (1995) in this volume clearly illustrates the great potential for lengthening the record of ENSO-cycle variability from coral reconstructions.

Dendroclimatic reconstructions are of limited value in tropical regions because most tree species there do not form distinct annual rings, and growth is less susceptible to the interannual variability of climate (Bradley and Jones, 1992). Proxy indices of remote ENSO teleconnections can be used as indices of the ENSO cycle if the teleconnection is well correlated with the core-region variations, and the correlation remains stable over decades and centuries. Estimates of the Southern Oscillation Index time series have been made from western North American-northern Mexican tree-ring chronologies (Lough and Fritts, 1989; Michaelsen, 1985), which index variations in winter rainfall that are modestly correlated with ENSO core-region variability.

Annual variations in the amount and chemical composition of precipitation accumulated on polar and alpine glaciers produce laminations that allow precise dating of these stratigraphic sequences over periods of centuries. Of particular relevance to the ENSO cycle are the cores from the Quelccaya ice cap in the southern Peruvian Andes (10°S) (Thompson, 1992; Baumgartner et al., 1989). Interannual variability in those ice cores is correlated with changes in the water level of Lake Titicaca (Thompson, 1992), which are in turn modestly correlated with the ENSO cycle.

In summary, rapid progress is being made in paleo-ENSO studies, but attempts to quantify ENSO-cycle variability from indirect indicators give rise to a variety of problems. Much of the research to date has been directed toward calibration and establishing the reliability of the proxy record. Each record has its own limitations, and calibration is often hindered by the lack of appropriate instrumental records. The most productive approach would seem to involve extensive intercomparison of proxy records and synthesis of information from a variety of sources.

ASPECTS OF THE INSTRUMENTAL RECORD

Literature Review

Following the work of Gilbert Walker in the 1920s and 1930s, there was much debate as to whether the SO was a physical reality or a statistical artifact. Troup (1965) examined data from the decades following Walker's study and found a continuation of the SO relationships. However, the correlations between station pairs were consistently weaker during the period 1921-1950 than they had been in Walker's earlier data. Berlage (1957) had previously observed that the SO was less well developed after 1920. Troup also found smaller variance in the 1921-1950 seasonal means, suggesting a decrease in the amplitude of the ENSO cycle by comparison with the earlier period.

Later studies have indicated an increase in the ENSO cycle variability after the 1921-1950 period. Trenberth and Shea (1987) comment that "the prominence of the SO has varied throughout this century. The SO was very strong from 1900-1920 at the time it was first documented and named. But it was not conspicuous again until after the later 1930s." Elliott and Angell (1988) reached a more or less similar conclusion. They found that seasonal correlations between eastern equatorial Pacific SST and SLP at a number of index stations have "been greatest since World War II, and were relatively high prior to World War I." They suggest that there have been changes in the character of the cycle: i.e., prior to World War I the pressure difference between the eastern Pacific and Darwin was more representative of the SO, but since World War II the pressure difference between the central Pacific and Darwin has better reflected the SO.

Further Analysis

The view of what constitutes "climate noise" and what distinguishes climate change from climate variability is largely dependent on the time scales of interest and the spectrum of climate variability. The conceptual framework for viewing this question is considerably simplified if the climate signal of interest exhibits a significant concentration of variance that rises well above the higher-frequency "climate noise" background and is also separated to some extent from concentrations of lower-frequency variance by a "spectral gap". The lower-frequency variability can then be thought of as reflecting changes in the climate "base state" on which the higher-frequency variability is superimposed. The ENSO signal satisfies this ideal situation reasonably well over large parts of the core region.

We have adopted this conceptual framework for further analysis of a number of core-region time series. Results from four of the mean monthly series will be described.

1. SST* (1874-1990). This is the mean SST over the area 0°-10°S, 90°-180°W; it serves as an index of ENSO-cycle SST variations over the equatorial Pacific. The time series was provided by H. Diaz (personal communication, 1992). Values are recorded back to 1854, but we found little information content in the series prior to 1874.

2. Bombay mean sea level pressure (1847-1990) from Parthsarathy et al. (1991). This time series serves as one index of variations in the Indian Ocean-west Pacific pole of the ENSO-cycle pressure seesaw.

3. Darwin mean SLP (1882-1988), from the NCAR surface data tape. This is a commonly used index of variations in the Indian Ocean-west Pacific pole of the ENSO-cycle pressure seesaw. Pressures prior to 1898 have been corrected by + 1 mb, as recommended by Trenberth and Shea (1987).

4. Tahiti mean SLP (1876-1988), which is commonly used as an index of variations in the southeast Pacific pole of the ENSO-cycle pressure seesaw. The record prior to 1935, assembled by Ropelewski and Jones (1987), has several gaps, ranging from 3 to 28 months. Elliott and Angell (1988) filled the gaps by interpolating from the Apia time series. We simply filled the gaps with long-term mean

monthly values, since Elliott and Angell found that this did not materially affect their results.

Singular Spectrum Analysis

We performed singular spectrum analysis (SSA) of these and other long time series from the ENSO core region to obtain a general picture of the degree to which various time series reveal a consistent picture of ENSO cycle variability. SSA is a variant of the principal component analysis applied to a time series. (See Vautard et al. (1992), Keppenne and Ghil (1992), and Vautard and Ghil (1989) for detailed discussions of the analysis technique and its practical application.) Like conventional EOF analysis of a two-dimensional array, SSA decomposes a time series into an ordered set of eigenvectors (EVs) and principal components (PCs). SSA EVs define the dominant empirical modes of variability in the time series. Each EV is a data-adaptive filter for a particular empirical mode from which the corresponding PC time series is obtained by passing this filter over the original time series. The number of points in the EV (filter) is the "window size" that determines the maximum period of fluctuations that can be resolved by SSA. The PCs provide a time series for each EV that reveals variations in the amplitude and intermittency of the individual modes. Following Rasmusson et al. (1990), a window size of 61 months was adapted that resolves the variations with periods less than 10 years, which includes ENSO-cycle variability.

Results of SSA analyses of Bombay SLP and SST* time series are now described. We chose Bombay rather than Darwin as our primary SLP index series because of its considerably greater length and the remarkable consistency of its ENSO signal with that in the SST* series. Figures 1 and 2 show the first five EVs for SST* and Bombay SLP. The leading EV for each of the two series primarily reflects low-frequency variations with periods longer than those that can be resolved by the 61-month window. EV 1 is essentially a running mean filter, although it also contains some ENSO-cycle variability for SST*. PC 1 (Figures 3 and 4) is quite similar for the two series for time scales longer than a decade, the main difference being the higher-frequency "ENSO embroidery" that appears in PC 1 for SST*.

EV 2 and EV 3 reflect the low-frequency component of ENSO variability identified by Rasmusson et al. (1990) in equatorial SST and zonal wind fields. The corresponding EVs for the two series are almost identical; they exhibit a period of around 44 to 48 months. Figures 5 and 6 show PC 2 for the two series. The correspondence is truly remarkable, with close-to-perfect correspondence of the major swings. This comparison provides convincing evidence of the consistency of both time series, and therefore of their validity as indices of ENSO-cycle variability since 1875. It also suggests that the Bombay SLP series can be used to extend the record back to 1847.

EVs generally appear in pairs, one even and one odd function. The EV 4–EV 5 pair reflects the near-biennial mode of ENSO variability described by Rasmusson et al. (1990) and Ropelewski et al. (1992). However, EV 4 also includes a trend that is the low-frequency, odd-function companion of the EV 1 running mean. The close correspondence between EVs and PCs continues through EV 7, and to a lesser extent through EV 10 (not shown).

SSA modes 2 through 7 reflect ENSO cycle variability and account for a total of 49 percent of the Bombay SLP variance and 65 percent of the SST* variance.

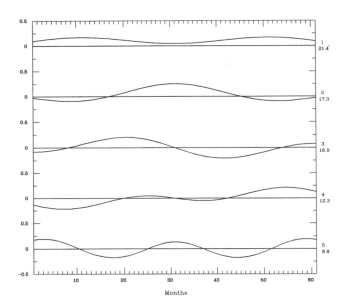

FIGURE 1 First five eigenvectors for SST* (see text). Percentage of variance explained by each eigenvector is shown on the right.

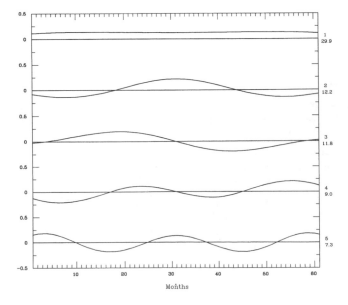

FIGURE 2 First five eigenvectors for Bombay SLP.

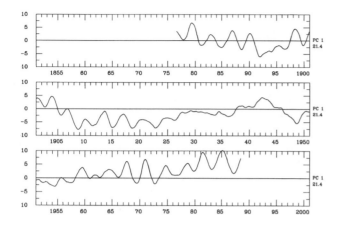

FIGURE 3 First principal component (PC1) for SST*.

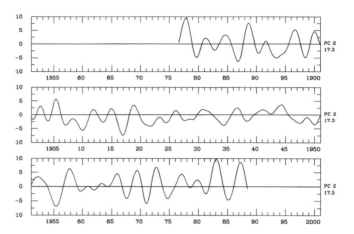

FIGURE 5 Second principal component (PC2) for SST*.

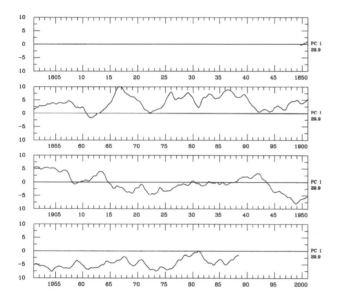

FIGURE 4 First principal component for Bombay SLP.

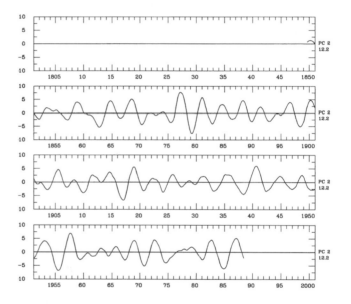

FIGURE 6 Second principal component for Bombay SLP.

We also used the SSA analysis as an aid in the somewhat arbitrary choice of what constitutes the appropriate base state for each series, and what averaging period should be used to illustrate the nature of multi-decadal to century time scales of variability. Following the initial SSA analysis, the low-frequency component of variability was removed from the anomaly series by applying a multi-stage low-pass filter that has been extensively used in the field of astronomy (Zeng and Dong, 1986). It is basically a data-adaptive recursive filter. We first used the low-pass filter to obtain a base state that has a very sharp cutoff near periods of 30 years (Figure 7). Periods longer than 33 years are passed essentially unattenuated, while periods shorter than 27 years are almost entirely removed. We show filtered values to the ends of the series (Figure 8). The confidence limits for these end values progressively broaden as the end of the series is approached, but are tighter than those for a conventional

linear filter. The general direction at the ends of the curves is more reliable than the absolute values.

SSA analyses were repeated after the series had been detrended by using this low-pass filter. The detrending removed EV 1 from each series, leaving the ENSO variability of each deterended series as the leading SSA modes—i.e., EV 2 of Figures 1 and 2 became EV 1 of the detrended series, EV 3 became EV 2, etc. We therefore judged this filter to be satisfactory for the purposes of this study, although further fine-tuning of the procedure for obtaining the base state may be desirable.

Base States

Figure 8 shows an example of the base-state curve for Bombay SLP, superimposed on the original time series. Figure 9 shows the base-state time series and 31-year mean-

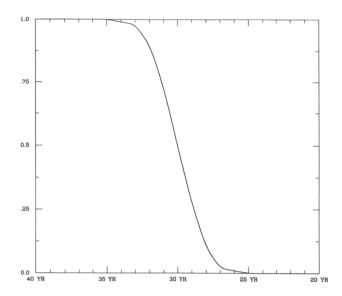

FIGURE 7   Response function for 30-year low-pass filter used in these analyses.

square departures (31MSD) for each of the four mean monthly time series. The 31-year mean-square departures are computed from the monthly values, i.e.,

$$(31MSD)_i = \frac{1}{373} \sum_{j=-186}^{186} X_{i+j}^2 , \qquad (1)$$

where X_i is a monthly value of the detrended series. For annual values, such as monsoon-season rainfall, the j index ranges from -15 to 15. The 31-year averaging period is a compromise between resolution and stability. It should be firmly kept in mind that there are only three or four independent running-mean values and therefore relatively few degrees of freedom in the base state and 31MSD series.

We first discuss briefly the base-state curves. The Bombay SLP and SST* base-state curves are smoothed representations of PC 1 for each untrended series (compare with Figures 3 and 4). The SST data are uncorrected for biases that may have arisen as a result of the change from bucket to intake temperature measurements since World War II,

but the broad-scale features of the SST* base-state curve are nevertheless similar to the global averages of SST, which we computed from several analyses of corrected SST data (Figure 10). The SST* base state shows an increase of more than 1°C since the 1910-to-1920 period, roughly twice that computed for the global average. The minor SST* maximum around 1940 and minimum around 1955 lead similar features in the global-average curves by around 15 years. The fall of SST* from the late nineteenth century to the minimum around 1915 is again around twice that computed for the global average. Thus, changes in tropical Pacific SST were generally of the same sense and larger than the century-scale changes in globally averaged SST.

Bombay exhibits a substantial century-time-scale trend toward lower pressure until 1955-1960, when values turned upward. Although of lesser amplitude, the same pattern is observed in the Madras, India, station pressure (not shown). The primary difference in the data from the two Indian stations occurs after the early 1970s, when the Madras upward trend flattens, while the upward trend at Bombay, as projected by the filter, continues to the end of the series.

There is no obvious relationship between the century-scale or decadal variations in SST* and Bombay SLP. Darwin does not show the high values prior to 1900 that are a feature of the Indian station data. However, the substantial correction applied to the Darwin pressures prior to 1898, together with the uncertainties of projecting the base-state curve to the beginning of the series (1882), leave the values prior to about 1910 in question. One could dismiss the Darwin data as biased, but the Tahiti base state also shows no high values during this period. Unfortunately, there are also uncertainties in the Tahiti data, since Ropelewski and Jones (1987) applied substantial corrections to the data prior to 1935.

After 1905, Darwin and Bombay base-state SLP curves follow a similar pattern of multi-decadal variations. Tahiti traces out the same pattern, lagging the opposite phase at Darwin by about 5 years. Except for the phase lag, the relationship between Darwin and Tahiti is similar to ENSO-cycle variations.

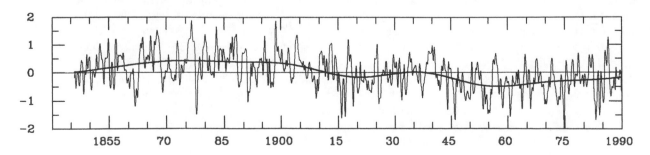

FIGURE 8   Base-state curve for Bombay SLP (heavy solid line). Lighter curve is the original time series after application of a light median filter (Rabiner et al., 1975) to eliminate isolated "noise points." Units: millibars.

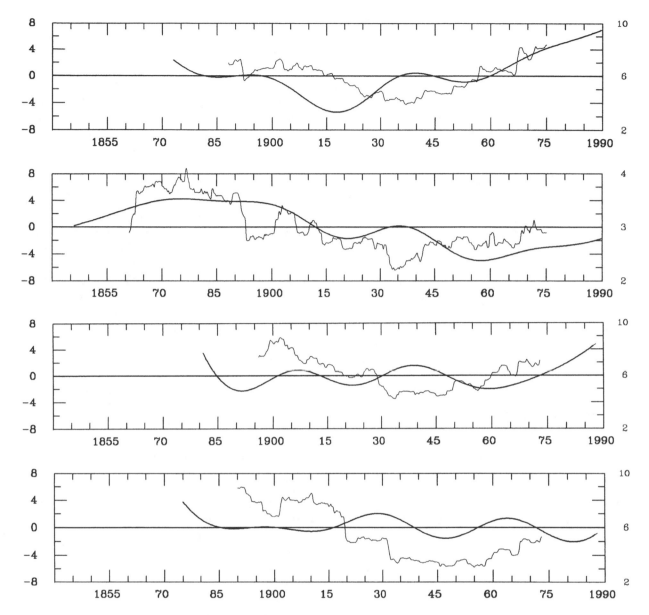

FIGURE 9   Time series of low-pass filtered base-state (solid curve) and 31-year mean-square departures for (top to bottom) SST*, Bombay SLP, Darwin SLP, and Tahiti LSP. Base-state units (left) are 10^{-1}°C or 10^{-1} mb; mean-square departure units (right) are 10^{-1}(°C)^2 or 10^{-1}(mb)^2.

Variability

The ENSO cycle is the dominant source of variability in the detrended time series. The first three EV pairs of the detrended series, which capture most of the ENSO-cycle variance, account for 68 percent, 75 percent, 79 percent, and 80 percent of the total Bombay, Tahiti, Darwin, and SST* variance, respectively.

The 31MSD series (Figure 9) exhibit a dominant century-scale variation with high variability late in the nineteenth century, the lowest variability during the third through fifth decades of the twentieth century, and increasing variability thereafter. The changes in the intensity of the ENSO cycle during the past century are substantial, with the 31-year mean maxima and minima differing by a factor of 2 or more.

The four 31MSD series of Figure 9 show a similar century-scale variation. The longer Bombay series shows that the period of high variability in the late nineteenth century probably extends back to around 1860. This is confirmed by the Madras station pressure series (not shown). The recovery from the twentieth-century minimum has been most pronounced in SST*, which has returned to the level observed prior to the sharp drop during the early 1890s. The question of whether this is at least partly due to inadequate sampling during the early decades needs to be investigated.

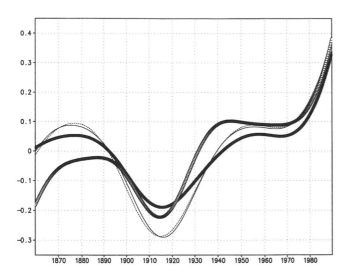

FIGURE 10 Time series of globally averaged SST. Heavy solid curve is obtained from the Comprehensive Ocean Atmosphere Data Set (COADS) (Woodruff et al., 1987) using optimum averaging (Vinnikov et al., 1990) with climatology as a first guess. The dashed curve is obtained from COADS using the box-averaging analysis technique (Bottomley et al., 1990), while the thin solid curve is obtained from COADS using optimum averaging with the previous month as the first guess. The open continuous dots are the curve obtained from the new SST data set and analyses described by Parker et al. (1995) in this volume. Units are °C.

The broadly similar character of the century-scale changes in SLP variability at the three stations shown suggests that the changes reflect primarily changes in the intensity of the ENSO cycle rather than shifts in one or both poles of the ENSO-cycle pressure seesaw. The 31MSD series for SLP shows greater irregularity than the SST* series. These irregularities are due to the contribution of individual large-amplitude swings, most notably that of 1877 to 1889. The presence of this feature maintained high 31MSD values until 1890, after which they dropped sharply as the 31MSD averaging window moved past this episode. To a great extent the great 1877-1878 ENSO cycle marked the climax of the late nineteenth-century period of high-amplitude ENSO swings in SLP; it was less of a singular event in the SST* series, although this could be an artifact of inadequate sampling.

The ENSO swings during 1915 to 1920 also had a pronounced effect as they moved into and then out of the 31-year averaging window. The singular character of some of the ENSO-cycle swings and the marginal stability of the SLP 31MSD series suggest caution in drawing conclusions concerning any variability other than century scale.

The drop in variability at Tahiti after the turn of the century lags a similar feature in the other three series. This is also true for the subsequent minimum, and the recovery during the final decades is relatively small. The delay in the dropoff of variability can be attributed to the Tahiti

ENSO swing of 1903-1907, which was much larger, in a relative sense, than that observed in the other series.

DISCUSSION

In the conceptual framework adopted for this study, ENSO-cycle variability that is concentrated in periods of two to six years is viewed as being superimposed on a varying base state that includes periods longer than around 30 years. The detrending of the time series by removal of the base state effectively "normalizes" the ENSO-cycle swings, so that they appear as more or less symmetric swings about a slowly changing base state. This picture of ENSO-cycle variability can be fundamentally different from what appears if the anomalies are simply computed as departures from long-term monthly means. Questions concerning the secular variability of the ENSO-cycle shift from "When and why were there periods when warm or cold "events" predominated and were strong?" to "When and why did the amplitude or frequency of the cycle change?" The difference in perspective is illustrated by Figure 11, which shows the original and detrended anomaly series for SST*. An upward trend in SST* can be seen in the period since the late 1950s. From the original series, one might hypothesize a "jump" to a warmer state in 1977 (Trenberth, 1990), with a subsequent predominance of the ENSO warm phase, but there is little or no evidence of such a change in the ENSO cycle appearing in the detrended series.

We examined the question of whether the results of these analyses could be linked in any quantitative way with the previously noted coastal El Niño chronology of Quinn et al. (1987). Hocquenghem and Ortilieb (1992) have questioned the intensity classification of Quinn et al., as well as the reality of some of their earliest events. Their doubts arise from the fact that the coastal El Niño rains are usually localized in northern Peru and southern Ecuador, and within that region the rainfall distribution is highly variable from event to event. This makes it difficult to determine intensity from information at a single point. Since some of the events listed in the Quinn et al. chronology were identified on the basis of rainfall or flooding reports from the central and southern coastal regions of Peru, they may have been unrelated to El Niño occurrences.

Hocquenghem and Ortilieb (1992) conclude that "it is probably still untimely to use the Quinn et al. sequence of paleo-El Niño events as a solid basis for recurrence studies of the phenomenon". In contrast, Enfield and Cid (1991), while skeptical of the intensity classification for the weaker events, believe that recurrence statistics for the two strongest event categories, strong (S) and very strong (VS), are reliable. Their recurrence-interval analysis showed no indication of a multi-century Little Ice Age signal in the S and VS events. There were, however, significant variations in the recurrence interval on shorter (century) time scales.

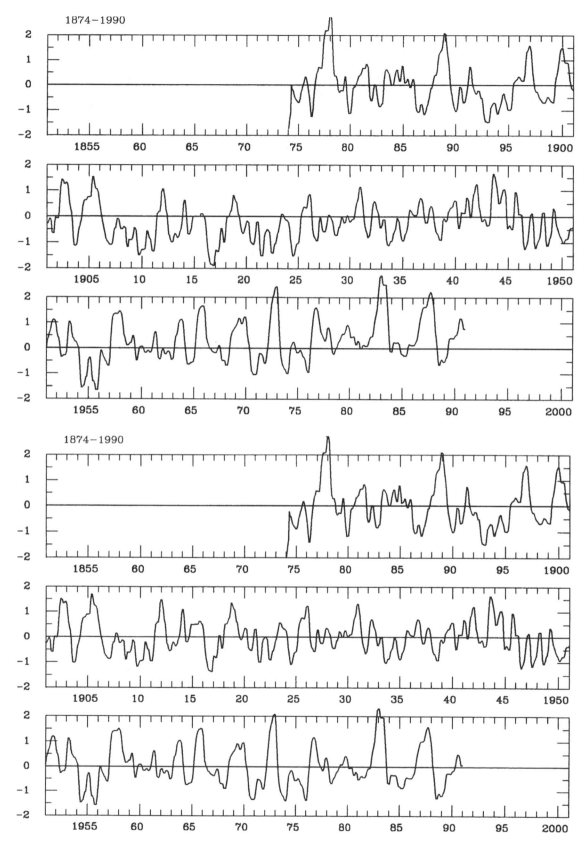

FIGURE 11 SST* anomalies. Upper panel, basic anomalies; lower panel, detrended series. A light median filter (Rabiner et al., 1975) has been applied to eliminate isolated "noise points." Units are °C.

The problems inherent in the subjective classification of El Niño or ENSO-cycle intensity, even from instrumental data, become apparent when we compare our results with the chronology of Quinn et al. During the periods spanned by the Bombay SLP and SST* series, there were, respectively, 14 and 12 S and VS events in the Quinn et al. chronology. The average recurrence interval is therefore around a decade. This immediately raises questions regarding the stability of recurrence statistics for periods shorter than the century time scale.

Comparisons with the original SLP and SST* series were meaningless, since 11 of the 14 highest peaks in the Bombay series occurred during the period of high-base-state SLP prior to 1900, and 13 of the 14 occurred before 1920. For the SST* series, none of the 12 highest peaks occurred during the cold period between 1907-1956. When the detrended series was used instead, 20 of the positive Bombay anomaly peaks reached approximately 1 mb. Only 7 of the 14 S and VS events in the Quinn et al. chronology were included among these 20 peaks. For the SST* series, only 7 of the 12 S and VS events were included among the 13 largest positive swings. Nine of the 13 highest SST* peaks coincided with one of the 15 1 mb positive peaks in Bombay SLP during this period, indicating problems of classification even when instrumental records were compared. Comparisons were also made using the relatively smooth low-frequency ENSO-mode PCs instead of the more irregular detrended series. The level of agreement was about the same, although there were changes in the list of events exceeding a given threshold value.

The problems of subjective classification arise because of regional differences in the response from cycle to cycle and the relatively continuous distribution of amplitudes of ENSO-cycle swings. There are no obvious discontinuities in the distribution that could serve to separate classes, and slight irregularities from swing to swing and station to station often shift events from one class to another. The poor correspondence between the Quinn et al. chronology and the instrumental time series may arise for several reasons, among them the inherent difficulty of meaningful subjective classification, the poor agreement between the

amplitude of ENSO-cycle swings and the amplitude of the regional El Niño response, and the difficulty of quantifying anecdotal information. Regardless of the reasons, we conclude from these simple comparisons that recurrence statistics derived from the Quinn et al. compilation of S and VS regional El Niño events cannot be considered a reliable index of basin-scale ENSO-cycle variability. Similar problems are likely to arise with any proxy record if ENSO fluctuations are classified by categories of intensity.

These analysis results provide no clear evidence of a relationship between base-state changes and changes in the statistics of the ENSO cycle. However, the data are admittedly inadequate for quantifying the base-state parameters of greatest interest (e.g., stratification, depth of the thermocline, strength of ocean-atmosphere coupling). Even if these parameters were available, the length of record might be far too short to reveal a relationship. For example, the ENSO cycle in the model used by Cane et al. (1995) for their study of the long-term behavior of the ENSO cycle results from low-order chaotic dynamics. It exhibits a pronounced regime-like behavior, and it would require a record many times the length of the observational time series to distinguish this regime-like behavior from changes caused by variations in the base state.

The question arises as to whether this pattern of ENSO-cycle secular variability over the past century can be related to secular variations in other aspects of global climate over the same period of time. In a recent review of large-scale precipitation variability, Rasmusson and Arkin (1993) cited three examples of pronounced secular variability in large-scale precipitation regimes: Indian monsoon-season rainfall, North American Great Plains drought, and Sahel rainfall. Figure 12 shows the base state and 31MSD series for all Indian monsoon-season rainfall (1871-1990) derived from the data published by Parthsarathy et al. (1991). The curves are similar to the 31-year running mean and variance shown in Rasmusson and Arkin (1993, Figure 19). The century-scale change in variability is broadly similar to that appearing in SST* and Bombay SLP. This is not surprising, since a significant correlation between Indian monsoon-season

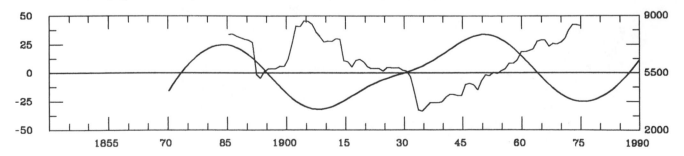

FIGURE 12 Time series of low-pass filtered base-state (solid curve) and 31-year mean-square departures (31MSD) for all-India monsoon-season (June-Sept.) rainfall. Base-state units (left) are mm, 31-year mean departure units (right) are mm².

rainfall and the SO has been well documented (e.g., Walker (1924), Shukla and Paolino (1983)).

Folland et al. (1991) have shown a correspondence between the twentieth-century variation in Sahel rainfall (Nicholson, 1985) and a pattern of SST variability that appears as the third global EOF of SST. The EOF3 coefficient time series and the rainfall values show a large amount of annual- and decadal-scale variability, but both exhibit a pronounced century-scale variation that has a vague similarity to the century-scale changes in the ENSO-cycle variability, particularity the rapid changes after 1960 in the SST* and Sahel rainfall series. The EOF3 coefficients imply an increase in tropical Pacific SST after the mid-1940s, and while this represents only a small part of the total SST variability, it is of the same sign as the trend in the SST* base state.

Finally, the time series of the percentage of the U.S. Great Plains experiencing severe or extreme drought (Figure 13) shows a dominant interdecadal pattern of variability (Rasmusson and Arkin, 1992). It also shows a pattern of increasingly intense drought episodes up to the great Dust Bowl droughts of the 1930s, followed by a decrease in intensity. This suggests a modulation of the interdecadal fluctuations on the century time scale, with the more severe drought epochs corresponding to the period of low ENSO-cycle variability.

It is of course hazardous to even suggest relationships

on the basis of a single realization of a "cycle." We are primarily calling attention to these "coincidences" as an area perhaps deserving of further investigation.

CONCLUSIONS

Because of the global-scale coherence of the ENSO signal, considerable information on ENSO-cycle variability over the past 100 to 150 years can be extracted from a small number of long instrumental time series available from the core region of the oscillation. The results from earlier studies have been extended using SLP data from Bombay, Darwin, and Tahiti, and a time series of SST averaged over the central and eastern equatorial Pacific. The major conclusions from these and earlier analyses can be summarized as follows.

- ENSO-cycle variance for 31-year periods has changed by a factor of two or more during the past century. The cycle was quite pronounced late in the nineteenth century, was relatively weak between 1920 to 1950, and has increased in intensity since then.
- The century-scale variation in equatorial Pacific SST was broadly similar to that in globally averaged SST.
- No obvious relationship could be detected between variations in the base-state parameters we analyzed (SST and SLP) and variations in the intensity of the ENSO cycle.
- Regional recurrence statistics, such as those derived from the Quinn et al. (1987) compilation of strong and very strong El Niño events in Peru, cannot be considered a reliable index of basin-scale ENSO-cycle variability.
- The century-scale variations in ENSO-cycle intensity broadly correspond to changes in all-India monsoon-season rainfall variability, to the modulation of the intensity of drought episodes over the U.S. Great Plains during the twentieth century, and, less clearly, to the century-scale variation in Sahel rainfall.

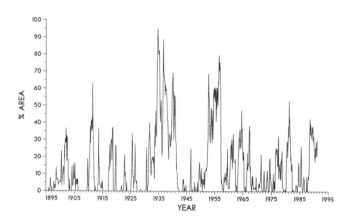

FIGURE 13 Area of U.S. Great Plains covered by severe and extreme drought. Data through March 1991. (Figure courtesy of the National Climatic Data Center, Asheville, N.C.)

ACKNOWLEDGMENTS

This work was partially supported by the NOAA Equatorial Pacific Ocean Climate Studies. The senior author was partially supported by NSF Grant ATM-9013435. Data for Figure 10 were kindly provided by Thomas M. Smith of the NWS/NOAA Climate Analysis Center.

Commentary on the Paper of Rasmusson et al.

YOCHANAN KUSHNIR
Lamont-Doherty Earth Observatory

Dr. Rasmusson has just given us an example of the "extended bellybutton" approach to data analysis. He used an a priori knowledge of the pattern of variability in the data to identify key time series useful in monitoring the behavior of the system even during periods when data coverage was sparse. This method could be used with proxy data, as we saw yesterday in Julia Cole's talk. This is a model to follow when using sparse data in the study of decadal variability. This approach is preferable to looking at a single time series, since it contains information on the spatial structure of the phenomenon.

Here we have an advantage because ENSO is so coherent over space. Even in that vast area where we do not have enough observations, we can still find meaningful information simply because the phenomenon itself is so coherent over such a large part of the tropics.

I should like to say something about the "basic-state" idea discussed in the paper. The paper emphasizes that it is an arbitrary choice, a compromise in terms of the length of the data and the length of the ENSO cycle. It is indeed arbitrary; one of the things that we see, as Dr. Rasmusson mentioned, is that the basic state variables relate to one another in the same way that ENSO relates in different locations. This may mean that the basic state is just one of the low-frequency, maybe non-linear manifestations of the ENSO cycle itself, and thus it may not be the true basic state for the phenomenon we are dealing with.

The reference to a 30-year interval as the basic state is a helpful approach, however, at least for following the change in the main spectral peaks of ENSO, the four- to five-year cycle and the biennial cycle. In this way, he can actually follow changes in the variance of the more intense part of the phenomenon. Although it is hard to refer to the results as statistically significant, it is interesting to note that they are so coherent over space. They show, in a quantitative way, that there is a change in the variance of ENSO over time scales of a century or so.

The connection that is made at the end of the paper to other phenomena like monsoons, changes in droughts in North America, and the Sahel drought is very interesting and is food for thought in the future. It is obviously very hard to tie these phenomena together with only one cycle of each event. They may be in phase during one cycle and then drift out of phase, as is the case with two cycles slightly different in period. This is a delicate subject, and needless to say, we have to approach it very carefully.

Discussion

TRENBERTH: I'd like to go on record with a couple of comments on the very long time scales. I'm a little concerned about the skimpiness of the pre-1950 SST data, as well as its homogeneity. I'm also worried about the fact that at least one correction needs to be made for every one of the Indian stations. For example, detrending can remove some very real signals. On a century time scale this begins to call into question what you can really say.

RASMUSSON: If the detrending had removed a real signal, it would have induced jumps at the ends of the detrended period. However, it corresponded magnificently with Henry Diaz's data, which greatly increased my confidence in both time series.

DIAZ: Because of the large spatial coherence of the ENSO signal in that area, you can actually get pretty good signals back to about the 1877-1878 event.

RASMUSSON: I think that event was really the beginning of the turn-down of the early twentieth century.

DIAZ: I'd like to see a study of the variations in the strength of the coupling between El Niño and Southern Oscillation events.

RASMUSSON: Yes. Quinn's long data set is very useful, but I don't think the categorization into strong, very strong, and so on is a good approach. The overlap with strong SST events is only half to two-thirds.

MYSAK: Gene, can you give us any insight into whether the slow interdecadal variations in the basic state in the Pacific are related to variations in the thermohaline circulation, as they appear to be in the Atlantic?

RASMUSSON: The correlations extend into the Atlantic, and particularly in the subtropics the Atlantic variability is definitely associated with the Pacific variability.

LEVITUS: That SOI/SST correlation seemed to be at an awfully low level. It's not clear to me that it could affect the thermohaline circulation at high latitudes.

RASMUSSON: That's for all the months; for the winter months the correlations get up to .5 or a little higher. Van Loon and Madden's pressure correlations extend out in exactly the same way. There's some sort of connection, and my guess is that it would affect things like the West Atlantic teleconnection pattern.

BRYAN: We've seen tremendous progress in proxy data sets. Gene, if we extend this research into the future, what would you like to see?

RASMUSSON: I'd like to see the coral people establish the long-term—300-to-500-year—character of the variations in what I call the core region, and then tie in with the tree-ring people at higher latitudes to see whether they could say anything about the stability of the teleconnnections. I am very much impressed with the coral data's potential for helping us reach a new level of understanding of secular variability.

KEELING: Julie's data show that same variance around 1920 to 1950. It seems to have hit a lot of different parameters. I think we need to keep an open mind about the possibility that something significant happened in that period.

RIND: Another unique thing about the 1920-to-1950 time frame is that there were no major volcanic eruptions.

CANE: We shouldn't forget that the variation during that period could also result from the internal dynamics of the system. There might be some way we could use the spatial patterns and multivariate changes to discriminate between what might be caused by some shift in the tropical Pacific and what might be caused by something else.

Decadal Climate Variations in the Pacific

KEVIN E. TRENBERTH AND JAMES W. HURRELL[1]

ABSTRACT

Considerable evidence has emerged of a substantial change in the North Pacific atmosphere and ocean lasting from about 1976 to 1988. Significant changes observed in the atmospheric circulation throughout the troposphere during that period show that a deeper, eastward-shifted Aleutian Low pressure system in the winter half-year advected warmer and moister air along the west coast of North America and into Alaska and colder air over the North Pacific. This advection caused substantial changes in sea surface temperatures over the North Pacific, as well as in coastal rainfall and streamflow, and in sea ice in the Bering Sea. Associated changes occurred in the surface wind stress, and, by inference, in the Sverdrup transport in the North Pacific Ocean. Changes in the monthly mean flow also imply substantial changes in the storm tracks and associated synoptic eddy activity, and in the sensible and latent heat fluxes at the ocean surface. In addition to the changes in the physical environment, large changes are found in the biology in the Northeast Pacific and in fish and other animal behavior.

It is suggested here that clues to possible causes of these changes lie in the close link between North Pacific changes on the decadal time scale and changes in the tropical Pacific and Indian Ocean, as well as the changes in frequency and intensity of El Niño versus La Niña events. A hypothesis is put forward outlining the tropical and extratropical relationships, which stresses the role of tropical forcing but includes important feedbacks in the extratropics that serve to emphasize the decadal more than interannual time scales. Whether the observed decadal variations are linked to "global warming" issues is an open question.

INTRODUCTION

Climate variations over the North Pacific and teleconnections downstream across North America have long been of interest; they have been particularly highlighted by the work of Namias (1959, 1963, 1969). Recently, considerable evidence has emerged of a substantial decade-long change in the North Pacific atmosphere and ocean that began about 1976. Changes in the atmospheric circulation throughout the troposphere at this time have been documented by Trenberth (1990; see Figure 1 below) and Nitta and Yamada (1989)

National Center for Atmospheric Research, Boulder, Colorado

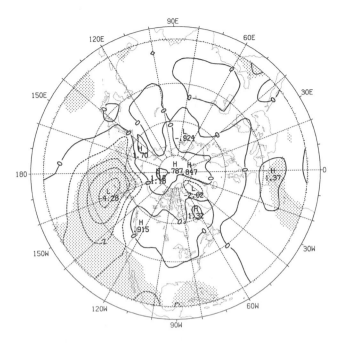

FIGURE 1 The difference between mean sea level pressures from 1977 to 1988 for November through March and those of 1924 to 1976 (mb). Stippling indicates statistical significance at 5 percent. (From Trenberth, 1990; published with permission of the American Meteorological Society.)

for the winter half-year. These changes were associated with changes in the surface wind stress, and, by inference, in the Sverdrup transport in the North Pacific Ocean (Trenberth, 1991) that have been directly measured in the Gulf of Alaska (Royer, 1989). Bakun (1990) has also noted changes in the alongshore wind stress off California that are probably related. Changes in the monthly mean flow also imply substantial changes in the storm tracks and associated synoptic eddy activity (e.g., Lau, 1988; Rogers and Raphael, 1992) and in the sensible and latent heat fluxes at the ocean surface (Cayan, 1992). The circulation changes were accompanied by changes in heat and moisture advection due to the quasistationary flow (see, e.g., Rogers and Raphael, 1992), so that there were substantial changes in the temperatures along the west coast of North America and in sea surface temperatures (SSTs) over the North Pacific (Trenberth, 1990), as well as in coastal rainfall and streamflow (Cayan and Peterson, 1989), and in sea ice in the Bering Sea (Manak and Mysak, 1987).

In addition to the changes in the physical environment, Venrick et al. (1987) observed associated large changes in the epipelagic ecosystem in the North Pacific, with increases in total chlorophyll in the water column and thus in phytoplankton. Mysak (1986) earlier noted that the changes in ocean currents and temperatures had altered the migration patterns of fish, in particular tuna and salmon, in the northeast Pacific (see also Hamilton, 1987). Ebbesmeyer et al.

(1991) found that the "step in Pacific climate" in 1976 had a profound effect on 40 environmental variables. They contrasted the periods 1968 to 1975 with 1977 to 1984 and noted that climate-related changes were found not only in the above parameters, but in the behavior of geese, salmon, and crabs in the Northeast Pacific, and in mollusk abundance, salinity, and water temperature in Puget Sound. Also accompanying the changes in the Pacific was a higher incidence of cold outbreaks across the plains of North America, which ultimately led to major freezes affecting the Florida citrus crop (Rogers and Rohli, 1991).

The change in 1976 is but one of several large changes that have occurred in the North Pacific. As noted above, Namias has documented decadal-scale variations in the past, with coherent signals in the atmospheric circulation and in the SSTs, and with teleconnections downstream across North America (see also Douglas et al., 1982; Namias et al., 1988). The fairly sluggish response of the mid-latitude ocean to changes in the ocean forcing through the surface momentum and heat fluxes effectively serves as a low-pass filter and emphasizes the longer time scales. More recent studies throw further light on these aspects and on the physical links between the atmosphere and ocean in the North Pacific, and will be discussed below. New evidence has also emerged on the teleconnections downstream across North America, particularly those associated with the Pacific-North American (PNA) teleconnection pattern.

Possible causes of the changes were discussed by Trenberth (1990); he noted the close link between North Pacific changes on the decadal time scale and those in the tropical Pacific and Indian Ocean, as well as the changes in frequency and intensity of El Niño versus La Niña events. That paper expanded on the similar link noted during El Niño events by Bjerknes (1969). These aspects will be pursued and quantified further below.

The time series in Trenberth (1990) ceased after the northern winter of 1987-1988, and it is quite interesting to see what the subsequent evolution has been. Accordingly, in this paper, we update the time series and carry out some more comprehensive correlation analyses with surface temperature and SST analyses, and examine further the links with the tropical Pacific and the Southern Oscillation. The time scales of the main circulation anomaly patterns contributing to the decadal variation and the links with the tropics and teleconnections elsewhere are also of interest. A more complete version of this paper is presented in Trenberth and Hurrell (1994).

NORTH PACIFIC LARGE-SCALE OBSERVED TRENDS

In the North Pacific, a close association between SST anomalies and the atmospheric circulation has been well recognized. Changes in surface temperatures arise from changes in temperature and moisture advection over the oceans by anomalous winds, and from the associated changes in surface fluxes and vertical mixing within the ocean. Anomalous northerly winds over the ocean are typically not only cold but also dry, so that large increases in surface fluxes of both sensible and latent heat into the atmosphere can be expected, which in turn cools the ocean. Convection and mechanical mixing in the ocean can spread those influences through considerable depth and may also entrain water from below the thermocline, giving the anomalies a finite lifetime (see, e.g., Frankignoul, 1985).

In the SSTs, a very distinctive pattern (cf. Figure 5) emerges as the dominant mode of an empirical orthogonal function (EOF) analysis (Davis, 1976; Lau and Nath, 1990) that is linked with a preferred mode of variability in winter in the Northern Hemisphere and is similar to the PNA teleconnection pattern of Wallace and Gutzler (1981). The PNA consists of four centers of action in the mid-tropospheric height field; they are of one sign near Hawaii and along the west coast of North America, and of opposite sign over the North Pacific and southeast United States. Wallace and Gutzler (1981) defined a PNA index using single-point values of 500 mb monthly mean geopotential height at each of the centers. This index does not appropriately weight the four centers of the PNA, and it is sensitive to errors in the analyses. Also, Northern Hemisphere upper-air analyses are available only after 1947. Wallace and Gutzler show that the surface signature of the PNA is mostly confined to the Pacific. An evaluation of sea level pressure charts (Trenberth and Paolino, 1980) shows them to be most reliable after 1924, and we therefore choose the area-weighted mean sea level pressure over the region 30°N to 65°N, 160°E to 140°W as a robust but simple measure of the circulation in the North Pacific. We refer to this as the NP index (for North Pacific). This is slightly smaller than the area used in Trenberth (1990), but corresponds better to the area participating in the decadal time-scale variations (see Figure 1). Trenberth (1990) noted that the correlation of the NP index with a PNA index based on all four PNA teleconnected centers at 700 and 500 mb is −0.92 for 1947 to 1987.

Figure 2 shows the NP index for all months of the year from 1924 through 1991. The first panel gives the total value of the index, and thus includes the mean annual cycle for three-month (seasonal) means that have been smoothed in time using a low-pass filter (with weights [1, 4, 8, 10, 8, 4, 1]/36, for which the half-power point is 11 years), which emphasizes the interdecadal fluctuations. The second panel shows the individual monthly anomalies (with the mean annual cycle removed) and is rather noisy, with the month-to-month variations prominent. The third panel shows the anomalies from the first panel, so that it is a smoothed version of the second panel, to show the most persistent anomalies. The figures emphasize that the period

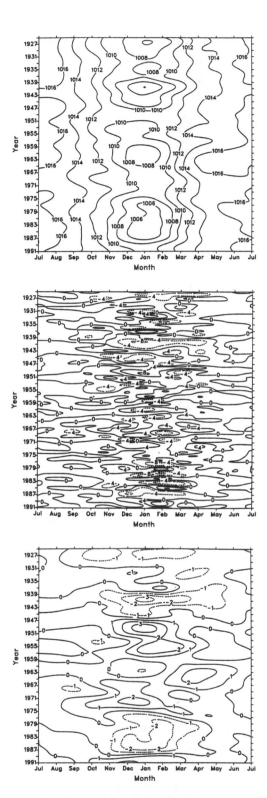

FIGURE 2 Time series of mean North Pacific sea level pressures averaged over 30°N to 65°N, 160°E to 140°W as a function of month and time. Top, total three-monthly mean values smoothed with a low-pass filter with seven weights ([1, 4, 8, 10, 8, 4, 1]/36) across years to emphasize the decadal time scales in mb. Center, monthly mean anomalies in mb. Bottom, seasonal (three-monthly) mean anomalies smoothed with the low-pass filter in mb.

November through March corresponds to the wintertime regime, with low pressures in the Pacific, while for the rest of the year the Pacific tends to be dominated more by the subtropical anticyclone. Also, the main variability occurs only in November through March. Standard deviations of the monthly anomaly times series range from about 1 mb in the summer months to more than 4.5 mb in January and February. Values exceed 2.8 mb only from November through March. The variability from one month to the next in the same winter season stems in part from westward-propagating planetary-scale waves with a 20-day period (Branstator, 1987; Madden and Speth, 1989). Averaging over the five winter months removes much of this kind of noise.

Accordingly, Figures 3 and 4 show the NP time series for the five-month wintertime average. In Figure 3 the 1946 to 1991 time series is broken up to emphasize the regime from 1976 to 1988, while in Figure 4 the time series from 1925 is shown with the low-passed curve to reveal, without the arbitrariness of deciding on a start and end for the regime, just how unusual the 1977-to-1988 period is; the

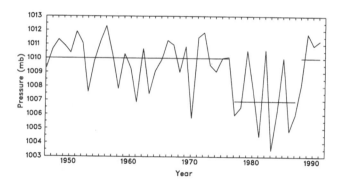

FIGURE 3 Time series of mean North Pacific sea level pressures averaged over 30°N to 65°N, 160°E to 140°W for the months November through March. Means for the combined periods 1946-to-1976 plus 1989-to-1991 and for 1977-to-1988 are indicated (1988 refers to the 1987-1988 winter).

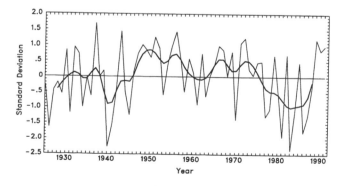

FIGURE 4 Time series of mean North Pacific sea level pressures for November through March, as in Figure 3, but beginning in 1925 and smoothed with the low-pass filter.

only previous time for which comparable values occurred was the much shorter interval from 1940 to 1941. Nevertheless, considerable interannual variability is present within the regime.

In the Aleutian Low from 1977 to 1988, for November through March, pressures were lower by 3.0 mb when averaged over the NP area of the North Pacific. Lower pressures were present individually in all five winter months and are highly statistically significant (Figure 1). No such change is present in any of the other months of the year (Figure 2). The wintertime changes correspond to the center of the low farther east and deeper on average by 4.3 mb for the five winter months, being deeper by 7 to 9 mb in January (Trenberth, 1991; see also Figure 1).

A climatology for surface wind stress based on the years 1980 to 1986 reveals changes in the North Pacific relative to a climatology based on ship data prior to 1977 that help confirm the reality of the sea level pressure changes (Trenberth, 1991), and so do analyses with independent data sets (Nitta and Yamada, 1989). Moreover, the associated changes in the curl of the wind stress and the corresponding Sverdrup transport in the ocean (Trenberth, 1991) over such a long period imply significant changes in the North Pacific Ocean currents.

The corresponding changes in surface temperatures are shown in Figure 5. The surface temperatures are taken from the updated Intergovernmental Panel on Climate Change (IPCC) data set (IPCC, 1990, 1992), which consists of land surface data from the University of East Anglia (Jones, 1988) blended with sea surface temperature data from the U.K. Meteorological Office (Bottomley et al., 1990; see also Trenberth et al., 1992). The surface temperature anomalies, expressed as departures from the 1951-to-1980 means, are shown for both the whole year and the five winter months November through March averaged over the 1977-to-1988 period. The temperature anomalies are strongly regional, and have both positive and negative signs. The 12-year period features very large North Pacific basin temperature anomalies, with warming of more than 1.5°C in Alaska and cooling of more than 0.5°C in the central North Pacific. The pattern in Figure 5 over the North Pacific is similar to the first EOF of SSTs (Davis, 1976; Lau and Nath, 1990). The annual mean anomaly clearly arises from the wintertime atmospheric anomaly, but over the North Pacific it is sufficiently persistent throughout the year that there is little difference between the annual and the five-month means. The wintertime pattern also reveals below-normal temperatures over the southeastern part of the United States, illustrating the PNA teleconnection. This is reflected in a higher-than-usual incidence of major freezes affecting the Florida citrus crop after 1977 (Rogers and Rohli, 1991).

The most compelling argument that the changes in sea level pressure are real is the physical consistency with the very large regional Pacific temperature anomalies for 1977

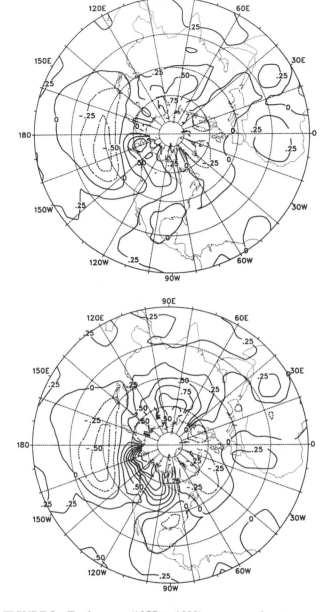

FIGURE 5 Twelve-year (1977-to-1988) average surface temperature or sea surface temperature anomalies as departures from the 1951-to-1980 mean. Contours every 0.25°C. Shown are the annual mean anomalies (upper panel) and the anomalies for the five winter months (November to March). Negative values are dashed.

to 1988, shown in Figure 5. The warming over Alaska and along the west coast of North America, together with the cooling in the central and western North Pacific, would be expected to accompany a stronger Aleutian Low, because of thermal advection (Rogers and Raphael, 1992) and increased ocean mixing and changes in the surface fluxes (Cayan, 1992; Alexander, 1992a,b). The increased southward gradient flow in the eastern North Pacific, revealed by the pressure pattern in Figure 1, would bring warmer and moister air

into Alaska and along the west coast of North America, while anomalous northerly winds would give rise to colder-than-normal conditions in the central and western North Pacific. Lower SSTs are a consequence of large sensible and latent heat fluxes into the atmosphere, combined with increased mixing in the ocean (Cayan, 1992). Cayan and Peterson (1989) found that increased stream flow in the coastal region of the northern Gulf of Alaska results from increased coastal rainfall associated with a deepened Aleutian Low and changes in the PNA.

To further illustrate the nature of the surface temperature changes associated with the NP index, Figure 6 shows the correlations for the November-to-March five-month average over 1935 to 1990, along with the corresponding departure pattern in degrees Celsius associated with a unit standard-deviation departure of the NP index. For each grid point, "seasonal" values were computed only when data existed for at least three of the five months defining the season. Correlations between variables were not computed if the two variables had fewer than 75 percent of the total number of seasons in common. Across the North Pacific and North American regions, these patterns show that the anomaly featured in Figure 5 is consistent with the whole record: Below-normal NP values are associated with below-normal temperatures over the North Pacific and southeast United States and above-normal surface temperatures along the West Coast, extending throughout Alaska and across most of Canada.

We have also investigated these relationships as a function of various lags. To objectively decide how much variance is explained by the correlations across an area, we have averaged the correlation coefficient squared for the region 140°E to 60°W, 30° to 65°N. The largest surface temperature variance explained by the NP index for this region occurs with NP leading by 1 to 2 months (r^2 values with NP leading by 3, 2, 1, and 0 months are 0.15, 0.19, 0.20, and 0.16). The pattern is similar to that at zero lag, but the magnitude of the correlation coefficients increase by about 0.1 (to >0.6 over the North Pacific, and to < −0.7 over British Columbia).

The above results are consistent with those of Davis (1976). The link between SST in the North Pacific and the overlying atmospheric circulation has become well established. The main relationship seems to be one where the changes in the atmospheric circulation are responsible for the SST changes, as shown by simultaneous and lagged correlations, for instance (Davis, 1976, 1978; Lanzante, 1984; Wallace et al., 1990). Nevertheless, there is the strong expectation that extratropical SST anomalies also influence and may reinforce the atmospheric circulation (Kushnir and Lau, 1992).

Further confirmation of the link between temperatures and the atmospheric circulation comes from correlations between the NP index time series in Figure 2 and tempera-

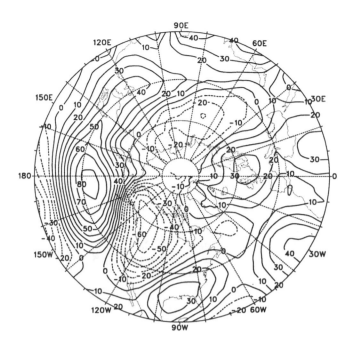

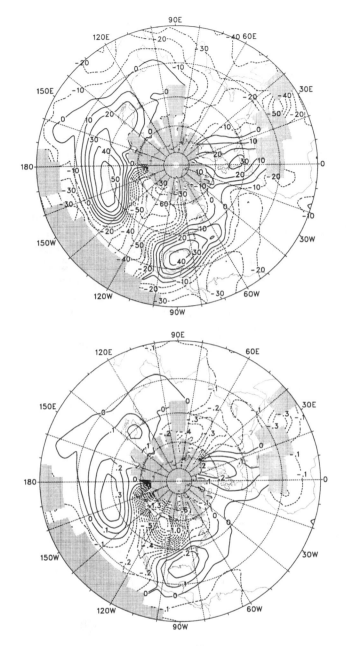

FIGURE 7 Correlations between the NP index in Figure 2 and 700 mb temperatures. Negative values are dashed. The 5 percent significance level is 0.30. Twenty percent of the variance is accounted for by the correlations over the 140°E to 60°W, 30°N to 65°N region.

FIGURE 6 Correlations between the NP index and surface temperatures for 1935 to 1990 (upper panel) and the departure pattern corresponding to a unit standard deviation of NP. Negative values are dashed. The 5 percent significance level is 0.27. Shaded areas indicate insufficient data.

tures at 700 mb (Figure 7). The pattern of correlations reveals the four centers of the PNA. Over the Pacific and North America this pattern (with opposite sign) bears a striking resemblance to the actual surface temperature anomalies for 1977 to 1988 (Figure 5), including the cooling over the eastern part of North America. As shown in Figure 6, the latter is thus revealed as part of the overall teleconnection pattern. Over most of the domain, the 700 mb correla-

tions are highest at zero lag. The exception is for the center over the southeastern United States, where correlations are 10 percent higher one month later, indicating that the development of the teleconnection downstream and the associated change in tracks of synoptic systems (see, e.g., Lau, 1988; Rogers and Rohli, 1991) may be somewhat delayed.

To complete the picture, Figure 8 shows the correlations of the NP index with the 500 mb height field, and the corresponding departure pattern. Once again, the PNA pattern emerges strongly, with all the PNA centers showing up, although the associated anomaly departure pattern clearly emphasizes the North Pacific.

CAUSES OF CHANGE IN THE NORTH PACIFIC

Examination of the possible causes of the various types of changes focuses attention on the association between the large-scale coherent climate variations and changes in atmospheric waves. The stationary planetary waves in the atmosphere are forced by orography and patterns of diabatic heating arising from the distribution of land and sea, both in the extratropics and in the tropics (see, e.g., Frankignoul, 1985). Therefore, in the Northern Hemisphere, changes in diabatic heating (for instance) can change the planetary waves and associated poleward heat fluxes. This mechanism does not operate in summer. In addition, when the temperature gradients on which the transients feed are altered

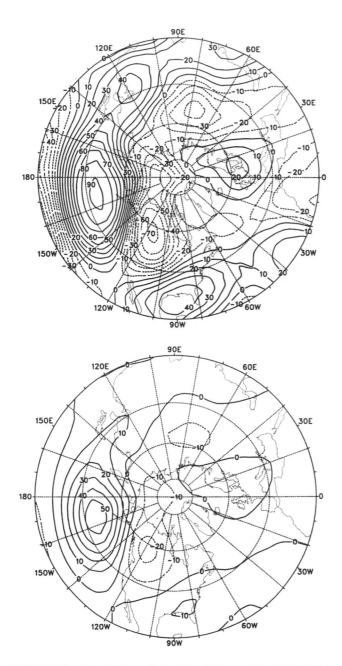

FIGURE 8   Correlations of the November to March NP index with 500 mb heights for 1948 to 1991 (upper panel) and the departure pattern corresponding to a unit standard deviation of NP. Negative values are dashed. Twenty-six percent of the variance is accounted for by the correlations over the 140°E to 60°W, 30°N to 65°N region.

through baroclinic instability, changes in the transient storm tracks result (van Loon, 1979; Lau, 1988).

When possible causes of changes are considered for the North Pacific, one prospect is in situ forcing through the influence of extratropical SST anomalies in the North Pacific on the circulation (Namias, 1959, 1963). It has been difficult to substantiate such influences either statistically (Davis,

1976, 1978) or with models (Ting, 1991; Kushnir and Lau, 1992). Recent modeling studies of SST anomalies in the Northern Hemisphere indicate that the changes in the storm tracks alter the eddy vorticity fluxes in the upper troposphere in such a way that they often reinforce and help maintain the circulation anomalies (Lau and Nath, 1990; Ting, 1991; Kushnir and Lau, 1992).

While the changes in eddy transports from the altered synoptic systems are one major complication, another is that the atmospheric heating effects may not be local. The sensible heat exchanged between the ocean and atmosphere is realized locally, but the latent heat lost by the ocean through evaporation is realized only as an increase in moisture, and the actual atmospheric heating is not realized until precipitation occurs, often far downstream. This latter aspect depends on the prevailing synoptic situation at the time, and varies with location according to the prevailing winds and background climatological flow. These nonlocal effects are therefore a sensitive function of position, and they add a large nondeterministic component to any forcing. This means that it is much more difficult to detect any systematic effects in both the real atmosphere and models. It also helps account for differences in results from many different model experiments, because inserted SST anomalies vary in location and intensity and the model climatologies vary. Placing "super SST anomalies" into a model will enhance the local effects so that results are more likely to appear as significant, but they are also much more likely to be unrealistic and inappropriate for the real atmosphere.

Another prospective cause of changes in the North Pacific comes from changes in teleconnections. The best-known examples of global impacts of local forcing are those involving changes in tropical SSTs, like the El Niño/Southern Oscillation (ENSO) phenomenon. Such changes in the atmosphere and the underlying ocean in the tropical Pacific affect higher latitudes (Bjerknes, 1969; Horel and Wallace, 1981).

LINKS WITH THE TROPICAL PACIFIC

The period of the deeper Aleutian Low regime extends from 1977 to 1988; during it there were three El Niño (warm) events in the tropical Pacific but no compensating La Niña (cold) events. Because of the El Niños, the tropical Pacific experienced above-normal SSTs and a persistently negative Southern Oscillation index (SOI) for that period (Figure 9). Modeling studies (e.g., Blackmon et al., 1983; Alexander, 1992b) confirm the causal link between SSTs in the tropics and the North Pacific circulation, with a deeper Aleutian Low resulting from El Niño conditions. Alexander (1992a,b) further shows that the observed changes in the North Pacific Ocean SSTs can be accounted for largely by the atmospheric changes, by means of the associated changes in surface fluxes and mixing through the upper layers of the ocean, and by the deepening of the mixed

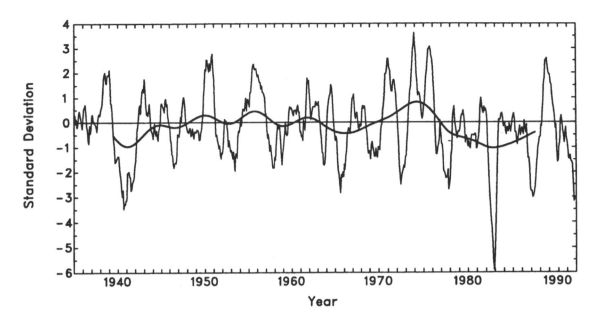

FIGURE 9 Time series of the normalized Southern Oscillation Index (Tahiti minus Darwin sea level pressure anomalies) monthly, filtered with a five-month running mean and with a low-pass filter with 109 weights that removes periods less than 10 years.

layer by entrainment. But the results obtained here are not simply due to the 1982-1983 and 1986-1987 El Niños; the Aleutian Low was also much deeper than normal in several other years, especially in the winter of 1980-1981. Note, however, from Figure 4 that the previous time when comparably low values occurred over the North Pacific was during the major 1939-to-1942 El Niño event.

We have examined the correlation of the NP index with the SOI for the period 1935 to 1991. The Tahiti-minus-Darwin normalized surface pressure index has been used (Trenberth, 1984); it is used only from 1935 on because the Tahiti record prior to then is poor. Correlations of the SOI with the Northern Hemisphere sea level pressures for the November-to-March winter months combined show the link with the North Pacific and the extension across North America (Figure 10). Note the values of opposite sign over North America, which are very important as part of the overall pattern. The anomalous wind flow accompanying this pattern is indeed one where stronger southerlies along the west coast of North America accompany a negative SOI (i.e., El Niño conditions).

We have examined correlations of the SOI with NP at several lags, using five-month running mean values of the SOI (Table 1). This smoothing is needed to make the SOI representative of the Southern Oscillation (Trenberth, 1984) and its scale is compatible with the time scales of the NP index. The highest cross-correlations of 0.53 occur at zero lag ±1 month. With the SOI leading by six months the correlation is 0.44, and with a lag of six months it is 0.16. An interpretation of this time/lag dependence is that the correlated changes are largely contemporaneous, and the

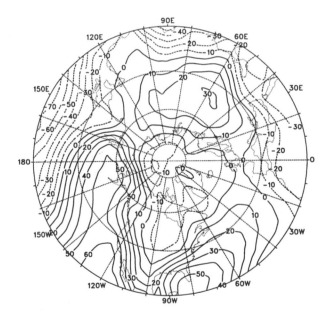

FIGURE 10 Correlations of the five-month mean (November to March) SOI with sea level pressures over the Northern Hemisphere for 1935 to 1991. The 1 percent significance level is 0.34.

persistence of the SOI is the factor contributing to the strong correlations when the SOI is leading. The lower values when the SOI is lagging are a reflection of the influence of the March-to-April time of year when the SOI tends to be weakest and preferentially changes sign (Trenberth, 1984).

However, this does not mean that there are not precursors in the tropics. On the contrary, it is well established that there is an evolution of the Southern Oscillation and the

TABLE 1   Correlations between the NP November-to-March Index, Indices of SST, and the SOI*

Index Lead	Niño 1 + 2	Niño 3	Niño 4	Niño 3 + 4	SOI
+6	−.39	−.47	−.46	−.47	.44
+4	−.44	−.49	−.50	−.51	.46
+2	−.48	−.50	−.49	−.51	.50
0	−.44	−.51	−.45	−.51	.52
−2	−.30	−.44	−.34	−.43	.47
−4	−.18	−.34	−.25	−.31	.39
−6	−.12	−.15	−.23	−.19	.16

*All values are five-month means. The period for the SOI and Niño 1 + 2 regions is 1935 to 1990 inclusive, so the one-tailed 1 percent significance level is 0.32. For the other Niño regions the period is 1951 to 1990, and the 1 percent significance level is 0.38. The lead (in months) refers to the Niño SST or SOI index with respect to the NP index. To be included in the computation of the area-average SSTs, at least half the points in an area were required. Maximum values are in italics.

SST fields in the tropical Pacific as El Niño events develop (Trenberth, 1976; Rasmusson and Carpenter, 1982; Trenberth and Shea, 1987; Wright et al., 1988). Trenberth (1976), Trenberth and Shea (1987), and Wright et al. (1988) noted that pressures in the South Pacific (e.g., at Easter Island) respond about a season earlier than the SOI does, and Barnett (1985) suggested that changes can often be seen over the southeast Asian region before the SOI responds. Barnett et al. (1989) further suggested that this evolution might be linked to snow cover over Asia.

We have therefore examined in more detail the relationships between SSTs in the tropics and the NP index. Problems with data coverage are severe in the tropics prior to 1951. To help summarize the results, we have computed correlations between the area-averaged SST anomalies for the tropical Pacific Niño regions—Niño 1 and 2 (0 to 10°S, 90 to 80°W), Niño 3 (5°N to 5°S, 150 to 90°W), and Niño 4 (5°N to 5°S, 160°E to 150°W)—with NP at several leads and lags (see Table 1). As larger areas are taken, the correlation coefficient increases in magnitude; for the Niño 3 and 4 regions combined, all correlations are larger, with maximum values of −0.52 at a 3-month lead by the SSTs. This shows that the changes in SST throughout much of the tropical Pacific lead the NP index by about three months, although the cross-correlation is not sharply defined and values are only slightly smaller at zero lag. Nevertheless, these results emphasize the involvement of the tropical SST variations in the atmospheric and surface temperature variations over the North Pacific and North America.

DISCUSSION AND CONCLUSIONS

The picture emerging from these empirical and modeling studies is not yet fully clear, but the evidence suggests the following hypothesis. In the tropics, coupled ocean-atmosphere interactions result in coupled modes, of which ENSO is the most prominent. This coupling results in large

interannual variability in the Pacific sector, with preferred time scales of 2 to 7 years, but with small-amplitude decadal variations. All these fluctuations have manifestations in higher latitudes through teleconnections within the atmosphere. In the North Pacific, ENSO variability is found in the PNA pattern (and the NP index), but is best seen when averages can be taken over the entire winter half-year, because the noise level associated with natural weather variability is high on monthly time scales. The deepened Aleutian low in ENSO events results in a characteristic SST anomaly pattern that, on average, is enhanced through positive feedback effects from effects of the extratropical SST anomaly itself and from changes in momentum (and vorticity) fluxes associated with changes in high-frequency storm tracks (see Kushnir and Lau, 1992). The same influences are present on long time scales, but whereas surface fluxes and mixed-layer processes are dominant in changing SSTs on interannual time scales, changes in ocean currents also become a factor on decadal time scales and would reinforce the SST changes. Moreover, the long time scale involved in changing the currents and the Sverdrup circulation adds further persistence to the extratropical system that, along with heat storage in the top 500 m of the ocean, serves to emphasize decadal over interannual time scales.

Aspects of the above hypothesis have appeared in the extensive works of Namias, but here we have emphasized much more the links with the tropics. A major but as yet unanswered question is whether either the intensity or the frequency of ENSO events might change as a result of global warming. A longer observational record than that given in Figure 9 reveals that the frequency and intensity of ENSO events have changed in the past (Trenberth and Shea, 1987), with strong ENSO fluctuations from about 1880 to 1920. Aside from the major event from 1939 to 1942, stronger and more regular ENSO events did not resume until the 1950s. However, the low-passed curve in Figure 9 indicates that the recent imbalance between the number of warm and cold events in the tropical Pacific is unprecedented.

Whether the unusual 1977-to-1988 imbalance can be ascribed in part to some identifiable contributor, or merely reflects natural variability, is a very difficult question to answer. The major change that occurred in March-to-April 1988, a transition from El Niño to a very strong La Niña (Figure 4), apparently ended the climate regime, although the underlying ocean currents and heat storage must be still perturbed and the pattern could reemerge. Indeed, the 1991- 1992 ENSO event was noted for its exceptionally warm water along the west coast of both North and South America in early 1992.

ACKNOWLEDGMENTS

We wish to especially thank Dennis Shea for preparing some of the figures. This research is partially sponsored by the Tropical Oceans/Global Atmosphere Project Office under grant NA86AANRG0100.

Commentary on the Paper of Trenberth and Hurrell

YOCHANAN KUSHNIR
Lamont-Doherty Earth Observatory

Dr. Trenberth has demonstrated in this study how complex the middle-latitude system is. We have to deal with seasonality (there is a nice diagram in his paper that actually emphasizes the seasonality in this very-low-frequency phenomenon that he discussed). We also have to deal with the fact that middle latitudes filter the signal that comes out of the tropics, if it does indeed come out of the tropics as suggested here. Actually, the Pacific may be a more complex environment in which to study mid-latitude interactions than the Atlantic, where the effect of ENSO or of the tropics seems to be weaker.

One of the biggest issues in mid-latitude interaction is whether the ocean and the atmosphere are really coupled. It has been suggested that the ocean is being forced by the atmosphere and no feedback is involved. Local interactions like mixing and heat exchange, or maybe some non-local interactions due to the currents and transports, could be responsible for the fact that the pattern is long-lived. But why is it seen also in summer when the forcing from the atmosphere disappears, as the paper emphasizes?

If there is feedback in mid-latitude interactions, can we learn about it from GCMs? As it turns out, we are dealing with a very confusing set of results presented in several papers. They are confusing not only because of the use of super-anomalies versus regular anomalies, but also because different kinds of models and different kinds of methodologies have been used to run the models. We have seen perpetual experiments run, and we have seen experiments of a more transient nature where the seasonal cycle varies and where the SSTs vary continuously.

The coupling between the mid-latitude ocean and the atmosphere is one of the unresolved challenges today. It remains to be seen whether it is really a two-way interaction, and whether feedback in the middle latitudes is involved.

Discussion

BRYAN: This all reminds me of the Hoskins and Karoly ideas of about 12 years ago about the connection between the tropics and mid-latitudes.

TRENBERTH: Well, that may be one of the ways in which the whole system is tied together, but the statistical results indicate that it's probably more than just Southern Oscillation. The transients in the middle latitudes seem to be adding a considerable chaotic component to any teleconnections, and the forcing itself is different for every ENSO.

DICKSON: I worry a bit that your picture doesn't take into account the overlap of the North Atlantic and the Southern Oscilla- tion signals that Rogers described in his 1984 paper. The biggest anomaly gradient of the lot sits on the boundary between them on the U.S. eastern seaboard, where there's such a strong land/ sea temperature contrast.

TRENBERTH: We can't treat those signals as linear and independent; I think the statistics indicate that there is a highly variable and sometimes strong connection between them. And I think there's a lot of feedback involved in the East Coast baroclinicity.

RASMUSSON: Yes, the correlation of all the oscillations in that area is fascinating. But the mean patterns in the North Pacific and the tropics do seem to move in concert when the convection shifts, and I think that linkage is clear.

MARTINSON: Peter Weyl's 1967 climate scenario—90 percent of which has since been corroborated, I'd say—started with precipitation change in the Caribbean and the exchange of moisture between the Pacific and the Atlantic.

KARL: Kevin, I was interested in your comments on the character of the variance in the southeastern United States. Could you elaborate a little on those figures?

TRENBERTH: A figure in Leathers and Palecki (J. Climate, 1992), shows a step-like discontinuity in 1957. But I'm convinced that some of it isn't real, since it's only this one particular point that never returns to the pre-1957 level. There's some evidence of changes in analysis procedures or radiosondes.

CANE: It seemed to me that one difference between what you're doing and the older work was the emphasis on changes that persisted for most of a decade and the idea that the ocean's circulation would change and reinforce the SST pattern. How might the surface wind changes affect the ocean, and how consistent would that be with the SST pattern?

TRENBERTH: There's a nice figure in my 1991 paper showing the annual change in the wind stress. It's directly related to the change in the Aleutian system, and contributes substantially to a change in both the subtropical and the polar gyres in the North Pacific. It seems to me it wouldn't be hard to find out the actual spinup time of a circulation change to a decent depth by using an ocean model; the typical number you hear is 10 to 30 years to really modify the basin-wide sverdrup transport.

Coupled Systems
Reference List

Aceituno, P. 1992. El Niño, the Southern Oscillation and ENSO: Confusing names for a complex ocean-atmosphere interaction. Bull. Am. Meteorol. Soc. 73:483-485.

Alexander, M. 1992a. Midlatitude atmosphere-ocean interaction during El Niño. Pt. I. The North Pacific Ocean. J. Climate 5:944-958.

Alexander, M. 1992b. Midlatitude atmosphere-ocean interaction during El Niño. Pt. II. The Northern Hemisphere atmosphere. J. Climate 5:959-972.

Anderson, R.Y. 1989. Solar-cycle modulation of ENSO: A mechanism for Pacific and global climate change. Sixth Pacific Climate Workshop, Asilomar, Calif.

Bakun, A. 1990. Global climate change and intensification of coastal ocean upwelling. Science 247:198-201.

Barnett, T.P. 1985. Variations in near global sea level pressure. J. Atmos. Sci. 42:478-501.

Barnett, T.P. 1991. The interaction of multiple time scales in the tropical climate system. J. Climate 4:269-285.

Barnett, T.P., N. Graham, M.A. Cane, S.E. Zebiak, S. Dolan, J.J. O'Brien, and D. Legler. 1988. On the prediction of the El Niño of 1986-1987. Science 241:192-196.

Barnett, T.P., L.D.U. Schlese, E. Roeckner, and M. Latif. 1989. The effect of Eurasian snow cover on regional and global climate variations. J. Atmos. Sci. 46:661-685.

Battisti, D.S. 1988. Dynamics and thermodynamics of a warming event in a coupled tropical atmosphere-ocean model. J. Atmos. Sci. 45:2889-2919.

Battisti, D.S. 1995. Decade-to-century time-scale variability in the coupled atmosphere-ocean system: Modeling issues. In Natural Climate Variability on Decade-to-Century Time Scales. D.G. Martinson, K. Bryan, M. Ghil, M.M. Hall, T.R. Karl, E.S. Sarachik, S. Sorooshian, and L.D. Talley (eds.). National Academy Press, Washington, D.C.

Battisti, D.S., and A.C. Hirst. 1989. Interannual variability in the tropical atmosphere/ocean system: Influence of the basic state and ocean geometry. J. Atmos. Sci. 46:1687-1712.

Baumgartner, T. R., J. Michaelsen, L.G. Thompson, G.T. Shen, A. Soutar, and R.E. Casey. 1989. The recording of interannual climatic change by high-resolution natural systems: Tree rings, coral bands, glacial ice layers, and marine varves. In Aspects of Climate Variability in the Pacific and Western Americas. D.H. Peterson (ed.). Geophysical Monograph 55, American Geophysical Union, Washington, D.C., pp. 1-14.

Berlage, H.P. 1957. Fluctuations in the general atmospheric circulation of more than one year, their nature and prognostic value. Mededelingen en Verhandelingen, Kon. Ned. Met. Inst. 69, 152 pp.

Bjerknes, J. 1964. Atlantic air-sea interaction. Adv. Geophys. 10:1-82.

Bjerknes, J. 1969. Atmospheric teleconnections from the equatorial Pacific. Mon. Weather Rev. 97:163-172.

Blackmon, M.L., J.E. Geisler, and E.J. Pitcher. 1983. A general circulation model study of January climate anomaly patterns associated with interannual variation of equatorial Pacific sea surface temperatures. J. Atmos. Sci. 40:1410-1425.

Blumenthal, M.B. 1991. Predictability of a coupled ocean-atmosphere model. J. Climate 4:766-784.

Blumenthal, M.B., Y. Xue, and M.A. Cane. 1992. Predictability of an ocean/atmosphere model using adjoint model analysis. In Proceedings of the Workshop on Predictability, 13-15 November 1991. European Centre for Medium-Range Weather Forecasts.

Bond, G. W. Broecker, J. McManus, and R. Lotti. 1992. The North Atlantic's record of millennial-to-century scale climate changes and their link to Greenland ice core climate records. In NOAA Climate and Global Change Program Special Report No. 2, pp. 55-62.

Bottomley, M., C.K. Folland, J. Hsiung, R.E. Newell, and D.E. Parker. 1990. Global Ocean Surface Temperature Atlas

(GOSTA). Joint project of the Meteorological Office, Bracknell, U.K., and the Massachusetts Institute of Technology. HMSO, London, 20+iv pp. and 313 plates.

Bradley, R.S. (ed.). 1991. Global Changes of the Past. UCAR/ Office for Interdisciplinary Earth Studies, Boulder, Colo., 514 pp.

Bradley, R.S., and P.D. Jones. 1992. Climate since A.D. 1500: Introduction. In Climate Since A.D. 1500. R.S. Bradley and P.D. Jones (eds.). Routledge, London, pp. 1-16.

Branstator, G., 1987. A striking example of the atmosphere's leading traveling mode. J. Atmos. Sci. 44:2310-2323.

Braudel, F. 1949. La Méditerranée et le monde Méditerranéen à l'époque de Philippe II. Colin, Paris, 1160 pp.

Bryan, F. 1986. High latitude salinity effects and inter-hemispheric thermohaline circulations. Nature 323:301-304.

Bryan, K. 1969. Climate and the ocean circulation. III: The ocean model. Mon. Weather Rev. 97:806-827.

Cane, M.A. 1991. Forecasting El Niño with a geophysical model. In ENSO Teleconnections Linking Worldwide Climate Anomalies: Scientific basis and societal impacts. M. Glantz, R. Katz, and M. Nicholls (eds.). Cambridge University Press, pp. 345-369.

Cane, M.A., S. Dolan, and S.E. Zebiak. 1986. Experimental forecasts of El Niño. Nature 321:827-832.

Cane, M.A., M. Münnich, and S.E. Zebiak. 1990. A study of self-excited oscillations of the tropical ocean-atmosphere system. Part I: Linear analysis. J. Atmos. Sci. 47:1853-1863.

Cane, M.A., S.E. Zebiak, and Y. Xue. 1995. Model studies of the long-term behavior of ENSO. In Natural Climate Variability on Decade-to-Century Time Scales. D.G. Martinson, K. Bryan, M. Ghil, M.M. Hall, T.R. Karl, E.S. Sarachik, S. Sorooshian, and L.D. Talley (eds.). National Academy Press, Washington, D.C.

Cayan, D.R. 1992. Latent and sensible heat flux anomalies over the northern oceans: The connection to monthly atmospheric circulation. J. Climate 5:354-369.

Cayan, D.R., and D.H. Peterson. 1989. The influence of North Pacific atmospheric circulation on streamflow in the West. Geophys. Monogr. 55:375-397.

Chao, Y., and S.G.H. Philander. 1993. On the structure of the southern oscillation. J. Climate 6:450-469.

Chatfield, C. 1989. The Analysis of Time Series: An Introduction. Chapman and Hall, London, 241 pp.

CLIMAP Project Members. 1981. Seasonal reconstruction of the Earth's surface at the last glacial maximum. Geol. Soc. Am. Map Chart Ser. MC-36.

Cole, J.E., R. Fairbanks, and G.T. Shen. 1995. Monitoring the tropical ocean-atmosphere using chemical records from long-lived corals. In Natural Climate Variability on Decade-to-Century Time Scales. D.G. Martinson, K. Bryan, M. Ghil, M.M. Hall, T.R. Karl, E.S. Sarachik, S. Sorooshian, and L.D. Talley (eds.). National Academy Press, Washington, DC.

Committee on Earth Sciences. 1989. Our Changing Planet: The FY 1990 Research Plan. The United States Global Change Research Program. OSTP FCCSET, Executive Office of the President.

Conover, W.J. 1980. Practical Non-Parametric Statistics. John Wiley and Sons, Inc., New York.

Crowley, T.J., and G.R. North. 1991. Paleoclimatology. Oxford

Monographs on Geology and Geophysics, No. 16. Oxford Press, Oxford, 339 pp.

Cubash, U., K. Hasselmann, H. Hock, E. Maier-Reimer, U. Mikolajewicz, B.D. Santer, and R. Sausen. 1991. Time dependent greenhouse warming computations with a coupled ocean-atmosphere model. Report 67, Max Planck-Institut für Meteorologie, Hamburg, 18 pp.

D'Arrigo, R.D., G.C. Jacoby, and E.R. Cook. 1992. Decadal-scale variability of the North Atlantic atmosphere-ocean system as revealed by dendroclimatic evidence from surrounding land areas. In NOAA Climate and Global Change Program Special Report No. 2, pp. 45-53.

Davis, R. 1976. Predictability of sea surface temperature and sea level pressure anomalies over the North Pacific Ocean. J. Phys. Oceanogr. 6:249-266.

Davis, R. 1978. Predictability of sea-level pressure anomalies over the North Pacific Ocean. J. Phys. Oceanogr. 8:233-246.

Delworth, T., S. Manabe, and R.J. Stouffer. 1993. Interdecadal variations of the thermohaline circulation in a coupled ocean-atmosphere model. J. Climate 6:1993-2011.

Delworth, T.L., S. Manabe, and R.J. Stouffer. 1995. North Atlantic interdecadal variability in a coupled model. In Natural Climate Variability on Decade-to-Century Time Scales. D.G. Martinson, K. Bryan, M. Ghil, M.M. Hall, T.R. Karl, E.S. Sarachik, S. Sorooshian, and L.D. Talley (eds.). National Academy Press, Washington, D.C.

Deser, C., and M. Blackmon. 1991. Analysis of decadal climate variations over the Atlantic basin. In Proceedings of the Fifth Conference on Climate Variations. American Meteorological Society, Boston, pp. 470-471.

Deser, C., and M.L. Blackmon. 1995. Surface climate variations over the North Atlantic ocean during winter: 1900-1989. In Natural Climate Variability on Decade-to-Century Time Scales. D.G. Martinson, K. Bryan, M. Ghil, M.M. Hall, T.R. Karl, E.S. Sarachik, S. Sorooshian, and L.D. Talley (eds.). National Academy Press, Washington, D.C.

Deser, C., and J.M. Wallace. 1987. El Niño events and their relationship to the Southern Oscillation: 1925-86. J. Geophys. Res. 92:14189-14196.

Diaz, H.F., and V. Markgraf (eds.). 1992. El Niño: Historical and Paleoclimatic Aspects of the Southern Oscillation. Cambridge University Press, 440 pp.

Douglas, A.V., D.R. Cayan, and J. Namias. 1982. Large-scale changes in North Pacific and North American weather patterns in recent decades. Mon. Weather Rev. 112:1851-1862.

Ebbesmeyer, C.C., D.R. Cayan, D.R. McLain, F.H. Nichols, D.H. Peterson, and K.T. Redmond. 1991. 1976 step in the Pacific climate: Forty environmental changes between 1968-1975 and 1977-1984. In Proc. Seventh Annual Pacific Climate (PACLIM) Workshop, April 1990. J.L. Betancourt and V.L. Sharp (eds.). Interagency Ecological Studies Program Tech. Rept. 26, California Department of Water Research, Sacramento, Calif., pp. 129-141.

Elliot, W.P., and J.K. Angell. 1988. Evidence for changes in Southern Oscillation relationships during the last 100 years. J. Climate 1:729-737.

Enfield, D.B. 1992. Historical and prehistorical overview of El Niño: Historical and paleoclimatic aspects of the Southern Oscillation. In El Niño: Historical and Paleoclimatic Aspects of the Southern Oscillation. H.F. Diaz and V. Markgraf (eds.). Cambridge University Press, Cambridge, U.K., pp. 95-117.

Enfield, D.B., and L. Cid. 1991. Low frequency changes in El Niño/Southern Oscillation. J. Climate 12:1137-1146.

Fairbanks, R.G., J. Cole, M. Moore, L. Wells, and G. Shen. 1992. The variance spectra of the Southern Oscillation under different climatological boundary conditions. In Paleo-ENSO Records International Symposium (Extended Abstracts). L. Ortlieb and J. Macharé (eds.). OSTROM-CONCYTEC, Lima, Peru, p. 101.

Folland, C., J. Owen, M.N. Ward, and A. Colman. 1991. Prediction of seasonal rainfall in the Sahel region using empirical and dynamical methods. J. Forecasting 10:21-56.

Frankignoul, C. 1985. Sea surface temperature anomalies, planetary waves, and air-sea feedbacks in middle latitudes. Rev. Geophys. 8:233-246.

Frankignoul, C., and K. Hasselmann. 1977. Stochastic climate models, Part II: Application to sea-surface temperature anomalies and thermocline variability. Tellus 29:289-305.

Glantz, M., R. Katz, and N. Nicholls. 1991. ENSO Teleconnections Linking Worldwide Climate Anomalies: Scientific basis and societal impacts. Cambridge University Press.

Gordon, C. 1989. Tropical-ocean-atmosphere interactions in a coupled model. Philos. Trans. Roy. Soc. London A329:207-223.

Gordon, A.L., S.E. Zebiak, and K. Bryan. 1992. Climate variability and the Atlantic Ocean. EOS 73:161-165.

Hamilton, K. 1987. Interannual environmental variation and North American fisheries. Bull. Am. Meteorol. Soc. 68:1541-1548.

Handler, P., and K. Andsager. 1990. Volcanic aerosols, El Niño, and the southern oscillation. J. Climatol. 10:413-424.

Hibler III, W.D., and K. Bryan. 1987. A diagnostic ice-ocean model. J. Phys. Oceanogr. 17:987-1015.

Hirst, A.C. 1988. Slow instabilities in tropical ocean basin-global atmosphere models. J. Atmos. Sci. 45:830-852.

Hocquenghem, A.M., and L. Ortlieb. 1992. Historical record of El Niño events in Peru (XVI-XVIII centuries): The Quinn et al. (1987) chronology revisited. In Paleo-ENSO Records International Symposium (Extended Abstracts). L. Ortlieb and J. Macharé (eds.). OSTROM-CONCYTEC, Lima, Peru, pp. 143-150.

Horel, J.D., and J.M. Wallace. 1981. Planetary-scale atmospheric phenomena associated with the Southern Oscillation. Mon. Weather Rev. 109:813-829.

IGBP. 1992. PAGES. J.A. Eddy (ed.). Global Change Report 19, International Geosphere-Biosphere Programme, Stockholm.

IPCC. 1990. Climate Change: The IPCC Scientific Assessment. J.T. Houghton, G.J. Jenkins, and J.J. Ephraums (eds.). Prepared for the Intergovernmental Panel on Climate Change by Working Group I. WMO/UNEP, Cambridge University Press, 365 pp.

IPCC. 1992. Climate Change 1992: The Supplementary Report to the IPCC Scientific Assessment. J.T. Houghton, B.A. Callander, and S.K. Varney (eds.). Prepared for the Intergovernmental Panel on Climate Change by Working Group I. WMO/UNEP, Cambridge University Press, 200 pp.

James, I.N., and P.M. James. 1989. Ultra-low frequency variability in a simple atmospheric model. Nature 342:53-55.

Jin, F.-F., and J.D. Neelin. 1992. Modes of interannual tropical

ocean-atmosphere interaction—a unified view. Part I: Numerical results. J. Atmos. Sci. 50:3477-3503.

Jones, P.D. 1988. Hemispheric surface air temperature variations: Recent trends and an update to 1987. J. Climate 1:654-660.

Keigwin, L.D., and E.A. Boyle. 1992. Centry and millennial-scale changes in North Atlantic surface and deep waters during the past 80,000 years. In NOAA Climate and Global Change Program Special Report No. 2, pp. 63-66.

Keppenne, C.L., and M. Ghil. 1992. Adaptive filtering and prediction of the Southern Oscillation Index. J. Geophys. Res. 97: 20449-20454.

Kushnir, Y. 1994. Interdecadal variations in North Atlantic sea surface temperature and associated atmospheric conditions. J. Climate 7: 141-157.

Kushnir, Y. and N.-C. Lau. 1992. The general circulation model response to a North Pacific SST anomaly: Dependence on time scale and pattern polarity. J. Climate 5:271-283.

Lanzante, J.R. 1984. A rotated eigenanalysis of the correlation between 700-mb heights and sea surface temperatures in the Pacific and Atlantic. Mon. Weather Rev. 112:2270-2280.

Latif, M., J. Biercamp, H. von Storch, and F.W. Zwiers. 1988. A ten-year climate simulation with a coupled ocean-atmosphere general circulation model. Report 21, Max Planck-Institut für Meteorologie, Hamburg.

Latif, M., A. Sterl, E. Maier-Reimer, and M.M. Junge. 1991. Climate variability in a coupled GCM. Part I: The tropical Pacific. J. Climate 6:5-21.

Latif, M., A. Sterl, E. Maier-Reimer, and M.M. Junge. 1993. Structure and predictability of the El Niño/Southern Oscillation phenomenon. J. Climate 6:700-708.

Lau, N.-C. 1988. Variability of the observed midlatitude storm tracks in relation to low frequency changes in the circulation pattern. J. Atmos. Sci. 45:2718-2743.

Lau, N.-C., and M.J. Nath. 1990. A general circulation model study of the atmospheric response to extratropical SST anomalies observed in 1950-79. J. Climate 3:965-989.

Lazier, J. 1988. Temperature and salinity changes in the deep Labrador Sea, 1962-1086. Deep-Sea Res. 35:1247-1253.

Leathers, D.J., and M.A. Palecki. 1992. The Pacific North American teleconnection pattern and United States climate: 29 temporal characteristics and index specification. J. Climate 5:707-716.

Levitus, S. 1989a. Interpentadal variability of temperature and salinity at intermediate depths of the North Atlantic Ocean, 1970-1974 versus 1955-1959. J. Geophys. Res. 94:6091-6131.

Levitus, S. 1989b. Interpentadal variability of salinity in the upper 150 m of the North Atlantic Ocean. J. Geophys. Res. 94:9679-85.

Lorenz, E.N. 1963. Deterministic nonperiodic flow. J. Atmos. Sci. 20:130-141.

Lough, J.M., and H.C. Fritts. 1985. The Southern Oscillation and tree rings: 1600-1961. J. Clim. Appl. Meteorol. 24:952-966.

Ma, C.-C., C.R. Mechoso, A. Arakawa, and J.D. Farrara. 1994. Sensitivity of a coupled ocean-atmosphere model to physical parameterizations. J. Climate 7:1883-1896.

Madden, R.A., and P. Speth, 1989. The average behavior of large-scale westward traveling disturbances evident in the Northern Hemisphere geopotential heights. J. Atmos. Sci. 46:3225-3239.

Manabe, S. 1969a. Climate and the ocean circulation. I: The atmospheric circulation and the hydrology of the earth's surface. Mon. Weather Rev. 97:739-774.

Manabe, S. 1969b. Climate and the ocean circulation. II: The atmospheric circulation and the effect of heat transfer by ocean currents. Mon. Weather Rev. 97:806-827.

Manabe, S., and R.J. Stouffer. 1988. Two stable equilibria of a coupled ocean-atmosphere model. J. Climate 1:841-866.

Manabe, S., R.J. Stouffer, M.J. Spelman, and K. Bryan. 1991. Transient response of a coupled ocean-atmosphere model to gradual changes of atmospheric CO_2. Part I: Annual mean response. J. Climate 4:785-818.

Manabe, S., M.J. Spelman, and R.J. Stouffer. 1992. Transient response of a coupled ocean-atmosphere model to gradual changes of atmospheric CO_2. Part II: Seasonal response. J. Climate 5:105-126.

Manak, D.K., and L.A. Mysak. 1987. Climatic atlas of arctic sea ice extent and anomalies, 1953-1984. Climate Research Group Report 87-8. McGill University, Montréal, 214 pp.

Mechoso, C.R., C.-C. Ma, J.D. Farrara, and J. Spahr. 1991. Simulations of interannual variaiblity with a coupled atmosphere-ocean GCM. In Proceedings of Fifth Conference on Climate Variations. American Meteorological Society, Boston, pp. J1-J4.

Mechoso, C.R., A.W. Robertson, N. Barth, M.K. Davey, P. Delecluse, P.R. Gent, S. Ineson, B. Kirtman, M. Latif, H. Le Treut, T. Nagai, J.D. Neelin, S.G.H. Philander, J. Polcher, P.S. Schopf, T. Stockdale, M.J. Suarez, L. Terray, O. Thual, and J.J. Tribbia. 1995. The seasonal cycle over the tropical Pacific in coupled ocean-atmosphere general circulation models. Mon. Weather Rev. 123(9):2825-2838.

Michaelsen, J. 1989. Long-period fluctuations in El Niño amplitude and frequency reconstructed from tree rings. In Aspects of Climate Variability in the Pacific and the Western Americas. D.H. Peterson (ed.). Geophysical Monograph 55, American Geophysical Union, Washington, D.C., pp. 69-74.

Mikolajewicz, U., and E. Maier-Reimer. 1990. Internal secular variability in an ocean general circulation model. Climate Dyn. 4:145-156.

Münnich, M., M.A. Cane, and S.E. Zebiak. 1991. A study of self-excited oscillations of the tropical ocean-atmosphere system. Part II: Nonlinear cases. J. Atmos. Sci. 48:1238-1248.

Mysak, L.A. 1986. El Niño interannual variability and fisheries in the northeast Pacific Ocean. Can. J. Fish. Aquatic Sci. 43:464-497.

Nagai, T., T. Tokioka, M. Endoh, and Y. Kitamura. 1991. El Niño/Southern Oscillation simulated in an MRI atmosphere-ocean coupled general circulation model. J. Climate 5:1202-1233.

Namias, J. 1959. Recent seasonal interactions between North Pacific waters and the overlying atmospheric circulation. J. Geophys. Res. 64:631-646.

Namias, J. 1963. Large-scale air-sea interactions over the North Pacific from summer 1962 through the subsequent winter. J. Geophys. Res. 68:6171-6186.

Namias, J. 1969. Seasonal interactions between the North Pacific Ocean and the atmosphere during the 1960s. Mon. Weather Rev. 97:173-192.

Namias, J., X. Yuan, and D.R. Cayan. 1988. Persistence of North Pacific sea surface temperature and atmospheric flow patterns. J. Climate 1:682-703.

Neelin, J.D., M. Latif, M.A.F. Allaart, M.A. Cane, U. Cubasch, W.L. Gates, P.R. Gent, M. Ghil, N.C. Lau, C.R. Mechoso, G.A. Meehl, J.M. Oberhuber, S.G.H. Philander, P.S. Schopf, K.R. Sperber, A. Sterl, T. Tokioka, J. Tribbia, and S.E. Zebiak. 1992. Tropical air-sea interaction in general circulation models. Climate Dynamics 7:73-104.

Nicholls, N. 1988. Low Latitude Volcanic Eruptions and the El Niño Oscillation. J. Climatol. 8:91-95.

Nicholson, S.E. 1985. Sub-Saharan rainfall 1981-84. J. Clim. Appl. Meteorol. 24:1388-1391.

Nitta, T., and S. Yamada. 1989. Recent warming of tropical sea surface temperature and its relationship to the Northern Hemisphere circulation. J. Meteorol. Soc. Japan 67:375-383.

NOAA. 1990. The Atlantic Climate Change Program Science Plan. NOAA Climate and Global Change Program Special Report No. 2, UCAR, Boulder, Colorado, 29 pp.

NOAA. 1992. Proceedings from the Principal Investigators' Meeting of the ACCP. NOAA Climate and Global Change Program Special Report No. 7, UCAR, Boulder, Colorado.

Ortlieb, L., and J. Macharé (eds.). 1992. Paleo-ENSO Records International Symposium (Extended Abstracts). OSTROM-CONCYTEC, Lima, Peru, 333 pp.

Pan, Y.H., and A.H. Oort. 1990. Correlation analyses between sea surface temperature anomalies in the eastern equatorial Pacific and the world ocean. Climate Dynamics 4:191-205.

Parker, D.E., C.K. Folland, A.C. Bevan, M.N. Ward, M. Jackson, and K. Maskell. 1995. Marine surface data for analysis of climatic fluctuations on interannual-to-century time scales. In Natural Climate Variability on Decade-to-Century Time Scales. D.G. Martinson, K. Bryan, M. Ghil, M.M. Hall, T.R. Karl, E.S. Sarachik, S. Sorooshian, and L.D. Talley (eds.). National Academy Press, Washington, DC.

Parthsarathy, B., K. Rupa Kumar, and A.A. Munot. 1991. Evidence of secular variations in Indian monsoon rainfall-circulation relationships. J. Climate 4:927-938.

Philander, S.G.H., W.J. Hurlin, and A.D. Seigel. 1987. Simulation of the seasonal cycle of the tropical Pacific Ocean. J. Phys. Oceanogr. 17:1986-2002.

Philander, S.G.H., R.C. Pacanowski, N.C. Lau, and M.J. Nath. 1992. A simulation of the Southern Oscillation with a global atmospheric GCM coupled to a high-resolution, tropical Pacific Ocean GCM. J. Climate 5:308-329.

Phillips, N.A. 1956. The general circulation of the atmosphere: A numerical experiment. Quart. J. Roy. Meteorol. Soc. 82:123-164.

Pitcher, E.J., M.L. Blackmon, G.T. Bates, and S. Muñoz. 1988. The effects of North Pacific sea surface temperature anomalies on the January climate of a general circulation model. J. Atmos. Sci. 45:173-188.

Quinn, W.H., V.T. Neal, and S. Antunez de Mayolo. 1987. El Niño occurrences over the past four and a half centuries. J. Geophys. Res. 92:14449-14461.

Rabiner, L.R., M.R. Sambar, and C.E. Schmidt. 1975. Applications of nonlinear smoothing algorithm to speech processing. IEEE Trans. Acoust. Speech Signal Process. 23:552-557.

Rasmusson, E.M., and P.A. Arkin. 1993. A global view of large-scale precipitation variability. J. Climate 6:1495-1522.

Rasmusson, E.M., and T.H. Carpenter. 1982. Variations in tropical sea surface temperature and surface wind fields associated with the Southern Oscillation/El Niño. Mon. Weather Rev. 110:354-384.

Rasmusson, E.M., X. Wang, and C.F. Ropelewski. 1990. The biennial component of ENSO variability. J. Mar. Syst. 1:70-96.

Rasmusson, E.M., X. Wang, and C.F. Ropelewski. 1995. Secular variability of the ENSO cycle. In Natural Climate Variability on Decade-to-Century Time Scales. D.G. Martinson, K. Bryan, M. Ghil, M.M. Hall, T.R. Karl, E.S. Sarachik, S. Sorooshian, and L.D. Talley (eds.). National Academy Press, Washington, D.C.

Rogers, J.C. 1984. The association between the North Atlantic oscillation and the southern oscillation in the northern hemisphere. Mon. Weather Rev. 112:1999-2015.

Rogers, J.C., and M.N. Raphael. 1992. Meridional eddy sensible heat fluxes in the extremes of the Pacific/North American teleconnection pattern. J. Climate 5:127-139.

Rogers, J.C., and R.V. Rohli. 1991. Florida citrus freezes and polar anticyclones in the Great Plains. J. Climate 4:1103-1113.

Ropelewski, C.F., and M.S. Halpert. 1987. Global and regional scale precipitation patterns associated with the El Niño/Southern Oscillation. Mon. Weather Rev. 115:1606-1626.

Ropelewski, C.F., and M.S. Halpert. 1989. Precipitation patterns associated with the high index phase of the Southern Oscillation. J. Climate 2:268-284.

Ropelewski, C.F., and P.D. Jones. 1987. An extension of the Tahiti-Darwin Southern Oscillation Index. Mon. Weather Rev. 115:2161-2165.

Ropelewski, C.F., M.S. Halpert, and X. Wang. 1992. Observed tropospheric biennial variability and its relationship to the Southern Oscillation. J. Climate 5:594-614.

Royer, T.C. 1989. Upper ocean temperature variability in the Northeast Pacific Ocean: Is it an indicator of global warming? J. Geophys. Res. 94:18175-18183.

Schopf, P.S., and M.J. Suarez. 1988. Vacillations in a coupled ocean-atmosphere model. J. Atmos. Sci. 45:549-566.

Semtner, A.J., Jr., and R.M. Chervin. 1988. A simulation of the global ocean circulation with resolved eddies. J. Geophys. Res. 93:15502-15522.

Shen, G.T., L.J. Linn, M.T. Price, J.T. Cole, R.G. Fairbanks, D.W. Lea, and T.A. McConnaughey. 1992. Paleochemistry of reef corals: Historical variability of the tropical Pacific. In Paleo-ENSO Records International Symposium (Extended Abstracts). L. Ortlieb and J. Macharé (eds.). OSTROM-CONCYTEC, Lima, Peru, pp. 287-294.

Shukla, J., and D.A. Paolino. 1983. The Southern Oscillation and long-range forecasting of the summer monsoon rainfall over India. Mon. Weather Rev. 111:1830-1837.

Stommel, H. 1961. Thermohaline convection with two stable regimes of flow. Tellus 13:224-230.

Tans, P., T. Conway, and T. Nakazawa. 1989. Latitudinal distribution of the sources and sinks of atmospheric carbon dioxide derived from surface observations and an atmospheric transport model. J. Geophys. Res. 94:5151-5172.

Thompson, L.G. 1992. Reconstructing the paleo-ENSO records from tropical and subtropical ice cores. In Paleo-ENSO Records International Symposium (Extended Abstracts). L. Ortlieb and J. Macharé (eds.). OSTROM-CONCYTEC, Lima, Peru, p. 311.

Thompson, P.D., E. Mosley-Thompson, and B.M. Brano. 1984.

El Niño-Southern Oscillation events recorded in the Quelccaya Ice Cap, Peru. Science 226:50-53.

Ting, M. 1991. The stationary wave response to a midlatitude SST anomaly in an idealized GCM. J. Atmos. Sci. 48:1249-1275.

Toggweiler, J.R., K. Dixon, and K. Bryan. 1989. Simulations of radiocarbon in a coarse-resolution world ocean model. II: Distributions of bomb-produced ^{14}C. J. Geophys. Res. 94:8243-8264.

Trenberth, K.E. 1976. Spatial and temporal variations of the Southern Oscillation. Quart. J. Roy. Meteorol. Soc. 102:639-653.

Trenberth, K.E. 1984. Signal versus noise in the Southern Oscillation. Mon. Weather Rev. 112:326-332.

Trenberth, K.E. 1990. Recent observed interdecadal climate changes in the Northern Hemisphere. Bull. Am. Meteorol. Soc. 71:988-993.

Trenberth, K.E. 1991. Recent climate changes in the Northern Hemisphere. In Greenhouse-Gas-Induced Climate Change: A Critical Appraisal of Simulations and Observations. M. Schlesinger (ed.). Elsevier, Amsterdam/Oxford/New York, pp. 377-390.

Trenberth, K.E., and J.W. Hurrell. 1994. Decadal atmosphere-ocean variations in the Pacific. Climate Dynamics 9:303-319.

Trenberth, K.E., and J.W. Hurrell. 1995. Decadal climate variations in the Pacific. In Natural Climate Variability on Decade-to-Century Time Scales. D.G. Martinson, K. Bryan, M. Ghil, M.M. Hall, T.R. Karl, E.S. Sarachik, S. Sorooshian, and L.D. Talley (eds.). National Academy Press, Washington, D.C.

Trenberth, K.E., and D.A. Paolino. 1980. The Northern Hemisphere sea-level pressure data set: Trends, errors, and discontinuities. Mon. Weather Rev. 108:855-872.

Trenberth, K.E., and D.J. Shea. 1987. On the evolution of the Southern Oscillation. Mon. Weather Rev. 115:3078-3096.

Trenberth, K.E., J.R. Christy, and J.W. Hurrell. 1992. Monitoring global monthly mean surface temperatures. J. Climate 5:1405-1423.

Troup, A.J. 1965. The "southern oscillation". Quart. J. Roy. Meteorol. Soc. 91:490-506.

van Loon, H. 1979. The association between latitudinal temperature gradient and eddy transport. Pt. I: Transport of sensible heat in winter. Mon. Weather Rev. 107:525-534.

Vautard, R., and M. Ghil. 1989. Singular spectrum analysis in nonlinear dynamics, with application to paleoclimatic time series. Physica 35D:395-424.

Vautard, R., P. Yiou, and M. Ghil. 1992. Singular spectrum analysis: A tool kit for short, noisy chaotic signals. Physica 58D:95-126.

Venrick, E.L., J.A. McGowan, D.A. Cayan, and T.L. Hayward. 1987. Climate and chlorophyll-a: Long-term trends in the central North Pacific Ocean. Science 238:70-72.

Vinnikov, K.Ya., P.Ya. Groisman, and K.M. Lugina. 1990. Empirical data on contemporary global climate changes (temperature and precipitation). J. Climate 3:662-677.

Walker, G.T. 1924. Correlation in seasonal variations of weather, IX: A further study of World Weather. Mem. Indian Meteorol. Dept. 24:275-332.

Walker, G.T., and E.W. Bliss. 1932. World Weather V. Mem. Roy. Meteorol. Soc. 4:53-84.

Wallace, J.M., and D.S. Gutzler. 1981. Teleconnections in the geopotential height field during the northern hemisphere winter. Mon. Weather Rev. 109:784-812.

Wallace, J.M., and Q. Jiang. 1987. On the observed structure of the interannual variability of the atmosphere-ocean climate system. In Atmospheric and Oceanic Variability. H. Cattle (ed.). Roy. Meteorol. Soc., Bracknell, Berkshire, pp. 17-43.

Wallace, J.M., C. Smith, and Q. Jiang. 1990. Spatial patterns of atmosphere-ocean interaction in the northern winter. J. Climate 3:990-998.

Wallace, J.M., C. Smith, and C.S. Bretherton. 1992. Singular value decomposition of wintertime sea surface temperature and 500-mb height anomalies. J. Climate 5:561-576.

Walsh, J.E., W.D. Hibler III, and B. Ross. 1985. Numerical simulation of Northern Hemisphere sea ice variability, 1951-1980. J. Geophys. Res. 90:4847-4865.

Washington, W.M., and G.A. Meehl. 1989. Climate sensitivity due to increased CO_2: Experiments with a coupled atmosphere and ocean general circulation model. Climate Dynamics 4:1-38.

Weaver, A.J., and E.S. Sarachik. 1991a. The role of mixed boundary conditions in numerical models of the ocean's climate. J. Phys. Oceanogr. 21:1470-1493.

Weaver, A.J., and E.S. Sarachik. 1991b. Evidence for decadal variability in an ocean general circulation model: An advective mechanism. Atmos.-Ocean 29:197-231.

Weaver, A.J., E.S. Sarachik, and J. Marotzke. 1991. Internal low frequency variability of the ocean's thermohaline circulation. Nature 353:836-838.

Winton, M., and E.S. Sarachik. 1993. Thermohaline oscillations induced by strong steady salinity forcing of ocean general circulation models. J. Phys. Oceanogr. 23:1389-1410.

Woodruff, S.D., R.J. Slutz, R.L. Jenne, and P.M. Steurer. 1987. A comprehensive ocean-atmosphere data set. Bull. Am. Meteorol. Soc. 68:1239-1250.

Wright, P.B., J.M. Wallace, T.P. Mitchell, and C. Deser. 1988. Correlation structure of the El Niño/Southern Oscillation phenomenon. J. Climate 1:609-625.

Zebiak, S.E. 1989. On the 30-60 day oscillation and the prediction of El Niño. J. Climate 2:1381-1387.

Zebiak, S.E., and M.A. Cane. 1985. A theory for El Niño and the Southern Oscillation. Science 228:1085-1087.

Zebiak, S.E., and M.A. Cane. 1987. A model El Niño-Southern Oscillation. Mon. Weather Rev. 115:2262-2278.

Zebiak, S.E., and M.A. Cane. 1991. Natural climate variability in a coupled model. In Greenhouse Gas-Induced Climatic Change: A critical appraisal of simulations and observations. M.E. Schlesinger (ed.). Elsevier, Amsterdam, pp. 457-470.

Zheng, D.-W., and D.-N. Dong. 1986. Realization of narrow band filtering of the polar motion data with multi-stage filter (translated). Acta Astron. Sin. 27:368-376.

5

PROXY INDICATORS OF CLIMATE

Proxy Indicators of Climate: An Essay

SOROOSH SOROOSHIAN AND DOUGLAS G. MARTINSON

INTRODUCTION

Establishing a baseline of natural climate variability over decade-to-century time scales requires a perspective that can be obtained only from a better knowledge of past variability, particularly that which precedes the pre-industrial era. Information revealing these past climate conditions is contained in historical records and "proxy" indicators. The historical records of climate (other than systematic weather observations, which began in the late 1800s), while invaluable because of their scope and often uniquely relevant perspective, are usually limited to the last several hundred years (see Chapter 2). The proxy indicators represent any piece of evidence that can be used to infer climate. Typically, proxy evidence includes the characteristics and constituent compositions of annual layers in polar ice caps, trees, and corals; material stored in ocean and lake sediments (including biological, chemical, and mineral constituents); records of lake levels; and certain historical documents.

Such proxy indicators can provide a wealth of information on past atmospheric compositions, tropospheric aerosol loads, volcanic eruptions, air and sea temperatures, precipitation and drought patterns, ocean chemistry and productivity, sea-level changes, former ice-sheet extent and thickness, and variations in solar activity—among other things. These records are particularly appropriate for detecting three manifestations of climate variability:

- Periodic or near-periodic variations (the latter are those that become evident only after examination of consider-

able data through which a clear statistical signal stubbornly emerges);
- Large and pronounced climate signals, such as severe and sustained droughts, drastically altered precipitation patterns, anomalously warm or cold periods, or floods; and
- Gradual trends, infrequent shifts, or other characteristics of natural variability that are difficult to recognize without the benefit of a long, continuous (or near-continuous) record.

Because the bulk of these proxy indicators are recorded naturally, their time span is potentially unlimited; their resolution and accuracy are limited only by the fidelity of the recorder itself. These "natural archives" are simply there for the taking—awaiting discovery, recovery, means of extraction, and interpretation. Consequently, proxy indicators represent a potential treasure trove of unique past climate information. The use of these data can present difficult problems of interpretation, particularly in light of the scanty spatial and temporal sampling, but can enhance our ability to reconstruct global changes. This chapter provides a state-of-the-art look at several aspects of this relatively new topic with respect to the climate of the last few millennia.

HISTORY OF PROXY INDICATORS

The use of proxy indicators in the study of modern climate draws on a broad range of disciplines and techniques. The potential of tree rings was recognized in the early 1900s, when the 11-year sunspot cycle was found to be recorded in the rings of trees in the southwestern United States. Major advances have been made since then, particularly within the last few decades. Many problems associated with interpretation of the rings (e.g., their causal relationship to climate) have been overcome, and the records have been extended from several hundred years to several thousand years or longer. Tree rings are now invaluable proxy indicators, because of their continuity and remarkable precision. In a similar vein, though more recently, the study of annual rings in massive corals is now yielding information regarding marine climates.

Many of the other proxy indicators, such as isotopes, fossil assemblages, and lake levels, had been used in the past to provide geological evidence in paleoclimate and paleoceanographic studies; they represent one of the standard tools in the geologist's arsenal. The relatively low resolution of the typical geological record had restricted their use chiefly to the study of long-time scale phenomena, such as glacial cycles. However, the desire to answer questions demanding higher-resolution data (for instance, whether the rapid climate excursions observed during the last deglaciation were anomalous, or were characteristic of a major climate transition) spurred improvements in the methods of recovery, analysis, and interpretation, which have yielded higher-resolution data. Improved collection techniques, such as long coring and better drilling methods, have made it possible to acquire the larger samples needed with the higher-resolution deposits; more natural data banks, such as ice caps and high deposition sediments, have been identified; and our understanding of both natural recorders and appropriate extraction techniques has increased, permitting higher precision with smaller samples.

All these developments are beginning to contribute to our knowledge of natural climate variability on decade-to-century time scales. Indeed, proxy indicators are now producing some of the most exciting and valuable records of variability to date. Furthermore, given the relatively embryonic state of the science, they have great potential for contributing to our understanding of the modern climate, particularly over longer time scales. New indicators are constantly being evaluated. For example, the use of biological ecosystems as proxy indicators of climate is demonstrated by McGowan (1995) and Reifsnyder (1995), both of which appear in this chapter. Numerous other proxy recorders—lake sediments, cave deposits, marshes—can now provide significant new insights. These are reviewed more thoroughly in Bradley and Jones (1992). In addition, the tremendous quantity of material in the historic records that is pertinent to climate is becoming increasingly useful, thanks to recent cataloguing that includes important metadata.

OCEANIC PROXY INSIGHTS

Numerous proxy indicators exist in the ocean, representing a wide spectrum of different oceanic and climate variables and spanning a wide range of time scales. Indicators residing in the sea-floor sediments may provide a nearly continuous record, often spanning tens of millions of years. These include pollen, faunal, and floral assemblages that have settled and accumulated, the isotopic compositions of skeletal material and tests from bottom-dwelling or floating organisms, and certain geological deposits and sediment types or compositions.

These ocean proxy records are accessible through sediment coring, drilling, and, in the case of certain bottom dwellers such as isolated corals, dredging. While the degrees of accuracy and resolution available vary, the information that can be extracted is staggering. For example (see the papers of Ruddiman and McIntyre, 1973; CLIMAP, 1981; and Imbrie et al., 1992 for more details), ocean proxy indicators have yielded regional information on the following characteristics: sea-surface and bottom temperature (from fossil assemblage composition), continental and landlocked ice volume and sea-level height (from oxygen isotope ratios), the partitioning of carbon between the land and oceans (from carbon isotope ratios), alkalinity of local water (from fossil preservation indices), deep-water circulation (from relative isotope compositions), surface productivity (from vertical isotope gradients), water-column stability (from radiolarian abundances), major front locations (from fossil, ice-rafted debris, and sediment-type distributions), deep-water temperature changes and surface salinity (from relative isotope compositions), vertical gradients of water-mass properties like temperature or salinity or of water-mass distributions (from analyses of sediments from different depths), deep-water velocity changes and source information (from sediment distributions near restricted passages), and predominant wind directions and intensities (from sediment compositions).

At typical sedimentation rates in the deep ocean basins (about 3 cm per 1000 years), sediment mixing serves as a low-pass filter, limiting the resolution of these proxy indicators to time scales of thousands of years. Regions in which the sediments accumulate faster offer the potential for resolving variations over significantly shorter time scales, but such areas often occur along continental margins where the interpretation of the sediment column is notoriously difficult, due to processes such as mass wasting. Consequently, until recently, these oceanic proxy indicators were used primarily to document and study climate variability on millennial or greater time scales. However, as Lehman (1995) explains in this chapter, areas with high deposition rates and relatively clean sediment records have now been located and sampled. These sample data are providing useful information about natural variability and rapid climate change on time scales as short as decadal.

The paper in this chapter by Cole et al. (1995) describes the shorter, but often high-quality, records that can be found in the chemical composition of corals. Coral records spanning hundreds of years can be pieced together from a single region to provide information regarding sea surface temperature, upwelling, rainfall, and winds. Because corals often grow at rates three orders of magnitude greater than the rate of sediment accumulation in the deep sea, their temporal resolution is exceptionally good, providing high-fidelity records of natural variability on time scales as short as seasonal.

Finally, in addition to the oceanic proxy indicators preserved through time that are discussed above, some researchers have proposed that the spatial and temporal distributions of plankton and fish populations are intimately linked to ocean and climate conditions. The paper in this chapter by McGowan (1995) reviews the current state of our knowledge and discusses the problems in using such data as proxies for ocean/climate variabilities. McGowan's view of the relationships between climate and various proxies contrasts with the findings of Dickson (1995) that appear in Chapter 3. Their differences highlight the uncertainties surrounding this relatively new endeavor and the challenges associated with using highly complex biologic indicators.

ATMOSPHERIC PROXY INSIGHTS

Scientists have been examining a number of noninstrumental atmospheric proxy data sources in search of clues to past climate conditions. Each type of record contains information on one or more aspects of climate. Among these are historical documents (almost all aspects of climate), tree rings (temperature, precipitation, pressure patterns, drought, and runoff), ice cores (temperature, precipitation, atmospheric aerosols, atmospheric composition, and more), lake levels (runoff and drought patterns), and varved sediments (temperature, precipitation, and solar radiation). With the exception of the historical documents, perhaps, the climate information in these proxy indicators must be isolated from the nonclimate portion of the signal and any accompanying noise.

Of the various sources of atmospheric proxy indicators, the greatest attention has so far been given to the use of historical documents, tree rings, and ice cores. (Of these, tree rings have been subjected to the most rigorous testing as sources of information on past climate.) A number of papers in this volume make reference to non-instrumental records, primarily tree rings (Jones and Briffa, 1995; Cook et al., 1995b, and Reifsnyder, 1995, all in this chapter; and Diaz and Bradley, 1995, in Chapter 2.) The uses of ice cores (Grootes, 1995, in this chapter), which reflect past temperatures, and of lake levels (Street-Perrott, 1995, in

this chapter), which reflect overall net moisture supply to the landscape, are presented as well.

The paper in this chapter by Jones and Briffa (1995) sets out to show the potential value of proxy records by looking at dendroclimate reconstructions of summer temperatures of four regions: northern Fennoscandia, the northern Urals, Tasmania, and northern Patagonia. In none of the thousand-year reconstructions did the twentieth century stand out as the warmest century, although it was among the warmest. Jones and Briffa also provide a useful discussion of the potential limitations of single-site dendroclimate reconstruction, and of the difficulties and uncertainties associated with comparison of multi-site records. Similar limitations can be expected to apply to other site-specific proxy indicators, so they represent a general caution about the interpretation of climate proxy records.

Despite the limitations, these long tree-ring records clearly document large and pronounced climate signals, such as the century-long cold period beginning in 1550. Tree-ring evidence is of critical importance in establishing the magnitude and duration of natural climate anomalies, since the instrumental records are too short to do so and also may reflect anthropogenic contamination. In addition, the tree-ring data can provide key information for the interpretation of gradual trends in climate. For example, they show that the period from 1880 to 1910 was anomalously warm throughout much of the Northern Hemisphere. Thus, the apparent magnitude of the present warming trend will depend on when the selected period begins.

The contribution to this chapter of Cook et al. (1995b) examines the power spectrum of a 2290-year reconstruction of warm-season Tasmanian temperatures to detect signs of periodic decadal-scale fluctuations. The paper suggests that the decadal-scale temperature anomalies over Tasmania during the twentieth century, both warm and cold, have been driven in part by long-term climate oscillations. It neither supports nor eliminates greenhouse warming as a possible contributor to the recent temperature increase in Tasmania. However, their study also investigates mechanisms of climate change that are testable (e.g., internal forcing related to deep-water formation, or external forcing by long-term solar variability) and provides specific explanations of synoptic-scale variability (e.g., the expansion and contraction of the circumpolar vortex).

Grootes (1995, in this chapter) discusses the potential role of ice-core records in reconstructing decade-to-century-scale climate variations. He seems confident that the new ice-core records being obtained from the summit of the Greenland ice sheet provide an accurate and remarkably detailed history of changes in climate over the North Atlantic basin as far back as 200,000 years ago. Indeed, there is little doubt as to the potential utility of the high-resolution, fast-response climate records from ice cores. At present, however, they can be obtained only from high-latitude or high-altitude locations, and few of the ice-core reconstructions of temperature have been explicitly calibrated against instrumental time series. The differences between ice-core records from various sites also make it clear that there is a pressing need to identify and differentiate the relative influences of local mesoscale and hemispheric atmospheric factors.

In her paper in this chapter, Street-Perrott (1995) discusses how fluctuations in the water level and surface area of relatively undisturbed lakes (i.e, there has been no human-induced change in the water budget) can provide a measure of climate variability on time scales of months to millions of years. The paper provides examples of the response of tropical lakes to variations in ocean-atmosphere interactions over the Pacific (Lake Pátzcuaro), the Indian Ocean (Lake Victoria), and the Atlantic (Lake Chad, Lake Malawi), and points to paleolimnological evidence for century-scale droughts in southern Africa and the tropical Americas.

Finally, Reifsnyder (1995, in this chapter) analyzes observational and paleoclimate records of temperature in an attempt to determine the realism of models' predicted global-warming rates. Some models predict a rate of warming that is 10 to 40 times faster than the natural warming that followed the last ice age. On the basis of his analysis, however, Reifsnyder argues that the climate change between now and the end of the next century, for a "business as usual" scenario, should take place at no more than twice the maximum rate estimated for changes since the last ice age. He further argues that, because models are known to overpredict by a factor of two the rate of change over the past century in response to the CO_2 increase, the rate of global warming over the next century is in fact unlikely to be higher than any estimated post-ice-age maximum rate—a position which certainly could generate a lot of debate, given the other forcings likely to be operating.

Soil and varved sediments, and historical documents, are two sources of atmospheric proxy data that are not covered in this volume. Sedimentary and soil cores can reveal significant paleoclimate information. For example, in a recently published paper in Science, Weiss et al. (1993) attributed the demise of the Akkadian culture (fl. 3000 B.C.) of southern Mesopotamia (present-day northeast Syria) in 2200 B.C. to an abrupt change in climate. The combined archaeological and soil-stratigraphic data for the area point to a shift in climate toward the arid, and a dry period persisting for about 300 years.

A variety of historical sources, such as ancient inscriptions, personal diaries and correspondence, scientific and quasi-scientific writings, government records, and public and private chronicles and annals record climate events that were seen as having some significance at the time. The information may include observations of weather phenomena, such as the occurrence of extreme rain- and snowfalls, droughts, floods, or lake and river freeze-ups and break-ups. Historical records, unfortunately, fail to give a complete picture of former climate conditions. They are often discontinuous observations, biased towards extreme events. The important long-term trends tend to go unremarked. A good source of discussions of documentary records of past climate is the recently published book by Bradley and Jones (1992).

CURRENT LIMITATIONS

While proxy indicators are extremely important, even crucial, to climate reconstruction, their limitations should be kept in mind. The most significant are:

- It is not always clear whether the signal they record reflects only local conditions, or is representative of regional or global conditions.
- Their accuracy is often unknown or untested.
- They often reflect more than one variable, making interpretation difficult.
- The absolute, and even the relative, timing of the record is often not certain.

Bradley and Eddy (1989) stressed the importance of the last of these limitations, and noted that without accurate dating it is impossible to determine whether certain events occurred synchronously or not. In the Vostok ice core of central East Antarctica, for example, variations in air temperature (Jouzel et al., 1987) and atmospheric CO_2 concentration (Barnola et al., 1987) have been reconstructed for the last 160,000 years. Within the dating uncertainties, these records show striking correspondence between high CO_2 concentrations and warm temperatures. However, whether the increases in CO_2 concentration precede or follow the temperature increases cannot be assessed accurately from these records.

Accurate dating is also a problem in attempting to overcome the first limitation listed above. That is, in order to reconstruct a global picture of climate at a given time, it is necessary to reconstruct climate variables from a number of different locations for that time. The accurate dating needed to establish the relationships of the various records is not always available. This problem plagues most of the ocean records, because distinct annual varves are rarely present. Dating is mostly accomplished through other techniques, again of limited accuracy, which constrains the degree to which comparisons can be made. Sowers et al. (1993) attempted to relate the variations in the Vostok ice record to those in the ocean by correlating supposedly globally synchronous changes in oxygen isotope concentrations. This technique has met with some success, but the range of error precludes comparisons over time scales less than millennial.

Other limitations arise because recording is often discontinuous. For instance, the recording mechanism may be disturbed during or after the recording, or the recovery may be incomplete. This introduces gaps or perturbations that disguise or break the general patterns. An excellent example of the type of inconsistencies that may occur from the different recording properties of different climate proxies is provided by Grootes's (1995) paper in this chapter. However, the promising news in this area is that more researchers are reporting long-term reconstructions from different proxy sources and different geographic locations. The latest is the study by Lara and Villalba (1993) of a 3620-year reconstructed temperature record from alerce trees (*Fitzroya cupressoides*) from southern Chile. Their work, which has shown the alerce tree as being the second longest-living tree after the bristlecone pine, expands the availability of long-term climate records for the Southern Hemisphere.

In Chapter 2 Diaz and Bradley (1995) provide a useful discussion of a number of potential sources of uncertainties and biases associated with climate variable reconstruction, even for the relatively well-understood tree rings. They warn that changes over time in the composition of the tree-ring network used for reconstruction are likely to affect the high-frequency variance and, to some extent, the low-frequency variance as well. Considerable effort has been made to reduce the uncertainties associated with the long-term reconstruction of climate variables from the composition of tree-ring samples. For instance, an index value (temperature, for instance) for a given site is obtained from samples of trees that may vary considerably in age or even be dead. Attention must therefore be given during the reconstruction analysis to getting the chronology right for every tree; removing the biological growth trend; identifying the nature of the climate signal and its strength; replication; and the way in which the analysis method deals with changes in site conditions over the recorded interval. As a result of this careful scrutiny, tree-ring records exhibit exceptionally high fidelity, with comparatively minor interpretational problems. Similar sets of considerations must be addressed for other proxy indicators, no doubt including some not yet identified.

THE FUTURE OF PROXY INDICATORS

It is clear that accurate dating is required to assess the rate at which past climate changes have occurred and to reconstruct globally synchronous records of climate, particularly for changes on time scales of less than a century. The duration of high-frequency, short-term events may be less than the normal error associated with most methods used in dating the proxy records available. The issue of dating thus merits continued attention. Similarly, a sustained effort must be made to identify problems and uncertainties in reconstructions based on proxy indicators and to assess the associated errors; this should come naturally with increased experience in these relatively novel techniques. Last, combining records that have been drawn from different areas and that use different types of indicators into a consistent picture will be crucial for the study and reconstruction of global climate variations.

Finally, as with climate modeling, the reconstruction of past climates requires the employment of a full range of different proxy indicators. These must be developed more fully so that they can be used to intercalibrate the reconstructed climates and cross-check the reconstructions; to provide multiple images of climate through different climate indicators; and to reduce the influence of noise through averaging. In this early stage, we are seeing only the tip of the iceberg; the results of the use of proxy indicators only hint at the climate insights yet to be won by means of these invaluable resources.

Monitoring the Tropical Ocean and Atmosphere Using Chemical Records from Long-Lived Corals

JULIA E. COLE[1], RICHARD G. FAIRBANKS[2], AND GLEN T. SHEN[3]

ABSTRACT

The tropical ocean and atmosphere constitute an active component of the global climate system on interannual-to-century time scales, yet instrumental climate records from this extensive region are scarce, short, and unevenly distributed. Geochemical records from corals offer a promising means of extending the record of tropical climate variability. Corals grow rapidly (more than 1 cm/yr), may live for several centuries, are datable by several independent means, and incorporate the signatures of key atmospheric and oceanic processes in their skeletal chemistry. Efforts to extend coral records beyond the instrumental period and to recover Holocene and deglacial-age samples are under way throughout the tropics. Coral-based reconstructions of the El Niño/Southern Oscillation (ENSO) phenomenon demonstrate how this approach can contribute to understanding past variability in tropical climate dynamics.

The abundance of atolls and coral reefs in the tropical Pacific enables coral studies to target specific features of ENSO, including variability in sea surface temperatures, upwelling, rainfall, and winds. Short records from three sites spanning the Pacific show coherent variations in these parameters associated with warm and cool ENSO extremes. In the Galapagos Islands, a tracer intercomparison study from a single coral head demonstrates high correlations with measured ENSO-related variability between 1936 and 1982, especially for the interannual periods characteristic of ENSO. Farther west, a coral record from Tarawa Atoll closely monitors central-western Pacific rainfall over the past century. Evolutionary spectral analysis of this record suggests a mode of variation between about 1930 and 1950 that differs from preceding or subsequent periods. Over the span of this coral record, variance at the annual cycle has been weakest during the past few decades. Between 1930 and 1950, however, the annual cycle reaches a magnitude comparable to that of the interannual variations, and a strong 3-year component virtually disappears. This shift occurs concurrently with an observed weakening of the Southern Oscillation and its teleconnections.

[1]Institute for Arctic and Alpine Research, University of Colorado, Boulder, Colorado
[2]Lamont-Doherty Earth Observatory and Department of Geological Sciences, Columbia University, New York, New York
[3]School of Oceanography, University of Washington, Seattle, Washington

INTRODUCTION

The tropical ocean and atmosphere constitute an active component of the global climate system, especially on sub-millennial time scales. Understanding the climatic processes that act on these time scales across this extensive region requires that we characterize their past variability. Yet instrumental records from the sparsely inhabited tropical oceans are scarce and short. Those records that extend beyond the past few decades often owe their locations to opportunistic, rather than climatic, rationales. Chemical records from the skeletons of long-lived corals provide one of the few high-quality means for extending the length of the climatic record in the tropical oceans beyond the brief period of instrumental coverage. Coral records offer an unusual opportunity to monitor the history of key tropical climate systems, especially since sites and tracers can be chosen explicitly for their climatic relevance. The development of proxy climate records from coral archives promises to offer important insights into the natural patterns and causes of tropical climate change, and into the sensitivity of low-latitude oceans to shifts in the global boundary conditions of climate. In the tropical Pacific, we have found that short records of isotopic and trace-metal variability in coral skeletons closely track specific ocean-atmosphere processes over the past few decades. An examination of multiple tracers within a single coral specimen demonstrates the varied utility of isotopic and trace-metal variations in tracking interannual sea surface temperature (SST) and upwelling changes. A 96-year record from the equatorial region west of the date line provides an index of ENSO variability comparable in quality to climatological records, and suggests that the spectrum of climate variability in this region has varied over the course of the present century.

Tropical climate variability may take many forms; describing variations in tropical ocean-atmosphere systems thus requires a multivariate approach capable of resolving specific processes. SST changes dramatically over annual- to-interannual periods in regions of upwelling such as the eastern Pacific, eastern Atlantic, and northern Indian oceans, while SST in the western basins of tropical oceans tends to vary much less. Changes in upwelling have important consequences for surface ocean chemistry as well as for temperature. Surface levels of nutrients and other dissolved species enriched in deep water also vary in accordance with the upwelling of deep water to the surface. Advective changes have site-dependent effects on both chemical and thermal distributions. Patterns of rainfall and wind may shift in response to changed SST distribution; they in turn alter the thickness and salinity of the ocean's surface mixed layer, which affect the degree of ocean-atmosphere interaction (Godfrey and Lindstrom, 1989; Lukas and Lindstrom, 1991). Significant changes in SST, surface upwelling, currents, rainfall, and winds may result from interactions within the tropical ocean-atmosphere system. Alternatively, such changes may be forced by continental systems, such as the Asian monsoon. Many chemical tracers in coral skeletons possess distinct sensitivities to specific environmental variables such as SST, rainfall, winds, and upwelling; as with instrumental data, many coral records monitor specific oceanic or atmospheric processes. Reconstructing the behavior of large-scale, multivariate climate systems such as the El Niño/Southern Oscillation (ENSO) will require the integration of many such coral indices over extensive regions (Figure 1).

CORAL RECORDS

Individual coral heads may live for several centuries, offering the potential for climate reconstructions from remote sites that predate even the longest instrumental records. Because they grow at rates of 1 to 2 cm/yr and can be sampled at intervals less than 1 mm, records at sub-monthly resolution are obtainable. Interpretation of such long, high-resolution records demands precise chronologies.

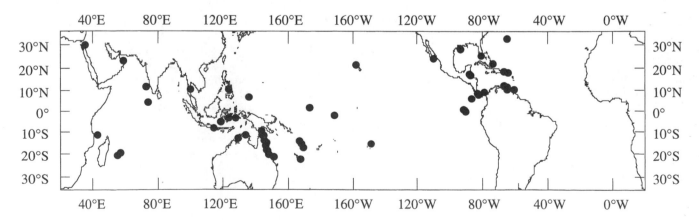

FIGURE 1 Map showing the distribution of sites where living coral heads have been cored for paleoclimatic analysis (after Dunbar and Cole, 1993). Sites with long records are mentioned in the text.

Corals can be dated over a wide range of time scales (from seasons to periods greater than 100,000 years) by a variety of independent means, including density and fluorescent banding, $^{230}Th/^{234}U$ ratio, ^{14}C content, stable-isotope fluctuations, and amino acid racemization (Dodge et al., 1974; Isdale, 1984; Edwards et al., 1988; Fairbanks, 1989, 1990; Bard et al., 1990; Cole and Fairbanks, 1990; Goodfriend et al., 1993). Recent work on monthly and even daily coral banding offers the promise for increasingly refined chronological determinations (Barnes and Lough, 1989; Risk and Pearce, 1992).

Corals precipitate a skeleton of aragonite ($CaCO_3$) that can incorporate several independent chemical tracers used to monitor variability in oceanic and atmospheric processes. The most consistently useful of these are the isotopic ratios of carbon and of oxygen in particular, and the concentration of certain lattice-bound trace metals that appear to substitute for skeletal calcium (expressed as Cd/Ca, Ba/Ca, Mn/Ca, and Sr/Ca ratios). In this section, we present an overview of the many tracers now in use for coral-based paleoclimatic reconstruction. Subsequent sections discuss examples of both short records from a variety of sites and longer records that yield insight into the recent history of tropical climate variability.

Isotopic Indicators

The stable isotopic content of coral aragonite is expressed in parts per thousand (‰) as $\delta^{18}O$ and $\delta^{13}C$, where the δ notation is defined in terms of the isotopic ratios R (either $^{13}C/^{12}C$ or $^{18}O/^{16}O$) in the sample relative to a standard:

$$\delta = [(R_{sample} - R_{standard})/R_{standard}] \times 1000$$

For the oxygen and carbon isotopic content of coralline aragonite, the conventional reference standard is the Pee Dee Belemnite, or PDB.

Coral skeletal $\delta^{18}O$ reflects a combination of local SST and the $\delta^{18}O$ of ambient seawater. The $\delta^{18}O$ of biogenic calcium carbonate precipitated in equilibrium with seawater decreases by about 0.22‰ for every 1°C rise in water temperature (Epstein et al., 1953). In corals, this effect is biologically mediated such that the $\delta^{18}O$ of the coral skeleton is offset below seawater $\delta^{18}O$. This offset is constant within a coral genus (Weber and Woodhead, 1972) for the rapidly growing central axis of the skeleton (Land et al., 1975; McConnaughey, 1989). Coral $\delta^{18}O$ records taken along the axis of maximum growth thus track ambient temperatures at subseasonal resolution (Fairbanks and Dodge, 1979; Dunbar and Wellington, 1981; Pätzold, 1984; McConnaughey, 1989). Coral skeletal $\delta^{18}O$ also records any variations in the $\delta^{18}O$ of the seawater (Epstein et al., 1953; Fairbanks and Matthews, 1979; Swart and Coleman, 1980). Such variations are small in most regions of the tropical ocean, but in some locations changes in evaporation, precipitation, or

runoff may cause pronounced $\delta^{18}O$ variability (Dunbar and Wellington, 1981; Cole and Fairbanks, 1990). In regions with fairly constant or well-known temperature histories, coral $\delta^{18}O$ provides a record of past variations in the hydrologic balance (Cole and Fairbanks, 1990). In many cases, coral skeletal $\delta^{18}O$ reflects a combination of thermal and hydrologic factors.

The interannual $\delta^{13}C$ signal in coral skeletons is often difficult to decipher in environmental terms, because of complicated interactions with biological processes that involve strong isotopic fractionation. Environment-related controls on skeletal $\delta^{13}C$ include (1) the isotopic composition of the ambient seawater (Nozaki et al., 1978), (2) coral geometry and growth rate (e.g., apex versus side of coral head) (Land et al., 1975; McConnaughey, 1989), and (3) photosynthesis of endosymbiotic dinoflagellates (Weber, 1974; Goreau, 1977; Fairbanks and Dodge, 1979; Swart, 1983; McConnaughey, 1989). This last parameter depends upon ambient light levels, as regulated by water depth and insolation. Coral skeletal $\delta^{13}C$ correlates positively with insolation in many contexts, from depth-dependent variation (Weber and Woodhead, 1970; Fairbanks and Dodge, 1979; McConnaughey, 1989) to annual cycles that reflect rainy (i.e., cloudy) seasons (Fairbanks and Dodge, 1979; Pätzold, 1984; McConnaughey, 1989; Cole and Fairbanks, 1990). However, shallow corals may experience reduced photosynthesis during brighter periods, while deeper corals may respond to increased light by increasing photosynthesis (Erez, 1978; McConnaughey, 1989). These responses can produce opposite skeletal $\delta^{13}C$ signatures in deep and shallow corals (McConnaughey, 1989). Environmental reconstruction from coral $\delta^{13}C$ records requires more information about growth conditions and physiological responses than is usually available in a paleoceanographic context.

Trace Metals

Specific environmental processes, including upwelling, advection, aeolian transport, and runoff, influence the surface-water concentrations of certain trace elements (Boyle, 1988; Martin et al., 1976; Shen and Boyle, 1988; Lea et al., 1989; Shen and Sanford, 1990; Shen et al., 1991, 1992b). Several such metals appear to substitute readily for Ca in the aragonite lattice of coral skeletons. Estimated distribution coefficients between corals and seawater allow the reconstruction of ambient seawater metal concentrations from metal concentrations in the coral skeleton. Trace metal records from corals thus yield a history of the processes that control the local distribution of trace metals.

The most useful metals for coral reconstructions of ENSO variability include Cd, Ba, Mn, and Sr. Many authors (Shen and Boyle, 1988; Lea et al., 1989; Linn et al., 1990; Shen and Sanford, 1990; Shen et al., 1987, 1991; Beck et al., 1992; de Villiers et al., 1993) describe these applications in greater detail, including specific techniques for sample cleaning and analysis. Studies of growth rate and species influences show very limited evidence of biological mediation of metal incorporation (de Villiers et al., 1993; G.T. Shen, unpublished results). However, SST may play a role in the incorporation of Ba (Lea et al., 1989), and the dependence of Sr incorporation on ambient SST (Smith et al., 1979) makes possible precise reconstruction of SST from coral Sr records (Beck et al., 1992).

Cadmium

The modern distribution of cadmium follows that of marine nutrients. Low levels in surface waters reflect biological removal, while higher levels at depth result from the regeneration of organic matter (Boyle et al., 1976; Martin et al., 1976). In coral records, the skeletal Cd content usually depends on the balance between Cd-rich upwelled deep water and Cd-poor oligotrophic surface waters. In Galapagos corals, Cd/Ca ratios directly reflect upwelling variations associated with both seasonal cycles (Linn et al., 1990; Shen and Sanford, 1990) and interannual ENSO variability (Shen et al., 1987, 1992a).

Barium

This trace metal exhibits nutrient-like behavior akin to cadmium's (Chan et al., 1977). Higher concentrations in both seawater and corals render coral Ba records less susceptible to contamination. Seasonal-resolution records of Ba/Ca from Galapagos corals reflect regional upwelling variability (Lea et al., 1989). Relative to Cd, Ba may exhibit greater sensitivity to periods of weak upwelling, possibly because the biological uptake of Cd occurs at rates comparable to the slow rate of supply during these times. Ba is also enriched in continental runoff waters, and Ba/Ca records from corals near continental margins may reflect this input (Shen and Sanford, 1990). However, a slight temperature effect may occur upon incorporation of Ba into the coral skeleton, which would complicate paleoclimatic reconstruction from coral Ba/Ca records (Lea et al., 1989).

Manganese

Unlike Cd and Ba, Mn reaches high concentrations in surface waters. A mid-depth maximum coincides with the local O_2-minimum zone, where particulate Mn oxides are reduced and solubilized; concentrations then diminish with increasing depth. Aeolian and fluvial input provide important Mn sources, as do reducing environments where particulate Mn is degraded, such as the O_2-minimum zone, shelf sediments, and lagoons. The interpretation of coral Mn/Ca records is thus usually site-specific, reflecting transport and mixing of water masses with varying Mn levels. In the

Galapagos, for example, seawater Mn levels reflect a combination of aerosol deposition on the surface waters and long-range advection of Mn-enriched surface water from the continental shelf (Shen et al., 1991). In localized reef settings, dissolved Mn in the water column can be augmented by local sediment fluxes, producing very high skeletal concentrations of Mn in corals from the Gulf of Panama and some Caribbean islands (Shen et al., 1991). Diagenetic Mn fluxes from lagoonal sediments offer useful indicators of climate variability at certain Pacific atoll sites far removed from continental sources of Mn (Shen et al., 1992b).

Strontium

Work by Smith et al. (1979) demonstrated that the Sr/Ca ratio of coral skeletons may provide a monitor of past temperature changes. However, the measurement technique used in that study was not sufficiently precise to offer detailed SST reconstructions at a useful resolution for the tropics. The recent application of thermal ionization mass spectrometry (TIMS) to these measurements has greatly improved the precision of Sr/Ca determinations (Beck et al., 1992); recent results indicate that monthly SST can be reconstructed with an apparent accuracy of better than 0.5°C (Beck et al., 1992; de Villiers et al., 1993). SST reconstructions based on skeletal Sr determinations are less susceptible to artifacts associated with hydrologic changes than $\delta^{18}O$-based reconstructions. The paired application of $\delta^{18}O$ and Sr/Ca data may thus allow the separation of SST changes from seawater $\delta^{18}O$ variability.

Paleoceanographic Applications

Many studies have documented the paleoclimatic utility of isotopic and trace-metal measurements by developing short (2-20 yr) data sets that correlate with nearby instrumental records (e.g., Fairbanks and Dodge, 1979; Pätzold, 1984; Shen et al., 1987, 1992b; Carriquiry et al., 1988; Lea et al., 1989; Cole and Fairbanks, 1990; Winter et al., 1991; Cole et al., 1992; Beck et al., 1992). With the growing interest in long records of natural climate variability, many groups are applying these techniques to the development of proxy records that extend beyond the period of local instrumental coverage. Efforts to develop long, high-resolution reconstructions of climate are in progress using coral records from sites throughout the low-latitude oceans. Several studies have focused explicitly on reconstructing interannual-to-decadal variability in the ENSO system; we detail current results from the equatorial Pacific in the following section. Other sites from which coral records extending beyond the present century have been developed include Florida Bay (Kramer et al., 1991), Cebu Island, Philippines (Pätzold and Wefer, 1986), Espiritu Santo Island (Quinn et al., 1992), the southern Great Barrier Reef (Druffel and

Griffin, 1992), Bermuda (Pätzold and Wefer, 1992), and Puerto Rico (A. Winter, pers. commun., 1992). Records from sites spanning the tropics are in progress at many institutions (Figure 1; see also Dunbar and Cole, 1993).

CORAL MONITORS OF ENSO VARIABILITY

The ENSO phenomenon consists of the large-scale oscillation of the tropical Pacific ocean-atmosphere system between two extreme states, characterized by alternately warm and cool SST anomalies in the eastern Pacific (Philander, 1990; Deser and Wallace, 1990). Both warm- and cool-phase ENSO conditions exhibit coherent patterns of coupled oceanic and atmospheric anomalies that leave distinct signatures in the isotopic and trace-metal content of coral skeletons (Figure 2). The influence of ENSO dominates interannual climate variability throughout the equatorial Pacific and the global tropics, and ENSO-related climate anomalies propagate to higher latitudes via mechanisms that include the displacement of upper atmospheric pressure patterns and the generation of troughs that penetrate into southwestern North America (van Loon and Rogers, 1981; Rasmusson and Wallace, 1983; Horel and Wallace, 1981).

The cool phase of ENSO is characterized by strong easterly trade winds at the surface, a convective maximum (the Indonesian Low) over Indonesia and northern Australia, and return flow aloft that brings dry subsiding air over most of the eastern and central Pacific. This zonal atmospheric pattern that spans the tropical Pacific is known as the Walker circulation. The strong trade winds during this phase of ENSO intensify the upwelling of cool, nutrient-rich waters in the eastern Pacific and transport surface waters westward, generating a strong zonal SST gradient and an eastward slope in sea level.

Transition to the warm phase of ENSO may occur when the trade winds relax or reverse west of the date line, depending on surface ocean conditions. As the warm phase of ENSO begins, the western Pacific warm pool spreads eastward (Lukas et al., 1984; McPhaden and Picaut, 1990), and the Indonesian Low convective system migrates to the region near the date line and the equator. In the eastern Pacific, surface waters become warm and oligotrophic, as the thermocline deepens and warm waters move in from the west. Trade winds remain weak and variable as a result of the diminished zonal SST gradient. Dramatic shifts in precipitation patterns occur across the entire tropical Pacific, from South America to southeast Asia, in response to Indonesian Low migration and the development of convection over newly warmed ocean regions.

Frequency-domain analyses of climate data that span the past four decades indicate that ENSO operates on three fundamental time scales. The annual cycle sets the phasing of ENSO anomalies; the evolution of ENSO extremes appears phase-locked to the annual cycle (Rasmusson and Carpenter, 1982). Several aspects of ENSO variability, including SST, winds, rainfall, and sea level pressure (SLP), also possess a quasi-biennial pulse that varies in intensity throughout the instrumental record (Trenberth, 1980; Rasmusson et al., 1990; Barnett, 1991; Ropelewski et al., 1992). Finally, ENSO warm extremes recur approximately every 3-7 years, lending a "low-frequency" beat to the spectrum of ENSO variability (Rasmusson and Carpenter, 1982; Rasmusson et al., 1990; Barnett, 1991; Ropelewski et al., 1992). Recent analyses of multi-decadal records of Pacific climate suggest that ENSO variability may also experience significant decadal-scale shifts (Elliott and Angell, 1988; Cooper et al., 1989; Trenberth, 1990). Such a shift is evident over the interval between 1977 and 1988, which exhibited no strong cool anomalies (Trenberth, 1990; Kerr, 1992).

The impacts of the Southern Oscillation both within and beyond the tropical Pacific fluctuate on a multi-decadal time scale (Trenberth and Shea, 1987; Elliott and Angell, 1988; Cole et al., 1993; Michaelsen and Thompson, 1992; Rasmusson et al., 1995, in this volume). During the first 20 years of this century, for example, the interannual pulse of the Southern Oscillation was strong and highly correlated with climatic records both within the tropical Pacific and in sensitive teleconnected sites. The subsequent three decades, however, witnessed a general decline in the correlations among Pacific climate variables and between Pacific climate and teleconnected sites. Elliott and Angell (1988) propose that these fluctuations may result from the migration of the centers of strongest coherent variability in the Southern Oscillation over the course of the present century.

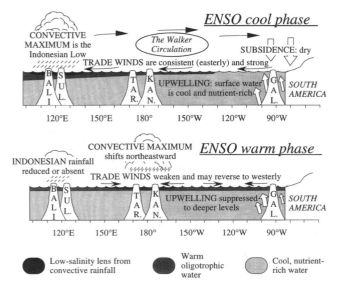

FIGURE 2 Schematic cross-section of equatorial Pacific showing major features of the warm and cool phases of ENSO in relation to selected sites where coral paleoclimatic reconstructions of ENSO are under way: Bali, Sulawesi, Tarawa, Kanton, and the Galapagos (after Cole, 1992).

Short Records Document Spatial Patterns of ENSO

Major dynamic features of ENSO variability include SST and upwelling anomalies in the eastern Pacific, and rainfall and wind anomalies in the western Pacific. Coral records can be targeted to address variability in each of these parameters through careful site and tracer selection. Figure 3 (from Cole et al., 1992) compares six short (20-30 yr) monthly-to-seasonal-resolution records from corals that grew at sites spanning the equatorial Pacific: the Galapagos Islands (1°S, 89°W), Tarawa (1°N, 172°E), and Bali (8°S, 115°E). In the eastern Pacific, Galapagos corals yield records of $\delta^{18}O$, Cd/Ca, and Ba/Ca that track variations in upwelling and SST (Shen et al., 1987; McConnaughey, 1989; Lea et al., 1989). In Tarawa corals, $\delta^{18}O$ primarily monitors monthly variability in local rainfall, a consequence of the displaced Indonesian Low (Cole and Fairbanks, 1990; Cole et al., 1993), and Mn/Ca variations reflect wind reversals associated with especially strong ENSO warm extremes (Shen et al., 1992b). In Bali, the migration of the Indonesian Low causes drought during ENSO warm extremes; coral $\delta^{18}O$ shows moderate

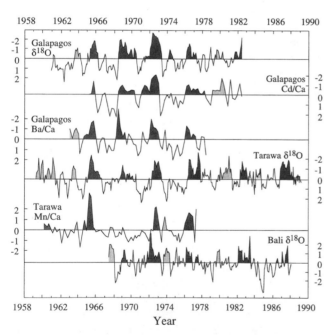

FIGURE 3   Comparison of short proxy records from corals that grew at three sites across the Pacific. (From Cole et al., 1992; reprinted with permission of Cambridge University Press.) All records are presented in units of standard deviation. Dark shading denotes recognized ENSO warm extremes (Quinn and Neal, 1992); lighter shading indicates conditions similar to the warm phase of ENSO at individual sites. These records target distinct components of the ENSO system; together they indicate both the Pacific-wide extent of ENSO extremes and the potential for the reconstruction of subtle patterns of spatial variability in ENSO. Periods of anomalies similar to ENSO warm phases but not recognized by Quinn and Neal (1992) occur in most records in 1963 and 1979-1980.

sensitivity to these fluctuations but probably includes a significant SST component as well.

Overall, these records monitor warm phases of ENSO consistently between sites. Cool phases of ENSO are best recorded by the Galapagos records and by the Tarawa $\delta^{18}O$ history. The Mn/Ca record at Tarawa appears to indicate changes in westerly wind intensity associated with strong warm phases, although it may not track variations in the intensity of easterly winds, and the Bali record does not show cool-phase anomalies consistent with the other coral records or with large-scale indices of Pacific climate.

Comparison of these records reveals patterns of spatial variability that are confirmed by concurrent instrumental records. For example, during 1979-1980, most of the Pacific experienced conditions similar to a weak ENSO warm anomaly. Although this year is not generally recognized as a major event (*sensu* Quinn et al., 1987), other studies have documented unusually warm SSTs and rainfall in the central Pacific during that time (Donguy and Dessiers, 1983). In 1977, the warm extreme of 1976 intensified in the western Pacific but decayed more rapidly into cool-phase conditions in the eastern Pacific. Variations in intensity of anomalies are also reflected in the coral data and generally confirmed by climatological data. In 1963 and 1969, for example, weak anomalies in the Tarawa $\delta^{18}O$ record are not concurrent with anomalies in the Mn/Ca record, which is consistent with observations of rainfall increases but no wind reversals at Tarawa. Cole et al. (1992) discuss these patterns in greater detail. This proxy intercomparison demonstrates that coral paleoclimatic tracers monitor large-scale ENSO anomalies across the entire Pacific basin, both warm and cool, and that they can also discern relatively subtle patterns of spatial variability and anomaly intensity in specific climatic and oceanographic parameters.

Calibration of Geochemical Tracers in a Galapagos Coral

In Galapagos corals, several tracers hold promise for reconstructing the SST and upwelling variability that define the eastern Pacific signature of ENSO. Shen et al. (1992a) measured parallel records of Cd/Ca, Ba/Ca, Mn/Ca, $\delta^{18}O$, and $\delta^{13}C$ at quarterly resolution from a coral that grew at Punta Pitt, Isla San Cristobal, on the eastern side of the Galapagos archipelago. These records span the period 1936 through mid-1982. With the exception of Mn/Ca, all tracers show high correlations with SSTs measured at Puerto Chicama, Peru; for seasonal records, R falls between 0.51 and 0.65, and for annual averages, R ranges from 0.61 to 0.73. Even higher correlations are found between the tracers and the shorter SST record from Academy Bay, Galapagos, with seasonal R values ranging from 0.57 to 0.79 and annual values falling between 0.72 and 0.90. These tracers monitor slightly different aspects of the eastern Pacific ENSO signal: $\delta^{18}O$ tracks SST variability, Ba and Cd reflect the surface concentrations (i.e., upwelling) of these nutrient-like tracers, and $\delta^{13}C$ probably responds to insolation variations, although this relationship is not constant for all coral $\delta^{13}C$ records from the Galapagos (McConnaughey, 1989).

Figure 4 compares the coral record of $\delta^{18}O$ anomaly from Punta Pitt with SST anomalies from Puerto Chicama and with an index of eastern-central Pacific SST anomalies (Wright, 1989). The $\delta^{18}O$ record captures most measured SST anomalies in these curves, both positive and negative. In certain cases where the $\delta^{18}O$ and Puerto Chicama records disagree (e.g., in 1980), the Wright SST index appears to corroborate the $\delta^{18}O$ record, suggesting that variability in the coastal SST record does not always represent conditions in nearby open ocean regions. Deser and Wallace (1987) have also noted that SSTs along the Peru coast occasionally become decoupled from large-scale variations in Pacific climate.

Cross-spectral analysis of the Punta Pitt coral records with records of Puerto Chicama SST indicate that correlations at the frequency bands associated with ENSO (about 2 and 3.8 years, in these records) reach values as high as 80 to 90 percent. Shen et al. (1992a) note that the observed degree of correlation, in both time and frequency domains, is truly remarkable, considering that these single proxy records derive from a site several hundred kilometers away from the coastal SST monitoring station at Puerto Chicama.

A Century of ENSO Variability in the Central-Western Pacific

In the western Pacific, ENSO variability is characterized by dramatic shifts in rainfall patterns associated with the migration of the Indonesian Low from northern Australia/ Indonesia to the region of the equator and the date line.

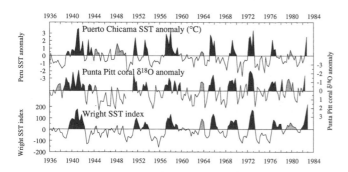

FIGURE 4   Plot of $\delta^{18}O$ anomaly from a Galapagos coral (Shen et al., 1992a) compared to SST anomalies measured at Puerto Chicama (Peru) and indexed over the eastern-central Pacific (Wright, 1989). Dark shading denotes recognized ENSO warm extremes (Quinn and Neal, 1992); lighter shading indicates conditions similar to the warm phase of ENSO at individual sites. Periods of anomalies similar to ENSO warm phases but not recognized by Quinn and Neal (1992) occur in all records in 1944, 1963, and 1979-1980.

Biweekly monitoring of surface ocean $\delta^{18}O$ at Tarawa Atoll demonstrates that at this site, rainfall causes a significant decrease in the $\delta^{18}O$ of the surface water in which corals grow (Cole, 1992). Coral $\delta^{18}O$ records from this site closely track these rainfall changes (Cole and Fairbanks, 1990; Cole et al., 1993). Our coral $\delta^{18}O$ record extends back to 1894 at monthly resolution, double the length of the instrumental record of Tarawa rainfall that began in 1946. Between 1976 and 1989, the Tarawa coral record suggests a period of increased rainfall that is unprecedented within this record. This apparent baseline shift is consistent with instrumental records from the Pacific (Trenberth, 1990; Kerr, 1992).

This $\delta^{18}O$ record correlates with large-scale indices of ENSO (Figure 5) at levels comparable to monthly instrumental records from individual Pacific stations, such as Tarawa rainfall and Darwin SLP. Monthly, seasonal, and annual $\delta^{18}O$ data linearly explain fractions of ENSO variance that are among the highest for any proxy record of ENSO (Cole, 1992; Cole et al., 1993). Despite the strong correlation between the coral and instrumental records of ENSO, the Tarawa $\delta^{18}O$ record does not capture the very strong ENSO warm anomaly of 1982-1983. Outgoing long-wave radiation measurements indicate that during that year the Indonesian Low moved much farther east than usual (Rasmusson and Wallace, 1983). Tarawa rainfall was not unusually intense at this time, and the coral $\delta^{18}O$ data reflect these conditions accurately. This anomaly that bypassed Tarawa is the only major ENSO extreme of the past century that the coral $\delta^{18}O$ record does not monitor, attesting both to the fidelity of the coral recorder and to the unusual nature of the 1982-1983 ENSO extreme. To reconstruct the full range of ENSO variability will require developing records from sites throughout the Pacific.

The coral record generally indicates warm and wet ENSO extremes during periods identified by Quinn et al. (1987) as "El Niño events" along the coast of South America. However, Figure 5 suggests exceptions to this correlation. In 1907, 1917, 1932, and 1943, Quinn et al. identify moderate or strong El Niño events, but neither the coral nor the instrumental indices of tropical Pacific climate indicate ENSO warm anomalies. In other years (e.g., 1946, 1963) the opposite pattern occurs; coral and instrumental indices demonstrate ENSO warm extremes that do not appear as moderate or strong in Quinn's most recent summaries (Quinn et al., 1987; Quinn and Neal, 1992), although these years were noted as weak anomalies in an earlier compilation (Quinn et al., 1978). Rasmusson et al. (1995, in this volume) have also noted, on the basis of instrumental climate records, that recurrence statistics derived from Quinn et al.'s "strong/very strong" events do not reflect basin-scale ENSO variability. Quinn et al. (1978, 1987, 1992) explicitly state that their index may not be entirely consistent with climatic records from across the tropical Pacific, such as the Southern Oscillation Index or central Pacific rainfall

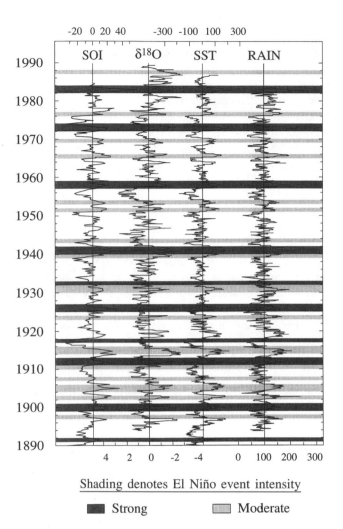

FIGURE 5 Comparison of the δ¹⁸O measured in a Tarawa Atoll coral with large-scale indices of Pacific climate from Wright (1989). Shaded bars denote El Niño events identified by Quinn and Neal (1992) from historical information; strongest events are shaded most darkly (from Cole, 1992). Strong and moderate El Niño events in 1907, 1917, 1932, and 1943 have no counterparts in the instrumental or coral records, indicating anomalies localized to coastal South America in those years. The coral record indicates a shift to wetter conditions at Tarawa between 1976-1989 that is unprecedented in the past century.

histories, especially with regard to anomaly intensity. However, their compilation of El Niño events remains widely (if perhaps inappropriately) used as a history of large-scale ENSO variability. Both climatological and coral data from across the Pacific indicate that the Quinn index of El Niño events in coastal South America does not consistently track the state of the Pacific-wide ENSO system during the past century (Rasmusson et al., 1995, in this volume; Cole et al., 1993).

Frequency-domain analysis of the Tarawa δ¹⁸O record suggests that interannual variance across periods of 1.9-2.5, 3.0, 3.6, and 5.6 years are highly coherent with large-scale

ENSO indices; at these periods 80 to 90 percent of the variance in the Tarawa δ¹⁸O record is linearly related to variance in the instrumental ENSO indices (Cole, 1992; Cole et al., 1993). These results, shown in Figure 6, fit into the general framework of studies that indicate biennial and low-frequency concentrations of variance in ENSO-related climatological data. With 96 years of coral δ¹⁸O data, we can address the issue of whether the spectral signature of ENSO has changed over the period of our record.

We examine the distribution of variance among dominant periods over the length of the record by performing spectral analysis on a series of 30-year windows of δ¹⁸O data, each shifted by 2 years from the previous. This evolutionary spectral analysis suggests that the variance spectrum of ENSO may have changed during the past century (Figure 7; Cole et al., 1993). Our 30-year intervals are too short to identify the specific frequencies noted above with statistical confidence, but this analysis suggests intriguing broad-scale changes in the variance spectrum of rainfall at Tarawa. Figure 7 maps the changing concentrations of variance at periods between 1 and 10 years over the past 96 years. In

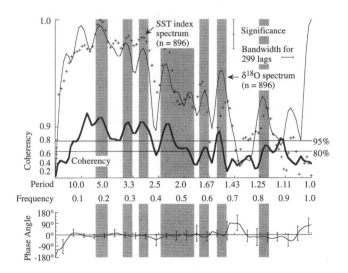

FIGURE 6 Cross-spectral comparison between SST index from Wright (1989) and Tarawa δ¹⁸O record (Cole, 1992). Upper panel shows the variance spectra of the Tarawa δ¹⁸O record (solid fine line) and the Wright SST index (plus symbols), plotted against frequency. The coherency between these records, which represents the degree of correlation between them as a function of frequency, is shown by the heavy line; the squared coherency gives the percent variance in common between the records. The lower panel shows the phasing of the records as a function of frequency. Shaded bars indicate frequency bands over which variance peaks are aligned; coherency is non-zero at more than 80-95 percent significance. At interannual periods characteristic of ENSO (here centered at 5.6, 3.5, 3.0, and 2.1 years), these records share at least 80 percent of their variance and are in phase. These results are consistent with cross-spectral analyses between the Tarawa record and other instrumental indices of ENSO (Cole, 1992).

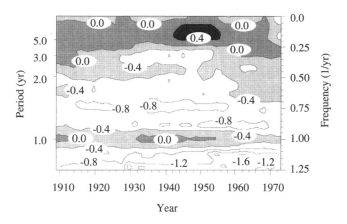

FIGURE 7 Evolutionary spectrum of variability in the Tarawa coral $\delta^{18}O$ record. (From Cole et al., 1993; reprinted with permission of the American Association for the Advancement of Science.) This figure maps the shifting distribution of variance (in units of log spectral density) in the Tarawa $\delta^{18}O$ record throughout the present century. The information in this figure derives from spectral analysis of overlapping 30-year intervals of monthly $\delta^{18}O$ data, offset by 2 years. Results are plotted against the midpoint of the analysis interval. The upper error bar on these values is 0.36 and the lower bar is 0.22. Overall, this figure suggests a change in the behavior of ENSO between about 1930 and 1950, with greater power at 5-to-6-year and annual periods and reduced power at 3-year periods.

the early part of this century, interannual variance at Tarawa is concentrated at periods longer than 2.5 years, and the annual cycle is moderately strong. In the interval between 1930 and 1950, annual variability intensifies and a minimum in variance occurs at the 3-year period. During this interval, in which the Southern Oscillation was thought to have generally weakened (Elliott and Angell, 1988), annual variance has approximately the same strength as low-frequency variability. Between 1955 and 1985, a strong variance peak centered at about 3.3 years reappears, and the annual cycle is at its weakest. Overall, these results imply a mode of ENSO behavior between about 1930 and 1950 that differs from preceding or subsequent periods.

Climatological studies that span only the most recent decades generally note that in the central-western Pacific, annual variability is secondary to interannual variance in many climatic parameters. Our coral record suggests that the strength of the annual cycle has varied at Tarawa, perhaps reflecting changes in the seasonal influence of either the southeast Asian monsoon or the eastern Pacific cool tongue. The annual cycle at Tarawa has been particularly weak since about 1950, coincidentally the time at which many instrumental records from the tropical Pacific begin. Evolutionary spectral analysis of the Tarawa $\delta^{18}O$ record highlights the limitations of using a short climatological data set to evaluate relative concentrations of variance at characteristic ENSO frequencies. At Tarawa, changes in the rela-

tive concentrations of annual and interannual variance have occurred over the past century; they may relate to overall shifts in the strength of the Southern Oscillation or to changes in the location of maximum expression of ENSO system variability. Our coral results require confirmation by further studies; additional records from the central and western tropical Pacific will allow us to draw more specific conclusions about the nature of these changes.

Urvina Bay Records: 400 Years of Oceanic Variability in the Eastern Pacific

A sub-aerially exposed Galapagos coral reaching nearly 400 years in age offers the opportunity for long paleoceanographic reconstructions from the eastern Pacific. In 1954 a portion of Urvina Bay, on the west coast of Isabela Island, was tectonically uplifted, exposing several square kilometers of pristine coral reef and marine shelf (Colgan and Malmquist, 1987). The oldest of the uplifted corals has been sampled for paleoenvironmental purposes. Dunbar et al. (1991, 1994) present a 371-year, annual-resolution record of coral $\delta^{18}O$ and growth rates from Urvina Bay that indicates significant decadal-scale variability in SST. Variance peaks at 11- and 22-year periods hint at sunspot modulation of eastern Pacific climate, although a direct link between sunspot activity and eastern Pacific upwelling remains to be established.

A 380-year annual record of Mn/Ca from the same coral has also been developed (Shen et al., 1991). Short segments of seasonal-resolution data throughout this record indicate the persistence of seasonal upwelling variability since the mid-seventeenth century. Upwelling intensity is difficult to quantify with this tracer, however, due to the large-scale spatial gradients in surface water Mn content. A long-term decreasing trend in Mn/Ca may indicate oceanic or atmospheric processes. Linn et al. (1990) suggest that higher Mn/Ca ratios in seventeenth-century samples from this Urvina Bay coral may be explained by either increased trade-wind strength or increased influence of Panama Basin water masses at that time. However, the possibility of an accumulating diagenetic Mn phase in this coral has not been eliminated (Shen et al., 1991).

Additional work in progress on this coral includes high-resolution $\delta^{18}O$ records and records of other trace elements. Results from this site will be placed in a regional perspective through ongoing studies of coral records that attain ages of 100-400 years from elsewhere in the eastern Pacific, including the Gulfs of Panama and Chiriquí, the Costa Rican islands of Caño and Cocos, and the southern Gulf of California (Linsley et al., 1991, 1994; Dunbar et al. 1993, 1994). Analysis of variability in the Pacific-wide ENSO system over the past several centuries will require the comparison of these eastern Pacific records with long reconstructions in progress from central and western Pacific sites such as Tarawa Atoll, Sulawesi, and Kanton Island.

CONCLUSIONS

Isotopic and trace-metal records from massive corals offer the opportunity for focused retrospective monitoring of tropical climate processes, including many of the key dynamic features of the ENSO system. Short records from sites throughout the tropics have shown the sensitivity of various chemical tracers in coral skeletons to environmental variability. Recent application of these methods is beginning to yield insight into the past behavior of specific components of the global climate system. For example, extensive work with tropical Pacific corals indicates that coral proxy records not only capture the large-scale ENSO anomalies that recur interannually, but also track subtle patterns of variation in the intensity and spatial patterns of anomalies. Statistical comparisons in the time and frequency domains suggest that many Pacific coral records approach the utility of single-station instrumental records reconstructing large-scale ENSO indices. Evolutionary spectral analysis of a proxy ENSO record from a western Pacific coral suggests that the relative concentrations of variance at annual, biennial, and low-frequency periods has varied over the course of this century, perhaps in relation to fluctuations in the overall strength of the Southern Oscillation or to changes in the locations of centers of action of this system. Thus, the development of coral chemical records that span centuries will improve the canonical description of ENSO variability that is currently based on the instrumental record of only the past few decades.

Paleoclimatic records from remote tropical sites can yield substantive information on natural climate variability, offering insights not available from the limited instrumental record or even from skilled numerical simulations. Paleoclimatic records from corals capture the full range of natural variability, including decadal-scale shifts and changes in the dominant frequency components of climate variability. Furthermore, calibration of coral tracers over recent periods will provide the basis for developing records of tropical climate that extend into times of altered boundary conditions; such records offer a logical means of evaluating the sensitivity of tropical systems to global climate change.

ACKNOWLEDGMENTS

This report benefited greatly from careful reviews by Rob Dunbar and Clara Deser. Many participants in the Dec-Cen workshop provided helpful commentary, and we thank the National Academy of Sciences/National Research Council for their sponsorship of this event. We also thank the NOAA Paleoclimatology Program and the National Science Foundation for their support of this research.

Commentary on the Paper of Cole et al.

CLARA DESER
University of Colorado

Dr. Cole et al. have demonstrated that selected chemical records from corals at key locations in the tropical Pacific can yield an accurate history of ENSO variability. The coefficients of the correlations between the coral records and measured environmental variables exceed 0.7 or so on seasonal and annual time scales. Their study has also shown that a network of coral sites across the equatorial Pacific can be used to investigate the spatial characteristics of individual ENSO cycles.

The long (approximately 100-year) coral record at Tarawa in the western equatorial Pacific reveals changes in the frequency characteristics of ENSO during this century. Further insight into the dependence of ENSO on surface boundary conditions may come when the coral records are extended back several more centuries.

There may also be other applications for the coral records, such as monitoring the atmospheric heat sources over Indonesia, South America, and Africa.

Discussion

SOCCI: If the mean temperature rose substantially, would you have to get a new set of correlations to predict the coral's response to variations?

COLE: No, we don't expect the relationships we've found to change. The temperature dependence of $\delta^{18}O$ and the biogenic carbonates is pretty well established, and corals appear to have a very consistent response over a range of time scales and temperature levels. Similarly, the rainfall signal in coral $\delta^{18}O$ should not change, but the proportion of the coral signal due to rainfall rather than temperature changes could vary with changes in the mean state.

KEELING: Enfield's paper says that El Niños tend to be spaced further apart for a while and then closer together for a while. Do you see anything similar to that, at least after 1890?

TRENBERTH: Dennis Shea and I analyzed the record for this century. We saw regular El Niño events early in the century, then a break with only the unusual one of 1939-1942, then since the 1950s more regular ones again, with a strong peak around 4 years.

RASMUSSON: I think these coral data are exciting stuff. I wanted to introduce a note of caution as regards Enfield's study, though. First, he didn't use his weaker data because he didn't trust them, but I wonder whether we can then trust his distinction between weaker and stronger. Second, he had only 8 to 10 events per century, which doesn't yield stable statistics. Third, when the events are compared over 120 years of basin-scale historical data, there is correspondence in only about half of them.

I have two questions for you, Julie: Will you be comparing Kanton and Tarawa, and when will you extend this record to 300 years?

COLE: Starting with the second question: as soon as I can find corals that old. We know they exist, and we'll be looking at less settled islands near Tarawa as soon as we get funding. As for Kanton corals, Rick Fairbanks has just collected some, and they should be analyzed soon.

MOREL: It will be fascinating to see what corals will show for very different climate conditions. When do you think it will be possible to have information reflecting, for example, ENSOs from past glacial ages?

COLE: Naturally, that will depend on the funds available; we have proposals out already. Rick drilled a series of cores off Barbados that go back about 30,000 years, and found many decade-to century-long time series. We need similar cores from locations in the Pacific to answer that kind of question.

TRENBERTH: There's a lot we can learn from corals on the longer time scales, though the absence of the 1982-83 El Niño from the Tarawa record underlines the importance of using more than one site. I was glad you looked at the coherence of the signal at different frequencies to determine the source of the relationship; I wish more paleo people would. It will be critical on the longer time scales.

CANE: Satellite data is too short-term to tell us much of anything on the time scales we're interested in, and even the instrumental record can't give us more than a few realizations of anything on time scales of a decade or more. The coral records are exciting, particularly since our ability to interpret what comes out of them as a geophysical signal seems to be much greater than for any other kind of proxy data available to us. Within the next few years I think we'll be able to get some ideas about what the natural variability has been like over the last few centuries, and even why ENSO is irregular.

MUNK: I confess I was disappointed in your 300-year limit; I was hoping to go back to the Devonian.

COLE: At present we can't go back to the Devonian continuously. Also, with the older material you have the problem of diagenesis, during which the oxygen in the skeleton is exchanged. But by patching together records we should be able to get some ideas about older time periods. For example, corals 88 and 125 thousand years old preserve a nice seasonal $\delta^{18}O$ cycle.

PHILANDER: Wasn't there a recent article in *Science* reporting coral records of much colder SSTs 9 or 10 thousand years ago?

COLE: Yes, by Warren Beck et al. They use an exciting high-precision technique, measurement of strontium/calcium ratios by thermal-ionization mass spectrometry. The strontium seems to be a close recorder of a pure temperature signal, so it can be used to separate the patterns of variability in the combined rainfall and temperature change we usually see in the tropics.

LEHMAN: Another possible application of this coral work is as an indicator of how much the tropical moisture pump spikes the surface of the ocean with rainwater, on time scales much longer than El Niño events. That will be very important.

COLE: In a 1960 paper Colin Ramage said that something like 50 percent of tropospheric water vapor comes from the equatorial Pacific and the Indonesian area. If this system is responding to conditions such as changes in sea level, small changes in this region might be amplified by positive water-vapor feedback.

LINDZEN: For the lower 2 or 3 km of the atmosphere the water vapor is largely determined by temperature, whereas the greenhouse in the tropics is almost entirely determined by water vapor above 3 km. That's only a few percent of the total, and decoupled from the boundary besides. I don't think the surface budget will give you much insight there.

COLE: But upper-level water vapor must originate from the surface initially. And large changes at lower levels may have significant impact.

RASMUSSON: Julie, did you have problems related to the 1982-83 coral mortality, or tectonic shifts?

COLE: In the Galapagos, 1982-83 caused a lot of coral damage and also made possible an amazing amount of bioerosion. There aren't many long coral records left there. The uplifts resulting from tectonic activity have given Rob Dunbar and Glen Shen some useful samples, one of them the longest existing coral paleoclimate record.

Natural Variability of Tropical Climates on 10- to 100-Year Time Scales: Limnological and Paleolimnological Evidence

F. ALAYNE STREET-PERROTT[1]

ABSTRACT

A lake is a passive hydrological storage. It varies in depth and surface area in response to changes in its atmospheric, land-surface, and subsurface fluxes of water. Provided that groundwater inflows and outflows are negligible, it can be modeled as a simple, low-pass signal filter with a characteristic time constant (*e*-folding time) τ_e. For closed lakes, which lack surface outlets and lose water solely by evaporation, τ_e varies from about 1.5 to at least 350 years. For open lakes regulated by surface outflow, τ_e is generally shorter, of the order of 10^{-2} to 5 years. Small, shallow closed lakes or large open lakes with τ_e equal to 5 to 10 years provide good records of both interannual-to-decadal and longer-than-century variations in climate, whereas large, closed lakes with τ_e greater than 50 years provide the best coverage over the intervening range of decade-to-century fluctuations. However, allowance must be made for characteristic time lags in the response of lakes to climatic fluctuations of period less than $2\tau_e$. This paper illustrates the response of tropical lakes to variations in ocean-atmosphere interaction over the Pacific, Indian, and Atlantic oceans during the period of limnological observations, and concludes with examples of paleolimnological evidence for century-scale droughts in southern Africa and the tropical Americas.

INTRODUCTION

In areas where humans have not significantly perturbed the water budget, fluctuations in the water level and surface area of lakes provide a physical indicator of climate on time scales of 10^{-1} to 10^6 yr. This paper has three aims: first, to summarize the general principles governing the response of lakes to variations in aridity, focusing on the decade-to-century time scale; second, to present examples of the behavior of tropical lakes over the last few centuries; and third, to comment on the links between the observed hydrological record and air-sea interaction over the tropical oceans, since the time scale highlighted here is longer than the response time of the atmosphere on its own. It concludes by emphasizing the need for a more coordinated approach to data collection and storage.

---

[1]Environmental Change Unit and School of Geography, University of Oxford, Oxford, U.K.; now at the University of Wales, Swansea.

THE RESPONSE OF LAKES TO VARIATIONS IN SURFACE WATER BALANCE

The general water balance of a lake can be written as

$$\frac{dV}{dt} = R - A_L(E_L - P_L) - D + G_i - G_o, \quad (1)$$

where V is the lake volume, R and D are the surface runoff and discharge rates into and out of the lake, E_L and P_L are the evaporation and precipitation rates over the lake surface (in units of depth per unit time), A_L is the lake-surface area, and G_i and G_o are the groundwater inflows and outflows from the lake.

For closed, sealed lakes, defined as those that lack a surface outflow and for which the subsurface fluxes are negligible, equation (1) reduces to

$$\frac{dV}{dt} = R - A_L(E_L - P_L). \quad (2)$$

At equilibrium, $dV/dt = 0$, and the equilibrium lake-surface area A_{Le} is determined by $(E_L - P_L)$ and R, which reflect the surface water balance over the lake and its surrounding catchment, respectively. Hence,

$$A_{Le} = \frac{R}{E_L - P_L}. \quad (3)$$

At equilibrium, the ratio of the lake-surface area A_{Le} to the total catchment area A_C (including the water surface) is given by

$$\frac{A_{Le}}{A_C} = C, \quad (4)$$

where C is a dimensionless index dependent only on R, E_L, and P_L (Mason et al., 1994):

$$C = \frac{R}{A_C(E_L - P_L)}. \quad (5)$$

On time scales longer than a year, R may be approximated by the expression

$$R = A_B(P_B - E_B), \quad (6)$$

where A_B is the area draining into the lake (excluding the water surface), and P_B and E_B are, respectively, the mean precipitation and evapotranspiration rates over the basin (in units of depth per unit time). Since by definition $A_C = A_L + A_B$, C can now be rewritten as follows:

$$C = \left[\left(\frac{E_L - P_L}{P_B - E_B} \right) + 1 \right]^{-1}. \quad (7)$$

C is an inverse measure of aridity, ranging from near 0 in extremely arid conditions (small P and large E) to a theoretical maximum of 1 in very humid climates.

So far, only the equilibrium case has been considered.

The response of the area of a closed, sealed lake to a perturbation in its inputs or outputs is determined by its characteristic response time (e-folding constant) τ_e, defined as the time taken to reach a fraction $(1 - 1/e)$, or 63 percent, of the total change in area. The response time τ_e is given by

$$\tau_e = \frac{A_L}{\frac{dA_L}{dL}(E_L - P_L)}, \quad (8)$$

where L is lake level (Mason et al., 1985, 1994). Values of τ_e calculated by Mason et al. (1994) for modern closed lakes vary from 1.5 to 350 years. τ_e is greatest for extensive, steep-sided lakes in relatively moist climates, such as Lake Malawi in the historical past (see below).

For an open lake (a lake possessing an outflow) with negligible groundwater fluxes,

$$\tau_e = A_L \frac{dL}{dD} \quad (9)$$

(Hutchinson, 1975). Calculated values of τ_e for open lakes are in general much smaller than those for closed lakes, varying from 10^{-2} to about 5 years (Mason et al., 1994).

The theoretical response of a closed, sealed lake to small perturbations in the aridity index C with time is summarized in Figure 1 (after Mason et al., 1994). Three simple types of change in C are illustrated: a step increase or decrease, a "spike" (short lived fluctuation), and a sinusoidal oscillation of period p. A step increase (shown in curve a) or decrease (curve b) in C results in an asymptotic approach of lake area (or level) to its new equilibrium value over a time span dependent on τ_e. A spike produces a rapid increase (curve c) or decrease (curve d) in lake area (or level), followed by a slower, asymptotic recovery. In contrast, a sinusoidal oscillation in climate produces an oscillating response with a phase shift dependent on the relative magnitudes of p and τ_e. For low-frequency (LF) variations in climate of period $p \gg 2\pi\tau_e$, the lake is approximately in equilibrium and exhibits negligible phase shift (curve e). For high-frequency (HF) variations of period $p \ll 2\pi\tau_e$, the lake lags the variations in climate with a phase shift of $-\pi/2$ (curve f); the maximum and minimum rates of increase in area (or level) correspond to the maxima and minima in C, respectively (Mason et al., 1994). For practical purposes, the limits of the LF and HF bands can be taken as $p > 20\tau_e$ and $p < 2\tau_e$. In theory, it should be possible to invert a time series of water-surface area or water level derived from a lake with known characteristics to obtain a record of variations in C (I. M. Mason, pers. commun.).

In summary, a lake acts as a simple, low-pass signal filter with a characteristic time constant τ_e. A wide range of lakes can act as climatic indicators on the decade-to-century time scale, although not necessarily over the whole range. Small, shallow, closed lakes or large, open lakes

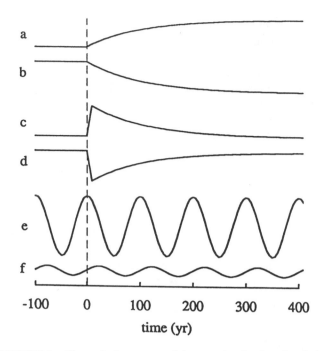

FIGURE 1 Theoretical response of the area A_L of a closed sealed lake (vertical axis) to simple variations in the aridity index C with time (from Mason et al., 1994). (a) and (b) show the response of a lake with equilibrium response time $\tau_e = 100$ yr to a 10 percent step increase and decrease in C, respectively. (c) and (d) show the response of the same lake to a "spike," a 100 percent change in C that lasts for 10 years and then reverts to the previous value. (d) and (e) show the response of the lake to a sinusoidal variation in C of amplitude 10 percent. Curve (e) is for a lake with $\tau_e = 1$ yr, showing negligible phase shift, and representing the equilibrium response in the LF band. Curve (f) is for a lake with $\tau_e = 100$ yr, giving a phase shift of $-\pi/2$, as an example of the response in the HF band.

with τ_e equal to 5 to 10 years potentially provide a good record of interannual to decadal fluctuations (their respective HF band). They should also exhibit negligible lag with respect to secular (LF) climatic variations. Large, closed lakes with τ_e greater than 50 years can provide good coverage of decade-to-century (HF) variations. Whenever a response in the HF band is considered, however, allowance needs to be made for a phase shift of $-\pi/2$ with respect to the climate signal.

EVIDENCE FOR PAST FLUCTUATIONS IN LAKE LEVEL AND SURFACE AREA

Time series of lake levels have traditionally been compiled from observational, historical, or paleolimnological (geological) data (Street-Perrott and Harrison, 1985). Few lakes, even in developed countries, possess instrumental (gauge-board) records extending back before A.D. 1875. In some cases, however, historical observations of the emergence or drowning of specific landmarks permit lake-level

time series to be extended for several centuries further back, at least in outline (de Terra and Hutchinson, 1934; Street-Perrott and Harrison, 1985; O'Hara, 1993).

In data-poor regions, or for the period before the start of reliable historical observations, high-resolution paleolimnological studies provide qualitative and, in some cases, quantitative information about past variations in lake depth and area, based on direct dating of former shorelines, or analyses of stratigraphical, geochemical, and paleoecological data from sediment cores and surface exposures (Street-Perrott and Harrison, 1985). The time resolution of these studies, however, is dependent on the dating framework. The most precise data should in principle be obtainable from lakes with finely laminated mud, such as Lake Turkana, Kenya (Halfman and Johnson, 1988); however, these are generally rare in the tropics.

Recent studies at the Mullard Space Science Laboratory, University College London, have established the feasibility of measuring variations in lake level by satellite radar altimetry (Mason et al., 1984, 1990) and variations in lake area with imaging instruments such as the NOAA Advanced Very High Resolution Radiometer (AVHRR) or the Earth Resources Satellite-1 (ERS-1) Along-Track Scanning Radiometer (Harris and Mason, 1989; Mason et al., 1990; Harris et al., 1992). The overall accuracy is expected to be around ± 0.25 m for water level and 1 percent for surface area (Mason et al., 1990). Figure 2 shows the variations in area of the closed Lake Abiyata, Ethiopia (7°N) (τ_e ca. 4 to 9 years) between 1985 and 1991, derived from AVHRR images (Harris et al., 1992).

EXAMPLES OF LAKE BEHAVIOR OVER THE LAST TWO MILLENNIA

The Influence of the El Niño/Southern Oscillation (ENSO) System

Lake Titicaca (16°S) is a large, open lake, 8,100 km² in area, situated at an altitude of 3,812 m above sea level in the Altiplano of Peru and Bolivia. The equilibrium response time τ_e of the lake is probably less than 5 years. It is fed mainly by tropical summer precipitation. Wet (dry) summers in the altiplano are characterized by a poleward (equatorward) displacement and weakening (strengthening) of the mid-latitude upper westerlies over South America (Kessler, 1974). High water levels were observed in the 1920s, 1930s, 1950s, 1960s, late 1970s, and late 1980s, with low levels in the 1910s and 1940s (very marked) and around 1970 (Figure 3) (Kessler, 1974; Künzel and Kessler, 1986; Martin et al., 1993).

The time series of the yearly rise (maximum minus previous minimum level) of Lake Titicaca between 1915 and 1981 was analyzed by Künzel and Kessler (1986) using maximum-entropy spectral analysis. They identified sig-

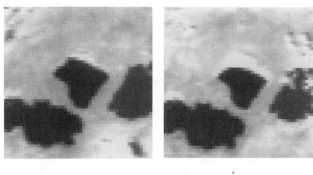

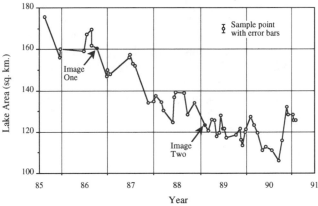

FIGURE 2   Water-level fluctuations of Lake Abiyata, A.D. 1984-1991, based on a time series of AVHRR imagery, together with sample images illustrating the appearance of the lake (center) in mid-1986 and late 1988 (From Harris et al., 1992; reprinted with permission of the European Space Agency.)

FIGURE 3   Measured water-level fluctuations of Lake Titicaca, A.D. 1912-1970. (Redrawn from Kessler, 1974; reprinted with permission of Dr. Kessler.)

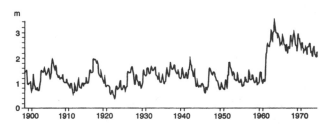

FIGURE 4   Measured water-level fluctuations of Lake Victoria, A.D. 1899-1974; (Redrawn from Vincent et al., 1979; reprinted with permission of Kluwer Academic Publishers.)

nificant spectral peaks with periods of 10.6 and 2.4 years. The former was assigned to the sunspot cycle and the latter to the quasi-biennial cycle. They also found a very weak period of about 4.7 years, which they ascribed to El Niño. More recent work suggests that both the quasi-biennial 2.4-year and lower-frequency 4.2-year cycles are associated with ENSO and the related variations in atmospheric angular momentum (Keppenne and Ghil, 1992; Dickey et al., 1992).

The Influence of Sea Surface Temperature Anomalies in the Indian Ocean

A lake that at least in part acts as an indicator of sea surface temperature (SST) anomalies in the Indian Ocean is Lake Victoria (1°S) in equatorial Africa. This is a very large, open lake (67,000 km^2), with an equilibrium response time τ_e of about 4.5 years (from data in Institute of Hydrology, 1985). Its water level curve for 1899 to 1978 shows the following features: oscillations about a stationary mean during the period 1899 to 1961; a step increase in lake level in 1961-1962; and an oscillating decline over the following 16 years (Figure 4) (Mörth, 1967; Vincent et al., 1979; Institute of Hydrology, 1985).

Spectral analysis of the lake-level record for 1899 to 1974 by Vincent et al. (1979) revealed only two significant peaks, at 10-13 years and 5-6 years, the first of which was attributed to solar variability in a famous paper by Brooks (1923). The record contains no power in the 2-to-3-year range; nor does it show any correlation with the Southern Oscillation Index (Vincent et al., 1979).

The sharp increase in lake level in 1961-1962 is an example of a "spike" (Figure 4), although a preliminary inversion of the lake-level curve suggests that the climate over the Victoria catchment has not fully returned to its pre-1961 state (I. M. Mason, pers. commun.). The sudden rise in water level was the result of a remarkable rainfall anomaly over East Africa in August 1961 to January 1962, covering approximately 2.1×10^6 km^2. The peak occurred in November 1961 when rainfall over the Lake Victoria basin was almost 400 percent of normal (Flohn, 1987). Flohn has summarized evidence suggesting that this was merely the western end of a large, positive anomaly of SST, cloud, and rainfall covering about 6×10^6 km^2 of the western Indian Ocean, which is an extent similar to that of a typical ENSO warm pool or of the South Atlantic warming event of 1984 (Philander, 1986).

The Influence of the Atlantic (and Global) Sea Surface Temperatures

Pronounced decade-to-century-scale fluctuations in water balance have affected Lake Chad, situated at 13°N in the West African Sahel (Figure 5). This is a shallow, closed lake in a semiarid climatic zone; its area has fluctuated between about 1,950 and 26,000 km^2 during the twentieth

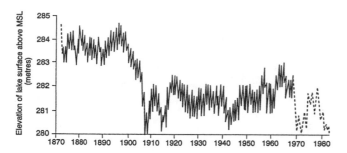

FIGURE 5   Water-level fluctuations of Lake Chad, A.D. 1870-1982, based on historical and instrumental data. (Adapted from Sikes, 1972, and updated by Thambyahpillay, 1983; reprinted with permission of Methuen and Co. and the University of Maiduguri.)

century (Thambyahpillay, 1983; G. Wells, pers. commun.). Over the past 40 years, its characteristic response time τ_e has varied from less than 1 year in 1985 to a maximum of 7.7 years in 1954-1969, when it covered an average of 21,520 km² (Mason et al., 1994). Note, however, that a small component of seepage loss, approximately 5 percent of the water budget (Roche, 1977), has been ignored in making these estimates. The lake-level curve from A.D. 1900 onward lags the instrumental record of sub-Saharan rainfall (Nicholson, 1985) by a year or two at most.

The reconstructed Lake Chad water-level curve for the last millennium (Figure 6) shows large fluctuations of 20 to more than 100 years' duration. Particularly high levels were recorded during the twelfth to fourteenth and the seventeenth centuries.

On the time scales of the last 100 and the last 10,000 years, fluctuations in Sahelian rainfall have been attributed to large-scale changes in SST fields. Following earlier work by Lamb (1978) and Lough (1986), Folland et al. (1986) and Parker et al. (1988) showed that twentieth-century droughts in the Sahel were significantly correlated with cold SST anomalies in the northern oceans, particularly the North Atlantic, and warm anomalies in the southern oceans, particularly the South Atlantic and the Indian Ocean. This

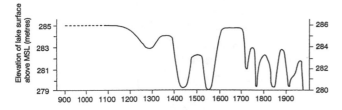

FIGURE 6   Water-level fluctuations of Lake Chad, A.D. 900-1980, reconstructed from a combination of stratigraphic, palynological, and historical evidence. (Redrawn from Maley, 1981; reprinted with permission of ORSTOM Editions.) Due to progressive sedimentation, two elevation scales are required; the left-hand one refers to the earlier part of the record and the right-hand one to the later part.

hemispherically asymmetrical anomaly pattern, which also characterized important low stands of Lake Chad on the millennial time scale, was ascribed by Street-Perrott and Perrott (1990) to a reduction in the northward heat transport in the Atlantic by the global thermohaline circulation. There is as yet no evidence to determine whether or not drought episodes within the last millennium can be explained by the same mechanism.

Examples of Droughts Lasting Several Centuries

The paleolimnological record provides examples of prolonged climatic fluctuations that have had profound significance for human communities in the tropics. For example, Lake Malawi (12°S), a large rift lake (28,750 km²) in equatorial Africa, has experienced variations in water level of at least 120 m vertical amplitude during the last few centuries (Owen et al., 1990). From A.D. 1860 onward, it fluctuated between 469 and 476 m above sea level. It was open throughout this period, with a response time τ_e of about 3 to 8 years (R. Crossley, pers. commun.), apart from 1915 to 1935 when it ceased to overflow (Beadle, 1974). Between A.D. 1390 and 1860, however, there was a prolonged period of very low lake levels, represented by a widely traceable erosional unconformity and significant changes in diatom (algal) assemblages. During this low stand, the lake must have been closed, probably with a response time τ_e much greater than 100 years. A rich oral tradition confirms that very low levels prevailed during the late eighteenth and early nineteenth centuries. Hydrological modeling by Owen et al. (1990) suggests that a reduction in rainfall of 40-50 percent, lasting for 100-150 years, could produce the required drop in lake level. The causes of such a prolonged anomaly remain uncertain, but probably involve large-scale oceanographic changes in the Indian and Atlantic Oceans.

An anomaly of similar duration is recorded by Lake Pátzcuaro (Mexico, 20°N) (O'Hara, 1992; Metcalfe et al., 1994). Lake Pátzcuaro is a small, closed lake with a modern τ_e value of 22 years (from data in O'Hara, 1992). Sediment cores from the lake testify to a prolonged low stand, dated A.D. 800-1200. Low levels affected at least two other climatically sensitive lakes in the central Mexican highlands between A.D. 650 and 1200. This suggests that drought stress may have been a factor in the collapse of the great city of Teotihuacán (Basin of Mexico) and possibly of the Classic Maya ceremonial centers as well (Metcalfe et al., 1994; Hodell et al., 1995).

In contrast, isotopic investigations of Wallywash Great Pond, Jamaica (Street-Perrott et al., 1993; Holmes et al., 1995) indicate that a prolonged period of heavy rainfall, possibly of temperate origin, was centered on A.D. 750. Wet conditions also prevailed at this time in Lake Miragoane, Haiti (Curtis and Hodell, 1993). Paleolimnological studies of small lakes in the equatorial upper Amazon

(Colinvaux et al., 1985) have revealed an episode of widespread flooding in the interval A.D. 650-1250. The overlap in time between these prolonged anomalies of differing sign in the West Indies, Mexico, and Ecuador, all of which were the most important of the last 3,000 years in their respective regions, suggests that they reflect a large-scale reorganization of the areas of moisture convergence and cloudiness over the tropical Americas, compatible with a slight southward displacement of the Bermuda High.

RECOMMENDATIONS FOR FUTURE MONITORING

Where lakes are concerned, unlike glacier termini and river flows, there is as yet no coordinated program of monitoring and data archiving on a global scale. Although monthly gauge-board measurements are very simple to make, they are commonly interrupted by large changes in water level that necessitate relocation of the monitoring site, and the data are stored, often in haphazard fashion, by a great diversity of regional or national agencies.

A lake integrates the water balance over its catchment area. In arid and semiarid regions in particular, lakes offer the potential of bridging enormous gaps in the network of conventional meteorological and hydrological stations. Moreover, significant changes in water level are predicted to occur over the next few decades in response to greenhouse warming (Cohen, 1986).

What is needed now is a global program of data collection and storage, making use of appropriate satellite technology to provide monthly coverage on a near-real-time basis. An effort in this direction has been initiated by the Mullard Space Science Laboratory, University College London, using the data transmitted by ERS-1 (I. M. Mason, pers. commun.). However, much more could be done to coordinate both the collection of conventional instrumental, historical, and paleolimnological data, and the development of automated data-acquisition systems for other satellites.

ACKNOWLEDGMENTS

I wish to thank Ian Mason, Rob Crossley, and Sarah O'Hara for their help in preparing this paper.

Decade-to-Century-Scale Variability of Regional and Hemispheric-Scale Temperatures

PHILIP D. JONES AND KEITH R. BRIFFA[1]

ABSTRACT

Variability is examined in three types of temperature data: mean hemispheric temperatures from the last 100 years, longer single-site temperature records, and millennium-long tree-ring reconstructions of summer temperature. The annual average warming, since the late nineteenth century, is of the order of 0.5°C for both hemispheres. The warming is more erratic in the Northern Hemisphere, with a slight cooling between 1940 and 1970. The Southern Hemisphere shows less variability, with a more monotonic warming trend in this century.

Longer European temperature records indicate that the warming of the twentieth century is not unusual compared to that of some decades in the late eighteenth century. In most European records the nineteenth century was cooler than the eighteenth, with the 1880s the coldest decade. Some of the warmth of the last 100 years, therefore, may reflect the unusually low starting point. Four millennium-long tree-ring reconstructions (Fennoscandia, northern Urals, Tasmania, and northern Patagonia) are studied. In only one of these is the twentieth century the warmest, but in the others it is one of the warmest. Warmer conditions have been reflected in three of the summer reconstructions in previous centuries, but never at the same time at all four locations.

INTRODUCTION

One of the main reasons put forward for our inability to detect the enhanced greenhouse effect is that the inherent natural variability of the climate system is sufficiently great to obscure the signal (Wigley and Barnett, 1990). Although the 0.5°C global warming that has occurred over the last 100 years is consistent with model predictions resulting from greenhouse-gas buildup over this time (Wigley and Raper, 1992), our poor understanding of natural climate variability on the decade-to-century time scale means that a large part, or even all, of the observed warming might have occurred anyway.

There are no climatic elements for which we have records long enough to fully define the characteristics of past decade-to-century climate variability. The longest and geographically most extensive measurements are of surface

---

[1]Climate Research Unit, University of East Anglia, Norwich, U.K.

temperature, precipitation, and pressure, but there are problems associated with even these records. Measurements have never been global in extent, nor as spatially uniform as required by statistical sampling theory (Madden et al., 1993; Trenberth et al., 1992). Furthermore, the expected signals of greenhouse-gas-induced changes (e.g., as predicted by modeling experiments) for precipitation and pressure are poorly understood for a variety of reasons (Barnett and Schlesinger, 1987; Wigley and Barnett, 1990).

In this paper we consider two aspects of temperature variability. First, we study hemispheric-mean temperature variability over the last 100 years. Second, in order to examine temperature variations over longer periods, we look at some long single-site and composite instrumental records, principally from Europe, and at the few millennium-long paleoclimatic reconstructions of summer temperature. Although imperfect, these long records currently represent one of our only methods of looking at century-time-scale variability with "real" data.

VARIATIONS IN HEMISPHERIC MEAN TEMPERATURE

Land Regions

Before about 140 years ago, instrumental temperature measurements were limited to Europe, parts of Asia and North America, and some coastal regions of Africa, South America, and Australasia. By the 1920s the only areas without instrumentation were some interior parts of Africa, South America, and Asia; Arctic coasts; and the whole of Antarctica.

Although incomplete, the coverage since 1850 allows the development of continental and hemispheric averages of temperature. Using compilations of homogeneous station records, Jones (1988) and Jones et al. (1986a,c) produced a gridded (5° latitude by 10° longitude) data set of surface air-temperature anomalies over land for each month since January 1851. The basic data were interpolated on a regular grid to mitigate the effects of uneven spatial density of the station network. Interpolating the station data in absolute degrees is not a viable option, since this would be affected by varying station numbers, different station elevations, and different formulae for calculating monthly averages. The use of anomaly values from a common reference period (1951-1970) overcomes these problems. A consequence of this procedure is that hemispheric-mean temperatures are expressed in relative rather than absolute terms.

Time series of hemispheric-mean seasonal and annual temperature anomalies are shown in Figure 1. The features exhibited by the two sets of curves have been discussed before (see, for example, Folland et al., 1990, 1992; Jones and Briffa, 1992; Jones et al., 1986a,c; Wigley et al., 1985, 1986). The differences in temperature between the second half of the record (1946-90) and the first half (1901-45) are listed in Table 1. The annual series for each hemisphere shows a warming of the order of 0.5°C since the late nineteenth century. The warming is considerably more erratic in the Northern Hemisphere, where a cooling of about 0.2°C clearly occurred between about 1940 and 1970 in all seasons except spring. In the Southern Hemisphere the warming is more monotonic, and there is no evidence of cooling after 1940.

The various seasonal curves in Figure 1 show considerable variation in periods of warming and cooling. The greater difference between seasonal trends over the last 140 years occurs over the Northern Hemisphere. In summer and autumn, the 1980s were barely warmer than the temperature levels of the 1930s and 1940s. In winter and spring the 1980s were clearly the warmest decade. All seasons except summer show the long-term warming evident in the annual data. In summer, the 1850s to 1870s were as warm as the most recent two decades. The cooling from the 1870s to the 1880s is the most pronounced feature of the summer

TABLE 1 Temperature Differences, 1946-1990 Average Minus 1901-1945 (land-only data)

	Whole Year	Jan	Feb	Mar	Apr	May	Jun	Jul	Aug	Sep	Oct	Nov	Dec
Northern Hemisphere													
$\Delta T(°C)$	0.14	0.14	0.22	0.20	0.28	0.24	0.16	0.07	0.06	0.07	0.06	0.11	0.21
$\sigma(1901-90)$	0.23	0.47	0.52	0.43	0.32	0.27	0.23	0.23	0.23	0.24	0.31	0.36	0.45
$\sigma(1901-45)$	0.22	0.46	0.54	0.36	0.31	0.29	0.24	0.26	0.25	0.26	0.35	0.38	0.50
$\sigma(1946-90)$	0.22	0.48	0.49	0.48	0.28	0.20	0.20	0.18	0.20	0.21	0.26	0.32	0.38
Southern Hemisphere													
$\Delta T(°C)$	0.21	0.20	0.17	0.21	0.15	0.28	0.23	0.30	0.29	0.19	0.17	0.20	0.19
$\sigma(1901-90)$	0.19	0.25	0.26	0.27	0.26	0.32	0.28	0.30	0.33	0.27	0.26	0.24	0.25
$\sigma(1901-45)$	0.15	0.22	0.27	0.23	0.22	0.26	0.27	0.26	0.30	0.22	0.25	0.22	0.23
$\sigma(1946-90)$	0.17	0.24	0.21	0.26	0.28	0.31	0.25	0.27	0.29	0.29	0.24	0.21	0.23

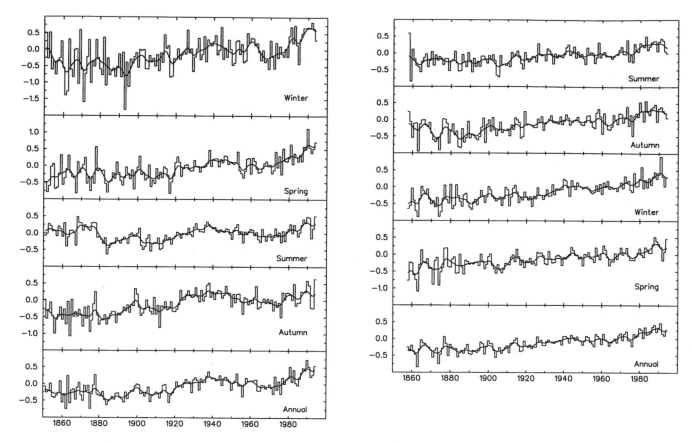

FIGURE 1 Surface air temperatures (land only) by season and year. Data are expressed as anomalies with respect to the 1951-1970 average. Smooth curves in this and subsequent plots were obtained by using a 10-year Gaussian filter. Left, Northern Hemisphere temperatures, 1851-1994; right, Southern Hemisphere temperatures, 1858-1994. Northern Hemisphere winters and Southern Hemisphere summers in this and subsequent figures are dated by the year in which the January occurs.

data. Over the Southern Hemisphere there is more agreement between the long-time-scale seasonal trends.

In both hemispheres, greater year-to-year variability is apparent during the nineteenth century. This increased variability is due to the sparser spatial coverage at that time. Although the estimates for individual years may be less reliable in the nineteenth than the twentieth century, the "frozen-grid" analyses undertaken by Jones et al. (1986a,c) indicate that the decadal-scale temperature fluctuations are reliably reproduced by the available data (for more discussion of this see Madden et al., 1993). The coolness of the 1880s compared with 1851 to 1880 in the Northern Hemisphere is, therefore, probably real. Thus, for the hemispheric temperature analyses of Hansen and Lebedeff (1987, 1988) and Vinnikov et al. (1990), who begin their analyses around 1880, warming rates measured over 1881 to 1990 are slightly greater than rates calculated over 1861 to 1990 (see also Folland et al. (1990, 1992) for more discussion of this point).

Land and Marine Regions

Since land represents only 29 percent of the area of the earth's surface, it is important to incorporate marine data

into hemispheric averages if we want to get the "best possible" global series. Merchant and naval ships have taken weather observations and measured the temperature of the sea surface since the beginning of the nineteenth century. In the last 20 years, major international efforts have been made to transfer all of the climate data contained in ships' log books into computer data banks. One such compilation is the Comprehensive Ocean-Atmosphere Data Set (COADS) produced by NOAA workers at Boulder, Colorado (Woodruff et al., 1987). COADS contains about 80 million non-duplicated sea surface temperature (SST) observations. Another similar data set has been assembled by the U.K. Meteorological Office (Bottomley et al., 1990). Much of the data is common to both sets, but comparisons are under way to isolate the unique observations in each.

Unfortunately, as with land data, marine records are affected by inhomogeneities and errors. Correction schemes have been devised to adjust both marine air temperatures and sea surface temperatures for biases attributable to the method of measurement, time of day, etc. (see, e.g., Bottomley et al., 1990; Folland and Parker, 1990, 1991; Folland et al., 1990, 1992; Jones and Wigley, 1990; and Jones et al., 1991).

Land and marine temperatures may be combined in a number of ways. We describe below the Climatic Research Unit's combined data set, comprising the land temperatures discussed earlier and SST anomalies corrected for changes in measurement technique. The main change is from uninsulated-bucket to engine-intake measurements (and some insulated buckets) around the start of the 1940s. The SST correction scheme is discussed in detail by Jones and Wigley (1990) and Jones et al. (1991). The data, which are given as a set of 5° by 5° grid-box values, are anomalies from the 1950-1979 reference period. Where co-located land and marine grid data occur, the resulting value is the average of the land and marine components.

Annual and seasonal time series of hemispheric-mean temperatures, based on the combined data set, are shown in Figure 2. Detailed discussions of the features exhibited by the two sets of curves may be found elsewhere (see, for example, Folland et al., 1990, 1992; Jones and Briffa, 1992; Jones and Wigley, 1990). The differences in temperature between the second half of the record (1946-90) and the first half (1901-45) are listed in Table 2. In many respects the year-to-year and decadal-scale variations in Figure 2 are damped versions of those seen in Figure 1. There are, however, subtle differences. The greater nineteenth-century variability prominent in the land-only data is not as apparent in the land-plus-marine data. The cooling between 1940

and 1970 evident in the Northern Hemisphere land-only seasonal series is hardly apparent in the combined data. It is replaced by a period of little trend in 1940-1960, followed by a slight cooling from 1960 to the mid-1970s. Variability in the combined data set is now similar between the two hemispheres, whereas in the land-only data the Northern Hemisphere variability was clearly greater.

The intercorrelation between the two hemispheric combined series is remarkably high (for annual values between 1901 and 1990, $r = 0.79$). Some of this high correlation is due to coincident long-term trends, and the remainder is due to common high-frequency (year-to-year) variability. Some of the interannual correspondence derives from a response to the common forcing of the El Niño/Southern Oscillation (ENSO) phenomenon (Philander, 1983). Many of the years that are warm relative to the filtered curves in Figure 2 have been shown to be El Niño years or "warm event" years (Bradley et al., 1987b; van Loon and Shea, 1985). In contrast, many cold years, relative to the filtered curves, correspond to La Niña or "cold event" years (Philander, 1985). Using the Southern Oscillation Index (SOI) (Allan et al., 1991; Ropelewski and Jones, 1987), Jones (1989) has removed the ENSO influence from the hemispheric and global temperature series. The SOI explains about 25 to 30 percent of the high-frequency (less than 20 years) variations in hemispheric temperature anomalies.

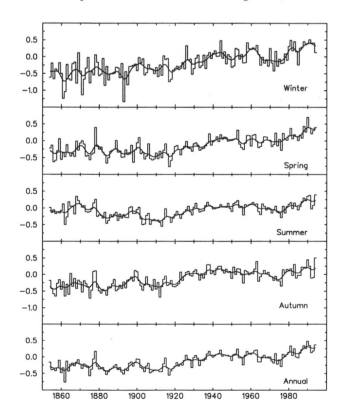
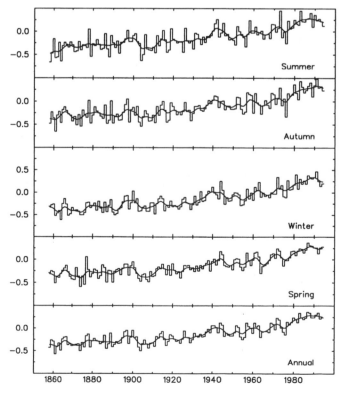

FIGURE 2   Northern Hemisphere (left) and Southern Hemisphere (right) surface air temperatures (land-plus-marine data) by season, 1854 to 1994. Data are expressed as anomalies from 1950 to 1979.

TABLE 2   Temperature Differences, 1946-1990 Average Minus 1901-1945 (land-plus-marine data)

	Whole Year	Jan	Feb	Mar	Apr	May	Jun	Jul	Aug	Sep	Oct	Nov	Dec
Northern Hemisphere													
$\Delta T(^\circ C)$	0.23	0.24	0.28	0.30	0.30	0.28	0.23	0.19	0.18	0.17	0.14	0.17	0.28
$\sigma(1901\text{-}90)$	0.21	0.32	0.36	0.32	0.26	0.23	0.21	0.20	0.20	0.21	0.24	0.26	0.35
$\sigma(1901\text{-}45)$	0.20	0.29	0.35	0.27	0.24	0.22	0.20	0.21	0.21	0.22	0.27	0.28	0.36
$\sigma(1946\text{-}90)$	0.16	0.31	0.32	0.30	0.19	0.14	0.15	0.13	0.15	0.16	0.19	0.21	0.28
Southern Hemisphere													
$\Delta T(^\circ C)$	0.24	0.25	0.23	0.26	0.22	0.23	0.22	0.26	0.25	0.25	0.24	0.24	0.22
$\Delta(1901\text{-}90)$	0.20	0.24	0.23	0.23	0.22	0.23	0.23	0.24	0.25	0.23	0.21	0.20	0.23
$\sigma(1901\text{-}45)$	0.14	0.21	0.21	0.19	0.16	0.17	0.18	0.19	0.19	0.15	0.18	0.16	0.19
$\sigma(1946\text{-}90)$	0.17	0.21	0.20	0.21	0.22	0.23	0.21	0.20	0.24	0.22	0.18	0.17	0.21
Global													
$\Delta T(^\circ C)$	0.23	0.25	0.26	0.28	0.26	0.26	0.23	0.22	0.22	0.21	0.19	0.20	0.25
$\sigma(1901\text{-}90)$	0.19	0.25	0.27	0.26	0.22	0.21	0.19	0.19	0.20	0.19	0.20	0.21	0.26
$\sigma(1901\text{-}45)$	0.16	0.22	0.24	0.20	0.19	0.18	0.17	0.17	0.18	0.16	0.21	0.20	0.24
$\sigma(1946\text{-}90)$	0.15	0.21	0.23	0.23	0.17	0.16	0.15	0.14	0.16	0.15	0.14	0.15	0.21

about 25 to 30 percent of the high-frequency (less than 20 years) variations in hemispheric temperature anomalies. Removal of the ENSO influence produces data series that may assist in the early detection of greenhouse-gas-related trends (Nicholls and Katz, 1991; Wigley and Jones, 1981).

VARIATIONS IN LONGER RECORDS

Study of the hemispheric time series since 1850 allows the variability on the 10-to-30-year time scale to be investigated. Century-time-scale variability, however, can be considered only with records that span at least several centuries. In this section we examine some of the longest instrumental records available (principally from Europe) and some of the longest annually resolved proxy climatic reconstructions.

Longer Instrumental Records

The most detailed compilation of long-term instrumental climate data (both temperature and precipitation) currently available is that of Bradley et al. (1985). This compilation extended and improved on the previously available data sets by searching meteorological and other archives for published and manuscript sources of early instrumental records. An important aspect of the resulting compilation is that it contains details of the sources of all of the station data sets and, where possible, details of their long-term homogeneity (see Bradley et al., 1985; Jones et al., 1985, 1986b). Clearly, if one is to study climatic change, it is vital to ensure homogeneity of time-series data.

Figure 3 shows annual time series for 12 stations (for seasonal time series see Jones and Bradley, 1992). The time series are smoothed using a 10-year Gaussian filter to emphasize variations on decadal and longer time scales, and expressed as anomalies from the 1901-1950 period. The longest time series are restricted to the Northern Hemisphere between 40° and 64°N. All the series, except for Toronto, were considered homogeneous by Jones et al. (1985). We have included Toronto here because the extension of its record to 1770 (Crowe, 1992) makes it the longest such time series from North America. Urban warming is clearly evident from the 1880s on.

For the European sites, the warming of the twentieth century is not unusual compared to the longer record. The period encompassing the late nineteenth and early twentieth centuries is one of the coldest for Europe. Temperatures in the late twentieth century are only marginally higher than those during parts of the eighteenth century. The North American and Asian records clearly show long-term warming, but all these records begin much later than in Europe. Evidence from tree-ring reconstructions during the seventeenth and eighteenth centuries in western North America indicates conditions as warm as today's (Briffa et al., 1992b; Fritts, 1991).

Within-Region Similarities

A comparison of the long European records shows consistency on the decadal time scale, with strongly similar cool and warm decades since 1700. Cool decades are evident

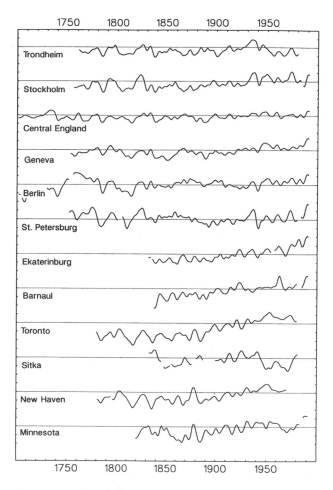

FIGURE 3 Low-frequency temperature variations at 12 selected sites having the longest records. Two records, central England and Minnesota, represent regional composite series rather than individual site records. The time series shown result from the application of a 10-year filter. Each tick mark on the vertical scale represents 1°C. Data are expressed as anomalies from 1901 to 1950.

during the 1740s, 1780s, 1800s, and 1810s, and the late 1830s and 1840s. Warm decades are limited to the 1820s. Similar decadal similarities are seen in other regions. The two Asian records show cool decades during the 1830s and 1840s as in Europe. The four North American records show relative cold during the 1780s, 1810s, and 1830s, again similar to the European data, although warmth is indicated during the 1800s, the 1820s, and to some extent during the 1840s. The cold-warm-cold oscillatory pattern between the 1810s and 1830s is particularly striking in the New Haven and Toronto records in North America and at Stockholm and St. Petersburg in Europe.

Between-Region Similarities

Since 1850 the best-known difference between European temperatures and those in the rest of the hemisphere occur-

red during the early 1940s. At this time Europe was cold while the rest of the hemisphere was warm, at a level exceeded in magnitude only during the 1980s (Jones et al., 1986a). North American station records show warmth during the 1800s and 1840s, while European stations show cool conditions. Although some correspondence appears between warm and cool decades over the northern landmasses, more comprehensive regional studies of twentieth-century temperature variations over the Northern Hemisphere have shown that no one region can be said to be fully representative of hemispheric-wide temperatures (Briffa and Jones, 1993; Jones and Kelly, 1983). There is, therefore, no reliable short-cut method of estimating hemispheric or global temperatures on the basis of relatively few station records, at least on the decadal time scale, as has been tried by Groveman and Landsberg (1979). (Lack of longer-time-scale data prevents reaching a similar conclusion about estimating temperatures on century time scales.)

Proxy Records

Many methods are used to reconstruct past climates (see Bradley, 1985). Only a few, however, have the potential for providing records that can be resolved to an annual or seasonal level. Of those that can, the most widely used sources are documentary historical data, dendroclimatic data, and ice-core records (see Bradley and Jones, 1992). Documentary reconstructions are mainly restricted to Europe and Asia and to the period since about A.D. 1400 (see, e.g., Mikami, 1992). Few documentary reconstructions extend continuously over long time periods, the proxy temperature series for Switzerland from 1525 (Pfister, 1992), for China from 1470 (Wang and Wang, 1991), and for Iceland from 1590 (Ogilvie, 1992) being the longest. Documentary evidence probably represents the most reliable source of past climatic information on short time scales. The ability of some documentary indicators to faithfully record low-frequency climatic variations on time scales longer than the human life span (in particular a century or more) must be questioned, however.

Ice-core evidence is limited to high-latitude and high-altitude locations. Furthermore, few ice-core reconstructions of temperature are explicitly calibrated against instrumental time series, and there may be some ambiguity in interpreting the seasonality of the climate variations (Bradley and Eischeid, 1985; Thompson et al., 1986). When isotope-temperature relationships have been tested, the best correlations on the interannual time scale explain only about 25 percent of the temperature variance (Jones et al., 1993; Peel, 1992). Notwithstanding these points, ice cores represent important environmental information, and may be the principal means of reconstructing highly resolved long temperature series in terrestrial tropical areas (albeit at high elevations).

To illustrate the potential value of proxy data further, we concentrate on dendroclimatic reconstructions. Some of the longest represent "warm-season" temperatures in northern Fennoscandia, the northern Urals, Tasmania, and northern Patagonia (Briffa et al., 1990, 1992a, 1995; Graybill and Shiyatov, 1992; Cook et al., 1991, 1992; Villalba, 1990; and Lara and Villalba, 1993, respectively). Details of some of these reconstructions are given in Table 3 and in the caption of Figure 4. As noted in an earlier section, summer (particularly in the Northern Hemisphere) is the season that appears from the instrumental record since 1850 to be most atypical in its long-term changes (see also Briffa and Jones, 1993). Another potential limitation of dendroclimatic reconstructions is that they may not fully represent very-long-time-scale variability because of the need to transform original ring-width measurements into indices in order to minimize any potential bias in the mean chronology resulting from different ages of the constituent trees (Cook et al., 1990, 1995a). The apparent character, in terms of long-time-scale variability, of the different chronologies (and hence their climate reconstructions) depends on the longevity of the trees (and hence the length of the measurement time series), the ecology of the site(s), and the data-processing method used to remove the 'age effect'.

Figure 4 shows the reconstructions filtered with 20- and 100-year low-pass filters. Each of the series has been rescaled to represent anomalies from the 1901-1960 period. The importance of the different chronology-production methods is demonstrated in the upper halves of both panels in Figure 4, which show alternate reconstructions for each of two northern high-latitude regions: northern Fennoscandia, produced using the same ring-width and density measurement data (Briffa et al., 1990, 1992a), and the northern Urals region, based on different data sets (ring width only for June-July (Graybill and Shiyatov, 1992) and ring widths and ring density for May-September (Briffa et al., 1995)). Differences in the way the original growth measurements for the many individual trees were amalgamated within these chronologies has markedly affected the appearance of the long-time-scale variability in the resulting climate

reconstructions. However, establishing the veracity, or true confidence levels, associated with the long-period fluctuations is extremely problematic.

Notwithstanding the methodological considerations, Figure 4 also shows that, compared to the southern series, the variance of the northern reconstructions is much greater. This may reflect the strong moderating influence of the southern oceans evident in the lower variance of the observational data in these regions. The explained variances in the southern reconstructions are also generally lower than those in the north (see Table 3). On the 20-year time scale, the temperatures for the two northern regions are largely out of phase between A.D. 1000 and 1200. In these comparisons, we use the two thicker curves for Fennoscandia and the Urals, which show the greater variability on decadal and longer time scales when Fennoscandia was warm and the Urals cold. After 1600, the two curves are more similar, both being cold through much of the seventeenth and the second half of the nineteenth centuries. The drop in temperatures in Fennoscandia began around 1580 and lasted to around 1750, whereas in the northern Urals the drop began earlier, the coldest period being from 1530 to 1670. Both series indicate cold summers around the end of the nineteenth century, although conditions were more severe in the Urals. The Fennoscandian series also shows protracted or relatively severe cold during the sixth, seventh, ninth, early twelfth, thirteenth, and early fourteenth centuries. On the 100-year time scale the Fennoscandian series shows generally continuous warmth exceeding that of the 1930s lasting from about 900 to 1100. Shorter warm periods also occurred during the 760s, around 1160, and notably in the early decades of the fifteenth century. The Urals series shows warmth in the thirteenth and fourteenth centuries, but on this time scale it does not exceed that of the twentieth century.

The nature of the Southern Hemisphere reconstructions is very different from that of those in the north. Low-frequency variability is conspicuously less. In Tasmania, the main feature of the entire reconstruction is a rapid increase in temperature since early this century. For northern Patagonia, the two series show less agreement with each

TABLE 3   Variance (r^2) Explained by Calibration of Different Summer Temperature Reconstructions Based on Tree-ring Data (see Figure 4)

Region	Reference	Calibration Period	r^2
Northern Fennoscandia	Briffa et al., 1990	1876-1975	0.51
Northern Fennoscandia	Briffa et al., 1992a	1876-1975	0.55
Northern Urals	Graybill and Shiyatov, 1992	1881-1969	0.60
Northern Urals	Briffa et al., 1995	1882-1990	0.68
Tasmania	Cook et al., 1992	1938-1989	0.37
Northern Patagonia	Villalba, 1990	1908-1984	0.42
Northern Patagonia	Lara and Villalba, 1993	1910-1987	0.36

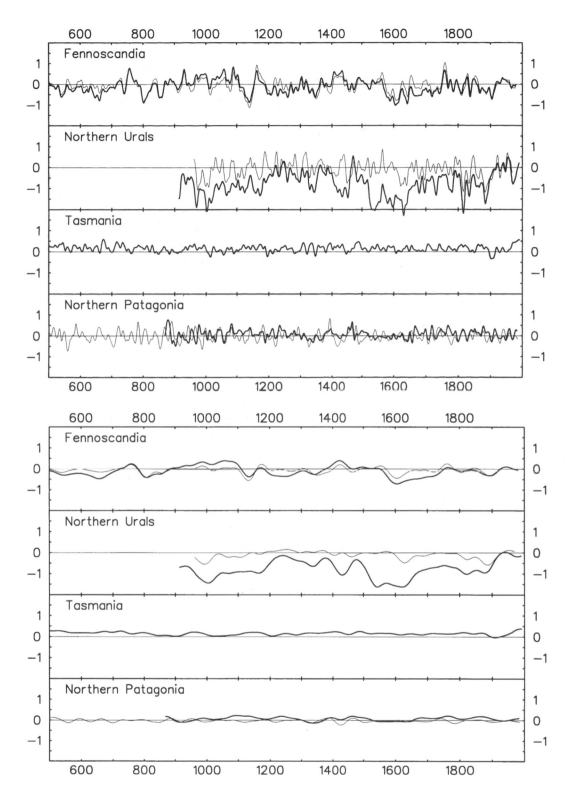

FIGURE 4 Variations in four of the longest tree-ring reconstructions of summer temperature. The reconstructions are for northern Fennoscandia (April to August; thick line, Briffa et al., 1992a; thin line, Briffa et al., 1990), northern Urals (May to September, thick line, Briffa et al., 1995; June to July, thin line, Graybill and Shiyatov, 1992), Tasmania (November to April; Cook et al., 1992), and northern Patagonia (December to February; thick line (Rio Alerce, Argentina), Villalba, 1990; thin line (Lenca, Chile), Lara and Villalba, 1993). Variations are shown in degrees Celsius after application of a 20-year low-pass filter (upper panel) and a 100-year low-pass filter (lower panel). For all series the zero line is 1901 to 1960.

other than might be expected. There is no evidence of any recent dramatic or unprecedented rise in temperature.

Taken together, the reconstructions for the two hemispheres show little coherence in century-time-scale variability. The most consistent features are the cool episodes from 1550 to 1650 and again between 1880 and 1910. The early period probably represents the coldest 100-year period of the last millennium, while the latter underscores the possibly exaggerated warming trends that may be estimated by calculating hemispheric trends beginning in the 1880s rather than the 1860s (see also Folland et al., 1990).

CONCLUSIONS

On the basis of the instrumental record, the 1980s was clearly the warmest decade. Warmth is evident in all seasons and in both hemispheres. There is evidence, clearer in the Northern Hemisphere, that the twentieth-century warming occurred in two phases, between 1920 and 1940 and since 1975. All seasonal curves, however, exhibit large decadal-time-scale variability superimposed on century-time-scale warming.

Longer single-site and composite instrumental records confirm the warming from the late nineteenth century on. The 250-year-long records from Europe show that the 1880s were the coldest decade, at least since 1700, so part of the

warming since then may reflect this unusually low starting point. In most European records the eighteenth century was warmer than the nineteenth century.

In only one (the Northern Urals, Briffa et al., 1995) of the thousand-year dendroclimatic reconstructions is the twentieth century clearly the warmest century. In all of the other reconstructions, however, this century is one of the warmest. Warmer (summer) conditions occurred in previous centuries, but never at the same time at all locations. Clearly, a greater geographical spread of long paleoclimatic reconstructions is required; types of proxy evidence that cover seasons other than summer must be employed before the apparently unprecedented warmth shown in instrumental records of the twentieth century can be placed in a longer-term context.

ACKNOWLEDGMENTS

The authors thank Don Graybill, Ed Cook, Antonio Lara, and Ricardo Villalba for making the results of their long dendroclimatic reconstructions available. This research has been supported by U.S. Department of Energy's Atmospheric and Climate Division (Grant No. DE-FG02-86ER60397), the U.K. Department of the Environment (Contract No. PECD/7/12/78), and the E.U. Environment Programme (Grant EV5V-CT94-0500; DG12 DTEE).

Commentary on the Paper of Jones and Briffa

THOMAS R. KARL
NOAA National Climatic Data Center

I very much enjoyed reading this paper. I should just like to call your attention to some questions that were raised in my mind as I read through it.

First, there is the question of whether or not we can really adequately measure global-scale decadal fluctuation prior to about 1900. I am primarily concerned about the spatial resolution and spatial comprehensiveness of the data. I want to show one diagram to give you an idea of the types of errors that occur when you attempt to calculate global temperature trends and do not have complete global coverage of the phenomena. Starting out with complete global coverage and systematically removing data to calculate a trend over 10, 50, 150 years—removing it in a way that matches the historical coverage of the observations over both land and sea, of course—on a 10-year time scale you get standard errors that are over 1°C. The standard errors come down to perhaps 0.4°C to 0.6°C for 50 years; when you go out to 100 years, the standard errors are much

smaller. There are also some significant biases that can creep into the record. This is cause for thought.

Another question I have is related to an issue Gene Rasmusson raised: Are we optimally using the data we have? Are 5- by 10-degree boxes the best way of calculating these global numbers? Is some kind of optimal averaging preferable? You get slightly different answers depending on how you go about the averaging.

My third point is the issue that Dr. Jones raised with respect to the very cool conditions during the 1880s, especially during the summer. You wonder not only about the spatial comprehensiveness of the data at that time, but also whether there could be some exposure problems.

Another question was raised in my mind by Dr. Jones's comment that decadal-scale temperature anomalies are not hemispherically consistent. One might ask whether there is any evidence to suggest that this may be less likely to occur at longer time scales, such as centuries. Can we get away with a less comprehensive data set when we go to longer time scales?

Because over two-thirds of the globe is covered by ocean, we also need to ask how confident we are about the SST corrections that have been applied to the data.

Dr. Jones mentioned the ice-core isotope-temperature relationships. He indicated that they explain only about 25 percent of the annual temperature variance in the observed record. The question here is: Is that a high enough proportion for these records to be useful indicators of larger-scale changes? Perhaps it is not if we are looking at annual scales, but we are really interested in century scales here.

Another issue is the sensitivity of the tree rings and ice cores to climate change. For instance, if climate is changing and the moisture source regions that contribute to tree growth or ice depth change, the signal may be contaminated. If they are responding to climate change, can they still adequately reflect what happened several hundred years in the past?

Dr. Jones also showed that there is little coherence among the several 1000-year tree-ring time series that have been assembled. That is very interesting; if it is true, putting together a global-scale multi-century time history of temperature change will be very challenging indeed! Another question about the analyses relates to the fact that the tree rings represent only certain portions of the year. For example, although the analyses all related to the summer, each was done for different months during the summer. Could part of the difference in the time series be due to the fact that they were not all measured in exactly the same seasons?

My last question is whether we can prioritize the work that needs to be done to improve our ability to estimate global temperature change on the century time scale. Should we even attempt to do this? Or should we let science work from the bottom up?

Discussion

SOCCI: Do the isotopes used to measure temperature account for only 25 percent of the variability because a large part of the signal is an ice-volume signal?

JONES: What I've done is to take an isotope record from the coastal portions of Greenland and the Antarctic peninsula and to compare the ^{18}O or deuterium values with year-to-year temperature values at a coastal site. That's where the 25 percent comes from. Of course, you might get more variance on longer time scales.

MYSAK: What's the state of the art in extracting, say, precipitation or runoff from tree-ring data?

JONES: Most of the tree-ring work we've dealt with is from Scandinavia, where the response is clearly to temperature. In southern Europe or the southern part of the United States you can get a good reconstruction of river flow or precipitation from trees, because there the trees are responding to moisture. In between, as in England or northern Germany, the trees are responding to a mix of the two, and it's hard to unravel the climate signal.

GROOTES: Remember that the isotopes are a measure only of the conditions when snow falls, and the season of greatest accumulation may vary considerably. The value of Phil's tree-ring records is that they have been carefully selected for their response to climate to get at the margins of the trees' limits.

BERGMAN: Would it be possible to take information from mountain-valley glaciers and deconvolve some of the problems with ice dynamics to get something useful?

JONES: Well, yes, but you wouldn't get the year-to-year reconstructions I can get from the tree-ring records. And remember that different source locations will give different response times, and changes to the source area can make dramatic differences to ice cores, as the formation of the Weddell Sea polynya did to Antarctic cores in the late 1970s.

LEHMAN: The Greenland isotope data suggests that higher elevations on an ice sheet are less likely to be contaminated by source variation effects. On a different topic, I was disturbed by how short the correlation decay lengths were for your Greenland records. It seems to me that they would have to be much longer for the larger temperature changes that characterize the deglaciation.

JONES: Yes, the ones I showed were calculated on decadal time scales from 90 years of data. But I don't think it has yet been shown, for the longer time scales, that you do have a much larger-scale signal. We also need to assess how representative our sites are of the time and space scales of interest.

KEELING: I was wondering how you took care of the heat-island effect. Also, those coastal Greenland stations you used for comparison are subject to very strong temperature inversions in the winter, and thus have very high variability. They would not be totally representative of the conditions on the top of the ice sheet.

JONES: We had only one station with a heat-island problem, Toronto.

RIND: We should remember that isotopes are an integrator over the path between the moisture source and the ice core, which will extend the length scale of a perturbation by comparison with a station record.

PARKER: You could estimate the representativeness of the

Greenland data by using something other than surface data, such as upper-air temperatures, though you'd have fewer years available. My other points relate to Tom's comments. First, the early instrumental temperature data seem to have been a little high for summer and a little low for winter, relatively speaking. As far as I can tell from nineteenth-century papers, this is fairly well documented for Europe, though less so for North America. Second, uncertainties in the corrections to instrumental SSTs may make a difference of about 0.1° C in estimates of nineteenth-century temperature. I suspect that the lack of spatial coverage then is a greater problem, however.

JONES: The important thing is that the difference between summer and winter in the 1850s and 1860s is markedly different from what it has been for most of the twentieth century.

KARL: David, from your work would you say that the cooling Phil showed in the 1880s is real, or might it be the result of the switch to using Stevenson screens?

PARKER: Perhaps both; you'd have to compare the changes with atmospheric circulation data from the same part of the world. The change of three- or four-tenths of a degree around 1880 seems a bit much to be simply instrumental.

KARL: The point I'd like to bring out is that some of the global analyses begin calculating trends and changes from 1880, so we may be starting at a low point.

RASMUSSON: Speaking of data problems, I recently saw that the Climate Analysis Center people found that in their data set the famous World War II SST average jump occurred between December 1941 and January 1942, which seems suspiciously coincidental.

JONES: David tells me that the data receipts prior to December 8, 1941 came from various nations' ships, whereas after that date they came mainly from U.S. ships.

ROOTH: When we look at this business of correlation lengths, we shouldn't forget that the regional meteorology can be used to help us interpret the data. For instance, Fennoscandia is very much influenced by intermittent invasions of Atlantic maritime air. A lot of the low-frequency variability there may reflect those maritime invasions as well as variability over longer time scales. The invasions don't really reach the Urals, and you see different frequency characteristics there.

Interdecadal Temperature Oscillations in the Southern Hemisphere: Evidence from Tasmanian Tree Rings Since 300 B.C.

EDWARD R. COOK[1], BRENDAN M. BUCKLEY[2], AND ROSANNE D. D'ARRIGO[1]

ABSTRACT

A 2290-year reconstruction of warm-season Tasmanian temperatures from tree rings has been analyzed for decadal-scale fluctuations. Spectral analyses indicate the existence of four oscillatory modes with mean periods of 31, 56, 79, and 204 years. The waveforms of these oscillations, estimated by singular spectrum analysis, appear to be reasonably stable through time, although each exhibits varying degrees of amplitude and phase modulation. The temperature oscillations, especially at periods of 31 and 56 years, may be related to the expansion and contraction of the circumpolar vortex around Antarctica and the compensating north-south movement of the subtropical high-pressure belt off eastern Australia. However, the cause of such sustained oscillatory behavior remains a mystery. Self-sustained internal forcing related to ocean circulation dynamics and deep-water formation is one possibility. Another possibility, related principally to the 79- and 204-year terms, is external forcing caused by long-term solar variation. In particular, the 79-year oscillation appears to have been phase-locked since 1700 with the envelope of solar cycle length associated with the sunspot numbers. Regardless of their cause(s), these decadal- and century-scale temperature oscillations seem to be important features of the climate system in the Tasman Sea Region that may need to be considered when searching for evidence of greenhouse warming.

INTRODUCTION

Compared to the Northern Hemisphere (e.g., Lamb, 1977), the Southern Hemisphere (SH) has relatively few high-resolution, instrumental climate records that extend back more than 100 years (Barry, 1978). To properly characterize and evaluate recent climatic trends and decadal-scale fluctuations apparent in SH instrumental records (e.g., Flet-cher et al., 1982; Jones et al., 1986c; Hansen and Lebedeff, 1987), however, much longer proxy climate records are needed. In this paper, we describe and analyze one such series: a 2290-year temperature reconstruction for Tasmania developed from a precisely dated annual tree-ring chronology. In so doing, we document the probable existence of interdecadal oscillations that appear to be important features of the SH climate system in the Tasmania-New Zealand sector.

[1]Lamont-Doherty Earth Observatory of Columbia University, Palisades, New York
[2]Institute of Antarctic and Southern Ocean Studies, University of Tasmania, Hobart

The existence of decadal-scale temperature and precipitation fluctuations in continental, hemispheric, and global instrumental records is well documented (e.g., Folland et al., 1984; Karl and Riebsame, 1984; Bradley et al., 1987a; Ghil and Vautard, 1991), if still poorly understood. Mechanisms believed to be responsible for some of the observed multi-year climatic fluctuations include explosive volcanic eruptions (Bradley, 1988), solar variability (Newell et al., 1989), lunar tidal effects (Currie, 1981), and the ocean thermohaline circulation (Stocker and Mysak, 1992). Indeed, the difficulty in determining the causes of decadal-scale climatic fluctuations has relegated the 10-to-100-year bandwidth of climatic variability to the "gray area of climatic change" (Karl, 1988), where physical theory is at present inadequate to explain the observations.

Understanding the causes of interdecadal climatic variability is a challenging task, made all the more difficult by the brief span of the instrumental climate records. Fluctuations on the time scale of 10 to 100 years are difficult to study if the instrumental data extend back only 80 to 100 years. Therefore, extending climate records back in time, particularly in poorly covered regions of the world like the SH, is critical to the study of interdecadal variability. One of the few regions of the SH that has the potential for millennia-long, high-resolution proxy climate time series is Tasmania, with its temperate rainforests of long-lived conifers. Recently, Cook et al. (1991) reported the development of a climatically sensitive Huon pine (*Lagarostrobos franklinii* C.J. Quinn) tree-ring chronology from western Tasmania that extended back to A.D. 900. The yearly variations in ring width were shown to correlate well with changes in warm-season (i.e., November to April) temperature. Cook et al. (1991) then went on to draw inferences about the significance of the present-day warming trend apparent in the series since 1965, and speculated about the occurrence of the Little Ice Age and Medieval Warm Period in Tasmania.

In a later paper, Cook et al. (1992) described more fully the development and characteristics of this tree-ring chronology and extended their analyses to the quantitative reconstruction of November-to-April average temperatures for Tasmania back to A.D. 900. In so doing, they produced the first well-verified estimates of past temperature change in the Australia-New Zealand sector of the SH that extend back through the Medieval Warm Period and the Little Ice Age as broadly defined by Lamb (1965) and Grove (1988), respectively. The Huon pine temperature estimates explained approximately 37 percent of the temperature variance and correlated well with actual data withheld from the regression-based reconstruction model. Cross-spectral analysis also indicated that the temperature reconstruction was especially good at estimating decadal-scale temperature fluctuations, with magnitude-squared coherencies between actual and estimated temperatures exceeding 60 percent for periods longer than 12 years.

Since the Cook et al. (1992) paper was written, the Tasmanian temperature reconstruction has been extended back to 300 B.C. through the inclusion of tree-ring series from several old Huon pine stumps found at the same site where the original series was developed. This extended reconstruction is shown in Figure 1. (Although no year zero actually exists for the B.C. to A.D. time scale, our series

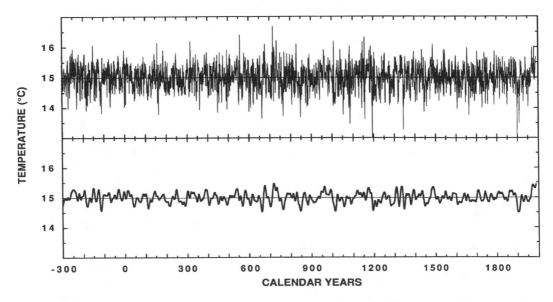

FIGURE 1   The extended November-to-April temperature reconstruction for Tasmania. The reconstruction covers the period 300 B.C. to A.D. 1989. The period from A.D. 900 is nearly identical to that previously published in Cook et al. (1992). The extension back to 300 B.C. is based on completely independent tree-ring data from old stumps found at the same site. The smooth curve superimposed on the annual values is a smoothing spline that acts as a 25-year low-pass filter.

for the sake of continuity includes a datum for that year.) Qualitatively, the extension prior to A.D. 900 looks similar to, and therefore homogeneous with, the later period. In the next section, we will describe the spectral properties of this record, which indicate the probable existence of interdecadal oscillatory modes in the SH climate system that have persisted over the past 2290 years. Later, we will take a detailed look at these oscillations through time and speculate on mechanisms that might be responsible for their existence.

SPECTRAL ANALYSES

Cook et al. (1992) examined the power spectrum of the Tasmanian temperature reconstruction estimated for the period A.D. 900-1989. Using both Blackman-Tukey and maximum-entropy spectral analysis (Jenkins and Watts, 1968; Marple, 1987), they found evidence for statistically significant (a priori $p < 0.05$) peaks in the spectrum at approximately 30, 56, 80, and 180 years. Cook et al. (1992) did not examine the properties of these apparent oscillations in the series. However, they speculated that the 30- and 56-year modes could be reflecting similar oscillations in the SH climate system associated with different wave-number flow regimes of the circumpolar zonal westerlies in the 40° to 50°S latitude zone (Enomoto, 1991).

The extended temperature reconstruction described earlier provides us with a rare opportunity to validate the existence of these apparent oscillations. Clearly, if they disappear from the early portion of the extended record, then the physical basis for the existence for these modes is doubtful. For this purpose, we split the series into two essentially independent 1145-year segments covering the intervals from 300 B.C. to A.D. 844 (early period) and A.D. 845 to 1989 (late period).

The Blackman-Tukey power spectra for these periods are shown in Figure 2 (estimates are shown only for periods greater than 10 years). As expected, the late-period spectrum

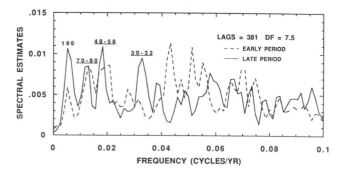

FIGURE 2   Blackman-Tukey spectra of the first and second halves of the Tasmanian temperature reconstruction. Each half is 1145 years long. The spectra are shown only for periods greater than 10 years. Note the similarity between the spectra in the low-frequency end.

has peaks very close in frequency to those found previously, the principal difference being that the longest-period oscillation is now 190 years. The early-period spectrum likewise has peaks in the same vicinity, although there are some differences. The 190-year peak is more subdued in the earlier period, and the 30-year peak has virtually disappeared. In contrast, the 56- and 80-year peaks found in the late period are present in the early period at closely associated frequencies and with similar power. If the bandwidths of the 56-, 80-, and 190-year peaks in the late-period spectrum are taken into account, the frequencies of the associated peaks in the early-period spectrum are fully covered. The same formal coverage applies for the power estimates in the early period if 80 percent confidence intervals are computed for the late-period peaks. Thus, in terms of both frequency and power, there is little that significantly distinguishes the spectra of the early and late periods at approximately 56, 80, and 190 years. In contrast, the 30-year peak appears to be less stable through time. This result covering the past 2290 years suggests that the putative oscillations are not statistical artifacts.

For completeness, the spectrum of the entire series is shown in Figure 3. The identified periods differ slightly due to the increased record length. All four peaks of interest here exceed the a priori 95 percent confidence level (dashed line), on the basis of a first-order Markov null-continuum model.

SINGULAR SPECTRUM ANALYSIS

The previous spectral analyses revealed the probable existence of oscillatory behavior in warm-season Tasmanian temperatures over the past 2290 years. However, changes in amplitude and phase are indicated, which necessitates a more thorough examination of these oscillations in the time domain. For this purpose we used singular spectrum analysis (SSA), a data-adaptive technique that is particularly well suited for isolating weak signals embedded in red noise (Vautard and Ghil, 1989). SSA decomposes a time series into signal and noise sub-spaces by applying principal-components analysis (PCA) to the autocorrelation function (ACF) of that process. Vautard and Ghil (1989) refer to the eigenvectors produced by PCA as empirical orthogonal functions (EOFs) and the eigenvalues as singular values, hence SSA. A pure oscillation embedded in red noise, whose average period is shorter than the length of the ACF, will be composed of an even-and-odd EOF pair. The EOF pair will resemble the shape of the oscillation and be in quadrature (i.e., 90° out of phase). The EOF loadings can be thought of as digital filter weights that are used to estimate the waveform. For this purpose, only the even (or symmetric) EOF, which preserves phase, is used here. Vautard and Ghil (1989) refer to these recovered waveforms as principal components (PCs). The final step of SSA is spectral analysis

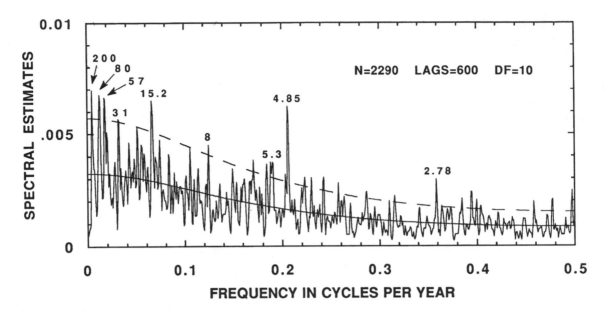

FIGURE 3   The Blackman-Tukey spectrum of the entire temperature reconstruction shown in Figure 1. A first-order Markov null spectrum was used for constructing the a priori 95 percent confidence level (dashed line). The four spectral periods of interest here are all statistically significant on the basis of these confidence levels, with the 31-year term being the weakest. These results are consistent with the split spectra shown in Figure 2.

of the PCs using high-resolution, maximum-entropy spectral analysis (Marple, 1987).

As in classical spectral analysis based on the Blackman-Tukey approach (Jenkins and Watts, 1968), the number of lags used in computing the ACF for SSA is somewhat arbitrary, being a tradeoff between resolution and stability. Initial experiments indicated that about 300 lags (13.1 percent of the series length) was the minimum needed to simultaneously resolve the 31-, 56-, 80-, and 200-year oscillations. In fact, for lag windows over the range 300 to 700, these four oscillations were always found in the four leading EOF pairs after suitable low-pass prefiltering to remove unwanted high-frequency variance. However, spectral analyses of their PCs indicated that about 400 lags (17.5 percent) were needed to cleanly separate the closely spaced 56- and 80-year oscillations. When only 300 lags were used, the spectra of these oscillations were somewhat contaminated by each other's power. In contrast, the 31-year term alone could be reasonably well resolved using only 100 lags (4.3 percent). For this reason, 100 lags were chosen to isolate the 31-year term, while 400 lags were used to isolate the longer oscillations.

Figure 4 shows the four even EOFs estimated as described above. The average period of each EOF is very similar to that obtained from the spectral analysis of its respective waveform. The EOFs are remarkably regular, given that they are based solely on the data and are not constrained to be sinusoidal. Figure 5 shows the normalized power spectra estimated by the maximum entropy method, which confirms the strongly periodic nature of these oscillations. These spectra are purely descriptive and cannot be

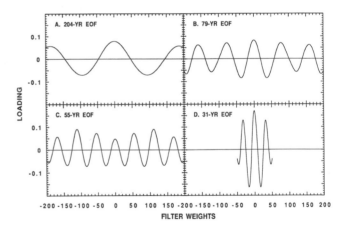

FIGURE 4   The four even empirical orthogonal functions (EOFs) estimated by singular spectrum analysis representing the oscillatory modes in the temperature reconstruction.

readily tested for statistical significance. The identified periods are very close to those found earlier (compare Figures 2 and 3 and Cook et al., 1992). Each spectrum has some degree of side-lobe power in the form of small secondary peaks, which may reflect a degree of amplitude and phase modulation in the waveform. However, they are small enough to be relegated to noise at this stage of analysis. The 31-year oscillation has the most complicated spectrum, with distinct secondary peaks of 29 and 35 years. This may reflect some drift or instability in the mean period over time.

The waveforms upon which the maximum entropy spectra are based are shown in Figure 6. The 204-year waveform

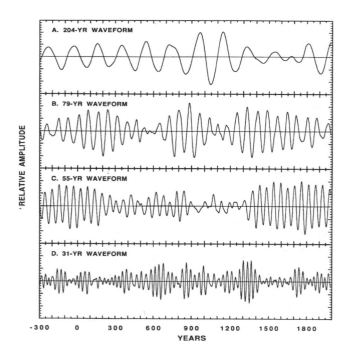

FIGURE 5 The spectra, estimated by maximum entropy, of the temperature waveforms shown in Figure 6. The spectra have all been scaled to unit peak heights, and the principal peaks are labeled.

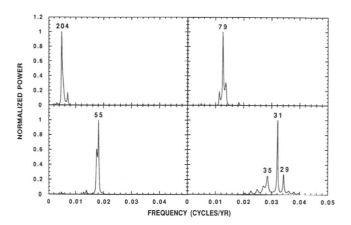

FIGURE 6 The waveforms of the temperature oscillations identified previously through standard spectral analysis. The waveforms were estimated using singular spectrum analysis, and their units are dimensionless.

is perhaps the most well behaved, with remarkable stability over the first half of the record. However, an increase in amplitude is indicated in the A.D. 900-1200 interval (during the Medieval Warm Period), followed by greatly reduced amplitude in the A.D. 1300-1800 period (during the Little Ice Age). Since about 1800, the 204-year waveform has increased in amplitude to a level not seen since A.D. 1300. The other waveforms are much more complicated, with each showing varying degrees of amplitude and phase modulation.

POSSIBLE LINKS TO THE CLIMATE SYSTEM AFFECTING TASMANIA

The statistical evidence for persistent decade-to-century-scale oscillations in warm-season Tasmanian temperatures is provocative and begs for a physical explanation. We do not have one yet, but some relationships between sea level pressures, sea surface temperatures, and land air temperatures over Tasmania for the past 100 years may provide some clues for where to look.

As noted earlier, Cook et al. (1992) suggested that the 31- and 55-year oscillations could be related to 20-to-30 and 40-to-60 year fluctuations in zonally averaged July sea level pressure (SLP) data in the 40° to 50°S latitude zone (Enomoto, 1991). Specifically, the 20-to-30-year SLP fluctuation was considered by Enomoto (1991) to be a standing oscillation of wave-number-zero structure related to the expansion and contraction of the circumpolar vortex. Inter-

estingly, one of the centers of maximum variance was located in the Tasman Sea region (130°E to 170°W). The 40-to-60-year fluctuation was considered by Enomoto (1991) to be an oscillation with wave-number-one structure, meaning an eccentricity of the circumpolar vortex. Each of these SLP fluctuations showed large amplitude changes in the 1880 to 1920 interval in the Tasmania-New Zealand sector, with anomalously low pressure being indicated for the 1890 to 1905 period. This implies an expansion of the circumpolar vortex at that time, with a greater tendency for cool, southwesterly winds and below-average temperatures over Tasmania.

A comparison of Hobart SLP data and Tasmanian actual and reconstructed temperatures for the November-to-April warm season supports this interpretation. Figure 7 shows the

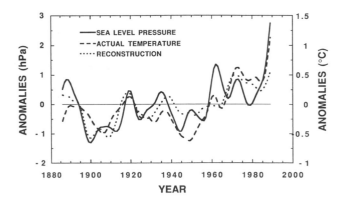

FIGURE 7 The relationship between warm-season sea level pressure (SLP) at Hobart and temperatures over Tasmania. The cold early-1900s period was characterized by anomalously low SLP, which suggests an expansion of the circumpolar vortex at that time. In contrast, the warm interval beginning in the 1960s was characterized by anomalously high SLP and an inferred expansion of the subtropical high off eastern Australia.

pressure and temperature anomalies since 1886, smoothed to highlight decadal-scale fluctuations. Each series shows persistently negative anomalies in the 1895 to 1920 period, indicating a clear link between anomalously low warm-season SLP and below-average temperatures over Tasmania, presumably related to an expansion of the circumpolar vortex at that time and the northward migration of the subtropical high-pressure belt off the eastern coast of Australia. Figure 7 also indicates a link between SLP and the recent warming in Tasmania. Since about 1960, both warm-season SLP and temperature have been anomalously high, which suggests a southward movement of the subtropical high-pressure belt and a greater frequency of warm, northeasterly winds over Tasmania. This inference is supported by Lough (1991), who showed that the high-pressure belt moved southward about 2° over the 1942-to-1981 period. It is also supported by Coughlan (1979), who showed that annual mean maximum temperatures over Tasmania were significantly correlated with the latitude of the subtropical high.

There is also a possible link between ocean circulation features in the Tasman Sea region and changing land and air temperatures over Tasmania. This link is suggested by changes in the correlation fields of warm-season (November to April) land and marine temperatures versus reconstructed Tasmanian temperatures for two 40-year periods (1888 to 1927 and 1950 to 1989) containing the cold and warm periods indicated in Figure 7. To generate the correlation fields, we obtained corrected land and marine monthly temperature data for 5° × 5° grid cells covering the southern Australia-Tasmania-New Zealand sector. These data were kindly provided by P.D. Jones and K.R. Briffa of the Climatic Research Unit, University of East Anglia, Norwich.

Figure 8 shows these correlation fields. The integer in each cell reflects the correlation coefficient rounded to the nearest tenth. For the cool early period (1888 to 1927), the correlation field is weighted toward the grid boxes off the west coast of Tasmania. This result implies that the inferred expansion of the circumpolar vortex coupled with increased southwesterly winds in the 1890 to 1920 period may have caused the cold West Wind Drift to be displaced northward from its normal position, resulting in anomalously cool temperatures over Tasmania (see Figure 7). In contrast, the correlation field of the late period (1950-1989) is strongly weighted toward the Tasman Sea east of Tasmania. Given the anomalously warm temperatures in Tasmania at that time (Figure 7), this result suggests concomitant warming in the Tasman Sea and more southerly penetration of the warm East Australian Current. In fact, the actual SSTs during this time were anomalously warm, in parallel with warm temperatures over Tasmania. This result is consistent with the southward shift of the subtropical high-pressure belt (Lough, 1991) and increased frequency of warm northeasterly winds over Tasmania.

None of the empirical relationships between Tasmanian

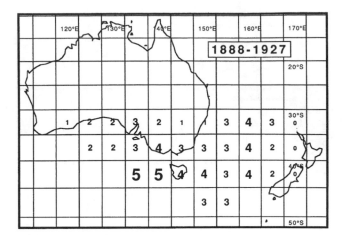

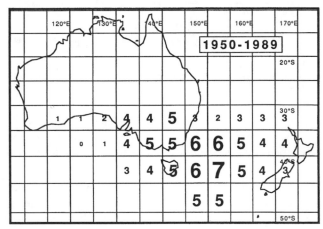

FIGURE 8   Correlation fields of 5° × 5° grid-cell temperatures versus Tasmanian temperatures for two contrasting cold and warm periods. Each integer reflects a correlation coefficient rounded to the nearest tenth. The 1888-to-1927 period contains the coldest period in the Tasmanian instrumental records, while the 1950-to-1989 period contains the warmest period. Note that the correlation field is weighted toward the western ocean cells during the cold period. During the later warm period, the field is strongly weighted toward the east in the Tasman Sea. The differences in these fields suggest changes in atmospheric and oceanic circulation during cold and warm phases of climate in Tasmania.

temperatures and large-scale atmosphere/ocean processes can explain why oscillations (which have apparently persisted over the past 2290 years) are present in the reconstruction. However, should a physical model be developed, it will probably involve interactions in the atmosphere/ocean/cryosphere system that influence the movements of the circumpolar vortex and subtropical high-pressure belt. Recently, Stocker and Mysak (1992) proposed that self-sustained oscillations, of similar order to those found here and in many other proxy climatic records, could be induced by internal changes of the ocean thermohaline circulation. Using a simple two-dimensional ocean model, they showed that internal nonlinear dynamics alone could produce self-sustained oscillations in ocean-to-atmosphere heat flux with periods of 19, 38, and 110 years for the Atlantic Ocean and 110 years for the Pacific Ocean. Similar results have been obtained from a GFDL coupled ocean-atmosphere model (Delworth et al., 1993). Figure 9 shows the power spectrum of detrended mean annual surface air temperatures for the Southern Hemisphere generated by the GFDL model (Delworth et al., 1993). There is a statistically significant (a priori $p < 0.05$) oscillation in simulated temperatures with a mean period of about 110 years, along with minor secondary peaks at about 56 and 29 years. Thus, there appears to be a reasonable model-based argument for the Tasmanian temperature oscillations' having an internal origin in the Southern Hemisphere ocean-atmosphere system.

With regard to possible solar forcing, we compared our 79- and 204-year waveforms with those related to the 80-to-90-year Gleissberg sunspot cycle and an approximately 200-year oscillation found in high-precision radiocarbon measurements from tree rings (Sonett, 1984; Stuiver and Braziunas, 1989), which Sonett and Suess (1984) have suggested could influence temperatures. These comparisons produced suggestive but equivocal results.

Figure 10 shows the waveform of the 79-year temperature oscillation and the envelope of the annual Wolf sunspot numbers (i.e., the Gleissberg cycle) estimated by SSA for the common period 1700 to 1987. There is a surprising degree of agreement between the two, with temperatures lagging sunspots by about 25 years over the central portion of the curves. However, the phasing apparently breaks down after 1900, with temperatures lagging sunspots by only 10 years. The phase drift may simply mean that the good relationship in the central portion is spurious. However, if sunspot cycle length is used instead of sunspot number as an index of solar activity (Friis-Christensen and Lassen, 1991), the phasing between that index and Tasmanian temperatures actually improves. Peaks in solar cycle length

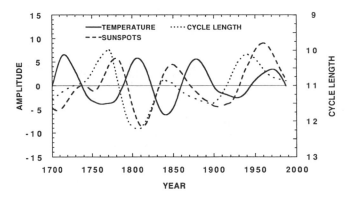

FIGURE 10   The waveform of the 79-year temperature oscillation compared to the envelopes of sunspot number (the Gleissberg cycle) and sunspot cycle length since 1700. The temperature and sunspot number waveforms were estimated using SSA, and are in dimensionless units. The cycle-length envelope is in years. There is an apparent lag of about 30 to 40 years between Gleissberg-cycle phase and temperature oscillations in Tasmania.

occurred around 1770, 1840, and 1940 (Friis-Christensen and Lassen, 1991). The corresponding peaks in the 79-year temperature waveform occurred around 1805, 1880, and 1970. In turn, solar-cycle-length minima occurred around 1805 and 1900, whereas temperature minima are found around 1840 and 1930. Thus, there appears to be a reasonably consistent lag relationship of 30 to 40 years between Tasmanian warm-season temperatures and solar cycle length since the early 1700s. In being nearly 180° out of phase, this putative relationship either implies that (1) the direct correlation between solar cycle length and Northern Hemisphere temperatures found by Friis-Christensen and Lassen (1991) is effectively opposite in the SH sector around Tasmania, (2) the thermal inertia of the SH oceans significantly delays the response of SH surface air temperatures to solar forcing, or (3) the results of Friis-Christensen and Lassen (1991) and those presented here are spurious. With only about three realizations of the solar-cycle-length envelope available from the sunspot record, it is difficult to make a strong case for either of the first two possibilities.

To compare our 204-year temperature oscillation with the approximately 200-year term in $\Delta^{14}C$, we obtained high-precision decadal radiocarbon measurements of dendro-chronologically dated wood from Stuiver and Becker (1986) covering the period 2500 B.C. to A.D. 1950. These measurements were detrended with a fifth-order polynomial to remove the geomagnetic dipole effect, and the data after 1885 were deleted to eliminate the industrial "Suess effect." Figure 11 shows the residuals from the polynomial curve since 300 B.C., which have also been smoothed to emphasize the century-scale fluctuations in $\Delta^{14}C$. Notable persistent departures are indicated in the A.D. 1100 to 1220 period (the Medieval Solar Maximum) and the A.D. 1450 to 1550 and 1650 to 1700 periods (the Spörer and Maunder Minima,

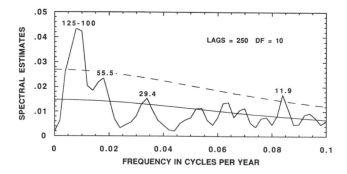

FIGURE 9   The power spectrum of Southern Hemisphere surface air temperatures generated by a GFDL coupled ocean-atmosphere model. The model was run for 1000 years. A first-order Markov null spectrum was used for constructing the a priori 95 percent confidence level (dashed line). An oscillation with a mean period of around 110 years is clearly indicated in the spectrum.

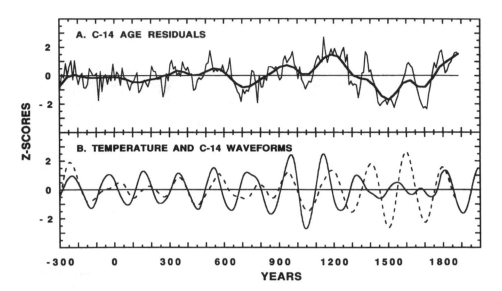

FIGURE 11   The ^{14}C age residuals from the polynomial trendline (A) and the approximately 200-year waveforms of temperature (solid curve) and ^{14}C residuals (dashed curve) (B). The residuals have also been smoothed to highlight long-term ^{14}C fluctuations particularly evident in the Medieval Warm Period and Little Ice Age. The waveforms are approximately in phase up to about A.D. 1050, after which the relationship breaks down until 1800.

respectively). Prior to A.D. 600, the series is very stable, with few significant long-term departures.

We applied SSA to the Δ^{14}C residuals to isolate the waveform of the approximately 200-year oscillation. This waveform is also shown in Figure 11, with the 204-year temperature waveform superimposed on it for comparison. Not surprisingly, this comparison produced mixed results. There is some degree of phase-locking during the first 1300 years, especially considering the error in the radiocarbon dates of about ± 10 years (Stuiver and Becker, 1986), after which it completely breaks down during the Little Ice Age period until about A.D. 1800. There is also a lack of agreement in the amplitude modulation apparent in each series. The Δ^{14}C oscillation has maximum amplitude through the period of the Spörer and Maunder Minima (also the Little Ice Age). In contrast, the temperature oscillation has maximum amplitude around the Medieval Solar Maximum (also the Medieval Warm Period).

The net result of this examination for a climate-sun link in Tasmanian temperatures at periods of about 80 and 200 years is mixed. If the relationship between solar cycle length and temperatures found by Friis-Christensen and Lassen (1991) is real, then there is some support for a climate-sun link in the temperature reconstruction at a period of 80 to 90 years. The link between the 204-year temperature oscillation and solar variability is more tenuous because of unstable phasing and the mismatch in amplitude modulation noted above. Given these somewhat contradictory results, the climate-sun link cannot be accepted at this time.

IMPLICATIONS FOR DETECTING THE GREENHOUSE EFFECT IN THE SOUTHERN HEMISPHERE

The decadal-scale natural oscillations found in warm-season Tasmanian temperatures indicate that the climate system in this part of the SH is to some degree internally, and perhaps also externally, forced. The level of forcing due to the four oscillations (as a percentage of the total yearly variance in temperatures) is comparatively small, amounting to only about 10 percent. However, when compared to the variance in the reconstruction due to decadal-scale temperature fluctuations alone (i.e., only variance at wavelengths >10 years), the oscillations account for approximately 46 percent of that low-frequency fraction. Given that only four oscillatory modes are necessary to explain a substantial fraction of the low-frequency variance, the following question is posed: To what extent have these natural oscillations contributed to the recent decadal-scale anomalous warming over Tasmania, as described in Cook et al. (1991)?

An examination of the waveform plots in Figure 6 suggests that their collective effect on Tasmanian temperatures since 1960 has been significant. Since that time, all four waveforms have either peaked or are still increasing. In terms of relative amplitude, the 204-year oscillation is also in its most active phase since the Medieval Warm Period. Therefore, it is plausible that the anomalous warming since 1960 in Tasmania is simply due to an unusual coincidence between the four oscillations and an increase in the amplitude of the 204-year term. It is also noteworthy that during the anomalously cold early 1900s, all of the oscillations were either decreasing or were at a minimum. These relationships are shown in Figure 12, where the four-waveform average is plotted with low-pass filtered actual and reconstructed warm-season temperatures since 1886. The similarity of the curves indicates that both unusually cool and unusually warm decadal-scale temperature anomalies over Tasmania during the twentieth century have been driven in part by long-term climatic oscillations.

This conclusion does not necessarily eliminate greenhouse warming as a possible contributor to the recent temperature increase in Tasmania. Indeed, temperatures since 1985 appear to have increased sharply even as the waveform average has decreased. However, it would be premature to claim that this departure is a manifestation of the greenhouse effect. The average waveform shown in Figure 12 can be used only qualitatively at this time; it does not necessarily contain all of the important terms related to multi-year temperature fluctuations. Nonetheless, its agreement with instrumental temperature data indicates that oscillatory modes with time constants of decades to centuries are an important part of the SH climate system. Until this behavior is better understood, the existence of greenhouse warming in the SH will be difficult to prove.

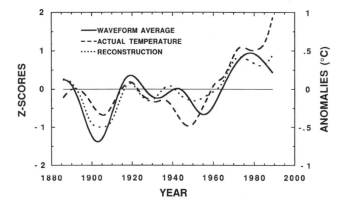

FIGURE 12 Comparison of the four-oscillation waveform average with actual and reconstructed Tasmanian temperatures for the period 1886 to 1989. The waveform average is expressed in standard normal deviates, while the temperatures are expressed as anomalies. The temperatures have been low-pass filtered to a level comparable to the waveform average. Note the coincidence between the waveform minimum around 1905 and cold temperatures, and between the waveform maximum around 1975 and warm temperatures.

ACKNOWLEDGMENTS

We gratefully acknowledge the assistance of Trevor Bird (Tasmanian Trades and Labor Council, Forestry Unit) and Mike Peterson (Tasmanian Forestry Commission), who truly made this research possible. In addition, Roger Francey (C.S.I.R.O. Division of Atmospheric Research) and Mike Barbetti (University of Sydney) actively assisted us in our field work. We also thank the Tasmanian Forestry Commission and the C.S.I.R.O. Division of Forest Research in Hobart for local logistical support. Peter Kelly dated and measured the Huon pine stump wood, Rob Allan (C.S.I.R.O. Division of Atmospheric Research) provided the Hobart SLP data, and Tom Delworth at GFDL provided the model-generated SH temperature data. This research is supported by the National Science Foundation, Division of Earth Sciences, Geologic Record of Global Change Program, Grant EAR 91-04922. Lamont-Doherty Earth Observatory Contribution No. 5215.

Commentary on the Paper of Cook et al.

WILLIAM E. REIFSNYDER

I should probably begin with the disclaimer that I am a forester, and by no means a statistician and number-cruncher in the sense that tree-ring people have become. But I strongly believe that one of our best sources of proxy data for the fairly recent past—several thousands of years—is tree-ring chronology. I am delighted to see people working hard and getting good sequences, and I really commend Ed Cook, Brendan Buckley, and Rosanne D'Arrigo for doing just that.

I always get nervous, however, at seeing these power spectra that have peaks with rather little power in them.

And I notice too that often when one gets more data many of the relationships either diminish or disappear. I wonder, then, whether this very real problem might be solved with other, quite different data sets. For example, Pieter Grootes told us about the ability of the ice-core analyzers to get very high resolution, on the order of single years, in data from good locations. So to open the discussion, I would like to throw this question out to all of you: Has anyone really attempted to use other data sets to see whether the same peaks will occur in those data?

Discussion

BUCKLEY: We hope to extend our ^{14}C chronology further, which may give us an answer to that question. Also, I expect to find similar temperature resolution at other sites in Tasmania next year.

GHIL: I can show the same peaks from an instrumental record [see Figure 1 in the essay introducing Atmospheric Modeling in Chapter 2]. On a power spectrum of a 300-year record of central England temperatures obtained by the singular spectrum analysis method you can clearly see two interdecadal and two interannual peaks. One is 25, near enough to your 30, and one 14.2, like your 15. The interannual peaks are 7.6, near your 8, and 5.2, matching your 5.2.

MYSAK: The 200-year time scale you passed over rather lightly seems to be the overturning time scale, at least for the Atlantic, and is what we found in our two-dimensional model.

JONES: Michael, it seems to me that your central England temperature record may show the Little Ice Age, but not the modern minimum. A sunspot feature, perhaps. And Brendan, relative to your Figure 10 with the solar cycle length, I wanted to point out that for that Southern Hemisphere pressure data set you're using there are absolutely no data in the zone between 40° and 50° south. Now, the dotted solar-cycle-length curve shows an enormous cycle about 1750 that Kristen and Lassen have postulated as forcing global temperatures. However, they used that curve from 1850 on while ignoring the record back to 1700.

RASMUSSON: These points are awfully important in critically examining what we really have here. But my question is about the treatment of the tree-ring data. Did you have to remove the growth trend, and if so, how did it affect the low-frequency variations?

JONES: The Tasmanian trees are so much longer-lived than the Scandinavian ones I was referring to that it's less of a problem to remove the growth trend.

BUCKLEY: Our paper does talk about bias, particularly at the end of the series, and describes Ed Cook's technique for dealing with it, which is different from Phil's. Also, yew and pine tend not to follow the normal exponential growth curve; they can be very slow-growing subcanopy trees for 200 or 300 years. Often we don't even use those 300 years because it's hard to measure that small growth with any accuracy.

KEELING: If I may, I should like to show for comparison the analysis we made at Scripps of the Jones-Quigley temperature record. I'm not forgetting, of course, that one tends to favor spectral analyses that confirm one's own position. But it seems to me that those spectral lines do come quite close to those we find in the temperature record. The 4.85 line is not far from one of the principal lines of the El Niño signal. Then there are some that may or may not correspond with others. But that 15-year line, that I mentioned was so hard to explain as solar, seems to be quite persistent. It's also half of the 31-year peak I mentioned in my presentation, which is a principal tidal line. It's not as striking as the 93-year cycle, but it's significant. In any case, whether it's the 80 and 200 you showed, and whether there is a Gleissberg effect or not, wherever you see evenly spaced patterns you should look to see what the beat frequency is, and whether it might relate to a long-period line too.

KUSHNIR: I think we need to remember that if we do a spectral analysis with such high resolution and are looking at the 95 percent significance level, there's always the chance that 5 percent of the peaks will be above the line.

MCWILLIAMS: Dave Keeling's tides have very clear frequencies. But in general, why are we looking for lines in the climate record? Why should we expect a priori to see periodicities rather than more distributive broad-band behavior?

GHIL: We like periodicities because they enhance predictability. But within the context of non-linear dynamics, even though what you observe is broad-band, that is a reflection of instabilities in the underlying phenomena you are trying to track. My suspicion is that each of those lines has a story behind it, and I hope that over the next 10 or 20 years we can begin to understand the interactions of the various phenomena.

REIFSNYDER: It's also true that some of those lines might disappear as more data are accumulated, though our need for cycles—circadian, budget—won't.

KAROLY: I didn't want to talk about cycles, but mechanisms. I think the mechanism you identified as associated with the pressure and temperature fluctuations in Tasmania suggests a structural change, possibly associated with the Antarctic current or the large-scale atmospheric circulation, that fits in quite well with the modulation of the strength of the zonal winds and the large-scale atmospheric circulation on interannual time scales. And it may well be a mode of the decadal and longer time scales as well, with a broad-band rather than a specific frequency.

REIFSNYDER: We should perhaps keep in mind that statistical significance does not equate with reality.

Variability of Atlantic Circulation on Sub-Millennial Time Scales

SCOTT J. LEHMAN[1]

ABSTRACT

New sediment records from the southeastern Norwegian Sea document a series of abrupt changes in poleward flow of warm surface waters during the last deglaciation (15,000-8,000 yr BP), some of which produced shifts in sea surface temperature of more than 5°C in less than 40 years. These findings confirm the timing, magnitude, and rates of circum-Atlantic climate change implied by ice-core data and predicted by numerical models, and support the theory that the intensity of thermohaline overturning controlled air temperatures around and downwind of the northern North Atlantic.

Unlike the deglacial interval, the last 8,000 years or so have been marked by relatively stable climate conditions in the North Atlantic region. However, as pointed out earlier by Broecker (1987), the conditions that conferred stability over this interval are not yet understood; thus, it cannot be assumed that such conditions will persist in the face of the gradual changes already forecast as a response to anthropogenic perturbation of the climate system.

INTRODUCTION

A key question confronting scientists and policy makers is the degree to which the 0.3°C to 0.6°C rise in globally averaged surface air temperatures observed since the mid-1800s (e.g., Jones and Wigley, 1986) can be ascribed to humankind's impact on atmospheric concentrations of CO_2 and other trace gases. Although the direction and magnitude of the observed temperature change is compatible with many model predictions, the extent to which it may be embraced or overprinted by natural climate variations is not yet known. Indeed, the need to document and understand natural climate variations in this context has motivated many of the contributions to this volume. This paper, on the other hand, views the issue of natural climate variation from a different perspective. Here I present evidence from the end of the last glacial period for natural shifts in circulation of the North Atlantic Ocean that produced regional changes in air and sea temperatures of 5 to 10°C in less than 50 years, which are much too large and sudden to be confused with gradual changes of the type seen in the instrumental record. Rather, the point is that comparable ocean-driven changes might be an unforeseen response to gradual anthropogenic forcing

[1]Woods Hole Oceanographic Institution, Woods Hole, Massachusetts; present address, INSTAAR and Department of Geological Sciences, University of Colorado, Boulder

of the ocean-atmosphere system in the future. Such changes are "unforeseen" because the vast majority of numerical models used to predict the climate response to CO_2 forcing treat the oceans as fixed, or restrict their interaction with the atmosphere to the mixed layer. However, ocean currents are crucial in determining global transports of heat and moisture, and the transient response of climate will thus depend critically on the response of the oceanic circulation. Although the importance of simulating such interactions is widely recognized, it is nevertheless hampered by the difficulty of simulating the general circulation of the oceans (i.e., "getting it right for the right reasons") and the extreme computational requirements of coupling three-dimensional models of the ocean and atmosphere.

Citing the geologic evidence for abrupt climate changes in the past, Broecker (1987) has already articulated the concern that there may be "surprises in the greenhouse" that cannot be adequately portrayed, let alone predicted, using existing models. I follow Broecker's lead here, adding new evidence that, as predicted on the basis of terrestrial indications of rapid climate change around the Atlantic region, ocean circulation patterns and sea surface temperatures varied with degree-per-decade rapidity. Understanding and modeling the mechanisms underlying these changes is of high priority. If, on the one hand, the climate response to increasing greenhouse-gas content of the atmosphere is expected to be gradual, then we may choose to pursue a strategy of adaptation. On the other hand, if we cannot rule out the possibility that the mechanisms that provoked the "surprises" seen in the geologic record may once again become active in the future, we may wish to adopt a policy of restraint, while at the same time trying to enhance our capacity for the recognition and detection of the premonitory signs of abrupt circulation and climate change.

THE CASE FOR "SURPRISES"

There is ample evidence from the geologic and instrumental record for changes in patterns of ocean circulation on a variety of time and space scales. These range from global-scale reorganizations of the thermohaline circulation on glacial to interglacial (10^5-yr to 10^4-yr) time scales (Boyle and Keigwin, 1987; Duplessy et al., 1988; Broecker and Denton, 1989; Raymo et al., 1990) to quasi-cyclic, interannual changes of regional scale in historical times, such as El Niño events (Cane, 1983). The spectrum of variability also appears to include possibly cyclic phenomena such as oceanic cooling and reduced ventilation associated with the North Atlantic's Great Salinity Anomaly during the 1960s to 1980s (Dickson et al., 1988b; Lazier, 1988; Mysak and Power, 1991). One of the most extreme examples of rapid ocean circulation change seen anywhere has been inferred from the geologic record of the last deglaciation (15,000

to 8,000 radiocarbon yr BP)[2] around the North Atlantic, when air temperatures shifted 5°C to 10°C in a few centuries or less. The proxy record of these temperature changes appears to be coherent over a broad region; it includes oxygen isotopic shifts in Greenland ice cores (Dansgaard et al., 1982, 1989), pollen and isotopic shifts in European lake sediments (Iversen, 1973; Siegenthaler et al., 1984; Ammann and Lotter, 1989), and changes in fossil assemblages of coleoptera in the British Isles (Coope, 1977; Atkinson et al., 1987) (Figure 1). Similar-looking changes characterize the Greenland ice-core record between 8,000 and 80,000 years ago; they thus appear to be a characteristic feature of glacial and deglacial climate in the North Atlantic region (Dansgaard et al., 1982).

The extraordinary rate of these changes was revealed by isotope studies of annually layered ice deposited approximately 10,500 to 10,000 yr BP on Greenland, when air temperatures over the ice sheet rose by 7°C in just 50 years (Dansgaard et al., 1989). As there is no known external (e.g., solar or orbital) climate-forcing agent of sufficient potency and frequency to explain the rapid and recurrent temperature changes seen in these records, the question remains: What process(es) drove the observed variations? The most likely answer, according to a variety of authors (e.g., Oeschger et al., 1984; Broecker et al., 1985), is that they were the result of sudden shifts in the strength and, therefore, in the heat-carrying capacity of the ocean's conveyor-belt circulation system—a global-scale overturning marked by sinking of relatively cold and salty water in the northern North Atlantic, generalized upwelling in the Indian and Pacific oceans, and the return of warm salty surface water to the Atlantic around Africa and South America (Gordon, 1986; Broecker, 1991a).

The impact of heat transport by the conveyor on the climate of the North Atlantic region is clearly evident in the large positive deviation from zonally averaged January air temperatures centered over and downwind of the northern North Atlantic and Nordic seas, as seen in Figure 2 (after Barry and Chorley, 1982). While there is little doubt that the conveyor contributes importantly to the warmth and habitability of the circum-North Atlantic region (especially northwestern Europe and Scandinavia), a key question from the climate-change point of view is whether the conveyor circulation is subject to significant variation. In an attempt to address this issue, the stability of the conveyor has been investigated in a large number of numerical models of ocean circulation, virtually all of which indicate that its operation is threatened by subtle changes in fresh-water balance at the surface near regions of convection in the northern North

---

[2]The abbreviation "yr BP" will be used hereafter to refer to ages in reservoir-corrected radiocarbon years. No attempt has been made to correct for differences between radiocarbon years and calendar years.

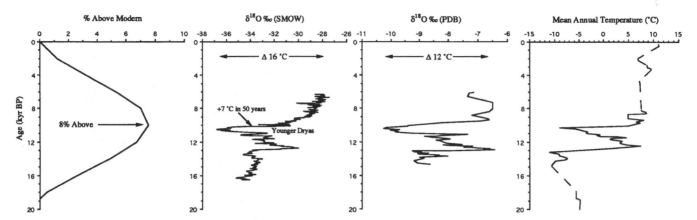

FIGURE 1 Comparison of insolation and various temperature proxy records for the last deglaciation (0-20 kyr BP). Beginning at the left, they are: the slow change in summer insolation for the Northern Hemisphere during that interval (Berger, 1978); oxygen isotope results from the Greenland ice core Dye 3 (Dansgaard et al., 1984); oxygen isotope results from Lake Gerzensee marls in Switzerland (Siegenthaler et al., 1984); and the mean annual temperature estimate for the British Isles based on fossil coleoptera (Atkinson et al., 1987). Note that summer insolation in the Northern Hemisphere, expressed as a percentage deviation from today's values (first panel), changed gradually over the interval 20 kyr to present. However, various climate proxy indicators from around the North Atlantic region indicate that air temperatures changed abruptly (remaining panels). Detailed studies of Dye 3 by Dansgaard et al. (1989) demonstrate that the large isotope shift at about 10,000 yr BP occurred within 50 years and corresponded to a temperature change of 7°C.

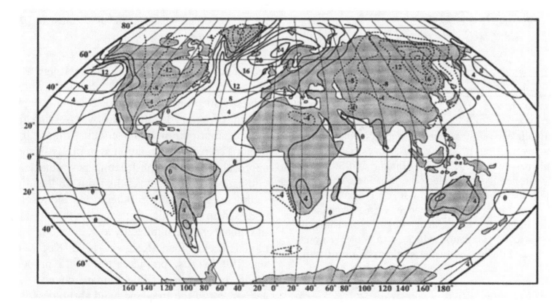

FIGURE 2 Deviation of local temperatures from the zonal average, for January, in °C. (From Barry and Chorley, 1982; reprinted with permission of Routledge.) The largest positive anomalies overlie the area of conveyor overturning in the northern Atlantic and Norwegian Sea.

Atlantic (see, e.g., Bryan, 1986; Manabe and Stouffer, 1988). In many cases the model circulation is observed to collapse within a few decades of surface freshening (Maier-Reimer and Mikalojewicz, 1989; Stocker and Wright, 1991).

As an example of the possible climate change arising from collapse of the conveyor circulation, Figure 3 shows the depression of (annual) surface ocean and air temperatures associated with the conveyor-off mode in the GFDL coupled atmosphere/ocean general-circulation model (Manabe and Stouffer, 1988). The main point to be made

here is that even though the sensitivity of the overturning circulation to fresh-water and temperature forcing at the surface in this (or any other) model cannot yet be validated, cooling of the surface ocean in response to weakened or eliminated conveyor circulation is capable of producing pronounced cooling of the air over and downwind of the North Atlantic. This is further borne out by the results of an earlier experiment in which Rind et al. (1986) used the NASA-GISS atmospheric general-circulation model with sea surface temperatures specified (rather than freely pre-

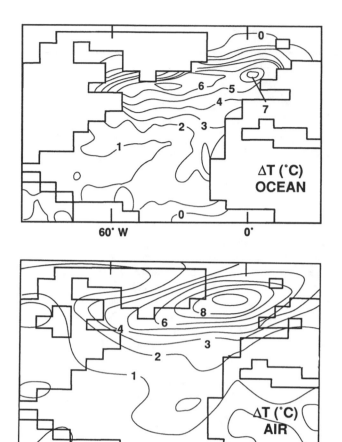

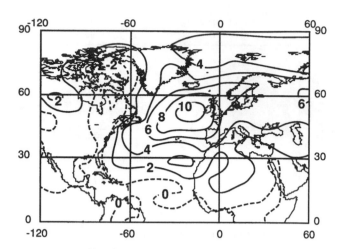

FIGURE 4   Difference in wintertime surface air temperatures between the control run and one with sea surface temperatures specified at glacial (LGM) values reconstructed by CLIMAP (1981). The results are similar to those seen for air temperatures in the GFDL coupled model (Figure 3, lower panel), although these are for wintertime and the area of maximum temperature difference is located somewhat further south, in accordance with the difference between the location of maximum ocean cooling as reconstructed for the LGM and simulated for the "conveyor off" state in the upper panel of Figure 3.

FIGURE 3   Difference in sea surface and surface air temperatures between the "conveyor on" and "conveyor off" states in the GFDL coupled atmosphere-ocean model (after Manabe and Stouffer, 1988; reprinted with permission of the American Meteorological Society). Although the sensitivity of the conveyor circulation in this model (and others) is debated, and cannot yet be empirically evaluated, there is little doubt that a weakening or elimination of the overturning conveyor circulation will lead to cooler sea and air temperatures in and downwind of the northern North Atlantic region.

record around the Atlantic region, through a review of the evidence for such changes in the ocean sediments themselves. Following up on the discussion brought forward in this introduction, I will begin by assessing the magnitude of heat transport into the circum-Atlantic region by the ocean's conveyor circulation system. I will then outline the approach used in reconstructing past changes in conveyor circulation and present results from studies of new Atlantic sediment records with sub-century-scale resolution and climate sensitivity sufficient to capture the amplitude and rates of climate change implied by ice-core studies and numerical models (Lehman and Keigwin, 1992; Koc-Karpuz and Jansen, 1992). These provide a test of the proposed conveyor model of circum-Atlantic climate change, as well as an assessment of the frequency and abruptness with which the conveyor circulation may vary.

HEAT TRANSPORT AND THE ATLANTIC CONVEYOR

The net transports of heat and salt in the world's oceans are characterized by a southward, trans-equatorial flow of relatively cold, salty North Atlantic Deep Water (NADW) formed by cooling and sinking in the northern North Atlantic, and an apparently compensatory, northward, trans-equatorial flow of warm, salty water in the Atlantic thermocline (Gordon, 1986; Broecker, 1991a). Cooling and sinking occurs predominantly in the Labrador and the Nordic seas (Figure 5). The dense products of surface cooling in the

dicted) at the cold, wintertime values reconstructed for the last glacial maximum (the LGM, which occurred about 18,000 yr BP) by CLIMAP (1981). As can be seen in Figure 4, a similar pattern of cooling (in this case, wintertime) was achieved. Although the presence of large ice sheets contributed to oceanic cooling during the LGM (Manabe and Broccoli, 1985), in both this and the coupled atmosphere/ocean experiment ice sheets were not introduced, and all other climate boundary conditions were those of today. Figures 3 and 4 provide an impression of the influence of a cold ("conveyor off") ocean alone.

The main purpose of the present paper is to assess the capacity of the ocean to undergo circulation changes of the magnitude and rapidity suggested by the climate proxy

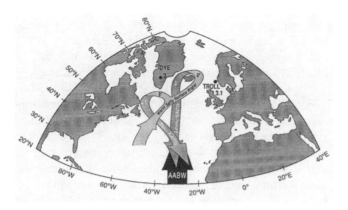

FIGURE 5 A schematic representation of the components of NADW that can be traced directly to source areas at the surface of the northern North Atlantic. As discussed in the text, these components may comprise as little as two-thirds of the total southward flux of NADW. The remainder may derive from entrainment of recirculated waters. Temperatures (in °C) refer to the surface temperatures just prior to convection, and to the average temperature of each of the convecting water masses that feed upper and lower NADW (cf. Clarke and Gascard, 1983; Aagaard et al., 1985; Bainbridge, 1981).

Labrador Sea descend directly to the interior of the open Atlantic, and are the dominant source of upper NADW (McCartney and Talley, 1984; Clarke and Gascard, 1983). While cooling and sinking occur at a variety of locations within the Nordic seas, the dense waters overflowing the Denmark Strait and crossing the Iceland-Faeroe Ridge into the open Atlantic appear to originate primarily from inter-mediate-depth convection within the Iceland Sea (Aagaard et al., 1985). These waters, which are denser than those formed in the Labrador Sea, are the dominant source of lower NADW (Worthington, 1976). Together these water masses comprise approximately two-thirds or more of the total southward, trans-equatorial transport of NADW. The remainder may result from entrainment of Antarctic Bottom Water (AABW) and Antarctic Intermediate Water (AAIW) (McCartney and Talley, 1984; Schmitz and McCartney, 1993). The southward-flowing deep waters upwell toward the surface at a variety of locations in the Pacific and Indian oceans. From these locations surface waters are returned to the Atlantic either via the Drake Passage (the "cold-water route") or via the Straits of Indonesia and the Aghullas retroflection around Cape Hope (the "warm-water route"). These return flows complete the Great Ocean Conveyor loop described by Gordon (1986) and Broecker (1991a). Estimates of the net deep-water export from the North Atlantic, based on a variety of different approaches (Broecker, 1991a; Schmitz and McCartney, 1993, and references therein; Wunsch, 1984), range between 14 and 20 sverdrups (1 Sv = 10^6 m^3/sec).

Broecker (1987, 1991b) made a rough calculation of the heat release to the atmosphere resulting from warm-to-cold water conversion by the conveyor, assuming average temperatures of 11°C for northward-flowing thermocline waters and 3°C for southward-flowing NADW. Using an estimated flux of NADW of 20 Sv, approximately 5 × 10^{21} calories are released to the northern atmosphere per year (i.e., about 0.7 petawatt). According to the circulation scheme of Schmitz and McCartney (1993), however, as much as 5 to 7 Sv of the total southward flow at depth is comprised of relatively old, cold, recirculated deep waters that have not been directly involved in warm-to-cold water conversion at the surface of the northern North Atlantic. Using this transport scheme, which also depicts a larger proportion of the compensatory northward flow at greater, colder depths, the average temperature of surface and near-surface waters subject to warm-to-cold conversion is 8°C, and the flux of NADW produced by convection in the northern North Atlantic is only 13 Sv, yielding an estimated heat release of about 2 × 10^{21} calories per year.

The two different estimates can probably be regarded as extremes that bracket the actual heat release to the atmosphere over the northern North Atlantic. (I suggest this because (1) in these calculations the larger estimate of deep-water flux is tied to the larger estimate of temperature difference between exported deep water and returning surface water, and vice versa, and (2) the smaller estimate of NADW flux would predict a nutrient content for the deep Atlantic at the high end of what can be accommodated by current observational estimates.) In any case, the impact of the conveyor, on the basis of heat release alone, is huge. Broecker (1987, 1991a) noted that his estimate of 5 × 10^{21} calories per year was equivalent to approximately 30 percent of the solar energy absorbed by the troposphere over the northern North Atlantic. The fingerprint of this oceanic heat source is seen in Figure 2. Its removal would result in temperature deficits similar to those seen in Figures 3 and 4. It follows from these examples that changes in the strength of the conveyor may exert a strong influence on the climate of the North Atlantic region.

THE PROXY RECORD OF SURFACE CIRCULATION

The proxies used here to assess past variations in the strength of the conveyor circulation are changes in fossil assemblages of planktonic foraminifera (carbonate-shelled protozoans) and diatoms (silica-shelled plants) with known temperature tolerances. Fossil assemblages of planktonic foraminifera, in particular, have been used routinely to chart the history of polar-front movements and related rates of surface-circulation change in the North Atlantic (McIntyre et al., 1976; Ruddiman et al., 1977; Ruddiman and McIntyre, 1981; Bard et al., 1987). The general premise is that north-ward penetrations of subtropical to subpolar assemblages occurred during periods when poleward flow of Atlantic surface waters was vigorous (as it is today), and that southward penetrations of polar assemblages occurred during periods of reduced poleward surface flow, such as the LGM (see, e.g., McIntyre et al., 1976; Kellogg, 1980). These states can be equated with what Broecker and colleagues have termed "conveyor on" and "conveyor off" modes, and with meridional versus zonal orientations of the North Atlantic polar front, respectively (see Figure 6, for example).

Synoptic reconstructions based on studies of a large number of Atlantic deep-sea sediment cores have outlined the general pattern of polar-front movement during the last deglaciation (Ruddiman and McIntyre, 1981). The front began to retreat from its LGM position approximately 13,000 yr BP, then advanced southward once again between about 11,000 and 10,000 yr BP during the well-known Younger Dryas cold interval, just prior to a final retreat to near its present position (see Figure 6). Initial attempts to calculate the rate of polar-front migration during the deglaci-

ation suggest that, in the fastest case, it swept between extreme positions in less than 400 years (Bard et al., 1987). This estimate is close to the maximum rate that can be discerned using ^{14}C dates from different sediment cores. The typical precision of ^{14}C ages between 8,000 and 15,000 years BP is ± 80 to ± 160 years at 1σ; the compound error associated with dating isochronous events in two different cores is therefore ± 160 to ± 320 at the 80 percent confidence level (1σ), and twice that at the 95 percent confidence level (2σ). Although the individual (and therefore compound) errors can be reduced by dating the same stratigraphic level repeatedly, even if there were no other stratigraphic uncertainties such as bioturbation (stirring by bottom-dwelling organisms), approximately 100 such dates would be required to reduce the error from several centuries to several decades (using $2\sigma/n^{1/2}$). It is therefore not surprising that such estimates of circulation response time greatly exceed those suggested by ice-core studies.

Rather than fight the errors involved in trying to chart the sweep rate of faunal boundaries between cores, one can alternatively measure the timing and rates of faunal change at a single site that is diagnostic of the overall behavior of the Atlantic circulation system. One such location is the region between the Faeroe Islands and southwestern Norway, which is effectively a gateway to the Nordic seas through which warm surface waters are drawn by deep-water formation and export across the Iceland-Faeroe Ridge and through the Denmark Strait. At this location, the degree and rate of warming or cooling indicated by faunal changes should be indicative of both the degree and the rate of change of poleward surface-water flow through the northern North Atlantic. Because conditions in the Norwegian Sea would be of uniformly polar character in the absence of Atlantic inflow, it is sufficient for the present purpose to measure the proportion of the single polar taxon, *Neogloboquadrina pachyderma* (sin), with respect to the sum of all planktonic foraminifera. *N. pachyderma* (sin) constitute more than 95 percent of the fauna at summer sea surface temperatures (SSTs) below 5°C or so (polar and Arctic water masses) and are extremely rare in waters with summer SSTs above about 10°C (Bé and Tolderlund, 1971; see Figure 7). Summer temperatures in this range are associated primarily with the region of mixing between Atlantic and polar waters within the basin today (Johannesson, 1986; Lee and Ramster, 1981). Past changes in *N. pachyderma* (sin) percentages between values corresponding to the 5°C and 10°C isotherms should record nearly the full range of SSTs associated with past changes in flow of Atlantic waters into the basin.

Measurements of *N. pachyderma* (sin) percentages were carried out in core Troll 3.1 from an area of high deposition rates in the western North Sea (see Figure 6b for core location). The seismostratigraphy, lithostratigraphy, and dating of the core site are described in Lehman et al. (1991).

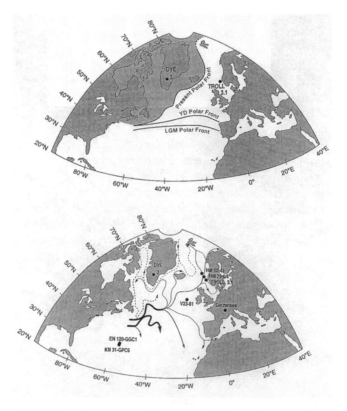

FIGURE 6   Above, map of the present, LGM, and Younger Dryas polar fronts and the approximate range of polar foraminifera today (after Ruddiman and McIntyre, 1981; printed with permission of Elsevier). Below, a schematic of the main surface transports in the northern North Atlantic (after Dietrich et al., 1975; reprinted with permission of John Wiley & Sons and Gebüder Borntraeger). Solid lines in the latter refer to warm-water transport and dashed lines to cold-water transport. Line thicknesses are roughly proportional to flux. The surface flux of warm waters toward areas of warm-to-cold conversion in the Labrador and Nordic seas is estimated to be that required to balance the export of new NADW production (see Figure 5).

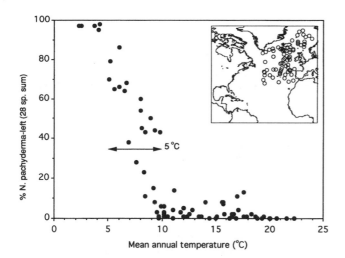

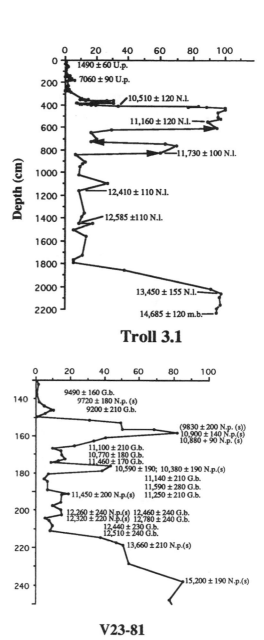

FIGURE 7   Percentage of *N. pachyderma* (sin) in core-top sediments from the northern North Atlantic and Norwegian Sea versus temperature of the overlying surface waters. *N. pachyderma* (sin) percentage appears to have a range of temperature sensitivity of about 5°C. The inset map shows the location of the sediment cores used in the calibration. Data used in this figure were supplied by David Anderson of the National Geophysical Data Center, NOAA.

Average rates of deposition at the core location were 5 m/kyr during deglaciation, some 20 to 50 times higher than at other high-deposition-rate locations in the open Atlantic. The core site is located a few degrees east of the main axis of Atlantic water inflow to the Norwegian Sea basin (Figure 6b). Atlantic water bleeding off the main axis of inflow and into the North Sea greatly influences SST and salinity over the core site, as is clearly apparent in satellite images of the region (see Figure 8, in the color well). Instrumental data suggest that changes in salinity and SST of the North Sea reflect similar changes in the open North Atlantic (Dickson et al., 1984; Heath et al., 1991). These observations suggest that the location Troll 3.1 provides a suitable monitor of surface water exchange between the open North Atlantic and the Nordic Seas.

The down-core variation in percentage of *N. pachyderma* (sin) and associated accelerator mass spectrometer (AMS) ^{14}C dates in Troll 3.1 are given in Figure 9. These match closely the faunal variations from another AMS ^{14}C-dated sediment core located off Ireland (core V23-81), just upstream of Troll 3.1 in the present-day circulation (Figure 6b; Ruddiman et al., 1977; Broecker et al., 1988). The agreement between these records provides additional evidence that faunal changes in the western North Sea were diagnostic of conditions in the open Atlantic during the last deglaciation. While the two records are complementary in this regard, the record in Troll 3.1 offers a key advantage in that rates of deposition are high enough to permit resolu-

FIGURE 9   Percentage of *N. pachyderma* (sin) against depth. The panel for core Troll 3.1 includes AMS ^{14}C dates from Lehman et al. (1991). Despite the large difference in depth scales, the downcore pattern is similar to that in core V23-81 panel, which is from Ruddiman et al. (1977, reprinted with permission of the Royal Society) and includes dates from Broecker et al. (1988, 1990b). The largest *N. pachyderma* (sin) peak in each core occurred during the Younger Dryas. Although dates of the *N. pachyderma* (sin) peak just prior to the Younger Dryas in V23-81 are discordant with those for a similar event in Troll 3.1, bracketing dates on subpolar foraminifera (*Globoratalia bulloides*) are similar. Detailed analyses of photographs of V23-81 indicate that out-of-sequence dates come from worm burrows (G. Bond, personal communication). Arrows mark faunal transitions in Troll 3.1 corresponding to SST changes of about 5°C in 40 years or less. (From Lehman and Keigwin, 1992; reprinted with permission of MacMillan Magazines.)

tion of the true rates of faunal change at the site, unmasked by bioturbation.[3]

The comparison of high-resolution faunal results for Troll 3.1 with other climate proxy records leads to the following conclusions concerning the variability and climatic impact of surface circulation during deglaciation:

1. In addition to initial deglacial warming from approximately 13,500 to 13,100 yr BP (20 to 18 m depth in core) and the Younger Dryas cold oscillation between about 11,200 and 10,500 yr BP (6 to 4 m), both of which have been recognized and dated previously in several cores from the northern North Atlantic, additional warm-cold-warm oscillations are documented at about 12,600 to 12,400 yr BP (15 to 10 m), 11,700 to 11,500 yr BP (8.5 to 7 m), and 9,700 yr BP (3.8 m) (see Figure 9). Although none of these oscillations was as severe as the Younger Dryas, they indicate that the surface limb of the conveyor was more variable in strength than previously imagined.

2. Estimates of the duration of single warm-to-cold and cold-to-warm transitions based on interpolation between ^{14}C dates indicate that several of these transitions (marked by arrows in Figure 8) occurred in 40 years or less. As these transitions span the full or nearly the full range of conditions recorded by N. pachyderma (sin), they represent swings between near-present-day levels and near-absence of Atlantic water within the basin, that is, between near-extreme states of the conveyor circulation. These changes appear to occur with an abruptness similar to that of the isotopic shifts recorded in Greenland ice cores (Dansgaard et al., 1989). (The transition to warmer water conditions at the close of the Younger Dryas appears to have taken around 100 years, but since deposition rates at the site of Troll 3.1 drop to about 1 m/kyr by this time, the apparently slower transition may be an artifact of the decrease in stratigraphic resolution. Nonetheless, absolute rates of SST change at this time may have been just as great as earlier ones. As we shall see below, this transition may have involved a warming of as much as 9°C; see, e.g., Figure 10.)

3. On the basis of the observed range of temperature sensitivity of N. pachyderma (sin), Lehman and Keigwin (1992) estimated that larger faunal changes in Troll 3.1 corresponded to changes in SST of 5°C or more. A new, independently dated, high-resolution record of SST changes in the southeastern Norwegian Sea based on diatom assemblages, which have a much broader range of temperature

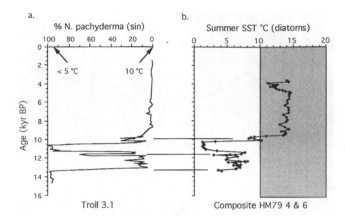

FIGURE 10   Comparison of N. pachyderma (sin) percentage in Troll 3.1 (panel a) with an estimate of summer SSTs based on diatom assemblages in an independently dated, composite record of two cores (HM79 6/4, panel b) from the southeastern Norwegian Sea (see Figure 6b for core locations). The diatom record is from Koc-Karpuz and Jansen (1992). The area of shading marks SSTs outside the ecological range of N. pachyderma (sin) indicated by data in Figure 7 and by Bé et al. (1971).

sensitivity than N. pachyderma (sin), confirms the earlier estimates from Troll 3.1 (Koc-Karpuz and Jansen, 1992). The comparison in Figure 10 shows that the range of SST that might be inferred from the percentage of N. pachyderma (sin) was truncated primarily at the warm end, as might be expected on the basis of present knowledge of its modern distribution (e.g., Bé and Tolderlund, 1971). The duration of major SST changes evident in Troll 3.1 and the magnitude of change confirmed by the diatom record suggest that the rates of SST change in the southeastern Norwegian Sea were approximately 1°C per decade, which is directly comparable to the rates of air-temperature change inferred from ice-core studies.

4. When plotted by age, the percentages of N. pachyderma (sin) variations in Troll 3.1 correspond closely with the history of mean annual air temperatures over the Greenland ice sheet deduced from changes in $\delta^{18}O$ of ice in Dye 3, which has the highest resolution of the long Greenland ice cores (Figure 11). This relationship is consistent with earlier suggestions that sudden changes in circum-Atlantic temperatures were governed by changes in strength of the conveyor circulation (Oeschger et al., 1984; Broecker et al., 1985). The strong similarity of temperature proxy records from Greenland, Britain, and Switzerland (Figure 1) indicate that conveyor circulation also affected air temperatures downwind of the northern North Atlantic at least as far as central Europe, in accordance with model predictions (e.g., Rind et al., 1986; see Figure 4).

---

[3]With an average deposition rate of 5 m/kyr, an average mixed-layer thickness of 10 cm would disturb only two years of sedimentation.

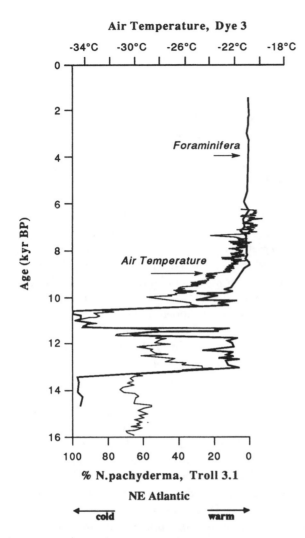

Air Temperature, Dye 3

Foraminifera →

Air Temperature →

% N.pachyderma, Troll 3.1

NE Atlantic

← cold   warm →

FIGURE 11 *N. pachyderma* (sin) percentage in Troll 3.1, plotted to age using the ^{14}C dates in Figure 9, and compared with mean annual air temperature based on δ^{18}O of ice from core taken at Dye 3, Greenland. The age-to-depth relationship in Dye 3 is roughly linear prior to approximately 10,000 yr BP, as suggested by correlation to dated variations at Lake Gerzensee. The correlation between Troll 3.1 and Dye 3.1 after about 10,000 yr BP is based on annual layer counting and ^{14}C-to-calendar year conversion, as discussed in Lehman and Keigwin (1992). As a result, the age model after 10,000 yr BP differs slightly from that used in Figure 1b of Hammer et al. (1986).

CONCLUSIONS

A series of abrupt changes in the poleward flow of warm surface waters has been documented in the northeastern North Atlantic during the last deglaciation (15,000 to 8,000 yr BP). They appear to correlate with evidence of atmospheric temperature changes over Greenland and downstream over Europe. Most important, they provide the first direct indication of the capacity of the conveyor circulation to shift abruptly in strength, as recorded by changes in temperature on the order of 1°C per decade across a range of more than 5°C at the surface of the northern North Atlantic. Numerical models of ocean circulation produce large changes in vertical and lateral ocean-heat fluxes in the North Atlantic as a result of changes in thermohaline circulation arising from fresh-water forcing at the ocean surface (Manabe and Stouffer, 1988; Maier-Reimer and Mikalojewicz, 1989; Stocker and Wright, 1991). Although the sensitivity of the different models to fresh-water forcing is debated and cannot yet be empirically evaluated, the rates and magnitude of the simulated changes now find support from high-resolution proxy records of ocean-circulation change. Owing to associated changes in heat transport, these non-linear aspects of circulation change must be considered a potent agent of climate variability.

Unlike the deglacial interval discussed in this paper, the last 8,000 years have been marked by relatively stable climate conditions in the North Atlantic region (Figure 11). But, as pointed out earlier by Broecker (1987), the conditions that conferred stability over this interval are not yet understood; thus, it cannot be assumed that such conditions will persist in the face of the gradual changes already forecast as a response to anthropogenic perturbation of the climate system.

ACKNOWLEDGMENTS

I wish to thank Delia Oppo and Bill Curry for stimulating discussions and their reviews of an early draft of this manuscript. Julia Cole also provided useful criticism. Ernest Joynt III and Fritz Heidi assisted with graphics. This work was supported by NSF grants OCE-90819660 and OCE-91816259.

Commentary on the Paper of Lehman

JULIA E. COLE
University of Colorado

In commenting on Dr. Lehman's paper, I would like to start by pointing out a few things that he has not emphasized, which make it especially relevant to observational data and model simulations.

Paleoclimate reconstructions play a unique role in the study of ocean-atmosphere dynamics, by offering a window onto climate during periods that we cannot observe first-hand. Dr. Lehman's paper provides strong evidence for rapid (that is, decadal) transitions in oceanic circulation during the last deglaciation, and highlights the information to be found in high-resolution studies of the marine sediment record.

Certain aspects of this study make it particularly complementary to ocean modeling studies. First, the site is sensitive to a specific aspect of the large-scale "conveyor belt" of ocean circulation. The core has extremely high, continuous sedimentation rates for the period of interest, which are critical to addressing rapid climate change. The response of the paleoceanographic proxy to temperature change, including its range of sensitivity, is well characterized in the modern ocean and provides quantitative estimates of SST changes. Chronologic control over this interval is extremely tight, which makes it possible to estimate rates of paleoceanographic changes. The rapid rates of circulation change inferred from the Troll core are in fact consistent with the high degree of sensitivity simulated by ocean GCMs for the ocean circulation conveyor.

This study raises many questions about past and future ocean-circulation dynamics. The Troll core provides a high-resolution window onto only a few thousand years, in the most recent deglaciation. Are rapid changes in the conveyor intensity characteristic of full glacial and interglacial as well as deglacial climates? New ice-core records from Greenland and Antarctica may help to shed light on this question. But high-resolution records from the ocean will be required as well, to help us understand the relationship between changing climate boundary conditions and the linear and nonlinear responses of large-scale ocean circulation.

This study challenges traditional ideas about the speed at which the oceanic conveyor starts and stops, and presents new observations on ocean sensitivity that concur with both ice-core evidence and model behavior. If rapid changes in the Troll core truly represent rapid changes in conveyor circulation intensity, they should be apparent in other North Atlantic records with short response time and high resolution. Such records will test and add detail to the proposed model of deglacial ocean circulation, in which distinct source regions vary in their contributions to northern-source deep water.

The recognition that ocean circulation has changed rapidly, and that we can reconstruct these changes in detail, should intensify efforts to focus paleoceanographic sampling and analysis strategies on decade-to-century-scale objectives. To reconstruct high-frequency oceanic variability, we need to target environments where rapid changes can be preserved by high sedimentation rates (where possible, in annually laminated sequences). We also need to encourage continuing support for appropriate coring techniques, rigorous dating, and proxy calibration. A new emphasis on high-resolution paleoceanography will help us to close the "information gap" that exists between annual and Milankovich time scales of climate variability, and will provide insight into how the climate system responds on societal time scales to changing boundary conditions.

Discussion

RIND: You called the oscillation seen during the last time period "characteristic of the system". It would be interesting to see whether you could find perturbations like this during the last interglacial.

LEHMAN: We intend to do just that at Bermuda Rise. I also think that it might be fruitful to begin to simulate these 3000-year oscillations that appear to arise from some sort of salinity forcing. But these variations during glacials are hard to relate to the kind of variability we seem to see during interglaciation, which may

be what the two different limbs of my double conveyor belt reflect. There might be a link between the spatial and temporal variability.

RIND: How do you square the subtropical gyre's apparent sensitivity to thermohaline circulation with the CLIMAP reconstruction of very little temperature change there at a time when we know that the thermohaline circulation was strongly restricted?

LEHMAN: Well, CLIMAP could be wrong. Maybe I should submit a proposal to test their conclusions by doing organic geochemical analysis. I have great faith in the high-amplitude variability we've found in our high-resolution records from the Bermuda Rise core. We have a very sensitive site and materials with a high deposition rate, and because the subtropical area isn't sensitive to front movement I really think it's diagnostic of a large part of the ocean.

GROOTES: I'd like to show some records of decadal-scale isotope fluctuations in the GISP2 core that complement what Scott has been showing us. Figure 4b in my paper shows these fluctuations for the last few hundred years, but they are even stronger during the glacial-to-interglacial transition about 14,000 years ago. If we use the same isotope-temperature gradient of 0.63 permil per °C, we calculate a change in mean annual temperature of about 10°C over a 5-year period. The electroconductivity of the ice, which reflects neutralization of acidity by alkaline dust, also indicates large and very rapid climate fluctuations—both interstadial Dansgaard-Oeschger events on a time scale of centuries, and periods of very high climate instability on a decadal time scale. We have to ask what caused the rapid changes in air circulation reflected by the dust and temperature changes at GISP 2.

COLE: Could they just be a function of change of source regions? Could you use deuterium excess to distinguish source and local climate changes?

GROOTES: Deuterium excess can be used to separate changes in local climate from changes in the source region of the water vapor. Even if the observed changes are a function of a change in source regions, we still need to find a cause for this latter change, one that can effect very rapid changes. We need to put together all our climate data—ice-core, ocean, terrestrial, instrumental—to determine the mechanism for these rapid climate fluctuations. It won't be easy, but we should be able to sort it out eventually.

NORTH: Could those large fluctuations relate to the extremely large ice sheet in North America at that time?

LEHMAN: Such an ice sheet ought to buffer the Pacific/Atlantic isotope source differences. We've got a lot to look at.

Ice Cores as Archives of Decade-to-Century-Scale Climate Variability

PIETER M. GROOTES[1]

ABSTRACT

Ice cores form a unique archive of past climatic and atmospheric conditions, because they can record these conditions continuously, with annual resolution, and may preserve them over very long times. The information that can be deduced from an ice core includes temperature fluctuations, derived from stable-isotope ratios of hydrogen and oxygen in the ice; chemistry and dust content, reflecting availability in the source area or biospheric activity, as well as atmospheric transport; and volcanic eruptions, recorded as high non-sea-salt sulfate concentrations and sometimes as volcanic ash. The composition of the atmosphere in the past is preserved in air bubbles trapped in the ice, while cosmogenic isotopes reflect the intensity of cosmic rays reaching the earth, and may be useful for dating the core.

The new ice cores from the summit area of the Greenland ice sheet show considerable variability on decadal time scales. Century-scale fluctuations are small during the Holocene, but frequent and large before that. A multiparameter core analysis allows a reconstruction of the changes in atmospheric circulation and in the biosphere (terrestrial and marine) that accompany climate change.

INTRODUCTION

Understanding decade-to-century-scale climate variability requires not only high-resolution climate records for at least the last 1000 years, but also a good definition of the background of slow climate changes on which shorter climate variations are superimposed. The ideal is to have long ($>10^5$ yrs), high-resolution records showing the full range of natural climate variability, from glacial/interglacial and stadial/interstadial (on 10^3 to 10^5 year time scales) to decadal. In addition to the long-term background of climate change, the long records also offer a better opportunity to ascertain what interactions in the climate system determine climate change, because the major changes in atmospheric circulation and composition and in land and ocean surface conditions that accompany glacial/interglacial climate change are much better defined than the comparatively minor changes of the last several thousand years. Yet the small climate

---

[1]Now at the Leibniz Laboratorium für Altersbestimmung und Isotopenforschung, Christian Albrechts Universität, Kiel, Germany

changes on a decade-to-century time scale are important for society, because these may significantly affect living conditions.

Reliable instrumental observations go back only about a century, so for the study of climate variability we are dependent on proxy records. Many different sources of proxy records have been used, such as ocean and lake sediments, peat bogs, tree rings, speleothems, loess deposits, and ice cores. Problems in the reconstruction of climate from proxy sources are that the climate information recorded depends on the response time and sensitivity of the recording systems, which differ from proxy to proxy, and that recording is often discontinuous and represents merely local conditions. Thus we need to combine records from different sources and areas into a consistent picture to reconstruct global climate variations.

An example of the difference in recording properties of three common climate proxies is given in Figure 1. Climate in northwest Europe during the last glacial-to-interglacial transition, reconstructed from pollen analyses (Figure 1a: Van der Hammen et al., 1967), showed two interstadial periods. The younger one contained the more warmth-loving pollen assemblages, and thus was believed to be the warmer

of the two. However, oxygen isotopes in ice cores (Figure 1b: Grootes, unpublished; also Dansgaard et al., 1982, 1984, 1989; Johnsen et al., 1992a) and in lake sediments (Oeschger et al., 1984) as well as beetle remains (Figure 1c: Coope, 1977; Coope and Brophy, 1972; Atkinson et al., 1987) indicate that warmer conditions prevailed during the older interstadial. In this case the pollen represents a distorted climate record (see also Kolstrup, 1990), because climate changed more rapidly than vegetation could respond, and the fluctuations were more faithfully recorded by the faster-reacting isotopes and beetles. Trees, which are frequently used as climate indicators in the arboreal part of the pollen spectrum, migrate particularly slowly into new territory in response to climate change. Reconstruction of decade-to-century-scale climate variations thus requires a fast-reacting climate recorder that preserves a high-resolution environmental record over long time periods. Ice is such a recorder.

ICE CORES AS ENVIRONMENTAL ARCHIVES

High-latitude and high-altitude cold ice masses contain a record of past climate and atmosphere that is unique, because it may be deposited as continuous, discrete annual

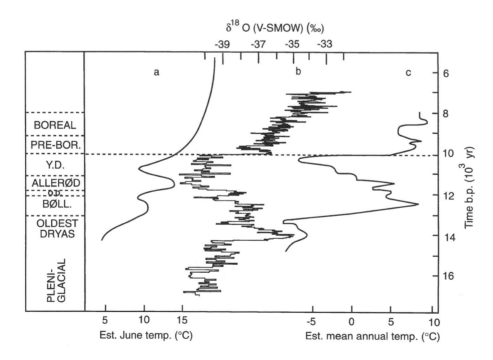

FIGURE 1   Climate change across the last glacial-to-interglacial transition as reconstructed from (a) northwest European pollen assemblages (after Van der Hammen et al., 1967), (b) the GISP2 oxygen isotope record, and (c) beetle remains in Great Britain (Coope and Brophy, 1972; Coope, 1977; Atkinson et al., 1987; modified from Atkinson et al., printed with permission of Macmillan Magazines Ltd.), each on its own relative time scale. The rapid warming from Younger Dryas to Preboreal has been used to line up the three curves at 10,000 yr before present (Mangerud et al., 1974). The latest results from the GISP2 and GRIP cores and from elsewhere indicate that the Younger Dryas-Preboreal transition really occurred around 11,600 years ago (Johnsen et al., 1992a; Alley et al., 1993). The δ¹⁸O values for Greenland are generally translated into temperature changes using 0.63 permil per °C (Dansgaard et al., 1973). The slow response of the pollen assemblages to the extremely rapid climate changes during the transition indicated by (b) and (c) led to an underestimate of temperatures in the Bølling, which was therefore considered to have been cooler than the Allerød.

layers of snow, and be preserved with little alteration over long periods of time ($>10^5$ years in the central areas of cold ice sheets). Moreover, the ice preserves a direct record of the atmospheric aerosols and gases. Over the past 30 years much research has been done on ice cores and on the climatic and environmental information they contain. Excellent summaries are given in *The Climatic Record in Polar Ice Sheets* (Robin, 1983) and *The Environmental*

Record in Glaciers and Ice Sheets (Oeschger and Langway, 1989). A review of global ice-core research is outside the scope of this paper, which aims to demonstrate that ice cores can provide detailed records of rapid climate fluctuations going back far in time. To provide access to the ice-core research literature, a limited selection of ice-core references, grouped under water isotopes, chemistry, dust, and gases, appears in Table 1.

TABLE 1   Ice-Core Research References

Core	$\delta^{18}O$, δD	Chemistry	Dust	Gases
Greenland				9,10,11
Camp Century	1,2,3,4	5,6,7	8	9,16,17,18,19,20,21
Dye-3	2,12,13,14	7,15	13	
Renland	22,23	22,24	22	
GRIP/GISP2	23,25,26,27,28	25,27,29,30,31,32,33		34,35
other[a]	36,37	5,6,30,31,38		39
Arctic				
Devon Island	40			
Agassiz	41			
Barnes	42	31,43		
Mt. Logan	43,44			
Antarctic				
Byrd Station	4	45,46	8	10,11,17,21,39,47,48
Dome C	49,50	51,52	8,51	20,39,53
Vostok	54,55,56,57	58,59	57,60,61,62	35,48,54,57,63,64,65,66
J-9	67,68			
Law Dome	69			48,70,71
other[b]	72	72,73		20,39,74,75,76
Other				
Quelccaya	77	77,78	79	
Dunde	80	80	80	

[a]includes Crête, North Central, Milcent, D-20
[b]includes South Pole, D-10, D-57, Siple Station
KEY:

1—Dansgaard et al., 1971; 2—Dansgaard et al., 1984; 3—Dansgaard et al., 1969; 4—Johnsen et al., 1972; 5—Hammer, 1977; 6—Hammer et al., 1980; 7—Herron and Langway, 1985; 8—Thompson and Mosley-Thompson, 1981; 9—Herron and Langway, 1987; 10—Neftel et al., 1982; 11—Raynaud and Delmas, 1977; 12—Dansgaard et al., 1982; 13—Dansgaard et al., 1989; 14—Oeschger et al., 1984; 15—Hammer et al., 1985; 16—Stauffer et al., 1984; 17—Stauffer et al., 1988; 18—Craig and Chou, 1982; 19—Craig et al., 1988a; 20—Craig et al., 1988b; 21—Leuenberger and Siegenthaler, 1992; 22—Johnsen et al., 1992b; 23—Johnsen et al., 1992a; 24—Hansson and Saltzman, 1993; 25—Taylor et al., 1993; 26—Dansgaard et al., 1993; 27 -GRIP Members, 1993; 28—Grootes et al., 1993; 29—Legrand et al., 1992; 30—Mayewski et al., 1990; 31—Mayewski et al., 1993a; 32—Mayewski et al., 1993b,c, 1994; 33—Whitlow et al., 1994; 34—Wahlen et al., 1991; 35—Bender et al., 1994; 36—Hammer et al., 1978; 37—Dansgaard et al., 1975; 38—Hammer, 1977; 39—Stauffer and Oeschger, 1985; 40—Paterson et al., 1977; 41—Fisher et al., 1983; 42—Hooke and Clausen, 1982; 43—Holdsworth and Peake, 1985; 44—Holdsworth et al., 1992; 45—Palais and Legrand, 1985; 46—Legrand and Delmas, 1988; 47—Neftel et al., 1988; 48—Raynaud et al., 1993; 49—Lorius et al., 1979; 50—Jouzel et al., 1982; 51—Boutron, 1980; 52—Petit et al., 1981; 53—Delmas et al., 1980; 54—Barnola et al., 1991; 55—Jouzel et al., 1987; 56—Lorius et al., 1985; 57—Jouzel et al., 1993; 58—Legrand et al., 1988; 59—Legrand et al., 1991; 60—De Angelis et al., 1987; 61—De Angelis et al., 1992; 62—Petit et al., 1990; 63—Barnola et al., 1987; 64—Chappelaz et al., 1990; 65—Raynaud et al., 1988; 66—Sowers et al., 1991; 67—Grootes and Stuiver, 1986; 68—Grootes and Stuiver, 1987; 69—Budd and Morgan, 1977; 70—Etheridge et al., 1988; 71—Pearman et al., 1986; 72—Mosley-Thompson et al., 1991; 73—Mulvaney and Peel, 1988; 74—Legrand and Feniet-Saigne, 1991; 75—Delmas et al., 1980; 76—Stauffer et al., 1985; 77—Thompson et al., 1986; 78—Thompson, 1992; 79—Thompson et al., 1984; 80—Thompson et al., 1989.

The isotopic composition (^{18}O/^{16}O and ^{2}H/^{1}H) of the ice matrix, expressed as the relative difference between the sample and Standard Mean Ocean Water (V-SMOW) in parts per thousand (permil, ‰), generally reflects local temperature. This parameter has been studied in a large number of intermediate and deep ice cores, and has been used to construct paleotemperature curves. Variations in other core properties, concurrent with those in the isotopes, then reflect the changes in atmospheric composition and circulation that accompanied, or led to, the temperature changes recorded by the isotopes. Variations in the concentrations of major cations and anions in ice cores have been interpreted as indications of changes in atmospheric transport and in terrestrial and oceanic biological productivity, or of volcanic activity. The dust and trace elements in the ice reveal variations in atmospheric aerosols due to changes in source areas and atmospheric transport, and in certain cases permit identification of these source areas. Samples of the atmosphere itself, obtained from air bubbles in the ice, show changes in the concentrations of greenhouse gases such as CO_2 and CH_4 during major glacial/interglacial climate changes, as well as variations in the $\delta^{18}O$ of atmospheric oxygen that are tied to global ice volume and mean ocean $\delta^{18}O$. Below the firm-ice transition this multi-parameter record is closed off from the outside world and preserved for a long time.

The quality of a reconstruction of the global environment from an ice core increases substantially when several core properties are used together. This is the basis for the multi-investigator Greenland Ice Sheet Project 2 (GISP2), which aims to obtain a detailed history of the climatic and environmental changes in the Northern Hemisphere for at least the last 200,000 years. The properties currently measured in the GISP2 ice core from the summit area of the Greenland ice sheet, and the information to be obtained from them, are given in Table 2. Because of the high time resolution possible, ice cores can be used to determine lag/lead relationships between various parameters of environmental change, and may thus reveal the dynamics and feedbacks of climate change.

I will use examples taken mainly from the new high-resolution GISP2 core from the Greenland summit and from my own field of stable-isotope research to demonstrate the importance of ice-core records for understanding climatic and environmental changes on decade-to-century time scales.

NON-POLAR ICE CORES

Ice-core research has been largely limited to the polar regions (Figure 2) because of the distribution of the global ice masses and the need to have little or no surface melting so that a good environmental record will be preserved in the ice. Over the last decade good ice-core records, one going back into the last ice age, have also been obtained

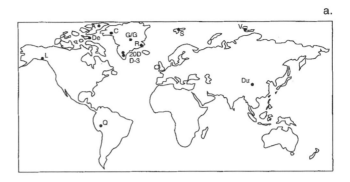

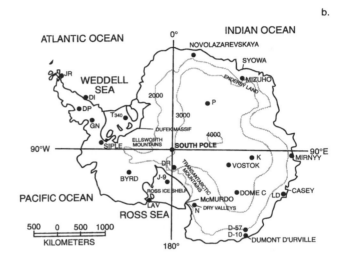

FIGURE 2 Global distribution of intermediate and deep ice cores and ice-core records. For the Arctic and lower latitudes: A—Agassiz ice cap, C—Camp Century, D-3—Dye-3, G/G—GISP2 and GRIP at the Greenland summit, De—Devon Island ice cap, Du—Dunde, China, L—Mt. Logan, Q—Quelccaya, Peru, R—Renland, S—Svalbard, V—Vavilov Dome, Severnaya Zemlya, 20-D—Greenland. For Antarctica: DI—Dolleman Island, DP—Dyer Plateau, DR—Dominion Range, D-10 and D-57—Dumont D'Urville-Dome C traverse, GN—Gomez Nunatak, JR—James Ross Island, J-9—Ross ice shelf, K—Komsomolskaya, LAV—Little America V, Ross ice shelf, LD—Law Dome, N—Newall Glacier, P—Plateau, T340—Ronne-Filchner ice shelf. (Source: World Data Center A for Glaciology (Snow and Ice), 1980, 1989.)

from a few low-latitude, high-altitude sites (Thompson et al., 1986, 1989; Thompson, 1992; Holdsworth and Peake, 1985; Holdsworth et al., 1992; Mayewski et al., 1993a). The tropical Quelccaya ice cap in the Peruvian Andes (13°56'S, 70°50'W, elevation 5670 m) provides a 1500-year record with annual resolution of variations in accumulation, oxygen isotopes, chemistry, and dust. In Quelccaya the period from about A.D. 1500 to A.D. 1900, corresponding with that of the Little Ice Age in northwest Europe, is characterized by more negative $\delta^{18}O$ values and higher liquid conductivity, while the accumulation was above average in the first part of the period and below average in the second (Figure 3a:

TABLE 2   Properties Measured in the GISP2 Core and the Environmental Information Obtained

Property	Environmental Information
Gas Content	
Total gas content	Paleo-elevation of the ice surface
CO_2 concentration in occluded air	Greenhouse gas; increases with temperature; positive climate feedback
CH_4 and other trace gases like N_2O, CH_3Cl and light hydrocarbons	Greenhouse gases as well as indicators of the oxidizing capacity of the atmosphere
O_2/Ar and N_2/Ar ratios	Fractionation of gases trapped in ice, Dole effect
Stable Isotopes	
In gases	
$^{13}C/^{12}C$ in CO_2	Relative size of carbon reservoirs
$^{18}O/^{16}O$ in O_2	O_2 cycle, global-ice volume indicator
$^{15}N/^{14}N$ in N_2	Fractionation of gases trapped in ice
$^{13}C/^{12}C$ and D/H in CH_4	Source distribution of atmospheric CH_4
In ice	
D/H and $^{18}O/^{16}O$ in ice	Primary paleotemperature indicators; deuterium excess *d* indicates ocean-surface conditions
Cosmogenic isotopes	
$^{14}C/C$ in CO_2	Dating climate and ice up to 30 to 40 kyr, carbon-reservoir changes
^{10}Be, ^{26}Al, ^{36}Cl in ice	Cosmic-ray production rate, snow accumulation rate, dating
He isotopes	Cosmic-ray production, crustal He
Chemistry	
Major anions and cations	Atmospheric chemistry and circulation, source areas, biogeochemical cycling, biomass burning, sea-ice extent, volcanic events, accumulation and its seasonal distribution
Trace metals (Pb, Ir)	Sources of trace metals and their variability
Methanesulfonic acid and iodide/iodate in ice	Biological productivity
H_2O_2	Oxidative atmospheric chemistry, seasonality
Particulate concentrations	Atmospheric circulation, global dust sources, volcanic events, annual layer counting, marker horizons, interhemispheric correlation
Electrical conductivity	Air circulation, dustiness, volcanic events, biomass burning events, seasonal cycles, chemical marker horizons
Physical and mechanical properties	Density, texture, fabric, ultrasound velocity, annual layers for accumulation and dating, stratigraphic continuity and flow deformation, core relaxation, firnification processes
Borehole and surface geophysics	Airborne and surface ice radar, internal layering, surface strain net for ice flow velocity, accumulation, flow modeling, temperature profile and history, borehole deformation and vertical strain
Atmospheric studies	Automatic weather stations, atmosphere-snow transfer functions for ions, particulates, isotopes, radionuclides, and gases

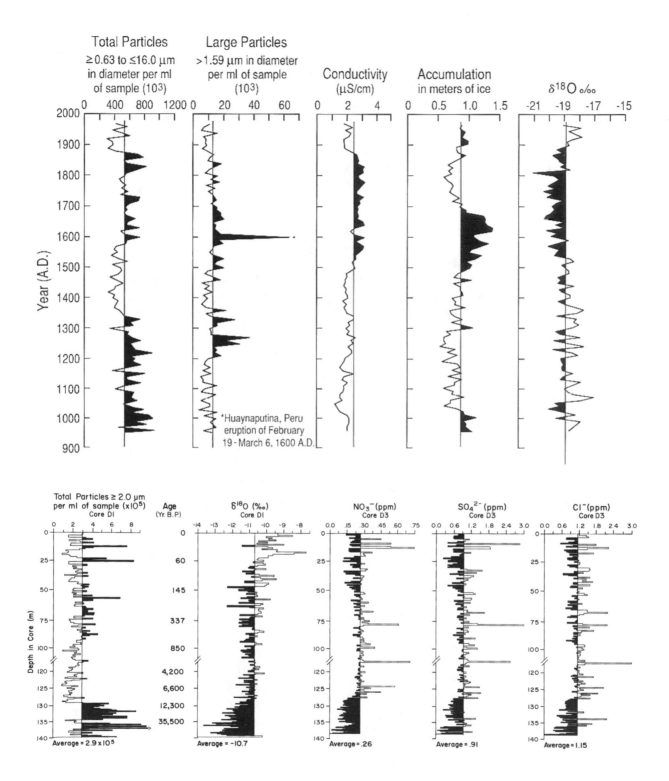

FIGURE 3 Ice-core records from low-latitude/high-altitude ice caps. Upper group: Dust, conductivity, accumulation, and δ¹⁸O of the upper part of the 1500-year ice-core record from Quelccaya, Peru (Thompson et al., 1986). An accurate time scale was derived from seasonal cycles in particle concentrations and verified against the historic Huaynaputina eruption of February 19—March 6, 1600 A.D. Lower group: Dust, δ¹⁸O, and chemistry in the Dunde core, Qinghai-Tibetan plateau, China. The time scale, based on seasonal cycles in particle concentrations and δ¹⁸O down to 70 m, on visible dust stratigraphy to 117 m, and on ice-flow modeling below 117 m, extends well into the last glacial period (Thompson et al., 1989). (Both figures reprinted with permission of the American Association for the Advancement of Science.)

Thompson et al., 1986). The Quelccaya record is particularly important because long climatic records are rare for the tropics. The accumulation appears to correlate with the recent El Niño/Southern Oscillation record (Thompson et al., 1984, 1992). The Dunde ice cap from the Qinghai-Tibetan plateau of China (38°06′N, 96°24′E, elevation 5325 m) provides the first non-polar ice-core record extending into the last glacial period (Figure 3b: Thompson et al., 1989). The variations in chemical concentrations and dust probably reflect aridity and windiness on the Qinghai-Tibetan plateau and in the Qaidam Basin.

The addition of ice-core records from lower latitudes to those of the polar regions is important for the understanding of global climate variability. First, it allows an evaluation in one medium (ice cores) of the latitude-dependent expression of global climate changes over a considerable range of latitudes. Second, ice cores outside the polar regions allow a comparison between the record of past climate changes in the ice and those in other media such as tree rings, lake sediments, loess, and peat bogs from the same geographic area, albeit at lower elevations. The potential for obtaining more non-polar ice core records exists in a number of the high mountain ranges of the world.

POLAR ICE CORES: THE GREENLAND SUMMIT

Many ice cores have been obtained in the Arctic and Antarctic over the past decades (see Table 1). Of particular interest for decade-to-century-type climate fluctuations are two ice-core projects that have been under way in the summit area of the Greenland ice sheet since 1989, viz., the U.S. Greenland Ice Sheet Project 2 (GISP2) and the European multinational projects Eurocore and GRIP (Greenland Ice Core Project). Basal, silty ice was reached by GRIP in summer 1992; the basal ice plus about 1.5 m of underlying bedrock was obtained by GISP2 in July 1993. Analyses of the two 3000+ m cores reveal a continuous, high-resolution record of a suite of climate-related ice-core properties (Table 2) in an area that, because of its latitude (72.6°N) and elevation (3200+ m), rarely experiences melting. The GISP2 and GRIP ice core records show in great detail frequent, rapid, and simultaneous changes in many climate-related core parameters, in both glacial and interglacial times (Alley et al., 1993; Taylor et al., 1993; Mayewski et al., 1993c; Grootes et al., 1993; Johnsen et al., 1992a; Dansgaard et al., 1993; GRIP members, 1993). These large and rapid changes are already forcing us to revise our ideas about climate stability. Over the next several years, these cores will contribute greatly to our knowledge and understanding of Northern Hemisphere and global climate change. Because the two cores are located at the current ice divide (GRIP/EUROCORE) and 28 km (about 10 ice thicknesses) west of the divide (GISP2), they are currently in different ice flow regimes (divide flow and flank flow respectively). Com-

parison of the two core records makes it possible to identify ice-flow effects and to verify the fluctuations in environmental parameters observed in each core. Such comparisons greatly increase the credibility of any reconstruction of climatic and environmental changes that is based on the core records. I will discuss some of the features of the GISP2 ice core record to illustrate the information obtainable from ice cores.

Recent Climate Fluctuations

The oxygen-isotope records from the Greenland summit area indicate a remarkably constant average climate during the Holocene (the last 10 kyr) (Dansgaard et al., 1993; Grootes et al., 1993; Grootes and White, unpublished data). Yet the $\delta^{18}O$ record shows significant short-term variability (Figure 4). The $\delta^{18}O$ values, obtained by calculating mean annual values from the 8 to 10 individual measurements over each annual layer, fluctuate over a range of about 6 permil around a fairly constant mean value of −35 permil

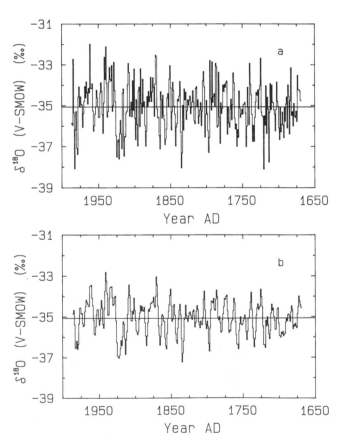

FIGURE 4   Oxygen isotope record for the upper 100 m at GISP2, Greenland summit. (a) Mean annual $\delta^{18}O$ values, calculated from about 10 samples per year, on the GISP2 time scale determined from seasonality in visible stratigraphy, ECM, and isotopes (Meese et al., 1994a). (b) The same record smoothed with a simple 3-year moving average.

(Figure 4a). Part of the variability appears to be related to year-to-year weather fluctuations that may be suppressed by smoothing with a simple three-point running mean. This brings out fluctuations on a decadal time scale that have a range of about 4 permil (Figure 4b). With a $\delta^{18}O$-temperature coefficient of 0.63 permil/°C (Dansgaard et al., 1973), a range of mean annual temperatures of about 6°C can be inferred. The rapidity and magnitude of the shifts indicate that the fluctuations do not reflect global climate changes, but are most likely attributable to changes in the atmospheric circulation over the summit of Greenland.

For the interpretation of the isotope record in the core it should also be realized that the isotopic composition of precipitation reflects primarily the temperature difference between the ocean surface at which the water vapor formed and the place of precipitation (air temperature at the condensation level some distance above the surface). Thus the isotopes do not provide a simple, direct local temperature record, although the precipitation temperature often dominates. Measurement of the pair of water isotope ratios ($^2H/^1H$ (or D/H) and $^{18}O/^{16}O$) in the ice permits the calculation of the so-called deuterium excess, $d \equiv \delta D - 8\delta^{18}O$, that is sensitive to sea surface temperature, relative humidity, and wind speed in the source area of the water vapor over the ocean. Thus it is possible to obtain from the ice-core record information about local temperature changes over Greenland as well as about changes in surface conditions over the Atlantic Ocean (Jouzel et al., 1982; Johnsen et al., 1989).

Application of longer smoothing filters eliminates the decadal fluctuations and reveals weaker, century-scale fluctuations (Figure 5). Interpretation of such filtered records requires great care and needs support from independent environmental evidence to ensure that the fluctuations have physical meaning. Instrumental records from coastal Greenland and the North Atlantic basin (e.g., World Weather Disc, 1990; Dickson et al., 1988b), which are available for about the last 100 years, offer the resolution required for the study of the decadal fluctuations in ice-core parameters shown in Figure 4b. When the rapid fluctuations in isotopes and other properties, observed in the GISP2 ice core, can be related to conditions at the ocean surface and over Greenland and Europe over the last century, then the ice core offers a long, continuous, and detailed record of such conditions. The instrumental record is too short to study century-scale climate variability, so the validation and calibration of the century-scale ice-core parameter fluctuations (Figure 5) will have to come from other proxy records of climate. Such validation is difficult to obtain for the Holocene, because the generally stable climate shows only relatively minor fluctuations that are difficult to correlate. Moreover, few records can match the continuity and detail of the ice cores. Long tree-ring sequences (e.g., Briffa et al., 1992a, b)

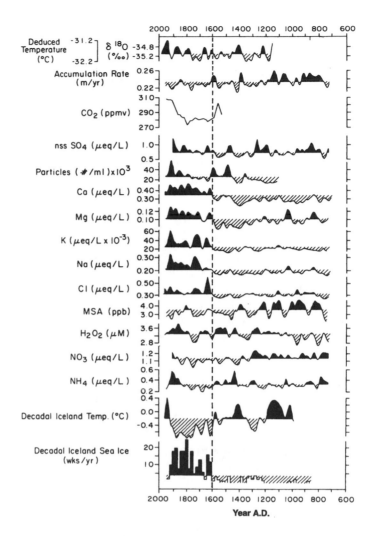

FIGURE 5   Century-scale variability over the last 1200 years in a suite of parameters measured in the GISP2 ice core from the summit of the Greenland ice sheet. The records have been subjected to a simple 50-year smoothing. Decadal Iceland sea-ice frequency and Iceland temperatures derived from sea ice (Bergthorsson, 1969) have been added for comparison. The time scale was derived by counting seasonal cycles in $\delta^{18}O$, visual stratigraphy, and electrical conductivity, and was verified against known volcanic events (Meese et al., 1994a). Accumulation rates were calculated from the annual layer thicknesses corrected for measured density, with a very small correction for ice-flow thinning (Alley et al., 1991). CO_2 results were obtained with dry extraction and laser absorption spectroscopy (Wahlen et al., 1991). Concentrations of cations, anions, and H_2O_2 were measured on biyearly samples (Mayewski et al., 1993b; Drummey, 1993). Methanesulfonic acid (MSA), an oxidation product of dimethylsulfide, indicates biological activity. The time interval includes the northwest European Medieval Warm Period, which extended from approximately A.D. 1000 to between A.D. 1350 and 1450, and the subsequent Little Ice Age, which lasted until approximately A.D. 1900. A significant change in chemistry shortly before A.D. 1600 correlates with Icelandic sea ice, but is not obvious in the $\delta^{18}O$ record. (From data shared by GISP2 investigators; figure used courtesy of the GISP2 Science Management Office.)

and varved sediments (e.g., Rozanski et al., 1992) offer possibilities.

Figure 5 shows a 1200-year multi-parameter record from GISP2 subjected to a 50-year smoothing. Records of decadal Iceland sea-ice frequency and of decadal Iceland temperatures derived from the sea-ice record (Bergthorsson, 1969) have been added for comparison. The period covered includes the Medieval Warm Period (MWP), from about A.D. 1000 to between A.D. 1350 and A.D. 1450, and the Little Ice Age (LIA), from about A.D. 1450 until about A.D. 1900, as defined in northwest Europe. The Icelandic record shows clear warm and cold periods corresponding to the northwest European MWP and LIA. The GISP2 $\delta^{18}O$ record, on the other hand, shows no clear, sustained evidence of the MWP and LIA. Rather, it shows a pattern of minor, century-scale fluctuations throughout this part of the record. During the period of the LIA the warm phase of these fluctuations appears to be more subdued, giving on average a colder climate. A similar observation was made at Crête, Greenland (Dansgaard et al., 1975). Accumulation, determined by identification of individual annual layers and flow modeling (Alley et al., 1993; Meese et al., 1994b), decreases after about A.D. 1200 and is, on average, lower during the LIA period. Terrestrial source indicators such as particles, Ca, Mg, and K, as well as marine indicators such as Na and Cl, are low during the early part of this period (MWP and early LIA) and increase around A.D. 1600, indicating increased transport by stronger winds since A.D. 1600 (see Mayewski et al., 1993b). Particles can also be of volcanic origin, in which case they are accompanied by elevated levels of non-sea-salt sulfate, and may include volcanic glass shards. The non-sea-salt sulfate, which reflects primarily volcanic input as plotted here, does not indicate that volcanism was a major factor in LIA climate (Mayewski et al., 1993b). Concentrations of nitrate (from lightning and soil exhalation, for example) and methanesulfonic acid (MSA, from oxidation of ocean-produced dimethylsulfide) decrease around A.D. 1350 and A.D. 1430, respectively, coinciding with the transition from the MWP to the LIA in Europe. Ammonium, which reflects primarily biomass burning as plotted here, shows high values during periods of climate (and, presumably, vegetation) change.

The multi-parameter ice-core data thus reveal a far richer and more detailed picture of climatic changes and the accompanying environmental alterations during the MWP and ensuing LIA than was available from other sources. This highly detailed record extends back through the glacial-to-interglacial transition well into the last glacial period. The significance for society of the relatively minor climate fluctuations observed in the recent GISP2 ice core record is dramatically illustrated by the fate of the Norse colonies on Greenland that thrived during the MWP from about A.D. 1000 to A.D. 1300, but lost contact with Iceland and disappeared when climate deteriorated after A.D. 1400.

The Glacial-to-Interglacial Transition

Figure 5 shows coherent patterns of change in various properties in the recent record. Yet it is difficult to study the dynamics of comparatively small century-scale climate changes in the presence of much larger short-term climate fluctuations that are not yet understood. A better situation exists across the last glacial-to-interglacial transition, which is apparent between 1670 and 1800 m of depth in the GISP2 core. There the ice has preserved in great detail a record of various climate parameters that show rapid century-scale fluctuations during this period of major climate change. A detailed time scale for the core is being constructed from the annual layering observed in visual stratigraphy, in the electrical conductivity (ECM), and in the particle concentrations of the ice (Alley et al., 1993; Meese et al., 1994a). This will provide the core with an internal time scale with annual resolution for determining the relative timing of the changes in climate-related core properties, such as stable isotopes, chemistry, dust, ice fabric, and atmospheric gas and isotope composition, during the transition from glacial to interglacial climate.

The $\delta^{18}O$ record of the glacial-to-interglacial transition (Figure 6) clearly shows the Younger Dryas as the last period of full glacial, low $\delta^{18}O$ values before the Holocene, lasting from about 11.6 to 12.9 kyr BP (Alley et al., 1993). The Younger Dryas is preceded until 14.7 kyr BP by a period of "warmer" but not fully interglacial climate, probably corresponding to the Allerød/Bølling interstadial complex described in Europe (Mangerud et al., 1974). The interstadial is interrupted by three brief (about a century) cooler episodes, of which the middle one is fairly weak. The ECM record (Taylor et al., 1993), which is measured continuously along the core with a resolution of 1 mm per sample, shows a pattern of highs and lows quite similar to those of the $\delta^{18}O$, except that ECM values drop to almost zero during periods of low $\delta^{18}O$. The second, weak, cold period of the Allerød/Bølling is not clearly developed in the ECM. The ECM records the presence or absence in the ice of alkaline dust that neutralizes the acid responsible for most of the electrical conductivity of the core (Taylor et al., 1992); it is thus strongly anti-correlated with the calcium concentrations (Mayewski et al., 1993c) in the ice.

The high-Ca, high-dust/low-ECM, low $\delta^{18}O$ combination in the Younger Dryas indicates a cold climate with strong winds. This finding, which is supported by other terrestrial and marine source indicators such as Mg, Na, and Cl (Mayewski et al., 1993c), fits the general concept of increased meridional temperature gradients and atmospheric circulation during glacial times (Manabe and Hahn, 1977). Comparison of the three Allerød/Bølling cool phases shows that while $\delta^{18}O$ and Ca concentrations assume intermediate values, the ECM drops to near zero in the two colder ones and shows little change in the third. This is as expected for

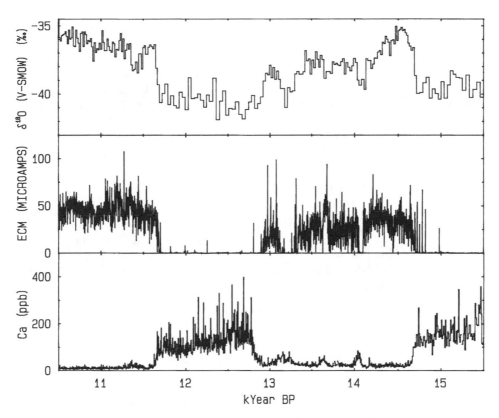

FIGURE 6 The last glacial-to-interglacial transition (11.6 to 14.7 kyr BP, Alley et al., 1993; Taylor et al., 1993) is indicated by δ¹⁸O, electrical conductivity (Taylor), and calcium concentrations (Mayewski) in the GISP2 core from the Greenland summit.

a neutralization/titration of acidity. The accumulation over this period, determined from visual stratigraphy (Alley et al., 1993), closely resembles the ECM and the δ¹⁸O records; it shows a drop to half its Holocene value in the Younger Dryas. Excellent agreement exists between the GISP2 records and the GRIP record (Johnsen et al., 1992a; Grootes et al., 1993).

Particularly interesting in all the records is the rapidity of the climatic changes. Transitions between the high (warm) and low (cold) ECM sections in the core are abrupt; they occur in less than 10 years, sometimes in as little as 3 years. Several transitions show an oscillatory behavior in which the ECM returns rapidly if briefly to pre-transition values (Taylor et al., 1993). Changes in accumulation (Alley et al., 1993) and in chemistry (Mayewski et al., 1993c) appear to be equally rapid. The isotopes have so far been measured only at 1 m (15- to 50-year) resolution, so they still lack this level of detail. Samples cut at 0.6 to 1.7 cm resolution are awaiting a detailed isotope study of these transitions.

The ice-core record of the glacial-to-interglacial transition thus shows an extremely dynamic climate system, with frequent alternations between warm and cold periods of varying length and with major changes in atmospheric circulation occurring over less than a decade (see Mayewski et

al., 1993c). Recent work on high-sedimentation rate ocean cores in the North Atlantic (Lehman and Keigwin, 1992; Lehman, 1995, in this chapter; Ruddiman et al., 1977; Sarnthein et al., 1992; Vogelsang, 1990; Veum et al., 1992; Karpuz and Jansen, 1992; Broecker et al., 1990a; Eglinton et al., 1992) shows corresponding changes in the ocean.

The deeper part of the GISP2 and GRIP records (Johnsen et al., 1992a; Dansgaard et al., 1993; GRIP members, 1993; Grootes et al., 1993) shows that major, rapid climate fluctuations were not unique to the last glacial-interglacial transition, but occurred throughout the glacial and even interglacial periods. The pattern of interstadial and stadial episodes in the glacial part of the core provides evidence for frequent changes between cold glacial and nearly interglacial conditions during a time when the deep-sea isotope records (Martinson et al., 1987; Imbrie et al., 1984) indicate the existence of large continental ice masses. This pattern has been observed, although not in the same detail, in all long ice-core records in Greenland (Johnsen et al., 1992a). Major rapid changes during this time are also documented in Atlantic ocean cores (Broecker et al., 1990a; Sarnthein and Tiedeman, 1990), albeit at lesser resolution. A question still to be answered is whether corresponding climate changes exist in Antarctica.

CONCLUSIONS

The new ice-core records being obtained from the summit of the Greenland ice sheet promise to provide a 200,000-year history of changes in climate and environmental conditions over the Greenland ice sheet and in the North Atlantic basin, in unprecedented detail. To utilize this information for the determination of global climate and its variability on all time scales, we need to find first the spatial variability in the ice-core properties, and second the relationship between the ice-core records and paleoclimatic records from ocean sediments and terrestrial sources elsewhere on the globe. For the first we must study ice cores from different parts of Greenland and the Arctic, from lower-latitude, high-altitude ice caps, and especially from Antarctica. For progress toward the second, active collaboration among the glaciologic, the oceanographic, and the terrestrial paleocli-mate research communities is needed. In view of the many and very rapid climate changes observed in the ice-core record, it is crucial to obtain accurate and detailed independent time scales for all paleoclimate records so that they can be used in intercomparisons. The detailed information on past changes in climate and environment will then allow us to study not only the major glacial-interglacial climate changes but also climate variability on decade-to-century time scales, and to develop an understanding of the causes and mechanisms of such changes.

ACKNOWLEDGMENTS

This work was supported by NSF grants DPP-8822073 and DPP-8915924, Division of Polar Programs. I thank my GISP2 co-principal investigators and L. G. Thompson for contributing their results, and P. A. Mayewski for critical comments.

Discussion

WALLACE: Are there cores from ice sheets in other Northern-Hemisphere locations, such as Alaska, comparable to these Greenland cores?

GROOTES: There are several cores from ice masses in other Northern Hemisphere locations, though none of them is as long and detailed as the Greenland cores. A core drilled by Holdsworth on Mount Logan in Canada provides a record of about 1000 years. Of the three lower-latitude cores drilled by Thompson, the tropical Quelccaya ice cap in Peru covers about 1500 years, and the two cores in China, Dunde and Guliya, go back well into the last glacial period. Also, there are cores in the Canadian Arctic on Devon Island and the Agassiz and Barnes ice caps, and others on Svalbard, Novaya Zemlya, and Severnaya Zemlya. Several of these cores come from lower elevations that have summer melting, so the climate record is less well preserved.

BRYAN: Do the earlier cores, like those Dansgaard analyzed, correspond with your findings?

GROOTES: On a large scale we found the $\delta^{18}O$ and conductivity of the ice to be quite close for the different cores. In details, though, you find quite different signatures for the various locations. To understand the details of atmospheric circulation, we need a multitude of cores from different locations and elevations so we can distinguish local phenomena and more general patterns. The Quelccaya records in the Southern Hemisphere go back about 1500 years, and I'd say there's the potential in the high Andes for going back to 2000 years.

Temporal Change in Marine Ecosystems

JOHN A. MCGOWAN[1]

ABSTRACT

Natural populations are highly variable in space and time, so biogeochemical processes and events must also be. Population and community biologists have shown that internal dynamics in biological systems, such as those due to competition and predation, can cause some of the population changes observed. It is generally conceded (but seldom demonstrated) that climatic variations also influence population growth or decline. In addition, harvesting, pollution, and habitat destruction are important agents of change. It is difficult to separate the influences of these several causal factors, so ascertaining the effects of climate change on organized living systems (ecosystems) is not straightforward.

The prevailing approach to the determination of the causes of population variability and ecosystem change is essentially reductionist. There have been many small-scale, local studies of the natural history, physiology, behavior, or growth of single species. But species populations in nature are constantly interacting with a complex system of other species and to broad spectra of physical and chemical variables. Many ecologists have therefore chosen to make the simplifying assumption that different species can be aggregated into functional groupings of species, the dynamics of which adequately represent the state of the biotic system. The group dynamics actually studied, however, are usually the rates of transfer or "flux" of carbon, nitrogen, or calories between a few aggregated categories.

Many ecosystem models assume a steady state. But the systems seldom seem to be in a steady state when field studies are done, and concurrent measurements of "fluxes" involve only a limited number of groups. Ecosystems are organized on very large spatial scales, and the heterogeneity of the ocean makes it difficult to extrapolate upward from small scales of study. Furthermore, the roles of external forcings that are independent of population density, such as climate, are seldom evaluated empirically.

As a consequence, we have little good conceptual insight into how climate affects populations or ecosystems, and almost no predictive capability. But large-scale, long-term monitoring of both individual species populations and aggregated groups has been done. In general, when harvestable populations have been monitored, the relationship between population variations and climate is vague at best. Monitoring of unharvested pelagic systems, however, clearly shows the relationship between population and low-frequency climatic variations.

---

[1]Scripps Institution of Oceanography, University of California San Diego, La Jolla, California

INTRODUCTION

Laymen always ask how climatic variations will affect commercial fish stocks or agriculture. Climatologists and oceanographers are seldom able to answer. And yet almost everyone believes that there is a strong linkage between climatic variability and biology, especially population biology. Establishing the nature of this linkage in the case of the ocean has proven to be very difficult, and at present we have little predictive ability. This inability is due partly to the approach biologists have taken to the problem, partly to the inherent difficulties in studying the complex, flexible webs of natural biological systems with their numerous feedback loops, and very much to problems of scale.

We are just beginning to understand that different population phenomena happen on different scales for different reasons (Haury et al., 1978). It is the larger spatial scales and low-frequency temporal scales that seem to be the most important, because most of the variability resides there

(Figure 1). But so few observations have been made on the larger scales that even this observation may be tenuous. It does seem true, however, that conclusions of studies of short-term, local changes or functional processes cannot be categorically extrapolated upward and outward to achieve large-scale understanding. The ocean and its populations are too rich in temporal and spatial variability for that to be so. Figure 1 shows typical time/space coverage of most studies designed to measure functional aspects of the ecosystem, such as the flux of energy or materials through selected categories of the biota. Sample separation on such studies may be on the order of hours to about 40 days—the maximum duration of most oceanographic cruises. Although such cruises may be repeated, there is seldom continuity between them, and there are large gaps in the frequency bands we wish to resolve. Variations on the decade-to-century scale cannot be resolved by studies done on smaller frequencies.

There have always been good reasons to study population

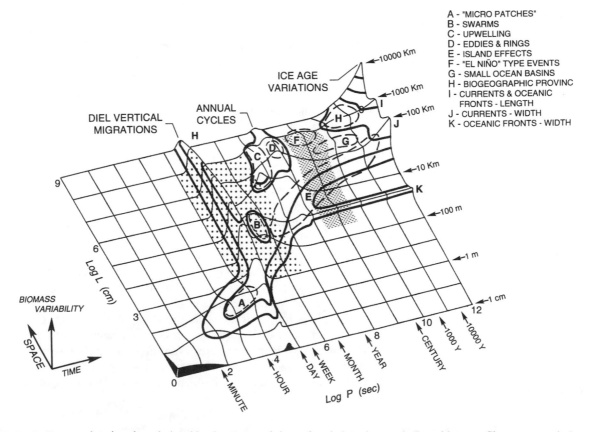

FIGURE 1   A diagram showing the relationships in space and time of variations in zooplankton biomass. Short-term variations generally take place on small spatial scales. Longer-lived variations are widespread. Different scales of variability almost always have different causes. For example, spawning aggregations, called "swarms," occur in euphausiids. These differ in kind from aggregations due to "upwelling centers" or Gulf Stream cold-core rings. The shaded area on the left indicates the time and space coverage of most process-oriented studies and the resolvable variance. The shaded area on the right is the decade-to-century time scale (data from Haury et al., 1978).

dynamics on the scale of generations, for it usually takes generations for detectable population growth to occur. Also, it has always been acknowledged that somehow or other the environment, including climate, must play a role in affecting population dynamics. Organisms can adapt and acclimate to changing environments. But now it seems that important climatic changes may occur much more rapidly than those experienced during the evolutionary history of most natural populations. If we are to make judgments about the seriousness of the threat of climate change, we will require a much better understanding of how relatively rapid climatic change affects populations, ecosystems, and biogeochemical cycling. It is the 10-to-100-year time scale over which significant changes in population growth rates

may be detected, so it is the one that should be studied for most macroscopic organisms.

Feedback loops exist between the biota and atmospheric gases (Figure 2). Rapid climatic change will almost certainly perturb the structure of marine communities and therefore the feedbacks, but for the most part we do not know what direction these perturbations will take or of what magnitude they will be. Carpenter et al. (1992) have shown how changes in the structure of plankton communities in lakes can have significant consequences for nutrient cycles and thus the stability, resilience, and resistance of ecosystems.

It seems essential to begin the process of acquiring the information that will lead us to a better understanding of the role of climate as it affects population and community

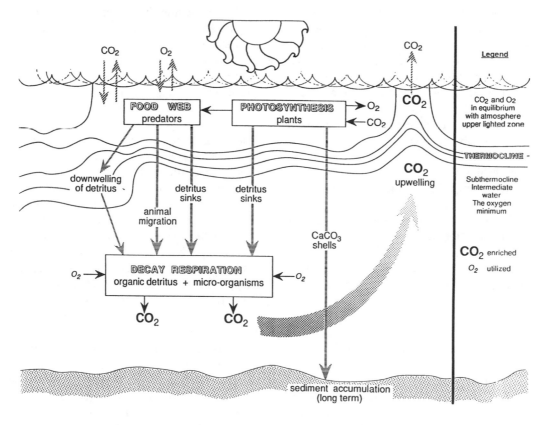

FIGURE 2   The ocean-atmosphere carbon cycle. Dissolved CO_2 and O_2 in the mixed layer are usually in equilibrium with the atmosphere. Photosynthesis in the upper, lighted zone removes dissolved CO_2 and "fixes" carbon into organic compounds. Some phytoplankton secrete calcium carbonate thecae which fall out of the mixed layer and either dissolve or form vast amounts of sediment. Plants also produce dissolved organic matter and particulate organic detritus which may sink. Much of the organic carbon produced is eaten by animals and passed through the food web (see Figures 3 and 4). Animals also produce dissolved and particulate organic matter. Some zooplankton engage in a diel vertical migration, feeding in the mixed layer and respiring and excreting in the sub-thermocline waters. Physical downwelling carries with it dissolved inorganic and organic carbon and detritus. Detritus is fed upon by microorganisms which through the process of respiration remove oxygen and add carbon dioxide to the sub-thermocline water. These waters are greatly oversaturated with CO_2 and undersaturated with O_2 everywhere. The sediment and the sub-thermocline waters of the world's ocean are the two greatest reservoirs of carbon on earth. Upwelling of oversaturated intermediate water may complete the cycle. The relative importance of the biological portions of this cycle is unknown.

biology. Time-series monitoring of critical but measurable components of large natural systems is needed to achieve this understanding. Limited efforts in this direction have already yielded important results.

BACKGROUND

Ecologists and other biologists want to understand why populations vary so widely in time and space. The scales of interest extend to geological time, which includes extinctions, speciation, and therefore evolution and the origins of diversity. Malthus in 1798 pointed out that all organisms could more than replace themselves during their lifetime, but that the resources (food, space, light) necessary to sustain such implied population-growth rates were in short supply. If essential resources are limited, not only individuals but species populations must compete for those resources. Observations of such competition led Darwin to propose a natural selection in favor of the fittest competitors. Ever since, competition theory has been a mainstay of the most fundamental concepts in ecology and evolution (Lotka, 1926; Volterra, 1926; May, 1973). This interplay between competitors can be shown in models and experiments to yield oscillations in population sizes.

One important outcome of these studies was the recognition that the relationships between the population growth rates of two or more competing species must be density dependent (i.e., non-linear) in order to stabilize the system. Stability allows for the continued presence of all of the species involved, and thus the persistence of diversity. The latter is an observed fact of nature. A very similar conceptual basis exists for predator-prey interactions with density-dependent, non-linear relationships between population growth rates required for dynamic equilibria of the two (or usually more) interacting populations.

Empiricists working with real, complex biological systems in nature often question the utility of such models. Not only species interactions (as predators, prey, and competitors) but demographic structure, genetic structure, stochastic factors, and in particular external non-density-dependent environmental forcing (i.e., climate) play very important roles. All exert their effects concurrently in complex, species-rich systems where everything seems to depend on everything else. This makes the models difficult to test. There are, unsurprisingly, few well-documented cases in nature showing competitive exclusions, or demonstrations of environmental variations' changing the course of a competitive contest. Some relatively short-term (in terms of the generation times of the subjects) experiments in removal or exclusion (of predators) have been done, but, as Pimm (1991) has pointed out, often not enough time has elapsed for effects, expected or otherwise, to appear. For these species-removal experiments to be useful for full validation of competition, predation, and community theory, the effects

on populations must be followed for the generations it takes for differences in population growth rate to be detected and to influence other species' population growth rates (e.g., through competition). For many marine organisms of interest, decades are required.

In the cases where adequate time-series measurements of population sizes of pelagic organisms have been obtained, the largest, and therefore the ecologically most important, changes have been found to be low frequency, mainly interannual. This result holds despite the organisms' intrinsic generation times, which may differ by orders of magnitude between species.

PERTURBATIONS AND CHANGE

The difficulties in applying competition/predation theory in the attempt to understand the configuration or kinetics of natural, multi-species communities have led to the development of an additional body of theory that depends on either external disturbances that perturb the system or the existence of patchy, non-random aggregations (Paine and Levin, 1981; Yodzis, 1978). In both cases the expectation of having a superior competitor present in the system, e.g., the competitive exclusion of others, is mitigated. This happens because the superior competitor's populations are episodically diminished by disturbances, thus allowing the inferior competitors to persist. The inferior ones may have special adaptations to allow them to reproduce at an early age or to disperse well, or have some other opportunistic trick that allows them to continue as populations while co-occurring with competitors. This may happen if density-independent perturbations occur that inhibit the dominant competitor (e.g., storms). Patchiness can also work, in theory, if there is constant or at least frequent migration between patches of species or groups of species with varying competitive abilities. This leads to a sort of contemporaneous disequilibrium (Richerson et al., 1970). Observations have been made, particularly in tropical forests and the intertidal zone, that are consistent with these theories and tend to support them. However, few sets of observations cover the necessary number of generations (particularly in forest trees). Furthermore, almost all of the studies have been done on sessile species, where space is clearly a strong limiting resource. There is evidence (McGowan and Walker, 1985; Venrick, 1990) that mobile oceanic populations and communities may not behave in this way.

This body of theory explicitly includes forcing from outside the biological system in a way that is independent of population density. The primary extrinsic force is climate, which is not generally thought to be affected by intrinsic biological processes, even though there are feedbacks from oceanic biota to the atmosphere as shown in Figure 2.

There is consensus that climatic variation is of central importance in the regulation of the availability of resources, the intensity and even direction of competitive contests, and the success of prey populations in overcoming the effects of predation. It is not clear why this outlook has developed. Perhaps it is because population cycles are clearly linked to the seasonal cycle, which is a well-understood climatic phenomenon. Other lower- (and higher-) frequency climatic events are also sporadically invoked as sort of a *deus ex machina*, although they are not often clearly shown to cause or to be linked to population changes in a predictable or quantifiable way. There is no well-developed body of theory of climatic effects, other than those of temperature on the fecundity or metabolic rate of individuals. But the study of the functional response of an individual to, say, temperature change cannot be extrapolated to yield information on the numerical response of entire populations, whose increase or decrease depends on the ratio of birth rates to death rates.

COMMUNITY STRUCTURE, TROPHIC STRUCTURE, AND CHANGE

Organized pelagic ecosystems do exist, and can be shown to have natural boundaries. They occupy regions that are physically distinctive. They are large, few in number, and relatively simple by comparison with land ecosystems because there are fewer constituent species (McGowan, 1971). But in spite of their relative "simplicity," they all contain hundreds of species populations of producers, herbivores, carnivores, omnivores, and degraders. Each of these species populations is thought to be genetically unique, having its own life-history pattern, physiology, and behavior. Since the species of these systems have spatial and temporal overlapping distributions, they all form nearly constant parts of each other's biotic environment. There is plenty of opportunity for interaction, so they are all more or less interconnected dynamically through nutrient exchange or competitive or predator-prey linkages.

There are significant logistic difficulties in studying the dynamics of such complex ecosystems on the basis of their individual species' population dynamics or physiology. Conceptual or mathematical models resulting from such ecosystem studies are often intractable (Hedgepeth, 1977). Ecologists, especially marine ecologists, have therefore resorted to the study of aggregated categories or functional groupings, such as "primary producers" or "herbivores" (see Figure 3), thus reducing the number of entities (or terms). Such approaches have little to do with species competition or predation theory, but nevertheless have theory of their own (Lindemann, 1942; Steele, 1974; Fasham, 1984): For example, trophic theory attempts to explain variations in abundance of entire trophic levels (primary producers, herbivores, pelagic fish). These are aggregated categories of species either shown, or more often assumed, to have the same or similar trophic (i.e., food) requirements. The rates of transfer between levels, either of energy or of

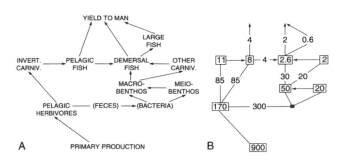

FIGURE 3 (a) A North Sea food web in which the details of the organization have been aggregated into major, trophic categories. (b) The values in the boxes indicate the "assimilation" in kcal m^{-2} yr^{-1}; the arrows show input or "flux", also in kcal m^{-2} yr^{-1}. In this model Steele has emphasized the efficiency of energy transfer between boxes and the stability of the system. He was concerned about the difficulty of making suitable observations at sea. (After Steele, 1974; reprinted with permission of Harvard University Press.)

some nutrient such as nitrogen, and the efficiency of these transfers are what is measured. But models based on nutrient flux usually include only one variable, and they differ conceptually from energy-flow models. Nutrients are retained and recycled, energy is not.

Pelagic ecosystems may be very far from equilibrium, and indeed they never seem to be in steady state when the measurements are made. The complete suite of pertinent fluxes within the food web have never been measured at the same time. Transfer rates must be non-linear and density dependent to stabilize the system. Determining a sufficient number of such rates to describe the state of the systems in nature has proven to be very difficult. Furthermore, ascertaining how extrinsic perturbations or variations in climate may change these rates seems as yet to be beyond current capabilities. The problems are likely to remain until the rates of change of biomass for the aggregated categories (i.e., the trophic levels) have been measured over a length of time sufficient to include some biologically "important" climatic events. Similar difficulties arise in attempts to model the role of the marine biosphere in the global carbon cycle. For this task one must understand how oceanic food webs withdraw CO_2 from the surface waters, package it as organic parcels, and export it to the subpycnocline waters (Figures 2 and 4). This process sequesters millions of gigatonnes of carbon (GtC) in the sediments and tens of thousands of Gt as dissolved inorganic and organic carbon in the intermediate and deep waters of the world's oceans. But our understanding of the rates of transfer between what are assumed to be the important aggregated categories (boxes in Figures 2 and 4) is very meager (Longhurst, 1991; Sundquist, 1993). In most cases we cannot estimate the error in these rates, and have little idea of which kinds of extrinsic perturbations will alter them.

Time series of concentrations of the most easily measured

Stocks as Gt C   *Flows as Gt C yr⁻¹*

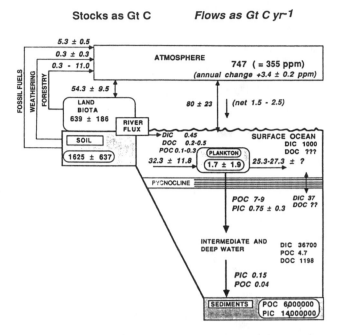

FIGURE 4   A diagram of the global carbon budget, showing what are thought to be the most important pathways between stocks or transient reservoirs. An attempt has been made to assign values to the amount of carbon in each reservoir (in GtC) and the flux of carbon between them (in GtC yr^{-1}). Varying levels of uncertainty are associated with these values; many are essentially unknown (Sundquist, 1993). Many or perhaps most of the fluxes are nonlinear, and few of the reservoir sizes have been followed through time. Some known mechanisms of flux are not included—for example, the advective sinking of particulate organic carbon shown in Figure 2. (After Longhurst, 1991; reprinted with permission of the American Society of Limnology & Oceanography.)

quantities, along with climatic variables, are needed to allow us to at least describe a system's response (or the lack of it) to perturbations, and to promote realistic modeling efforts and their validation. We do not, at present, really know what types of atmospheric or hydrographic perturbations affect biological systems and what types do not. There are many kinds of disturbances and perturbations, ranging from microscale turbulence to El Niños, almost all of which have been implicated one way or another. Some of these seem to represent severe disturbances, whereas others do not. Which ones, then, are which? How, and in what directions, do the systems respond to different kinds of climatic events (Wiebe et al., 1987)?

FISHERY DATA

Some marine populations have been followed very closely over many generations. A great many detailed studies have been made of the population dynamics of commercial and sport fish, because of their obvious importance to man. These studies typically focus on populations of single

species that were targeted because of economic or political pressures, and because the catch itself constituted a convenient and useful population-sampling tool. The almost universal result, where long-term studies were done, is that individual species' population sizes vary enormously through time, frequently by factors of ten or more, and that they do not often correlate with each other or with anomalies in climatic proxy measurements such as sea surface temperature (Garrod and Colebrook, 1978; MacCall and Prager, 1988; McGowan, 1989; Skud, 1982; Steele and Henderson, 1984; Sund and Norton, 1990; see Figures 5, 6, and 7). But Kostlow (1984) has pointed out that if the number of fish larvae and juvenile survivorship (hence recruitment to adult stocks) depend on large-scale environmental variations, there should be correlations in recruitment among species of fishes whose larvae have overlapping large-scale distributions, since they would be expected to respond in similar ways to perturbations. Kostlow examined time series of recruitment in 14 stocks in the northwest Atlantic; out of 91 comparisons in his correlation matrix, fewer than one-half were significant and 35 were negative. He also examined relationships between stock and recruitment and found no evidence for density dependence, and on this basis concluded that large-scale extrinsic processes were likely to be responsible for those correlations observed. However, no physical or climatic data were used in the study, and apparently no corrections were made for multiple testing in his correlation matrix.

Hollowed et al. (1987) examined pairwise correlations of 35 years of recruitment data for all possible pairs of 59 species or stocks from the northeastern Pacific. When they applied a correction for multiple testing, very few of the pairwise correlations were significant. One of their conclusions was that the principal value of the correlation patterns seen in their study lay in the possible use of the patterns for generating hypotheses about factors influencing recruitment. Garrod and Colebrook (1978) looked at recruitment patterns of many species over large sectors of the North Atlantic using principal-component analyses, but failed to find meaningful general patterns. Sund and Norton (1990) have examined the time series of California fish landings for the period 1928 to 1985. They tested the utility of the data set in relating variability in landings to climate variability for warm-water species (Figure 5). Only one species of the seven was associated with temperature variations other than seasonal. Sund and Norton conclude that "data concerning any resource highly impacted by man's economic or other activities are likely to have variability in amplitudes and frequencies that are greater than those induced by non-anthropogenic factors. Declines in fish landings as a possible result of high exploitation during warm water periods and high abundance is thought to be a major factor which confuses simple interpretation of the data." Steele and Henderson (1984) point out that some of the most dramatic population

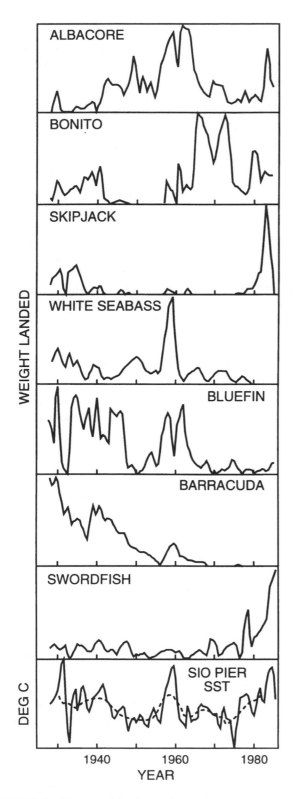

FIGURE 5   The annual landings of some important warm-water commercial and sports fish in the California Current, along with sea surface temperature (and its 5-year running mean) at Scripps Institution of Oceangraphy's pier. Each panel is scaled to give the maximum weight landed at full scale. (From Sund and Norton, 1990; reprinted with permission.)

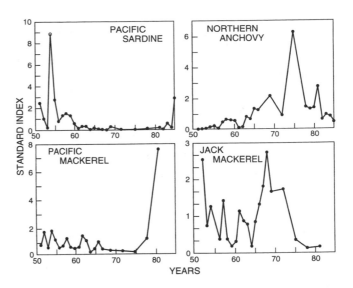

FIGURE 6   Variations in abundance indices of the larvae of four important schooling fish off California. These fish (Pacific sardine, northern anchovy, Pacific mackerel, and jack mackerel) have broadly overlapping distributions and very similar diets. (After MacCall and Prager, 1988; reprinted with permission of the California Cooperative Oceanic Fisheries Investigation.)

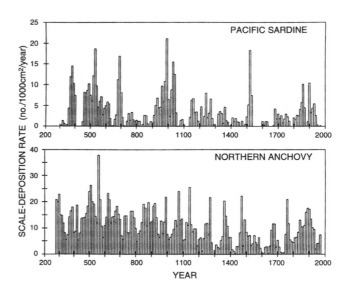

FIGURE 7   A composite time series of fish-scale deposition rate in the Santa Barbara basin.

fluctuations occur in pelagic fish stocks, but there are no obvious or simple general relationships between these and contemporaneous variations in climate or physical changes in the ocean. They claim that abundance changes are rapid, but tend to occur following 50-year periods of relative stability. Steele and Henderson suggest (through the use of a model) that fish populations are resistant to environmental pressures until they reach a certain threshold, at which point they flip over to a new relationship with their environment; in other words, they show multiple steady states.

Although multiple steady states may exist, many factors besides population size influence catch data; these include fishing-gear changes, the amount of fishing effort, and market-demand factors. Furthermore, most commercial species show long-term, secular trends of decreasing abundance (catch) that are not associated with climatic trends. There are strong indications that human harvesting of populations has introduced such a large added source of mortality at such an unusual point in the age structure that natural and climate-induced variability may be masked. Thus, the use of fisheries catch data (or other estimates of population change) to test competition and predation theory, or to assess the role of extrinsic, density-independent factors such as climate, is questionable. In this case anthropogenic and natural variability cannot be separated, and long-term studies of unharvested, co-occurring species of fish are clearly called for.

FINE-SCALE SEDIMENTARY RECORDS

In an attempt to assess anthropogenic influence on fish populations, Soutar and Isaacs (1974) utilized a time-series of fish-scale counts from cores of the annually varved sediments from the anoxic Santa Barbara basin off Southern California. They looked chiefly at the Pacific sardine and the northern anchovy, two pelagic, schooling species with overlapping distributions in the ocean and almost identical food habits. Judging from 40 years of catch records (meager for the anchovy in the early days) and from egg and larval surveys, the abundances of these two fish had apparently oscillated out of phase. Soutar and Isaacs assumed that the fish-scale count from a core (7.6 cm in diameter) provided a reliable estimate of the abundance of fish in the overlying waters, and that the region of the Santa Barbara basin was representative of the much larger area occupied by these two fish. Their results, which covered at least 1800 years in 10-year increments, showed large oscillations in abundance of both fish as far back as A.D. 400; these variations had a red spectrum, and the two species may have varied in abundance with respect to one another before man began harvesting them. Subsequent studies (e.g., Baumgartner et al., 1992) from this core plus a second one, however, have shown that the two species' abundances are most often positively correlated with one another (Figure 7). Furthermore, attempts to calibrate the sedimentary record, by using 5-year blocks from much larger box cores of sediment from only the past 50 years or so that can be matched to actual in vivo population estimates, have not yielded very convincing results. Baumgartner et al. attempted to overcome some of the sampling problems by time-averaging the data into 10-year blocks and by using only four categories of abundance; they were thus able to resolve only variations of 50-year periodicity. Since both species have generation times of two to three years, these cases have not been very useful

for tests of the importance of climate in theories of competition and predation. If those estimates of the lower-frequency variations can be validated, however, they should be examined with respect to the climatic-change indices that can also be found in the cores. That comparison would seem especially apt, since at low frequencies the two species appear to be positively correlated. Such a correlation would be expected if these two ecologically similar fish were sensitive to climatic changes.

Both Lange and Schimmelmann have examined high-frequency signals of several kinds found in Santa Barbara cores, with the goal of recovering climate-related information from the varved sediments in the Santa Barbara basin that have year-to-year resolution (Lange et al., 1990; Schimmelmann and Tegner, 1991; Schimmelmann et al., 1990; Schimmelmann et al., 1992). While in some cases there seems to be low-frequency (several decades) correspondence of climate indicators and biogenic flux, the existence of higher-frequency relationships is not clear.

TIME SERIES OF DIRECT MEASUREMENTS

There is a third body of biological research that has no very structured theory behind it, nor is it concerned with the details of the intermediary mechanisms responsible for population changes. It is empirical and observational. Its main aim, which does not depend on preconceived ideas of how systems function, is to describe complex ecosystems by asking what the temporal scales of variability of basic properties such as climate, hydrography, nutrients, and biological functional groups are. In other words, what are the frequency spectra? With such information, hypotheses may be tested. Are some frequencies more important—i.e., of greater amplitude—than others? Are there regular patterns of succession over time (Wiebe et al., 1987)? Are there connections (i.e., correlations) between climate change and ecosystem change?

The assumption behind the studies that constitute this third body of research is that various components of the physical-chemical-biological systems interact to influence each other's magnitude or concentration over time. If these interactions happen in a patterned or regular way, there should be detectable statistical relationships between them in spite of a large amount of noise. Most of the researchers in this field want to know whether there are cross-correlations and, if so, at which frequencies. Since both density-dependent and density-independent forcing are probable, the fact that this approach avoids the necessity of discriminating between them is in its favor. It does, however, presuppose a series of measurements long enough and frequent enough to sample adequately all of the potentially important frequencies of variation. Since at present there are not many cases in which we know what these are, we must begin with frequent sampling over long periods of time to capture the essential features of the variations and to lessen the possibility of aliasing. Furthermore, since the ocean is so heterogeneous horizontally, it is critical to separate local variations caused by local, in situ processes or events from variations caused by the advection of water past the sampling site—water that may have had a different (and unmeasured) history. While the number of moderately high-frequency, long-time-series studies of biology of the ocean is small, the number of these in which spatial averaging is possible is even smaller. Spatial averaging can help to ameliorate the effects of patchiness and advective events. A spatially arrayed time series can also be used to determine the correlation length scales of variations, and may also allow discrimination between anthropogenic and natural variations (which will be discussed later).

A few such series of long duration have been examined in this way. In all cases practical considerations have kept the sampling frequencies to monthly or, at best, biweekly intervals. However, two of these, the continuous plankton recorder (CPR) surveys of the United Kingdom and the California Cooperative Oceanic Fisheries Investigations (CalCOFI), are of over 40 years' duration, and both have spatial dimensions.

Continuous Plankton Recorder Surveys

The continuous plankton recorder surveys of zooplankton and phytoplankton species' abundance in the North Atlantic between Scotland and Iceland began in 1939. After World War II, ship routes to and from the Atlantic Ocean weather stations were added to the survey. It was extended progressively until by 1965 a large part of the North Atlantic north of 45°N was sampled monthly by 20 merchant ships and 16 weather ships. In that year, 112,000 miles of continuous sampling was done.

Three early, major results of this effort have been: the preparation of composite biogeographic maps of spatial variations in species abundance and of patterns of diversity in the northern North Atlantic, based on thousands of observations; the derivation of species assemblages and their spatial and temporal trends (i.e., the outlines of community structure); and the analysis of temporal variations in plankton populations. The power of this last analysis is greatly enhanced by the availability of so many samples and the ability to spatially average the data. Much of the recent work on temporal changes has been done for a single area (10°E to 20°W, 45°N to 65°N) comprising part of the northeastern Atlantic, the North Sea, and the English Channel (CPR Survey Team, 1992). This effort has emphasized the relationships between environmental variations and plankton abundance (Figures 8 and 9).

The earlier studies of the CPR group were impeded by the lack of physical and chemical data on commensurate scales of sampling. As time-series information accumulated,

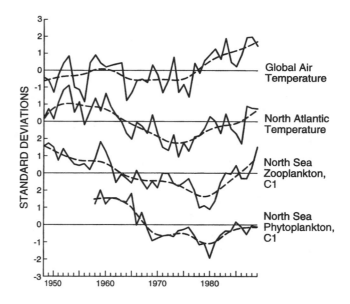

FIGURE 8 Low-pass-filtered Continuous Plankton Recorder and climate data from the Northeast Atlantic. The data shown are anomalies in terms of standard deviations from the mean of a long-term data set. The zooplankton and phytoplankton first-principle components of abundance fluctuations were used here. (From the CPR Survey Team, 1992; reprinted with permission of the International Council for the Exploration of the Sea.)

it became apparent that it was the larger time and space scales that showed the greatest amplitudes of variation; interannual variations were evident, and in some cases dominated the variability. One of the more spectacular of the long-period events was the progressive delay of the spring bloom for phytoplankton species (Figure 9) and a shortening of the seasonal cycles for many zooplankton. Dickson et al. (1988a) relate the declining trend in abundance of both zooplankton and phytoplankton between the mid-1950s and 1980 (the trend may have reversed during the 1980s) to a long-term increase in the northerly wind component over the eastern North Atlantic and European seaboard between 1950 and 1980. They suggest that the consequence of a delay in the onset of the spring bloom of phytoplankton and a consequent shortening of the production season reduced the carrying capacity for zooplankton. The increase in spring winds should have resulted in increased wind mixing, thus changing the critical depth in the sense of Sverdrup (1953). This would mix the phytoplankton beyond the depth at which light is sufficient for photosynthesis to exceed respiration, and thus delay the initiation of net production (i.e., population growth). But Colebrook (1985) associates these trends with variations in west winds. It seems highly likely that in both cases the same basic pattern of climatic change is involved. What is fairly certain, however, is that the changes in the phytoplankton are not a direct response to changes in temperature, and that the

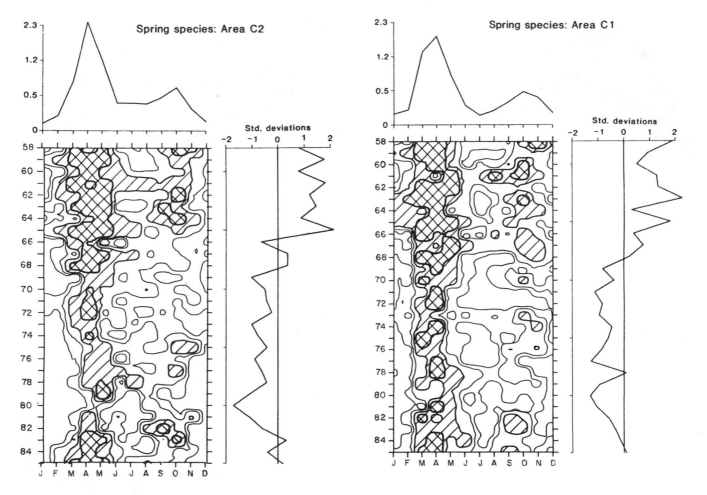

FIGURE 9   Month/year contour plots, long-term monthly means, and standard deviations of 12 species of phytoplankton in CPR standard areas C2 (west-central North Sea) and C1 (east-central North Sea), 1958-1985. (From Dickson et al., 1988a; reprinted with permission of Oxford University Press.)

pattern of change in the North Atlantic should be considered, from the point of view of plankton, simply as an index of climate change (CPR Survey Team, 1992).

Additional studies of the CPR data (Aebischer et al., 1990; Taylor et al., 1992) have confirmed the very close relationship between large-scale, low-frequency variations in plankton and climate changes. These relationships indicate a climate connection spanning the complete width of the North Atlantic, and support the view that low-frequency, large-amplitude planktonic trends are driven externally (Taylor et al., 1992). Radach (1984) reviewed much of the CPR information to discover how much of the observed change in production cycles could be explained by climate. Spectral analyses imply that at least 50 percent can be attributed to the physical environment. Perhaps density-dependent biological processes can account for the rest, but the CPR sampling scheme and sample-processing methods, integrating through time and space as they do, prevent the detection of smaller-scale processes and events that may contribute to the process of change.

The California Cooperative Oceanic Fisheries Investigations

The CalCOFI program is not primarily a fisheries study; rather, it is concerned with all of the elements of the pelagic ecosystem that can be measured routinely. It was the continuing decline of the Pacific sardine that led to the establishment of this large environmental study of the California Current in 1949. The CalCOFI study resembles the CPR work in some important ways. Both allow for the spatial averaging of time series, cover large areas, and are of over 40 years' duration, thus allowing frequencies of change from monthly to decadal to be resolved. Both target the larger, more efficiently sampled zooplankton. They differ in that CalCOFI measures physical and chemical properties in addition to biological ones, and has done so over a greater depth range (500 m), thus adding another dimension to the study.

As in the CPR study, species' biogeographic patterns and seasonal cycles have been described by CalCOFI. There is a strong north-south gradient in biomass and an offshore, midstream maximum in the California Current. Most of the species present are not endemic, but have much larger populations outside of this current system. Elements of four different, large faunal provinces mix together in the central sector, which results in very high species diversity. Time-series studies (Bernal and McGowan, 1981; Chelton et al., 1982) have yielded important results. Low-frequency inter-annual variations dominated the spectra of zooplankton abundance, temperature at 10 m, salinity at 10 m, and mass transport from the north. Furthermore, these variables were coherent and well correlated throughout the system (Figures 10 and 11). The changes in temperature and salinity were consistent with the changes in transport from the north, since the North Pacific and Aleutian currents are sources of cool, low-salinity water. These waters also tend to be high in plant nutrients, and their movement down the coast of North America and into the California Current would certainly have a fertilizing effect that should lead to higher

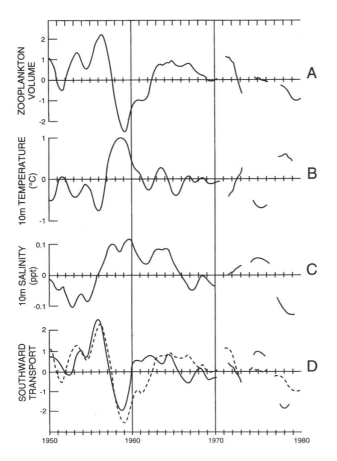

FIGURE 10 The space-averaged, non-seasonal, low-pass-filtered (double 13-month running mean) anomalies of zooplankton bio-mass, temperature, salinity, and mass transport of the California Current. The dashed line in panel (d) is the zooplankton data, repeated from panel (a). (From Chelton et al., 1982; reprinted with permission of Kline Geology Laboratory.)

biomass. But there is also a normal north-to-south gradient of plankton; during times of enhanced southward transport, this could shift southwards and give the appearance locally of a temporal change in population growth. In either event, this change in growth is clearly associated with whatever climatic changes cause such large-scale shifts in the mass transport of the California Current system.

Bernal and McGowan (1981) found that the CalCOFI plankton anomalies were uncorrelated with anomalies in the most popular index of coastal upwelling. This unexpected result has led to new interpretations of the regulation of biomass and new concepts of which physical events are important and which are not. Standing out in this study are the profound effects of the 1958-1959 California El Niño. The largest negative plankton anomalies of the time series from 1949 to 1980 occurred then. This anomaly was repeated in 1983-1984. Further studies (Roemmich and McGowan, 1995) have brought this time series up to date. In their longer series, covering 1950-1993, a clear warming trend, a negative salinity anomaly, and a 70% decrease in zooplankton biomass are all seen on the interdecadal scale (Figure 11). But there was no such long-term trend in mass transport from the north. During this period, temperature differences across the thermocline increased. The resulting increase in stratification inhibited wind-driven upwelling from sources rich in the inorganic nutrients needed for new production, so zooplankton populations decreased. This proposed mechanism for the interdecadal variations in plankton abundance is quite different from the one proposed for interannual changes (Chelton et al., 1982). A model designed to "explain" interannual variations would be inappropriate for the interdecadal scale, since quite different mechanisms seem to be involved.

The observed rate of decrease in zooplankton biomass is 6×10^{-10} s^{-1} over the 43 years of the Roemmich and McGowan (1995) study. On such long time scales, zoo-plankton biomass is controlled by the net of decreases due to an excess of death rates over birth rates, and increases through advection by currents from the north. Advective input is estimated to be $\sim 5 \times 10^{-8}$ s^{-1}, which is far larger than the long-term trend. The trend is a small residual of much larger terms; it cannot be isolated by case-history studies of currents, reproduction, or mortality. It nevertheless is a dramatically large signal when accumulated over 43 years. The Roemmich and McGowan study shows that climate-ecosystem studies dominated by short-term, process-oriented experiments cannot simply be extrapolated to decadal time scales, where the balance of terms is different from monthly or seasonal balances.

Taken together, the CPR and CalCOFI studies have shown clearly and unambiguously that biotic variability in the ocean has a red spectrum, and that the largest anomalies are very closely related to climatic variations. These studies have even allowed us to focus on some intermediary mecha-

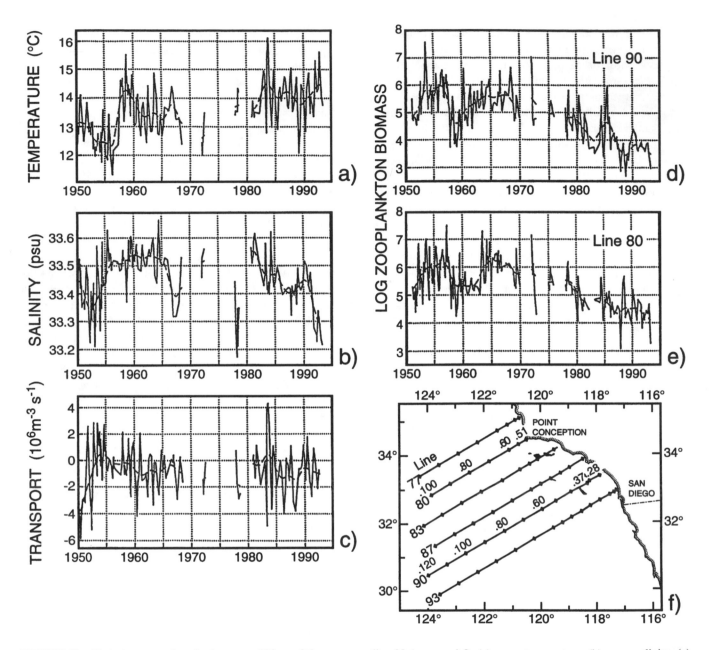

FIGURE 11   Variations over time in the upper 100 m of the ocean on line 90 (see panel f). (a) mean temperature, (b) mean salinity, (c) transport (in sverdrups, with a 500 m reference level), and (d) zooplankton biomass. Panel (e) shows biomass at line 80 for comparison. (After Roemmich and McGowan, 1995; reprinted with permission of the American Association for the Advancement of Science.)

nisms that seem to be important, and to reject others that seem less so.

Allen's Scripps Pier Data

During the years 1919 through 1939, W.E. Allen took daily samples of phytoplankton from the pier at Scripps Institution of Oceanography, and from 1920 to 1939 a similar time series was taken at Port Hueneme, some 120 miles to the north (Allen, 1941). Other five-year high-frequency series were also obtained at various distances from La Jolla

until either they were shown to resemble La Jolla so closely that effort was abandoned, or logistic and financial difficulties arose. Allen tracked the variations in abundance of 98 species of diatoms and over 25 species of dinoflagellates. Population changes of several orders of magnitude occurred both within years and between years. Although he was able to demonstrate a clear seasonal change in both total dinoflagellates and total diatoms, interannual differences obviously dominated the series. Allen lacked the statistical tools to investigate this further, however. He did not analyze the phytoplankton as a community or ask questions about species diversity shifts with respect to hydrography, for such concepts had not yet been well developed. Tont (1981, 1987, 1989), using Allen's 1930-1939 data, looked at seasonality and seasonal anomalies of 23 of the most abundant species of diatoms. He found negative correlations between a few of the species pairs; some of these appear to be significant, whereas others do not. He also cross-correlated seasonal anomalies of total diatoms and dinoflagellates with sea-surface and air-temperature anomalies. All of these yielded negative coefficients, some of which appear to be significant. Tont (1987) states ". . . at very best, however, climatic change, as defined by the variables used in this study, explains only 36 percent of the total variance. . . ." It is clear from his figures that cold years at the SIO pier are more productive than warm ones.

Sugihara and May (1990) have also used Allen's data (along with those for the incidence of chickenpox and measles in New York City from 1949-1972) to test their approach to making short-term predictions about the trajectories of chaotic dynamical systems. They show that "apparent noise associated with deterministic chaos can be distinguished from sampling error and other sources of externally induced environmental noise." They conclude that the method gives sensible answers with the observed time series of measles (deterministic chaos), chickenpox (seasonal cycles with additive noise), and diatoms (a mixture of chaos and additive noise). The dynamics of diatoms near shore reflect at least partial governance by a chaotic attractor. This result is new and valuable information that could only have come from such a time series. It is an indication of the degree to which the variability of plankton populations is driven by external, density-independent forces rather than by the ecosystem's internal nonlinear dynamics.

NATURAL AND ANTHROPOGENIC VARIABILITY

Most of the biological time-series data come from populations that are strongly influenced by humans, either directly by their harvesting forests, fish, and game, or indirectly through pollution or habitat disruption. Our inability to separate the effects of natural processes from anthropogenic ones has proven to be a serious impediment to progress in environmental management, the testing of theories of ecosystem behavior, or determining the role of climate in population variability. Conversi and McGowan (1994) attempted to make this separation through the use of spectral and time-series analyses of three geographically separated sites with similar kinds of sewage discharge and similar water-quality monitoring programs. Her approach was simple: to examine the relationships between the frequency spectra of sewage discharge and those of water transparency, particle concentration, and temperature within and between sites. Each site (Los Angeles City's wastewater treatment plant at Santa Monica Bay, Los Angeles County's at Palos

Verdes, and San Diego County's at Point Loma) had a somewhat different 17-year history of discharge and different mass-emission curves. Conversi found that these were, in each case, uncorrelated with the local water-quality monitoring data. But the water quality did vary with time, especially at the lower frequencies. Most of this varation was seasonal (which the discharge was not), and the rest was correlated between sites and with temperature. There are known natural low-frequency coherent temperature variations in the area of the Southern California Bight. It appears that most of the variations in water clarity (a prime water-quality variable) were natural.

Radach et al. (1990) studied a long time series of temperature, salinity, plant nutrients, and phytoplankton samples taken 3 to 5 days per week at a single station in the German Bight in the southeastern North Sea. They observed conspicuous changes in the annual cycles of nutrients, a general increase in phytoplankton biomass, a shift in the ratio of flagellates to diatoms, and other evidence for a strong systematic change in the ecosystem. There is no evidence for concurrent changes in meteorological or hydrographic conditions. They attribute the observed changes to the anthropogenic increase in nutrient loads of the rivers that discharge into the coastal zone. On the basis of the changes over time in the magnitude, proportions, and pulse-like nature of the nutrient inputs, they suggested anthropogenic mechanisms for the observed shift in the Bight's carrying capacities for different kinds of phytoplankton.

CONCLUSIONS

Relatively few of the time series in the open ocean are of sufficient length and frequency to permit the exploration of important questions of population and community dynamics or the role of climate in influencing these. Many of the series have such large gaps that aliasing is likely, and some frequencies cannot be determined. Others track populations so affected by human activities that it seems unlikely that climate-induced variations can be separated from anthropogenic changes.

However, there are some marine populations for which the sampling has been long-term and of relatively high frequency, as with the CalCOFI and CPR programs. In both cases long-term, space-averaged data have shown that zooplankton variance has a red spectrum and that it is the interannual and interdecadal changes in abundance that are most important, although both study programs showed clear seasonality as well. The low-frequency variability was coherent over large geographic areas in both the Atlantic and Pacific. Both programs have shown significant correlations of plankton-biomass changes with large-scale climatic variations. There is strong evidence in both cases of comprehensive ecosystem response. The effects of the 1958-1959 and 1983-1984 Californian El Niños are unambiguous and large.

The sampling and data-processing methods of the two groups of researchers were different, as were their choices of climatic indicators, but they came to the same conclusion: The largest changes in plankton abundance are clearly associated with climatic changes. Apparently smaller climatic variations have lesser or no detectable effects. We are now in the position of asking how these large variations work to change the carrying capacities of entire ecosystems. What structural changes take place? How is biogeochemical cycling affected?

It is difficult to see how short-term, ad hoc studies of the flux of energy or an element can be extrapolated upward and outward in space and time to get answers. Many such studies are driven by concepts rather than empirically derived inverse models. Realistic models of reasonably complete systems must consist of partial differentials and multiple variables which themselves are functions of other extrinsic variables (Fisher, 1988). If these are to mimic nature, close coupling between parts with vastly different intrinsic rates of natural increase (i.e., genetically programmed generation times) will be needed, including numerous feedback loops that allow external forcings to cause switches between states.

There are two cases in which time series have apparently resolved the question of anthropogenic versus natural variations. Radach et al. (1990) showed that changes in phytoplankton production and community structure were very likely due to the observed enhanced nutrient input to the German Bight from rivers, chiefly the Elbe: "Elbe freshwater discharges and nitrate/nitrite concentrations are running fairly parallel to each other." Conversi and McGowan (1994) have managed to separate low-frequency natural variability from anthropogenic, and in their study the percentage of the variance due to each may be estimated. Neither of these analyses would have been possible without time-series monitoring. Sugihara and May's 1990 analysis of Allen's magnificent time series shows that even very noisy time series have predictive value, and that given the proper set of time-series data there is a real chance of separating intrinsic, biologically driven variations from extrinsic environmental ones.

Because multiple, non-linear interactions exist in complex bio-geochemical marine ecosystems, they never seem to be in steady state when we measure them. Reductionist, experimental micro-research may therefore never provide sufficient understanding of their structure and function on the critical frequency scales of decades to centuries, much less answer the questions of what keeps ecosystems stable or changes them. Even the few existing time series of observations, crude though they are, have already provided us with valuable insight into the behavior of some of these systems and how climate affects them. More observational evidence is badly needed if we are to understand the natural biogeochemical variability of the oceans and assess the consequences of anthropogenic climate change. Natural variability is the reference against which such change must be measured. There may still be time to acquire the needed evidence.

Commentary on the Paper of McGowan

ROBERT R. DICKSON
U.K. Ministry of Agriculture, Fisheries, and Food

I am very glad to have the chance to review Dr. McGowan's paper, for many reasons. High among them is my gratification that the National Academy of Sciences has been so broad-minded as to include plankton and biology in this decadal symposium. The changes in the planktonic ecosystem are arguably the most important changes that this workshop addresses, because they have the greatest socioeconomic impact.

Although the maze of interactions in the planktonic ecosystem that Dr. McGowan showed us is absolutely correct, he was a bit more pessimistic than I would be about their use in understanding what is going on, and also a bit more disparaging about the contributions that can be made by detailed biological studies. We agree entirely on the efficiency and cost-effectiveness of planktonic monitoring, however. I have been soliciting funding for the continuous plankton recorder for the last five years, and Dr. McGowan has been doing the same for many more years for the CalCOFI data set. This issue is of key importance to this symposium because of the need to assess whether an observed change in the plankton is due to climate or to man, by which we usually mean anthropogenic nutrient inputs or eutrophication.

Perhaps you think I exaggerate the importance of answering this question. Signatories to the Ministerial North Sea Conference—and that is everyone around the North Sea—are obliged to reduce nutrient use in agriculture in their countries by 50 percent if their coastal waters are or are likely to be subject to eutrophication. One of the signs of eutrophication adopted by the Paris Convention is an increased prevalence of dinoflagellate blooms in shelf water.

In Århus, Denmark, a university team is working out the cost of remeandering previously straightened rivers in order to allow greater and more effective denitrification by geochemical processes as these rivers flow to the Kattegat. Elsewhere in Europe a team is working on a fast analytical observing system for the detection of toxicity in all catches of fish, in case one day algal toxins are found in fish flesh (they have not been to date).

Such simple approaches as the "precautionary principle," which relies on minimizing inputs in lieu of understanding, do not solve this problem of algal blooms. When municipal authorities reduce their nutrient inputs as much as they can, they always find it far easier to reduce phosphorus than nitrogen inputs, and a lowering of the P/N ratio can be shown to create algal toxicity in previously nontoxic species in the laboratory.

A final problem—and the most difficult of the lot—is that an increasing prevalence of toxic dinoflagellate blooms is the expected result not only of anthropogenic eutrophication via increased nutrient inputs, but also of climatic global warming, because of the increase in stability due to surface warming, increased melt water, fresh-water accession, or whatever. The attribution of a cause to any trend is going to be difficult, yet crucial to the application of effective countermeasures.

Finally, I should like to show some CPR-derived illustrations related to the examples Dr. McGowan presented, just to make the point that with monitoring you can find out rather quickly what is going on.

Figure 1 is the spatial decomposition of the picture of post-war zooplankton trends in seas around the British Isles, which Dr. McGowan showed as an overall mean in his Figure 8. It does not really matter what the numbers are. B4, B5, C5, and D5 are actually in the eastern Atlantic, while the other statistical rectangles are on the European shelf. We find the same trend of change in all parts of this system (except for the continental coastal strip, where trends in primary and secondary production are different), so this long-period downturn is demonstrably not due to anthropogenic effects; it is most likely a climatic response.

The reason I bring up this point is that to react appropriately we need to know whether the cause of the trend is natural or man-made. The actual cause is a change in the effectiveness of wind mixing, which controls the time of initiation of primary production in spring. It also determines the so-called "production ratio" (D_c/D_m) that regulates the subsequent development of the spring bloom. A peculiar thing seems to be going on in our part of the world. There was a long-term increase in the frequency of watches reporting gale force or greater winds from light vessels in the North Sea, at a time when the long downturn in plankton that Dr. McGowan showed was going on. Supporting this increase in windiness, we also have a very remarkable change, from 1960 to 1990, in the significant wave height at the Seven Stones Lightvessel in the western Channel. In their 1991 and 1993 papers Bacon and Carter reported that the Northeast Atlantic was getting increasingly rougher. Their findings (see Figure 2) have now been backed up and supplemented with GEOSAT data, and indeed there has been a 2 percent increase per year in significant wave height.

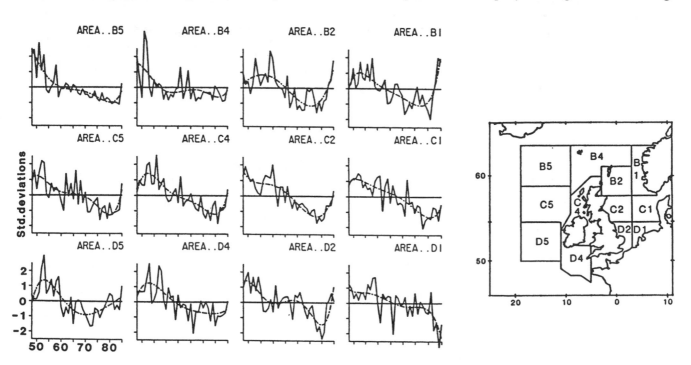

FIGURE 1   Trends of zooplankton abundance in 12 CPR Standard Areas (see map) around the British Isles since 1948.

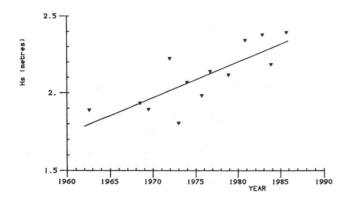

FIGURE 2   Increase in significant wave height (H$_s$) from ship-borne wave recorder data at the Seven Stones Lightvessel (50°N, 6°W) between 1962 and 1986. (From Bacon and Carter, 1991; reprinted with the permission of the Royal Meteorological Society.)

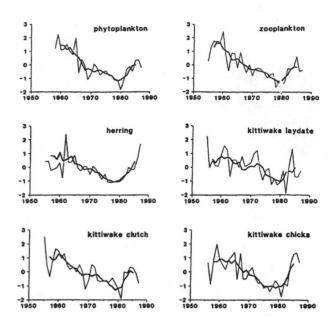

FIGURE 3   Standardized (zero mean, unit variance) time series and 5-year running means for abundances of phytoplankton, zoö-plankton, and herring, for kittiwake laying date, clutch size and chick production, and for frequency of westerly weather, from 1955 to 1987. Ordinates in standard deviation units. In absolute terms, 5-year running means for frequency of westerly weather ranged from 60 to 82 days per year, corresponding to a maximum decrease of 27 percent.

Andy Bakun is also showing something similar in ship observations off Spain. However, Hammond has published a measure of U.K. windiness that goes quite the other way. I do not believe that these things are irreconcilable, but they do show that looking for more and more evidence may turn up something contradictory.

But the final point I should like to make is one that Dr. McGowan and I agree on: Long-term monitoring of large numbers of biological components will eventually yield "system" and understanding. Figure 3 shows that we have for the first time an indication that the long trend in the plankton, which we think is associated with wind speed, is actually apparent in four different trophic levels in the northeast Atlantic: phytoplankton; zoöplankton (the same trend you saw before); herring; and the lay date, clutch size, and chick survival of the kittiwake, our only true marine gull. It would be a terrible shame if we went on monitoring climate and the environment without knowing about the changes going on underneath the sea surface.

Discussion

KUSHNIR: I'd like to add to John's comment about the recent increase in storminess in the North Atlantic. This figure is a different version of Bob's Figure 3. Here you can see the differences between the 15 warm years in the North Atlantic, 1950 to 1964, and the 15 cold years, 1970 to 1984, for both SST and sea-level pressure and winds. In the atmosphere the change from warm to cold conditions corresponded to an increase in sea-level pressure in the central North Atlantic and a decrease north of Iceland — which in turn means an increase in the strength of the westerlies, and hence also storminess, north of about 50° N.

REIFSNYDER: Let me second what John just said about the daunting complexity of systems. Many of the ecosystem models, or even just predator/prey relationships, are predicated on a stable climate. As we recognize that climate is not stable, we call into question even more than usual the predictive value of ecosystem models. We have very little hard information on climate and microclimate effects to feed into what is already something of a black-box approach.

MCGOWAN: I agree. I'm convinced that the only way we can understand how climatic variability—both magnitude and frequency—affects ecosystems is to do lots of monitoring. The reductionist studies some biologists have done are useless for prediction.

TALLEY: Bob Dickson showed some correlation of cod and haddock with events. Have you found catch data useful, John?

MCGOWAN: Not really. They correlate well with events for a few species, but mostly they don't seem to correlate with climate, or even each other. Garrod and Colebrook (1978) and other teams did a massive multi-species cross-correlation of recruitment to the North Atlantic fishery—that's the number of young fish entering per unit of time—and found a vast number of non-significant correlations. If climate is affecting the ecosystems, most of the commercially harvested fish just aren't responding. On the other hand, phytoplankton and zooplankton from those same areas track low-frequency climatic events very well.

DICKSON: Plankton data can be extremely useful in telling you which seas are dominated by eutrophication and which by climate.

RIND: John, would you care to comment on land ecosystem models? How about the agricultural ones?

MCGOWAN: Well, models for managed monocultures do rather well at predicting. They have a great deal of observational and experimental data to work with. But when you start talking about an entire complex of wild stocks, or about ecosystems, you have some complex dynamics.

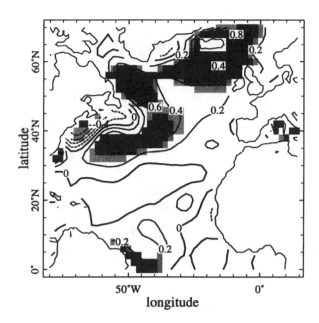

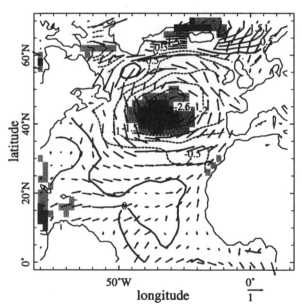

FIGURE 1   The difference between the average wintertime (December-April) conditions during the interval 1950 to 1964 and average wintertime conditions during the interval 1970 to 1984. Upper panel: sea-surface temperature distribution. Countour interval is 0.2°C, negative contours are dashed. Bottom panel: sea-level pressure (contours) and winds (arrows). Contour interval is 0.5 mb; arrow scale is shown at the bottom. Distribution of the t-variable corresponds to sea-surface temperature; sea-level pressure differences are denoted by three levels of gray: light for 2-2.5, medium for 2.5-3, and dark for 3-3.5. (From Kushnir, 1994; reprinted with permission of the American Meteorological Society.)

Maximum Rates of Projected and Actual Increases in Global Mean Temperature as Compared with Bioclimatic Fluctuations

WILLIAM E. REIFSNYDER[1]

ABSTRACT

Predictions have been made and promulgated by authoritative world bodies that the rate of model-predicted global climate warming will be 10 to 40 times that observed following the end of the last ice age (IPCC, 1990b; Schneider, 1989). If this rate of global warming were to actually occur, the effects on earth ecosystems could be dramatic, perhaps catastrophic.

A survey of published rates of increase in global mean temperature for various periods of rapid warming during the past 850,000 years shows that there appears to be an upper limit represented by the relationship $\Delta T = 0.38\ \Delta y^{0.27}$, where Δy is the period during which ΔT shows sustained warming.

Analysis shows that the actual rate of warming during the millennium marking the end of the last ice age was approximately 2.5°C. Furthermore, the IPCC-I "most likely" model-estimated rate of global temperature increase in the past century (IPCC, 1990b) is 2 to 3 times the observed rate for the same period. My analysis shows that for a predicted "business-as-usual" emissions scenario, the climate change that can be expected by the end of the next century is at most twice the maximum rates observed since the last ice age. But since the models used by IPCC appear to over-predict global climate change for the past century by a factor of at least 2, one can assume that in actuality the rate of global warming over the next century will be no more than past maximum rates.

Last, contemporary measured temporal and spatial variations of climate and microclimate are shown to be several orders of magnitude greater than the secular increase of temperature predicted to result from anthropogenic global warming. This relationship may have important consequences for prediction of ecological change in the next century.

---

[1]Professor Emeritus, School of Forestry and Environmental Studies, Yale University, New Haven, Connecticut

INTRODUCTION

It is supposed, and oft stated, that the increase in the rate of global warming due to anthropogenic influences may be so great that terrestrial ecosystems will not be able to adapt, and catastrophic change or collapse will ensue. The IPCC impacts assessment group stated (IPCC, 1990b):

> Such warming would not only be greater than recent natural fluctuation, but it would occur 15 to 40 times faster than past natural changes. Such a rate of change may exceed the ability of many species to adapt or disperse into more favorable regions and many plant and animal species may become extinct.

This statement was based on Schneider (1989), which states that the global warming typically projected is "10 to 60 times as fast as the natural average rate of temperature change that occurred from the end of the last Ice Age to the present warm period (that is, 2°C to 6°C warming in a century from human activities compared to an average natural warming of 1°C to 2°C per millennium from the waning of the Ice Age to establishment of the present interglacial epoch)."[2]

Ecologists and foresters are properly concerned that the predicted changes might produce catastrophic ecological disruption. In this context, it is appropriate to examine the climatic record to determine the magnitudes of past climatic fluctuations, in particular, temperature variability on various time and space scales. Since our object of concern is ecological systems, we need to find what conditions plants and animals have been exposed to in the course of their existence. This will not tell us how any particular plant, species, or ecosystem will respond, but it will provide a useful background against which we can assess or predict response to global warming.

The objectives of the analysis below are threefold: (1) to analyze records of past climates in order to determine actual rates of warming so these rates can be compared with model predictions; (2) to investigate the variability of climate and microclimate on time scales ranging from a few minutes to a half-year; and (3) to take a brief look at the spatial variation of microclimates—the climate that plants actually live in—over space scales ranging from a meter or so to about one kilometer. I will concentrate on temperature, even though it is only one of many factors affecting plant behavior, because it is the most readily available type of observational data.

---

[2]It should be noted that the ratio of rates of warming depends on the rate chosen for the denominator. If one chooses a period of very small temperature increase, the ratio can be made very large, producing an obviously absurd comparison.

RATES OF INCREASE OF GLOBAL MEAN TEMPERATURE

Climatic modeling has emphasized global mean temperature and its generalized latitudinal variation (IPCC, 1990a; hereinafter referred to as IPCC-I). Similarly, I intend to focus on global mean temperature, with only a cursory examination of the Northern Hemisphere and the temperate latitudes, even though according to IPCC-I "land surfaces warm more rapidly than the ocean, and high northern latitudes warm more than the global mean in winter."

Only in the past 150 years or so have the instrumental observations of temperature been sufficiently good to permit the calculation of global means. Nevertheless, there are many surrogate, or proxy, observations that may be suitable for estimating global mean temperatures prior to the era of instrumental observations. While there are many uncertainties as to their applicability, such surrogate observations have been widely used for that purpose (see, for example, Chapter 7 in IPCC-I). Since most of the concern over global temperature increase appears to relate to the rate of increase rather than to absolute magnitude, I will examine records of historical climate for examples of such rates, and attempt to clarify the related uncertainties.

The IPCC Scientific Assessment Panel (IPCC-I) has made an evaluation of global temperature since 1861. For the period from 1861 to 1989, they relied on the data of Jones (1988), which show a value of 0.45°C per hundred years, or 0.58°C for the 128 years of record. Given the scantiness of the earlier observations, a more reliable estimate might be made using data from 1881 to date. Such an analysis yields a value of 0.53°C per century, or 0.57°C for the 108 years of record (see Figure 1). (Note that the abscissa is for the appropriate time interval, Δy, *not* for time from the present.) Since my investigation spans several orders of magnitude in the time scales and temperature scales, it is appropriate to plot the data on a log/log scale.

Balling (1992), using data limited to a more recent interval, has calculated a linearized increase in global mean temperature of 0.45°C, slightly less than the 0.53°C of the IPCC-I report. Using the relationship between the global-temperature anomalies of Jones et al. (1986) and a stratospheric dust index (estimated from solar radiation records from a high-elevation station in Austria; see Wu et al., 1990), Balling estimated the effect of dust loading on global temperature. He found that one-third of the global-temperature trend of the past 100 years disappeared, bringing that portion of the global-temperature rise that might be attributed to atmospheric carbon dioxide increase to about 0.30°C. Although it is perhaps not directly comparable, Wu et al. (1990) have estimated the increase in global night marine air temperature to be 0.49°C for the period 1856-1988, and 0.29°C for the period 1888-1988. When the variability caused by two additional factors, solar irradiance

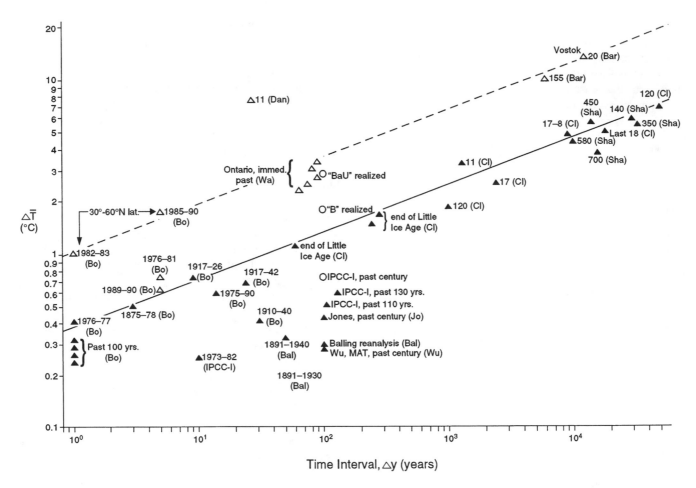

FIGURE 1   Rates of global warming during periods of most rapid warming in the geologic past, gleaned from various sources. Filled triangles represent global mean temperatures as inferred from measurement data. Open triangles indicate north temperate zone measurements. Open circles represent modeled or predicted values. Numbers near each point indicate the period of warming (past century) or thousands of years before present (prior to the past century) when the period of warming began. Letters in parentheses indicate source of data: Balling, 1992; Barnola et al., 1987; Boden et al., 1992; Clark, 1982; IPCC, 1990a,b; Shackleton and Opdyke, 1973; Wu et al., 1990. Solid line marks approximate upper limit of the periods of rapid global warming. Broken line indicates approximate upper limit of warming in the north temperate zone.

and the Southern Oscillation Index, is removed, the increase in global marine air temperature attributable to CO_2 increase is reduced to 0.24°C.

How does this recent global temperature change compare with changes found in other periods in the geological past during which temperatures have increased rather more than the average? I have searched historical records for maximum rates of warming for various periods of time.[3] From these I have selected periods during which there was sustained warming, and have calculated a warming rate for each. Data

from these periods of maximum rates of warming are plotted on Figure 1 as filled triangles. They include the sharp 60-year recovery from the Little Ice Age; the 1.5°C increase over the 250 years at the end of the Little Ice Age; and the average for the 17,000 years since the start of warming at the end of the last glaciation. For comparison, I show several periods of rapid warming in the past century or so, including the largest year-to-year increases.

Shackleton and Opdyke (1973; quoted in Balling, 1992) used oxygen-isotope data from deep-sea cores as a reflection of global ice-volume changes. The ice-volume change can be interpreted in terms of global mean temperature changes over the past 850,000 years. Temperature increases for the periods that show the most rapid warming (interpreted from Balling, 1992, Figure 1) have also been plotted on Figure 1. In all of these cases, I have chosen those periods exhibiting the most rapid temperature increases.

---

[3] I have used secondary sources for most of the data for two reasons: For this first analysis, I preferred to use data that have been accepted by others as representing global mean temperature; and my lack of access to a major scientific library severely limits the availability of primary data sources.

There appears to be an upper bound to these data, represented by the straight line on the plot. The only "outliers" are the Vostok ice-core points and a group of points representing recent ground-surface temperatures inferred from borehole temperatures in Ontario, Canada (Wang and Lewis, 1992).[4] The line (fitted by eye) represents a decrease of the time-rate of warming in the form of a power law, $\Delta T = 0.38 \Delta y^{0.27}$, where 0.38 is the value of the function at $\Delta y = 1$ year. I propose that this represents the maximum rate of increase of global mean temperature observed over the past 850,000 years, no matter what the cause, or whether it arises from internal or external forcings. It seems appropriate to use this function as the standard against which we should compare model predictions of rates of global temperature increase.

Also plotted on Figure 1 (as open triangles) are several points representing Northern Hemisphere or northern temperate-zone temperature increases. These include the data of Angell (Boden et al., 1992) for the 30°-60°N latitude zone; Ontario (Canada) borehole temperatures (Wang and Lewis, 1992)[5]; and information from the Vostok ice cores (Barnola et al., 1987). From the Vostok data, I have chosen the two periods of most sustained and rapid warming and calculated the temperature difference for those periods.

For these data, I have drawn an upper-limit curve parallel to the global curve, a reasonable first guess.[6] The equation for this line is $\Delta T = 1.05 \Delta y^{0.27}$. As expected, mid-latitude temperature increases are greater than global increases. The upper curve indicates rates of increase about 2.75 times the global curve. Also, the rates of increase in Northern Hemisphere temperature are only slightly greater (about 10 percent) than the whole-globe rates.

The first thing to note on this graph is that the "maximum" observed global rate of increase is greater by a factor

of 2 or 3 than the temperature increases observed in the past century. Why should this be so? If greenhouse-gas increases over the past century (estimated to be an increase of 50 percent in the equivalent CO_2 concentration (IPCC, 1990a)) have resulted in global warming, we might expect the rates of increase in global mean temperature to be at or above the "maximum" rate indicated by the solid line in Figure 1. The fact that they are well below this line implies that there may be some unknown climatic mechanism partially compensating for any warming caused by increases in greenhouse gases, or even that the observed warming is a natural trend reflecting little or none of the increase in greenhouse gases.

Another possibility is that model predictions of the effect of past increases of greenhouses gases are too high. It is thus appropriate to compare model-projected rates of increase for the past century with the observational record.

COMPARISON WITH MODEL PROJECTIONS

IPCC-I modeled the past century's observed climate, using observed increases in greenhouse gases, for various "climate sensitivities" (Figure 2). Their "best estimate" is a global mean temperature increase of 0.73°C (plotted on Figure 1 as the lowest open circle). Since the observed increase has been only 0.45°C, or even less if one accepts Balling's (1992) reanalysis, the IPCC-I model predictions are high by at least 60 percent, and perhaps as much as 150 percent.

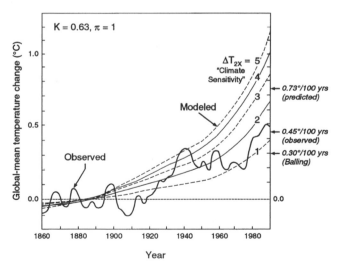

FIGURE 2 Global mean temperature change for the past century as modeled by IPCC-1, for various values of climate sensitivity (temperature change for a doubling of greenhouse gases) and the recommended ocean thermal diffusivity (K) and downwelling parameter (π). Observed global mean temperature is superimposed on the modeled curves. (Adapted from IPCC, 1990a, Figure 8.1a; reprinted with permission of the Intergovernmental Panel on Climate Change.)

---

[4]Kirk Bryan (personal communication, 1992) has suggested that there are two spectral regimes in the temperature record, one for the ice age with relatively large oscillations, and another, relatively calmer, for the last 8,000 years, in which oscillations have been less rapid. However, it is likely that the physical causes limiting the rate of temperature rise would be the same during both periods; only the magnitudes of the temperature changes would be expected to be different in the two periods.

[5]The average ground-surface temperature was inferred from the temperature profile in the boreholes and the thermal conductivity of the subsurface material. This inferred temperature increase over the past century is much greater than the measured increase in global temperature, perhaps because the rate of increase is larger at high latitudes than at low latitudes. The borehole data could (and should) be checked against standard meteorological observations from nearby weather stations. I scaled the data off rather small, perhaps not perfectly accurate, diagrams.

[6]I have ignored one outlier, the temperature increase marking the end of the Younger Dryas as inferred from Greenland ice cores by Dansgaard, et al. (1989). This point implies a rate of warming twice that of the high-latitude curve of Figure 1.

We can also compare predictions of future warming (as modeled with IPCC-I projections of increases in greenhouse gases in the next 100 years) with the "maximum" warming indicated in Figure 1. The IPCC-I business-as-usual (BaU) scenario for "realized" temperature is plotted on Figure 1 (open circle). The "realized" temperature for their "B" scenario, representing a substantial reduction in greenhouse-gas emissions, is also plotted for comparison. If the IPCC-I model predictions are taken to be high by a factor of 2, as indicated by their overprediction of the past climate, then the predicted rates of increase in global temperature will be no greater than the observed increases over the recent geologic past. This further implies a climate sensitivity (increase in mean temperature for a doubling of greenhouse gases) closer to 1.25 than to 2.5, the value chosen by IPCC-I.[7]

MICROCLIMATIC REQUIREMENTS OF PLANTS

But suppose that global temperature does increase at a rate greater than that observed in the past? What effect might that have on plants and ecosystems? It is appropriate in this context to compare the variations in temperature to which plants are actually exposed and the microclimatic temperature requirements of plants.

Clearly, long-term trends in climate have been reflected in ecosystem function and species ranges, and will continue to be in the future. It is important to note, however, that insofar as the physiological response of a particular plant is climatically controlled (whether it lives or dies, flowers, reproduces, etc.), it is the local energy balance and microclimate that are controlling. Indeed, the global energy balance and climate are controlled in the first instance by the sum of local climates. Thus, it is not that global climate change influences plants and ecosystems, but the other way around: Local climates, summed over the globe, constitute global climate.

There are, of course, the questions of how, and how much, local climates are modified by changes in the global energy fluxes produced by changes in such exogenous variables as the carbon dioxide concentration of the atmosphere. It is clearly reasonable to expect the predicted changes in the surface radiation balance to result in higher surface temperatures. What is not so clear is what the time and space scales of such increases might be.

It is my contention that any secular increase in the "average" temperature of a particular location would be produced by the slow change in the energy balance concomitant with an enhanced greenhouse effect, and would appear as a superimposition on the existing climate and microclimate. It is important, therefore, to investigate the actual climatic fluctuations to which plants are exposed, in both time and space, for comparison with those likely to result from a slow secular warming. It should be noted that a warming rate of 3°C per hundred years (the BaU "realized" temperature scenario) translates to a rate of 0.03°C per year for the globe, with perhaps 2 to 3 times that rate for high latitudes. But there is little empirical evidence that such secular temperature increases will be accompanied by an increase in the variability of climatic or microclimatic temperatures (IPCC-I; NAS, 1992). An increase in the mean temperature of, say, 0.03°C might thus be expected to be accompanied by a similar increase in the mean maximum temperature. Information on the specific microclimatic requirements of various species of plants provides a perspective for assessing the significance of the changes that are expected by, for example, IPCC-I.

Information on plant species' microclimatic requirements is available in the published literature, although there is less than one might expect, given the importance of environmental conditions to the growth and survival of plants. An example of this information is presented in Figure 3, which

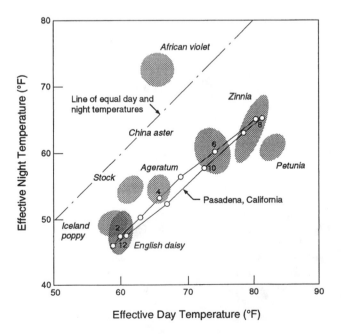

FIGURE 3 Optimal growing conditions for various flowers. "Effective day temperature" is the temperature during active photosynthesis; "effective night temperature" is the temperature during the dark period. Dotted circles indicate the range within which growth is optimum. Climagram for Pasadena is shown by the solid line connecting numbered months. (Source: Went, 1957, as reported in Brooks, 1958.)

---

[7]I have assumed that the warming trend in the last century is the result of increasing atmospheric carbon dioxide and other greenhouse gases. This seems the most reasonable assumption in the absence of contrary theory. Balling (1992) states that the 95 percent confidence interval of the trend line is between 0.37°C and 0.53°C.

shows the optimum range of day and night temperatures for flower and seed production of a number of ornamentals. Lowry (1969) provides a compendium of what is known about specific physiological requirements of plants in a microclimatic context. For example, Lowry discusses in detail the "heat units" concept, which relates the development of a plant to the thermal environment to which it is exposed. Briefly, the rate of development of a plant increases linearly with temperature above a certain threshold. It would be expected, therefore, that any increase in the temperature climate to which a plant is exposed would result in its more rapid development, other factors being equal. A relevant question in this context is what contribution a secular increase in temperature due to an augmented greenhouse effect would make to the development of a particular plant or plant association.

CLIMATIC FLUCTUATIONS ON SHORT TIME SCALES

Plants are exposed to variations in climate and microclimate that span time scales ranging from a few seconds to years or decades. They survive this variation in a number of ways, ranging from inherent flexibility and adaptation to dispersal to more favorable microclimates. It is not my purpose to create a dictionary of strategies that plants adopt. Rather, it is to look at the range of microclimates in which plants, especially, exist. In a rough way, we can compare the tempurature variability of their microclimates with the increase in temperature we presume will result from an enhanced greenhouse effect. This predicted increase is of the order of 0.02°C to 0.03°C per year in the global average, and perhaps 50 percent to 100 percent greater than the global average in the high northern latitudes (IPCC-I). A location might thus experience a secular increase of 0.02°C to 0.06°C per year.

Such increases would clearly have some impact on specific plants; indeed, some would not be able to survive. Maximum microclimatic temperatures might increase to the point where a particular plant could not reproduce or would suffer fatal heat injury. An assembly of plants with similar microclimatic requirements would either die out or, alternatively, invade an area with a more suitable microclimate. It would thus be of interest to see what plants are actually exposed to in terms of climatic variability, against a background of annual increases that might be the result of global warming.

A wealth of climatic and microclimatic data is available in various compendia, notably those of Geiger (1965), Yoshino (1975), and Landsberg (1958), and in many standard texts on climatology. Data from these and other sources are plotted in Figure 4. (Again, this is a log-log plot to encompass the many orders of magnitude involved.) Data were selected not only to represent "maximum" microcli-

matic fluctuations, but also to be representative of various climates and microclimates. In general, these are air temperatures measured at various heights from near-surface to screen height. Unless otherwise indicated on the figure, measurements are presumed to be at heights of 1.25 to 2.0 meters above ground level, the standard specified by the World Meteorological Organization (WMO, 1983).[8]

The annual range of mean daily temperatures varies from about 1°C in the tropics to about 67° in the extreme continentality of Siberia. Most mid-latitude values range from about 20° to 30°. Diurnal temperature ranges are somewhat smaller. The largest I could find for an agricultural surface was about 30°, at ground level beneath an orchard. (I have ignored diurnal fluctuations on bare ground surfaces as being less relevant to plants. The greatest ground-surface diurnal range I could find was about 50°, in the Sahara (Geiger, 1965).)

Information on shorter-period fluctuations is scarce in the scientific literature; a few representative values are given in Figure 4. It is apparent that many plants and plant surfaces are subjected to short-period fluctuations on the order of 1°C to 30°C.

Does this variability have any significance for how well plants and ecosystems would survive a secular warming on the order of 0.06°C per year? My own experience leads me to believe that such warming would, in general, lead to slow ecological changes of the same order of magnitude as those that have always been a part of ecosystem behavior. In terms of human perceptibility, these changes are usually scarcely noticeable. Thus, the forest-prairie boundary may move back and forth in response to climatic change, but the rate may be measured in centimeters or meters per year. Looking back in geologic time tends to produce a foreshortened view. The thousand-year-long warming at the end of the last ice age may appear to us to have had ecological consequences more drastic than any observer on the scene at the time would have noticed.

SPACE VARIATIONS IN MICROCLIMATE

As I noted previously, a plant that is experiencing a secular change in the microclimate, slow though it may be, may find that it can no longer exist or reproduce and reestablish itself in that changed microclimate. It may be subject to ravages of disease or pests that did not exist previously. However, its reproductive processes may permit the plant to become established in a nearby microclimate that is more favorable to its survival. Indeed, this is one of the ways that species distribution changes in response to

---

[8]It should be noted that temperatures at the ground surface and on plant surfaces are frequently more extreme than those observed in a standard weather shelter.

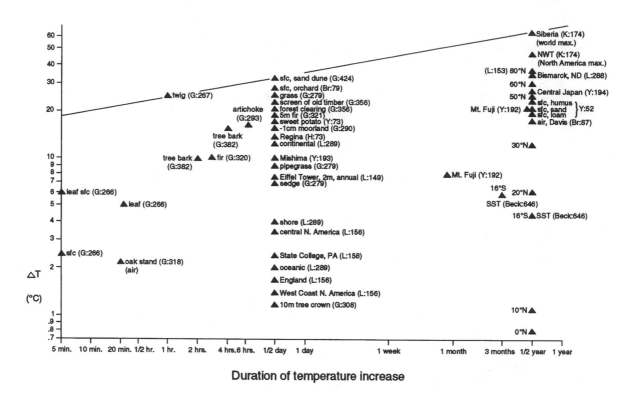

Duration of temperature increase

FIGURE 4   Temperature change as a function of time interval in various bioclimates. It is the time during which temperature at the particular location was increasing. Data sources: Brooks, 1958; Geiger, 1965; Hare and Hay, 1974; Koeppe and De Long, 1958; Landsberg, 1958; Yoshino, 1975.

climatic trends. So it is of interest to see how microclimate (and mesoclimate) varies over space.

Figure 5 presents data from a number of sources on the variations in temperature that may be found within distances ranging from a meter or so to one kilometer. Most of these are for maximum daytime temperatures, and have been chosen to illustrate typical conditions.

Large differences in temperature—on the order of 10°C to 15°C—can be found in close proximity, as on the sunny side of a small mound or furrow as compared with the opposite side. Significant but less extreme temperature differences can be found over distances ranging upwards of 10 meters. Microclimatic extremes are greater than mesoclimatic extremes. Indeed, a negative exponential decrease in the temperature differences as a function of distance between measurement locations appears in this sample.

DISCUSSION

It seems to me that this considerable variability has significance for predictions of ecological change in response to "global warming". Any long-term increase (or decrease) in the temperature to which a plant or ecosystem is exposed will have some effect. Plants in specific locations within an ecosystem may not be able to survive. But plants at unfavorable boundaries in the same ecosystem may repro-

duce in more favorable conditions nearby; their geographical range may thus be altered. Whether this is a "catastrophic" response depends on human definitions.[9]

Science (1988, 242:1010) asks, "Is there life after climate change?" and answers "Yes, but the world will be a different place, with an abundance of male alligators, migrating trees, and a plethora of parasites." It is absurd to ask what will happen *after* climate change; climate is changing all the time, at all possible scales of time and space. There is no *before* and *after*. Thompson Webb is quoted in this article as saying that we are moving into a new biological world. In actuality, we are *forever* moving into new biological worlds. The relevant question for public policy is whether the climatic changes that occur as the result of increases in greenhouse gases will have consequences that are obviously and dramatically disruptive of ecosystem behavior and human response.

We might ask ourselves whether, if we had never heard about "global warming," would we have inferred that such a process was at work, on the basis of observations of plant and animal behavior (or even the climatic record) during the past century. I contend that we could not. Long-term climatic change—from whatever cause—is masked by the

---

[9]This point is elaborated in Reifsnyder (1994).

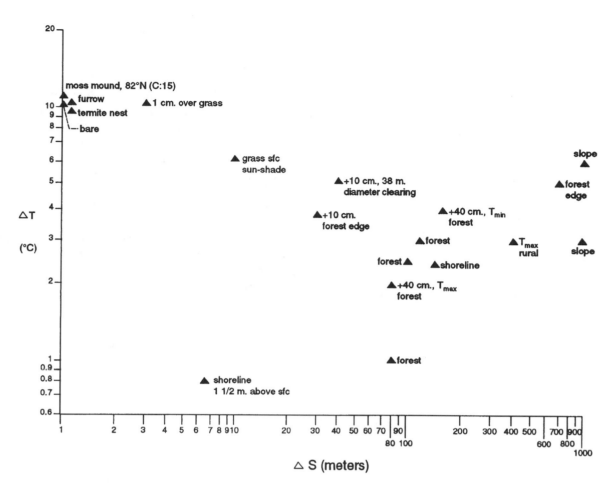

FIGURE 5 Variation in temperature over various distances; locations as indicated. Data sources: Corbet, 1972; Fowler, 1971; Geiger, 1965; Landsberg, 1958; Shitara, 1971; Yoshino, 1975.

shorter-term fluctuations, which are several orders of magnitude greater. It is difficult to see how such long-term secular changes could be catastrophic to the world's biological systems.

This view is consistent with the "final words" of the Panel on Policy Implications of Greenhouse Warming of the National Academy of Sciences (NAS/NAE/IOM, 1992):

> So far as we can reason from the assumed gradual changes in climate, their impacts will be no more severe, and adapting to them will be no more difficult, than for the range of climates already on earth and no more difficult than for other changes humanity faces. However, because we cannot rule out major changes in ocean currents, atmospheric circulations, or other natural or social surprises, we need to be alert to any substantial probability or signs of such changes that would require responses not considered in our report.

From a scientific point of view, long-term changes and interconnections are of great interest and worthy of continued attention. For many users of climatic information, however, predictions of short-term climate would be of more value than predictions of the slow secular changes expected to result form anthropogenic greenhouse-gas increases or of likely average conditions for the end of the next century. For example, farmers would find an accurate forecast of the next growing season's climate to be very valuable, but a forecast of a slow climate trend even over the next decade would be of marginal value to them. Foresters are in a similar situation. Despite the long life of a tree or a forest ecosystem, most forest-management decisions consider a time frame of less than ten years.

Now it may be, as many have argued, that it will never be possible for modelers to predict the next year's climate, or even the next decade's, because of the inherent randomness of climatic processes. But where do endogenous variables end and exogenous variables take over? What is the scale of ultimate believability of climatic forecasts? Decades? Centuries? Millennia? This is not a trivial question. Furthermore, as I have argued elsewhere (Reifsnyder, 1989), society must always make important decisions in the face of climatic uncertainty. We may plan for global warming, but there is always some non-zero chance that we will be faced with global cooling instead.

SUMMARY AND CONCLUSIONS

My findings are these:

1. There is a well-defined limit to temporal rates of increase in global mean temperature for periods of warming during the past 850,000 years. It can be expressed in the relationship $\Delta T = 0.38\ \Delta y^{0.27}$, where Δy is the number of years during which sustained warming has occurred.
2. Natural fluctuations of climate are several orders of magnitude larger than both the secular increase of temperature predicted by current climate models and the probable actual increase over the next century.
3. The results of comparisons of predicted rates of global temperature increase with past observed rates of increase depend greatly on the time periods chosen. Unless the time periods are carefully selected to represent comparable periods (of rapid warming, for example), comparisons can be so skewed as to be misleading or meaningless.

Even if the IPCC-1 model predictions were correct, the rate of global temperature increase over the next century would probably be no more than the maximum rates of increase experienced at various times in the past 8,000 years. At worst, they would be double those rates. Even these predicted increases in temperature will be so small by comparison with natural climatic fluctuations, and spread over such a long time scale, that their short-term impact on human behavior and welfare is likely to be minimal and probably not measurable. It thus seems reasonable to conclude that the long-term secular temperature change will not produce dramatic and disruptive changes in the earth's biological systems.

Does this mean that we as scientists should not be interested in the scientific issues entailed by enhanced greenhouse effects and their potential impacts? Not at all: What happens to the earth's climate because of human activities, and what happens to the earth's ecosystems as a result, should be foci of intense scientific scrutiny. We have only one earth to lose, and we should not lose it by default. Let us put our effort into devising productive research approaches, not waste it in chasing chimeras.

Commentary on the Paper of Reifsnyder

SOROOSH SOROOSHIAN
University of Arizona

In his analysis, Dr. Reifsnyder has used temperature as the sole indicator of climate variability, in both space and time. I should like to bring up two points related to this approach.

First, the data Dr. Reifsnyder has used in his analysis come from the published literature and from analyses made by numerous researchers. Some of them are instrumental observations; others result from the analysis of proxy records. They come from many sources, various time periods, and locations worldwide. Not only does each type carry an individual potential for error, but combining them may risk an apples-and-oranges comparison. When data from varied sources are used, as Dr. Reifsnyder and some others have done, one must ask, "To what extent do such uncertainties affect the results and the conclusions drawn from them?"

Even if we ignore the temporal and spatial uncertainties associated with the temperature data, data can still be biased because of changes in instrument technology or in the manner in which measurements are taken. A fair degree of confidence can be placed in data that have been corrected for sources of error; when we are talking about deducing several degrees of temperature change over hundreds or thousands of years from proxy records, we need to acknowledge the magnitude of the associated uncertainties. Let me adduce a similar situation with precipitation data. In collaboration with the Agricultural Research Service, we at Arizona have been monitoring precipitation amounts over a small, well-instrumented experimental watershed. We have a very high density of sophisticated rain gauges, yet the analysis of our data shows that if the measurement errors are not taken into account, the information value may be limited. Conclusions drawn from those data need to be questioned as well.

Dr. Reifsnyder's paper also deals with the impact of temperature change on vegetation. I will not take issue with his position that vegetation will have the ability to tolerate the changes in global mean temperature estimated by various models. I will, however, argue that not only temperature is involved. Even small changes in mean global temperature can have an impact on the general circulation of the atmosphere and the oceans. A temperature change not only might increase the severity of droughts and floods but might expand the regions of the globe affected by them as well. Trees might tolerate a temperature change, but could they also survive a sustained drought?

I should like to endorse a third point that Dr. Reifsnyder makes, although indirectly: Scientists need to be responsible in conveying the results of their studies to decision- and policy-makers. The use of data with a high degree of uncertainty, or the results of models based on questionable assumptions, places the credibility of our science at risk.

Discussion

RIND: I find myself in the strange position of having to defend Steve Schneider. I'm not sure it's reasonable to compare the glacial-to-interglacial warming over 5000 years with the projected warming over the next 100 years or so, when the time spans are so different.

REIFSNYDER: Oh, I admit that's comparing apples with oranges. If you extrapolate the IPCC values for 1000 years they become absurd. You do have to be careful how you choose the periods for comparison. The longer the time period, the slower the rate of change.

MOREL: Allow me in turn to defend the IPCC report. For a U.N.-sponsored six-month exercise, I would say it's remarkably respect-worthy. We were trying to give the reader a reasonable idea of what could happen, working from uncertainties quite as large as one finds in nature. After all, we can't even link some substantial ecosystem changes, such as the desertification in Africa, with any changes in global climate indicators.

REIFSNYDER: I quite agree. However, that report has been used by non-scientific people as the basis for so many outrageous suggestions that I think it's important to call attention to its inconsistencies and limitations.

LEHMAN: I hesitate to add fuel to the fire, but we did find that during the deglaciation sea-surface temperatures in the North Atlantic changed by as much as 5°C in less than 40 years, in both negative and positive directions. Ice cores show warming of 7° in 50 years. These regional modulations probably relate to some fairly significant global changes.

TRENBERTH: It seems to me that the models essentially represent the anthropogenic change. Why should they be relevant to natural climate change? Conversely, how can one deduce the rate at which man can change the climate from that of natural change?

REIFSNYDER: My point was more a matter of what plants have been exposed to, whatever the origin of the change. The slow secular temperature increase ascribable to increased carbon dioxide in the atmosphere is several orders of magnitude less than that of the climate to which a plant is exposed in the course of its life.

JONES: I think we should keep in mind that most of the points to the right of the 100-year band on your Figure 1 are regional. A 10°C change in Vostok, or even the entire Antarctic region, probably means a global change on the order of 4° or 5°. Also, I'd like to emphasize that small changes can make a significant difference. In the 1920s and 1930s there were some very warm summers in northern Scandinavia, meaning 1° or 2°C warmer on average. As a result, the tree line moved many kilometers further north in a short space of time.

REIFSNYDER: That's not what's at issue. My question is how much an increase due to global warming will affect such natural changes.

GROISMAN: I'd just like to add another reminder that the rate of mean temperature changes at high latitudes is not the same as the global rate, either in the last hundred years' observations or in the model results, whatever the IPCC report may say.

KEELING: I think this is just the kind of healthy challenge the IPCC report needs to counter the word-bite publicity it's been getting. I'd like to make a few points here. First, I don't see how we can be sure that the reason for the temperature increase you cited is global warming. Even worse, it may be that there are several anthropogenic signals whose effects are partially canceling each other out; Mike Schlesinger brought that possibility up at a recent meeting.

Second, those fluxes bother me. The fluxes involved in the CO_2 increase over the last 100 years are dwarfed by the natural fluxes in the carbon cycle, and I don't think they have anything to do with the accumulation of CO_2 in the atmosphere. Third, I don't think we need to worry so much about what level of temperature we reach as about how long the increased temperature will persist. Some trees can handle a passing fire; that doesn't mean they could survive 5 years of heat and drought, followed by another 5. That sort of possible anthropogenic effect is what we need to know about.

REIFSNYDER: Indeed. But in response to your first point, let me say that we really have no better hypothesis then a CO_2 increase for what caused the warming over the last 100 years. My position is that it's our best estimate.

GROOTES: A propos of climate's societal impacts, let us not forget the Norse colonies in Greenland. They lived there happily for 500 years before being driven out by the effects of the Little Ice Age. But I have a question for you, Bill. I understood you to say that the concern about global warming should and probably would disappear, and we should concentrate our efforts elsewhere. What are your suggestions?

REIFSNYDER: Oh, there's a great deal more we need to know about anthropogenic global warming. I just think it will die down as a social-policy issue, since we can't yet relate it to the warming during this century. As for Greenland, remember that the Little

Ice Age temperature changes were significantly more severe than the 0.02° or 0.04°C per year that presumably can be attributed to anthropogenic carbon dioxide.

PERRY: You seem to be advancing a new biological principle here: Changes in the mean state over long periods are insignificant as long as they are small in relation to the total range of conditions experienced by the individual organism. Surely you don't mean you would be happy if the temperatures you survived on your drive here from New Mexico became your long-term environment?

REIFSNYDER: No, that would certainly qualify as a profound change to my local ecosystem. But my point, really, is that the changes we have seen on the scale of a lifetime cannot be ascribed to anthropogenic global warming.

KARL: I'd like to raise a question related to your Figure 1, Bill, that I think we should be considering at this workshop: Can you get by with a much smaller spatial resolution as the time interval increases with respect to the time frame you're dealing with? Can you get away with not having a complete global average if you're dealing with 1000 years instead of 10,000?

MCGOWAN: Very few ecologists I know of have ever thought that the effects of temperature per se on photosynthesis or animal behavior were the issue. For instance, the California Current, which is very heterogeneous in both space and time, has lots of 2°C variations on many time scales. Despite this exposure to temperature shifts, when the mean temperature of the upper 100 meters of the Current rose a little less than 2° during the 1958 and 1983 El Niños, the effects on both plants and animals were catastrophic. Of course it wasn't just the temperature, which I think is a rather poor proxy measurement of what's really happening; it was the fact that the mixed layer collapsed. There were no nutrients coming in, there was advective movement of boundaries, and so on.

REIFSNYDER: Temperature just happens to be the one thing we have data on, and I think it may be important as a proxy for other variations. Certainly ecosystems respond to all kinds of things, often catastrophically. It seems to me that your example demonstrates my point that it's these short-term excursions from the mean, which as far as we know are normal climate behavior, that produce catastrophic ecosystem changes, not the warming of 0.05°C or less per year ascribable to anthropogenic CO_2.

Proxy Indicators of Climate Reference List

Aagaard, K., J.H. Swift, and E.C. Carmack. 1985. Thermohaline circulation in the Arctic Mediterranean seas. J. Geophys. Res. 90(C3):4833-4846.

Aebischer, N.J., J.C. Coulson, and J.M. Colebrook. 1990. Parallel long-term trends across four marine trophic levels and weather. Nature 347:753-755.

Aharon, P. 1985. Carbon isotope record of late Quaternary coral reefs: Possible index of sea surface paleoproductivity. In The Carbon Cycle and Atmospheric CO_2: Natural Variations Archean to Present. E.T. Sundquist and W.S. Broecker (eds.). Geophysical Monograph 32, American Geophysical Union, Washington, D.C., pp. 343-355.

Allan, R.J., N. Nicholls, P.D. Jones, and I.J. Butterworth. 1991. A further extension of the Tahiti-Darwin SOI, early ENSO events and Darwin pressure. J. Climate 4:743-749.

Allen, W.E. 1941. Twenty years' statistical studies of marine plankton dinoflagellates off southern California. Amer. Mid. Nat. 26:603-635.

Alley, R.B., P.M. Grootes, D. Meese, A.J. Gow, K. Taylor, and K.M. Cuffey. 1991. Climate at the Greenland Summit: Little Ice Age to modern. EOS 72:66.

Alley, R.B., D.A. Meese, C.A. Shuman, A.J. Gow, K.C. Taylor, P.M. Grootes, J.W.C. White, M. Ram, E.D. Waddington, P.A. Mayewski, and G.A. Zielinski. 1993. Abrupt increase in Greenland snow accumulation at the end of the Younger Dryas event. Nature 362:527-529.

Ammann, B., and A.F. Lotter. 1989. Late glacial radiocarbon- and palynostratigraphy on the Swiss Plateau. Boreas 18:109-126.

Atkinson, T.C., K.R. Briffa, and G.R. Coope. 1987. Seasonal temperatures in Britain during the past 22,000 years, reconstructed using beetle remains. Nature 325:587-592.

Bacon, S., and D.J.T. Carter. 1991. Wave climate changes in the North Atlantic and North Sea. Int. J. Climatol. 11:545-558.

Bacon, S., and D.J.T. Carter. 1993. A connection between mean wave height and atmospheric pressure gradient in the North Atlantic. Int. J. Climatol. 13:423-436.

Bainbridge, A. (ed.). 1981. GEOSECS Atlantic Expedition Hydrographic Data. Vol. 1. U.S. Government Printing Office, Washington, D.C.

Balling, R.C., Jr. 1992. The Heated Debate: Greenhouse Predictions versus Climate Reality. Pacific Research Institute for Public Policy, San Francisco.

Bard, E., M. Arnold, P. Maurice, J. Duprat, J. Moyes, and J.-C. Duplessy. 1987. Retreat velocity of the North Atlantic polar front during the last deglaciation determined by ^{14}C accelerator mass spectrometry. Nature 328:791-794.

Bard, E., B. Hamelin, R.G. Fairbanks, and A. Zindler. 1990. Calibration of the 14-C timescale over the past 30,000 years using mass spectrometric U-Th ages from Barbados corals. Nature 345:405-410.

Barnes, D.J., and J.M. Lough. 1989. The nature of skeletal density banding in scleractinian corals: Fine banding and seasonal patterns. J. Exp. Mar. Bio. 126:119-134.

Barnett, T.P. 1991. The interaction of multiple time scales in the tropical climate system. J. Climate 4:269-285.

Barnett, T.P., and M.E. Schlesinger. 1987. Detecting changes in global climate induced by greenhouse gases. J. Geophys. Res. 92:14772-14780.

Barnola, J.M., D. Raynaud, Y.S. Korotkevich, and C. Lorius. 1987. Vostok ice core provides 160,000-year record of atmospheric CO_2. Nature 329:408-414.

Barnola, J.M., P. Pimienta, D. Raynaud, and Y.S. Korotkevich. 1991. CO_2-climate relationship as deduced from the Vostok ice core: A re-examination based on new measurements and on a re-evaluation of the air dating. Tellus 43B:83-90.

Barry, R.G. 1978. Climate fluctuations during the periods of historical and instrumental record. In Climatic Change and Variability: A Southern Hemisphere Perspective. A.B. Pittock, L.A. Frakes,

D. Jenssen, J.A. Peterson, and J.W. Zillman (eds.). Cambridge University Press, London, pp. 150-166.

Barry, R.G., and R.J. Chorley. 1982. Atmosphere, Weather, and Climate (6th ed.). Routledge, New York, 387 pp.

Baumgartner, T.R., A. Soutar, and V. Ferreira-Bartrina. 1992. Reconstruction of the history of Pacific sardine and northern anchovy populations over the past two millennia from sediments of the Santa Barbara Basin, California. Calif. Coop. Oceanic Fisheries Invest. Rpts. 33:24-40.

Bé, A.W.H., and D.S. Tolderlund. 1971. Distribution and ecology of living planktonic foraminifera in surface waters of the Atlantic and Indian oceans. In Micropaleontology of Oceans. B.M. Funnell and W.R. Riedel (eds.). Cambridge University Press, London, pp. 105-149.

Beadle, L.C. 1974. The Inland Waters of Tropical Africa: An Introduction to African Limnology. Longman, London.

Beck, J.W., R.L. Edwards, E. Ito, F.W. Taylor, J. Recy, F. Rougerie, P. Joannot, and C. Henin. 1992. Sea-surface temperature from coral skeletal strontium/calcium ratios. Science 257:644-647.

Bender, M., T. Sowers, M.-L. Dickson, J. Orchardo, P. Grootes, P.A. Mayewski, and D.A. Meese. 1994. Climate correlations between Greenland and Antarctica during the past 100,000 years. Nature 372: 663-666.

Berger, A.L. 1978. Long-term variations of daily insolation and Quaternary climate changes. J. Atmos. Sci. 35:2362-2367.

Bergthorsson, P. 1969. An estimate of drift ice and temperature in Iceland in 1000 years. Jokull 19:94-101.

Bernal, P.A., and J.A. McGowan. 1981. Advection and upwelling in the California Current. In Coastal Upwelling. F.A. Richards (ed.). American Geophysical Union, Washington, D.C., pp. 381-399.

Boden, T.A., R.J. Sepanski, and F.W. Stoss (eds.). 1992. Trends '91: A Compendium of Data on Global Change. Publ. No. ORNL/CDIAC-49, Carbon Dioxide Information Analysis Center, Oak Ridge National Laboratory, Tennessee.

Bottomley, M., C.K. Folland, J. Hsiung, R.E. Newell, and D.E. Parker. 1990. Global Ocean Surface Temperature Atlas (GOSTA). H.M.S.O., London, 20 pp., 313 plates.

Boutron, C. 1980. Respective influence of global pollution and volcanic eruptions on the past variations of the trace metals content of Antarctic snows since 1880s. J. Geophys. Res. 85:7426-7432.

Boyle, E.A. 1988. Cadmium: Chemical tracer of deepwater paleoceanography. Paleoceanography 3:471-489.

Boyle, E.A., and L. Keigwin. 1987. North Atlantic thermohaline circulation during the past 20,000 years linked to high-latitude surface temperature. Nature 330:35-40.

Boyle, E.A., F. Sclater, and J.M. Edmond. 1976. On the marine geochemistry of cadmium. Nature 263:42-44.

Bradley, R.S. 1985. Quaternary Paleoclimatology: Methods of Paleoclimatic Reconstruction. Allen and Unwin, London and Boston, 472 pp.

Bradley, R.S. 1988. The explosive volcanic eruption signal in northern hemisphere continental temperature records. Climatic Change 12:221-243.

Bradley, R.S., and J.A. Eddy. 1989. Records of past global changes.

In Global Changes of the Past. R.S. Bradley (ed.). UCAR/Office for Interdisciplinary Earth Studies, Boulder, Colorado, pp. 5-9.

Bradley, R.S., and J.K. Eischeid. 1985. Aspects of the precipitation climatology of the Canadian high Arctic. In Glacial, Geologic and Glacio-Climatic Studies in the Canadian High Arctic. R.S. Bradley (ed.). Dept. of Geology and Geography, University of Massachusetts at Amherst, pp. 250-271.

Bradley, R.S., and P.D. Jones (eds.). 1992. Climate Since A.D. 1500. Routledge, New York, 679 pp.

Bradley, R.S., P.M. Kelly, P.D. Jones, C.M. Goodess, and H.F. Diaz. 1985. A Climatic Data Bank for Northern Hemisphere Land Areas, 1851-1980. Technical Report TR017, U.S. Dept. of Energy, Carbon Dioxide Research Division, 335 pp.

Bradley, R.S., H.F. Diaz, J.K. Eischeid, P.D. Jones, P.M. Kelly, and C.M. Goodess. 1987a. Precipitation fluctuations over northern hemisphere land areas since the mid-19th century. Science 237:171-175.

Bradley, R.S., H.F. Diaz, G. Kiladis, and J.K. Eischeid. 1987b. ENSO signal in continental temperature and precipitation records. Nature 327:497-501.

Brassel, S.C., G. Eglinton, I.T. Marlowe, U. Pflaumann, and M. Sarnthein. 1986. Molecular stratigraphy: A new tool for climatic assessment. Nature 320:129-133.

Briffa, K.R., and P.D. Jones. 1993. Global surface air temperature variation over the twentieth century: Part 2, Implications for large-scale palaeoclimatic studies of the Holocene. Holocene 3:82-93.

Briffa, K.R., T.S. Bartholin, D. Eckstein, P.D. Jones, W. Karlén, F.H. Schweingruber, and P. Zetterberg. 1990. A 1400-year tree-ring record of summer temperature. Nature 346:434-439.

Briffa, K.R., P.D. Jones, T.S. Bartholin, D. Eckstein, F.H. Schweingruber, W. Karlén, P. Zetterberg, and M. Eronen. 1992a. Fennoscandian summers from A.D. 500: Temperature changes on short and long time scales. Climate Dynamics 7:111-119.

Briffa, K.R., P.D. Jones, and F.H. Schweingruber. 1992b. Tree-ring density reconstructions of summer temperature patterns across western North America since A.D. 1600. J. Climate 5:735-754.

Briffa, K.R., P.D. Jones, F.H. Schweingruber, S.G. Shiyatov, and E.R. Cook. 1995. Unusual twentieth-century summer warmth in a 1,000-year temperature record from Siberia. Nature 376:156-159.

Broecker, W.S. 1987. Unpleasant surprises in the greenhouse? Nature 328:123-126.

Broecker, W.S. 1991a. The great ocean conveyor. Oceanography 4:79-89.

Broecker, W.S. 1991b. Strength of the Nordic heat pump. In The Last Deglaciation: Absolute and Radiocarbon Chronologies. E. Bard and W.S. Broecker (eds.). NATO ASI Series I, Vol. 2, Springer-Verlag, pp. 173-182.

Broecker, W.S., and G.H. Denton. 1989. The role of ocean-atmosphere reorganizations in glacial cycles. Geochim. Cosmochim. Acta 53:2465-2501.

Broecker, W.S., D.M. Peteet, and D. Rind. 1985. Does the ocean-atmosphere system have more than one stable mode of operation? Nature 315:21-25.

Broecker, W.S., M. Andree, W. Wolfli, H. Oeschger, G. Bonani, J. Kennett, and D. Peteet. 1988. The chronology of the last deglaciation: Implications for the cause of the Younger Dryas event. Paleoceanography 3:1-19.

Broecker, W.S., G. Bond, M. Klas, G. Bonani, and W. Wölfli. 1990a. A salt oscillator in the glacial Atlantic? 1. The concept. Paleoceanography 4:469-477.

Broecker, W.S., M. Klas, E. Clark, S. Trumbore, G. Bonani, W. Wolfli, and S. Ivy. 1990b. Accelerator mass spectrometric radiocarbon measurements on foraminifera shells from deep-sea cores. Radiocarbon 32:119-133.

Brooks, F.A. 1958. An Introduction to Physical Microclimatology. University of California, Davis.

Brooks, C.E.P. 1923. Variations in the Levels of the Central African Lakes, Victoria and Albert. Geophysical Memoir No. 20. U.K. Meteorological Office, London.

Bryan, F. 1986. High-latitude salinity effects and interhemisphere thermohaline circulations. Nature 323:301-304.

Budd, W.F., and V.I. Morgan. 1977. Isotopes, climate and ice sheet dynamics from core studies on Law Dome, Antarctica. IAHS No. 118:312-325.

Cane, M.A. 1983. Oceanographic events during El Niño. Science 222:1189-1195.

Carpenter, S.R., K.L. Cottingham, and D.E. Schindler. 1992. Biotic feedback in lake phosphorus cycles. Trends Ecol. Evol. 7(10):332-336.

Carriquiry, J.D., M.J. Risk, and H.P. Schwarcz. 1988. Timing and temperature record from the stable isotopes of the 1982-1983 El Niño warming event in eastern Pacific corals. Palaios 3:359-364.

Chan, L.H., D. Drummond, J.M. Edmond, and B. Grant. 1977. On the barium data from the Atlantic GEOSECS Expedition. Deep-Sea Res. 24:613-649.

Chappelaz, J., J.M. Barnola, D. Raynaud, Y.S. Korotkevich, and C. Lorius. 1990. Ice-core record of atmospheric methane over the past 160,000 years. Nature 345:127-131.

Chelton, D.B., P.A. Bernal, and J.A. McGowan. 1982. Large-scale interannual physical and biological interaction in the California Current. J. Mar. Res. 40:1095-1125.

Clark, W.C. (ed.). 1982. Carbon Dioxide Review 1982. Oxford University Press, Oxford.

Clarke, R.A., and J.C. Gascard. 1983. The formation of Labrador Sea water. Part 1: Large-scale processes. J. Phys. Oceanogr. 13:1764-1778.

CLIMAP Project Members. 1981. Seasonal reconstruction of the earth's surface during the last glacial maximum. Map and Chart Series No. MC-36, Geological Society of America, Boulder, Colo., 18 pp.

Cohen, S.J. 1986. Impacts of CO_2-induced climatic change on water resources in the Great Lakes Basin. Climatic Change 8:135-153.

Cole, J.E. 1992. Interannual-Decadal Variability in Tropical Climate Systems: Stable Isotope Records and General Circulation Model Experiments. Ph.D. thesis, Columbia University.

Cole, J.E., and R.G. Fairbanks. 1990. The Southern Oscillation recorded in the oxygen isotopes of corals from Tarawa Atoll. Paleoceanography 5:669-683.

Cole, J.E., G.T. Shen, R.G. Fairbanks, and M. Moore. 1992. Coral monitors of El Niño/Southern Oscillation dynamics across the equatorial Pacific. In El Niño: Historical and Paleoclimatic Aspects of the Southern Oscillation. H. Diaz and V. Markgraf

(eds.). Cambridge University Press, Cambridge, U.K., pp. 349-375.

Cole, J.E., R.G. Fairbanks, and G.T. Shen. 1993. The spectrum of recent variability in the Southern Oscillation: Results from a Tarawa Atoll coral. Science 262:1790-1793.

Cole, J.E., R.G. Fairbanks, and G.T. Shen. 1995. Monitoring the tropical ocean and atmosphere using chemical records from long-lived corals. In Natural Climate Variability on Decade-to-Century Time Scales. D.G. Martinson, K. Bryan, M. Ghil, M.M. Hall, T.R. Karl, E.S. Sarachik, S. Sorooshian, and L.D. Talley (eds.). National Academy Press, Washington, D.C.

Colebrook, J.M. 1985. Sea surface temperatures and zooplankton, North Sea 1948 to 1983. J. Cons. Int. Explor. Mer 42:179-185.

Colgan, M.W., and D. Malmquist. 1987. The Urvina Bay uplift: A dry trek through a coral community. Oceanus 30:61-66.

Colinvaux, P.A., M.C. Miller, K.-B. Liu, M. Steinitz-Kannan, and I. Frost. 1985. Discovery of permanent Amazon lakes and hydraulic disturbance in the upper Amazon Basin. Nature 313:42-45.

Conversi, A. 1992. Variability of water quality data collected near three major southern California sewage outfalls. Ph.D. dissertation. Scripps Institution of Oceanography, Univ. Calif., San Diego, 109 pp.

Conversi, A., and J.A. McGowan. 1994. Natural versus human-caused variability of water clarity in the Southern California Bight. Limnol. Oceanogr. 39(3):632-648.

Cook, E.R., K.R. Briffa, S. Shiyatov, and V. Mazepa. 1990. Tree-ring standardization and growth-trend estimation. In Methods of Dendrochronology: Applications in the Environmental Sciences. E.R. Cook and L.A. Kairiukstis (eds.). Kluwer, Dordrecht, pp. 104-123.

Cook, E., T. Bird, M. Peterson, M. Barbetti, B. Buckley, R. D'Arrigo, R. Francey, and P. Tans. 1991. Climatic change in Tasmania inferred from a 1089-year tree-ring chronology of subalpine Huon pine. Science 253:1266-1268.

Cook, E., T. Bird, M. Peterson, M. Barbetti, B. Buckley, R. D'Arrigo, and R. Francey. 1992. Climatic change over the last millennium in Tasmania reconstructed from tree rings. Holocene 2(3):205-217.

Cook, E.R., K.R. Briffa, D.M. Meko, D.A. Graybill, and G. Funkhouser. 1995a. The 'segment-length curse' in long tree-ring chronology development for paleoclimatic studies. Holocene 5:229-237.

Cook, E.R., B.M. Buckley, and R.D. D'Arrigo. 1995b. Interdecadal temperature oscillations in the Southern Hemisphere: Evidence from Tasmanian tree rings since 300 B.C. In Natural Climate Variability on Decade-to-Century Time Scales. D.G. Martinson, K. Bryan, M. Ghil, M.M. Hall, T.R. Karl, E.S. Sarachik, S. Sorooshian, and L.D. Talley (eds.). National Academy Press, Washington, D.C.

Coope, G.R. 1977. Fossil coleopteran assemblages as sensitive indicators of climatic changes during the Devensian (last) cold stage. Phil. Trans. Roy. Soc. London B 280:313-340.

Coope, G.R., and J.A. Brophy. 1972. Late glacial environmental changes indicated by a coleopteran succession from North Wales. Boreas 1:97-142.

Cooper, N.S., K.D.B. Whysall, and G.R. Bigg. 1989. Recent decadal climate variations in the tropical Pacific. Int. J. Climatol. 9:221-242.

Corbet, P.S. 1972. The Microclimate of Arctic Plants and Animals, on Land and in Fresh Water. Acta Arctica 18. 43 pp.

Coughlan, M.J. 1979. Recent variations in annual-mean maximum temperatures over Australia. Quart. J. Roy. Meteorol. Soc. 105:707-719.

CPR Survey Team. 1992. Continuous plankton records: The North Sea in the 1980s. ICES Mar. Sci. Symp. 195:243-248.

Craig, H., and C.C. Chou. 1982. Methane: The record in polar ice cores. Geophys. Res. Lett. 9:1221-1224.

Craig, H., C.C. Chou, J.A. Welhan, C.M. Stevens, and A.E. Engelkemeir. 1988a. The isotopic composition of methane in polar ice cores. Science 242:1535-1539.

Craig, H., Y. Horibe, and T. Sowers. 1988b. Gravitational separation of gases and isotopes in polar ice caps. Science 242:1675-1678.

Crowe, R.B. 1992. Extension of Toronto temperature time-series from 1840 to 1778 using various United States and other data. In The Year Without a Summer? World Climate in 1816. C.R. Harington (ed.). Canadian Museum of Nature, Ottawa, pp. 145-161.

Currie, R.G. 1981. Evidence for 18.6 year M_N signal in temperature and drought conditions in North America since A.D. 1800. J. Geophys. Res. 86:11055-11064.

Curtis, J.H., and D.A. Hodell. 1993. An isotopic and trace-element study of ostracods from Lake Miragoane, Haiti: A 10,500 year record of paleosalinity and paleotemperature. In Climate Change in Continental Isotopic Records. P.K. Swart, K.C. Lohmann, J. McKenzie, and S. Savin (eds.). AGU Geophys. Monogr. 78:135-152.

Dansgaard, W., S.J. Johnsen, J. Møller, and C.C. Langway, Jr. 1969. One thousand centuries of climatic record from Camp Century on the Greenland ice sheet. Science 166:377-381.

Dansgaard, W., S.J. Johnsen, H.B. Clausen, and C.C. Langway, Jr. 1971. Climatic record revealed by the Camp Century ice core. In The Late Cenozoic Glacial Ages. K.K. Turekian (ed.). Yale University Press, pp. 37-56.

Dansgaard, W., S.J. Johnsen, H.B. Clausen, and N. Gundestrup. 1973. Stable isotope glaciology. Medd. Grønland 197:1-53.

Dansgaard, W., S.J. Johnsen, N. Reeh, N. Gundestrup, H.B. Clausen, and C.U. Hammer. 1975. Climatic changes, Norsemen and modern man. Nature 255:24-28.

Dansgaard, W., H.B. Clausen, N. Gundestrup, C.U. Hammer, S.F. Johnsen, P.M. Kristinsdottir, and N. Reeh. 1982. A new Greenland deep ice core. Science 218:1273-1277.

Dansgaard, W., S.J. Johnsen, H.B. Clausen, D. Dahl-Jensen, N. Gundestrup, C.U. Hammer, and H. Oeschger. 1984. North Atlantic climate oscillations revealed by deep Greenland ice cores. In Climate Processes and Climate Sensitivity. J.E. Hansen and T. Takahashi (eds.). Geophys. Monograph 29 (Maurice Ewing Series), American Geophysical Union, Washington, D.C., pp. 288-298.

Dansgaard, W., J.W.C. White, and S.J. Johnsen. 1989. The abrupt termination of the Younger Dryas climate event. Nature 339:532-534.

Dansgaard, W., S.J. Johnsen, H.B. Clausen, D. Dahl-Jensen, N.S. Gundestrup, C.U. Hammer, C.S. Hvidberg, J.P. Steffensen, A.E.

Sveinbjørnsdottir, J. Jouzel, and G. Bond. 1993. Evidence for general instability of past climate from a 250-kyr ice-core record. Nature 364:218-220.

De Angelis, M., N.I. Barkov, and V.N. Petrov. 1987. Aerosol concentrations over the last climatic cycle (160 kyr) from an Antarctic ice core. Nature 325:318-321.

De Angelis, M., N.I. Barkov, and V.N. Petrov. 1992. Sources of continental dust over Antarctica during the last glacial cycle. J. Atmos. Chem. 14:233-244.

Delmas, R.J., J.-M. Ascencio, and M. Legrand. 1980. Polar ice evidence that atmospheric CO_2 20,000 yr B.P. was 50% of present. Nature 284:155-157.

Delworth, T., S. Manabe, and R.J. Stouffer. 1993. Interdecadal variations of the thermohaline circulation in a coupled ocean-atmosphere model. J. Climate 6:1993-2011.

Deser, C., and J.M. Wallace. 1987. El Niño events and their relation to the Southern Oscillation. J. Geophys. Res. 92:14189-14196.

Deser, C., and J.M. Wallace. 1990. Large-scale atmospheric circulation features of warm and cold episodes in the tropical Pacific. J. Climate 3:1254-1281.

de Terra, H., and G.E. Hutchinson. 1934. Evidence of recent changes shown by Tibetan highland lakes. Geogr. J. 84:311-320.

de Villiers, S., G.T. Shen, and B.K. Nelson. 1993. Sr/Ca thermometry: Coral skeletal uptake and surface ocean variability in the eastern equatorial Pacific upwelling area. Geochim. Cosmochim. Acta 58:197-208.

Dickey, J.O., S.L. Marcus, and R. Hide. 1992. Global propagation of interannual fluctuations in atmospheric angular momentum. Nature 357:484-488.

Dickson, R.R., S.-A. Malmberg, S.R. Jones, and A.J. Lee. 1984. An investigation of the earlier great salinity anomaly of 1910-14 in waters west of the British Isles. C.M. 1984/Gen:4, International Council for the Exploration of the Sea, Copenhagen, 15 pp. + 15 figs. (mimeo).

Dickson, R.R., P.M. Kelly, J.M. Colebrook, W.S. Wooster, and D.H. Cushing. 1988a. North winds and production in the eastern North Atlantic. J. Plank. Res. 10:151-169.

Dickson, R.R., J. Meincke, S.-A. Malmberg, and A.J. Lee. 1988b. The "Great Salinity Anomaly" in the northern North Atlantic 1968-1982. Prog. Oceanogr. 20:103-151.

Dietrich, G., W. Kalle, W. Krauss, and G. Siedler. 1975. General Oceanography (second edition). John Wiley and Sons, New York, 626 pp.

Dodge, R.E., R.C. Aller, and J. Thomson. 1974. Coral growth related to resuspension of bottom sediments. Nature 247:574-577.

Donguy, J.-R., and A. Dessier. 1983. El Niño-like events observed in the tropical Pacific. Mon. Weather Rev. 111:2136-2139.

Druffel, E.R.M., and S. Griffin. 1993. Large variations of surface ocean radiocarbon: Evidence of circulation changes in the southwestern Pacific. J. Geophys. Res. 98:20249-20259.

Drummey, S.M. 1993. From snow and ice: A study of the H_2O_2 content of polar snow and ice from Greenland and Antarctic ice sheets. M.Sc. thesis, University of New Hampshire, Durham.

Dunbar, R.B., and J.E. Cole. 1993. Coral Records of Ocean-Atmosphere Variability: Report from the Workshop on Coral Paleoclimate Reconstruction. NOAA Climate and Global

Change Program Special Report No. 10, University Corporation for Atmospheric Research, Boulder, Colo., 38 pp.

Dunbar, R.B., and G.M. Wellington. 1981. Stable isotopes in a branching coral monitor seasonal temperature variation. Nature 293:453-455.

Dunbar, R.B., G.M. Wellington, M.W. Colgan, and P.W. Glynn. 1991. Eastern tropical Pacific corals monitor low-latitude climate of the past 400 years. In Proceedings of the Seventh Annual Pacific Climate (PACLIM) Workshop. J.L. Betancourt and V.L. Tharp (eds.). California Department of Water Resources, Sacramento, Calif., pp. 183-198.

Dunbar, R.B., B.K. Linsley, W.A. Jones, D.A. Mucciarone, G.M. Wellington. 1993. Eastern equatorial Pacific climate variability during the past several centuries: Results from stable isotopes in corals from Mexico, Costa Rica, Panama, and Ecuador. EOS, Trans. AGU 74:373 (abstract).

Dunbar, R.B., G.M. Wellington, M.W. Colgan, and P.W. Glynn. 1994. Eastern Pacific sea surface temperature since 1600 A.D.: The $\delta^{18}O$ record of climate variability in Galapagos corals. Paleoceanography 9:291-316.

Duplessy, J.C., N.J. Shackleton, R.G. Fairbanks, L. Labeyrie, D. Oppo, and N. Kallel. 1988. Deepwater source variations during the last climatic cycle and their impact on the global deepwater circulation. Paleoceanography 3:343-360.

Edwards, R.L., F.W. Taylor, and G.J. Wasserburg. 1988. Dating earthquakes with high-precision thorium-230 ages of very young corals. Earth and Planet. Sci. Lett. 90:371-381.

Eglinton, G., S.A. Bradshaw, A. Rosell, M. Sarnthein, U. Pflaumann, and R. Tiedemann. 1992. Molecular record of secular sea surface temperature changes on 100-year time scales for glacial terminations I, II, and IV. Nature 356:423-426.

Elliott, W.P., and J.K. Angell. 1988. Evidence for changes in Southern Oscillation relationships during the last 100 years. J. Climate 1:729-737.

Enomoto, H. 1991. Fluctuations of snow accumulation in the Antarctic and sea level pressure in the southern hemisphere in the last 100 years. Climatic Change 18:67-87.

Epstein, S., R. Buchsbaum, H.A. Lowenstam, and H.C. Urey. 1953. Revised carbonate-water isotopic temperature scale. Bull. Geol. Soc. Am. 64:1315-1326.

Erez, J. 1978. Vital effect on stable-isotope composition seen in foraminifera and coral skeletons. Nature 273:199-202.

Etheridge, D.M., G.I. Pearman, and F. de Silva. 1988. Atmospheric trace-gas variations as revealed by air trapped in an ice core from Law Dome, Antarctica. Ann. Glaciol. 10:28-33.

Fairbanks, R.G. 1989. A 17,000-year glacio-eustatic sea level record: Influence of glacial melting rates on the Younger Dryas event and deep-ocean circulation. Nature 342:637-643.

Fairbanks, R.G. 1990. The age and origin of the Younger Dryas climate event in Greenland ice cores. Paleoceanography 5:937-948.

Fairbanks, R.G., and R.E. Dodge. 1979. Annual periodicity of the O-18/O-16 and C-13/C-12 ratios in the coral *Montastrea annularis*. Geochim. Cosmochim. Acta 43:1979.

Fairbanks, R.G., and R.K. Matthews. 1979. The marine oxygen isotope record in Pleistocene coral, Barbados, West Indies. Quat. Res. 10:181-196.

Fasham, M.J.R. (ed.). 1984. Flows of Energy and Materials in Marine Ecosystems: Theory and Practice. Plenum Press, New York, 733 pp.

Fisher, A. 1988. One model to fit all. National Science Foundation, Mosaic 19 (3/4): 52-59.

Fisher, D.A., R.M. Koerner, W.S.B. Paterson, W. Dansgaard, N. Gundestrup, and N. Reeh. 1983. Effect of wind scouring on climatic records from ice-core oxygen-isotope profiles. Nature 301:205-209.

Fletcher, J.O., U. Radok, and R. Slutz. 1982. Climatic signals of the Antarctic Ocean. J. Geophys. Res. 87:4269-4276.

Flohn, H. 1987. East Africa rains of 1961/62 and the abrupt change of the White Nile discharge. Palaeoecol. Africa 18:3-18.

Folland, C.K., and D.E. Parker. 1990. Observed variations of sea surface temperature. In Climate-Ocean Interaction. M.E. Schlesinger (ed.). Kluwer Academic Press, Dordrecht, pp. 31-52.

Folland, C.K., and D.E. Parker. 1991. Worldwide surface temperature trends since the mid 19th century. In Greenhouse-Gas-Induced Climatic Change: A critical appraisal of simulations and observations. M.E. Schlesinger (ed.). Elsevier, Amsterdam, pp. 173-194.

Folland, C.K., D.E. Parker, and F.E. Kates. 1984. Worldwide marine temperature fluctuations, 1856-1981. Nature 310:670-673.

Folland, C.K., T.N. Palmer, and D.E. Parker. 1986. Sahel rainfall and worldwide sea temperatures, 1901-1985. Nature 320:602-607.

Folland, C.K., T.R. Karl, and K.Ya. Vinnikov. 1990. Observed climate variations and change. In Climate Change: The IPCC Scientific Assessment. J.T. Houghton, G.J. Jenkins, and J.J. Ephraums (eds.). Prepared for the Intergovernmental Panel on Climate Change by Working Group I. WMO/UNEP, Cambridge University Press, pp. 195-238.

Folland, C.K., T.R. Karl, N. Nicholls, B.S. Nyenzi, D.E. Parker, and K.Ya. Vinnikov. 1992. Observed climate variability and climate. In Climate Change 1992: The Supplementary Report to the IPCC Scientific Assessment. J.T. Houghton, B.A. Callander, and S.K. Varney (eds.). Prepared for the Intergovernmental Panel on Climate Change by Working Group I. WMO/UNEP, Cambridge University Press, pp. 135-170.

Fowler, W.B. 1971. Measurement of seasonal air temperatures near the soil surface. J. Range Mgmt. 24(2):158-160.

Friis-Christensen, E., and K. Lassen. 1991. Length of the solar cycle: An indicator of solar activity closely associated with climate. Science 254:698-700.

Fritts, H.C. 1991. Reconstructing Large-scale Climatic Patterns from Tree-ring Data. University of Arizona Press, Tucson, 286 pp.

Garrod, C.J.R., and J.M. Colebrook. 1978. Biological effects of variability in the North Atlantic ocean. Rapp. P.-v. Réun. Cons. Int. Explor. Mer 173:128-144.

Geiger, R. 1965. The Climate near the Ground. Harvard University Press, Cambridge, Mass.

Ghil, M., and R. Vautard. 1991. Interdecadal oscillations and the warming trend in global temperature time series. Nature 350:324-327.

Godfrey, J.S., and E.J. Lindstrom. 1989. The heat budget of the equatorial western Pacific surface mixed layer. J. Geophys. Res. 94:8007-8017.

Goodfriend, G.A., P.E. Hare, and E.R.M. Druffel. 1993. Aspartic acid racemization and protein diagenesis in corals over the last 350 years. Geochim. Cosmochim. Acta 56:3847-3850.

Gordon, A.L. 1986. Interocean exchange of thermocline water. J. Geophys. Res. 91:5037-5046.

Goreau, T.J. 1977. Carbon metabolism in calcifying and photosynthetic organisms: Theoretical models based on stable isotope data. In Proceedings, Third International Coral Reef Symposium. University of Miami, pp. 395-401.

Graybill, D.A., and S.G. Shiyatov. 1992. Dendroclimatic evidence from the northern Soviet Union. In Climate Since A.D. 1500. R.S. Bradley and P.D. Jones (eds.). Routledge, New York, pp. 393-414.

Greenland Ice-core Project (GRIP) Members. 1993. Climate instability during the last interglacial period recorded in the GRIP ice core. Nature 364:203-207.

Grootes, P.M. 1995. Ice cores as archives of decade-to-century-scale climate variability. In Natural Climate Variability on Decade-to-Century Time Scales. D.G. Martinson, K. Bryan, M. Ghil, M.M. Hall, T.R. Karl, E.S. Sarachik, S. Sorooshian, and L.D. Talley (eds.). National Academy Press, Washington, D.C.

Grootes, P.M., and M. Stuiver. 1986. Ross ice shelf oxygen isotopes and West Antarctic climate history. Quat. Res. 26:49-67.

Grootes, P.M., and M. Stuiver. 1987. Ice sheet elevation changes from isotope profiles. In The Physical Basis of Ice Sheet Modelling. E.D. Waddington and J.S. Walder (eds.). IAHS No. 170, pp. 269-281.

Grootes, P.M., M. Stuiver, J.W.C. White, S.J. Johnsen, and J. Jouzel. 1993. Comparison of the oxygen isotope records from the GISP2 and GRIP Greenland ice cores. Nature 366:552-554.

Grove, J.M. 1988. The Little Ice Age. Methuen, London, 498 pp.

Groveman, B., and H.C. Landsberg. 1979. Simulated Northern Hemisphere temperature departures 1579-1880. Geophys. Res. Lett. 6:767-769.

Halfman, J.D., and T.C. Johnson. 1988. High-resolution records of cyclic climatic change during the past 4 ka from Lake Turkana, Kenya. Geology 16:496-500.

Hammen, T. van der, G.C. Maarleveld, J.C. Vogel, and W.H. Zagwijn. 1967. Stratigraphy, climatic succession and radiocarbon dating of the Last Glacial in the Netherlands. Geologie en Mijnbouw 46:79-95.

Hammer, C.U. 1977. Past volcanism revealed by Greenland ice sheet impurities. Nature 270:482-486.

Hammer, C.U. 1980. Acidity of polar ice cores in relation to absolute dating, past volcanism, and radio-echoes. J. Glaciol. 25:359-372.

Hammer, C.U., H.B. Clausen, W. Dansgaard, N. Gundestrup, S.J. Johnsen, and N. Reeh. 1978. Dating of Greenland ice cores by flow models, isotopes, volcanic debris and continental dust. J. Glaciol. 20:3-26.

Hammer, C.U., H.B. Clausen, and W. Dansgaard. 1980. Greenland ice sheet evidence of post-glacial volcanism and its climatic impact. Nature 288:230-235.

Hammer, C.U., H.B. Clausen, W. Dansgaard, A. Neftel, P. Kristinsdottir, and E. Johnson. 1985. Continuous impurity analysis along the Dye-3 deep core. In Greenland Ice Core: Geophysics, Geochemistry, and the Environment. C.C. Langway, Jr., H.

Oeschger, and W. Dansgaard (eds.). Geophys. Monograph 33, American Geophysical Union, Washington, D.C., pp. 90-94.

Hammer, C.U., H.B. Clausen, and H. Tauber. 1986. Ice-core dating of the Pleistocene/Holocene boundary applied to a calibration of the ^{14}C time scale. Radiocarbon 28:284-291.

Hansen, J.E., and S. Lebedeff. 1987. Global trends of measured surface air temperature. J. Geophys. Res. 92:13345-13372.

Hansen, J.E., and S. Lebedeff. 1988. Global surface temperatures: Update through 1987. Geophys. Res. Lett. 15:323-326.

Hansson, M.E., and E.S. Saltzman. 1993. The first Greenland ice core record of methanesulfonate and sulfate over a full glacial cycle. Geophys. Res. Lett. 20:1163-1166.

Hare, F.K., and J.E. Hay. 1974. The climate of Canada and Alaska. In Climates of North America. R.A. Bryson and F.K. Hare (eds.). World Survey of Climatology 11, Elsevier, Amsterdam, pp. 49-192.

Harris, A.R., and I.M. Mason. 1989. Lake area measurement using AVHRR: A case study. Int. J. Remote Sens. 10:885-895.

Harris, A.R., I.M. Mason, C.M. Birkett, and J.A.D. Mansley. 1992. Lake remote sensing for global climate research. In Proceedings of the Central Symposium, "International Space Year" Conference, Münich, 30 March—4 April 1992. Publication ESA SP-341, European Space Agency, Nordwijk, pp. 173-178.

Haury, L.R., J.A. McGowan, and P.H. Wiebe. 1978. Patterns and processes in the time-space scales of plankton distribution. In Spatial Patterns in Plankton Communities. J.H. Steele (ed.). Plenum Press, New York, pp. 277-327.

Heath, M.R., E.W. Henderson, and G. Slesser. 1991. High salinity in the North Sea. Nature 352:116.

Hedgepeth, J.W. 1977. Models and muddles: Some philosphical observations. Helgo. Wiss. Meers. 30:92-104.

Herron, M.M., and C.C. Langway, Jr. 1985. Chloride, nitrate, and sulfate in the Dye-3 and Camp Century, Greenland ice cores. In Greenland Ice Core: Geophysics, Geochemistry and the Environment. C.C. Langway, Jr., H. Oeschger, and W. Dansgaard (eds.). Geophys. Monograph 33, American Geophysical Union, Washington, D.C., pp. 77-84.

Herron, S.L., and C.C. Langway, Jr. 1987. Derivation of paleoelevations from total air content of two deep Greenland ice cores. In The Physical Basis of Ice Sheet Modelling. E.D. Waddington and J.S. Walder (eds.). IAHS No. 170, pp. 283-295.

Hodell, D.A., J.H. Curtis, and M. Brenner. 1995. Possible role of climate in the collapse of Classic Maya civilisation. Nature 375:391-394.

Holdsworth, G., and E. Peake. 1985. Acid content of snow from a mid-troposphere sampling site on Mount Logan, Yukon Territory, Canada. Ann. Glaciol. 7:153-160.

Holdsworth, G., H.R. Drouse, and M. Nosal. 1992. Ice core climate signals from Mount Logan, Yukon, A.D. 1700-1987. In Climate since A.D. 1500. R.S. Bradley and P.D. Jones (eds.). Routledge, London, pp. 517-548.

Hollowed, A.B., K.M. Bailey, and W.S. Wooster. 1987. Patterns of recruitment of marine fishes in the northeast Pacific ocean. Biol. Oceanogr. 5:99-131.

Holmes, J.A., F.A. Street-Perrott, M. Ivanovich, and R.A. Perrott. 1995. A late Quaternary palaeolimnological record from Jamaica based on trace-element chemistry of ostracod shells. Chem. Geol. 124:143-160.

Hooke, R.LeB., and H.B. Clausen. 1982. Wisconsin and Holocene δ¹⁸O variations, Barnes ice cap, Canada. Geol. Soc. Amer. Bull. 93:784-789.

Horel, J.D., and J.M. Wallace. 1981. Planetary-scale atmospheric phenomena associated with the Southern Oscillation. Mon. Weather Rev. 109:813-829.

Hughes, M.K., and Brown, P.M. 1992. Drought frequency in central California since 101 B.C. recorded in giant sequoia tree rings. Climate Dynamics 6:161-167.

Hutchinson, G.E. 1975. A Treatise on Limnology, 2nd ed. Vol. 1: Geography, Physics and Chemistry. John Wiley and Sons, New York.

Imbrie, J., J.D. Hays, D.G. Martinson, A. McIntyre, A.C. Mix, J.J. Morley, N.G. Pisias, W. Prell, and N.J. Shackleton. 1984. The orbital theory of Pleistocene climate: Support from a revised chronology of the marine ¹⁸O record. In Milankovitch and Climate. A. Berger, J. Imbrie, J. Hays, G. Kukla, and B. Saltzman (eds.). D. Reidel, Hingham, Mass., pp. 269-305.

Imbrie, J., E. Boyle, S. Clemens, A. Duffy, W. Howard, G. Kukla, J. Kutzbach, D. Martinson, A. McIntyre, A. Mix, B. Molfino, J. Morley, L. Peterson, N. Pisias, W. Prell, M. Raymo, N. Shackleton, and J. Toggweiler. 1992. On the structure and origin of major glaciation cycles, 1: Linear responses to Milankovich forcing. Paleoceanography 7(6): 701-738.

Institute of Hydrology. 1985. Further Review of the Hydrology of Lake Victoria. Unpublished report to the UK Overseas Development Administration. Institute of Hydrology, Wallingford, U.K.

IPCC. 1990a. Climate Change: The IPCC Scientific Assessment. J.T. Houghton, G.J. Jenkins, and J.J. Ephraums (eds.). Prepared for the Intergovernmental Panel on Climate Change by Working Group I. WMO/UNEP, Cambridge University Press, 365 pp.

IPCC. 1990b. Climate Change: The IPCC Impacts Assessment. W.J.McG. Tegart, G.W. Sheldon, and D.C. Griffiths (eds.). Prepared for the Intergovernmental Panel on Climate Change by Working Group II. WMO/UNEP, Australian Government Publishing Service, Canberra, 270 pp.

IPCC. 1991. Climate Change: The IPCC Response Strategies. Prepared for the Intergovernmental Panel on Climate Change by Working Group III. WMO/UNEP, Island Press, Washington, D.C., 272 pp.

Isdale, P. 1984. Fluorescent bands in massive corals record centuries of coastal rainfall. Nature 310:578-579.

Iversen, J. 1973. The Development of Denmark's Nature since the Last Glaciation. Danmarks Geologiska Undersokelse, Vol. 7c, Copenhagen, 126 pp.

Jenkins, G.M., and D.G. Watts. 1968. Spectral Analysis and Its Applications. Holden-Day, Inc., San Francisco, 525 pp.

Johannessen, O. 1986. A brief overview of the physical oceanography. In The Nordic Seas. B.G. Hurdle (ed.). Springer-Verlag, New York, pp. 103-127.

Johnsen, S.J., W. Dansgaard, H.B. Clausen, and C.C. Langway, Jr. 1972. Oxygen isotope profiles through the Antarctic and Greenland ice sheets. Nature 235:429-434.

Johnsen, S.J., W. Dansgaard, and J.W.C. White. 1989. The origin of Arctic precipitation under present and glacial conditions. Tellus 41B:452-468.

Johnsen, S.J., H.B. Clausen, W. Dansgaard, K. Fuhrer, N. Gundestrup, C.U. Hammer, P. Iversen, J. Jouzel, B. Stauffer, and J.P. Steffensen. 1992a. Irregular glacial interstadials recorded in a new Greenland ice core. Nature 359:311-313.

Johnsen, S.J., H.B. Clausen, W. Dansgaard, N.S. Gundestrup, M. Hansson, P. Jonsson, J.P. Steffensen, and A.E. Sveinbjørnsdottir. 1992b. A "deep" ice core from East Greenland. Medd. Grønland Geosci. 29:3-22.

Jones, P.D. 1988. Hemispheric surface air temperature variations: Recent trends and an update to 1987. J. Climate 1:654-660.

Jones, P.D. 1989. The influence of ENSO on global temperatures. Climate Monitor 17:80-89.

Jones, P.D., and R.S. Bradley. 1992. Climate since the period of instrumental records. In Climate Since A.D. 1500. R.S. Bradley and P.D. Jones (eds.). Routledge, New York, pp. 246-268.

Jones, P.D., and K.R. Briffa. 1992. Global surface air temperature variations over the twentieth century: Part 1. Spatial, temporal and seasonal details. Holocene 2:174-188.

Jones, P.D., and K.R. Briffa. 1995. Decade-to-century-scale variability of regional and hemispheric-scale temperatures. In Natural Climate Variability on Decade-to-Century Time Scales. D.G. Martinson, K. Bryan, M. Ghil, M.M. Hall, T.R. Karl, E.S. Sarachik, S. Sorooshian, and L.D. Talley (eds.). National Academy Press, Washington, D.C.

Jones, P.D., and P.M. Kelly. 1983. The spatial and temporal characteristics of Northern Hemisphere surface air temperature variations. J. Climatol. 3:243-252.

Jones, P.D., and T.M.L. Wigley. 1986. Global temperature variations between 1861 and 1984. Nature 322:430-434.

Jones, P.D., and T.M.L. Wigley. 1990. Global warming trends. Sci. Amer. 263:84-91.

Jones, P.D., S.C.B. Raper, B.D. Santer, B.S.G. Cherry, C.M. Goodess, P.M. Kelly, T.M.L. Wigley, R.S. Bradley, and H.F. Diaz. 1985. A Grid Point Surface Air Temperature Data Set for the Northern Hemisphere. Technical Report TR022, U.S. Dept. of Energy, Carbon Dioxide Research Division, 251 pp.

Jones, P.D., S.C.B. Raper, R.S. Bradley, H.F. Diaz, P.M. Kelly, and T.M.L. Wigley. 1986a. Northern Hemisphere surface air temperature variations: 1851-1984. J. Clim. Appl. Meteorol. 25:161-179.

Jones, P.D., S.C.B. Raper, B.S.G. Cherry, C.M. Goodess, and T.M.L. Wigley. 1986b. A Grid Point Surface Air Temperature Data Set for the Southern Hemisphere, 1851-1984. Technical Report TR027, U.S. Dept. of Energy, Carbon Dioxide Research Division, 73 pp.

Jones, P.D., S.C.B. Raper, and T.M.L. Wigley. 1986c. Southern Hemisphere surface air temperature variations: 1851-1984. J. Clim. Appl. Meteorol. 25:1213-1230.

Jones, P.D., T.M.L. Wigley, and G. Farmer. 1991. Marine and land temperature data sets: A comparison and a look at recent trends. In Greenhouse-Gas-Induced Climatic Change. M.E. Schlesinger (ed.). Kluwer Academic Publishers, Dordrecht, pp. 153-172.

Jones, P.D., R. Marsh, T.M.L. Wigley, and D.A. Peel. 1993. Decadal time scale links between Antarctic Peninsula ice core oxygen-18 and deuterium and temperature. Holocene 3:14-26.

Jouzel, J., L. Merlivat, and C. Lorius. 1982. Deuterium excess in an East Antarctic ice core suggests higher relative humidity at the oceanic surface during the last glacial maximum. Nature 299:688-691.

Jouzel, J., C. Lorius, J.R. Petit, C. Genthon, N.I. Barkov, V.M. Kotlyakov, and V.M Petrov. 1987. Vostok ice core: A continuous isotope temperature record over the last climatic cycle (160,000 years). Nature 329:403-407.

Jouzel, J., N.I. Barkov, J.M. Barnola, M. Bender, J. Chappellaz, C. Genthon, V.M. Kotlyakov, V. Lipenkov, C. Lorius, J.R. Petit, D. Raynaud, G. Raisbeck, C. Ritz, T. Sowers, M. Stievenard, F. Yiou, and P. Yiou. 1993. Vostok ice cores: Extending climatic signal over the penultimate glacial period. Nature 364:407-412.

Karl, T.R. 1988. Multi-year fluctuations of temperature and precipitation: The gray area of climate change. Climatic Change 12:179-197.

Karl, T.R., and W.E. Riebsame. 1984. The identification of 10- to 20-year temperature and precipitation fluctuations in the contiguous United States. J. Clim. Appl. Meteorol. 23:950-966.

Karpuz, N.K., and E. Jansen. 1992. A high-resolution diatom record of the last deglaciation from the SE Norwegian sea: Documentation of rapid climatic changes. Paleoceanography 7:499-520.

Keigwin, L.D., and G.A. Jones. 1989. Glacial-Holocene stratigraphy, chronology, and paleoceanographic observations on some North Atlantic sediment drifts. Deep-Sea Res. 36:845-867.

Keigwin, L.D., G.A. Jones, S.J. Lehman, and E.A. Boyle. 1991. Deglacial meltwater discharge, North Atlantic deep circulation, and abrupt climate change. J. Geophys. Res. 96:16811-16826.

Kellogg, T.B. 1980. Paleoclimatology and paleo-oceanography of the Norwegian and Greenland seas: Glacial-interglacial contrasts. Boreas 9:115-137.

Keppenne, C.L., and M. Ghil. 1992. Extreme weather events. Nature 358:547.

Kerr, R.A. 1992. Unmasking a shifty climate system. Science 255:1508-1510.

Kessler, A. 1974. Atmospheric circulation anomalies and level fluctuations of Lake Titicaca (South America). Preprints of the International Tropical Meteorology Meeting, January 31-February 7, 1974, Nairobi, Kenya, Part I. American Meteorological Society, Boston, pp. 90-91.

Koc-Karpuz, N., and E. Jansen. 1992. A high resolution diatom of the last deglaciation from the SE Norwegian Sea: Documentation of rapid climatic changes. Paleoceanography 7:499-520.

Koeppe, C.E., and G.C. De Long. 1958. Weather and Climate. McGraw-Hill, New York.

Kolstrup, E. 1990. The puzzle of Weichselian vegetation types poor in trees. Geologie en Mijnbouw 69:253-262.

Koslow, J.A. 1984. Recruitment patterns in Northwest Atlantic fish stocks. Can. J. Fish. Aquat. Sci. 41:1722-1729.

Kramer, P.A., J.J. Leder, P.K. Swart, and H.D. Hudson. 1991. A 100-year climatic reconstruction of Florida Bay waters based on C and O isotopic analysis of a coral skeleton (abstract). Geological Society of America Abstracts with Programs 23(A105).

Künzel, F., and A. Kessler. 1986. Investigation of level changes of Lake Titicaca by maximum entropy spectral analysis. Arch. Meteor. Geophys. Bioklimat. B36:219-227.

Kushnir, Y. 1994. Interdecadal variations in North Atlantic sea surface temperature and associated atmospheric conditions. J. Climate 7:141-157.

Lamb, H.H. 1965. The medieval warm epoch and its sequel. Palaeogeog. Palaeoclim. Palaeoecol. 1:13-37.

Lamb, H.H. 1977. Climate: Present, Past, and Future. Vol. 2, Climate History and the Future. Methuen, London, 835 pp.

Lamb, P.J. 1978. Large-scale tropical Atlantic surface circulation patterns associated with sub-Saharan weather anomalies: 1967 and 1968. Tellus 30:240-251.

Land, L.S., J.C. Lang, and D.J. Barnes. 1975. Extension rate: A primary control on the isotopic composition of West Indian (Jamaican) scleractinian reef coral skeletons. Mar. Biol. 33:221-233.

Landsberg, H. 1958. Physical Climatology. Gray Printing Co., Dubois, Penn.

Lange, C.B., S.K. Burke, and W.H. Berger. 1990. Biological production off southern California is linked to climatic change. Climatic Change 16:319-329.

Lara, A., and R. Villalba. 1993. A 3620-year temperature record from Fitzroya cupressoides tree rings in southern South America. Science 260:1104-1106.

Lazier, J. 1988. Temperature and salinity changes in the deep Labrador Sea 1962-1986. Deep-Sea Res. 35:1247-1253.

Lea, D.W., E.A. Boyle, and G.T. Shen. 1989. Coralline barium records temporal variability in equatorial Pacific upwelling. Nature 340:373-376.

Lee, A.J., and J.W. Ramster. 1981. Atlas of the Seas Around the British Isles. U.K. Ordnance Survey, Southampton, England.

Legrand, M.R., and R.J. Delmas. 1988. Soluble impurities in four Antarctic cores over the last 30,000 years. Ann. Glaciol. 10:116-120.

Legrand, M., and C. Feniet-Saigne. 1991. Methanesulfonic acid in South Polar snow layers: A record of strong El Niño? Geophys. Res. Lett. 18:187-190.

Legrand, M.R., C. Lorius, N.I. Barkov, and V.N. Petrov. 1988. Vostok (Antarctica) ice core: Atmospheric chemistry changes over the last climatic cycle (160,000 yr). Atmos. Environ. 22:317-331.

Legrand, M., C. Feniet-Saigne, E.S. Saltzman, C. Germain, N.I. Barkov, and V.N. Petrov. 1991. Ice core record of oceanic emissions of dimethylsulfide during the last climate cycle. Nature 350:144-146.

Legrand, M., M. de Angelis, T. Staffelbach, A. Neftel, and B. Stauffer. 1992. Large perturbations of ammonium and organic acids content in the Summit-Greenland ice core. Fingerprint from forest fires? Geophys. Res. Lett. 19:473-475.

Lehman, S.J. 1995. Variability of Atlantic circulation on sub-millennial time scales. In Natural Climate Variability on Decade-to-Century Time Scales. D.G. Martinson, K. Bryan, M. Ghil, M.M. Hall, T.R. Karl, E.S. Sarachik, S. Sorooshian, and L.D. Talley (eds.). National Academy Press, Washington, D.C.

Lehman, S.J., and L.D. Keigwin. 1992. Sudden changes in North Atlantic circulation during the last deglaciation. Nature 356:757-762.

Lehman, S.J., G.A. Jones, L.D. Keigwin, E.S. Andersen, G. Butenko, and S.-R. Østmo. 1991. Initiation of Fennoscandian ice-sheet retreat during the last deglaciation. Nature 349:513-516.

Leuenberger, M., and U. Siegenthaler. 1992. Ice-age atmospheric concentration of nitrous oxide from an Antarctic ice core. Nature 360:449-451.

Lindeman, R.L. 1942. The trophic-dynamic aspect of ecology. Ecology 23:399-418.

Linn, L.J., M.L. Delaney, and E.R.M. Druffel. 1990. Trace metals in contemporary and seventeenth century Galapagos coral: Records of seasonal and annual variations. Geochim. Cosmochim. Acta 54:387-394.

Linsley, B.K., R.B. Dunbar, D.A. Mucciarone, and G.M. Wellington. 1991. Seasonal to decadal-scale oceanographic variability recorded by eastern Pacific corals over the last 200+ years (abstract). EOS, Trans. AGU (supplement) 72:150.

Linsley, B.K., R.B. Dunbar, G.M. Wellington, and D.A. Mucciarone. 1994. A coral-based reconstruction of Intertropical Convergence Zone variability over Central America since 1707. J. Geophys. Res. 99:9977-9994.

Longhurst, A.R. 1991. Role of the marine biosphere in the global carbon cycle. Limnol. Oceanogr. 36:1507-1526.

Lorius, C., L. Merlivat, J. Jouzel, and M. Pourchet. 1979. A 30,000 yr isotope climatic record from Antarctic ice. Nature 280:644-648.

Lorius, C., J. Jouzel, C. Ritz, L. Merlivat, N.I. Barkov, Y.S. Korotkevich, and V.M. Kotlyakov. 1985. A 150,000-year climatic record from Antarctic ice. Nature 316:591-596.

Lotka, A.J. 1926. Elements of Physical Biology. Williams and Wilkins, Baltimore, Md. (1956 Dover reprint available, 465 pp.)

Lough, J.M. 1986. Tropical Atlantic sea surface temperatures and rainfall variations in sub-Saharan Africa. Mon. Weather Rev. 114:561-570.

Lough, J.M. 1991. Rainfall variations in Queensland, Australia: 1891-1986. Intl. J. Climatol. 11:745-768.

Lowry, W.P. 1969. Weather and Life: An Introduction to Biometeorology. Academic Press, New York.

Lukas, R., and E. Lindstrom. 1991. The mixed layer of the western equatorial Pacific ocean. J. Geophys. Res. 96 (supplement):3343-3357.

Lukas, R., S.P. Hayes, and K. Wyrtki. 1984. Equatorial sea level response during the 1982-1983 El Niño. J. Geophys. Res. 89:10425-10430.

MacCall, A.D., and M.H. Prager. 1988. Historical changes in abundance of six fish species off southern California based on CalCOFI egg and larva samples. CalCOFI Rept. 29:91-101.

Madden, R.A., D.J. Shea, G.W. Branstator, J.J. Tribbia, and R. Weber. 1993. The effects of imperfect spatial and temporal sampling on estimates of the global mean temperature: Experiments with model and satellite data. J. Climate 6:1057-1066.

Maier-Reimer, E., and U. Mikolajewicz. 1989. Experiments with an OGCM on the cause of the Younger Dryas. Report No. 39, Max Planck Institut für Aeronomie, Hamburg, 13 pp.

Maley, J. 1981. Études Palynologiques dans le Bassin du Tchad et Paléoclimatologie de l'Afrique Nord-tropicale de 30,000 Ans à l'Époque Actuelle. Travaux et Documents de l'ORSTOM 129. Éditions de l'Office de la Recherche Scientifique et Technique Outre-Mer, Paris.

Manabe, S., and T. Broccoli. 1985. The influence of ice sheets on the climate of an Ice Age. J. Geophys. Res. 90(C2):2167-2190.

Manabe, S., and D.G. Hahn. 1977. Simulation of the tropical climate of an ice age. J. Geophys. Res. 82:3889-3911.

Manabe, S., and R.J. Stouffer. 1988. Two stable equilibria of a coupled ocean-atmosphere model. J. Climate 1:841-866.

Mangerud, J., S.T. Andersen, B.E. Berglund, and J.J. Donner. 1974. Quaternary stratigraphy of Norden: A proposal for terminology and classification. Boreas 3:109-127.

Marple, S.L. 1987. Digital Spectral Analysis. Prentice-Hall, Inc., Englewood Cliffs, N.J., 492 pp.

Martin, J.H., K.B. Bruland, and W.W. Broenkow. 1976. Cadmium transport in the California Current. In Marine Pollutant Transfer. H.L. Windom and R.A. Duce (eds.). D.C. Heath, Lexington, Mass., pp. 159-184.

Martin, L., M. Fournier, P. Mourguiart, A. Sifeddine, B. Turcy, M.L. Absy, and J.-M. Flexor. 1993. Southern Oscillation signal in South American paleoclimatic data of the last 7,000 years. Quat. Res. 39:338-346.

Martinson, D.G., N.G. Pisias, J.D. Hays, J. Imbrie, T.C. Moore, Jr., and N.J. Shackleton. 1987. Age dating and the orbital theory of the Ice Ages: Development of a high-resolution 0 to 300,000-year chronostratigraphy. Quat. Res. 27:1-29.

Mason, I.M., C.G. Rapley, F.A. Street-Perrott, and S.P. Harrison. 1984. Satellite altimetric measurements of lake levels. Proceedings of the Workshop on ERS-1 Radar Altimeter Data Products (Frascati, Italy, 8-11 May, 1984). Publication ESA SP-221, European Space Agency, Paris, pp. 165-169.

Mason, I.M., C.G. Rapley, F.A. Street-Perrott, and M. Guzkowska. 1985. ERS-1 observations of lakes for climate research. Proceedings of the EARSeL/ESA Symposium "European Remote Sensing Opportunities" (Strasbourg, 31 March-3 April 1985). Publication ESA SP-233, European Space Agency, Paris, pp. 235-241.

Mason, I., A. Harris, C. Birkett, W. Cudlip, and C. Rapley. 1990. Remote sensing of lakes for the proxy monitoring of climatic change. In Remote Sensing and Global Change (Proceedings of the 16th Annual Conference, Swansea), Remote Sensing Society, Nottingham, U.K., pp. 314-324.

Mason, I.M., M.A.J. Guzkowska, C.G. Rapley, and F.A. Street-Perrott. 1994. The response of lake levels and areas to climatic change. Climatic Change 27:161-197.

May, R.M. 1973. Stability and Complexity in Model Ecosystems. Princeton University Press, Princeton, N.J., 235 pp.

Mayewski, P.A., W.B. Lyons, M.J. Spencer, M.S. Twickler, C.F. Buck, and S. Whitlow. 1990. An ice-core record of atmospheric response to anthropogenic sulphate and nitrate. Nature 346:554-556.

Mayewski, P.A., G. Holdsworth, M.J. Spencer, S. Whitlow, M. Twickler, M.C. Morrison, K.K. Ferland, and L.D. Meeker. 1993a. Ice core sulfate from three Northern Hemisphere sites: Source and temperature forcing implications. Atmos. Environ. 27A(17/18):2915-2919.

Mayewski, P.A., L.D. Meeker, M.C. Morrison, M.S. Twickler, S.I. Whitlow, K.K. Ferland, D.A. Meese, M.R. Legrand, and J.P. Steffensen. 1993b. Greenland ice core "signal" characteristics: An expanded view of climate change. J. Geophys. Res. 98(D7): 12839-12847.

Mayewski, P.A., L.D. Meeker, S. Whitlow, M.S. Twickler, M.C.

Morrison, R.B. Alley, P. Bloomfield, and K. Taylor. 1993c. The atmosphere during the Younger Dryas. Science 261:195-197.

Mayewski, P.A., L.D. Meeker, S.I. Whitlow, M.S. Twickler, M.C. Morrison, P. Bloomfield, G.C. Bond, R.B. Alley, A.J. Gow, P.M. Grootes, D.A. Meese, M. Ram, K.C. Taylor, and W. Wumkes. 1994. Changes in atmospheric circulation and ocean ice cover over the North Atlantic during the last 41,000 years. Science 263:1747-1751.

McCartney, M.S., and L.D. Talley. 1984. Warm-to-cold water conversion in the northern North Atlantic. J. Phys. Oceanogr. 14:922-935.

McConnaughey, T.A. 1989. C-13 and O-18 isotopic disequilibria in biological carbonates: I. Patterns. Geochim. Cosmochim. Acta 53:151-162.

McGowan, J.A. 1971. Oceanic biogeography of the Pacific. In The Micropaleontology of Oceans. B.M. Funnell and W.R. Riedel (eds.). Cambridge University Press, pp. 3-74.

McGowan, J.A. 1989. Pelagic ecology and Pacific climate. In Aspects of Climate Variability in the Pacific and the Western Americas. D.G. Peterson (ed.). Geophysical Monograph 55, American Geophysical Union, Washington, D.C., pp. 141-150.

McGowan, J.A. 1995. Temporal change in marine ecosystems. In Natural Climate Variability on Decade-to-Century Time Scales. D.G. Martinson, K. Bryan, M. Ghil, M.M. Hall, T.R. Karl, E.S. Sarachik, S. Sorooshian, and L.D. Talley (eds.). National Academy Press, Washington, D.C.

McGowan, J.A., and P.W. Walker. 1985. Dominance and diversity maintenance in an oceanic ecosystem. Ecol. Mono. 55:103-118.

McIntyre, A., N.G. Kipp, A.W.H. Bé, T. Crowley, T. Kellogg, J.V. Gardner, W. Pressl, and W.F. Ruddiman. 1976. Glacial North Atlantic 18,000 years ago: A CLIMAP reconstruction. In Investigation of Late Quaternary Paleoceanography and Paleoclimatology. R.M. Cline and J.D. Hays (eds.). Memoir 145, Geological Society of America, Boulder, Colo., pp. 43-76.

McPhaden, M.J., and J. Picaut. 1990. El Niño-Southern Oscillation displacements of the western equatorial Pacific warm pool. Science 250:1385-1388.

Meese, D.A., R.B. Alley, A.J. Gow, P. Grootes, P.A. Mayewski, M. Ram, K.C. Taylor, E. Waddington, and G. Zielinski. 1994a. Preliminary depth-age scale of the GISP2 ice core. Special Report 91-01, U.S. Army Cold Regions Research and Engineering Laboratory, Hanover, N.H., 66 pp.

Meese, D.A., A.J. Gow, P. Grootes, P.A. Mayewski, M. Ram, M. Stuiver, K.C. Taylor, E.D. Waddington, and G.A. Zielinski. 1994b. The accumulation record from the GISP2 core as an indicator of climate change throughout the Holocene. Science 266:1680-1682.

Meko, D.M., M.K. Hughes, and C.W. Stockton. 1991. The science of global change and climate variability: The paleo record. In Managing Water Resources Under Conditions of Climate Uncertainty. National Research Council, Washington, D.C., pp. 71-100.

Meko, D.M., E.R. Cook, D.W. Stahle, C.W. Stockton, and M.K. Hughes. 1993. Spatial patterns of tree-growth anomalies in the United States and southeastern Canada. J. Climate 6:1773-1786.

Metcalfe, S.E., F.A. Street-Perrott, S.L. O'Hara, P.E. Hales, and R.A. Perrott. 1994. The paleolimnological record of environmental change: Examples from the arid frontier of Mesoamerica.

In Effects of Environmental Change in Drylands: Biogeographical and Geomorphological Perspectives. A.C. Millington and K. Pye (eds.). John Wiley and Sons, Chichester, pp. 131-145.

Michaelsen, J., and L.G. Thompson. 1992. A comparison of proxy records of El Niño/Southern Oscillation. In El Niño: Historical and Paleoclimatic Aspects of the Southern Oscillation. H.F. Diaz and V. Markgraf (eds.). Cambridge University Press, Cambridge, 440 pp.

Mikami, T. (ed.). 1992. Proceedings of the International Symposium on the Little Ice Age Climate. Department of Geography, Tokyo Metropolitan University, 342 pp.

Mörth, H.T. 1967. Investigations into the meteorological aspects of the variations in the level of Lake Victoria. Vol. 4, East African Meteorology Department Memoirs, 10 pp.

Mosley-Thompson, E., J. Dai, L.G. Thompson, P.M. Grootes, J.K. Arbogast, and J.F. Paskievitch. 1991. Glaciological studies at Siple Station (Antarctica): Potential ice-core paleoclimatic record. J. Glaciol. 37:11-22.

Mulvaney, R., and D.A. Peel. 1988. Anions and cations in ice cores from Dolleman Island and the Palmer Land plateau, Antarctic Peninsula. Ann. Glaciol. 10:121-125.

Mysak, L.A., and S.B. Power. 1991. Greenland sea ice and salinity anomalies and interdecadal climate variability. Climatol. Bull. 25:81-91.

NAS. 1975. Understanding Climatic Change: A Program for Action. U.S. Committee for the Global Atmospheric Research Program. National Academy of Sciences, Washington, D.C., 239 pp.

NAS/NAE/IOM. 1992. Policy Implications of Greenhouse Warming: Mitigation, Adaptation, and the Science Base. Committee on Science, Engineering, and Public Policy; National Academy of Sciences, National Academy of Engineering, and Institute of Medicine. National Academy Press, Washington, D.C., 918 pp.

Neftel, A., H. Oeschger, J. Schwander, B. Stauffer, and F. Zumbrunn. 1982. Ice core sample measurements give atmospheric CO_2 content during the past 40,000 yr. Nature 295:220-223.

Neftel, A., H. Oeschger, T. Staffelbach, and B. Stauffer. 1988. CO_2 record in the Byrd ice core 50,000-5,000 years BP. Nature 331:609-611.

Newell, N.E., R.E. Newell, J. Hsiung, and Z. Wu. 1989. Global marine temperature variation and the solar magnetic cycle. Geophys. Res. Lett. 16:311-314.

Nicholls, N., and R.W. Katz. 1991. Teleconnections and their implications for long-range forecasts. In Teleconnections Linking Worldwide Climate Anomalies. M.H. Glantz, R.W. Katz, and N. Nicholls (eds.). Cambridge University Press, pp. 511-525.

Nicholson, S.E. 1985. Sub-Saharan rainfall 1981-84. J. Clim. Appl. Meteorol. 24:1388-1391.

Nozaki, Y., D.M. Rye, K.K. Turekian, and R.E. Dodge. 1978. A 200 year record of carbon-13 and carbon-14 variations in a Bermuda coral. Geophys. Res. Lett. 5:825-828.

O'Hara, S.L. 1992. Stratigraphic Evidence for Anthropogenic Soil Erosion in Central Mexico. Unpublished D. Phil. thesis, University of Oxford, U.K.

O'Hara, S.L. 1993. Historical evidence of fluctuations in the level of Lake Pátzcuaro, Michoacán, Mexico, over the last 600 years. Geogr. J. 159:51-69.

Oeschger, H., and C.C. Langway, Jr. (eds.). 1989. The Environmental Record in Glaciers and Ice Sheets. In Proceedings of the Dahlem Conference, Berlin, 1988. John Wiley & Sons, Chichester, 400 pp.

Oeschger, H., J. Beer, U. Siegenthaler, B. Stauffer, W. Dansgaard, and C.C. Langway, Jr.. 1984. Late-glacial climate history from ice cores. In Climate Processes and Climate Sensitivity. J.E. Hansen and T. Takahashi (eds.). Geophys. Monograph 29 (Maurice Ewing Series), American Geophysical Union, Washington, D.C., pp. 299-306.

Ogilvie, A.E.J. 1992. Documentary evidence for changes in the climate of Iceland, A.D. 1500 to 1800. In Climate Since A.D. 1500. R.S. Bradley and P.D. Jones (eds.). Routledge, New York, pp. 92-117.

Owen, R.B., R. Crossley, T.C. Johnson, D. Tweddle, I. Kornfield, S. Davison, D.H. Eccles, and D.E. Engstrom. 1990. Major low levels of Lake Malawi and their implications for speciation rates in cichlid fishes. Proc. Roy. Soc. London B240:519-553.

Paine, R.T., and S.A. Levin. 1981. Intertidal landscapes: Disturbance and the dynamics of pattern. Ecol. Monogr. 51:145-178.

Palais, J.M., and M. Legrand. 1985. Soluble impurities in the Byrd Station ice core, Antarctica: Their origin and sources. J. Geophys. Res. 90:1143-1154.

Parker, D.E., C.K. Folland, and M.N. Ward. 1988. Sea-surface temperature anomaly patterns and prediction of seasonal rainfall in the Sahel region of Africa. In Recent Climatic Change. S. Gregory (ed.). Belhaven Press, London, pp. 166-178.

Paterson, W.S.B., R.M. Koerner, D. Fisher, S.J. Johnsen, H.B. Clausen, W. Dansgaard, P. Bucher, and H. Oeschger. 1977. An oxygen-isotope record from the Devon Ice Cap, Arctic Canada. Nature 266:508-511.

Pätzold, J. 1984. Growth rhythms recorded in stable isotopes and density bands in the reef coral Porites lobata (Cebu, Philippines). Coral Reefs 3:87-90.

Pätzold, J. 1986. Temperature and CO_2 Changes in Tropical Surface Waters of the Philippines During the Past 120 Years: Record in the Stable Isotope Content of Corals. Berichte-Reports, Geol.-Paläont. Inst. Kiel, 12, 92 pp. (in German).

Pätzold, J., and G. Wefer. 1992. Bermuda coral reef record of the past 1000 years. Abstracts of the Fourth International Conference on Paleoceanography, Kiel, pp. 224-225.

Pearman, G.I., D. Etheridge, F. de Silva, and P.J. Fraser. 1986. Evidence of changing concentrations of atmospheric CO_2, N_2O, and CH_4 from air bubbles in Antarctic ice. Nature 320:248-250.

Peel, D.A. 1992. Ice core evidence from the Antarctic Peninsula. In Climate Since A.D. 1500. R.S. Bradley and P.D. Jones (eds.). Routledge, New York, pp. 549-571.

Petit, J.R., M. Briat, and A. Royer. 1981. Ice age aerosol content from East Antarctic ice core samples and past wind strength. Nature 293:391-394.

Petit, J.R., L. Mounier, J. Jouzel, Y.S. Korotkevich, V.I. Kotlyakov, and C. Lorius. 1990. Paleoclimatological and chronological implications of the Vostok core dust record. Nature 343:56-58.

Pfister, C. 1992. Monthly temperature and precipitation in central Europe from 1525-1979: Quantifying documentary evidence on weather and its effects. In Climate Since A.D. 1500. R.S. Bradley and P.D. Jones (eds.). Routledge, New York, pp. 118-142.

Philander, S.G.H. 1983. El Niño Southern Oscillation phenomena. Nature 302:295-301.

Philander, S.G.H. 1985. El Niño and La Niña. J. Atmos. Sci. 42:2652-2662.

Philander, S.G.H. 1986. Unusual conditions in the tropical Atlantic Ocean in 1984. Nature 322:236-238.

Philander, S.G.H. 1990. El Niño, La Niña, and the Southern Oscillation. Academic Press, San Diego, Calif., 293 pp.

Pimm, S.L. 1991. The Balance of Nature? Ecological Issues in the Conservation of Species and Communities. University of Chicago Press, 434 pp.

Prahl, F.G., and S.G. Wakeham. 1987. Calibration of unsaturation patterns in long-chain ketone compositions for paleotemperature assessment. Nature 330:367-369.

Quinn, T., F. Taylor, and T. Crowley. 1992. A coral record from the tropical South Pacific (abstract). EOS, Trans. AGU (supplement) 73:150.

Quinn, W.H., and V.T. Neal. 1992. The historical record of El Niño events. In Climate Since A.D. 1500. R.S. Bradley and P.D. Jones (eds.). Routledge, London, pp. 623-648.

Quinn, W.H., D.O. Zopf, K.S. Short, and K.T.W. Yang. 1978. Historical trends and statistics of the Southern Oscillation, El Niño, and Indonesian droughts. Fisheries Bull. 76:663-678.

Quinn, W.H., V.T. Neal, and S.E. Antunez de Mayolo. 1987. El Niño occurrences over the past four and a half centuries. J. Geophys. Res. 92:14449-14461.

Radach, G. 1984. Variations in plankton in relation to climate. Hydrobiological variability in the North Atlantic and adjacent seas. Rapp. P.-V. Réun. Cons. Int. Explor. Mer 185:234-254.

Radach, G., J. Berg, and E. Hagmeier. 1990. Long-term changes of meteorological, hydrographic, nutrient and phytoplankton time-series at Helgoland and L V ELBE 1 in the German Bight. Cont. Shelf Res. 10(4):305-328.

Ramage, C.S. 1960. Role of a tropical "maritime continent" in the atmospheric circulation. Mon. Weather Rev. 96:365-370.

Rasmusson, E.M., and T.H. Carpenter. 1982. Variations in tropical sea surface temperature and surface wind fields associated with the Southern Oscillation/El Niño. Mon. Weather Rev. 110:354-383.

Rasmusson, E.M., and J.M. Wallace. 1983. Meteorological aspects of the El Niño/Southern Oscillation. Science 222:1195-1202.

Rasmusson, E.M., X. Wang, and C.F. Ropelewski. 1990. The biennial component of ENSO variability. J. Mar. Syst. 1:71-96.

Raymo, M.E., W.F. Ruddiman, N.J. Shackleton, and D.W. Oppo. 1990. Evolution of Atlantic-Pacific $\delta^{13}C$ gradients over the last 2.5 m.y. Earth Planet. Sci. Lett. 97:353-368.

Raynaud, D., and R. Delmas. 1977. Composition des gaz contenues dans la glace polaire. In Isotopes and Impurities in Snow and Ice. IAHS No. 118:377-381.

Raynaud, D., J. Chappelaz, J.M. Barnola, Y.S. Korotkevich, and C. Lorius. 1988. Climatic and CH_4 cycle implications of glacial-interglacial CH_4 change in the Vostok ice core. Nature 333:655-657.

Raynaud, D., J. Jouzel, J.M. Barnola, J. Chappelaz, R.J. Delmas, and C. Lorius. 1993. The ice record of greenhouse gases. Science 259:926-934.

Reifsnyder, W.E. 1989. A tale of ten fallacies: The skeptical enquirer's view of the carbon dioxide/climate controversy. Agric. For. Meteorol. 47:349-371.

Reifsnyder, W.E. 1994. Global warming: a catastrophe for plants and ecosystems? In Proceedings of the 13th International Congress of Biometeorology (12-18 September 1993, Calgary, Alberta). Biometeorology, Part 2, Vol. I:68-87.

Reifsnyder, W.E. 1995. Maximum rates of projected and actual increases in global mean temperature as compared with bioclimatic fluctuations. In Natural Climate Variability on Decade-to-Century Time Scales. D.G. Martinson, K. Bryan, M. Ghil, M.M. Hall, T.R. Karl, E.S. Sarachik, S. Sorooshian, and L.D. Talley (eds.). National Academy Press, Washington, D.C.

Richerson, P., R. Armstrong, and C.R. Goldman. 1970. Contemporaneous disequilibrium: A new hypothesis to explain the paradox of the plankton. Proc. Nat. Acad. Sci. 67:1710-1714.

Rind, D., D. Peteet, W.S. Broecker, A. McIntyre, and W. Ruddiman. 1986. The impact of cold North Atlantic sea surface temperatures on climate: Implications for the Younger Dryas cooling (11-10 K). Climate Dynamics 1:3-33.

Risk, M.J., and T.H. Pearce. 1992. Interference imaging of daily growth bands in massive corals. Nature 358:572-573.

Robin, G. de Q. (ed.). 1983. The Climatic Record in Polar Ice Sheets. Cambridge University Press, Cambridge, 212 pp.

Roche, M.A. 1977. Lake Chad: A subdesertic terminal basin with fresh waters. In Desertic Terminal Lakes. D.C. Greer (ed.). Utah Water Research Laboratory, Logan, pp. 213-223.

Roemmich, D., and J. McGowan. 1995. Climate warming and the decline of zooplankton in the California Current. Science 267:1324-1326.

Ropelewski, C.F., and P.D. Jones. 1987. An extension of the Tahiti-Darwin Southern Oscillation Index. Mon. Weather Rev. 115:2161-2165.

Ropelewski, C.F., M.S. Halpert, and X. Wang. 1992. Observed tropospheric biennial variability and its relationship to the Southern Oscillation. J. Climate 5:594-614.

Rozanski, K., T. Goslar, M. Dulinski, T. Kuc, M.F. Pazdur, and A. Walanus. 1992. The late Glacial-Holocene transition in Central Europe derived from isotope studies of laminated sediments from Lake Gosciaz (Poland). In The Last Deglaciation: Absolute and Radiocarbon Chronologies. E. Bard and W.S. Broecker (eds.). NATO ASI Series, Vol. 12, Springer-Verlag, Berlin, pp. 69-80.

Ruddiman, W.F., and A. McIntyre. 1973. Time-transgressive deglacial retreat of polar waters from the North Atlantic. J. Quat. Res. 3(1):117-130.

Ruddiman, W.F., and A. McIntyre. 1981. The North Atlantic Ocean during the last deglaciation. Palaeogeogr. Palaeoclim. Palaeoecol. 35:145-214.

Ruddiman, W.F., C.D. Sancetta, and A. McIntyre. 1977. Glacial-interglacial response rate of subpolar North Atlantic waters to climatic change: The record in oceanic sediments. Phil. Trans. Roy. Soc. London B 280:119-142.

Sarnthein, M., and R. Tiedemann. 1990. Younger Dryas-style cooling events at glacial terminations I-VI: Associated benthic $\delta^{13}C$ anomalies at ODP Site 658 constrain meltwater hypothesis. Paleoceanography 5:1041-1055.

Sarnthein, M., E. Jansen, M. Arnold, J.C. Duplessy, H. Erlenkeuser, A. Flatoy, T. Veum, E. Vogelsang, and M.S. Weinelt. 1992. $\delta^{18}O$ time-slice reconstruction of meltwater anomalies at termination I in the North Atlantic between 50 and 80°N. In The Last Deglaciation: Absolute and Radiocarbon Chronologies.

E. Bard and W.S. Broecker (eds.). NATO ASI Series, Vol. 12, Springer-Verlag, Berlin, pp. 183-200.

Schimmelmann, A., and M.J. Tegner. 1991. Historical oceanographic events reflected in $^{13}C/^{12}C$ ratio of total carbon in laminated Santa Barbara basin sediments. Glob. Biogeochem. Cycles 5:173-188.

Schimmelmann, A., C.B. Lange, and W.H. Berger. 1990. Climatically controlled marker layers in Santa Barbara basin sediments and fine-scale core to core correlation. Limnol. Oceanogr. 35:165-173.

Schimmelmann, A., C.B. Lange, W.B. Berger, A. Simon, S.K. Burke, and R.B. Dunbar. 1992. Extreme climatic conditions recorded in Santa Barbara Basin laminated sediments: The 1835-1840 Macoma Event. Mar. Geol. 106:279-299.

Schmitz, Jr., W.J., and M.S. McCartney. 1993. On the North Atlantic circulation. Rev. Geophys. 31:29-49.

Schneider, S.H. 1989. The greenhouse effect: Science and policy. Science 243:771-781.

Shackleton, N.J., and N.D. Opdyke. 1973. Oxygen isotope and paleomagnetic stratigraphy of equatorial Pacific core V28-238. Oxygen isotope temperatures and ice volumes on a 10^5 and 10^6 year scale. Quat. Res. 3:39-55. (Quoted in Balling, 1992.)

Shen, G.T., and E.A. Boyle. 1988. Determination of lead, cadmium, and other trace metals in annually banded corals. Chem. Geol. 67:47-62.

Shen, G.T., and C.L. Sanford. 1990. Trace element indicators of climate variability in reef-building corals. In Global Ecological Consequences of the 1982-83 El Niño-Southern Oscillation. P.W. Glynn (ed.). Elsevier, New York, pp. 255-284.

Shen, G.T., E.A. Boyle, and D.W. Lea. 1987. Cadmium in corals as a tracer of historical upwelling and industrial fallout. Nature 328:794-796.

Shen, G.T., T.M. Campbell, R.B. Dunbar, G.M. Wellington, M.W. Colgan, and P.W. Glynn. 1991. Paleochemistry of manganese in corals from the Galapagos Islands. Coral Reefs 10:91-101.

Shen, G.T., J.E. Cole, D.W. Lea, L.J. Linn, T.A. McConnaughey, and R.G. Fairbanks. 1992a. Surface ocean variability at Galapagos from 1936-1982: Calibration of geochemical tracers in corals. Paleoceanography 7:563-588.

Shen, G.T., L.J. Linn, T.M. Campbell, J.E. Cole, and R.G. Fairbanks. 1992b. A chemical indicator of trade wind reversal in corals from the western tropical Pacific. J. Geophys. Res. 97:12689-12698.

Shitara, H. 1971. Thermal influence of the Lake Inawashiro on the local climate in summer daytime. Science Reports of the Tohoku University, 7th series (Geography), 18(2):213-220.

Siegenthaler, U., U. Eicher, H. Oeschger, and W. Dansgaard. 1984. Lake sediments as continental $\delta^{18}O$ records from the glacial-post-glacial transition. Ann. Glaciol. 5:149-152.

Sikes, S.K. 1972. Lake Chad. Eyre Methuen, London.

Skud, B.E. 1982. Dominance in fishes: The relation between environment and abundance. Science 216(9):144-149.

Smith, S.V., R.W. Buddemeier, R.C. Redalje, and J.E. Houcke. 1979. Strontium-calcium thermometry in coral skeletons. Science 204:404-407.

Sonett, C.P. 1984. Very long solar periods and the radiocarbon record. Rev. Geophys. Space Phys. 22:239-254.

Sonett, C.P., and H.E. Suess. 1984. Correlation of bristlecone pine ring widths with atmospheric ^{14}C variations: A climate-sun relation. Nature 307:141-143.

Soutar, A., and J.D. Isaacs. 1974. Abundance of pelagic fish during the 19th and 20th centuries as recorded in anaerobic sediment off the Californias. Fish. Bull. 72:257-273.

Sowers, T., M. Bender, D. Raynaud, Y.S. Korotkevich, and J. Orchardo. 1991. The δ^{18}O of atmospheric O_2 from air inclusions in the Vostok ice core: Timing of CO_2 and ice volume changes during the penultimate deglaciation. Paleoceanography 6:679-696.

Sowers, T., M. Bender, L. Labeyrie, D.G. Martinson, J. Jouzel, D. Raynaud, J.J. Pichon, Y.S. Korotkevich. 1993. A 135,000 year Vostok-SPECMAP common temporal framework. Paleoceanography 8:737-766.

Stahle, D.W. and M.K. Cleaveland. 1992. Reconstruction and analysis of spring rainfall over the southeastern U.S. for the past 1000 years. Bull. Am. Meteorol. Soc. 73:1947-1961.

Stauffer, B., and H. Oeschger. 1985. Gaseous components in the atmosphere and the historic record revealed by ice cores. Ann. Glaciol. 7:54-59.

Stauffer, B., H. Hofer, H. Oeschger, J. Schwander, and U. Siegenthaler. 1984. Atmospheric CO_2 concentration during the last glaciation. Ann. Glaciol. 5:160-164.

Stauffer, B., G. Fischer, A. Neftel, and H. Oeschger. 1985. Increase of atmospheric methane recorded in Antarctic ice core. Science 229:1386-1388.

Stauffer, B., E. Lochbronner, H. Oeschger, and J. Schwander. 1988. Methane concentration in the glacial atmosphere was only half that of the pre-industrial Holocene. Nature 332:812-814.

Steele, J.H. 1974. The Structure of Marine Ecosystems. Harvard University Press, Cambridge, 128 pp.

Steele, J.H., and Henderson. 1984. Modelling long-term fluctuations in fish stocks. Science 224:985-987.

Stocker, T.F., and L.A. Mysak. 1992. Climatic fluctuations on the century time scale: A review of high-resolution proxy data and possible mechanisms. Climatic Change 20:227-250.

Stocker, T.F., and D.G. Wright. 1991. Rapid transitions of the ocean's deep circulation induced by changes in surface water fluxes. Nature 351:729-732.

Street-Perrott, F.A. 1995. Natural variability of tropical climates on 10- to 100-year time scales: Limnological and paleolimnological evidence. In Natural Climate Variability on Decade-to-Century Time Scales. D.G. Martinson, K. Bryan, M. Ghil, M.M. Hall, T.R. Karl, E.S. Sarachik, S. Sorooshian, and L.D. Talley (eds.). National Academy Press, Washington, D.C.

Street-Perrott, F.A., and S.P. Harrison. 1985. Lake levels and climate reconstruction. In Paleoclimate Analysis and Modeling. A.D. Hecht (ed.). John Wiley and Sons, New York, pp. 291-340.

Street-Perrott, F.A., and R.A. Perrott. 1990. Abrupt climate fluctuations in the tropics: The influence of Atlantic Ocean circulation. Nature 343:607-612.

Street-Perrott, F.A., P.E. Hales, R.A. Perrott, J.Ch. Fontes, V.R. Switsur, and A. Pearson. 1993. Late Quaternary palaeolimnology of a tropical marl lake: Wallywash Great Pond, Jamaica. J. Paleolimnology 9:3-22.

Stuiver, M., and B. Becker. 1986. High-precision decadal calibration of the radiocarbon time scale, A.D. 1950-2500 B.C. Radiocarbon 28:863-910.

Stuiver, M., and T.F. Braziunas. 1989. Atmospheric ^{14}C and century-scale solar oscillations. Nature 338:405-408.

Sugihara, G., and R.M. May. 1990. Nonlinear forecasting as a way of distinguishing chaos from measurement error in time series. Nature 344:734-741.

Sund, P.N., and J.G. Norton. 1990. Interpreting long-term fish landings records: Environment and/or exploitation? In Proc. Sixth Annual Pacific Climate (PACLIM) workshop (Asilomar, California). J.L. Betancourt and A.M. Mackay (eds.). Tech. Rept. No. 23, Interagency Ecological Studies Program for the Sacramento-San Joaquin Estuary, California Dept. of Water Resources, Sacramento, pp. 71-75.

Sundquist, E.T. 1993. The global carbon dioxide budget. Science 259:934-941.

Sverdrup, H.V. 1953. On conditions for the normal blooming of phytoplankton. J. Cons. Exp. Mer 18:287-295.

Swart, P.K. 1983. Carbon and oxygen isotope fractionation in scleractinian corals: A review. Earth-Sci. Rev. 19:51-80.

Swart, P.K., and M.L. Coleman. 1980. Isotopic data for scleractinian corals explain their palaeotemperature uncertainties. Nature 283:557-559.

Taylor, A.H., J.M. Colebrook, J.A. Stephens, and N.G. Baker. 1992. Latitudinal displacements of the Gulf Stream and the abundance of plankton in the northeast Atlantic. J. Mar. Biol. Assoc. 72:919-921.

Taylor, K., R. Alley, J. Fiacco, P. Grootes, G. Lamorey, P. Mayewski, and M.J. Spencer. 1992. Ice-core dating and chemistry by direct-current electrical conductivity. J. Glaciol. 38:325-332.

Taylor, K.C., G.W. Lamorey, G.A. Doyle, R.B. Alley, P.M. Grootes, P.A. Mayewski, J.W.C. White, and L.K. Barlow. 1993. The "flickering switch" of late Pleistocene climate change. Nature 361:432-436.

Thambyahpillay, G.G.R. 1983. Hydrogeography of Lake Chad and environs: Contemporary, historical, and paleoclimatic. Annals of Borno 1:105-145.

Thompson, L.G. 1992. Ice core evidence from Peru and China. In Climate Since A.D. 1500. R.S. Bradley and P.D. Jones (eds.). Routledge, London, pp. 517-548.

Thompson, L.G., and E. Mosley-Thompson. 1981. Microparticle concentration variations linked with climatic change: Evidence from polar ice cores. Science 212:812-815.

Thompson, L.G., E. Mosley-Thompson, and B.M. Arnao. 1984. El Niño-Southern Oscillation events recorded in the stratigraphy of the tropical Quelccaya ice cap, Peru. Science 226:50-53.

Thompson, L.G., E. Mosley-Thompson, W. Dansgaard, and P.M. Grootes. 1986. The Little Ice Age as recorded in the stratigraphy of the tropical Quelccaya Ice Cap. Science 234:361-364.

Thompson, L.G., E. Mosley-Thompson, M.E. Davis, J.F. Bolzan, J. Dai, T. Yao, N. Gundestrup, X. Wu, L. Klein, and Z. Xie. 1989. Holocene-Late Pleistocene climatic ice core records from Qinghai-Tibetan Plateau. Science 246:474-477.

Tont, S.A. 1981. Temporal variations in diatom abundance off southern California in relation to surface temperature, air temperature and sea level. J. Mar. Res. 39:191-201.

Tont, S.A. 1987. Variability of diatom species populations: From days to years. J. Mar. Res. 45:985-1006.

Tont, S.A. 1989. Climatic change: Response of diatoms and dinoflagellates. In Aspects of Climate Variability in the Pacific and the Western Americas. D. Peterson (ed.). Geophysical Monograph 55, American Geophysical Union, Washington, D.C., pp. 161-163.

Trenberth, K.E. 1980. Atmospheric quasi-biennial oscillations. Mon. Weather Rev. 108:1370-1377.

Trenberth, K.E. 1990. Recent observed interdecadal climate changes in the Northern Hemisphere. Bull. Am. Meteorol. Soc. 71:988-993.

Trenberth, K.E., and D.J. Shea. 1987. On the evolution of the Southern Oscillation. Mon. Weather Rev. 115:3078-3096.

Trenberth, K.E., J.R. Christy, and J.W. Hurrell. 1992. Monitoring global monthly mean surface temperatures. J. Climate 5:1405-1423.

van Loon, H., and J.C. Rogers. 1981. The Southern Oscillation. Part II: Associations with changes in the middle troposphere in the northern winter. Mon. Weather Rev. 109:1163-1168.

van Loon, H., and D.J. Shea. 1985. The Southern Oscillation. Part IV: The precursors south of 15°S to the extremes of the oscillation. Mon. Weather Rev. 115:370-379.

Vautard, R., and M. Ghil. 1989. Singular spectrum analysis in nonlinear dynamics, with applications to paleoclimatic time series. Physica 35D:395-424.

Venrick, E.L. 1990. Phytoplankton in an oligotrophic ocean: Species structure and interannual variability. Ecology 71:1547-1563.

Veum, T., E. Jansen, M. Arnold, I. Beyer, and J.-C. Duplessy. 1992. Water mass exchange between the North Atlantic and the Norwegian Sea during the past 28,000 years. Nature 356:783-785.

Villalba, R. 1990. Climatic fluctuations in northern Patagonia during the last 1000 years as inferred from tree-ring records. Quat. Res. 34:346-360.

Vincent, C.E., T.D. Davies, and A.K.C. Beresford. 1979. Recent changes in the level of Lake Naivasha, Kenya, as an indicator of equatorial westerlies over East Africa. Climatic Change 2:175-189.

Vinnikov, K.Ya., P.Ya. Groisman, and K.M. Lugina. 1990. The empirical data on modern global climate changes (temperature and precipitation). J. Climate 3:662-677.

Vogelsang, E. 1990. Paläo-Ozeanographie des Europäischen Nordmeeres an Hand stabiler Kohlenstoff- und Sauerstoffisotope. Ph.D. Dissertation, Berichte Sonderforschungsbereich 313, University of Kiel, 136 pp.

Volterra, V. 1926. Fluctuations in the abundance of a species considered mathematically. Nature 118:558-560.

Wahlen, M., D. Allen, and B. Deck. 1991. Initial measurements of CO_2 concentrations in air in the GISP2 ice core. Geophys. Res. Lett. 18:1457-1460.

Wang, J.-Y. 1963. Agricultural Meteorology. Pacemaker Press, Milwaukee, Wisconsin.

Wang, K., and T.J. Lewis. 1992. Geothermal evidence for a cold period before recent climatic warming. Science 256:1003-1005.

Wang, S.-W., and R. Wang. 1991. Little Ice Age in China. Chinese Sci. Bull. 36:217-220.

Weber, J.N. 1974. C-13/C-12 ratios as natural tracers elucidating calcification processes in reef-building and non-reef-building corals. In Proceedings of the Second International Coral Reef Symposium 2. Great Barrier Reef Commission, pp. 289-298.

Weber, J.N., and P.M.J. Woodhead. 1970. Carbon and oxygen isotope fractionation in the skeletal carbonate of reef-building corals. Chem. Geol. 6:93-117.

Weber, J.N., and P.M.J. Woodhead. 1972. Temperature dependence of oxygen-18 concentration in reef coral carbonates. J. Geophys. Res. 77:463-473.

Weiss, H., M.A. Courty, W. Wetterstrom, F. Guichard, L. Senior, R. Meadow, and A. Curnow. 1993. The genesis and collapse of third millennium North Mesopotamian civilization. Science 261(5124):995-1003.

Went, F.W. 1957. Most favorable day and night temperatures for some garden flowers. Nurserymen's Institute, Agricultural Extension Service. (Quoted in Brooks, 1958.)

Whitlow, S.I, P.A. Mayewski, G. Holdsworth, M.S. Twickler, and J.E. Dibb. 1994. An ice core based record of biomass burning in North America, 1750-1980. Tellus 46B:239-242.

Wiebe, P.H., C.B. Miller, J.A. McGowan, and R.A. Knox. 1987. Long time-series study of oceanic ecosystems. EOS 60:1178-1190.

Wigley, T.M.L., and T.P. Barnett. 1990. Detection of the greenhouse effect in the observations. In Climate Change: The IPCC Scientific Assessment. J.T. Houghton, G.J. Jenkins, and J.J. Ephraums (eds.). Prepared for the Intergovernmental Panel on Climate Change by Working Group I. WMO/UNEP, Cambridge University Press, pp. 239-255.

Wigley, T.M.L., and P.D. Jones. 1981. Detecting CO_2-induced climatic change. Nature 292:205-208.

Wigley, T.M.L., and P.D. Jones. 1988. Do large-area-average temperature series have an urban-warming bias? Climatic Change 12:313-319.

Wigley, T.M.L., and S.C.B. Raper. 1992. Implications for climate and sea level of revised IPCC emissions scenarios. Nature 357:293-300.

Wigley, T.M.L., J.K. Angell, and P.D. Jones. 1985. Analysis of the temperature record. In Detecting the Climatic Effects of Increasing Carbon Dioxide. M.C. MacCracken and F.M. Luther (eds.). Report No. DOE/ER-0235, U.S. Dept. of Energy, Carbon Dioxide Research Division, pp. 55-90.

Wigley, T.M.L., P.D. Jones, and P.M. Kelly. 1986. Empirical climate studies: Warm world scenarios and the detection of climatic change induced by radiatively active gases. In The Greenhouse Effect, Climatic Change, and Ecosystems. B. Bolin, B.R. Döös, J.Jäger, and R.A. Warrick (eds.). Report No. 29, SCOPE series, John Wiley & Sons, New York, pp. 271-323.

Winter, A., C. Goenaga, and G.A. Maul. 1991. Carbon and oxygen isotope time series from an 18-year Caribbean reef coral. J. Geophys. Res. 96:16673-16678.

WMO. 1983. Guide to Meteorological Instruments and Methods of Observations. Fifth edition. WMO Report No. 8, World Meteorological Organization, Geneva.

Woodruff, S.D., R.J. Slutz, R.J. Jenne, and P.M. Steurer. 1987. A Comprehensive Ocean-Atmosphere Data Set. Bull. Am. Meteorol. Soc. 68:1239-1250.

World Data Center A for Glaciology (Snow and Ice). 1980. Glaciological Data: Ice Cores. Report GD-8, INSTAAR, University of Colorado, Boulder, 139 pp.

World Data Center A for Glaciology (Snow and Ice). 1989. Glaciological Data: Ice Core Update 1980-1989. Report GD-23, CIRES, University of Colorado, Boulder, 105 pp.

World Weather Disc. 1990. Weather-disc Associates, Inc., Seattle, Washington.

Worthington, L.V. 1976. On the North Atlantic Circulation. Johns Hopkins Oceanographic Studies, Vol. 6, Johns Hopkins Press, Baltimore, Maryland, 110 pp.

Wright, P.B. 1989. Homogenized long-period Southern Oscillation indices. Int. J. Climatol. 9:33-54.

Wu, Z., R.E. Newell, and J. Hsiung. 1990. Possible factors controlling global marine temperature variations over the past century. J. Geophys. Res. 95(D8):11799-11810.

Wunsch, C. 1984. An eclectic Atlantic circulation model. Part 1: The meridional flux of heat. J. Phys. Oceanogr. 14:1712-1733.

Yodzis, P. 1978. Competition for Space and the Structure of Ecological Communities. Lecture Notes in Biomathematics 25, Springer-Verlag, Berlin, 191 pp.

Yoshino, M. 1975. Climate in a Small Area. University of Tokyo Press.

6 CONCLUSIONS

The 42 papers presented in this volume span the field of natural climate variability on decade-to-century time scales. Together with the essays, commentaries, and discussions, they show that impressive progress has been made toward the goals of describing, understanding, and modeling the spatial and temporal structure, the magnitude, and the patterns of natural variability. Taken as a whole, they have provided the Climate Research Committee with the perspective needed to draw the conclusions that are discussed in this chapter. These conclusions suggest the research directions and priorities most likely to yield useful insights and further progress; they also were valuable as points of departure for the committee discussions that yielded the recommendations in Chapter 7.

The relatively short instrumental record of climate (the last 50 to 100 years), which reflects anthropogenic change as well as natural variations, does not represent a stationary or steady record. Instead, climate fluctuations over the past few millennia or so will need to be analyzed to establish a baseline of natural variability against which future (and present) variations can be gauged. Many of the papers in this volume contribute data or insights toward this end; they show that this natural propensity for change has manifested itself through all the possible modes of change shown in Figure 1 of the Introduction—periodic variations, sudden shifts, gradual changes, and changes in variability.

• Periodic variations (Figure 1a) are apparent, for example, in the 2290-year-long Tasmanian tree-ring record presented by Cook et al. in Chapter 5. The rings register temperature swings over periods averaging 31, 56, 79, and 204 years, and since 1700 the 79-year fluctuation has been marching nearly in step with a similar variation related to the 11-year sunspot cycle. Nearly periodic fluctuations in estimated mean annual temperature are apparent in the last few thousand years of the Greenland ice-core records (see Grootes's paper in Chapter 5). Similarly, periodic fluctuations are apparent in the 90-year recorded relationship between North Atlantic surface wind and air temperature (see Deser and Blackmon's paper in Chapter 2) and in 130 years of global surface-air temperature data (see Keeling and Whorf's paper, also in Chapter 2, and Figure 1 in Michael Ghil's essay introducing atmospheric modeling in Chapter 3). Quasi-

decadal periodicities in North Atlantic ocean properties have been documented at the surface (by Deser and Blackmon, e.g.), where there is a dipole of opposing tendencies with centers east of Newfoundland and off the southeastern United States. At depth in the North Atlantic, significant changes in salinity and temperature with decadal and longer time scales have also been observed (see Lazier's and Levitus's papers in Chapter 3).

• Sudden regional shifts or jumps (Figure 1b) of several degrees in mean annual temperature, sometimes in just a few years, can be seen in Grootes's Greenland ice-core data; they may exceed 10° in a century. The Northern Hemisphere land-temperature records used in the analyses of Jones and Briffa (Chapter 5) show regional jumps of autumn temperature of more than 0.5°C during the 1920s (the higher levels persisted for 20 years or so), and the temperature data used in Karl et al. (Chapter 2) shows a jump in the variation of diurnal temperature range of about 0.3° in the 1950s. Decadal-scale variability is by no means limited to temperature fluctuations, as is clearly demonstrated by Figure 1 in Thomas Karl's essay introducing the atmospheric observations section, and in the papers by Nicholson, by Shukla, and by Groisman and Easterling in that section. The precipitation in the United States was 5 to 10 percent higher in the 1970s than in the 1930s or 1950s. In the Sahel, precipitation abruptly decreased by more than 50 percent during the period 1968-93, and has persisted at that reduced level for the past few decades. Precipitation over southern Canada has been shown to have increased substantially (over 10 percent) during the 1970s and 1980s. Jumps in regional ocean temperatures and salinity have also been documented (see, for example, the papers by Mysak, Dickson, and Levitus in Chapter 3). Sudden shifts occurred in the surface properties of, and atmospheric circulation over, the North Pacific in 1976 and 1988 (see Cayan's paper in Chapter 3), and numerous researchers such as Dickson (see Chapter 3) have reported marked changes in ocean properties in various parts of the North Atlantic in response to the passage of a surface salinity anomaly.

• Gradual climate changes (Figure 1c) are apparent in a variety of records. For example, the mean hemispheric air temperature records analyzed by Jones and Briffa (Chapter 5) show a gradual warming of approximately

0.5°C over the last century (which occurred mostly between 1910 and 1940, and again since the mid-1970s). A gradual warming of about 0.3°C at mid-depths of the subpolar North Atlantic has been observed in the last 30 years, as is noted in Levitus's paper in Chapter 3.

• Changes in the variability of climate (Figure 1d) have also been documented in some of the data sets presented. For example, the interannual variance of winter temperatures in the United States increased by about 150 percent during the period 1975-1985, as described in Karl's essay. In Chapter 2, Diaz and Bradley provide evidence of a twentieth-century increase in large-scale Northern Hemisphere interannual temperature variability. Similarly, the diurnal temperature range has varied over the past 40 or so years (see Karl et al., also in that chapter). In this case, the minimum (night-time) daily temperature has risen nearly three times faster than the maximum (day-time) temperature. However, since the records in question begin in the 1950s, this change may reflect a significant anthropogenic component as well as natural variability.

We are not yet certain why these changes in climate occur. Various types of models must be used to test our hypotheses and to increase our understanding of the climate system. Models of the atmosphere, the ocean, and the coupled atmosphere-ocean-land-cryosphere system are beginning to yield insights into the causes of natural climate variations. We are beginning to realize their potential for:

• Identifying the responses of key components of the climate system to changes in internal parameters, and to changes in the external forcing such as insolation or volcanic eruptions.

• Explaining the sensitivity and climate signature of each of these components, internal modes of variability, and the interaction between system components (e.g., how a perturbation propagates through the system, or is attenuated or amplified by feedbacks).

• Clarifying how different climate variables respond to the same change, and how certain components of the climate system influence particular time and space scales of variability.

Progress in these areas is reflected in many of the papers appearing in the earlier chapters of this volume. The Climate Research Committee considers the following results particularly noteworthy.

• Recent modeling studies suggest that significant changes in the deep-water circulation may occur over time scales of decades to centuries, and that these changes may critically affect climate. The thermohaline circulation is fairly sensitive to local climate conditions in the high-latitude oceans, particularly air/sea/ice exchange in the sub-polar regions of the North and South Atlantic oceans (in Chapter 3, see the papers by Mysak and by McDermott and Sarachik; in Chapter 4, see Delworth et al.), where the deep water exchanges heat and salt with the atmosphere. Thus, relatively small changes in the climate or environment of these source regions may have profound impacts on the thermohaline circulation.

• A second ocean-model finding is that the thermohaline circulation can oscillate between quasi-steady 'equilibrium' modes. This effect is apparent in both ocean models (see the papers in Chapter 3 by Barnett et al., by McDermott and Sarachik, and by Weaver) and preliminary coupled ocean-atmosphere models (see Delworth et al. in Chapter 4). Multiple equilibria might contribute to rapid climate transitions, while sustained oscillatory changes might contribute to more-or-less regular fluctuations.

Atmospheric models have traditionally led the way in modeling the earth's climate system, and satisfactory simulations of the present atmospheric circulation do exist. Earlier simulations were restricted to fixed lower-boundary conditions; current models are starting to include interaction with the underlying ocean and land surface processes. Complementary progress is being made with ocean models, and coupled models are beginning to advance as well. Despite the encouraging results described above and throughout the volume, it should be noted that satisfactory simulation of the present-day climate does not guarantee that the sensitivity of the models to prescribed changes is realistic. To properly address model sensitivity, and the realism of models in simulating decade-to-century-scale climate change, a hierarchy ranging from simple, mechanistic models through detailed process-oriented models to fully coupled ocean-atmosphere-land-cryosphere-biosphere models is needed. Many models of different construction and complexity are required to test against each other, develop better understanding of the system components, and improve parameterizations and computational efficiency.

Systematically combining observations and models, and ensuring the long-term continuity and sufficient quality of the data, will be critical to the assessment of climate variability and of the models that are used for climate simulation and prediction. The observations permit us to initialize, force, and diagnose models, providing reassurance that we are simulating the real world. As was shown in the NRC's 1991 report on four-dimensional model assimilation of data, models not only serve as the measure of our understanding and a means of prediction, but are now good enough to help guide observation, monitoring, and data-management programs. Combining theoretical and empirical evidence from models and observing systems will permit us to focus our field experiments and observational programs on those components of the climate system that are most susceptible to change or most likely to provide early warning of impending change. Together, observations and models offer the possibility of differentiating between natural variability and anthropogenic change. They have already provided a sketchy description of the timing and character of natural climate variability, and tentatively identified some explanatory mechanisms. The results highlighted above, which suggest that decade-to-century-scale climate fluctuations over the last few millennia have been as varied and extensive as many of those observed over the last few decades, emphasize the need to distinguish between natural and anthropogenic signals.

Separation of natural climate fluctuation and anthropogenic change will require additional evidence involving a well-chosen combination of modeling studies and observational studies. As models improve, a better data base—one with a longer time span, broader spatial representation, and more climate variables—will permit the verification of the distinct signatures within, or key relationships between the specific components of the climate system that the models reveal. (The absence of such data sets for initialization, diagnosis, and validation already hinders modeling progress in some areas.) Increased collaboration among the designers of observation systems, the data analysts, and the modelers is needed, as well as interaction between modelers working on different scales of space and time.

Sophisticated validation methodologies can bridge the gap between observations and the simulations of statistical and dynamical models. For example, sparse data are typically processed (if only by simple interpolation and gridding) to facilitate comparison with the model output; when model output is subjected to the same processing, one model's results can be compared with another's, or with observational data, within a common statistical framework.

Additional data are needed to supplement and expand the currently sparse and sporadic record of past natural climate variability. Such additional data would ideally reflect an assortment of variables and represent a broad range of collection strategies. In some cases they are already available, but have been under-utilized (for example, the indirect or proxy data described in Chapter 5); in others they need to be obtained through special programs or refinement of existing collection programs. In either case, greater sensitivity to consistent data quality, continuity, and uniform data-management practices will be key.

• *Proxy Data.* A critical source of natural-variability information that can augment current instrumental records and model results is the various proxy indicators of the climate of the past several millennia. Tree rings, corals, ice cores, and ocean and lake sediments (see Chapter 5) are proving invaluable in supplying information over long periods of time at annual or even seasonal resolution. These data are particularly relevant to studies

of natural climate variability, because (unlike modern observations) they represent records of climate prior to significant human interference. Because these proxy records' utility to the study of modern climate was discovered relatively recently, we still need to identify new indicators, improve our understanding of existing ones, and hone our skill in collecting them. Increased acquisition, processing, and archiving of such valuable climate data will also be important.

Other proxy indicators that merit more active study and collection are traditional paleoclimate data, geochemical tracer data, and biological data.

— Traditional paleoclimate data (e.g., deep-sea sediment records, ice cores) typically encompass tens of thousands of years or longer; while their resolution is a few millennia to hundreds of years at best, they constitute an excellent data set against which to test a model's ability to simulate climate under conditions significantly different from today's. Successful simulation is critical to establish confidence in the fundamental physics of the models, while eliminating uncertainties related to parameters calibrated against modern conditions.

— Geochemical tracer data provide not only insight into the ocean circulation, but unique information on the rates of gas, heat, and momentum exchange between the ocean and atmosphere.

— Proxy data from modern-day biological indicators—for example, the marine life discussed in Dickson's (Chapter 3) and McGowan's (Chapter 5) papers—often reveal information on distinctive aspects of climate, because of their sensitivity to integrated climate conditions and to nonclimate variables such as species interactions or predation. Despite their complexity, they warrant additional study to see whether their climate signals can be successfully extracted.

• *Historical Records.* Historical records (e.g., travelers' journals, ships' logs, newspapers) offer important information on a variety of climate indicators over hundreds of years. Since much of it remains undiscovered, "archeological" data searches and associated data administration (e.g., reprocessing and quality control) are required to recover it. Besides providing unique, if sometimes subjective, information on past climate variability, historical data serve as a basis for evaluating records of proxy indicators.

• *Operational Data.* Only operational sources can provide the wide-ranging data sets required to initialize, force, diagnose, and validate models with the long-term consistency and coverage required for climate-change prediction. Some such sets are available from present-day monitoring networks, but they have not always been collected sytematically enough to yield the required quality, consistency, and uniformity. For example, the data currently gathered for weather forecasting often lack the accuracy, precision, and continuity in instrument characteristics and processing methods needed to permit the resolution of small but significant longer-time-scale climate variations that can be buried within the much larger diurnal or seasonal signals. Other such data have yet to be monitored. Particularly important are external forcings (e.g., solar variability) and critical variables (e.g., water vapor and moisture and energy fluxes).

Over the last few decades, satellites have provided some of the most helpful operational data. These data are especially useful for increasing spatial coverage and determining otherwise difficult-to-observe climate variables such as snow and ice distribution, which has been shown to be a major influence on the climate and the sensitive thermohaline circulation.

• *Research Data.* The aforementioned types of operational data will not be sufficient alone; they will need to be supplemented with focused data collected specifically for climate-oriented research. These latter data sets will be critical for formulating, improving, and validating specific processes in theoretical and numerical models that are necessary to our understanding of the climate system and the specific mechanisms that control it. For instance, the ocean is severely undersampled with respect to time except in specific coastal and island locations. Establishing and maintaining globally distributed observational systems to provide spatial and time-series measurements of velocity, temperature, and salinity will be central to producing a legacy for monitoring, understanding, and predicting future climate variability.

• *Model Simulations.* Decade-to-century-scale modeling activities are going on at universities, government laboratories, and other research centers in many countries. The complementary and overlapping results obtained at these institutions offer excellent opportunities for intercomparison among model results and validation against instrumental data sets. Proper documentation, archiving, and data-management procedures for model simulations are essential, however, since the quality and continuity are a concern with model-derived information just as with observational data.

The workshop and the papers in this volume show that the earth's climate is always changing, and that gradual changes, periodic variations, and sudden shifts are all characteristic of this natural propensity for change. The Climate Research Committee concludes that the climate fluctuations of the last few millennia were as varied and extensive as any of those observed over the last few decades, though modern climate seems to be intriguingly close to the warmest limits of our poorly documented record of natural variability over the past several thousand years (see Jones and Briffa's paper in Chapter 2). Modeling studies do suggest that climate is very sensitive to relatively small perturbations in key locations. For example, subtle changes in the air/sea/ice interaction of the high-latitude North Atlantic may lead to abrupt changes in ocean circulation and Northern Hemisphere climate. If one of these disproportionate responses were triggered by an anthropogenic effect, climate could be altered more rapidly than the past record would lead us to expect.

The magnitude of the climate changes indicated by records of natural variability, and the rapidity with which they have taken place, suggest that society should expect significant climate change even if anthropogenic influence is minimized. We need to understand and be able to predict this change so that we can adapt to it or modify our contribution to it. Given our recent advances in documenting past climate change, monitoring modern climate processes, and improving process and coupled atmosphere-ocean models, the study of natural climate variability represents an area of great scientific opportunity—one that is important not only for guiding policy, but for understanding how our present biological and geochemical environment evolved and learning to predict how it may respond to natural variations or anthropogenic changes.

7

RECOMMENDATIONS FOR CLIMATE-VARIABILITY RESEARCH PRIORITIES

Anthropogenic changes in climate have been the topic of many publications, and have received considerable public attention. An important part of the research strategy for understanding such changes depends on defining the natural state and variability of earth's climate, but natural climate changes are not yet well enough understood to constitute a baseline against which we might realistically measure human-induced effects. A broad spectrum of observations, including both instrumental records and paleoclimate data (the former possibly contaminated by anthropogenic change, the latter not) has revealed substantial variability in the earth's climate on time scales of decades to centuries. This natural variability alone has considerable socio-economic impact, particularly as it affects agriculture, fisheries, and water resources. To discover possible anthropogenic effects, we must assess variations of the modern climate, anticipate those of the future, and identify the regions or variables that are indicative of change.

The participants in the Dec-Cen workshop presented papers in a range of areas of climate research. These papers bear witness to substantial progress in our ability to describe, understand, and model the spatial and temporal structure, the magnitude, and the patterns of natural variability. However, it is clear that considerable effort is still required both to establish a real baseline of climate variability (natural and anthropogenically forced) and to determine the mechanisms controlling it. To achieve this dual objective, particularly in a time of limited available funding, will require careful consideration of proposed research efforts. On several occasions since the workshop, the CRC has discussed goals, community concerns, and available tools and information. We have identified four fundamental scientific questions that can give direction to work in the field:

1. How can data, theory, and models best be combined to permit us to separate natural climate variability from anthropogenic change on the decade-to-century time scale?

2. Once we can distinguish between natural variability and anthropogenic change, can we understand their interaction, and thus decade-to-century-scale variability?

3. To what extent is the climate state predictable on the decade-to-century time scale?

4. What types of observations would be most helpful in achieving the goal of understanding decade-to-century-scale variability and forecasting climate change on that scale?

To obtain a clear picture of the causes of climate variability, both modeling and real-world observations must be employed, and the traces left by past changes must be uncovered. National and international efforts to explore and document climate variability on decade-to-century time scales have begun to provide the necessary foundation for examining human influences on climate. In addition to inspiring the fundamental questions above, the workshop's presentations and wide-ranging discussions gave the CRC the basis for formulating the following set of recommendations for the conduct of future research.

Criteria must be established to ensure that key variables are identified and observations are made in such a way that their results will yield the most useful data base for future studies of climate variability on decade-to-century time scales. For instance:

- Minimal quality standards that exceed those required for measuring diurnal and seasonal cycles must be implemented.
- The quality and continuity of data acquisition must be maintained over time. (Converting successful, appropriate research programs to sustained operational ones is one way to provide this continuity.)
- Critical forcings and internal climate variables must be monitored as well, to complete the data base.
- Multiple quantities should be monitored simultaneously, to permit cross-checking and provide statistical control.
- Models should be consulted to help design optimal sampling strategies for monitoring systems.

Modeling studies must be actively pursued in order to improve our skill in simulating and predicting the climate state. The use of many types of models, and closer links between models and observational studies (such as data assimilation), will be necessary. Specific recommendations are:

- Different model types must be intercalibrated to establish levels of confidence.
- Models of the individual climate system components must be improved in order to facilitate the development of better coupled models, which integrate these components.
- Known weaknesses in both models and existing data

sets must be targeted for study to improve our understanding of the non-linear global dynamics of the complex interactive climate system—for instance, the interactions between the wind-driven and thermohaline circulations are poorly understood.

• To permit accurate data/model comparisons, both observational data and the model output must be subjected to the same processing.

• Easier access to existing computers must be available to facilitate wider participation in modeling efforts, and higher-speed, larger-memory computers must be developed to make possible more highly resolved models, longer simulations, and more careful sensitivity studies.

Records of past climate change, particularly those reflecting the pre-industrial era, must be actively addressed as a source of valuable new data on the natural component of climate variability. The following approaches are recommended:

• The coverage provided by existing proxy indicators must be expanded so that they yield regional, even global, information.

• Interpretations of proxy indicators currently in use must be continually evaluated for possible improvements; the associated uncertainties and limitations must be assessed and problems identified.

• New proxy indicators of climate must be developed to permit cross-checking of data derived from proxy records currently in use.

• Efforts to locate and fully exploit the wealth of information contained in historical records must be supported.

Climate data must be properly archived, and made readily and freely available to researchers worldwide. Exchange of model-derived information, data from in situ observations, and proxy-record knowledge is an overarching concern. Solutions to the most challenging and important research problems depend greatly on the integration of data from all these sources.

In order to further explore the research priorities that follow from the four key science questions noted earlier, the CRC has established a panel on decade-to-century-scale natural climate variability. The DEC-CEN panel will explore critical scientific questions and issues relating to ocean-atmosphere-land-cryosphere-biosphere interactions and long-term climate change, both natural and anthropogenic. The panel will examine national and international long-range policy, plans, and progress in all these areas. DEC-CEN will then develop not only a basis for determining U.S. research priorities and applications, but a strategy for addressing them in the context of current and planned programs designed to obtain climate-related information.

List of Participants

In The Climate Research Committee's Workshop on Natural Climate Variability on Decade-to-Century Time Scales

Tim P. Barnett
Climate Research Division (0224)
UCSD/SIO
9500 Gilman Drive
La Jolla, CA 92093-0224

David S. Battisti
Department of Atmospheric Sciences
University of Washington
Box 351640
Seattle, WA 98195-1640

Brad Berger
Department of Mechanical Engineering
University of California, Irvine
Irvine, CA 92717-3975

Kenneth Bergman
NASA Headquarters (SED)
300 E Street SW, 5th Flr, Q26
Washington, DC 20546

Doris E. Bouadjemi (HA 466)
Board on Atmospheric Sciences and Climate
National Research Council
2101 Constitution Ave.
Washington, DC 20418

Kirk Bryan
Atmospheric and Oceanic Sciences Program
Sayre Hall
Princeton University
P.O. Box 308
Princeton, NJ 08544-0710

Brendan M. Buckley
Institute of Antarctic and Southern Ocean Studies
University of Tasmania
Hobart, Tasmania 7001, AUSTRALIA

Mark A. Cane
Lamont-Doherty Earth Observatory of
    Columbia University
P.O. Box 1000
Palisades, NY 10964

Daniel R. Cayan
Code A-024
UCSD Scripps Institution of Oceanography
La Jolla, CA 92093

Paola Cessi
Physical Oceanography Research Division,
    Mail Code 0230
Scripps Institution of Oceanography
University of California, San Diego
La Jolla, CA 92093-0230

Robert J. Charlson
Department of Atmospheric Sciences, AK-40
University of Washington
Seattle, WA 98195

Ming Chu
Department of Physics
Chinese University
Shatin, N.T., HONG KONG

Ralph J. Cicerone
Earth System Science
220 Physical Sciences Building
University of California, Irvine
Irvine, CA 92717

Julia E. Cole
INSTAAR
Campus Box 450
University of Colorado
Boulder, CO 80309-0450

Thomas L. Delworth
NOAA Geophysical Fluid Dynamics Laboratory
Princeton University
P.O. Box 308
Princeton, NJ 08542

Clara Deser
Cooperative Institute for Research in Environmental
   Sciences (CIRES)
Campus Box 449
University of Colorado
Boulder, CO 80309-0449

Henry F. Diaz
NOAA Environmental Research Laboratories
325 Broadway
Boulder, CO 80303-3328

Robert R. Dickson
Fisheries Laboratory
Ministry of Agriculture, Fisheries, and Food
Pakefield Road
Lowestoft, Suffolk NR33 0HT, UNITED KINGDOM

Bruce C. Douglas
Department of Geography
LeTrak Hall
University of Maryland, College Park
College Park, MD 20742

Amelito Enriquez
Canada College
4200 Farm Hill Blvd.
Redwood City, CA 94061-1099

Michael Ghil
Institute of Geophysics and Planetary Physics
Slichter Hall, Room 3839
University of California, Los Angeles
405 Hilgard Avenue
Los Angeles, CA 90024-1567

David M. Goodrich
NOAA, Office of Global Programs
1100 Wayne Avenue, Suite 1225
Silver Spring, MD 20910-5603

Nicholas P. Graham
UCSD/SIO, CRD, 0224
9500 Gilman Drive
La Jolla, CA 92093-0224

Pavel Ya. Groisman
Dept. of Geosciences
Morrill Science Center
University of Massachusetts
Amherst, MA 01003

Pieter M. Grootes
Leibniz Labor. für Altersbestimmerung und
   Isotropenforschung
C-14 Labor., Leibnizstrasse 19
Christian Albrechts Universität
D 2300 Kiel, GERMANY

Mohan Gupta
Earth System Science
University of California, Irvine
Irvine, CA 92717-3100

David Hunter
Graduate Department, 0208
UCSD/SIO
La Jolla, CA 92093-0208

Philip D. Jones
Climate Research Unit
University of East Anglia
Norwich, NR4 7TJ UNITED KINGDOM

Thomas R. Karl
National Climatic Data Center
Federal Building
15 Patton Ave., Room 120
Asheville, NC 28801-5001

David J. Karoly
Cooperative Research Centre for Southern
    Hemisphere Meteorology
Mathematics Building
Monash University
Clayton, Victoria 3168, AUSTRALIA

Charles D. Keeling
UCSD/SIO, Geological Research Division
2314 Ritter Hall, MC 0220
La Jolla, CA 92093-0220

Yochanan Kushnir
Lamont-Doherty Earth Observatory of
    Columbia University
Palisades, NY 10964

John R. N. Lazier
Bedford Institute of Oceanography
Box 1006
Dartmouth, NS B2Y 4A2

Scott J. Lehman
Institute of Arctic and Alpine Research
Campus Box 450
University of Colorado
Boulder, CO 80309

Sydney Levitus
NOAA/NODC, E/OC05
1825 Connecticut Avenue NW, Rm 424
Washington, DC 20235

Richard S. Lindzen
Center for Meteorology and Physical
    Oceanography (54-1416)
Massachusetts Institute of Technology
Cambridge, MA 02139

Douglas G. Martinson
Lamont-Doherty Earth Observatory of
    Columbia University
P.O. Box 1000
Palisades, NY 10964-8000

David McDermott
Department of Atmospheric Sciences
University of Washington
Box 351640
Seattle, WA 98195-1640

John A. McGowan
Marine Life Research Group, A-028
UCSD/SIO
8602 La Jolla Shores Drive
La Jolla, CA 92093

James C. McWilliams
Institute of Geophysics and Planetary Physics
Slichter Hall
University of California, Los Angeles
405 Hilgard Avenue
Los Angeles, CA 90024-1567

Pierre Morel
NASA Office of Mission to Planet Earth
300 E St., SW
Washington, DC 20024

Michele Morris
Code A-030
UCSD/SIO
La Jolla, CA 92093

Walter H. Munk
UCSD/SIO, 0225
La Jolla, CA 92093-0225

Lawrence A. Mysak
Centre for Climate and Global Change Research
McGill University (C²GCR)
805 Sherbrooke Street West
Montreal, Quebec H3A 2K6, CANADA

Sharon E. Nicholson
Department of Meteorology
Florida State University
Tallahassee, FL 32306

Gerald R. North
Director, Climate Research Program
Department of Meteorology
Texas A&M University
College Station, TX 77843-3150/3148

David E. Parker
Hadley Centre for Climate Prediction and Research
    (E Div, Rm H-102)
U.K. Meteorological Office
London Road
Bracknell, Berks RG12 2SY, UNITED KINGDOM

John S. Perry (FO 2080)
Board on Sustainable Development
National Research Council
2101 Constitution Ave.
Washington, DC 20418

S. George H. Philander
Atmospheric and Oceanic Sciences Program
P.O. Box CN710, Sayre Hall
Princeton University
Princeton, NJ 08544-0710

Eugene M. Rasmusson
Department of Meteorology
2213 Computer & Space Science Building
University of Maryland
College Park, MD 20742-2425

William E. Reifsnyder
Lama Star Route, Box 3
Questa, NM 87556

Ellen F. Rice (HA 466)
Board on Atmospheric Sciences and Climate
National Research Council
2101 Constitution Ave.
Washington, DC 20418

David Rind
NASA/Goddard Institute for Space Studies
2880 Broadway
New York, NY 10025

David A. Robinson
Department of Geography
Lucy Stone Hall
Rutgers University
New Brunswick, NJ 08903

Claes H. Rooth
School of Marine & Atmospheric Science
University of Miami (RSMAS)
4600 Rickenbacker Causeway
Miami, FL 33149

Edward S. Sarachik
Department of Atmospheric Sciences
University of Washington
Box 351640
Seattle, WA 98195-1640

Robert A. Schiffer
NASA Headquarters (ER)
300 E Street, SW
Washington, DC 20546

Peter Schlosser
Lamont-Doherty Earth Observatory of
   Columbia University
Palisades, NY 10964

Jagadish Shukla
Center for Ocean-Land-Atmosphere Studies
Institute of Global Environment and Society
4041 Powder Mill Rd., Suite 302
Calverton, MD 20705-3106

Soroosh Sorooshian
Department of Hydrology & Water Resources
University of Arizona
Tucson, AZ 85721

William A. Sprigg (HA 466)
Board on Atmospheric Sciences and Climate
National Research Council
2101 Constitution Ave.
Washington, DC 20418

F. Alayne Street-Perrott
Department of Geography
University of Wales Swansea
Swansea SA2 8PP, Wales, UNITED KINGDOM

James H. Swift
UCSD/SIO, ODF 0214
9500 Gilman Drive
La Jolla, CA 92093-0214

Lynne D. Talley
Physical Oceanography Research Division,
   Mail Code 0230
Scripps Institution of Oceanography
University of California, San Diego
La Jolla, CA 92093-0230

Kevin E. Trenberth
National Center for Atmospheric Research
P.O. Box 3000
Boulder, CO 80307-3000

John M. Wallace
Department of Atmospheric Sciences, AK-40
University of Washington
Seattle, WA 98195

John E. Walsh
Department of Atmospheric Sciences
102 Atmospheric Sciences Bldg.
University of Illinois at Urbana-Champaign
105 South Gregory Avenue
Urbana, IL 61801

Andrew J. Weaver
School of Earth and Ocean Sciences
E Hut
University of Victoria
P.O. Box 1700
Victoria, BC V8W 2Y2, CANADA

Xiaojun Yuan
Lamont-Doherty Earth Observatory of
  Columbia University
P.O. Box 1000
Palisades, NY 10964-8000

Stephen E. Zebiak
Lamont-Doherty Earth Observatory of
  Columbia University
P.O. Box 1000
Palisades, NY 10964

Index